ENCYCLOPEDIA OF STATISTICAL SCIENCES

VOLUME 5

Lindeberg Condition to Multitrait-Multimethod Matrices

ENCYCLOPEDIA OF STATISTICAL SCIENCES

VOLUME 5

LINDEBERG CONDITION
to MULTITRAIT-MULTIMETHOD MATRICES

A WILEY-INTERSCIENCE PUBLICATION
John Wiley & Sons
NEW YORK · CHICHESTER · BRISBANE · TORONTO · SINGAPORE

Library of Congress Cataloging in Publication Data:
Main entry under title:
Encyclopedia of statistical sciences.

"A Wiley-Interscience publication."
Includes bibliographies.
Contents: v. 1. A to Circular probable error—v. 3. Faà di Bruno's formula to Hypothesis testing—[etc.]—v. 5. Lindeberg condition to multitrait-multimethod matrices.
1. Mathematical statistics—Dictionaries.
2. Statistics—Dictionaries. I. Kotz, Samuel.
II. Johnson, Norman Lloyd. III. Read, Campbell B.
QA276.14.E5 1982 519.5'03'21 81-10353
ISBN 0-471-05552-2 (v. 5)

Printed in the United States of America

10 9 8 7 6 5 4

CONTRIBUTORS

R. J. Adler, *Technion, Haifa, Israel.* Local Time; Markov Random Fields

M. C. Agrawal, *University of Delhi, Delhi, India.* Mean Reciprocal Values

I. A. Ahmad, *University of Maryland, Catonsville, Maryland.* Matusita's Distance

P. Albrecht, *Universität Mannheim, Federal Republic of Germany.* Mixed Poisson Processes

F. B. Alt, *University of Maryland, College Park, Maryland.* Matching Problems

I. Amara, *University of North Carolina, Chapel Hill, North Carolina.* Log-Rank Scores, Statistics and Tests

C. E. Antle, *Pennsylvania State University, University Park, Pennsylvania.* Lognormal Distribution

A. A. Assad, *University of Maryland, College Park, Maryland.* Management Science, Statistics in

J. K. Baksalary, *Academy of Agriculture in Poznan, Poznan, Poland.* Milliken's Estimability Criterion

Y. Bard, *IBM, Cambridge, Massachusetts.* Maximum Entropy Principle, Classical Approach

D. J. Bartholomew, *The London School of Economics, London, England.* Manpower Planning

C. Benedetti, *Università degli Studi di Roma, Rome, Italy. Metron*

J. O. Berger, *Purdue University, West Lafayette, Indiana.* Minimax Estimation

U. N. Bhat, *Southern Methodist University, Dallas, Texas.* Markov Processes

B. B. Bhattacharyya, *North Carolina State University, Raleigh, North Carolina.* Multinomial Probit: Multinomial Logit

H. M. Blalock, Jr., *University of Washington, Seattle, Washington.* Multiple Indicator Approach

D. Boos, *North Carolina State University, Raleigh, North Carolina.* L-Statistics

K. O. Bowman, *Union Carbide Corporation, Oak Ridge, Tennessee.* Method of Moments

B. M. Brown, *The University of Tasmania, Hobart, Tasmania, Australia.* Median Estimates and Sign Tests

M. W. Browne, *University of South Africa, Pretoria, South Africa.* Multitrait-Multimethod Matrices

R. J. Buehler, *University of Minnesota, Minneapolis, Minnesota.* Minimum Variance Unbiased Estimation

O. Bunke, *Humboldt-Universität zu Berlin, Berlin, German Democratic Republic. Mathematische Operationsforschung und Statistik, Series Statistics*

J. A. Cadzow, *Arizona State University, Tempe, Arizona.* Maximum Entropy Spectral Analysis

S. Cambanis, *University of North Carolina, Chapel Hill, North Carolina.* Measure Theory in Probability and Statistics

V. Chew, *University of Florida, Gainesville, Florida.* Multiple Range and Associated Test Procedures

P. L. Cornelius, *University of Kentucky, Lexington, Kentucky.* Linear Models with

Crossed-Error Structure; Linear Models with Nested-Error Structure

J. A. Cornell, *University of Florida, Gainesville, Florida.* Mixture Experiments

R. G. Cornell, *University of Michigan, Ann Arbor, Michigan.* Most Probable Number

P. R. Cox, *Mayfield, Sussex, England.* Marriage; Migration

E. M. Cramer, *University of North Carolina, Chapel Hill, North Carolina.* Multicollinearity

C. Dagum, *University of Ottawa, Ottawa, Canada.* Lorenz Curve

E. B. Dagum, *Statistics Canada, Ottawa, Canada.* Moving Averages

J. N. Darroch, *The Flinders University, Bedford Park, South Australia.* Marginal Symmetry

M. H. DeGroot, *Carnegie-Mellon University, Pittsburgh, Pennsylvania.* Minimax Decision Rules; Multiple Decision Procedures

J. L. Denny, *University of Arizona, Tucson, Arizona.* Markovian Dependence

D. J. De Waal, *University of the Orange Free State, Bloemfontein, South Africa.* Matrix-Valued Distributions

C. A. Drossos, *University of Patras, Patras, Greece.* Metamathematics of Probability

R. M. Dudley, *Massachusetts Institute of Technology, Cambridge, Massachusetts.* Manifolds

S. Edwards, *University of North Carolina, Chapel Hill, North Carolina.* Logistic Regression

B. S. Everitt, *University of London, London, England.* Mixture Distributions

P. C. Fishburn, *Bell Laboratories, Murray Hill, New Jersey.* Mean-Variance Analyses

M. A. Fligner, *Ohio State University, Columbus, Ohio.* Location Tests

M. R. Frankel, *National Opinion Research Center, New York, New York.* Master Samples

G. H. Freeman, *National Vegetable Research Station, Wellesbourne, Warwick, England.* Magic Square Designs

J. Gani, *University of Kentucky, Lexington, Kentucky.* Literature, Statistics in; Meteorology, Statistics in

S. I. Gass, *University of Maryland, College Park, Maryland.* Linear Programming

J. L. Gastwirth, *George Washington University, Washington, DC.* Measures of Uniformity

S. Geman, *Brown University, Providence, Rhode Island.* Method of Sieves

J. E. Gentle, *IMSL, Houston, Texas.* Monte Carlo Methods

B. L. Golden, *University of Maryland, College Park, Maryland.* Management Science, Statistics in

J. C. Gower, *Rothamsted Experimental Station, Harpenden, Herts, England.* Measures of Similarity, Dissimilarity and Distance

P. E. Green, *University of Pennsylvania, Philadelphia, Pennsylvania.* Marketing, Statistics in

S. T. Gross, *Baruch College, New York, New York.* Median Estimation, Inverse

F. E. Grubbs, *U.S. Army, Aberdeen Proving Ground, Maryland.* Mann–Grubbs Method

W. C. Guenther, *University of Wyoming, Laramie, Wyoming.* Modified (or Curtailed) Sampling Plans

R. C. Gupta, *University of Maine, Orono, Maine.* Modified Power Series Distribution

P. Hackl, *Wirtschaftsuniversität Wien, Vienna, Austria.* Moving Sums (MOSUM)

P. Hall, *Australian National University, Canberra, Australia.* Martingales

E. J. Hannan, *Australian National University, Canberra, Australia.* Linear Systems, Statistical Theory

H. L. Harter, *Wright State University, Dayton, Ohio.* Method of Averages; Method of Group Averages; Method of Least Absolute Values; Method of Least p-th Powers; Minimax Method; Most Approximative Method

D. M. Hawkins, *National Research Institute for Mathematical Sciences, Pretoria, South Africa.* Masking and Swamping

T. P. Hettmansperger, *Pennsylvania State*

University, University Park, Pennsylvania. Mann's Test for Trend; Mathisen's Median Test

C. C. Heyde, *University of Melbourne, Parkville, Victoria, Australia.* Multidimensional Central Limit Theorem

R. R. Hocking, *Texas A & M University, College Station, Texas.* Linear Regression

C. Hsiao, *University of Toronto, Toronto, Ontario, Canada.* Minimum Chi Square

R. Hultquist, *Pennsylvania State University, University Park, Pennsylvania.* Multiple Linear Regression

J. S. Hunter, *Princeton University, Princeton, New Jersey.* Measurement Errors

C. M. Jarque, *Secretaria de Programacion y Presupuesto, Mexico City, Mexico.* Multistratified Sampling

N. T. Jazairi, *York University, Downsview, Ontario, Canada.* Log-Change Index Numbers; Marshall–Edgeworth–Bowley Index

E. H. Jebe, *Ann Arbor, Michigan.* Multiphase Sampling

D. E. Johnson, *Kansas State University, Manhattan, Kansas.* Messy Data

B. L. Joiner, *University of Wisconsin, Madison, Wisconsin.* MINITAB

K. G. Jöreskog, *University of Uppsala, Uppsala, Sweden.* LISREL

V. M. Joshi, *University of Western Ontario, London, Ontario, Canada.* Linear Sufficiency

K. Kafadar, *Hewlett-Packard, Palo Alto, California.* Monte Carlo Swindle

G. Kallianpur, *University of North Carolina, Chapel Hill, North Carolina.* Mahalanobis, Prasantha Chandra

N. Keyfitz, *Harvard University, Cambridge, Massachusetts.* Malthus, Thomas Robert; Malthusian Parameter

G. G. Koch, *University of North Carolina, Chapel Hill, North Carolina.* Logistic Regression; Log-Rank Scores, Statistics and Tests; Multiple-Record Systems

K. Kocherlakota, *University of Manitoba, Winnipeg, Canada.* Multinormal Distributions

S. Kocherlakota, *University of Manitoba, Winnipeg, Canada.* Multinormal Distributions

H. C. Kraemer, *Stanford University Medical Center, Stanford, California.* Measures of Agreement

R. A. Kronmal, *University of Washington, Seattle, Washington.* Marsaglia's Table Method; Mixture Method

S. Kullback, *George Washington University, Washington, D.C.* Minimum Discrimination Information Estimation

H. E. Kyburg, *University of Rochester, Rochester, New York.* Logic in Statistical Reasoning

N. M. Laird, *Harvard University, Cambridge, Massachusetts.* Missing Information Principle

D. Lambert, *Carnegie-Mellon University, Pittsburgh, Pennsylvania.* Minimax Tests; Most Stringent Test

L. R. LaMotte, *University of Houston, Houston, Texas.* Linear Estimators, Bayes

C. Leach, *University of Newcastle upon Tyne, Newcastle upon Tyne, England. Mathematical and Statistical Psychology, British Journal of*

E. A. Lew, *Punta Gorda, Florida.* Medico-Actuarial Investigations

K. McConway, *The Open University, Milton Keynes, England.* Marginalization

H. MacGillivray, *University of Queensland, St. Lucia, Queensland, Australia.* Mean, Median, Mode and Skewness

H. J. Malik, *University of Guelph, Ontario, Canada.* Logistic Distribution

B. F. J. Manly, *University of Otago, Dunedin, New Zealand.* Manly–Parr Estimators

N. R. Mann, *University of California, Los Angeles, California.* Mann–Fertig Statistic; Mann–Grubbs Method

K. V. Mardia, *The University of Leeds, Leeds, England.* Mardia's Test of Multivariate Normality

D. G. Marx, *University of the Orange Free State, Bloemfontein, South Africa.* Matric-t Distribution

P. W. Mielke, *Colorado State University, Fort Collins, Colorado.* Mood's Dispersion Test; Multi-Response Permutation Procedures

R. G. Miller, *Stanford University, Stanford, California.* Multiple Comparisons

G. A. Milliken, *Kansas State University, Manhattan, Kansas.* Messy Data

L. E. Moses, *Stanford University, Stanford, California.* Matched Pairs; Matched Pairs *t* Test

G. S. Mudholkar, *University of Rochester, Rochester, New York.* Multiple Correlation Coefficient

B. Natvig, *University of Oslo, Oslo, Norway.* Multistate Coherent Systems

D. G. Nel, *University of the Orange Free State, Bloemfontein, South Africa.* Matric-*t* Distribution

K. O'Brien, *East Carolina University, Greenville, North Carolina.* Mantel–Haenszel Statistic

J. K. Ord, *Pennsylvania State University, University Park, Pennsylvania.* Linear-Circular Correlation

T. Papaioannou, *University of Ioannina, Ioannina, Greece.* Measures of Information

W. C. Parr, *University of Florida, Gainesville, Florida.* Minimum Distance Estimation

E. Parzen, *Texas A & M University, College Station, Texas.* Multiple Time Series

G. P. Patil, *Pennsylvania State University, University Park, Pennsylvania.* Linear Exponential Family; Logarithmic Series Distribution

A. V. Peterson, *University of Washington, Seattle, Washington.* Marsaglia's Table Method; Mixture Method

A. N. Pettitt, *University of Technology, Loughborough, Leicester, England.* Mann–Whitney–Wilcoxon Statistic

E. C. Pielou, *University of Lethbridge, Alberta, Canada.* Line Intercept Sampling; Line Intersect Sampling; Line Transect Sampling

K. C. S. Pillai, *Purdue University, West Lafayette, Indiana.* Mahalanobis *D* Square

J. H. Pollard, *Macquarie University, North Ryde, New South Wales, Australia.* Mathematical Theory of Population

D. Pregibon, *Bell Laboratories, Murray Hill, New Jersey.* Link Tests

R. L. Prentice, *The Fred Hutchinson Cancer Research Center, Seattle, Washington.* Linear Rank Tests

C. P. Quesenberry, *North Carolina State University, Raleigh, North Carolina.* Model Construction, Selection of Distributions

D. Raghavarao, *Temple University, Philadelphia, Pennsylvania.* Main Effects

C. R. Rao, *University of Pittsburgh, Pittsburgh, Pennsylvania.* Matrix Derivatives, Applications in Statistics; MINQE

M. V. Ratnaparkhi, *Wright State University, Dayton, Ohio.* Liouville–Dirichlet Distribution; Multinomial Distributions

C. B. Read, *Southern Methodist University, Dallas, Texas.* Linear Hazard Rate Distribution; Markov Inequality; Mathematical Functions, Approximations to; Maxwell, James Clerk; Mean Deviation; Mean Squared Error; Median Unbiased Estimators; Midranges; Mills' Ratio; Morbidity

J. Rissanen, *IBM, San Jose, California.* Minimum-Description-Length Principle

T. Robertson, *University of Iowa, Iowa City, Iowa.* Monotone Relationships

V. K. Rohatgi, *Bowling Green State University, Bowling Green, Ohio.* Moment Problem

D. Ruppert, *University of North Carolina, Chapel Hill, North Carolina.* M-Estimators

B. F. Ryan, *Pennsylvania State University, University Park, Pennsylvania.* MINITAB

T. A. Ryan, *Pennsylvania State University, University Park, Pennsylvania.* MINITAB

D. Sankoff, *University of Montréal, Montréal, Canada.* Linguistics, Statistics in

F. W. Scholz, *Boeing Computer Services Company, Tukwila, Washington.* Maximum Likelihood Estimation

H. T. Schreuder, *U.S. Rocky Mountain Forest and Range Experiment Station, Fort Collins, Colorado.* Mickey's Unbiased Ratio and Regression Estimators

S. J. Schwager, *Cornell University, Ithaca, New York.* Mean Slippage Problems

H. Seal*, *École Polytechnique Federale de*

* Deceased

Lausanne, Lausanne, Switzerland. Multiple Decrement Tables

A. R. Sen, *Calgary, Alberta, Canada.* Location Parameter, Estimation of

P. K. Sen, *University of North Carolina, Chapel Hill, North Carolina.* Locally Optimal Statistical Tests; Log-Rank Scores, Statistics and Tests; Multidimensional Contingency Tables

E. Seneta, *University of Sydney, Sydney, New South Wales, Australia.* Markov, Andrei Andreevich; Montmort, Pierre Rémond de

G. Shafer, *University of Kansas, Lawrence, Kansas.* Miller's Paradox; Moral Certainty

M. Shaked, *University of Arizona, Tucson, Arizona.* Majorization and Schur Convexity

A. Shapiro, *University of South Africa, Pretoria, South Africa.* Minimum Rank Factor Analysis; Minimum Trace Factor Analysis

L. R. Shenton, *University of Georgia, Athens, Georgia.* Method of Moments

D. M. Shoemaker, *U.S. Department of Education, Washington, D.C.* Multiple Matrix Sampling

G. Simons, *University of North Carolina, Chapel Hill, North Carolina.* Measure Theory in Probability and Statistics

B. Singer, *The Rockefeller University, New York, New York.* Longitudinal Data Analysis

N. D. Singpurwalla, *The George Washington University, Washington, D.C.* Military Standards for Fixed-Length Life Tests; Military Standards for Sequential Life Testing

H. Solomon, *Stanford University, Stanford, California.* Military Statistics

G. W. Somes, *East Carolina University, Greenville, North Carolina.* Mantel–Haenszel Statistic; McNemar Statistics; Miettinen's Statistic

A. P. Soms, *University of Wisconsin, Madison, Wisconsin.* Lindstrom–Madden Method

F. W. Steutel, *Technisch Hogeschool Eindhoven, Eindhoven, The Netherlands.* Log-Concave and Log-Convex Distributions

G. W. Stewart, *University of Maryland, College Park, Maryland.* Linear Algebra, Computational; Matrix, Ill-Conditioned

S. Stigler, *University of Chicago, Chicago, Illinois.* Mansfield, Merriman

D. J. Strauss, *University of California, Riverside, California.* Luce's Choice Axioms and Generalizations

R. Syski, *University of Maryland, College Park, Maryland.* Multi-Server Queues

H. Thomas, *Pennsylvania State University, University Park, Pennsylvania.* Measurement Structures and Statistics

D. M. Titterington, *University of Glasgow, Glasgow, Scotland.* Medical Diagnosis, Statistics in

J. W. Tolle, *University of North Carolina, Chapel Hill, North Carolina.* Mathematical Programming

R. Trader, *University of Maryland, College Park, Maryland.* Moments, Partial

R. C. Tripathi, *University of Texas, San Antonio, Texas.* Modified Power Series Distribution

W. Uhlmann, *Institut für Angewandte Mathematik und Statistik, Würzburg, Federal Republic of Germany. Metrika*

T. S. Wallsten, *University of North Carolina, Chapel Hill, North Carolina.* Measurement Theory

E. J. Wegman, *Office of Naval Research, Arlington, Virginia.* Military Statistics

L. Weiss, *Cornell University, Ithaca, New York.* Maximum Probability Estimation

D. J. White, *Manchester University, Manchester, England.* Markov Decision Processes

H. White, *University of California, La Jolla, California.* Misspecification, Tests for

Y. Wind, *University of Pennsylvania, Philadelphia, Pennsylvania.* Marketing, Statistics in

D. B. Wolfson, *McGill University, Montreal, Canada.* Lindeberg–Feller Theorem; Lindeberg–Lévy Theorem

F. W. Young, *University of North Carolina, Chapel Hill, North Carolina.* Multidimensional Scaling

M. Zeleny, *Fordham University at Lincoln Center, New York, New York.* Multiple Criteria Decision Making

L *continued*

LINDEBERG CONDITION *See* LINDEBERG–FELLER THEOREM

LINDEBERG–FELLER THEOREM

HISTORY

The Lindeberg–Feller central limit theorem gives necessary and sufficient conditions for the convergence in distribution of a sequence of suitably standardized sums of independent random variables with finite variance, to a standard normal random variable. An important forerunner to the Lindeberg–Feller theorem was proved by Liapunov* in 1900, who established sufficient conditions for convergence to normality assuming the existence of moments of order $2 + \delta$ for some $\delta > 0$. Lindeberg [5] strengthened Liapunov's theorem* by assuming the existence of the second moments while Feller [1] established the necessity of Lindeberg's condition for convergence to normality, provided that the uniformly asymptotically negligible condition is satisfied. The theorem given below is for a single sequence of random variables. A more general version for double arrays of random variables may be found in most textbooks on probability theory (see, e.g., Gnedenko and Kolmogorov [4]).

STATEMENT OF THE LINDEBERG–FELLER THEOREM

Let $X_1, X_2, \ldots,$ be a sequence of independent random variables and let the distribution function of X_k be $F_k(x)$. Suppose that $\text{var}(X_k) = \sigma_k^2 < \infty$ for all k. Let $S_n = \sum_{k=1}^{n}(X_k - E[X_k])/s_n$, where $s_n^2 = \sum_{k=1}^{n}\text{var}(X_k)$. Then in order that as $n \to \infty$,

$$\text{(a)} \quad \Pr[S_n \leqslant x] \to (1/\sqrt{2\pi}) \int_{-\infty}^{x} e^{-t^2/2}\, dt \qquad \text{for all} \quad x > 0,$$

$$\text{(b)} \quad \lim_{n \to \infty} \max_{1 \leqslant k \leqslant n} \Pr[|(X_k - E[X_k])/s_n| > \epsilon] = 0 \qquad \text{for all} \quad \epsilon > 0,$$

it is necessary and sufficient that for each $\eta > 0$, we have

$$\text{(L)} \quad (1/s_n^2) \sum_{k=1}^{n} \int_{|x| > \epsilon s_n} x^2\, dF_k(x + E[X_k]) \to 0 \qquad \text{for each} \quad \epsilon > 0.$$

IDEA OF PROOF: SUFFICIENCY

Form a sequence of truncated bounded random variables and using the Lindeberg condition together with a limit result due to Liapunov, establish sufficiency for this truncated sequence. Finally, show that the standardized sums arising from the truncated random variables behave the same way in the limit as do the standardized sums arising from the original random variables.

An alternative approach using convolution operators may be found in Feller [2].

NECESSITY

The proof is technical and involves manipulations with characteristic functions.*

REMARKS

1. Condition (b), sometimes called the uniformly asymptotically negligibility (UAN) condition, ensures that the relative contribution of each of the summands of S_n tends to zero as $n \to \infty$.
2. Condition (L) is referred to the Lindeberg condition. Again, it is a condition on the relative smallness of each summand, and may be shown to imply that $\max_{1 \leqslant j \leqslant n} \sigma_j / s_n \to 0$ as $n \to \infty$.
3. Condition (L) may be easily shown to imply the Liapunov condition.*
4. As in the case of the Lindeberg–Lévy theorem,* there exist various extensions of the Lindeberg–Feller theorem to sums of random vectors [3] in $\mathbb{R}^k$ and to sums of dependent random variables. (*See* LIMIT THEOREM, CENTRAL.)

References

[1] Feller, W. (1935). *Math. Zeit.*, **40**, 521–559.

[2] Feller, W. (1966). *An Introduction to Probability Theory and Its Applications*, Vol. 2. Wiley, New York.

[3] Gnedenko, B. V. (1944). *Uspekhi Mat. Nauk*, **10**, 230 (in Russian).

[4] Gnedenko, B. V. and Kolmogorov, A. N. (1954). *Distributions for Sums of Independent Random Variables*. Addison-Wesley, Reading, Mass.

[5] Lindeberg, J. W. (1922). *Math. Zeit.*, **15**, 221.

(LIMIT THEOREM, CENTRAL
LINDEBERG–LÉVY THEOREM)

D. WOLFSON

LINDEBERG–LÉVY THEOREM

HISTORY

The Lindeberg–Lévy central limit theorem is widely known more simply as "the" central limit theorem, although it is, strictly speaking, a single representative of central limit theorems in general. It refers to the convergence in distribution of a suitably standardized sum of independent, identically distributed random variables with finite variance, to a normal random variable with zero mean and unit variance. The form given below is due, independently, to Lévy [1] and Lindeberg [2].

STATEMENT OF THE LINDEBERG–LÉVY CENTRAL LIMIT THEOREM

Let $X_1, X_2, X_3, \ldots$ be a sequence of independent and identically distributed random variables with $E[X_i] = \mu$ and $0 < \text{var}(X_i) = \sigma^2 < \infty$. Then

$$\Pr\left[\sum_{i=1}^{n} (X_i - \mu)/(\sqrt{n}\,\sigma) \leqslant x \right] \to (1/\sqrt{2\pi}) \int_{-\infty}^{x} e^{-t^2/2}\, dt$$

for all $x \in (-\infty, \infty)$, as $n \to \infty$.

IDEA OF PROOF

The characteristic function* of

$$\sum_{i=1}^{n} (X_i - \mu)/(\sqrt{n}\,\sigma)$$

is expanded using Taylor's theorem and the

assumed existence of a second moment. This expanded characteristic function (ch.f) is then shown to converge pointwise to $e^{-t^2/2}$, the ch.f. of the standard normal distribution. An application of the continuity theorem then leads to the stated result.

EXTENSIONS AND GENERALIZATIONS

The Lindeberg–Lévy theorem may be extended to include random vectors:

Random Vector Lindeberg–Lévy Theorem. Let $\mathbf{X}_n = (X_{n1}, X_{n2}, \ldots, X_{nk})$ be independent, identically distributed random vectors with mean vector $\boldsymbol{\mu} = (\mu_1, \mu_2, \ldots, \mu_k)$, where $\mu_j = E[X_{nj}]$, $j = 1, 2, \ldots, k$. Suppose further that $E[X_{nj}^2] < \infty$ and that the covariance matrix * of $\mathbf{X}_n$ is $\boldsymbol{\Sigma}$. Then $\sum_{j=1}^{n}(\mathbf{X}_j - \boldsymbol{\mu})/\sqrt{n}$ converges in distribution to a multivariate normal* random vector having mean zero and covariance matrix $\boldsymbol{\Sigma}$.

A second extension, to random sums of random variables, is exemplified by the following theorem:

Random Sums Lindeberg–Lévy Theorem. Let $X_1, X_2, \ldots$ be a sequence of independent, identically distributed random variables with $E[X_i] = \mu$ and $\operatorname{var}(X_i) = \sigma^2$. Let $\{\nu_n, n \geqslant 1\}$ be a sequence of random variables taking on only strictly positive integer values such that $\nu_n/n \rightarrow \alpha$ in probability, where α is a constant satisfying $0 < \alpha < \infty$. Then

$$\Pr\left[\sum_{i=1}^{\nu_n} (X_i - \mu)/\left(\sigma\sqrt{\nu_n}\right) \leqslant x\right] \rightarrow \left(1/\sqrt{2\pi}\right)\int_{-\infty}^{x} e^{-t^2/2}\,dt$$

for all $x \in (-\infty, \infty)$ as $n \rightarrow \infty$.

Related to the central limit theorem but requiring more delicate analyses are the laws of the iterated logarithm, * which describe the transient behavior of the standardized sums as n increases. These results complement those describing the limiting behavior given by the central-limit-type theorems.

Other generalizations of the Lindeberg–Lévy theorem may be found in LIMIT THEOREMS, CENTRAL, which includes examples of central-limit-type theorems for dependent random variables; also discussed are the concepts of infinite divisibility* and stability as well as the famous Berry–Esseen theorem on rate of convergence to normality (*see* ASYMPTOTIC NORMALITY).

APPLICATIONS

Either in its original form or in one of its extended forms, the Lindeberg–Lévy theorem lies at the heart of most of the asymptotic theory of statistics. The construction of confidence intervals, indeed much of the theory and practice of statistical inference* and hypothesis testing, * is facilitated by the availability of the central limit theorem. The frequent occurrence of sums of random variables in statistics is the reason for the wide applicability of this important theorem.

References

[1] Lévy, P. (1925). *Calcul des Probabilités*. Gauthier-Villars, Paris.

[2] Lindeberg, J. W. (1922). *Math. Zeit.*, **15**, 221.

Further Reading

In addition to the references just given, the following texts contain many of the topics referred to in this article.

Billingsley, P. (1979). *Probability and Measure*. Wiley, New York.

Feller, W. (1966), *An Introduction to Probability Theory and Its Applications*, Vol. 2. Wiley, New York.

Gnedenko, B. V. and Kolmogorov, A. N. (1954). *Limit Distributions for Sums of Independent Random Variables*. Addison-Wesley, Reading, Mass.

All of these books are rigorous in spirit and require some mathematical sophistication.

(LIMIT THEOREM, CENTRAL)

D. WOLFSON

LINDLEY'S EQUATION

Let W_n be the waiting time of the nth customer in a queue, S_n be the corresponding service time, and X_{n+1} be the length of time between the nth and $(n+1)$th arrivals. Then, for a general $G/G/1$ queueing system (with one customer),

$$W_{n+1} = \max(0, W_n + S_n - X_n).$$

This basic result, due to Lindley [2], implies that the distribution function

$$F_n(x) = \Pr(W_n \leqslant x)$$

of the Ws converges as $n \to \infty$ to some limit function F. For more details see Lindley [2] and Cohen [1].

References

[1] Cohen, J. W. (1969). *The Single Server Queue*. North-Holland, Amsterdam, London.

[2] Lindley, D. V. (1952). *Proc. Cambridge Phil. Soc.*, **48**, 177–184.

(QUEUEING THEORY)

LINDSTROM–MADDEN METHOD

The Lindstrom–Madden method is a method for constructing approximate lower confidence limits* on the reliability of a series system, given Bernoulli subsystem data. This is a fundamental problem in reliability theory* and is discussed in ref. 6, together with the Lindstrom–Madden method. A series system of independent components functions only if all the components function. More precisely, let Y_i, $i = 1, 2, \ldots, k$, be independent binomial random variables with parameters (n_i, p_i), $n_1 \leqslant n_2 \leqslant \cdots \leqslant n_k$, where p_i is the probability that the ith subsystem will function, and let the observed values be $y_1, y_2, \ldots, y_k$, with $x_i = n_i - y_i$, $i = 1, 2, \ldots, k$. The reliability of the system is $\prod_{i=1}^k p_i$. A general method of constructing a lower $1-\alpha$ level confidence limit for $\prod_{i=1}^k p_i$ was given in ref. 1. However, this is very difficult to implement in practice. Lipow and Riley [5] constructed the exact lower confidence limit for $\prod_{i=1}^k p_i$ for specific values of x_i, n_i, $k = 2, 3$, and Lloyd and Lipow [6] noted that the tabulated values were close to an approximation, the Lindstrom–Madden method, described below. Consider building systems by randomly selecting without replacement a single test result (success or failure) from each of the k subsystem data. Then there are n_1 systems and the expected number of failures is $z_1 = n_1 q_0$, $q_0 = 1 - \prod_{i=1}^k ((n_i - x_i)/n_i)$. Let

$$I_p(r,s) = \frac{1}{\beta(r,s)} \int_0^p t^{r-1}(1-t)^{s-1}\,dt;$$

i.e., $I_p(r,s)$ is the incomplete beta function. Then if y is an integer, $y < n$, we have

$$\sum_{i=0}^{y} \binom{n}{i} p^{n-i} q^i = I_p(n-y, y+1).$$

A complete discussion of the above can be found in ref. 4. For $0 \leqslant y < n$, real, define $u(n, y, \alpha)$ by $\alpha = I_{u(n,y,\alpha)}(n-y, y+1)$. Thus, for integer values of y, $u(n, y, \alpha)$ is a $100(1-\alpha)\%$ lower confidence limit for p. Then the Lindstrom–Madden method consists of using $u(n_1, z_1, \alpha)$ as an approximation to the exact lower confidence limit b and reduces to the usual method for putting a lower confidence limit on the success probability if z_1 is an integer.

Denote by $[x]$ the integral part of x, x real. Sudakov [7] showed that $u(n_1, z_1, \alpha) \leqslant b \leqslant u(n_1, [z_1], \alpha)$, and hence if z_1 is an integer, then the Lindstrom–Madden method is exact. The special case when only x_1 is nonzero (in this case $z_1 = x_1$) was proved by Winterbottom [8]. Sudakov's results were simplified and generalized by Harris and Soms [2]. Also, Harris and Soms [3] improved the lower bound $u(n_1, z_1, \alpha)$ by the use of a short FORTRAN program, whose listing they provide. We now give two examples.

Example 1. Let $\alpha = 0.05$, $(n_1, n_2, n_3) = (5, 10, 20)$, and $(x_1, x_2, x_3) = (1, 0, 0)$. Then $z_1 = 1$, $u(5, 1, 0.05) = 0.343$, so $b = 0.343$.

Example 2. Let $\alpha = 0.05$, $(n_1, n_2, n_3, n_4) = (10, 15, 20, 25)$, and $(x_1, x_2, x_3, x_4) = (1, 3, 2, 4)$. Then $z_1 = 4.557$, $u(10, 4.557, 0.05) = 0.257$, $u(10, 4, 0.05) = 0.304$, so $0.257 \leqslant b \leqslant 0.304$. Using the program in Harris and Soms [3], this can be improved to $0.291 \leqslant b \leqslant 0.304$.

Addendum—Added in Proof An error has been discovered in the proof of one of the lemmas employed in the proof of Sudakov's inequality. This is documented in "The Theory of Optimal Confidence Limits for Systems Reliability with Counterexamples for Results on Optimal Confidence Limits for Series Systems", B. Harris and A. P. Soms, Tech. Rep. 643, Dept. of Statistics, University of Wisconsin-Madison, Madison, Wis. Numerical evidence indicates that for confidence levels of practical interest the inequality is still valid.

References

[1] Buehler, R. J. (1957). *J. Amer. Statist. Ass.*, **52**, 482–493. (The fundamental paper on one-sided confidence limits on parametric functions.)

[2] Harris, B. and Soms, A. P. (1980). Bounds for Optimal Confidence Limits for Series Systems. *Tech. Rep. No. 2093*, Mathematics Research Center, University of Wisconsin–Madison, Madison, Wis. (Generalizes and simplifies the results of ref. 7.)

[3] Harris, B. and Soms, A. P. (1981). Improved Sudakov-Type Bounds for Optimal Confidence Limits on the Reliability of Series Systems. *Tech. Rep. No. 637*, Dept. of Statistics, University of Wisconsin–Madison, Madison, Wis. (Further sharpening of the lower bound for the optimal lower confidence limit on system reliability.)

[4] Johnson, N. L. and Kotz, S. (1970). *Distributions in Statistics: Discrete Distributions*. Wiley, New York, Chap. 3. (Provides an excellent reference for the binomial distribution.)

[5] Lipow, M. and Riley, J. (1959). *Tables of Upper Confidence Bounds on Failure Probability of 1, 2, and 3 Component Serial Systems*. Vols. 1 and 2. Space Technology Laboratories, Los Angeles, Calif. (Often used as a standard of comparison for various approximate methods; contains only equal sample sizes.)

[6] Lloyd, D. K. and Lipow, M. (1962). *Reliability: Management, Methods, and Mathematics*. Prentice-Hall, Englewood Cliffs, N. J., Chap. 9. (An early and often cited text on all aspects of reliability.)

[7] Sudakov, R. S. (1974). *Eng. Cybern.*, **12**, 55–63. (Contains the first theoretical justification of the Lindstrom–Madden method; difficult to read.)

[8] Winterbottom, A. (1974). *J. Amer. Statist. Ass.*, **69**, 782–788. (Proves a special case of ref. 7 and gives a survey of different methods.)

Bibliography

Harris, B. (1977). In *Theory and Applications of Reliability: With Emphasis on Bayesian and Nonparametric Methods*, C. P. Tsokos and I. N. Shimi, eds., Vol. 2. Academic Press, New York, pp. 275–297. (Comprehensive survey paper of the many approaches used on this problem.)

Harris, B. and Soms, A. P. (1981). Recent Advances in Statistical Methods for System Reliability Using Bernoulli Sampling of Components. *Tech. Rep. No. 643*, Dept. of Statistics, University of Wisconsin–Madison, Madison, Wis. (An update of the previous paper.)

Johns, M. V., Jr. (1975). Reliability Assessment for Highly Reliable Systems. *Tech. Rep. No. 1*, Dept. of Statistics, Stanford University, Stanford, Calif. (Uses the Poisson approximation and has tables for $k = 2$.)

(RELIABILITY
SYSTEM RELIABILITY ANALYSIS)

ANDREW P. SOMS

LINEAR ALGEBRA, COMPUTATIONAL

Computational linear algebra may be divided into three parts. The first part deals with discrete computations, such as manipulating the graph of a matrix or performing exact calculations on matrices with integer or rational elements. The second part is concerned with constrained optimization problems that are defined in terms of matrices—for example, linear and quadratic programming problems. The third part is often called numerical linear algebra. It deals with the solution in floating-point arithmetic of such problems as solving linear systems of equations, linear least-squares problems, the computation of eigenvalues, etc.

Because of limitations of space this article will be restricted to numerical linear algebra.

It should be appreciated, however, that even this restriction is insufficient. Specifically, a matrix may be too large to fit into the high-speed memory of the computer in question, and it is a matter of considerable algorithmic ingenuity to solve such problems. Most of the recent research in this area has been devoted to *sparse matrices*, most of whose elements are zero. No attempt will be made to survey sparse matrix computations here; instead, this article will treat matrix problems that can be held in the high-speed memory of the computer or at least can be solved with one or two passes of the data through memory.

It is hardly necessary to argue the importance of numerical linear algebra in statistics. This is obvious for areas like multivariate analysis, * where matrix notation has become standard. It is less well appreciated that a significant amount of matrix computations is essential for maximum likelihood estimation, * robust estimation, * and other essentially nonlinear statistical problems. This is because nonlinear problems are frequently solved by solving a sequence of linearized approximations.

Modern numerical linear algebra has five principal features:

1. The systematic use of matrix decompositions
2. The use of updating methods to recompute decompositions of slightly altered matrices
3. The use of backward rounding-error analysis to assess the stability of algorithms
4. The use of perturbation theory to assess the accuracy of computed solutions
5. The implementation of matrix algorithms in high-quality mathematical software

In selecting or developing methods for solving a problem, it is necessary to have an understanding of each of these five features. Accordingly, they will be discussed separately in the next five sections. The treatment is primarily didactic; technical comments and references will be found in the section "Bibliographical Notes."

DECOMPOSITIONS

The method of Gaussian elimination for solving the linear equation

$$Ax = b \tag{1}$$

of order p will serve to introduce the use of matrix decompositions in numerical linear algebra. The result of performing Gaussian elimination on the system is to produce a lower triangular matrix L and an upper triangular matrix U such that

$$A = LU. \tag{2}$$

Given this *LU decomposition*, one may solve the system (1) by first solving the upper triangular system

$$Uy = b$$

(this is equivalent to performing the elimination operations on the vector b). The upper triangular system

$$Lx = b$$

is then solved for x (this is usually called the back-substitution phase of the algorithm); *see* GAUSS – JORDAN ELIMINATION.

From the foregoing it is evident that the LU decomposition (2) and its computation are distinct from the problem (1) and its solution. Once an LU decomposition has been computed [an $O(p^3)$ process], it may be used again and again to solve systems of the form (1) at a cost of $O(p^2)$ work per solution. Moreover, the same LU decomposition can be used to solve the system $A^Tx = b$. Finally, if the LU decomposition of A is available, it is unnecessary to form A^{-1} to solve (1), even though the formula $x = A^{-1}b$ suggests that it is. In fact, the calculation of A^{-1} is not only unnecessary, but it is expensive [an extra $O(p^3)$ work] and undesirable on grounds of numerical stability.

The foregoing comments about the LU decomposition hold generally for the decompositional approach to numerical linear algebra. A decomposition may be regarded as a computational platform from which a variety of problems may be solved. It may be relatively expensive to compute; but once it is available, it can be used repeatedly at little additional cost. Moreover, a decomposition may make it unnecessary to compute such notationally convenient but numerically tricky objects as inverses or generalized inverses*. Finally, although it is not evident from the LU decomposition, a well-chosen decomposition may aid in the mathematical analysis of a statistical process.

There are a large number of decompositions that are used in numerical linear algebra, and it can be a matter of some delicacy to select the right one. Accordingly, the rest of this section is devoted to a survey of the most frequently used matrix decompositions.

Pivoted *LU* Decomposition

The LU decomposition (2) need not exist, and even when it does it may be impossible to compute it in a stable manner. The cure for this situation is to interchange rows and columns of A during the computations in order to ensure numerical stability, a process that is called *pivoting*. The result is a decomposition of the form

$$P_r A P_c = LU,$$

where P_r and P_c are the permutation matrices corresponding to the row and column interchanges. There are two common strategies for choosing interchanges. At the kth step of the algorithm, *partial pivoting* interchanges rows to bring the largest element in column k into the (k, k)-position. *Complete pivoting* interchanges both rows and columns to place the largest element in the matrix in the (k, k)-position. Although one can prove stronger theorems about the stability of complete pivoting, in practice the simpler partial-pivoting strategy is just as stable and is consequently the method of choice.

The principal application of the LU decomposition is to the solution of linear equations and, where it is required, the computation of inverses. It is computed by Gaussian elimination or one of its variants, such as the algorithms of Crout or Doolittle. It requires $O(p^3)$ work.

Cholesky Decomposition

If A is a $p \times p$ positive-definite matrix (which here implies symmetry), it can be factored uniquely in the form

$$A = R^T R, \tag{3}$$

where R is upper triangular with positive-diagonal elements. This factorization is called the *Cholesky decomposition.*

There is an important variant in which pivoting is used to produce a decomposition of the form

$$P^T A P = R^T R,$$

where P is a permutation matrix. The interchanges may be chosen so that the elements of R satisfy

$$r_{kk}^2 \geqslant \sum_{i=k}^{j} r_{ij}^2 \qquad (j = k+1, k+2, \ldots, p).$$

This implies that the diagonals of R are decreasing. In particular, if A is near a semidefinite matrix, then a trailing principal submatrix of A will be small.

The Cholesky decomposition is used to solve linear systems involving positive-definite matrices. In statistical applications the matrix is often a correlation matrix. The Cholesky decomposition of the augmented cross-product matrix $(X\ y)^T(X\ y)$ is used in the analysis of the linear model $y = XB + e$. The Cholesky decomposition is also used in the solution of symmetric eigenvalue problems of the type $Ax = \lambda Bx$, a problem that occurs frequently in statistical applications. The algorithm for computing the decomposition is called variously the Cholesky algorithm or the square-root method.[1] It is simply an adaptation of Gaussian elimination and requires $O(p^3)$ work.

QR Decomposition

Given any $n \times p$ $(n > p)$ matrix X there is an $n \times n$ orthogonal matrix Q such that

$$Q^T X = \begin{bmatrix} R \\ O \end{bmatrix},$$

where R is upper triangular with nonnegative diagonal elements. If Q is partitioned in the form

$$Q = (\overset{p}{Q_X} \quad \overset{n-p}{Q_\perp}),$$

then

$$X = Q_X R, \tag{4}$$

an expression that is sometimes called the *QR factorization*. The QR decomposition is closely related to the Cholesky factorization, as the relation

$$X^T X = R^T R \tag{5}$$

shows.

The QR decomposition is extraordinarily versatile. To cite one example, if X is of rank p, $Q_\perp Q_\perp^T$ is the projection onto the orthogonal complement of the column space of X. Thus $Q_\perp Q_\perp^T y$ is the residual vector $y - X\hat{b}$ of linear regression. As another example, the generalized inverse $X^\dagger$ of X is given by $X^\dagger = R^{-1}X^T$. Thus operations involving $X^\dagger$ can be replaced by a multiplication by X^T followed by the solution of upper triangular systems.

The QR decomposition may be computed by three distinct algorithms: the Golub–Householder algorithm, the method of plane rotations, and the Gram–Schmidt* method with reorthogonalization. They each have their own advantages and drawbacks. Each requires $O(np^2)$ work.

Spectral Decomposition

It is well known that a symmetric matrix A of order p has a set of orthonormal eigenvectors $v_1, v_2, \ldots, v_p$ satisfying

$$Av_i = \lambda_i v_i \qquad (i = 1, 2, \ldots, p). \tag{6}$$

If $V = (v_1, v_2, \ldots, v_p)$, then V is orthogonal and it follows from (6) that

$$V^T A V = \Lambda = \operatorname{diag}(\lambda_1, \lambda_2, \ldots, \lambda_p). \tag{7}$$

The decomposition (7) is called the *spectral decomposition* of A.

The spectral decomposition is widely used both inside and outside of statistics. For example, the spectral decomposition of a sample correlation matrix gives estimates of the principal components (*see* COMPONENTS ANALYSIS).

All methods for computing eigenvalues and eigenvectors are necessarily iterative. For the spectral decomposition the procedure is to perform a direct reduction to a tridiagonal matrix* followed by the iterative QR algorithm (not to be confused with the QR decomposition). The amount of work varies, but it is always $O(p^3)$. This algorithm supersedes the older Jacobi algorithm, which, unfortunately, is still to be found in some statistical programs.

Singular Value Decomposition

Given any $n \times p$ $(n \geqslant p)$ matrix X, there is an $n \times n$ orthogonal matrix U and a $p \times p$ orthogonal matrix V such that

$$U^T X V = \begin{bmatrix} \Psi \\ 0 \end{bmatrix}.$$

where

$$\Psi = \operatorname{diag}(\psi_1, \psi_2, \ldots, \psi_p)$$

with

$$\psi_1 \geqslant \psi_2 \geqslant \cdots \geqslant \psi_p \geqslant 0.$$

The numbers ω_i are called the *singular values* of X, the columns of U the *left singular vectors*, and the columns of V the *right singular vectors*. The singular value decomposition is related to the spectral decomposition of $X^T X$ in much the same way as the QR and Cholesky decompositions are related; compare the relation

$$V^T X^T X V = \Psi^2$$

with (7).

Since there are relatively few computational tasks that require a singular value decomposition and since the decomposition is expensive to compute, it is less used than the other decompositions. On the other hand, it has many interesting properties that make it of theoretical interest to both numerical analysts and statisticians. For example, the matrix $\overline{X}$ of rank k nearest X in the least-squares sense may be obtained by setting $\overline{\Psi} = \text{diag}(\psi_1, \psi_2, \ldots, \psi_k, 0, \ldots, 0)$ and forming

$$\overline{X} = U\begin{bmatrix} \overline{\Psi} \\ 0 \end{bmatrix} V^T.$$

The customary way of calculating the singular value decomposition is by an initial reduction to bidiagonal form followed by a variant of the QR algorithm. When $n > p$, a great deal of work can be saved by first calculating the QR decomposition of X and then the singular value decomposition of R, from which the singular value decomposition of X can be reconstituted.

Schur Decomposition

It is natural to attempt to generalize the spectral decomposition of a symmetric matrix to nonsymmetric matrices by asking for a nonsingular matrix W such that $W^{-1}AW$ is diagonal, in which case the columns of W are eigenvectors of A. Unfortunately, such a decomposition need not exist, and even when it does, the matrix W may be so near a singular matrix as to be computationally useless. If W is restricted to be unitary, then it can be determined so that

$$W^H A W = T,$$

where T is an upper triangular matrix. The diagonal elements of T, which are eigenvalues of A, can be made to appear in any order, although usually they are thought of as appearing in descending order of magnitude. Whatever the order, the decomposition is called a *Schur decomposition* and the columns of W are called *Schur vectors*. The first k Schur vectors span the invariant subspace corresponding to the first k eigenvalues of A as they appear in T.

When A is real, it is desirable on grounds of computational efficiency to remain in the real field, even when A has complex eigenvalues. It can be shown that there is an orthogonal matrix W such that W^TAW is quasi-triangular; i.e., W^TAW is block upper triangular with at most 2×2 blocks. The 1×1 blocks are eigenvalues of A, while the 2×2 blocks contain complex conjugate pairs of eigenvalues. The decomposition is sometimes called a real Schur decomposition.

The Schur decomposition is used in the computation of eigenvectors and other objects related to the spectrum of A. In fact, many problems that would seem to require eigenvectors for their solution actually require no more than the Schur decomposition. The decomposition is computed by an initial reduction to Hessenberg form followed by the QR algorithm at a cost of $O(p^3)$ work.

Generalized Schur Decomposition

Although a theory of canonical forms exists for the generalized eigenvalue problem $Ax = \lambda Bx$, an attempt to compute one of these forms may lead to numerical difficulties. If one restricts oneself to unitary equivalences of the form $(A, B) \rightarrow (Y^H AW, Y^H BW)$, then one can find unitary matrices Y and W such that $Y^H AW$ and $Y^H BW$ are both upper triangular. This decomposition may be called a *generalized Schur decomposition*. The eigenvalues of the problem are ratios of the corresponding diagonal elements of $Y^H AW$ and $Y^H BW$, and they may be made to appear in any order. If B is nonsingular, this decomposition is related to the Schur decomposition of $B^{-1}A$, since $W^H B^{-1} AW$ is upper triangular. When A and B are real, there is a variant decomposition in which B is quasi-triangular.

The generalized Schur decomposition is used principally to solve the generalized eigenvalue problem. It is computed by a variant of the QR algorithm.

For square matrices, the total work required to compute any of the decompositions described here is $O(p^3)$; however, the order constants vary widely. The following is a list of the decompositions in order of work required to decompose a square matrix:

Cholesky
LU
QR
Spectral
Schur
Singular value
Generalized Schur

As a rule the last three decompositions should be used only when there is an explicit need.

It is important to note that all the decompositions, except the *LU* and Cholesky, are based on orthogonal transformations, from which they derive their extraordinary stability (in the sense to be defined in the sequel). Pivoting is necessary for the stability of the *LU* decomposition, and positive definiteness for the Cholesky.

UPDATING

The term "updating" is used broadly to mean the recomputation of a numerical object when the data defining it have been altered. A very simple example of this process is a running mean, which is updated each time a new data point is added. There are also algorithms for updating standard deviations and correlation coefficients.

Perhaps the oldest example of an updating method for matrices is given by the well-known formula

$$(A + uv^T)^{-1} = A^{-1} - \frac{1}{1 + v^T A^{-1} u} A^{-1} uv^T A^{-1}, \tag{8}$$

which expresses the inverse of A after a rank-one modification in terms of A^{-1}. The critical point is that the computation of $(A + uv^T)^{-1}$ from A^{-1} requires $O(p^2)$ work, whereas its ab initio computation requires $O(p^3)$ work. This saving of computation is one of the principal reasons for using update techniques.

Unfortunately, the formula (8) has the disadvantage that if in a sequence of updates a nearly singular matrix A is encountered, then every subsequent update will be inaccurate. For example, consider the sequence

$$\begin{aligned} A &= \begin{bmatrix} \pi & 0 \\ \pi & e \end{bmatrix} \to \begin{bmatrix} \pi & 0 \\ \pi & e \end{bmatrix} + \begin{bmatrix} e+\epsilon \\ 0 \end{bmatrix} \begin{bmatrix} 0 & 1 \end{bmatrix} \\ &= \begin{bmatrix} \pi & e+\epsilon \\ \pi & e \end{bmatrix} \to \begin{bmatrix} \pi & e+\epsilon \\ \pi & e \end{bmatrix} \\ &\quad - \begin{bmatrix} e+\epsilon \\ 0 \end{bmatrix} \begin{bmatrix} 0 & 1 \end{bmatrix} = A, \end{aligned}$$

where $\epsilon = 5.10^{-7}$. The matrix $B = A^{-1}$ is given

$$B = \begin{bmatrix} \pi^{-1} & 0 \\ -e^{-1} & e^{-1} \end{bmatrix}$$

and if the updating formula (8) is applied twice to the sequence above, starting with B, the result should be again B. But if this is done numerically in 10-digit, decimal, floating-point arithmetic, the result is the sequence

$$\begin{bmatrix} 0.3183098861 & 0 \\ 0.3678794412 & 0.3678794412 \end{bmatrix}$$
$$\to \begin{bmatrix} -1{,}729{,}945.033 & 1{,}729{,}945.351 \\ 1{,}999{,}344.789 & -1{,}999{,}344.789 \end{bmatrix}$$
$$\to \begin{bmatrix} 0.3180000000 & 0 \\ 0.3680000000 & 0.3680000000 \end{bmatrix},$$

and the first and last members agree to only three significant figures.

For reasons illustrated by the example above, there is a tendency in modern numerical linear algebra to avoid the formation of explicit matrix inverses whenever possible. In the updating game this means that what one updates is a matrix decomposition. Methods for updating *LU* decompositions find their greatest application in mathematical programming* and related areas, and for that reason they will not be treated here. For the statistician, the most important update techniques are those related to the Cholesky and *QR* decompositions.

Let A be a positive-definite matrix with Cholesky decomposition (3). There are order

p^2 algorithms for computing from R the Cholesky decompositions

(a) $A + xx^T$ (x a p-vector)

(b) $A - xx^T$ (assumed positive definite)

(c) P^TAP (P a permutation matrix obtained by interchanging two rows of the identity).

(9)

If x is regarded as a regression observation or an observation of a multivariate random variable, the first update amounts to adding an observation to the data, while the second amounts to deleting one. The third amounts to rearranging the order of the variables and is used in adding and deleting variables.

If X is an $n \times p$ matrix with QR factorization (4), then $O(n \times p)$ updates exist for the following modifications of X:

(a) Add a row to X.

(b) Delete a row from X.

(c) Interchange two columns of X.

(d) Add a column to X.

(e) Delete a column from X.

(f) The general rank-one modification $X + uv^T$.

(10)

The relation $X^TX = R^TR$ shows that the first three updates in (10) correspond to the three in (9). However, it is important to note that deleting a row of X is stable, while the update (9b) has numerical problems when $A - xx^T$ is nearly singular.

There are techniques for updating the other decompositions; unfortunately, the gains are less spectacular, since they require $O(p^3)$ work for square matrices. An outstanding problem, whose solution would be of wide applicability, is to design an updating algorithm for rank-one modifications of symmetric tridiagonal matrices under orthogonal similarity transformations.

Although it is not, strictly speaking, an update technique, the SWEEP operator* should be mentioned here. Essentially a variant of Gauss–Jordan elimination,* it is widely used in linear regression* and multivariate analysis because it provides at once regression coefficients, error sum of squares, and inverse correlation matrix for an arbitrary subset of the variables. This, along with the simplicity of the operator itself, makes it the method of choice when a short code yielding easily accessible results is required (e.g., the author has coded a version for his TI-59, a hand-held calculator).

However, the operator has the serious drawback that it proceeds by forming the explicit inverse and therefore has the same limitations as the formula (8). One implication of this is that the SWEEP operator cannot be used to determine the residual sum of squares for a nearly collinear set of variables without an expensive reinversion. Another difficulty with the operator is that no satisfactory rounding-error analysis of it has been published. This is probably of less importance, since in practice it seems to work well when collinearities are avoided. However, in view of the fact that everything done by the SWEEP operator can also be done by updating Cholesky or QR decompositions, the latter is to be preferred in constructing portable software.

BACKWARD ROUNDING-ERROR ANALYSIS[2]

When a numerical algorithm is used to solve a problem in floating-point arithmetic, the answer may be very accurate, completely inaccurate, or anywhere in between. Unless the computed solution is quite close to the correctly rounded answer, there arises the question of what is causing the inaccuracies. Numerical linear algebra answers this question by distinguishing between good and bad problems on the one hand and good and bad algorithms on the other.

Loosely speaking, a bad problem is one that is sensitive to small perturbations in the data. The problem of measuring the sensitivity, or, as it is usually called, the *ill-conditioning* of a specific problem, is discussed in the next section. What is important here is that it is unreasonable to expect any algorithm to perform well on an ill-conditioned problem. In the first place the data are seldom known exactly, and even when they are they must be rounded to be expressed in the finite arithmetic of the computer in question. Thus the algorithm never sees the original problem—only a perturbed version whose answer may differ greatly from that of the original.

If an algorithm cannot be expected to solve an ill-conditioned problem accurately, it should at least introduce no large errors of its own. This is the key to the notion of stability in numerical linear algebra. An algorithm is said to be stable if it computes the exact solution of a perturbed problem, where the perturbation is of the same order as perturbation made by rounding the data. The aptness of the definition is seen from the following statements. If a stable algorithm is applied to a well-conditioned problem, it must compute an accurate solution, since the backward perturbation accounting for the effects of rounding error cannot alter the answer by much. On the other hand, if a stable algorithm is applied to an ill-conditioned problem and the backward perturbation is smaller than the errors in the original data, then it is the data, not the algorithm, that is to blame for inaccuracies in the answers. In this case computing in higher precision will not help unless more accurate data are furnished.

The backward error analysis of the Golub–Householder method for computing the QR factorization of an $n \times p$ matrix X is typical. It can be shown that the computed Q and R satisfy exactly

$$Q^T(X+E) = \begin{bmatrix} R \\ 0 \end{bmatrix}, \qquad (11a)$$

where

$$\|E\|_F \leqslant \phi(n, p)\|A\|_F \epsilon_M . \qquad (11b)$$

Here $\|E\|_F^2 = \text{trace}(E^TE)$, $\phi(n, p)$ is a low-degree polynomial in n and p, and ϵ_M is the rounding unit—i.e., $\epsilon_M = 2^{-t}$ if t-digit binary floating-point arithmetic is used in the calculation. Similar results hold for all the decompositions described above.

The notion of numerical stability is useful in explaining a number of phenomena. For purposes of illustration two will be briefly discussed: the undesirability of forming matrix inverses and the drawbacks of the normal equations*.

A common procedure for solving the linear system (1) is to compute an "inverse" B of A and set $\bar{x} = Bb$. This rather natural procedure is unstable; specifically, it can happen that $\bar{x}$ satisfies $(A + E)\bar{x} = b$ for *no* small E. This has the important implication that the residual $b - A\bar{x}$ can become very large; that is, the computed solution need not satisfy the original equations, even approximately. For this reason, it is preferable to generate a solution directly from an LU or Cholesky decomposition rather than from an explicit inverse. The same comments hold for generalized inverses.

Concerning the normal equations, from (5) it is seen that the triangular part R of the QR factorization of a matrix X carries the same information as the cross-product matrix X^TX. The practical difference is that R has a backward error analysis, provided that it is computed properly, whereas X^TX does not. In fact, the correctly rounded X^TX can be indefinite and therefore not equal to $(X + E)^T(X + E)$ for any E. Since X^TX is the matrix of the normal equations, the foregoing gives strong reason for preferring computational methods based on the QR decomposition in regression calculations.

It is important to maintain perspective on the comments above. In the section on updating, we saw the catastrophic effects of forming the explicit inverse of a nearly singular matrix, and this, coupled with the known instability of matrix inversion, has led numerical analysts to exercise considerable algorithmic ingenuity in avoiding the use of inverses. This is largely justified, but a dogma should not be made of it. In statistical applications, there are many matrices whose inverses can be written down analytically, and if these matrices are well conditioned, there is no reason why these formulas should not be used. Moreover, in some applications there is information to be gained from inspecting the elements of an inverse matrix. In this case, one should not hesitate to compute and output the inverse.

Much the same may be said of the assertion that the normal equations should be avoided. It is true that at a fixed level of precision methods based on the QR decomposition will solve a wider range of problems than methods based on the normal equations, and other things being equal, the former should be preferred to the latter. However, there are problems—unbalanced analysis of variance* is an example—where convenience or efficiency may shift the balance to the normal equations. Unfortunately, there is no good survey in the literature aimed at the statistician who must make such a choice.

PERTURBATION THEORY

As powerful as backward rounding-error analysis is in confirming the essential goodness of an algorithm, it cannot guarantee the accuracy of the computed solution or even distinguish between accurate and inaccurate solutions. The necessary supplement is perturbation theory, which asks the following questions. Given a function $f(A)$ of a matrix A and a perturbation $A + E$ of A, how does the difference $f(A + E) - f(A)$ behave as a function of E? When E comes from a backward error analysis of an algorithm for computing f, the analysis gives some idea of how accurate the computed solution is.

An example will make clear the use of perturbation theory in numerical linear algebra. Consider the perturbation

$$(A + E)\tilde{x} = b \tag{12}$$

of the problem (1). It can be shown that

$$(A + E)^{-1} = A^{-1} - A^{-1}EA^{-1} + O(\|E\|^2),$$

where $\|\cdot\|$ is any matrix norm. Hence, since $x = A^{-1}b$,

$$\tilde{x} \simeq x - A^{-1}Ex \tag{13}$$

or

$$\frac{\|\tilde{x} - x\|}{\|x\|} \lesssim \|A^{-1}E\| \lesssim \kappa(A)\frac{\|E\|}{\|A\|}, \tag{14}$$

where

$$\kappa(A) \equiv \|A\|\,\|A^{-1}\| \geqslant 1.$$

The interpretation of a bound like (14) goes as follows. The right-hand side can be interpreted as a relative error in $\tilde{x}$. For example, if the Euclidean norm is used, the components of x are roughly equal, and $\|\tilde{x} - x\|/\|x\| \simeq 10^{-k}$, then x and $\tilde{x}$ will agree to roughly k decimal digits. Similarly, the factor $\|E\|/\|A\|$ represents a relative error introduced by the perturbation of A. Thus $\kappa(A)$ is a constant associated with the problem (1) that indicates how relative errors in A will be magnified in the solution. It is called the *condition number of A* (with respect to inversion).

The analysis of the linear system (1) is typical of the general case. First an error expression, such as (13), is developed (for simplicity we did not treat terms of order $\|E\|^2$ or higher, but they are included in a complete analysis). Then norms are used to isolate the features that make a problem well- or ill-conditioned. It requires some care, both in the derivation of the error expression and its dissection by norms, to ensure that the final bound tells a true story. When it does, it produces condition numbers, which can be monitored to tell when a problem is ill-behaved.

The advantages of being able to summarize in a single number the sensitivity of a problem to perturbations are obvious. As an example, suppose that pivoted Gaussian elimination in t-digit decimal arithmetic is used to solve (1). Then the backward error analysis shows that the computed solution $\tilde{x}$ satisfies (12) where $\|E\|/\|A\| = O(10^{-t})$. If $\kappa(A) = 10^l$, then the right-hand side of (14) is $O(10^{l-t})$, which implies that x and $\tilde{x}$ agree to about $t - l$ significant features, *provided that the components of x are roughly equal.* This justifies a widely quoted rule of thumb: If $\kappa(A) = 10^l$, then expect a loss of about l significant figures in solving a linear system involving A. We have already seen an example of this rule of thumb in the section on updating. The matrix there has a condition number of about 10^7, and a loss of seven digits occurred in the computations.

It is important to remember that reducing a complex perturbation problem to a single number necessarily involves a loss of information. In particular, a bound of the form (14) is useful only when the quantities involved are properly scaled. A reasonable problem that is poorly scaled may have an unreasonable condition number, a situation that is sometimes called artificial ill-conditioning. It may even happen that no appropriate scaling exists; for example, the proviso italicized above may be at odds with some natural scaling of A. In this case one must seek a more delicate analysis, presumably beginning with (13); *see also* MATRIX, ILL-CONDITIONED.

SOFTWARE QUALITY

The standards by which the quality of a computer program is judged have risen greatly over the past decade. The design, implementation, and testing of mathematical software is a subject in itself worthy of an entire article. Here we can only sketch some of the goals and methods (*see also* STATISTICAL SOFTWARE).

A piece of mathematical software should be robust, transportable, and legible. A robust program is one that performs properly for all legitimate input. This is not to say that a robust program solves all problems; that can hardly be expected of any program. However, when a robust program fails, it tells the user so. For example, a program to compute a Cholesky decomposition should, on encountering a negative diagonal element, return with an error flag, instead of causing a program abort by attempting to take the square root of a negative number.

Ensuring robustness is primarily a design problem, and a failure in robustness can usually be traced to unconscious, false assumptions about the computation. An example, which is particularly relevant to statistical computations, occurs when one tries to calculate sums of squares, say

$$\nu = (x^2 + y^2)^{1/2}. \qquad (15)$$

If 10^{99} is the largest number representable in the computer in question and either x or y is greater than 10^{50}, then the direct use of (15) will result in an overflow abort, even though the final answer may be perfectly representable. Thus the assumption that a computer has an infinite range results in programs that will fail on reasonable problems. The cure in this case is simple: Compute $\nu = \mu[(x/\mu)^2 + (y/\mu)^2]^{1/2}$, where $\mu = |x| + |y|$. This will work on any system that sets underflows to zero, provided x and y are not both zero (another potential failure in robustness).

From the point of view of the user, the ideal program is *portable*; that is, it can be taken from one computer to another without modification. A surprising number of algorithms from numerical linear algebra can be made effectively portable. However, programs of greater complexity, especially those involving 1/0 and string manipulation, will generally require changes if they are to work on a new system. A program is *transportable* if these changes are minimal and routine.

The two chief problems in moving programs are the differences in computer architectures and the vagaries of programming languages. The cure for the first problem is to isolate and label the machine-dependent features of a program so that they may be changed easily. For example, machine-dependent constants should be specified in a fixed place at the beginning of the program. The answer to the second problem is to work well within the standard of the programming language being used. The successful writer of transportable code is as linguistically conservative as a prescriptive grammarian of the old school; he or she shuns innovation until extensive usage has made it traditional.

Legibility has two interrelated aspects: the readability of the code and the existence of good documentation. Anyone who has had to modify someone else's program is aware of the need for clean code. It is achieved by tricks of the trade that are described in any current introduction to programming. They include uniform layout of the programs, suggestive nomenclature, suitable comments, careful choice of data structures, and, especially, the use of a small number of control structures to prevent the program from becoming a can of worms.

Documentation must serve both the technician, who will have to maintain and perhaps modify the program, and the naive user. Technical documentation can often be achieved by running comments in the code itself; however, matrix algorithms do not lend themselves to this kind of treatment. What is required is an external description of the algorithm in standard mathematical prose and a discussion of the implementation details. The program-naming conventions should be consistent with the description, and well-chosen comments should indicate the main subdivisions of the algorithm.

User documentaion will depend on the sophistication of the audience, although experience suggests that it is hard to be too simple-minded. It is highly desirable to include a compressed but complete description of the usage in the program comments, since programs have a way of becoming separated from their external documentation.

It is sometimes objected that the exacting standards of quality software slow down program development. On the contrary, they encourage precisely the techniques that make programs easy to write and debug. This observation is particularly important now, since the availability of cheap computers will turn many people into de facto software engineers. Before the hard-won advances in software engineering are dismissed as time consuming and irrelevant, such critics would do well to recall who won the race between the tortoise and the hare.

BIBLIOGRAPHICAL NOTES

J. H. Wilkinson is correctly regarded as the founder of modern numerical linear algebra, and his two books [58, 59]—one slim, one thick, and both hard-going—are required for anyone who would claim mastery of the field. Other important advanced treatises are by Householder [32], Faddeev and Faddeeva [18], and Parlett [40]. The book by Lawson and Hanson [35], on solving least-squares problems will be of particular interest to regression analysts.

Treatments of elementary numerical linear algebra may be found in most introductory numerical analysis texts. The books of Fox [21] and Stewart [51] are devoted exclusively to numerical linear algebra. Forsythe and Moler [20] give a fine introduction to the subject. A software-oriented treatment is given in ref. 44.

The diffusion of the techniques of modern numerical linear algebra into the statistical community has been slow. This is in part due to incompatibilities of notation, and partly because numerical analysts, with some notable exceptions, have not been aware of what statisticians need to compute. The most comprehensive treatment of numerical linear algebra in statistics is given by Kennedy and Gentle [33]. Seber [48] gives an excellent summary of methods for solving regression problems, and Chambers [9] a useful survey.

The importance of numerical linear algebra in nonlinear problems may be appreciated from a perusal of ref. 37.

The most important source of sparse matrices in statistics are design matrices from unbalanced analysis of variance [47, 50] and the analysis of categorical data* [29]. For more on sparse matrices the reader is referred to the proceedings of various conferences on the subject [8, 16, 43, 45]. Of particular interest are the recent developments in the solution of large sparse least-squares problems; e.g., see refs. 6 and 24.

Decompositions

Although matrix decompositions have long been used in numerical analysis, the mature decompositional point of view emerged only in the mid-1960s with the publication of the books by Householder [32] and Wilkinson

[59]. In particular, Householder used matrix decompositions to show that seemingly different algorithms were actually minor variants of the same algorithm.

Decompositions are less frequently used in statistics. The spectral decomposition occurs naturally in principal component analysis* and has been used in the analysis of ridge regression*. The approximation $\overline{X}$ is but one of several related to the singular value decomposition that are important in statistical application. Rao [42] gives a good survey. A systematic exposition of linear regression theory is possible in terms of the *QR* decomposition. This has yet to be written, but pieces may be found in the book by Weisberg [56] and especially in a paper by Golub and Stayan [26].

FORTRAN subroutines for computing the decompositions described above appear in EISPACK [23, 49] and LINPACK [15], two packages of high-quality mathematical software. Algol programs are contained in a series edited by Wilkinson and Reinsch [62]. Rice [44] surveys some of the major program libraries. Incidentally, these references, together with the above-mentioned books by Householder and Wilkinson, are good places to start in tracing the history of a particular decomposition.

Not all possible decompositions have been listed here; some are too specialized and others are unstable. Rather complicated, but stable decompositions exist for symmetric indefinite matrices [1, 7]. The most important of the unstable decompositions are spectral (eigenvalue–eigenvector) decomposition of a general matrix, the Jordan canonical form [34], the generalized spectral decomposition of a definite generalized eigenvalue problem [40], and the generalized singular value decomposition of van Loan [55] (however, the variant proposed by Paige and Saunders [39] may be stable). This is not to say that these decompositions should never be used. On the contrary, one of these may be the decomposition of choice for a particular problem. But it should not be used without an analysis to show that its potential instability will not affect the results adversely.

The decompositions described here represent the end of the line, in the sense that they produce the simplest form attainable under a fixed class of transformations. There is currently considerable interest is using intermediate decomposition, such as the Hessenberg, tridiagonal, and bidiagonal forms mentioned above. These often can be used in place of, respectively, the Schur, spectral, and singular value decompositions, at a considerable saving in cost.

Although the decompositional approach has been stressed in this article, there are some dense problems for which it is inappropriate. For example, an estimate of the largest eigenvalue of a positive definite matrix may be obtained cheaply by means of the power method [36].

Updating

There has been a recent interest in updating formulas for means and standard deviations (ref. 10 has recent references with good bibliographies). This literature is apparently not read by the manufacturers of hand-held calculators, whose $\sum +$ buttons represent a low point in the art of numerical computation. A natural generalization of these updating formulas is the algorithm for updating a covariance matrix [57].

Some statements by Plackett [41] suggest that the formula (8) may have been known to Gauss.* The history of this formula, its extensions, and its variation have been surveyed in refs. 30 and 38. It is implicit in the simplex method for linear programming* [14], where the limitations of updating inverses first became a serious problem. The crux of the matter is illustrated by the example following (8). For some years the standard solution for this problem was to reinvert the matrix periodically. In 1969, Bartels and Golub [2] suggested the alternative of updating the *LU* decomposition, a suggestion that sparked the systematic investigation of updating algorithms for matrix decompositions.

Algorithms for performing the updates (9) of the Cholesky decomposition are given in LINPACK [15]. A detailed analysis of the algorithm for removing observations ("downdating," as it is sometimes called) has been given in ref. 54. Cholesky updating algorithms that are suitable for large, sparse matrices are described in ref. 25. Algorithms for updating a QR factorization are found in refs. 13 and 28.

The formulas for the SWEEP operator may be found in Fisher [19], although Beaton [3], who coined the term SWEEP, appears to be responsible for its current popularity. Goodnight [27] gives an able exposition of its properties and application. The SWEEP operator is at its worst in applications such as all subsets regression [22], where the attempt to compute a residual sum of squares for a nearly collinear set of variables will force an expensive reinversion. Berk [5] shows how collinearities may be detected before a disastrous SWEEP is performed.

Backward Rounding-Error Analysis

A readable history of modern rounding-error analysis has been given by Wilkinson [61], who is largely responsible for the notions of ill-conditioning and stability. Simple examples of backward rounding-error analysis may be found in almost any numerical analysis text.

The result (11) concerning the stability of the QR decomposition is established in ref. 60. It is interesting to note that the specific bound is rather seldom used in practice, since it is usually pessimistic. In general, rounding-error analyses are less valued for their final bounds than for the insight they provide about a numerical algorithm.

Perturbation Theory

Although the condition number $\kappa(A)$ appears in one of the earliest analyses of Gaussian elimination, the idea of perturbation theory as an independent supplement to rounding-error analysis had to await the development of backward rounding-error analysis. Since then the technique has been applied to many different problems, and the literature has grown so voluminous that it cannot be surveyed here. Of greatest interest to the statistician will be the results on least squares (ref. 52 is a good survey) and the symmetric eigenvalue problem [40].

Since the definition of $\kappa(A)$ involves A^{-1}, it would appear that a matrix inversion is required to compute it. However, it turns out that $\kappa(A)$ can be reliably estimated from an LU decomposition of A [11], and the LINPACK codes [15] contain such a condition estimater.

The role of row and column scaling in matrix computations is still imperfectly understood. This is because it can affect both the computational properties of the algorithm in question and the interpretation of the perturbation analysis. Stewart [53] gives a discussion of scaling in Gaussian elimination. A more general discussion is given in the introduction to LINPACK [15].

Since perturbation theory applies to errors from any source, its use extends beyond the analysis of the effects of rounding error. For example, the perturbation theory for the least-squares problem can be used to generate a set of sensitivity coefficients that are useful in assessing the effects of errors in the variables on regression coefficients. (See Weisberg [56, p. 68].) For other closely related regression diagnostics*, see ref. 4.

It does not seem to be widely appreciated that perturbation expansions such as (13) can be used to approximate the distribution of the result when E is assumed to be random and small. For example, in (13) let $E = \sigma F$, where f_{ij} are independent random variables distributed $N(0, 1)$. Let $\dot{x} = x - A^{-1}Ex$. Then it is easy to show that as $\sigma \to 0$, the random variable $\sigma^{-1}(\tilde{x} - \dot{x})$ approach zero surely. This implies that as σ becomes small $\tilde{x}$ has approximately the distribution of $\dot{x}$, which is $N[x, \sigma^2\|x\|^2(A^TA)^{-1}]$. This approach has been used by Hodges and Moore [31] in an analysis of a stochastic model of errors in regression variables.

Software Quality

High-quality mathematical software has only begun to emerge as a separate discipline. Consequently, most of its literature is contained in conference proceedings or even technical reports. A recent bibliography is contained in ref. 17 and a somewhat older survey in ref. 12. Further references will be found in the pages of the Association for Computing Machinery's *Transactions on Mathematical Software*.

NOTES

1. There is a variant of the algorithm that avoids square roots. However, prejudice against square roots is a relic of the days of hand computation and has no place in the world of high-speed computation, where the time to take a square root may be comparable to the division time and is much less than the time to take a logarithm.

2. Throughout this article the term "rounding error" will refer to errors introduced by computer arithmetic, not errors made in recording data with a fixed number of significant figures. Although the two are closely related, the latter will usually be much larger than the former.

References

[1] Aasen, J. O. (1971). *BIT (Nord. Tidskr. Inf.—Behandl)* **11**, 233–242.

[2] Bartels, R. H. and Golub, G. H. (1969). *Commun. AMC*, **12**, 266–268.

[3] Beaton, A. E. (1964). The Use of Special Matrix Operators in Statistical Calculus. *Bull. No. 64–51*, Educational Testing Service Research, Princeton, N.J.

[4] Belsley, D. A., Kuh, E., and Welsch, R. E. (1980). *Regression Diagnostics: Identifying Influential Data and Sources of Collinearity*. Wiley, New York.

[5] Berk, K. N. (1977). *J. Amer. Statist. Ass.*, **72**, 863–866.

[6] Björk, A. and Duff, I. S. (1980). *Linear Algebra Appl.*, **34**, 43–68.

[7] Bunch, J. R. and Kaufman, L. (1977). *Math. Comp.*, **31**, 163–179.

[8] Bunch, J. R. and Rose, D. J., eds. (1976). *Sparse Matrix Computations*. Academic Press, New York.

[9] Chambers, J. M. (1977). *Computational Methods for Data Analysts*. Woley, New York.

[10] Chan, T. F. and Lewis, J. G. (1979). *Commun. ACM*, **22**, 526–531.

[11] Cline, A. K., Moler, C. B., Stewart, G. W., and Wilkinson, J. H. (1979). *SIAM J. Numer. Anal.*, **16**, 368–375.

[12] Cody, W. J. (1974). *SIAM Rev.*, **16**, 36–46.

[13] Daniel, J., Gragg, W. B., Kaufman, L., and Stewart, G. W. (1976). *Math. Comp.*, **30**, 772–795.

[14] Dantzig, G. B. (1963). *Linear Programming and Extensions*. Princeton University Press, Princeton, N.J.

[15] Dongarra, J. J., Bunch, J. R., Moler, C. B., and Stewart, G. W. (1979). *LINPACK Users' Guide*. SIAM, Philadelphia.

[16] Duff, I. S. and Stewart, G. W., eds. (1979). *Sparse Matrix Proceedings 1978*. SIAM, Philadelphia.

[17] Einarsson, B. (1979). *J. Comp. Appl. Math.*, **5**, 145–159.

[18] Faddeev, D. K. and Faddeeva, V. N. (1963). *Computational Methods of Linear Algebra*. W. H. Freeman, San Francisco.

[19] Fisher, R. A. (1958). *Statistical Methods for Research Workers*, 13th ed. Hafner, New York.

[20] Forsythe, G. and Moler, C. B. (1967). *Computer Solution of Linear Algebraic Systems*. Prentice-Hall, Englewood Cliffs, N.J.

[21] Fox, L. (1965). *An Introduction to Numerical Linear Algebra*. Oxford University Press, New York.

[22] Furnival, G. and Wilson, R. (1974). *Technometrics*, **16**, 499–511.

[23] Garbow, B. S., Boyle, J. M., Dongarra, J. J., and Moler, C. B. (1977). *Matrix Eigensystem Routines—EISPACK Guide Extension*. Lect. Notes Computer Sci., Springer-Verlag, New York.

[24] George, A. and Heath, M. T. (1980). *Linear Algebra Appl.*, **34**, 69–84.

[25] Gill, P. E., Golub, G. H., Murray, W., and Saunders, M. A. (1974). *Math. Comp.*, **28**, 505–535.

[26] Golub, G. H. and Styan, G. P. (1973). *J. Statist. Comp. Simul.*, **2**, 253–274.

[27] Goodnight, J. H. (1979). *Amer. Statist.*, **33**, 149–158.

[28] Gragg, W. B., Leveque, R. J., and Trangenstein, J. A., (1979). *J. Amer. Statist. Ass.*, **74**, 161–168.

[29] Haberman, S. J. (1974). *The Analysis of Frequency Data*. University of Chicago Press, Chicago.

[30] Henderson, H. V. and Searle, S. R. (1981). *SIAM Rev.*, **23**, 53–60.

[31] Hodges, S. D. and Moore, P. G. (1972). *Appl. Statist.*, **21**, 185–195.

[32] Householder, A. S. (1964). *The Theory of Matrices in Numerical Analysis*. Ginn/Blaisdell, Waltham, Mass.

[33] Kennedy, W. J. and Gentle, J. E. (1980). *Statistical Computing*. Marcel Dekker, New York.

[34] Kugstrom, B. and Ruhe, A. (1980). *Trans. Math. Software*, **6**, 398–419.

[35] Lawson, C. L. and Hanson, R. J. (1974). *Solving Least Squares Problems*. Prentice-Hall, Englewood Cliffs, N.J.

[36] O'Leary, D. P., Stewart, G. W., and Vandergraft, J. S. (1979). *Math. Comp.*, **33**, 1289–1292.

[37] Ortega, J. M. and Rheinboldt, W. C. (1970). *Iterative Solution of Nonlinear Equations in Several Variables*. Academic Press, New York.

[38] Ouelette, D. V. (1981). *Linear Algebra Appl.*, **36**, 187–295.

[39] Paige, C. C. and Saunders, M. A. (1981). *SIAM J. Numer. Anal.* (in press).

[40] Parlett, B. N. (1980). *The Symmetric Eigenvalue Problem*. Prenctice-Hall, Englewood Cliffs, N.J.

[41] Plackett, R. L. (1950). *Biometrika*, **37**, 149–157.

[42] Rao, C. R. (1980). In *Multivariate Analysis V*, P. R. Krishnaiah, ed. North-Holland, Amsterdam, 3, p. 22.

[43] Reid, J. K., ed. (1971). *Large Sparse Sets of Linear Equations*. Academic Press, New York.

[44] Rice, J. R. (1981). *Matrix Computations and Mathematical Software*. McGraw-Hill, New York.

[45] Rose, D. J. and Willoughby, R. A., eds. (1972). *Sparse Matrices and Their Applications*. Plenum Press, New York.

[46] Scheffé, H. (1959). *The Analysis of Variance*. Wiley, New York.

[47] Searle, S. R., Speed, F. M., and Henderson, H. V. (1981). *Amer. Statist.*, **35**, 16–33.

[48] Seber, G. A. F. (1977). *Linear Regression Analysis*. Wiley, New York.

[49] Smith, B. T., Boyle, J. M., Dongarra, J. J., Garbow, B. S., Ikebe, Y., Klema, V. C., and Moler, C. B. (1976). *Matrix Eigensystem Routines—EISPACK Guide*, 2nd ed. Lect. Notes Computer Sci., Springer-Verlag, New York.

[50] Speed, F. M., Hocking, R. R., and Hackney, O. P. (1978). *J. Amer. Statist. Ass.*, **73**, 105–112.

[51] Stewart, G. W. (1974). *Introduction to Matrix Computations*. Academic Press, New York.

[52] Stewart, G. W. (1977). *SIAM Rev.*, **19**, 634–662.

[53] Stewart, G. W. (1977). In *Mathematical Software III*, J. Rice, ed. Academic Press, New York, pp. 1–14.

[54] Stewart, G. W. (1979). *J. Inst. Math. Appl.*, **23**, 203–213.

[55] VanLoan, G. F. (1976). *SIAM J. Numer. Anal.*, **13**, 76–83.

[56] Weisberg, S. (1980). *Applied Linear Regression*. Wiley, New York.

[57] Welford, B. P. (1962). *Technometrics*, **4**, 419–420.

[58] Wilkinson, J. H. (1963). *Rounding Errors in Algebraic Processes*. Prentice-Hall, Englewood Cliffs, N.J.

[59] Wilkinson, J. H. (1965). *The Algebraic Eigenvalue Problem*. Oxford University Press, London.

[60] Wilkinson, J. H. (1965). In *Error in Digital Computation*, Vol. 2, L. B. Rall, ed. Wiley, New York, pp. 77–101.

[61] Wilkinson, J. H. (1971). *SIAM Rev.*, **13**, 548–568.

[62] Wilkinson, J. H. and Reinsch, C., eds. (1971). *Handbook for Automatic Computation*, Vol. 2: *Linear Algebra*. Springer-Verlag, New York.

(ALGORITHMS, STATISTICAL
COMPUTERS IN STATISTICS
GAUSS–JORDAN ELIMINATION
GAUSS–SEIDEL ITERATION
GENERALIZED INVERSES
MATRIX, ILL-CONDITIONED
MULTICOLLINEARITY
NUMERICAL ANALYSIS
STATISTICAL SOFTWARE)

G. W. STEWART

LINEAR–CIRCULAR CORRELATION

Suppose that the random variable X is interval scaled, whereas the variate Θ is a circular variable ($0 \leqslant \theta \leqslant 2\pi$). For example, X may refer to temperature and Θ to wind direction. To develop a measure of correlation* between these variates, we write the regression* equation for X given θ as

$$\begin{aligned} X &= \beta_0 + \beta\cos(\theta + \alpha) + \epsilon \\ &= \beta_0 + \beta_1\cos\theta + \beta_2\sin\theta + \epsilon, \end{aligned}$$

where ϵ denotes an independent error term with variance σ^2; the variance of X is σ_X^2, $\beta_1 = \beta\cos\alpha$, and $\beta_2 = -\beta\sin\alpha$.

The squared multiple correlation* coefficient for X, given $\cos\theta$ and $\sin\theta$, is

$$\begin{aligned} \rho_{X\Theta}^2 &= 1 - \left(\sigma^2/\sigma_X^2\right) \\ &= \left(\rho_{12}^2 + \rho_{13}^2 - 2\rho_{12}\rho_{13}\rho_{23}\right)/\left(1 - \rho_{23}^2\right), \end{aligned}$$

where $\rho_{12} = \text{corr}(X, \cos\Theta)$, $\rho_{13} = \text{corr}(X, \sin\Theta)$, and $\rho_{23} = \text{corr}(\cos\Theta, \sin\Theta)$. This coefficient was introduced in [2], albeit in a different form. The sample quantity $R_{x\theta}^2$ is obtained on replacing the ρ_{ij} by the sample coefficients r_{ij}.

Given a sample of n pairs of observations and assuming that the errors are normally distributed, it follows that $R_{x\theta}^2$ has an F-

distribution* with $(2, n-3)$ degrees of freedom under H_0: $\rho^2_{X\Theta} = 0$. When $\rho^2_{X\Theta} > 0$, the distribution usually is intractable, but progress on special cases has been made in [1] and [3]. Explicit results are available where the θ-values are equally spaced around the circle.

References

[1] Liddell, I. G. and Ord, J. K. (1978). *Biometrika*, **65**, 448–450.

[2] Mardia, K. V. (1976). *Biometrika*, **63**, 403–405.

[3] Mardia, K. V. and Sutton, T. W. (1978). *J. R. Statist. Soc. B*, **40**, 229–233.

(CORRELATION
DIRECTIONAL DATA ANALYSIS
DIRECTIONAL DISTRIBUTIONS
MULTIPLE CORRELATION)

J. K. ORD

LINEAR CONTAMINATED INDEPENDENCE *See* DEPENDENCE, CONCEPTS OF

LINEAR ESTIMATORS, BAYES

A Bayes linear estimator is one that has minimum average total mean squared error among estimators which are linear in the response variable **Y**. The average is a weighted average over points in the set of parameter values. The term "best linear estimator" is used with the same meaning [2, 4].

When estimating a vector parameter $\boldsymbol{\tau}$ by an estimator $\hat{\boldsymbol{\tau}}$, the total squared error is $(\hat{\boldsymbol{\tau}} - \boldsymbol{\tau})'(\hat{\boldsymbol{\tau}} - \boldsymbol{\tau})$. The risk function is then the expected value of total squared error over the distribution of the response variable; this risk function is called total mean squared error.*

Bayes linear estimators are defined more rigorously below. Such an estimator is Bayesian in the sense that, among linear estimators, it minimizes the expected value, over some prior distribution on the parameter values, of risk when the loss function is taken to be total squared error. However, that only linear estimators are considered means that a Bayes *linear* estimator may not be a formal Bayesian estimator because other nonlinear estimators may achieve lesser expected risk.

As a simple example, consider estimating the mean μ using n independent observations from a population with mean μ and variance σ^2. Let **Y** denote the vector of observations and consider a linear estimator $\mathbf{l}'\mathbf{Y}$ of μ. Total mean squared error is

$$\text{TMSE}_l(\sigma^2, \mu^2) = \sigma^2 \sum l_i^2 + \mu^2 \sum (l_i - 1/n)^2.$$

Let γ and δ denote expected values of σ^2 and μ^2 over a probability distribution on (σ^2, μ). The expected value of TMSE_l over this same distribution is then

$$\text{ETMSE}_l(\gamma, \delta) = \gamma \sum l_i^2 + \delta \sum (l_i - 1/n)^2.$$

If $\gamma > 0$, then the unique estimator which minimizes this average risk is

$$\mathbf{l}'\mathbf{Y} = \frac{n\delta}{\gamma + n\delta}\,\bar{y},$$

where $\bar{y}$ is the average of the observations. If the parameter set is such that all nonnegative values of (γ, δ) are possible, then all scalar multiples $a\bar{y}$ of $\bar{y}$, with $0 \leqslant a \leqslant 1$, are Bayes linear estimators. But if the parameter set is restricted so that, for some $m > 0$, $\delta \leqslant m\gamma$ for all points in the parameter set, then $\bar{y}$ is a Bayes linear estimator only at $(\gamma, \delta) = (0, 0)$.

Interest in Bayes linear estimators derives mainly from two consideration. First, in linear models in which the set of variance–covariance matrices of **Y** spans two or more dimensions there is, except in special cases [5], no uniformly minimum variance unbiased linear estimator of fixed-effects $\boldsymbol{\beta}$ in the mean vector $\mathbf{X}\boldsymbol{\beta}$. Examples are mixed analysis of variance models*, heteroscedastic regression models, regression models with random coefficients* and linear models for quadratics in multivariate normal random variables [2, 3]. In these models there is no clear-cut optimal linear estimator and the problem becomes one of identifying and comparing admissible linear estimators in order to choose an estimator suitable for a particular application.

The second consideration leading to Bayes linear estimators is the fact that, in a regression model, there are biased linear estimators which may have less mean squared error*, depending upon the parameter values, than the least-squares* estimator. This fact provides the motivation for ridge regression* estimators, fractional rank estimators [1], and a large body of literature investigating these and other biased estimation procedures for linear regression models. Estimators resulting from these procedures are commonly Bayes linear estimators.

The importance of the class of Bayes linear estimators lies in the fact that it contains all admissible linear estimators. To state this fact more rigorously, suppose that $E(\mathbf{Y}) = \boldsymbol{\mu}$ and $\mathrm{var}(\mathbf{Y}) = \boldsymbol{\Sigma}$, with $(\boldsymbol{\mu}, \boldsymbol{\Sigma})$ contained in an appropriate set Ω of parameter points. Different linear models for $\mathbf{Y}$ correspond to different sets Ω: e.g., $\Omega = \{(\mathbf{X}\boldsymbol{\beta}, \sigma^2\mathbf{I}) : \boldsymbol{\beta} \in R^p, \sigma^2 \geqslant 0\}$ describes a standard regression model, while $\Omega = \{(\mathbf{X}\boldsymbol{\beta}, \gamma_1\mathbf{V}_1 + \cdots + \gamma_k\mathbf{V}_k) : \boldsymbol{\beta} \in R^p,$ all $\gamma_i \geqslant 0\}$ represents a mixed analysis of variance model. In estimating specified linear functions $\mathbf{C}'\boldsymbol{\mu}$ of $\boldsymbol{\mu}$ by linear functions $\mathbf{L}'\mathbf{Y}$ of $\mathbf{Y}$, total mean squared error is

$$\begin{aligned}\mathrm{TMSE}_L(\boldsymbol{\Sigma}, \boldsymbol{\mu}\boldsymbol{\mu}') &= E\left[(\mathbf{L}'\mathbf{Y} - \mathbf{C}'\boldsymbol{\mu})'(\mathbf{L}'\mathbf{Y} - \mathbf{C}'\boldsymbol{\mu})\right] \\ &= \mathrm{tr}\left[\mathbf{L}'\boldsymbol{\Sigma}\mathbf{L} + (\mathbf{L} - \mathbf{C})'\boldsymbol{\mu}\boldsymbol{\mu}'(\mathbf{L} - \mathbf{C})\right].\end{aligned}$$

A weighted average of TMSE_L over points in Ω is equivalent to TMSE_L evaluated at the same weighted average $(\mathbf{V}, \mathbf{M})$ of points $(\boldsymbol{\Sigma}, \boldsymbol{\mu}\boldsymbol{\mu}')$. Because a weighted average is a convex combination, weighted averages of $\{(\boldsymbol{\Sigma}, \boldsymbol{\mu}\boldsymbol{\mu}') : (\boldsymbol{\mu}, \boldsymbol{\Sigma}) \in \Omega\}$ are points in the minimal closed convex cone Π containing $\{(\boldsymbol{\Sigma}, \boldsymbol{\mu}\boldsymbol{\mu}') : (\boldsymbol{\mu}, \boldsymbol{\Sigma}) \in \Omega\}$. The linear estimator $\mathbf{L}'\mathbf{Y}$ is a *Bayes linear estimator*, or a *best linear estimator*, at $(\mathbf{V}, \mathbf{M}) \in \Pi$ if no other linear estimator has less TMSE at $(\mathbf{V}, \mathbf{M})$ than $\mathrm{TMSE}_L(\mathbf{V}, \mathbf{M})$. A necessary and sufficient condition that $\mathbf{L}'\mathbf{Y}$ be Bayes at $(\mathbf{V}, \mathbf{M})$ is that $\mathbf{L}$ satisfy $(\mathbf{V} + \mathbf{M})\mathbf{L} = \mathbf{M}\mathbf{C}$.

If $\mathbf{L}'\mathbf{Y}$ is Bayes at $(\mathbf{V}, \mathbf{M}) \in \Pi$ and $\mathbf{V} + \mathbf{M}$ is positive definite, then $\mathbf{L}'\mathbf{Y}$ is the only linear estimator best at $(\mathbf{V}, \mathbf{M})$. The existence of another linear estimator with uniformly (over Ω) less or equal TMSE than $\mathbf{L}'\mathbf{Y}$ then contradicts this uniqueness, so $\mathbf{L}'\mathbf{Y}$ is admissible among linear estimators. On the other hand, some linear estimators which are Bayes at $(\mathbf{V}, \mathbf{M})$ with $\mathbf{V} + \mathbf{M}$ singular may not be admissible: trivially, at $\mathbf{V} = \mathbf{0}$, $\mathbf{M} = \mathbf{0}$, all linear estimators are Bayes.

The fundamental result relating admissible linear estimators and Bayes linear estimators is due to Olsen et al. [2]. Although their result is restricted to unbiased linear estimators, it is readily extended as follows: If $\mathbf{L}'\mathbf{Y}$ is admissible among linear estimators of $\mathbf{C}'\boldsymbol{\mu}$, then there exists a *nonzero* $(\mathbf{V}, \mathbf{M}) \in \Pi$ such that $\mathbf{L}'\mathbf{Y}$ is Bayes at $(\mathbf{V}, \mathbf{M})$.

References

[1] Marquardt, D. W. (1970). *Technometrics*, **12**, 591–612. (Following the ridge regression papers of A. E. Hoerl and R. W. Kennard, proposes other estimators which also have reduced mean squared error. Notes that in some restricted parameter spaces, such estimators may be uniformly better than the least-squares estimators.)

[2] Olsen, A., Seely, J., and Birkes, D. (1976). *Ann. Statist.*, **4**, 878–890. (A landmark paper, showing the connection between bestness and admissibility for unbiased linear estimators. Significant also for applying results on linear estimation to invariant unbiased quadratic estimation of variance components.)

[3] Pukelsheim, F. (1976). *J. Multivariate Anal.*, **6**, 626–629. (Uses linear models in quadratics to investigate variance component estimators.)

[4] Rao, C. R. (1976). *Ann. Statist.*, **4**, 1023–1037. (Characterizes admissible linear estimators in linear models where the dispersion matrix varies in one dimension. The approach is different from Olsen et al. [2], and does not connect bestness and admissibility.)

[5] Zyskind, G. (1967). *Ann. Math. Statist.*, **38**, 1092–1109. (Methods and approaches in this paper presage later work on admissibility among linear estimators.)

Further Reading

See the following works, as well as the references just given, for more information on the topic of Bayes linear estimators.

Bibby, J. and Toutenberg, H. (1979). *Prediction and Improved Estimation in Linear Models*. Wiley, New York. (A different approach, based on "improvement regions." Contains an extensive bibliography on efforts to describe improved linear regression estimators.)

LaMotte, L. R. (1978). *Technometrics*, **20**, 281–290. (A comprehensive treatment of Bayes linear estimators in linear regression models.)

(ADMISSIBILITY
BAYESIAN INFERENCE
ESTIMATION, POINT
JAMES–STEIN ESTIMATORS
REGRESSION ANALYSIS (BAYESIAN APPROACH)
RIDGE REGRESSION
SHRINKAGE ESTIMATORS)

L. R. LAMOTTE

LINEAR EXPONENTIAL FAMILY

A linear exponential family of distributions is defined by probability densities of the form

$$f(x;\omega) = \exp(\omega x)/q(\omega) \qquad (1)$$

with respect to a σ-finite measure μ over a Euclidean space for x.

In conventional terminology,

$$f(x;\omega) = a(x)\exp(\omega x)/q(\omega), \qquad (2)$$

where $q(\omega) = \sum a(x)\exp(\omega x)$ or $\int a(x)\exp(\omega x)$ as the case may be. It is known [4] that the set of parametric points ω such that $\int \exp(\omega x)\,d\mu(x) < \infty$ is an interval, finite or not.

The random variable x of (1) may be an s-vector, say $\mathbf{x} = (x_1, x_2, \ldots, x_s)$. In that case, ω will also be an s-vector, $\boldsymbol{\omega} = (\omega_1, \omega_2, \ldots, \omega_s)$, and $\boldsymbol{\omega}\mathbf{x}$ will be interpreted as a scalar product. In this setup, the family (1) will be called a multivariate linear exponential family.

Sometimes the linear exponential family has been discussed as the exponential-type families or distributions. The discrete analog of the linear exponential family is the power series distribution*. For the general exponential family, *see* EXPONENTIAL FAMILIES.

The linear exponential family provides a perceptive class and form for several distributions that are important in statistical and stochastic literature. Various characteristic properties of this family help develop a unified picture about its varied members. Some of these results are given below. For an overall review, see Wani and Patil [9].

UNIVARIATE LINEAR EXPONENTIAL FAMILY

Property 1 (Patil [5]). Let a probability density function $f(x;\omega)$ have analytic cumulant generating function*. Then f is linear exponential in ω if and only if its cumulants κ_r' satisfy the relation

$$\kappa_{r+1} = \frac{d\kappa_r}{d\omega}, \qquad r = 1, 2, \ldots. \qquad (3)$$

Property 2 (Bildikar and Patil [2]). Let $\kappa_1(\omega)$ and $\kappa_2(\omega)$ denote the first two cumulants* of a linear exponential family. Then the relation of variance to mean ratio $\kappa_2(\omega)/\kappa_1(\omega) = (1 + de^{\omega})^{-1}$, where d is some real number, holds true if and only if the corresponding distribution is either binomial*, Poisson*, or negative binomial* according as d is positive, zero, or negative.

Property 3 (Wani and Patil [9]). Equality of mean and standard deviation characterizes the exponential distribution* within the linear exponential family. This is a continuous analog similar to the characterization of the Poisson distribution in the linear exponential family by the equality of its mean and the variance.

Property 4 (Bolger and Harkness [3]). A linear exponential family is normal if and only if $\kappa_3(\omega) \equiv 0$, under some weak regularity conditions.

MULTIVARIATE LINEAR EXPONENTIAL FAMILY

Property 5 (Patil [6]). Let a multivariate probability density function $f(\mathbf{x}, \boldsymbol{\omega})$ have analytic cumulant generating function. Then f

is multivariate linear exponential in ω if and only if

$$\kappa_{\mathbf{r}+\mathbf{e}_i}(\omega) = \frac{\partial \kappa_{\mathbf{r}}(\omega)}{\partial \omega_i}, \qquad r = 1, 2, \ldots, s$$

where $\mathbf{r} = (r_1, r_2, \ldots, r_s)$ with $r_i \geqslant 0$ and $\sum r_i \geqslant 1$ and $\mathbf{e}_i$ is the ith unit basis vector.

Property 6 (Bildikar and Patil [2]). The component variables of the multivariate linear exponential family are mutually independent if and only if they are pairwise independent. Further, the subvectors of the multivariate linear exponential family are mutually independent if and only if the covariance of an arbitrary member of one subvector with an arbitrary member of another subvector is zero.

Property 7 (Bildikar and Patil [2]). A multivariate linear exponential family is multivariate normal if and only if all the cumulants of order 3 are zero. Also, if and only if, (a) the regression of one of the variables on the rest is linear, and (b) the rest of $(s-1)$ variables are distributed normally in pairs.

Property 8 (Bildikar [1]). A multivariate linear exponential family is multiple Poisson if and only if (a) the univariate marginals are Poisson, and (b) the distribution of the sum of any two random variables is univariate Poisson.

Property 9 (Bildikar [1]). A multivariate linear exponential family is multivariate normal if and only if the variables are distributed normally in pairs; also, if and only if (a) the regression of every variable on any other is linear, and (b) the sum of any two variables is univariate linear.

CONCLUDING REMARKS

Within the context of statistical estimation, Patil and Shorrock [8] have examined the structural properties of the linear exponential families under three headings: (a) mean value function, (b) a characterization of the gamma family, and (c) equality of the first two Bhattacharya bounds*. Of equal interest are the results that arise in size-biased sampling and related form-invariant weighted distributions. *See* Patil and Ord [7].

References

[1] Bildikar, S. (1968). *Sankhyā B*, **31**, 35–42.

[2] Bildikar, S. and Patil, G. P. (1968). *Ann. Math. Statist.*, **39**, 1316–1326.

[3] Bolger, E. M. and Harkness, W. L. (1966). *Pacific J. Math.*, **16**, 5–11.

[4] Lehmann, E. L. (1959). *Testing Statistical Hypotheses*. Wiley, New York.

[5] Patil, G. P. (1963). *Biometrika*, **50**, 205–207.

[6] Patil, G. P. (1965). In *Classical and Contagious Discrete Distributions*, G. P. Patil, ed. Statistical Publishing Society, Calcutta, India.

[7] Patil, G. P. and Ord, J. K. (1976). *Sankhyā B*, **38**, 48–61.

[8] Patil, G. P. and Shorrock, R. (1965). *J. R. Statist. Soc. B*, **27**, 94–99.

[9] Wani, J. K. and Patil, G. P. (1975). In *Statistical Distributions in Scientific Work*, Vol. 3, G. P. Patil, S. Kotz, and J. K. Ord, eds. D. Reidel, Dordrecht, Holland, pp. 423–431.

Bibliography

See the following works, as well as the references just given, for more information on the topic of linear exponential families.

Barndorff-Nielsen, O. (1977). *Information and Exponential Families in Statistical Theory*. Wiley, New York.

Doss, D. C. (1969). *Ann. Math. Statist.*, **40**, 1721–1727.

Patil, G. P. and Seshadri, V. (1964). *J. R. Statist. Soc. B*, **26**, 286–292.

Patil, G. P., Boswell, M. T., and Ratnaparkhi, M. V (1984). *A Modern Dictionary and Classified Bibliograph of Statistical Distributions*, Vol. 2: *Continuous Univaria Models*. International Co-operative Publishing Hous Fairland, Md.

Patil, G. P., Boswell, M. T., Ratnaparkhi, M. V., a Roux, J. J. J. (1984). *A Modern Dictionary and Classifi Bibliography of Statistical Distributions*, Vol. 3: *Mu variate Models*. International Co-operative Publish House, Fairland, Md.

Wani, J. K. (1968). *Proc. Camb. Philos. Soc.*, **64**, 4 483.

(EXPONENTIAL FAMILIES)

G. P. PAT

LINEAR FILTERING *See* PREDICTION, LINEAR

LINEAR HAZARD RATE DISTRIBUTION

Let X be a random variable with probability density function (PDF) $f(t)$ given by

$$f(t) = (c + kt)\exp\left[-\left(ct + \tfrac{1}{2}kt^2\right)\right], \quad t > 0. \quad (1)$$

Then

$$\Pr(X > t) = 1 - F(t) = \exp\left[-\left(ct + \tfrac{1}{2}kt^2\right)\right], \quad t > 0,$$

and the hazard function (failure rate, conditional mortality rate), $h(t)$, defined by

$$h(t) = f(t)/\left[1 - F(t)\right] = c + kt,$$

is linear in t; X is then said to have a *linear hazard rate distribution* (see HAZARD PLOTTING). Kodlin [5] has obtained properties of the distribution, which is useful in studying survival times of certain human populations; it is an increasing hazard rate distribution if and only if $k > 0$ (*see* FAILURE RATE CLASSIFICATION OF DISTRIBUTIONS).

The case $c = 0$, $k > 0$ gives the Rayleigh distribution*, a member of the family of Weibull distributions*; where $c > 0$ and $k = 0$, the exponential distribution* results. The mean μ, variance σ^2, and median (or 50% survival time) $\tilde{m}$ are given [2, (pp. 29–30), 5] by

$$\mu = \exp\left[c^2/(2k)\right]\sqrt{2\pi/k}\left[1 - \Phi\left(c/\sqrt{k}\right)\right],$$

where Φ is the standard normal distribution* function; $\sigma^2 = 2(1 - c\mu)/k - \mu^2$; and $\tilde{m} = [(c/k)^2 + 2(\ln 2)/k]^{1/2} - (c/k)$.

The term *Rayleigh distribution* is also defined to mean the PDF (1); see Gross and Clark [3, Section 4.8], who further define *generalized Rayleigh distributions* in terms of failure times having polynomial hazard rates. Other writers, however, have described the noncentral chi-square distribution* as generalized Rayleigh (see Johnson and Kotz [4, p. 131]).

Kodlin [5] and Bain [1] discuss evaluation of maximum likelihood* (ML) estimates of c, k, μ, etc.; see also Gross and Clark [3]. Kodlin derives a useful property of (1). Suppose that X is observed only from some time $t_1 > 0$ onward, and let $Z = X - t_1$. Then the conditional distribution of Z, given that $Z > 0$ (i.e., that $X > t_1$) has PDF (1) with c replaced by $c + kt_1$.

Bain [1] compares the ML and least squares* estimators of c and k; he provides a detailed study with Monte Carlo* simulation of tests of hypotheses * when nuisance parameters are present, examining properties of tests of the quantities $\beta = k/c^2$ and $\alpha = c/\sqrt{k}$, in which c and $\sqrt{k}$, respectively, become scale parameters.

References

[1] Bain, L. J. (1974). *Technometrics*, **16**, 551–559. (This paper also discusses general and polynomial hazard rate distributions and has further references.)

[2] Bain, L. J. (1978). *Statistical Analysis of Reliability and Life-Testing Models*. Dekker, New York.

[3] Gross, A. and Clark, V. (1975). *Survival Distributions: Reliability Applications in the Biomedical Sciences*. Wiley, New York. [There is an excellent discussion of estimation of parameters in (1), with examples, in Sec. 4.8.]

[4] Johnson, N. L. and Kotz, S. (1970). *Distributions in Statistics: Continuous Univariate Distributions*, Vol. 2. Wiley, New York.

[5] Kodlin, D. (1967). *Biometrics*, **23**, 227–239. [The author develops properties of the distribution (1) and discusses some examples.]

(HAZARD PLOTTING
LIFE TESTING
RAYLEIGH DISTRIBUTION
SURVIVAL ANALYSIS)

CAMPBELL B. READ

LINEARIZATION

Any reduction to a linear form, this term can be applied in various contexts. Sometimes a formula can be made linear in parameter(s) by appropriate reparametrization, or a regression* can be made linear by

appropriate choice of dependent and/or independent variables.

Approximate linearization is often effected by using the initial terms of Taylor expansions. An example of such approximate linearization is replacement of

$$f(\xi_1 + x_1, \xi_2 + x_2, \ldots, \xi_m + x_m)$$

by

$$f(\xi_1, \xi_2, \ldots, \xi_m) + \sum_{j=1}^{m} \frac{\partial f(\xi_1, \ldots, \xi_m)}{\partial \xi_j} \cdot x_j .$$

LINEAR LEAST SQUARES *See* LEAST SQUARES

LINEAR MODELS WITH CROSSED-ERROR STRUCTURE

This article and a companion article, LINEAR MODELS WITH NESTED-ERROR STRUCTURE, deal with linear models that have particular patterns of correlated errors in which generalized least-squares (GLS) analysis is desirable and feasible.

The distinction between crossed-error structure and nested-error structure is basically the same as the distinction between one-way and two-way classifications, between split-plots* and strip plots*, or between one-restrictional and two-restrictional lattice designs*. Crossed-error structure arises when sampling is stratified in two directions or, in the case of designed experiments, when blocking is done by two cross-classified criteria. Nested-error structure arises primarily as a result of subsampling from primary sampling units or, in the case of designed experiments, by incomplete blocking (*see* INCOMPLETE BLOCK DESIGNS) by a single blocking criterion. However, if either rows or columns are considered fixed (or ignored), the crossed-error model reduces to one having the same variance–covariance structure as a nested-error model.

Crossed-error models have been used in econometric problems which combine cross-sectional and time-series data [4, 7, 10, 11], e.g., annual data from a number of geographical regions over a series of years, and it is desired to estimate some demand or production function. This is an example of data structured as one crossed-error group. Another case of crossed-error structure is that of the two-restrictional lattice designs* in which each replicate is an independent crossed-error group. In this case the x's in the model consist of indicator variables for treatments and possibly also "covariates" (*see* ANALYSIS OF COVARIANCE). Strip plots* also have crossed-error structure, but in this case GLS analysis is ordinarily unnecessary because it yields the same estimates of treatment effects as does ordinary least squares (OLS), and the analysis of variance provides variance estimates for treatment comparisons of interest. However, the GLS analysis might be used if data are unbalanced owing to missing observations or covariates in the analysis.

Let us now construct a mathematical linear model for data with one crossed-error group. First consider the random-effects model for a two-way classification with one observation per cell and without interaction, $y_{jk} = \mu + e_{1j} + e_{2k} + e_{3jk}$, where the row, column, and residual effects (e_{1j}, e_{2k}, and e_{3jk}, respectively) are all uncorrelated and have variances σ_r^2, σ_c^2, and σ_e^2, respectively. Now suppose that for each observation, y_{jk}, of the response variable, we have measured or observed (or otherwise known) values of p explanatory variables $x_{jk1}, x_{jk2}, \ldots, x_{jkp}$, and that the interest is in estimating how the explanatory variables influence or relate (linearly) to the response. Variation due to row and column effects is of no particular interest per se, but represents nuisance variation with which we must deal in drawing inference from the data. We can now write the model as $y_{jk} = \sum_{m=1}^{p} x_{jkm}\beta_m + \epsilon_{jk}$, where the β's are the parameters to be estimated and $\epsilon_{jk} = e_{1j} + e_{2k} + e_{3jk}$. (A constant term may be included by defining $x_{jk1} = 1$ for every observation.) Thus the variance of an observed y_{jk} is $\sigma_r^2 + \sigma_c^2 + \sigma_e^2$ and two observations are correlated with covariance σ_r^2 if they occur in the same row or σ_c^2 if they are in the same column, but are uncorrelated otherwise.

Now suppose that there are n crossed-error groups, possibly differing in size, the ith of which can be arrayed as a rectangular $r_i \times c_i$ array with levels of two cross-classified random factors as rows and columns. We write the linear model (*see* GENERAL LINEAR MODEL) in the usual matrix form $\mathbf{y} = \mathbf{X}\boldsymbol{\beta} + \boldsymbol{\epsilon}$, where $\mathbf{y}$ contains all of the measurements of the response variable arrayed as a vector, $E(\mathbf{y}) = \mathbf{X}\boldsymbol{\beta}$ represents the fixed effects, and an element of $\boldsymbol{\epsilon}$ is of the form $\epsilon_{ijk} = e_{1ij} + e_{2ik} + e_{3ijk}$, $i = 1, \ldots, n$; $j = 1, \ldots, r_i$; $k = 1, \ldots, c_i$. The ϵ's within each group have crossed-error structure as we have previously defined, but are uncorrelated if in different groups. Thus the covariance matrix, $\mathbf{V}(\boldsymbol{\epsilon})$, is the block diagonal matrix,

$$\operatorname{diag}\left[\sigma_c^2 \mathbf{I}_{r_i c_i} + \sigma_r^2 \mathbf{R}_i \mathbf{R}_i' + \sigma_c^2 \mathbf{C}_i \mathbf{C}_i'\right], \quad i = 1, \ldots, n, \quad (1)$$

where $\mathbf{R}_i$ and $\mathbf{C}_i$ are design matrices[1], of dimensions $r_i c_i \times r_i$ and $r_i c_i \times c_i$, for rows and columns, respectively, within the ith group. We shall suppose that $\mathbf{X}$ is of full rank. We also suppose that $\boldsymbol{\beta}$ includes parameters for group effects. However, these may be omitted if group effects are orthogonal to the other fixed effects in the model, but are themselves of no particular interest (e.g., in two-restrictional lattices if data are complete and there are no covariates in the analysis).

The GLS estimator is then $\hat{\boldsymbol{\beta}} = (\mathbf{X}'\mathbf{V}^{-1}\mathbf{X})^{-1}\mathbf{X}'\mathbf{V}^{-1}\mathbf{y}$, where $\mathbf{V}$ is equal or proportional to $\mathbf{V}(\boldsymbol{\epsilon})$. If the variance components σ_e^2, σ_r^2, and σ_c^2 are known, $\hat{\boldsymbol{\beta}}$ is the best linear unbiased estimator (BLUE), but if the variance components must be estimated in order to obtain $\mathbf{V}$, then apparently situations may occur where either OLS or the "covariance estimator" is more efficient [10]. Usually, however, the estimated GLS estimator will be preferred and only GLS estimation will be further considered here. If variance components must be estimated, the estimates are simply substituted for the unknown values in the formulas to follow. Fuller and Battese [4] suggest estimating the variance components by the "fitting-of-constants" method (*see* VARIANCE COMPONENTS). They also discuss properties of the resulting estimated GLS estimator of $\boldsymbol{\beta}$.

Let us choose $\mathbf{V}$ such that $\mathbf{V}(\boldsymbol{\epsilon}) = \sigma_e^2 \mathbf{V}$. Then

$$\mathbf{V}^{-1} = \operatorname{diag}\left[\mathbf{I}_{r_i c_i} - \mathbf{A}_i \mathbf{H}_i \mathbf{A}_i'\right], \quad (2)$$

where

$$\mathbf{A}_i = \left[\mathbf{R}_i, \mathbf{C}_i\right]$$

and

$$\mathbf{H}_i = \left[\operatorname{diag}(\theta_r \mathbf{I}_{r_i}, \theta_c \mathbf{I}_{c_i}) + \mathbf{A}_i \mathbf{A}_i'\right]^{-1}$$

$$= \begin{bmatrix} (\theta_r + c_i)\mathbf{I}_{r_i} & \mathbf{J}_{r_i \times c_i} \\ \mathbf{J}_{c_i \times r_i} & (\theta_c + r_i)\mathbf{I}_{c_i} \end{bmatrix}^{-1} \quad (3)$$

$$= \begin{bmatrix} (\theta_r + c_i)^{-1}\mathbf{I}_{r_i} & \mathbf{0} \\ \mathbf{0} & (\theta_c + r_i)^{-1}\mathbf{I}_{c_i} \end{bmatrix} + \psi_i \begin{bmatrix} c_i(\theta_r + c_i)^{-1}\mathbf{J}_{r_i \times r_i} & -\mathbf{J}_{r_i \times c_i} \\ -\mathbf{J}_{c_i \times r_i} & r_i(\theta_c + r_i)^{-1}\mathbf{J}_{c_i \times c_i} \end{bmatrix}, \quad (4)$$

where $\theta_r = \sigma_e^2/\sigma_r^2$, $\theta_c = \sigma_e^2/\sigma_c^2$, $\mathbf{J}_{r_i \times c_i}$ is an $r_i \times c_i$ matrix of 1's, and $\psi_i = [(\theta_r + c_i)(\theta_c + r_i) - r_i c_i]^{-1}$. For the case of $n = 1$, $\mathbf{H}_i$ reduces to Henderson's [5, p. 400] $(\mathbf{Z}'\mathbf{Z} + \mathbf{D}^{-1})^{-1}$.

DIRECT SOLUTION OF THE GLS EQUATIONS

$\mathbf{A}_i \mathbf{H}_i \mathbf{A}_i'$ in (2) can be written as

$$(\theta_r + c_i)^{-1}\mathbf{R}_i \mathbf{R}_i' + (\theta_c + r_i)^{-1}\mathbf{C}_i \mathbf{C}_i' - \phi_i \mathbf{J}_{r_i c_i \times r_i c_i},$$

where $\phi_i = \psi_i[\theta_r(\theta_r + c_i)^{-1} + \theta_c(\theta_c + r_i)^{-1}]$ [11]. Consequently, the weighted sum of squares of the lth X-variable is

$$\sum_i \left[\sum_j \sum_k x_{ijkl}^2 - (\theta_r + c_i)^{-1} \sum_j x_{ij\cdot l}^2 - (\theta_c + r_i)^{-1} \sum_k x_{i\cdot kl}^2 + \phi_i x_{i\cdot\cdot l}^2 \right], \quad (5)$$

where the dot notation indicates sums over the particular subscripts. Elements of $\mathbf{X}'\mathbf{V}^{-1}\mathbf{y}$ or the off-diagonal elements of $\mathbf{X}'\mathbf{V}^{-1}\mathbf{X}$ may be computed by analogy to (5), replacing the squares with appropriate cross products. Having thus obtained the sums of squares and cross products it is a simple task to obtain $\hat{\boldsymbol{\beta}}$ and its covariance matrix $\mathbf{V}(\hat{\boldsymbol{\beta}}) = \sigma_e^2(\mathbf{X}'\mathbf{V}^{-1}\mathbf{X})^{-1}$. Formula (5) with x's replaced by y's will give the weighted total sum of squares, $\mathbf{y}'\mathbf{V}^{-1}\mathbf{y}$. The weighted regression sum of squares is given by $\hat{\boldsymbol{\beta}}'\mathbf{X}'\mathbf{V}^{-1}\mathbf{y}$ or, defining $\hat{\mathbf{y}} = \mathbf{X}\hat{\boldsymbol{\beta}}$, by $\hat{\mathbf{y}}'\mathbf{V}^{-1}\hat{\mathbf{y}}$, which also may be computed by analogy to (5). The residual mean square estimates σ_e^2 and may be used as a denominator for F-statistics for testing linear hypotheses. However, if variance components are estimated from the data by the fitting-of-constants method, the denominator of F, for purposes of obtaining critical values (*see* HYPOTHESIS TESTING) of F, should be assigned only as many degrees of freedom as the original data provided for estimating σ_e^2. For unbalanced data from strip plots, this allocation of degrees of freedom is probably too liberal for tests of main effects of treatment factors applied to rows and columns. More conservative tests may be done by considering each of these F-statistics to have only as many denominator degrees of freedom as the original data provide for estimation of the appropriate error component for the effects to be tested.

SOLUTION BY COMPUTING A CORRECTION TO THE OLS SOLUTION

Let $\mathbf{A}$ be the direct sum, $\bigoplus_{i=1}^n \mathbf{A}_i$. (Note that by this definition $\mathbf{A}$ is a rearrangement (by groups) of the columns of $[\mathbf{R}, \mathbf{C}]$, where $\mathbf{R}$ and $\mathbf{C}$ are design matrices for the row and column effects in the error structure.) Also let $\mathbf{H}^{-1} = \text{diag}(\mathbf{H}_i^{-1})$ and

$$\mathbf{W} = \left[\mathbf{H}^{-1} - \mathbf{A}'\mathbf{X}(\mathbf{X}'\mathbf{X})^{-1}\mathbf{X}'\mathbf{A}\right]^{-1}. \quad (6)$$

$\mathbf{H}_i^{-1}$ is obtained from (3) by dropping the exponent of -1. Then

$$(\mathbf{X}'\mathbf{V}^{-1}\mathbf{X})^{-1} = (\mathbf{X}'\mathbf{X})^{-1} + (\mathbf{X}'\mathbf{X})^{-1}\mathbf{X}'\mathbf{AWA}'\mathbf{X}(\mathbf{X}'\mathbf{X})^{-1}, \quad (7)$$

$$\hat{\boldsymbol{\beta}} = \tilde{\boldsymbol{\beta}} - (\mathbf{X}'\mathbf{X})^{-1}\mathbf{X}'\mathbf{AWA}'(\mathbf{y} - \tilde{\mathbf{y}}), \quad (8)$$

where $\tilde{\boldsymbol{\beta}} = (\mathbf{X}'\mathbf{X})^{-1}\mathbf{X}'\mathbf{y}$ is the OLS estimate and $\tilde{\mathbf{y}} = \mathbf{X}\tilde{\boldsymbol{\beta}}$ is the vector of "predicted" values obtained by OLS. The last term in (8) may be considered to be a correction for row and column effects, with $\mathbf{A}'(\mathbf{y} - \tilde{\mathbf{y}})$ being a vector of appropriately ordered row and column totals of deviations from OLS regression, and $\mathbf{W}$ providing the appropriate weighting to be given to these deviations in computing the correction. Except for the definitions of $\mathbf{A}$ and $\mathbf{W}$, formulae (7) and (8) are the same as (3) and (4) in LINEAR MODELS WITH NESTED-ERROR STRUCTURE. Furthermore, provided that both σ_r^2 and σ_c^2 are positive, $\mathbf{W}^{-1}$ is positive definite. Consequently, a computational algorithm involving Cholesky decompositions, similar to that suggested for the onefold nested-error model, will work here as well (*see* LINEAR ALGEBRA, COMPUTATIONAL).

Interestingly, (6) to (8) also hold if $\mathbf{A}$ and $\mathbf{H}$ are replaced by $\mathbf{A}^*$ and $\mathbf{H}^*$, where $\mathbf{A}^*$ is obtained in the same manner as $\mathbf{A}$ except that each $\mathbf{A}_i$ is augmented with a column of ones (an indicator variable for the group effect) and $\mathbf{H}_i^* = \text{diag}[(\theta_r + c_i)^{-1}\mathbf{I}_{r_i}, (\theta_c + r_i)^{-1}\mathbf{I}_{c_i}, -\phi_i]$. This has the nice property that $\mathbf{H}^*$ is diagonal, but it increases the dimensions of the matrix which must be inverted to obtain $\mathbf{W}$.

SOLUTION BY LINEAR TRANSFORMATION OF THE MODEL

Fuller and Battese [4] suggest a linear transformation of the model in the case where

$n = 1$. With slight modification, it is applicable to cases of $n > 1$ as well. The transformed model is given by $\mathbf{y}^* = \mathbf{X}^*\boldsymbol{\beta} + \boldsymbol{\epsilon}^*$, where

$$y_{ijk}^* = y_{ijk} - \alpha_{1i}\bar{y}_{ij\cdot} - \alpha_{2i}\bar{y}_{i\cdot k} + \alpha_{3i}\bar{y}_{i\cdot\cdot},$$

$$\alpha_{1i} = 1 - \left[1 + c_i\theta_r^{-1}\right]^{-1/2},$$

$$\alpha_{2i} = 1 - \left[1 + r_i\theta_c^{-1}\right]^{-1/2},$$

$$\alpha_{3i} = \alpha_{1i} + \alpha_{2i} - 1 + \left[1 + c_i\theta_r^{-1} + r_i\theta_c^{-1}\right]^{-1/2},$$

and $\mathbf{X}^*$ is obtained by the analogous transformation made on each column of the original $\mathbf{X}$ matrix. The transformed errors ϵ_{ijk}^* are uncorrelated and have variance σ_e^2. Thus, after transformation, $\hat{\boldsymbol{\beta}}$ can be obtained by OLS regression of $\mathbf{y}^*$ on $\mathbf{X}^*$.

MULTIWAY CROSSED-ERROR STRUCTURE AND COMBINATIONS OF NESTED- AND CROSSED-ERROR STRUCTURES

The theorem used by Fuller and Battese [3] to obtain transformations for estimation of linear models with nested- and crossed-error structures can be readily applied to some models with more complicated error structure as well. For example, Battese and Fuller [1] use it to obtain a transformation for estimation in a linear model applied to data from a split-plot design (nested error) cross-classified with a set of years regarded as random. Other potential applications are GLS estimation in lattice designs which involve combinations of lattice square and split-plot principles [2, p. 412; 6, p. 504], and other incomplete block designs with nested rows and columns [9]. Searle and Henderson [8] give a general method for deriving $\mathbf{V}^{-1}$ for error structures which consist of combinations of balanced crossed and/or nested classifications. The GLS estimate may also be obtained by the method described by Henderson [5].

NOTE

1. We define a "design matrix" for a treatment or blocking factor to be a matrix, consisting of 0's and 1's, having as many rows as there are observations and as many columns as there are levels of the treatment or blocking factor. The 1's indicate which observations belong to the various treatments or blocks. Thus here $\mathbf{R}_i$ has as many rows as there are observations in the ith group, namely, r_ic_i, and r_i columns. The (jk)th element of $\mathbf{R}_i$ is 1 if the jth observation is obtained from the kth row and 0 otherwise.

References

[1] Battese, G. E. and Fuller, W. A. (1972). *Biometrics*, **28**, 781–792. (Generalized least-squares estimation of a linear model with a combination of crossed- and nested-error structures.)

[2] Federer, W. T. (1955). *Experimental Design*. Macmillan, New York.

[3] Fuller, W. A. and Battese, G. E. (1973). *J. Amer. Statist. Ass.*, **68**, 636–642. (Gives a basic theorem useful for deriving transformations for estimation of linear models with various error structures. Application to nested-error models.)

[4] Fuller, W. A. and Battese, G. E. (1974). *J. Econometrics*, **2**, 67–78. (Transformation for GLS estimation of linear models with crossed-error structures. Also discusses other aspects of the GLS estimation problem with crossed-error structure.)

[5] Henderson, C. R. (1971). *Econometrica*, **39**, 397–401. (Gives a method, with application to the crossed-error case, of obtaining GLS estimates in a general mixed model.)

[6] Kempthorne, O. (1952). *Design and Analysis of Experiments*. Wiley, New York.

[7] Nerlove, M. (1971). *Econometrica*, **39**, 383–396. (Eigenvalues and eigenvectors of the crossed-error covariance matrix and their relationship to the GLS problem.)

[8] Searle, S. R. and Henderson, H. V. (1979). *J. Amer. Statist. Ass.*, **74**, 465–470. [Methods for deriving eigenvalues, determinant, and inverse of $\mathbf{V}(\boldsymbol{\epsilon})$ in error structures consisting of balanced combinations of crossed and/or nested classifications.]

[9] Singh, M. and Dey, A. (1979). *Biometrika*, **66**, 321–326. (Incomplete block designs with nested rows and columns.)

[10] Swamy, P. A. V. B. and Arora, S. S. (1972). *Econometrica*, **40**, 261–275. (Shows that it is not possible to choose an estimator of $\boldsymbol{\beta}$, in the

crossed-error model, on the basis of asymptotic efficiency. Discusses efficiency of an estimated GLS estimator compared to OLS and "covariance estimators.")

[11] Wallace, T. D. and Hussain, A. (1969). *Econometrica*, **37**, 55–72. [Crossed-error regression problem with one crossed-error group. Discusses relative merits (based primarily on asymptotic efficiency) of GLS estimation as compared to the covariance estimator.]

Acknowledgment

This article is published as Journal Article No. 81-3-104 of the Kentucky Agricultural Experiment Station with the approval of the Director.

(ANALYSIS OF COVARIANCE
ANALYSIS OF VARIANCE
F-TESTS
GENERAL LINEAR MODEL
HYPOTHESIS TESTING
LATTICE DESIGNS
LEAST SQUARES
LINEAR MODELS WITH NESTED-ERROR STRUCTURE
LINEAR REGRESSION
STRIP PLOTS
TWO-WAY CLASSIFICATION
VARIANCE COMPONENTS)

P. L. CORNELIUS

LINEAR MODELS WITH NESTED-ERROR STRUCTURE

Data sets with one-fold nested error structure occur in situations where there has been some form of subsampling from primary sampling units. Data from split-plot*, one-restrictional lattice* and incomplete block designs* and cases of more than one "measurement" taken on each "subject" are examples. The two-way mixed model with one observation per cell may also be considered as a case of nested-error structure (see, e.g., Hussain [7]).

Consider, for a moment, the random effects model (*See* FIXED-, RANDOM-, and MIXED-EFFECTS MODELS) for a one-way classification*, $y_{ij} = \mu + e_{1i} + e_{2ij}$, where μ is the overall mean, e_{1i} the "effect" of the ith group, and e_{2ij} the residual deviation associated with the jth individual in the ith group; $j = 1, \ldots, k_i$; $i = 1, \ldots, a$. We suppose that the e_{1i}'s are random variables with variance σ_1^2, the e_{2ij}'s are random variables with variance σ_2^2, and all e_{1i}'s and e_{2ij}'s are uncorrelated. Then each observation, y_{ij}, has variance $\sigma_1^2 + \sigma_2^2$ and two observations in the same group are correlated with covariance σ_1^2. In the case of an incomplete block design y_{ij} represents the jth plot (or experimental unit) in the ith block, or in the case of measurements within subjects, y_{ij} is the jth measurement on the ith subject.

Now, as we did in the case of linear models with crossed error structure*, let us suppose that, for each observation, we have values of p explanatory variables x_{ij1}, $x_{ij2}, \ldots, x_{ijp}$ which, we postulate, will explain some of the observed variation among the y_{ij}'s. Such explanatory variables may consist of indicator variables for treatment effects, numerical regressor variables, or both. The model now becomes $y_{ij} = \sum_{m=1}^{p} x_{ijm}\beta_m + e_{1i} + e_{2ij}$. Letting $\epsilon_{ij} = e_{1i} + e_{2ij}$, write the linear model in the usual matrix notation, $\mathbf{y} = \mathbf{X}\boldsymbol{\beta} + \boldsymbol{\epsilon}$ (*see* GENERAL LINEAR MODEL). The covariance matrix $\mathbf{V}(\boldsymbol{\epsilon})$ has block diagonal structure with $\sigma_1^2 + \sigma_2^2$ on the diagonal and off-diagonal elements σ_1^2 for two observations in the same group, and zero otherwise. Assume that the model is parametrized such that $\mathbf{X}$ is of full rank.

Some data analysis problems with nested-error structure, e.g., balanced split-plot designs, have nice solutions in which best linear unbiased estimates (BLUEs) of parameters are simply ordinary least squares* (OLS) estimates and appropriate mean squares can be partitioned in the analysis of variance* to provide estimates of error for various parameter estimates and hypotheses to be tested. However, unbalanced analyses in data sets of this type often present rather messy problems (*see* MESSY DATA). In the general case, the BLUE of $\boldsymbol{\beta}$ is given by the generalized least squares (GLS) estimate $\hat{\boldsymbol{\beta}} = (\mathbf{X}'\mathbf{V}^{-1}\mathbf{X})^{-1}\mathbf{X}'\mathbf{V}^{-1}\mathbf{y}$, where $\mathbf{V}$ is equal or proportional to the covariance matrix of the ϵ_i's, $\mathbf{V}(\epsilon)$. Fuller and Battese [3] suggested a transformation of the model, viz.,

$$y_{ij} - \alpha_i \bar{y}_{i.} = \sum_{m=1}^{p} (x_{ijm} - \alpha_i \bar{x}_{i.m})\beta_m + \epsilon_{ij}^*, \quad (1)$$

where

$$\alpha_i = 1 - \left[1 + k_i \theta^{-1}\right]^{-1/2}, \qquad \theta = \sigma_2^2/\sigma_1^2.$$

We denote (1) in matrix notation as

$$\mathbf{y}^* = \mathbf{X}^* \boldsymbol{\beta} + \boldsymbol{\epsilon}^*. \quad (2)$$

The ϵ_{ij}^*'s are uncorrelated and all have variance σ_2^2. Thus $\boldsymbol{\beta}$ can be estimated by OLS using Model (2), i.e., $\boldsymbol{\beta} = (\mathbf{X}^{*\prime}\mathbf{X}^*)^{-1}\mathbf{X}^{*\prime}\mathbf{y}^*$. This is indeed the same as the GLS estimator for the original model before transformation. The covariance matrix of $\hat{\boldsymbol{\beta}}$ is given by $\sigma_2^2(\mathbf{X}^{*\prime}\mathbf{X}^*)^{-1}$, and the residual mean square from the transformed regression analysis has expectation σ_2^2.

In practice, the variance components* σ_1^2 and σ_2^2 are usually unknown and must be estimated from the data. Fuller and Battese suggest they be estimated by the "fitting-of-constants" method (*see* VARIANCE COMPONENTS), and give formulae for the estimates. When estimated variance components are used to compute the transformation, it seems appropriate, in obtaining critical values for F-tests* of hypotheses in the subsequent regression analysis, to assign to the denominator of F only as many degrees of freedom as were originally available for estimating σ_2^2. For unbalanced data from split-plot or repeated-measurements* designs, even this allocation of degrees of freedom may be too liberal for tests concerning the main-plot factor. A more conservative test may be done by considering the F-statistic for the main-plot factor to have only as many denominator degrees of freedom as the original data provide for estimation of main-plot error.

With some algebraic manipulation the GLS solution can be obtained by computing a correction applied to the OLS solution of the original problem. The inverse matrix

$$(\mathbf{X}^{*\prime}\mathbf{X}^*)^{-1} = (\mathbf{X}'\mathbf{X})^{-1} + (\mathbf{X}'\mathbf{X})^{-1}\mathbf{X}'\mathbf{A}\mathbf{W}\mathbf{A}'\mathbf{X}(\mathbf{X}'\mathbf{X})^{-1}, \quad (3)$$

where $\mathbf{A}$ is a design matrix* for the nesting factor ("main plots," "blocks," "subjects," etc., as the case may be) and $\mathbf{W} = [\operatorname{diag}(\theta + k_i) - \mathbf{A}'\mathbf{X}(\mathbf{X}'\mathbf{X})^{-1}\mathbf{X}'\mathbf{A}]^{-1}$. Also,

$$\hat{\boldsymbol{\beta}} = \tilde{\boldsymbol{\beta}} - (\mathbf{X}'\mathbf{X})^{-1}\mathbf{X}'\mathbf{A}\mathbf{W}\mathbf{A}'(\mathbf{y} - \tilde{\mathbf{y}}), \quad (4)$$

where $\tilde{\boldsymbol{\beta}} = (\mathbf{X}'\mathbf{X})^{-1}\mathbf{X}'\mathbf{y}$ is the OLS estimator of $\boldsymbol{\beta}$ and $\tilde{\mathbf{y}} = \mathbf{X}\tilde{\boldsymbol{\beta}}$ is the vector of "predicted" values obtained by OLS. Furthermore, after applying (1), the total sum of squares is $\mathbf{y}'\mathbf{y} - \mathbf{y}'\mathbf{A}\mathbf{H}\mathbf{A}'\mathbf{y} = \sum_i \sum_j y_{ij}^2 - \sum_i (\theta + k_i)^{-1} y_{i.}^2$, $\mathbf{H} = \operatorname{diag}[(\theta + k_i)^{-1}]$, and $y_{i.} = \sum_j y_{ij}$. The regression sum of squares is $\hat{\mathbf{y}}'\hat{\mathbf{y}} - \hat{\mathbf{y}}'\mathbf{A}\mathbf{H}\mathbf{A}'\hat{\mathbf{y}}$ [or, equivalently, $\hat{\boldsymbol{\beta}}'(\mathbf{X}'\mathbf{y} - \mathbf{X}'\mathbf{A}\mathbf{H}\mathbf{A}'\mathbf{y})$, where $\hat{\mathbf{y}} = \mathbf{X}\hat{\boldsymbol{\beta}}$].

Equation (4) is interesting in that, if we consider the nesting factor as "blocks," $\mathbf{A}'(\mathbf{y} - \tilde{\mathbf{y}})$ is a vector of block totals of deviations from OLS regression, and $\mathbf{W}$ applies the proper weighting to convert these block deviations into corrections to be applied to individual observations within blocks. Premultiplying such corrections by $(\mathbf{X}'\mathbf{X})^{-1}\mathbf{X}'$ gives corrections to be applied to $\tilde{\boldsymbol{\beta}}$. Recovery of interblock information* in one-restrictional lattice and incomplete block designs follows from (4) as a special case. Furthermore, the general framework described here provides a feasible method for handling unbalanced data problems owing to incomplete data or covariates in the analysis in those designs and in split-plot designs as well. If we allow σ_1^2 to be negative, the method also provides a GLS analysis for cases with negative intraclass correlation, provided that the estimated covariance matrix $\hat{\mathbf{V}}(\epsilon)$ is positive definite, i.e., provided $\hat{\sigma}_2^2 + k_{\max}\hat{\sigma}_1^2 > 0$, where $k_{\max} = \max(k_i)$.

COMPUTATION

One advantage of solution by using (3) and (4) is that if variance components are estimated by the fitting-of-constants method, the OLS solution is easily obtained in the process. If $\hat{\sigma}_1^2 > 0$, then $\mathbf{W}^{-1} = \text{diag}(\theta + k_i) - \mathbf{A'X}(\mathbf{X'X})^{-1}\mathbf{X'A}$ is positive definite. Since $\mathbf{X'X}$ is also positive definite, we can, by Cholesky decomposition, find upper triangle matrices $\mathbf{Z}$ and $\mathbf{Q}$ such that $\mathbf{W}^{-1} = \mathbf{Z'Z}$ and $\mathbf{X'X} = \mathbf{Q'Q}$. The following is a suggested computational strategy:

1. Form $[\mathbf{X'X} \,|\, \mathbf{X'A} \,|\, \mathbf{X'y}]$, an augmented matrix, and perform a Cholesky decomposition (*see* LINEAR ALGEBRA, COMPUTATIONAL), giving $[\mathbf{Q} \,|\, \mathbf{B} \,|\, \mathbf{C}]$, where $\mathbf{B} = (\mathbf{Q}^{-1})'\mathbf{X'A}$ and $\mathbf{C} = (\mathbf{Q}^{-1})'\mathbf{X'y}$. This step may be performed as a part of the OLS and variance component estimation stage of the analysis. In fact, if instead we perform the Cholesky decomposition on
$$\begin{bmatrix} \mathbf{X'X} & \mathbf{X'A} & \mathbf{X'y} \\ \mathbf{A'X} & \mathbf{A'A} & \mathbf{A'y} \end{bmatrix},$$
getting
$$\begin{bmatrix} \mathbf{Q} & \mathbf{B} & \mathbf{C} \\ \mathbf{0} & \mathbf{R} & \mathbf{S} \end{bmatrix}$$
as the result, the sum of squares owing to the nesting factor adjusted for $\mathbf{X}$ is $\mathbf{S'S}$. The expectation of $\mathbf{S'S}$ is $\nu\sigma_2^2 + [\sum_i k_i - \text{tr}(\mathbf{B'B})]\sigma_1^2$, where ν is the degrees of freedom for the nesting factor adjusted for $\mathbf{X}$ and $\text{tr}(\mathbf{B'B})$ is the trace of $\mathbf{B'B}$ (which is equal to the sum of the squares of all elements of $\mathbf{B}$). Also, σ_2^2 is estimated by $(\mathbf{y'y} - \mathbf{C'C} - \mathbf{S'S})/(\sum k_i - p - \nu)$.
2. Obtain the solutions $\mathbf{D}$ and $\tilde{\boldsymbol{\beta}}$ to the upper triangular system of equations $\mathbf{Q}[\mathbf{D} \,|\, \tilde{\boldsymbol{\beta}}] = [\mathbf{B} \,|\, \mathbf{C}]$ [11, Ch. 3] and form a second augmented matrix
$$\left[\mathbf{W}^{-1} \,|\, \mathbf{A'X}(\mathbf{X'X})^{-1} \,|\, \mathbf{A'}(\mathbf{y} - \tilde{\mathbf{y}})\right]$$
$$= \left[\text{diag}(\theta + k_i) - \mathbf{B'B} \,|\, \mathbf{D'} \,|\, \mathbf{A'y} - \mathbf{B'C}\right].$$
Perform a Cholesky decomposition giving $[\mathbf{Z} \,|\, \mathbf{E} \,|\, \mathbf{F}]$ where $\mathbf{E} = (\mathbf{Z}^{-1})'\mathbf{A'X}(\mathbf{X'X})^{-1}$, $\mathbf{F} = (\mathbf{Z}^{-1})'\mathbf{A'}(\mathbf{y} - \tilde{\mathbf{y}})$.
3. The correction to be added to $(\mathbf{X'X})^{-1}$ in (3) is now $\mathbf{E'E}$ and the correction to be subtracted from $\tilde{\boldsymbol{\beta}}$ in (4) is $\mathbf{E'F}$. Also, $(\mathbf{X'X})^{-1} = \mathbf{Q}^{-1}(\mathbf{Q}^{-1})'$.

The user should be warned that solution for $\hat{\boldsymbol{\beta}}$ by this algorithm will inevitably be somewhat more sensitive to ill-conditioning of the $\mathbf{X'X}$ matrix than the OLS solution $\tilde{\boldsymbol{\beta}} = \mathbf{Q}^{-1}\mathbf{C}$ (*see* MATRIX, ILL-CONDITIONED). Therefore, introduction of orthogonality properties into the $\mathbf{X}$-matrix (by centering and/or orthogonalizing $\mathbf{X}$-variables, for example) is a good idea.

For the case of negative intraclass correlation* ($\hat{\sigma}_1^2 < 0$), provided $\tilde{\mathbf{V}}(\boldsymbol{\epsilon})$ is positive definite, $\mathbf{W}^{-1}$ is negative definite. Thus the computational procedure above will work in that case also if we replace $\mathbf{W}^{-1}$ with $-\mathbf{W}^{-1}$ and change the sign of the corrections at step 3.

Patterson and Thompson [9] suggest a method of estimating the variance components*, known as restricted maximum likelihood* (REML) (Harville [4]), and give a set of equations to be solved iteratively. The REML solution also can be obtained (apparently with less computation) by the following modification of the algorithm suggested above.

In step 1, complete only enough of the decomposition to obtain $[\mathbf{Q} \,|\, \mathbf{B} \,|\, \mathbf{C}]$ and the adjustments $\mathbf{A'A} - \mathbf{B'B} = \mathbf{M}$ (say) and $\mathbf{A'y} - \mathbf{B'C} = \mathbf{T}$. Obtain the eigenvalues of $\mathbf{M}$, $\lambda_1 \geqslant \lambda_2 \geqslant \cdots \geqslant \lambda_\nu > \lambda_{\nu+1} = \cdots = \lambda_a = 0$, and an $a \times \nu$ matrix $\mathbf{P}$, the columns of which are the ordered normalized eigenvectors associated with the ν nonzero eigenvalues. Compute $\mathbf{U} = \mathbf{P'T}$ and $s_i^2 = u_i^2/\lambda_i$, where u_i is the ith element of $\mathbf{U}$. The s_i^2 values (each with one degree of freedom and $E(s_i^2) = \sigma_2^2 + \lambda_i\sigma_1^2$) together with the fitting-of-constants estimate of σ_2^2 (with $\sum k_i - p - \nu$ degrees of freedom) give a test of $\nu + 1$ uncorrelated mean squares from which the REML variance component estimates can be obtained by iterated weighted least squares*, as described by Hayman [5] and Cornelius and Byars [2]. The fitting-of-constants estimates, which can be computed substituting $\mathbf{S'S} = \sum s_i^2$ and $\sum k_i - \text{tr}(\mathbf{B'B}) = \sum \lambda_i$ in the formulas in step 1 (but note that the elements of $\mathbf{S}$ are not s_i), provide values for computing initial estimates of the weights. Steps 2 and 3 are computed as before with θ obtained from the REML variance estimates.

NUMERICAL EXAMPLE

We illustrate the computational method with a subset of data from the hypothetical 3×3 simple lattice design in LATTICE DESIGNS, obtained by deleting two observations from block 6 and one from block 3, so that treatment 9 is entirely missing. This data set is too small to provide reliable estimates of the variance components, but it provides a simple example. We will fit the treatment-means and replication-effects model, $y_{tlij} = \mu_t + \rho_l + \epsilon_{ij}$, where t, l, i, and j index the treatment, replication, block, and plot within the block, respectively; $t = 1, 2, \ldots, 8$; $l = 1, 2$; $i = 1, \ldots, 6$; $j = 1, \ldots, k_i$, where $k_1 = k_2 = k_4 = k_5 = 3$, $k_3 = 2$, and $k_6 = 1$. With respect to the replication effects ρ_l, let us impose the condition $\rho_2 = -\rho_1$, so that the model is of full rank. The replication, block, and treatment indices, and columns of $\mathbf{X}$ and $\mathbf{y}$ are shown in Table 1.

$\mathbf{X}'\mathbf{X}$ obtained from these data has values 2, 2, 2, 2, 2, 1, 2, 2, and 15 on the diagonal, and the only nonzero off-diagonal elements are the (6, 9)th and (9, 6)th, both equal to 1. $\mathbf{A}'\mathbf{A} = \text{diag}(k_i) = \text{diag}(3, 3, 2, 3, 3, 1)$, $\mathbf{X}'\mathbf{y} = [29.8, 11.0, 22.5, 16.2, 24.8, 6.2, 17.9, 28.4, 7.6]'$, $\mathbf{A}'\mathbf{y} = [34.5, 25.1, 22.6, 29.9, 36.2, 8.5]'$, and

$$\mathbf{X}'\mathbf{A} = \begin{bmatrix} 1 & 0 & 0 & 1 & 0 & 0 \\ 1 & 0 & 0 & 0 & 1 & 0 \\ 1 & 0 & 0 & 0 & 0 & 1 \\ 0 & 1 & 0 & 1 & 0 & 0 \\ 0 & 1 & 0 & 0 & 1 & 0 \\ 0 & 1 & 0 & 0 & 0 & 0 \\ 0 & 0 & 1 & 1 & 0 & 0 \\ 0 & 0 & 1 & 0 & 1 & 0 \\ 3 & 3 & 2 & -3 & -3 & -1 \end{bmatrix}.$$

$\mathbf{Q}$ has values $\sqrt{2}$, $\sqrt{2}$, $\sqrt{2}$, $\sqrt{2}$, $\sqrt{2}$, 1, $\sqrt{2}$, $\sqrt{2}$, and $\sqrt{14}$ on the diagonal, and the only nonzero off-diagonal element is $Q_{69} = 1$. $\mathbf{B}$ has the same pattern of zeroes and nonzeroes as $\mathbf{X}'\mathbf{A}$; rows 1, 2, 3, 4, 5, 7, and 8 have the 1's replaced by $1/\sqrt{2}$, row 6 is unchanged, and the ninth row is $(14)^{-1/2}[3, 2, 2, -3, -3, -1]$. $\mathbf{C} = [21.0718, 7.7782, 15.9099, 11.4551, 17.5362, 6.2000, 12.6572, 20.0818, 0.3742]'$.

Table 1

Block	Replication	Treatment	Column of **X** Associated with Parameter									
			μ_1	μ_2	μ_3	μ_4	μ_5	μ_6	μ_7	μ_8	ρ_1	y
1	1	1	1	0	0	0	0	0	0	0	1	16.3
1	1	2	0	1	0	0	0	0	0	0	1	4.2
1	1	3	0	0	1	0	0	0	0	0	1	14.0
2	1	4	0	0	0	1	0	0	0	0	1	8.6
2	1	5	0	0	0	0	1	0	0	0	1	10.3
2	1	6	0	0	0	0	0	1	0	0	1	6.2
3	1	7	0	0	0	0	0	0	1	0	1	9.1
3	1	8	0	0	0	0	0	0	0	1	1	13.5
4	2	1	1	0	0	0	0	0	0	0	−1	13.5
4	2	4	0	0	0	1	0	0	0	0	−1	7.6
4	2	7	0	0	0	0	0	0	1	0	−1	8.8
5	2	2	0	1	0	0	0	0	0	0	−1	6.8
5	2	5	0	0	0	0	1	0	0	0	−1	14.5
5	2	8	0	0	0	0	0	0	0	1	−1	14.9
6	2	3	0	0	1	0	0	0	0	0	−1	8.5

$\mathbf{S} = [2.7543, -0.7425, 0, -2.3826, 4.0805, 0]'$, and $\mathbf{S}'\mathbf{S} = 30.465$. The residual sum of squares after OLS fitting of **X** and **A** is $\mathbf{y}'\mathbf{y} - \mathbf{C}'\mathbf{C} - \mathbf{S}'\mathbf{S} = 2.165$, with $\sum k_i - p - \nu = 2$ degrees of freedom, giving $\hat{\sigma}_2^2 = 1.0825$.

$$\mathbf{B}'\mathbf{B} = (1/14) \times \begin{bmatrix} 30 & 6 & 6 & -2 & -2 & 4 \\ 6 & 32 & 4 & 1 & 1 & -2 \\ 6 & 4 & 18 & 1 & 1 & -2 \\ -2 & 1 & 1 & 30 & 9 & 3 \\ -2 & 1 & 1 & 9 & 30 & 3 \\ 4 & -2 & -2 & 3 & 3 & 8 \end{bmatrix}.$$

$\mathbf{S}'\mathbf{S}$ has $\nu = 4$ degrees of freedom and $\mathrm{tr}(\mathbf{B}'\mathbf{B}) = 148/14 = 10.5714$. Thus $\hat{\sigma}_1^2 = [30.465 - 4(1.0825)]/(15 - 10.5714) = 5.9015$, and $\theta = 1.0825/5.9015 = 0.18343$. The diagonal elements of $\mathbf{Q}^{-1}$are the reciprocals of the diagonal elements of **Q** and the only nonzero off-diagonal element of $\mathbf{Q}^{-1}$ is then the (6, 9)th, which is equal to $-1/\sqrt{14}$. In step 2, **D** has rows 1, 2, 3, 4, 5, 7 and 8 equal to 0.5 times the corresponding rows of $\mathbf{X}'\mathbf{A}$, row 6 is equal to $(1/14)[-3, 12, -2, 3, 3, 1]$, and row 9 is equal to $(1/14)[3, 2, 2, -3, -3, -1]$; $\tilde{\boldsymbol{\beta}} = [14.90, 5.50, 11.25, 8.10, 12.40, 6.10, 8.95, 14.20, 0.10]'$. We also have

$$\mathbf{B}'\mathbf{C} = [31.95, 26.90, 23.35, 31.65, 31.80, 11.15]',$$

and thus

$$\mathbf{A}'\mathbf{y} - \mathbf{B}'\mathbf{C} = [2.55, -1.80, -0.75, -1.75, 4.40, -2.65]'.$$

Now, adding $\theta = 0.183843$ to the diagonal of $\mathbf{A}'\mathbf{A}$ and subtracting $\mathbf{B}'\mathbf{B}$ gives $\mathbf{W}^{-1}$ (Table 2a). Augmenting $\mathbf{W}^{-1}$ with $\mathbf{D}'$ and $\mathbf{A}'\mathbf{y} - \mathbf{B}'\mathbf{C}$ and performing the second Cholesky decomposition gives **E** (Table 2b). Also

$$\mathbf{F} = [2.4988, -0.8829, -0.2765, -2.1013, 3.4526, -1.6388]'.$$

$(\mathbf{X}'\mathbf{X})^{-1} = \mathbf{Q}^{-1}(\mathbf{Q}^{-1})'$ has values 1/2, 1/2, 1/2, 1/2, 1/2, 15/14, 1/2, 1/2, 1/14 on the diagonal and the only nonzero off-diagonal elements are the (6, 9)th and (9, 6)th, which

Table 2

$$\mathbf{W}^{-1} = (10^{-4}) \begin{bmatrix} 10406 & -4286 & -4286 & 1429 & 1429 & -2857 \\ -4286 & 8977 & -2857 & -714 & -714 & 1429 \\ -4286 & -2857 & 8977 & -714 & -714 & 1429 \\ 1429 & -714 & -714 & 10406 & -6429 & -2143 \\ 1429 & -714 & -714 & -6429 & 10406 & -2143 \\ -2857 & 1429 & 1429 & -2143 & -2143 & 6120 \end{bmatrix}$$

(a)

$$\mathbf{E} = (10^{-4}) \begin{bmatrix} 4902 & 4902 & 4902 & 0 & 0 & -2101 & 0 & 0 & 2101 \\ 2425 & 2425 & 2425 & 5888 & 5888 & 9054 & 0 & 0 & 2721 \\ 5184 & 5184 & 5184 & 4916 & 4916 & 4014 & 7670 & 7670 & 5818 \\ 4470 & -481 & -481 & 5192 & 241 & 2672 & 5192 & 241 & -2191 \\ 3198 & 5475 & -1047 & 4768 & 7045 & 5811 & 4768 & 7045 & -4764 \\ 4866 & 4866 & 9152 & 3616 & 3616 & 3693 & 3616 & 3616 & -4526 \end{bmatrix}$$

(b)

$$(\mathbf{X}^{*\prime}\mathbf{X}^{*})^{-1} = (10^{-3}) \begin{bmatrix} 1607 & 958 & 958 & 958 & 810 & 810 & 958 & 810 & 0 \\ 958 & 1607 & 958 & 810 & 958 & 810 & 810 & 958 & 0 \\ 958 & 958 & 1919 & 654 & 654 & 589 & 654 & 654 & 117 \\ 958 & 810 & 654 & 1716 & 1067 & 1280 & 1005 & 856 & -58 \\ 810 & 958 & 654 & 1067 & 1716 & 1280 & 856 & 1005 & -58 \\ 810 & 810 & 589 & 1280 & 1280 & 2642 & 857 & 857 & -138 \\ 958 & 810 & 654 & 1005 & 856 & 857 & 1716 & 1067 & -58 \\ 810 & 958 & 654 & 856 & 1005 & 857 & 1067 & 1716 & -58 \\ 0 & 0 & 117 & -58 & -58 & -138 & -58 & -58 & 1008 \end{bmatrix}$$

(c)

Table 3

$$
(\mathbf{X}^* \mid \mathbf{y}^*) = (10^{-3}) \begin{bmatrix}
747 & -253 & -253 & 0 & 0 & 0 & 0 & 0 & 240 & 7560 \\
-253 & 747 & -253 & 0 & 0 & 0 & 0 & 0 & 240 & -4540 \\
-253 & -253 & 747 & 0 & 0 & 0 & 0 & 0 & 240 & 5260 \\
0 & 0 & 0 & 747 & -253 & -253 & 0 & 0 & 240 & 2242 \\
0 & 0 & 0 & -253 & 747 & -253 & 0 & 0 & 240 & 3942 \\
0 & 0 & 0 & -253 & -253 & 747 & 0 & 0 & 240 & -158 \\
0 & 0 & 0 & 0 & 0 & 0 & 645 & -355 & 290 & 1075 \\
0 & 0 & 0 & 0 & 0 & 0 & -355 & 645 & 290 & 5475 \\
747 & 0 & 0 & -253 & 0 & 0 & -253 & 0 & -240 & 5926 \\
-253 & 0 & 0 & 747 & 0 & 0 & -253 & 0 & -240 & 26 \\
-253 & 0 & 0 & -253 & 0 & 0 & 747 & 0 & -240 & 1226 \\
0 & 747 & 0 & 0 & -253 & 0 & 0 & -253 & -240 & -2370 \\
0 & -253 & 0 & 0 & 747 & 0 & 0 & -253 & -240 & 5330 \\
0 & -253 & 0 & 0 & -253 & 0 & 0 & 747 & -240 & 5730 \\
0 & 0 & 394 & 0 & 0 & 0 & 0 & 0 & -394 & 3346
\end{bmatrix}
$$

are both equal to $-1/14$. Adding $\mathbf{E}'\mathbf{E}$ to $(\mathbf{X}'\mathbf{X})^{-1}$ gives $(\mathbf{X}^{*\prime}\mathbf{X}^*)^{-1}$ (Table 2c). Finally, computing $\mathbf{E}'\mathbf{F}$ and subtracting it from $\tilde{\boldsymbol{\beta}}$ gives $\hat{\boldsymbol{\beta}} = [14.66, 3.44, 12.14, 8.79, 11.26, 6.69, 9.20, 12.62, 0.42]'$.

If the linear transformation method is used, we have $\alpha_1 = \alpha_2 = \alpha_4 = \alpha_5 = 0.75996$, $\alpha_3 = 0.71016$, $\alpha_6 = 0.60630$, and, to three-digit accuracy, $(\mathbf{X}^* \mid \mathbf{y}^*)$ as in Table 3. OLS regression of $\mathbf{y}^*$ on $\mathbf{X}^*$, of course, gives the solution we already have obtained.

If the REML solution is computed, the nonzero eigenvalues of $\mathbf{M}$ are $\lambda_1 = \lambda_2 = 1.5$, $\lambda_3 = 1$, and $\lambda_4 = 3/7$. The s_i^2 values are 0.00272, 24.61728, 0.55125, and 5.29375, and REML variance estimates $\hat{\sigma}_1^2 = 5.8099$, $\hat{\sigma}_2^2 = 1.1127$, and $\hat{\boldsymbol{\beta}} = [14.66, 3.45, 12.12, 8.79, 6.70, 9.20, 12.63, 0.42]'$. (Note: s_1^2 and s_2^2 are not unique since they depend on the choice of eigenvectors associated with an eigenvalue of multiplicity 2. The sum $s_1^2 + s_2^2 = 24.62000$ is unique.)

TWOFOLD AND MULTIFOLD NESTED-ERROR MODEL

In the twofold nested-error model an element of $\boldsymbol{\epsilon}$ is of the form $\epsilon_{ijk} = e_{1i} + e_{2ij} + e_{3ijk}$; $i = 1, \ldots, a$; $j = 1, \ldots, b_i$; $k = 1, \ldots, n_{ij}$, where e_{1i}, e_{2ij}, and e_{3ijk} are errors associated with the first, second, and third stages of nesting, with variances σ_1^2, σ_2^2, and σ_3^2, respectively, all e_{1i}, e_{2ij}, and e_{3ijk} being uncorrelated. A transformation analogous to (1) has been developed by Fuller and Battese [3] for the case of n_{ij}'s all equal for a given i. (The n_{ij}'s may differ for differing i.) A transformation, equivalent to theirs under that restriction, but appropriate for arbitrary n_{ij}'s, has been developed by Cornelius [1]. Cornelius [1] also gives the formulae analogous to (3) and (4) for the arbitrary twofold nested-error model.

Presumably analogous methods for more-than-twofold nested-error models can be derived. For error structures that consist of balanced nested classifications, this would follow by straightforward application of the theorems of Fuller and Battese [3] and Searle and Henderson [10]. For unbalanced nested classifications, LaMotte's [8] method of obtaining $\mathbf{V}^{-1}$ may be helpful. The method described by Henderson [6] might also be adapted to this problem.

References

[1] Cornelius, P. L. (1981). Unpublished Report. (Gives a transformation for generalized least squares estimation of linear models with arbitrary twofold nested-error structure.)

[2] Cornelius, P. L. and Byars, J. (1977). *Biometrics*, **33**, 375–382. (Describes procedures for computing iterated weighted least squares estimates of variance components from sets of uncorrelated mean squares or mean square and cross product matrices.)

[3] Fuller, W. A. and Battese, G. E. (1973). *J. Amer. Statist. Ass.*, **68**, 626–632. (Transformations for

generalized least squares estimation of linear models with onefold and a restricted case of two-fold nested-error structure. Also considers some properties of the estimator of β when estimates of variance components must be used.)

[4] Harville, D. A. (1977). *J. Amer. Statist. Ass.*, **72**, 320–338. (Primarily a review article concerning methods for estimating variance components, but the problem of joint maximum likelihood estimation (MLE) of fixed effects and variances of random effects in linear models is considered. The MLE estimates of the fixed effects are equivalent to GLS estimates obtained when the MLE estimates of the variance components are used.)

[5] Hayman, B. I. (1960). *Biometrics*, **16**, 369–381. (Iterated least squares estimation of variance components from uncorrelated mean squares.)

[6] Henderson, C. R. (1971). *Econometrica*, **39**, 397–401. (Gives a method of obtaining GLS estimates in a general mixed model.)

[7] Hussain, A. (1969). *Biometrika*, **56**, 327–336. (Considers a model which is essentially a two-way mixed model with covariates in its analysis.)

[8] LaMotte, L. R. (1972). *Ann. Math. Statist.*, **43**, 659–662. [Inverse and determinant of $\mathbf{V}(\boldsymbol{\epsilon})$ in a multifold nested model.]

[9] Patterson, H. D. and Thompson, R. (1971). *Biometrika*, **58**, 545–554. [Solution of the onefold nested-error problem by methods which have become known as restricted maximum likelihood (REML).]

[10] Searle, S. R. and Henderson, H. V. (1979), *J. Amer. Statist. Ass.*, **74**, 465–470. [Eigenvalues, determinant and inverse of $\mathbf{V}(\boldsymbol{\epsilon})$ in models with error structures consisting of balanced combinations of crossed and/or nested classifications.]

[11] Stewart, G. W. (1973). *Introduction to Matrix Computations*. Academic Press, New York. (Algorithm 1.3 on page 108 may be used to solve the upper triangular system $\mathbf{QD} = \mathbf{B}$ and also $\mathbf{Q}\tilde{\beta} = \mathbf{C}$ without having to invert $\mathbf{Q}$. $\mathbf{Q}^{-1}$ may be obtained by Algorithm 1.4 or 1.5 on page 110.)

Acknowledgment

This article is published as Journal Article No. 81-3-102 of the Kentucky Agricultural Experiment Station with the approval of the Director.

(ANALYSIS OF COVARIANCE
ANALYSIS OF VARIANCE
BLOCKS, BALANCED INCOMPLETE
COMPUTERS IN STATISTICS
FIXED-, RANDOM-, AND MIXED-EFFECTS MODELS
GENERAL BALANCE
GENERAL LINEAR MODEL
INCOMPLETE BLOCK DESIGNS
INTERBLOCK INFORMATION
LATTICE DESIGNS
LEAST SQUARES
LINEAR ALGEBRA, COMPUTATIONAL
LINEAR MODELS WITH CROSSED-ERROR STRUCTURE
NESTING AND CROSSING IN DESIGN
PARTIALLY BALANCED DESIGNS
REPEATED MEASUREMENTS
SPLIT PLOTS
VARIANCE COMPONENTS)

P. L. CORNELIUS

LINEAR PLATEAU MODELS *See* PLATEAU MODELS, LINEAR

LINEAR PREDICTION *See* PREDICTION, LINEAR

LINEAR PROGRAMMING

Programming problems are concerned with the efficient use or allocation of limited resources to meet desired objectives. These problems are characterized by the large number of solutions that satisfy the basic conditions of each problem. The selection of a particular solution as the best solution to a problem depends on some aim or overall objective that is implied in the statement of the problem. A solution that satisfies both the conditions of the problem and the given objective is termed an "optimum solution." A typical example is that of a manufacturer who must determine what combination of available resources will enable the manufacture of products in a way that not only satisfies a production schedule, but also maximizes profit. This problem has as its basic conditions the limitations of the available resources and the requirements of the production schedule, and as its objective, the desire of the manufacturer to maximize gain.

We consider only a very special subclass of programming problems called *linear programming problems.* A linear programming problem differs from the general variety in that a mathematical model or description of the problem can be stated using relationships which are called "straightline" or linear. Mathematically, these relationships are of the form

$$a_1x_2 + a_2x_2 + \cdots + a_jx_j + \cdots + a_nx_n = a_0,$$

where the a_j's are known coefficients and the x_j's are unknown variables. The mathematical statement of a linear programming problem includes a set of simultaneous linear equations and/or inequalities which represent the conditions of the problem and a linear function which expresses the objective of the problem.

DEFINITIONS AND BASIC THEOREMS

The *linear programming problem* is as follows: Find a set of numbers $x_1, x_2, \ldots, x_n$ that minimizes (or maximizes) the *linear objective function*

$$c_1x_1 + c_2x_2 + \cdots + c_jx_j + \cdots + c_nx_n \quad (1)$$

subject to the *linear constraints*

$$\begin{array}{ccccccccc} a_{11}x_1 + a_{12}x_2 + \cdots + a_{1j}x_j + \cdots + a_{1n}x_n = a_{10} \\ \vdots \\ a_{i1}x_1 + a_{i2}x_2 + \cdots + a_{ij}x_j + \cdots + a_{in}x_n = a_{i0} \\ \vdots \\ a_{m1}x_1 + a_{m2}x_2 + \cdots + a_{mj}x_j + \cdots + a_{mn}x_n = a_{m0} \end{array} \quad (2)$$

and the nonnegativity constraints

$$x_j \geqslant 0 \qquad (j = 1, \ldots, n). \quad (3)$$

In words, in a linear programming problem one wishes to find a nonnegative solution to a set of linear constraints which optimizes, i.e., minimizes or maximizes, a linear objective function. The c_j are called *cost coefficients*, the a_{ij} are termed *technological coefficients*, and the a_{i0} are called *right-hand-side coefficients*. The linear constraints can be equations, as noted above, or linear inequalities of the form

$$a_{i1}x_1 + \cdots + a_{in}x_n \leqslant a_{i0} \quad (4)$$

or

$$a_{i1}x_1 + \cdots + a_{in}x_n \geqslant a_{i0}. \quad (5)$$

Inequalities such as (4) and (5) can be transformed to equations by the suitable addition or subtraction of a nonnegative *slack variable*. For (4) we would have

$$a_{i1}x_1 + \cdots + a_{in}x_n + x_{n+1} = a_{i0}, \quad x_{n+1} \geqslant 0 \quad (6)$$

and for (5) we would have

$$a_{i1}x_1 + \cdots + a_{in}x_n - x_{n+1} = a_{i0}, \quad x_{n+1} \geqslant 0. \quad (7)$$

Each inequality has associated with it a different slack variable. A slack variable measures the difference between the left- and right-hand sides of the given inequality. As the form of a linear programming problem requires that all variables be nonnegative, we note that a variable not restricted to be nonnegative can always be expressed as the difference between two nonnegative variables; i.e., $x_j = x_j' - x_j''$, where $x_j' \geqslant 0$ and $x_j'' \geqslant 0$.

A linear programming problem can contain any mixture of linear constraints. For computational purposes, the basic constraints (2) of a linear programming problem must always be expressed in terms of equations such that the number of equations (m) is less than the number of variables (n). This causes (2) to be an *undetermined* set of linear equations which has many possible solutions. As each equation can be considered as a *hyperplane* in n-dimensional space, the *solution space* of the set of linear equations is, in general, a *convex polyhedron*. The computation algorithms of linear programming determine from all possible solutions one that optimizes the objective function. Since the solution space may be unbounded in the direction of optimization, the optimum value of the objective function can also be unbounded. As maximizing a linear function is equal to minimizing the negative of the linear function, we only consider the minimizing case.

In matrix notation the linear programming problem is given by

$$\begin{array}{ll} \text{minimize} & cX \\ \text{subject to} & AX = b,\ X \geqslant 0, \end{array}$$

where $c = (c_1, c_2, \ldots, c_n)$ is a row vector; $X = (x_1, x_2, \ldots, x_n)$ is a column vector; A is the $m \times n$ matrix of coefficients; $b = (a_{10}, a_{20}, \ldots, a_{m0})$ is a column vector; and $0 = (0, 0, \ldots, 0)$ is a n-rowed column vector. As it is convenient to consider the columns of A as points in m-dimensional space, the linear programming problem can be written as

$$\begin{array}{ll} \text{minimize} & cX \\ \text{subject to} & x_1P_1 + x_2P_2 + \cdots + x_nP_n = P_0, \; x_j \geqslant 0, \end{array}$$

where $P_j = (a_{1j}, a_{2j}, \ldots, a_{mj})$ is a column vector for $j = 0, 1, \ldots, n$.

A *feasible solution* to the linear programming problem is a vector $X = (x_1, x_2, \ldots, x_n)$ which satisfies conditions (2) and (3).

A *basic solution* to equations (2) is a solution obtained by setting $n - m$ variables equal to zero and solving for the remaining m variables, provided that the determinant of the coefficients of these m variables is nonzero. The m variables are called *basic variables*.

A *basic feasible solution* is a basic solution which also satisfies constraints (3); i.e., all basic variables are nonnegative.

A *nondegenerate basic feasible solution* is a basic feasible solution with exactly m positive x_j.

A *minimum feasible solution* is a feasible solution which also minimizes (1).

A *basis* is a linearly independent set of vectors. A *feasible basis* for the linear programming problem is a square matrix B composed of a linearly independent set of vectors selected from the rectangular matrix $A = (P_1P_2 \cdots P_n)$ such that, for the square set of equations, $BX_0 = P_0$, $X_0 = B^{-1}P_0 \geqslant 0$. For example, if $B = (P_1P_2 \cdots P_m)$, we would have $X_0 = (x_{10}, x_{20}, \ldots, x_{m0}) \geqslant 0$. Here the solution to the given problem is $X = (x_{10}, x_{20}, \ldots, x_{m0}, 0, \ldots, 0)$, where the last $n - m$ terms of X are zero. Finding a feasible basis corresponds to selecting a determined square set of equations from the given underdetermined rectangular set and letting those variables not in the square set be equal to zero.

A *convex combination* of the vectors $U_1, U_2, \ldots, U_n$ is a vector

$$U = \alpha_1U_1 + \alpha_2U_2 + \cdots + \alpha_nU_n,$$

where the α_i are all scalars, $\alpha_i \geqslant 0$, and $\sum \alpha_i = 1$.

A subset of points C of Euclidean space is a *convex set* if and only if, for all pairs of points U_1 and U_2 in C, any convex combination

$$U = \alpha_1U_1 + \alpha_2U_2 = \alpha U_1 + (1 - \alpha)U_2, \quad 1 \geqslant \alpha \geqslant 0,$$

is also in C. A convex set is one that contains the straight line joining any two points in the set.

A point U in a convex set C is called an *extreme point* if U cannot be expressed as a convex combination of any other two distinct points in C.

Theorem 1. The set of all feasible solutions to the linear programming problem is a convex set.

Theorem 2. The objective function (1) assumes its minimum at an extreme point of the convex set C generated by the set of feasible solutions to the linear programming problems. If it assumes its minimium at more than one extreme point, it takes on the same value for every convex combination of those particular points.

Theorem 3. If a set of $k \leqslant m$ vectors $P_1, P_2, \ldots, P_k$ can be found that are linearly independent and such that

$$x_1P_1 + x_2P_2 + \cdots + x_kP_k = P_0$$

and all $x_j \geqslant 0$, then the point $X = (x_1, x_2, \ldots, x_k, 0, \ldots, 0)$ is an extreme point of the convex set of feasible solutions. Here X is an n-dimensional vector whose last $n - k$ elements are zero.

Theorem 4. If $X = (x_1, x_2, \ldots, x_n)$ is an extreme point of C, then the vectors associated with positive x_j form a linearly independent set. From this it follows that, at most, m of the x_j are positive.

Theorem 5. $X = (x_1, x_2, \ldots, x_n)$ is an extreme point of C if and only if the positive x_j are coefficients of linearly independent vectors P_j in

$$\sum_{j=1}^{n} x_j P_j = P_0.$$

Theorem 6. If a feasible solution exists, then a basic feasible solution exists.

Theorem 7. If the objective function possesses a finite minimum, then at least one optimal solution is a basic feasible solution.

These theorems enable us to restrict the search for an optimum solution to extreme points of the convex set C of all possible solutions. A geometric discussion of the above concepts is given below in the following section.

If the linear programming problem is stated as the following *primal problem*:

$$\begin{array}{ll} \text{minimize} & cX \\ \text{subject to} & AX \geqslant b,\ X \geqslant 0, \end{array}$$

then the corresponding *dual problem* is given by

$$\begin{array}{ll} \text{maximize} & Wb \\ \text{subject to} & WA \leqslant c,\ W \geqslant 0, \end{array}$$

where $W = (w_1, w_2, \ldots, w_m)$ is the row vector of unknowns for the dual problem.

Duality Theorem. If either the primal or the dual problem has a finite optimum solution, then the other problem has a finite optimum solution and the extremes of the linear functions are equal, i.e., $\min cX = \max Wb$. If either problem has an unbounded optimum solution, then the other problem has no feasible solutions.

The concept of duality and the duality theorem have much significance in the theoretical and computational aspects of linear programming.

LINEAR PROGRAMMING TECHNIQUES

Simplex Method

The basic computational procedure for solving any linear programming problem is the *simplex method*. With the simplex method, we can, once a first basic (extreme-point) feasible solution has been determined, obtain a minimum basic feasible solution in a finite number of steps. These steps, or *iterations*, consist of finding a new basic feasible solution whose corresponding value of the objective function is less than (or, at worst, equal to) the value of the objective function for the preceding solution. This process is continued until a minimum solution with either a finite or infinite value of the objective function has been reached. The name *simplex* was given to this procedure as a geometric version of this technique contained the inequality $x_1 + x_2 + \cdots + x_n \leqslant 1$, which defines a simplex (generalized tetrahedron) with unit intercepts in n-dimensional space. A description of the standard simplex algorithm follows.

Assume that all coefficients of vector P_0 are nonnegative. We can always multiply an equation by -1 to make the corresponding $a_{i0} \geqslant 0$. Let B_1 be a feasible basis; i.e., a basic feasible solution has been found from the equations $B_1 X_{01} = P_0$, where $X_{01} = B_1^{-1} P_0 \geqslant 0$. In practice, B_1 is usually a unit matrix of order m, and the corresponding first feasible solution is readily obtained since the inverse of a unit matrix is a unit matrix. For this case, $X_{01} = P_0$. In those situations where a suitable unit matrix is not given as part of the problem, an *artificial unit basis* is attached to the problem to start off. This device is discussed below.

Assuming a unit basis for the first feasible solution and reordering the vectors of the basis so that $B_1 = (P_1 P_2 \cdots P_m)$ (this step is

Table 1

Basis	P_0	$P_1 \cdots P_l \cdots P_n$	$P_{m+1} \cdots P_j \cdots P_k \cdots P_n$
P_1	x_{10}	$1 \cdots 0 \ldots 0$	$x_{1,m+1} \cdots x_{1j} \cdots x_{1k} \cdots x_{1n}$
$\vdots$	$\vdots$	$\vdots \quad \vdots \quad \vdots$	$\vdots \quad \vdots \quad \vdots \quad \vdots$
P_l	x_{l0}	$0 \cdots 1 \cdots 0$	$x_{lm+1} \cdots x_{lj} \cdots x_{lk} \cdots x_{ln}$
$\vdots$	$\vdots$	$\vdots \quad \vdots \quad \vdots$	$\vdots \quad \vdots \quad \vdots \quad \vdots$
P_m	x_{m0}	$0 \cdots 0 \cdots 1$	$x_{1,m+1} \cdots x_{mj} \cdots x_{mk} \cdots x_{mn}$
	x_{00}	$0 \cdots 0 \cdots 0$	$x_{0,m} \cdots x_{0j} \cdots x_{0k} \cdots x_{0n}$

not necessary, but is done here to aid in the discussion), the *simplex tableau* (computational tableau) takes the form in Table 1, where $x_{ij} = a_{ij}$ for $i = 1, \ldots, m$ and $j = 0, 1, \ldots, n$; the basic feasible solution is $X_{01} = (x_{10}, \ldots, x_{l0}, \ldots, x_{m0}) = B_1^{-1}P_0$; and in general, we can define $X_{j1} = (x_{1j}, x_{2j}, \ldots, x_{mj}) = B_1^{-1}P_j$. The value of the objective function is $x_{00} = \sum_{i \text{ in basis}} c_i x_{i0}$. The numbers x_{0j} for $j = 1, \ldots, n$ are defined by $x_{0j} = \sum_{i \text{ in basis}} c_i x_{ij} - c_j$. The summation term is called the indirect cost and is sometimes denoted by $z_j = \sum_{i \text{ in basis}} c_i x_{ij}$. Note that $x_{0j} = 0$ for any j in the basis. The following theorems indicate the need and the use of the x_{0j}.

Theorem 8. If, for any j, the condition $x_{0j} > 0$ holds, then a set of feasible solutions can be constructed, such that $x'_{00} < x_{00}$ for any member of the set, where the lower bound of x'_{00} is either finite or infinite. (x'_{00} is the value of the objective function for a particular member of the set of feasible solutions.)

Case 1. If the lower bound is finite, a new feasible solution consisting of exactly m positive variables can be constructed whose value of the objective function is less than the value of the preceding solution; i.e., $-\infty < x'_{00} < x_{00}$.

Case 2. If the lower bound is infinite, a new feasible solution consisting of exactly $m + 1$ positive variables can be constructed whose value of the objective function can be made arbitrarily small.

Theorem 9. If for any basic feasible solution $X = (x_{10}, x_{20}, \ldots, x_{m0})$ the conditions $x_{0j} \leqslant 0$ hold for all $j = 1, 2, \ldots, n$, then the solution is a minimum feasible solution.

Theorem 8 assumes nondegeneracy; i.e., all basic feasible solutions to the problem are strictly positive (all $x_{i0} > 0$). The assumption is required from a theoretical point of view as it enables one to prove that the simplex method will converge in a finite number of steps. If a particular problem can have degenerate basic feasible solutions, then there is the possibility that the procedure will *cycle*, i.e., return to the same basis after a finite number of steps, and hence not converge to the optimum solution. Although examples have been constructed which do cycle, it is quite uncommon, and a problem, degenerate or not, usually converges. Although not usually employed, computational devices exist which will guarantee the convergence of any problem.

To determine a new basic feasible solution, the following steps are carried out. These steps change the basis one vector at a time until a stop condition is encountered.

1. Compute all x_{0j}.
2. Are all $x_{0j} \leqslant 0$ for $j = 1, 2, \ldots, n$? (This set of inequalities is called the *optimality criterion*.) If so, the current solution is an optimum solution, and the procedure stops. For any $x_{0j} = 0$ with P_j not in the optimum basis, an alternate optimum solution can be obtained by introducing this vector into the basis. If not, select

for the vector to be introduced into the new solution the vector P_k whose $x_{0k} = \max_j x_{0j} > 0$. If ties occur, select any one.

3. To ensure feasibility of the new solution, the vector to be eliminated from the basis is the vector P_l corresponding to

$$\frac{x_{l0}}{x_{lk}} = \min_{x_{ik}>0} \frac{x_{i0}}{x_{ik}} .$$

If ties occur, select any one. If all $x_{ik} \leqslant 0$, then the problem has an unbounded optimum solution and the procedure stops. If the ratio $x_{l0}/x_{lk} \geqslant 0$ happens to equal zero (this implies the degenerate case with $x_{l0} = 0$), the value of the objective function for the new solution will be the same as before. The element x_{lk} is called the *pivot element*.

4. Determine the new solution and new simplex tableau by applying the following formulas (Gaussian elimination):

$$x'_{ij} = x_{ij} - \frac{x_{lj}x_{ik}}{x_{lk}}, \qquad i \neq l,$$

$$x'_{lj} = \frac{x_{lj}}{x_{lk}} .$$

These formulas hold for $i = 0, 1, \ldots, m$ and $j = 0, 1, \ldots, n$. The x'_{ij} for $j = 0$ is the new basic feasible solution; x'_{00} is the new value of the objective function; x'_{0j} are the new indirect-minus-direct cost numbers. This transformation is equivalent to determining a new feasible basis B_2 such that the new solution vector is $X_{02} = B_2^{-1}P_0$ and the $X_{j2} = B_2^{-1}P_j$.

The steps above are repeated for the data in the new tableau. Note that the transformation will cause the unit matrix of the initial tableau to be transformed to the inverse of the current basis.

If a unit basis is not explicitly contained in the original statement of the problem, a set of *artificial nonnegative variables* is attached to the system, one new variable for each equation. In some instance a full set of m artificial variables will not be required, as the original problem contains a partial set of unit vectors. The costs coefficients for the artificial variables are assumed to be infinite, and hence, if a minimum feasible solution to the original problem exists, the simplex method will drive the values of the artificial variables to zero. If the original problem does not have any feasible solutions, the simplex method will terminate with artificial variables in the optimum solution at a positive level. The computational tableau and steps of the simplex method can be readily modified to take care of the infinite costs.

To illustrate the above, consider the following linear programming problem:

$$\begin{array}{ll} \text{maximize} & x_1 + 2x_2 \\ \text{subject to} & -x_1 + 3x_2 \leqslant 10 \\ & x_1 + x_2 \leqslant 6 \\ & x_1 - x_2 \leqslant 2 \\ & x_1 + 3x_2 \geqslant 6 \\ & 2x_1 + x_2 \geqslant 4 \\ & x_1 \geqslant 0 \\ & x_2 \geqslant 0. \end{array}$$

We convert the problem to minimization and to equalities by adding the slack variables x_3, x_4, x_5, x_6, and x_7. As a full unit basis is not available, we add two artificial variables, x_8 and x_9. The problem is then to

$$\text{minimize } -x_1 - 2x_2 + wx_8 + wx_9$$

subject to

$$\begin{array}{l} -x_1 + 3x_2 + x_3 = 10 \\ x_1 + x_2 + x_4 = 6 \\ x_1 - x_2 + x_5 = 2 \\ x_1 + 3x_2 - x_6 + x_8 = 6 \\ 2x_1 + x_2 - x_7 + x_9 = 4, \\ x_j \geqslant 0. \end{array}$$

The w represents the infinite artificial cost. The first basic feasible solution is given by $x_3 = 10$, $x_4 = 6$, $x_5 = 2$, $x_8 = 6$, $x_9 = 4$; the value of the objective function is $10w$, step 0 of Table 2. As the artificial part of the objective function and the indirect cost can be separated from the real part of these numbers, an additional row is added to the tableau, as shown. The five tableaus represent the complete solution to the problem

Table 2

	Basis	P_0	P_1	P_2	P_3	P_4	P_5	P_6	P_7	P_8	P_9
0	P_3	10	-1	3	1						
	P_4	6	1	1		1					
	P_5	2	1	-1			1				
	P_8	6	1	(3)				-1		1	
	P_9	1	2	1					-1		1
		0	1	2							
	w	10	3	4				-1	-1		
1	P_3	4	-2		1			1			
	P_4	4	$\frac{2}{3}$			1		$\frac{1}{3}$			
	P_5	4	$\frac{4}{3}$				1	$-\frac{1}{3}$			
	P_2	2	$\frac{1}{3}$	1				$-\frac{1}{3}$			
	P_9	2	($\frac{5}{3}$)					$\frac{1}{3}$	-1		1
		-4	$\frac{1}{3}$					$\frac{2}{3}$			
	w	2	$\frac{5}{3}$					$\frac{1}{3}$	-1		
2	P_3	$\frac{32}{5}$			1			($\frac{7}{5}$)	$-\frac{6}{5}$		
	P_4	$\frac{16}{5}$				1		$\frac{1}{5}$	$\frac{2}{5}$		
	P_5	$\frac{12}{5}$					1	$-\frac{3}{5}$	$\frac{4}{5}$		
	P_2	$\frac{8}{5}$		1				$-\frac{2}{5}$	$\frac{1}{5}$		
	P_1	$\frac{6}{5}$	1					$\frac{1}{5}$	$-\frac{3}{5}$		
		$-\frac{22}{5}$						$\frac{3}{5}$	$\frac{1}{5}$		
	w	0									
3	P_6	$\frac{32}{7}$			$\frac{5}{7}$			1	$-\frac{6}{7}$		
	P_4	$\frac{16}{7}$			$-\frac{1}{7}$	1			($\frac{4}{7}$)		
	P_5	$\frac{36}{7}$			$\frac{3}{7}$		1		$\frac{2}{7}$		
	P_2	$\frac{24}{7}$		1	$\frac{2}{7}$				$-\frac{1}{7}$		
	P_1	$\frac{2}{7}$	1		$-\frac{1}{7}$				$-\frac{3}{7}$		
		$-\frac{50}{7}$			$-\frac{3}{7}$				$\frac{5}{7}$		
4	P_6	8			$\frac{1}{2}$	$\frac{3}{2}$		1			
	P_7	4			$-\frac{1}{4}$	$\frac{7}{4}$			1		
	P_5	4			$\frac{1}{2}$	$-\frac{1}{2}$	1				
	P_2	4		1	$\frac{1}{4}$	$\frac{1}{4}$					
	P_1	2	1		$-\frac{1}{4}$	$\frac{3}{4}$					
		-10			$-\frac{1}{4}$	$-\frac{5}{4}$					

given above. Note that as long as artificial variables are in the solution, the vector to be introduced into the basis corresponds to the vector with the maximum artificial part of the indirect cost. As artificial variables would never be allowed to reenter a basis, they are dropped from the tableau once they are eliminated. The elements in the circles are the pivot elements. From the last (fourth) iteration, we have that the optimal solution is $x_1 = 2$; $x_2 = 4$; $x_5 = 4$; $x_6 = 8$; $x_7 = 4$; all other variables are equal to zero, and the maximum value of the objective function is $+10$.

By plotting the original inequalities in two-dimensional space, as in Fig. 1, a geo-

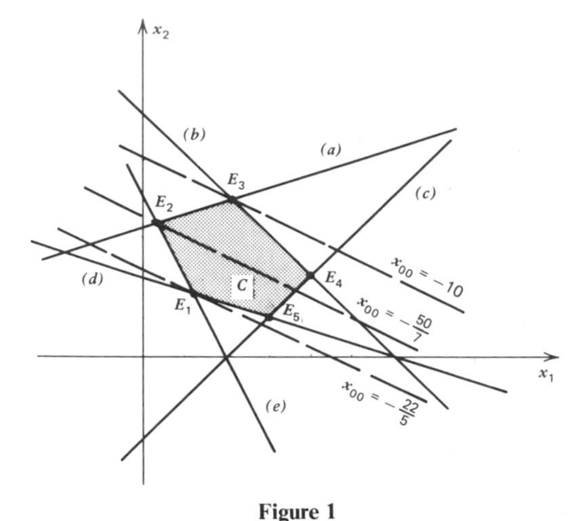

Figure 1

metric picture of the simplex method can be given. The shaded area C is the convex set of feasible solutions and the points E_i are extreme points. For this problem, the simplex method moves from the first basic feasible solution corresponding to E_1, to E_2, and finally to the optimum solution, E_3. The trace of the objective function through each of the extreme-point solutions is shown in the figure. In a sense, the simplex method moves the trace of the objective function from one extreme point to an adjacent extreme point until the optimum is reached. Note that if the trace of the objective function was parallel to the line joining E_3 and E_4, multiple optimum solutions would exist.

The kernel of the computational aspects of linear programming resides in the calculation of the inverse of the current basis. Given the inverse, all quantities required in a simplex iteration can be determined. This observation leads to the *revised simplex method*. Letting B_p be the feasible basis of the pth iteration, we have

$$X_{0p} = B_p^{-1} P_0$$

$$X_{jp} = B_p^{-1} P_j$$

$$\pi_p = c_p B_p^{-1}$$

$$\pi_p P_j = c_p B_p^{-1} P_j$$

$$\pi_p P_0 = c_p X_{0p} = c_p B_p^{-1} P_0,$$

where X_{0p} is the solution vector for the pth basic feasible solution; X_{jp} are the vectors that express the given vectors P_j as linear combinations of the basic vectors; c_p is the row vector of cost coefficients of the vectors in the pth basis; the elements of the row vector π_p are called the *simplex multipliers*; $\pi_p P_j$ is the indirect cost of the P_j vector; and $\pi_p P_0$ is the value of the objective function for the pth basis. From a computational point of view, using the explicit representation of the inverse and the simplex multipliers has a number of advantages. These include the reduction of the amount of computation and the reduction of the amount of information that has to be recorded per iteration. Whereas the standard simplex method transforms and records the complete simplex tableau, the revised procedure needs only to record the new inverse solution vector. Note that the revised procedure uses the original data in each step and that if, as is the case in many problems, the data contains many zeros, computation time can be saved.

The *product form* of the revised simplex method uses the fact that the inverse of the feasible basis, which starts out as an identity matrix, can be expressed as the product of elementary tranformation matrices. Each such matrix, which differs from a unit matrix in the lth column (l corresponds to the row position of the vector eliminated from the basis) contains the information necessary to determine the row inverse. For the pth iteration let

$$E_p^l = \begin{bmatrix} 1 & \cdots & y_{1l} & \cdots & 0 \\ \vdots & & \vdots & & \vdots \\ 0 & \cdots & y_{ll} & \cdots & 0 \\ \vdots & & \vdots & & \vdots \\ 0 & \cdots & y_{ml} & \cdots & 1 \end{bmatrix},$$

where

$$y_{il} = -x_{ik}/x_{lk}, \qquad i \neq l$$

$$y_{ll} = 1/x_{lk}.$$

The inverse for the pth basis is given by

$$E_p^l E_{p-1}^l \cdots E_2^l E_1 = B_p^{-1}$$

with $E_1 = I$. With this condensed form only

a limited amount of information needs to be recorded. It has been demonstrated that for most linear programming problems, the product form of the revised simplex method is the most efficient. Computational experience has shown that the number of iterations required to find an optimum solution can vary between m and $3m$. The number appears to be more a function of the number of equations than the number of variables. The number varies with the algorithm used, the method of finding the first feasible solution, and the criterion used to select a vector to go into the new basis.

Integer Programming

Many programming problems require the solution to be in terms of integers, e.g., whole units to be manufactured. The simplex algorithm does not guarantee an integer solution for the general linear programming problem. However, variations of the simplex method exist which find an integer optimum solution, if one exists. These integer programming* procedures add, in a systematic fashion, new constraints, or cutting planes, to the original set of constraints. The new constraints change the convex set of solutions so that it will contain as an optimum extreme point a point with integer coordinates. Other specialized computational procedures such as *branch-and-bound** or *enumerative algorithms* have been successful in solving problems of moderate size. Integer programming computer codes do exist, but as the amount of computation required to solve a particular problem appears to rest heavily on the structure and data of the problem, the utilization of these codes in an operational environment is open to question.

Upper-Bound Conditions

In many instances, the variables of a linear programming are bounded from above, i.e., $x_j \leqslant u_j$. A slight variation of the basic simplex method can solve the bounded problem without the explicit representation of the upper bound constraints. This procedure can be used when all or some of the variables are bounded. The situation when the x_j are bounded below is easily handled by direct substitution, i.e., if $d_j \leqslant x_k$. Let $x_j = d_j + x_j'$ and substitute $d_j + x_j'$ for the corresponding x_j, with $x_j' \geqslant 0$.

Quite a few linear programming applications in transportation, production scheduling, and resource allocation contain constraints that bound distinct subsets of the variables; i.e., they contain linear sums of the form $\sum_{j \in J} x_j = 1$, where J refers to a subset of the indices $j = 1, 2, \ldots, n$, and each j can appear at most once in some J. These equations are termed *generalized upper bounds* (*GUB*) *constraints*. A special form of the simplex algorithm has been developed that treats the GUB constraints implicitly, thus enabling problems with tens of thousands of such additional restrictions to be solved routinely.

Sensitivity Analysis

Sensitivity analysis in linear programming deals with the investigation of how the optimum solution varies with changes in the original data. For example, we are interested in how much a particular cost coefficient, c_j, can vary before the computed optimum is no longer optimal, or how much a right-hand-side coefficient, a_{i0}, can vary before the solution is no longer feasible, or, finally, what is the effect on the optimum solution if a change is made in an a_{ij}. Techniques are available for performing these analyses and are usually included in linear programming codes. Additional procedures exist for handling the *parametric programming* cases where each cost coefficient is a linear function of the same parameter, e.g., $c_j = d_j + \lambda d_j'$. This procedure yields sets of optimum solutions corresponding to ranges of the parameter.

Mathematical Programming Systems

Problems of over 4000 constraints and 10,000 variables are common and are solved routinely using present-day computers and extremely efficient computer codes called *mathematical programming systems* (*MPSs*).

The basic computational tool of the MPSs is the simplex algorithm, but it has been extended, refined, and honed to be an exceptional numerical analysis* tool. Many of these extensions take explicit advantage of the structure exhibited by just about all large-scale linear programming problems. This structure is usually due to the problem's multitime period basis and/or its consideration of interrelated production and distribution activities. Some special *decomposition algorithms* exist for particular structures. These include the *Dantzig–Wolfe decomposition algorithm* for block angular structures, and *Bender's partitioning algorithm*, which has been used to solve mixed-integer and other nonlinear problems. Most computer manufacturers support and maintain an MPS that includes provisions to solve GUB, integer, mixed-integer, parametric, and certain nonlinear programming* problems. *See* also MATHEMATICAL PROGRAMMING.

SELECTED LINEAR PROGRAMMING APPLICATIONS

Transportation Problems

BASIC TRANSPORTATION PROBLEM. A homogeneous product is to be shipped in the amounts $a_1, a_2, \ldots, a_m$, respectively, from each of m shipping origins and received in amounts $b_1, b_2, \ldots, b_n$, respectively, by each of n shipping destinations. The cost of shipping a unit amount from the ith origin to the jth destination is c_{ij} and is known for all combinations (i, j). The problem is to determine the amounts x_{ij} to be shipped over all routes (i, j) to minimize the total cost of transportation. The a_i are availabilities, and the b_j are requirements.

To ensure the consistency of the mathematical model, we restrict $\sum a_i = \sum b_j$; i.e., the sum of the availabilities equals the sum of the requirements. This is not a restriction that must be adhered to for a particular application in that if $\sum a_i > \sum b_j$, a dummy destination (e.g., a warehouse facility) is added to the problem with a requirement of $\sum a_i - \sum b_j$. If $\sum a_i < \sum b_j$, a dummy origin (e.g., purchases from a competitor) is added to the problem with an availability of $\sum b_j - \sum a_i$. Letting $x_{ij} \geqslant 0$ be the unknown amount to be shipped from origin i to destination j (i.e., the set of variables) the linear programming model for $m = 2$ and $n = 3$ is then

minimize

$$c_{11}x_{11} + c_{12}x_{12} + c_{13}x_{13} + c_{21}x_{21} + c_{22}x_{22} + c_{23}x_{23}$$

subject to

$$\begin{aligned}
x_{11} + x_{12} + x_{13} \qquad\qquad\qquad\qquad &= a_1\\
x_{21} + x_{22} + x_{23} &= a_2\\
x_{11} \qquad\qquad + x_{21} \qquad\qquad &= b_1\\
x_{12} \qquad\qquad + x_{22} \qquad &= b_2\\
x_{13} \qquad\qquad + x_{23} &= b_3,\\
x_{ij} &\geqslant 0.
\end{aligned}$$

As any one of the equations above can be deduced from the remaining $(m + n - 1)$, the rank of the system is $(m + n - 1)$. Any feasible basis is of order $(m + n - 1)$ and the corresponding matrix is triangular. This model has wide application with the sources and destinations taking on a variety of interpretations, e.g., warehouse, stores, ports, and the cost coefficients representing distances, times, dollars, etc. The model can be modified, for example, to supply information that would enable management to select from a set of possible new origins the one that is the best with respect to transportation costs. Here a set of problems would be solved, one for each possibility, with data reflecting the anticipated costs between the new origin and old destinations and the anticipated availability at the new origin and requirements for the destinations. A similar analysis would be done to aid in the selection of a new destination, or in the selection of an origin or destination to be closed down. Special simplex algorithms are available to solve transportation problems.

TRANSSHIPMENT PROBLEM. This is a transportation problem in which the origins and destinations can act as an intermediate shipping points from which the goods are transshipped to their final destination. This problem can be transformed into a basic transportation problem.

CAPACITATED TRANSPORTATION PROBLEM. This is a basic transportation problem in which the possible shipments are bounded from above, i.e., $x_{ij} \leqslant u_{ij}$. This problem can be solved with slight modification to the computational procedure for the basic problem.

GENERALIZED TRANSPORTATION PROBLEM AND MACHINE-ASSIGNMENT PROBLEM. In this problem, the constraint set is of the form

$$\text{minimize} \quad \sum_i \sum_j c_{ij} x_{ij}$$

subject to

$$\sum_{j=1}^{n} a_{ij} x_{ij} = a_i, \quad i = 1, \ldots, m, \tag{8}$$

$$\sum_{i=1}^{m} b_{ij} x_{ij} = b_j, \quad j = 1, \ldots, n, \; x_{ij} \geqslant 0. \tag{9}$$

Problems that fit this model can be found in the transportation field and in problems of machine assignment. For the latter, the $b_{ij} = 1$, and equation (8) becomes an inequality ($\leqslant$); the a_{ij} represent the time it takes to process 1 unit of product j on machine i; x_{ij}, the number of units of j produced on machine i; a_i, the time available on machine i; b_j, the number of units, j, that must be completed; and c_{ij}, the cost of processing 1 unit of product j on machine i.

Allocation or Assignment Problems

In this problem we have a number of individuals, machines, etc., to be assigned to perform a set of jobs. Each individual i has a given rating c_{ij} which measures his or her effectiveness in doing job j. An individual can be assigned to only one job. If x_{ij} represents the assignment of the ith individual to the jth job, the linear programming formulation is then

$$\text{maximize} \quad \sum_i \sum_j c_{ij} x_{ij}$$

$$\text{subject to} \quad \sum_j x_{ij} = a_i, \quad i = 1, \ldots, m,$$

$$\sum_i x_{ij} = b_j, \quad j = 1, \ldots, n,$$

$$x_{ij} \geqslant 0,$$

where a_i is the number of persons of type i available and b_j is the number of jobs of type j available. We assume that $\sum a_i = \sum b_j$. In many instances all a_i and b_j equal 1 and $m = n$. Mathematically, the assignment problem is a transportation problem.

Network-Flow Problems

Given a network (road, railway, pipeline, etc.) consisting of a single source (origin), a single sink (destination), and intermediate nodes (transfer points), let x_{ij} be the flow between any point i to point j of the network. A point is either the source, sink, or any node. We assume that the network is directed and that there is a capacity restriction $f_{ij} \geqslant 0$ on each arc (i, j). We wish to determine the flow f through the network which maximizes the flow from source to sink.

For a general network, with $i = 0$ the source node and $i = m$ the sink node, we have the following formulation of the *maximal flow problem*:

$$\text{maximize} \quad f$$

subject to

$$\sum_j (x_{0j} - x_{j0}) - f = 0 \tag{10}$$

$$\sum_j (x_{ij} - x_{ji}) = 0 \qquad (i \neq 0, m) \tag{11}$$

$$\sum_j (x_{mj} - x_{jm}) + f = 0, \tag{12}$$

$$0 \leqslant x_{ij} \leqslant f_{ij}. \tag{13}$$

Equations (10) to (12) represents the *conservation of flow* at the nodes; i.e., what flows in must flow out, and the summations are taken over those j which form arcs for the given i. The objective function maximizes the flow out of the source or the flow into the sink, (10) and (12).

A related problem is the *minimum-cost network problem.* Here we assume a linear cost function with cost coefficients c_{ij}, which represents the cost of sending 1 unit of flow across arc (i, j). There is a known amount of goods F at the source node that must be sent across the network to arrive at the sink node at minimum cost. The corresponding formulation is the following:

$$\text{minimize} \quad \sum_{(i,j)} c_{ij}x_{ij}$$

$$\text{subject to} \quad \sum_j (x_{0j} - x_{j0}) = F$$

$$\sum_j (x_{ij} - x_{ji}) = 0 \qquad (i \neq 0, m)$$

$$\sum_j (x_{mj} - x_{jm}) = -F,$$

$$0 \leqslant x_{ij} \leqslant f_{ij}.$$

If $F = 1$ and each c_{ij} is interpreted as the distance between the corresponding node i and node j, then the minimum-cost network problem can be interpreted as finding the shortest route from node 0 to node m. Special algorithms exist for solving all these network problems, many of which are not based on the simplex algorithm. The transportation problem is a minimum-cost network problem.

Activity-Analysis Problems

A manufacturing company has at its disposal fixed amounts of a number of different resources. These resources, such as raw material, labor, and equipment, can be combined to produce any one of several different commodities or combination of commodities. The manufacturer knows how much of resource i it takes to produce 1 unit of commodity j. The firm also knows how much profit it makes for each unit of commodity j produced. The manufacturer desires to produce that combination of commodities that will maximize the total profit. For this problem, we define the following:

m = number of resources

n = number of commodities

a_{ij} = number of units of resource i required to produce 1 unit of the commodity j

b_i = maximum number of units of resource i available

c_j = profit per unit of commodity j produced

x_j = level of activity (the amount produced) of the jth commodity.

The a_{ij} are sometimes called *input-output coefficients*.

The total amount of the ith resource that is used is given by the linear expression

$$a_{i1}x_1 + a_{i2}x_2 + \cdots + a_{in}x_n.$$

Since this total amount must be less than or equal to the maximum number of units of the ith resource available, we have, for each i, a linear inequality of the form

$$a_{i1}x_1 + a_{i2}x_2 + \cdots + a_{in}x_n \leqslant b_i.$$

Since a negative x_j has no appropriate interpretation, we require that all $x_j \geqslant 0$. The profit derived from producing x_j units of the jth commodity is given by c_jx_j. This formulation yields the problem of maximizing the profit function

$$c_1x_1 + c_2x_2 + \cdots + c_nx_n$$

subject to the conditions

$$\begin{aligned} a_{11}x_1 + a_{12}x_2 + \cdots + a_{1n}x_n &\leqslant b_1 \\ a_{21}x_1 + a_{22}x_2 + \cdots + a_{2n}x_n &\leqslant b_2 \\ \vdots \qquad \vdots \qquad\quad \vdots \quad &\quad \vdots \\ a_{m1}x_1 + a_{m2}x_2 + \cdots + a_{mn}x_n &\leqslant b_m, \\ x_j &\geqslant 0. \end{aligned}$$

Diet Problem

Here we are given the nutrient content of a number of different foods. For example, we might know how many milligrams of phosphorus or iron are contained in 1 ounce of each food being considered. We are also given the recommended daily allowance (RDA) for each nutrient. Since we know the cost per ounce of food, the problem is to determine the diet that satisfies the RDAs and is also the minimum-cost diet. Define

m = number of nutrients

n = number of foods

a_{ij} = number of milligrams of the ith nutrient in 1 ounce of the jth food

b_i = RDA in milligrams of the ith nutrient

c_j = cost per ounce of the jth food

x_i = number of ounces of the jth food to be purchased ($x_j \geqslant 0$).

The total amount of the ith nutrient contained in all the purchased food is given by

$$a_{i1}x_1 + a_{i2}x_2 + \cdots + a_{in}x_n .$$

Since this total amount must be greater than or equal to the RDA of the ith nutrient, this linear programming problem involves minimizing the cost function

$$c_1x_1 + c_2x_2 + \cdots + c_nx_n$$

subject to the conditions

$$\begin{aligned}
a_{11}x_1 + a_{12}x_2 + \cdots + a_{1n}x_n &\geqslant b_1\\
a_{21}x_1 + a_{22}x_2 + \cdots + a_{2n}x_n &\geqslant b_2\\
\vdots \qquad \vdots \qquad\qquad \vdots \quad &\quad \vdots\\
a_{m1}x_1 + a_{m2}x_2 + \cdots + a_{mn}x_n &\geqslant b_m ,\\
x_j &\geqslant 0.
\end{aligned}$$

Although questionable when applied directly to diets for human consumption, this formulation has proved highly satisfactory for the evaluation of diets for cattle and chickens. Special menu planning models are used for human diets.

Blending Problem

Blending problems refer to situations where a number of components are mixed together to yield one or more products. There are restrictions on the available quantities of raw materials, restrictions on the quality of the product, and restrictions on the quantities of the products to be produced. There is usually a very large number of different ways in which the raw materials can be blended to form the final products while satisfying the various constraints. It is desired to carry out the blending operation so that the given objective function can be optimized. Blending problems are found in the making of petroleum (blending of gasolines), paint, steel, etc.

Production Scheduling

A manufacturing firm knows it has to produce $r_i (i = 1, \ldots, n)$ items of a certain commodity during the next n months. They can be produced either in regular time, subject to a ceiling of a_i per month, or in overtime, subject to a ceiling of b_i per month. The cost of one item produced in the ith month is c_i on regular time and d_i on overtime. The variation of cost with time and also the capacity restrictions might make it more economical to produce in advance of the period when the items are acually needed. Storage cost is assumed to be s_i per item in each month. We wish to determine the production schedule that minimizes the sum of production and storage costs. This problem can be formulated as a standard linear programming model or as a transportation problem.

Trim-Loss Problem

Paper mills produce rolls of a given, standard width. Customers require rolls of various widths, and hence the rolls of standard width must be cut. In general, some waste occurs at the end of the cutting process, i.e., trim loss. A mill wishes to cut the rolls as ordered by its customers and to minimize the total trim loss. This application applies to similar manufacturing situations in which a standard roll, sheet, etc., must be cut with resulting trim loss.

Caterer's Problem

A catering firm knows that, in connection with the meals it has arranged to serve during the next n days, it will need r_j fresh napkins on the jth day, with $j = 1, 2, \ldots, n$. Laundering normally takes p days; i.e., a soiled napkin sent for laundering immediately after use on the jth day is returned in time to be used again on the $(j + p)$th day. However, the laundry also has a higher-cost service which returns the napkins in $q < p$ days (p and q integers). Having no usable napkins on hand or in the laundry, the caterer will meet its early needs by purchasing napkins at a cents each. Laundering expenses are b and c cents per napkin for the normal and high-cost services, respectively. How does the firm arrange matters to meet its needs and minimize its outlays for the n days?

Critical-Path Planning and Scheduling

A characteristic of many projects is that all work must be performed in some well-defined order; e.g., in construction work, forms must be built before concrete can be poured. This formulation concerns the scheduling of the jobs that combine to make a project. The analysis requires a graphical representation of a project for which the cost and starting and ending times for each job of the project are known. The linear programming formulation provides a means of selecting the least costly schedule for desired and feasible project-completion time.

Zero-Sum Two-Person Games

Any zero-sum two-person game can be represented as a linear programming problem, and that linear program can also be transformed into a zero-sum two-person game. For a game, the dual of the linear-programming formulation for one player represents the linear programming formulation for the second player. The computational procedure from one area can be applied to solve a problem from the other area. *See* also GAME THEORY.

LINEAR PROGRAMMING AND STATISTICAL ANALYSIS

Linear programming theory usually assumes that the model's coefficients (i.e., the a_{ij}, b_i, c_j) are determined. However, as many of these data are measured and/or estimated, most analysts will conduct sensitivity studies and/or parametric analyses on selected data to obtain information on the stability of the optimum solution. A proper extension of such concerns is to study the situation when some or all of the model's coefficients are assumed to be random variables. This type of generalization is treated under the (usually equivalent) headings of *stochastic programming, programming under uncertainty*, or *probabilistic programming.*

The *stochastic programming model* can be considered to

$$\text{minimize} \qquad x_0 = \sum_{j=1}^{n} c_j x_j$$

$$\text{subject to} \qquad \sum_{j=1}^{n} a_{ij} x_j \geqslant b_i \qquad (i = 1, \ldots, m),$$

$$x_j \geqslant 0,$$

where the coefficients are random variables having known distributions. If only the cost coefficients c_j are randomly distributed independently of the x_j, then the minimum expected total cost solution is obtained by minimizing the objective function x_0 in which the expected costs $\bar{c}_j$ are substituted for the c_j. If c is a random vector with mean vector $\bar{c} = (\bar{c}_1, \ldots, \bar{c}_n)$ and covariance matrix V, then x_0 will be a random vector with mean $\bar{c}X$ and variance $X'VX$. If the vector c has a joint normal distribution, then x_0 is also normally distributed.

A basic stochastic problem known as the *two-stage model with recourse* is to obtain

$$\begin{array}{ll} \min_X & E\left[cX + \min_Y (dY)\right] \\ \text{subject to} & AX + BY = b \\ & X \geqslant 0 \\ & Y \geqslant 0, \end{array}$$

where E is expectation with respect to the random vector b, with c and A also random variables. The matrix B is a diagonal matrix with plus or minus ones, and d is a nonrandom penalty cost vector. Here a decision maker chooses an $X \geqslant 0$, then observes the random matrix A and random vector b and compares AX with b. The chosen X may or may not be feasible. In either event, the decision maker is allowed to make another decision Y, after the fact, to correct for any differences between AX and b. This correction is made at a penalty cost of dY. Explicit expressions for the objective function have been determined for certain distributions. In general, the minimum is convex in X and in A and b, but concave in d.

Another probabilistic model is that of *chance-constrained programming*. The basic model is

$$\begin{array}{ll} \text{minimize} & \sum_{j=1}^{n} c_j x_j \\ \text{subject to} & \Pr\left\{ \sum_{j=1}^{n} a_{ij} x_j \geqslant b_i \right\} \geqslant \alpha_i \\ & \qquad (i = 1, \ldots, m), \\ & x_j \geqslant 0. \end{array}$$

where the α_i are given probabilities not equal to zero, and, in general, the a_{ij}, b_i, and c_j are random variables. Depending on the form of the associated distributions, an approach to solving this problem is to convert it to a set of deterministic constraints.

Two statistical problems of interest can be formulated and solved by standard linear programming procedures. They are the L_1 absolute deviation regression problem and the L_∞ Chebyshev problem; their formulations follow.

The problem of minimizing the sum of absolute deviations can be handled as follows. Let a_{ij}, $i = 1, 2, \ldots, k$ and $j = 1, 2, \ldots, p$ denote a set of k observational measurements on p independent variables, and b_i, $i = 1, 2, \ldots, k$, denote the associated measurements on the dependent variable. The problem is to find the regression coefficients x_j that minimize $\sum_i |\sum_j a_{ij} x_j - b_i|$; that is, we wish to find values of the regression coefficients such that the sum of the absolute differences $|\sum_j a_{ij} x_j - b_i|$ is a minimum. Since for any set of x_j the expression $b_i - \sum_j a_{ij} x_j$ can be positive or negative, we represent this difference as the difference of two nonnegative numbers; i.e., we let $z_i' - z_i'' = b_i - \sum_j a_{ij} x_j$, $z_i' \geqslant 0$, $z_i'' \geqslant 0$. We now rewrite the problem as follows:

$$\begin{array}{ll} \text{minimize} & \sum_{i=1}^{k} (z_i' + z_i'') \\ \text{subject to} & \sum_{j=1}^{p} a_{ij} x_j + z_i' - z_i'' = b_i \\ & \qquad (i = 1, \ldots, k), \\ & z_i' \geqslant 0, \\ & z_i'' \geqslant 0, \end{array}$$

with the variables x_j unrestricted as to sign. Each x_j can be written as the difference of two nonnegative variables, to put the problem into the standard linear programming format. Since in a basic feasible solution z_i' and z_i'' cannot both be positive, the optimum basic solution will select a set of x_j that minimizes the sum of the absolute differences.

In a similar vein, the Chebyshev problem for a set of observations is to find a set of coefficients x_j such that we can obtain $\min_{x_j} \{\max_i |\sum_j a_{ij} x_j - b_i|\}$; that is, we wish to find a set of x_j such that the maximum deviation between the corresponding observations is a minimum. Let $z \geqslant |\sum_j a_{ij} x_j - b_i|$

for all i. The variable z is nonnegative and we wish to have z a minimum. This inequality in absolute terms can be rewritten for each i as two inequalities; i.e., we have that the value of $\sum_j a_{ij}x_j - b_i$ can lie between z and $-z$, or $-z \leqslant \sum_j a_{ij}x_j - b_i \leqslant z$. The Chebyshev problem is now as follows:

$$\text{minimize} \quad z$$

$$\text{subject to} \quad \sum_{j=1}^{p} a_{ij}x_j - b_i - z \leqslant 0 \qquad (i = 1, \ldots, k)$$

$$\sum_{j=1}^{p} a_{ij}x_j - b_i + z \geqslant 0,$$

$$z \geqslant 0,$$

and x_j unrestricted as to sign.

LITERATURE

The original papers describing the linear programming problem and the simplex algorithm, due to Dantzig, are contained in the volume edited by Koopmans [15]; see also Dantgiz's book [9], which includes a history of the subject. The books by Dantzig [9], and Charnes and Cooper [7] encompass most of linear-programming (and extensions) through the early 1960s. The more recent texts by Gass [12, 13], Bazaraa and Jarvis [3], and Bradley et al. [5] cover, collectively, most of the important material of the field through the 1970s. The field is a most active one; linear programming is now included in the more general area of mathematical programming. Many research papers appear in the journals *Mathematical Programming*, *Operations Research*, and *Management Science*.

The relationship between linear programming and economics is described in the books by Baumol [2] and Gale [10]. Vajda's book [21] covers the basic concepts that relate the fields of probability, statistics, and mathematical programming; see also the paper by Walkup and Wets [22] and the early papers by Charnes and Cooper [6], Dantzig [8], Madansky [17], Tintner [19, 20], and Williams [23].

References

[1] Appa, A. and Smith, C. (1973). *Math. Programming*, **5**, 73–87.

[2] Baumol, W. J. (1965) *Economic Theory and Operations Analysis*, 2nd ed. Prentice-Hall, Englewood Cliffs, N.J.

[3] Bazaraa, M. S. and Jarvis, J. (1977). *Linear Programming and Network Flows*. Wiley, New York.

[4] Benders, J. F. (1962). *Numer. Math.* **4**, 238–252.

[5] Bradley, S., Hax, A., and Magnanti, T. (1977). *Applied Mathematical Programming*. Addison-Wesley, Reading, Mass.

[6] Charnes, A. and Cooper, W. W. (1959). *Manag. Sci.*, **6**, 73–79.

[7] Charnes, A. and Cooper, W. W. (1960). *Management Models and Industrial Applications of Linear Programming*, 2 vols. Wiley, New York.

[8] Dantzig, G. B. (1955). *Manag. Sci.* **1**, 197–206.

[9] Dantzig, G. B. (1963). *Linear Programming*. Princeton University Press, Princeton, N.J.

[10] Gale, D. (1960). *The Theory of Linear Economic Models*. McGraw-Hill, New York.

[11] Garvin, W. W. (1960). *Introduction to Linear Programming*. McGraw-Hill, New York.

[12] Gass, S. I. (1970). *An Illustrated Guide to Linear Programming*. McGraw-Hill, New York.

[1`] Gass, S. I. (1975). *Linear Programming: Methods and Applications*, 4th ed. McGraw-Hill, New York.

[14] Hansotia, B. J. (1980). *Decis. Sci.*, **11**, 151–168.

[15] Koopmans, T. C., ed. (1951). *Activity Analysis of Production and Allocation*. Cowles Comm. Monogr. *No.* 13. Wiley, New York.

[16] Luce, R. D. and Raiffa, H. (1957). *Games and Decisions*. Wiley, New York.

[17] Madansky, A. (1963). In *Recent Advances in Mathematical Programming*, R. L. Graves and P. Wolfe, eds. McGraw-Hill, New York, pp. 103–120.

[18] Sengupta, J. K. (1972). *Stochastic Programming*. North-Holland, Amsterdam, Netherlands.

[19] Tintner, G. (1955). In *Proc. 2nd Symp. Linear Programming*, H. A. Antosiewicz, ed. U.S. Air Force, Washington, D.C., pp. 197–228.

[20] Tintner, G. (1960). *Econometrica*, **28**, 490–495.

[21] Vajda, S. (1972). *Probabilistic Programming*. Academic Press, New York.

[22] Walkup, D. W. and Wets, R. J.-B. (1967). *SIAM J. Appl. Math.* **15**, 1299–1314.

[23] Williams, A. C. (1965). *SIAM J. Appl. Math.* **13**, 927–940.

(BRANCH-AND-BOUND METHOD
GAME THEORY
INTEGER PROGRAMMING
MATHEMATICAL PROGRAMMING
NETWORK ANALYSIS
NONLINEAR PROGRAMMING
OPERATIONS RESEARCH
OPTIMIZATION IN STATISTICS)

SAUL I. GASS

LINEAR RANK TESTS

Linear rank statistics are concerned with testing the null hypothesis that random variables $Y_1, Y_2, \ldots, Y_n$ are independent and identically distributed, versus alternatives in which the distribution of Y_i, $i = 1, \ldots, n$, depends in some manner on a corresponding "regression" p vector, $\mathbf{x}_i' = (x_{1i}, \ldots, x_{pi})$, where the prime denotes vector transpose. Linear rank tests are of the form

$$\mathbf{S} = \sum_{i=1}^{n} \mathbf{x}_i a_n(R_i), \tag{1}$$

where R_i is the number of Y's equal to or less than Y_i and $a_n(i)$ is a "score" to be associated with the ith smallest of the Y's. A simple and useful scoring scheme is given by

$$a_n(R_i) = R_i, \qquad i = 1, \ldots, n, \tag{2}$$

in which case $\mathbf{S}$ is usually referred to as a Wilcoxon-type linear rank statistic [19, 29]. Other scoring systems include the van der Waerden [28] scores

$$a_n(R_i) = \Phi^{-1}[R_i/(n+1)], \quad i = 1, \ldots, n, \tag{3}$$

where Φ is the normal distribution function, and the Savage [27] scores

$$a_n(R_i) = n^{-1} + (n-1)^{-1} + \cdots + (n - R_i + 1)^{-1}, \qquad i = 1, \ldots, n. \tag{4}$$

The study of tests of the form (1) is largely motivated by a desire to avoid unnecessary distributional assumptions. The paper of Hotelling and Pabst [10] on rank correlation* is often credited as the beginning of study of nonparametric methods, although some earlier work can be identified. The work of Wilcoxon, cited above, appears to have sparked interest in the use and evaluation of rank statistics. The results of Chernoff and Savage [2] stimulated research on theoretical properties of linear rank tests, much of which is summarized in Hájek and Šidák [8]. There are a number of recent introductory and not-so-introductory books on nonparametric statistical methods that summarize the extensive literature on linear rank tests. Some of these are referenced below.

Under the null hypothesis each of the $n!$ possible rank vectors $(R_1, \ldots, R_n)$ is equally probable. It follows that the null hypothesis distribution of $\mathbf{S}$ is entirely independent of the (common) distribution of $Y_1, \ldots, Y_n$, whence $\mathbf{S}$ earns the adjective nonparametric, or distribution-free.

Simple permutation arguments can be used to calculate the moments of $\mathbf{S}$ under the null hypothesis. For example, in exactly $(n-1)!$ of the $n!$ possible rank vectors a particular $\mathbf{x}_i$ will multiply $a_n(j)$, for each $j = 1, \ldots, n$ in $\mathbf{S}$. It follows that

$$E(\mathbf{S}) = \sum_{i=1}^{n} \mathbf{x}_i (n-1)! [a_n(1) + \cdots + a_n(n)]/n!$$
$$= n^{-1} \sum_{i=1}^{n} \mathbf{x}_i \sum_{j=1}^{n} a_n(j) = n \bar{\mathbf{x}} \bar{a}_n. \tag{5}$$

Similarly,

$$\operatorname{Var} \mathbf{S} = (n-1)^{-1} \mathbf{X}'\mathbf{X} \sum_{1}^{n} [a_n(i) - \bar{a}_n]^2, \tag{6}$$

where the $p \times p$ matrix $\mathbf{X}'\mathbf{X}$ has (j, k)th element

$$\sum_{1}^{n} (x_{ji} - \bar{x}_j)(x_{ki} - \bar{x}_k).$$

The significance level of the test against "two-sided" (usually linear location shift) alternatives can be calculated by summing the probability that the distance between **S** and its null hypothesis expectation $E(\mathbf{S})$ is equal to or greater than the distance between **s**, the observed **S**-value, and $E(\mathbf{S})$. With moderate to large sample sizes this probability is, quite generally, well approximated by comparing

$$[\mathbf{s} - E(\mathbf{S})]'(\operatorname{Var}\mathbf{S})^{-1}[\mathbf{s} - E(\mathbf{S})] \quad (7)$$

with χ_p^2 tables, where it has been assumed that $\operatorname{var}\mathbf{S}$ is of full rank p.

Attractive features of linear rank tests, relative to standard parametric analysis of variance and regression tests, include freedom from distributional assumptions and associated efficiency robustness under departures from standard distributional assumptions. As illustrated below, linear rank tests also have a computational advantage in certain special cases. Before considering further the propertiies of linear rank statistics, some important special cases of (1) will be discussed.

RANK TESTS FOR THE COMPARISON OF TWO POPULATIONS

Suppose that $Y_1, \ldots, Y_m$ (sample 1) are independent and identically distributed (i.i.d.) with distribution function F_1, while $Y_{m+1}, \ldots, Y_n$ (sample 2) are i.i.d. with distribution function F_2. A linear rank statistic of the form (1) can be used to test the null hypothesis $F_1 = F_2$ versus, for example, a two-sided alternative in which $F_1 \geqslant F_2$ [i.e., $F_1(y) \geqslant F_2(y)$ for all y] or $F_1 \leqslant F_2$ but $F_1 \neq F_2$. To do so, define $X_i = 0$ for $i = 1, \ldots, m$ and $x_i = 1$ for $i = m+1, \ldots, n$ (any two values for the x's will do). The linear rank statistic S is then simply the sum of scores, $a_n(R_{m+1}) + \cdots + a_n(R_n)$, in the second sample.

As a simple illustration, consider the Wilcoxon rank-sum test with scores (2). Suppose that $m = 5$, $n = 9$, and that the first sample's y-values are 2.1, 4.7, 6.8, 7.9, and 8.6, while the second sample's y-values are 7.5, 8.9, 9.2, and 9.3. Summing the ranks in the second sample gives $s = 4 + 7 + 8 + 9 = 28$. The null hypothesis expectation of S is, from (5), $9(4/9)(10/2) = 20$. The significance level of the (two-sided) test is then

$$\Pr[|S - 20| \geqslant 8] = \Pr[S \leqslant 12] + \Pr[S \geqslant 28].$$

Tail area probabilities for the Wilcoxon rank-sum or the equivalent Mann–Whitney test* are tabulated in many places. For example, the tables of Lehmann [18, pp. 408–410] give

$$\Pr[S \geqslant 28] = \Pr[U \geqslant 18] = 1 - 0.9683 = 0.0317,$$

where $U = S - \frac{1}{2}(n-m)(n-m+1)$ has been standardized to have a minimum value of zero and $P[U \leqslant 17]$ is obtained by table lookup. Since the null hypothesis distribution of S is symmetric about its mean [whenever $a_n(i)$, $i = 1, \ldots, n$, are symmetric about $\bar{a}_n$] it follows that $\Pr[S \leqslant 12] = 0.0317$, so that the desired significance level is 0.0634. This method gives a distribution-free, exact significance test that is computationally simple and is even amenable to hand calculation in small or moderate size samples.

In the illustration above the significance level calculation is simple enough that a table need not be utilized. Specifically, $S \geqslant 28$ will occur if and only if the ranks $\{R_6, R_7, R_8, R_9\}$ in the second sample are one of the sets $\{4, 7, 8, 9\}$, $\{5, 7, 8, 9\}$, $\{6, 7, 8, 9\}$, or $\{5, 6, 8, 9\}$, so that

$$\Pr[S \geqslant 28] = 4(4!\,5!/9!) = 0.0317,$$

where the 4! and 5! factors arise since the ranks within samples can be permuted without changing the value of S.

If neither m nor $n - m$ is small (e.g., m, $n - m \geqslant 8$), one can use the χ^2 approximation to (7) to obtain the significance level. In the illustration above one obtains

$$\operatorname{var} S = (n-1)^{-1}[m(n-m)/n] \times [n(n+1)(n-1)/12] = m(n-m)(n+1)/12 = 5/3,$$

so that (7) has value $8(50/3)^{-1}8 = 3.84$ with corresponding significance level 0.05 from χ_1^2 tables, in good agreement with the exact calculation in spite of the small sample sizes.

A number of other linear rank statistics have been proposed for the two-sample problem. For example, the normal scores test* (e.g., Fisher and Yates [5]) arises if $a_n(i)$ is the ith order statistic in a standard normal sample of size n. The test based on van der Waerden scores (3) is asymptotically equivalent to the normal scores test. The easily computed sign (median) test is given by scores $a_n(i) = -1$ if $i < \frac{1}{2}(n+1)$, $a_n(i) = 0$ if $i = \frac{1}{2}(n+1)$, and $a_n(i) = 1$ if $i > \frac{1}{2}(n+1)$. In contrast to these special cases, the Savage or exponential scores* (4) are not symmetric about their mean, so the tail areas must be computed separately for a two-sided test. The reader may wish to verify that the significance level for the Savage test is 0.0317 based on the small data set given above.

MORE GENERAL LINEAR RANK TESTS

Comparison of the distributions of $p+1$ populations ($p > 1$) can be based on a linear rank test of the form (1) upon defining $\mathbf{x}_i' = [x_{1i}, \dots, x_{pi}]$ such that $x_{ji} = 1$ if the ith random variable, Y_i, arises from the jth population and $x_{ji} = 0$ otherwise (indicator variables corresponding to any p of the $p+1$ samples would give an equivalent test). Any of the specific scoring schemes $\{a_n(i), i = 1, \dots, n\}$ mentioned above for the two-sample problem could be used in the $p+1$ sample problem. With Wilcoxon scores, $a_n(i) = i$, the statistic (7) is referred to as the Kruskal–Wallis* [16, 17] statistic. After some algebra this statistic can be written

$$\frac{12}{n(n+1)} \sum_{j=1}^{p+1} n_j \left[R_{j\cdot} - \frac{1}{2}(n+1) \right]^2, \quad (8)$$

where $R_{j\cdot}$ is the average of the ranks in the jth sample and n_j is the size of the jth sample, $j = 1, \dots, p+1$. If each of the n_j are moderate to large (e.g., $n_j \geqslant 8$, all j), then the (null hypothesis) distribution of (8) will be well approximated by a χ_p^2 distribution.

Suppose now that random variables $W_1, \dots, W_n$ arise from a linear regression* model

$$W_i = \alpha + \boldsymbol{\beta}\mathbf{x}_i + \sigma\epsilon_i, \quad (9)$$

where $\boldsymbol{\beta} = (\beta_1, \dots, \beta_p)$ is a row p-vector of coefficients, α and $\sigma > 0$ are (unknown) parameters, and the ϵ_i, $i = 1, \dots, n$, are i.i.d. with distribution function F. The linear rank statistic (1) can be used to test the hypothesis $\boldsymbol{\beta} = \boldsymbol{\beta}_0$ upon setting

$$Y_i = W_i - \boldsymbol{\beta}_0\mathbf{x}_i = \alpha + (\boldsymbol{\beta} - \boldsymbol{\beta}_0)\mathbf{x}_i + \sigma\epsilon_i, \quad i = 1, \dots, n. \quad (10)$$

Under this hypothesis $Y_i, \dots, Y_n$ are i.i.d. with mean and variance given by (5) and (6). In many applications the sample size and regression vectors will be such that a χ_p^2 approximation to (7) will be appropriate (asymptotic results are discussed below). Exact significance levels are, however, tabled (e.g., Lehmann [18, p. 433]) for the equally spaced special case $x_i = i$, $i = 1, \dots, n$, and $a_n(i) = i$ for $n \leqslant 11$. Such tables usually give significance levels corresponding to (n, D), where

$$D = \sum_{i=1}^{n} (R_i - i)^2 = n(n+1)(2n+1)/3 - 2S. \quad (11)$$

The statistic D in (11) can be used as a test for trend* between Y_i and x_i upon replacing x_i by i in the sequence $x_1 < x_2 < \cdots < x_n$.

In some applications x_i-values will arise as independent realizations of a random variable X. The linear rank test (1) then will test the hypothesis of independence of the conditional distribution of Y given $X = x$ on x, that is, the hypothesis of independence of Y and X. A test that is symmetric in Y and X would replace x_i values by their ranks in the sequence $\{x_1, \dots, x_n\}$. If the data are then recorded so that $x_1 < x_2 < \cdots x_n$, the resulting statistic, with $a_n(i) = i$, is again $\sum_1^n iR_i$ so that the tables for the equivalent statistic (11) can be used to test independence. The corresponding statistic (7) is a multiple of Spearman's rank correlation* coefficient. This same statistic can be used in the $p+1$ sample problem to test the hypothesis of equal distributions in each sample, versus an ordered alternative in which, for example, $F_1 \geqslant F_2 \geqslant \cdots \geqslant F_{p+1}$, with at least one inequality. If the study design suggests such a natural ordering among groups this statistic can, of course, be expected to be more powerful than the corresponding Kruskal–Wallis test, particularly if the number of groups is at all large. A variety of tests could be generated for these purposes corresponding to possible choices of the scores $\{a_n(i)\}$ and possible choices of the "weights" associated with the ordered independent variables $x_1 < \cdots < x_n$.

It is perhaps worth mentioning that another popular nonparametric statistic, the Wilcoxon signed-rank test*, is not a linear rank test as defined in (1). This statistic is used to test the hypothesis that random variables $Z_1, \ldots, Z_n$ are i.i.d. from a distribution that is symmetric about 0, as may be of interest, for example, in the analysis of differences between response pairs in a matched-pair experiment. The Wilcoxon signed-rank test can be written

$$\sum_{i=1}^{n} X_i R_i, \tag{12}$$

where $X_i = 0$ or 1 according to whether $Z_i < 0$ or $Z_i \geqslant 0$ and $Y_i = |Z_i|$ (R_i is the rank of Y_i as usual). As such the statistic resembles the Wilcoxon rank-sum statistic. The distribution theory, however, differs since, for example, the number of X_i's that take value 1 is not fixed by the null hypothesis (see, e.g., Lehmann [18, p. 123] for further discussion of this statistic). Other scoring systems could, of course, be substituted for Wilcoxon scores in (12).

GENERATION AND ASYMPTOTIC PROPERTIES OF LINEAR RANK TESTS

Consider the linear regression model given by (10) with error density f. The probability of the rank vector $\mathbf{R} = (R_1, \ldots, R_n)$ for the Y's can be written

$$\begin{aligned}\Pr[\mathbf{R} : \boldsymbol{\gamma}] &= \int \cdots \int_{A_n} \prod_1^n f[\{Y_i - \alpha \\ &\qquad - (\boldsymbol{\beta} - \boldsymbol{\beta}_0)\mathbf{x}_i\}/\sigma]\, dY_i/\sigma \\ &= \int \cdots \int_{B_n} \prod_1^n f(Z_i - \boldsymbol{\gamma}\mathbf{x}_i)\, dZ_i,\end{aligned} \tag{13}$$

where $(Y_{(1)}, \ldots, Y_{(n)})$ is the order statistic for the Y's,

$$Z_i = (Y_i - \alpha)/\sigma, \qquad i = 1, \ldots, n,$$

with corresponding order statistic $(Z_{(1)}, \ldots, Z_{(n)})$, $Y_{(1)} < \cdots < Y_{(n)}$ in A_n, $Z_{(1)} < \cdots < Z_{(n)}$ in B_n, and $\boldsymbol{\gamma} = (\boldsymbol{\beta} - \boldsymbol{\beta}_0)/\sigma$. The rank probability therefore only depends on the parameter $\boldsymbol{\gamma}$, and the hypothesis $\boldsymbol{\beta} = \boldsymbol{\beta}_0$ is equivalent to $\boldsymbol{\gamma} = \mathbf{0}$. A locally most powerful rank test is provided by the statistic

$$\mathbf{S} = \partial \log \Pr[\mathbf{R} : \boldsymbol{\gamma}]/\partial \boldsymbol{\gamma}|_{\boldsymbol{\gamma}=\mathbf{0}},$$

which, assuming that differentation and integration can be interchanged, can be shown to be

$$\mathbf{S} = \sum_{i=1}^{n} \mathbf{x}_i E\{\phi_f(u_i)\}, \tag{14}$$

where $\phi_f(u) = -f'[F^{-1}(u)]/f[f^{-1}(u)]$ for $0 < u < 1$, [note that $f'(y) = df(y)/dy$] and u_i is the ith order statistic in a uniform sample of size n. Another statistic that is quite generally asymptotically equivalent to (14) (see, e.g., Hájek and Šidák [8, p. 163]) is given by

$$\mathbf{S} = \sum_{i=1}^{n} \mathbf{x}_i \phi_f\{i/(n+1)\}. \tag{15}$$

Rank statistics (1), with some known optimality properties therefore can be generated by selecting an error density f and computing the scores $a_n(i) = E\{\phi_f(u_i)\}$ in (14) or the scores $a_n(i) = \phi_f\{i/(n+1)\}$ in (15). For example, logistic*, normal*, and exponential* densities f give, respectively, Wilcoxon, normal scores, and Savage-type scores in (14). Similarly, logistic, normal, and double exponential* densities give, respectively, Wilcoxon, van der Waerden, and sign* (median) scores in (15).

Standard versions of the central limit theorem do not apply immediately to the statistics (1) since the random variables being summed are dependent. Hájek and Šidák [18, p. 216] give some flexible sufficient conditions on the regression variables, x_i, and the scores, $a_n(i)$, under which (1), appropriately standardized, will be asymptotically normal. For example, with scalar x the major conditions are that scores be generated according to (14) or (15) with slight restrictions on f and that

$$\sum_1^n (x_i - \bar{x})^2 \Big/ \max_{1 \leqslant i \leqslant n} (x_i - \bar{x})^2 \to \infty$$

as $n \to \infty$. The latter conditions ensure that each observation, Y_i, make an asymptotically trivial contribution to the test statistic. Derivation of the asymptotic distribution of statistics of the form (1) under a variety of alternatives has stimulated the development of new weak convergence methods and results of great interest (see, e.g., Pyke [25] for a discussion of this topic). Some particularly important papers are Chernoff and Savage [2], Pyke and Shorack [26], and Hájek [7]. Results with contiguous alternatives (e.g., Hájek and Šidák [8, p. 268]) show that the (Pitman) efficiency of tests of the form (14) or (15) within the regression model (10) will quite generally be given by

$$\frac{\left\{ \int_0^1 \phi_f(u) \phi_{f_0}(u)\, du \right\}^2}{\int_0^1 \phi_f^2(u)\, du}, \tag{16}$$

where f_0 is the "true" sampling density in (10) and the Fisher information* terms in the denominator are assumed to be finite. Expression (16) indicates the (asymptotic) efficiency of such tests against linear regression alternatives to be a property of the sampling and "score generating" densities only and not of the regression matrix. One can also readily note that the efficiency is one if f and f_0 agree up to location scaling. Some interesting special cases of (16) show that the efficiency of a linear rank test with normal scores will, under mild conditions on f_0 (e.g., Puri and Sen [24, p. 118]), be equal to or greater than that of the corresponding least squares test, with equality only if f_0 is normal. Similarly, if f_0 is normal, Wilcoxon-type liner rank statistics will have efficiency 0.955 relative to the optimal least-squares procedure and its efficiency is equal to or greater than 0.864 under any sampling distribution with finite variance [9]. On the other hand, the efficiency of the least-squares procedure relative to the Wilcoxon-type linear rank test can be very low, particularly if f_0 has one or more heavy tails (e.g., efficiency is zero under Cauchy sampling). Good efficiency properties under a range of distributional assumptions is one of the principal attractions of linear rank test procedures.

CONFIDENCE REGIONS AND RANK REGRESSION ESTIMATORS

Consider again the linear regression model (9). A $100(1 - \alpha)\%$ confidence region for $\boldsymbol{\beta}$ may be obtained by grouping together all $\boldsymbol{\beta}_0$ values such that the corresponding linear rank statistic based on (10) is not significant at level α (see, e.g, Bickel and Doksum [1, p. 179] for a discussion of such test "inversion"). Similarly, one may define an estimator of $\boldsymbol{\beta}$ as the $\boldsymbol{\beta}_0$ value for which the corresponding linear rank test is least significant in some sense (i.e., is as close as possible to a zero vector). Such rank regression estimators, with score generating density f, will quite generally have efficiency given by (16) relative to the optimal maximum likelihood estimator based on the actual (unknown) sampling density f_0 (see Jurečková [13, 14] and Jaeckel [12]). The rank regression estimator generated by a sampling density f can, in fact, be thought of asymptotically as an adaptive M-estimator* (e.g., Huber [11]) in which the M-estimator ψ function is given by $\psi(r) = \phi\{F_0(r)\}$, where

F_0 is the true (unknown) sampling distribution function (see Jurečková [15]). Rank regression estimators have the same attractive efficiency properties relative to parametric maximum likelihood estimators as do the generating rank tests relative to the corresponding parametric tests.

RANK TESTS WITH RIGHT-CENSORED DATA

A good deal of recent work has been directed toward the specification of linear rank tests when the response variable is subject to arbitrary right censorship (by means of an "independent" censoring mechanism). Linear rank statistics that have been proposed for this problem take the form

$$\mathbf{S} = \sum_{i=1}^{k} \{\mathbf{x}_i a(R_i) + \mathbf{z}_i A(R_i)\}, \quad (17)$$

where $Y_1, \ldots, Y_k$ are the k (random) uncensored response variables with corresponding order statistic $(Y_{(1)}, \ldots, Y_{(k)})$ and rank statistic $(R_1, \ldots, R_k)$, $\mathbf{z}_i$ is the sum of regression values ($\mathbf{x}$'s) corresponding to random variables that are censored in $[Y_{(i)}, Y_{(i+1)}]$, $i = 1, \ldots, k$, $(Y_{(k+1)} = \infty)$ and $A(R_i)$ is a score to be associated with these censored observations. A special case of particular interest is given by scores

$$a(i) = n_1^{-1} + n_2^{-1} + \cdots + n_i^{-1} - 1,$$

$$A(i) = n_1^{-1} + \cdots + n_i^{-1}, \qquad i = 1, \ldots, k, \quad (18)$$

which arises as a score test for $\boldsymbol{\beta} = \mathbf{0}$ in the partial likelihood* function from Cox's proportional hazards failure time regression model* [3,4]. Note that n_j is the number of Y-values (censored or uncensored) that are equal to or greater than $Y_{(j)}$, $j = 1, \ldots, k$. In the $p + 1$ sample problem the statistic given by (18) is referred to as the log-rank* statistic [20,21]. It is a censored data* generalization by the Savage statistic mentioned above.

A second interesting special case is given by the scores

$$a(i) = 1 - 2\prod_{j=1}^{i} \left(\frac{n_j}{n_j + 1}\right),$$

$$A(i) = 1 - \prod_{j-1}^{i} \left(\frac{n_j}{n_j + 1}\right), \qquad i = 1, \ldots, k, \quad (19)$$

which provides a censored data generalization of the Wilcoxon linear rank statistic [21,22], which is to be preferred as a Wilcoxon test generalization to that proposed earlier by Gehan [6] (*see* GEHAN–GILBERT TEST, and Prentice and Marek [23] for a discussion of this point).

Censored data rank tests of the type (17) can be generated from a linear regression model in a manner similar to that described above upon replacing (13) by the sum of probabilities of all possible underlying rank vectors given the set of (censored and uncensored) Y-values [22]. An extreme minimum error density generates the log-rank scores (18), while a logistic density gives the generalized Wilcoxon scores (19).

Significance level calculations for statistics (14) usually will need to be based on a χ^2 approximation to a statistic of the form (7) (Peto and Peto [21] discuss the use of higher-order moments for $\mathbf{S}$ to improve the approximation). If the censoring is independent of the regression values (the $\mathbf{x}$'s), formulae (5) and (6) will apply, provided one conditions on the set of scores (i.e., conditional on the censoring "pattern"). In fact, under quite general censorship the censored data rank statistics given by (18) or (19) will have mean zero and variance estimators as given in Prentice [22]. In the $p + 1$ sample problem both of these statistics have an alternate representation as a weighted sum over the k distinct "failure" times of the difference between the vector of observed numbers of failures in each sample minus a vector of conditionally expected numbers of failures [21]. This representation is very useful for the communication of findings and it naturally incorporates tied uncensored Y-values (failure times). It also leads to a convenient variance estimator that is the sum over distinct failure times of hypergeometric variances.

ADDITIONAL TOPICS

The previous paragraph mentioned tied Y-values for the first time. Such tied values can be accommodated by defining the linear rank statistic to be the sum (or average) of rank statistics corresponding to possible (untied) rankings. This is equivalent, with uncensored data, to replacing the score for a set of tied Y-values by the average of their possible scores. Formulae (5) and (6) apply to this rank statistic provided that one conditions on the set of averaged scores (i.e., on the "pattern" of ties) and that the mechanism giving rise to the tied data (e.g., grouping an underlying continuous random variable) does not depend on the values $\mathbf{x}_i$, $i = 1, \ldots, n$, of the regression variable. If the latter condition is violated, a variance estimator can still be developed, in the manner of Prentice [22], as the information matrix corresponding to a likelihood function that is the sum of rank probabilities for possible underlying (untied) rank vectors, given the data.

As a final, rather important, topic suppose that the distribution of the random variables $Y_1, \ldots, Y_n$ may depend not only on the corresponding p-vectors $\mathbf{x}_1, \ldots, \mathbf{x}_n$, but also on auxiliary variables or characteristics. For example, the $\mathbf{x}_i$ may specify treatment group, but treatments may be assigned within blocks in a randomized block experiment. A simple means of accommodating such additional sources of heterogeneity in testing the hypothesis of no dependence of the distribution of Y or x involves dividing of the sample into strata that are homogeneous, or nearly so, with respect to the auxiliary data. Upon selecting a scoring scheme $\{a_n(i)\}$, linear rank statistics of the form (1) may be formed in each stratum. A summary of linear rank tests then can be formed as the sum over strata of the specific tests, while the corresponding null hypothesis mean and variance are simply sums over strata of (5) and (6). Approximate significance levels can be computed using these summary calculations in (7). Such a test procedure can be developed readily as the score test based on a likelihood function that is the product of rank probabilities (13) over strata. The stratified version of the log-rank test (18) is often referred to as the Mantel–Haenszel* test.

References

[1] Bickel, P. J. and Doksum, K. A. (1977). *Mathematical Statistics: Basic Ideas and Selected Topics*. Holden-Day, San Francisco.

[2] Chernoff, H. and Savage, I. R. (1958). *Ann. Math. Statist.*, **29**, 972–994.

[3] Cox, D. R. (1972). *J. R. Statist. Soc. B*, **34**, 187–220.

[4] Cox, D. R. (1975). *Biometrika*, **62**, 269–276.

[5] Fisher, R. A. and Yates, F. (1963). *Statistical Tables for Biological, Agricultural and Medical Research*, 6th ed. Oliver & Boyd, Edinburgh.

[6] Gehan, E. A. (1965). *Biometrika*, **52**, 203–223.

[7] Hájek, J. (1968). *Ann. Math. Statist.*, **39**, 325–346.

[8] Hájek, J. and Šidák, Z. (1967). *Theory of Rank Tests*. Academia, Prague.

[9] Hodges, J. L. and Lehmann, E. L. (1956). *Ann. Math. Statist.*, **27**, 324–335.

[10] Hotelling, H. and Pabst, M. R. (1936). *Ann. Math. Statist.*, **7**, 29–43.

[11] Huber, P. J. (1973). *Ann. Statist.*, **1**, 799–821.

[12] Jaeckel, L. A. (1972). *Ann. Math. Statist.*, **43**, 1949–1958.

[13] Jurečková, J. (1969). *Ann. Math. Statist.*, **40**, 1889–1900.

[14] Jurečková, J. (1971). *Ann. Math. Statist.*, **42**, 1328–1338.

[15] Jurečková, J. (1977). *Ann. Statist.*, **5**, 464–472.

[16] Kruskal, W. H. and Wallis, W. A. (1952). *J. Amer. Statist. Ass.*, **47**, 583–612.

[17] Kruskal, W. H. and Wallis, W. A. (1953). *J. Amer. Statist. Ass.*, **48**, 907–911.

[18] Lehmann, E. L. (1975). *Nonparametrics: Statistical Methods Based on Ranks*. Holden-Day, San Francisco.

[19] Mann, H. B. and Whitney, D. R. (1947). *Ann. Math. Statist.*, **18**, 50–60.

[20] Mantel, N. (1966). *Cancer Chemother. Rep.*, **50**, 163–170.

[21] Peto, R. and Peto, J. (1972). *J. R. Statist. Soc. A*, **135**, 185–206.

[22] Prentice, R. L. (1978). *Biometrika*, **65**, 167–179.

[23] Prentice, R. L. and Marek, P. (1979). *Biometrics*, **35**, 861–867.

[24] Puri, M. L. and Sen, P. K. (1971). *Nonparametric Methods in Multivariate Analysis*. Wiley, New York.

[25] Pyke, R. (1970). In *Nonparametric Techniques in Statistical Inference*, M. L. Puri, ed. Cambridge University Press, Cambridge, pp. 21–37.

[26] Pyke, R. and Shorack (1968). *Ann. Math. Statist.*, **39**, 755–771.

[27] Savage, I. R. (1956). *Ann. Math. Statist.*, **27**, 590–615.

[28] van der Waerden, B. L. (1953). *Math. Ann.*, **126**, 93–107.

[29] Wilcoxon, F. (1945). *Biometrics Bull.*, **1**, 80–83.

Acknowledgment

This work was supported by grants GM-28314, GM-24472, and CA-15704 from the National Institutes of Health.

(ADAPTIVE TESTS
CENSORED DATA
DISTRIBUTION-FREE METHODS
FISHER–YATES TESTS
KRUSKAL–WALLIS TEST
LOG-RANK SCORES, STATISTICS, AND TESTS
MANN–WHITNEY–WILCOXON TEST
M-ESTIMATORS
NORMAL SCORES TESTS
ORDER STATISTICS
RANK CORRELATIONS
RANK ORDER STATISTICS
RANKING PROCEDURES
ROBUSTNESS
WILCOXON SIGNED-RANK TEST)

ROSS L. PRENTICE

LINEAR REGRESSION

In almost all fields of study, the researcher is frequently faced with the problem of trying to describe the relation between a response variable and a set of one or more input variables. Given data on input (predictor, independent) variables labeled $x_1, x_2, \ldots, x_p$ and the associated response (output, dependent) variable y, the objective is to determine an equation relating output to input. The reasons for developing such an equation include the following:

1. To predict the response from a given set of inputs
2. To determine the effect of an input on the response
3. To confirm, refute, or suggest theoretical relations

To illustrate, the simplest situation is that of a single input for which a linear relation is assumed. Thus, if the relation is exact, it is given for appropriate values of β_0 and β_1 by

$$y = \beta_0 + \beta_1 x. \qquad (1)$$

The determination of β_0 and β_1 in this case is easy, requiring only two distinct pairs of observations (x_1, y_1) and (x_2, y_2).

In general, the problem is more complex in that the response is not given exactly by (1). This may be true because, although the relation is theoretically given by (1), the observations are not measured without error. Alternatively, there may be no theoretical justification for an exact linear relation but it is used as an approximation.

A model, commonly used in both cases, is

$$y = \beta_0 + \beta_1 x + e. \qquad (2)$$

Here e denotes the measurement error* or other random fluctuations in y which cause the response to depart from (1). (In this discussion, it is assumed that the input variables are either specified by the user or measured without error.)

The appropriate analysis of (2) is dictated by the assumptions made on the distribution of errors. Typically, it is assumed that the errors have mean zero and variance σ^2 and that the errors associated with distinct observations are uncorrelated. That is, if a very large number of pairs (x_i, y_i) were observed for a situation modeled by (2), then (a) the errors

$$e_i = y_i - \beta_0 - \beta_1 x_i \qquad (3)$$

would average to zero, (b) the error associated with one observation would in no way influence any other error, and (c) the mean of the squares of the errors would be σ^2.

Based on n pairs of observations (x_i, y_i), $i = 1, 2, \ldots, n$, the objective of the analyst is to estimate β_0, β_1, and σ^2 and to make inferences about these parameters. In addition, it may be desirable to indicate the precision of a prediction obtained for a given input when the estimates b_0 and b_1 of β_0 and β_1 are used in (1). These inferences require further specification of the distribution of the errors. The classical results are developed assuming a Gaussian (or normal) distribution.

The general linear regression model is now given as

$$y = \beta_0 + \beta_1 x_1 + \beta_2 x_2 + \cdots + \beta_p x_p + e. \tag{4}$$

Here the assumption on the errors is the same as given above and the analysis is to be based on n, $(p + 1)$-tuples $(x_{1i}, x_{2i}, \ldots, x_{pi}, y_i)$, $i = 1, \ldots, n$.

The sense in which (4) is a linear model must be emphasized. As written, the average response in (2) is a linear function of x and in (4) is a linear (planar) function of $x_1, \ldots, x_p$, but this is not the essential linearity. The critical feature is that the average response is a linear function of the coefficients $\beta_0, \beta_1, \ldots, \beta_p$. The variables indicated by y and x_i, $i = 1, \ldots, p$, may represent functions of the variables which are actually observed as long as these functions do not depend on unknown parameters. For example, the model

$$\log z = \beta_0 + \beta_1 / w + e \tag{5}$$

does not represent z as a linear function of w, but by letting $y = \log z$ and $x = 1/w$ this model is seen to be equivalent to (2). Similarly, the polynomial model

$$y = \beta_0 + \beta_1 x + \beta_2 x^2 + e \tag{6}$$

is a special case of (4) with $x_1 = x$ and $x_2 = x^2$. The important point is that the analysis of (4) is not complicated by such modifications.

In the next section, the classical least-squares analysis is summarized. In the section "Detection of Departures from Assumptions," some of the departures from the basic assumptions are discussed and methods are given for detecting them. Finally, in the section "Alternatives to Classical Least Squares," some alternatives to least squares, designed to cope with these problems, are indicated.

CLASSICAL LEAST-SQUARES ANALYSIS

The estimation of the unknown parameters in the general linear regression model is most frequently achieved by the method of least squares*. Given n observations (or cases) $(x_{1i}, x_{2i}, \ldots, x_{pi}, y_i)$, $i = 1, \ldots, n$, let the ith residual be $e_i = y_i - \beta_0 - \sum_{j=1}^{p} \beta_j x_{ji}$. The method of least squares determines values, b_j, as estimates of β_j so as to minimize the sum of squared residuals. The estimated regression function (predicted value) for the ith set of inputs, $\hat{y}_i$, and the estimated residual r_i are given by

$$\hat{y}_i = b_0 + \sum_j b_j x_{ji}$$
$$r_i = y_i - \hat{y}_i. \tag{7}$$

There are essentially two major advantages of this method. The first is computational, since the method only requires the solution of a system of linear equations. The second is statistical in that the estimates possess desirable small sample properties. In particular, the b_j are unbiased estimates of the β_j which have minimum variance in the class of estimators which are unbiased. Further, the assumption of normality allows for simple inferences on the β_j. The estimate of σ^2 is also unbiased and minimum variance.

With the advent of high-speed computing the computational advantage is less compelling than in the past. This has encouraged a study of alternatives to least squares, some of which are indicated in the section "Alternatives to Classical Least Squares."

Since all but the most elementary regression computations are performed by computer, the discussion will be directed at understanding the analysis in terms of a typical computer output. This will minimize the need for complex formulas and algebra.

The requirement of the user is to specify the data matrix, that is, the rectangular array

$$[X \mid Y] = \left[\begin{array}{cccccc|c} 1 & x_{11} & x_{21} & \cdots & x_{p1} & & y_1 \\ 1 & x_{12} & x_{22} & \cdots & x_{p2} & & y_2 \\ \vdots & \vdots & \vdots & & \vdots & & \vdots \\ 1 & x_{1n} & x_{2n} & \cdots & x_{pn} & & y_n \end{array}\right]. \tag{8}$$

Here each row represents a particular case and each column a variable. Implicit in this display is the assumption of model (4). Note that the column of 1's in (8) is usually inserted automatically by the computer unless the user specifies that the constant term, β_0, is not in the model. The computer then operates on the array (8) to obtain two new arrays,

$$[S \mid d] \tag{9}$$

and

$$[C \mid b]. \tag{10}$$

[In particular, (9) consists of sums of squares and cross products of the columns in (8) with $S = X'X$ and $d = X'Y$. Then $C = S^{-1}$ and $b = S^{-1}d$.] In addition, most programs report an additional array,

$$[R \mid r_y]. \tag{11}$$

Here $R = (r_{ij})$ is the matrix of correlation coefficients among the x_i and r_y is the vector of correlations between x_i and y, $i = 1, \ldots, p$.

Note that R is analogous to S in that it is a sum of squares and cross-products matrix, but it is constructed from standardized data. This standardization is achieved by modifying the elements in (8) according to

$$x_{ji}^* = \frac{x_{ji} - \bar{x}_{j\cdot}}{\left[\sum_i (x_{ji} - \bar{x}_{j\cdot})^2\right]^{1/2}}. \tag{12}$$

Here $\bar{x}_{j\cdot}$ is the mean of the elements in a column of (8). Note that the standardized data x_{ji}^* have mean zero and unit sum of squares.

The parameter estimates and other relevant statistics can now be defined in terms of these arrays as follows.

Estimates

The regression coefficients $\beta_0, \beta_1, \ldots, \beta_p$ are estimated by the $p + 1$ elements of b in (10).

The residual variance σ^2 is estimated by

$$\begin{aligned} s^2 &= \sum_{i=1}^{n} r_i^2 \Big/ (n - p - 1) \\ &= \left(\sum_{i=1}^{n} y_i^2 - \sum_{i=1}^{p+1} b_i d_i \right) \Big/ (n - p - 1). \end{aligned} \tag{13}$$

Variance of b

The variance–covariance matrix of b is given by

$$V(b) = C\sigma^2. \tag{14}$$

In particular, the variance of b_i is given by $C_{ii}\sigma^2$, where C_{ii} is the ith diagonal element of C. The estimated variance of b is obtained by replacing σ^2 by s^2 from (13).

Tests of Hypotheses

For testing the hypothesis $\beta_i = \beta_i^*$, the test statistic is

$$t_i = \frac{b_i - \beta_i^*}{\sqrt{C_{ii}s^2}}, \tag{15}$$

which is distributed as Student's t with $n - p - 1$ degrees of freedom. The t-statistics for $\beta_i^* = 0$ are usually included in the computer output. For more general hypotheses of the form $H\beta = \delta$, where H is $q \times (p + 1)$ of rank q, the test statistic is

$$F = \frac{(Hb - \delta)'(HCH')^{-1}(Hb - \delta)}{qs^2}, \tag{16}$$

which is distributed as F with q and $n - p - 1$ degrees of freedom.

Confidence Regions

From the test statistic (15), a 100 $(1 - \alpha)$% confidence interval for β_i is given by

$$b_i \pm t_{\alpha/2}\sqrt{C_{ii}s^2}, \tag{17}$$

where $t_{\alpha/2}$ is the $\alpha/2$ percentile of Student's t.

More generally, joint $(1 - \alpha)$ confidence regions on linear functions, $H\beta$, of the coefficients are given by the interior of the ellipsoid

$$(H\beta - Hb)'(HCH')^{-1}(H\beta - Hb) \leqslant qs^2F(\alpha, q, n - p - 1). \quad (18)$$

Prediction

One of the more common uses of the fitted regression equation is in the prediction of the response to a new input x^*. The predicted value is given by

$$y^* = b_0 + \sum_{j=1}^{p} b_j x_j^* \quad (19)$$

and a $100(1 - \alpha)\%$ prediction interval is

$$y^* \pm t_{\alpha/2}\sqrt{s^2(1 + x^{*\prime}Cx^*)}. \quad (20)$$

Quality of the Fitted Equation

It is natural to seek a measure of how effective the inputs are in describing the response. One suggestion is to compare the residual sum of squares

$$RSS = \sum_{i=1}^{n} (y_i - \hat{y}_i)^2 \quad (21)$$

with the (total) sum of squared deviations about $\bar{y}$, that is,

$$TSS = \sum_{i=1}^{n} (y_i - \bar{y})^2. \quad (22)$$

A popular measure is the squared multiple correlation coefficient

$$R^2 = 1 - RSS/TSS. \quad (23)$$

Here, $0 \leqslant R^2 \leqslant 1$, where R^2 close to 1 suggests a good fit and R^2 close to zero indicates a poor fit.

It has been argued that (21) and (22) are not directly comparable since one parameter is estimated in (22) and $p + 1$ parameters are estimated to obtain (21). The alternative is to compare the mean squares

$$RMS = RSS/(n - p - 1) \quad (24)$$

and

$$TMS = TSS/(n - 1). \quad (25)$$

Analogous to (23), the "adjusted" R^2 is defined by

$$R_a^2 = 1 - RMS/TMS. \quad (26)$$

DETECTION OF DEPARTURES FROM ASSUMPTIONS

The classical analysis assumes that the model is correct and that the data are good. In practice, this is rarely the case and it is essential that the violations be detected and evaluated. In this section, several of the problems are cited together with means of diagnosing them and some suggested remedies.

Problems

1. Incorrect functional form for the regression function. Additional variables and/or different functions of the variables may be required.
2. Violations of assumptions of independence, constant variance, and normality of errors.
3. Outliers* and extreme points. The former are observations in which the response is abnormally large or small and the latter are cases in which the inputs are different from the rest of the data.
4. Multicollinearity* among the input variables, that is, nearly exact linear relations among subsets of the input variables. This includes the case where one of the inputs is nearly constant.

One or more of these problems may completely invalidate the analysis and, unfortunately, the standard statistics provide little indication of their presence. Several additional indicators have been proposed, including the following.

Diagnostics

1. **Bivariate Plots.** Plots of x_i against x_j and x_i against y for all i and j may help to identify extreme points and suggest model changes.
2. **Residual Plots.** Plots of the residuals, r_i, against x_i and $\hat{y}$ as well as normal plots of the residuals may suggest model changes and may reveal departures from variance homogeneity and normality of errors.
3. **The Hat Matrix*.**

$$H = (h_{ij}) = XCX'. \tag{27}$$

Equation (27) indicates the influence of the jth response on the ith predicted value. Large diagonal elements (greater than $2(p+1)/n$) suggest highly influential observations.

4. **Studentized Residual.**

$$t_i = r_i \sqrt{s^2(1 - h_{ii})}\ . \tag{28}$$

These combine the magnitude of the residual and the measure of influence.

5. **Cook's Distance.**

$$D_i = \frac{t_i^2}{p+1}\left(\frac{h_{ii}}{1 - h_{ii}}\right). \tag{29}$$

This formula measures the squared distance from $\hat{Y}$, the predicted response vector using all the data, to $\hat{Y}(i)$, the predicted response if the ith case is deleted when estimating β. Large values of D_i [greater than $F(0.1, p+1, n-p-1)$] suggest influential observations which merit further study.

6. **Multcollinearities.** These are relations among the input variables which may be revealed by examining the correlation matrix, R. Some of the indicators are
   (a) Large correlations, r_{ij}.
   (b) Large variance inflation factors. $\text{VIF}_i = i$th diagonal element of $V = R^{-1}$. [$\text{VIF}_i < 10$ is acceptable.]
   (c) Small eigenvalues, λ_i, of R indicate multicollinearities and the associated eigenvector defines the relation among the standardized variables.

Unfortunately, there are no guaranteed solutions to any of the problems cited above. The following remedies are typical but should be used with caution.

Remedies

1. Nonuniform residual plots may suggest nonlinear functions. Individual points which are outstanding may suggest other variables that could be included, especially cateogorical variables defining subgroups of cases.
2. The most common cause of variance inhomogeneity is that the variance is proportional to one of the inputs. Division of the equation by this variable, or some power of it, will help. Normality may be achieved by transformations.
3. Outliers and extreme points may be deleted from the analysis but care must be taken, as these may be valid, informative observations. Alternatively, one of the robust procedures in the next section might be used.
4. The eigenvector may identify the multicollinearity, but the action to be taken depends on the cause. If the linear relation is inherent in the system being modeled and the relation is strong, it may be appropriate to eliminate one of the variables in the relation. If the apparent linear relation is due to the peculiarities of the particular sample, then, if possible, additional data should be taken which are more uniformly spread over the sample space. Alternatively, one might simulate this by using ridge regression* or one of the related methods discussed in the next section.

ALTERNATIVES TO CLASSICAL LEAST SQUARES

Since least-squares analysis is vulnerable to departures from the basic assumptions, several alternatives have been suggested. The following is a brief discussion of some of the recent developments.

Robust Regression

Least squares is known to be equivalent to the method of maximum likelihood* when the errors are normally distributed, and this leads to minimization of the sum of squared residuals. Other distributional assumptions lead to the minimization of the sum of other functions of the residuals, that is,

$$\text{minimize} \sum_{i=1}^{n} f(e_i). \qquad (30)$$

This includes least squares when $f(e) = e^2$ but allows for more general functions. For example, $f(e) = |e|$ yields the absolute value estimator. A compromise between the two is given by

$$f(r) = \begin{cases} e^2/2, & |e| \leqslant h, \\ h|e| - h^2/2, & |e| > h, \end{cases} \qquad (31)$$

where h is to be determined. The idea here is that observations with large residuals are given less weight and hence are less influential. Numerous other functions have been proposed.

Ridge Regression and Related Methods

When multicollinearities are present, least-squares estimates of the coefficients may be abnormally large or even have the wrong sign. A method that effectively adjoins fictitious data is known as ridge regression. The ridge estimator is the solution to

$$(R + kI)b^* = r_y, \qquad (32)$$

where $k \geqslant 0$ is to be determined. (Note that these coefficients are for the standardized data.)

Dramatic evidence of the effect of multicollinearity is given by the ridge trace, a plot of the solutions of (32) as a function of k. The choice of k is usually based on the data and complicates the properties of the estimates. Clearly, this choice should be based on the acceptable degree of multicollinearity. Note that the eigenvalues of $R + kI$ are $\lambda_i + k$.

Closely related to ridge regression are the methods known as (a) principal components (*see* COMPONENTS ANALYSIS), (b) latent root*, (c) generalized inverse, and (d) generalized ridge regression.

Variable Elimination

One of the first modifications of least squares was that of eliminating variables. This has been a confusing and controversial topic primarily because it has been applied indiscriminately to data that have not been subjected to the above diagnostics. Variable elimination only should be applied after the data have been examined for extremes, outliers, and multicollinearities, and appropriate action has been taken. Variables which are then not contributing to the description of the response may be eliminated. *See also* ELIMINATION OF VARIABLES.

SUMMARY

Regression analysis is a useful and powerful tool of the data analyst if carefully applied. The precautions noted here may prevent incorrect analyses.

Bibliography

General Texts

Belsley, D. A., Kuh, E., and Welsch, R. E. (1980). *Regression Diagonstics*. Wiley, New York. (Recent developments in outlier and multcollinearity detection.)

Daniel, C. and Woods, F. S. (1971). *Fitting Equations to Data*. Wiley, New York. (An excellent resource book with interesting diagnostics.)

Draper, N. R. and Smith, H. (1966). *Applied Regression Analysis*. Wiley, New York. (A classic text with a good discussion of nonlinear regression.)

Gunst, R. F. and Mason, R. L. (1980). *Regression Analysis and Its Applications*. Marcel Dekker, New York. (A thorough text with good examples.)

Hocking, R. R. (1984). *Technometrics*, **25**, 219–229. (Developments in linear regression methodology since 1959 are summarized. Discussion and response follow on pp. 230–249.)

Neter, J. and Wasserman, W. (1974). *Applied Linear Statistical Models*. Richard D. Irwin, Homewood, Ill. (Complete and easy-to-read basic text.)

Research Papers

REGRESSION DIAGNOSTICS

Cook, R. D. (1977). *Technometrics*, **19**, 15–18.

Hoaglin, D. C. and Welsch, R. E. (1978). *Amer. Statist.*, **32**, 17–22.

RIDGE REGRESSION AND RELATED METHODS

Hocking, R. R. (1976). *Biometrics*, **32**, 1–49. (A comprehensive review paper.)

Hocking, R. R., Speed, F. M., and Lynn, M. J. (1976). *Technometics*, **18**, 425–437. (Unified presentation of ridge and other methods.)

Hoerl, A. E. and Kennard, R. W. (1970). *Technometrics*, **12**, 55–67. (The basic paper on ridge regression.)

ROBUST ESTIMATION

Andrews, D.F. (1974). *Technometics*, **16**, 523–532. (Discusses reweighted least squares with several weight functions.)

Huber, P. J. (1973). *Ann. Statist.* **1**, 799–821. (A basic paper on robust regression.)

(GENERAL LINEAR MODEL
LEAST SQUARES
MULTCOLLINEARITY
MULTIPLE LINEAR REGRESSION
REGRESSION COEFFICIENTS
REGRESSION DIAGNOSTICS
REGRESSION VARIABLES, SELECTION OF
RIDGE REGRESSION
STEPWISE REGRESSION)

RONALD R. HOCKING

LINEAR STRUCTURAL RELATIONSHIPS *See* LISREL

LINEAR SUFFICIENCY

The concept of linear sufficiency has been introduced by Barnard [1] in connection with the Gauss–Markov theorem* on estimation by least squares and provides an additional optimal property—besides that of uniformly minimum variance—of the least-squares estimator.

Let the sample space consist of the observed values of a set of random variables (RVs) $X_1, X_2, \ldots, X_n$, whose joint probability distribution involves an unknown parameter $\boldsymbol{\theta} = (\theta_1, \theta_2, \ldots, \theta_k)$, $k \geqslant 1$. For defining linear sufficiency we restrict ourselves to statistics that are linear functions of the X's of the form $T = l_1X_1 + l_2X_2 + \cdots + l_nX_n + c$. We shall refer to such statistics as "linear statistics." Barnard's definition of linear sufficiency is essentially as follows:

Definition. A random vector $\mathbf{T} = (T_1, T_2, \ldots, T_k)$, $k \geqslant 1$, where each component T_i is a linear statistic, is *linearly sufficient* for $\boldsymbol{\theta}$ if, for any other linear statistic, V, that is uncorrelated with each component T_i of $\mathbf{T}$, it implies that $E_{\boldsymbol{\theta}} V$ is independent of $\boldsymbol{\theta}$.

The Gauss–Markov theorem considers a setup satisfying the following conditions:

$$E_{\boldsymbol{\theta}} X_i = \sum_{i=1}^{k} a_{ir}\theta_r, \quad (1)$$

$$\operatorname{var}_{\boldsymbol{\theta}} X_i = \sigma^2 > 0 \quad (2)$$

and

$$\operatorname{cov}_{\boldsymbol{\theta}}(X_i X_j) = 0 \quad (3)$$

for $i, j = 1, 2, \ldots, n$, $i \neq j$. Here a_{ir} are known constants and σ^2 may be known or unknown. The theorem proves that for any linear function ϕ of $\boldsymbol{\theta}$, given by

$$\phi = \sum_{i=1}^{k} \lambda_i \theta_i,$$

the estimator having uniformly minimum variance (UMV) in the class of linear unbiased estimators is

$$T = \sum_{i=1}^{k} \lambda_i T_i,$$

where $T_1, T_2, \ldots, T_k$ are linear statistics. Thus the random vector $\mathbf{T} = (T_1, T_2, \ldots, T_k)$ is the optimal linear estimator of $\boldsymbol{\theta} = (\theta_1, \theta_2, \ldots, \theta_k)$. Barnard [1] shows that $\mathbf{T}$ is also linearly sufficient for $\boldsymbol{\theta}$, according to the foregoing definition. Conversely, it can be shown that **T** is the only linear estimator of $\boldsymbol{\theta}$ which is linearly sufficient.

Remark. Barnard [1] also introduces the notion of the "true value" of the observed random variable X_i, which may differ from the mean value $E_{\boldsymbol{\theta}}X_i$ by a (possibly unknown) constant, c_i (not depending on $\boldsymbol{\theta}$), representing a systematic error of observation, and defines linear sufficiency for $\boldsymbol{\theta}$ in terms of the independence on $\boldsymbol{\theta}$ of the true value of V. This, however, is only a minor generalization. Since the constants c_i do not involve $\boldsymbol{\theta}$, they do not affect the independence on $\boldsymbol{\theta}$ of the mean value $E_{\boldsymbol{\theta}}V$.

A simple example of the foregoing result is obtained when θ is a real parameter, i.e., $k = 1$ and the condition (1) is replaced by the condition $E_\theta X_i = \theta$. An elementary calculation shows that in this case the sample mean $\overline{X} = (1/n)\sum_{i=1}^{n} X_i$ is the UMV estimator of θ in the linear unbiased class and that $\overline{X}$ is linearly sufficient for θ and also is the only unbiased linear estimator which is linearly sufficient for θ.

If the variables X_i in the Gauss–Markov theorem have a joint normal distribution, $\mathrm{cov}_{\boldsymbol{\theta}}(T, V) = 0$ for any two linear statistics T, V implies that T and V are distributed independently, and $E_{\boldsymbol{\theta}}V$ is independent of $\boldsymbol{\theta}$ only if V is distributed independently of $\boldsymbol{\theta}$. Thus, in this particular case, the notion of linear sufficiency coincides with the usual notion of sufficiency. Hence the former notion is significant only when normality in the model cannot be assumed.

LINEAR SUFFICIENCY FOR SURVEY SAMPLING

A concept of linear sufficiency for finite populations has been formulated by Godambe [2]. This, however, is entirely unrelated either to least squares or to UMV estimation. Let $\mathbf{x} = (x_1, x_2, \ldots, x_N)$ be the characteristic numbers associated with the units of a finite population labeled by integers $1, 2, \ldots, N$. A linear estimator is of the form $b(s, \mathbf{x}) = \sum_{i \in s} b_{si} x_i$ where s denotes a sample (a set of distinct units). A linear parametric function is of the form $G(\mathbf{x}) = \sum_{i=1}^{N} g_i x_i$. For a given sampling design, estimators $b_1(s, \mathbf{x})$, $b_2(s, \mathbf{x})$ are said to be independent if $\sum_{i \in s} b_{b1,s_i} b_{2,si} = 0$ for all samples s having positive probability, and an estimator $b(s, \mathbf{x})$ and a linear parametric function $G(\mathbf{x})$ are said to be independent if $\sum_{i \in s} b_{si} g_i = 0$ for all such samples.

A linear estimator $b(s, \mathbf{x})$ is defined as linearly sufficient for $G(\mathbf{x})$ if any other linear estimator $b_1(s, \mathbf{x})$ is independent of $G(\mathbf{x})$. As an application of this notion, Godambe shows that for the population total $T(\mathbf{x}) = \sum_{i=1}^{N} x_i$, any estimator $b(s, \mathbf{x}) = \kappa(x) \sum_{i \in s} x_i$ is linearly sufficient (*see* SURVEY SAMPLING for more detailed explanation of the terminology).

References

[1] Barnard, G. A. (1963). *J. R. Statist. Soc. B*, **25**, 124–127.

[2] Godambe, V. P. (1966). *J. R. Statist. Soc. B*, **28**, 310–319.

(ESTIMATION, POINT
GAUSS–MARKOV THEOREM
LABELS
LEAST SQUARES
SUFFICIENCY
SURVEY SAMPLING)

V. M. JOSHI

LINEAR SYSTEMS, STATISTICAL THEORY

The idea of a dynamic system is an old one, dating back at least to Poincaré and his investigations relating to celestial mechanics. Its development has been associated with problems of the control of mechanical systems (such as a steam engine) and especially of systems in electrotechnology (e.g., communications systems). World War II and, later, problems of system control associated with space vehicles led to the field's very rapid development. The work of R. E. Kalman played a major part in this growth, and it is his state-space description of linear systems with which we mainly deal here. In sum: The state of the system at time t is described by a vector $x(t)$, of dimension n, let us say. This state is determined by previous values of itself and of input vectors, $u(t)$, of m components, that are observed and, indeed, may constitute the variables under the control of the systems engineer. There are effects from unobserved stochastic disturbances, $\xi(t)$, also. Thus

$$x(t+1) = Fx(t) + Gu(t) + \xi(t+1). \quad (1)$$

Here and below we restrict ourselves to the case where F, G, and other matrices of parameters are not time dependent. The underlying state vectors may not be observed but rather an output vector, $y(t)$, of s components which has $x(t)$ and $u(t)$ as inputs together with a further stochastic disturbance, $\eta(t)$. Thus

$$y(t) = Hx(t) + Du(t) + \eta(t). \quad (2)$$

The autocovariance properties of $\xi(t)$ and $\eta(t)$ are prescribed as follows:

$$E\{\xi(s)\xi(t)'\} = \delta_{st}Q,$$
$$E\{\eta(s)\eta(t)'\} = \delta_{st}R,$$
$$E\{\xi(s)\eta(t)'\} = \delta_{st}S.$$

Such a system may always be put in prediction error form wherein $x(t)$ is replaced by its best linear estimate, $x(t\,|\,t-1)$, from the past of $y(t)$ and $u(t)$. Then

$$x(t+1\,|\,t) = Fx(t\,|\,t-1) + Gu(t) + K\epsilon(t)$$
$$y(t) = Hx(t\,|\,t-1) + Du(t) + \epsilon(t),$$
$$E\{\epsilon(s)\epsilon(t)'\} = \delta_{st}\Sigma.$$

Here $\epsilon(t)$ is the innovation sequence and is the vector of errors of linear prediction for $y(t)$ from the past of the $y(t)$ and $u(t)$ sequences. In turn, this prediction error form may be transformed into ARMAX form, i.e., ARMA models (*see* AUTOREGRESSIVE-MOVING AVERAGE (ARMA) MODELS) with exogenous components,

$$\sum_{0}^{p} A(j)y(t-j) = \sum_{0}^{r} D(j)u(t-j) + \sum_{0}^{q} B(j)\epsilon(t-j),$$
$$A(0) = B(0). \quad (3)$$

This form is that in which models of econometric systems have usually been built, no state vector being explicitly introduced. In econometrics* the words "input" and "output" are translated into "exogenous" and "endogenous." However, the descriptions are equivalent in that they define the same family of stochastic processes*, $y(t)$.

We explicitly deal now only with the stable case where (1) may be chosen so that all eigenvalues of F are of less than unit modulus. A third description of a linear system is the transfer function* form, where

$$y(t) = \sum_{0}^{\infty} L(j)u(t-j) + \sum_{0}^{\infty} K(j)\epsilon(t-j),$$
$$K(0) = I_s, \quad (4)$$

assuming the system to have been initiated indefinitely far in the past. The characteristics of the system are now described by the transfer functions

$$k(z) = \sum_{0}^{\infty} K(j)z^{-j}, \qquad l(z) = \sum_{0}^{\infty} L(j)z^{-j},$$
$$k(0) = I_s.$$

These are related to (1)–(3) as follows. Put

$$a(z) = \sum_{0}^{p} A(j)z^{p-j}, \qquad b(z) = \sum_{0}^{q} B(j)z^{q-j},$$
$$d(z) = \sum_{0}^{r} D(j)z^{r-j}.$$

Then

$$k(z) = a(z)^{-1}b(z) = I_s + H(zI_n - F)^{-1}K,$$
$$l(z) = a(z)^{-1}d(z) = D + H(zI_n - F)^{-1}G. \quad (5)$$

Moreover, if $y(t)$ in (4) is to be of the form in (1), (2), or (3), then $k(z)$ and $l(z)$ must be such rational functions of z. In the case

$s = 1$, when $y(t)$ is scalar, such rational transfer function systems are sometimes called Box–Jenkins models*. Then $A(j)$ and $B(j)$ are scalars and the $D(j)$ are vectors.

As was pointed out at an early stage by M. G. Kendall in relation to (3), a major problem when $s > 1$ is that of introducing parameters in a satisfactory manner. Until this is done, any theory of estimation will be deficient. The problem arises because there are many equivalent descriptions [(1) and (2)] with different $x(t)$, F, G, H, and $\xi(t)$. Similarly, the $A(j)$, $B(j)$, and $D(j)$ in (3) are not uniquely prescribed. The latter problem was recognized by econometricians, who spoke of it as the problem of identification*. (In the control literature the word *identification* stands for the whole process of ascribing values of F, G, H, etc., using observations for $t = 1, 2, \ldots, T$). However, $k(z)$ and $l(z)$ are uniquely determined if it is required that they be composed of functions analytic for $|z| \geqslant 1$ and that the determinant of $k(z)$ be nonzero for $|z| > 1$. The first condition is related to the possibility of representing $y(t)$ in the form (4), while the latter is related to the fact that $\epsilon(t)$ depends only on $y(s)$ and $u(s)$, $s \leqslant t$. Some element of uniqueness can be achieved in (5) by requiring a, b, and d to be left coprime so that if $a(z) = u(z)a_1(z)$, $b(z) = u(z)b_1(z)$, and $d(z) = u(z)d_1(z)$, where u, a_1, b_1, and d_1 are all matrices of polynomials, then the determinant, $\det u(z)$, must be a nonzero constant. This requirement amounts to the removal of redundant factors from the left of a, b, and d. For (1) and (2) a certain element of uniqueness can be achieved by taking the state vector to be of minimal dimension. This again amounts to the removal of a certain form of redundancy. If, for every $x(0 \mid -1)$, $x(t \mid t-1)$ can, by a choice of a suitable sequence $u(t)$, $\epsilon(t)$, $t \geqslant 1$, be changed into any vector of n dimensions, then the system is said to be *controllable* (reachable). If that is not so, the system is to an extent redundantly specified and (in the stationary case; see below) a new state vector can be introduced so that the system is controllable. If, when $u(t)$ and $\epsilon(t)$ are zero for all t, then $x(0 \mid -1)$ can be uniquely determined from $y(0)$, $y(1)$, $\ldots, y(t)$, for some t, then the system is observable. If not, again a redefinition of $x(t \mid t-1)$ will produce an observable system. If the system is observable and controllable, then it is minimal. Thus we can, for inference purposes, always assume that the system is minimal. The minimal dimension, n, is a key integer parameter and is called the "order" or the "McMillan degree." It is the degree of $\det a(z)$ in a left coprime factorization, (5). Call $M(n)$ the set of all systems of order n. The problem of fitting a rational transfer function model can be thought of as that of determining n and then the "system parameters" that determine k and l, given n, as well as Σ. To do this it is necessary to introduce coordinates into $M(n)$. This is not trivial since there are many left coprime factorizations [because of the arbitrariness of $u(z)$] and many minimal representations [(1) and (2)] for the same system. Let $y_k(t+j \mid t)$ be the best linear predictor of $y_k(t+j)$ from $u(s)$ and $y(s)$, $s \leqslant t$. Then the rational transfer linear systems of order n are those for which the $y_k(t+j \mid t)$; $k = 1, \ldots, s$; $j = 1, 2, \ldots$ constitute a set of random variables of rank n. We may always choose a linearly independent set of the form $y_k(t+j \mid t)$, $j = 1, 2, \ldots, n_k$, $\sum n_k = n$, for some suitable n_k. Given the partition $n = \sum n_k$, call $U(\{n_k\})$ the subset of $M(n)$ for which these $y_k(t + j \mid t)$ constitute a basis. Then $y_k(t + n_k + 1 \mid t)$ can be linearly expressed in terms of these. These expressions provide ns numbers. We obtain $n(s+m)$ further ones by taking the elements of the kth row of $[K(j) \vdots L(j)]$ for $j = 1, \ldots, n_k$. This gives $n(2s+m)$ numbers in all. If we add the sm element of D, we get $d(n) = n(2s+m) + ms$ numbers that serve to coordinatize these "neighborhoods," $U(\{n_k\})$. There are $\binom{n+s-1}{s-1}$ partitions of n and the union of all of these $U(\{n_k\})$ is $M(n)$. If n_k and n_k' are two partitions, then in $U(\{n_k\}) \cap U(\{n_k'\})$ the coordinates in the first neighborhood are functions of those in the second, and conversely. They are analytic functions, indeed finite rational functions, so that $M(n)$ is an analytic manifold. Each $U(\{n_j\})$ is dense in $M(n)$. Thus the coordinatization of the set of all rational transfer function systems is satisfactorily completed.

Such descriptions have, however, been little used because of the complexity of the problem of optimizing (say) a Gaussian likelihood over $M(n)$, partly because of the multiplicity of coordinate systems and partly because $d(n)$ is so large. Indeed, in econometrics values of $s = 30$ and $m = 10$ would not be large. Since n/s may be interpreted as the average lag in an ARMAX model, a value of $n = 60$ is not large. But then $d(n) = 60(2.30 + 10) + 10.30 = 4500$. As a consequence, when s is not small, prior constraints must be used. Of course, the use of such information introduces risks. One form of prior constraints is to examine only a particular $U(\{n_k\})$, for example, that for which $n_1 = n_2 \cdots = n_a = n_{a+1} + 1 = \cdots = n_s + 1$, $n = sn_a + a$. (This neighborhood has special features.) Alternatively, we might proceed as follows. Call $M(p,q,r)$ the set of systems (3) for which $A(o) = B(o) = I_s$ and the partitioned matrix, $[A(p) \vdots B(q) \vdots D(r)]$, is of rank s. This may be coordinatized by the elements of the $A(j)$, $B(j)$, and $D(j)$. Also, $M(p, p, p)$ is just the $U(\{n_k\})$ for $n_k \equiv p$ and is dense in $M(ps)$, which is dense in its topological closure, which is the union of all $M(n)$, $n \leqslant ps$. Thus in examining all $M(p, p, p)$, and hence in examining all $M(p,q,r)$, we are examining a set dense, in a suitable sense, in the set of all rational transfer function systems. In the case of models built in state-space form it is very likely that the parameters specifying D, F, G, H, and K will be heavily constrained a priori and n will be known so that the problem is reduced to reasonable proportions. In econometrics there will also be prior constraints on the parameters for $M(p,q,r)$ that further reduce the scale of the problem.

The asymptotic theory for the estimation of such systems as we have discussed exists mainly for the case where $y(t)$ is stationary [except possibly for deterministic behavior of $u(t)$].

The case $s = 1$ is of special importance and the problem of inference is then greatly simplified as there is only one coordinate system. Now also $n = \max(p,q,r)$, assuming that $A(p)$, $B(q)$, and $D(r)$ are not zero. The coordinates are the coefficients in the ARMAX form [with $A(o) = B(o) = 1$]. Often the ARMAX form is reexpressed, particularly for $s = 1$, as

$$\sum_0^p A(j)y(t-j) - \sum_0^r D(j)u(t-j) = \zeta(t),$$

$$\sum_0^v \Gamma(j)\zeta(t-j) = \sum_0^q B(j)\epsilon(t-j).$$

This is sometimes said to be the transfer function form or, when $s = 1$, the Box–Jenkins form. It has some advantages since the problem of estimation is to a certain extent split into two halves. (If all ARMAX systems are to be listed in this way, two integer parameters are needed rather than one.)

The methods of estimation used (once n or p,q,r is specified) have been almost entirely maximum likelihood*, with the likelihood constructed as if the $\epsilon(t)$ are Gaussian, or some close approximation to such a likelihood that gives asymptotically equivalent estimators. Other estimators have been suggested, but these have usually been to initiate an iterative solution of the likelihood equations. When $s = 1$, many programs are available for these calculations that have been widely used and that appear to work well. For $s > 1$ such programs as are available seem to relate only to special, heavily constrained forms of the problem, for the reasons stated above. One fashion in which the likelihood may be constructed is via the Kalman filter*, which constructs the linear prediction errors for $y(t)$ given $y(t-1)$, $\ldots, y(1), u(t), \ldots, u(1)$. These will be Gaussian and orthogonal, if the $\epsilon(t)$ are Gaussian, and the Kalman filter apparatus also defines their covariance matrix (which depends on t). This technique is especially useful if there are missing observations. The technique also relates closely to methods for estimating the system parameters recursively, i.e., so that the parameter vector from data to time t is obtained from that to time $t-1$ by a simple updating procedure.

To estimate the integer parameters, for example n, Akaike introduced a criterion which in the present context takes the form

$$Q(n) = \log \det \hat{\Sigma}(n) + d(n)C(T)/T, \quad (6)$$

which is to be minimized with respect to n. Here $\hat{\Sigma}$ is the estimate of Σ by maximum likelihood where n is the order and $C(T)$ is a specified sequence, e.g., $C(T) = 2$ or $C(T) = \log nT$. Similarly, p, q, r might be estimated by minimizing

$$Q(p,q,r) = \log \det \hat{\Sigma}(p,q,r) + \{(p+q)s^2 + mrs\}C(T)/T.$$

There is a substantial body of theory justifying these procedures, asymptotically. The essential requirement of the theory is much weaker than a Gaussian assumption for the $\epsilon(t)$ and, in fact the central requirement is that the best linear predictor of $y(t)$, in the least-squares sense, be the best predictor in that sense. Without this requirement the system could hardly be called linear. The asymptotic theory of the subject is remarkably complete and self-contained. The main problem is not this theory but the construction of good algorithms.

Bibliography

Akaike, H. (1969). *Ann. Inst. Statist. Math. (Tokyo)*, **21**, 243–247. [The method of estimating n via (6) is introduced.]

Akaike, H. (1976). In *System Identification: Advances and Case Studies*, R. K. Mehra and D. G. Lainiotis, eds. Academic Press, New York, pp. 27–96. (A technique for estimation in case $s > 1$ is discussed. Presently, it is not certain that this technique is the best to use.)

Akaike, H., Arahata, E., and Ozaki, T. (1975, 1976). *Computer Sci. Monogr. TIMSAC-74*, (1), (2), Institute of Statistical Mathematics, Tokyo. [Many standard packages (e.g., GENSTAT and GLIM) contain programs for ARMA systems for $s = 1$. A wide variety of programs, covering $s > 1$ also, are discussed. See also TIMSAC-78, 1979.]

Box, G. E. P. and Jenkins, G. M. (1970). *Time Series Analysis*. Holden-Day, San Francisco. (The importance of the case $s = 1$ is emphasized.)

Brockett, R. W. (1970). *Finite Dimensional Linear Systems*. Wiley, New York. (An account of some of the theory associated with that of linear differential equations.)

Byrnes, C. I. (1980). *Geometrical Methods for the Theory of Linear Systems*, C. I. Byrnes and L. F. Martin, eds. D. Reidel, Dordrecht, Holland. (Algebraic geometric aspects of linear systems.)

Hannan, E. J. (1979). In *Developments in Statistics*, Vol. 2, P. R. Krishnaiah, ed. Academic Press, New York, pp. 83–121. (A survey in a statistical context; with references.)

Hannan, E. J. (1980). *Ann. Statist.*, **8**, 1071–1081. [The asymptotic properties of the method of estimating n via (6) are discussed. Later work avoids many of the restrictions herein (e.g., independence for the $\epsilon(t)$) and extends the results to the vector case.]

Hannan, E. J., Dunsmuir, W. T. M., and Deistler, M. (1980). *J. Multivariate Anal.*, **10**, 275–295. (A discussion of the asymptotic properties of estimation, given n or p, q, r, is given.)

Hannan, E. J. and Heyde, C. C. (1972). *Ann. Math. Statist.*, **43**, 2058–2066. (The importance of the requirement that the becst linear predictor be the best predictor was first emphasized.)

Jakeman, A., Steele, L. P., and Young, P. C. (1980). *IEEE Syst. Man. Cybern.*, **SMC-10**, 593–602. [The importance of the transfer function form is stressed. There are many references to earlier work by Young and co-workers here and the work related to the recursive methods mentioned above (6).]

Kailath, T. (1980). *Linear Systems*. Prentice-Hall, Englewood Cliffs, N.J. (A recent book giving great detail concerning the algebraic and topological structure of linear systems.)

Kalman, R. E. (1969). *Proc. First IFAC*. Butterworth, London, pp. 481–492. (One of Kalman's earliest publications that sets out the theory on celestial mechanics.)

Kendall, M. G. and Stuart, A. (1966). *The Advanced Theory of Statistics*., Vol. 3. Hafner, New York. (For the observation of Kendall, see Sec. 50.29.)

Koopmans, T. C. (1950). *Statistical Inference in Dynamic Economic Models*, T. C. Koopmans, ed. Wiley, New York. (The identification problem for econometric systems is considered extensively.)

MacFarlane, A. G. J. (1980). *IEEE Trans. Aut. Control*, **AC-24**, 250–265. [An excellent survey of the history of dynamical systems in relation to automatic control. The terminology "McMillan degree" is derived from an extension to rational functions of a theorem on the reduction to diagonal form of a matrix of polynomials (see p. 259).]

Poincaré, H. (1905–1910). *Leçons de Mécanique Celeste*. Gauthier-Villars, Paris. (Poincaré's work on celestial mechanics.)

Rosenbrock, H. H. (1970). *Multivariable and State Space Theory*. Wiley, New York. (An earlier, briefer work. The ideas presented are all for discrete-time systems.)

(AKAIKE'S CRITERION
AUTOREGRESSIVE-MOVING AVERAGE (ARMA) MODELS
BLACK BOX
BROWNIAN PROCESSES
ECONOMETRICS
GLIM
KALMAN FILTERING
LAG MODELS
LINEAR REGRESSION)

E. J. HANNAN

LINE INTERCEPT SAMPLING

Line intercept sampling is a method used by plant ecologists for estimating the "cover" of a plant species in a given area. The cover of a species is the area occupied on the ground by the projection onto it of aboveground parts of the plant. Thus for a low-growing species that forms part of the mosaic of ground vegetation, its cover is the sum of the areas of its patches; for a shrub or tree species, the cover is the sum of the areas vertically beneath the canopies of all individuals of the species.

The method consists in laying a sampling line, of length L, across the area to be sampled (see Fig. 1); one then measures the length of every intercept along the line where it lies across a patch (or canopy projection) of the species of interest. Then an estimate of the proportion, say C, of the total area covered by the species is given by the intuitively obvious estimator

$$\hat{C} = \frac{1}{L} \sum_{i=1}^{m} y_i .$$

Here y_i is the length of the ith patch intercept and m is the number of patches cut by the sampling line. Kendall and Moran [1, p. 67] prove that this estimator is unbiased. An unbiased estimator of $\text{var}(\hat{C})$ is [2]

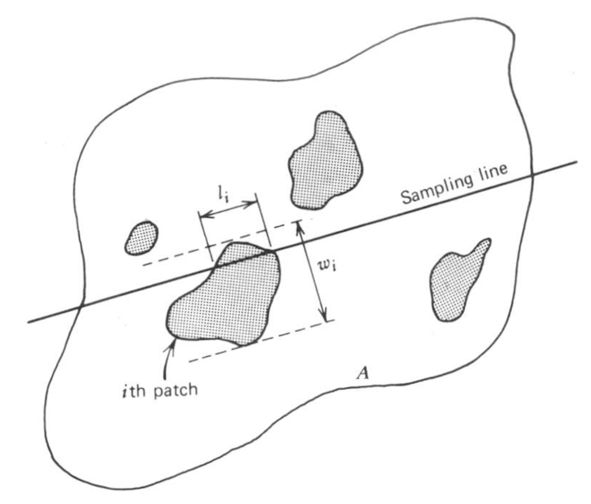

Figure 1

$$\text{var}(\hat{C}) = \frac{1}{L^2} \sum_{i=1}^{m} \left(y_i^2 - \frac{L\pi}{3A} y_i^3 \right) \simeq \frac{1}{L^2} \sum_{i=1}^{m} y_i^2,$$

since the term in y_i^3 is usually negligibly small.

McIntyre [3] was the originator of the method. He, and also Lucas and Seber [2] (and see Seber [4]), adapted the method so that it could be used to measure the mean number, say $\bar{P}$, of patch centers per unit area (a patch center is the center of gravity of a patch). When $\bar{P}$ is to be estimated, it is necessary to measure w_i, the maximum width, at right angles to the sampling line, of the ith patch ($i = 1, \ldots, m$). Then [2]

$$\hat{\bar{P}} = \frac{1}{L} \sum_{i=1}^{m} w_i^{-1},$$

$$\text{var}(\hat{\bar{P}}) = \frac{1}{L^2} \left(\sum_{i=1}^{m} w_i^{-2} \right) - \frac{\hat{\bar{P}}}{A},$$

where A is the area of land being sampled.

Line intercept sampling should not be confused with line intersect sampling* or line transect sampling*. The first two, which for clarity and brevity might conveniently be called "intercept sampling" and "Buffon sampling," respectively, resemble each other in that, in both methods, population members are selected as sample members if, *and* *only if*, they are cut by a randomly placed sampling line. In "intercept sampling," the length of the intercept cut off along the line by a sample member (a "patch") is a variate that is measured and used for the estimation of cover. In "Buffon sampling" a sample member is treated as an indivisible unit; i.e., the measurement made on it is unaffected by the length of its intercept on the sampling line. Therefore, the method of estimating $\overline{P}$ described in this article, although it was derived from intercept sampling, is, strictly, an example of "Buffon sampling." Line transect sampling consists in counting the organisms seen (usually mammals or birds) from a "transect," or straight path, though the area of interest.

References

[1] Kendall, M. G. and Moran, P. A. P. (1963). *Geometrical Probability*. Charles Griffin, London.

[2] Lucas, H. A. and Seber, G. A. F. (1977). *Biometrika*, **64**, 618–622.

[3] McIntyre, G. A. (1953). *J. Ecol.*, **41**, 319–330.

[4] Seber, G. A. F. (1979). In *Sampling Biological Populations*, R. M. Cormack, G. P. Patil, and D. S. Robson, eds. International Co-operative Publishing House, Fairland, Md., pp. 183–192.

Further Reading

See the following works, as well as the references just given, for more information on the topic of line intercept sampling.

Eberhardt, L. L. (1978). *J. Wildl. Manag.*, **42**, 1–31.

Gates, C. E. (1979). In *Sampling Biological Populations*, R. M. Cormack, G. P. Patil, and D. S. Robson, eds. International Co-operative Publishing House, Fairland, Md., pp. 71–154.

Strong, C. W. (1966). *Ecology*, **47**, 311–313.

(ECOLOGICAL STATISTICS
FISHERIES, STATISTICS IN
LINE INTERSECT SAMPLING
LINE TRANSECT SAMPLING
QUADRAT SAMPLING
STATISTICS IN FORESTRY)

E. C. PIELOU

LINE INTERSECT SAMPLING

Line intersect sampling is a method of sampling a population whose members are shaped, approximately, as line segments of various lengths. The sampling is length biased, and there is no sampling frame. The method is much used by foresters and forest ecologists to estimate such quantities as the volume of wood in logging residues in tracts of clear-felled land. It is occasionally used in other contexts, for example in industry, for estimating the mean lengths of microscopic fibers. Because it is typically a forestry method, however, the account below assumes a forest context.

To estimate the mean wood volume per unit area, in downed logs, it is impractical to collect and measure a random sample of individual logs; it is impractical, also, to measure the volumes of the logs in a sample of quadrats (small sampling plots) since many logs will lie across quadrat boundaries; *see* QUADRAT SAMPLING.

Line intersect sampling is a way of overcoming these difficulties. The method consists in laying down at random a straight sampling line of length L in the tract of area A to be sampled. Then every log intersected by the sampling line is examined. The volume, v, of each log is determined, usually by measuring its length, l, and diameter, d, and assuming the log to be cylindrical, so that $v = \pi d^2 l/4$.

Suppose that the line intersects n logs. Then the total volume of the logs examined is $\sum_n v_i$, where v_i is the volume of the ith log.

Let the total number of logs in the whole tract be N; N is unknown. The population mean volume of wood per unit area, say $\overline{V} = (1/A)\sum_N v_i$, is the quantity to be estimated.

Now envisage a strip of land, of width W, within the tract, so placed that the sampling line L bisects it longitudinally (see Fig. 1); let $W > 2l(\text{max})$, where $l(\text{max})$ is the length of the longest log.

The probability that the center of the ith log lies in the strip is obviously LW/A. The unconditional probability, p_i, that the ith log

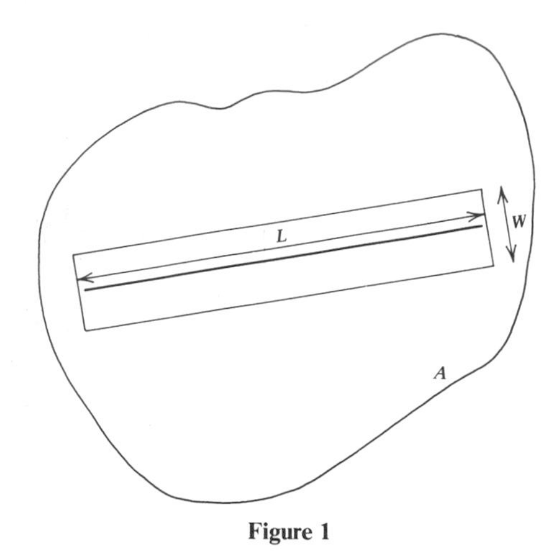

Figure 1

(from a total of N logs), which is of length l_i, is cut by the sampling line is then (using the well-known result of Buffon's needle problem*)

$$p_i = \frac{LW}{A}\frac{2l_i}{\pi W} = \frac{2Ll_i}{\pi A}.$$

Therefore, the expected total volume of the measured logs is

$$E\sum_n v_i = \sum_N p_i v_i = \frac{2L}{\pi A}\sum_N v_i l_i.$$

Hence

$$E\sum_n \left(\frac{v_i}{l_i}\right) = \frac{2L}{\pi A}\sum_N v_i = \frac{2L}{\pi}\bar{V},$$

so that

$$\hat{\bar{V}} = \frac{\pi}{2L}\sum_n \left(\frac{v_i}{l_i}\right) \quad \text{or} \quad \frac{\pi^2}{8L}\sum d_i^2$$

if it is assumed that the logs are cylinders.

De Vries [3] has shown that the variance of $\hat{\bar{V}}$ is estimated by

$$\text{var}(\hat{\bar{V}}) = \left(\frac{\pi}{2L}\right)^2 \sum_n \left(\frac{v_i}{l_i}\right)^2.$$

The method was first applied to the estimation of wood volumes by Warren and Olsen [5]. Since then it has been extended and elaborated. De Vries [4] described multistage line intersects sampling. Brown [1] developed "planar intersect sampling," in which logs that intersect a plane, rather than a line, are sampled and measured. Brown and Roussopoulos [2] showed how to correct for the biases caused by the simplifying assumptions that (1) the measured logs are true cylinders; and (2) all logs lie in one horizontal plane.

Line intersect sampling* should not be confused with line intercept sampling*. These sampling methods bear a superficial resemblance to each other but have quite different purposes. Neither of them resemble line transect sampling* in either method or purpose.

References

[1] Brown, J. K. (1971). *Forest Sci.*, **17**, 96–102.

[2] Brown, J. K. and Roussopoulos, P. J. (1974). *Forest Sci.*, **20**, 350–356.

[3] de Vries, P. G. (1973). *A General Theory on Line Transect Sampling, with Application to Logging Residue Inventory*. Mededelingen Landbouw Hogeschool, Wageningen, The Netherlands, Rep. No. 73(11).

[4] de Vries, P. G. (1979). In *Sampling Biological Populations*, R. M. Cormack, G. P. Patil, and D. S. Robson, eds. International Co-operative Publishing House, Fairland, Md., pp. 1–70.

[5] Warren, W. G. and Olsen, P. F. (1964). *Forest Sci.*, **13**, 267–276.

Further Reading

See the following works, as well as the references just given, for more information on the topic of line intersect sampling.

de Vries, P. G. and van Eijnsbergen, A. C. (1973). *Line Intersect Sampling over Populations of Arbitrarily Shaped Objects*. Mededeligen Landbouw Hogeschool, Wageningen, The Netherlands, Rep. No. 73(19).

van Wagner, C. E. (1968). *Forest Sci.*, **14**, 20–26.

(ECOLOGICAL STATISTICS
FISHERIES, STATISTICS IN
LINE INTERCEPT SAMPLING
LINE TRANSECT SAMPLING
QUADRAT SAMPLING
STATISTICS IN FORESTRY)

E. C. PIELOU

LINEO-NORMAL DISTRIBUTION

A distribution with cumulative distribution function (CDF)

$$F_X(x) = \frac{\sqrt{2}}{\sqrt{\pi}\,\sigma} \int_{-\infty}^{x} \int_0^1 t^{1/2} \times \exp\left[-\tfrac{1}{2} y^2 (t\sigma^2)^{-1} \right] dt\, dy.$$

The distribution is symmetrical about zero; its variance is $2\sigma^2/3$. It belongs to a class of modified normal distributions* constructed by Romanowski [1, 2].

References

[1] Romanowski, M. (1964). *Bull. Géod.*, **73**, 195–216.

[2] Romanowski, M. (1965). *Metrologia*, **4**, 84–86.

(EQUINORMAL DISTRIBUTION
MODIFIED NORMAL DISTRIBUTIONS
QUADRI-NORMAL DISTRIBUTION
RADICO-NORMAL DISTRIBUTION)

LINE TRANSECT SAMPLING

Line transect sampling is a method used by wildlife ecologists for estimating population sizes, or densities, of active, fast-moving terrestrial animals, chiefly mammals and birds.

The field data are collected as follows. An observer walks along a straight line (the transect) of known length that has been placed at random in the area of interest. There is no sampling frame. The observer counts the number of animals, of the species concerned, that are seen; they may be to the right or left of the transect line. The exact point on the ground occupied by each animal at the moment it was flushed or first sighted is marked and the perpendicular distance x from this point to the transect is measured. Two additional measurements are occasionally made, as well as, or instead of, x. These are the radial distance (flushing distance) r from the observer to the point, and the flushing angle θ, which is the angle between the transect and the line joining observer and flushing point. Thus $x = r \sin\theta$.

In arriving at an estimate of D, the density of the animals within the area of interest, it is assumed that:

1. The animals are scattered over the area independently of one another according to a stochastic process with rate parameter D.
2. No individual animal is recorded more than once.
3. Each animal behaves independently of all others (i.e., the flushing of one does not cause the flushing of others).
4. The probability density function (PDF) of x, say $f(x)$, or of r, say $g(r)$, is known; $f(x)$ is the probability of noticing (or flushing) an animal whose perpendicular distance from the transect is x, and analogously for $g(r)$.

Numerous estimators have been derived, and are compared in Gates [2]. They differ because of different assumptions as to the form of the functions $f(x)$ or $g(r)$. It is usually assumed that $f(x)$ and $g(r)$ are monotone decreasing functions of x and r, respectively, and that $f(0) = g(0) = 1$ (equivalently, that an animal on the transect is certain to be observed). Gates [1] and many others assume that x has a negative exponential distribution*. Sen et al. [3] assume r to have a Pearson Type III distribution (*see* PEARSON'S DISTRIBUTIONS).

Sometimes only those animals are recorded for which x is less than some preassigned maximum admissible value, say w. In this case, only animals occurring within a belt, or strip transect, of width $2w$ are tallied.

Line transect sampling should not be confused with line intersect sampling* or line intercept sampling*. These are three entirely different sampling methods.

References

[1] Gates, C. E. (1969). *Biometrics*, **25**, 317–328.

[2] Gates, C. E. (1979). In *Sampling Biological Populations*, R. M. Cormack, G. P. Patil, and D. S. Robson, eds. International Cooperative Publishing House, Fairland, Md., pp. 71–154.

[3] Sen, A. T., Tourigny, J., and Smith, G. E. J. (1974). *Biometrics*, **30**, 329–340.

Further Reading

See the following works, as well as the references just given, for more information on the topic of line transect sampling.

Burnham, K. P. and Anderson, D. R. (1976). *Biometrics*, **32**, 325–336.

Eberhardt, L. L. (1978). *J. Wildl. Manag.*, **42**, 1–31.

Hemingway, P. (1971). In *The Scientific Management of Animal and Plant Communities for Conservation*, E. Duffy and A. S. Watts, eds. Blackwell, Oxford, pp. 405–411.

Kovner, J. L. and Patil, S. A. (1974). *Biometrics*, **30**, 225–230.

Pollock, K. H. (1978). *Biometrics*, **34**, 475–478.

(ECOLOGICAL STATISTICS
FISHERIES, STATISTICS IN
LINE INTERCEPT SAMPLING
LINE INTERSECT SAMPLING
STATISTICS IN ANIMAL SCIENCE
STATISTICS IN FORESTRY)

E. C. PIELOU

LINGUISTICS, STATISTICS IN

Linguistics is unique among scientific disciplines in that its practitioners generally do not require statistical methodology and are not constrained by statistical criteria of validity. Most linguists concur that the grammatical structure of a language consists, in large measure, of discrete entities or categories whose relationships and co-occurrence constraints are qualitative in nature, identical from speaker to speaker within the speech community, and relatively little affected by error or variability in behavior. These structures can thus be deduced by analyzing and comparing test utterances elicted from, or intuited by, any native speaker of the language (most frequently linguists serve as their own data source), without need for repeated trials, random sampling, estimation*, hypothesis testing*, or other statistical apparatus.

It is only since the ground-breaking work of William Labov in the late 1960s that any concerted attempt has been made, within the discipline of linguistics, to investigate questions of central interest to current linguistic theory using a conventionally statistical empirical approach. We shall deal with this in the ensuing sections, but first we note that outside the narrowly defined domain of linguistics, with its focal concern of identifying the formalizing grammatical structure, statistics have long played an important role in the study of languages, linguistic behavior, and communication theory*. Not only has standard statistical methodology been an integral part of such fields as psycholinguistics, infant language acquisition, language education, acoustic phonetics, and linguistic demography, but many original and important developments in applied probability and statistics have taken place in language-related fields.

For example, the transmission of information through written and spoken language has been the prototype for models of communication in information theory*. The study of word frequency* has led to original contributions to sampling theory, distribution theory, and stochastic processes (*see also* LITERATURE, STATISTICS IN). The development of spatial representations of data has at times been closely associated with language-related problems (e.g., the "semantic differential" of Osgood et al. [13]; and multidimensional scaling*, as in Shepard [28]).

Of particular statistical interest in historical linguistics is "glottochronology," Swadesh's [30] method for estimating the date at which two historically related languages split apart. This is based on an analogy with radiocarbon dating* whereby the historical replacement of a word by a synonym can be hypothesized to resemble a Poisson process*. The wider field of lexicostatistics*, an approach to studying genetically related languages, was put on a mathematically firm footing starting in the late 1960s when theory and methods were elaborated strikingly similar to those being worked out in numerical taxonomy*, cluster analysis, and classification*.

In comparative linguistics, the quantitative study of language typology pioneered by Greenberg [6] has culminated in Hawkins' [7] statistical demonstration, in a worldwide sample of languages, of how the modifiers within major sentence components (verb phrase, noun phrase, etc.) all tend to co-occur in the same relative position (i.e., before or after) with respect to the main element (verb, noun, etc.), in the characteristic word-order pattern of a language.

VARIATION THEORY

As mentioned at the outset, the formal models of grammatical theory have discrete structures of an algebraic, algorithmic, and/or logical nature. Such structures often involve sets of two or more alternate components, such as synonyms, paraphrases, or "allophones," which can carry out identical or similar linguistic functions. By allowing a degree of randomness into the choice between such alternates, the grammatical formalisms are converted into probabilistic models of linguistic performance susceptible to statistical study [20].

The most appropriate source of data for this type of study is natural speech. The construction of a sample of natural speech is very different from sampling for sociological questionnaire administration, for psychological experimentation, or for educational testing, since one generally cannot predict when or how often the linguistic phenomenon under study will occur in the flow of conversation. Hence the sample usually involves relatively few speakers (20 to 120), carefully chosen to represent the diversity of linguistic behavior within the community being studied, with a large volume of material tape-recorded from each speaker [12, 26]. This material is transcribed, nowdays often in computer-accessible form [27], and is systematically scanned for occurrences of the words, sounds, or grammatical structures of interest. Usually, the same data set (the "corpus") can be used for many different studies, since it is representative of all the structures and usages of natural speech.

The key concept underlying statistical linguistics is the "linguistic variable." An example from spoken English is the copula verb *be*, which occurs as the contracted variant (*John's a doctor*, *we're coming*, *I'm at home*) or the full variant (*John is . . .* , *we are . . .* , *I am . . .*). Another example involves the sounds written "th," is in *this* and *think*, which usually have an interdental "fricative" pronunciation, but which are also pronounced at least occasionally by speakers of most varieties of English as "stops" (*dis*, *tink*). A third example is the alternation of future and "periphrastic future" tenses (*You will hear about it* versus *you are going to hear about it*).

The choice of one variant or another of a linguistic variable can be heavily influenced by a wide range of factors, including the phonological and syntactic context in which it occurs, the topic of conversation, the degree of situational or contextual formality, idiosyncratic tendencies of the speaker, and the identity of the hearer(s). These factors, however, usually cannot account for all the variability in the data, and so a probabilistic model is set up to evaluate their influence:

$$\log \frac{p(\mathbf{x})}{1 - p(\mathbf{x})} = \mu + \sum \alpha_i x_i, \qquad (1)$$

where $p(\mathbf{x})$ is the probability that a particular one of the two variants will be chosen and $1 - p(\mathbf{x})$ the probability of the other, in the context represented by the vector $\mathbf{x} = (x_1, x_2, \ldots)$. In this "logistic-linear" model, the α_i represent the effects on the choice of variant of the x_i—the latter are often 0–1 indicator variables for the absence or presence of the ith linguistic or sociolinguistic feature in the context of the variable. In the example of copula contraction given above, if x_1 and x_2 indicate that the sound preceding the copula is a vowel or consonant, respectively, while x_3 and x_4 indicate that the grammatical category following the copula is a verb or adjective, and x_5 and x_6 indicate informal and formal speaking styles, then α_1, α_3, and α_5 will be high (approximately 1.0) while α_2, α_4, and α_6 will be low (about -1.0), and $\mu = 0$. Thus the *is* in *John is not aware* as uttered in a formal context is far less susceptible to contraction ($p = 0.05$) than the *am* in *I am coming*, spoken informally ($p = 0.95$).

For statistical analysis, the speech sample is scanned for occurrences of one or the other variant of a variable, and each occurrence is recorded along with $\mathbf{x}$, representing the features or factors present in the context. Each observed vector $\mathbf{x}$ is considered to define a data cell for the analysis, where the number $\mathbf{R}(\mathbf{x})$ of occurrences of one of the variants, compared to the total occurrences $N(\mathbf{x})$ in the cell, is assumed to be a binomial random variable with parameters $N(\mathbf{x})$ and $p(\mathbf{x})$.

Because the distribution of the data among the cells cannot be controlled when a natural speech sample is used, and because many different factors x_i may influence the probability $p(\mathbf{x})$, the final "design" is often a high-dimensional array with many or even most of the possible cells empty ($N(\mathbf{x}) = 0$), and the data distributed very unevenly among the others. Estimation methods based on sum-of-squares approximations are thus inappropriate and the parameters must be estimated using exact maximum likelihood* methods. Many statistical computing packages have the capability of carrying out this type of analysis, although most of the linguistic work has made use of one or other version of the "variable rule" program [3, 16]. Elimination of statistically irrelevant influences can be assured by a multiple regression type of analysis with a stepwise* selection of significant factors [21]. For example, in expressing future time in the French spoken in Montreal, if a verb is negated this has a statistically very significant effect in reducing the use of the periphrastic future (discussed above). Elevated socioeconomic status of the speaker is also a significant factor, while neither the nature of the subject of the verb nor the age of the speaker has a significant effect.

DETECTING HETEROGENEITY

A major preoccupation in statistical linguistics is the question of the homogeneity of the speech community. Do all speakers in the community share a common model of type (1)—possibly involving a single parameter to account for individual differences—or might different individuals or different segments of the community each have a substantively different model of type (1)? Rousseau [15] has developed a way of answering this question based on the dynamic clustering of Diday (e.g., Diday et al. [5]), which generalizes the k-means algorithm* (*see* CLASSIFICATION).

An initial (random) partition into k groups of the speakers in the speech sample is made, followed by an estimation of the parameters in k versions of model (1), a separate model for each group. Speakers are then reassigned to groups according to which model they "fit" best, using the likelihood criterion. Further iterations are carried out of the estimation and reassignment procedures until convergence. The significance of the analyses for each of $k = 2, 3, \ldots$ can be tested based on the increase in likelihood with the increase in k, compared to the number of additional parameters estimated.

Thus, using the data of Laberge [11], Rousseau found that in expressing an indefinite referent, Montreal French speakers fell into two groups according to how they varied between *on*, "one," and *tu*, "you." In one group, speakers had a high rate of *on* usage in conveying proverb-like sentiments and in a certain class of syntactic constructions, while the other group shared the former but not the latter (syntactic) effect.

IMPLICATIONAL SCALES

The principles of Guttman scaling were developed independently for use in linguistics by DeCamp [4], and have been applied extensively (e.g., Bailey [1] and Bickerton [2]). The data on a linguistic variable are typically given as a two-dimensional array Y, where each row i represents a different speaker or speech variety and each column j represents a different linguistic or sociolinguistic context. The entry y_{ij} represents the pair (R_{ij}, N_{ij}), the successes and total trials of a binomial experiment (or the uses of one of the variants and the total occurrences of the variable). The problem is to find row and column permutations (or relabelings) such that the relabeled matrix, R_{ij}/N_{ij} is nondecreasing in both i and j. This, of course, is not always possible, so various somewhat arbitrary measures of scaling have been proposed, to assess to what extent a data set is "scalable" or forms an "implicational scale" based on the minimum possible number of "scaling errors," i.e., cases where $R_{ij}/N_{ij} > R_{hk}/N_{hk}$ but $i < h$ or $j < k$.

It is of particular linguistic interest when for many or most of the cells $R_{ij}/N_{ij} = 0$ or $R_{ij}/N_{ij} = 1$. These correspond to nonvariable ("categorical") usage of one variant rather than the other, by speaker i in context j. When such data scales well, as in Fig. 1, linguists consider the pattern of variable and nonvariable behavior in the community to be well characterized and easily interpretable; see Fig. 1.

At first glance a model such as (2) would not seem capable of giving this type of characterization since p_{ij} could not be 0 or 1 in for any i as j as long as μ, the α_i and the β_j are finite.

$$\log \frac{p_{ij}}{1 - p_{ij}} = \mu + \alpha_i + \beta_j . \qquad (2)$$

In other words, model (2) could only give an approximate account for nonvariable behavior (i.e., when $R_{ij}/N_{ij} = 1$ or $R_{ij}/N_{ij} = 0$). It has been shown, however, that for arrays such as that in Fig. 1, maximum likelihood estimation, of the parameters in (2) become singular, but that the estimates of p_{ij} for each i and j remain well defined, and *can* take on values 0 and 1 where the data predicts nonvariable behavior [17]. This forms the basis for an integrated logistic-linear/implicational scale analysis, including a principled basis for rejecting data which cause "scaling errors," since these data turn out to be outliers* in terms of their extremely low likelihood under a maximum likelihood analysis of the data set [25].

In data where few of the R_{ij}/N_{ij} are 0 or 1, there is little linguistic interest in searching for implicational scales, and it becomes appropriate to resort to more general methods such as principal components analysis*

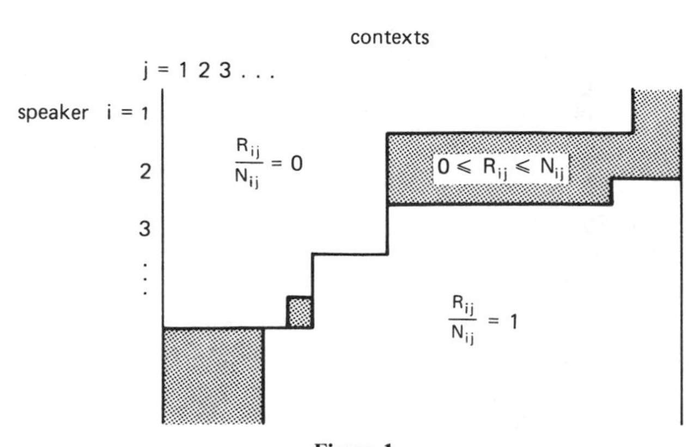

Figure 1

and multidimensional scaling* to understand the relationships among contexts and/or speakers (e.g., Poplack [14] and Sankoff and Cedergren [23]).

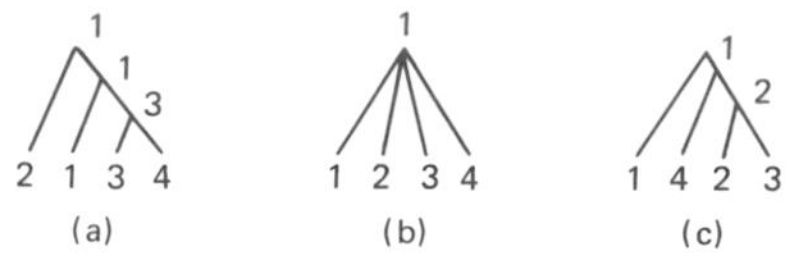

(a) (b) (c)

Figure 2

MULTIPLE VARIANTS

Many linguistic variables have more than two variants. For example, in the clauses *the hat that I bought*, *the hat which I bought*, and *the hat I bought*, the three variants of the "complementizer" are *that*, *which*, and zero (or no complementizer). Phonological variables may commonly have five or ten variants. Model (1) generalizes easily as follows. If $p^{(1)}(\mathbf{x}), p^{(2)}(\mathbf{x}), \ldots, p^{(m)}(\mathbf{x})$ are probabilities of each of the variants under the conditions represented by $\mathbf{x}$, then

$$\begin{aligned} \log \frac{p^{(1)}(\mathbf{x})}{p^{(2)}(\mathbf{x})} &= \mu^{(1)} + \sum \alpha_i^{(1)} x_i, \\ \log \frac{p^{(2)}(\mathbf{x})}{p^{(3)}(\mathbf{x})} &= \mu^{(2)} + \sum \alpha_i^{(2)} x_i, \qquad (3) \\ &\vdots \\ \log \frac{p^{(m-1)}(\mathbf{x})}{p^{(m)}(\mathbf{x})} &= \mu^{(m-1)} + \sum \alpha_i^{(m-1)} x_i, \end{aligned}$$

where the parameters may be estimated from the number of occurrences $R^{(1)}(\mathbf{x}), R^{(2)}(\mathbf{x}), \ldots, R^{(m)}(\mathbf{x})$ of each variant for each context $\mathbf{x}$.

When there are three or more variants of a phonological variable, the linguistic problem of "rule ordering" is raised. It is thought that one of the variants, say variant 1, is the "underlying form" and the rest are generated from it by a series of "rules," or random experiments, which have the same form in each cell $\mathbf{x}$.

For example, suppose that there are N occurrences of the variable in a cell. These are all thought to have been originally cases of variant 1. Then a first binomial experiment created a number M of variant 2, leaving $N - M$ cases of variant 1. Another experiment then created a number L of variant 3, leaving $N - M - L$ cases of variant 1. A final experiment transformed K out of L cases of variant 3 into variant 4, leaving $L - K$ cases of variant 3. Then $R^{(1)} = N - M - L$, $R^{(2)} = M$, $R^{(3)} = L - K$, and $R^{(4)} = K$.

Each of the three experiments in this example can be analyzed according to the independent model of type (1), the first using the values (N, M) from each cell, the second using the values $(N - M, L)$, and the third (L, K). The likelihood of the entire "rule order scheme" is just the product of the maximum likelihood found in the three individual analyses.

The schema can be represented as a tree as in Fig. 2a, where each nonterminal vertex represents a binomial experiment. The same data set could, however, have been generated as a 4-nomial as in Fig. 2b (where the single nonterminal vertex represents the 4-nomial experiment) and analyzed using model (3). Alternatively, it could have been generated as in Fig. 2c and analyzed using one model of type (3) and one of type (1). In fact, there are a total of 184 different schemata. Each schema has exactly the same number of parameters to be estimated on the basis of the same amount of data, and the schema having maximum likelihood would seem, all other considerations being equal, to be the most reasonable solution to the rule-ordering problem.

Examining Fig. 2a more closely, it is not hard to show that all the schemata in Fig. 3 have the same likelihood as that in Fig. 2a. This subset of rule schemata contains all and only those which have the same topological tree structure as Fig. 2a and the same labeling of the terminal vertices. The implications of this equivalence for the problem

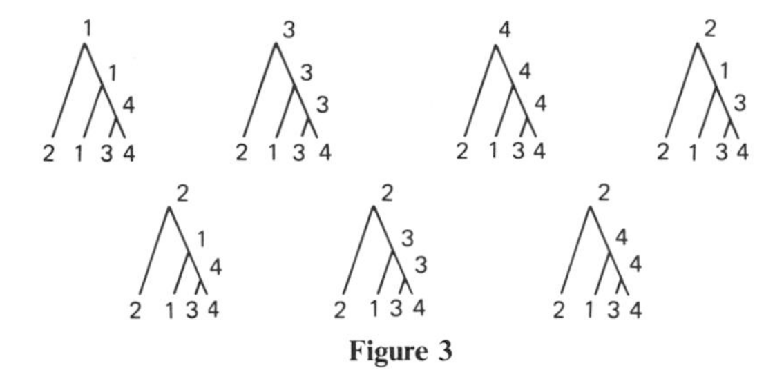

Figure 3

of rule ordering is that this problem can be decomposed into two aspects. One is the identity of the "underlying form" or, more generally, which variants give rise to which other, and the second aspect is the most likely arrangement of the variants into a treelike, or hierarchial classification,* of which there are only 26 different possibilities compared to the 184 schemata in the example. The data on variant occurrences in context do not bear at all on the first aspect, but do allow us to use statistical means, namely the comparison of the likelihoods of the different schemata, to infer the hierarchy.

This methodology has been used most extensively in analyzing the extreme variation in the pronunciation of the word-final consonants *s*, *n*, and *r* that characterize most varieties of Caribbean Spanish [22].

PROBABILISTIC GRAMMARS

Grammars may be thought of as consisting of a finite set of categories, e.g., $\{S, V, N\}$ including one distinguished element, S, a finite vocabulary, e.g., $\{v, n\}$, and a finite set of "rewrite rules," e.g., $\{S \to NV, N \to n, N \to nS, V \to v, V \to vN\}$. The "language" generated by the grammar is the set, generally infinite, of sentences (or strings, or finite sequences) of vocabulary terms which can be constructed as follows.

The first step is to write down the symbol for the distinguished category S. A rule that rewrites S must then be chosen from the set of rules; since there is only one such, $S \to NV$, this is chosen and the S is erased and replaced by NV. For the next step, any term in the current string which is a category and not a vocabulary term is "rewritten." Suppose it is the V which is to be rewritten; then it must be replaced by the right-hand side of some rule which rewrites V, either $V \to v$ or $V \to vN$. Suppose that the latter is chosen; the revised string is thus NvN. Similarly, at each successive step one category term X in the current string is replaced by the right-hand side of some rule in the grammar which rewrites X. If and when the string contains only vocabulary terms, the process stops and we have obtained a sentence. Some of the sentences generated by the grammar above are *nv*, *nvn*, *nnvvn*, etc.

Grammars of natural languages generally contain at most a few dozen categories, e.g., noun, noun phrase, verb phrase, . . . , a vocabulary with thousands of elements, and several dozen rules.

Grammars can be probabilized as formal models of linguistic behavior. This is most easily achieved by making each step in the derivation of a sentence independent of the others, and assigning probabilities to the various rules which can rewrite a given category, e.g.,

$$p(S \to NV) = 1, \qquad p(N \to n) = \theta,$$

$$p(N \to nS) = 1 - \theta,$$

$$p(V \to v) = \tau, \qquad p(V \to vN) = 1 - \tau.$$

Such probabilistic grammars are special cases of multitype Galton–Watson branching processes*. Conditions on the probabilities (e.g., that θ and τ not be too small) are known which ensure that the probability is one that the sentence derivation process is finite (i.e., terminates after a finite number of steps).

Inference based on probabilistic grammars has been used in the study of style [9], infant language learning [29], second language acquisition [10], ambiguous grammars [18, 19], discourse effects on noun phrase structure [8], and bilingual speakers' code switching between one language and another [24].

DISCUSSION

At present, statistics is only beginning to have an impact in linguistics, despite its importance in such subfields or related fields as lexicology and psycholinguistics. The further acceptance by linguists of quantitative methodology and the statistical mode of thinking will depend upon the degree to which this approach can contribute to the understanding of problems of interest to the practitioners of the discipline. This must involve the development or adaptation of methods prompted by specific concerns of linguistic theory rather than the naive application of preexisting methodology to language-based data, interesting as this may be. The progress that has been made has been in applications to phonology and morphology, as well as some fragmentary or drastically simplified syntactic or semantic analyses. It is in these latter two domains, with their complex algebraic and logical representation of language, that the introduction of statistical methods is the most difficult, but potentially the most fruitful.

References

[1] Bailey, C.-J. N. (1973). *Variation and Linguistic Theory*. Center for Applied Linguistics, Arlington, Va.

[2] Bickerton, D. (1975). *Dynamics of a Creole System*. Cambridge University Press, Cambridge, England.

[3] Cedergren, H. J. and Sankoff, D. (1974). *Language*, **50**, 333–355.

[4] De Camp, D. (1971). In *Pidginization and Creolization of Languages*, D. Hymes, ed. Cambridge University Press, Cambridge, England, pp. 349–370.

[5] Diday, E. et al. (1979). *Optimisation en Classification Automatique*. INRIA, France.

[6] Greenberg, J. H. (1960). *Int. J. Amer. Linguistics*, **26**, 178–194.

[7] Hawkins, J. A. (1980). *J. Linguistics*, **16**, 193–235.

[8] Hindle, D. (1981), In *Variation Omnibus*, D. Sankoff and H. J. Cedergren, eds. Linguistic Research Inc., Edmonton, Alberta, pp. 349–357.

[9] Klein, S. (1965). *Language*, **41**, 619–631.

[10] Klein, W. and Dittmar, N. (1979). *Developing Grammars*. Springer-Verlag, Berlin.

[11] Laberge, S. (1977). Etude de la variation des pronoms sujets définis et indéfinis dans le français parlé à Montréal. Ph. D. dissertation, Université de Montréal, Quebec, Canada.

[12] Labov, W. (1966). *The Social Stratification of English in New York City*. Center for Applied Linguistics, Arlington, Va.

[13] Osgood, C. E., Suci, G. J., and Tannenbaum, P. H. (1957). *The Measurement of Meaning*. Urbana, Ill.

[14] Poplack, S. (1979). Process and Function in a Variable Phonology. Ph.D. dissertation, University of Pennsylvania.

[15] Rousseau, P. (1978). Analyse de données binaries. Ph.D. dissertation, Université de Montréal.

[16] Rousseau, P. and Sankoff, D. (1978). In *Linguistic Variation: Models and Methods*, D. Sankoff, ed. Academic Press, New York, pp. 57–69.

[17] Rousseau, P. and Sankoff, D. (1978). *Biometrika*, **65**, 603–608.

[18] Sankoff, D. (1971). *J. Appl. Prob.*, **8**, 233–240.

[19] Sankoff, D. (1972). *Linear Algebra Appl.*, **5**, 277–281.

[20] Sankoff, D. (1978). *Synthese*, **37**, 217–238.

[21] Sankoff, D. (1979). VARBRUL 2S, Appendix B. (in Poplack (1979), pp. 252–257).

[22] Sankoff, D. (1980). *Tech. Reps. 931 and 965*, Centre de Recherche de Mathématiques Appliquées, Université de Montréal, Quebec, Canada.

[23] Sankoff, D. and Cedergren, H. J. (1976). *Language*, **52**, 163–178.

[24] Sankoff, D. and Poplack, S. (1981). *Papers Linguistics*, **14**, 3–46.

[25] Sankoff, D. and Rousseau, P. (1979). In *Papers from the Scandinavian Symposium on Syntactic Variation*, S. Jacobson, ed. Almqvist & Wiksell, Stockholm, Sweden, pp. 7–22.

[26] Sankoff, D. and Sankoff, G. (1973). In *Canadian Languages in Their Social Context*, R. Darnell, ed. Linguistic Research Inc., Edmonton, Alberta, pp. 7–64.

[27] Sankoff, D., Lessard, R., and Nguyen, B. T. (1978). *Computers Humanities*, **11**, 185–191.

[28] Shepard, R. N. (1972). In *Human Communication: A Unified View*, E. E. David and P. B. Denes, eds. McGraw Hill, New York, pp. 67–113.

[29] Suppes, P. (1970). *Synthese*, **22**, 95–116.

[30] Swadesh, M. (1952). *Proc. Amer. Philos. Soc.*, **96**, 452–463.

Bibliography

Bishop, M. M. Fienberg, S. E. and Holland, P. W. (1975). *Discrete Multivariate Analysis: Theory and Practice*. MIT Press, Cambridge, Mass. (Contains a good treatment of logistic-linear and related models.)

Bloom, L. (1978). *Readings in Language Development*. Wiley, New York. (Contains articles illustrating quantitative methods in the study of language acquisition.)

David, J. and Martin, P. (1977). *Études de Statistique Linguistique*. Klincksieck, Paris. (Includes an annotated bibliography of statistical linguistics in the "Anglo-Saxon" countries from 1970 to 1976, by C.-M. Charpentier.)

Doležel, L. and Bailey, R. W. (1969). *Statistics and Style*. Elsevier, New York. (A diverse and excellent collection of studies of literary language.)

Herdan, G. (1964). *Quantitative Linguistics*. Butterworth, London. (One of several major works on the topic by this author. His ideas, although often original and profound, have had little impact among linguists.)

Journal of Child Language. Cambridge University Press, Cambridge, England. (Current acquisition research, much of it statistical.)

Journal of the Acoustical Society of America, New York. (Includes articles on phonetics using multivariate statistical techniques.)

Journal of Verbal Learning and Verbal Behavior. Academic Press, New York. (The leading psycholinguistics journal, exemplifying rigorous criteria of statistical validity.)

Labov, W. (1969). *Language*, **45**, 715–762. (The key article that led to the development of the statistical theory of linguistic variation.)

Language and Speech. Kingston Press Services, Hampton Hall, Middlesex, England. (Quantitative psycholinguistic research.)

Lieberson, S. (1981). *Language Diversity and Language Contact*. Stanford University Press, Stanford, Calif. (A collection of articles on linguistic demography.)

Muller, C. (1977). *Principes et Méthodes de Statistique Lexicale*. Hachette, Paris. (Most recent of a series of textbooks by the leading French statistical linguist.)

Oller, J. W. Jr. (1979). *Language Texts at School*. Longman, New York. (Contains statistical treatment of data on educational linguistics.)

Prideaux, G., in collaboration with Derwing, B. L. and Baker, W. J. (1980). *Experimental Linguistics*. E. Story-Scientia, Ghent, Belgium. (Includes several sophisticated statistical treatments of psycholinguistic and stylistic data.)

(CLASSIFICATION
LEXICOSTATISTICS
LITERATURE, STATISTICS IN
MULTIDIMENSIONAL SCALING
WORD FREQUENCY)

DAVID SANKOFF

LINKAGE CLUSTERING *See* CLASSIFICATION

LINKED BLOCK DESIGNS

These designs were introduced by Youden [4] as a subclass of incomplete block designs* and were classified by Roy and Laha [3]. They can be obtained by dualizing (in the sense of plane projective geometry) incomplete block designs.

A design is said to be a linked block design if every pair of sets has exactly the same number, μ, say, in common.

Roy and Laha's classification subdivides linked block designs into three categories: (1) symmetrical balanced incomplete block designs, (2) partially balanced incomplete block designs, and (3) "irregular" (not belonging to any of "known" types).

Adhikary [1] extended the idea of linked block designs to sets of different sizes by introducing multiple linked block designs. See Raghavarao [2] for more details.

References

[1] Adhikary, B. (1965). *Bull. Calcutta Statist. Ass.*, **14**, 36–64.

[2] Raghavarao, D. (1971). *Constructions and Combinatorial Problems in Design of Experiments*. Wiley, New York.

[3] Roy, J. and Laha, R. G. (1956). *Sankhyā*, **17**, 115–132.

[4] Youden, W. G. (1951). *Biometrics*, **7**, 124 (abstract).

(DESIGN OF EXPERIMENTS
INCOMPLETE BLOCK DESIGNS)

LINK INDEX

An index that is constructed with the immediately preceding period as a base.

(INDEX NUMBERS
LINK RELATIVES)

LINK RELATIVES

In a series of indices over time the base period may be changed with successive calculations. This results in indices that are link relatives. Frequently, link relatives are used because data are not available from the same elements for every time period, only for the present and preceding periods. A series of link relatives can be transformed into a series with common base which results in chain indices.

(INDEX NUMBERS)

LINK TESTS

The purpose of many statistical modelling procedures is to explain the variation in a response variable y by a function of some explanatory variables $x_1, x_2, \dots, x_k$. The most frequently used model in this context is the linear regression* model which relates the mean (μ) of y as a linear function of the explanatory variables, *viz.*

$$\mu = x_1\beta_1 + x_2\beta_2 + \cdots + x_k\beta_k .$$

In the analysis of count data and multiway contingency tables* the common model assumes the log-linear form of the mean, *viz.*

$$\log(\mu) = x_1\beta_1 + x_2\beta_2 + \cdots + x_k\beta_k .$$

Similarly in the analysis of dose-response experiments a commonly used model relating the probability of response (μ) and dosage is given by the logit-linear model, *viz.*

$$\log[\mu/(1-\mu)] = x_1\beta_1 + x_2\beta_2 + \cdots + x_k\beta_k .$$

All of the above models share a common feature, namely, that a function $g = g(\mu)$ of the mean is assumed to be linearly related to the explanatory variables. Since the mean μ implicitly depends on the stochastic behavior of the response, and the x's are assumed fixed, the function g provides the link between the random (stochastic) component and the systematic (deterministic) component of the response variable y. For this reason Nelder and Wedderburn [8] refer to $g(\mu)$ as a *link function*. This article provides a method of testing the adequacy of a particular hypothesized link function.

METHODOLOGY

The difference between an hypothesized link $g_0(\mu)$ and the correct (but unknown) link $g_*(\mu)$ can be statistically assessed in exactly the same manner as the difference between an hypothesized value for β, say β_0, and the correct (but unknown) value β_*. This can be done by embedding $g_0(\mu)$ and $g_*(\mu)$ in a family of link functions and using local linear expansions to model their differences.

Consider the parametric family of link functions $g(\mu;\gamma)$ such that $g_0(\mu) = g(\mu;\gamma_0)$ and $g_*(\mu) = g(\mu;\gamma_*)$. The parameter γ can be vector-valued although in the development which follows we deal exclusively with the scalar case.

The relationship between μ and the explanatory variables can be modelled by

$$g(\mu;\gamma) = x_1\beta_1 + x_2\beta_2 + \cdots + x_k\beta_k . \quad (1)$$

Two common methods of testing an hypothesis concerning γ are the likelihood ratio and scoring methods. The former requires fitting the model with γ constrained to be equal to γ_0 and comparing the maximum of the log-likelihood obtained with that of the unconstrained model. For nonlinear models,* iterative methods are needed to fit each of these models. The scoring method requires a fit of the constrained model plus some auxiliary calculations, which in nonlinear models amount to a single cycle in the usual iterative model fitting scheme.

The likelihood ratio method is primarily used in cases where estimation is the primary objective. For diagnostic testing purposes however, the score test is more appropriate since it requires far less computation. Moreover, local linear expansions of $g(\mu;\gamma)$ allow the score test to be cast in a framework analogous to hypothesis tests concerning individual regression coefficients β_j.

The correct relationship between μ and

the explanatory variables is given by

$$g(\mu;\gamma_*) = x_1\beta_1 + x_2\beta_2 + \cdots + x_k\beta_k. \tag{2}$$

For γ_* close to γ_0,

$$g(\mu;\gamma_*) \simeq g(\mu;\gamma_0) + (\gamma_* - \gamma_0)g'(\mu;\gamma_0),$$

$$g'(\mu;\gamma_0) = \frac{d}{d\gamma}\, g(\mu;\gamma)|_{\gamma=\gamma_0}.$$

Upon substitution into (2) the correct relationship between μ and the explanatory variables is given (approximately) by

$$g(\mu;\gamma_0) = x_1\beta_1 + x_2\beta_2 + \cdots + x_k\beta_k + z\delta, \tag{3}$$

where $z = g'(\mu;\gamma_0)$ and $\delta = \gamma_0 - \gamma_*$. A test of the hypothesis that δ is equal to zero corresponds to a test of the hypothesis that $g_0(\mu)$ is the correct link function. Since z depends on μ which itself must be estimated, Pregibon [9], following Andrews [1], suggested replacing μ by its estimated value from an initial fit of the model $g_0(\mu) = x_1\beta_1 + x_2\beta_2 + \cdots + x_k\beta_k$. With this substitution equation (3) becomes

$$g(\mu;\gamma_0) = x_1\beta_1 + x_2\beta_2 + \cdots + x_k\beta_k + \hat{z}\delta. \tag{4}$$

The interesting feature of this substitution is that the score test of the hypothesis $\delta = 0$ in (4) is identical to the score test of the hypothesis $\gamma = \gamma_0$ in (1). The asymptotic optimality of the procedure therefore follows from the optimality properties of the score test.

To summarize, the link test requires the following steps:

- fit the hypothesized model $g_0(\mu) = x_1\beta_1 + x_2\beta_2 + \cdots + x_k\beta_k$
- construct the derived variable $\hat{z} = g'(\hat{\mu};\gamma_0)$
- use the score test to determine if augmenting the model with $\hat{z}$ significantly improves the fit.

If the final step results in a significant finding, another member of the link family should be used to model the relationship between μ and the explanatory variables. A judicious choice of link family can be most helpful in this regard. Some possibilities are given below.

GRAPHICAL PRESENTATION

An advantage of modelling link inadequacy by a derived variable $\hat{z}$ is that familiar methods of graphical analysis are generally applicable. See for example, Cook and Weisberg [3, Chap. 2]. These methods will be especially important when outliers* and/or high leverage* points are suspected as the calculated value of the test statistic can be unduly influenced by such atypical points. The preferred graphical display is a scatter plot of residuals* versus an adjusted version of the derived variable. For linear models this adjustment amounts to orthogonally projecting $\hat{z}$ onto the residual space defined by $\mathbf{X}$. For nonlinear models a nonorthogonal projection is required. Details of this procedure are given in the example below.

COMMONLY USED LINK FAMILIES

In most modelling contexts a particular link function is routinely assumed. Some of these were outlined in the introduction. This section introduces some commonly used families of link functions which can be used to assess the adequacy of a particular hypothesized link function.

A particularly simple link family applicable to linear models is the polynomial family [11]:

$$g(\mu;\gamma) = \mu + \gamma\cdot\mu^2.$$

This simplifies to the identity link at $\gamma = 0$. The appropriate derived variable is $\hat{\mu}^2$, which corresponds to Tukey's celebrated one degree of freedom for nonadditivity (*see* TUKEY'S ADDITIVITY TEST). This family is easily generalized to other than linear models by replacing μ by $g_0(\mu)$ on the right-hand side of the equation.

A well-studied family of link functions applicable to positive responses is the power family [12]:

$$g(\mu;\gamma) = \begin{cases} \mu^{\gamma} & \text{if} \quad \gamma \neq 0, \\ \log(\mu) & \text{if} \quad \gamma = 0. \end{cases}$$

This family has the identity, log, and reciprocal links all as special cases. The corresponding derived variables are, respectively, $\hat{\mu}\log(\hat{\mu})$, $\log^2(\hat{\mu})$, and $\log(\hat{\mu})/\hat{\mu}$.

A family of links which is not restricted to positive responses is the exponential family* [7]:

$$g(\mu;\gamma) = \begin{cases} \exp(\mu\gamma) & \text{if} \quad \gamma \neq 0, \\ \mu & \text{if} \quad \gamma = 0. \end{cases}$$

This family is most useful for linear models since the identity link is a limiting case. The appropriate derived variable is $\hat{\mu}^2$, which is identical to the derived variable for the polynomial link family. This reflects the local similarity between exponential and polynomial link families.

For dose-response experiments and similar problems where one observes r "successes" at each of n independent "trials", commonly used link functions include the probit (inverse normal CDF), logit* (inverse logistic CDF), log-log (inverse maximal extreme value CDF), and complementary log-log (inverse minimal extreme value CDF). Prentice [10], Van Montfort and Otten [13], and Pregibon [9] provide generalized link families which include a subset of the above as special cases. In the example which follows this section we will use Pregibon's family:

$$g(\mu;\gamma) = \begin{cases} \mu^{\gamma} + (1-\mu)^{-\gamma} & \text{if } \gamma \neq 0, \\ \log(\mu) - \log(1-\mu) & \text{if } \gamma = 0. \end{cases} \tag{5}$$

This family of link functions is most useful in assessing asymmetric alternatives to the logit function. The corresponding derived variable is $\log^2(\hat{\mu}) + \log^2(1-\hat{\mu})$.

EXAMPLE

Lindsey [6, Table 5] presents data from an experiment to determine how salinity and temperature of sea water affect the proportion of eggs of English sole hatching. There are 72 observations covering all combinations of three salinity levels and three temperature settings. A second-order response surface* is used to model the effects of the experimental factors.

Since the response variable represents the number of eggs hatching out of n, the binomial distribution will be used to model the stochastic behavior of the data. The initial link specification will be the logit function, logit $(\mu) = \log[\mu/(1-\mu)]$. Thus if μ denotes the proportion of eggs hatching at salinity level s and temperature setting t, the initial model is

$$\text{logit}(\mu) = \beta_1 + \beta_2 s + \beta_3 t + \beta_4 s^2 + \beta_5 t^2 + \beta_6 st \tag{6}$$

where β_1 is the overall mean effect.

The maximum likelihood fit of this model yields a chi-squared goodness-of-fit* statistic of 2195 with 66 degrees of freedom. A link test will be used to determine if a significant proportion of this value can be attributed to deviations from logit linearity. Pregibon's link family (5) will be used since it economically models asymmetric alternatives to the logit function.

The value of the score statistic* corresponding to the hypothesis that δ is zero is 235.8. If the data were exactly binomial, this value should be compared to percentage points of the $\chi^2(1)$ distribution. When the data exhibit super-binomial variation, a "heterogeneity factor" [4] must be applied. In the present case, we use the value of the chi-squared goodness-of-fit statistic from model (6) divided by its degrees of freedom. This leads to a significance level of the observed score statistic of 0.01. Thus there is rather strong evidence that the logit link function is inadequate.

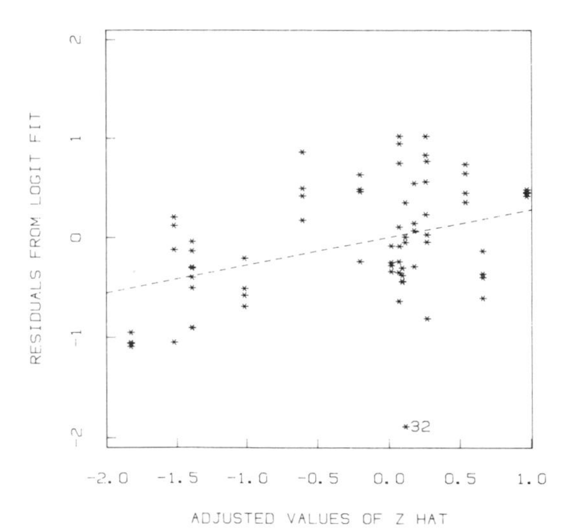

Figure 1 Added variable plot.

A scatter plot of residuals from model (6) versus the adjusted derived variable $\hat{z}_{\text{adj}}$ is displayed in Fig. 1. The adjustment was accomplished by projecting $\hat{z}$ onto the residual space defined by **X** with inner product $W = \text{diag}\{n_i\hat{\mu}_i(1-\hat{\mu}_i)\}$. The dashed line represents an estimate of δ derived from the augmented model

$$\begin{aligned}\text{logit}(\mu) &\\ &= \beta_1 + \beta_2 s + \beta_3 t + \beta_4 s^2 \\ &\quad + \beta_5 t^2 + \beta_6 st + \hat{z}\delta.\end{aligned}$$

A clear dependence of residuals on adjusted derived variable is evident. An outlier (#32) does not seem to unduly affect the overall dependence of residuals on $\hat{z}_{\text{adj}}$.

RELATIONSHIP OF LINK MODIFICATION AND DATA TRANSFORMATION

There is a large body of statistical literature dealing with transforming a reponse variable y to $g(y)$ so that $g(y) \sim N(\mu_g, \sigma_g^2)$, where $\mu_g = x_1\beta_1 + x_2\beta_2 + \cdots + x_k\beta_k$ and $\sigma_g^2 = constant$. This model should not be confused with the corresponding link modification model $y \sim N(\mu, \sigma^2)$ where $g(\mu) = x_1\beta_1 + x_2\beta_2 + \cdots + x_k\beta_k$ and $\sigma^2 = constant$. Since $\mu_g = Eg(y) \simeq g(Ey) = g(\mu)$, both of these methods model linearity on the transformed scale g. The important difference therefore is how the two methods model the stochastic behavior of the response. The data transformation model assumes normality on the transformed scale, whereas the link modification method assumes normality on the original scale.

Theoretical comparisons between the data transformation and link modification methods are not entirely relevant because of the different models being considered. The applicability of the methods can only be judged when applied to particular data analysis problems. Even though experience has indicated that data is usually collected on a scale such that transformations which improve linearity also tend to improve variance homogeneity and normality, examples to the contrary are not difficult to find (e.g., Fisher [5], and Box and Hill [2]).

References

[1] Andrews, D. F. (1971). *Biometrika*, **58**, 249–254.

[2] Box, G. E. P. and Hill, W. J. (1974). *Technometrics*, **16**, 385–389.

[3] Cook, R. D. and Weisberg, S. (1982). *Residuals and Influence in Regression*, Chapman and Hall, London.

[4] Finney, D. J. (1971). *Statistical Method in Biological Assay*, 2nd Edition, Charles Griffin, London.

[5] Fisher, R. A. (1949). *Biometrics* **5**, 300–316.

[6] Lindsey, J. K. (1975). *Appl. Statist.*, **24**, 1–16.

[7] Manly, B. F. J. (1975). *The Statistician*, **25**, 37–42.

[8] Nelder, J. A. and Wedderburn, R. W. M. (1972). *J. R. Statist. Soc. A.*, **135**, 370–384.

[9] Pregibon, D. (1978). *Appl. Statist.*, **29**, 15–24.

[10] Prentice, R. L. (1976). *Biometrics*, **32**, 761–768.

[11] Tukey, J. W. (1949). *Biometrics*, **5**, 232–242.

[12] Tukey, J. W. (1957). *Ann. Statist.*, **38**, 602–632.

[13] Van Montfort, M. A. J. and Otten, A. (1976). *Biometrische Zeitschr.*, **18**, 371–380.

(GENERALIZED LINEAR MODELS RESIDUALS)

DARYL PREGIBON

LIOUVILLE–DIRICHLET DISTRIBUTION

The Liouville–Dirichlet family of distributions, recorded by Marshall and Olkin [4], is an application of Liouville's extension [2] of the Dirichlet integral. This family includes the Dirichlet*, inverted Dirichlet*, multivariate unit-gamma, and multivariate gamma*-type distributions.

DEFINITION AND PROPERTIES

The Liouville–Dirichlet distribution (LDD) is defined by the joint probability density function (PDF) given by

$$f_{X_1,X_2,\ldots,X_s}(x_1,x_2,\ldots,x_s) = Ch\left(\sum x_i\right)\prod_{i=1}^{s} x_i^{\alpha_i-1} \Big/ \int_a^b u^{\alpha-1}h(u)\,du, \tag{1}$$

$x_i > 0$, $a < \sum x_i < b$, $\alpha_i > 0$, $i = 1, 2, \ldots, s$; where $C = \Gamma(\alpha)/\prod_{i=1}^{s}\Gamma(\alpha_i)$, $\alpha = \sum_1^s \alpha_i$, and $h(u)$ is a nonnegative continuous function defined on (a, b).

Property 1. Let $X_1, X_2, \ldots, X_s$ have the LDD with PDF given by (1). Then the marginal distribution of $X_1, X_2, \ldots, X_n$ $(n < s)$ is LDD with corresponding parameters.

Property 2: Preservation of Schur Convexity. The family of s-dimensional distribution functions $\{F_X(\mathbf{x})\}$ with parameters $\boldsymbol{\alpha}' = (\alpha_1, \alpha_2, \ldots, \alpha_s)$ is said to be *parametrized to preserve Schur convexity* if

$$g(\boldsymbol{\alpha}) = E[k(\mathbf{x})] = \int k(\mathbf{x})\,dF_{\mathbf{X}}(\mathbf{x})$$

is Schur convex for $\boldsymbol{\alpha} \in \mathbf{A}$, where $\mathbf{A}$ is a subset of s-dimensional Euclidean space.

The LDD possesses the property of preservation of Schur convexity (*see* MAJORIZATION AND SCHUR CONVEXITY). For details see Cheng [1], Marshall and Olkin [4], and Nevius et al. [5].

SOME SPECIAL CASES OF LDD

1. **Dirichlet Distribution.** In the PDF of the LDD given by (1), if $a = 0$, $b = 1$, and $h(u) = (1 - u)^{\alpha_0 - 1}$, $0 < u < 1$, $\alpha_0 > 0$, we get the Dirichlet distribution with PDF

$$f_{X_1,X_2,\ldots,X_s}(x_1,x_2,\ldots,x_s) = \frac{\Gamma(\alpha+\alpha_0)}{\prod_{j=0}^{s}\Gamma(\alpha_j)}\left(\prod_1^s x_i^{\alpha_i-1}\right)\left(1-\sum_1^s x_i\right)^{\alpha_0-1},$$

where $\alpha = \sum_1^s \alpha_i$.

2. **Inverted Dirichlet Distribution.** The inverted Dirichlet distribution is defined by the joint PDF

$$f_{X_1,X_2,\ldots,X_s}(x_1,x_2,\ldots,x_s) = C\prod_1^x x_i^{\alpha_i-1}\Big/\left(1+\sum_1^s x_i\right)^{\alpha+\alpha_0}, \tag{2}$$

$x_i > 0$, $\alpha_i > 0$, $i = 0, 1, 2, \ldots, s$; where

$$C = \Gamma(\alpha+\alpha_0)\Big/\prod_{j=0}^{s}\Gamma(\alpha_j), \qquad \alpha = \sum_{i=1}^{s}\alpha_i.$$

For $a = 0$, $b = \infty$, and

$$h(u) = (1+u)^{-\alpha+\alpha_0}, \qquad u \geqslant 0, \quad \alpha_0 > 0,$$

the PDF (1) reduces to (2).

3. **Multivariate Gamma Distribution(s).**
   (a) If $a = 0$, $b = \infty$, and $h(u) = u^{\alpha_0}e^{-u}$ in (1), we get

$$f_{X_1,X_2,\ldots,X_s}(x_1,x_2,\ldots,x_s) = \frac{\Gamma(\alpha)}{\Gamma(\alpha+\alpha_0)}\left(\sum x_i\right)^{\alpha_0}\prod_{i=1}^{s}\frac{x_i^{\alpha_i-1}}{\Gamma(\alpha_i)}e^{-x_i}, \ldots, \tag{3}$$

   the joint PDF of correlated gamma variables.

   (b) In case 3(a), if $h(u) = e^{-u}$, (3) reduces to the joint PDF of independent gamma variables.

4. **Multivariate Unit-Gamma-Type Distribution.** In the PDF given by (1), if $a = 0$, $b = 1$, and $h(u) = (-\log u)^{k-1}$, $0 < u < 1$, $k > 0$, we get a multivariate distri-

bution with PDF of the form

$$f_{X_1,X_2,\ldots,X_s}(x_1,x_2,\ldots,x_s)$$
$$\propto \left(\prod_1^s x_i^{\alpha_i-1}\right)\left(-\log\left(\sum_1^s x_i\right)\right)^{k-1},$$
$$0 < x_i < 1, \quad 0 < \sum_1^s x_i < 1.$$

This distribution is an s-dimensional extension of the unit-gamma distribution discussed in refs. 3 and 6 and hence could be called the multivariate unit-gamma distribution.

References

[1] Cheng, K. W. (1977). Majorization: Its Extensions and Preservation Theorems. *Tech. Rep. No. 121*, Dept. of Statistics, Stanford University, Stanford, Calif.

[2] Edwards, J. (1922). *A Treatise on the Integral Calculus*, Vol. 2. Macmillan, New York, pp. 160–162.

[3] Grassia, A. (1977). *Aust. J. Statist.*, **19**, 108–114.

[4] Marshall, A. W. and Olkin, I. (1979). *Inequalities: Theory of Majorization and Its Applications*. Academic Press, New York, Chap. 11.

[5] Nevius, S. E., Proschan, F., and Sethuraman, J. (1977). *Ann. Statist.*, **5**, 263–273.

[6] Ratnaparkhi, M. V. (1981). In *Statistical Distributions in Scientific Work*, Vol. 4, C. Taillie, G. P. Patil, and B. A. Baldessari, eds. D. Reidel, Dordrecht, Holland, pp. 389–400.

(DIRCHLET DISTRIBUTION
INVERTED DIRICHLET DISTRIBUTION
MULTIVARIATE GAMMA
DISTRIBUTIONS)

M. V. RATNAPARKHI

LISREL

Structural equation models* have been useful in attacking many substantive problems in the social and behavioral sciences. Such models have been used in the study of macroeconomic policy formation, intergeneration occupational mobility, racial discrimination in employment, housing, earnings, studies of antecedents and consequences of drug use, scholastic achievement, evaluation of social action programs, and many other areas. In methodological terms, the models have been referred to as simultaneous equation systems, linear causal analysis, path analysis*, structural equation models, dependence analysis, cross-lagged panel correlation technique, etc.

LISREL (LInear Structural RELationships) is a general model for studying a set of linear structural equations. The variables in the equation system may be either directly observed variables or unmeasured latent variables (hypothetical construct variables). The latter are not observed but are related to observed variables. The general model is particularly designed to handle models with latent variables, measurement errors*, and reciprocal causation (simultaneity, interdependence). In its most general form the model assumes that there is a causal structure among a set of latent variables. The LISREL model consists of two parts: the measurement model and the structural equation model. The measurement model specifies how the latent variables are measured in terms of the observed variables and the structural equation model specifies the causal relationships of the latent variables.

In addition to models of functional and structural relationships, the LISREL model covers a wide range of models useful in the social and behavioral sciences: e.g., exploratory and confirmatory factor analysis* models, path analysis models, econometric* models for time-series* data, recursive and nonrecursive models for cross-sectional and longitudinal data*, and covariance structure models.

Associated with the LISREL model is a computer program (LISREL VI) which may be used to analyze data from a single sample or from several samples simultaneously. This program provides three kinds of estimates of the parameters in the LISREL model using the instrumental variables* method, the method of least squares*, or the maximum likelihood* method. Several measures of goodness of fit* of the model to the data are provided and structural hypotheses about parameters in the model may be tested by means of a likelihood ratio χ^2 statistic.

The LISREL model was first introduced by Jöreskog [5] and was further developed and described by Jöreskog [6–8]. A complete description of the LISREL model and the uses of the computer program LISREL VI is given by Jöreskog and Sörbom [9]. Applications of LISREL to psychological, educational, sociological, and other social science problems are discussed in Kenny [10], Munck [12], Bohrnstedt and Borgatta [3], and Bagozzi [1]. The robustness of the estimation procedure has been studied by Boomsma [4]. Alternative model formulations and estimation procedures are considered by McDonald [11], Bentler and Weeks [2], and Wold [13].

LISREL MODEL

Consider random vectors $\boldsymbol{\eta}' = (\eta_1, \eta_2, \ldots, \eta_m)$ and $\boldsymbol{\xi}' = (\xi_1, \xi_2, \ldots, \xi_n)$ of latent dependent and independent variables, respectively, and the system of linear structural relations

$$\boldsymbol{\eta} = \mathbf{B}\boldsymbol{\eta} + \boldsymbol{\Gamma}\boldsymbol{\xi} + \boldsymbol{\zeta}, \tag{1}$$

where $\mathbf{B}(m \times m)$ and $\boldsymbol{\Gamma}(m \times n)$ are parameter matrices and $\boldsymbol{\zeta}' = (\zeta_1, \zeta_2, \ldots, \zeta_m)$ is a random vector of residuals (errors in equations, random disturbance terms). The elements of $\mathbf{B}$ represent direct causal effects of η-variables on other η-variables and the elements of $\boldsymbol{\Gamma}$ represent direct causal effects of ξ-variables on η-variables. It is assumed that $\boldsymbol{\zeta}$ is uncorrelated with $\boldsymbol{\xi}$ and that $\mathbf{I} - \mathbf{B}$ is nonsingular. (Although the computer program LISREL VI can deal with location parameters as well, it is assumed here, for simplicity of presentation, that all variables, observed as well as unobserved, have means zero.) The vectors $\boldsymbol{\eta}$ and $\boldsymbol{\xi}$ are not observed but instead vectors $\mathbf{y}' = (y_1, y_2, \ldots, y_p)$ and $\mathbf{x}' = (x_1, x_2, \ldots, x_q)$ are observed, such that

$$\mathbf{y} = \boldsymbol{\Lambda}_y\boldsymbol{\eta} + \boldsymbol{\epsilon} \tag{2}$$

and

$$\mathbf{x} = \boldsymbol{\Lambda}_x\boldsymbol{\xi} + \boldsymbol{\delta}, \tag{3}$$

where $\boldsymbol{\epsilon}$ and $\boldsymbol{\delta}$ are vectors of errors of measurement in $\mathbf{y}$ and $\mathbf{x}$, respectively. The matrices $\boldsymbol{\Lambda}_y(p \times m)$ and $\boldsymbol{\Lambda}_x(q \times n)$ are regression matrices of $\mathbf{y}$ on $\boldsymbol{\eta}$ and of $\mathbf{x}$ on $\boldsymbol{\xi}$, respectively. It is convenient to refer to $\mathbf{y}$ and $\mathbf{x}$ as the *observed variables* and $\boldsymbol{\eta}$ and $\boldsymbol{\xi}$ as the *latent variables*. The errors of measurement are assumed to be uncorrelated with $\boldsymbol{\eta}$, $\boldsymbol{\xi}$, and $\boldsymbol{\zeta}$ but may be correlated among themselves.

Let $\boldsymbol{\Phi}(n \times n)$ and $\boldsymbol{\Psi}(m \times m)$ be the covariance matrices of $\boldsymbol{\xi}$ and $\boldsymbol{\zeta}$, respectively, and let $\boldsymbol{\Theta}_\epsilon$ and $\boldsymbol{\Theta}_\delta$ be the covariance matrices of $\boldsymbol{\epsilon}$ and $\boldsymbol{\delta}$, respectively. Then it follows, from the assumptions above, that the covariance matrix $\boldsymbol{\Sigma}[(p+q) \times (p+q)]$ of $(\mathbf{y}', \mathbf{x}')$ is

$$\boldsymbol{\Sigma} = \begin{bmatrix} \mathbf{C}_y(\boldsymbol{\Gamma}\boldsymbol{\Phi}\boldsymbol{\Gamma}' + \boldsymbol{\Psi})\mathbf{C}_y' + \boldsymbol{\Theta}_\epsilon & \mathbf{C}_y\boldsymbol{\Gamma}\boldsymbol{\Phi}\boldsymbol{\Lambda}_x \\ \boldsymbol{\Lambda}_x\boldsymbol{\Phi}\boldsymbol{\Gamma}'\mathbf{C}_y' & \boldsymbol{\Lambda}_x\boldsymbol{\Phi}\boldsymbol{\Lambda}_x + \boldsymbol{\Theta}_\delta \end{bmatrix}, \tag{4}$$

$$\mathbf{C}_y = \boldsymbol{\Lambda}_y(\mathbf{I} - \mathbf{B})^{-1}.$$

The elements of $\boldsymbol{\Sigma}$ are functions of the elements of $\boldsymbol{\Lambda}_y$, $\boldsymbol{\Lambda}_x$, $\mathbf{B}$, $\boldsymbol{\Gamma}$, $\boldsymbol{\Phi}$, $\boldsymbol{\Psi}$, $\boldsymbol{\Theta}_\epsilon$, and $\boldsymbol{\Theta}_\delta$. In applications some of these elements are fixed and equal to assigned values. In particular, this is so for elements of $\boldsymbol{\Lambda}_y$, $\boldsymbol{\Lambda}_x$, $\mathbf{B}$, and $\boldsymbol{\Gamma}$, but it is possible to have fixed values in the other matrices also. For the remaining nonfixed elements of the eight parameter matrices, one or more subsets may have identical but unknown values. Thus the elements in $\boldsymbol{\Lambda}_y$, $\boldsymbol{\Lambda}_x$, $\mathbf{B}$, $\boldsymbol{\Gamma}$, $\boldsymbol{\Phi}$, $\boldsymbol{\Psi}$, $\boldsymbol{\Theta}_\epsilon$, and $\boldsymbol{\Theta}_\delta$ are of three kinds:

1. *Fixed parameters* that have been assigned given values
2. *Constrained parameters* that are unknown but equal to one or more other parameters
3. *Free parameters* that are unknown and not constrained to be equal to any other parameter

This specification makes the model very flexible so that several classes of submodels are contained within the general model

The two most common submodels are the factor analysis model,

$$\mathbf{x} = \mathbf{\Lambda}_x \boldsymbol{\xi} + \boldsymbol{\delta}, \tag{5}$$

which involves only $x-$, $\xi-$, and $\delta-$ variables, and the structural equation model for directly observed y- and x-variables,

$$\mathbf{y} = \mathbf{B}\mathbf{y} + \mathbf{\Gamma}\mathbf{x} + \boldsymbol{\zeta}. \tag{6}$$

The last model is obtained from the general model by setting $m = p$, $n = q$, $\mathbf{\Lambda}_y = \mathbf{I}$, $\mathbf{\Theta}_\epsilon = \mathbf{0}$, $\mathbf{\Lambda}_x = \mathbf{I}$, and $\mathbf{\Theta}_\delta = \mathbf{0}$.

ESTIMATION

Equations (1) to (3) represent a model for a population of individuals (observational units). This population is characterized by the mean vector $\boldsymbol{\mu}$ and the independent parameters $\boldsymbol{\theta}$ in $\mathbf{\Lambda}_y$, $\mathbf{\Lambda}_x$, $\mathbf{B}$, $\mathbf{\Gamma}$, $\mathbf{\Phi}$, $\mathbf{\Psi}$, $\mathbf{\Theta}_\epsilon$, and $\mathbf{\Theta}_\delta$ generating the covariance matrix $\mathbf{\Sigma}$ in (4). In practice, $\boldsymbol{\mu}$ and $\boldsymbol{\theta}$ are unknown and must be estimated from data. It is assumed that the data is a random sample of independent observations from the population.

It is assumed that the distribution of the observed variables is sufficiently well described by the moments of first and second order, so that information contained in moments of higher order may be ignored. In particular, this will hold if the distribution is multivariate normal. Since the mean vector is unconstrained, the distribution of the observed variables is described by the parameters $\boldsymbol{\theta}$, which are to be estimated.

Let $\mathbf{S}$ be the sample covariance matrix computed from the sample data. The approach to estimation taken is to fit $\mathbf{\Sigma}$ to $\mathbf{S}$, using either the unweighted least-squares (ULS) method or the maximum likelihood (ML) method.

The fitting function for ULS is

$$F = \tfrac{1}{2}\operatorname{tr}(\mathbf{S} - \mathbf{\Sigma})^2 \tag{7}$$

and that for ML is

$$F = \log|\mathbf{\Sigma}| + \operatorname{tr}(\mathbf{S}\mathbf{\Sigma}^{-1}) - \log|\mathbf{S}| - (p + q). \tag{8}$$

Both fitting functions are regarded as a function of the parameters $\boldsymbol{\theta}$ and are to be minimized with respect to these. Both fitting functions are nonnegative and equal to zero only when there is a perfect fit, i.e., when the fitted $\mathbf{\Sigma}$ equals $\mathbf{S}$. The fitting function for ML is derived from the maximum likelihood principle based on the assumption that the observed variables have a multinormal distribution. The fitting functions are minimized by an iterative procedure starting with a set of consistent initial estimates computed by instrumental variables* techniques (see Jöreskog and Sörbom [9]).

In ML, $(2/N)$ times the inverse of the information matrix $\mathbf{E}$ evaluated at the minimum of F provides an estimate of the asymptotic covariance matrix of the estimators $\hat{\boldsymbol{\theta}}$ of $\boldsymbol{\theta}$. The square root of the diagonal elements of $(2/N)\mathbf{E}^{-1}$ are large-sample estimates of the standard errors of the parameter estimates.

ASSESSMENT OF FIT

With the ML method, the validity of the model may be tested by a likelihood ratio test. The logarithm of the likelihood ratio is simply $(N/2)$ times the minimum value of the function F. Under the model this is distributed, in large samples, as a χ^2 distribution with degrees of freedom equal to

$$d = \tfrac{1}{2}(p + q)(p + q + 1) - t, \tag{9}$$

where t is the total number of parameters estimated.

For both ULS and ML, two other measures of fit, the goodness-of-fit index and the root-mean-square error, have been constructed, which are less sensitive to departures from normality than the χ^2-measure. A more detailed assessment of fit may be obtained by inspection of normalized residuals and modification indices. For information about these and other details, see Jöreskog and Sörbom [9]. For a more comprehensive description of the mathematical-statistical theory of LISREL, see Jöreskog [6, 8].

TESTS OF STRUCTURAL HYPOTHESES

Once the validity of the model has been established, various structural hypotheses within the model may be tested. One can test hypotheses of the forms:

1. That certain θ's are fixed equal to assigned values
2. That certain θ's are equal in groups

Each of these two types of hypotheses leads to a covariance structure $\boldsymbol{\Sigma}(\boldsymbol{v})$, where $\boldsymbol{v}$ is a subset of $u < t$ elements of $\boldsymbol{\theta}$. Let F_v be the minimum of F under the structural hypothesis and let F_θ be the minimum of F under the general model. Then $(N/2)(F_v - F_\theta)$ is approximately distributed as χ^2 with $t - u$ degrees of freedom.

References

[1] Bagozzi, R. P. (1980). *Causal Models in Marketing*. Wiley, New York.

[2] Bentler, P. M. and Weeks, D. G. (1980). *Psychometrika*, **45**, 298–308.

[3] Bohrnstedt, G. W. and Borgatta, E. F., eds. (1981). *Social Measurement*. Sage, Beverly Hills, Calif.

[4] Boomsma, A. (1982). On the Robustness of LISREL against Small Sample Size and Nonnormality. Ph.D. dissertation, University of Groningen.

[5] Jöreskog, K. G. (1973). In *Structural Equation Models in the Social Sciences*, A. S. Goldberger and O. D. Duncan, eds. Seminar Press, New York, pp. 85–112.

[6] Jöreskog, K. G. (1977). In *Applications of Statistics*, P. R. Krishnaiah, ed. North-Holland, Amsterdam, pp. 265—287.

[7] Jöreskog, K. G. (1978). *Psychometrika*, **43**, 443–477.

[8] Jöreskog, K. G. (1981). *Scand. J. Statist.*, **8**, 65–92.

[9] Jöreskog, K. G. and Sörbom, D. (1984). *Analysis of Linear Structural Relationships by Maximum Likelihood and Least Squares Methods*. Scientific Software, Inc., Chicago.

[10] Kenny, D. A. (1979). *Correlation and Causality*. Wiley, New York.

[11] McDonald, R. P. (1978). *Brit. J. Math. Statist. Psychol.*, **31**, 59–72.

[12] Munck, I. M. E. (1979). *Model Building in Comparative Education*. Almqvist & Wiksell, Stockholm.

[13] Wold, H. (1982). In *Systems under Indirect Observation: Causality, Structure and Prediction*, Part 2, K. G. Jöreskog and H. Wold, eds. North-Holland, Amsterdam, pp. 1–32.

(ECONOMETRICS
FACTOR ANALYSIS
GENERAL LINEAR MODEL
INSTRUMENTAL VARIABLE ESTIMATION
LATENT STRUCTURE ANALYSIS
PATH ANALYSIS
PSYCHOLOGY, STATISTICS IN
SOCIOLOGY, STATISTICS IN
STRUCTURAL EQUATION MODELS)

KARL G. JÖRESKOG

LITERATURE AND STATISTICS

HISTORICAL INTRODUCTION

Statistical considerations are important both in the spoken and written aspects of language. Concise dictionaries, for example, include words with the greatest frequency of usage, as Kučera [20] has pointed out. Similarly, statistical counts of the frequencies of letters in alphabetic languages must clearly have entered into the development of the Braille and Morse codes in the 1830s and the design of typewriter keyboards in the 1870s. However, these can hardly be considered to have initiated statistical studies in literature. Although there is no clear dividing line between statistical linguistics* and statistics in literature, this article will concentrate primarily, although not exclusively, on the literary aspects of statistics.

Williams [35] reports that the Masoretes, Jewish scribes concerned with the exact preservation of the Old Testament text between AD 500 and 1000, counted the numbers of letters and words in each book, as well as the repetitions of certain words (usually names) of special significance. More recently, statistical computer-based studies of the Old Testament have been carried out by such Hebrew scholars as Radday [30]. In the nineteenth century, several English works were studied intensively, especially Shakespeare (see Clarke [3] and Fleay [6]); both repetitions of words and variations of meter in verse were counted. The concern of this school of authors was to characterize literary texts quantitatively, and to solve problems of chronology and authorship using only elementary mathematical methods. In the same tradition, Spevack [34] has provided a complete modern *Concordance* of Shakespeare's works.

Perhaps the earliest statistical linguistic study in a literary context was Markov's [22] paper on the sequence of vowels and consonants in the poem "Evgeni Onegin" by Pushkin; it was this which led to the definition of the simple Markov chain*, where

$$P = \begin{pmatrix} p_{cc} & p_{cv} \\ p_{vc} & p_{vv} \end{pmatrix}$$

denotes the matrix of transition probabilities, with, for example,

$$p_{cv} = \Pr\{\text{consonant is followed by a vowel}\}.$$

In any language, one can study whether phonemes or sequences of words in sentences form a Markov chain (see Mandelbrot [21].) In fact, it can be proved that language cannot be a Markov process* of finite order, although Good [11] has shown that fair approximations to English can be obtained with a third-order Markov chain of words. Colleagues who were asked to provide the next word in a sentence independently, given only the previous three, produced as one of their results: "The best film on television tonight is there no-one here who had a little bit of fluff."

Among the pioneers of the statistical study of literature are Mendenhall [24, 25], Zipf [38], Yule [37], Herdan [16–19], Guiraud [13], Morton and MacGregor [27], Mosteller and Wallace [28], Muller [29], Williams [35], Brainerd [1, 2], and Hantrais [15]. These authors have used statistical methods to date literary works, identify their likely sources, discriminate between authors or between the different genres of a single author, compare vocabularies and literary structures, and make comparative analyses of different languages. In most modern studies, the computer has played an important if not indispensable role.

Although different authorities may give varying weights to the several aspects of the statistical study of literary texts, all are agreed that redundancy, word length, sentence length, and vocabulary as measures of style, and comparisons of different languages are important topics in the subject. We shall therefore consider each of these in turn.

REDUNDANCY

All languages are redundant: the information conveyed in speech or writing is repetitious, but such redundancy avoids error and misunderstanding. Shannon [32], among others, was interested in the entropy* of English, and the possibility of predicting the next letter when the preceding text was known. For a sequence of symbols a_1, $a_2, \ldots$ such that the probability that $a_r = i$, is p_i, where $i = 1, \ldots, t$ are the t letters of a generalized alphabet, we can define measures of heterogeneity

$$c_{rs} = \sum_i p_i^r(-\log p_i)^s. \tag{1}$$

The simplest of these is the repeat rate

$$c_{20} = \sum_i p_i^2, \tag{2}$$

which is of importance in cryptanalysis.

The entropy

$$c_{11} = -\sum_i p_i \log p_i \tag{3}$$

arises in the study of the expected weight of evidence per character for the two multinomial hypotheses $H_1 : p_i$ as against $H_0 : q_i = 1/t$; in the first, each symbol is assumed to have a different probability p_i, as against the second, where each has equal probability $1/t$. In the latter case, the expected weight per character for a long text is $\log t - c_{11}$, and can be shown to approximate the value

$$\log t - c_{11} = \sum_i p_i \log(tp_i) \sim \tfrac{1}{2}(tc_{20} - 1), \quad (4)$$

when the repeat rate c_{20} is close to its minimum. A relation clearly exists between the weight of evidence, entropy, and the repeat rate.

Good [11] has pointed out that in the development of Braille, where $t = 41$, the principle used intuitively was to maximize the transmission of information per symbol, without loss of information; the same appears to be true of Morse code. But for speech, this would not be possible, and a more sensible principle is probably to maximize the transmission of information per symbol. For details, the reader is referred to Good [10] and Good and Toulmin [12] as well as the article referred to previously.

WORD AND SENTENCE LENGTH

An important problem considered as early as 1851 by Augustus de Morgan was that of distinguishing between different authors by word lengths in their texts. Mendenhall [25] used the same criterion to discriminate between the writings of Shakespeare, Bacon, and Marlowe. Mendenhall reported that Shakespeare used more four-letter words, and Bacon more three-letter words, than any others; there was a close resemblance, however, between the word-length distributions of Shakespeare and Marlowe. Williams [36], reviewing Mendenhall's analysis, noted that Shakespeare's and Marlowe's works consisted of verse, whereas Bacon wrote prose. The differences in word lengths could thus be attributed to the different styles of composition. Word-length counts in the works of Mill, Dickens, and Shakespeare have also shown significant differences, and these can be used to identify each author.

Another criterion discussed by Yule [37] was the frequency of sentence length as an individual characteristic of authors. This showed a wider variation than word length and is not considered entirely reliable in view of known changes within an author's work during his lifetime. In the works of Bacon, Coleridge, and Macaulay, for example, the peak frequencies of sentence length were in the 31–35, the 21–25, and the 11–15 word groups, respectively. The empirical distribution of sentence length tends to be skewed, with a peak at the lower end, and long tail at the upper. Williams [35] suggested that it might be fitted by a lognormal distribution, and analyzed works by Shaw, Wells, and Chesterton to test his hypothesis. He was able to obtain a reasonable fit in each case, with clear differences among the three writers. There are also differences in sentence length in the work of the same author, not only temporally but also depending on the genre of writing: e.g., according to whether the prose is descriptive or consists of dialogue. These parallel the differences in word-length distributions in the prose and verse of the same author, which Williams [36] has documented for Sir Philip Sidney.

VOCABULARY

The vocabulary of an author is often recognized by his readers; this is sometimes a question of the frequency with which certain words are used. The actual frequencies of words in particular works are of interest, and have been investigated by Zipf [38], Yule [37], and Good [9], among others. Zipf had conjectured that the correct frequency distribution would be based on a compound Poisson model; Sichel [33] was able to fit successfully a family of compound Poisson distributions to the word frequencies of a number of authors and works in different languages.

Mosteller and Wallace [28] used word count criteria, in the main, to determine the authorship of *The Federalist* papers; Hamilton wrote "while" whereas Madison preferred "whilst," and used "upon" roughly 18 times as frequently per 1000 words as did Madison. The word "enough" also emerged in their study as a Hamilton marker. In their very interesting book, Mosteller and Wallace outlined four different approaches to the study of discrimination between authors on the basis of word frequency* counts. The first and principal approach was Bayesian*, the second relied on a classical linear discrimination procedure (*see* DISCRIMINANT ANALYSIS), the third was a robust Bayesian analysis (*see* BAYESIAN INFERENCE), while the fourth consisted of a simplified rate study on classical lines. Evidence was adduced to show that the 12 disputed *Federalist* papers were written by Madison.

Another interesting statistical problem is the estimation* of the author's total latent vocabulary from a sample of his work. This is referred to as the type (new word) and token (sample word) problem: a sample of 1000 tokens may in fact contain only 400 to 450 types. Yule estimated that the *Shorter Oxford Dictionary* of 1933 contained 58,000 nouns, 27,000 adjectives, and 13,500 verbs; there would also be about 1000 adverbs, and these all together form the substantive words usually classed in group I. There are, in addition, some 500 pronouns, prepositions, conjunctions, auxiliary verbs, articles, and interjections, which form the auxiliary words of group II. An educated person might be expected to know at least 20,000 or so words.

There are distinct differences in the size of authors' vocabularies and also in the different parts of speech which they commonly use. There are also differences in the use of words in the different works of a single author. For example, in Shakespeare's plays, the vocabulary size is larger for the historical plays than for the comedies [8]. A new measure of vocabulary richness devised by Ratkowsky et al. [31] indicates that *Macbeth* has the richest vocabulary and *Much Ado About Nothing* the poorest.

The total size of the vocabulary used by Shakespeare is 31,534, of which 14,376 types appear once, 4343 twice, and so on up to 846 types which appear more than 100 times; for details, the reader is referred to Spevack [34], in whose concordance 884,647 occurrences of words are listed. Earlier estimates based on extracts from every tenth page of Clark's *Concordance* [3] yielded 1082

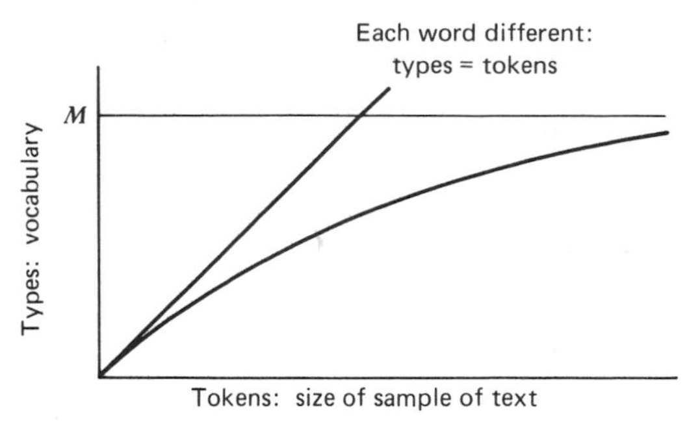

Figure 1 Relation between types and tokens.

words used once only, 1111 used 2 to 20 times, and 293 used over 20 times. This gave 2486 words with total usages of 30,900: hence the rather lower estimate of a 25,000-word vocabulary for Shakespeare, based on the total of 310,000 words listed in this earlier and less complete concordance.

Various models, some based on Fisher's work on the abundance of species, others on a stochastic sampling* scheme, have been proposed for the rate of increase of an author's vocabulary with the increase in sample size, i.e., the relationship between types and tokens. The relation clearly differs depending on whether one groups all words together or separates them into group I and group II. Roughly speaking, if the total vocabulary of the author is M, the graph of the relation may be illustrated by Fig. 1. The first token will clearly be a type (a new word), but gradually the slope of the curve decreases as it approaches M as an asymptote. For further details, see Efron and Thisted [5], who have given a lower estimate of about 66,000 words for Shakespeare's total vocabulary, as well as McNeil [23], Gani [7], and Gani and Saunders [8].

COMPARISON OF DIFFERENT LANGUAGES

Morton [26] studied the relative use of the Greek "and" in various works by ancient Greek authors such as Herodotus and Thucydides, as well as in the different Greek Epistles of the New Testament attributed to

St. Paul. The range of total usage in the latter is from about 3 to 7%, and the number of sentences containing "and" ranges from 29 to 67% of the total. The conclusion drawn by Morton was that some of the Epistles were likely to be the work of writers other than St. Paul. Computerized concordances of both Hebrew and Greek books of the Bible have also been compiled in *The Computer Bible Series* (see, e.g., Radday [30]) to assist in the detailed linguistic analysis of their language.

The usage of words containing different numbers of letters differs in different languages. Mendenhall compared Latin, Italian, Spanish, French, and German, and found differences in their frequency distributions for words of different lengths. Italian, Latin, and Spanish had a bimodal distribution of word length, German had greater frequencies of long words, and French had a very high peak for two-letter words, more marked than that which existed in Spanish, Italian, and Latin.

Brainerd [2], among others, has considered some models for testing the affinities between various languages of unknown origin. This is concerned with the relationship between two languages which may both arise from a common source, or which have borrowed extensively from each other, or for whose resemblance there is no fully understood cause. The method used involves finding the probability that a single pair of words selected at random from the two languages share certain characteristics, when the null hypothesis* assumes the languages to be independent. For example, in two Austronesian languages studied by Cowan [4], the probability of four or more agreements of initial consonants in nine possible comparisons of pronoun subject prefixes of the verb was 0.001; this indicated that the null hypothesis of no causal relationship between the two languages was untenable.

CONCLUDING REMARKS

Statistics in literature is an interesting and fairly recent area of study, which has its place as a quantitative procedure for characterizing language and style. It is of value in a wide set of problems concerned with differentiating between authors, and helping to determine the exact chronology and authorship of literary texts. The methods used are clearly also relevant to the creation of artificial languages such as those used in computer programming. It is possibly in this area that the principles laid down for statistics in literature may find a valuable new application in future (see Halstead [14]).

References

Letters at the end of reference entries denote one of the following categories:

G: general, can be read by the layman
L: literary
Sp: specialist
St: statistical
T: technical, or including some technical sections

[1] Brainerd, B. (1970). *Introduction to the Mathematics of Language Study*. Elsevier, New York. (T, St; very readable, particularly by statisticians.)

[2] Brainerd, B. (1974). *Weighing Evidence in Language and Literature: A Statistical Approach*. University of Toronto Press, Toronto. (T, St; very readable, particularly by statisticians.)

[3] Clarke, M. C. (1845). *Complete Concordance to Shakespeare*. Knight, London. (L)

[4] Cowan, H. K. J. (1962). *Studia Linguistica*, **16**, 57–96. (Sp, St)

[5] Efron, B. and Thisted, R. (1976). *Biometrika*, **63**, 435–447. (Sp, St)

[6] Fleay, F. G. (1876). *Shakespeare Manual*. London. (L)

[7] Gani, J. (1975). In *Perspectives in Probability and Statistics*, J. Gani, ed. Academic Press, London, pp. 313–323. (Sp, St)

[8] Gani, J. and Saunders, I. W. (1976). *Sankhyā B*, **38**, 101–111. (Sp, St)

[9] Good, I. J. (1953). *Biometrika*, **40**, 237–264. (Sp, St)

[10] Good, I. J. (1963). *Ann. Math. Statist.*, **34**, 911–934 . (Sp, St)

[11] Good, I. J. (1969). Statistics of Language. In *Encyclopaedia of Linguistics, Information and Control*. Pergamon Press, Oxford, pp. 567–581. (G, T)

[12] Good, I. J. and Toulmin, G. H. (1968). *J. Inst. Math. Appl.* **4**, 94–105. (T, St)

[13] Guiraud, P. (1959). *Problèmes et Méthodes de la Statistique Linguistique*. D. Reidel, Dordrecht, Holland. (G, T)

[14] Halstead, M. H. (1977). *Elements of Software Science*. Elsevier, New York. (T)

[15] Hantrais, L. (1976). *Le Vocabulaire de Georges Brassens*. Klincksieck, Paris. (L)

[16] Herdan, G. (1960). *Type Token Mathematics*. Mouton, The Hague. (G, T)

[17] Herdan, G. (1961). *Quantitative Linguistics*. Butterworth, London. (G, T)

[18] Herdan, G. (1962). *The Calculus of Linguistic Observations*. Humanities Press, The Hague. (G, T)

[19] Herdan, G. (1966). *The Advanced Theory of Language as Choice and Chance*. Springer-Verlag, Berlin. (G, T)

All of Herdan's books repay reading. The layman can benefit from them, particularly if he or she is numerate.

[20] Kučera, H. (1980). Computers in Language Analysis and Lexicography. In *The American Heritage Dictionary of the English Language*. Houghton Mifflin, Boston, pp. xxxviii–xl. (G)

[21] Mandelbrot, B. (1961). In *Structure of Language and Its Mathematical Aspects. Proc. Symp. Appl. Math.*, Vol. 12, R. Jakobson, ed. American Mathematical Society, Providence, R.I., pp. 190–219. (T, St)

[22] Markov, A. A. (1913). *Bull. Acad. Imp. Sci. St. Pétersbourg*, **7**, 153–162 (in Russian). (Sp, St)

[23] McNeil, D. (1973). *J. Amer. Statist. Ass.*, **68**, 92–96.

[24] Mendenhall, T. C. (1887). *Science*, **11**, 237–249 (supplement, Mar. 1887). (G)

[25] Mendenhall, T. C. (1901). *Popular Sci. Monthly*, **60**, 97–105. (G).

[26] Morton, A. Q. (1965). *J. R. Statist. Soc. A*, **128**, 169–233. (G, T; can be read by the layman, but will also appeal to the statistician.)

[27] Morton, A. Q. and MacGregor, G. H. C. (1964). *The Structure of Luke and Acts*. Hodder and Stoughton, London. (G, T; can be read by the layman, but will also appeal to the statistician.)

[28] Mosteller, F. and Wallace, D. L. (1964). *Inference and Disputed Authorship: The Federalist*. Addison-Wesley, Reading, Mass. (G, St; can be read for general interest, but is more appropriate for statisticians.)

[29] Muller, C. (1967). *Étude de Statistique Lexicale: le Vocabulaire du Théâtre de Pierre Corneille*. Larousse, Paris. (G, T)

[30] Radday, Y. T. (1979). *An Analytical Linguistic Key-Word-in-Context Concordance to Genesis*. The Computer Bible, Vol. 18, J. A. Baird and D. N. Freedman, eds. Biblical Research Associates, Wooster, Ohio. (G, T)

[31] Ratkowsky, D. A., Halstead, M. H., and Hantrais, L. (1980). *Glottometrika*, **2**, 125–147. (Sp, St)

[32] Shannon, C. E. (1951). *Bell. Syst. Tech. J.*, **30**, 50–64.

[33] Sichel, H. S. (1975). *J. Amer. Statist. Ass.*, **70**, 542–547. (Sp, St)

[34] Spevack, M. (1968). *A Complete and Systematic Concordance to the Works of Shakespeare*, 6 vols. George Olms, Hildesheim, West Germany. (L)

[35] Williams, C. B. (1970). *Style and Vocabulary: Numerical Studies*. Charles Griffin, London. (G, T; can be read by the layman, particularly if he or she is numerate.)

[36] Williams, C. B. (1975). *Biometrika*, **62**, 207–212. (Sp, St)

[37] Yule, G. U. (1944). *The Statistical Study of Literary Vocabulary*. Cambridge University Press, Cambridge. (G, T; can be read by the layman, particularly if he or she is numerate.)

[38] Zipf, G. K. (1932). *Selected Studies of the Principle of Relative Frequency in Language*. Harvard University Press, Cambridge, Mass. (G, T)

(CLASSIFICATION
DISCRIMINANT ANALYSIS
LINGUISTICS, STATISTICS IN
WORD FREQUENCY)

J. GANI

LLOYD DAM *See* DAM THEORY

LOCALLY MOST POWERFUL RANK TESTS *See* RANK ORDER STATISTICS

LOCALLY OPTIMAL STATISTICAL TESTS

In the classical Neyman–Pearsonian setup for testing a simple null hypothesis against a simple alternative, the *optimal*, i.e., *most powerful* (MP), *test* is prescribed by the Neyman–Pearson Lemma* (NPL) (*see* HYPOTHESIS TESTING). This simple prescription may stumble into obstacles when these hypotheses are not simple or when the alternative hypothesis* may belong to a given class (so that one would naturally like to have the optimality property extended to this class as well). Under some regularity conditions, the MP test prescribed by the NPL may remain uniformly MP (UMP) for a given class of alternatives. However, these conditions may not hold in all situations and, moreover, the class of alternatives pertaining to such UMP tests may be somewhat restricted.

To make this point clear, we consider two examples, treated already in HYPOTHESIS TESTING (see Vol. 3, pp. 717–719). In Ex. 2 (p. 717), which pertains to the hypothesis on the mean of a normal distribution (variance known), the MP test remains UMP for one-sided alternatives. However, in Ex. 3 (p. 718), which pertains to the location parameter of a Cauchy distribution (scale parameter known), the MP test is not UMP even for such one-sided alternatives. Further, these UMP tests (when they exist) are necessarily *unbiased* for the given class of alternatives (*see* UNBIASEDNESS), but they may fail to be so for other alternatives not belonging to this class (viz., one-sided UMP tests against two-sided alternatives), so that if one wants to set such a wider class of alternatives, the restricted alternative UMP tests may not perform well over the entire class. For this reason, one may like to confine interest to the class of tests that are unbiased for the broader class of alternatives and, within this class, a UMP test, whenever it exists, is termed a UMP unbiased (UMPU) test; the generalized NPL (*see* NEYMAN – PEARSON LEMMA) provides such UMPU tests.

Again, a UMPU test may not always exist. In the normal distributional example cited before it does, but for the Cauchy distribution it does not. Further, when the hypotheses are composite, these MP, UMP, or UMPU tests may not exist and one may be restricted to the class of similar regions* (where the influence of the nuisance parameters* on the size of the test is eliminated); within this class of similar regions, one may then choose an optimal one. Such an optimal similar region may not always exist. Finally, in the multiparameter case, even for the simple null vs. simple alternative problems, a UMP (or UMPU) test may not be available. In such a case, over different subspaces of the parameter space (under the alternative hypotheses), one may have different MP tests. Hence the concepts of best average power, maximin power, most stringency, etc. have been employed to identify some tests having some optimality properties. However, in general, it may be difficult to construct such an optimal test for a composite hypothesis testing problem (e.g., the location parameter of the Cauchy distribution when the scale parameter is unknown).

The developments in the theory of testing statistical hypotheses in this general setup have taken place in two broad avenues. First, for the general exponential family* of densities, the regularity conditions pertaining to the existence and construction of such optimal tests have been verified, and the optimality properties are studied in detail; Lehmann [5] is an excellent source of reference. Second, for densities not necessarily belonging to such exponential families (viz., Cauchy), special attention has been paid to the development of the theory of optimal tests when the alternative hypotheses are "close" to the null ones. Such tests for "local alternatives" may retain their optimality under less restrictive regularity conditions. Also, from the practical point of view, such local alternatives are quite appropriate in the large sample case (where for any fixed alternative away from the null one, a consistent test has a power converging to one, so that for a meaningful study of the asymptotic power of tests, one would naturally confine oneself to the locality of the null hypothesis for which the power functions may not converge to 1). With such shrinking domains in mind, local optimality concepts for various statistical tests have been developed, and are discussed here. We remark that a locally optimal test may not perform that well for nonlocal alternatives, particularly when the sample size is not large.

LOCALLY MOST POWERFUL TESTS

Let E be a random element defined on a probability space $(\Omega, \mathscr{B}, P)$, where P belongs to a class $\mathscr{P}$. Consider the null hypothesis $H_0: P \in \mathscr{P}_0 \subset \mathscr{P}$, against an alternative $H: P \in \mathscr{P}_\Delta \subset \mathscr{P} \backslash \mathscr{P}_0$, where the index $\Delta(>0)$ is used to define a (nonnegative) distance Δ

$= d(\mathscr{P}_0, \mathscr{P}_\Delta)$, between the two measures, in a meaningful way. For example, if $P(= P_\theta)$ is characterized by a single parameter θ, $\mathscr{P}_0 = P_{\theta_0}$ and $\mathscr{P}_\Delta = P_{\theta_\Delta}$, then $\Delta = \theta_\Delta - \theta_0$ (for the one-sided case) or $\Delta = |\theta_\Delta - \theta_0|$ (for the two-sided case). Thus $\mathscr{P}_\Delta \to \mathscr{P}_0$ as $\Delta \to 0$.

Consider a test function $\phi(E)$ such that $E_{H_0}\phi(E) \leqslant \alpha$, the level of significance. Then, $\phi(E)$ is termed a *locally most powerful* (LMP) test function for H_0 against $\Delta > 0$ at some level of significance α, if it is UMP at level α for H_0 against $\mathscr{H}_\delta = \{H_\Delta : 0 < \Delta \leqslant \delta\}$, for some $\delta > 0$. Thus an LMP test is locally UMP. Whereas the global UMP test may not exist, such LMP tests exist under less stringent regularity conditions.

Consider first the simplest case where the power function of every test for H_0 against H_Δ solely depends on Δ and is continously differentiable (from the right) at $\Delta = 0$. Then, by the local expansion of the power function, one may conclude that the LMP test for $\Delta = 0$ against $\Delta > 0$ exists and is defined by the fact that it maximizes the (right-hand) derivative of the power function (with respect to Δ) at $\Delta = 0$ among all level α tests of H_0. In particular, for P_θ characterized by a single parameter θ, $H_0 : \theta = \theta_0$ against $H : \theta > \theta_0$, the LMP test corresponds to the critical region for which the first derivative of the (log) density at θ_0 is a maximum. In the case of a composite hypothesis, we may define a LMP similar region in the same manner; we need to confine ourselves to the class of level α similar regions, and within this class a UMP test for local alternatives, whenever existent, will be the LMP similar region. In this context, see the early works of Neyman and Pearson [8, 9].

For the two examples cited earlier, for the normal distribution, the LMP test (for one-sided alternatives) and LMP similar region (when the variance is unknown) both work out nicely. However, for the Cauchy distribution, in the case of known scale parameter, a LMP one-sided test exists, but the corresponding LMP similar region (when the scale parameter is unknown) does not work out. In the case of a single parameter, an LMP test, besides being locally unbiased, maximizes the minimum power whenever its power function is bounded away from the level of significance, for all other alternatives which are away from the null one.

Consider next the more common (multiparameter) case where the index $\Delta(= \Delta(\boldsymbol{\theta}))$ depends on a vector $\boldsymbol{\theta} = (\theta_1, \ldots, \theta_q)$ of parameters. If tests for $H_0 : \Delta = 0$ against $\Delta > 0$ have constant powers of the contours $\Delta(\boldsymbol{\theta}) = c(\geqslant 0)$, then one may reduce the testing problem to a uniparameter case, and the previous treatments apply. However, in the multiparameter case, not all tests (of the same size α) may have the same family of equipower contours, and hence the picture may well depend on the direction cosines of the θ_j, even in the locality of $\boldsymbol{\theta}_0$. In such a case, one possibility is to assign a distribution to $\boldsymbol{\theta}$ on such contours (usually, the uniform one) and to have a test having locally the best average power. Alternatively, one may also consider a locally maximin* power test by maximizing the minimum power locally over such a contour. A locally most stringent test* may also be defined by minimizing (over the contour) the difference between the actual and the envelope power functions locally. Such locally optimal statistical tests may be conveniently derived in an asymptotic setup where E $(= E_n)$ is based on a sample of size n so that for testing $H_0 : \Delta = 0$, an alternative $\Delta = \Delta_n$ is chosen in such a way that $\Delta_n \to 0$ as $n \to \infty$, but asymptotically the power functions of the competing tests are bounded away from 1. Typically, Δ_n depends on n and converges to 0 at the rate of $n^{-1/2}$. In this setup, whenever we allow $n \to \infty$, we have asymptotically a local setup and hence the corresponding optimal tests are termed locally asymptotically optimal. Optimal $C(\alpha)$ tests* considered by Neyman (1959) deserve special mention in this context.

LOCALLY MOST POWERFUL UNBIASED TESTS

We have observed that LMP tests are locally unbiased. There are some situations (e.g., two-sided alternatives in a single parametric case, or global alternatives in the multiparameter case) where some (e.g., one-sided) tests may not be unbiased, even locally. In such a case, we confine ourselves to the class of tests that are at least locally unbiased and, within this class, we choose a locally most powerful one, whenever it exists. Such a test is termed as *locally most powerful unbiased* (LMPU). Thus, an LMPU test is uniformly most powerful among the locally unbiased ones (for local alternatives). In the case of a single parameter θ (two-sided alternatives), an LMPU test has been termed a *Type A test* (see [6, 8, 9]). The critical region w_A of a Type A test satisfies two conditions:

$$(\partial/\partial\theta)\log P\{E \in w_A \mid \theta\}|_{\theta_0} = 0 \quad \text{(local unbiasedness)}, \quad (1)$$

$$(\partial^2/\partial\theta^2)\log P\{E \in w_A \mid \theta\}|_{\theta_0} \quad \text{is a maximum (LMP)}. \quad (2)$$

For a composite hypothesis (involving some nuisance parameter(s)), one needs to confine oneself to the class of similar regions and obtain an LMPU similar region. Such tests are termed *Type B tests*. For similar regions, we need tests with Neyman structure*, where completeness and sufficiency play a fundamental role. Given a complete, sufficient statistic T, one needs to have $P_0\{E \in w \mid T\} = \alpha$ a.e., where P_0 denotes the probability under H_0.

In the multiparameter case, there are several generalizations of Type A and B tests [4, 7, 9]. If w is a critical region with a power function locally twice differentiable with respect to the elements of $\boldsymbol{\theta}$, and the null hypothesis is $H_0: \boldsymbol{\theta} = \boldsymbol{\theta}_0$, then the local unbiasedness of w entails that the vector of first-order partial derivatives of $P\{E \in w \mid \boldsymbol{\theta}\}$ (at $\boldsymbol{\theta}_0$) is null and the second-order partial derivative matrix (at $\boldsymbol{\theta}_0$), $\mathbf{Q}_0(w)$, is positive semidefinite (p.s.d.). Hence, locally,

$$\begin{aligned} P\{E \in w \mid \boldsymbol{\theta}\} &= P\{E \in w \mid \boldsymbol{\theta}_0\} \\ &\quad + \tfrac{1}{2}(\boldsymbol{\theta} - \boldsymbol{\theta}_0)'\mathbf{Q}_0(w)(\boldsymbol{\theta} - \boldsymbol{\theta}_0) \\ &\quad + o(\|\boldsymbol{\theta} - \boldsymbol{\theta}_0\|^2). \end{aligned} \quad (3)$$

Thus, for any two critical regions, w_1 and w_2, of this local unbiased type, having the same size α, we have by (3), locally,

$$\begin{aligned} P\{E \in w_1 \mid \boldsymbol{\theta}\} - P\{E \in w_2 \mid \boldsymbol{\theta}\} &= \tfrac{1}{2}(\boldsymbol{\theta} - \boldsymbol{\theta}_0)'\mathbf{Q}_0(w_1)(\boldsymbol{\theta} - \boldsymbol{\theta}_0)Z(\boldsymbol{\theta}, \boldsymbol{\theta}_0) \\ &\quad + o(\|\boldsymbol{\theta} - \boldsymbol{\theta}_0\|^2), \end{aligned} \quad (4)$$

$$Z(\boldsymbol{\theta}, \boldsymbol{\theta}_0) = 1 - \frac{(\boldsymbol{\theta} - \boldsymbol{\theta}_0)'\mathbf{Q}_0(w_2)(\boldsymbol{\theta} - \boldsymbol{\theta}_0)}{(\boldsymbol{\theta} - \boldsymbol{\theta}_0)'\mathbf{Q}_0(w_1)(\boldsymbol{\theta} - \boldsymbol{\theta}_0)}.$$

By the Courant theorem on the ratio of two quadratic forms (see, [10, p. 122]), $Z(\boldsymbol{\theta}, \boldsymbol{\theta}_0)$ will be nonnegative for all $\boldsymbol{\theta}$, if the largest characteristic root, $\text{ch}_1[\mathbf{Q}_0(w_2)][\mathbf{Q}_0(w_1)]^{-1}$, is less than or equal to 1. Hence among the class of locally unbiased critical regions, w_1 will be LMP if for any other w_2, $\text{ch}_1[\mathbf{Q}_0(w_2)][\mathbf{Q}_0(w_1)]^{-1}$ is $\leqslant 1$. For Type C regions, Neyman assumed the proportionality of the matrices $\mathbf{Q}_0(w)$, while for Type D regions, Isaacson [4] considered the maximization of the Gaussian curvature of the power function at $\boldsymbol{\theta}_0$, leading to the maximization of the determinant of $\mathbf{Q}_0(w)$. For composite hypotheses, again, one needs to confine oneself to the class of locally unbiased similar regions and, granted the Neyman-structure, the results follow on the same lines.

Tests with Neyman structure may not generally hold (particularly for densities not belonging to the exponential families). In such a case, one may again consider the local asymptotic optimality criteria (in the sense described earlier), and in the light of these, asymptotically (locally) optimal $C(\alpha)$-tests may be formulated. In this local asymptotic optimality setup, the classical likelihood ratio tests* and the associated Wald (1943) W-tests (based on maximum likelihood* estimators) deserve special mention. These tests are asymptotic in character (in the sense that they may attain the level of significance only asymptotically) and in that way they may not need the Neyman-structure required for similar regions. On the other hand, under fairly general regularity conditions [14], these tests possess the total optimality properties for large sample sizes.

However, small sample local optimality properties of these tests may not follow in a general case.

LOCALLY MOST POWERFUL INVARIANT TESTS

In many situations (particularly in the nonparametric case), the problem of testing the null hypothesis against an alternative remains invariant under a group (G) of transformations on the sample space (and the induced group in the parameter space); *see* INVARIANCE CONCEPTS IN STATISTICS. In such a case, it is natural to have the test procedure also invariant under G. The maximal invariant with respect to the group G, say T, plays a fundamental role in this context. An invariant test function depends on the sample point E through the maximal invariant T only. Hence, for the testing problem, it suffices to consider the density $p^*(T;\theta)$ of this maximal invariant. Once we have this density, we may appeal to the LMP or LMPU test function based on it. In particular, for a single parameter (one-sided alternative), the LMP invariant test function is solely based on $(\partial/\partial\theta)\log p^*(T\mid\theta)|_{\theta_0}$. For the case of rank tests (in the two-sample as well as the simple regression model), for the hypothesis of randomness*, the group G relates to the class of all monotone functions and the maximal invariant T is the vector of the ranks of the observations. In such a case, the LMPR (LMP Rank) test statistics are all linear rank statistics with appropriate scores [3, 13]. A more general treatment of LMPR tests for various hypotheses of invariance arising in nonparametric problems is due to Hájek and Šidák [2, pp. 64–71]. There are certain situations (viz., censored/truncated data) where one partitions T as (T_1, T_2) and may desire to have a LMP test based on T_1 alone. If we denote the LMP invariant test statistic (based on T) by $L(T)$, then an LMP test statistic based on T_1 alone may simply be obtained by letting

$$L_1(t_1) = E_0\{L(T_1, T_2)\mid T_1 = t_1\}, \quad (5)$$

where E_0 denotes the expectation under the null hypothesis [1].

UI-LMP (INVARIANT) TESTS

In the multiparameter case, we have discussed several possibilities of deriving some locally optimal tests. Roy [11] considered a heuristic approach, the *Union-Intersection* (UI) *principle**, which is quite flexible and may be incorporated easily in deriving suitable (and sometimes optimal) tests for the multiparameter case. The UI principle has been incorporated in the LMPR testing problem by Sen [12]. Basically, the theory of UI tests may be adapted to general LMP tests as follows. Write $\boldsymbol{\theta} = \Delta\boldsymbol{\gamma}$, and for a specified $\boldsymbol{\gamma}$, find the LMP (or LMPU or LMP Invariant or LMPR, etc.) test statistic, which we denote by $L(\boldsymbol{\gamma})$. Normalize $L(\boldsymbol{\gamma})$ in such a way that under the null hypothesis it has mean 0 and unit variance. Then, take the supremum over $\boldsymbol{\gamma} \in \Gamma$ (the parameter space under the alternative), and take this as the test statistic. For global alternatives, such a UI test statistic becomes asymptotically (and sometimes universally) equivalent to the likelihood ratio test statistic (or a variant version of it) and shares the asymptotic local optimality properties of the likelihood ratio tests. This UI–LMP test also can be adapted when we do not have a global alternative (e.g., ordered alternatives in analysis of variance*, orthant alternatives in the multiparameter location model, etc), though in such a case the (asymptotic) local optimality properties have not yet been fully explored. For such restricted alternatives, UI-LMP test statistics are computationally much simpler than the corresponding likelihood ratio test statistics, and for either type the (asymptotic) local optimality property may not be generally true.

References

[1] Basu, A. P., Ghosh, J. K., and Sen, P. K. (1983). *J. Roy. Statist. Soc., Ser. B*, **45**, 384–390.

[2] Hájek, J. and Šidák, Z. (1967). *Theory of Rank Tests*. Academic Press, New York.

[3] Hoeffding, W. (1951). *Proc. 2nd Berkeley Symp., Math. Statist. Prob.*, pp. 83–92.

[4] Isaacson, S. L. (1951). *Ann. Math. Statist.*, **22**, 217–234.

[5] Lehmann, E. L. (1959). *Testing Statistical Hypotheses*. Wiley, New York.

[6] Neyman, J. (1935). *Bull. Soc. Math. France*, **63**, 246–266.

[7] Neymann, J. (1937). *Skand. Actuar Tidskr.*, **20**, 149–161.

[8] Neyman, J. and Pearson, E. S. (1936). *Statist. Res. Mem.*, **1**, 1–37.

[9] Neyman, J. and Pearson, E. S. (1938). *Statist. Res. Mem.*, **2**, 25–57.

[10] Puri, M. L. and Sen, P. K. (1971). *Nonparametric Methods in Multivariate Analysis*. Wiley, New York.

[11] Roy, S. N. (1953). *Ann. Math. Statist.*, **24**, 220–238.

[12] Sen, P. K. (1982). *Coll. Math. Soc. János Bolyai*, **32**, 843–858.

[13] Terry, M. E. (1952). *Ann. Math. Statist.*, **23**, 346–366.

[14] Wald, A. (1943). *Trans. Amer. Math. Soc.*, **54**, 426–482.

(HYPOTHESIS TESTING
INVARIANCE PRINCIPLES IN STATISTICS
LIKELIHOOD RATIO TEST
MOST STRINGENT TESTS
NEYMAN–PEARSON LEMMA
NEYMAN STRUCTURE, TESTS WITH
OPTIMAL $C(\alpha)$ TESTS
UNBIASEDNESS
UNION-INTERSECTION PRINCIPLE)

P. K. SEN

LOCAL TIME

To define the notion of local time we start with a function $f(t)$, with $0 \leqslant t \leqslant 1$. From this we can obtain a function $L(x,t)$ that describes the amount of time f spends *below* the level x, $-\infty < x < \infty$, during $[0,t]$. This is given by $L(x,t) = \int_0^t I(f(t) - x)\,dt$, where $I(y)$ is the indicator function $I(y) = 1$ if $y \leqslant 0$, and $I(y) = 0$ if $y > 0$. Figure 1 illustrates an example of this when $f(t) = t$. Here $L(x,t)$ is graphed as a function of x with t fixed at $1/2$.

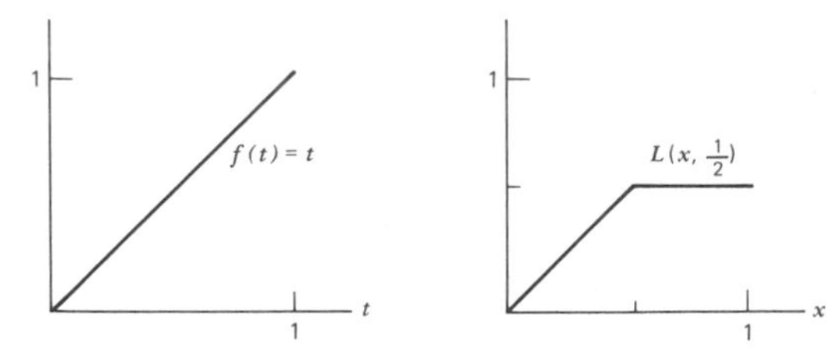

Figure 1 Function $f(t) = t$ and its occupation distribution function at $t = \frac{1}{2}$.

The function L is called the *occupation time distribution* of the function f, and for fixed t has all the properties of a distribution function in x, except that $L(\infty,t)$ may be less than 1. Thus it is not unreasonable to expect that, for fixed t, $L(\cdot,t)$ can be differentiated to yield a function, $l(\cdot,t) = \partial L(\cdot\, t)/\partial x$, which serves as a sort of "density function" for $L(\cdot,t)$. This density is known as the *local time function* for f, and the quantity $l(x,t)$ is called the *local time* of f at x in $[0,t]$. In essence, $l(t,x)$ describes the amount of time f spends *at* the level x during $[0,t]$.

Although virtually every function f gives rise to a corresponding occupation time distribution, many do not possess a properly defined local time function. The function depicted in Fig. 1 provides an example of this, since the fact that the function $L(x,\frac{1}{2})$ is not differentiable at the point $x = \frac{1}{2}$ implies that $l(\frac{1}{2},\frac{1}{2})$ cannot be properly defined. This particular example is indicative of a general underlying principle in the theory of local times that, basically, goes as follows: If a function f is smooth, then either it will not possess a local time function, or, at best, the latter function will be extremely badly behaved in the sense that it will oscillate very wildly. Conversely, if f itself is a highly erratic function, it is likely not only to possess a local time function, but this function will be smooth and well behaved.

It is because of this relationship between a function and its local time that this concept has proven to be so useful in probability theory. Many of the stochastic processes* studied by probabilists exhibit extremely erratic behavior. A prime example is the widely studied Brownian motion*, which is continuous at every point t, but differentiable nowhere. In accordance with the relationship above, such processes tend to have very smooth local time functions. Thus since it is generally easier to study smooth rather than erratic functions, it often turns out that the mathematically most convenient way to study many of the properties of these processes is via their local times. For example, local time turns out to be an indispensible tool for studying problems related to the Hausdorff dimension* of random sets generated by erratic stochastic processes.

Although local time, as described above, is essentially a tool of ordinary real analysis, it was first introduced by Lévy in 1948 [1] in a stochastic setting related to Brownian motion. He called it "le mesure du voisinage." As is true of so many of Lévy's concepts, its further study has been highly demanding mathematically, requiring an esoteric mixture of probability, measure theory, and Fourier analysis.

Much of the recent literature is concerned with the local time of vector-valued stochastic processes defined on N-dimensional spaces.

References

[1] Lévy, P. (1948). *Processus Stochastiques et Mouvement Brownien.* Gauthier-Villars, Paris.

Further Reading

See the following works, as well as the reference just given, for more information.

Adler, R. J. (1981). *The Geometry of Random Fields*. Wiley, London. (This contains one chapter that includes a discussion of the local time of Gaussian processes, essentially a simplified version of the following entry.)

Geman, D. and Horowitz, J. (1980). *Ann. Prob.*, **8**, 1–67. (This is an extremely wide ranging survey covering virtually every aspect of local time. With its bibliography of over 150 items, it provides an excellent starting point for further reading. However, as is the case with essentially every article on this topic, it is written for the probability specialist.)

(BROWNIAN MOTION
FRACTALS
HAUSDORFF DIMENSION)

R. J. ADLER

LOCATION, MEASURES OF *See* MEAN, MEDIAN, MODE AND SKEWNESS; MEDIAN ESTIMATION, INVERSE

LOCATION PARAMETER, ESTIMATION OF

Let $F(x;\theta)$ be the distribution function (d.f.) of a family of one-dimensional distributions depending on a parameter θ. Let Ω be the set of admissible values of θ. We shall call F a *location parameter* family of distributions if for all $x \in R = (-\infty, +\infty)$ and for all $\theta \in \Omega$, $F(x,\theta)$ is of the form

$$F(x;\theta) = F_1(x-\theta), \tag{1}$$

where $F_1(x-\theta)$ is a function of $x-\theta$.

If the density function of F exists and is of the form

$$\begin{aligned} f(x,\theta) &= f_1(x-\theta) \\ &= \exp\{c_1(x-\theta)^2 + c_2(x-\theta) + c_3\}, \\ &\qquad (-\infty < x < \infty, -\infty < \theta < \infty, \\ &\qquad c_1 < 0, -\infty < c_2 < \infty) \cdots . \end{aligned} \tag{2}$$

then the statistic $\bar{x}$ (sample mean) is sufficient for θ (*see* SUFFICIENCY). A statistic $T = T(x)$ is said to be sufficient for x if the conditional distribution of x, given $T = t$, is independent of θ for all t. If $T = t$ is any unbiased estimator of θ, the totality of unbiased estimators is given by $T = t - U$, where

$$E_\theta(U) = 0 \quad \text{for all} \quad \theta \in \Omega.$$

The estimator T^* which minimizes the variance of T among all unbiased estimators is defined as the uniformly minimum variance (UMVUE) estimator among all unbiased estimators of θ. If we put $c_2 = 0$ in (2) then the sample mean $\bar{x}$ is UMVUE; *see* MINIMUM VARIANCE UNBIASED ESTIMATION.

Let $x_1, \dots, x_n$ denote a random sample from the density function $g(\Omega; \theta)$, where θ is a location parameter and Ω is the real line. The estimator

$$t(x_1, \dots, x_n) = \frac{\int \theta \prod_{i=1}^{n} f(x_i; \theta)\, d\theta}{\int \prod_{i=1}^{n} f(x_i; \theta)\, d\theta} \tag{3}$$

is the estimator of θ which has uniformly smallest mean-squared-error* within the class of location-invariant estimators, i.e., $t(x_1, \dots, x_n)$ is the best location invariant estimator (BLIV), due to Pitman [9] (*see* PITMAN ESTIMATORS). As far as we know, $t(x_1, \dots, x_n) = \bar{x}$ holds only when $x_1, \dots, x_n$ come from a normal population.

If $x_1, \dots, x_n$ come from the family of distributions with density functions belonging to (2), then $\bar{x}$ is the maximum likelihood* estimate (MLE) of θ. (*See also* ARITHMETIC MEAN.)

The sample median as an estimator of θ exhibits less sensitivity to heavy tails than the $\bar{x}$. In fact, the asymptotic relative efficiency* (ARE) of the median with respect to the mean increases as the tail of the distribution gets heavier. For example, the ARE of the median is somewhat higher ($\pi^2/12 \sim 0.82$) for the logistic (with heavier tail than the normal) than for the normal ($2/\pi \sim 0.637$).

Although the median is less sensitive than the mean to outlying observations, it goes too far in discarding observations. This led to the construction of estimators whose behavior is more like that of the mean when x is close to θ. A class of estimators providing such a compromise and including both the mean and the median is called the *trimmed mean**. Denote the resulting estimator by $\bar{x}_\alpha$, $0 \leqslant \alpha < 1/2$, where

$$\bar{x}_\alpha = \frac{1}{n - 2[n\alpha]} \{x_{([n\alpha]+1)} + \cdots + x_{(n-[n\alpha])}\}, \tag{4}$$

where $[n\alpha]$ is the largest integer $\leqslant n\alpha$ and $x_{(1)}, \dots, x_{(n)}$ are the ordered observations. For distributions with heavy tails, the performance of a trimmed mean with moderate amount of trimming (α) can be much better than $\bar{x}$, as shown in Lehmann [8, pp. 361–362], with practically no loss in efficiency at the normal. The disadvantages of estimating θ by $\bar{x}_\alpha$ is that we do not know the value of α to choose. On the basis of numerical results Lehmann [8] suggests a value of about 0.1. This would vary with situations and should be carefully determined on the basis of individual cases. Another disadvantage is that the trimmed mean tends to be very inefficient compared to $\bar{x}$ for distributions with thin tails when the location model may not hold. However, in situations where the model is applicable, heavy-tailed distributions* are generally encountered.

Now we will briefly discuss the case where it is required to estimate the mean θ based on a sample $y_1, \dots, y_n$ from an infinite population, when either the form of the parent distribution (F) is unknown or it is known but involves nuisance parameters. For details see FIXED-WIDTH AND BOUNDED-LENGTH CONFIDENCE INTERVALS.

Consider first the problem of obtaining a confidence interval* (L_n, U_n) for θ with the properties that $P\{L_n \leqslant \theta \leqslant U_n\}$ is greater than some preassigned confidence coefficient $1 - \alpha$ and the width $d_n = U_n - L_n$ is $\leqslant 2d$, for some specified $d > 0$. If the underlying F is normal and the variance σ^2 were known, then setting $n^0 = n^0(d, \sigma) = \min[n : n^{-1/2}\sigma Z_{\alpha/2} \leqslant d]$ (where Z_ϵ is the upper $100\epsilon\%$ point of the standard normal distribution*), we see that for every $n \geqslant n^0$, $(\bar{y}_n - d, \bar{y}_n + d)$ has the desired properties. However, as is generally the case, if σ is unknown, the quantity $n^0(d, \sigma)$ is also so, so that no fixed n will be $\geqslant n^0(d, \sigma)$, simultaneously for all σ. Dantzig [3] showed it is impossible to get an estimate with the required properties by means of samples of fixed size. The number of observations must depend in some way on the observations themselves, i.e., some form of sequential or multistage procedures must be used. Stein [15] showed that a two-stage procedure satisfies the requirements mentioned above.

Chow and Robbins [2] suggested a sequential procedure for interval estimation of θ based on an updated estimate of the variance at each stage, and for a much wider class of F, for which the variance σ^2 is finite.

ESTIMATION WHEN CV IS KNOWN

For many distributions, particularly in the physical and biological sciences, it is not uncommon to find that the population standard deviation is proportional to the population mean, i.e., the coefficient of variation* (CV) is a constant. In such cases it is possible to obtain a more efficient estimator of the population mean than the sample mean, assuming that the CV is known.

Assume that a random sample $y = (y_1, y_2, \ldots, y_n)$ of fixed sized $n \geqslant 2$ is taken from the normal distribution $N(\theta, a\theta^2)$, $(\theta > 0, a > 0)$ where the coefficient of variation $\sqrt{a}$ is known. Let

$$d_1 = \bar{y} = n^{-1}\sum_{i=1}^{n} y_i;$$

$$d_2 = c_n\left[\sum_{i=1}^{n}(y_i - \bar{y})^2\right]^{1/2} = c_n\sqrt{n}\, s, \tag{5}$$

where $c_n = (1/\sqrt{2a})\Gamma((n-1)/2)/\Gamma(n/2)$. Both d_1 and d_2 are unbiased estimates of θ.

Khan [7] showed that the estimator

$$\hat{\theta}_1 = \alpha d_2 + (1 - \alpha)d_1, \qquad 0 \leqslant \alpha \leqslant 1, \tag{6}$$

where $\alpha = a\{a + n[ac_n^2(n-1) - 1]\}^{-1}$, has the smallest variance, uniformly in θ among all unbiased estimators of θ that are linear in $\bar{y}$ and s. Khan further showed that $\hat{\theta}_1$ is a BAN estimator (*see* LARGE-SAMPLE THEORY) and its asymptotic efficiency relative to the MLE of θ given by

$$\hat{\theta}_n = \left[\left\{4as^2 + (1 + 4a)\bar{y}^2\right\}^{1/2} - \bar{y}\right]/(2a) \tag{7}$$

is 1.

Sen [14] obtained a biased but consistent estimator

$$\hat{\theta}_2 = \alpha_1 d_1 + (1 - \alpha_1)\frac{d_2'}{\sqrt{a}}, \qquad 0 \leqslant \alpha_1 \leqslant 1, \tag{8}$$

where $d_2' = s$, which has minimum expected squared error among all convex combinations of $\bar{y}$ and s. Gleser and Healy [5] considered a more general class of estimators of θ for the normal distribution which are linear in $\bar{y}$ and s, but not necessarily unbiased. They showed that in this class the estimator which has minimum expected squared error is

$$\hat{\theta}_3 = (d_n d_1 + n^{-1}ad_2)/(d_n + n^{-1}a + n^{-1}ad_n). \tag{9}$$

For a random sample from the normal distribution with known CV, the minimal sufficient statistic includes an ancillary statistic*. Hinkley [6] discussed the effects of conditioning on the ancillary statistic and the degree to which it affects inference about θ.

Searls [11] obtained an estimator

$$\hat{\theta}_4 = W\sum_{i=1}^{n} y_i, \tag{10}$$

where W is chosen so that the mean squared error (MSE) $E(\hat{\theta}_4 - \theta)^2$ is a minimum. This leads to

$$W = \frac{1}{n + a}, \qquad \text{MSE}(\hat{\theta}_4) = \frac{\sigma^2}{n + a}. \tag{11}$$

The relative efficiency of $\hat{\theta}_4$ with respect to the sample average $\bar{y}$ is given by $(n + a)/n$, which tends to be high for small values of n and large values of a. Searls [12] extended the result to the case when an estimate of the CV is available.

For large samples and for a wider class of distributions (F) Sen [13] obtained an estimator which has minimum expected squared error among all convex combinations of $\bar{y}$ and s.

For the normal parent the estimator is given by (4). Exact expressions for bias and MSE were obtained. However, for other distributions, use was made of the Taylor's expansion of s around Es and retaining terms to the order of n^{-1}. It was shown that the relative efficiency of the estimator with respect to the sample average $\bar{y}$ is approximately

$$1+\frac{\left(\sqrt{\beta_1}-2\sqrt{a}\right)^2}{\beta_2-\beta_1-1}, \qquad (12)$$

where β_1 and β_2 are shape parameters of the population. For the normal and near normal populations for which $\sqrt{\beta_1}<2\sqrt{a}$, the efficiency is high, is equal to 1 for gamma and exponential populations for which $\sqrt{\beta_1}=2\sqrt{a}$, and is less than 1 for lognormal* and inverse gaussian* populations for which $\sqrt{\beta_1}>2\sqrt{a}$. Sen extended the theory when an estimate of the CV is available, and discussed the case when the form of the distribution is not known. The results were illustrated with data from biological populations. This estimator differs in form from Searls' estimator and is applicable to large samples.

Other work in this area includes that of Rutemiller and Bowers [10]. Let $\{Y_j\}$ be a set of normally distributed random variables with parameters μ_j, σ_j, and

$$E(Y_j)=\theta_j=\beta_o+\sum_{i=1}^{p}\beta_i x_{ij},$$

$$\sqrt{V(Y_j)}=\sigma_j=\gamma_o+\sum_{i=1}^{p}\gamma_i x_{ij},$$

$$j=1,2,\ldots,n. \qquad (13)$$

The regression model allows the expected value of the dependent variable and its standard deviation to be different linear functions of the independent variables x_1, $\ldots,x_p$. The method is applicable to large samples and yields an approximation of the MLEs of $2p+2$ regression coefficients, which may be used to construct confidence intervals for the parameters of the normal distribution and tolerance intervals for the individual Y.

Later, Amemiya [1] considered the regression model where the variance of the dependent variable is proportional to the square of its expectation. The model is:

$\{Y_j\}$ is independent with

$$E(Y_j)=\theta_j=\sum_{i=1}^{p}\beta_i x_{ij},$$

$$V(Y_j)=\sigma^2 Y_j=a\left(\sum_{i=1}^{p}\beta_i x_{ij}\right)^2,$$

$$j=1,2,\ldots,n, \qquad (14)$$

where $\beta_i (i=1,2,\ldots,p)$ are unknown parameters, x_{ij} are known quantities, and a is a scalar unknown parameter. Although the model can be conceived without specifying the distribution of Y_j, it naturally arises when Y_j follows either a lognormal or a gamma distribution, provided that $\sum_{i=1}^{p}\beta_i x_{ij}>0$. Amemiya obtained a measure of the asymptotic efficiency of the weighted least squares estimator as compared to the MLE when Y_j follow (1) a normal, (2) a lognormal, and (3) a gamma distribution. It was shown that the asymptotic efficiency was 1 when Y_j has a gamma distribution, but is less than 1 for normal and lognormal Y_j's. When Y_j is normal, the author recommends the use of model (13) which includes (14) as a special case.

Gerig and Sen [4] derived MLEs for the parameters of two normal populations with CVs equal but unknown. The model is:

$(Y_{h\eta_h})$ is distributed normally with sample size η_h,

$$E(Y_{hi})=\theta_h, \qquad (15)$$

$$V(y_{hi})=\theta_h^2\eta_h^2, \qquad h=1,2.$$

They assumed that the two populations have a common coefficient of variation, $\eta_1=\eta_2=\eta$. MLEs were used to estimate θ_1, θ_2, and η. The relative efficiency of the proposed mean estimator with respect to the sample mean is shown to be greater than 1. The effect of departures from the assumptions of normality and equal CVs on the relative efficiency was studied. Monte Carlo* simulation was used to deal with the former case. An example was provided from wildlife populations to illustrate the proposed methods. Although model (14) (proposed by Amemiya) appears to include (15), the two models serve two distinct purposes. Also, model (14) is at a level of generality that does not allow explicit solution of the likelihood equations and detailed analysis of the relative efficiency considered in the model (15).

References

[1] Amemiya, T. (1973). *J. Amer. Statist. Ass.*, **68**, 928–934.

[2] Chow, Y. S. and Robbins, H. (1965). *Ann. Math. Statist.*, **36**, 457–462.

[3] Dantzig, G. B. (1940). *Ann. Math. Statist.*, **11**, 186–192.

[4] Gerig, T. M. and Sen, A. R. (1980). *J. Amer. Statist. Ass.*, **75**, 704–708.

[5] Gleser, L. J. and Healy, J. D. (1976). *J. Amer. Statist. Ass.*, **71**, 977–981.

[6] Hinkley, D. V. (1977). *Biometrika*, **64**, 105–108.

[7] Khan, R. A. (1968). *J. Amer. Statist. Ass.*, **63**, 1038–1041.

[8] Lehmann, E. L. (1983). *Theory of Point Estimation*. Wiley, New York.

[9] Pitman, E. J. G. (1939). *Biometrika*, **30**, 391–421.

[10] Rutemiller, H. C. and Bowers, D. A. (1968). *J. Amer. Statist. Ass.*, **63**, 552–557.

[11] Searls, D. T. (1964). *J. Amer. Statist. Ass.*, **59**, 1225–1226.

[12] Searls, D. T. (1967). *Amer. Statist.*, **21**, 20–21.

[13] Sen, A. R. (1978). *Commun. Statist. A*, **7**, 657–672.

[14] Sen, A. R. (1979). *Biometrische Zeit.*, **21**, 131–137.

[15] Stein, C. (1945). *Ann. Math. Statist.*, **16**, 243–258.

(ARITHMETIC MEAN
COEFFICIENT OF VARIATION
FIXED-WIDTH AND BOUNDED-LENGTH CONFIDENCE INTERVALS
MAXIMUM LIKELIHOOD ESTIMATION
MINIMUM VARIANCE UNBIASED ESTIMATION
PITMAN ESTIMATORS
TRIMMED MEAN)

A. R. SEN

LOCATION-SCALE FAMILIES

All random variables $(x_1, \ldots, x_m)$ that have a joint (CDF) which can be expressed as

$$g\left(\frac{x_1 - \theta}{\tau}, \frac{x_2 - \theta}{\tau}, \ldots, \frac{x_m - \theta}{\tau}\right) \quad (\tau > 0)$$

with the *same* function $g(\cdot)$ are said to belong to a location-scale family. Different $g(\cdot)$'s correspond to different families. The parameter θ is the *location parameter*; τ is the *scale parameter*.

For any fixed τ we have a subfamily which is a *location family* (with parameter θ); for $\theta = 0$ we have a *scale family* (with parameter τ).

The member of the family with $\theta = 0$ and $\tau = 1$ is called the "standard"; the corresponding CDF $g(x_1, x_2, \ldots, x_m)$ is the "standard CDF" of the family.

All CDFs in the family have the same shape. If the standard deviation is σ, then, in the general case, the exponential value is $\theta + \tau\sigma$ and the standard deviation is $\tau\sigma$. (Very often, $\mu = 0$ and $\tau = 1$, so the goal expected value is θ, and the standard deviation is τ, but this is not necessarily the case. It is possible, for example, that μ and/or σ may not exist, as in the case of the Cauchy distribution*.)

LOCATION-SCALE PARAMETER

A two-dimensional parameter (μ, σ) with $\sigma > 0$ is called a location-scale parameter for the distribution of a random variable X if the cumulative distribution function $F_X(x \mid \mu, \sigma)$ is a function only of $(x - \mu)/\sigma$. In this case

$$F_X(x \mid \mu, \sigma) = F((x - \mu)/\sigma),$$

where F is a distribution function. If the distribution of X is absolutely continuous with the density $f_X(x \mid \mu, \sigma)$, then (μ, σ) is a location-scale parameter for the distribution of X if (and only if)

$$f_X(x \mid \mu, \sigma) = (1/\sigma) f((x - \mu)/\sigma)$$

for some density $f(x)$.

(INVARIANCE PRINCIPLE IN STATISTICS
LOCATION-SCALE FAMILIES)

LOCATION TESTS

A location parameter for a population is generally chosen as a measure of central tendency and in this case describes the size of a "typical" observation drawn at random from the population. Two commonly used location parameters are the mean and the median. It is also useful to have some idea as to how close observations will tend to be to this typical value, and this information is usually contained in a suitable scale parameter for the population such as variance (*see* SCALE TESTS).

When comparing populations, it is often of interest to determine which population will tend to produce the largest measurements, as in comparing the yield of several varieties of a crop or the strength of two alloys. If the variability in the observations from the populations of interest are about equal, a comparison of the location parameters of the populations will generally give a definitive answer as to which crop is better or which alloy is stronger.

In this entry, location tests with varying assumptions on the underlying populations are discussed in the one-sample, two-sample, and k-sample settings. The following section contains a definition of a location parameter family, with several examples of location parameters. The section "One-Sample Problem" discusses tests for some common location parameters in the one-sample problem. The section "Two-Sample Problem" provides tests for the equality of location parameters in the two-sample problem, and the section "k-Sample Problem" deals with testing for equality of location parameters in the k-sample setting. The section "Numerical Example" contains a numerical example to illustrate the various procedures, and the final section provides a discussion of some related problems.

LOCATION PARAMETERS

Let the random variable ϵ have continuous cumulative distribution function (CDF) $F(t) = P(\epsilon \leqslant t)$. Then the shifted random variable, $X = \eta + \epsilon$, has CDF given by $P(X \leqslant x) = P(\epsilon \leqslant x - \eta) = F(x - \eta)$. For fixed F and varying η, the distributions $F(x - \eta)$ constitute a location family. The distributions in this family all have the same shape and differ only in the value of the parameter η, which is called a location parameter.

The location parameter η is not fully specified without making further assumptions on F. For example, if it is assumed that $F(0) = 1/2$ and X has a distribution of the form $F(x - \eta)$, then η represents the median of this distribution, since

$$P(X \leqslant \eta) = F(\eta - \eta) = 1/2.$$

Similarly, if $E(\epsilon) = 0$, i.e., $\int_{-\infty}^{\infty} t\, dF(t) = 0$, then η represents the mean of X. Some other common location parameters are the first or third quartiles, or their average.

In order to make inferences about η when X has CDF $F(x - \eta)$, it is important to specify which location parameter η is to represent. In the next section, hypothesis tests about the mean and the median are discussed.

ONE-SAMPLE PROBLEM

Let $X_1, \ldots, X_n$ denote a random sample of size n from a population with continuous CDF $F(x - \eta)$. In this setting, the appropriate procedure for testing hypotheses about η depends upon the assumptions made regarding F.

The classical assumption is that F is a normal CDF with mean 0 and variance σ^2, or equivalently, that $X_1, \ldots, X_n$ is a random sample from a normal population with mean η and variance σ^2. Hypothesis tests about η are then based on the one-sample t-statistic. For example, to test $H_0 : \eta = \eta_0$ vs. $H_1 : \eta \neq \eta_0$, the appropriate procedure is to reject H_0 whenever

$$|\sqrt{n}\,(\bar{X} - \eta_0)/S_x| \geqslant t_{1-\alpha/2}(n-1),$$

$\bar{X} = \sum X_i/n$, $S_x^2 = \sum(X_i - \bar{X})^2/(n-1)$, and $t_{1-\alpha/2}(n-1)$ is the $100(1-\alpha/2)$ percentile of Student's t-distribution with $n-1$ degrees of freedom. If the assumption of normality is violated, the procedure above provides only an approximate test about the population mean, with the approximation tending to improve as the sample size becomes large.

A weaker assumption than normality is symmetry of the distribution about the median η, namely $1 - F(x) = P(X_i > \eta + x) = P(X_i < \eta - x) = F(-x)$ for all x. Under this assumption the most commonly used procedure for testing hypotheses about η is based on the Wilcoxon signed-rank* statistic [19] (*see* DISTRIBUTION-FREE METHODS). In order to test $H_0 : \eta = \eta_0$ vs. $H_1 : \eta \neq \eta_0$, first compute $|X_1 - \eta_0|, \ldots, |X_n - \eta_0|$. Then rank these absolute deviations from η_0 from smallest to largest, letting R_i^+ denote the rank of $|X_i - \eta_0|$ in this ranking. The Wilcoxon signed-rank statistic is given by $T^+ = \sum \psi_i R_i^+$, where $\psi_i = 1$ if $X_i > \eta_0$ and $\psi_i = 0$ otherwise. The null hypothesis is rejected if T^+ is either too large or too small. Tables of the null distribution for small samples, as well as a large-sample approximation can be found in Ref. 8.

Finally, the weakest assumption considered here is that F is any continuous CDF with $F(0) = 1/2$; i.e., η represents the median of the distribution. The sign test* statistic [3], $S = \sum \psi_i$ with ψ_i defined above, can be used to test $H_0 : \eta = \eta_0$ vs. $H_1 : \eta \neq \eta_0$, with H_0 being rejected for large or small values of s. Under the null hypothesis S has the binomial distribution with n trials and success probability $1/2$, and the appropriate rejection region can be obtained easily. Note that the sign test is an inference procedure about the population median and must not be used to test hypotheses about the population mean, *unless* these parameters are known to coincide, as in a symmetric population.

In computing the signed-rank statistic or the sign statistic, some of the X's may be equal to the hypothesized median η_0. It is generally recommended that these observations be discarded and n reduced accordingly. For the signed-rank statistic it may also happen that some of the $|X_i - \eta_0|$ are tied and this creates a difficulty in determining their ranks. In this case, all observations within a tied group are assigned the midrank, which is the average of the ranks that these tied observations would have received had they not been tied.

The sign and signed-rank tests are valid under very general assumptions concerning the type of population sampled and are usually referred to as nonparametric procedures. Discussion of these and related one-sample nonparametric procedures can be found in ref. 5, 8, or 12, together with many references. The confidence intervals associated with these tests are discussed in refs. 8 and 12.

Finally, when observations are collected in pairs (X_i, Y_i), $i = 1, \ldots, n$, where X and Y may measure the response of a subject to two different treatments, it is often of interest to determine if either treatment tends to produce larger measurements. This question can often be answered by considering the differences, $Z_i = Y_i - X_i$, $i = 1, \ldots, n$, and testing hypotheses about the mean or median of the distribution of Z.

TWO-SAMPLE PROBLEM

Let $X_1, \ldots, X_m$ and $Y_1, \ldots, Y_n$ be two independent random samples from populations with continuous CDFs $F(x - \eta_1) = P(X \leqslant x)$ and $F(y - \eta_2) = P(Y \leqslant y)$. In this problem interest is usually focused on $\Delta = \eta_2 - \eta_1$, the difference in the location parameters. Note that the size of this difference does not depend on which location parameter is represented by η_1 and η_2. When $\Delta > 0$ (< 0), the Y observations tend to be larger (smaller) then the X observations, while the two distributions are identical when $\Delta = 0$.

The classical assumption is that F is a normal CDF with mean 0 and variance σ^2, so that the X and Y samples are from normal populations with means η_1 and η_2, respectively, and common variance σ^2. The two-sample t-statistic,

$$t = \left[(nm)^{1/2}(\overline{X} - \overline{Y})\right] / \left[S_p(n+m)^{1/2}\right],$$

$S_p^2 = [(m-1)S_x^2 + (n-1)S_y^2]/(m+n-2)$, is used to test $H_0 : \Delta = 0$ vs. $H_1 : \Delta \neq 0$, with H_0 being rejected whenever $|t| \geqslant t_{1-\alpha/2}(m+n-2)$.

When F is allowed to be any continuous CDF, not necessarily a normal, the two distributions have the same shape but may differ in the value of their location parameters. In this setting, the two-sample t-statistic provides an approximate test for equality of the location parameters (provided that the population means exist), the approximation improving as the sample sizes become larger. As an alternative, an exact nonparametric procedure based on the Mann–Whitney–Wilcoxon* [15, 19], statistic can be used. To compute the statistic, first rank the combined sample of X's and Y's from smallest to largest, letting R_i denote the rank of Y_i in this joint ranking. In the case of tied observations, all observations within a tied set are assigned the midrank as in the signed-rank statistic. The Mann–Whitney–Wilcoxon statistic is given by $W = \sum_{i=1}^{n} R_i$, with $H_0 : \Delta = 0$ being rejected in favor of $H_1 : \Delta \neq 0$ for large or small values of W. Tables of the small sample null distribution of W as well as an approximation to the null distribution of W in large samples can be found in ref. 8. Although the two-sample t-statistic provides a more powerful procedure for testing hypotheses about Δ when a populations are normal, the test based on the Mann–Whitney–Wilcoxon statistic tends to be more powerful, for distributions having heavier tails than the normal.

Another commonly used nonparametric test statistic for this problem is due to Mood [16] and simply counts the number of Y's larger than the combined sample median (*see* BROWN–MOOD MEDIAN TEST). Further discussion of these and other nonparametric tests for the two-sample problem can be found in ref. 5, 8, or 12. A confidence interval procedure for Δ associated with the Mann–Whitney–Wilcoxon test is described in ref. 8 or 12.

k-SAMPLE PROBLEM

Let $X_{i1}, \ldots, X_{in_i}$ denote a random sample from a population with continuous CDF $F(x - \eta_i)$, $i = 1, \ldots, k$. There are k populations of interest with n_i observations drawn from the ith population, and it is assumed that the observations from population i are independent of those from population j for each pair i and j. The populations all have the same shape and may differ only in the values of their location parameters.

The classical theory assumes that F is a normal CDF with mean 0 and variance σ^2, or equivalently, that the k populations are normal with common variance and possibly unequal means. The procedure for testing $H_0 : \eta_1 = \cdots = \eta_k$ vs. $H_1 : \eta_i$ not all equal is often referred to as a one-way analysis of variance*. Let $N = \sum_{i=1}^{k} n_i$, $\overline{X}_{i\cdot} = \sum_{j=1}^{n_i} X_{ij}/n_i$, and $\overline{X}_{\cdot\cdot} = \sum_{i=1}^{k}\sum_{j=1}^{n_i} X_{ij}/N$. The appropriate test statistic is $F = \text{MST}/\text{MSE}$, where $\text{MST} = \sum_{i=1}^{k} n_i(\overline{X}_{i\cdot} - \overline{X}_{\cdot\cdot})^2/(k-1)$ and $\text{MSE} = \sum_{i=1}^{k}\sum_{j=1}^{n_i}(X_{ij} - \overline{X}_{i\cdot})^2/(N-k)$, with H_0 being rejected whenever $F \geqslant F_{1-\alpha}(k-1, N-k)$, the upper αth percentile of the F-distribution with $k-1$ and $N-k$ degrees of freedom.

The nonparametric setting for this problem replaces the normality assumption with the assumption that F is any continuous CDF. The procedure most often used is due to Kruskal and Wallis [11] (*see* KRUSKAL–WALLIS TEST) and requires that all the N observations first be jointly ranked from smallest to largest, with R_{ij} denoting the rank of X_{ij} in this ranking. Ties are again broken using midranks. The test statistic is given by

$$H = 12 \sum_{i=1}^{k} n_i\left(\overline{R}_{i\cdot} - \overline{R}_{\cdot\cdot}\right)^2 / \{N(N+1)\},$$

$\bar{R}_{i\cdot} = \sum_{j=1}^{n_i} R_{ij}/n_i$, $R_{\cdot\cdot} = (N+1)/2$, with $H_0: \eta_1 = \cdots = \eta_k$ being rejected in favor of $H_1: \eta_i$ not all equal, whenever H is sufficiently large. Fairly extensive tables of the small sample null distribution of H can be found in ref. 9, while for large sample sizes H_0 is rejected whenever $H \geqslant \chi^2_{1-\alpha}(k-1)$, where $\chi^2_{1-\alpha}(k-1)$ is the upper αth percentile of the χ^2 distribution with $k-1$ degrees of freedom.

A significantly large value of F or H indicates that at least some of the η_i's are different. In order to determine the exact nature of these differences, multiple comparison procedures (*see* MULTIPLE COMPARISONS) are usually performed.

When the treatments are ordered in some way and the experimenter anticipates a deviation from $H_0: \eta_1 = \cdots = \eta_k$ in a particular direction, it is advantageous to test H_0 against some type of restricted alternative. For example, an extension of the one-sided hypothesis in the two-sample setting would be the ordered alternative $H_1: \eta_1 \leqslant \eta_2 \leqslant \cdots \leqslant \eta_k$, with at least one strict inequality. A procedure for testing H_0 against the ordered alternative H_1 which assumes normality is given by Bartholomew in ref. 2, while a nonparametric competitor can be found in Jonckheere* [10]. For further reading on ordered or restricted alternatives, see ref. 1, 7, or 14.

NUMERICAL EXAMPLE

In order to test a hypothesis, the computed value of the test statistics would need to be compared with the critical value obtained from an appropriate table of its null distribution. In practice it is not recommended that all test statistics be computed as is done here for demonstration purposes. Instead, only those test statistics which are appropriate to the particular problem should be used.

Hald [6] gave an example in which samples from a given product were taken and assigned at random to several groups (of predetermined size). The groups were stored under different conditions and after storage the water content of each sample was obtained. A subset of the data is given below, showing the water content of the samples for each of three methods of storage.

1	2	3
7.8	5.4	8.1
8.3	7.4	6.4
7.6	7.1	
8.4		
8.3		

In order to illustrate the computation of the one-sample statistics restrict attention to the five observations using storage method 1. To test the null hypothesis that the mean water content using storage method 1 is 8, the appropriate test statistic, under the assumption that the observations are normally distributed, is $t = \sqrt{n}(\bar{X} - 8)/S_x = \sqrt{5}(8.08 - 8)/0.356 = 0.50$. For testing the null hypothesis that the median water content is 8, the Wilcoxon signed-rank statistic or the sign statistic can be used, depending on whether the observations can be assumed to come from a symmetric population. The following table is useful in illustrating the computation of these test statistics.

Observation	1	2	3	4	5
X_i	7.8	8.3	7.6	8.4	8.3
$\lvert X_i - 8\rvert$	0.2	0.3	0.4	0.4	0.3
R_i^+	1	2.5	4.5	2.5	4.5
ψ_i	0	1	0	1	1

The midrank $R_i^+ = 2.5$ is obtained since two values of $|X_i - 8|$ are tied for ranks 2 and 3, while the midrank of 4.5 occurs since two of the $|X_i - 8|$ are tied for ranks 4 and 5. The value of the Wilcoxon signed-rank statistic is $T^+ = \sum \psi_i R_i^+ = 9.5$, and the sign statistic is $S = \sum \psi_i = 3$.

The computation of the two-sample statistics can be illustrated by a comparison of storage methods 1 and 2. Following the notation of the section "Two-Sample Problem" let the X's correspond to method 1 observations and the Y's to method 2 observations. The following table gives each observation followed by its rank (in parentheses) when the X and Y observations are ranked jointly.

Method 1(X's)	7.8(5)	8.3(6.5)	7.6(4)	8.4(8)	8.3(6.5)
Method 2(Y's)	5.4(1)	7.4(3)	7.1(2)		

To test the null hypothesis that the difference between the location parameters of the two storage methods is 0, one obtains $t = (nm)^{1/2}(\bar{X} - \bar{Y}) / [S_p(n + m)^{1/2}] = (15)^{1/2}(8.08 - 6.633)/0.687(8)^{1/2} = 2.88$ as the appropriate test statistic if the observations from both methods are normally distributed with the same variance. In this case the t-statistic is testing the equality of the two population means. If the observations from both methods are not assumed to be normally distributed, the Mann–Whitney–Wilcoxon test statistic, $W = \sum R_i = 1 + 3 + 2 = 6$ can be used provided that the populations have the same shape.

Finally, to test the hypothesis of equality of the location parameters of all three storage methods, the k-sample location tests can be used. If the observations taken on each method are normally distributed with common variance, the appropriate test statistic is $F = \text{MST}/\text{MSE} = 2.03/0.611 = 3.32$. When the observations are not normally distributed, the Kruskal–Wallis statistic, $H = 12\sum n_i(\bar{R}_i. - \bar{R}..)^2/N(N+1) = 5.25$, can be used.

RELATED PROBLEMS

A location family of distributions can be broadened by allowing the members of the family to differ in their scale parameters as well as their location parameters. For a fixed CDF F, the distributions $F[(x - \eta)/\sigma]$ form a location and scale family as η and σ are varied. Increases (decreases) in the location parameter η cause the probability distribution to be shifted to the right (left), while increases (decreases) in the scale parameter σ cause the probability distribution to become less (more) concentrated about η.

In a two-sample setting, the location problem is generalized to a location and scale problem by letting $X_1, \ldots, X_m$ and $Y_1, \ldots, Y_n$ denote independent random samples from two populations having CDFs $F[(x - \eta_1)/\sigma_1]$ and $F[(x - \eta_2)/\sigma_2]$, respectively. Tests for equality of the location parameters η_1 and η_2, under the assumption $\sigma_1 = \sigma_2$, were discussed in the section "Two-Sample Problem." When the assumption of equal scale parameters is dropped, testing for equality of the location parameters becomes more difficult. When both populations are normal but the scale parameters (variances) are unequal, the problem of testing for equality of the means is known as the Behrens–Fisher problem*, and an approximate solution is given by Welch [18]. If the populations are not normal, the problem of testing for equality of the location parameters, for example the medians, without assuming that the populations have the same shape would be a nonparametric generalization of the Behrens–Fisher problem. Some possible solutions can be found in refs. 4, 7, or 17, although there as yet appears to be no "standard" solution as in the parametric case.

Another important hypothesis in the two-sample location and scale problem, testing for equality of the scale parameters, is discussed in the entry SCALE TESTS. Finally, it may sometimes be of interest to test for location or scale differences simultaneously. Lepage considered this problem of testing $H_0: \eta_1 = \eta_2,\ \sigma_1 = \sigma_2$ vs. $H_1: \eta_1 \neq \eta_2$, or $\sigma_1 \neq \sigma_2$ in ref. 13.

References

[1] Barlow, R. E., Bartholomew, D. J., Bremmer, J. M., and Brunk, H. D. (1972). *Statistical Inference under Order Restrictions*. Wiley, New York.

[2] Bartholomew, D. J. (1959). *Biometrika*, **46**, 36–48.

[3] Dixon, W. J. and Mood, A. M. (1946). *J. Amer. Statist. Ass.*, **41**, 557–566.

[4] Fligner, M. A. and Policello, G. E. P. (1981). *J. Amer. Statist. Ass.*, **76**, 162–168.

[5] Hájek, J. and Sidák, Z. (1967). *Theory of Rank Tests*. Academic Press, New York.

[6] Hald, A. (1952). *Statistical Theory with Engineering Application*. Wiley, New York.

[7] Hettmansperger, T. P. and Malin, J. S. (1975). *Biometrika*, **62**, 527–529.

[8] Hollander, M. and Wolfe, D. A. (1973). *Nonparametric Statistical Methods*. Wiley, New York.

[9] Iman, R. L., Quade, D., and Alexander, D. A. (1975). *Selected Tables in Mathematical Statistics*, Vol. 3, D. B. Owen and R. E. Odeh, eds. American Mathematical Society, Providence, R.I., pp. 329–384.

[10] Jonckheere, A. R. (1954). *Biometrika*, **41**, 133–145.

[11] Kruskal, W. H. and Wallis, W. A. (1952). *J. Amer. Statist. Ass.*, **47**, 583–621.

[12] Lehmann, E. L. (1975). *Nonparametrics: Statistical Methods Based on Ranks*. Holden-Day, San Francisco.

[13] Lepage, Y. (1971). *Biometrika*, **58**, 213–217.

[14] Mack, G. A. and Wolfe, D. A. (1981). *J. Amer. Statist. Ass.*, **76**, 175–181.

[15] Mann, H. B. and Whitney, D. R. (1947). *Ann. Math. Statist.*, **18**, 50–60.

[16] Mood, A. M. (1954). *Ann. Math. Statist.*, **25**, 514–522.

[17] Potthoff, R. F. (1963). *Ann. Math. Statist.*, **34**, 1596–1599.

[18] Welch, B. L. (1937). *Biometrika*, **29**, 350–362.

[19] Wilcoxon, F. (1945). *Biometrics*, **1**, 80–83.

(ARITHMETIC MEAN
BROWN–MOOD MEDIAN TEST
DISTRIBUTION-FREE METHODS
FISHER–YATES TESTS
HYPOTHESIS TESTING
INFERENCE, STATISTICAL
KRUSKAL–WALLIS TEST
LOCATION-SCALE FAMILIES
MANN–WHITNEY–WILCOXON TESTS
NORMAL SCORES TESTS
ORDER STATISTICS
RANK TESTS
t-TESTS
WILCOXON SIGNED-RANK TEST)

Michael A. Fligner

LOGARITHMIC NORMAL DISTRIBUTION *See* LOGNORMAL DISTRIBUTION

LOGARITHMIC SERIES DISTRIBUTION

Partly because of its rather long tail on the right, the logarithmic series distribution (LSD) has been found useful in the analysis of various kinds of data. See, for example, Patil et al. [19] and Williams [24]. Broadly speaking, these applications can be divided into two main types:

1. Random collection of individuals or units (e.g., moths or insects) and then classified into groups (species). This collection gives rise to the distribution of groups (species): e.g., groups with one individual, two individuals, three individuals, and so on, which may follow an LSD.
2. Random collection of groups (rats) and the number of individuals or units (fleas) in each group is counted. This gives rise to the distribution of groups with (no individual) one individual, two individuals, etc., which again may follow an LSD.

The LSD was introduced by Fisher et al. [6] while studying abundance and diversity for insect trap data. Let various species of insects be trapped. Suppose that for a given species the number of individuals trapped or represented in the sample is a random variable, X, having a Poisson distribution* with parameter $E[X] = \omega t$. The parameter t is a characteristic of the trap and ω is a characteristic of the species. Further suppose that the characteristic ω varies from species to species in such a manner that a species chosen at random has characteristic Ω which possesses a gamma distribution with parameters $1/\lambda$ and k. Then the number of representatives, X, has a negative binomial distribution* with parameters k and $1/(1 + \lambda t)$. Fisher, using his great intuition, notes that one cannot observe the number of species with no representation. Thus he does something like taking the limit as k goes to zero of the zero-truncated negative binomial and obtains the LSD.

Boswell and Patil [1] discuss various chance mechanisms generating the LSD used in the analysis of number of species and individuals. See Ehrenberg [4] for applications in consumer behavior in repeat buying, Engen [5] for abundance models, Ord et al. [11] for ecological applications, and Grassle et al. [7] and Patil and Taillie [15] for a recent discussion in the context of diversity.

DEFINITION AND STRUCTURE

The LSD is defined by the probability function given by

$$p(x) = \Pr\{X = x\} = p(x;\theta) = \alpha\theta^x/x, \qquad x = 1, 2, \ldots, \infty, \quad (1)$$

where $\alpha = -1/\log(1-\theta), 0 < \theta < 1$. Clearly, an LSD is a power series distribution* (PSD) with series function $f(\theta) = -\log(1-\theta)$, and inherits the properties of a PSD. Specific properties follow:

Property 1.

$$E[X] = \mu = \alpha\theta/(1-\theta),$$

$$V(X) = \mu\left(\frac{1}{1-\theta} - \mu\right).$$

Property 2.

i. The probability $e(\theta)$ of the LSD with parameter θ giving rise to an even observation is less than $1/2$ irrespective of the value of θ.

ii. The conditional distribution of $Y = \frac{1}{2}X$ given that Y has assumed an integral value is also an LSD but with parameter θ^2. See Patel and Wani [16].

Property 3. The limit of a zero-truncated negative binomial distribution with parameters k and $1-\theta$ is a LSD with parameter θ.

Property 4. A randomly stopped sum $\sum_{i=1}^{N} X_i$ of LSD RVs X_i with parameter θ has a negative binomial distribution with parameters k and $1-\theta$ if the stopping RV N is Poisson with parameter $-k\log(1-\theta)$. This is useful as a clustering model (see, e.g., Douglas [3]).

Property 5. Let $I_k = \{k, 2k, 3k, \ldots; k \text{ an integer}\}$ be the set of integral multiples of integers. Then the conditional distribution of X/k, given that X/k is an integer, is also an LSD but with parameter θ^k (Patil and Seshadri [14]).

Property 6. Let X_i, $i = 1, 2, \ldots, n$, be a random sample of size n drawn from the population of the LSD with parameter θ. Then the sample total $Z = \sum_{i=1}^{n} X_i$ is complete sufficient for θ. Further, Z follows what Patil and Wani [16] have called the first type Stirling distribution* with parameters θ and n defined by

$$p(z;\theta,n) = b(z,n)\theta^z/f_n(\theta),$$

$$z = n, n+1, \ldots, \qquad n = 1, 2, \ldots,$$

$f_n(\theta) = \theta^{-n}$ and $b(z,n) = n!\ s(n,z)/z!$ with $s(n,z)$ being the absolute value of the Stirling number of the first kind with arguments n and z.

MINIMUM VARIANCE UNBIASED (MVU) ESTIMATION*

Based on a random sample $x_1, x_2, \ldots, x_n$ of size n from the LSD defined by (1), the unique MVU estimator of θ is given by $h(z,n) = zs(n, z-1)/s(n,z)$ if $z = \sum_{i=1}^{n} x_i > n$, and $= 0$ if $z = n$, where $s(n,z)$ is the absolute value of the Stirling number of the first kind with arguments n and z. For large z,

$$h(z,n) \approx \frac{z}{z-1}\left[\frac{\log(z-1)+c}{\log z + c}\right]^{n-1},$$

where c is Euler's constant. A recurrence relation is available for $h(z,n)$ given by

$$h(z+1,n) = \frac{(z+1)h(z,n-1)}{zh(z,n-1)-(z-1)h(z,n)+z},$$

with $h(z,n) = z(z-1)$ for $n = 1$ and $h(z,n) = 0$ for $z = n$.

Tables for the MVU estimator $h(z,n)$ are available to four decimal places for values of z and n with the following ranges:

$$n = 3(1)12, \quad z \leqslant 128;$$

$$n = 14(2)22(3)37, \quad z \leqslant 133,$$

$$n = 40(5)70(10)100, \quad z \leqslant 156.$$

The MVU estimator of the variance of the MVU estimator of θ is then obtained as $h(z,n)[h(z,n) - h(z-1,n)]$.

For details, see Patil and Bildikar [13] and relevant references in Patil et al. [19].

MAXIMUM LIKELIHOOD ESTIMATION*

Based on a random sample $x_1, x_2, \ldots, x_n$ of size n from the LSD defined by (1), the likelihood equation for estimating θ is given by the moment equation $\bar{x} = \mu(\hat{\theta}) = \hat{\mu}$, where $\bar{x}$ is the sample mean and $\mu(\theta)$ is the mean of the LSD. Since $\mu(\theta)$ is an increasing function of θ, the likelihood equation has a unique solution $\hat{\theta}$ given by its inversion. Patil and Wani [17] give $\hat{\theta}$ to four decimal places for $\bar{x} = 1.02(0.02), 2.00(0.05), 4.00(0.1), 8.0(0.2), 14.0(0.5), 30.0(2), 40(5), 60(10), 140(20), 200$. They also tabulate the bias and standard error of $\hat{\theta}$ in a few cases. For more details and references, see Johnson and Kotz [8], Patil [12], and Patil et al. [19].

TABULATION

Williamson and Bretherton [25] provide tables for both the probability function and the cumulative distribution function. The argument of the table is the mean value μ and not θ. Entries are given to five decimal places for $\mu = 1.1(0.1), 2.0(0.5), 5.0(1), 10$, until the cumulative probability exceeds 0.999. A key for conversion from θ to μ is also given.

For extensive tables, see Patil et al. [18]. These give values of the probability function and the cumulative function to six decimal places for $\theta = 0.01(0.01), 0.70(0.005), 0.900(0.001), 0.999$.

CONCLUDING REMARKS

For more details and related references concerning structural and inferential properties of the LSD and related versions that are truncated, modified, etc., see Boswell et al. [2], Douglas [3], Johnson and Kotz [8], Kempton [9], Ord [10], Patil et al. [19], Rao [20], and Wani and Lo [21, 22, 23]. In their latest paper, Wani and Lo compare in terms of shortness of intervals Crow's system of confidence intervals with Clopper and Pearson's and the corresponding randomized counterparts.

Kempton [9] considers a generalized form of the LSD based on the beta distribution of the second kind in place of the usual gamma in the classical development of Fisher's LSD model. He has proposed the generalized form for fitting species abundance data with exceptionally long tails.

References

[1] Boswell, M. T. and Patil, G. P. (1971). In *Statistical Ecology*, Vol. 1, G. P. Patil, E. C. Pielou, and W. E. Waters, eds. Pennsylvania State University Press, University Park, Pa., pp. 99–130.

[2] Boswell, M. T., Ord, J. K., and Patil, G. P. (1979). In *Statistical Distributions in Ecological Work*, J. K. Ord, G. P. Patil, and C. Taillie, eds. International Co-operative Publishing House, Fairland, Md., pp. 3–156.

[3] Douglas, J. B. (1979). *Analysis with Standard Contagious Distributions*. International Co-operative Publishing House, Fairland, Md.

[4] Ehrenberg, A. S. C. (1972). *Report Buying—Theory and Applications*. North-Holland Elsevier, New York.

[5] Engen, S. (1978). *Stochastic Abundance Models*. Chapman & Hall, London/Wiley, New York.

[6] Fisher, R. A., Corbet, A. S. and Williams, C. B. (1943). *J. Anim. Ecol.*, **12**, 42–58.

[7] Grassle, J. F., Patil, G. P., Smith, W. K., and Taillie, C., eds. (1979). *Ecological Diversity in Theory and Practice*. International Co-operative Publishing House, Fairland, Md.

[8] Johnson, N. L. and Kotz, S. (1970). *Distributions in Statistics: Discrete Distributions*. Wiley, New York.

[9] Kempton, R. A. (1975). *Biometrika*, **62**, 29–38.
[10] Ord, J. K. (1972). *Families of Frequency Distributions*. Charles Griffin, London.
[11] Ord, J. K., Patil, G. P., and Taillie, C., eds. (1979). *Statistical Distributions in Ecological Work*. International Co-operative Publishing House, Fairland, Md.
[12] Patil, G. P. (1962). *Biometrics*, **18**, 68–75.
[13] Patil, G. P. and Bildikar, S. (1966). *Sankhyā A*, **28**, 239–250.
[14] Patil, G. P. and Seshadri, V. (1975). In *Statistical Distributions in Scientific Work*, Vol. 1, G. P. Patil, S. Kotz, and J. K. Ord, eds. D. Reidel, Dordrecht, Holland, pp. 83–86.
[15] Patil, G. P. and Taillie, C. (1979). *Bull. Int. Statist. Inst.*, **44**, 1–15.
[16] Patil, G. P. and Wani, J. K. (1965). *Sankhyā A*, **27**, 271–280.
[17] Patil, G. P. and Wani, J. K. (1965). In *Classical and Contagious Discrete Distributions*, G. P. Patil, ed. Statistical Publishing Society, Calcutta/Pergamon Press, Elmsford, N.Y., pp. 398–409.
[18] Patil, G. P., Kamat, A. R., and Wani, J. K. (1964). Certain Studies on the Structure and Statistics of the Logarithmic Series Distribution and Related Tables. *Tech. Rep.*, Aerospace Research Laboratory, Wright-Patterson Air Force Base, Ohio.
[19] Patil, G. P., Boswell, M. T., Joshi, S. W., and Ratnaparkhi, M. V. (1984). *A Modern Dictionary and Classified Bibliography of Statistical Distributions*, Vol. 1: *Discrete Models*. International Co-operative Publishing House, Fairland, Md.
[20] Rao, C. R. (1971). In *Statistical Ecology*, Vol. 1, G. P. Patil, E. C. Pielou, and W. E. Waters, eds. Pennsylvania State University Press, University Park, Pa., pp. 131–142.
[21] Wani, J. K. and Lo, H. P. (1975). *Canad. J. Statist.*, **3**, 277–284.
[22] Wani, J. K. and Lo, H. P. (1975). *Biometrics*, **31**, 771–775.
[23] Wani, J. K. and Lo, H. P. (1977). *Canad. J. Statist.*, **2**, 153–158.
[24] Williams, C. B. (1964). *Patterns in the Balance of Nature*. Academic Press, London.
[25] Williamson, E. and Bretherton, M. H. (1964). *Ann. Math. Statist.*, **35**, 284–297.

(COMPOUND DISTRIBUTIONS
DISCRETE DISTRIBUTIONS
MIXTURE DISTRIBUTIONS
MULTIVARIATE LOGARITHMIC SERIES DISTRIBUTIONS
POWER SERIES DISTRIBUTIONS)

G. P. PATIL

LOGARITHMIC TRANSFORMATION

Transformation on the variable (or on sample data) of the type $Y' = \log Y$ or $Y'' = a + b \log Y$ is commonly used in ANOVA* if true effects are multiplicative (or proportional). See, e.g., Bartlett [1] for an earlier discussion or any textbook on ANOVA and design of experiments.* It is also used extensively for normalization purposes (see, e.g., Whitmore and Yalovsky [2] as an example, where the logarithmic normalizing transformation is found to be superior to the competing square-root* transformation).

The logarithmic transformation of a gamma random variable yields a variable whose (approximate) standard deviation is independent of the shape parameter and which is more nearly normally distributed than the original gamma variable.

References

[1] Bartlett, M. S. (1947). *Biometrics*, **3**, 39–51.
[2] Whitmore, G. A. and Yalovsky, M. (1978). *Technometrics*, **20**, 207–208.

(ANGULAR TRANSFORMATION
ANOVA
SQUARE-ROOT TRANSFORMATION)

LOG-CHANGE INDEX NUMBERS

Index numbers* are measures of relative change, which are usually based on ratios. Log-change index numbers are based on the natural logarithms of those ratios. For example, a change in the price of a commmodity from \$10 to \$11 is equal to 10 *percent* in terms of conventional index numbers $[((11/10) - 1) \times 100]$, and to 9.531 *log-percent* in terms of log-change index numbers $[(\ln(11/10) \times 100]$. For a set of commodities, log-change index numbers are weighted arithmetic means* of the natural logarithms of the price or quantity ratios. Log-change index numbers possess a number of desirable properties. For example, some versions of these indexes are exact for certain classes of utility and production functions, and satisfy the commodity reversal test, the unit of measurement test, some forms of the proportionality test, and the time and factor reversal tests.

GENERAL FORM OF LOG-CHANGE INDEX NUMBERS

Let the set of commodities under consideration be $A = \{a_1, a_2, \dots, a_n\}$, observed in two situations, 0 and 1, assumed here to be time periods. The prices of a_i in the two periods are p_{i0} and p_{i1}, the corresponding quantities are q_{i0} and q_{i1}, the money values are $v_{i0} = p_{i0}q_{i0}$ and $v_{i1} = p_{i1}q_{i1}$, and the value shares are then $w_{i0} = v_{i0}/\sum_{j=1}^{n} v_{j0}$ and $w_{i1} = v_{i1}/\sum_{j=1}^{n} v_{j1}$.

The log-change price and quantity index numbers in period 1 relative to period 0, $\ln(P_1/P_0)$ and $\ln(Q_1/Q_0)$, are of the following general form:

$$\ln(P_1/P_0) = \sum_{i=1}^{n} \overline{w}_i \ln(p_{i1}/p_{i0}),$$

$$\ln(Q_1/Q_0) = \sum_{i=1}^{n} \overline{w}_i \ln(q_{i1}/q_{i0}),$$

where the weights w_i are some function of the money values. There are many proposals found in the literature regarding the form of the weight function. Walsh [6] proposed the following weights:

$$\overline{w}_i \sqrt{v_{i1}v_{i0}} \bigg/ \sum_{j=1}^{n} \sqrt{v_{j1}v_{j0}}\,.$$

Tornqvist [4] used a set of weights defined by

$$\overline{w}_i = \tfrac{1}{2}(w_{i1} + w_{i0}),$$

which is also found in Fisher [1]. Theil [3] proposed instead a kind of average of the Walsh–Tornqvist weights, that is,

$$\overline{w}_i = \frac{\sqrt[3]{\tfrac{1}{2}(w_{i1} + w_{i0})w_{i1}w_{i0}}}{\sum_{j=1}^{n} \sqrt[3]{\tfrac{1}{2}(w_{j1} + w_{j0})w_{j1}w_{j0}}}\,.$$

None of the formulas defined by these weight functions, however, satisfies the factor reversal test.

VARTIA INDEXES

Two important weight functions have been proposed recently by Vartia [5]. They are defined by

$$\hat{w}_i = \frac{L(v_{i0}, v_{i1})}{L(\sum_j v_{j0}, \sum_j v_{j1})}, \qquad i = 1, 2, \dots, n,$$

$$\tilde{w}_i = \frac{L(w_{i0}, w_{i1})}{\sum_j L(w_{j0}, w_{j1})}, \qquad i = 1, 2, \dots, n,$$

where $L(x, y) = L(y, x)$ denotes the *logarithmic mean*, defined by

$$L(x, y) = \begin{cases} (y - x)/\ln(y/x), & y \neq x, \\ x, & y = x. \end{cases}$$

The two indexes defined by the weight functions $\hat{w}_i$ and $\tilde{w}_i$ are called, respectively, Vartia Index I and Vartia Index II.

The function $L(x, y)$ is an average in the sense that $\min(x, y) \leqslant L(x, y) \leqslant \max(x, y)$, and that $L(ax, ay) = aL(x, y)$ for $a > b$. From the definition of $L(x, y)$ we have the identity $\ln(y/x) = (y - x)/L(x, y)$, which shows that $\ln(y/x)$, and hence Vartia indexes, are measures of relative change with respect to the logarithmic mean.

Vartia indexes satisfy various tests of index numbers, including in particular the factor reversal test. The satisfaction of this test can be proved for the weights $\hat{w}_i$ as follows:

$$\ln\left(\frac{\sum_i v_{i1}}{\sum_i v_{i0}}\right) = \frac{\sum_i v_{i1} - \sum_i v_{i0}}{L(\sum_j v_{j1}, \sum_j v_{j0})}$$

$$= \sum_i \left[\frac{L(v_{i1}, v_{i0})}{L(\sum_j v_{j1}, \sum_j v_{j0})} \ln\left(\frac{v_{i1}}{v_{i0}}\right)\right]$$

$$= \sum_i \hat{w}_i \ln\left(\frac{p_{i1}}{p_{i0}}\right) + \sum_i \hat{w}_i \ln\left(\frac{q_{i1}}{q_{i0}}\right).$$

The two Vartia indexes differ in that, on the one hand, the weights of Vartia Index II add up to 1, while those of Vartia Index I add up to 1 at most; and, on the other, Vartia Index I is consistent in aggregation, whereas Vartia Index II is not.

Consistency in Aggregation

Vartia Index I is consistent in aggregation* in the sense that if the set A which contains n commodities was divided into m disjoint subsets each containing n_k commodities ($k = 1, 2, \dots, m$; and $\sum_k n_k = n$), then the value of the index is the same whether computed for A directly, or indirectly as the weighted sum of the equivalent indexes of the m subsets. This is demonstrated by equating the total price index with the weighted sum of the corresponding subindexes as follows:

$$\ln\left(\frac{P_1}{P_0}\right)$$

$$= \sum_{i=1}^{n}\left[\frac{L(v_{i0}, v_i\exp)}{L(\sum_{j=1}^{n} v_{j0}, \sum_{j=1}^{n} v_{j1})}\ln\left(\frac{p_{i1}}{p_{i0}}\right)\right]$$

$$= \sum_{k=1}^{m}\sum_{i=1}^{n_k}\left[\frac{L(v_{i0}, v_{i1})}{L(\sum_{j=1}^{n} v_{j0}, \sum_{j=1}^{n} v_{j1})}\ln\left(\frac{p_{i1}}{p_{i0}}\right)\right]$$

$$= \sum_{k=1}^{m}\left(\frac{L(\sum_{j=1}^{n_k} v_{j0}, \sum_{j=1}^{n_k} v_{j1})}{L(\sum_{j=1}^{n} v_{j0}, \sum_{j=1}^{n} v_{j1})}\right.$$

$$\left.\times\left[\sum_{j=1}^{n_k}\frac{L(v_{i0}, v_{i1})}{L(\sum_{j=1}^{n_k} v_{j0}, \sum_{j=1}^{n_k} v_{j1})}\ln\left(\frac{p_{i1}}{p_{i0}}\right)\right]\right).$$

The expression in brackets is the Vartia Index I for the kth subset, and the weighted sum of these subindexes, where the weights are of the type $\hat{w}_i$, is equal to the total Vartia Index I computed from the first line directly for the whole set.

References

[1] Fisher, I. (1922). *The Making of Index Numbers*. Houghton Mifflin, Boston.

[2] Sato, K. (1976). *Rev. Econ. Statist.*, **58**, 223–228. (Develops Vartia Index II; few references.)

[3] Theil, H. (1973). *Rev. Econ. Statist.*, **55**, 498–502. (Uses different weight function; few references.)

[4] Tornqvist, L. (1936). *Bank Finland Monthly Bull.*, **10**, 1–8.

[5] Vartia, Y. O. (1976). *Scand. J. Statist.*, **3**, 121–126. (Few references.)

[6] Walsh, C. M. (1901). *The Measurement of General Exchange Value*. Macmillan, New York.

Bibliography

Diewert, W. E. (1978). *Econometrica*, **46**, 883–900. (Discusses economic properties of Vartia Index I; intermediate; references.)

Hulten, C. R. (1973). *Econometrica*, **41**, 1017–1025. (Approximates Divisia index numbers by log-change indexes; economic theory; intermediate; few references.)

Jevons, W. S. (1865). *J. R. Statist. Soc. A*, **28**, 294–320. (Of purely historical interest.)

Lau, L. J. (1979). *Rev. Econ. Statist.*, **61**, 73–82. (Discusses economic properties of Vartia Index II; mathematical; few references.)

Sato, K. (1974). *Rev. Econ. Statist.*, **56**, 649–652. (Improves on Theil's weights; few references.)

Stuvel, G. (1957). *Econometrica*, **25**, 123–131. (Proposes an index numbers formula which, like Vartia Index I, satisfies the factor reversal test and is consistent in aggregation.)

Theil, H. (1965). *Econometrica*, **33**, 67–87. (Theory and application of log-change index numbers.)

Theil, H. (1974). *Rev. Econ. Statist.*, **56**, 552–554. (Further discussion of Theil's formula.)

Vartia, O. Y. (1976). *Relative Changes and Index Numbers*. Ser. A4. The Research Institute of the Finnish Economy, Helsinki. (Contains detailed discussion of various measures of relative change, including the notion of the logarithmic mean, Vartia indexes, and Divisia indexes; mostly elementary; extensive references.)

(DIVISIA INDICES
INDEX NUMBERS)

NURI T. JAZAIRI

LOG-CONCAVE AND LOG-CONVEX DISTRIBUTIONS

A probability distribution on (a, b), $-\infty \leqslant a < b \leqslant \infty$, is called log-concave (log-convex) if it has a density $f > 0$ on (a, b) such that

$$\log f \text{ is concave (convex) on } (a, b), \quad (1)$$

i.e., if f is log-concave (log-convex); for a slightly more general definition, see Ibragimov [1].

If f has a second derivative on (a, b), then (1) is equivalent to

$$f''(x)f(x) - \{f'(x)\}^2 \leqslant 0 \quad (\geqslant 0)$$
$$(x \in (a, b)); \quad (2)$$

other, practically equivalent, conditions are that f'/f is nonincreasing (nondecreasing), and $f(x+h)/f(x)$ is nonincreasing (nondecreasing) in x for each $h > 0$.

The most important properties of log-concave densities are given in the following theorem due to Ibragimov [1].

Theorem 1. Let $\mathscr{U}$ be the set of log-concave densities, and let $*$ denote convolution*.

1. If $f_1 \in \mathscr{U}$ and $f_2 \in \mathscr{U}$, then $f_1 * f_2 \in \mathscr{U}$.
2. If $f \in \mathscr{U}$ and g is unimodal, then $f * g$ is unimodal.

Property (2) of f is called *strong unimodality**; it is equivalent to log concavity (see Ibragimov [1]). Examples of log-concave distributions are the normal, gamma (shape parameter $\geqslant 1$), and uniform. Related classes of distributions are the Pólya frequency functions or PF_r distributions; according to Karlin [2], the class PF_2 is contained in $\mathscr{U}$. For definitions see Karlin [2] and PÓLYA TYPE 2 FREQUENCY DISTRIBUTIONS.

For distributions (p_n) on the integers, log concavity (log convexity) can be defined by the analog to (2):

$$p_{n+1}p_{n-1} - p_n^2 \leqslant 0 \quad (\geqslant 0);$$

these distributions have analogous properties (see Keilson [3]).

Log concavity and log convexity of $\bar{F}(x) = \int_x^\infty f(u)\,du$ are important in reliability* theory; these properties are equivalent to the failure rate, being increasing and decreasing, respectively. Log concavity (log convexity) of f implies log concavity (log convexity) of $\bar{F}$ (see Keilson [3, p. 74]).

If f is log-convex (on its support), then it is convex and infinitely divisible*; log concavity of f does not imply its concavity. If f is completely monotone, then it is log-convex. Log-convex densities, being convex, are necessarily restricted to a half-line. Most practical examples are completely monotone (i.e., mixtures of exponential densities). Mixtures* of log-convex densities are again log-convex.

For the occurrence of log concavity and log convexity in birth–death processes*, and for the associated moment inequalities, see Karlin [2], Keilson [3] and the references therein.

References

[1] Ibragimov, I. A. (1956). *Theory Prob. Appl.*, **1**, 255–260.

[2] Karlin, S. (1968). *Total Positivity*. Stanford University Press, Stanford, Calif.

[3] Keilson, J. (1979). *Markov Chain Models, Rarity and Exponentiality*. Springer-Verlag, New York.

(INFINITE DIVISIBILITY
TOTAL POSITIVITY
UNIMODALITY)

F. W. STEUTEL

LOGICAL TESTS *See* EDITING STATISTICAL DATA

LOGIC OF STATISTICAL REASONING

Statistical reasoning has always characterized human cognition, and in all likelihood antedates deductive reasoning. Until very recently, however, this form of reasoning has remained informal and intuitive. The probabilistic foundations of statistical reasoning were laid only in the sixteenth and seventeenth centuries, and statistical *inference** itself became a subject of formal study only in the present century.

The impetus for this study was the realization that in many areas of empirical inquiry in which it proved very difficult to account for individual phenomena, it nevertheless proved possible to obtain important and useful information concerning mass phenomena. The study of statistical distributions turned out to be immensely fruitful in a wide variety of scientific disciplines.

The contrast between the study of the statistical characteristics of populations in biology, the social sciences, meteorology, and the like, and the study of individual phenomena in celestial mechanics and engineering, however, should not be taken too strictly. Measurement and manufacture involve error and approximation, and error and approximation can only be understood statistically. Thus even in the most precise and deterministic of sciences, statistical inference lies at the interface between theory and experience.

The logic of statistical reasoning may take many forms. What form it takes depends in part on what the outcome of that reasoning is taken to be—whether it is the choice of an action, the acceptance of a statement, the rejection of a hypothesis, the assignment of a probability, or whatever—and also on what interpretation of probability is taken to be involved in statistical reasoning—subjective, logical, empirical, or other. Furthermore, the foundations of statistical reasoning are highly controversial. While most processes of statistical reasoning yield conclusions that are relatively uncontroversial, there is not only a great deal of disagreement about what form of reasoning warrants those uncontroversial conclusions, but there are many instances in which the conclusions themselves are regarded as controversial. It is quite possible that what is a sound inference or procedure according to one approach to statistical reasoning may be regarded as unsound, and its conclusion unwarranted, from the point of view of another approach to statistical reasoning. The nature of these controversies is reflected in the essays collected in ref. 4.

STATISTICAL TESTING

The general idea behind statistical testing is that proposed by many philosophers for scientific inference in general. This idea is that you must have a specific hypothesis in mind to begin with; that you then deduce some consequence of this hypothesis; and that you then put this consequence to the test. If the test yields a negative result, you reject the hypothesis.

R. A. Fisher* [3] argues that the force of the reasoning behind significance tests* is that of a disjunction: either the hypothesis is false, or something very unusual has occurred. He uses the example of throwing four dice, and obtaining four sixes, with the implication that were we not convinced to start with that the dice were fair, we would be inclined to reject the null hypothesis of fairness.

But on the null hypothesis* any particular outcome of the throw of the four dice would be equally unlikely, and the outcome of four sixes can only incline us, more than some other outcome, to be doubtful of the fairness of the dice if we have some other hypothesis in mind which would serve to explain the observed outcome.

Significance testing is relatively quick and easy—therein lies its advantage for the ordinary investigator—but if the explicit reasoning goes no further than the significance level, it is very difficult to assess the evidential import of the data. This fact has led some writers (e.g., Fisher) to claim that significance testing is of use only as a vague, heuristic tool to suggest areas of inquiry worthy of serious investigation. Critical discussions of significance testing are to be found in refs. 8 and 12.

HYPOTHESIS TESTING

In classical hypothesis testing, as expounded in ref. 7, for example, we take account of the exact character of the alternative hypotheses. We devise tests that not only yield a small probability of falsely rejecting the null hypothesis, but which provide the maximum possible power* for discriminating between the null hypothesis and its alternatives, which are efficient and unbiased, or which have other desirable long-run properties. The fact that the alternative hypotheses are considered explicitly allows for much more explicit treatment both of the properties of the tests employed and the limitations and assumptions involved in the application of the test. It is still true, however, that the size of the test—its p-value*—reflects no more than the long-run frequency with which the null hypothesis will be erroneously rejected, and thus may or may not be appropriate to the epistemological characterization of a particular instance of statistical reasoning, and will almost certainly not reveal the whole story.

The logic of the "justification" of both significance and hypothesis testing* deserves two further remarks. Assuming that all the appropriate conditions are satisfied, it may be the case that the significance level associated with a particular investigation—its p-value—provides some epistemic guide to the cogency of its results. But if we consider a large class of such investigations, we can be quite sure that a twentieth of those whose significance level is 0.05 represent erroneous rejections of the null hypothesis. More important, if we look at *published* results, from which most of the results that have failed to achieve significance have been weeded, we are looking at a highly biased sample of applications of significance testing, in which we may be quite sure that the frequency of erroneous rejections of the null hypothesis is very much greater than the significance level cited.

CONFIDENCE METHODS

The usual reconstruction of confidence interval* reasoning proceeds as follows. Let P be a population, and X a random quantity defined on that population. We assume that we know the forn of the distribution of X—e.g., that it is normal. Let μ be an unknown parameter. We define an interval-valued function of sample statistics with the property that: Whatever be the distribution of X of the assumed form, the long-run frequency with which samples will yield values of the function covering the parameter μ will be at least $1 - p$. This part of the reasoning is strictly analytic and deductive. We now draw a sample from the population, and compute the corresponding interval. We assert, with confidence $1 - p$, that the interval computed covers μ.

On the frequency interpretation of probability, this confidence coefficient is not a probability. We cannot say that the probability is $1 - p$ that μ lies in the specified interval. The parameter μ either lies in that interval or does not, and the frequency with which it does so is either 0% of 100%.

The interpretation of the confidence coefficient $1 - p$ is therefore open to question. Why should we prefer large coefficients to small ones? One conventional answer is that in the long run of cases in which the assumptions of the method are satisfied, the relative frequency of correct assertions will be about $1 - p$. That says nothing about the case before us, but it may make us feel better. Neyman [9] takes the bull by the horns and claims that what is the outcome of the inference is not the acceptance or rejection of a statement or hypothesis, but the decision to act in a certain way—to act as if μ were in the specified interval—and that this pattern of inductive *behavior* can be (and can only be) justified in terms of its long-run properties. The concept of inductive behavior is the link between the long-run properties that are amenable to statistical treatment and the particular instances about which we must come to practical decisions. Confidence methods and their interpretation are extensively discussed in ref. 7.

FIDUCIAL INFERENCE

Although the reasoning involved in fiducial inference* [3] yields results that are sometimes much like those yielded by the confidence interval approach, the logic involved is different. The outcome of a fiducial inference is not the acceptance or rejection of a hypothesis, not a decision, but a probability distribution, albeit a probability distribution of a special kind. Again, let P be a population, and let the random quantity X have a normal distribution* in that population with unknown mean μ and unit variance. It follows that the difference between the mean of X and an observed value of X will have a normal distribution with unit variance, and *known* mean equal to 0.

Now we observe a value of X. From here on the reasoning is quite different from that previously discussed. Whatever be the value of μ, the distribution of $X - \mu$ is normal (0, 1). In particular, then, the *fiducial* distribution of μ is normal with variance 1 and mean equal to the observed value. We can use this distribution to obtain statements of fiducial probability* about the parameter μ. These probabilities are not frequencies, but they are *based* on frequencies.

Fiducial inference is not well understood, nor often explicitly employed. Yet in simple cases, the logic of the fiducial analysis seems to conform to both common sense and the demands of practicality.

BAYESIAN INFERENCE*

The reasoning underlying Bayesian inference is strictly probabilistic. It is based on Bayes' theorem*: $P(H/E) = P(H) * P(E/H) * (P(E))^{-1}$. The crucial and controversial feature of Bayesian reasoning is that the prior probability of H is always assumed to exist. For the personalistic or subjectivistic Bayesian, for whom probability represents a personal degree of belief rather than any sort of empirical frequency, Bayes' theorem is always applicable. References 6 and 10 provide classical statements of the Bayesian position; a short critical discussion is to be found in ref. 1.

The usual result of a Bayesian inference is an assertion of the probability of the hypothesis in question, or, if we are concerned with a parametrized family of hypotheses, a distribution of probability over that family. The result is not the acceptance or rejection of a hypothesis, but the attribution to the hypothesis of its appropriate degree of probability. An alternative way of viewing the outcome of Bayesian reasoning is that it combines data, prior probabilities, and utilities to yield a decision or an action (*see* DECISION THEORY).

One may raise the question of whether it is essential for the purposes of science to *accept* scientific hypotheses. The firm subjectivistic Bayesian will deny this, on the grounds that all we need from science is the ability to make informed decisions, based on utilities and probabilities, and thus that we need never *accept* statistical hypotheses. Some methodologists and philosophers have argued that the acceptance of a scientific hypothesis can itself be construed as an action, to which the canons of Bayesian decision theory can be applied by employing *epistemic utilities*, in short, by taking account of the relative values of truth and error and of knowledge and ignorance.

GENERAL CONSIDERATIONS

The structure one attributes to statistical reasoning depends on both the interpretation of probability one accepts, and on what one takes to be the outcome of a statistical inference. If one adopts a frequency or empirical interpretation of probability (*see* FREQUENCY INTERPRETATION IN PROBABILITY AND STATISTICAL INFERENCE), then the structure of statistical reasoning is embodied in the long-run characteristics of a particular pattern of reasoning. The logic of hypothesis testing, for example, depends on the frequency with which the hypothesis tested will be erroneously rejected, as well as on other long-run properties of the test—power, unbiasedness, efficiency, etc. Since it is the long run that is important, and not the particular case at hand, whatever contributes to desirable long-run properties is relevant to the logic of statistical reasoning: random sampling, randomized or mixed tests, etc.

When probability is interpreted as empirical frequency or measure, we can never assign a probability to a single event, e.g., the event of a particular confidence region covering a parameter. The outcome of statistical reasoning is not the assignment of a probability to a statistical hypothesis, simple or complex, or of a probability distribution to the possible values of a parameter. It must therefore be given some other interpretation. Among the common interpretations of the result of statistical reasoning are: that it is the rejection of a (null) hypothesis, in significance testing and in the testing of hypotheses; that it is the acceptance of a hypothesis, under some interpretations of confidence methods; that it is a decision to perform an action such as to reject a shipment of items or to shut down a certain machine for adjustment; that it is an epistemic decision to accept a hypothesis, to suspend judgment about it, or to reject it.

On the subjective interpretation of probability, Bayes' theorem can always be applied. The prior probabilities required for the application of the theorem need represent no more than prior subjective opinion, and thus are always available. The outcome of statistical reasoning may therefore always be the assignment of a probability to a statistical hypothesis, or the assignment of a posterior distribution to the parameters characterizing a population.

The outcome may also be construed as a decision or a choice, by taking account of the relevant utilities. Since the long run plays no role, on this view, neither random sampling, nor randomized or mixed tests have a place in the logic of statistical reasoning. If random sampling is employed at all, it is simply for practical and psychological reasons.

It should be noted that among Bayesians it is common to use the term "subjective probability*" or "subjective prior distribution" even when there are perfectly good statistical grounds for accepting the probability or the distribution. But there appears to be no way, within the theory, of distinguishing between the cases in which there are good statistical grounds for accepting a prior distribution, and cases in which the prior distribution reflects merely ungrounded personal opinion.

Probability may be given a logical interpretation in two ways. According to one approach, probabilities are assigned to the sentences of a language, the prior probabilities of statistical hypotheses (or the prior distributions of statistical parameters) are well defined, and Bayes' theorem may be applied to the problem of decision or the problem of belief. This is the approach of Carnap [2]. The outcome of statistical reasoning may be construed as a posterior distribution of probability, or as a decision, just as in the case of the subjective interpretation.

The other logical approach, described in ref. 5, gives rise to epistemological probability. On this approach probabilities apply to statements, as they do on the subjectivistic interpretation, but all probabilities are based upon known empirical frequencies. Since these frequencies (or distributions) may be known only approximately, there are conditions under which all we can say of a prior probability is that it lies in the closed interval $[0, 1]$. Under these circumstances, the confidence approach may apply. Under other circumstances, the limits on the prior distribution of probability, combined with evidence through the mechanism of Bayes' theorem, yield useful posterior probabilities.

There is a close connection between the approach of Fisher and the approach of the frequency-based logical interpretation of probability. This connection has been explored in depth in ref. 11.

What one takes to be the logic of statistical reasoning depends both on what view one has of probability—frequency, logical, epistemological, subjective, fiducial—and on what one takes to be the conclusion of statistical reasoning—the acceptance or rejection of a statistical hypothesis, the assignment of a probability to a hypothesis or a probability distribution to a parameter, or a course of action. Both views of probability and views about the result of statistical reasoning are controversial. The controversy is not academic, since what counts as good statistical reasoning on one view will be rejected as inadequate from another. The best advice for the ordinary practitioner or consumer of statistics is simply to be aware that the controversies exist, that they do have a bearing on the conduct of statistical reasoning, and that, fortunately, the reasoning involved in most practical applications of statistics admits of analysis and justification within each of the variety of approaches to statistics.

References

[1] Barnard, G. A. and Cox, D. R., eds. (1962). *The Foundations of Statistical Inference*. Wiley, New York. (A discussion of the subjectivistic approach to statistical reasoning among adherents of diverse viewpoints.)

[2] Carnap, R. (1950). *The Logical Foundations of Probability*. University of Chicago Press, Chicago. (The best known attempt to provide an interpretation of probability as logical measure.)

[3] Fisher, R. A. (1956). *Statistical Methods and Scientific Inference*. Hafner, New York. (An exposition of the logic of fiducial inference, together with more general remarks on statistical reasoning.)

[4] Godambe, V. P. and Sprott, D. A., eds. (1971). *Foundations of Statistical Inference*. Holt, Rinehart and Winston, Toronto, Ontario. (A collection of papers, from diverse points of view, concerning the foundations and nature of statistical reasoning.)

[5] Kyburg, H. E., Jr. (1974). *The Logical Foundations of Statistical Inference*. Reidel, Dordrecht, Holland. (A technical exposition of epistemological probability and its applications in statistics, with criticisms of alternative points of view.)

[6] Kyburg, H. E., Jr. and Smokler, E., eds. (1980). *Studies in Subjective Probability*. Krieger, Huntington, N.Y. (A collection of the early papers advocating the subjective interpretation of probability.)

[7] Lehmann, E. L. (1959). *Testing Statistical Hypotheses*. Wiley, New York. (The classic exposition of the main outlines of the frequentist approach to hypothesis testing and confidence methods.)

[8] Morrison, E. and Henkel, E. (1970). *The Significance Test Controversy*. Aldine, Chicago. (An extensive collection of papers concerning the pros, cons, and caveats of significance testing, particularly in the social sciences.)

[9] Neyman, J. (1957). *Bull. Int. Statist. Inst.*, **25**, 7–22. (A concise statement of Neyman's behavioristic approach to statistical inference.)

[10] Savage, J. (1954). *The Foundations of Statistics*. Wiley, New York. (The first thorough effort to establish statistical subjectivistic principles.)

[11] Seidenfeld, T. (1979). *Philosophical Problems of Statistical Inference*. Reidel, Dordrecht, Holland. (A meticulous logical analysis of fiducial reasoning, together with remarks on other approaches.)

[12] Spielman, S. (1974). *Philos. Sci.*, **41**, 211–226. (A critical analysis of tests of significance as general inductive procedures; complementary to Morrison and Henkel [8].)

(BAYESIAN INFERENCE
CHANCE
CONFIDENCE INTERVALS
DECISION THEORY
DEGREES OF BELIEF
FIDUCIAL INFERENCE
FIDUCIAL PROBABILITY
FOUNDATIONS OF PROBABILITY
FREQUENCY INTERPRETATION IN PROBABILITY AND STATISTICAL INFERENCE
HYPOTHESIS TESTING
INFERENCE, STATISTICAL
SIGNIFICANCE TESTS, HISTORY AND LOGIC
SUBJECTIVE PROBABILITIES
WEIGHT OF EVIDENCE)

H. Kyburg

LOGISTIC CURVE

The graph of the function

$$g(t) = (a + be^{-ct})^{-1}$$

with a, b, c each positive is a *logistic curve*. It increases with t (has a positive slope); it has an upper asymptote at

$$\lim_{t \to \infty} g(t) = a^{-1}$$

and a lower asymptote (as $t \to -\infty$) of zero.

The curve can be used, with t denoting time, as a growth curve*, to represent the size of an increasing population (human or otherwise). In such a context, the upper limit a^{-1} is called the "ultimate" or "saturation" population.

From data of population sizes at a number of times $(t_1, t_2, \ldots)$ values of the parameters a, b, c can be estimated. The fitted logistic curve may then be used to predict size at future times (in particular, the "ultimate" size). If the available observations are only for relatively early times, estimates of the parameters and consequently of future population sizes are rather volatile (i.e., are substantially affected by small changes in observed sizes). This feature has tended to discredit the use of logistic curves for prediction. When the times of observation include values greater than the point of inflection $[t = c^{-1}\log(b/a)$ giving $g''(t) = 0]$, estimates are more stable. There is a good discussion in Leach [1].

The cumulative distribution function* of a logistic distribution* is represented by a logistic curve with $a = 1$. For additional information, see Pearl and Reed [2].

References

[1] Leach, D. (1981). *J. R. Statist. Soc. A*, **144**, 94–105.

[2] Pearl, R. and Reed, L. J. (1920). *Proc. Nat. Acad. Sci. U.S.A.*, **6**, 275–288. (A "classic" on this topic.)

(DEMOGRAPHY
GROWTH CURVES
LOGISTIC DISTRIBUTION
LOGIT
MATHEMATICAL THEORY OF POPULATION
STOCHASTIC DEMOGRAPHY)

LOGISTIC DISTRIBUTION

The use of the logistic distribution as a growth curve was first given by Verhulst [42]. The logistic function as a growth curve can be based on the differential equation

$$\frac{dF}{dx} = c\left[F(x) - A\right]\left[B - F(x)\right], \quad (1)$$

where c, A, and B are constants with $c > 0$, $B > A$. For the solution of (1) and an excellent account of the logistic distribution, see Johnson and Kotz [25, Chap. 22]. Equation (1) can be put in the form

$$F(x) = \left[1 + D^{-1}e^{-x/c}\right]^{-1}, \quad (2)$$

where D is a constant. Equation (2) is the cumulative distribution function of the logistic distribution. The logistic distribution has been widely used by Berkson [2–4] as a model for analyzing bioassay and other experiments involving quantal response.* Pearl and Reed [31] used this in studies connected with population growth. Plackett [32, 33] has considered the use of this distribution with life test data. Gupta and Shah [21] have applied this distribution to biochemical data. Tsutakawa [41] uses the one-parameter logistic distribution in one- and two-stage bioassay designs. The shape of the logistic distribution is similar to the normal distribution. The simple explicit relationships between X, $f_X(x)$, and $F_X(x)$ render much of the analysis of the logistic distribution attractively simple and many authors, e.g., Berkson [3], prefer it to the normal distribution.

The logistic density function in its reduced form is defined by

$$f_X(x) = e^{-x}/(1 + e^{-x})^2, \quad -\infty < x < \infty. \quad (3)$$

This distribution is symmetrical about the mean zero and has a variance $\sigma^2 = \pi^2/3$. The cumulative distribution function (CDF) corresponding to the density function (3) is given by

$$F_X(x) = (1 + e^{-x})^{-1}. \quad (4)$$

Equation (3) may be written in terms of $F_X(x)$ as

$$f_X(x) = F_X(x)(1 - F_X(x)). \quad (5)$$

The moment generating function* (MGF) of the density (3) is

$$M_X(t) = \Gamma(1 + t)\Gamma(1 - t) = \pi t/\sin \pi t, \quad |t| < 1. \quad (6)$$

We can express (6) in terms of Bernoulli numbers*, (see Tarter and Clark [39]) as

$$M_X(t) = \sum_{j=0}^{\infty} (-1)^{j-1}\left[2(2^{2j} - 1)/(2j)!\right] \times B_{2j}(\pi t)^{2j}. \quad (7)$$

The standard logistic distribution, with mean zero and variance unity, sometimes denoted by $L(0,1)$, is defined by the equation

$$f_X(x) = ge^{-gx}/(1+e^{-gx})^2, \quad -\infty < x < \infty, \quad (8)$$

where $g = \pi/\sqrt{3}$. The distribution of the density function given by (8) is symmetrical about the mean zero.

The standard logistic distribution belongs to the first class of initial distributions of the exponential type and has a shape similar to the standard normal distribution. The density curve of the logistic distribution crosses the density curve of the normal between 0.68 and 0.69. The inflection points of the standard logistic are ± 0.53 (approx.), whereas the inflection points of the standard normal are ± 1.00.

The parametric form of the cumulative distribution function* (CDF) is

$$F_X(x) = \left[1 + \exp\left\{-\left(\frac{x-\alpha}{\beta}\right)\right\}\right]^{-1}. \quad (9)$$

The corresponding probability density function* (PDF) is

$$f_X(x) = \beta^{-1}\left[\exp\{-(x-\alpha)/\beta\}\right] \times \left[1 + \exp\{-(x-\alpha)/\beta\}\right]^{-2}. \quad (10)$$

This distribution is sometimes called the *sech-square(d) distribution* (see Harkness and Harkness [23]). Using (10), it can be shown that (see Johnson and Kotz [25, Chap. 22])

$$E(X) = \alpha; \operatorname{var}(X) = \beta^2\pi^2/3.$$

The probability density function of X can be expressed in terms of $E(X) = \mu$ and $\operatorname{var}(X) = \sigma^2$ as

$$f_X(x) = \left(\pi/(\sigma\sqrt{3})\right)e^{-\pi(x-\mu)/(\sigma\sqrt{3})} \times \left[1 + e^{-\pi(x-\mu)/(\sigma\sqrt{3})}\right]^{-2}, \quad (11)$$

where $-\infty < x < \infty$, $-\infty < \mu < \infty$, $\sigma > 0$.

Gumbel [15] has shown that the asymptotic distribution of the midrange* of exponential-type initial distributions is logistic. Dubey [12] proved that the logistic distribution defined by (3) is a compound extreme-value distribution* with an exponential distribution as a compounder. Gumbel and Keeney [19] showed that a logistic distribution is obtained as the limiting distribution of an appropriate multiple of the "extremal quotient," i.e., (largest value)/(smallest value). Talacko [37] has shown that the logistic is the limiting distribution (as $n \to \infty$) of the standardized variable corresponding to $\sum_{j=1}^{n} j^{-1}X_j$, where the X_j's are independent random variables each having a type I extreme-value distribution. Galambos and Kotz [14] give the characterization of the logistic distribution.

ORDER STATISTICS

Let $X_{1:n} \leqslant X_{2:n} \leqslant \cdots \leqslant X_{n:n}$ denote the order statistics* of a sample of size n from the standard logistic distribution defined by the density function (8). The distribution of the rth order statistics $X_{r:n}$ (see Gupta and Shah [21]), is

$$h_r(x) = \frac{1}{\beta(r, n-r+1)} \frac{ge^{-g(n-r+1)x}}{(1+e^{-gx})^{n+1}}, \quad -\infty < x < \infty,$$

and the CDF of $X_{r:n}$ is

$$H_r(x) = \frac{1}{\beta(r, n-r+1)} \sum_{j=0}^{n-r} (-1)^j \binom{n-r}{j} \times \left[(j+r)(1+e^{-gx})^{j+r}\right]^{-1}. \quad (12)$$

The percentage points of the rth order statistic in a sample of size n from the standard logistic distribution (8) for $n = 1(1)(25)$ have been computed by Gupta and Shah [21].

The MGF of the rth order statistic (see Birnbaum and Dudman [8], Plackett [33], and Gupta and Shah [21]) is

$$M_r(t) = \frac{\beta(r + t/g, n - r + 1 - t/g)}{\beta(r, n-r+1)}. \quad (13)$$

A number of recurrence relations exist among the moments of order statistics and for details the reader is referred to Shah [35]. Exact moments of the rth order statistic in a sample of size n from the logistic distribution (8) are tabulated by Gupta and Shah

[21] for $n = 1(1)(10)$ and the covariances of the rth and sth order statistics also have been computed by Shah [35], for $n = 1(1)(25)$.

The kth cumulant of the rth order statistic is (see Gupta and Shah [21] and Birnbaum and Dudman [8])

$$K_k(r,n) = \left[(k-1)!(-1)^k/g^k\right]\left[\sum_{j=1}^{\infty}(j+r-1)^{-k} + (-1)^k\sum_{j=1}^{\infty}(j+n-r)^{-k}\right], \qquad k \geqslant 2,$$

$$K_1(r,n) = -\frac{1}{g}\left[\frac{1}{r} + \frac{1}{r+1} + \cdots + \frac{1}{n-r}\right], \qquad n - r \geqslant r - 1.$$

Plackett [32] has evaluated $K_k(r,n)$ for $k = 1,2,3,4$ when $n - r < r - 1$.

The moment generating function, variance, and the cumulant generating function* of the median ($X_{(k+1)}$) from a sample of size $n = 2k + 1$ have been derived by Tarter and Clark [39]. The relative efficiency of the median in estimating the mean of the logistic distribution has been compared with that in the normal distribution and it was found that it is greater in the logistic distribution.

The distribution of the range* from the standard logistic distribution (8) (see Gupta and Shah [21]), is

$$P(w) = n\sum_{j=0}^{n-1}\binom{n-1}{j}(-1)^j a^{j+1}A(j,n), \qquad 0 < w < \infty,$$

where

$$A(j,n) = -\frac{1}{(1-a)^n}\left[\binom{n-1}{j+1}(-1)^{j+1}\log a + \sum_{\alpha \neq j+1}^{n-1}\binom{n-1}{\alpha}(-1)^{\alpha}\frac{a^{\alpha-j-1}-1}{\alpha-j-1}\right]$$

and $a = e^{-gw}$. Malik [28] obtained an exact formula for the cumulative distribution function of the rth quasi-range.

ESTIMATION

A great variety of methods have been used to estimate the parameters of the logistic distribution. Pearl and Reed [31] use a Taylor series expansion to obtain a least squares* solution by successive approximations. Berkson [6] used the method of maximum likelihood* for estimation for complete samples. A major difficulty with the method of maximum likelihood is the inability to solve explicitly for the estimates. The use of iterative numerical methods and extensive computer work is generally required. Plackett [32] used the likelihood equations for doubly censored samples and used a Taylor's series expansion to obtain linearized maximum likelihood estimators for samples which have been singly censored from above. Berkson and Hodges [7] found a minimax* estimator for the logistic function. Gupta et al. [22] computed the best linear unbiased estimates of μ and σ with a minimum variance based on ordered observations for complete samples as well as censored samples from the logistic distribution (11) for $n = 2(1)25$. Chernoff et al. [11] provided weight functions to compute asymptotically efficient linear estimators from complete logistic samples. Harter and Moore [24] used iterative procedures for solving the likelihood equations for doubly censored samples. Schafer and Sheffield [34] studied the moment estimators.

The maximum likelihood estimators of μ and σ in (11) based on a mutually independent set of random variables $X_1, X_2, \ldots, X_n$ each having the distribution (11) satisfy the equations

$$n^{-1}\sum_{i=1}^{n}\left[1 + e^{\pi(x_i-\hat{\mu})/(\hat{\sigma}\sqrt{3})}\right]^{-1} = \frac{1}{2}, \qquad (14)$$

$$n^{-1}\sum_{i=1}^{n}\left(\frac{x_i-\hat{\mu}}{\hat{\sigma}}\right)\frac{1-e^{\pi(x_i-\hat{\mu})/(\hat{\sigma}\sqrt{3})}}{1+e^{\pi(x_i-\hat{\mu})/(\hat{\sigma}\sqrt{3})}} = \frac{\sqrt{3}}{\pi}. \qquad (15)$$

Equations (14) and (15) must be solved by trial and error.

For large n:

$$n\,\mathrm{var}(\hat{\mu}) \doteqdot (9/\pi^2)\sigma^2 \doteqdot 0.91189\sigma^2$$

$$n\,\mathrm{var}(\hat{\sigma}) \doteqdot 9/(3+\pi^2)\sigma^2 \doteqdot 0.69932\sigma^2.$$

The parameters μ and σ may also be estimated by taking the sample mean $\overline{X}$ and the sample standard deviation S as the initial values of $\hat{\mu}$ and $\hat{\sigma}$, respectively, and then solving the likehood equations (14) and (15) by iterative proceedures. The asymptotic efficiency of $\overline{X}$ is 91.2%; that of S is 87.4%. Gupta and Shah [21] show that the actual efficiency of $\overline{X}$, as the estimator of μ, and of S, as an estimator of σ, is greater than the asymptotic efficiency when the sample size is small. These estimators are, however, less efficient than the best linear unbiased estimators.

MULTIVARIATE LOGISTIC DISTRIBUTIONS

Gumbel [18] extended the logistic distribution to the bivariate case. He proposed two types of bivariate logistic distributions for which the marginals are also logistic. The first type is defined by

$$F(x,y) = (1 + e^{-x} + e^{-y})^{-1},$$
$$-\infty < x < \infty, \quad -\infty < y < \infty \tag{16}$$

or

$$f(x,y) = e^{-(x+y)}/(1 + e^{-x} + e^{-y})^2,$$
$$-\infty < x < \infty, \quad -\infty < y < \infty. \tag{17}$$

Since the function (17) cannot be split into the product of the marginals

$$f(x) = e^{-x}/(1 + e^{-x})^2,$$
$$f(y) = e^{-y}/(1 + e^{-y})^2,$$

the variables X and Y are not independent. The MGF of the distribution (17) is

$$M_{X,Y}(t_1, t_2) = \Gamma(1 - t_1)\Gamma(1 - t_2) \times \Gamma(1 + t_1 + t_2). \tag{18}$$

The correlation coefficient between X and Y, denoted by ρ, has the fixed value $\frac{1}{2}$.

The largest observations of the components X and Y, following a bivariate logistic distribution (17), are asymptotically independent. But the smallest values are not asymptotically independent and this is due to the asymmetry of the bivariate logistic distribution (17).

A second type of bivariate logistic distribution, defined by Gumbel [18], is

$$F(x,y) = (1 + e^{-x})^{-1}(1 + e^{-y})^{-1} \times \left[1 + \frac{\alpha e^{-(x+y)}}{(1 + e^{-x})(1 + e^{-y})}\right],$$
$$|x| < \infty, \quad |y| < \infty, \quad |\alpha| \leqslant 1, \tag{19}$$

or

$$f(x,y) = \frac{e^{-(x+y)}}{(1 + e^{-x})^2(1 + e^{-y})^2} \times \left[1 + \frac{\alpha(1 - e^{-x})(1 - e^{-y})}{(1 + e^{-x})(1 + e^{-y})}\right]. \tag{20}$$

The coefficient of correlation between X and Y is linked to the parameter α by the relation

$$\rho = 3\alpha/\pi^2. \tag{21}$$

For $\alpha \neq 0$ the extreme values are asymptotically independent.

Malik and Abraham [29] derived a p-variate logistic distribution. The joint distribution function of $X_1, X_2, \ldots, X_p$ is

$$F(x_1, \ldots, x_p) = \left\{1 + \sum_{k=1}^{p} \exp(-x_k)\right\}^{-1},$$

with density function

$$f(x_1, \ldots, x_p) = p!\exp\left\{-\sum_{k=1}^{p} x_k\right\} \times \left[1 + \sum_{k=1}^{p} \exp(-x_k)\right]^{-(p+1)},$$
$$-\infty < x_k < \infty, \quad k = 1, 2, \ldots, p. \tag{22}$$

The density function (23) is defined as the p-variate multivariate logistic density. When $p = 2$ we obtain the bivariate logistic distribution obtained by Gumbel [18]. Gumbel's [18] bivariate logistic distribution of type II may also be extended to a p-variate case. A

p-variate distribution with logistic marginal CDFs $F(x_1), \ldots, F(x_p)$ may be obtained (see Malik and Abraham [29]) from a system

$$F(\mathbf{x}) =$$

$$\prod F(x_i)\Big[1 + \sum b_{12}\bar{F}(x_1)\bar{F}(x_2) + \sum b_{123}\bar{F}(x_1)\bar{F}(x_2)\bar{F}(x_3) + \cdots + b_{123\ldots p}\bar{F}(x_1)\bar{F}(x_2)\ldots\bar{F}(x_p)\Big], \quad (23)$$

where $\bar{F} = 1 - F$ and $\sum b_{12} = \sum_{i<j} b_{ij}$, etc. The system (23) seems to be Johnson and Kotz's [26, 27] multivariate *F-G-M* distribution (*see* FARLIE–GUMBEL–MORGENSTERN DISTRIBUTIONS).

TABLES

Values of $f_X(x)$ and $F_X(x)$ as given by (3) and (4) are included in the collection of tables by Owen [30]. For many calculations associated with the use of the logistic distribution, standard tables of hyperbolic functions (sinh, cosh, tanh, etc.) can be used with facility.

References

[1] Abraham, B. (1970). Study of Some Distributional Problems from a Logistic Distribution. M.Sc. thesis, University of Guelph, Guelph, Ontario.

[2] Berkson, J. (1944). *J. Amer. Statist. Ass.*, **39**, 357–365.

[3] Berkson, J. (1950). *Biometrics*, **6**, 327–339.

[4] Berkson, J. (1953). *J. Amer. Statist. Ass.*, **48**, 565–599. (A simple method of estimating the bioassay and quantal response based on the logistic function.)

[5] Berkson, J. (1955). *J. Amer. Statist. Ass.*, **50**, 130–162.

[6] Berkson, J. (1957). *Biometrics*, **13**, 28–34.

[7] Berkson, J. and Hodges, J. L. (1961). *Proc. 4th Berkeley Symp. Math. Statist. Prob.*, Vol. 4. University of California Press, Berkeley, Calif., pp. 77–86.

[8] Birnbaum, A. and Dudman, J. (1963). *Ann. Math. Statist.*, **34**, 658–663.

[9] Chan, L. K. Chan, N. N., and Mead, E. R. (1971). *J. Amer. Statist. Ass.*, **66**, 889–892.

[10] Chan, L. K. and Cheng, S. W. (1974). *J. Amer. Statist. Ass.*, **69**, 1027–1030.

[11] Chernoff, H., Gastwirth, J., and Johns, M. (1967). *Ann. Math. Statist.*, **38**, 52–72.

[12] Dubey, S. D. (1969). *Naval Res. Logist. Quart.*, **16**, 37–40.

[13] Engelhardt, M. (1970). *J. Amer. Statist. Ass.*, **70**, 899–902.

[14] Galambos, J. and Kotz, S. (1978). Characterizations of Probability Distributions. *Lect. Notes Math.*, **675**. Springer-Verlag, Berlin.

[15] Gumbel, E. J. (1958). *Statistics of Extremes*. Columbia University Press, New York. (The classical reference for extreme-value distributions.)

[16] Gumbel, E. J. (1958). Multivariate Distributions with Given Margins. *Tech. Rep.*, Columbia University, New York.

[17] Gumbel, E. J. (1960). *Ann. Math. Statist.*, **31**, 1210.

[18] Gumbel, E. J. (1961). *J. Amer. Statist. Ass.*, **56**, 335–349.

[19] Gumbel, E. J. and Keeney, R. D. (1950). *Ann. Math. Statist.*, **21**, 523–538.

[20] Gupta, S. S. and Gnanadesikan, M. (1966). *Biometrika*, **53**, 656–670.

[21] Gupta, S. S. and Shah, B. K. (1965). *Ann. Math. Statist.*, **36**, 907–920.

[22] Gupta, S. S., Qureishi, A. S., and Shah, B. K. (1967). *Technometrics*, **9**, 43–56.

[23] Harkness, W. L. and Harkness, M. L. (1968). *J. Amer. Statist. Ass.*, **63**, 329–337.

[24] Harter, H. L. and Moore, A. H. (1967). *J. Amer. Statist. Ass.*, **62**, 675–683.

[25] Johnson, N. L. and Kotz, S. (1970). *Distributions in Statistics: Continuous Univariate Distributions*, Vol. 2. Wiley, New York. (An excellent treatment of the logistic distribution is given in Chap. 22.)

[26] Johnson, N. L. and Kotz, S. (1975). *Commun. Statist.*, **4**, 414–428.

[27] Johnson, N. L. and Kotz, S. (1977). *Commun. Statist.*, **6**, 485–496.

[28] Malik, H. J. (1980). *Commun. Statist.*, **9**, 1527–1534.

[29] Malik, H. J. and Abraham, B. (1973). *Ann. Statist.*, **1**, 588–590.

[30] Owen, D. B. (1962). *Handbook of Statistical Tables*. Addison-Wesley, Reading, Mass.

[31] Pearl, R. and Reed, L. J. (1920). *Proc. Nat. Acad. Sci. USA*, **6**, 275–288.

[32] Plackett, R. L. (1958). *Ann. Math. Statist.*, **29**, 131–142.

[33] Plackett, R. L. (1959). *Technometrics*, **1**, 9–19.

[34] Schafer, R. E. and Sheffield, T. S. (1973). *Biometrics*, **29**, 449–455.

[35] Shah, B. K. (1966). *Ann. Math. Statist.*, **37**, 1002–1010.

[36] Shah, B. K. (1970). *Ann. Math. Statist.*, **41**, 2150–2152.

[37] Talacko, J. (1956). *Trab. Estadist.*, **7**, 159–174.

[38] Tarter, M. E. (1966). *J. Amer. Statist. Ass.*, **66**, 514–525.

[39] Tarter, M. E. and Clark, V. A. (1965). *Ann. Math. Statist.*, **36**, 1779–1786.

[40] Tiku, M. L. (1968). *Aust. J. Statist.*, **10**, 64–74.

[41] Tsutakawa, R. K. (1972). *J. Amer. Statist. Ass.*, **67**, 864–885.

[42] Verhulst, P. F. (1945). *Acad. Brux.*, **18**, 1–38.

(EXTREME-VALUE DISTRIBUTIONS)

H. J. MALIK

LOGISTIC REGRESSION

Logistic regression is a statistical method for analysis of the relationship between an observed proportion or rate and a set of explanatory variables. It is based upon the fitting of the linear logistic model

$$\pi(\mathbf{x}) = \{1 + \exp(-\eta - \mathbf{x}'\boldsymbol{\beta})\}^{-1}, \quad (1)$$

where $\pi(\mathbf{x})$ is the expected value of a randomly obtained proportion $p(\mathbf{x})$ from the subpopulation corresponding to the vector $\mathbf{x} = (x_1, x_2, \dots, x_t)'$ of t explanatory variables; and η and $\boldsymbol{\beta}$ are, respectively, the unknown constant term and the vector of t regression coefficients to be estimated. Also, $\boldsymbol{\beta}$ can be interpreted as a set of measures for the extent to which the *logit* transformation*

$$\psi(\mathbf{x}) = \log_e\{\pi(\mathbf{x})/[1 - \pi(\mathbf{x})]\} = \eta + \mathbf{x}'\boldsymbol{\beta} \quad (2)$$

increases (or decreases) relative to $\mathbf{x}$. An important property of the model (1) is that the range of possible values for η and the β_k is unrestricted, since all values in $(-\infty, \infty)$ yield $\pi(\mathbf{x})$ in the $(0, 1)$ interval for all $\mathbf{x}$. Otherwise, the logistic model relationship for $\pi(\mathbf{x})$ in (1) is S-shaped with respect to increasing $\mathbf{x}'\boldsymbol{\beta}$ in $(-\infty, \infty)$, and is essentially linear for $\mathbf{x}$ such that $0.2 \leqslant \pi(\mathbf{x}) \leqslant 0.8$.

Most applications of logistic regression are concerned with situations where the observed proportions $p(\mathbf{x})$ pertain to a binary response variable from a sampling process for which the product binomial probability distribution can be assumed; e.g., stratified simple random sampling, the strata being the distinct vectors $\mathbf{x}$. These include *quantal bioassay experiments* concerned with dose–response relationships (see Finney [17] and BIOASSAY, STATISTICAL METHODS IN), *paired comparison* studies* involving *Bradley-Terry models** [8], and epidemiologic investigations concerned with certain health outcomes ([2], [9], [26]). Also, since logistic models are a special case of log-linear models (*see* CONTINGENCY TABLES), they can be used for the analysis of binary* response data from general multiway cross-classifications; for further discussion here, see refs. 1, 6, 7, 11, 16, 18, 20, 25, 31, and 33.

When the product binomial distribution is assumed for binary response data, the parameters η and $\boldsymbol{\beta}$ in (1) are usually estimated by maximum likelihood* methods as discussed by Walker and Duncan [40] and the previously cited references for log-linear models. More specifically, let $i = 1, 2, \dots, s$, index the set of subpopulations that corresponds to the distinct vectors $\mathbf{x}_i = (x_{i1}, x_{i2}, \dots, x_{it})'$ of the t explanatory variables where $t < s$; and let $\mathbf{n}_i = (n_{i1}, n_{i2})'$ denote the sample distribution of the $n_{i+} = (n_{i1} + n_{i2})$ observations from the ith subpopulation. The product binomial likelihood for the data can then be written as

$$\phi(\mathbf{n} \mid \boldsymbol{\pi}) = \prod_{i=1}^{s} \{n_{i+}!\pi_i^{n_{i1}}(1 - \pi_i)^{n_{i2}}/(n_{i1}!n_{i2}!)\}, \quad (3)$$

where the $\pi_i = \pi(\mathbf{x}_i)$ denote the respective probabilities of the first category of the binary response for randomly obtained observations from the subpopulation corresponding to $\mathbf{x}_i$; $\mathbf{n} = (\mathbf{n}_1', \mathbf{n}_2', \dots, \mathbf{n}_s')'$ denotes the concatenated vector of all frequencies of the two levels of the binary response across the s subpopulations, and $\boldsymbol{\pi} = (\pi_1, \pi_2, \dots, \pi_s)'$ is a similarly arranged vector of the $\{\pi_i\}$. The maximum likelihood (ML) estimates $\hat{\boldsymbol{\beta}}_A = (\hat{\eta}, \hat{\boldsymbol{\beta}}')'$ for the linear logistic model (1) can be expressed as the solution of the nonlinear equations obtained from substituting the model counterparts from (1) for the π_i into the likelihood (3) for ϕ, differentiating

$\log_e \phi$ with respect to $\boldsymbol{\beta}_A = (\eta, \boldsymbol{\beta}')'$, and equating the result to 0. After some simplifications, these equations have the form

$$\sum_{i=1}^{s} (n_{i1} - n_{i+}\hat{\pi}_i)\mathbf{x}_{iA} = \mathbf{0}; \tag{4}$$

here $\hat{\pi}_i = \hat{\pi}(\mathbf{x}_i, \hat{\boldsymbol{\beta}}_A) = \{1 + \exp(-\hat{\eta} - \mathbf{x}_i'\hat{\boldsymbol{\beta}})\}^{-1}$is the model-predicted ML estimate of the π_i based on the ML estimates $\hat{\boldsymbol{\beta}}_A$ of $\boldsymbol{\beta}_A$ and $\mathbf{x}_{iA} = [1, \mathbf{x}_i']'$. Since the equations (4) are generally nonlinear, iterative procedures are required for the computation of $\hat{\boldsymbol{\beta}}_A$. For this purpose, one useful approach is the *Newton–Raphson method** (or, effectively, *iterative (re-)weighted least squares**) as described in Nelder and Wedderburn [31]. Its use involves adjusting an lth step estimate $\hat{\boldsymbol{\beta}}_{Al}$ to an $(l+1)$th step estimate $\hat{\boldsymbol{\beta}}_{A,l+1}$ via

$$\hat{\beta}_{A,l+1} = \hat{\boldsymbol{\beta}}_{Al} + [\mathbf{V}(\hat{\boldsymbol{\beta}}_{Al})][\mathbf{F}(\hat{\boldsymbol{\beta}}_{Al})], \tag{5}$$

$$\mathbf{F}(\hat{\boldsymbol{\beta}}_{Al}) = \sum_{i=1}^{s} \{n_{i1} - n_{i+}\hat{\pi}_{i,l}\}\mathbf{x}_{iA},$$

$$\mathbf{V}(\hat{\boldsymbol{\beta}}_{Al}) = \left\{\sum_{i=1}^{s} n_{i+}\hat{\pi}_{i,l}(1 - \hat{\pi}_{i,l})\mathbf{x}_{iA}\mathbf{x}_{iA}'\right\}^{-1};$$

here $\hat{\pi}_{i,l} = \hat{\pi}(\mathbf{x}_i, \hat{\boldsymbol{\beta}}_{A,l})$ and $\mathbf{V}(\hat{\boldsymbol{\beta}}_{Al})$ is the lth step estimate for the asymptotic covariance matrix for $\hat{\boldsymbol{\beta}}_A$. Such adjustments are terminated after a convergence criterion is reached (e.g., maximum distance between two successive sets of values $\leqslant 0.001$) or a specified maximum number of iterations is undertaken (e.g., $l \leqslant 10$). In practice, convergence usually occurs rapidly for situations where the model is appropriate in the sense of providing a good fit to the data, and nonredundant in the sense that the matrices $\{\mathbf{X}_j^*\}$ whose rows are the vectors $[1, \mathbf{x}_i']$ for the subpopulations with $\{n_{ij} \geqslant 1\}$ both have full rank $(1 + t)$ for $j = 1, 2$; see Silvapulle [37] for more formal conditions. Otherwise, the computations in (5) are initiated with a preliminary estimate $\hat{\boldsymbol{\beta}}_{A0}$ which is provided externally. Here, two general methods are potentially applicable and are given in (7) and (8).

If the data structure under analysis primarily corresponds to the responses of individual subjects in the sense that most of the n_{i+} are 1's, then *discriminant function* methods (*see* DISCRIMINANT ANALYSIS), as discussed by Cornfield [10] and Truett *et al.* [38], could be used to obtain $\hat{\boldsymbol{\beta}}_{A0}$. The rationale is that if the conditional distribution of the explanatory variables $\mathbf{x}$ given the jth level of the binary response is multivariate normal, $N(\boldsymbol{\mu}_j, \boldsymbol{\Sigma})$, then *Bayes' theorem** implies that the conditional probability of $j = 1$ given $\mathbf{x}$ has the form shown in (1) with

$$\eta = \log_e\left(\frac{1-\theta}{\theta}\right) - \tfrac{1}{2}(\boldsymbol{\mu}_1 - \boldsymbol{\mu}_2)'\Sigma^{-1}(\boldsymbol{\mu}_1 + \boldsymbol{\mu}_2),$$

$$\boldsymbol{\beta} = \boldsymbol{\Sigma}^{-1}(\boldsymbol{\mu}_1 - \boldsymbol{\mu}_2), \tag{6}$$

where θ is the unconditional probability that $j = 1$. It follows that a preliminary estimate $\hat{\boldsymbol{\beta}}_{A0} = (\hat{\eta}_0, \hat{\boldsymbol{\beta}}_{*0})'$ is

$$\hat{\eta}_0 = -\log_e(n_{+2}/n_{+1}) - \tfrac{1}{2}\hat{\boldsymbol{\beta}}_{*0}'(\bar{\mathbf{x}}_1 + \bar{\mathbf{x}}_2),$$

$$\hat{\boldsymbol{\beta}}_{*0} = \mathbf{S}^{-1}(\bar{\mathbf{x}}_1 - \bar{\mathbf{x}}_2), \tag{7}$$

where $\bar{\mathbf{x}}_j$ is the mean vector for the explanatory variables for the $n_{+j} = \sum_{i=1}^{s} n_{ij}$ subjects with the jth level of the response, and $\mathbf{S}$ is the pooled within-response-level sample covariance matrix. The estimates $\hat{\boldsymbol{\beta}}_{A0}$ from (7) can be used in their own right without iteration when the data have been obtained from a simple random sample of the overall population and the multivariate normal assumptions stated prior to (6) are realistic. Otherwise they can be used to initiate the iterative process in (5); for this, no assumptions concerning $\mathbf{x}$ are required, so implementation with respect to both continuous and discrete explanatory variables is possible. For a comparative discussion of the discriminant function estimates $\hat{\boldsymbol{\beta}}_{A0}$ and the ML estimates $\hat{\boldsymbol{\beta}}_A$, see refs. 14, 21, and 35.

Alternatively, if the data structure under analysis primarily corresponds to groups of subjects with the same vectors of explanatory variables in the sense that many of the n_{i+} are $\geqslant 10$, then the *minimum logit chi-square* method discussed by Berkson [4, 5] (*see* MINIMUM CHI-SQUARE) can be used to obtain preliminary estimates $\bar{\boldsymbol{\beta}}_{A0} = (\bar{\boldsymbol{\eta}}, \bar{\boldsymbol{\beta}}_*')'$ via the weighted least squares computational procedures outlined in Grizzle *et al.* [19]. In this regard, let $f_i = \log_e(n_{i1}/n_{i2})$ and $v_i = (n_{i1}^{-1} + n_{i2}^{-1})$ denote the sample logit and its linear Taylor series (δ-method)-based estimated variance with 0 frequencies being replaced by 0.5. Let $\mathbf{f} = (f_1, f_2, \ldots, f_s)'$ and let $\mathbf{D}_v$ be the diagonal matrix with $\mathbf{v} = (v_1, v_2, \ldots, v_s)'$ on the diagonal. Then the minimum logit chi-square estimates $\bar{\boldsymbol{\beta}}_{A0}$ are obtained via

$$\bar{\boldsymbol{\beta}}_{A0} = \begin{bmatrix} \bar{\eta} \\ \bar{\boldsymbol{\beta}}_* \end{bmatrix} = \left[\mathbf{X}_A'\mathbf{D}_v^{-1}\mathbf{X}_A\right]^{-1}\left[\mathbf{X}_A'\mathbf{D}_v^{-1}\mathbf{f}\right], \tag{8}$$

where $\mathbf{X}_A$ is the matrix whose rows are the vectors $[1, \mathbf{x}_i']$ for the respective subpopulations. For situations involving moderately large samples (e.g., most $n_{ij} \geqslant 5$ and few $\leqslant 3$), the estimates $\bar{\boldsymbol{\beta}}_{A0}$ can be used in their own right for $\boldsymbol{\beta}_A$ by virtue of their asymptotic equivalence to the maximum likelihood estimates $\hat{\boldsymbol{\beta}}_A = (\hat{\eta}, \hat{\boldsymbol{\beta}}')'$. In this regard, the $\bar{\boldsymbol{\beta}}_{A0}$ are *minimum modified chi-square* estimates as discussed in Cramer [12], Neyman [32], and in CHI-SQUARE TESTS. The analogous goodness-of-fit* statistic is the *minimum logit chi-square statistic*

$$\begin{aligned} Q_W &= \left(\mathbf{f} - \mathbf{X}_A\bar{\boldsymbol{\beta}}_{A0}\right)'\mathbf{D}_v^{-1}\left(\mathbf{f} - \mathbf{X}_A\bar{\boldsymbol{\beta}}_{A0}\right) \\ &= \sum_{i=1}^{s}\left(f_i - \bar{\eta} - \mathbf{x}_i'\bar{\boldsymbol{\beta}}_*\right)^2 / v_i, \end{aligned} \tag{9}$$

which has the approximate chi-square distribution with $(s - t - 1)$ degrees of freedom (DF) when the model (1) applies. Also, the statistic Q_W is asymptotically equivalent to the log-likelihood ratio chi-square statistic

$$Q_L = \sum_{i=1}^{s}\sum_{j=1}^{2} 2n_{ij}\left[\log_e(n_{ij}/\hat{m}_{ij})\right], \tag{10}$$

where $\hat{m}_{i1} = n_{i+}\hat{\pi}_i$ and $\hat{m}_{i2} = n_{i+}(1 - \hat{\pi}_i)$ are the ML estimates for the expected values of the n_{ij} under (1); and to the Pearson chi-square statistic

$$Q_P = \sum_{i=1}^{s}\sum_{j=1}^{2}(n_{ij} - \hat{m}_{ij})^2/\hat{m}_{ij}. \tag{11}$$

For large samples, the choice among Q_L, Q_P, and Q_W is mostly a matter of personal preference, since they provide essentially the same information concerning the goodness of fit of the model (1).

When many of the n_{ij} are small, some caution is necessary in the use of Q_L, Q_P, or Q_W and the related estimates $\bar{\boldsymbol{\beta}}_{A0}$ or $\hat{\boldsymbol{\beta}}_A$. In this regard, the estimates $\hat{\boldsymbol{\beta}}_A$ are considered preferable, since they are functions of linear statistics in (4) to which asymptotic arguments are more readily applicable than the sample logits $\mathbf{f}$ in (8). At the same time, the computation of $\bar{\boldsymbol{\beta}}_{A0}$ in (8) is still useful for the purpose of obtaining a straightforward preliminary estimate to initiate the iterative process in (5); also, convergence is usually rapid when this is done. Among Q_L, Q_P, and Q_W, numerical studies (e.g., [30]) seem to suggest that Q_P is preferable in the sense of the applicability of its approximate chi-square distribution with $(s - t - 1)$ DF to a broad range of small sample situations, particularly those with most $\hat{m}_{ij} > 2$ and few < 1.

As noted in reference to (6) and (7), logistic regression is often applied to data for which most of the n_{ij} are 0 or 1. For these situations, a different strategy for the evaluation of goodness of fit is necessary since the asymptotic rationale for Q_L, Q_P, or Q_W in (9) to (11) is not realistic. One useful approach is to verify that various expanded models $[\mathbf{X}_A, \mathbf{W}]$ with full rank $(w + t + 1)$ for w additional explanatory variables $\mathbf{W}$ can be reduced to the model (1) of original interest via corresponding differences of *likelihood ratio* statistics*. Alternatively, as noted in Breslow and Day [9, pp. 205–210] and elsewhere, the significance of the expansion $\mathbf{W}$ can be equivalently assessed without the fitting of $[\mathbf{X}_A, \mathbf{W}]$ by the use of the *score statistic*

$$Q_S = \mathbf{G}'\mathbf{V}_{\mathbf{G}}^{-1}\mathbf{G}, \tag{12}$$

$$\mathbf{G} = \mathbf{W}'\left(\mathbf{n}_{*1} - \hat{\mathbf{m}}_{*1}\right),$$

$$\mathbf{V}_{\mathbf{G}} = \mathbf{W}'\mathbf{D}_{\hat{v}}^{-1}\left[\mathbf{D}_{\hat{v}} - \mathbf{X}_{\mathbf{A}}(\mathbf{X}_{\mathbf{A}}'\mathbf{D}_{\hat{v}}^{-1}\mathbf{X}_{\mathbf{A}})^{-1}\mathbf{X}_{\mathbf{A}}'\right]\mathbf{D}_{\hat{v}}^{-1}\mathbf{W};$$

here $\mathbf{n}_{*1} = (n_{11}, n_{21}, \ldots, n_{s1})'$, $\hat{\mathbf{m}}_{*1} = (\hat{m}_{11}, \hat{m}_{21}, \ldots, \hat{m}_{s1})$, and $D_{\hat{v}}$ is the diagonal matrix of elements $\{\hat{v}_i = [n_{i+}\hat{\pi}_i(1 - \hat{\pi}_i)]^{-1}\}$. Given that $\boldsymbol{\pi}$ is compatible with the model $\mathbf{X}_A$, Q_S has an approximate chi-square distribution with w DF. For additional discussion of methods for assessing the goodness of fit of

Table 1 Numbers of Dead and Alive Animals at 96 Hours, Cross-classified by Operation and Dose

	Fraction of Spleen Removed									
	Sham (None)		One-fourth		One-half		Three-Fourths		Entire Spleen	
Bacterial Dose	Dead	Alive	Dead	Alive	Dead	Alive	Dead	Alive	Dead	Alive
1.2×10^3	0	5	0	5	0	5	0	6	4	2
1.2×10^4	1	4	0	5	1	4	2	4	5	1
1.2×10^5	0	5	2	3	5	1	5	0	4	1
1.2×10^6	0	5	4	2	6	0	5	0	5	0
1.2×10^7	4	2	5	1	4	1	5	0	5	0
1.2×10^8	5	1	5	0	5	0	5	0	5	0

logistic regression models, see refs. 15, 24, 26, and 39.

Aspects of the application of logistic regression to binary response data are illustrated by an example. The data in Table 1 are from an experiment to assess the survival at 96 hours of animals given a bacterial challenge subsequent to the removal of a specific fraction of their spleens. A model of interest for these data is the parallel line logistic with respect to $\log_e$(dose). It can be expressed as

$$\pi(\mathbf{x}_{hi}) = \left\{1 + \exp\left(-\eta - \sum_{k=1}^{4} \tau_k x_{hik} - \xi x_{hi5}\right)\right\}^{-1}, \qquad (13)$$

where $\pi(x_{hi})$ is the probability of death for the animals with operation $h = 1, 2, 3, 4, 5$ and dose $i = 1, 2, 3, 4, 5, 6$; also, η is the reference intercept for a 1 unit dose to the sham operation group; the τ_k are treatment effect parameters relative to indicator variables x_{hik}, which are 1 for the $(k+1)$th operation and 0 otherwise; and ξ is the slope parameter for $x_{hi5} = \log_e[(\text{dose})_i]$. A rationale for the model (13) is the assumption that the animals with the hth operation have a logistic tolerance distribution for the value of the $\log_e$ (dose) which would cause death at equal or larger doses; *see* QUANTAL RESPONSE ANALYSIS for further discussion.

Since many frequencies in Table 1 are small, maximum likelihood methods are used to estimate the parameters. The resulting estimates and their estimated standard errors are given in (14).

Parameter	η	τ_1	τ_2	τ_3	τ_4	ξ
ML estimate	− 10.9	1.9	3.5	4.2	6.1	0.68
Estimated SE	1.9	0.8	0.9	1.0	1.2	0.11

(14)

These were obtained by the Newton–Raphson method* in (5), which was initiated via the minimum logit chi-square estimates (8); related computer programs are documented in refs. 3, 15, and 22. The likelihood ratio and Pearson goodness of fit statistics for the model (13) were $Q_L = 22.62$ and $Q_P = 35.29$ with 24 DF. Although both are nonsignificant at the $\alpha = 0.05$ type I error level, they should be viewed cautiously because many of the estimated expected frequencies are $\leqslant 1$. For this reason, the expanded model with cross-product variables $x_{hi1}x_{hi5}$, $x_{hi2}x_{hi5}$, $x_{hi3}x_{hi5}$, and $x_{hi4}x_{hi5}$ for operation $\times$ [linear $\log_e$(dose)] interaction added to (13) was considered. For it, $Q_{LE} = 15.79$ with 20 DF, and so $(Q_L - Q_{LE}) = 6.83$ with 4 DF is nonsignificant ($\alpha = 0.10$); similarly, from (12), $Q_S = 4.65$ with 4 DF is also supportive and has the computational advantage of not requiring the fitting of the expanded model. Thus the data in Table 1 are judged to be compatible with the model (13). Additional justification for this conclusion is the objective of having a model provide a parsimonious overall representation of the data even though there may be a few worrisome discrepancies between observed frequencies and their estimated expected values.

Since the $\hat{\tau}_k$ in (14) tend to increase in an essentially linear manner with respect to increasing percentages of the amount of spleen removed, it is of interest to test the hypothesis

$$H_0 : (\tau_2 - 2\tau_1) = (\tau_3 - 3\tau_1) = (\tau_4 - 4\tau_1) = 0 \tag{15}$$

by fitting the corresponding reduced model

$$\pi(x_{hi}) = \left\{1 + \exp\left(-\eta - \tau \sum_{k=1}^{4} kx_{hik} - \xi x_{hi5}\right)\right\}^{-1}, \tag{16}$$

where τ is the slope parameter for the combined explanatory variable $(\sum_{k=1}^{4} kx_{hik})$ for the number of quarters of spleen removed. The ML estimates (and their estimated standard errors) for this model are $\hat{\eta} = -10.5$ (1.8), $\hat{\tau} = 1.4(0.3)$, and $\hat{\xi} = 0.68(0.11)$. The likelihood ratio statistic for the model reduction hypothesis H_0 in (15) is $Q_{LC} = 24.06 - 22.62 = 1.44$ with 3 DF, which is nonsignificant ($\alpha = 0.10$). Otherwise, $\exp(\hat{\tau}) = 4.1$ can be interpreted as the extent to which the odds $\pi(\mathbf{x}_{hi})]/[1 - \pi(\mathbf{x}_{hi})]$ of death vs. survival increases per quarter of spleen removed; and $\exp(\hat{\xi}) = 2.0$, as the extent to which it increases per unit of $\log_e$(dose).

For further discussion of alternative methods for testing hypotheses and fitting of reduced models, see Imrey *et al.* [25], and CHI-SQUARE TESTS and CHI-SQUARE TESTS: NUMERICAL EXAMPLES. Additional examples illustrating the application of logistic regression are given in refs. 9, 23, 26, and 27.

Finally, the concepts included in this entry can be extended in several directions. These can be summarized in two ways:

1. Methods for response variables with more than two outcomes which may or may not be ordinally scaled (see refs. 25, 28, and 34).
2. Methods which pertain to a sampling framework different from the product binomial distribution in (3); e.g., conditional logistic regression for *relative risk analyses* of case control studies in epidemiology [9, 26]; models for proportions based on data from cluster samples or to which random effects considerations are applicable [13, 29, 36]; and a variety of heuristic strategies including ordinary least squares with an assumption-free nature for situations where the basic observations are either percentage scores or ratios constrained to the interval (0, 1).

References

[1] Andersen, E. B. (1980). *Discrete Statistical Models with Social Science Applications*. North-Holland, Amsterdam.

[2] Anderson, S., Auquier, A., Hauck, W. W., Oakes, D., Vandaele, W., and Weisberg, H. I. (1980). *Statistical Methods for Comparative Studies*. Wiley, New York.

[3] Baker, R. J. and Nelder, J. A. (1978). *The GLIM System Manual (Release 3)*. The Numerical Algorithms Group/Royal Statistical Society, Oxford.

[4] Berkson, J. (1953). *J. Amer. Statist. Ass.*, **48**, 565–599.

[5] Berkson, J. (1955). *J. Amer. Statist. Ass.*, **50**, 130–162.

[6] Bishop, Y. M. M., Fienberg, S. E., and Holland, P. W. (1975). *Discrete Multivariate Analysis: Theory and Practice*. MIT Press, Cambridge, Mass.

[7] Bock, R. D. (1975). *Multivariate Statistical Methods in Behavioral Research*. McGraw-Hill, New York.

[8] Bradley, R. A. (1976). *Biometrics*, **32**, 213–233.

[9] Breslow, N. E. and Day, N. E. (1980). *Statistical Methods in Cancer Research. Vol. 1: The Analysis of Case-Control Studies*. IARC, Lyon, France.

[10] Cornfield, J. (1962). *Fed. Proc.*, **21**, 58–61.

[11] Cox, D. R. (1970). *The Analysis of Binary Data*. Methuen, London.

[12] Cramer, H. (1946). *Mathematical Methods of Statistics*. Princeton University Press, Princeton, N.J.

[13] Dempster, A. P. and Tomberlin, T. J. (1980). *Proc. 1980 Conf. Census Undercount*, pp. 88–94.

[14] Efron, B. (1975). *J. Amer. Statist. Ass.*, **70**, 892–898.

[15] Engelman, L. (1981). In *BMDP Statistical Software*, W. J. Dixon *et al.*, eds. University of California Press, Los Angeles, Chap. 14.5.

[16] Fienberg, S. E. (1980). *The Analysis of Cross-classified Categorical Data*. (2nd ed.) MIT Press, Cambridge, Mass.

[17] Finney, D. J. (1971). *Statistical Methods in Biological Assay*, 2nd Edition. Hafner, New York.

[18] Gokhale, D. V. and Kullback, S. (1978). *The Information in Contingency Tables*. Dekker, New York.

[19] Grizzle, J. E., Starmer, C. F., and Koch, G. G. (1969). *Biometrics*, **25**, 489–504.

[20] Haberman, S. J. (1978). *Analysis of Qualitative Data*, Vols. 1 and 2. Academic Press, New York.

[21] Halperin, M., Blackwelder, W. C., and Verter, J. I. (1971). *J. Chron. Dis.*, **24**, 125–158.

[22] Harrell, F. E., Jr. (1983). LOGIST. *SUGI Supplemental Library User's Guide*. SAS Institute, Inc., Cary, N.C., pp. 181–202.

[23] Higgins, J. and Koch, G. G. (1977). *Amer. Statist. Ass., Proc. Social Statist. Sect.*, pp. 974–979.

[24] Hosmer, D. W. and Lemeshow, S. (1980). *Commun. Stat. A*, **9**, 1043–1071.

[25] Imrey, P. B., Koch, G. G., Stokes, M. E., and collaborators (1981, 1982). *Int. Statist. Rev.*, **49**, 263–283 and **50**, 35–63.

[26] Kleinbaum, D. G., Kupper, L. L., and Chambless, L. E. (1982). *Commun. in Statist.*, **11**, 485–547.

[27] Koch, G. G., Amara, I. A., Davis, G. W., and Gillings, D. B. (1982). *Biometrics*, **38**, 563–595.

[28] Koch, G. G., Gillings, D. B., and Stokes, M. E. (1980). *Ann. Rev. Public Health*, **1**, 163–225.

[29] Laird, N. (1975). Ph.D. dissertation, Department of Statistics, Harvard University, Cambridge, Mass.

[30] Larntz, K. (1978). *J. Amer. Statist. Ass.*, **73**, 253–263.

[31] Nelder, J. A. and Wedderburn, R. W. (1972). *J. R. Statist. Soc. A*, **135**, 370–384.

[32] Neyman, J. (1949). *Proc. First Berkeley Symp. Math. Statist. Prob.*, J. Neyman, ed. University of California Press, Berkeley, pp. 230–273.

[33] Plackett, R. L. (1981). *The Analysis of Categorical Data*, 2nd ed. Charles Griffin, London.

[34] Prentice, R. L. and Pyke, R. (1979). *Biometrika*, **66**, 403–411.

[35] Press, S. J. and Wilson, S. (1978). *J. Amer. Statist. Ass.*, **73**, 699–705.

[36] Segreti, A. C. and Munson, A. E. (1981). *Biometrics*, **37**, 153–156.

[37] Silvapulle, M. J. (1981). *J. R. Statist. Soc. B*, **43**, 310–313.

[38] Truett, J., Cornfield, J., and Kannel, W. (1967). *J. Chronic Dis.*, **20**, 511–524.

[39] Tsiatis, A. A. (1980). *Biometrika*, **67**, 250–251.

[40] Walker, S. H. and Duncan, D. B. (1967). *Biometrika*, **54**, 167–179.

Acknowledgment

This research was supported in part by the U.S. Bureau of the Census (JSA-80-19). The authors would like to thank P. Bradshaw and C. Thomas for providing the data in Table 1.

(BAYES THEOREM
BIOASSAY, STATISTICAL METHODS IN
BRADLEY-TERRY MODEL
CATEGORICAL DATA
CHI-SQUARE TESTS
CHI-SQUARE TESTS: NUMERICAL EXAMPLES
CONTINGENCY TABLES
DISCRIMINANT ANALYSIS
LIKELIHOOD RATIO TESTS
LOGIT
MINIMUM CHI-SQUARE
NEWTON-RAPHSON METHOD
PAIRED COMPARISONS
RELATIVE RISK)

GARY G. KOCH
SUZANNE EDWARDS

LOGIT

If p is a number between zero and 1, then logit (p) is defined as

$$\log(p/(1-p)).$$

See BIOASSAY, STATISTICAL METHODS IN, for applications of this transformation.

(QUANTAL RESPONSE ANALYSIS
TOLERANCE DISTRIBUTION)

LOG-LAPLACE DISTRIBUTION

If a random variable $Z = \gamma + \delta \log X$ has a standard Laplace distribution* with density function $\frac{1}{2}e^{-|z|}$, the random variable $X = \exp((Z-\gamma)/\delta)$ is said to have a log-Laplace distribution with parameters γ and δ. The density function of X is

$$\begin{array}{ll} \frac{1}{2}\delta e^{\gamma}x^{\delta-1} & (0 \leqslant x \leqslant e^{-\gamma/\delta}), \\ \frac{1}{2}\delta e^{-\gamma}x^{-\delta-1} & (x \geqslant e^{-\gamma/\delta}), \end{array}$$

with the corresponding cumulative distribution function

$$\begin{array}{ll} \frac{1}{2}e^{\gamma}x^{\delta} & \text{for } 0 \leqslant x \leqslant e^{-\gamma/\delta} \\ 1-\frac{1}{2}e^{-\gamma}x^{-\delta} & \text{for } x \geqslant e^{-\gamma/\delta}. \end{array}$$

Any power of a log-Laplace variable also has a log-Laplace distribution (see, e.g., Johnson [1]).

The rth moment of X about zero is $[1-(r/\delta)^2]^{-1}\exp(-r\gamma\delta^{-1})$ $(r<\delta)$. The shape of the frequency curve depends on δ, but not on γ. It is unimodal with a cusped mode at $x = \exp(-\gamma/\delta)$. Uppuluri [2] presents a characterization of the distribution based on properties of dose–response curves at low doses.

References

[1] Johnson, N. L. (1954). *Trab. Estadíst.*, **5**, 283–291.

[2] Uppuluri, V. R. R. (1980). *Some Properties of the Log-Laplace Distribution*. Manuscript No. 103, International Summer School on Statistical Distributions in Scientific Work, Trieste, Italy, July–Aug. 1980.

(JOHNSON SYSTEM OF DISTRIBUTIONS
LAPLACE DISTRIBUTION
LOG-NORMAL DISTRIBUTION
TRANSFORMATIONS)

LOG-LINEAR MODELS *See* CONTINGENCY TABLES

LOGNORMAL DISTRIBUTION

If for some value of a, $\ln(X-a)$ has a normal distribution with mean μ and variance σ^2, then X has a lognormal distribution with parameters a, μ, and σ. It is convenient to let $b = \exp(\mu)$ so that a, b, and σ are the location, scale, and shape parameters for this distribution. The probability density function for the lognormal may be written as

$$f_X(x) = \frac{1}{\sigma(x-a)\sqrt{2\pi}} \exp\left[-\frac{1}{2\sigma^2}\left\{\log\left(\frac{x-a}{b}\right)\right\}^2\right], \quad x > a.$$

Hereafter the notation $X \sim \mathrm{LN}(a,b,\sigma)$ will indicate that X has a lognormal distribution with location, scale, and shape parameters a, b, and σ, respectively.

IMPORTANT PROPERTIES

1. If $X_1 \sim \mathrm{LN}(a_1, b_1, \sigma_1)$ and $X_2 \sim \mathrm{LN}(a_2, b_2, \sigma_2)$ are independent random variables, then the product $(x_1 - a_1)(x_2 - a_2) \sim \mathrm{LN}(0, b_1 b_2, \sqrt{\sigma_1^2 + \sigma_2^2})$. This multiplicative property for independent lognormal random variables follows directly from the additive properties for normal random variables.
2. $$\begin{aligned} E(X) &= a + b\exp(\sigma^2/2), \\ \mathrm{var}(X) &= b^2\exp(\sigma^2)\left[\exp(\sigma^2)-1\right], \\ \mathrm{median}(X) &= a+b, \\ \mathrm{mode}(X) &= a + b\left[\exp(-\sigma^2)\right], \\ E\left[(X-a)^r\right] &= b^r\exp(r\sigma^2/2). \end{aligned}$$
3. For very small values of σ (less than 0.3), it is very difficult to distinguish between the lognormal and the normal distributions. (See Klimko, et al. [12] and Kotz [13].)
4. The moment generating function for the lognormal distribution does not exist.
5. The hazard function for the lognormal increases at first and then decreases. (See Bury [5].) It would appear that this would preclude its use in time to failure studies. However, a mixture of heterogeneous items (such as cancer patients) might indeed produce such a hazard function. Oberhoffer, et al. [15] use the

lognormal distribution as a model for the time from diagnosis to death for patients with lymphatic leukemia.

6. The lognormal model has been used as model for the concentrations of air contaminants (see, e.g., the paper by Owen and DeRouen [16] and its references). It is also often used as a model for the distribution of income and wealth. It should be considered as a possible model whenever a model with positive skewness is needed.

ESTIMATION OF PARAMETERS

When the location parameter (a) is known, the methods of statistical inference for the normal are generally applied to the transformed (logs) data. When the location parameter is unknown, the problem is more complicated. Hill [10] notes that the likelihood function is unbounded at $a = \min X_i$ and hence unrestricted maximum likelihood estimation* is not appropriate. He suggests that a local maximum is to be considered and offers a Bayesian argument to justify its use in a problem of interest to him. Harter and Moore [9] also advocate the use of local maximum likelihood estimation for the three parameter lognormal.

The problems of maximum likelihood estimation with the three parameter lognormal are not necessarily overcome by restrictions on the parameter space. It is noted in Klimko et al. [12] that when the third sample moment of the observations was negative they were unable to find a relative maximum for the likelihood function. This condition occurs fairly often with small sample sizes and small values of σ as noted in this article. They suggest that in such cases some other model may be more appropriate since the lognormal distribution always has positive skewness.

In an unpublished paper, Monlezum et al. [14] have shown that, when the shape parameter (σ) is known, the maximum likelihood estimators of a and b always exist and are unique. Thus if either the shape or location parameter is known, maximum likelihood estimation of the unknown parameters is well defined.

Tiku [21] suggests alternatives to the maximum likelihood estimators which may be useful, especially with censored samples. Cohen [6] also suggests a slight modification of the usual maximum likelihood estimators. Cohen and Whitten [7] present several methods for estimation of the parameters in three parameter lognormal models and give the results of a Monte Carlo evaluation of these methods.

Aitchison and Brown [1] describe the early development of the lognormal distribution. Their extensive list of references would be helpful to anyone interested in the early applications. They describe applications in the fields of astronomy, biology, small-particle statistics, economics, sociology, and physical and industrial processes. They give special consideration to the role of the lognormal for the distribution of income.

Sartwell [19] used the lognormal as the distribution function for the period of incubation of certain diseases. Gross and Clark [8] suggest the lognormal distribution for the time to recover from surgery. Boag [4] considers the lognormal for the distribution of survival times for patients with certain types of cancer. Additional references may be found in the books by Bain [2], Patel et al. [17], Johnson and Kotz [11] and Patil et al. [18].

Sums of independent lognormal random variables have received some attention in the literature in recent years. Interested readers are referred to R. Barakat [3] and S. C. Schwartz and Y. S. Yeh [20] for detailed derivations and discussions.

References

[1] Aitchison, J. and Brown, J. A. C. (1957). *The Lognormal Distribution*. Cambridge University Press, Cambridge, England.

[2] Bain, L. J. (1978). *Statistical Analysis of Reliability and Life-Testing Models*. Marcel Dekker, New York.

[3] Barakat, R. (1976). *J. Opt. Soc. Amer.*, **66**, 211–216.

[4] Boag, J. W. (1949). *J. R. Statist. Soc. B*, **15**. (Suggests the use of the log-normal model for the relapse time of cancer patients.)

[5] Bury, K. B. (1975). *Statistical Models in Applied Science*. Wiley, New York.

[6] Cohen, A. C., Jr. (1951). *J. Amer. Statist. Ass.*, **46**, 206–212.

[7] Cohen, A. D., Jr. and Whitten (1980). *J. Amer. Statist. Ass.*, 399–404. (Reviews procedures for estimation of the parameters in the three-parameter log normal.)

[8] Gross, A. T. and Clark, V. A. (1975). *Survival Distributions: Reliability Applications in the Biometrical Sciences*. Wiley, New York.

[9] Harter, H. L. and Moore, A. H. (1966). *J. Amer. Statist. Ass.*, **61**, 842–851. (Provides a good review of the maximum likelihood estimation of the parameters in the three-parameter lognormal. Discusses the asymptotic variances and provides some Monte Carlo evaluations.)

[10] Hill, B. M. (1963). *J. Amer. Statist. Ass.*, **58**, 72–84.

[11] Johnson, N. L. and Kotz, S. (1970). *Distributions in Statistics: Continuous Univariate Distributions*, Vol. 1. Wiley, New York.

[12] Klimko, L. A., Rademaker, A., and Antle, C. E. (1975). *Commun. Statist.*, **4**, 1009–1019.

[13] Kotz, S. (1973). *Commun. Statist.*, **1**, 113–132.

[14] Monlezum, C. J., Antle, C. E., and Klimko, L. A. (1975). Unpublished manuscript. (Concerning maximum likelihood estimation of the parameters in the lognormal model.)

[15] Oberhoffer, G., Schmitz-Draeger, H. G., and Thurn, P. (1957). *Strahlinthuapie*, **108**, 325–355.

[16] Owen, W. J. and DeRouen, T. A. (1980). *Biometrics*, **36**, 707–719. (Suggest and evaluate methods of estimating the mean of a lognormal population from data containing zeros or left-censored data.)

[17] Patel, J. K., Kapadia, C. H., and Owen, D. B. (1976). *Handbook of Statistical Distributions*. Marcel Dekker, New York.

[18] Patil, G. P., Boswell, M. T., and Ratnaparkhi, M. V. (1981). *A Modern Dictionary and Classified Bibliography of Statistical Distributions*. International Co-operative Publishing House, Fairland, Md.

[19] Sartwell, P. E. (1950). *Amer. J. Hyg.*, **51**, 310–318.

[20] Schwartz, S. C. and Yeh, Y. S. (1982). *Bell System Tech. J.*, **61**, 1442–1462.

[21] Tiku, M. L. (1968). *J. Amer. Statist. Ass.*, **63**, 134–140.

(APPROXIMATIONS TO DISTRIBUTIONS
JOHNSON SYSTEM OF DISTRIBUTIONS
NORMAL DISTRIBUTION
QUANTAL RESPONSE ANALYSIS)

CHARLES E. ANTLE

LOG-RANK SCORES, STATISTICS, AND TESTS

Log-rank scores are a set of values which are used in nonparametric test procedures suggested by Mantel [15], Cox [2], and Peto and Peto [20] for comparing the survival experiences of two or more groups in the presence of right censorship. More simply, for data that do not involve any censoring* and can be ranked without ties, they have the form

$$a_{j,N} = 1 - \sum_{k=1}^{j} (N - k + 1)^{-1}$$
$$= 1 - E\{T_{j:N}\}, \qquad (1)$$

where $j = 1, 2, \ldots, N$ indexes the ordering of the observations from smallest to largest and $T_{j:N}$ denotes the jth order statistic from the unit exponential distribution* (i.e., scale parameter $\lambda = 1$); see David [5, p. 39] concerning the equality of the two expressions for the $\{a_{j,N}\}$ in (1). In this setting, the use of the scores $\{(1 - a_{j,N})\}$ has been previously discussed by Savage [23] and Hájek [8] as providing nonparametric tests with good power properties for data from exponential distributions. More generally, such log-rank tests are particularly advantageous for comparing distributions in the *proportional hazards family* (e.g., Weibull distributions with common shape parameter) in the sense of being the *locally* most powerful rank-invariant* (*LMPR*) *tests* against *Lehmann alternatives**. Thus appropriate areas of application include not only survival data, but also data for measures of productivity or work load (e.g., distance covered in exercise studies), time (e.g., sentence length or disposition time in criminal justice studies), extent of activity (e.g., severity of disease in medical studies of pain or injury), or other phenomena involving mostly L-shaped* distributions with positive skewness.

An impression as to why the $\{a_{j,N}\}$ are called log-rank scores can be gained by noting that the series approximation

$$\log_e c = \left(1 + \frac{1}{2} + \frac{1}{3} + \cdots + \frac{1}{c}\right) - 0.5772 - \frac{1}{2c} + O(c^{-2}) \qquad (2)$$

(see Cramer [3, p. 125]) implies their approximate similarity to the quantities

$$a^*_{j,N} = 1 + \log_e\{1 - (j-1)/N\}$$
$$= 1 + \log_e(N - j + 1) - \log_e(N), \quad (3)$$

where the values of $(N - j + 1)$ are the reverse ranks for $j = 1, 2, \ldots, N$. Otherwise, since the transformation $T = -\log_e(1 - P)$ of the random variable P with the uniform distribution on $(0, 1)$ has the unit exponential distribution, it follows that

$$a_{j,N} = 1 + E\{\log_e(1 - P_{j:N})\}, \quad (4)$$

where $P_{j:N}$ denotes the jth order statistic from the unit uniform distribution on $(0, 1)$; so the $\{a^*_{j,N}\}$ are the rank analogs of the $\{a_{j,N}\}$ because they involve the replacement of order statistics with ranks under the logarithm.

The scores $\{a_{j,N}\}$ may be used for the purpose of comparing two or more groups (or subpopulations) via tests of randomness* as discussed in CHI-SQUARE TESTS and RANDOMIZATION TESTS. The resulting *randomization log-rank statistic* Q_N for the comparison of s groups has the general one-way analysis-of-variance form

$$Q_N = [(N-1)/(Nv_N)] \sum_{i=1}^{s} n_i(\bar{a}_i - \mu_N)^2, \quad (5)$$

where $\bar{a}_i = \{\sum_{j=1}^{N} U_{j,i} a_{j,N}/n_i\}$ is the sample mean for the n_i log-rank scores pertaining to the ith group as defined via the indicator random variables

$$U_{j,i} = \begin{cases} 1 & \text{if observation } j \text{ belongs to} \\ & \text{the } i\text{th group,} \\ 0 & \text{otherwise,} \end{cases} \quad (6)$$

with $i = 1, 2, \ldots, s$; $\mu_N = \sum_{j=1}^{N}(a_{j,N}/N) = 0$ and $v_N = \sum_{j=1}^{N}(a^2_{j,N}/N)$ are the pooled groups mean and variance of the $\{a_{j,N}\}$; also, when there is neither censoring nor ties so that (1) applies, $v_N = \{1 - N^{-1}\sum_{j=1}^{N} j^{-1}\}$ as given in Hájek [8, p. 84]. Under the hypothesis H_0 that the s groups are equivalent in the sense of a random distribution of the $\{a_{j,N}\}$ among them, Q_N approximately has the chi-square distribution* with DF $= (s - 1)$ when the sample sizes $n_1, n_2, \ldots, n_s$ are moderately large (e.g., all $n_i \geqslant 10$) via *randomization central limit theory* as discussed in Hájek and Sidak [9, pp. 160–164]. Thus the rejection of H_0 for $Q_N \geqslant \chi^2_{1-\alpha}(s-1)$ constitutes a log-rank test with significance level α. Alternatively, for small-sample situations, the permutation distribution of Q_N relative to the $(N!/\prod_{i=1}^{s} n_i!)$ possible allocations of observations to groups can be used to obtain a log-rank test of H_0 via the exact probability (or p-value*) for possible values at least as large as the observed value. For the cases of $s = 2$ groups and sample sizes $6 \leqslant n_1 \leqslant n_2 \leqslant 10$ with neither censoring nor ties, tables for the exact distribution of the *Savage* (*or exponential scores*) *statistic* $S = n_1(1 - \bar{a}_1)$ are given in Hájek [8, pp. 170–171].

The randomization log-rank statistic (5) can be extended to situations involving a multivariate set of response variables and /or covariables via the methods discussed in CHI-SQUARE TESTS. Similarly, for situations where patients are stratified according to one or more explanatory variables (e.g., indicators for demographic or pretreatment status), average partial association log-rank tests can be formulated via the methods discussed in CHI-SQUARE TESTS. Briefly, these involve quadratic forms* in which the within-stratum sums of log-rank scores are added across strata and are compared relative to the correspondingly summed covariance matrix.

There are two approaches to formulating log-rank scores for data in which the ranking involves ties. One of these is to form the scores (1) as if there were no ties and then to assign to ties the average of the values for the corresponding ranks; i.e., it involves average log ranks. As such, it is appealing for essentially continuous data for which ties are anticipated to be rare by maintaining the connection between log-rank scores and the order statistics of the unit exponential distribution. The other approach to handling ties is oriented more to discrete data and can be interpreted as providing log-rank scores for average ranks; it also indicates the basis for the *conditional log-rank statistic* $Q_{C,N}$ in (14) suggested in Mantel [15]. To see its nature, consider the $(2 \times r)$ contingency table shown in (7) for the comparison of $s = 2$ groups with respect to the r categories of an ordinally scaled response variable.

Group	Response Category						Total
	1	2	3	4	...	r	
1	n_{11}	n_{12}	n_{13}	n_{14}	...	n_{1r}	n_1
2	n_{21}	n_{22}	n_{23}	n_{24}	...	n_{2r}	n_2
Total	n_{+1}	n_{+2}	n_{+3}	n_{+4}	...	n_{+r}	N

(7)

$\{n_{ij}\}$ denotes the number of subjects in the ith treatment group with the jth response category, and the $\{n_{+j} = (n_{1j} + n_{2j})\}$ denote the total number of subjects in the pooled groups with the jth response category. Under the hypothesis H_0 that the $s = 2$ groups are equivalent in the sense of being a random partition of a common population, the $\{n_{ij}\}$ have the multiple hypergeometric distribution* (8) where the $\{n_i\}$ and $\{n_{+j}\}$ are considered fixed (by virtue of either finite population randomization and/or conditional multinomial distribution arguments.)

$$\Pr(\{n_{ij}\} \mid H_0) = \prod_{i=1}^{2} n_i! \prod_{j=1}^{r} n_{+j}! \bigg/ \left(N! \prod_{i=1}^{2} \prod_{j=1}^{r} n_{ij}! \right). \tag{8}$$

Moreover, (8) can be written as the product (9) of $(r-1)$ hypergeometric distributions for (2×2) contingency tables corresponding to the successive columns of (7) vs. the corresponding sums of remaining columns.

$$\Pr(\{n_{ij}\} \mid H_0) = \prod_{j=1}^{(r-1)} \left\{ \frac{n_{1j}^*! n_{2j}^*! n_{+j}! N_{j+1}^*!}{N_j^*! n_{1j}! n_{2j}! n_{1,j+1}^*! n_{2,j+1}^*!} \right\}, \tag{9}$$

where the $N_j^* = \sum_{k=j}^{r} n_{+k}$ and the $n_{ij}^* = \sum_{k=j}^{r} n_{ik}$. From (9), it follows that the conditional expectation of n_{1j} given n_{+j}, N_j^*, and n_{1j}^* and the hypothesis H_0 is

$$m_{1j} = (n_{+j} n_{1j}^* / N_j^*), \tag{10}$$

where the n_{1j}^* are considered fixed in the sense of successively fixing n_{1k} for $k < j$ and then rewriting it as $n_{1j}^* = (n_1 - \sum_{k=1}^{(j-1)} n_{1k})$. Let

$$\begin{aligned} g_{1+} &= \sum_{j=1}^{(r-1)} (n_{1j} - m_{1j}) \\ &= \sum_{j=1}^{r} \left\{ n_{1j} - n_{+j} \left[\sum_{k=j}^{r} n_{1k} \bigg/ \sum_{k=j}^{r} n_{+k} \right] \right\} \\ &= \sum_{j=1}^{r} a_{j,N} n_{1j}. \end{aligned} \tag{11}$$

For survival studies, g_{1+} can be interpreted as the sum of the differences between the observed numbers of deaths or failures in successive time intervals for group 1 and their conditional expected values relative to the corresponding numbers of subjects at risk. The coefficients

$$a_{j,N} = \left\{ 1 - \sum_{k=1}^{j} \left[n_{+k} \bigg/ \sum_{l=k}^{r} n_{+l} \right] \right\} \tag{12}$$

are the log-rank scores for the ordered categories $j = 1, 2, \ldots, r$; if there are no ties so that $r = N$ and $n_{+j} = 1$ for $j = 1, 2, \ldots, N$, it can be verified that the $\{a_{j,N}\}$ in (12) are identical to those from (1). Thus the $\{a_{j,N}\}$ in (12) can be used in the randomization log-rank statistic Q_N in (5) as described previously.

Alternatively, the differences $g_{1j} = (n_{1j} - m_{1j})$ are, under H_0, conditionally uncorrelated with conditional variances

$$v_{1j} = n_{1j}^* n_{2j}^* n_{+j} N_{j+1}^* / \{(N_j^*)^2 (N_j^* - 1)\}. \tag{13}$$

Thus a *conditional log-rank statistic* for testing H_0 relative to (9) is

$$\begin{aligned} Q_{C,N} &= \left(g_{1+}^2 \bigg/ \sum_{j=1}^{(r-1)} v_{1j} \right) \\ &= \left[(n_1 \bar{a}_1)^2 \bigg/ \sum_{j=1}^{(r-1)} v_{1j} \right]. \end{aligned} \tag{14}$$

In Mantel [15], this formulation of $Q_{C,N}$ was expressed as the *Mantel–Haenszel statistic** [17] for the $(r-1)$ conditioned (2×2) contingency tables in (9); as such, it is analogous to the average partial association statistics in CHI-SQUARE TESTS except that it involves pseudo-strata for the successive response categories as opposed to actual strata for subsets of distinct subjects. For moderately large samples, $Q_{C,N}$ has an approximate chi-square distribution with DF = 1. When sample sizes are small (e.g., $n_1, n_2 \leqslant 10$), the exact methods reviewed in Gart [7] for sets of (2×2) contingency tables can be applied via the computer procedures discussed in Thomas [26] and Thomas et al. [27]. Otherwise, for the comparison of s groups, $Q_{C,N}$ can be extended to a test statistic with an approximate chi-square distribution with DF $= (s-1)$ by replacing the g_{1j} and v_{1j} by their vector counterparts for the first $(s-1)$ groups.

For situations such as survival studies, the data are often right-censored in the sense that the observed values are lower bounds for the corresponding true values (which are not observed); e.g., subjects who remain alive throughout a follow-up period have censored values for time to death. Since such censored data* do not have a strict ordinal relationship to each other or all noncensored data, the formulation of log-rank scores in (7) to (11) needs to be modified when they occur. For this purpose, let the $\{n_{ij}\}$ in (7) be written as $n_{ij} = (n_{ij0} + n_{ij1})$, where n_{ij0} corresponds to the number of right-censored observations with true values $\geqslant j$, and n_{ij1} to the number of observed values with j. From arguments similar to those in (8) to (12) for

$$g_{1+r} = \sum_{j=1}^{r} (n_{1j1} - m_{1j1}) = \sum_{j=1}^{r} \sum_{j'=0}^{1} a_{jj',N} n_{1 1j'}, \tag{15}$$

where $m_{ij1} = (n_{1j1} + n_{2j1}) n_{1j}^{*} / N_j^{*}$, it follows that the log-rank scores are

$$a_{j0,N} = -\sum_{k=1}^{j} \left(n_{+k1} \Big/ \sum_{l=k}^{r} n_{+l} \right), \quad \text{where} \quad j = 1, 2, \ldots, r,$$

$$a_{j1,N} = (1 + a_{j0,N}), \quad \text{where} \quad j = 1, 2, \ldots, r, \tag{16}$$

$$a_{r1,N} = a_{r0,N}.$$

A computer algorithm for calculating the $\{a_{jj',N}\}$ is documented in Peto [19]. Given that there is no association between the nature of right censorship and the groups to be compared, these scores can be used in the randomization log-rank statistic Q_N in (5). Similarly, with the modification

$$v_{1j} = n_{1j}^{*} n_{2j}^{*} n_{+j1} (N_{j+1}^{*} + n_{+j0}) / \left\{ (N_j^{*})^2 (N_j^{*} - 1) \right\}, \tag{17}$$

the conditional log-rank statistic $Q_{C,N}$ in (14) is applicable.

Under the assumptions of random and equal censorship and no ties, Q_N can be viewed as a locally most powerful rank-invariant test against Lehmann alternatives; see Peto [18] and Crowley [4] for discussion. Other useful methods for comparisons involving right-censored data are the *Gehan–Gilbert test* statistic* and a modified Wilcoxon procedure described in Mantel [16]. In this regard, the Gehan–Gilbert test is more sensitive to differences between groups for the lower categories, while the log-rank test is more sensitive relative to the higher categories. A statistic utilizing the maximum of the Gehan–Gilbert statistic and the log-rank statistic is described in Tarone [25] and applied to numerical examples. For other discussion of statistical test procedures for censored data, *see* COX'S REGRESSION MODEL and SURVIVAL ANALYSIS; see also references dealing with survival data methodology such as Breslow [1], Cox [2], Elandt-Johnson and Johnson [6], Holford [11], Kalbfleisch and Prentice [12], Laird and Olivier [13], and Lee [14].

Some aspects of the application of log-rank statistics are illustrated through two examples. In this regard, the data in Table 1 pertain to an evaluation of the effectiveness of an active treatment for an overdose of a particular drug relative to a placebo control treatment. They are from a randomized experiment for which a cage of five mice was the primary experimental unit with $n_1 = 8$ cages being assigned to the control treatment and $n_2 = 8$ cages to the active treatment. A response variable of interest for this study is a measure of liver damage known as SGOT, for which the upper limit of the

Table 1 Minimum and Median SGOT Values, Ranks, and Log-rank Scores for Cages of Five Mice in Experiment to Evaluate Treatment for a Type of Drug Overdose

	Observed Values		Ranks		Log-Rank Scores	
Treatment	Minimum SGOT	Median SGOT	Minimum SGOT[a]	Median SGOT	Minimum SGOT	Median SGOT
Control	62	145	2	6	0.871	0.548
Control	62	171	3	8	0.799	0.337
Control	124	1244	10	11	0.069	−0.097
Control	124	2613	11	14	−0.097	−0.881
Control	140	1276	13	12	−0.547	−0.297
Control	149	3873	14	16	−0.881	−2.381
Control	684	1742	15	13	−1.381	−0.547
Control	1555	2800	16	15	−2.381	−1.381
Mean	362.5	1733	10.5	11.9	−0.443	−0.587
Active	25	25	1	1	0.938	0.938
Active	68	105	4	2	0.722	0.871
Active	83	156	5	7	0.639	0.448
Active	93	187	6	9	0.548	0.212
Active	93	482	7	10	0.448	0.069
Active	100	109	8	3	0.337	0.799
Active	112	124	9	4	0.212	0.722
Active	139	139	12	5	−0.297	0.639
Mean	89.1	166	6.5	5.1	0.443	0.587

[a]The ties have been broken here via the ranking for Median SGOT so the tables for exact distributions of Wilcoxon and Savage statistics could be used.

normal range is 40, but very large values are possible. The corresponding data for the mice within each cage have been summarized in terms of the minimum value and the median value. These quantities are given in Table 1 for the $N = (n_1 + n_2) = 16$ cages, together with their ranks and log-rank scores. From this information, it follows that the Wilcoxon rank sum statistics with respect to the control group are $W^{(1)} = 84$ for minimum SGOT and $W^{(2)} = 95$ for median SGOT. The former approaches significance with one-sided exact $p = 0.052$ while the latter is clearly significant with one-sided exact $p = 0.001$. In contrast, the log-rank scores-based Savage statistic provides a stronger result $S^{(1)} = 11.54$ for minimum SGOT with one-sided exact $p \leqslant 0.025$, and a somewhat weaker result $S^{(2)} = 12.70$ for median SGOT with one-sided exact $p \leqslant 0.005$. Thus the log-rank test performs better for minimum SGOT for which there is some overlap between the two groups for smaller values and little overlap for larger values, but the Wilcoxon test performs better for median SGOT for which there is little overlap between the two groups for both smaller and larger values.

The data in Table 2 are from a North Carolina Highway Safety Research Center study concerned with the comparison of driver injury severity in utility vehicles and pickup trucks for the set of all multivehicle accidents in North Carolina during 1973–1978. Since the data are discrete, log-rank scores are obtained via (12); they are shown at the bottom of Table 2. The log-rank statistic is obtained via (5) as $Q_N = 7.07$, which is significant ($p \leqslant 0.01$) relative to the $\chi^2(1)$ distribution. Similarly, the conditional log-rank statistic $Q_{C,N} = 6.82$ is also significant ($p \leqslant 0.01$). The Wilcoxon statistic counter-

Table 2 Frequency Distribution for Driver Injury in Utility Vehicles and Pickup Trucks from the Set of All Multivehicle Accidents, in North Carolina, 1973–1978

	Driver Injury Severity					Total
Vehicle Class	Not Injured	Minor Injury	Moderate Injury	Serious Injury	Fatality	Number of Accidents
Utility vehicles	2708	165	156	48	11	3088
Pickup trucks	59953	3785	2711	973	135	67557
Total	62661	3950	2867	1021	146	70645
$\{\text{Rank}/(N+1)\}$ scores	0.4435	0.9149	0.9632	0.9907	0.9990	
Log-rank scores[a]	0.1130	−0.3817	−1.0924	−1.9673	−2.9673	

[a] Log-rank scores are obtained via (12).

part of the log-rank statistic is obtained via (5) with respect to rank scores as $Q_N = 3.77$, which is almost significant ($p = 0.052$) relative to the $\chi^2(1)$ distribution. Thus the log-rank statistic provides a more sensitive result. In addition, its scores can be interpreted as providing a more appropriate framework for the analysis of injury severity by involving greater distances between the more severe categories than rank scores for which the greater distances are between the less severe categories. Otherwise, it should be noted that the difference between utility vehicles and pickup trucks which has been detected may be due to characteristics of the accident environment, such as speed at impact or impact site, rather than the nature of the vehicles.

References

[1] Breslow, N. E. (1975). *Int. Statist. Rev.*, **43**, 45–58.

[2] Cox, D. R. (1972). *J. R. Statist. Soc. B*, **34**, 187–220.

[3] Cramer, H. (1946). *Mathematical Methods of Statistics*. Princeton University Press, Princeton, N.J.

[4] Crowley, J. (1974). *Biometrika*, **61**, 533–538.

[5] David, H. A. (1970). *Order Statistics*. Wiley, New York.

[6] Elandt-Johnson, R. C. and Johnson, N. L. (1980). *Survival Models and Data Analysis*. Wiley, New York.

[7] Gart, J. E. (1971). *Int. Statist. Rev.*, **39**, 148–161.

[8] Hájek, J. (1969). *Nonparametric Statistics*. Holden-Day, San Francisco, Calif.

[9] Hájek, J. and Sidak, Z. (1967). *Theory of Rank Tests*. Academic Press, New York.

[10] Haybittle, J. L. and Friedman, L. S. (1979). *Statistician*, **28**, 199–208. (London.)

[11] Holford, T. R. (1980). *Biometrics*, **36**, 299–306.

[12] Kalbfleisch, J. D. and Prentice, R. L. (1980). *The Statistical Analysis of Failure Time Data*. Wiley, New York.

[13] Laird, N. and Olivier, D. (1981). *J. Amer. Statist. Ass.*, **76**, 231–240.

[14] Lee, E. T. (1981). *Statistical Methods for Survival Data Analysis*. Lifetime Learning Publications, Belmont, Calif.

[15] Mantel, N. (1966). *Cancer Chemother. Rep.*, **50**, 163–170.

[16] Mantel, N. (1981). *Amer. Statist.*, **35**, 244–247.

[17] Mantel, N. and Haenszel, W. (1959). *J. Natl. Cancer Inst.*, **22**, 719–748.

[18] Peto, R. (1972). *Biometrika*, **59**, 472–474.

[19] Peto, R. (1973). *Appl. Statist.*, **22**, 112–118.

[20] Peto, R. and Peto, J. (1972). *J. R. Statist. Soc. A*, **135**, 185–207.

[21] Prentice, R. L. and Gloecker, L. A. (1978). *Biometrics*, **34**, 57–67.

[22] Prentice, R. L. and Marek, P. (1979). *Biometrics*, **35**, 861–867.

[23] Savage, I. R. (1956). *Ann. Math. Statist.*, **27**, 590–616.

[24] Tarone, R. E. (1975). *Biometrika*, **62**, 679–682.

[25] Tarone, R. E. (1981). *Biometrics*, **37**, 79–85.

[26] Thomas, D. G. (1975). *Computer Biomed. Res.*, **8**, 423–446.

[27] Thomas, D. G., Breslow, N., and Gart, J. J. (1977). *Computer Biomed. Res.*, **10**, 373–381.

Acknowledgment

This research was supported in part by the U.S. Bureau of the Census (JSA-80-19). The authors would like to thank K. H. Donn and G. D. Rudd for providing the data in Table 1, and D. W. Reinfurt for providing the data in Table 2.

(CHI-SQUARE TESTS
CLINICAL TRIALS
COX'S REGRESSION MODEL
DISTRIBUTION-FREE METHODS
MANTEL–HAENSZEL STATISTIC
ORDER STATISTICS
SURVIVAL ANALYSIS)

GARY G. KOCH
P. K. SEN
INGRID AMARA

LOMAX DISTRIBUTION

Also referred to as the Pareto distribution* of the second kind. It is given by the cumulative distribution function

$$F_X(x) = 1 - \frac{K_1}{(x+c)^a}, \qquad K_1 > 0, \quad a > 0.$$

This is a special form of Pearson type VI* distribution.

(PARETO DISTRIBUTION
PEARSON'S DISTRIBUTIONS)

LONGITUDINAL DATA ANALYSIS

DOMAIN OF LONGITUDINAL DATA ANALYSIS

Longitudinal data analysis is a subspecialty of statistics in which individual histories—interpreted as sample paths, or realizations, of a stochastic process—are the primary focus of interest. A wide variety of scientific questions can only be addressed by utilizing longitudinal data together with statistical methods which facilitate the detection and characterization of regularities across multiple individual histories. Some examples are:

1. **Persistence.** Do persons who vote according to their political party identification in one presidential election (e.g., 1956) tend to vote this way in a subsequent election [14]? Do persons or firms who are repeatedly victimized by criminals always tend to be victimized in the same way [15, 35]?
2. **Structure of Individual Time Paths.** Are there simple functions—e.g., low-order polynomials—which characterize individuals' changes in systolic blood pressure with increasing age in various male cohorts [45]? This question arises in the study of factors associated with the onset of coronary heart disease [13].
3. **Interaction among Events.** Are West African villagers infected with one species of malaria parasite (e.g., *Plasmodium falciparum*, Pf) more resistant to subsequent infection with another species (*Plasmodium malariae*, Pm) than if Pf was not already present in their peripheral blood [11, 32]?
4. **Stability of Multivariate Relationships.** For neurologists making prognoses about the recovery of patients from nontraumatic coma, are the same neurological indicators useful at admission to a hospital, 24 hours later, 3 days, and 7 days, or does the list of key prognostic indicators change over time—and in what manner [27]?

In this article we present some examples of analytical strategies that can assist research workers in answering questions such as these. Our aim is to exhibit the general flavor of longitudinal data analysis, as well as to illustrate how the idiosyncrasies of a scientific problem suggest different methodologies.

SOME HISTORY

Statistical methods which are especially suited to the quantitative study of individual histories have their roots in John Graunt's first attempt to construct a life table (*see* LIFE TABLES). However, the development of a diverse set of techniques for measuring the dynamics of vector-valued stochastic processes*—as opposed to a single positive random variable (e.g., waiting time until death) —is a phenomenon of the twentieth century, primarily stimulated by the large longitudinal field studies initiated in the 1920s. Among the most influential of these investigations were:

1. L. M. Terman's follow-up study of California school children who scored in the top 1% of the national IQ distribution. Initiated in 1921, a principal aim of the study was to follow the original sample of 857 boys and 671 girls into adult life to assess whether high IQ was a good predictor of success in later life. For details on the design and early analyses of this data, see Terman et al. [42] and Terman and Oden [41].
2. E. Sydenstricker's Hagerstown morbidity* study, initiated in December 1921, with follow-up* on almost 2000 households through March 1924, had a central aim of assessing sickness incidence over a sufficiently long period of time to distinguish it from sickness prevalence. In this connection, see Sydenstricker [39]. The Hagerstown sample was subsequently followed up in 1941, and some straightforward descriptive analyses suggested the important relation that in this population chronic illness led to poverty rather than the reverse implication.

The natural conceptual framework in which to consider alternative analyses of these and other longitudinal data is the theory of stochastic processes. However, in the 1920s this subject was very much in its infancy, and the first attempts to utilize process models in longitudinal analyses did not occur until the late 1940s and early 1950s. In the social sciences, Tinbergen [43] presented a graphical caricature of a finite-state process with intricate causal relations among the states and across time which triggered P. Lazarsfeld's 1954 utilization of this framework to study voting behavior and opinion change in election campaigns. This work can be viewed as a precursor to Blumen et al.'s [9] classic study of interindustry job mobility utilizing Markov and simple mixtures of Markov models to describe the individual dynamics. For a superb and up-to-date review of subsequent utilization of stochastic process models for the study of intraindividual dynamics in the social sciences, the reader should consult Bartholomew [5]. A useful and comprehensive presentation of longitudinal analysis methods and associated substantive problems in economics are the papers in *Annals de l'INSEE* [4]. Despite this extensive development most longitudinal analyses in the social sciences from the 1920s through the present involve the estimation and interpretation of correlation coefficients in linear models relating a multiplicity of variables to each other and to change over time. For critiques of this technology, see Rogosa [36] and Karlin et al. [22]. For an imaginative application, see Kohn and Schooler [25]. A balanced appraisal of path analysis* by its originator and insights appear in Wright [50].

In medicine and public health, the Hagerstown morbidity study anticipated the more recent major longitudinal data collections such as the Framingham study* of atherosclerotic disease [13], the University Group Diabetes Program evaluations of oral hypoglycaemic agents [18], and the World Health Organization field surveys of malaria in Nigeria [31]. Analyses exploiting the availability of individual histories, or portions of them, in these and other studies have primarily been of two types:

1. Estimation of age-dependent incidence rates; i.e., the expected number of occurrences of a given event per unit time per individual at risk of the event (see, e.g., Bekessey et al. [6].
2. Survival analysis*, where the waiting time until occurrence of an event is the

primary dependent variable of interest. For details on the analysis of survival data, which can be viewed as representing the duration of one episode in a single state in what may be a multistate stochastic process; *see also* COMPETING RISKS. Insightful examples appear in Crowley and Hu [12] and Menken et al. [30].

Analyses where the modeling of continuous functions of time are of interest, or where the goal is to characterize transition rates between discrete states and assess the influence of a variety of covariates on them, are of relatively recent vintage. Particularly important in facilitating such analyses is the literature on growth curves*—e.g., Rao [34] and the references in Ware and Wu [45], Foulkes and Davis [17], and McMahon [29] —and the recently developed nonparametric methods for the analysis of counting processes (see, e.g., Aalen et al. [2].

DESIGNS AND THEIR IMPLICATIONS FOR ANALYSIS

In empirical applications, testing whether specific classes of processes describe the occurrence of events or the evolution of a continuous variable is best facilitated by observing, in full, many realizations of the underlying process $X(t)$ for all t in an interval $[T_1, T_2]$. Examples of such data are the work histories in the Seattle and Denver Income Maintenance Experiments [44], the fertility histories in the Taichung IUD experiment [28], and the job vacancy histories for ministers in Episcopalian churches in New England [46]. In most substantive contexts, however, ascertaining the exact timing of each occurrence of an event for each individual is either impossible, economically infeasible, or both. Observations usually contain gaps and censoring* relative to a continuously evolving process. Some examples of this situation are:

1. In the Framingham study of atherosclerotic disease, individuals were examined once every two years at which times symptoms of illness, hospitalizations, or other events occurring between examinations were recorded (retrospective information). In addition, a physical examination, some blood studies, and other laboratory work (current information) were completed. One topic of considerable interest is the intraindividual dynamics of systolic blood pressure. This is a continuous-time and continuous-state process which can only be modeled using the biennial samples, i.e., measurements made at the examinations. Such data represent fragmentary information about the underlying process. Lazarsfeld and Fiske [26] introduced the terminology, "panel" study, to refer to this kind of data collection*. An associated body of statistical techniques is frequently referred to as *methods of panel analysis*.
2. The WHO field surveys of malaria in Nigeria—the Garki Project [31]—involved the collection of a thick blood film every 10 weeks for $1\frac{1}{2}$ years from individuals in eight village clusters. The blood films were examined for presence or absence of any one or more of three species of malaria parasite together with an estimate of the density of *P. falciparum* parasites if they were present. Data on the mosquito vectors, including person-biting rates measured via human bait, was collected in some of the villages every five weeks in the dry season and every two weeks in the wet season. Thus the dynamics of intraindividual infection and of parasite transmission between humans and mosquitos can only be modeled from partial information about a continuously evolving process.
3. In Taeuber et al.'s [40] residence history study, observations are taken retrospectively on current residence, first and second prior residence, and birthplace of individuals in particular age cohorts. Analyses in which duration of residence is a dependent variable of interest must accommodate censoring on the right for

> current residence. Furthermore, characterization of the pattern of adult residence histories is complicated by the fact that initial conditions are unknown for persons who have occupied more than three residences beyond, for example, age 18.

A feature of modeling with such fragmentary data is that algebraic characterizations of the data sets which can possibly be generated by given continuous-time models are frequently very difficult to obtain. On the other hand, these characterizations are, of necessity, the basis of tests for compatibility of the data with proposed models. In addition, estimation of quantities such as rates of occurrence of events per individual at risk of the event at a given time is made complicated by the fact that some of the occurrences are unobserved. This necessitates estimation of rates that have meaning within stochastic process models which are found to be compatible with the observed data.

ANALYTICAL STRATEGIES—EXAMPLES

It is our view that the flavor of longitudinal data analysis is best conveyed by a variety of examples. It is to be understood, however, that the issues raised in each example are applicable to a much wider range of studies.

NonParametric Estimation of Integrated Incidence Rates and Assessment of Possible Relationships between Events

Aalen et al. [2] utilize retrospective data on 85 female patients at the Finsen Institute in Copenhagen to assess whether hormonal changes in connection with menopause or similar artifically induced changes in ovarian function might affect the development of a chronic skin disease, *pustulosis palmoplantaris*. They propose a stochastic compartment model of the possible disease dynamics and mortality which can be summarized by the directed graph in Fig. 1.

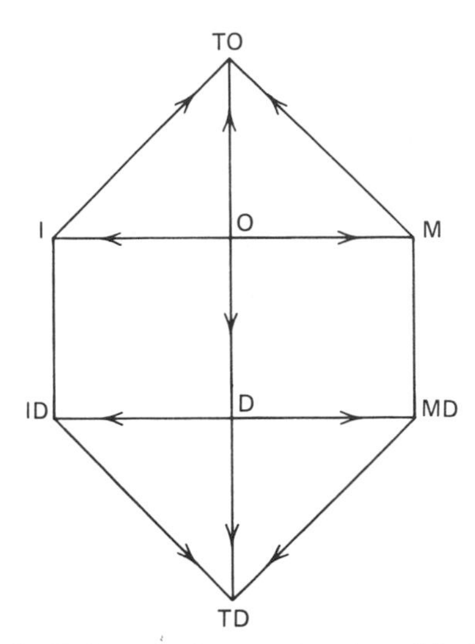

Figure 1 Caricature of compartment model of skin disease, menopause, and mortality. O, no event has occurred; M, natural menopause has occurred; D, disease has been detected; I, induced menopause has occurred; TO, dead without disease; TD, dead with disease.

Since the etiology of *pustulosis palmoplantaris* is unknown, there is no defensible basis for proposing a very restrictive parametric family of stochastic process models as candidates to describe the movement of persons among nodes on the graph in Fig. 1. However, the question of possible influence of natural and induced menopause on the outbreak of disease only requires comparisons of the age-dependent rates of transition per person at risk, $r_{O,D}(t), r_{M,MD}(t), r_{I,ID}(t)$, for the transitions $O \to D$, $M \to MD$, and $I \to ID$, respectively. To this end we first define the integrated rate of transition from a state labeled i to a state labeled j in a general stochastic compartment model as $A_{ij}(s,t) = \int_s^t r_{ij}(u)\,du$. Then we bring in the nonparametric estimator

$$A_{ij}(s,t) = \sum_{k\,:\,s \leqslant t_k^{(i,j)} < t} \left[Y_i\left(t_k^{(i,j)}\right) \right]^{-1}, \quad (1)$$

where $Y_i(t)$ = number of individuals in state i at age t,
$t_k^{(i,j)}$ = age of kth transition from state i to state j with $0 < t_1^{(i,j)} < t_2^{(i,j)} < \cdots$.

As verified in a remarkable paper by Aalen [1], (1) is an unbiased, consistent,

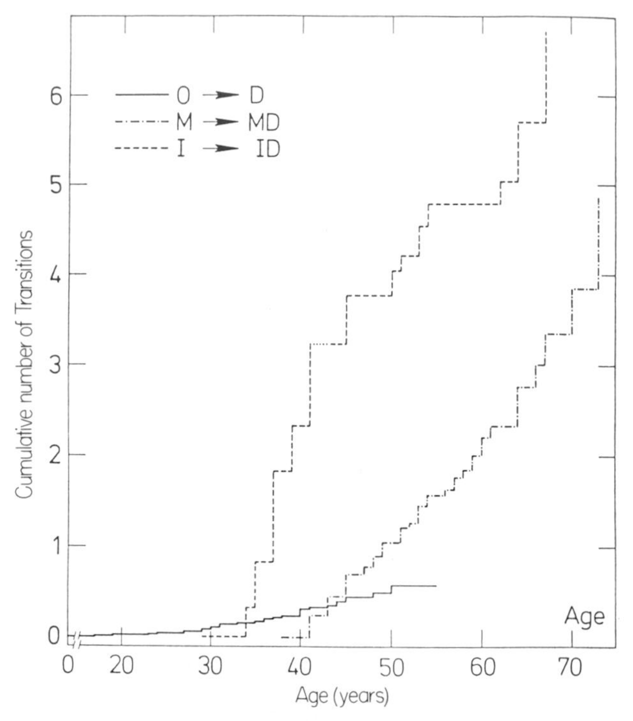

Figure 2 Estimated integrated "conditional" intensities for outbreak of pustulosis palmo-plantaris before and after natural and induced menopause. From Aalen et al. [2].

asymptotically normal estimator of $A_{ij}(s,t)$ for quite general discrete-state, continuous-time stochastic processes. In particular, no strong assumptions about dependencies across time—such as the Markov property—are necessary to validate the desirable statistical properties of (1). A good estimator of var $\hat{A}_{ij}(s,t)$ is given by

$$\sum_{k\,:\,s \leqslant t_k^{(i,j)} < t} \left[Y_i\left(t_k^{(i,j)}\right) \right]^{-2}. \qquad (2)$$

Applying (1) to the Finsen Institute data yields the estimates of the integrated rates $A_{O,D}(0,t)$, $A_{M,MD}(0,t)$ and $A_{I,ID}(0,t)$ shown in Fig. 2.

The graph suggests, and more formal tests (see Aalen et al. [2] for details) support, the conclusion that induced and natural menopause increase the chance of appearance of pustulosis palmo-plantaris. Additional analyses, based on $A_{D,MD}(0,t)$ and $A_{O,M}(0,t)$ utilizing data from the Finsen Institute and a Norwegian study allowing for estimation of the age distribution for natural menopause, indicated that the outbreak of the disease does *not* influence the occurrence of the natural menopause.

The implications of this example for longitudinal data analysis generally are:

1. If the individual level dynamics are interpretable as realizations of a discrete-state, continuous-time stochastic process and there is no well-developed substantive theory to guide the modeling, then estimation of quantities such as integrated transition rates is best carried out using nonparametric methods such as (1) and (2). This at least provides unbiased estimates of basic descriptive quantities—i.e., the integrated rates—which must constrain any subsequent modeling based on proposed subject-matter theory.
2. Graphical displays of integrated rates such as Fig. 2 are a particularly useful guide to parametric specification of the transition rates $r_{ij}(t)$. Figure 2 suggests

that curves of the form ct^{α} with $c > 0$ and $\alpha > -1$ would be appropriate for the Finsen Institute data.

3. If the underlying process is actually Markovian—see Anderson and Goodman [3] for formal tests—then $\{r_{ij}(t)\}$ may be interpreted as entries in a one-parameter family of intensity matrices, $R(t)$, governing the Kolmogorov forward and backward differential equations for the transition probabilities of continuous-time inhomogeneous Markov chains:

$$\frac{\partial P(s,t)}{\partial t} = P(s,t)R(t)$$

$$\frac{\partial P(s,t)}{\partial s} = -R(s)P(s,t),$$

where $P(t,t) = I$ and $P(s,t)$ has entries interpreted as $p_{ij}(s,t) = Pr(X(t) = j \mid X(s) = i)$. Here $(X(t), t \geqslant 0)$ is a realization of the Markov chain and i and j index states.

4. Nonparametric procedures such as (1) and (2) require observations on a process over an interval of time, i.e., knowledge of $(x(t,l),\ 0 \leqslant t \leqslant T_l) =$ (observed realization of the underlying process for individuals labeled $l = 1, 2, \ldots$, etc.). The interval T_l may, in many studies, vary across individuals. Some mild grouping of events—e.g., in the Aalen et al. [2] study, we only know that menopause occurred in some small age range—still allows for reasonable estimates of integrated rates. However, appropriate analogues of these procedures for estimation in panel designs where there are multiple unobserved transitions in a discrete-state, continuous-time process remain to be developed.

Malaria Parasite Interaction in a Common Human Host: An Example of Panel Analysis to Assess Specific Theoretical Proposals

As part of a field study of malaria in Nigeria—see Molineaux and Gramiccia [31]—blood samples from persons in eight clusters of villages were collected on eight occasions, each pair of surveys being separated in time by approximately 10 weeks. For each person at each survey, a blood sample is examined to assess whether that individual is infected with either one or both of two species of malaria parasite called *Plasmodium falciparum* (Pf) and *Plasmodium malariae* (Pm), respectively. We define a four-state stochastic process for an individual's infection status where the states are:

State	Pf	Pm
1	−	−
2	−	+
3	+	−
4	+	+

and (−) means absence of parasites, (+) means presence of parasites.

A question of considerable importance is whether the presence of Pf in an individual's blood makes him (her) more or less resistant to infection by Pm than if there is no Pf. In principle, this question should be answerable using the methods from Aalen's theory of counting processes described in the previous example. Especially, we should first assess whether the possibly time-dependent transition rates per individual at risk of a transition, $r_{12}(t)$ and $r_{34}(t)$, satisfy $r_{12}(t) \equiv r_{34}(t)$. If this null hypothesis is rejected and our estimates suggest that $r_{12}(t) > r_{34}(t)$, then the evidence would favor an interpretation of parasite interaction in which the presence of Pf inhibits the acquisition of Pm. Similarly, a competitive effect would be suggested if $r_{43}(t) > r_{21}(t)$—i.e., the presence of Pf promotes the loss of Pm.

Because data on infection status is collected only every 10 weeks, there can be multiple unobserved transitions among the four states in the continuous-time infection status process. In particular, the times of occurrence of transitions, which are necessary for utilization of (1) and (2), cannot be determined. One response to this problem of missing data relative to a continuously evolving process is to estimate the rates $r_{ij}(t)$ within a class of models which are at least compatible with the observed data. To this end we introduce a time series* of time-homogeneous Markov chains, each of which is a candidate to describe the unobserved dynamics between a pair of successive surveys. These models are constrained by the requirement that the transition probabilities of the chain are—to within sampling variability—coincident with the conditional probabilities $Pr(X(k\Delta) = i_k \mid X((k-1)\Delta) = i_{k-1})$ for $k = 0, 1, 2, \ldots, 7$. Here $X(t), 0 \leqslant t \leqslant 7\Delta$, is the four-state infection status process; $0, \Delta, 2\Delta, \ldots, 7\Delta$ are the survey dates, $\Delta = 10$ weeks, and i_k may be any one of the states $(1, 2, 3, 4)$ in which an individual is observed at time $k\Delta$.

We illustrate this strategy and its implications in detail in a 4×4 table of transition counts from the WHO survey of malaria in Nigeria mentioned above (see Table 1). The entries in the table are denoted by $N_{ij}(3\Delta, 4\Delta) =$ (number of persons in state i at time 3Δ and in state j at time 4Δ)—e.g., $N_{31}(3\Delta, 4\Delta) = 77$. If these observations are generated by a continuous-time homogeneous Markov chain, there must be a 4×4 stochastic matrix $\mathbf{P}(3\Delta, 4\Delta)$ representable as $\mathbf{P}(3\Delta, 4\Delta) = e^{\Delta\mathbf{R}}$ and such that $N_{ij}(3\Delta, 4\Delta) \approx N_{i+}(3\Delta, 4\Delta)(e^{\Delta\mathbf{R}})_{ij}$. Here $N_{i+}(3\Delta, 4\Delta) = \sum_{j=1}^{4} N_{ij}(3\Delta, 4\Delta) =$ (number of persons in state i at time 3Δ) and $\mathbf{R}$ is a matrix whose entries satisfy $r_{ij} \geqslant 0$ for $i \neq j$ and $\sum_{j=1}^{4} r_{ij} = 0$. The off-diagonal entries, r_{ij}, are the transition rates per person at risk, and they are constrained by the model to be constant during the 10-week interval Δ.

Table 1 Transitions in Infection Status from Survey 4 to Survey 5 (i.e., $k = 3$ and 4) for All Individuals Aged 19–28 Years Present at Both Surveys

State at Survey 4	State at Survey 5			
	1	2	3	4
1	340	14	171	7
2	21	2	9	0
3	77	3	103	13
4	16	2	20	4

Source. Cohen and Singer [11].

We also introduce the more restricted class of models where $r_{14} = r_{23} = r_{32} = r_{41} = 0$. These zero elements on the minor diagonal of $\mathbf{R}$ exclude the possibility that both Pf and Pm would either be gained or be lost simultaneously.

Introducing the measure of goodness-of-fit*

$$G^2 = -2\sum(\text{observed frequency}) \times \log\left(\frac{\text{frequency predicted by the model}}{\text{observed frequency}}\right) = -2\sum_{i,j} N_{ij} \log\left(\frac{N_{i+}(e^{\Delta\mathbf{R}})_{ij}}{N_{ij}}\right),$$

we calculate $\mathbf{R}$, which minimizes this quantity subject to the constraint $r_{14} = r_{23} = r_{32} = r_{41} = 0$. For the data in Table 1, we find that the constrained Markovian model fits the data well and that

$$\hat{\mathbf{R}} = \begin{bmatrix} -0.751 & 0.116 & 0.635 & 0 \\ 3.351 & -3.351 & 0 & 0 \\ 0.764 & 0 & -0.970 & 0.206 \\ 0 & 0.621 & 1.946 & -2.567 \end{bmatrix}.$$

The surprising feature of this matrix is that

$$0.116 = \hat{r}_{12} < \hat{r}_{34} = 0.206$$

$$1.946 = \hat{r}_{43} < \hat{r}_{21} = 3.351.$$

This suggests that contrary to expectations based on previous literature [10], the presence of Pf *promotes* the acquisition and *reduces* the loss of Pm. Thus there is a cooperative rather than a competitive effect. The same calculations applied to many other 4×4 tables indicated that in each instance a constrained, time-homogeneous Markov model fits the data; and the empirical regularity $\hat{r}_{12} < \hat{r}_{34}$ holds in this population regardless of season or age of the individuals. On the other hand, $\hat{r}_{43} < \hat{r}_{21}$ tends to hold for younger persons, but for individuals over age 44 we typically find $\hat{r}_{43} > \hat{r}_{21}$. In this modeling strategy time variation in the transition rates is measured only by their variation across different pairs of successive surveys. It is important to emphasize, however, that this does not necessarily imply that the infection histories across all eight surveys are representable as a time-inhomogeneous Markov chain. In fact, tests of the Markov property on this data reveal that there is dependence in the infection statuses across several surveys.

A next step in the study of parasite interactions should be the estimation of the transition rates $r_{ij}(t)$ within a model which is compatible with the frequencies

$n_{i_0,i_1,\ldots,i_7}$ = (number of individuals with infection status i_0 at time 0, i_1 at time Δ, i_2 at time $2\Delta, \ldots$, etc.).

Furthermore, there should be an assessment of whether the important qualitative conclusions about cooperative, as opposed to competitive, effects still hold up in a model which accounts for non-Markovian dependence. It is at this stage that we need procedures analogous to (1) and (2) tailored to data where there are gaps in the observation relative to a continuously evolving process. This is an important but currently unresolved research problem.

The strategy employed in this example is based on a philosophy about modeling longitudinal microdata which is in sharp contrast to the methodology utilized in the example in the preceding section. In particular, we have here adopted the view that one should:

1. Begin with very simple, somewhat plausible classes of models as candidates to describe some portion of the observed data and within which the unobservable dynamics are well defined—e.g., the time series of time-homogeneous Markov chains where each separate model only describes unobserved dynamics between a pair of successive surveys and fits the observed transitions.
2. Estimate and interpret the parameters of interest—e.g., the transition rates r_{ij}—within the simplified models, and then assess whether these models can, in fact, account for finer-grained detail such as the sequence frequencies $n_{i_0,i_1,\ldots,i_7}$.
3. Typically, the original proposed models—they are usually first-order Markovian across a wide range of subject matter contexts—which may adequately represent data based on pairs of consecutive surveys will not account for higher-order dependencies. Such dependencies tend to be the rule rather than the exception in longitudinal microdata. We then look for structured residuals from the simple models to guide the selection of more realistic and interpretable specifications (see, e.g., Singer and Spilerman [38] for a discussion of this kind of strategy in a variety of sociology and economics investigations).

The repeated fitting of models and then utilizing structured residuals to guide successively more realistic model selection is a strategy which, on the surface, seems to be very reasonable. However, the process frequently stagnates after only one or two stages because the possible explanations for given structured residuals are usually too extensive to be helpful by themselves. One really needs, in addition, specific subject-matter theories translated into mathematics to guide the model selection process. Unfortunately, in most fields where analysis of longitudinal microdata is of interest, the development of substantive theory is quite weak.

The potential danger of the foregoing strategy, even in the use of transition rate estimates, such as in the malaria example, is that parameter estimates may be biased simply as a result of model misspecification. The biases, in turn, can lead to incorrect conclusions about relationships between events. However, if a process is observed continuously over an interval—producing what is frequently referred to as event history data—then this possibility can be avoided by utilizing methods such as (1) and (2) which are not based on strong, substantively indefensible assumptions about dependencies in the underlying process. In this connection, an important research problem is to provide guidance about the class of situations for which rates from a time series of Markovian models are good approximations to the corresponding rates estimated within quite general counting processes as in Aalen's [1] theory.

Growth Curves, Polynomial Models, and Tracking

There is frequently a sharp distinction between what one can learn from repeated cross-sectional surveys as opposed to prospective longitudinal designs. An instance of this arises if one asks whether the time trend of population means from repeated cross sections in any way reflects the structure of individual time paths. That the answer is often negative is illustrated by a comparison of the time trend in systolic blood pressure (SBP) for two male cohorts in the Framingham study with the pattern of individual serial measurements. As mentioned previously, data in the Framingham study were collected biennially on the same individuals over a 12-year period. Averaging across persons' SBP measurements at each examination yields the linear trend in Fig. 3.

This pattern could also be obtained if independent random samples—cross-sectional data collection—had in fact been utilized. However, an analysis of individual change in SBP over time indicates that these serial measurements are best represented as

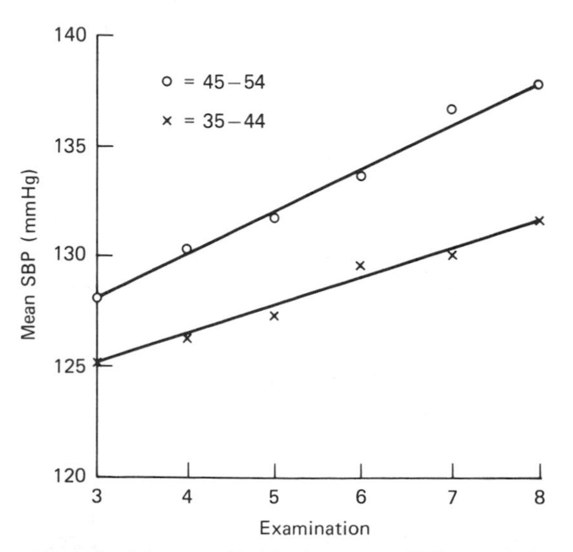

Figure 3 Mean systolic blood pressure (SBP) at examinations 3–8 in Framingham Heart Study: men aged 35–44 and 45–54 at examination 2. From Ware and Wu [45].

cubic functions of time. Thus the points in the linear pattern in Fig. 3 may be interpreted as averages across cubic functions sampled at discrete times. The cubic relationships, characterizing individual SBP dynamics, are not retrievable from repeated cross sections.

A systematic strategy for ascertaining the foregoing relationships proceeds according to the following steps:

1. If $t_1 < t_2 < \cdots < t_K$ are the examination dates, then introduce the $K \times K$ matrix $\mathbf{\Phi}$ whose rows $\boldsymbol{\phi}_l'$, $l = 0, 1, \ldots, K-1$ are the orthogonal polynomials of degree l on $(t_1, \ldots, t_K)$. Let

$$\mathbf{Y}_i = \begin{bmatrix} y_i(t_1) \\ \vdots \\ y_i(t_K) \end{bmatrix}$$

denote the vector of observations for the ith individual and consider the family of models for $\mathbf{Y}_i$ where

$$y_i(t_j) = \sum_{l=0}^{L} \beta_{li}\phi_l(t_j) + \epsilon_{ij}, \qquad L = 0, 1, \ldots, K-1,$$

and ϵ_{ij} is a residual to be interpreted as a value of a normally distributed random variable with mean 0 and variance–covariance structure satisfying the assumptions

$$\mathbf{Y}_i \mid \boldsymbol{\beta}_i \sim N(\mathbf{\Phi}\boldsymbol{\beta}_i, \sigma^2\mathbf{I}) \tag{3}$$

$$\boldsymbol{\beta}_i \sim N(\boldsymbol{\beta}, \mathbf{\Lambda}). \tag{4}$$

Here $N(\boldsymbol{\mu}, \mathbf{\Sigma})$ denotes the multivariate normal distribution with mean vector $\boldsymbol{\mu}$ and covariance matrix $\mathbf{\Sigma}$. We define $b_{li} = \boldsymbol{\phi}_l'\mathbf{Y}_i$ and determine the degree L of the polynomial describing the time trend in the population as the largest l for which $Eb_{li} \neq 0$. (See Ware and Wu [45, p. 429] for formal tests.) We then estimate $E\mathbf{Y}$ using $\mathbf{\Phi}_L'\bar{b}$, where $\bar{b}$ is the vector of sample means of the first L orthogonal polynomial coefficients. This produced the linear relationship shown in Fig. 3, based on the Framingham data.

2. The specification (3) and (4) implies that $\mathbf{Y}_i \sim N(\mathbf{X}\boldsymbol{\beta}, \mathbf{X}\boldsymbol{\Lambda}\mathbf{X}^T + \sigma^2\mathbf{I})$, where $\mathbf{X}$ is a design matrix whose columns are orthonormal polynomials on $(t_1, \ldots, t_K)$. This suggests that in order to estimate polynomial representations for individuals we should test the data for compatibility with the covariance structure $\mathbf{X}\boldsymbol{\Lambda}\mathbf{X}^T + \sigma^2\mathbf{I}$ against the alternative of an arbitrary covariance matrix, $\boldsymbol{\Sigma}$. Here $\boldsymbol{\Lambda}$ and σ^2 are estimated by maximum likelihood. By finding the polynomial of lowest degree for which a likelihood ratio test* fails to reject $\mathbf{X}\boldsymbol{\Lambda}\mathbf{X}^T + \sigma^2\mathbf{I}$ as an adequate model, we choose a polynomial model for the individual curves. For details on this kind of procedure, see Rao [34]. When applied to the male cohorts—ages 35–44 and 45–54, respectively, at examination 2—we obtain cubic functions of age characterizing SBP measurements.

An important application of polynomial growth curve models arises in the problem of "tracking." The papers by Ware and Wu [45], Foulkes and Davis [17], and McMahon [29] should each be consulted for alternative views about the concept of tracking*. Here one is interested in ascertaining whether an initial ordering of individual observations persists over a prescribed interval of time. For example, in order to carry out early identification of persons at risk of cardiovascular disease it is necessary to know whether children with high blood pressure also tend to have high blood pressure as they grow older. Alternatively, in developmental psychology there is a substantial literature dealing with individual differences and their possible stability with increasing age for persons in the same birth cohort. L. Terman's classic follow-up study of gifted children raises the question of stability of individual differences in terms of performance on tests. This is precisely the question of tracking as set forth in the biometry literature and which has received in-depth consideration in the context of longitudinal data analysis only in the past few years.

For an assessment of tracking in a time interval $[T_1, T_2]$ we introduce the index

$$\begin{aligned}\gamma(T_1, T_2) = \Pr(f(t, \boldsymbol{\beta}_j) &\geqslant f(t, \boldsymbol{\beta}_i) \\ &\text{for all } t \text{ in } [T_1 T_2] \\ &\text{or } f(t, \boldsymbol{\beta}_j) < f(t, \boldsymbol{\beta}_i) \\ &\text{for all } t \text{ in } [T_1, T_2]),\end{aligned}$$

where $f(t, \boldsymbol{\beta}_i)$ and $f(t, \boldsymbol{\beta}_j)$ are time paths of two randomly chosen individuals. With polynomial specifications, $f(t, \boldsymbol{\beta}_i)$ may be written in the form

$$f(t, \boldsymbol{\beta}_i) = \beta_{i0} + \beta_{i1}t + \cdots + \beta_{iL}t^L.$$

Foulkes and Davis [17] propose the rule that no tracking will be said to occur if $\gamma(T_1, T_2) < \frac{1}{2}$. Then for estimated values in the interval $(\frac{1}{2}, 1)$ γ may be interpreted as a measure of the extent of tracking.

In an interesting application of this idea, Foulkes and Davis [17] utilize data assembled by Grizzle and Allen [20] to assess the quantity of coronary uric potassium (in milliequivalents per liter) following a coronary occlusion in three groups of dogs. The assessment is based on measurements taken at 2-minute intervals during the first 13 minutes following coronary occlusion. The three populations consist of (a) 9 control dogs, (b) 10 dogs with extrinsic cardiac denervation three weeks prior to coronary occlusion, and (c) 8 dogs subjected to extrinsic cardiac denervation immediately prior to coronary occlusion. They find that for individuals in each population, the amount of potassium is representable as a cubic function of time. Furthermore, the tracking index $\gamma(1, 13)$ is estimated to be 0.444, 0.711, and 0.500 in groups (a) to (c), respectively, with corresponding standard errors given by 0.059, 0.060, and 0.094. This suggests that only the group (b) dogs track—i.e., those with extrinsic cardiac denervation three weeks prior to coronary occlusion.

Repeat Victimization: Detecting Regularities in Turnover Tables

One of the original aims of the National Crime Survey [33] (*see* JUSTICE STATISTICS, BUREAU OF) was the measurement of annual

Table 2 Repeat Victimization Data for Eight Major Crime Categories[a]

First Victimization in Pair	Second Victimization in Pair								
	Rape	Assault	Robbery	Purse Snatching/ Pocket Picking	Personal Larceny	Burglary	Household Larceny	Motor Vehicle Theft	Totals
Rape	26	50	11	6	82	39	48	11	273
Assault	65	2,997	238	85	2,553	1,083	1,349	216	8,586
Robbery	12	279	197	36	459	197	221	47	1,448
Purse snatching/ pocket picking	3	102	40	61	243	115	101	38	703
Personal larceny	75	2,628	413	229	12,137	2,658	3,689	687	22,516
Burglary	52	1,117	191	102	2,649	3,210	1,973	301	9,595
Household larceny	42	1,251	206	117	3,757	1,962	4,646	391	12,372
Motor vehicle theft	3	221	51	24	678	301	367	269	1,914
Total	278	8,645	1,347	660	22,558	9,565	12,394	1,960	57,407

Source. *Reiss* [35].

[a] Reported crimes by households with two or more victimizations while in survey July 1, 1972, to December 31, 1975.

change in crime incidents for a limited set of major crime categories. However, longitudinal analysis is facilitated by the fact that the basic sample is divided into six rotation groups of approximately 10,000 housing units each. The occupants of each housing unit are interviewed every six months over a three-year period. For individuals victimized at least once in a given six-month interval, a detailed record of their victimization history in that period is collected. This retrospective information forms the basis of individual victimization histories. For a detailed critique of the NCS design and of the measurement of criminal victimization generally, see Penick and Owen [33] and Fienberg [15].

A question of considerable interest and importance is whether persons, or households, that are victimized two or more times within a three-year period tend to be victimized in the same or a similar manner. A useful first cut at this kind of question can be developed by preparing a turnover table of the frequency with which a particular succession of crimes is committed on the same individual or household. To this end, Table 2 lists the number of successive victimizations for eight major crime categories in households with two or more victimizations from July 1, 1972, to December 31, 1975.
Although there is no natural order relation among these categories, similar types of crime are listed, to the extent possible, in adjacent rows and columns. In particular, crimes of personal violence are in rows 1 to 3, while those involving theft without personal contact are in rows 5 to 7.

Informal examination of the table suggests that in repeatedly victimized households there is a strong propensity for persons to be victimized in the same way two or more times. Formal support for this claim, together with some refinements to crimes of similar character, can be obtained by the following strategy.

1. First test whether the transition probabilities $p_{ij} = \Pr$ (second victimization is of type $j \mid$ first victimization is of type i) are such that the row proportions in Table 2 are homogeneous. This hypothesis is, as you would expect, clearly rejected.
2. Prepare a new table of counts in which the diagonal entries in Table 2 are deleted and again test the hypothesis of homogeneous row proportions. This constrained specification is much closer to satisfying the baseline model of homogeneous row proportions than the original data.
3. Delete cells from the original table in diagonal blocks in each of which there are crimes of similar type. Then test the hypothesis of homogeneous row proportions on the reduced table. A pattern of deletions which yields a table consistent with this hypothesis is shown in Fig. 4. See Fienberg [15] for more details on this strategy as applied to victimization data.

What this analysis suggests is that relative to a table with homogeneous row proportions, there is elevated repeat victimization involving crimes of similar type. This example is the prototype of strategies which use residuals from simple baseline models—quasi-independence or quasi-homogeneity—to detect special regularities in turnover tables.

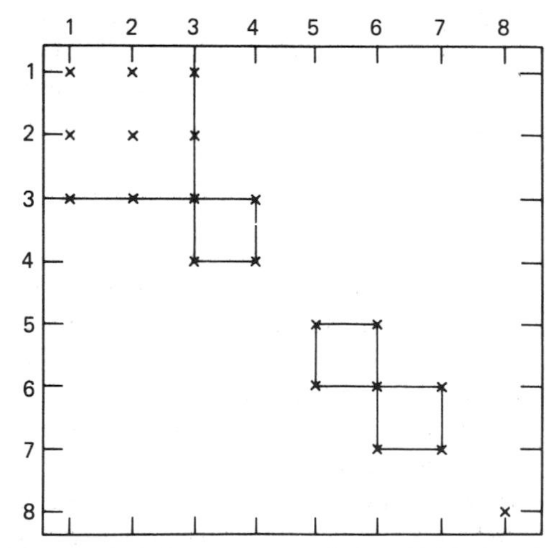

Figure 4 Deletion pattern applied to Table 2 which yields a table with transition probabilities such that row proportions are homogeneous. ×, cell deleted.

BRIEF GUIDE TO OTHER LITERATURE

The remarks and examples in the two preceding sections can, at best, convey a rudimentary impression of the issues involved in analyses of longitudinal data. Some important topics—e.g., strategies for incorporating measurement error* into process models and the introduction of continuous-time, continuous-state processes as covariates in survival analyses—were hardly mentioned at all. Thus it seems prudent to mention some of the literature that an interested reader could follow up to become acquainted with other aspects of longitudinal data analysis.

1. For an enlightening discussion of biases due to measurement error in panel surveys, see Williams and Mallows [48] and Williams [47]. A nice discussion of measurement models and unobservables with particular emplasis on the psychology literature appears in Bentler [7].
2. Trade-offs between data collection designs, particularly prospective vs. retrospective studies, are important to understand, and very nicely discussed in Schlesselman [37]. The impact of different designs on analytical strategies is lucidly treated in Hoem and Funck-Jensen [21] with particular emphasis on demography. See also Goldstein [19]. Experimental studies in biology, medicine, and psychology called repeated measurement* designs represent an important area discussed in the review papers of Koch et al. [23] and Koch et al. [24].
3. The introduction of time-varying covariates in survival models, particularly as stochastic process models, is an important and quite recent development. For an insightful discussion emphasizing problems in labor economics, see Flinn and Heckman [16]. In the context of medical epidemiology the analysis of Woodbury et al. [49] is quite illuminating.

References

[1] Aalen, O. (1978). *Ann. Statist.*, **6**, 701–726.

[2] Aalen, O., Borgan, Ø., Keiding, N., and Thormann, J. (1980). *Scand. J. Statist.*, **7**(4), 161–171.

[3] Anderson, T. W. and Goodman, L. (1957). *Ann. Math. Statist.*, **28**, 89–110.

[4] *Annals de l'INSEE* (1978). The Econometrics of Panel Data. INSEE, Paris, pp. 30–31.

[5] Bartholomew, D. J. (1982). *Stochastic Models for Social Processes*, 3rd ed. Wiley, New York.

[6] Bekessy, A., Molineaux, L., and Storey, J. (1976). *Bull. WHO*, **54**, 685–693.

[7] Bentler, P. (1980). *Annu. Rev. Psychol.*, **31**, 419–456.

[8] Berelson, B., Lazarsfeld, P., and McPhee, W. (1954). *Voting*. University of Chicago Press, Chicago.

[9] Blumen, I., Kogan, M., and McCarthy, P. J. (1955). *The Industrial Mobility of Labor as a Probability Process*. Cornell Stud. Ind. Labor Relations, Vol. 6. Cornell University Press, Ithaca, N.Y.

[10] Cohen, J. E. (1973). *Quart. Rev. Biol.*, **48**, 467–489.

[11] Cohen, J. E. and Singer, B. (1979). In *Lectures on Mathematics in the Life Sciences*, Vol. 12, S. Levin, ed. American Mathematical Society, Providence, R.I., pp. 69–133.

[12] Crowley, J. and Hu, M. (1977). *J. Amer. Statist. Ass.*, **72**, 27–36.

[13] Dawber, T. (1980). *The Framingham Study*. Harvard University Press, Cambridge, Mass.

[14] Duncan, O. D. (1981). In *Sociological Methodology*, S. Leinhardt, ed. Jossey-Bass, San Franscisco, Calif.

[15] Fienberg S. (1980). *Statistician*, **29**, 313–350.

[16] Flinn, C. J. and Heckman, J. J. (1982). *Adv. Econometrics*, **1**, 35–95.

[17] Foulkes, M. A. and Davis, C. E. (1981). *Biometrics*, **37**, 439–446.

[18] Gilbert, J. P., Meier, P., Rümke, C. L., Saracci, R., Zelen, M., and White, C. (1975). *J. Amer. Med. Ass.*, **231**, 583–608.

[19] Goldstein, H. (1979). *The Design and Analysis of Longitudinal Studies*. Academic Press, New York.

[20] Grizzle, J. and Allen, D. (1969). *Biometrics*, **25**, 357–381.

[21] Hoem, J. and Funck-Jensen (1982). In *Multidimensional Mathematical Demography*, K. Land and A. Rogers, eds. Academic Press, New York.

[22] Karlin, S., Cameron, E., and Chakraborty, R. (1983). *Amer. J. Hum. Genet.*, **35**, 695–732.

[23] Koch, G., Amara, I., Stokes, M., and Gillings, D. (1980). *Int. Statist. Rev.*, **48**, 249–265.

[24] Koch, G., Gillings, D., and Stokes, M. (1980). *Annu. Rev. Public Health*, **1**, 163–225.

[25] Kohn, M. and Schooler, C. (1978). *Amer. J. Sociol.*, **84**, 24–52.

[26] Lazarsfeld, P. F. and Fiske, M. (1938). *Public Opinion Quart.*, **2**, 596–612.

[27] Levy, D. E., Bates, D., Caronna, J., Cartlidge, N., Knill-Jones, R., Lapinski, R., Singer, B., Shaw, D., and Plum, F. (1981). *Ann. Intern. Med.*, **94**, 293–301.

[28] Littman, G. and Mode, C. J. (1977). *Math. Biosci.*, **34**, 279–302.

[29] McMahon, C. A. (1981). *Biometrics*, **37**, 447–455.

[30] Menken, J., Trussell, T. J., Stempel, D., and Balakol, O. (1981). *Demography*, **18**, 181–200.

[31] Molineaux, L. and Gramiccia, G. (1980). *The Garki Project: Research on the Epidemiology and Control of Malaria in the Sudan Savanna of West Africa*. WHO, Geneva.

[32] Molineaux, L., Storey, J., Cohen, J. E., and Thomas, A. (1980). *Amer. J. Tropical Med. Hyg.*, **29**, 725–737.

[33] Penick, B. K. and Owens, M. E. B., eds. (1976). *Surveying Crime* (Report of Panel for the Evaluation of Crime Surveys.) National Academy of Sciences, Washington, D.C.

[34] Rao, C. R. (1965). *Biometrika*, **52**, 447–458.

[35] Reiss, A. J. (1980). In *Indicators of Crime and Criminal Justice: Quantitative Studies*, S. Fienberg and A. Reiss, eds. U.S. Government Printing Office, Washington, D.C.

[36] Rogosa, D. (1980). *Psychol. Bull.*, **88**, 245–258.

[37] Schlesselman, J. J. (1982). *Case Control Studies*. Oxford University Press, Oxford, England.

[38] Singer, B. and Spilerman, S. (1976). *Ann. Econ. Soc. Meas.*, **5**, 447–474.

[39] Sydenstricker, E. (1927). A Study of Illness in a General Population Group. *Hagerstown Morbidity Stud. No. 3, Public Health Rep.*, p. 32.

[40] Taeuber, K. E., Chiazze, L., Jr., and Haenszel, W. (1968). *Migration in the United States*. U.S. Government Printing Office, Washington, D.C.

[41] Terman, L. M. and Oden, M. H. (1959). *The Gifted Group at Mid-Life*. Genetic Studies of Genius, Vol. 5. Stanford University Press, Stanford, Calif.

[42] Terman, L. M., Burks, B. S., and Jensen, D. W. (1930). *The Promise of Youth*. Genetic Stud. Genius, Vol. 3. Stanford University Press, Stanford, Calif.

[43] Tinbergen, J. (1940). *Rev. Econ. Stud.*, 73–90.

[44] Tuma, N., Hannan, M., and Groeneveld, L. (1979). *Amer. J. Socio.*, **84**, 820–854.

[45] Ware, J. H. and Wu, M. C. (1981). *Biometrics*, **37**, 427–437.

[46] White, H. C. (1970). *Chains of Opportunity*. Harvard University Press, Cambridge, Mass.

[47] Williams, W. H. (1978). In *Contributions to Survey Analysis and Applied Statistics*, H. A. David, ed. Academic Press, New York, pp. 89–112.

[48] Williams, W. H. and Mallows, C. L. (1970). *J. Amer. Statist. Ass.*, **65**, 1338–1349.

[49] Woodbury, M. A., Manton, K. G., and Stallard, E. (1979). *Biometrics*, **35**, 575–585.

[50] Wright, S. (1983). *Amer. J. Hum. Genet.*, **35**, 757–768.

(BIOSTATISTICS
EPIDEMIOLOGICAL STATISTICS
FOLLOW-UP
FRAMINGHAM: AN EVOLVING LONGITUDINAL STUDY
MORBIDITY
PANEL DATA
SURVIVAL ANALYSIS
TRACKING)

BURTON SINGER

LORD'S TEST STATISTICS

These are "quick" test statistics proposed by Lord [1].

1. A one-sample analog of the t-test with test statistic of the form

$$t_W = \frac{\overline{X} - \mu}{W},$$

where W is the sample range*.

This test is often used in industrial applications, although it is not resistant to outliers*. The efficiency of this test was studied by Pillai [3]. Tables of critical values for various significance levels are presented in Lord [1], Pillai [3], and Snedecor and Cochran [4].

2. A two-sample analog of the t-test for testing the equality of two means based on two independent samples from normal populations. The test statistic is of the form

$$t_w = \frac{|\overline{X}_2 - \overline{X}_1|}{\overline{W}},$$

where $\overline{W} = (W_1 + W_2)/2$ and $W_i (i = 1, 2)$ are the sample ranges.

The test is no more robust under nonnormality than the t-test and it is more vulnerable to extreme sample values. Moore [2] has extended this test for the case of independent samples of unequal size. Tables of critical values for significance levels $\alpha = 0.10$, 0.05, 0.02, and 0.01 are presented in Lord [1] (for the equal-sample-size case).

An abridged version of Lord's tables appears in Snedecor and Cochran [4].

References

[1] Lord, E. (1947). *Biometrika*, **34**, 41–67.

[2] Moore, P. G. (1957). *Biometrika*, **34**, 482–489.

[3] Pillai, K. C. S. (1951). *Ann. Math. Statist.*, **22**, 469–472.

[4] Snedecor, G. W. and Cochran, W. G. (1980). *Statistical Methods*, 7th ed. Iowa State University Press, Ames, Iowa.

(STUDENTIZED RANGE TESTS
STUDENT'S t-TESTS)

LORENZ CURVE

Lorenz [19] was dissatisfied with, and rightly critical of, the methods that had been used to assess whether an income distribution* (ID) is becoming more or less unequal. The main purpose of the methods proposed was the comparison of income distributions of a country at different periods in time or of different countries at the same period in time. "We wish to be able to say at what point a community is to be placed between the two extremes, equality, on the one hand, and the ownership of all wealth by one individual on the other" [19, p. 209]. This objective was hereafter called the welfare ordering of income distributions, which is closely related to another modern concept, that of stochastic dominance. To accomplish his objective, Lorenz introduced a new approach, later termed the Lorenz curve, which simultaneously takes into account the changes in income and in population, thus putting any two communities of the most diverse conditions on a comparable basis.

LORENZ CURVE: DEFINITION AND PROPERTIES

Consider a sample of n individuals, and let x_i be the income of individual i, $i = 1, 2, \ldots, n$, such that $x_1 \leqslant x_2 \leqslant \cdots \leqslant x_n$. The sample Lorenz curve is the polygon joining the points $(h/n, L_h/L_n)$, $h = 0, 1, \ldots, n$, where $L_0 = 0$ and $L_n = \sum_{i=1}^{h} x_i$ is the total income of the poorest h individuals. Hence the Lorenz curve $q = L(y)$ has as its abscissa the cumulative proportion of *income receivers*, arrayed by increasing size of their incomes, and as its ordinate the corresponding proportion of income received (Fig. 1). Its formal representation is

$$L(y) = \int_0^y x\,dF(x)/E(Y), \qquad (1)$$

where Y is a nonnegative income variable for which the mathematical expectation $\mu = E(Y)$ exists, and $p = F(y)$ is the cumulative distribution function (CDF) of the population of income receivers. When all

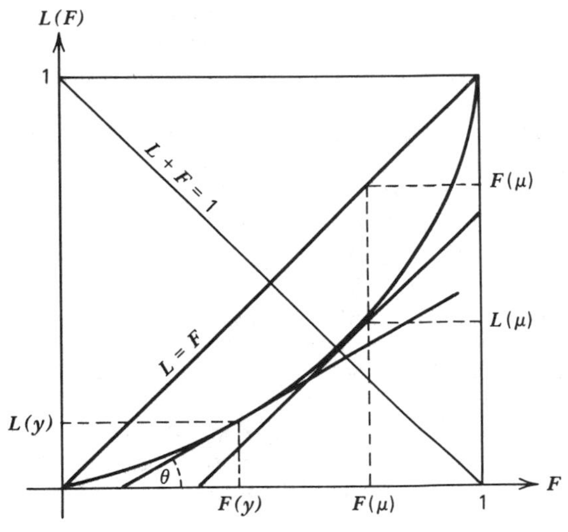

Figure 1 Lorenz curve:

$$L(y) = \int_0^y x\,dF(x)/E(Y)$$

$$F(\mu) - L(\mu) = \max_{\{y\}}[F(y) - L(y)]$$

$$F(\mu) + L(\mu) > 1, \qquad dL/dF|_{y=\mu} = 1$$

$$dL/d\mu = \tan\theta = y/E(Y).$$

members of the population receive the same income, the Lorenz curve is the equidistribution or identity function $F = L$ (Fig. 1). As the distribution becomes more unequal, the Lorenz curve bends downward and to the right within the unit square. It follows from (1) that the Lorenz curve is the first moment distribution function of $F(y)$.

Since the CDF of all specified models of ID are strictly increasing and continuously differentiable functions, $y = F^{-1}(p)$ is well defined. Replacing it in (1), we deduce that

$$L(p) = \int_0^p F^{-1}(\Theta)\,d\Theta / E(Y).$$

Some properties of the Lorenz curve for the class of continuously differentiable ID models with finite mathematical expectation are:

(a) the Lorenz curve is a CDF with $L(0) = 0$ and with $L(1) = \lim_{p \to 1-} L(p) = \lim_{y \to \infty}(F(y)) = 1$;

(b) $dL/dF = y/E(Y)$, thus the differential is also a strictly increasing and continuously differentiable function, taking the value of 1 at $y = E(Y)$;

(c) $L(F)$ is convex;

(d) $E(|X - \mu|)/(2\mu) = F(\mu) - L(\mu) = \max_{\{y\}}[F(y) - L(y)]$ (Fig. 1);

(e) $E(|X - m|)/(2\mu) = F(m) - L(\frac{1}{2})$, where m is the median income;

(f) the Lorenz curve is symmetric if and only if (iff) $1 - p = L(1 - q)$, in which case the diagonal function $F + L = 1$ in the unit square bisects the equidistribution function $F = L$ and intersects the Lorenz curve at the point with coordinates $(F(\mu), L(\mu))$;

(g) the Lorenz curve is left (right) asymmetric iff $F(\mu) + L(\mu) > 1$ (< 1) (Fig. 1).

Taking the identity function $L = F$ as the abscissa, the asymmetry of the Lorenz curve can be more effectively studied. This transformation determines the new coordinate system $(u, v) = ((F + L)/\sqrt{2}, (F - L)/\sqrt{2})$, and was first introduced and extensively analyzed by Gini [12]. More recently, Kakwani and Podder [17] and Kakwani [15] have used the same transformation.

Table 1 presents the mathematical form of the Lorenz curve for some well-known ID models (*see* INCOME DISTRIBUTION MODELS). As an illustration, we deduce it for the Pareto* type I and the gamma distribution* functions.

LORENZ CURVE FOR THE PARETO TYPE I MODEL. The CDF is

$$F(y) = 1 - \alpha(y/y_0)^{-\alpha},$$

$$y \geqslant y_0 > 0, \quad \alpha > 1.$$

Table 1

ID Model	Lorenz Curve[a]
Pareto type I	$L(F) = 1 - (1 - F)^{(\alpha-1)/\alpha}$
Lognormal[b]	$L(F(y)) = \Lambda(y;\, \mu + \sigma^2, \sigma^2)$, or $N^{-1}(L) = N^{-1}(F) - \sigma$
Gamma	$(F, L) = (\Gamma(y; \lambda, \alpha), \Gamma(y; \lambda, \alpha + 1)$
Log-logistic[c]	$L(F) = B(F; 1 + 1/\alpha, 1 - 1/\alpha), \alpha > 1$
Singh–Maddala[c]	$L(F) = B[1 - (1 - F)^{\beta-1}; 1 + 1/(\delta + 1), \beta - 1 - 1/(\delta + 1)], \beta < 2 + \delta$
Dagum, type I[c]	$L(F) = B(F^{\beta}; 1/\beta + 1/\alpha, 1 - 1/\alpha), \alpha > 1$

[a] For the derivation of the Lorenz curves shown in this table, see Dagum [6].

[b] Λ symbolizes the lognormal CDF with parameters $\mu + \sigma^2$ and σ^2, and N^{-1} is the inverse of the standardized normal, such that $\mu = E(\log Y)$ and $\sigma^2 = \mathrm{var}(\log Y)$.

[c] $B(\cdot)$ is the beta cumulative distribution function. Making $\beta = 2$ in Singh and Maddala's model, and $\beta = 1$ in Dagum's type I model, we obtain the log-logistic model [5, 6].

Its probability density function (PDF) is $f(y) = \alpha y_0^{\alpha} y^{-\alpha-1}$ and $E(Y) = \alpha y_0/(\alpha - 1)$. Taking these results into (1), we obtain

$$L(F(y)) = (a-1) y_0^{\alpha-1} \int_{y_0}^{y} x^{-\alpha}\, dx$$
$$= 1 - (y/y_0)^{-\alpha+1}. \qquad (2)$$

It follows from (2) and Pareto's CDF that $L(F) = 1 - (1 - F)^{(\alpha-1)/\alpha}$, as given in Table 1.

LORENZ CURVE FOR THE GAMMA MODEL. The gamma PDF is $\gamma(y;\alpha,\lambda) = (\lambda^{\alpha}/\Gamma(\alpha)) y^{\alpha-1} e^{-\lambda y}$, $y > 0$, $(\alpha,\lambda) > 0$, and zero for $y \leqslant 0$. The CDF is

$$\Gamma(y;\alpha,\lambda) = \int_0^y \gamma(x;\alpha,\lambda)\, dx,$$

and $E(Y) = \alpha/\lambda$. The corresponding Lorenz curve is

$$L(F(y)) = \frac{\lambda^{\alpha+1}}{\alpha\Gamma(\alpha)} \int_0^y x^{\alpha} e^{-\lambda x} dx$$
$$= \Gamma(y, \alpha, \lambda + 1), \qquad (3)$$

the gamma CDF with parameters α and $\lambda + 1$.

APPLICATIONS OF THE LORENZ CURVE

The Lorenz curve proved to be a powerful tool for the analysis of a variety of scientific problems: e.g., (a) to measure the income inequality within a population of income receivers, (b) as a criterion to perform a partial ordering of social welfare states, (c) to assess the progressiveness of a tax system, (d) to extend the concept of the Lorenz curve to functions of income or other variables, (e) to study the stochastic properties of the sample Lorenz curve, and (f) to derive goodness-of-fit tests for exponential distribution functions, as well as upper and lower bounds for the Gini ratio.

LORENZ CURVE AND INCOME INEQUALITY. Gini [10] introduced a measure of income inequality as a function of Gini's mean difference*. In 1914 he proved the important theorem that relates the Gini mean difference to the area of concentration [11], i.e., the area between the equidistribution function $L = F$ and the Lorenz curve (*see* INCOME INEQUALITY MEASURES).

LORENZ CURVE AND SOCIAL WELFARE ORDERING OF INCOME DISTRIBUTIONS. For this analysis, the equidistribution function is used both as a benchmark and as the most preferred social welfare state. For the usual assumptions that the utility function for all income receivers is an identical strictly increasing and concave function of income, it can be proved that the subset of Lorenz curves *that do not intersect* can be ordered from left to right (in the Lorenz diagram) by decreasing order of preference [1,22]. Given three nonintersecting Lorenz curves L_1, L_2, and L_3, where $L_1(L_3)$ has the minimum (maximum) distance to the equidistribution function, then $L_1 \overset{L}{>} L_2 \overset{L}{>} L_3$, where the symbol $\overset{L}{>}$ stands for "strictly Lorenz superior to" or "strictly preferred to." Therefore, $L_1(y) > L_2(y) > L_3(y)$, or equivalently, $F_1(y) < F_2(y) < F_3(y)$, for all y in the open interval $(0, \infty)$, i.e., for all $F \in (0, 1)$. The definition of "Lorenz superior to" uses the symbol "$\geqslant$," where the strict inequality

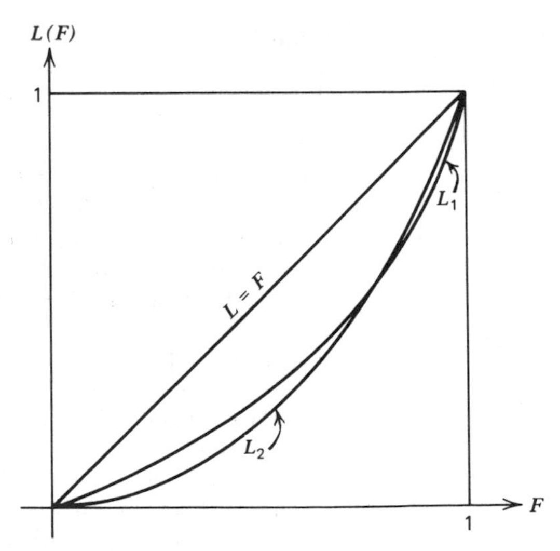

Figure 2 Intersecting Lorenz curves: before (L_1) and after (L_2) tax.

holds for at least one value of y. The same ordering of Lorenz curves can be obtained by applying the first order stochastic dominance (SD). For the ordering of intersecting Lorenz curves (Fig. 2), the second- and higher-order SD are relevant. For surveys on SD, see Bawa [2] and Whitmore and Findlay [24]. In statistics, Lehmann [18] proved the first-order SD and Blackwell [3] the second-order SD. See also Marshall and Olkin [21].

LORENZ CURVE AND TAX PROGRESSIVENESS. One important comparison of income distributions is between the ID of a given population of taxpayers *before* and *after* tax, in order to assess if the tax system is progressive, regressive, or proportional. A tax system is progressive (regressive) when to increasing levels of income correspond increasing (decreasing) percentages of taxes. Hence the income elasticity of taxes is greater (less) than 1. If it is equal to 1, the tax system is proportional. [*Note*: the elasticity of a continuously differentiable function $Q(x)$ is, by definition, $d\log Q(x)/d\log x$).] These are particular cases of the concentration curves discussed below. It follows that the after-tax Lorenz curve is to the left (right) of the before-tax curve when the tax system is progressive (regressive); when it is proportional, it will leave the Lorenz curve unchanged. Should the after-tax curve (L_2 in Fig. 2) intersect from below (above) the before-tax (L_1) Lorenz curve, we conclude that the tax system is regressive (progressive) for the lower-income group and progressive (regressive) for the upper-income group.

LORENZ AND CONCENTRATION CURVES. The Lorenz curve has been successfully extended to the analysis of phenomena other than the distribution of income and wealth. This extension is applied to a variety of economic, sociodemographic, biological, and physical data, and takes the generic name of concentration curve, containing the Lorenz curve as a particular case. Gini [11] and a group of his collaborators applied the concentration curve to the study of several economic and sociodemographic variables; Mahalanobis [20] applied it to consumer demand; Chandra and Singpurwalla [4] applied it to the total time on the test transform in reliability theory, introduced the Lorenz process, and studied its weak convergence to functionals of a Brownian motion* process. Thompson [23] applied the Lorenz curve to the study of the distribution of the number of fish caught by a population of fishermen.

The definition of a *concentration curve* follows the definition of the Lorenz curve given in (1). Thus, if the two variables are related by $y = g(x)$, $x > 0$, $g(x) > 0$, and $E[g(x)]$ exists, the concentration curve of $g(x)$ is, by definition, the cumulative share of $g(x)$ arrayed by increased size of x, that is,

$$L(g(x)) = \int_0^x g(t)\,dF(t)/E[g(X)], \quad (4)$$

which is also a CDF of x.

The properties stated above for the Lorenz curve also apply to the concentration curve, provided that $g'(x) > 0$ for all $x > 0$, and $g(0) = 0$. For $g'(x) < 0$, the concentration curve is strictly concave; therefore $L(g(x)) > F(x)$, and the elasticity of $g(x)$ is negative. For $g'(x) = 0$, i.e., $g(x) = c$, we have $L = F$ and the elasticity of a constant is equal to zero. For $g(x) = x$, the concentration curve is the standard Lorenz curve and the elasticity is equal to 1. Hence the identity (or equidistribution) function $L = F$ and the standard Lorenz curve $L(g(x)) = L(x)$ divide the unit square in three regions (Fig. 3). In between these two curves, the elasticity of $g(x)$ is positive and less than 1, above negative, and below positive and greater than 1.

Let $h(x)$ be another function of x such that $h(x) > 0$ for all $x > 0$ and $E[h(X)]$ exists; then $L(h(x))$ is the concentration curve of $h(x)$. If $g'(x) > 0$ and $h'(x) > 0$ for all x, then $L(g)$ and $L(h)$ would be convex functions of $F(x)$. It can be proved [15] that $L(g)$ will be above (below) $L(h)$ iff the elasticity of $g(x)$ is less (greater) than the elasticity of $h(x)$ for all $x > 0$.

The *Gini concentration ratio* for the variable $g(x)$ with CDF $F(x)$ is equal to twice

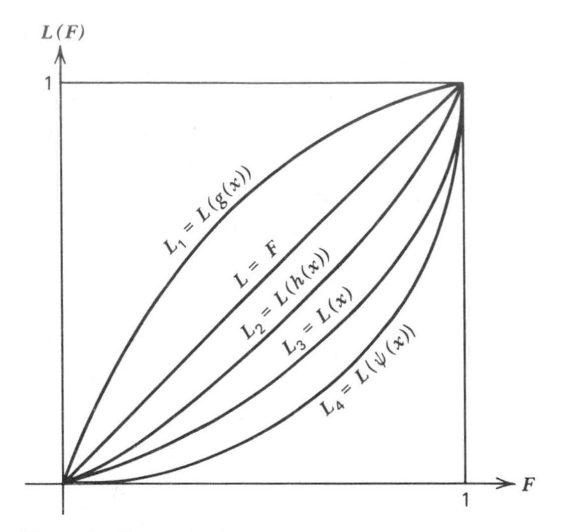

Figure 3 Concentration curves:

$$\epsilon(\theta(x), x) = d \log \theta / d \log x.$$

$$L_1 = L(g(x)), \quad \epsilon(g(x), x) < 0;$$

$$L = F, \quad \epsilon(c, x) = 1;$$

$$L_2 = L(h(x)), \quad 0 < \epsilon(h(x), x) < 1;$$

$$L_3 = L(x), \quad \epsilon(x, x) = 1;$$

$$L_4 = L(\psi(x)), \quad \epsilon(\psi(x), x) > 1.$$

the area between the equidistribution function $L(g) = F$ and the concentration curve $L = L(g)$.

LORENZ CURVE, PROBABILITY THEORY, AND STATISTICAL INFERENCE. The Lorenz curve and the Gini ratio stimulated contributions in the fields of probability theory* and statistical inference*. Gastwirth [9] reviewed some basic properties of the Lorenz curve and proceeded to derive upper and lower bounds to the Gini mean difference and the Gini ratio from data which were grouped in intervals, the mean income in each interval being known. These bounds are sharpened when the underlying density function is locally decreasing (or increasing), or when the density has a decreasing hazard rate in the last interval.

Kakwani and Podder [16] specified several mathematical forms for the Lorenz curve, offering neither a probability base nor an empirical foundation, and discussed alternative methods of estimation from grouped data*.

Thompson [23] proved the important theorem whereby each distribution with a finite mean uniquely determines its Lorenz curve (up to a scale transformation) and conversely. This theorem allows one to perform goodness-of-fit* tests either on the sample distribution function or on the sample Lorenz curve. Gail and Gastwirth [7] developed a scale-free goodness-of-fit test for the Laplace distribution* based on the Lorenz curve. They deduced the asymptotic relative efficiency* of the sample Lorenz curve against Weibull and gamma alternatives, and compared the power of the Lorenz statistics evaluated at the median with that of 10 other goodness-of-fit tests for exponentiality (Laplace distribution) against seven alternatives. Gail and Gastwirth [8] extended these results, including the Gini ratio as a scale-free goodness-of-fit test of exponentiality. Under this parametric hypothesis, the exact distribution, mean, and variance of the sample Gini ratio are deduced and the asymptotic convergence of its standard deviation to the standard distribution is established.

Goldie [14] and Gail and Gastwirth [7] proved that the sample Lorenz curve converges, with probability 1, uniformly to the population Lorenz curve. The same result holds for its inverse. The central limit theorem* is then proved under the condition that the probability density function is continuous and has finite variance.

References

[1] Atkinson, A. B. (1970). *J. Econ. Theory*, **2**, 244–263.

[2] Bawa, V. S. (1975). *J. Financ. Econ.*, **2**, 95–121.

[3] Blackwell, D. (1951). *Proc. 2nd Berkeley Symp. Math. Statist. Prob.*, J. Neyman and L. M. LeCam, eds. University of California Press, Berkeley, Calif., pp. 93–102.

[4] Chandra, M. and Singpurwalla, N. D. (1978). The Gini Index, the Lorenz Curve, and the Total Time on Test Transforms. *Res. Paper T-368*, George Washington University, Washington, D.C.

[5] Dagum, C. (1977). *Econ. Appl.*, **30**, 413–436.

[6] Dagum, C. (1980). *Econ. Appl.*, **33**, 327–367.

[7] Gail, M. H. and Gastwirth, J. L. (1978). *J. Amer. Statist. Ass.*, **73**, 787–793. (A formal probabilistic and statistical discussion of the Lorenz curve.)

[8] Gail, M. H. and Gastwirth, J. L. (1978). *J. R. Statist. Soc. B*, **40**, 350–357. (An excellent study of the exact and asymptotic distribution of the Gini statistic and its use in a scale-free test of exponentiality.)

[9] Gastwirth, J. L. (1972). *Rev. Econ. Statist.*, **54**, 306–316.

[10] Gini, C. (1912). *Studi Economicogiuridici*, Università di Cagliari, Cagliari, Italy, III, 2a. (In Gini [13, pp. 211–382].)

[11] Gini, C. (1914). Sulla misura della concentrazione e della variabilità dei caratteri. *Atti R. Ist. Veneto Sci. Lett. Arti.* (In Gini [13, pp. 411–459].)

[12] Gini, C. (1932). *Metron*, **9** (3–4). (In Gini [13, pp. 651–724].)

[13] Gini, C. (1955). *Variabilità e concentrazione.* Memorie di Metodologia Statistica, Vol. 1. Libreria Eredi Virgilio Veschi, Rome, Italy.

[14] Goldie, C. M. (1977). *Adv. Appl. Prob.*, **9**, 765–791. (An excellent study on the asymptotic sampling theory of sample Lorenz curves and their inverses.)

[15] Kakwani, N. C. (1980). *Income Inequality and Poverty*. Oxford University Press, Oxford.

[16] Kakwani, N. C. and Podder, N. (1973). *Int. Econ. Rev.*, **14**, 278–292.

[17] Kakwani, N. C. and Podder, N. (1976). *Econometrica*, **44**, 137–148.

[18] Lehmann, E. L. (1955). *Ann. Math. Statist.*, **26**, 399–419.

[19] Lorenz, M. O. (1905). *J. Amer. Statist. Ass.* (N.S.), No. 70, 209–219.

[20] Mahalanobis, P. C. (1960). *Econometrica*, **28**, 325–351.

[21] Marshall, A. W. and Olkin, I. (1979). *Inequalities: Theory of Majorization and its Applications*. Academic Press, New York.

[22] Sen, A. K. *On Economic Inequality*. Oxford University Press, Oxford, England.

[23] Thompson, W. A., Jr. (1976). *Biometrics*, **32**, 265–271.

[24] Whitmore, G. A. and Findlay, M. C., eds. (1978). *Stochastic Dominance: An Approach to Decision-Making under Risk*. D. C. Heath, Lexington, Mass.

(ECONOMETRICS
FISHERIES RESEARCH, STATISTICS IN
GOODNESS-OF-FIT
INCOME DISTRIBUTION MODELS
INCOME INEQUALITY MEASURES
LIMIT THEOREM, CENTRAL)

CAMILO DAGUM

LOSS OF INFORMATION

This term can have a variety of meanings in different contexts. It usually refers to loss (or gain) resulting from the use of a particular form of design of experiment*. The "information" is defined more or less arbitrarily; it is not always the Fisher "information*."

When used in connection with confounding* it refers (roughly) to the proportion of the data from which no contribution to the estimation of a specified main effect* or contrast* can be obtained.

(CONFOUNDING
MEASURES OF INFORMATION)

LOT TOLERANCE PERCENT DEFECTIVE (LTPD) *See* ACCEPTANCE SAMPLING

LOWER CONTROL LIMIT *See* CONTROL CHARTS

LOWER PROBABILITIES *See* NONADDITIVE PROBABILITIES

LOWSPREAD

A term used in exploratory data analysis denoting the distance between the median and the lowest value in a distribution of a variable quantity.

(FIVE-NUMBER SUMMARIES
MIDSPREAD)

L-STATISTICS

If a sample $X_1, \ldots, X_n$ is rearranged in ascending order of magnitude, the resulting ordered random variables are denoted

$$X_{1:n} \leqslant X_{2:n} \leqslant \cdots \leqslant X_{n:n}$$

and called the *order statistics**. Here we are interested in linear combinations of these random variables, that is, statistics of the form

$$T_n = \sum_{i=1}^{n} c_{in} X_{i:n}, \qquad (1)$$

where $c_{1n}, \ldots, c_{nn}$ are chosen constants. Mosteller [27] used the term "systematic statistics" to describe any function of the order statistics, and T_n is still often called a "linear systematic statistic." Modern usage, however, favors the term "linear combination of order statistics," or the shortened versions "L-statistics" or "L-estimators." Simple examples include the ith order statistic $X_{i:n}$ (choose $c_{in} = 1$ and $c_{jn} = 0$, $j \neq i$), the mean $\bar{X} = n^{-1}\sum X_i = n^{-1}\sum X_{i:n}$, the range* $X_{n:n} - X_{1:n}$, and the median $X_{k+1:n}$ for $n = 2k + 1$. L-statistics are attractive because they allow one to choose between the competing demands of computational ease, efficiency* (statistical accuracy), and robustness* (insensitivity to minor errors in assumptions). In *censored* samples where only a subset of the ordered values are available, use of L-statistics is often the natural way to proceed.

Introductions to the general subject of order statistics including L-statistics may be found in Sarhan and Greenberg [29], David [14], Johnson and Kotz [21], and Kendall and Stuart [23]. Serfling [30] gives an excellent survey of asymptotic results for T_n.

There are two basic versions of form (1). The first uses constants that approximate a smooth function J, $nc_{in} \approx J(i/(n+1))$, and the second uses "discrete" constants that select only some of the order statistics.

WEIGHTS OF THE CONTINUOUS TYPE

The mean $nc_{in} = 1$ is of this type. Other important examples are:

Example 1. The trimmed mean,

$$T_n = \frac{1}{n-2r} \sum_{i=r+1}^{n-r} X_{i:n},$$

where the first r and the last r order statistics have been "trimmed" from the sample. If r is chosen to be $[n\alpha]$, where $[x]$ is the greatest integer $\leqslant x$ and $0 < \alpha < \frac{1}{2}$, then T_n is often called the α-trimmed mean.

Example 2. The asymptotically efficient estimate of location for the logistic distribution*,

$$T_n = \frac{1}{n} \sum_{i=1}^{n} 6\left(\frac{i}{n+1}\right)\left(1 - \frac{i}{n+1}\right) X_{i:n}.$$

Example 3. Gini's mean difference*,

$$T_n = \frac{2}{n(n-1)} \sum_{i=1}^{n} (2i - n - 1) X_{i:n}.$$

Example 4. The asymptotically efficient estimate of scale for the normal distribution,

$$T_n = \frac{1}{n} \sum_{i=1}^{n} \Phi^{-1}\left(\frac{i}{n+1}\right) X_{i:n},$$

where Φ^{-1} is the inverse of the standard normal distribution function.

Examples 2 and 4 have the form

$$T_n' = \frac{1}{n} \sum_{i=1}^{n} J\left(\frac{i}{n+1}\right) X_{i:n} \qquad (2)$$

with $J(t) = 6t(1-t)$ and $J(t) = \Phi^{-1}(t)$, respectively. Examples 1 and 3 are very close to this form with $J(t) = 1/(1-2\alpha)$ for $\alpha < t < 1 - \alpha$ and $J(t) = 0$ elsewhere for Example 1 and $J(t) = 4(t - \frac{1}{2})$ for Example 3. The weight function $J(t)$ defined on (0, 1) should be piecewise continuous and (if integrable) such that $\int_0^1 J(t)\,dt = 1$ when estimating location and $\int_0^1 J(t)\,dt = 0$ when estimating scale.

Bennett [2] and Jung [22] introduced the form (2) for estimation of location and scale parameters and showed how to derive J functions which make T_n' asymptotically equivalent to the best linear unbiased estimators (BLUEs) given by Lloyd [26]. The latter estimators are computationally difficult since they require tables of the expected values and covariance matrix of $(X_{1:n}, \ldots, X_{n:n})$ for each n and each family of distributions. Some of these tables, however, are available in refs. 29 and 32. Further results about optimal J functions were given by Chernoff et al. [12] using the asymptotic normal distribution of T_n'.

The finite distributions of T_n' are in general unknown (see ref. 13), but most of the classical asymptotic results for $\bar{X}$ have been generalized to T_n'. These include the law of large numbers* (strong consistency), the central limit theorem* (asymptotic normality*), the law of the iterated logarithm*, the Berry–Esseen theorem, and Edgeworth expansions. See Serfling [30] for references and also Helmers [18]. Each of these results allows trade-offs between restrictions on J and restrictions on F, the distribution function of the data.

Robustness of T_n' to outliers* requires that $J(t)$ be zero in neighborhoods of 0 and 1. This keeps the extreme order statistics of the sample from dominating T_n'. In the four examples above, only the trimmed mean is robust in this sense.

WEIGHTS OF THE DISCRETE TYPE

Mosteller [27] suggested the use of linear combinations of *selected* order statistics having the form

$$T_n'' = \sum_{i=1}^{k} a_i X_{n_i : n}, \tag{3}$$

where $a_1, \ldots, a_k$ are nonzero constants and $(n_1, \ldots, n_k)$ is a subset of $(1, \ldots, n)$. We are really interested in the case where $\lim_{n\to\infty} n_i/n = p_i$, $0 < p_i < 1$. Then $X_{n_i : n}$ is called a sample percentile (approximately $100p_i\%$ of the sample values lie below $X_{n_i : n}$) and we shall denote it by $\hat{\xi}_{p_i}$. When np_i is not an integer, $\hat{\xi}_{p_i} = X_{[np_i]+1 : n}$, but when np_i is an integer there is no agreement as to the definition of $\hat{\xi}_{p_i}$. For convenience we take $\hat{\xi}_{p_i} = X_{np_i : n}$ when np_i is an integer. Note that the *usual* definition of the median when n is even is the average of the middle two order statistics rather than $\hat{\xi}_{1/2}$. Also, the range is of form (3), but $n_1/n = 1$ and $n_2/n \to 0$ do not lie between 0 and 1. Other examples include:

Example 5. The interquartile range,

$$T_n'' = \hat{\xi}_{3/4} - \hat{\xi}_{1/4}.$$

Example 6. An estimator of location suggested by Gastwirth [15],

$$T_n'' = 0.3\hat{\xi}_{1/3} + 0.4\hat{\xi}_{1/2} + 0.3\hat{\xi}_{2/3}.$$

Mosteller [27] derived the joint asymptotic normal distribution of $(\hat{\xi}_{p_1}, \ldots, \hat{\xi}_{p_k})$ and thereby the asymptotic distribution of T_n'' when $n_i/n \to p_i$. He used the term "inefficient" in describing T_n'' because the estimators generally have variances larger than the best estimators. Although the smallest variance is not attainable, Ogawa [28] showed how to choose the p_i optimally (given k) for the normal location and scale parameters. Since that time a huge literature has developed for estimating parameters from different families using only a few order statistics. Many of these articles are referenced in ref. 21. Recent papers include refs. 11 and 25. The estimators T_n'' are typically quick to compute and robust against outliers* unless the p_i are very close to 0 and 1, and are often used as starting points in the iterative calculation of more complicated estimators.

MIXTURE OF TYPES

Sometimes a combination of forms (2) and (3) is used. The best known example is

Example 7. The α-Winsorized mean (see ref. 33)

$$T_n = \frac{1}{n}\left(rX_{r:n} + \sum_{i=r+1}^{n-r} X_{i:n} + rX_{n-r:n} \right),$$

where $r = [n\alpha]$. T_n is the mean of a sample for which each of the smallest r order statistics has been replaced by $X_{r:n}$ and each of the r largest order statistics has been replaced by $X_{n-r:n}$. T_n can also be viewed as approximately $(1 - 2\alpha)$ times the α-trimmed mean of Ex. 1 plus $\alpha(\hat{\xi}_\alpha + \hat{\xi}_{1-\alpha})$.

Other examples occur naturally in *censored* situations (see Chernoff et al. [12, pp. 68–69]

Table 1. Efficiencies Relative to $\overline{X}$ for $n = 20$

Estimator	Uniform	Normal	$0.9N(0,1)+0.1N(0,9)$	Laplace	t_3
$T_{0.05}$	0.86	0.98	1.30	1.19	1.67
$T_{0.10}$	0.75	0.95	1.39	1.32	1.86
$T_{0.25}$	0.55	0.83	1.32	1.58	1.97
T_{adapt}	—	0.90	1.36	1.42	1.94
Median	0.41	0.67	1.10	1.53	1.73
Gastwirth	0.53	0.81	1.29	1.56	1.93

for a general theory). Consider type II censoring* on the right from the exponential distribution*, $F(x) = 1 - \exp(-x/\sigma)$, $x \geqslant 0$. Here, only the first $n - r$ order statistics $X_{1:n} \leqslant \cdots \leqslant X_{n-r:n}$ are available for inference and the maximum likelihood estimator of σ is the L-estimator

$$\hat{\sigma} = \left[\sum_{i=1}^{n-r} X_{i:n} + rX_{n-r:n} \right] \Big/ (n-r).$$

If $r = [n\alpha]$, then $\hat{\sigma}$ is just a one-sided trimmed mean plus approximately $\gamma\hat{\xi}_{1-\alpha}$, where $\gamma = \alpha/(1-\alpha)$.

COMPARISON OF SOME POPULAR ESTIMATORS

Table 1 lists the ratio of the variance of the mean $\overline{X}$ to the variance of six L-estimators for samples of size $n = 20$. The estimators are three α-trimmed means T_α, $\alpha = 0.05$, 0.10, and $\alpha = 0.25$, a trimmed mean due to Jaeckel [20] where α is chosen by the sample (T_{adapt}), the median, and the estimator of Example 6 (Gastwirth). The distributions are uniform, normal, contaminated normal ($F(x) = 0.9\Phi(x) + 0.1\Phi(x/3)$), Laplace ($f(x) = \frac{1}{2}\exp(-|X|)$), and the t distribution with 3 degrees of freedom. All of the estimators outperform $\overline{X}$ for the last three distributions, and all except the median perform well at the normal. The results for the uniform were taken from ref. 9, and all the rest were taken from ref. 1.

Table 2 compares four scale estimators to the sample standard deviation $S = [(n-1)^{-1}\sum(X_i - \overline{X})^2]^{1/2}$. Each entry is the ratio of the asymptotic variance of $\log S$ to the asymptotic variance of the log of each estimator. This measure was suggested in ref. 6 and adjusts for differences in the parameter being estimated. The estimators may all be defined in terms of the *quasi-ranges* $w_i = X_{n+1-i:n} - X_{i:n}$, $i = 1, [n/2]$. GINI is the estimator of Example 2 defined in terms of the w_i by

$$\text{GINI} = \frac{2}{n(n-1)} \sum_{i=1}^{[n/2]} (n - 2i + 1)w_i.$$

ADMED is the average deviation from the sample median given by

$$\text{ADMED} = \frac{1}{n} \sum_{i=1}^{n} |X_i - \text{median}|$$

$$= \frac{1}{n} \sum_{i=1}^{[n/2]} w_i.$$

Table 2. Standardized Asymptotic Efficiencies Relative to the Sample Standard Deviation

Scale Estimator	Uniform	Normal	Logistic	Laplace	Exponential
GINI	1.000	0.978	1.127	1.205	1.500
ADMED	0.600	0.876	1.124	1.250	1.566
$W_{0.0694}$	1.241	0.652	0.752	0.786	1.010
$W_{0.25}$	0.200	0.368	0.543	0.601	0.910

Note that GINI gives relatively more weight to the extreme quasi-ranges, whereas ADMED gives equal weight to each quasi-range. The last two estimators are individual quasi-ranges $W_\alpha = w_{[n\alpha]}$. $W_{0.0694}$ is the quasi-range which is optimal for normal data (see ref. 27). $W_{0.25}$ is the interquartile range (Example 5). Some of the results in Table 2 were taken from refs. 7 and 27. GINI and ADMED both do well relative to S. In fact, if GINI were compared to the maximum likelihood estimators of scale for the normal, logistic, and Laplace, the standardized asymptotic efficiencies would be 0.978, 0.985, and 0.964, respectively. $W_{0.0694}$ performs fairly well except at the (skewed) exponential. The interquartile range cannot be recommended except for simplicity.

L-STATISTICS AS FUNCTIONALS

A slight variation of form (2) is to replace $n^{-1}J(i/(n+1))$ by $c_{in} = n\int_{(i-1)/n}^{i/n} J(u)\,du$. Then (2) becomes $\int_0^1 F_n^{-1}(t)J(t)\,dt$, where $F_n(x)$ is the *empirical distribution function** of the data and $F_n^{-1}(t) = \min\{x : F_n(x) \geqslant t\}$. $F_n(x)$ is a step function with jumps of size $1/n$ at the order statistics, that is, the distribution function of a random variable Y_n which takes the value $X_{i:n}$ with probability equal to $1/n$. On the other hand, the inverse $F_n^{-1}(t)$ is equal to $X_{i:n}$ for $(i-1)/n < t \leqslant i/n$ so that $\hat{\xi}_{p_i} = F_n^{-1}(p_i)$. A graph is the easiest way to see these relationships. One version of (1) which contains both types of weights is then

$$T(F_n) = \int_0^1 F_n^{-1}(t)J(t)\,dt + \sum_{i=1}^{k} a_i F_n^{-1}(p_i). \tag{4}$$

Here we use the notation $T(\cdot)$ to indicate that (4) is a *functional* of F_n. This representation has two important advantages. First, since $F_n \to F$ ($F(x) = P(X_i \leqslant x)$) in a very strong stochastic sense, $T(F)$ is identified as the target parameter which $T(F_n)$ is estimating. For sample percentiles, it may be obvious that $\hat{\xi}_p$ is an estimator of the population percentile ξ_p defined by $F(\xi_p) = p$ or $\xi_p = F^{-1}(p)$. In general, the target of T_n may not be as obvious until written in the form (4). Second, the asymptotic analysis of $T(F_n)$ is greatly aided by studying smoothness properties of $T(\cdot)$. In particular, a special Gâteaux derivative of $T(\cdot)$, called by Hampel [17] the *influence curve**, is useful for computing asymptotic distributions and assessing the robustness of $T(F_n)$. If we denote the influence curve of $T(\cdot)$ at F by $\mathrm{IC}_{T,F}(x)$, then typically one can show that

$$n^{1/2}\left(T(F_n) - T(F) - \frac{1}{n}\sum_{i=1}^{n} \mathrm{IC}_{T,F}(X_i)\right) \xrightarrow{p} 0. \tag{5}$$

When (5) holds and $\operatorname{var} \mathrm{IC}_{T,F}(X_1) < \infty$, asymptotic normality of $T(F_n)$ follows from the central limit theorem applied to $n^{1/2}\sum \mathrm{IC}_{T,F}(X_i)$. From (5) we can also see that $\mathrm{IC}_{T,F}(X_i)$ is asymptotically the contribution (or influence) of X_i to the error approximation $T(F_n) - T(F)$. In the latter context Hampel [17] has suggested ways of using $\mathrm{IC}_{T,F}(x)$ to measure the robustness of $T(F_n)$.

Von Mises [34] pioneered the functional approach in statistics. Serfling [30, Chap. 6] gives the development since that time. Other sources include refs. 1, 4, 5, 6, and 8. *See also* STATISTICAL FUNCTIONALS.

EXTENSIONS

1. *L*-statistics have been generalized to $\sum c_{in} g(X_{i:n})$, where g is a suitable function (see ref. 12 or 31).
2. The c_{in} may be chosen by the data, in which case T_n is called adaptive (see ref. 19 and ADAPTIVE METHODS).
3. The form (1) does not easily generalize to multivariate data or regression. For multivariate trimming, see ref. 16. For regression, see refs. 3, 9 and 24.

References

[1] Andrews, D. F., Bickel, P. J., Hampel, F. R., Huber, P. J., Rogers, W. H., and Tukey, J. W. (1972). *Robust Estimates of Location.* Princeton

University Press, Princeton, N.J. (Includes a wealth of results but is poorly indexed.)

[2] Bennett, C. A. (1952). *Asymptotic Properties of Ideal Linear Estimators*. Ph.D. thesis, University of Michigan, Ann Arbor, MI.

[3] Bickel, P. J. (1973). *Ann. Statist.*, **1**, 597–616.

[4] Bickel, P. J. and Lehmann, E. L. (1975). *Ann. Statist.*, **3**, 1038–1044.

[5] Bickel, P. J. and Lehmann, E. L. (1975). *Ann. Statist.*, **3**, 1045–1069.

[6] Bickel, P. J. and Lehmann, E. L. (1976). *Ann. Statist.*, **4**, 1139–1158.

The above three papers suggest criteria for judging measures of location and scale defined as functionals of F.

[7] Boos, D. D. (1978). Gini's Mean Difference as a Nonparametric Measure of Scale. *Inst. Statist. Mimeo Ser. No. 1166*. North Carolina State University, Raleigh, N.C.

[8] Boos, D. D. (1979). *Ann. Statist.*, **7**, 955–959.

[9] Carroll, R. J. and Ruppert, D. (1980). *J. Amer. Statist. Ass.*, **75**, 828–838.

[10] Carroll, R. J. and Wegman, E. J. (1979). *Commun. Statist. A*, **6**, 795–812.

[11] Chan, L. K. and Rhodin, L. S. (1980). *Technometrics*, **22**, 225–237.

[12] Chernoff, H., Gastwirth, J. L., and Johns, M. V., Jr. (1967). *Ann. Math. Statist.*, **38**, 52–72. (Ground-breaking paper on asymptotic normality of L-statistics. Fairly technical.)

[13] Cicchitelli, G. (1976). *Metron*, **34**, 219–228.

[14] David, H. A. (1981). *Order Statistics*, 2nd ed. Wiley, New York.

[15] Gastwirth, J. L. (1966). *J. Amer. Statist. Ass.*, **61**, 929–948.

[16] Gnandesikan, R. and Kettenring, J. R. (1972). *Biometrics*, **28**, 81–124.

[17] Hampel, F. R. (1974). *J. Amer. Statist. Ass.*, **64**, 383–397.

[18] Helmers, R. (1980). *Ann. Statist.*, **8**, 1361–1374.

[19] Hogg, R. V. (1974). *J. Amer. Statist. Ass.*, **69**, 909–923. (Very readable.)

[20] Jaeckel, L. A. (1971). *Ann. Math. Statist.*, **42**, 1540–1552.

[21] Johnson, N. L. and Kotz, S. (1970). *Distributions in Statistics: Univariate Distributions*, 2 vols. Wiley, New York.

[22] Jung, J. (1955). *Ark. Mat.*, **3**, 199–209.

[23] Kendall, M. G. and Stuart, A. (1977). *The Advanced Theory of Statistics*, Vols. 1 and 2. Macmillan, New York, Chaps. 14, 19, 31.

[24] Koenker, R. and Bassett, G., Jr. (1978). *Econometrica*, **46**, 33–50.

[25] Kubat, P. and Epstein, B. (1980). *Technometrics*, **22**, 575–581.

[26] Lloyd, E. H. (1952). *Biometrika*, **39**, 41–67.

[27] Mosteller, F. (1946). *Ann. Math. Statist.*, **17**, 377–408.

[28] Ogawa, J. (1951). *Osaka Math. J.*, **3**, 175–213.

[29] Sarhan, A. E. and Greenberg, B. G., eds. (1962). *Contributions to Order Statistics*. Wiley, New York.

[30] Serfling, R. J. (1980). *Approximation Theorems of Mathematical Statistics*. Wiley, New York. (Chapter 8 deals with L-statistics. Chapter 6 introduces the functional approach.)

[31] Shorack, G. R. (1972). *Ann. Math. Statist.*, **43**, 412–427.

[32] Tietjen, G. L., Kahaner, D. K., and Beckman, R. J. (1977). In *Selected Tables in Mathematical Statistics*, Vol. 5, D. B. Owen and R. E. Odeh, eds. American Mathematical Society, Providence, R.I., pp. 1–73.

[33] Tukey, J. W. (1962). *Ann. Math. Statist.*, **33**, 1–67.

[34] von Mises, R. (1947). *Ann. Math. Statist.*, **18**, 309–348.

(CENSORING
EDF STATISTICS
GINI'S MEAN DIFFERENCE
INFLUENCE FUNCTIONS
ORDER STATISTICS
RANGES
TRIMMING AND WINSORIZATION)

DENNIS D. BOOS

LUBBOCK'S FORMULA

This formula expresses the sum of $(mn + 1)$ terms $u_0, u_1, \ldots, u_{mn}$ in terms of the sum of $(m + 1)$ terms $u_0, u_n, \ldots, u_{mn}$ with correction terms depending upon the forward differences* $\{\Delta^r u_0, \Delta^r u_{mn}\}$. The formula is

$$\sum_{j=0}^{mn} u_j = n \sum_{j=0}^{m} u_{nj} - \frac{1}{2}(n-1)(u_0 + u_{mn})$$
$$- \frac{n^2-1}{12n}(\Delta u_{mn} - \Delta u_0)$$
$$+ \frac{n^2-1}{24n}(\Delta^2 u_{mn} - \Delta^2 u_0)$$
$$- \frac{(n^2-1)(19n^2-1)}{720n^3}(\Delta^3 u_{mn} - \Delta^3 u_0)$$
$$+ \frac{(n^2-1)(9n^2-1)}{480n^3}(\Delta^4 u_{mn} - \Delta^4 u_0)$$
$$- \cdots .$$

If the mathematical form of u_x is known, and its derivatives can be calculated, it is usually more convenient to use Woolhouse's formula*.

LUCE'S CHOICE AXIOM AND GENERALIZATIONS

Suppose that we are to choose one out of three possible restaurants for dinner tonight, and that we are pairwise indifferent between them. That is, if our choice is restricted to any two, our choice probabilities are both $\frac{1}{2}$. Is it true that when all three restaurants are offered our choice probabilities are all $\frac{1}{3}$? The answer is no, in general, although it would be yes if the *choice axiom* of R. D. Luce holds.

Choice behavior long has been a major topic of discussion in both economics and psychology*. The economic literature mostly has been concerned with algebraic, or nonprobabilistic choice theories, with emphasis on formal mathematical properties of various axioms rather than on data fitting and statistical inference. In psychology there has been a similar emphasis on formalism, but following the classic work of Thurstone [13], it has been concerned mostly with probabilistic conditions. In his seminal book, *Individual Choice Behavior* [7], the psychologist R. D. Luce proposed his axiom for choice probabilities and explored its implications for the theories of learning, utility, and psychometrics.

Let $a, b, c, \ldots$ be elements of a set S of choice alternatives, and for any $a \in R \subseteq S$ write $P_R(a)$ for the chance that alternative a is chosen when a choice is to be made from R. For convenience we denote $P_{\{a,b\}}(a)$ by $P(a,b)$. Luce's choice axiom asserts that

A1 There is no pair $a, b \in S$ such that $P(a,b) = 0$.

A2

$$P_R(a) = P_S(a)/P_S(R). \tag{1}$$

The second condition, which is the more substantive part, says that the choice probabilities for a choice set R are identical to those for the choice set S conditional on R having been chosen. Notice that (1) would be a trivial identity if $P_R(a)$ were defined as a conditional probability. The setting here is quite general, however, and it is even possible that, for example, $P_R(a) < P_S(a)$. An example, from Corbin and Marley [2], is of a lady choosing between a red hat, a, and a blue hat, b. Here $R = \{a, b\}$. If the milliner produces an identical blue hat b', the lady may be more likely to choose a, since in choosing b she risks the embarrassment that another lady will appear attired in b'. Here $S = \{a, b, b'\}$ and $P_R(a) < P_S(a)$.

Simple consequences of the axiom are:

1. $P_R(a)/P_R(b)$ is independent of R. This may be regarded as a probabilistic formulation of K. J. Arrow's well-known principle of independence of irrelevant alternatives; it is also known as the constant ratio rule.
2. $P(a, b)\,P(b, c)\,P(c, a) = P(a, c)\,P(c, b)\,P(b, a)$, the product rule.
3. There exists a ratio scale v over S such that for all $a \in R \subseteq S$,

$$P_R(a) = v(a)/\sum_{b \in R} v(b). \tag{2}$$

[For example, take $v(a) = P_S(a)$.] This formulation is closely related to logit* analysis; economists often call it the strict utility model. For paired comparisons—i.e., two-element choice sets—the axiom is equivalent to the Bradley–Terry model*.

An important and illuminating counterexample to the Luce model was suggested by Debreu [3] in a review of Luce [7]. Let a, b, c be three alternatives such that the pairwise choice probabilities are all $\frac{1}{2}$, and suppose that a and b are much more similar to each other than to c. For instance, let a and b be two bicycles and c be a pony. Write $S = \{a, b, c\}$. According to Luce's axiom we must have $P_S(a) = P_S(c) = \frac{1}{3}$; yet in practice we would surely find $P_s(a)$ and $P_S(b)$ both close to $\frac{1}{4}$ and $P_S(c)$ close to $\frac{1}{2}$.

RELATION TO THE THURSTONE MODEL

According to the Thurstone model [13], there are random variables $\{U_a : a \in S\}$ such that for all $R \subseteq S$

$$P_R(a) = P\left(U_a = \max_{b \in R}\{U_b\}\right). \quad (3)$$

The maximizing variable is assumed to be unique with probability 1. The $\{U_a\}$ are taken to be independent, and identically distributed except for location shifts. *See* RANDOM UTILITY MODELS. Yellott [15] and others have shown that, *for the Thurstone model, the axiom holds if the common distribution is the extreme value distribution**,

$$P(U \leqslant x) = \exp(-e^{-(ax+b)}) \quad (4)$$

for some real a, b. Conversely, for $|S| \geqslant 3$, (4) is necessary for the choice axiom to hold. Yellott also gives the following interesting characterization:

> Let each element of a choice set be replicated k times—think of k identical cups of tea, k cups of coffee, etc., giving rise to k utility variables for each $a \in S$—and suppose that the probability of choosing a particular element (such as a cup of tea) is independent of k.

For the Thurstone model this condition implies that the variables have an extreme value distribution, and hence that the choice probabilities satisfy the choice axiom.[1]

TESTS OF THE CHOICE AXIOM

A wide variety of experimental tests have been performed. Some are based on the product or constant ratio rules, some on likelihood ratio tests. Domains include psychophysics (judgments of weights, sound intensities, colors), sociology* (prestige of occupations), animal studies, etc. Luce [8] discusses the tests and gives references. No very clear pattern emerges; sometimes the model fits well, other times not. In the case of animal studies (*see* STATISTICS IN ANIMAL SCIENCE) for example, food preferences of rats gave good agreement with the product rule, but the proportions of times spent by rats on alternative activities did not. Likelihood ratio tests carried out by Hohle [5] on various sets of data—lifted weights, preferences among vegetables, samples of handwriting, etc.—tended to show significant departures from the model. The model definitely fails in studies where some of the alternatives are similar, as in Debreu's example.

GENERALIZATIONS AND RELATED WORK

Because Luce's choice axiom fails to hold empirically in many situations, later theorists have proposed more general choice models. Much attention has been paid to Tversky's Choice By Elimination [14]. In this there are *aspects* according to which each choice alternative is either satisfactory or unsatisfactory. When presented with a choice set R the subject selects, by a well-defined procedure, some desirable aspect and rejects those elements of R which are unsatisfactory with respect to that aspect. If a subset T remains with $|T| > 1$, the subject restricts attention to T, selects another aspect, and so on until a single choice alternative remains. Tversky's model was not intended as a data analytic tool, and seems never to have been used as such; indeed, the technical problems of fitting it to data are formidable. Instead, interest has centered on axiomatic properties; for example, the model is a random utility model, generalizes Luce's and satisfies *weak stochastic transitivity*:

$$P(a,b) > \tfrac{1}{2},\ P(b,c) > \tfrac{1}{2} \Rightarrow P(a,c) > \tfrac{1}{2}.$$

A feature of Tversky's model (and, indeed, of many of the subsequent generalizations of Luce's model) is that the notion of a utility scale, with v-values as in (2) giving the strengths or intensities, is lost. Indow [6] gives a useful discussion and comparison of the models of Tversky and Luce. Corbin and Marley [2] propose a generalization of Tversky's model, based on a random utility model for which there is positive probability that the maximizing variable is not unique.

The idea of *correlated* utility variables, which seems to arise naturally from examples such as Debreu's, has been developed by McFadden [10], Strauss [12], and perhaps others. They consider Gumbel's type B extreme-value distribution

$$P(X_1 \leqslant x_1, X_2 \leqslant x_2) = \exp\left\{\left[e^{-\alpha x_1} + e^{-\alpha x_2}\right]^{1/\alpha}\right\}, \qquad \alpha > 1, \tag{5}$$

for which $\operatorname{corr}(X_1, X_2) = 1 - 1/\alpha^2$. In the independence case $\alpha = 1$, (5) reduces to the product of two distribution functions of form (4). If $v_i = e^{m_i}$, we have

$$P(X_1 + m_1 > X_2 + m_2) = \frac{v_1^\alpha}{v_1^\alpha + v_2^\alpha}. \tag{6}$$

Thus as α increases we are increasingly likely to pick the "better" alternative. The form (5) thus captures the notion of similarity, accounting for the well-known examples such as Debreu's and

$$P(\text{trip to Paris} + \$1, \text{trip to Paris}) = 1.$$

Strauss [12] extends this in the "choice by features" model, which allows several sets of correlated variables of form (5), each corresponding to a different "feature" of the choice set. According to this the subject takes each feature in turn, notes the maximum utility with respect to its utility model [a k-variate version of (5)], and finally selects the alternative whose maximizing utility with respect to one of the features proved to be the overall maximum. See also McFadden [10].

Models such as these raise questions about the general random utility model, in which the variables need not be independent. Strauss [11] gives some results relating various axiomatic and distributional properties; e.g., the choice probabilities

$$P(i, R) = P\left(X_i + m_i = \max_{j \in R}\{X_j + m_j\}\right)$$

are shown to satisfy the choice axiom for all sets $R \subseteq S$ and all real $\{m_i\}$ if and only if there is an $\alpha > 0$ such that

$$P(X_i + m_i = \max\{X_j + m_j\}) = \frac{e^{\alpha m_i}}{e^{\alpha m_i} + e^{\alpha m_j}}.$$

Compare with (6). Distribution families of this type include Gumbel's type B extreme-value distribution (5), and the multivariate logistic

$$F(x_1, \ldots, x_k) = \left\{1 + \sum_{i=1}^{k} e^{-\alpha x_i}\right\}^{-1}$$

In the paired comparison* case, sometimes called the Bradley–Terry model*, the representation

$$p(a, b) = \frac{v(a)}{v(a) + v(b)}$$

is clearly not unique. We can, for instance, choose a new scale w from S to the reals such that $p(a, b) = \{1 + \exp[w(b) - w(a)]\}^{-1}$ for all a, b. Colonius [1] characterizes the class of scales equivalent for a set of paired-comparison probabilities.

Finally, there is a considerable literature on the relationships of the choice axiom to conditions on the ranking of alternatives, discard mechanisms for choice, and other axioms. See Luce [8] and Luce and Suppes [9]. As an illustration, suppose that subjects can rank the elements of a choice set, and write $P(a, b, c, \ldots)$ for the probability that the rank order of choice is a followed by b followed by $c, \ldots$. One typical theorem asserts that for a random utility model the *decomposition* condition

$$P(a, b, c, \ldots, k) = P(a, \{a, b, c, \ldots, k\}) \times P(b, c, \ldots, k) \tag{7}$$

is sufficient for the choice axiom to hold, while the choice axiom is sufficient for (7) restricted to sets of three elements [11]. See Georgescu-Roegen [4] for some related results.

NOTE

1. Note, however, that in his proof Yellott assumes the k utility variables for the k replicates are independent. It seems much more natural to suppose that they are instead identical, but in that case Yellott's condition becomes trivially true.

References

[1] Colonius, H. (1980). *Brit. J. Math. Statist. Psychol.*, **33**, 99–110.

[2] Corbin, R. and Marley A. A. J. (1974). *J. Math. Psychol.*, **11**, 274–299.

[3] Debreu, G. (1960). *Amer. Econ. Rev.*, **50**, 186–188.

[4] Georgescu-Roegen, N. (1969). *Econometrica*, **37**, 728–730.

[5] Hohle, R. H. (1966). *J. Math. Psychol.*, **3**, 174–183.

[6] Indow, T. (1975). *Behaviormetrika*, **2**, 13–31.

[7] Luce, R. D. (1959). *Individual Choice Behavior*. Wiley, New York. (The seminal work on the choice axiom.)

[8] Luce, R. D. (1977). *J. Math. Psychol.*, **15**, 215–233. (Probably the best general review article on the axiom.)

[9] Luce, R. D. and Suppes, P. (1965). In *Handbook of Mathematical Psychology*, Vol. 3, R. D. Luce, R. R. Bush, and E. Galanter, eds. Wiley, New York, pp. 249–410. (A major reference article on the earlier axiomatic work on choice.)

[10] McFadden, D. (1981). In *Structure and Analysis of Discrete Data*, D. McFadden and C. Mansky, eds. MIT Press, Cambridge, Mass.

[11] Strauss, D. J. (1979). *J. Math. Psychol.*, **20**, 35–52.

[12] Strauss, D. J. (1981). *Brit. J. Math. Statist. Psychol.*, **34**, 50–61.

[13] Thurstone, L. L. (1927). *Psychol. Rev.*, **34**, 273–286.

[14] Tversky, A. (1972). *J. Math. Psychol.*, **9**, 341–367.

[15] Yellott, J. I. (1977). *J. Math. Psychol.*, **15**, 109–144. (A well-written and important paper on the relationships between the axiom, the random utility model, and the double exponential distribution.)

(CHANCE
PAIRED COMPARISONS
PSYCHOLOGY, STATISTICS IN
RANDOM UTILITY MODEL
UTILITY)

DAVID J. STRAUSS

LURKING VARIABLES

In observing data from some real-world process, we are unable to measure all the variables that affect it, a fact we tacitly admit as soon as we postulate a model containing an error term. The error term is a catchall for all the *lurking variables* that affect the process but are not observed, far less known to exist.

In correlation studies a lurking variable W may produce a significant correlation between two variables X and Y which themselves are not causally related. This happens when a change in W produces a change in each of X and Y.

The term "lurking variables" was coined by Box, Hunter, and Hunter [1]. The concept, however, goes back to the concept of error in the eighteenth century, to the practice of taking the arithmetic mean* as the "best" measurement of the physical quantity of interest; see LAWS OF ERROR I and II.

Reference

[1] Box, G. E., Hunter, W. G., and Hunter, J. S. (1978). *Statistics for Experimenters*. Wiley, New York.

(CAUSATION
SPURIOUS CORRELATION)

LYAPUNOV *See* LIAPUNOV

M

MacDONALD DISTRIBUTION

The MacDonald function*—also known as the modified Bessel function* of the second kind (or the modified Hankel function—is explicitly defined by

$$K_n(z) = \pi(2\sin n\pi)^{-1}\left[I_{-n}(z) - I_n(z)\right],$$

where

$$I_n(z) = \exp(-n\pi i/2)J_n(z(\exp(\pi i/2))$$

is the modified Bessel function of the first kind and

$$J_n(z) = \sum_{m=0}^{\infty}(-1)^m(z/2)^{2m+n} \times\left[\Gamma(m+1)\cdot\Gamma(m+n+1)\right]^{-1}$$

is the Bessel function of the first kind of order n.

The "type $x^n K_n(x)$" MacDonald density function with scale parameter b is given by:

$$f(x) = \pi^{-1/2}2^{-n+1}\left[\Gamma(n+\tfrac{1}{2})b\right]^{-1} \times (x/b)^n K_n(x/b), \qquad x \geqslant 0, \quad (1)$$

when n can be viewed as a shape parameter.

For this distribution the expected value is

$$2b\pi^{-1/2}\Gamma(n+1)\left[\Gamma(n+\tfrac{1}{2})\right]^{-1}$$

and the variance is

$$b^2\left\{2n+1-\frac{4}{\pi}\left[\frac{\Gamma(n+1)}{\Gamma(n+\frac{1}{2})}\right]^2\right\}.$$

In applications the symmetrized MacDonald density

$$f(x) = \pi^{-1/2}2^{-n}\left[\Gamma(n+\tfrac{1}{2})b\right]^{-1} \times (|x|/b)^n K_n(|x|/b) \qquad (2)$$

with the expected value 0 and variance $(2n+1)b^2$ is used (see, e.g., McGraw and Wagner [4]).

Other types of MacDonald's distribution are $x^{n+p}K_n(x)$, $p \geqslant 1$ (see Kropáč [3] for more details).

The distribution (1) approaches a folded normal distribution (*See* FOLDED DISTRIBUTIONS) as $n \to \infty$ while (2) approaches a normal distribution. The MacDonald distribution of $x^{n+1}K_n(x)$ type approaches the Rayleigh distribution* as $n \to \infty$. (For $n = 0.5$ the distribution (1) is a (negative) exponential distribution*.)

A MacDonald distribution of the form $(\pi\sigma_x\sigma_y)^{-1}K_0[z/(\sigma_x\sigma_y)]$ arises as the distribution of the product XY where X and Y are independent normal random variables with mean values 0 and standard deviations σ_x and σ_y, respectively.

If the mean values are *not* zero the corresponding density can be expressed as an infinite series of terms containing products of K_n and I_{2n}.

A MacDonald distribution also arises as the distribution of the product XY of two independent gamma* or generalized gamma variables, and in the distribution of mean values from random samples from exponential and Laplace populations. (See Kropáč [3] for more details.) It also arises as the compound* (mixture*) distribution Normal Λ_{σ^2} Gamma (α, β) or Λ Gamma (α, β).

These are *elliptically symmetric distributions** which also can be viewed as generalized Laplace distributions.* (See Kropáč [2] and Johnson and Kotz [1] for more details and interrelations.)

References

[1] Johnson, N. L. and Kotz, S. (1972). *Continuous Multivariate Distributions*. Wiley, New York, pp. 296–97.

[2] Kropáč, O. (1981). *Kybernetika*, **17**, 401–412.

[3] Kropáč, O. (1982). *Aplikace Matematiky*, **27**, 285–302.

[4] McGraw, D. K. and Wagner, J. F. (1968). *IEEE Trans. Inf. Theory*, **14**, 110–120.

Bibliography

MacDonald, H. M. (1899). *Proc. London Math. Soc.*, **30**, 165–179.

(BESSEL FUNCTIONS)

MACDONALD FUNCTION

The MacDonald function is defined for noninteger ν by

$$K_\nu(z) = \frac{\pi\left[I_{-\nu}(z) - I_\nu(z)\right]}{2\sin\nu\pi}$$

where

$$I_\nu(z) = \left(\frac{1}{2}z\right)^\nu \sum_{j=0}^{\infty}\left(\frac{1}{4}z^2\right)^j \Big/ \left(j!\,\Gamma(\nu + j + 1)\right)$$

is a modified Bessel function*. ($K_\nu(z)$ is also called a modified Bessel function of the third kind.) For integer values n we have

$$K_n(z) = \lim_{\nu\to n} K_\nu(z) \quad \left[n = 0, \pm 1, \pm 2, \ldots\right].$$

See BESSEL FUNCTIONS for more details.

MACLAURIN SERIES

A particular form of Taylor series, obtained by expanding an infinitely differentiable function in powers of x:

$$f(x) = f(0) + \frac{x}{1!}f'(0) + \frac{x^2}{2!}f''(0) + \cdots + \frac{x^n}{n!}f^{(n)}(0) + R_n,$$

where $f^{(i)}(0)$ is the ith derivative of $f(x)$ at $x = 0$. The remainder term can be expressed as

$$R_n = \frac{x^{n+1}}{(n+1)!}f^{(n+1)}(\theta x) = \frac{1}{n!}\int_0^x (x - t)^n f^{(n+1)}(x)\,dt,$$

where $0 < |\theta| < 1$. The convergence of a Maclaurin series can be determined by investigating the remainder directly or by establishing the radius of convergence of the series. It is possible that the series may converge as $n \to \infty$, but the sum $S(x)$ may not be equal to $f(x)$; i.e., $R_n \to f(x) - S(x)$. In probability theory Maclaurin series are used in connection with moment generating functions* and their relation to moments of a random variable.

Bibliography

Apostol, T. M. (1957). *Mathematical Analysis*. Addison-Wesley, Reading, Mass.

Courant, R. (1937). *Differential and Integral Calculus*, Vol. 1, 2nd. ed. Blackie and Sons, London, England.

Mullins, E. R., Jr. and Rosen, D. *Probability and Calculus*. Bogden and Quigley, Tarrytown-on-Hudson, N.Y.

Parzen, E. (1960). *Modern Probability Theory and Its Applications*. Wiley, New York.

(GENERATING FUNCTIONS
MOMENT PROBLEM
TAYLOR EXPANSION)

MacQUEEN'S CLUSTERING ALGORITHM

This was the original version of the k-means algorithm* (see MacQueen [1]).

Reference

[1] MacQueen, J. B. (1967). *Proc. Fifth Berkeley Sym. Math. Statist. and Prob.*, **1**, 281–297, University of California Press, Berkeley.

(k-MEANS ALGORITHMS)

MAD

MAD is an analysis of variance* program for unbalanced designs. It is a collection of Fortran subroutines implementing the procedure described by Bryce and Carter. [1]. See also Bryce [2] for a general description of the program and its implementation.

References

[1] Bryce, G. R. and Carter, M. W. (1974). COMSTAT 1974 Proc. in Comp. Statist., G. Bruckman et al., eds., Physica Verlag, Vienna, Austria.

[2] Bryce, G. R. (1975). *Appl. Statist.*, **24**, 350–352.

(STATISTICAL SOFTWARE)

MAGIC SQUARE DESIGNS

A *magic square* is a set of integers arranged in the form of a square so that the sums of the integers in each row, column and diagonal of the square are the same. If the integers are the first n^2 natural numbers, the square is said to be of the nth order, and it is easy to see that each sum is $\frac{1}{2}n(n^2+1)$. Such squares have been known to mathematicians for hundreds (or possibly thousands) of years, and whole books have been written about them; for example ref. 1. There is also a chapter in the book by Rouse Ball and Coxeter [4], and this, in one edition or another, has been used as the source by most of those recently interested in the topic.

There is no magic square of order 2, and only one basic magic square of order 3:

8	1	6
3	5	7
4	9	2

This can be presented in eight forms by reflection and rotation. The numbers of basic squares of higher order increase rapidly, there being 880 of order 4, and more than 15 million of order 5.

In the square

15	10	3	6
4	5	16	9
14	11	2	7
1	8	13	12

,

not only do the rows, columns and diagonals sum to $34 = \frac{1}{2}n(n^2+1)$, but so also do the six *broken diagonals* 15, 9, 2, 8; 10, 16, 7, 1; 10, 4, 7, 13; 3, 5, 14, 12; 3, 9, 14, 8; 6, 4, 11, 13. A square with this property is called *pandiagonal*, and such a square can be found for any value of $n > 3$, such that $n \neq 4m+2$. There are 48 pandiagonal squares of order 4, 3600 of order 5, more than 38 million of order 7 and more than 6.5×10^{12} of order 8.

If a magic square is such that the numbers in the cells symmetrically related to the center of the square sum to n^2+1, then the square is said to be *symmetrical*. All the squares shown here, except the pandiagonal one above, have this property.

Many constructions of magic squares are known. The simplest for order 4 consists of writing the numbers 1 to 16 in their natural order; then, if the number p occurs on either the diagonal from top left to bottom right, or the diagonal from top right to bottom left, replace it by $(17-p)$; otherwise, leave it alone. The resulting square is:

16	2	3	13
5	11	10	8
9	7	6	12
4	14	15	1

The same principle applies to other squares

of order $4m$ by similarly changing the numbers in those cells on the diagonals of component blocks of 16 cells.

For odd order the simplest construction is to start with 1 in the middle of the top row and then put the numbers in order in a diagonal line sloping up to the right, except that (a) when the top row is reached, the next number is written in the bottom row as if it came immediately above the top row; (b) when the right-hand column is reached, the number is written in the left-hand column as if it were to the right of the right-hand column; (c) when a cell already filled or the top right-hand corner is reached, the next number is put directly below the previous one. This method was used to construct the square of order 3 given above and gives the following square of order 5:

17	24	1	8	15
23	5	7	14	16
4	6	13	20	22
10	12	19	21	3
11	18	25	2	9

The method gives rise to a Graeco-Latin square*; for order 5 with numbers going from 00 to 44 in the scale of 5 this is

31	43	00	12	24
42	04	11	23	30
03	10	22	34	41
14	21	33	40	02
20	32	44	01	13

A slightly more complicated rule is required for constructing magic squares of order $4m + 2$, details being given in ref. 4.

There only appears to have been one use of magic squares for the design of an experiment, the principles of which are explained by Phillips [3]. The idea is that in psychological experiments balancing for trend is sometimes necessary, and this can be achieved by presenting a set of factorial* treatments in the order given by the numbers in a symmetrical magic square. Thus, the square of order 4 could represent four treatments each at two levels in a 2^4 factorial design. If fewer factors are needed, then the omitted factor can be used to measure order effects. A pandiagonal square can be used to balance a Latin square of order $n > 3$, $n \neq 4m + 2$, or a Graeco-Latin square of odd order.

An entirely different type of experimental design has been described as a *magic Latin square* [2]. This is a Latin square* of side n, not prime, with the additional restriction that a complete replicate of the treatments is found not only in each row and each column but also in $nr \times s$ rectangles, where $r \times s = n$. An example is as follows:

A	*B*	*C*	*D*	*E*	*F*
D	*E*	*F*	*A*	*B*	*C*
C	*A*	*B*	*E*	*F*	*D*
F	*D*	*E*	*C*	*A*	*B*
B	*C*	*A*	*F*	*D*	*E*
E	*F*	*D*	*B*	*C*	*A*

The randomization of the treatments in such a design requires more care than that for an ordinary Latin square, and there is an extra term in the analysis of variance* of the results. Such designs have been used for practical experiments only rarely, and are not really recommended since the additional stratification will not usually control any considerable amount of variation.

There seems to be no reason why magic square designs, as originally defined, should not be used if the experimental requirements are satisfied, particularly because of the relation with Latin and Graeco-Latin squares. However, magic squares have tended to remain the province of recreational mathematicians and professional magicians.

References

[1] Andrews, W. S. (1917). *Magic Squares and Cubes*. Dover, New York.

[2] Federer, W. T. (1955). *Experimental Designs: Theory and Application*. Macmillan, New York, Chap. 15.

[3] Phillips, J. P. N. (1964). *Appl. Statist.*, **13**, 67–73.

[4] Rouse Ball, W. W. and Coxeter, H. S. M. (1974). *Mathematical Recreations and Essays*, 12th ed., University of Toronto Press, Toronto, Ontario, Chap. 7.

(GRAECO-LATIN SQUARES
LATIN SQUARES)

G. H. FREEMAN

MAHALANOBIS, PRASANTA CHANDRA

Born: June 29, 1893, in Calcutta, India.

Died: On the eve of his seventy-ninth birthday, June 28, 1972, in Calcutta, India.

Contributed to: Statistics, economic planning.

Prasanta Chandra Mahalanobis's family originally belonged to the landed aristocracy of Bengal, but had moved to Calcutta by the middle of the last century. Prasantha was educated first in Calcutta, where he received the B.Sc. honors degree in Physics, and later at Cambridge University, where he passed Part I of the Mathematical Tripos in 1914 and the Natural Science Tripos, Part II, in 1915, and was elected a senior research scholar. A visit to India on a holiday turned out to be a permanent homecoming because of an opportunity to teach at the Presidency College in Calcutta and pursue his growing interest in statistical problems. The latter began to absorb his interest so completely that he abandoned his intention of pursuing a career of research in physics and decided to devote himself to statistics, although he continued to teach physics until 1948.

A man of great originality, Mahalanobis's contribution to statistical thought goes far beyond his published work. He considered as artificial any distinction between "theoretical" and "applied" statistics. His own work in statistics was always associated with, and arose out of, some field of application. His investigations on anthropometric problems led him to introduce the D^2—statistic, known later as Mahalanobis's distance, and used extensively in classification problems (*see* MAHALANOBIS D^2). This early work yielded a wealth of theoretical problems in multivariate analysis*, many of which were later solved by his younger colleagues. Among other major areas in which he worked during the 1930s and 1940s are (a) meteorological statistics, (b) operations research* (to give it its modern name), (c) errors in field experimentation, and (d) large-scale sample surveys*.

It is characteristic of Mahalanobis's research that fundamentally new ideas are introduced in the course of studying some immediate, practical problem. Thus his concept of pilot surveys was the forerunner of sequential analysis*, anticipating it by a decade. The 1944 memoir on sample surveys in the *Philosophical Transactions of the Royal Society* established the theory, gave the estimating procedures, and, at the same time, raised basic questions concerning randomness*, as to what constitutes a random sample. He also introduced the concepts of optimum survey design and interpenetrating* network of subsamples. His development of cost and variance functions in the design of sampling* (originating in the work of J. Neyman*) may be regarded as an early use of operational research techniques.

Mahalanobis's work in field experimentation (1925), carried out in ignorance of the design of experiments introduced earlier by R. A. Fisher*, brought the two men together and started a close professional and personal friendship that lasted until Fisher's death in 1962. Mahalanobis and Fisher held similar views on the philosophy of statistics as well as its methodological aspects. Both regarded statistics as a "key technology" for "increasing the efficiency of human efforts in the widest sense."

Mahalanobis's personal contributions to statistics have, to some extent, been obscured by his two monumental achievements —the founding of the Indian Statistical Institute* (ISI) and the creation of the National Sample Survey (NSS). The ISI, established in 1931, not only has produced a generation of statisticians of world stature but has played a part equal to that of western countries in making statistics the highly developed, precise science that it is today. (For a brief account of the history and activities of the ISI, see INDIAN STATISTICAL INSTITUTE.)

In 1950 Mahalanobis established the NSS as a division within the ISI. The NSS rapidly grew into an agency (for which there hardly was a parallel elsewhere in the world) noted for its use of continuing sample surveys for the collection of socioeconomic and demographic data, which covered an entire country. It came to play so vital a role during the economic five-year plans in India that the NSS was taken over by the government and now continues to function as an integral part of the Ministry of Planning.

If Professor Mahalanobis had retired in 1947, these achievements alone would have assured him an enduring place in the history of statistics. However, during the last 20 years of his life he became intensely interested in the applications of statistics to problems of planning. He came up with a two-sector and later, a four-sector model for economic growth. (In doing so, he had no pretensions to making a contribution to economic theory.) The resources of the ISI were fully mobilized to assist the Planning Commission, of which Mahalanobis was himself a member for the years 1955–1967.

Under his leadership, the ISI prepared the draft frame of the Indian second Five-Year Plan. The importance of this aspect of Mahalanobis's activity—though far removed from the world of academic statistics—can hardly be exaggerated. It provided the government with a scientific data base for industrialization as well as a cadre of highly skilled statisticians at the very outset of the Five-Year Plans. If, in the course of the last 30 years, India has been transformed from a country with virtually no industrial base into one ranking among the first 10 countries of the world with the largest industrial output, some measure of credit belongs to Professor Mahalanobis and the statistical institutions created by him.

Mahalanobis's interests were not confined to statistics and the physical sciences. He had more than a casual interest in ancient Indian philosophy, particularly in those aspects relating to multivalued logic. Bengali literature was his second love having, in his younger days, been a protégé of the great Indian poet Rabindranath Tagore. In the midst of an extremely busy life he found time to take a prominent part in the activities of the Brahmo Samaj, a religious reform movement which spearheaded a renaissance in the intellectual life of Bengal in the nineteenth and early twentieth centuries.

From 1949 until his death, Professor Mahalanobis was the Statistical Advisor to the government of India. He received many awards and honors. Among the ones which he treasured most were perhaps the Fellowship of the Royal Society and one of his country's highest civilian awards, the *Padma Vibhushan*.

Professor Mahalanobis published over 200 scientific papers in addition to numerous articles in Bengali and English in nontechnical journals. A bibliography of his scientific publications and a more detailed account of his work can be found in the biographical memoir written by C. R. Rao for the Royal Society [1].

Reference

[1] Rao, C. R. (1973). *Biog. Mem. Fellows of the R. Society*, **19**, 455–492.

(INDIAN STATISTICAL INSTITUTE
MAHALANOBIS D^2)

G. KALLIANPUR

MAHALANOBIS D^2

INTRODUCTION

Mahalanobis D^2, which is the two-sample version of Hotelling's T^2*, is a generalization of the two-sample t^2, and a special case of Hotelling's T_0^2 for MANOVA* (*see* HOTELLING'S TRACE) when the number of samples $l = 2$. The statistic is defined by

$$D^2 = \mathbf{d}'\mathbf{S}^{-1}\mathbf{d},$$

where $\mathbf{d} = \overline{\mathbf{X}}_1 - \overline{\mathbf{X}}_2$, $\mathbf{S} = (n_1\mathbf{S}_1 + n_2\mathbf{S}_2)/n$, $n = n_1 + n_2$, and where $\overline{\mathbf{X}}_i, \mathbf{S}_i$ are the p-dimensional column vector of means and $p \times p$ covariance matrix with $n_i = N_i - 1$ degrees of freedom (DF), respectively, computed from a sample of size N_i $(N_i > p)$

drawn from a p-variate normal population $N_p(\boldsymbol{\mu}_i, \boldsymbol{\Sigma})$ with mean $\boldsymbol{\mu}_i$ and symmetric positive definite covariance matrix $\boldsymbol{\Sigma}(i = 1, 2)$. The two samples are drawn independently. $\mathbf{A}', \mathbf{A}^{-1}$ denote the transpose and inverse, respectively, of the matrix $\mathbf{A}$. Mahalanobis [10, 11] suggested $\boldsymbol{\delta}'\boldsymbol{\Sigma}^{-1}\boldsymbol{\delta}$, $\boldsymbol{\delta} = \boldsymbol{\mu}_1 - \boldsymbol{\mu}_2$, as a measure of the distance between the two populations. The statistic can be derived through the likelihood ratio or union–intersection principles* (*see* HOTELLING'S T^2) for testing the hypothesis $H_0: \boldsymbol{\mu}_1 = \boldsymbol{\mu}_2$ versus $H_1: \boldsymbol{\mu}_1 \neq \boldsymbol{\mu}_2$.

DISTRIBUTIONS

Mahalanobis D^2

The method of derivation of the distribution of D^2 is exactly the same as that of Hotelling's T^2 and hence $[(n - p + 1)/(pn)]cD^2$ is distributed as a noncentral F* with p and $n - p + 1$ DF and noncentrality parameter $c\boldsymbol{\delta}'\boldsymbol{\Sigma}^{-1}\boldsymbol{\delta}$, where $c = N_1N_2/N$, $N = N_1 + N_2$. If $\boldsymbol{\delta} = \mathbf{0}$ then $[n - p + 1)/(pn)]cD^2$ is distributed as central $F_{p,n-p+1}$ (see Rao [19], Bose and Roy [3] for earlier derivation, and references in HOTELLING'S T^2).

Related Criteria

Consider $N_{p+q}(\boldsymbol{\mu}_i, \boldsymbol{\Sigma})$, $i = 1, 2$. Let

$$\boldsymbol{\delta} = \begin{matrix} p \\ q \end{matrix}\underset{1}{\begin{pmatrix} \boldsymbol{\delta}_1 \\ \boldsymbol{\delta}_2 \end{pmatrix}}, \qquad \boldsymbol{\Sigma} = \begin{matrix} p \\ q \end{matrix}\underset{\quad p \qquad\quad q}{\begin{pmatrix} \boldsymbol{\Sigma}_{11} & \boldsymbol{\Sigma}_{12} \\ \boldsymbol{\Sigma}_{12}' & \boldsymbol{\Sigma}_{22} \end{pmatrix}}.$$

To test the hypothesis $H_0: (\boldsymbol{\delta}_1 - \boldsymbol{\Sigma}_{12}\boldsymbol{\Sigma}_{22}^{-1}\boldsymbol{\delta}_2) = \mathbf{0}$ given $\boldsymbol{\delta}_2 \neq 0$, i.e., the mean vectors in the two populations are equal after adjustment for the covariable (see Subrahmaniam and Subrahmaniam [22]), the following test criteria have been suggested (see Rao [15, 16]; Cochran and Bliss [4]; Subrahmaniam and Subrahmaniam [21]):

$$T_1 = c\left(D_{p+q}^2 - D_q^2\right)/\left[n + cD_q^2\right] \quad \text{and}$$

$$T_2 = c\left(D_{p+q}^2 - D_q^2\right)/n,$$

where D_r^2 denotes the Mahalanobis D^2 for r variables. The hypothesis H_0 considered here reformulates the hypotheses H_{01}, H_{02}, and H_{03} of Rao [18] into a single hypothesis. T_1 is a conditional test while T_2 is unconditional on the basis of the distributions involved. The nonnull distribution of $[(n - p - q + 1)T_1/p]$ conditional on D_q^2 is a noncentral F with p and $n - p - q + 1$ DF and noncentrality parameter $\lambda/(1 + u)$, where

$$\lambda = c\boldsymbol{\delta}_{1\cdot 2}'\boldsymbol{\Sigma}_{11\cdot 2}^{-1}\boldsymbol{\delta}_{1\cdot 2},$$

$$\boldsymbol{\delta}_{1\cdot 2} = \boldsymbol{\delta}_1 - \boldsymbol{\Sigma}_{12}\boldsymbol{\Sigma}_{22}^{-1}\boldsymbol{\delta}_2,$$

$$\boldsymbol{\Sigma}_{11\cdot 2} = \boldsymbol{\Sigma}_{11} - \boldsymbol{\Sigma}_{12}\boldsymbol{\Sigma}_{22}^{-1}\boldsymbol{\Sigma}_{12}',$$

and $u = cD_q^2/n$. When H_0 is true, $(n - p - q + 1)T_1/p$ is distributed as a central $F_{p,n-p-q+1}$ and does not involve D_q^2. Hence in this case the unconditional distribution is also a central $F_{p,n-p-q+1}$, whether $\boldsymbol{\delta}_2$ is zero or not. The nonnull distribution of $T_2 = T_1(1 + u)$ has been obtained by Subrahmaniam and Subrahmaniam [22] using those of T_1 and u, the latter being a constant times D_q^2.

OPTIMUM PROPERTIES

The optimum power properties are the same as those of Hotelling's T^2. Also the properties discussed for Hotelling's trace* for MANOVA apply to D^2 as a special case when $l = 2$.

Further, in regard to robustness* against nonnormality also, the properties stated for Hotelling's T^2 and Hotelling's trace are relevant in this case, i.e., D^2 is reasonably robust against the kurtosis* aspect of nonnormality. In addition, it may be pointed out from the study of Ito and Schull [6] that, in large samples, if $N_1 = N_2$, the unequal covariance matrices have no effect on the size of the type I error probability or the power function. Further, Rao [18] has shown that, in small samples, the power can decrease with an increase in the number of variables unless the corresponding true D^2 is of certain order.

As regards optimal properties of T_1 and T_2, power studies including numerical tabulations were made [21, 22] assuming $\boldsymbol{\delta}_2 = \mathbf{0}$, and $\boldsymbol{\delta}_2 \neq \mathbf{0}$. Based on these studies it has been observed that, for testing $\boldsymbol{\delta}_1 = \mathbf{0}$ in the presence of a set of covariates, T_1 has generally better power than T_2 for small values of D_q^2, but the reverse is true for large values of D_q^2.

APPLICATIONS

A few applications are discussed to illustrate the usefulness of the test criteria defined previously.

Discriminant Function

The linear discriminant function between two populations $N_p(\boldsymbol{\mu}_i, \boldsymbol{\Sigma})$, $i = 1, 2$ could be defined as $\mathbf{l}'\mathbf{X}$, $\mathbf{l}' = (l_1, \ldots, l_p)$ and $\mathbf{X}' = (X_1, \ldots, X_p)$, the vector of p variables, for which $(\mathbf{l}'\boldsymbol{\delta})^2/(\mathbf{l}'\boldsymbol{\Sigma}\mathbf{l})$ is a maximum with respect to $\mathbf{l}$ (*see* DISCRIMINANT ANALYSIS). If $\mathbf{l}'\boldsymbol{\delta} = \mathbf{l}'\boldsymbol{\Sigma}\mathbf{l}$, the maximum is attained when $\mathbf{l} = \boldsymbol{\Sigma}^{-1}\boldsymbol{\delta}$. The linear discriminant function due to R. A. Fisher [5] is obtained when $\boldsymbol{\delta}$ is estimated by $\mathbf{d}$ and $\boldsymbol{\Sigma}$ by $\mathbf{S}$, such that $\hat{\mathbf{l}} = \mathbf{S}^{-1}\mathbf{d}$. In order to test the significance of the discriminant function, the test employed is that of the hypothesis $H_0: \boldsymbol{\mu}_1 = \boldsymbol{\mu}_2$ versus $H_1: \boldsymbol{\mu}_1 \neq \boldsymbol{\mu}_2$ using D_p^2, and the significance of the discriminant function is concluded if

$$[(n-p+1)/pn]cD_p^2 > F_{p,n-p+1,1-\alpha}.$$

Value of Additional Information

Consider the hypothesis

$$H_0: l_{r+1} = \cdots = l_p = 0,$$

i.e., $\mathbf{l}_2 = \mathbf{0}$ versus H_1: (not all $l_{r+1}, \ldots, l_p$ equal zero), where $\mathbf{l}'(1 \times p) = (\mathbf{l}_1', \mathbf{l}_2')$ is the coefficient vector of the linear discriminant function for p variables, $\mathbf{l} = \boldsymbol{\Sigma}^{-1}\boldsymbol{\delta}$. Here $\mathbf{l}_2 = \boldsymbol{\Sigma}_{22\cdot 1}^{-1}\boldsymbol{\delta}_{2\cdot 1}$ where $\boldsymbol{\delta}$ as well as $\boldsymbol{\Sigma}$ are partitioned according to r, $p - r$. Hence H_0: $(\mathbf{l}_2 = \mathbf{0})$ can be tested using

$$T_1 = c(D_p^2 - D_r^2)/[n + cD_r^2],$$

noting that $(n - p + 1)T_1/(p - r)$ is distributed as a central $F_{p-r,n-p+1}$. If $F_{p-r,n-p+1} > F_{p-r,n-p+1,1-\alpha}$, conclude in favor of H_1.

Assigned Discriminant Function

Let the hypothesis be H_0: ($\mathbf{a}'\mathbf{X}$ discriminates well between the two populations) versus H_1: ($a'\mathbf{X}$ does not), where $\mathbf{a}$ is a given vector. The T_1 statistic could be used for the test of H_0 since a discriminant function is invariant under a linear transformation, and a nonsingular linear transformation $\mathbf{Z}(p \times 1) = \mathbf{AX}$ with $z_1 = \mathbf{a}'\mathbf{X}$ as the first component of $\mathbf{Z}$ would leave the coefficients of $z_2, \ldots, z_p$ in the discriminant function all zero. Hence $T_1 = c(D_p^2 - D_1^2)/[n + cD_1^2]$ and the central $F_{p-1,n-p+1} = (n - p + 1)T_1/(p - 1)$ are the suitable test criteria, where D_p^2 is based on all the components of $\mathbf{Z}$. If $F_{p-1,n-p+1} > F_{p-1,n-p+1,1-\alpha}$, conclude in favor of H_1.

Discriminant Function Coefficients

Consider the hypothesis $H_0: (l_j/l_i) = \rho$ (specified) versus $H_1: (l_j/l_i) \neq \rho$. The hypothesis could be tested using $T_1 = c(D_p^2 - D_{p-1}^2)/[n + cD_{p-1}^2]$ and $F_{1,n-p+1} = (n - p - 1)T_1$ where D_{p-1}^2 is computed using the $p - 1$ variables $X_1, \ldots, X_{i-1}, X_{i+1}, \ldots, X_{j-1}, X_{j+1}, \ldots, X_p, X_i + \rho X_j$. Then if $F_{1,n-p+1} > F_{1,n-p+1,1-\alpha}$, conclude in favor of H_1.

The test has been extended to the case for the hypothesis H_0: the ratios of coefficients of $X_1, \ldots, X_s$ are $\rho_1, \ldots, \rho_s$, respectively (see Rao [19]). The test criteria in this case are obtained by replacing D_{p-1}^2 above in T_1 by D_{p-s+1}^2 based on the variables $\rho_1 X_1 + \cdots + \rho_s X_s, X_{s+1}, \ldots, X_p$, and further replacing 1 DF in the F-statistic by $s - 1$ DF. In the discriminant function the ratios of coefficients are unique, unlike the individual coefficients. (See Kshirsagar [8], Rao [19] for more details on the above tests.)

Profile Analysis

The profile of a group is defined as the graph joining successively the points (i, μ_i), $i = 1, \ldots, p$, where μ_i is the expected score in the ith test in a battery of p tests. The profiles are similar in two groups if

$\mathbf{C}\boldsymbol{\mu}_1 = \mathbf{C}\boldsymbol{\mu}_2$, where

$$\mathbf{C}((p-1)\times p) = \begin{bmatrix} 1 & -1 & 0\cdots 0 & 0 & 0 \\ 0 & 1 & -1\cdots 0 & 0 & 0 \\ \cdot & \cdot & \cdots\cdots & \cdot & \cdot \\ 0 & 0 & 0\cdots 0 & 1 & -1 \end{bmatrix}.$$

Given p-variate samples of sizes N_1 and N_2, respectively, of the two independent groups, similarity of profiles could be tested by the hypothesis

$$H_0: \mathbf{C}\boldsymbol{\mu}_1 = \mathbf{C}\boldsymbol{\mu}_2 \quad \text{versus} \quad H_1: \mathbf{C}\boldsymbol{\mu}_1 \neq \mathbf{C}\boldsymbol{\mu}_2.$$

The test will reject H_0 if $F_{p-1,n-p+2} > F_{p-1,n-p+2,1-\alpha}$, where

$$F_{p-1,n-p+2} = [(n-p+2)/((p-1)n)]cD^2_{p-1},$$

$$D^2_{p-1} = \mathbf{d}'\mathbf{C}'(\mathbf{CSC}')^{-1}\mathbf{Cd}.$$

Confidence Bounds for δ

A $100(1-\alpha)\%$ joint confidence region for $\boldsymbol{\delta}$ is given by

$$(\mathbf{d}-\boldsymbol{\delta})'\mathbf{S}^{-1}(\mathbf{d}-\boldsymbol{\delta}) \leqq D^2_{p,n-p+1,1-\alpha},$$

$$D^2_{p,n-p+1,1-\alpha} = [pn/(c(n-p+1))]F_{p,n-p+1,1-\alpha}.$$

Further, for all nonnull $\mathbf{a}(p\times 1)$ (see Morrison [14], Roy [20])

$$\mathbf{a}'\mathbf{d} - [\mathbf{a}'\mathbf{Sa}D^2_{p,n-p+1,1-\alpha}]^{1/2} \leqq \mathbf{a}'\boldsymbol{\delta} \leqq \mathbf{a}'\mathbf{d} + [\mathbf{a}'\mathbf{Sa}D^2_{p,n-p+1,1-\alpha}]^{1/2}$$

gives $100(1-\alpha)\%$ simultaneous confidence* bounds on all linear functions of $\boldsymbol{\delta}$.

Cluster Analysis

If $\mathbf{x}_{ij}$, $j = 1, \ldots, n_i, i = 1, \ldots, k$ be k p-dimensional random samples, cluster analysis aims at grouping the k samples into g homogeneous classes where g is unknown, $g \leqq k$. If attention is restricted to $N_p(\boldsymbol{\mu}_i, \boldsymbol{\Sigma})$ populations, $i = 1, \ldots, k$, the samples being independently drawn from the respective populations, given $\boldsymbol{\Sigma}$, the problem implies that there are only g distinct means in the set $\mu_1, \ldots, \mu_k$. Let λ be the set of identifying labels and let $\mathscr{C}_m$ be the set of $\bar{\mathbf{x}}_r$ assigned to the mth group by λ, $m = 1, \ldots, g$, where $\bar{\mathbf{x}}_r$ is the sample mean vector of the rth sample. The maximum likelihood* estimate of λ is the grouping that minimizes (see Mardia et al. [13])

$$W_g^2 = \sum_{m=1}^{g} (1/(2N_m)) \sum_{\mathscr{C}_m} n_r n_s D^2_{rs},$$

$$D^2_{rs} = (\bar{\mathbf{x}}_r - \bar{\mathbf{x}}_s)'\boldsymbol{\Sigma}^{-1}(\bar{\mathbf{x}}_r - \bar{\mathbf{x}}_s),$$

$$N_m = \sum_{\bar{x}_r \in \mathscr{C}_m} n_r.$$

When $\boldsymbol{\Sigma}$ is unknown it may be estimated by

$$\mathbf{S} = \frac{n_1\mathbf{S}_1 + \ldots + n_k\mathbf{S}_k}{n_1 + \ldots + n_k - k}.$$

Mahalanobis D^2 also is used as a distance function in other methods in cluster analysis. See Mardia [12] for a study of D^2 when the common covariance matrix $\boldsymbol{\Sigma}$ may be singular, and for applications of D^2 to (a) factor analysis*, showing that Bartlett's scores are better in separating out individuals than Thomson's, in view of the fact that the D^2 between the estimated factors for two individuals in Bartlett's method is greater than or equal to that in Thomson's (see Lawley and Maxwell [9] for the two methods); (b) genetic difference, proving an extended version of Rao's [17] inequality involving D^2; (c) growth curves*, removing the requirements that the inverses of the matrices in the D^2's involved there be nonsingular; and (d) other related problems. Also see Jain and Waller [7] for a study of the pattern recognition* system in which they relate p_{opt} to the number of training samples and the Mahalanobis distance between the two p-variate normal populations, where p_{opt} is the number of variables (features) for which the average probability of misclassification using Anderson's W statistic attains a minimum. In addition, see Bartkowiak [1] for an abbreviated method of calculating D^2 by a procedure called *linres*.

EXAMPLE

A numerical example is given below to illustrate the use of Mahalanobis D^2 in discriminant analysis.

Table 1 Sample Mean Vectors and Sample Pooled Covariance Matrix from Four Physical Measurements of 23 and 22 Reserve Officers of the Armed Forces of the Philippines from Two Regions

				$\mathbf{S} = (s_{ij})$,	$i, j = 1, 2, 3, 4$		
i	$\bar{\mathbf{x}}_1$	$\bar{\mathbf{x}}_2$	$\mathbf{d} = \bar{\mathbf{x}}_1 - \bar{\mathbf{x}}_2$	1	2	3	4
1	64.130	64.182	− 0.052	3.7647	8.1590	1.2587	0.7250
2	125.478	127.682	− 2.204		153.9189	15.1768	20.6117
3	32.565	33.545	− 0.980			3.1420	1.9090
4	28.609	29.000	− 0.391				4.8716

Table 1 gives the mean vectors and pooled covariance maxtrix **S** computed from data on four physical measurements, namely, height (inches), weight (pounds), chest (inches), and waist (inches) of 23 and 22 male reserve officers in civilian status of the Armed Forces of the Philippines (see Ventura [23]) hailing from two different regions of the Philippine Islands but all within the age interval 29 to 31. (The assumption of equality of covariance matrices was found to be justified in view of earlier tests.) From Table 1, $\hat{\mathbf{l}}' = (0.0796,\ 0.0378,\ -0.4900,\ -0.0602)$, $D_4^2 = 0.4162$ and $F_{4,40} = 1.088 < 2.61 = F_{4,40,0.95}$. Hence the discriminant function is not significant, in view of the fact that the test concludes that $\boldsymbol{\mu}_1 = \boldsymbol{\mu}_2$.

References

[1] Bartkowiak, A. (1976). *Zast. Matematyki, Applic. Mathemat.*, **15**, 215–220.

[2] Bose, R. C. (1936). *Sankhyā*, **2**, 379–384.

[3] Bose, R. C. and Roy, S. N. (1938). *Sankhyā*, **4**, 19–38.

[4] Cochran, W. G. and Bliss, C. I. (1948). *Ann. Math. Statist.*, **19**, 151–176.

[5] Fisher, R. A. (1936). *Ann. Eugenics*, **7**, 179–188.

[6] Ito, K. and Schull, W. J. (1964). *Biometrika*, **51**, 71–82.

[7] Jain, A. K. and Waller, W. G. (1978). *Patt. Recog.*, **10**, 365–374.

[8] Kshirsagar, A. M. (1972). *Multivariate Analysis*. Dekker, New York.

[9] Lawley, D. N. and Maxwell, A. E. (1963). *Factor Analysis as a Statistical Method*. Butterworth, London.

[10] Mahalanobis, P. C. (1930). *J. Asiat. Soc. Beng.*, **26**, 541–588.

[11] Mahalanobis, P. C. (1936). *Proc. Nat. Inst. Sci. India*, **12**, 49–55.

(Although the concept of the measure of divergence between two p-variate normal populations originated in 1925 in the Presidential Address given by Mahalanobis* at the Anthropological Section of the Indian Science Congress, the pioneering study of classical D^2, i.e., D^2 with known parameters, was carried out by him in 1930 (see [10]), evaluating its first four moments through approximate methods which were found to be exact by Bose [2], who obtained the exact distribution. Further, in 1936 Mahalanobis (see [11]) proposed the Studentized D^2 for two samples, i.e., D^2 as defined in the Introduction, whose distribution was later derived by Bose and Roy [3]. However, consideration of the Studentized D^2 in the uncorrelated case had arisen in the earlier papers of Mahalanobis since 1925 (see references in [2].)

[12] Mardia, K. V. (1977). In *Multivariate Analysis IV*, P. R. Krishnaiah, ed., North-Holland, New York, "Mahalanobis distances and angles."

[13] Mardia, K. V., Kent, T., and Bibby, M. (1979). *Multivariate Analysis*, Academic Press, New York.

[14] Morrison, D. F. (1976). *Multivariate Statistical Methods*, 2nd ed. McGraw-Hill, New York.

[15] Rao, C. R. (1946). *Sankhyā*, **7**, 407–413.

[16] Rao, C. R. (1949). *Sankhyā*, **9**, 343–366.

[17] Rao, C. R. (1954). *Bull. Int. Statist. Inst.*, **34**, 90–97.

[18] Rao, C. R. (1966). In *Multivariate Analysis*, P. R. Krishnaiah, ed. Academic Press, New York, "Covariance adjustment and related problems in multivariate analysis."

[19] Rao, C. R. (1973). *Linear Statistical Inference and its Applications*, 2nd ed. Wiley, New York.

[20] Roy, S. N. (1957). *Some Aspects of Mutlivariate Analysis*. Wiley, New York.

[21] Subrahmaniam, K. and Subrahmaniam, K. (1973). *Sankhyā (B)*, **35**, 51–78.

[22] Subrahmaniam, K. and Subrahmaniam, K. (1976). *J. Multivariate Anal.*, **6**, 330–337.

[23] Ventura, S. R. (1957). "On the extreme roots of a matrix in multivariate analysis and associated tests." Unpublished thesis. The Statistical Center, University of the Philippines, Manila.

(DISCRIMINANT ANALYSIS
HOTELLING'S T^2
MULTIVARIATE ANALYSIS OF VARIANCE)

K. C. S. PILLAI

MAIL SURVEYS *See* SURVEY SAMPLING

MAIN EFFECTS

When several factors affect the response of a variable in an experiment, it is a common practice to take two or more levels of each factor and use all or a subset of all treatment combinations in the experiment (*see* FACTORIAL EXPERIMENTS). In such cases the response will consist of a contribution solely due to the level of the factor used, in addition to the other parameters. The specific contribution of the level of the factor used in the model is called the *main effect* of the factor and one can form $s - 1$ independent contrasts* of the main effects of the levels of the factor if it occurs at s levels in the experiment. Such contrasts of the main effects of the levels are called *differential main effects*. A differential main effect measures the extent to which the response for two levels within the same factor differ.

To illustrate the concept of main effects and differential main effects, let us consider an experiment with one factor a at s levels and let the ith level be used on n_i experimental units. Let the random variables Y_{ij} denote the observation on the jth experimental unit receiving the ith level of the factor a. We assume the model

$$E(Y_{ij}) = \mu + a_i; \qquad j = 1, 2, \ldots, n_i; \quad i = 1, 2, \ldots, s,$$

where $E(Y_{ij})$ is the expected value of the random variable Y_{ij}, μ is the general mean, and a_i is the effect of the ith level of the factor a. Here a_i is the ith level main effect of the factor a, and $a_i - a_{i'}$ for $i \neq i'$ is one of the differential main effects of factor a. This differential main effect is also known in the literature as an elementary contrast of the main effects of the levels of the factor a. One can form $s - 1$ independent differential main effects of the form $a_1 - a_i$ for $i = 2, 3, \ldots, s$.

In 2^n experiments, that is, the experiments with n factors each at two levels, the differential main effects are simply called main effects, denoted by the corresponding capital letters, and they can be estimated by using all 2^n treatment combinations or a suitable fractional plan. Yates' algorithm* provides a nice tool for calculating main effects in a 2^n experiment using all treatment combinations or a suitable $1/2^k$ fraction of all treatment combinations.

When the levels of a factor are greater than two, one may form linear, quadratic, and cubic contrasts of main effects of the levels instead of forming the differential main effects.

In factorial experiments, one likes to draw inferences about the interactions* of the factors along with the main effects.

It is not always true that the important effects for the experimenter are the main effects. There exist situations where the experimenter will be interested in the interactions of the factors rather than the main effects.

The linear model in a factorial experiment using only the main effects of the factors involved and ignoring the interaction of the factors is called an *additive model*; this is the model commonly used in block designs.

In the presence of significant interactions, one gets erroneous interpretations of the main effects of the factors. Inferences on main effects can be made meaningfully only if all interactions in the experiment are negligible. Cochran and Cox [1] described a sugar cane experiment in which three varieties (V) of sugar cane were compared in combination with three levels of nitrogen (N) (150, 210, and 270 lbs. per acre, respectively), giving the following summary table of the treatment totals (tons):

	n_0	n_1	n_2	Total	SE
v_1	266.1	275.9	303.8	845.8	
v_2	245.8	250.2	281.7	777.7	± 23.0
v_3	274.4	258.1	231.6	764.1	
Total	786.3	784.2	817.1	2387.6	
SE		± 23.0			

In this experiment the *VN* interaction is highly significant. Although the differential main effect of nitrogen is significantly larger for the second level of nitrogen over the zeroth level of nitrogen, a closer scrutiny of the table indicates that such is the case for varieties 1 and 2, while for variety 3, n_2 gives a smaller yield than n_0. In such cases the differential main effects of a factor have to be interpreted for a given combination of levels of the other factors used in the experiment rather than interpreting overall differential main effects.

The variance of the main effects of the levels of a factor measures the overall differential main effects of that factor and it can be estimated in many experiments from the analysis of variance* table.

The fractional factorial* plans used only to estimate main effects are *main effect plans*; they are *saturated* if the number of treatment combinations is just enough to estimate all the differential main effects in the experiment. Using saturated main effect plans to draw inference on main effects, one needs to replicate the plan in the experiment. Unsaturated main effect plans provide the estimates of all differential main effects and also leave some degrees of freedom for error.

Reference

[1] Cochran, W. G., and Cox, G. M. (1957). *Experimental Designs*. 2nd ed. Wiley, New York.

(FACTORIAL EXPERIMENTS
INTERACTION)

DAMARAJU RAGHAVARAO

MAJORIZATION AND SCHUR CONVEXITY

The concept of majorization is hinted at by the famous Lorenz curve*, introduced in 1905 to describe the distribution of wealth. Mathematically, the development of the theory of majorization started in the 1920s with Schur and Dalton and later with Hardy, Littlewood, and Pólya (see references and a historical discussion in Marshall and Olkin [32], Ch. 1).

Schur convexity is the property of functions that preserve the order of majorization; thus both of these concepts will be defined and discussed in this article. An early account of majorization and Schur functions can be found in the book by Hardy, Littlewood, and Pólya [17]. Since then the theory, and mainly its useful application in almost every part of mathematics, have been developed extensively. A complete account of the theory of majorization can be found in Marshall and Olkin [32]. Excellent surveys on the applications of majorization and Schur functions in probability and statistics are Proschan [39], Marshall and Olkin [32; Chps. 11–13], and Tong [55, Chap. 6]. For a review of these books see Kemperman [27]; *see also* INEQUALITIES ON DISTRIBUTIONS: BIVARIATE AND MULTIVARIATE.

The basic definitions and key results will be presented. The main theorems which yield many of the applications in probability and statistics then will be described. Almost all of the specific applications that were discussed in the literature up to 1979 can be found in at least one of the three works mentioned above or in references cited there. For this reason early applications will not be emphasized. On the other hand, a description of some recent research will be used to illustrate the power of the general theory.

Finally, some references will be listed. The bibliographies of the works of Proschan, Marshall and Olkin, and Tong are extensive and essentially complete. Here we avoid duplicating these; the bibliography consists mainly of papers written or published after Marshall and Olkin [32] and Tong [55]. The bibliography also lists some papers that were listed as technical reports in these books, but which have since been published.

GENERAL THEORY

The applications of majorization and Schur convexity in probability and statistics are centered around a few basic results of Marshall, Proschan, Olkin, and Sethuraman which will be described in the following section. These results, and numerous others, link the general theory of majorization and Schur convexity with applications in ranking and selection*, sampling*, reliability*, experimental designs, entropy*, unbiasedness*, order statistics*, species diversity, and random numbers generation*. In order to make full use of the results of the following section, some basic definitions and properties are needed; most of the following results can be found in [32] and [55].

Conceptually, a vector $\mathbf{y} = (y_1, \dots, y_n)$ majorizes a second vector $\mathbf{x} = (x_1, \dots, x_n)$ if the components of $\mathbf{y}$ are "more diverse" than the components of $\mathbf{x}$. Formally, let $x_{[1]} \geqslant x_{[2]} \geqslant \cdots \geqslant x_{[n]}$ and $y_{[1]} \geqslant y_{[2]} \geqslant \cdots \geqslant y_{[n]}$ denote the decreasing rearrangements of $\mathbf{x}$ and $\mathbf{y}$; then, $\mathbf{y}$ is said to *majorize* $\mathbf{x}$ (in symbols, $\mathbf{x} \prec \mathbf{y}$) if

$$\sum_{i=1}^{j} y_{[i]} \geqslant \sum_{i=1}^{j} x_{[i]}, \qquad j = 1, \dots, n-1 \quad (1)$$

and

$$\sum_{i=1}^{n} y_{[i]} = \sum_{i=1}^{n} x_{[i]}. \quad (2)$$

For example, if $\mathbf{x}$ is a vector such that $x_i \geqslant 0$, $i = 1, \dots, n$ and $\sum_{i=1}^{n} x_i = s$, then $(s/n, \dots, s/n) \prec \mathbf{x} \prec (s, 0, \dots, 0)$.

Conditions which are equivalent to majorization may be useful in theory and in applications. One of them is the following. Let $\mathbf{x}$ and $\mathbf{y}$ be two vectors; then $\mathbf{x} \prec \mathbf{y}$ *if and only if there exists an* $n \times n$ *doubly stochastic matrix** $\mathbf{P}$ (i.e., $\mathbf{P}$ has nonnegative entries, p_{ij}, and $\sum_{i=1}^{n} p_{ij} = 1$, $j = 1, \dots, n$, $\sum_{j=1}^{n} p_{ij} = 1$, $i = 1, \dots, n$) *such that* $\mathbf{x} = \mathbf{yP}$.

To obtain a majorization relation between two vectors it is often sufficient to consider only one pair of coordinates at a time. This follows from the fact that $\mathbf{x} \prec \mathbf{y}$ *if and only if there exist* $n-2$ *vectors* $\mathbf{z}^{(1)}, \dots, \mathbf{z}^{(n-2)}$ *such that*

$$\mathbf{x} \prec \mathbf{z}^{(1)} \prec \cdots \prec \mathbf{z}^{(n-2)} \prec \mathbf{y}, \quad (3)$$

where any two adjacent vectors in (3) *differ by at most two coordinates*. For example $(11, 8, 4, 3) \prec (14, 6, 4, 2)$, so in (3) one can take $\mathbf{z}^{(1)} = (14, 5, 4, 3)$ and $\mathbf{z}^{(2)} = (14, 6, 3, 3)$.

Not every two vectors can be ordered by majorization, but the majorization relation is a partial ordering in $\mathbb{R}^n$. A *Schur function* is a function which is monotone with respect to this partial ordering. The function ϕ is *Schur convex* if it is nondecreasing with respect to majorization, that is, if

$$\mathbf{x} \prec \mathbf{y} \quad \text{implies} \quad \phi(\mathbf{x}) \leqslant \phi(\mathbf{y}), \quad (4)$$

and ϕ is *Schur concave* if

$$\mathbf{x} \prec \mathbf{y} \quad \text{implies} \quad \phi(\mathbf{x}) \geqslant \phi(\mathbf{y}). \quad (5)$$

Chapter 3 of Marshall and Olkin [32] provides an excellent discussion on Schur functions, with proofs of the following claims.

When ϕ *is differentiable and symmetric* (that is, permutation invariant) *a necessary and sufficient condition for* ϕ *to be Schur convex is that* $\partial\phi/\partial x_i(\mathbf{x})$ *is nonincreasing in* $i = 1, \dots, n$ *for all* $\mathbf{x}$ *such that* $x_1 \geqslant x_2 \geqslant \cdots \geqslant x_n$, or, equivalently,

$$(x_1 - x_2)\left(\frac{\partial\phi}{\partial x_1}(\mathbf{x}) - \frac{\partial\phi}{\partial x_2}(\mathbf{x})\right) \geqslant 0 \quad \text{for all } \mathbf{x}. \quad (6)$$

Similarly, *if* ϕ *is differentiable and symmetric then a necessary and sufficient condition for it to be Schur concave is that* (6) *holds with the inequality reversed.*

Actually, if ϕ is symmetric and differentiable on $\mathscr{D} = \{\mathbf{x} : x_1 \geqslant x_2 \geqslant \cdots \geqslant x_n\}$ then, to show that ϕ is Schur convex (or Schur concave), it is enough to show that (6) holds (or that (6) holds with the inequality reversed) on $\mathscr{D}$. Thus, for example, $\phi(\mathbf{x}) = \max(x_1, \dots, x_n)$ is Schur convex, or more generally, for $l = 1, 2, \ldots, n$,

$$\phi(\mathbf{x}) = \sum_{i=l}^{n} x_{(i)} \tag{7}$$

is Schur convex (where $x_{(1)} \leqslant x_{(2)} \leqslant \cdots \leqslant x_{(n)}$ is the increasing rearrangement of $x_1, \ldots, x_n$) because ϕ is symmetric and it satisfies (6) on $\mathscr{D}$.

Identification of Schur functions is useful because then (4) and (5) can be applied to obtain a host of inequalities. Many examples and procedures for identifying Schur functions are described in Chapter 3 of [32]. The most common Schur functions are of the form

$$\phi(x_1, \ldots, x_n) = \sum_{i=1}^{n} g(x_i) \tag{8}$$

where g is a univariate convex or concave function. Actually, whenever ϕ is symmetric and convex or concave (or, merely, even if ϕ is symmetric and quasi-convex or quasi-concave), then ϕ is a Schur function. It follows from (8) that functions of the form

$$\phi(x_1, \ldots, x_n) = \prod_{i=1}^{n} h(x_i) \tag{9}$$

are Schur functions whenever h is logconvex or logconcave.

Often it is necessary to compare vectors $\mathbf{x}$ and $\mathbf{y}$ which do not have the same coordinates sum. Then they cannot be ordered by majorization because (2) fails. The partial ordering defined by weak majorization, as follows, can then be useful. Two versions of weak majorization have been extensively studied.

Let $\mathbf{x}$ and $\mathbf{y}$ be two vectors. The vector x is said to be *weakly submajorized* by $\mathbf{y}$, in symbols $\mathbf{x} \prec_{w} \mathbf{y}$, if

$$\sum_{i=1}^{j} x_{[i]} \leqslant \sum_{i=1}^{j} y_{[i]}, \qquad j = 1, \ldots, n. \tag{10}$$

The vector $\mathbf{x}$ is said to be *weakly supermajorized* by $\mathbf{y}$, in symbols $\mathbf{x} \prec^{w} \mathbf{y}$, if

$$\sum_{i=j}^{n} x_{[i]} \geqslant \sum_{i=j}^{n} y_{[i]}, \qquad j = 1, \ldots, n. \tag{11}$$

Thus, roughly $\mathbf{x} \prec_{w} \mathbf{y}$ if the components of $\mathbf{y}$ tend to be more diverse or larger than those of $\mathbf{x}$, whereas $\mathbf{x} \prec^{w} \mathbf{y}$ if the components of $\mathbf{y}$ tend to be more diverse or smaller than those of $\mathbf{x}$. The vector $\mathbf{x}$ is said to be *weakly majorized* by $\mathbf{y}$ if either (10) or (11) holds.

Some results that deal with majorization have weak majorization analogs. For example, $\mathbf{x} \prec_{w} \mathbf{y}$ *if and only if there exists some doubly substochastic matrix* $\mathbf{P}$ *such that* $\mathbf{x} = \mathbf{y}\mathbf{P}$. Similarly, $\mathbf{x} \prec^{w} \mathbf{y}$ *if and only if* $\mathbf{x} = \mathbf{y}\mathbf{P}$ *for some doubly superstochastic matrix* P.

The weak majorization analogs of (3) are the following:

(a) $\mathbf{x} \prec_{w} \mathbf{y}$ *if and only if there exist* $k < \infty$ *vectors* $\mathbf{z}^{(1)}, \ldots, \mathbf{z}^{(k)}$ *such that*

$$\mathbf{x} \leqslant \mathbf{z}^{(1)} \prec \mathbf{z}^{(2)} \prec \cdots \prec \mathbf{z}^{(k)} = \mathbf{y}, \tag{12}$$

where any two adjacent $\mathbf{z}^{(i)}$*'s differ by at most two coordinates.*

(b) $\mathbf{x} \prec^{w} \mathbf{y}$ *if and only if there exist* $n-1$ *vectors* $\mathbf{z}^{(1)}, \ldots, \mathbf{z}^{(n-1)}$ *such that*

$$\mathbf{x} \geqslant \mathbf{z}^{(1)} \prec \mathbf{z}^{(2)} \prec \cdots \prec \mathbf{z}^{(n-1)} = \mathbf{y} \tag{13}$$

where any two adjacent $\mathbf{z}^{(i)}$*'s differ by at most two coordinates.*

Clearly, every function which preserves any weak majorization ordering must be a Schur function. It turns out that

$$\mathbf{x} \prec_{w} \mathbf{y} \quad \textit{implies} \quad \phi(\mathbf{x}) \leqslant \phi(\mathbf{y}) \tag{14}$$

if and only if ϕ *is nondecreasing and Schur convex* and that

$$\mathbf{x} \prec^{w} \mathbf{y} \quad \textit{implies} \quad \phi(\mathbf{x}) \leqslant \phi(\mathbf{y}) \tag{15}$$

if and only if ϕ *is nonincreasing and Schur convex*. The counterpart of (6) says that ϕ *satisfies* (14) *if and only if*

$$\frac{\partial \phi}{\partial x_1}(\mathbf{x}) \geqslant \frac{\partial \phi}{\partial x_2}(\mathbf{x}) \geqslant \cdots \geqslant \frac{\partial \phi}{\partial x_n}(\mathbf{x}) \geqslant 0$$

whenever $x_1 \geqslant x_2 \geqslant \cdots \geqslant x_n$, and that (15) *holds if and only if*

$$0 \geqslant \frac{\partial \phi}{\partial x_1}(\mathbf{x}) \geqslant \frac{\partial \phi}{\partial x_2}(\mathbf{x}) \geqslant \cdots \geqslant \frac{\partial \phi}{\partial x_n}(\mathbf{x})$$

whenever $x_1 \geqslant x_2 \geqslant \cdots \geqslant x_n$.

APPLICATIONS

Most of the important applications of majorization and Schur functions in probability and statistics are of the following format: In some instances one has a distribution function, $F_{\boldsymbol{\theta}}$, which depends on the parameter vector $\boldsymbol{\theta} = (\theta_1, \ldots, \theta_n)$, and one is interested in a real functional of $F_{\boldsymbol{\theta}}$, taking on the values $\psi(\boldsymbol{\theta})$. In other instances one has a univariate or multivariate distribution function F and one considers a real functional of F which depends on an n-dimensional vector $\boldsymbol{\theta}$; the value that the functional takes on will be denoted by $\psi(\boldsymbol{\theta})$. The objective in either case is to show that ψ is a Schur function of $\boldsymbol{\theta}$. In many instances ψ is also monotone. A host of inequalities can be obtained then by picking two vectors $\boldsymbol{\theta}^{(1)}$ and $\boldsymbol{\theta}^{(2)}$ such that $\boldsymbol{\theta}^{(1)} \prec \boldsymbol{\theta}^{(2)}$ or $\boldsymbol{\theta}^{(1)} \prec_{w} \boldsymbol{\theta}^{(2)}$ or $\boldsymbol{\theta}^{(1)} \prec^{w} \boldsymbol{\theta}^{(2)}$, and deriving $\psi(\boldsymbol{\theta}^{(1)}) \leqslant \psi(\boldsymbol{\theta}^{(2)})$ or $\psi(\boldsymbol{\theta}^{(1)}) \geqslant \psi(\boldsymbol{\theta}^{(2)})$. Often the resulting inequalities have various statistical applications. In the following the most useful results and some typical applications will be described. A more detailed discussion can be found in Marshall and Olkin [32, Chaps. 11–13] and Tong [55, Chap. 6].

The first result seems to be that of Marshall and Proschan [33]. They show that *if $X_1, \ldots, X_n$ are exchangeable random variables* (*see* EXCHANGEABILITY) and, in particular, if $X_1, \ldots, X_n$ are independent and identically distributed random variables*, *and if ϕ is a symmetric convex function, then*

$$\psi(\theta_1, \ldots, \theta_n) = E\phi(\theta_1 X_1, \ldots, \theta_n X_n) \tag{16}$$

is symmetric and convex and hence Schur convex. Similarly,

$$\tilde{\psi}(\theta_1, \ldots, \theta_n) = E\phi(X_1 - \theta_1, \ldots, X_n - \theta_n) \tag{17}$$

is symmetric and convex [32, p. 286].

Marshall and Olkin [32, p. 289 and pp. 319–321] also derived versions of (16) and (17) which are fit for weak majorization.

For an illustration of the use of (16) see Shaked and Tran [49]. For further applications see Marshall and Olkin [32, pp. 290–295].

The second result of the type discussed above was obtained by Marshall and Olkin [31] and generalized by Hollander et al. [18]. The basic theorem in that paper states that *the convolution* of two n-variate Schur-concave densities is Schur concave* (for a simpler proof than that of Marshall and Olkin [31], see [32], p. 100). The corollaries which follow are useful for applications in probability and statistics. One of these [32, pp. 300] states that *if $X_1, \ldots, X_n$ have a joint density, f, which is Schur concave* (in particular, if the X_i's are independent and identically distributed with a common logconcave density) and if *$A \subset \mathrm{R}^n$ is a Lebesgue measurable set satisfying*

$$(\mathbf{y} \in A \text{ and } \mathbf{x} \prec \mathbf{y}) \quad \text{imply} \quad \mathbf{x} \in A, \tag{18}$$

then $\int_{A+\boldsymbol{\theta}} f(\mathbf{x})\, d\mathbf{x} = P(\mathbf{X} \in A + \boldsymbol{\theta})$ is a Schur-concave function of $\boldsymbol{\theta}$. From this fact it is not difficult to show (see, e.g., [31]) that *if $X_1, \ldots, X_n$ have a joint Schur concave density, then the distribution function $F(\mathbf{x}) = P(X_1 \leqq x_1, \ldots, X_n \leqslant x_n)$* is a Schur-concave function; *see* LOG-CONCAVE AND CONVEX DISTRIBUTIONS.

An immediate application is the inequality

$$F(x_1, \ldots, x_n) \leqslant F(\bar{x}, \bar{x}, \ldots, \bar{x}) \tag{19}$$

where $\bar{x} = n^{-1}(x_1 + \cdots + x_n)$. This inequality provides an upper bound on $F(\mathbf{x})$. Numerically, the upper bound can sometimes be found using existing statistical tables.

In many statistical applications bounds on the distribution function of $(|X_1|, \ldots, |X_n|)$ are more useful than the bound (19). Tong [56] has proven that if $X_1, \ldots, X_n$ have a Schur-concave joint density then $\gamma(\mathbf{x}) = P(|X_1| \leqslant x_1, \ldots, |X_n| \leqslant x_n)$ *is a Schur concave-function of* $\mathbf{x} = (x_1, \ldots, x_n)$. Various applications of the result are discussed in Tong [56].

Another corollary of the main theorem of Marshall and Olkin [32, p. 299] states that *if $X_1, \ldots, X_n$ have a Schur-concave joint density, then $E[\phi(\mathbf{X} + \boldsymbol{\theta})]$ is Schur convex in $\boldsymbol{\theta}$*

whenever ϕ is a Schur-convex function such that the expectation exists.

The third result of the type discussed above has been obtained by Proschan and Sethuraman [40]. *Suppose that for every* $\theta \in \Theta$ (where Θ is an interval) *the function* $\alpha(\theta, x)$ *is a density* (continuous or discrete) *of a non-negative random variable such that* α *is totally positive of order* 2 (see Karlin [20]) *and* α *satisfies the semigroup property*

$$\int_0^x \alpha(\theta_1, x)\alpha(\theta_2, y - x)\, d\nu(x)$$
$$= \alpha(\theta_1 + \theta_2, y)$$

for some measure ν *on* R. *Let* $X_1, \ldots, X_n$ *be independent random variables such that* X_i *has the density* $\alpha(\theta_i, \cdot)$ *for some* $\theta_i \in \Theta$. *Then*

$$\psi(\boldsymbol{\theta}) = E\phi(\mathbf{X}) \qquad (20)$$

is Schur convex whenever ϕ *is Schur convex.*

For example, the families of Poisson* distributions, the binomial* and the negative binomial* distributions and the family of the gamma* distributions have densities which satisfy the above conditions. Thus, many useful inequalities can be derived for these families of distributions (see Proschan and Sethuraman [40] and Nevius et al. [36]). For example, if ϕ is the indicator function of a set A which satisfies (18), then $-\phi$ is Schur convex and thus

$$\psi(\boldsymbol{\theta}) = P_{\boldsymbol{\theta}}(\mathbf{X} \in A) \qquad (21)$$

is Schur concave.

Nevius et al. [37] have obtained some weak majorization extensions of the above results. See also Marshall and Olkin [32, pp. 321–324).

The preceding results have numerous applications in various areas of statistics. One typical usage is to prove unbiasedness* of statistical tests. Let $\mathbf{X}$ be a vector of observations whose distribution depends upon the parameter vector $\boldsymbol{\theta} = (\theta_1, \ldots, \theta_n)$. Very often a statistical procedure for testing the null hypothesis $H_0: \theta_1 = \theta_2 = \cdots = \theta_n$ versus some alternative is of the form "reject H_0 if $\phi(\mathbf{X}) > c$" where ϕ is some function and c is a constant. The power of the test then is $\beta(\boldsymbol{\theta}) = P_{\boldsymbol{\theta}}(\phi(\mathbf{X}) > c)$.

Using the previous results it is often possible to show that β is a Schur-convex function of $\boldsymbol{\theta}$. For example, if ϕ is Schur convex then the complement A of the set $\{\mathbf{x} : \phi(\mathbf{x}) > c\}$ satisfies (18); thus, by (21), $\beta(\boldsymbol{\theta})$ is Schur convex.

In many applications it is further possible to show that $\beta(\theta, \theta, \ldots, \theta) = \alpha$ does not depend upon θ. It follows then that for every $\boldsymbol{\theta}$, $\beta(\boldsymbol{\theta}) \geqslant \alpha$. Hence the test procedure is unbiased. In a similar manner one can show that the power function is monotone (in the right sense). Marshall and Olkin [32, pp. 386–392] list explicit applications. Further applications can be found in Perlman [38], Alam and Mitra [1], and Shaked [47].

Other statistical areas mentioned in Chapter 13 of Marshall and Olkin [32] and in which majorization is a useful mathematical tool are the following:

Optimal linear estimators can be found using majorization [32, pp. 392–395] and Tong [56].

In ranking and selection* studies, numerous inequalities which give bounds on the probability of correct selection can be derived using Schur convexity ([32, pp. 395–402], Huang and Panchapakesan [19], Gupta and Panchapakesan [16, pp. 276–278, and references there], Berger [3, 4], Bjornstad [5], and Tong [57]).

In reliability theory* majorization has been used quite frequently in recent years. A collection of results can be found in Marshall and Olkin [32, pp. 402–405]. See also El-Neweihi et al. [12], Ross et al. [44] and El-Neweihi [11].

Majorization and Schur convexity are also useful mathematical tools in sampling* [32, pp. 331–343]. Various inequalities involving order statistics* can be obtained using majorization [32, pp. 348–355; 51, 52]. Brown and Solomon [6] discuss techniques to combine random number generators using majorization (*see* GENERATION OF RANDOM VARIABLES) Karlin and Rinott [23, 24] apply majorization to obtain entropy* inequalities and Solomon [50] and Kempton [28] use it to define species diversity. For further applications see Marshall and Olkin [32, Chaps. 12, 13].

An area in which Schur convexity is starting to play a role is that of optimal experimental designs. Assume that we are to compare v varieties (labeled $1, 2, \ldots, v$) via b blocks of size $k(< v)$. A design, d, in this case is a $k \times v$ array with blocks as columns and varieties as entries. If the effects of blocks and varieties are additive, then, corresponding to each design d, associate the information matrix $\mathbf{C}_d \equiv \text{diag}(r_1, \ldots, r_v) - (1/k)\mathbf{NN}'$, where r_i is the number of replications of variety i in d and a typical element n_{ij} of $\mathbf{N}$ is the number of times variety i occurs in block j, $i = 1, \ldots, v$; $j = 1, \ldots, k$.

Various optimality criteria have been introduced with the objective of obtaining an optimal design. The optimal designs according to various criteria are often defined as those which minimize functions of various sets of variances of families of some estimators. Many can be shown to yield optimal designs by minimizing functions of the nonzero eigenvalues of the information matrix $\mathbf{C}_d$. Giovagnioli and Wynn [15], Cheng [7], and Constantine [8] have noticed, independently, that some of the functions which are to be minimized are Schur convex and nonincreasing. This observation has various applications:

In searching for an optimal design, d^*, Constantine [8] looked for the vector $\boldsymbol{\lambda}_{d^*}$ of eigenvalues which is weakly majorized (in the sense $\overset{w}{\prec}$) by the vector $\boldsymbol{\lambda}_d$ corresponding to any other design d in a given class of designs.

For the case of four varieties, $v = 4$, Cheng [7] found optimal designs according to various criteria, most of them based on minimizing various Schur-convex functions of the eigenvalues $\lambda_{d1} \geqslant \lambda_{d2} \geqslant \lambda_{d3}$.

Giovagnioli and Wynn [15] have found the design d and the corresponding vector $\boldsymbol{\lambda}_{\tilde{d}}$, which is the highest in majorization (in the sense $\prec$) among all vectors $\boldsymbol{\lambda}_d$ corresponding to the designs d in a particular class of designs. This result enables them to obtain upper limits on various quantities of interest without the complete knowledge of the underlying design, provided it belongs to the particular class of designs.

Other appearances of majorization in design problems can be found in Rinott and Santner [41], Kiefer and Wynn [29] and Gaffke [14].

A continuous version of majorization is dilation (*see* GEOMETRY IN STATISTICS: CONVEXITY); for a somewhat elementary discussion see Karlin and Novikoff [21] or Kemperman [26]. DeGroot and Fienberg [10], Karlin and Rinott [22], Karr and Pittenger [25], Meilijson and Nadas [34], Ruschendorf [45], Shaked [46, 48], Spiegelman [53], Torgensen [58], Vasicek [59], Vickson [60], and Whitt [61] further study this concept.

Majorization also appears in Beck [2] Dauer and Krueger [9], Felzenbaum and Tamir [13], Leon and Proschan [30], Tamir [54], Ross [43], and Ruschendorf [45].

By requiring (1) and (2) to hold with the subscript i replacing $[i]$, one obtains a partial ordering in $\mathbb{R}^n$ which is closely related to majorization; see Narayana [35] and Robertson and Wright [42].

References

[1] Alam, K. and Mitra, A. (1981). *J. Amer. Statist. Ass.*, **76**, 107–109.

[2] Beck, E. (1977). *Monats. Math.*, **83**, 177–189.

[3] Berger, R. L. (1980). *J. Statist. Plan. Inference*, **4**, 391–402.

[4] Berger, R. L. (1982). In *Statistical Decision Theory and Related Topics, III*, Vol. 1, S. S. Gupta and J. O. Berger, eds. Academic Press, New York. "A minimax and admissible subset selection rule for the least probable multinomial cell", pp. 143–156.

[5] Bjornstad, J. E. (1981). *Ann. Statist.*, **9**, 777–791.

[6] Brown, M. and Solomon, H. (1979). *Ann. Statist.*, **7**, 691–695.

[7] Cheng, C-S. (1979). *Sankhyā A*, **41**, 1–14.

[8] Constantine, G. M. (1980). On Schur optimality. Tech. Rep., Department of Mathematics, Indiana University, Bloomington, IN.

[9] Dauer, J. P. and Krueger, R. J. (1978). *J. Optim. Theory. Appl.*, **25**, 361–373.

[10] DeGroot, M. H. and Fienberg, S. E. (1982). In *Statistical Decision Theory and Related Topics*, III, Vol. 1, S. S. Gupta and J. O. Berger, eds. Academic Press, New York, "Assessing probability assessors: calibration and refinement", pp. 291–314.

[11] El-Neweihi, E. (1980). *Commun. Statist. A*, **9**, 399–414.

[12] El-Neweihi, E., Proschan, F., and Sethuraman, J. (1978). *Adv. Appl. Prob.*, **10**, 232–254.

[13] Felzenbaum, A. and Tamir, A. (1979). *Lin. Alg. Appl.*, **27**, 159–166.

[14] Gaffke, N. (1981). *Ann. Statist.*, **9**, 893–898.

[15] Giovagnioli, A. and Wynn, H. P. (1980). *J. Statist. Plan. Inference*, **4**, 145–154.

[16] Gupta, S. S. and Panchapakesan, S. (1979). *Multiple Decision Procedures: Theory and Methodology of Selecting and Ranking Populations*. Wiley, New York.

[17] Hardy, G. H., Littlewood, J. E., and Polya, G. (1952). *Inequalities*. Cambridge University Press, London. (Contains an early account of the basic main results of majorization and Schur convexity.)

[18] Hollander, M., Proschan, F., and Sethuraman, J. (1981). *J. Multivariate Anal.*, **11**, 50–57.

[19] Huang, D.-Y. and Panchapakesan, S. (1976). *Commun. Statist. A*, **5**, 621–633.

[20] Karlin, S. (1968). *Total Positivity*. Stanford University Press, Stanford, Calif.

[21] Karlin, S. and Novikoff, A. (1963). *Pacific J. Math.*, **13**, 1251–1279.

[22] Karlin, S. and Rinott, Y. (1980). *J. Multivariate Anal.*, **10**, 467–498.

[23] Karlin, S. and Rinott, Y. (1981a). *Adv. Appl. Prob.*, **13**, 93–112.

[24] Karlin, S. and Rinott, Y. (1981b). *Adv. Appl. Prob.*, **13**, 325–351.

[25] Karr, A. F. and Pittenger, A. O. (1979). *Stoch. Proc. Appl.*, **9**, 35–53.

[26] Kemperman, J. H. B. (1973). *Nederl. Akad. Wetensch. Prod. Ser. A*, **76** (= *Indag. Math.*, **35**), 149–164, 165–180, and 181–188.

[27] Kemperman, J. H. B. (1981). *Bull. Amer. Math. Soc.*, **5**, 319–324.

[28] Kempton, R. A. (1979). *Biometrics*, **35**, 307–321.

[29] Kiefer, J. and Wynn, H. P. (1981). *Ann. Statist.*, **9**, 737–757.

[30] Leon, R. V. and Proschan, F. (1979). *J. Math. Anal. Appl.*, **69**, 603–606.

[31] Marshall, A. W. and Olkin, I. (1974). *Ann. Statist.*, **2**, 1189–1200.

[32] Marshall, A. W. and Olkin, I. (1979). *Inequalities: Theory of Majorization and Its Applications*. Academic Press, New York. (A complete account of the theory of majorization and Schur functions.)

[33] Marshall, A. W. and Proschan, F. (1965). *J. Math. Anal. Appl.*, **12**, 87–90.

[34] Meilijson, I. and Nadas, A. (1979). *J. Appl. Prob.*, **16**, 671–677.

[35] Narayana, T. V. (1979). *Lattice Path Combinatorics with Statistical Applications*. University of Toronto Press, Toronto, Ontario.

[36] Nevius, S. E., Proschan, F., and Sethuraman, J. (1977a). *Ann. Statist.*, **5**, 263–273.

[37] Nevius, S. E., Proschan, F., and Sethuraman, J. (1977b). In *Statistical Decision Theory and Related Topics, II*. S. S. Gupta and D. S. Moore, eds. Academic Press, New York, pp. 281–296.

[38] Perlman, M. D. (1980). *Ann. Statist.*, **8**, 247–263.

[39] Proschan, F. (1975). In *Reliability and Fault Tree Analysis*, R. E. Barlow, J. B. Fussell, and N. D. Singpurwalla, eds. SIAM, Philadelphia, pp. 237–258. (An updated survey on the use of majorization and Schur convexity in reliability theory.)

[40] Proschan, F. and Sethuraman, J. (1977). *Ann. Statist.*, **5**, 256–262.

[41] Rinott, Y. and Santner, T. J. (1977). *Ann. Statist.*, **5**, 1228–1234.

[42] Robertson, T. and Wright, F. T. (1982). *Ann. Statist.*, **10**, 1234–1245.

[43] Ross, S. M. (1981). *J. Appl. Prob.*, **18**, 309–315.

[44] Ross, S. M., Shashahani, M., and Weiss, G. (1980). *J. Amer. Statist. Ass.*, **75**, 663–666.

[45] Ruschendorf, L. (1981). *Ann. Prob.*, **9**, 276–283.

[46] Shaked, M. (1980). *J. R. Statist. Soc. B*, **42**, 192–198.

[47] Shaked, M. (1981). Unbiasedness of some discordancy tests for outliers in samples from normal and from other distributions with logconcave densities. Tech. Rep., Department of Mathematics, University of Arizona, Tucson, AZ.

[48] Shaked, M. (1982). *J. Appl. Prob.*, **19**, 310–320.

[49] Shaked, M. and Tran, L. T. (1982). *J. Amer. Statist. Ass.*, **77**, 196–203.

[50] Solomon, D. L. (1979). In *Ecological Diversity in Theory and Practice*. J. F. Grassie, G. P. Patil, W. Smith, and C. Taillie, eds., pp. 24–35.

[51] Smith, N. L. and Tong, Y. L. (1980a). Inequalities for functions of order statistics under an additive model. Techn. Rep., Department of Mathematics and Statistics, University of Nebraska, Lincoln, NE.

[52] Smith, N. L. and Tong, Y. L. (1980b). Inequalities for functions of order statistics under a multiplicative model. Techn. Rep., Department of Mathematics and Statistics, University of Nebraska, Lincoln, NE.

[53] Spiegelman, C. (1982). *J. Res. National Bureau Standards*, **87**, 71–74.

[54] Tamir, A. (1980). *SIAM J. Control and Optimization*, **18**, 282–287.

[55] Tong, Y. L. (1980). *Probability Inequalities in Multivariate Distributions*. Academic Press, New York. (Contains a thorough account on the use of majorization and Schur convexity in deriving probability inequalities. Some discussion about the use of Schur convexity in statistical selection problems can also be found in this reference.)
[56] Tong, Y. L. (1982a). *Ann. Statist.*, **10**, 637–642.
[57] Tong, Y. L. (1982b). *Biometrics*, **38**, 333–339.
[58] Torgensen, E. N. (1981). *Ann. Statist.*, **9**, 638–657.
[59] Vasicek, O. A. (1977). *Oper. Res.*, **25**, 879–884.
[60] Vickson, R. G. (1977). *Math. Oper. Res.*, **2**, 244–252.
[61] Whitt, W. (1980). *J. Appl. Prob.*, **17**, 1062–1071.

(GEOMETRY IN STATISTICS: CONVEXITY
INEQUALITIES ON DISTRIBUTIONS:
BIVARIATE AND MULTIVARIATE)

MOSHE SHAKED

MAKEHAM–GOMPERTZ DISTRIBUTION

This is a continuous univariate distribution with hazard rate* function $h_X(x)$ given by

$$h_X(x) = A + bc^x,$$

where A is arbitrary and b and $c > 0$.

This distribution has been widely used in actuarial work.

(ACTUARIAL STATISTICS—LIFE
GOMPERTZ DISTRIBUTION
LIFE INSURANCE
LIFE TABLES
SURVIVAL ANALYSIS)

MALLOWS C_p *See* C_p STATISTICS

MALTHUS, THOMAS ROBERT

Born: February 17, 1766, in Guildford, Surrey, England.

Died: December 23, 1834, in Bath, England.

Contributed to: population studies, political economy.

Malthus, still today the best-known name in the field of population*, stated issues and dilemmas around which research and controversy have revolved since the first *Essay* [1] was published in 1798. People are the agents of production; how can there be too many of them? Rousseau and others saw population as the test of government, with population increase a sign that things were well. Yet increased population can make the survival and livelihood of individuals more precarious. Misery abounds in the world; what part of it is due to population pressure?

Thomas Robert Malthus (1766–1834) lived half his life in the eighteenth century and half in the nineteenth. His father was Daniel Malthus, a liberal of independent means and a friend of Hume and Rousseau. Robert (as he liked to be called) Malthus had three children, and led the life of a scholar and teacher, being for some time on the faculty of the East India College at Haileybury. His famous *Essay* went through seven editions, but the last six had little relation to the first.

In one sentence, the Malthusian principle is that population tends to grow until it presses against the means of subsistence. That was what Darwin took from Malthus and applied with such success to animals and plants. Malthus explained the matter in terms of geometric increase, of which a human population is capable, and arithmetic increase, which is the best that can be expected of the plants and animals on which humans subsist. The experience of America showed that where there were no environmental constraints humans could double each 25 years, a rate of nearly 3 % per year; one could visualize an equal increase of subsistence for one generation, but beyond that the most that could be hoped for was an absolute increase equal to that of the initial generation.

As often as the arithmetic and geometric increase have been quoted, they are not necessary to the Malthusian proposition. Malthus was careful in his phrasing—he never said that population grows geometrically, but only that it *tends* so to grow. And in fact any tendency of population to grow appreciably faster than the means of subsistence, along any curve whatever, would equally support his theory.

The public outcry in response to the first *Essay* came from both the perfectionists to his left and the pious churchgoers to his right. Malthus spent much of his life gathering data that would test his principle. Censuses of Canada and the United States, the Scandinavian countries, and a few others were already in existence, and Malthus used what he could find. The first census of England was taken just three years after the *Essay* appeared. Statisticians may note that he led the way both in the assembly of data and its interpretations. Malthus was more than aware that the means of subsistence are conventional rather than physiological. If people set higher standards of subsistence they will cease their increase sooner than they would with a lower standard of consumption.

The first edition of 1798 made much of the positive check of death. It spoke of the misery that is provoked by population growth beyond the means of subsistence and the harsh control through starvation exercised by nature, with phrases that sometimes seem to imply that equilibrium in the face of the passion between the sexes can be secured only by mortality.

But even in the first edition Malthus saw that something better than this was possible: the preventive check of birth limitation. An equilibrium based on control of births is infinitely preferable to one based on death. The more Malthus studied population the more hope he had for the preventive check. In the words of the last pages of the last edition (ref. 1): "The evils resulting from the principle of population have diminished rather than increased," and population considerations "by no means preclude that gradual and progressive improvement in human society."

Ignorant people cannot be expected to control the size of their families, and Malthus placed great stress on education. In an age when even liberals thought that educating the masses was dangerous, Malthus was wholly in favor of it. Educated people would apply the preventive check of fewer births; in the Malthusian perspective this was to be accomplished by later marriage*, which did indeed occur among the educated classes of England in his time. With increased income and education, people would stop having children before they drove down the standard of living.

Malthus saw dilemmas and difficulties arising in the economy as well as in the population. His work on political economy showed how insufficient demand could prevent full employment, anticipating Keynes, as Keynes himself notes, by more than a century. For Malthus the world was no easy place; yet difficulties had been placed in it not to create despair but to incite to activity. This sense that life is difficult attracts us to Malthus today, after the complacently optimistic Victorian age had seemingly refuted him once and for all.

Reference

[1] Malthus, T. R. (1914). *An Essay on Population.* J. M. Dent, London, Vol. 2, p. 261. (Everyman's Library.)

Bibliography

James, P. (1979). *Population Malthus: His Life and Times*. Routledge and Kegan Paul, London. (A well-researched biography.)

Keyfitz, N. (1982). In *Malthus: Past and Present*, J. Dupaquier and A. Fauve-Chamoux, eds. Academic Press, London. (The paper discusses the evolution of Malthus's thought. The book is a selection of papers presented at the Malthus Conference of IUSSP.)

Malthus, T. R. (1960). In *Three Essays on Population.* New American Library/Mentor, New York. ("A Summary View of the Principle of Population," first published in 1830.)

Peterson, W. (1979). *Malthus*. Harvard University Press, Cambridge, Mass. (Especially good on Malthus's doctrine, on how it contrasted with the views of his contemporaries, and how it has been treated in the century and a half since his time.)

Simpkins, D. W. (1972). In *Dictionary of Scientific Biography*, Vol. IX, p. 69. Scribners, New York.

(MATHEMATICAL THEORY OF
POPULATION
POPULATION PROJECTIONS)

NATHAN KEYFITZ

MALTHUSIAN PARAMETER

Malthusian parameter is the name used by R. A. Fisher* [2] and others for the rate of increase that a population would ultimately attain if its observed age-specific birth and death rates were to continue indefinitely. Alfred J. Lotka [3] and other demographers called the same entity the *intrinsic rate of natural increase*.

The Malthusian parameter is not either the actual statistically counted rate of increase of a population or in any sense a forecast of its future rate. It is rather an interpretation of an observed age-schedule of mortality and fertility*. If there are two populations, one of which has suffered migration* losses of individuals past childbearing, but with the same life tables* and the same rates of childbearing age for age, then the two will have different crude birth, death, and natural increase rates, but they will have the same Malthusian parameter.

The Malthusian parameter occurs as a constant in the solution for $B(t)$ of the integral equation*

$$B(t) = \int_0^\infty B(t-x)p(x)f(x)\,dx,$$

where $B(t)$ is the births at time t, $p(x)$ is the probability that a child just born will survive to age x, and $f(x)\,dx$ is the probability that an individual that has survived to age x will give birth in the succeeding period of time and age dx. (The starting population of the process has to be specified for the complete solution, but it does not affect the Malthusian parameter.) The right way to solve the equation is by the Laplace transform, as Willy Feller [1] showed, but a more elementary approach is to suppose that the births will increase exponentially.

Substituting e^{mt} for the unknown birth function $B(t)$ gives an equation of which the unknown is a number m, the Malthusian parameter, rather than the function $B(t)$:

$$1 = \int_0^\infty e^{-mx}p(x)f(x)\,dx.$$

This last is readily solved once the life table survivorship $p(x)$ and the birth function $f(x)$ are available.

References

[1] Feller, W. (1941). *Ann. Math. Statist.*, **12**, 243–267.

[2] Fisher, R. A. (1958). *The Genetical Theory of Natural Selection*. Dover, New York. (First pub. 1930.)

[3] Lotka, J. (1939). *Théorie Analytique des Associations Biologiques*. Part II. *Analyse Démographique avec Application Particulière à l'Espèce Humaine*. Hermann & Cie., Paris.

Bibliography

Keyfitz, N. (1977). *Applied Mathematical Demography*. Wiley, New York.

(DEMOGRAPHY
MATHEMATICAL THEORY OF POPULATION
POPULATION PROJECTIONS
VITAL STATISTICS)

NATHAN KEYFITZ

MANAGEMENT SCIENCE, STATISTICS IN

Since the inception of management science as an independent field, statistics always has been an integral part of the management scientist's training and kit of tools. As a model builder, the management scientist encounters statistical issues ranging from data collection* to model calibration, testing, and validation, in all phases of research and implementation. Over the years, certain areas of fruitful collaboration between statistics and management science have emerged; examples include forecasting*, simulation*, decision analysis, and statistical quality control*. Since each area constitutes a vast field in its own right, this article does not attempt to address them in detail; optimization*and PERT* are also excluded. Nonetheless they continue to provide fertile grounds for research in management science.

As an indication of their importance, one only needs to consult the leading professional journals of management science where research articles on these topics appear frequently. A classification of all articles appearing in *Management Science* over the period 1970–1982 in which articles are categorized according to the statistical tools employed, yielded the following analysis:

Regression	53
Forecasting/Time Series	42
Simulation	38
Bayesian Statistics	13
Multivariate Statistics	9
Statistical Quality Control	8
Correlation Analysis	8
ANOVA	7
Sampling	5

Given the extent and the diversity of management science studies that employ statistical methodology, this article does not aim to be comprehensive. Instead, it attempts to discuss a small sample of studies that brings out the vital interplay of statistics and management science. First we discuss the role of regression-based modeling; we begin with a celebrated application of regression analysis in which this tool is used to elicit the decision-making rules of experienced managers. This is followed by examples of the use of regression in the estimation of unavailable data and the construction of explanatory models. Analysis of variance is then discussed in an important context related to the evaluation of algorithms. Finally, sampling theory and three other statistical topics are briefly covered.

USING REGRESSION TO ESTIMATE MANAGERIAL DECISION RULES

A well-known problem faced by many manufacturing firms is that of production planning, where a firm has to determine the levels of production and workforce in each period of a given time horizon in such a way as to minimize costs associated with production, labor, and inventory while meeting projected customer demands. Working with a quadratic cost function, Holt et al. [22] were able to derive the optimal decision rules below for setting production and workforce levels P_t and W_t in a given period t.

$$\left.\begin{aligned} P_t &= c_1 + c_2 W_{t-1} - c_3 I_{t-1} + \sum_{i=0}^{m} \gamma_{t+i} S_{t+i}, \\ W_t &= d_1 + d_2 W_{t-1} - d_3 I_{t-1} + \sum_{i=0}^{m} \delta_{t+i} S_{t+i}, \end{aligned}\right\} \qquad (1)$$

where S_t is the projected sales in period t, I_t is the inventory at the end of period t, and m indicates the duration of the planning horizon. These results are called *Linear Decision Rules* (LDR).

To follow the optimal policy, the decision maker has to determine the coefficients (c, d γ, and δ) from the problem cost data. The optimal rules imply that past workforce and inventory levels as well as future sales must be taken into account in setting the production and workforce for the current period.

Bowman [9] raised a natural question motivated by the form of the LDR model: To what extent do managers actually follow a decision rule in the practice of production planning? He suggested that "managers and/or their organizations can be *conceived* of as decision rule coefficient estimators." (Bowman's approach is hence also called the *Management Coefficients theory*.) If his theory is accurate, then past behavior of the manager can be used to formulate a decision rule similar to (1) for which the coefficients may be estimated statistically. Such a rule would then specify the average behavior of the manager from which his actual decisions may deviate due to occasional erratic behavior.

To test this, Bowman used regression* analysis on data from different firms to obtain the form and coefficients of the decision rule that best fitted the observed past decisions. While coefficients of the relations in (1) may be obtained in this fashion, he found that the following model yielded superior results:

$$P_t = a + b_1 W_t + b_2\left[(\overline{W}/\overline{S})SA_t - W_t\right] + b_3\left[(\bar{I}/\overline{S})S_t - I_{t-1}\right]. \quad (2)$$

Here, SA_t equals the average sales over the next three periods; $\overline{W}$, $\bar{I}$, and $\overline{S}$ represent the averages of workforce, inventory, and sales, respectively, over the entire planning horizon. The ratios $\overline{W}/\overline{S}$ and $\bar{I}/\overline{S}$ essentially normalize the differences occurring within brackets in (2). A similar relation is obtained for W_t. Bowman found that such statistically estimated rules, if followed, would outperform the actual past performance of the firm.

Bowman's theory has led to a number of more detailed studies all of which use regression to elicit decision rules from past data on managerial performance. Kunreuther [28] studied an electronics firm where the production planning process involved drawing up an initial plan that was subsequently revised as more reliable information on future sales became available. He found that the estimated rules outperformed the actual past decisions for the initial planning process. Kunreuther also provided more detailed information on the statistical significance of the estimated coefficients. Hurst and McNamara [24] used Bowman's approach for the shorter-term problem of production scheduling in a textile mill and obtained acceptable results.

Moskowitz and Miller [31] focused on the verification of Bowman's theory that regression models accurately specify an individual's decision-making process. They considered six different decision-making environments corresponding to three levels of forecast errors and two different forecasting horizons (for sales). Regression was then used to obtain decision rules similar to (1) based on observed performance of over 80 experimental subjects. Statistical measures of goodness-of-fit* such as R^2, F, and t statistics were used to validate the regression models. Moreover, since the variants of the decision rules in (1) estimated in this study were such that P_t and W_t shared the same predictor variables involving I_{t-1} and sales, it was possible to eliminate these predictors to relate P_t, W_t, and W_{t-1} directly in a linear relationship whose coefficients could then be estimated independently. This provided the authors with an additional check on the validity of the regression models. Overall, the authors confirmed Bowman's assumptions and obtained an average R^2 of 0.93 across all subjects. More recently Remus [32] tested the Management Coefficients theory within a competitive environment by studying the decision-making behavior of subjects in an executive management game. He found that the variance in subjects' decision making was linearly related to their final ranking with respect to their competitors. Thus a player who followed his decision rules more consistently also ranked higher in performance.

The Management Coefficients theory is an interesting example of the fruitful interplay between model building in management science and statistical methodology; *see also* STATISTICAL MODELING. Here, the statistical model (regression) is motivated by the form of the optimal policy in (1), itself derived analytically. The optimal rules not only suggest the predictor variables in the regression but also justify the use of linear forms for such variables. Thereafter the statistical model becomes an independent schema in its own right which can be tested against actual decision-making patterns of experienced managers.

The underlying idea of using regression to calibrate management science models whose forms are derived theoretically occurs frequently in applied work. To give a simple example, Kolesar and Blum [27] in their work on response times of emergency service units (such as fire engines) used a theoretical probabilistic model to show that "the average response distance in a region is inversely proportional to the square root of the number of locations" from which such units are available to respond. Denoting the response time by R and the number of locations by

N, this suggests the form $R = aN^b$. The authors used simulation* data to find the parameters a and b; the resulting values were $(a,b) = (7.49, -0.597)$. Note that a value for b close to $-1/2$ supports the applicability of the theoretical result. More complicated regression-based models for inventory* control follow the same approach of using empirical or simulated data to estimate parameters (see, e.g., Ehrhardt and Wagner [15]).

USING REGRESSION FOR INPUT DATA IN MODELING

In some models, regression has been used as a means of providing input data required and utilized by the model. For example, in many network models of transportation and distribution activities, one needs the actual distance and travel time between nodes of the network. If the distance between two sites i and j is denoted by d_{ij}, then a matrix of distances specifying d_{ij} for all pairs of sites i and j is required. If the number of sites is n, roughly n^2 pieces of data are necessary. Since network models typically deal with a very large number of sites, this presents a major data collection* problem. To reduce this burden, one may instead choose to relate d_{ij} to explanatory variables that are significantly easier to measure. One example is given by

$$d_{ij} = a + bs_{ij} + cv_{ij} + dz_{ij}$$

where s_{ij} and v_{ij} are, respectively, the straight-line (Euclidean) and right-angle distances between sites i and j, and z_{ij} refers to the geographical zones in which i and j are found. The coefficients a, b, c, and d then may be obtained by regression by using data sampled from only a small percentage of all possible n^2 pairs (i, j).

The appeal of such a relation is that s_{ij} and v_{ij} may be computed readily by using the geographical (x, y) coordinates of the sites i and j. If such a model results in a good fit for the data, the effort required for data collection is reduced significantly. Cook and Russell [11] as well as other researchers have used this approach with some success. A similar approach was used by Hausman and Gilmour [19] to estimate the length of an optimal traveling salesman* tour through n customer sites. Since the exact computation of such tour lengths is too time-consuming, the authors suggested predicting such distances using explanatory variables related to the dispersion (standard deviation) of the x and y coordinates of the customers.

To take an example from a different problem area, one may cite Abernathy's work [1] on network scheduling. In project management network models, the duration of individual activities within the project must be estimated and provided as input data to the network scheduling model. Subjective estimates for such durations often have been found to be in error. Abernathy obtained a regression model relating the size of the error to several explanatory variables, including the error of the estimate in a previous period. Such models may be used to correct the subjective estimates that serve as input data.

CONSTRUCTION OF CAUSAL MODELS BY REGRESSION

Management scientists frequently use regression models to obtain explicit relationships between two sets of variables. The term "causal," as in causal models, refers to the ability of variables in explaining the behavior of the independent variable to be predicted. Many management science models rely heavily on explanatory models of this kind. For example, in marketing* science, market response to selling effort is related to time spent by salesmen, prior market share, and the manager's or salesman's experience, as well as other variables, by means of a multiplicative model. Such relationships then provide the basic framework for performing allocation studies. Beswick [4] solves the problem of allocating selling effort and setting the sales force size by dynamic programming* and by using response functions calibrated through regression analysis. Bass

[3] addresses the question of estimating geographical market potentials which may be used for setting sales quotas or allocating marketing effort. His work is based on a modified regression model to predict sales at a disaggregate level when only aggregate information on sales is available.

In the case of research and development (R&D) Birnbaum [6] uses stepwise regression* analysis to identify the problem characteristics and phases (stages) of a research process which makes an R&D effort most suitable for interdisciplinary research. In this study the independent variables refer to the problem characteristics and the stage of the research while the dependent variables measure the output (articles, patents, etc.) of the research effort. Lucas [30] develops regression models to determine the relationship between the use of an information system and the performance of the sales force. Searle and Udell [36] discuss regression models that include dummy variables in situations where cardinal measurements of variables are not readily available. Their discussion is illustrated with an application dealing with the attributes of span of control in a study of organizational behavior.

Teece [37] uses regression analysis to characterize time-cost tradeoffs in international technology-transfer projects. In particular, he investigates the determinants of the elasticity of project cost with respect to changes in the project's duration. Schendel and Patton [34] use an econometric approach to provide a simultaneous equation model of corporate strategy. The model uses two-stage* and three-stage least squares to obtain a system of equations relating the firm's performance goals with each other as well as with controllable and exogenous variables; *see* ECONOMETRICS.

STATISTICAL ANALYSIS OF ALGORITHMIC PERFORMANCE DATA

An important and interesting application of statistical data analysis may be found in the literature on algorithm performance. In management science one often finds several different algorithms for solving the same optimization* problem, and then comparison of the available algorithms becomes necessary. In the past, empirical work on such comparisons involved applying each algorithm to a battery of test problems and reporting the results. Recently, however, management scientists have become increasingly aware of the statistical issues arising in such comparisons and have started to focus on the methodology underlying their testing procedures; the following briefly describes two examples.

Zanakis [43] examined the performance of three heuristic methods for solving 0–1 integer programming* problems. One common performance measure for an algorithm is its computation time or speed. For heuristic algorithms, another performance criterion is accuracy, measured in terms of absolute or relative deviation of the heuristic solution from the optimum. Both these measures vary from one problem to the next and are influenced by problem characteristics such as the number of variables (V) or the number of constraints (C) in the problem. Another problem characteristic singled out by Zanakis is the degree of constraint slackness (S). Roughly speaking, a problem is expected to be more difficult to solve if its constraints are tight, i.e., if it has a low value of slackness. To test the influence of these problem characteristics on algorithmic performance, Zanakis used a factorial design* with three levels for each of the characteristics V, C, and S and five problems within each cell of the design. Analysis of variance* (ANOVA) was then used to test for interaction* effects between problem characteristics and algorithmic performance. Zanakis also was successful in using stepwise regression to develop predictive equations for each algorithm's computation time with V, C, S and some interaction variables such as VC and VS as predictor variables. Clearly, such predictive relations are of practical interest, since they provide information as to how changes in, say, problem size could impact an algorithm's performance.

More recently Lin and Rardin [29] gave a more comprehensive treatment of experimental design methodology for the testing and comparison of integer programming algorithms. They focused on designs that would help separate nuisance effects from the algorithm effect. The former effects include problem characteristics as well as effects due to the efficiency of the optimization algorithm under study. Since the two algorithms tested by the authors were exact, accuracy was not an issue and solution time was the only performance measure at play. Seven problem characteristics, including V and C, were chosen and controlled at high or low levels resulting in 2^7 combinations. Two test problems (replications) were used for each of these combinations. The results of the two exact algorithms on these problems then were used to perform an ANOVA. The authors also examined the data to see if the ANOVA assumptions were sufficiently satisfied. In short, the work of Lin and Rardin [29] is distinguished by its close consideration of statistical aspects of empirical tests of algorithmic performance. Other researchers such as Dembo and Mulvey [14] also are utilizing statistical tools when reporting empirical tests of algorithms. We foresee this area emerging as an important example of the interplay between statistics and algorithmic design.

SAMPLING THEORY

Management scientists frequently encounter sampling* methodology questions in constructing and using models. Sampling issues arise most frequently in connection with simulation models* and experiments. When using a simulation model, one is interested in the sample sizes required to ensure a desired level of accuracy for the measured simulation output characteristics, and in ways of reducing sample sizes. As early as 1953, Kahn and Marshall [25] discussed methods for reducing sample sizes in simulation experiments. One, called importance sampling*, involved distorting the underlying distribution of the sampled variable. This method was later described by Clark [10], who comments upon its prevalence among users of simulation. Today there is a vast literature on Variance Reduction Methods, of which importance sampling is a member. Since this topic requires a lengthy discussion, we refer the reader to the text by Rubinstein [33], where the heavily statistical flavor of the subject is brought out.

Multiple Ranking Procedures is an approach for analyzing simulations. An applications-oriented discussion is provided by Kleijnen et al. [26].

In certain areas where simulation is heavily used, there have been studies of the sample size issue. For example, in network scheduling (PERT*), the classic paper by Van Slyke [40] was a forerunner of such studies. Similarly, in the area of inventory control, studies of sample sizes required to guarantee stated precision and confidence levels in the estimation of key inventory-related measures have been made (see, e.g., refs. 16 and 17).

Apart from uses related to simulation models, stratified sampling* is used by management scientists in data analysis activities. An early example [41] involved collecting information from railroad waybills to estimate the percentage shares of different railroads that share shipment revenues. Another early paper [42] discussed a stratified sampling strategy for situations where sampling has a deterrent effect in that it reduces the probability of having defectives within the sampled stratum. More recently, Bitran, Hax, and Valor-Sabatier [7] have discussed appropriate stratified sampling techniques for the analysis of inventory* systems.

In some applications, a minimum-cost or maximum-revenue sampling plan must be determined. For example, Boockholdt and Finley [8] derive the optimal sample size for audit tests that minimizes a total cost function composed of sampling costs and cost of Type I error at a fixed level of Type II error. Heiner and Whitby [21] discuss a problem where medical insurance claims are audited to detect overpayment, and the amount of overpayment recovered is related to the length of the confidence interval for the amount of wrong payment. Trading off the expected amount to be recovered and the cost of sampling leads to an interesting problem in setting optimal sample sizes for the audit. Finally, Turley and Bui [39] discuss a sampling scheme for identifying excessive consumers of energy. While many other applications of sampling policies may be cited, the preceding examples point to the range of situations where sampling is of interest to the management scientist.

OTHER STATISTICAL TOOLS

The following is a brief list of examples of studies within management science that rely upon other statistical tools.

1. *Correlation Analysis* has been used as a method to analyze the relation between two sets of variables under study. For example, Cravens [12] uses correlation* analysis to determine how the amount of information processed by a decision maker to perform a particular task is influenced by the characteristics of the decision maker, the task, and the interaction between the two. Ginzberg [18] employs correlation analysis to investigate the relation between the success of an MIS system and the degree of realism in the expectations of users of such a system.
2. *Discriminant Analysis* is utilized to determine to what extent various groups overlap with or diverge from one another. Bariff and Lusk [2] utilize it to distinguish between two groups of users of an MIS system based on their cognitive styles and personality traits. This information then is used to generate different types of user reports based on the user's profile. Schwartz and Vertinsky [35] use discriminant analysis* together with multiple regression* to determine the impact of various project attributes (risk, payback, etc.) on the probability of funding for R&D projects.
3. Techniques of *Bayesian Statistics* are used primarily in the context of decision analysis; *see* BAYESIAN INFERENCE. Bierman and Hausman [5] address a credit-granting decision problem where the probability of collection is revised as collection experience becomes available. The paper combines Bayesian revision (updating) of probabilities with a dynamic programming solution technique. Hayes [20] uses a Bayesian approach to estimate parameters of a demand distribution when the objective is to minimize the expected total operating cost of inventory control policies; this objective leads to estimators that are different from the classical ones. Cunningham and Frances [13] use Bayesian decision theory* to determine how further data should be obtained on the cost coefficients of a linear programming* problem.

References

[1] Abernathy, W. J. (1971). *Manag. Sci.*, **18**, B80–B88.

[2] Bariff, M. L. and Lusk, E. J. (1977). *Manag. Sci.*, **23**, 820–829.

[3] Bass, F. M. (1971). *Manag. Sci.*, **17**, B485–B494.

[4] Beswick, C. A. (1977). *Manag. Sci.*, **23**, 667–678.

[5] Bierman, H. and Hausman, W. H. (1970). *Manag. Sci.*, **16**, B519–B523.

[6] Birnbaum, P. H. (1981). *Manag. Sci.*, **27**, 1279–1293.

[7] Bitran, G. R., Hax, A. C., and Valor-Sabatier, J. (1981). Diagnostic Analysis of Inventory Systems: A Statistical Approach. Technical Report #10, MIT Sloan School of Management.

[8] Boockholdt, J. L. and Finley, D. R. (1980). *Decision Sci.*, **11**, 702–713.

[9] Bowman, E. H. (1963). *Manag. Sci.*, **9**, 310–321.

[10] Clark, C. E. (1961). *Operat. Res.*, **9**, 603–620.

[11] Cook, T. M. and Russell, R. A. (1978). *Decision Sci.*, **9**, 673–687.

[12] Cravens, D. W. (1970). *Manag. Sci.*, **16**, B656–B670.

[13] Cunningham, A. A. and Frances, D. M. (1976) *Manag. Sci.*, **22**, 1074–1080.

[14] Dembo, R. S. and Mulvey, J. M. (1978). On the analysis and comparison of mathematical programming algorithms and software. In *Computers*

and *Mathematical Programming*, W. W. White, ed. National Bureau of Standards Publication 502, Washington, D.C.

[15] Ehrhardt, R. and Wagner, H. M. (1982). Inventory models and practice. In *Advanced Techniques in the Practice of Operations Research*, H. J. Greenberg, F. H. Murphy, and S. H. Shaw, eds. North-Holland, New York.

[16] Geisler, M. A. (1964). *Manag. Sci.*, **10**, 261–286.

[17] Geisler, M. A. (1964). *Manag. Sci.*, **10**, 709–715.

[18] Ginzberg, M. J. (1981). *Manag. Sci.*, **27**, 459–478.

[19] Hausman, W. H. and Gilmour, P. (1967). *Transportation Res.*, **1**, 349–357.

[20] Hayes, R. H. (1969). *Manag. Sci.*, **15**, 686–701.

[21] Heiner, K. W. and Whitby, O. (1980). *Interfaces*, **10**, No. 4, 46–53.

[22] Holt, C. C., Modigliani, F., Muth, J. F., and Simon, H. A. (1960). *Planning Production, Inventories, and Work Force*. Prentice-Hall, Englewood Cliffs, N. J.

[23] Horsky, D. (1977). *Manag. Sci.*, **23**, 1037–1049.

[24] Hurst, E. G. and McNamara, A. B. (1967). *Manag. Sci.*, **14**, B182–B203.

[25] Kahn, H. and Marshall, A. W. (1953). *Operat. Res.*, **1**, 263–278.

[26] Kleijnen, J. P., Naylor, T. H., and Seaks, T. G. (1972). *Manag. Sci.*, **18**, B245–B257.

[27] Kolesar, P. and Blum, E. H. (1973). *Manag. Sci.*, **19**, 1368–1378.

[28] Kunreuther, H., (1969). *Manag. Sci.*, **15**, B415–B439.

[29] Lin, B. W. and Rardin, R. L. (1980). *Manag. Sci.*, **25**, 1258–1271.

[30] Lucas, H. C. (1975). *Manag. Sci.*, **21**, 908–919.

[31] Moskowitz, H. and Miller, J. G. (1975). *Manag. Sci.*, **22**, 359–370.

[32] Remus, W. E., (1978). *Manag. Sci.*, **24**, 827–835.

[33] Rubinstein, R. Y. (1981). *Simulation and the Monte Carlo Method*. Wiley, New York.

[34] Schendel, D. and Patton, G. R. (1978). *Manag. Sci.*, **24**, 1611–1621.

[35] Schwartz, S. L. and Vertinsky, I. (1977). *Manag. Sci.*, **24**, 285–301.

[36] Searle, S. R. and Udell, J. G. (1970). *Manag. Sci.*, **16**, B397–B409.

[37] Teece, D. (1977). *Manag. Sci.*, **23**, 830–837.

[38] Tintner, G. and Rama Sastry, M. V. (1972). *Manag. Sci.*, **19**, 205–210.

[39] Turley, R. E. and Bui, M. (1981). *Interfaces*, **11**, No. 1, 62–66.

[40] Van Slyke, R. M. (1963). *Operat. Res.*, **14**, 839–860.

[41] Van Voorhis, W. R. (1953). *Operat. Res.*, **1**, 259–262.

[42] Whittle, P. (1954). *Operat. Res.*, **2**, 197–203.

[43] Zanakis, S. H. (1977). *Manag. Sci.*, **24**, 91–104.

(STATISTICAL MODELING)

A. A. ASSAD
B. L. GOLDEN
E. WASIL

MANIFOLDS

The best known manifolds are smoothly curved surfaces of dimension $d \leqslant k$ in Euclidean spaces R^k. (In general, a manifold can, in the neighborhood of each of its points, be represented by such a surface.) Examples are spheres $S^{k-1} = \{(x_1, \dots, x_k) : x_1^2 + \cdots + x_k^2 = 1\}$.

After a brief introduction to differential geometry, it will be shown how manifolds enter into statistics both as sample spaces*, in which observations take values, and as parameter spaces for families of laws (= probability measures).

DIFFERENTIAL GEOMETRY

For fuller expositions see, e.g., Boothby [4] or Helgason [11]. Most commonly studied manifolds M have a *Riemannian structure* or *metric*: the distance between two points is the length of the shortest arc joining them through M. Around each point p of M there is a chart, or set of local coordinates $x = (x_1, \dots, x_d)$, mapping some neighborhood of p in M smoothly, and with smooth inverse, onto an open set in R^d. In coordinates the Riemannian distance element ds is

$$ds^2 = \sum_{i,j=1}^{d} g_{ij}(x)\, dx_i\, dx_j$$

for some functions g_{ij}.

VOLUMES AND UNIFORMITY. A Riemannian manifold M has an element of "volume" or "surface area"

$$dV = (\det g_{ij})^{1/2} |dx_1 \dots dx_d|.$$

Then V is an analogue of Lebesgue or uniform measure, with respect to which many laws have densities. If M is compact, V can be normalized, giving the uniform probability law.

Homogeneous Spaces

Often there is a group G of one-to-one transformations of M onto itself, preserving ds^2 and hence dV, and such that for every p and q in M there is a g in G with $g(p) = q$. For example let $M = S^{k-1}$ where G is the group of all rotations of R^k.

Most often G will itself be a manifold consisting of matrices (a Lie group); see Helgason [11]. Then for p in M, let $K = K_p$ be the subgroup of all h in G with $h(p) = p$. For g in G let $gK = \{gh : h \in K\}$ (a coset). The G/K denotes the set of all such cosets. The map $g \to g(p)$ from G onto M gives a natural 1-1 map of G/K onto M.

Let $\mathscr{P}$ be the set of all laws on M preserved by all elements of K. Suppose K is compact. Then in $\mathscr{P}$, under rather general conditions, one can define convolution* and study infinite divisibility* of laws on M, finding among others analogues of Gaussian or normal* laws [10]. In S^1, for example, the "wrapped normal" laws retain some but not all of the useful properties of normal laws on R^k (*see* DIRECTIONAL DISTRIBUTIONS and WRAPPED NORMAL DISTRIBUTION).

Curvature

A Riemannian manifold has an intrinsic curvature. A two-dimensional manifold is positively curved if, as in S^2, the area and circumference of a disk of radius r are smaller than they are in R^2. A surface is negatively curved if it is saddle-shaped, so that the area and circumference of a disk are larger than in R^2. Curvature is also defined in higher dimensions.

General relativity theory explains gravitation through positive curvature induced by matter. Curvature is quite appreciable over the distances to the farthest observable galaxies and quasars; *see* STATISTICS IN ASTRONOMY.

MANIFOLDS AS SAMPLE SPACES

For the most studied cases, the circle S^1 and sphere S^2, *see* DIRECTIONAL DISTRIBUTIONS *and* DIRECTIONAL DATA ANALYSIS.

THE PROJECTIVE PLANE. The set of all lines, or *axes*, through the origin in R^{k+1} forms a *projective space* P^k with a naturally induced Riemannian structure. Here P^k is S^k with antipodal points identified. While P^1 is a circle, P^2 differs from S^2. Laws on P^2 often are called *axial distributions*.

For data in S^2, a 3×3 matrix M_{ij} gives the moments of inertia of the sample around any axis; M_{ij} is unchanged if any subset of the data points are replaced by their antipodal points. Thus M_{ij} is defined for samples in P^2. Procedures based on M_{ij} have good invariance properties (see Watson [16], Chap. 5). Anderson and Stephens [2] gave a test of uniformity on S^2 or P^2, using M_{ij}, which also indicates kinds of nonuniformity, whether a mode (maximum), or a "girdle" along a great circle.

HYPERBOLOIDS. The sheet $t > 0$ of the hyperboloid $t^2 - x^2 - y^2 - z^2 = 1$ occurs, e.g., in relativistic physics as the (negatively curved) manifold of "4-velocities" for particles of positive rest mass. Jensen [14] treats a natural family of distributions on hyperboloids, analogous to von Mises–Fisher distributions on spheres (*see* DIRECTIONAL DISTRIBUTIONS). Neither distribution is that of a Brownian diffusion [16, p. 99; 15].

MATRICES. The manifold M may be a set of k-tuples of vectors, i.e., matrices. For example, sets of possible sample covariance matrices form manifolds on which there are interesting families of laws, e.g., Wishart distributions*. On distributions of matrix argument generally see James [12, 13], Downs [6], and MATRIX-VALUED DISTRIBUTIONS.

MANIFOLDS AS PARAMETER SPACES

LOCATION FAMILIES. A group G acting on M and one law P on M give a location family $\mathscr{P} = \{P \circ g^{-1} : g \in G\}$ of laws on M. If $H = \{g \in G : P \circ g^{-1} = P\}$ then the coset space $\Theta = G/H$ indexes $\mathscr{P} = \{P_\theta : \theta \in \Theta\}$ in a one-to-one way.

Equivariant estimators* of location in M are those transformed by g in G if all the observations are. Likewise, an *invariant test* of uniformity gives the same result when the observations are all transformed by a g in G. *See* INVARIANCE CONCEPTS IN STATISTICS.

AFFINE GROUP AND SCALE. If $M = R^1$, the affine group G of all transformations $x \to ax + b$, where $a > 0$ and $-\infty < b < \infty$, gives changes of scale* as well as location (not preserving ds^2 for $a \neq 1$).

MULTINOMIAL DISTRIBUTIONS. Let a sample space X be decomposed into cells $A_1, \dots, A_k$. Any family $\{P_\theta : \theta \in \Theta\}$ of laws on X gives probabilities $\{\{P_\theta(A_i)\}_{i=1}^k : \theta \in \Theta\}$ forming a subset M of the collection of all multinomial distributions* on k points. In the most interesting cases, M is a manifold, of dimension less than $k - 1$. Differentiability of the surface M is of use in the asymptotic theory of tests of the composite hypothesis* $\{P_\theta : \theta \in \Theta\}$ such as chi-square tests* of fit [7, 8].

Geometry of densities

Let a family $\{P_\theta : \theta \in \Theta\}$ of laws on a space X all have densities $\{f_\theta : \theta \in \Theta\}$ with respect to one σ-finite measure μ on X. Then distances between points of Θ can be measured by various distances between densities in spaces of functions integrable (to various powers) with respect to μ. The resulting geometry of Θ may have little to do with the geometry of X, unlike the cases of location and scale families.

Čencov [5] extensively treats the differential geometry of parametric families in a framework of homological algebra (categories and functors).

If $\theta = (\theta_1, \dots, \theta_k)$ is a local coordinate system in Θ, the *Fisher information* matrix*

$$g_{ij}(\theta) = \int (\partial \log f_\theta(x)/\partial\theta_i) \times (\partial \log f_\theta(x)/\partial\theta_j)\, dP_\theta(x)$$

defines a natural Riemannian metric on Θ when it exists.

EXPONENTIAL FAMILIES. If each $f_\theta > 0$ almost everywhere for μ, and all functions $\log f_\theta$ belong to a finite-dimensional vector space F of measurable functions on X, then $\{P_\theta : \theta \in \Theta\}$ is included in an exponential family*. The set of all f in F such that $\int e^f d\mu < \infty$ is convex, and has a natural flat Euclidean geometry. On its curved submanifolds see, e.g., Amari [1]. Here curvature measures the extent to which a family is not itself exponential: *see* Efron [9] and STATISTICAL CURVATURE.

HELLINGER DISTANCE. Let P and Q be any two probability laws, both absolutely continuous with respect to a finite measure μ, such as $P + Q$, with $f = dP/d\mu$, $g = dQ/d\mu$. Then the *Hellinger distance**, defined by

$$d_H(P, Q) = \left(\int (f^{1/2} - g^{1/2})^2 \, d\mu \right)^{1/2}$$

does not depend on the choice of μ. If f is a nonparametric estimator of the unknown density (*see* DENSITY ESTIMATION), minimizing $d_H(f, f_\theta)$, $\theta \in \Theta$, gives an estimate of θ with good efficiency* and robustness* properties [3].

References

[1] Amari, S. (1982). *Ann. Statist.*, **10**, 357–385.

[2] Anderson, T. W. and Stephens, M. A. (1972). *Biometrika*, **59**, 613–621.

[3] Beran, R. (1977). *Ann. Statist.*, **5**, 445–463.

[4] Boothby, W. M. (1975). *An Introduction to Differentiable Manifolds and Riemannian Geometry*. Academic Press, New York.

[5] Čencov, N. N. (1972). Statistical Decision Rules and Optimal Inference. *Transls. of Math. Monographs*, Amer. Math. Soc., **53** (1982).

[6] Downs, T. D. (1972). *Biometrika*, **59**, 665–676.

[7] Dudley, R. M. (1976). Probabilities and Metrics. *Aarhus Univ. Mat. Inst. Lecture Note Series*, **45**.

[8] Dudley, R. M. (1979). In *Probability Theory, Banach Center Publs.*, **5**. Polish Scientific Publishers, Warsaw, pp. 75–87.

[9] Efron, B., and discussants (1975). *Ann. Statist.*, **3**, 1189–1242.

[10] Gangolli, R. (1964). *Acta Math.*, **111**, 213–246.

[11] Helgason, S. (1962). *Differential Geometry and Symmetric Spaces*. Academic Press, New York.

[12] James, A. T. (1964). *Ann. Math. Statist.*, **35**, 475–501.

[13] James, A. T. (1975). In *Theory and Applications of Special Functions*, R. A. Askey, ed. Academic Press, New York, pp. 497–520.

[14] Jensen, J. L. (1981). *Scand. J. Statist.*, **8**, 193–206.

[15] Karpelevič, F. I., Tutubalin, V. N., and Šur, M. G. (1959). *Theor. Prob. Appl.*, **4**, 399–404.

[16] Watson, G. S. (1983). Statistics on Spheres. *Univ. Arkansas Lecture Notes Math. Sci.*, **6**. Wiley, New York.

(DIRECTIONAL DATA ANALYSIS
DIRECTIONAL DISTRIBUTIONS
HELLINGER DISTANCE
INVARIANCE CONCEPTS IN STATISTICS
MEASURE THEORY IN STATISTICS AND PROBABILITY
SAMPLE SPACES
STATISTICAL CURVATURE)

R. M. DUDLEY

MANLY–PARR ESTIMATORS

The Manly–Parr estimator of population size can be calculated from data obtained by the capture-recapture method* of sampling an animal population. The assumption made by Manly and Parr [5] is that an animal population is sampled on a series of occasions in such a way that on the ith occasion all the N_i animals then in the population have the same probability p_i of being captured. In that case the expected value of n_i, the number of captures, is $E(n_i) = N_i p_i$, so that

$$N_i = E(n_i)/p_i. \quad (1)$$

Manly and Parr proposed that p_i be estimated by the proportion of the animals known to be in the population at the time of the ith sample that are actually captured at that time. For example, in an open population (where animals are entering through births and leaving permanently through deaths and emigration) any animal seen before the time of the ith sample and also after that time was certainly in the population at that time. If there are C_i individuals of this type, of which c_i are captured in the ith sample, then $\hat{p}_i = c_i/C_i$ is an unbiased estimator of p_i. Using equation (1) the Manly–Parr estimator of N_i is then

$$\hat{N}_i = n_i/\hat{p}_i = n_i C_i/c_i. \quad (2)$$

Based upon a particular multinomial* model for the capture process, Manly [4] gives the approximate variance

$$\text{Var}(\hat{N}_i) \simeq N_i(1 - p_i)(1 - \theta_i)/(p_i\theta_i), \quad (3)$$

where θ_i is the probability of one of the N_i animals being included in the class of C_i animals known to certainly be in the population at the time of the ith sample. This variance can be estimated by

$$\widehat{\text{Var}}(\hat{N}_i) = \hat{N}_i(C_i - c_i)(n_i - c_i)/c_i^2. \quad (4)$$

Manly and Parr also proposed estimators for survival rates and birth numbers in the open population situation. Let r_i denote the number of animals that are captured in both the ith and the $(i+1)$th samples. The expected value of this will be $E(r_i) = N_i s_i p_i p_{i+1}$, where s_i is the survival rate over the period between the two samples for the population as a whole. This relationship together with equation (1) suggests the Manly–Parr survival rate estimator

$$\hat{s}_i = r_i/(n_i\hat{p}_{i+1}), \quad (5)$$

where $\hat{p}_{i+1} = c_{i+1}/C_{i+1}$.

At the time of the $(i+1)$th sample the population size N_{i+1} will be made up of the survivors from the previous sample time, $N_i s_i$, plus the number of new entries B_i to the population (the births). On this basis the Manly–Parr estimator of the number of births is

$$\hat{B}_i = \hat{N}_{i+1} - \hat{s}_i\hat{N}_i. \quad (6)$$

Variance formulae for $\hat{s}_i$ and $\hat{B}_i$ have not been developed at the present time.

Table 1 Example Data to Illustrate the Manly–Parr Estimators[a]

Capture–Recapture Pattern				Number of Animals
1	1	1	1	2
1	1	1	0	9
1	1	0	0	14
1	0	1	1	5
1	0	1	0	2
1	0	0	1	2
1	0	0	0	22
0	1	1	1	1
0	1	1	0	8
0	1	0	1	3
0	1	0	0	15
0	0	1	1	4
0	0	1	0	20
0	0	0	1	14
$n_i = 56$	52	51	31	
$c_i =$ —	11	8	—	
$C_i =$ —	20	13	—	
$r_i = 25$	20	12	—	

[a]There are four samples and, for example, the capture–recapture pattern "1 0 1 1" indicates that the animals concerned were captured in the first, third, and fourth samples only. There are five animals with this pattern.

As an example of the use of the Manly–Parr equations consider the data shown in Table 1. The values of n_i, c_i, C_i, and r_i that are needed for equations (2), (4), (5), and (6) are shown at the foot of the table. Using these, eq. (2) produces the population size estimates $\hat{N}_2 = 94.5$ and $\hat{N}_3 = 82.9$. The square roots of the estimated variances from equation (4) then give the estimated standard errors $S\hat{e}(\hat{N}_2) = 16.9$ and $S\hat{e}(\hat{N}_3) = 16.7$. Eq. (5) gives the estimated survival rates $\hat{s}_1 = 0.812$ and $\hat{s}_2 = 0.625$. Finally, Eq. (6) produces the estimated birth number $\hat{B}_2 = 23.8$. The data in this example are part of the illustrative data used by Manly and Parr [5].

Seber [7], Southwood [8], and Begon [1] discuss the Manly–Parr method in the context of capture–recapture methods in general. There are two principal competitors for the analysis of data from open populations, the Jolly–Seber method [3, 6] and the Fisher–Ford method [2]. The main theoretical advantage of the Manly–Parr approach is that it does not require the assumption that the probability of survival is the same for animals of all ages.

References

[1] Begon, M. (1979). *Investigating Animal Abundance*. Edward Arnold, London. (An introduction to capture–recapture methods aimed mainly at biologists. There is a lengthy discussion on the relative merits of different methods of analysis for capture–recapture data.)

[2] Fisher, R. A. and Ford, E. B. (1947). *Heredity*, **1**, 143–174.

[3] Jolly, G. M. (1965). *Biometrika*, **52**, 225–247.

[4] Manly, B. F. J. (1969). *Biometrika*, **56**, 407–410.

[5] Manly, B. F. J. and Parr, M. J. (1968). *Trans. Soc. Brit. Ent.*, **18**, 81–89.

[6] Seber, G. A. F. (1965). *Biometrika*, **52**, 249–259.

[7] Seber, G. A. F. (1982). *The Estimation of Animal Abundance and Related Parameters*, 2nd. ed. Charles Griffin, London. (This is the standard reference for statistical and mathematical aspects of capture–recapture methods.)

[8] Southwood, T. R. E. (1978). *Ecological Methods*. Chapman and Hall, London. (This is a standard text for ecologists. One chapter gives a good survey of both statistical and practical aspects of capture–recapture methods.)

(CAPTURE–RECAPTURE METHODS
ECOLOGICAL STATISTICS
STATISTICS IN ANIMAL SCIENCE)

B. F. J. Manly

MANN–FERTIG STATISTIC

The Mann–Fertig [1] statistic was designed to test that a set of data came from a two-parameter Weibull distribution* with unknown parameters, under an alternative hypothesis that the distribution is three-parameter Weibull. Thus the test is designed for data which, because of theory (perhaps a weakest-link or other extreme-value consideration) are assumed to have been selected randomly from a Weibull distribution with distribution function given by

$$F_Z(z) = 1 - \exp\{-[(z-\lambda)/\delta]^{\beta}\}.$$

Here $\lambda \geqslant 0$ is a threshold before which, in failure analysis, failure occurs with probability zero, $\delta > 0$ is a scale parameter, and $\beta > 0$ is a shape parameter.

The Mann–Fertig goodness-of-fit* test essentially tests the hypothesis $H_0: (\lambda = 0)$ versus the alternative $\lambda > 0$ by determining whether the left tail is too short. A variation of the test statistic can be used to obtain a median unbiased* estimate of λ.

The statistic for testing H_0 is a slight modification of the S statistic of Mann et al. [2], designed earlier to test the hypothesis of a two-parameter Weibull against some more general alternatives. It is based on an asymptotic result of Pyke [3] concerning the approximate independence and approximate exponentiality of spacings*, i.e., differences of successive order statistics, from continuous distributions.

If Z is a Weibull variate with $\lambda = 0$ and if $X = \ln Z$, then X has a Type I distribution of the smallest extreme having distribution function

$$F_X(x) = 1 - \exp\{-\exp[(x-\eta)/\xi]\}$$

with location parameter η and scale parameter $\xi = \beta^{-1}$. The S statistic is based on normalized spacings from the distribution of X,

$$l_i = (X_{i+1,n} - X_{i,n})/[E(X_{i+1,n}) - E(X_{i,n})],$$
$$i = 1, \dots, m-1. \quad (1)$$

Here E is the expectation operator, $X_{i,n}$ is the ith smallest observation of X in a sample of size n, and $X_{m,n}$ with $m \leqslant n$, is the largest value of X observed. It was van Montfort [5] who observed that for a Type I extreme value distribution* the asymptotic result of Pyke pertaining to continuous distributions applies for very small samples. For samples as small as three or four the result applies nearly exactly at the lower end of the distribution and roughly at the upper end.

Since the Type I extreme-value distribution is a member of the location-scale parameter* family, one can observe that l_i is independent of the unknown location parameter η. If $E(Y_{i,n})$, the expected value of the reduced parameter-free ith order statistic, is used in l_i, $i = 1, \dots, m-1$, in place of the unknown $E(X_{i,n})$, $i = 1, \dots, m$, then each l_i is proportional to the unknown scale parameter ξ. Or,

$$\begin{aligned} E(X_{i+1,n}) - E(X_{i,n}) &= \xi E\{(X_{i+1,n} - \eta)/\xi\} \\ &\quad - \xi E\{(X_{i,n} - \eta)/\xi\} \\ &= \xi\{E(Y_{i+1,n}) - E(Y_{i,n})\}. \end{aligned}$$

One determines a value for k as discussed presently and forms the test statistic

$$P_{k,m} = \sum_{i=k+1}^{m-1} l_i \Big/ \sum_{i=1}^{m-1} l_i, \quad (2)$$

as the ratio of sums of l_i's. $P_{k,m}$ is therefore parameter free, even though the $E(Y_{i,n})$'s are used in place of the $E(X_{i,n})$'s. Moreover, the distribution of each $2l_i$, as defined in (1), is approximately chi-square with 2 degrees of freedom. Hence the distribution of $P_{k,m}$ is essentially Beta* with parameters $m-k-1$ and k when H_0 is true.

The S statistic of Mann et al. is of the form $P_{k,m}$ with $k = [m/2]$ and $[r]$ the greatest integer less than or equal to r. A Monte Carlo* generation of critical values of S resulted essentially in percentiles of appropriate Beta distributions. When k is smaller than $[m/2]$, as it tends to be for testing a three-parameter Weibull, the Beta approximation is enhanced. For $P_{k,m}$ large, H_0 is rejected.

The appropriate values for k for small sample sizes are given in Mann and Fertig [1]. If the sample size n is 15 or larger, k equal to the integer nearest to $m/3$ tends to be optimal for significance levels ranging from 0.01 to 0.25.

Before using the goodness-of-fit test, it is useful and, in fact, important to plot the data on Weibull probability paper. It may be, for example, that the data, rather than exhibiting a threshold, are from a mixture of two Weibull distributions reflecting two different phenomena operating at low and high levels of the random variable, respectively. Thus, early and late failures may result from a mixture of two or more types of hardware. In this case, the goodness-of-fit test would tend to reject the hypothesis of a two-parameter Weibull, but it would be incorrect to blindly accept the alternative of a three-parameter Weibull. A probability plot of such data on Weibull paper will ordinarily reveal two or more intersecting straight lines, rather than a smooth curve, concave downward, exhibited by three-parameter Weibull data.

CONFIDENCE BOUNDS AND POINT ESTIMATES

The theory used to develop the Mann–Fertig statistic $P_{k,m}$ can be used also to obtain iteratively confidence bounds and a median unbiased estimate of a threshold parameter λ. Note that if Z has a three-parameter Weibull distribution with threshold parameter λ, then $X_{i,n} = \ln(Z_{i,n} - \lambda)$ is the ith Type I extreme-value distribution order statistic and

$$P^*_{k,m}(\lambda) = \sum_{i=k+1}^{m-1} l^*_i \Big/ \sum_{i=1}^{m-1} l^*_i ,$$

with

$$l^*_i = \frac{\ln(T_{i+1,n} - \lambda) - \ln(T_{i,n} - \lambda)}{E(Y_{i+1,n}) - E(Y_{i,n})}, \qquad i = 1, \dots, m-1,$$

has the distribution of $P_{k,n}$ (defined in (2)) under H_0. Mann and Fertig demonstrate that $P^*_{k,m}(\lambda)$ is monotonically decreasing in λ for $0 < \lambda < Z_{1,n}$. Therefore a lower confidence bound at level γ can be determined for λ if H_0 is rejected at significance level α and $\gamma \leqslant 1 - \alpha$.

Somerville [4] discovered the necessary condition of rejection of H_0. He also determined that if $\alpha = 0.5$, as in obtaining a median unbiased estimate of λ, then the optimal k is the integer nearest to $m/5$. Somerville's results demonstrate that the power of a test based on $P_{k,m}$ is bounded above, even though λ may be greatly larger than zero. This bound increases as m and the significance level α increase and as the shape parameter $\beta = 1/\xi$ decreases. Thus, the maximum power of a test of H_0 based on $P_{k,m}$ is 1.00 if $m = 10$, $\beta = 0.5$, and $\alpha = 0.5$ or if $m = 25$, $\beta = 1.0$, and $\alpha = 0.10$. If $n = 25$ and $\beta = 2$, then the maximum power is 0.70 if $\alpha = 0.05$.

Example

As an example of the use of this statistic to obtain a median unbiased estimate of λ, we consider a sample of 96 magnitudes of largest California earthquakes occurring in six-month periods over 48 years. Reference to EXTREME-VALUE DISTRIBUTIONS shows that if magnitude M has a Type III distribution of largest values with upper threshold μ, then $-(M - \mu) = -M - (-\mu) = \mu - M$ has a Weibull distribution with lower threshold zero. Or, $-M$ has a Weibull distribution with lower threshold $\lambda = -\mu$. The presence of an upper threshold is corroborated by physical theory. Hence, a Type III distribution is appropriate for the analysis of the data.

A moment estimate of λ is $\hat{\lambda} = -12.3$. Using $\hat{\lambda}$ to calculate $P^*_{k,m}$ with $k = 19$ yields 0.803. The 50th percentile of a Beta distribution with parameters 76 and 19 is 0.802 and the 10th percentile is 0.745. The very close agreement of the two independent estimates, the moment estimate and the median unbiased estimate, of $\lambda = -\mu$ gives statistical corroboration of the Type III distribution and the upper threshold specified by physical theory. A plot of the data on Weibull probability paper yields a straight line for $Z = 12.3 - M$.

References

[1] Mann, N. R. and Fertig, K. W. (1975). *Technometrics*, **17**, 237–245.

[2] Mann, N. R., Scheuer, E. M., and Fertig, K. W. (1973). *Commun. Statist.*, **2**, 383–400.

[3] Pyke, R. (1965). *J. R. Statist. Soc. B*, **27**, 395–449.

[4] Somerville, P. N. (1977). In *Theory and Applications of Reliability*, Vol. I, P. Tsokos and I. N. Shimi, eds.

[5] van Montfort, M. A. J. (1970). *J. Hydrology*, **11**, 421–427.

Acknowledgment

Research supported by the Office of Naval Research, Contract N00014-82-K-0023, Project 047-204.

(WEIBULL DISTRIBUTION)

NANCY R. MANN

MANN–GRUBBS METHOD

Consider a series system made up of k independent subsystems, e.g., electronic components, each having exponentially distributed failure time T with either no censoring* of the subsystem data or only Type II censoring. Testing is without replacement. For this model, system reliability* $R_s(t_m)$ at time $t_m > 0$ is equal to

$$\prod_{i=1}^{k} \exp(-t_m\lambda_j) = \exp\left(-t_m \sum_{j=1}^{k} \lambda_j\right), \quad (1)$$

with $\lambda_j > 0, j = 1, \ldots, k$, the hazard rate for the jth subsystem. One can determine an upper confidence bound on the system hazard rate $\phi = \sum_{j=1}^{k}\lambda_j$, and from this, of course, can then obtain a lower confidence bound on $R_s(t_m) = \exp(-\phi t_m)$.

The Mann–Grubbs [7] method for the determination of confidence bounds for system reliability has been adapted to several situations, but was derived originally for this particular exponential series-system model with Type II censoring. Here the method yields a lower confidence bound that very closely approximates the lower confidence bound that is most accurate (has the highest probability of being close to the true system reliability) for all values of system reliability, among exact confidence bounds that are unbiased. The restriction of unbiasedness* is necessary here because of the nuisance parameters $\lambda_1, \ldots, \lambda_k$, the hazard rates for the k independent subsystems. For binomial models and exponential models with censoring by time, the Mann–Grubbs [8] method uses an approach similar to that used for the exponential model with Type II censoring. However, because of discreteness of the data in the binomial case, optimal bounds are more difficult to approximate. For all these situations, however, the method is quite simple to implement.

For the model given by equation (1) with $k = 2$, Lentner and Buehler [2] derived the uniformly most accurate unbiased lower confidence bound on $R_s(t_m)$. Generalization to $k \geqslant 2$ was made by El Mawaziny [1] in his doctoral thesis. As noted above, their results depend upon the assumption that for the jth subsystem n_j prototypes have been tested until r_j failures occur, $1 \leqslant r_j \leqslant n_j$, $j = 1, \ldots, k$. For the jth subsystem, one observes the ith smallest failure times $t_{i,j}$, $i = 1, \ldots, r_j$. One then computes the total time on test for the jth subsystem, $w_j = \sum_{i=1}^{r_j} t_{i,j} + (n_j - r_j)t_{r_j,j}$, $j = 1, \ldots, k$. Calculation of a lower confidence bound on $R_s(t_m)$ based on the w_j's and the r_j's by El Mawaziny's method must be performed iteratively by means of a computer. If both the number of subsystems and the total number of failures are large, problems of loss of precision will result; see Mann [4].

The Mann–Grubbs (M–G) method eliminates the need for a computer and the problem of loss of precision resulting from the lengthy calculations. The approach is based on the fact, demonstrated by Mann and Grubbs [7], that the conditional distribution of $\phi = \sum_{j=1}^{k}\lambda_j$, given the data, is that of a sum of weighted noncentral chi squares*. This sum can be well approximated for present purposes by a single weighted chi-square variate with mean m and variance v. Thus using a two-moment fit to chi-square, one assumes that $2m\phi/v$ is a chi-square* variate with $2m^2/v$ degrees of freedom (see Patnaik [10]).

The expressions for the conditional mean m and variance v of the system hazard rate derived by Mann and Grubbs for this model have been simplified by Mann [8] and are given by

$$m = \sum_{j=1}^{k} \frac{(r_j - 1)}{w_j} + w_{(1)}^{-1} \qquad (2)$$

and

$$v = \sum_{j=1}^{k} \frac{(r_j - 1)}{w_j^2} + w_{(1)}^{-2}, \qquad (3)$$

where $w_{(1)}$ is the smallest of the w_j's.

Once the hazard-function moments have been calculated, the Wilson–Hilferty* transformation of chi-square to normality can be used to facilitate the calculations, since the number of degrees of freedom $\nu = 2m^2/v$ for the approximate chi-square variate $2m\phi/v$ is not generally an integer. To approximate the uniformly most accurate lower confidence bound $\underline{R}_s(t_m)$ on series-system reliability $R(t_m)$ at time t_m and at confidence level $1 - \alpha$ (incorporating the Wilson–Hilferty transformation), one calculates

$$\underline{R}_s(t_m) = \exp\left[-t_m m \left(1 - \frac{v}{9m^2} + \frac{z_{1-\alpha} v^{1/2}}{3m} \right)^3 \right], \qquad (4)$$

where z_γ is the 100γth percentile of a standard normal distribution. The Wilson–Hilferty transformation yields an approximation to chi-square in this context that, for 3 or more degrees of freedom, is accurate to within a unit or two in the second significant figure.

For an example of calculation of an approximate confidence bound $\underline{R}_s(t_m)$ on series system reliability, we consider an independent series system containing three subsystems, each with exponential failure time. For each of the three subsystems, prototypes have been life tested with Type II censoring, resulting in total-times-on test $w_1 = 42.753$, $w_2 = 45.791$, and $w_3 = 31.890$, with $r_1 = 4$, $r_2 = 3$, and $r_3 = 2$. To obtain $\underline{R}_s(t_m)$ using (2), (3) and (4), one forms

$$m = \frac{4-1}{42.753} + \frac{3-1}{45.791} + \frac{2-1}{31.890} + \frac{1}{31.890}$$
$$= 0.17656$$

and

$$v = \frac{4-1}{(42.753)^2} + \frac{3-1}{(45.791)^2} + \frac{2-1}{(31.890)^2} + \frac{1}{(31.890)^2} = 0.00456.$$

Thus, an approximate 90% lower confidence bound at $t_m = 1$ is calculated as

$$\underline{R}_{.90}(1) = \exp\left\{ -.17656 \left[1 - \frac{.00456/.17656^2}{9} + 1.282\sqrt{.00456}/.52968 \right]^3 \right\}$$
$$= 0.766.$$

The El Mawaziny optimal exact lower confidence bound on $\underline{R}_{.90}(1)$ is .772. See Mann and Grubbs [7] or Mann et al. [9].

Schoenstadt [11] uses simulation* to compare an exact procedure of Lieberman and Ross [3] and this approximate method for obtaining lower confidence bounds on system reliability under the model specified by equation (1). He concludes that the results of the simulation runs "seem to demonstrate the superiority of the M–G bounds in those instances that can be considered of practical importance," and also demonstrates that the M–G bounds are exact to the accuracy of his simulation.

For binomial data, the parameter of interest for which a posterior mean and variance (conditional on the failure data) are calculated is $\xi = -\ln R$ or, for parallel systems, $\xi = -\ln(1 - R)$, with $\ln(\cdot)$, the natural logarithm. The parameter ξ has been demonstrated by Mann [6] to have a posterior distribution that is approximately proportional to a chi-square variate, as does ϕ in the exponential model. The expressions for the conditional mean and variance of ξ resemble the expressions (2) and (3) only for models for which *randomized* bounds, commonly used in binomial models, are obtained.

Expressions for obtaining approximate lower confidence bounds on R_s or R_p (either randomized or *nonrandomized*) for the binomial model and for an exponential model with fixed censoring times can be found in Sect. 10.4 of Mann et al. [9]. Comparisons of results with those of other methods are also given there for all applicable models, as well as methods for combining results to obtain bounds for models more general than simple series or parallel systems. The latter are also discussed in Mann and Grubbs [8].

References

[1] El Mawaziny, A. H. (1967). Ph.D. Thesis, Iowa State University, Ames, Iowa.

[2] Lentner, M. M. and Buehler, R. J. (1963). *J. Amer. Statist. Ass.*, **58**, 670–677.

[3] Lieberman, J. and Ross, S. (1971). *J. Amer. Statist. Ass.*, **66**, 837–840.

[4] Mann, N. R. (1970). *Naval Res. Logist. Quart.*, **17**, 41–54.

[5] Mann, N. R. (1974). *J. Amer. Statist. Ass.*, **69**, 492–495.

[6] Mann, N. R. (1974). *IEEE Trans. Rel.*, **R-22**, 293–304.

[7] Mann, N. R. and Grubbs, F. E. (1972). *Biometrika*, **59**, 191–204.

[8] Mann, N. R. and Grubbs, F. E. (1974). *Technometrics*, **16**, 335–347.

[9] Mann, N. R., Schafer, R. E., and Singpurwalla, N. D. (1974). *Methods for Statistical Analysis of Reliability and Life Data.* Wiley, New York.

[10] Patnaik, P. B. (1949). *Biometrika*, **36**, 202–243.

[11] Schoenstadt, A. L. (1980). *J. Amer. Statist. Ass.*, **75**, 212–216.

[12] Wilson, E. B. and Hilferty, M. M. (1931). *Proc. Nat. Acad. Sci.* (U.S.), **17**, 684–688.

Acknowledgment

Nancy Mann's research was supported by the Office of Naval Research, Contract N00014-82-K-0023, Project 047-204.

(RELIABILITY)

NANCY R. MANN
FRANK GRUBBS

MANN'S TEST FOR TREND

In 1938 M. G. Kendall [2] introduced his new measure of rank correlation* for assessing the degree of agreement in a sequence of paired observations $(X_1, Y_1), \dots, (X_n, Y_n)$. The numerator S of Kendall's τ^* can be written in various ways, for example,

$$S = \sum_{i<j} \operatorname{sign}(X_j - X_i)\operatorname{sign}(Y_j - Y_i)$$

$$= P - Q,$$

where P is the number of pairs (X_i, Y_i), (X_j, Y_j) for which $X_j > X_i$ and $Y_j > Y_i$ or $X_j < X_i$ and $Y_j < Y_i$, called the number of concordances, and Q is the number of discordances, defined in a similar way; $\operatorname{sign}(z) = 1, 0, -1$ as $z > 0, = 0, < 0$, respectively.

In 1945 H. B. Mann [4] recognized that S could be used to test for the presence of trend in a sequence of numbers. In Mann's test for trend $X_i \equiv i$, $i = 1, 2, \dots, n$, and the test statistic is simply

$$S^* = \sum_{i<j} \operatorname{sign}(Y_j - Y_i).$$

To test the null hypothesis that $Y_1, \dots, Y_n$ are independent and identically distributed against the alternative hypothesis that they are stochastically increasing, Mann suggested rejecting the null hypothesis for large values of S^*. We say Y_j is *stochastically larger than* Y_i if $F_i(t) > F_j(t)$ for all t, where F_i and F_j are the cumulative distribution functions. Under the null hypothesis, $ES^* = 0$ and the variance, corrected for ties, is given by

$$\operatorname{Var} S^* = \frac{1}{18}\Big[n(n-1)(2n+5) - \sum u(u-1)(2u+5)\Big],$$

where u denotes the multiplicity of tied values.

Mann extended Kendall's table of critical values of the test. He also provided an alternative proof of the limiting normality of S^* and discussed conditions under which S^* provides a consistent and an unbiased test.

To illustrate, we consider the water pollution data of Gerstein [1], also discussed by Lehmann [3, p. 290]. The Y values are the number of odor periods in a year near Lake Michigan for the years from 1950 through

1964. The observations are

10, 20, 17, 16, 12,
15, 13, 18, 17, 19,
21, 23, 23, 28, 28.

The value of S^* is 68. We have 15 observations and there are three sets of ties; 17, 23, 28, each with multiplicity 2. Hence, $\operatorname{Var} S^* = 405.3$, $S^*/(\operatorname{Var} S^*)^{1/2} = 3.38$, the approximate p-value of the test is .0004, and Mann's test supports the hypothesis of an upward trend in the number of odor periods per year, at any reasonable significance level.

References

[1] Gerstein (1965). *Amer. Water Works Ass. J.*, **57**, 841–857.

[2] Kendall, M. G. (1938). *Biometrika*, **30**, 81–93.

[3] Lehmann, E. L. (1975). *Nonparametrics: Statistical Methods Based on Ranks*. Holden-Day, San Francisco.

[4] Mann, H. B. (1945). *Econometrica*, **13**, 245–259.

(KENDALL'S TAU
RANKING PROCEDURES)

THOMAS P. HETTMANSPERGER

MANN–WHITNEY–WILCOXON STATISTIC

The Wilcoxon two-sample rank sum statistic, W, is defined as follows. Combine samples $X_1, \ldots, X_n$ and $Y_1, \ldots, Y_m$ and order the resulting sample of size $N = (n + m)$. Let r_j be the rank of X_j in the combined sample, that is, X_j is the r_jth smallest in the combined sample; then $W = \sum_{j=1}^{n} r_j$, the sum of ranks of the X's. This definition is given by Wilcoxon [35] in a paper which also proposed the one-sampled signed-rank statistic (*see* WILCOXON SIGNED RANK TEST). The statistic had been proposed earlier, for example by G. Deuchler in 1914 [10]; for further historical references see Kruskal [23]. The statistic W is used to test the null hypothesis H that the $\binom{N}{n}$ possible assignments of ranks to the X's are equally likely.

Mann and Whitney [25] introduced their statistic, U, and defined it to be the total number of times an X precedes a Y in the ordered combined sample, giving the relationship

$$U = nm + \tfrac{1}{2}n(n+1) - W. \tag{1}$$

They found a recurrence relationship for the distribution of U under H and tabulated significance points for small values of n, m. For larger n, m they proposed a normal distribution approximation, showing the large sample convergence of the distribution of U by consideration of high-order moments. They also showed the consistency of the U test against a certain class of alternatives.

The mean of W under H is $\frac{1}{2}n(N+1)$ and its variance is $(12)^{-1}mn(N+1)$; W takes integer values between $\frac{1}{2}n(n+1)$ and $\frac{1}{2}N(N+1) - \frac{1}{2}m(m+1)$, with its null distribution being symmetric about its mean. For n, m as small as 10, the normal approximation to the distribution of W is reasonable.

The statistics U, W are both referred to as the Mann–Whitney–Wilcoxon statistic (MWW) because of the relationship (1). As an illustration, suppose that the X sample is given by 60, 61, 63, 65, 58 and the Y sample by 75, 68, 59, 72, 64, 67, giving the X, Y observations in the ordered sample as $XYXXXYXYYYY$. Thus $W = 1 + 3 + 4 + 5 + 7 = 20$, and $U = 6 + 3 \times 5 + 4 = 25$, confirming (1) with $n = 5$, $m = 6$. It is found that $\Pr(W \leqslant 20 \mid H) = 0.0411$.

The hypothesis H can be true if one of the following holds. (a) N subjects are randomly assigned to different treatments, n to the first, m to the second treatment; if there is no difference between treatments H is true. (b) Random samples are drawn from two populations, n from the first, m from the second population; if there is no difference between populations, H is true.

In both cases it is assumed there are no tied data. Lehmann [24] considers the large sample distribution of U under H and other hypotheses using Hoeffding's [20] theory of U-statistics*, since the Mann–Whitney version is a U-statistic. This can be seen by writing $\text{sgn}(x) = -1$ if $x < 0$, 0 if $x = 0$, $+1$ if $x > 0$, for then

$$2U - mn = \sum_{j=1}^{n} \sum_{i=1}^{n} \text{sgn}(Y_i - X_j),$$

the "U-statistic," assuming there are no ties in the data. An alternative approach is to use the theorem of Chernoff and Savage [7] applied to the Wilcoxon version of the statistic (*see* CHERNOFF–SAVAGE THEOREM).

OPTIMALITY RESULTS

For the population model (b) above, let the $X(Y)$ population have continuous distribution function $F_X(F_Y)$. The important location shift alternative for this model is where $F_Y(y) = F_X(y - \theta)$. The MWW statistic is unbiased for a test of $H_0 : \theta \leqslant 0$ against $H_A : \theta > 0$. The small sample power of the test can be found numerically using the results of Milton [26] for normal F_X, F_Y.

The MWW statistic is an asymptotically locally most powerful* statistic for the location shift model when F_X, F_Y have the logistic density; Cox and Hinkley [8, Ch. 6.3, iv]. The asymptotic Pitman efficiency* of the MWW statistic relative to the t-test* is bounded below by 0.864 [18], equal to $3\pi^{-1} \sim 0.955$ for normal, 1.10 for logistic, and 1.50 for double-exponential F_X and F_Y.

FURTHER INFERENCES

If $X_1, \ldots, X_n$ and $Y_1 - \delta, \ldots, Y_m - \delta$ are all assumed to have the same distribution, where δ represents the treatment effect, then δ can be estimated by $\hat{\delta}$, where $\hat{\delta}$ is chosen so that W, calculated from $X_1, \ldots, X_n$ and $Y_1 - \delta, \ldots, Y_m - \delta$, is as close to its null mean value, $\frac{1}{2}n(N + 1)$, as possible. This is known as the Hodges–Lehmann estimator*. Confidence intervals* for δ also can be found.

The statistic $U(nm)^{-1}$ is an unbiased estimator of $\Pr(X < Y)$, and Green [16] considers a test of the hypothesis $\Pr(X < Y) = p_0$ for general F_X, F_Y. Bounds for the variance of U are given. Stedl and Fox [32] consider estimating $\Pr(X < Y)$ when F_X is related to F_Y by a Lehmann alternative*. Delong and Sen [9] truncate the MWW statistic to estimate $\Pr(X < Y)$.

EXTENSIONS AND MORE RECENT RESULTS

1. *Tied and grouped data*. Bradley [5, Ch. 3, §3.3] discusses the general problems of tied data when using rank statistics. From a practical point of view, the method of midranks provides a unique solution; theoretical aspects are considered by Behnen [2] and others. Hochberg [17] considers an estimate for the variance of the MWW statistic with grouped data*.
2. *Censored data**. Gehan [15] proposed a version of the MWW statistic when observations are randomly censored on the right. This modification can, however, give misleading results; see Prentice and Marek [31]. Basu [1] proposed a version for use when the smallest r observations are only available, and also considered a sequential scheme.
3. *Dependent observations*. Like the t-statistic, the distribution of the MWW statistic is sensitive to dependencies between observations. When $F_X = F_Y$, serial correlation* between observations (see Box [4]) and within-sample dependence* (see Pettitt and Siskind [30]) perturb greatly the distribution of the MWW statistic from its distribution under H.
4. *Likelihood-based inference*. Kalbfleisch [22] considered a marginal likelihood based on ranks. The MWW statistic is equivalent to a score statistic when the observations are assumed, after some arbitrary monotone transformation, to have the logistic distribution*, with the samples differing only in their location parameters. Bayesian inference* for this model using the MWW statistic is considered by Brooks [6] and Pettitt [29].

5. *Miscellaneous results*. Eplett [11, 13] considered extensions to the MWW statistic to enable data on the circle to be analyzed. Berchtold [3] modified the statistic by changing the ranks of the smallest and largest 10% of the observations in the combined sample. An increase in asymptotic Pitman efficiency resulted. Fligner et al. [14] gave details of other test procedures having the same null distribution as the MWW statistic. Pettitt [27, 28] used the MWW statistic to test for and estimate a change-point. Iman [21] and Ury [34] considered small sample approximations to the null distribution of the MWW statistic while Thorburn [33] considered a local limit theorem. Eplett [12] studied an influence curve for the MWW statistic (*see* INFLUENCE FUNCTIONS).

DISCUSSION

The MWW statistic provides a simple procedure to compare two independent samples. The U version admits an estimate of something meaningful, i.e., $\Pr(X < Y)$, and an estimate of location, i.e., the Hodges–Lehmann estimate. Probably the various refinements of the MWW statistic and other rank statistics offer few *practical* advantages over the MWW procedure, where the advantages of, say, increased asymptotic efficiency, have to be weighed (usually) against greatly increased complexities in use. Rank-based procedures tend to be very sensitive to violations of independence or exchangeability*, assumptions which can, to a certain extent, be guarded against by randomization* in experimental design.

References

[1] Basu, A. P. (1977). In *Th. Appl. Reliability*, Vol. 1. Tsokos and Shimi, eds. Academic Press, New York, pp. 131–150.

[2] Behnen, K. (1976). *Ann. Statist.*, **4**, 157–174.

[3] Berchtold, H. (1979). *Biom. J.*, **21**, 649–655.

[4] Box, G. E. P. (1976). *J. Amer. Statist. Ass.*, **71**, 791–799.

[5] Bradley, J. V. (1968). *Distribution-Free Statistical Tests*. Prentice-Hall, Englewood Cliffs, N.J.

[6] Brooks, R. J. (1978). *J. Roy. Statist. Soc. B*, **40**, 50–57.

[7] Chernoff, H. and Savage, I. R. (1958). *Ann. Math. Statist.*, **29**, 972–994.

[8] Cox, D. R. and Hinkley, D. (1974). *Theoretical Statistics*. Chapman & Hall, London.

[9] Delong, E. R. and Sen, P. K. (1981). *Commun. Statist. A*, **10**, 963–982.

[10] Deuchler, G. (1914). *Zeit. f. Pädagogische Psychologie u. Experimentelle Pädagogik*, **15**, 114–131, 145–159, 229–242.

[11] Eplett, W. J. R. (1979). *Ann. Statist.*, **7**, 446–453.

[12] Eplett, W. J. R. (1980). *J. Roy. Statist. Soc. B*, **42**, 64–70.

[13] Eplett, W. J. R. (1982). *J. Roy. Statist. Soc. B*, **44**, 270–286.

[14] Fligner, M. A., Hogg, R. V., and Killeen, T. J. (1976). *Commun. Statist. A*, **5**, 373–376.

[15] Gehan, E. A. (1965). *Biometrika*, **52**, 203–223.

[16] Green, J. R. (1979). *Biometrika*, **66**, 645–653.

[17] Hochberg, Y. (1981). *Commun. Statist. A*, **10**, 1719–1732.

[18] Hodges, J. L. and Lehmann, E. L. (1956). *Ann. Math. Statist.*, **27**, 324–335.

[19] Hodges, J. L. and Lehmann, E. L. (1963). *Ann. Math. Statist.*, **34**, 598–611.

[20] Hoeffding, W. (1948). *Ann. Math. Statist.*, **19**, 293–325.

[21] Iman, R. L. (1976). *Commun. Statist. A*, **5**, 587–598.

[22] Kalbfleisch, J. D. (1978). *J. Amer. Statist. Ass.*, **73**, 167–170.

[23] Kruskal, W. J. (1957). *J. Amer. Statist. Ass.*, **52**, 356–360.

[24] Lehmann, E. L. (1951). *Ann. Math. Statist.*, **22**, 165–179.

[25] Mann, H. and Whitney, D. R. (1947). *Ann. Math. Statist.*, **18**, 50–60.

[26] Milton, R. C. (1970). *Rank Order Probabilities: Two-Sample Normal Shift Alternatives*. Wiley, New York.

[27] Pettitt, A. N. (1979). *Appl. Statist.*, **28**, 126–135.

[28] Pettitt, A. N. (1980). *J. Statist. Comput. Simul.*, **11**, 261–272.

[29] Pettitt, A. N. (1982). *J. Roy. Statist. Soc. B*, **44**, 234–243.

[30] Pettitt, A. N. and Siskind, V. (1981). *Biometrika*, **68**, 437–441.

[31] Prentice, R. L. and Marek, P. (1979). *Biometrics*, **35**, 861–867.

[32] Stedl, J. and Fox, K. (1978). *Commun. Statist. B*, **7**, 151–161.

[33] Thorburn, D. (1977). *Ann. Prob.*, **5**, 926–939.

[34] Ury, H. K. (1977). *Commun. Statist. B*, **6**, 181–197.

[35] Wilcoxon, F. (1945). *Biometrics*, **1**, 80–83.

Bibliography

Lehmann, E. L. (1975). *Nonparametrics*. Holden-Day, San Francisco. (Chapter 2 gives an extensive coverage of the MWW statistic and a long bibliography. Table B gives complete null distribution of MWW statistic for $n = 3(1)10$ and $n \leqslant m \leqslant 10$.)

Puri, M. L. and Sen, P. R. (1971). *Nonparametric Methods in Multivariate Analysis*. Wiley, New York. (Very theoretical account.)

Singer, B. (1979). *Br. J. Math. Statist. Psychol.*, **32**, 1–60. (Bibliography of nonparametric, distribution-free techniques. See Sect. D1 for MWW statistics.)

(DISTRIBUTION-FREE METHODS
HODGES–LEHMANN ESTIMATORS
RANKING PROCEDURES
RANK TESTS)

A. N. PETTITT

MANOVA *See* MULTIVARIATE ANALYSIS OF VARIANCE

MANPOWER PLANNING

Manpower (or human resource) planning may be defined as the matching of the supply of people with the work available for them to do. It embraces forecasting* both supply and demand and is concerned with control techniques to bring them into balance. The need for statistical techniques arises both because of the variability of individual behaviour and the uncertainty of the social environment. This is of course only one aspect of the total management process and the statistician's role is therefore as one of a team concerned with the collection and analysis of relevant data. It is common to distinguish different levels at which planning may take place. At one extreme there is national manpower planning for a whole economy and at the other there is planning for the individual firm. The most successful work appears to have been carried out at the senior levels of large firms or public institutions. Computer packages for a wide range of modelling exercises are now available.

STATISTICAL ANALYSIS OF WASTAGE

Wastage, or attrition, is a key variable because it is the source of major uncertainty and is difficult to control. It is therefore important to have an adequate statistical description of the process for incorporation into more comprehensive models. In earlier work following the actuarial tradition, propensity to leave was expressed in terms of rates. Now it is usually approached via the *completed length of service* (CLS) distribution. The form of this frequency distribution can yield useful insights into the nature of the wastage process and is a basic tool for forecasting. The estimation of the distribution is rarely a simple matter as the data available are usually incomplete. Cohort* data, which give the lengths of service of a sample of individuals recruited at about the same time, are usually heavily censored on the right. More often we have census* or current data which gives a record of events over an interval of time—often a year. Lengths of service then may be censored at both ends but also there is information about the numbers exposed to risk of leaving. Estimation methods fall into two groups. Distribution-free methods* seek to estimate the distribution function directly. Curve-fitting* methods involve estimating the parameters of a family of distributions known to be successful in graduating CLS distributions. The latter procedure is greatly helped by the fact that the forms of CLS distributions are remarkably stable between industries, skill levels, and countries. In a seminal paper Lane and Andrew [5] demonstrated that CLS distributions in the British Steel industry could be graduated by lognormal distributions*. This feature has been repeatedly observed since and even when the fit is poor as judged by conventional statistical tests it is often adequate for practical purposes. The fact that the logarithm of CLS has an approximately normal distribution has considerable statistical advantages.

No entirely convincing model to explain the widespread occurrence of the lognormal distribution in this context has yet been forthcoming. In fact several other distributions which result from plausible models are, in practice, difficult to distinguish from the lognormal. These are the various mixed exponential distributions first proposed by Silcock [7], and the inverse Gaussian distribution* suggested much more recently by Bartholomew [2] and Whitmore [12]. Both of these distributions have advantages for various purposes and it is convenient that such a wide choice is available.

MARKOV MODELS FOR GRADED SYSTEMS

The typical manpower system consists of a set of grades through which individuals progress. The system is fed by recruitment and subject to loss by leaving. If the system is hierarchical, progression will be by means of promotion (or demotion). The grades can be formed on the basis of skill or salary level but also they may be based on age, sex, location, or any combination of such factors. In practice it often is found that the flow rates from the grades are fairly constant over a period of time. This suggests that the system should be modelled by postulating constant probabilities for each kind of flow. Let the grades be numbered $1, 2, \ldots, k$. Then we may define probabilities p_{ij} $(i, j = 1, 2, \ldots, k)$ and w_i $(i = 1, 2, \ldots, k)$ such that p_{ij} is the probability that an individual in i moves to j in some specified interval of time; w_i is the probability of leaving from that grade. Note that the probabilities do not depend on states occupied before i; it is assumed that people move independently. Such a model is called a Markov chain (*see* MARKOV PROCESSES). In the terminology of that theory it has k transient states and one absorbing state. Obviously $w_i = 1 - \sum_{j=1}^{k} p_{ij}$. For such a system we can calculate various quantities of interest for planning purposes. For example the ith row sum of the fundamental matrix $(\mathbf{I} - \mathbf{P})^{-1}$ (where $\mathbf{P} = \{p_{ij}\}$) is the expected length of stay in the system of a person entering grade i. Of greater practical interest is the prediction of future grade numbers. If $\mathbf{n}(T)$ denotes the vector of expected grade sizes the following recurrence relation enables it to be predicted:

$$\mathbf{n}(T + 1) = \mathbf{n}(T)\mathbf{P} + R(T)\mathbf{p}_0, \qquad (1)$$

where $R(T)$ denotes the total number recruited at time T and $\mathbf{p}_0$ is a vector giving the proportions of recruits allocated to the various grades. Particular interest attaches to the steady state structure which exists if $R(T) \to R$ as $T \to \infty$. In that case the steady state vector satisfies

$$\mathbf{n} = R(\mathbf{I} - \mathbf{P})^{-1}\mathbf{p}_0. \qquad (2)$$

This is the ideal structure for a given $\mathbf{P}$ and $\mathbf{p}_0$ in the sense that it can be maintained over a period of time.

The converse problem has received a good deal of attention recently; a goal structure $\mathbf{n}^*$ is given (or a sequence of such goals) and the question arises as to how to choose the values of those parameters which can be controlled. One can map out the region in the space of possible structures which can be maintained if, for example, it is only possible to control the proportions in which recruits are allocated to grades. Many such control problems can be expressed in terms of linear or quadratic programs and solved by routine methods. A limitation of many of these methods is that they treat the system deterministically by assuming that the actual grade structure will conform to its expectations as given by (1). Adaptive strategies have been devised which take into account the variations arising from the random fluctuations of the real process.

RENEWAL MODELS FOR GRADED SYSTEMS

In many organizations the assumption of the Markov model that grade sizes can fluctuate in response to the operation of fixed transition rates is unrealistic. Instead the grade sizes are fixed or at least subject to tight control. In such cases a Markov model is inappropriate. The process is now a renewal process in which movements can only occur when vacancies arise. A vacancy can occur when someone leaves, is transferred, or when a new position is created. The main stochastic mechanism involved is that governing loss which is usually specified by a CLS distribution. A second point at which randomness* may enter is in the selection of individuals to fill vacancies. There are two approaches to the modelling of such systems. One is to use standard renewal theory* suitably extended to cope with several grades. By this means one can predict the flow rates and such things as the age or length of service distribution of members of the system. In practice the full analysis of the models is difficult and one usually resorts to simulation* models.

The second approach, due to White [11], involves modelling the flows of vacancies. When a person moves from X to Y a vacancy moves in the opposite direction. White found empirical justification for assuming the vacancies to flow according to a Markov chain. Such a model can then be used to forecast stocks of vacancies in exactly the same way as stocks and flows of people are predicted in the usual Markov model.

LITERATURE

The literature of manpower planning is widely scattered but in recent years several textbook accounts have appeared. Apart from the papers referred to in the text only books are included in the following list. Bartholomew and Forbes [3] is specifically concerned with practical statistical methods and contains a list of references which is virtually complete up to 1978. Vajda [9], Grinold and Marshall [4], and Niehaus [6] provide a complementary viewpoint from an operational research/computing angle. Smith [8] is an account of applications on both the supply and demand side. Bartholomew [2] includes an account of most of the stochastic models used in manpower planning. The book of readings (Bartholomew [1]) ranges more widely than the statistical aspects; it also contains a classified bibliography. Verhoeven [10] is a case study based on the Netherlands air force using an interactive manpower planning system known as FORMASY.

References

[1] Bartholomew, D. J., ed. (1976). *Manpower Planning*. Penguin Modern Management Readings, Penguin Books, Harmondsworth, Middlesex, England.

[2] Bartholomew, D. J. (1982). *Stochastic Models for Social Processes*, 3rd ed. Wiley, Chichester, England.

[3] Bartholomew, D. J. and Forbes, A. F. (1979). *Statistical Techniques for Manpower Planning*. Wiley, Chichester. (Detailed numerical examples of both Markov and renewal models, including vacancy chain models, are given in Chapters 4 and 5. This book also contains an interactive computer program written in BASIC which enables the user to simulate either kind of model.)

[4] Grinold, R. C. and Marshall, K. T. (1977). *Manpower Planning Models*. North-Holland, New York and Amsterdam.

[5] Lane, K. F. and Andrew, J. E. (1955). *J. Roy. Statist. Soc. A*, **118**, 296–323.

[6] Niehaus, R. J. (1979). *Computer Assisted Human Resources Planning*. Wiley Interscience, New York.

[7] Silcock, H. (1954). *J. Roy. Statist. Soc.*, **A117**, 429–440.

[8] Smith, A. R., ed. (1976). *Manpower Planning in the Civil Service*. Civil Service Studies No. 3, HMSO, London.

[9] Vajda, S. (1978). *Mathematics of Manpower Planning*. Wiley, Chichester, England.

[10] Verhoeven, C. J. (1982). *Techniques in Corporate Manpower Planning Methods and Applications*. Kluwer-Nijhoff Publishing, The Hague.

[11] White, H. C. (1970). *Chains of Opportunity*. Harvard University Press, Cambridge, Mass.

[12] Whitmore, G. A. (1979). *J. Roy. Statist. Soc. A*, **142**, 468–478.

(FORECASTING
MANAGEMENT SCIENCE, STATISTICS IN)

D. J. BARTHOLOMEW

MANTEL–HAENSZEL STATISTIC

Traditional epidemiological studies, either prospective or retrospective, investigate the degree of association between the presence or absence of a suspected risk factor and the occurrence of a disease (*see* EPIDEMIOLOGICAL STATISTICS). A potential problem with such studies is the possibility of obtaining a spurious association due to a lack of control of "nuisance" variables which are related either to the disease or to the factor under study.

One method to control for nuisance variables at the design stage is to employ a matching case method. An example of a retrospective matched design consists of matching each individual case with a control on factors suspected of influencing the relationship between the disease and the factor under study (*see* MATCHED PAIRS). An alternative method of controlling for nuisance variables is assigning individuals (say cases and controls) to various strata according to their values on the nuisance variables and then analyzing within each strata as well as a summary analysis across all strata. Mantel and Haenszel [3] developed a statistic for the latter situation and also demonstrated its applicability to the matched sample design as a special case.

Mantel and Haenszel, in their paper introducing the statistic, address only the analysis of data from retrospective studies*, and spend a great deal of time deriving summary relative risk formulae for data subclassified by nuisance variables. Discussion is limited here to the Mantel–Haenszel statistic and one (common) summary relative risk estimator. The statistic is primarily used for obtaining inferences by randomization* arguments which do not require any assumed general probability structure (*see* CHI-SQUARE TESTS and INFERENCE, DESIGN BASED VS. MODEL BASED). The test is directed at alternatives of average partial association between the response variable and the factor under study in the sense that it is based on an across-strata summary measure (*see* CHI-SQUARE TESTS). Bishop et al. [2] point out that the Mantel–Haenszel statistic is not the preferred statistic when higher-order interaction of strata with the disease and factor under study is present.

To derive the statistic assume, for convenience, a case-control study design where cases and controls are classified as having or not having a suspected risk factor. In addition, the study subjects are further classified into strata by their respective values on nuisance variables that are to be controlled as categorical variables. Continuous nuisance variables must be categorized. An analysis then is performed on each of the s strata as well as an overall analysis testing for differences in exposure rates between the cases and controls.

If n_{ijk} represents the number of study subjects having condition i ($i = 1$(case) or 2(control)), exposure j ($j = 1$(exposed) or 2(not exposed)), and falling in stratum k ($k = 1, 2, \ldots, s$) then utilizing the dot notation for summation over a subscript we have:

$n_{1.k}$ as the number of cases in stratum k,

$n_{2.k}$ as the number of controls in stratum k,

$n_{.1k}$ as the number of exposed in stratum k,

$n_{.2k}$ as the number of unexposed in stratum k,

$n_{..k}$ as the number of study subjects in stratum k.

The statistic used in testing for no relationship between disease occurrence and exposure rates within the kth stratum is the square of the difference between any one of the n_{ijk} and its expected value divided by its variance. The expected value and the variance of the chosen (pivotal cell) n_{ijk} is derived assuming fixed marginal totals $n_{1.k}$, $n_{2.k}$, $n_{.1k}$, and $n_{.2k}$. For the situation presented here the data may be presented in a 2×2 contingency table*, and for such a situation the test statistic is invariant as to which of the four cells (n_{11k}, n_{12k}, n_{21k}, and n_{22k}) is selected as the pivotal cell. The within-stratum statistic corrected for continuity may be given (using n_{11k} as pivotal) as

$$\frac{(|n_{11k} - n_{1.k}n_{.1k}/n_{..k}| - 1/2)^2}{n_{1.k}n_{2.k}n_{.1k}n_{.2k}/\{n_{..k}^2(n_{..k} - 1)\}}$$

This test statistic, which is asymptotically $\chi^2(1)$ under the null hypothesis, reduces to

$$\frac{(|n_{11k}n_{22k} - n_{12k}n_{21k}| - n_{..k}/2)^2(n_{..k} - 1)}{n_{1.k}n_{2.k}n_{.1k}n_{.2k}}$$

for the 2×2 contingency table discussed here. This test statistic is different from the ordinary corrected chi-square statistic for 2×2 contingency tables by a factor of $n_{..k}/(n_{..k} - 1)$.

The statistic proposed by Mantel and Haenszel for the test of significance across all strata for the above situation may be presented as

$$\frac{\left(\left|\sum_{k=1}^{s} n_{11k} - \sum_{k=1}^{s} n_{1.k}n_{.1k}/n_{..k}\right| - 1/2\right)^2}{\sum_{k=1}^{s} n_{1.k}n_{2.k}n_{.1k}n_{.2k}/\{n_{..k}^2(n_{..k} - 1)\}}.$$

This statistic is asymptotically $\chi^2(1)$ under the null hypothesis of no relationship between disease occurrence and level of exposure. It is also optimal for the alternative hypothesis stating a constant partial association across all strata between the condition and exposure (see Birch [1]).

For testing the above null hypothesis of no relationship between disease occurrence and level of exposure over all strata one also may use the Pearson chi-square statistic

$$\sum_{ijk} \frac{(n_{ijk} - n_{i.k}n_{.jk}/n_{..k})^2}{n_{i.k}n_{.jk}/n_{..k}}$$

(*see* CHI-SQUARE TESTS).

This statistic, which is asymptotically $\chi^2(s)$ under the null hypothesis of the present situation, has an advantage over the Mantel–Haenszel statistic in that it is appropriate for all alternatives to the null hypothesis. Consequently, however, it is less powerful in detecting a consistent difference that shows up in most of the individual strata. Another disadvantage of the Pearson statistic is that the $n_{..k}$ must be sufficiently large (for each stratum) so that the distribution within each stratum is asymptotically $\chi^2(1)$. However, the asymptotic distribution of the Mantel–Haenszel statistic depends only on the overall sample size being large; thus the statistic may be valid even with small stratum samples. In fact, as few as two observations per stratum are sufficient for the statistic to be valid as long as the number of strata is large. Such a situation occurs when using the Mantel–Haenszel statistic for analyzing matched-pairs designs. For this design there are two observations per stratum and n strata. In this case, the Mantel–Haenszel statistic is identical with McNemar's [14] statistic (*see* MCNEMAR STATISTICS). For the R-to-one matching design the Mantel–Haenszel statistic becomes Miettinen's statistic* [15]. So, in fact, these two test statistics are special cases of the Mantel–Haenszel statistic. However, the Mantel–Haenszel statistic has an added advantage over both since it allows the number of matched controls per case to vary from stratum to stratum. Another test statistic which assumes a logistic model and looks very much like the Mantel–Haenszel statistic for comparing s 2×2 tables was proposed by Cochran [5]. However, Cochran's statistic differs from that of Mantel and Haenszel by a correction for continuity and by a factor $(n_{..k} - 1)/n_{..k}$ in the variance term of the pivotal cell. Gart [8] showed how to calculate an exact p-value for Cochran's statistic and hence the Mantel–Haenszel statistic.

Estimates of an overall measure of relative risk* were also given in the original paper by Mantel and Haenszel [13]. Here we present one such estimator of overall relative risk which has received attention in the statistical literature. The reader is referred to the original paper [13] if interested in the other methods.

Using the notation developed previously, Mantel and Haenszel give

$$\hat{\Psi}_k = n_{11k}n_{22k}/(n_{21k}n_{12k})$$

as an estimate of relative risk within the kth stratum, and the overall estimate they suggested is

$$\hat{\Psi} = \sum_{k=1}^{2} \frac{n_{11k}n_{22k}}{n_{..k}} \Big/ \sum_{k=1}^{s} \frac{n_{21k}n_{12k}}{n_{..k}}.$$

Hauck [9], Breslow [3], Breslow and Liang [4], and Ury [16] have all proposed and/or investigated large sample variance estimators of this statistic.

As pointed out by Mantel [1, 2] the Mantel–Haenszel statistic is appropriate for other than retrospective studies; and, in fact, its use has not been limited to epidemiology. It may be used for comparing life tables* from two populations where the test statistic is based on the difference between the observed number of deaths and the expected number of deaths accumulated across time intervals. It may be shown that for this problem the test statistic is identical to the conditional logrank statistic (*see* LOGRANK SCORES, STATISTICS, AND TESTS). The Mantel–Haenszel statistic is also appealing for *quantal* bioassay experiments* as opposed to *logistic regression** since the product binomial distribution need not be assumed. Recently Darroch [6] has discussed the similarity (equivalence) of the Mantel–Haenszel test with one observation for each treatment in a stratum to tests of marginal symmetry* in contingency tables.

As an example of the Mantel–Haenszel test consider the fictitious data in Table 1 for organic particulates in the air (high and low) and bronchitis. The nuisance variable is age. Illustrative computations for the first age group are:

Table 1. Number of Cases of Bronchitis by Level of Organic Particulates in the Air and by Age

		Bronchitis		
Age	Level	Yes	No	Total
0–14	High	62	915	977
	Low	7	442	449
15–24	High	20	382	402
	Low	9	214	223
25–39	High	10	172	182
	Low	7	120	127
40	High	12	327	339
	Low	6	183	189

$$E(n_{111}) = \frac{n_{1.1}n_{.11}}{n_{..1}} = \frac{977 \times 69}{1426} = 47.27,$$

$$\mathrm{Var}(n_{111}) = \frac{n_{1.1}n_{.11}n_{2.1}n_{.21}}{n_{..1}^2(n_{..1} - 1)}$$

$$= \frac{977 \times 69 \times 499 \times 1357}{(1426)^2(1425)}.$$

The composite test statistic value is:

$$Q = \frac{|104 - 87.48| - 1/2}{28.43} = 9.03$$

Hence the null hypothesis of no relationship between level of organic particulates in the air and bronchitis is not supported by these data. The relative risk estimator $\hat{\Psi}$ is 1.94.

For the more general situation Mantel and Haenszel give a numerical example for comparing cases and controls when there are three categories of exposure. They then talk through an analysis for the general situation of r categories of exposure, but never actually derive the statistic. To derive the statistic let n_{ijk} represent the number of observations having condition i(1 or 2), exposure j(1, 2, . . . , r), and falling in strata k(1, 2, . . . , s). If we again utilize the dot notation for summation over a subscript we have $n_{.jk}$ as the number of observations with exposure j in stratum k, $n_{i.k}$ as the number of observations with condition i in stratum k, and $n_{..k}$ as the number of observations in stratum k. For r exposure categories and two conditions per strata we choose $r - 1$ pivotal cells for which we need expected values and the corresponding variance-covariance matrix. If we let

$$\underline{a}'_k = (n_{11k}, n_{12k}, \ldots n_{1(r-1)k})_{1 \times (r-1)}$$

be the $r - 1$ pivotal cells for the kth stratum then the expected value of $\underline{a}'_k$ is

$$\underline{e}'_k = \frac{n_{1.k}}{n_{..k}}(n_{.1k}, n_{.2k}, \ldots, n_{.(r-1)k})_{1 \times (r-1)},$$

with corresponding variance–covariance matrix

$$\underline{V}_k =$$

$$\frac{n_{2.k}}{n_{..k} - 1}\left(\mathrm{Diag}\ \underline{e}_k - \frac{1}{n_{1.k}}\underline{e}_k\underline{e}'_k\right)_{(r-1)\times(r-1)},$$

under the null hypothesis of no relationship between disease occurrence and level of exposure. If $\underline{A} = \sum_{k=1}^{s} \underline{a}_k, \underline{E} = \sum_{k=1}^{s} \underline{e}_k$ and $\underline{V} = \sum_{k=1}^{s} \underline{V}_k$, the Mantel–Haenszel statistic may be expressed as

$$(\underline{A} - \underline{E})' \underline{V}^{-1} (\underline{A} - \underline{E});$$

it is asymptotically $\chi^2(r-1)$ under the null hypothesis of no difference in exposure rates between cases and controls.

The Mantel–Haenszel statistic also can be expanded to the multivariate situation of t disease states, r exposure rates, and s strata; the statistic then involves $(t-1)(r-1)$ pivotal cells, and is asymptotically $\chi^2((t-1)(r-1))$ under the null hypothesis of no relationship between exposure rates and disease states. Mantel [11] also developed a procedure for testing equality of cases and controls when the factor under study is ordinally scaled. He then extrapolated to the situation where the disease states also are ordinally scaled.

The best article to read for further information on the Mantel–Haenszel statistic is the original [13]. It deals only with retrospective studies but is an excellent epidemiological article dealing extensively with the concept of relative risk. Other excellent resources dealing with comparisons of different categorical data* statistics and their applications are the article by Landis et al. [10] and the book by Fleiss [7]. The entry CHI-SQUARE TESTS: NUMERICAL EXAMPLES also is a fine reference along with entries previously indicated.

References

[1] Birch, M. W. (1964). *J. Roy. Statis. Soc. B.*, **26**, 313–324.

[2] Bishop, Y. M. M., Fienberg, S. E., and Holland, P. W. (1975). *Discrete Multivariate Analysis: Theory and Practice*. MIT Press, Cambridge, Massachusetts.

[3] Breslow, N. (1981). *Biometrika*, **68**, 73–84.

[4] Breslow, N. and Liang, K. (1982). *Biometrics*, **38**, 943–952.

[5] Cochran, W. G. (1954). *Biometrics*, **10**, 417–451.

[6] Darroch, J. N. (1981). *Interntl. Statis. Rev.*, **49**, 285–307.

[7] Fleiss, J. L. (1983). *Statistical Methods for Rates and Proportions*. Wiley, New York.

[8] Gart, J. J. (1971). *Internatl. Statis. Rev.*, **39**, 148–169.

[9] Hauck, W. W. (1979). *Biometrics*, **35**, 817–819.

[10] Landis, J. R., Heyman, E. R., and Koch, G. G. (1978). *Internatl. Statis. Rev.*, **46**, 237–254.

[11] Mantel, N. (1963). *J. Amer. Statis. Ass.*, **58**, 690–700.

[12] Mantel, N. (1966). *Cancer Chemother. Rep.*, **50**, 163–170.

[13] Mantel, N. and Haenszel, W. (1959). *J. Nat. Cancer Inst.*, **22**, 719–748.

[14] McNemar, Q. (1947). *Psychometrika*, **12**, 153–157.

[15] Miettinen, O. S. (1969). *Biometrics*, **25**, 339–355.

[16] Ury, H. K. (1982). *Biometrics*, **38**, 1094–1095.

(CHI-SQUARE TESTS
CHI-SQUARE TESTS: NUMERICAL EXAMPLES
EPIDEMIOLOGICAL STATISTICS
INFERENCE, DESIGN-BASED VS. MODEL-BASED
LOGRANK SCORES, STATISTICS AND TESTS
MCNEMAR STATISTICS
MIETTINEN'S STATISTIC)

GRANT W. SOMES
KEVIN F. O'BRIEN

MANY-SERVER QUEUES *See* MULTI-SERVER QUEUES

MARCINKIEWICZ THEOREM

If a characteristic function* $\phi(t)$ is of the form $\exp(P(t))$ where $P(t)$ is a polynomial with complex coefficients, then the degree of this polynomial cannot exceed 2.

(The theorem implies, for example, that the function e^{t^4} cannot be a characteristic function.) For more details and bibliography, *see* CHARACTERISTIC FUNCTIONS.

MARDIA'S TEST OF MULTINORMALITY

MOTIVATION

In multivariate analysis*, the only distribution leading to tractable inference is the multivariate normal and therefore it is more important to test the hypothesis of multinormality than to test normality in the univariate case. (*See also* MULTIVARIATE NORMALITY, TESTING FOR.)

The univariate measures, skewness*, β_1, and kurtosis*, β_2, are given by

$$\beta_1 = \mu_3^2/\mu_2^3, \qquad \beta_2 = \mu_4/\mu_2^2,$$

where μ_i is the ith moment about the mean. These are constructed so that

1. They are invariant under change of scale and origin.
2. β_1 is a function of μ_3, the lowest central moment measuring symmetry.
3. β_2 is a function of μ_4, the lowest central moment measuring 'peakedness.'

These properties can be extended to the multivariate case but they do not lead to unique measures. However, β_1 appears in $\text{corr}(\overline{X}, S^2)$ for large n and β_2 appears in the approximation to Pitman's permutation test, and when these properties are extended, we are led to unique estimators (Mardia [4]). Further, in Davis [1] these statistics appear predominantly in his expansion of Hotelling's T^2* statistic under non-normality.

DEFINITION OF MULTIVARIATE SKEWNESS AND KURTOSIS

Consider the random vectors **X** and **Y** which are distributed identically and independently with $E(\mathbf{X}) = \boldsymbol{\mu}$ and $\text{Cov}(\mathbf{X}) = \boldsymbol{\Sigma}$. Then the population measures of p-variate skewness and kurtosis proposed by Mardia [4] are respectively (see also, Mardia [5])

$$\beta_{1,p} = E\left\{(\mathbf{X} - \boldsymbol{\mu})'\boldsymbol{\Sigma}^{-1}(\mathbf{Y} - \boldsymbol{\mu})\right\}^3,$$

$$\beta_{2,p} = E\left\{(\mathbf{X} - \boldsymbol{\mu})'\boldsymbol{\Sigma}^{-1}(\mathbf{X} - \boldsymbol{\mu})\right\}^2.$$

Both these measures are invariant under nonsingular transformations. Also $\beta_{1,p}$ is a function of the third central moment of **X** and $\beta_{2,p}$ is a function of the fourth central moment of **X**.

Sample measures may also be defined. Let **S** be the sample covariance matrix; then

$$b_{1,p} = \frac{1}{n^2}\sum_{i,j=1}^{n} r_{ij}^3, \qquad b_{2,p} = \frac{1}{n}\sum_{i=1}^{n} r_{ii}^2,$$

where

$$r_{ij} = (\mathbf{X}_i - \overline{\mathbf{X}})'\mathbf{S}^{-1}(\mathbf{X}_j - \overline{\mathbf{X}}).$$

We note from Mardia [4] that if $\text{var}(\overline{X})$, $\text{var}(S^2)$, and $\text{cov}(\overline{X}, S^2)$ are all taken to order n^{-1} and the fourth population cumulant is assumed negligible, then approximately

$$\text{corr}(\overline{X}, S^2) \sim (\tfrac{1}{2}\beta_1)^{1/2}.$$

In the multivariate case, if we additionally assume that cumulants* of order higher than 3 of X are negligible, then approximately

$$\beta_{1,p} = n^2 \sum_{i,j,k=1}^{p} \left\{\text{cov}(\overline{X}_i, S_{jk})\right\}^2,$$

where

$$\mathbf{S} = (S_{ij}), \qquad \overline{\mathbf{X}}' = (X_1, \ldots, X_p).$$

It can be shown (Mardia, [6]) that if the sample points are uniformly distributed on a p-dimensional hypersphere we have $b_{1,p} \simeq 0$. However, when abnormal clustering occurs, $b_{1,p}$ will be large (see Figs. 1*a* and 1*b*). The value of $b_{2,p}$ will be large for abnormal clustering but can take a range of values under symmetry. In Fig. 1*a*, $b_{2,p}$ is close to normality but if the long vectors are excluded it will become abnormally large. Thus, both measures are needed to obtain a full view of any departure from normality.

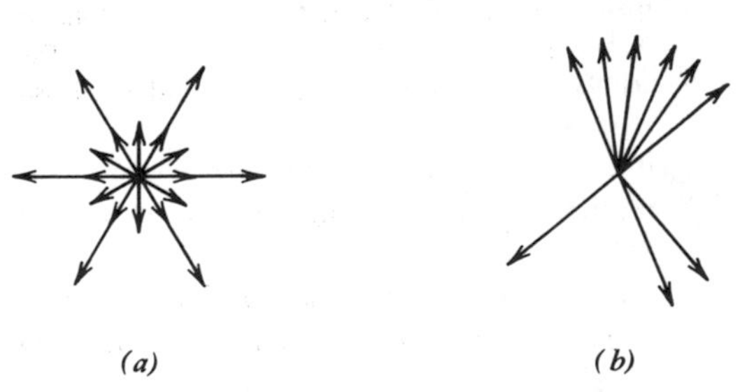

Figure 1 (a) Spherical symmetric case ($b_{1,p} = 0$, $b_{2,p} \simeq$ normal). (b) Abnormal clustering ($b_{1,p} \neq 0$, $b_{2,p} \neq$ normal).

EXACT MOMENTS OF $b_{1,p}$ AND $b_{2,p}$ FOR THE MULTINORMAL CASE

Mardia [5] has shown that if **X** is assumed to be a random sample from $N(\boldsymbol{\mu}, \boldsymbol{\Sigma})$,

$$E(b_{1,p}) = \frac{p(p+2)}{(n+1)(n+3)} \times \{(n+1)(p+1) - 6\}.$$

(The exact variance of $b_{1,p}$ is not known.) We can show that

$$E(b_{2,p}) = p(p+2)(n-1)/(n+1),$$

and

$$\text{var}(b_{2,p}) = \frac{8p(p+2)(n-3)}{(n+1)^2(n+3)(n+5)} \times (n-p-1)(n-p+1).$$

Mardia and Foster [8] have shown the correlation between $b_{1,p}$ and $b_{2,p}$ to be

$$\text{corr}(b_{1,p}, b_{2,p}) \doteq \frac{3(8p^2 - 13p + 23)}{(p+2)\{6(p+1)\}^{1/2}} n^{-1/2}.$$

Thus it is possible for the correlation to be substantial even for moderately large n. From Mardia [5],

$$\beta_{1,p} = 0, \qquad \beta_{2,p} = p(p+2).$$

The estimators are biased. They are, however, asymptotically unbiased and consistent as estimators of $\beta_{1,p}$ and $\beta_{2,p}$.

INDIVIDUAL TESTS OF MULTINORMALITY AND APPROXIMATION TO PERCENTAGE POINTS

We can take $b_{1,p}$ and $b_{2,p}$ as test statistics of multinormality where the null hypothesis is rejected when their values are large.

The calculation of exact percentage points is extremely difficult and so approximations have been developed using the above moments. We have approximately (Mardia [4]) that

$$A = \frac{n}{6} b_{1,p} \sim \chi_f^2, \qquad f = \frac{1}{6} p(p+1)(p+2).$$

More accurately (Mardia [5]) if

$$A' = \frac{n}{6} K b_{1,p},$$

$$K = \frac{(p+1)(n+1)(n+3)}{n\{(n+1)(p+1) - 6\}},$$

then A' also is approximately χ_f^2 with $E(A') = f$ for all n. Similarly, $B' = U/V$ is asymptotically $N(0, 1)$, where

$$U = \{(n+1)b_{2,p} - p(p+2)(n-1)\} \times \{(n+3)(n+5)\}^{1/2},$$

$$V = \{8p(p+2)(n-3)(n-p-1) \times (n-p+1)\}^{1/2}.$$

Also

$$B = \{b_{2,p} - p(p+2)\}/\{8p(p+2)/n\}^{1/2}$$

is asymptotically $N(0, 1)$.

Both $b_{1,p}$ and $b_{2,p}$ are related to the χ^2-variable (see Mardia and Kanazawa [9]) and hence the Wilson–Hilferty approximation can be used to define transformed measures, from which

$$\left\{\frac{nb_{1,p}}{6f_1}\right\}^{1/3} \sim N\left(1 - \frac{2}{9f_1}, \frac{2}{9f_1}\right),$$

$$f_1 = 6 + \{8p(p+2)(p+8)^{-2}\}^{1/2} n^{1/2} \times \left[\{\tfrac{1}{2}p(p+2)\}^{1/2}(p+8)^{-1}n^{1/2} + \{1 + \tfrac{1}{2}p(p+2)(p+8)^{-2}n\}^{1/2}\right].$$

It therefore is possible to define a transformed variable $W(b_{1,p})$ as

$$W(b_{1,p}) = \{12p(p+1)(p+2)\}^{-1/2} \times \left[\{27np^2(p+1)^2(p+2)^2 b_{1,p}\}^{1/3} - 3p(p+1)(p+2) + 4\right].$$

Asymptotically, $W(b_{1,p})$ will have a standard normal null distribution. Also, Mardia and Kanazawa have defined a transformation of $b_{2,p}$, approximating $b_{2,p}$ by $A'' + B''/\chi_C^2$, where χ_C^2 is a chi-square variable with C

degrees of freedom. The constants A'', B'' and C are chosen so that the first three moments of $b_{2,p}$ correspond to those of $A'' + B''/\chi^2_C$. See ref. 9 for details. Use of this transformation, with $x = (b_{2,p} - \mu)/\mu_2^{1/2}$, gives $W(b_{2,p}) \sim N(0, 1)$, where

$$W(b_{2,p}) = \left(\frac{9C}{2}\right)^{1/2}\left[\left(\frac{1-2/C}{x[2/(C-4)]^{1/2}+1}\right)^{1/3} + \frac{2}{9C} - 1\right].$$

Mardia [5] examined the adequacy of approximations A and A' and B and B' for $b_{1,p}$ and $b_{2,p}$ respectively. For $p = 2$, all approximations were inadequate for moderately small values of n although combinations of the measures produced reasonable results. For $p > 2$ and $n \geqslant 50$ it was found that:

1. For the upper 5% points of $b_{1,p}$ use A'.
2. For the lower 2.5% points of $b_{2,p}$ use $b_{2,p}$ treated as normal with mean $p(p+2)(n+p+1)/n$ and variance equal to $8p(p+2)/(n-1)$ when $50 \leqslant n \leqslant 400$ and use B for $n > 400$.
3. For the upper 2.5% points of $b_{2,p}$ use B.

Mardia [5] gives tables of the critical values of $b_{1,2}$ and $b_{2,2}$ for several common values of n and α.

Mardia and Kanazawa [9] compared the approximation $W(b_{2,p})$ with the normal approximation and exact values found by simulation. For $p = 2$, and $\alpha = 0.01$ or 0.05, $W(b_{2,p})$ gives a close fit. The normal approximation is better for $p = 4$ and $\alpha = 0.05$ or $\alpha = 0.95$. Otherwise, the average of the normal approximation and $W(b_{2,p})$ gives the closest estimate. To facilitate calculation, a statistical algorithm has been developed by Mardia and Zemroch [10].

OMNIBUS TESTS

Many omnibus tests, i.e., tests involving both $b_{1,p}$ and $b_{2,p}$, have been formulated and are summarized in Mardia and Foster [8]. The most important criteria for deciding which of the many are best are the closeness of the statistic to the χ^2_2 distribution in its null distribution and its power, for given size, when both symmetric and asymmetric margins are considered. The most worthwhile statistics are given by

$$S_W^2 = W^2(b_{1,p}) + W^2(b_{2,p})$$

and

$$C_W^2 = (W(b_{1,p}), W(b_{2,p}))'\mathbf{W}^{-1}(W(b_{1,p}), W(b_{2,p})),$$

where $\mathbf{W}$ is a 2×2 matrix with 1's in the diagonal and $\mathrm{Cov}\{W(b_{1,p}), W(b_{2,p})\}$ in the off-diagonal is given by

$$\mathrm{Cov}\{W(b_{1,p}), W(b_{2,p})\} = \frac{1}{4}\left(\frac{f_1}{f}\right)^{1/2}\left\{-\frac{40}{9}\left(1-\frac{2}{f_1}\right)\frac{1}{f_1-4} + \frac{n}{3\sigma}\left(1-\frac{2}{f_1}\right)^{1/3}\left(\frac{2}{f_1-4}\right)^{1/2}\right\} \times \mathrm{Cov}(b_{1,p}, b_{2,p}) + \cdots,$$

where f_1 is defined in Section 4 and $\sigma^2 = \mathrm{var}(b_{2,p})$.

POWER COMPARISON

Mardia [7] describes various other multinormality tests including the statistics b_1^*, b_2^* (skewness and kurtosis statistics derived from the union-intersection principle*), W^* (a generalization of the Shapiro–Wilk statistic*), $\mathbf{d}_\alpha$ (directional normality procedure), CM^* (multivariate Cramér–von Mises statistic*), K^* (multivariate Kolmogorov–Smirnov statistic*), $L_{\max}(\boldsymbol{\lambda})$ (obtained from Box–Cox* type transformations), $Q_{2,1}$ and $Q_{1,2}$ (Student-t approach), C, C_β, (maximum correlation tests), $\log \hat{\eta}^2_{\max}$ (maximum curvature), and radius, angle, and graphical techniques. For appropriate references, see Mardia [7]. Further, Small [12] constructed skewness and kurtosis statistics Q_1 and Q_2

on the basis of the univariate measures obtained from the marginals and also defined an omnibus statistic Q_3 based on Q_1 and Q_2.

Giorgi and Fattorini [3] compared the power of Mardia's tests with W^*, CM^*, K^* and two-dimensional criteria obtained on applying Shapiro–Wilk's statistics to $\mathbf{d}_1$ and $\mathbf{d}_{-1}$, whereas Foster [2] compared W^*, CM^*, $\mathbf{d}_\alpha$, Q_1, Q_2, $W(b_{1,p})$, $W(b_{2,p})$, some omnibus tests based on $b_{1,p}$ and $b_{2,p}$ and Small's omnibus test Q_3 with Mardia's tests. With a null hypothesis of multinormality for $p = 2$, 3, and 4, between the two studies a wide range of alternative hypotheses has been considered including both skew and symmetric distributions and various mixtures of distributions.

For symmetric alternatives, Q_2, Q_3, C_W^2 and S_W^2 have approximately the same power as $b_{2,p}$ and rather more than $b_{1,p}$. For skew alternatives, C and C_β are superior to all other statistics but Q_1, Q_2, Q_3 and S_W^2 are similar to $b_{2,p}$ and better than $b_{1,p}$. C_W^2 is somewhat worse than S_W^2. W^* is slightly better than $b_{1,p}$ and $b_{2,p}$ for alternatives consisting of mixtures of normals. With the test statistic W^*, however, estimated percentage points do not seem to be particularly reliable. Although specific statistics are more suited to specific alternative distributions, the study suggests that S_W^2 is an overall good test to use, with C_W^2 slightly worse. Schwager and Margolin [11] show that $b_{2,p}$ gives a locally best invariant test of multinormality against the presence of outliers from a mean slippage model.

References

[1] Davis, A. W. (1980). *Biometrika*, **67**, 419–427.

[2] Foster, K. J. (1981). Tests of Multivariate Normality. Ph.D. thesis, University of Leeds, U.K.

[3] Giorgi, G. M. and Fattorini, L. (1976). *Quaderni Dell'Instituto Di Statistica*, **20**, 1–8.

[4] Mardia, K. V. (1970). *Biometrika*, **57**, 519–530.

[5] Mardia, K. V. (1974). *Sankhyā B*, **36**, 115–128.

[6] Mardia, K. V. (1975). *Appl. Statist.*, **24**, 163–171.

[7] Mardia, K. V. (1980). Tests of Univariate and Multivariate Normality. In *Handbook of Statistics, Vol. 1* P. R. Krishnaiah, ed. North-Holland.

[8] Mardia, K. V. and Foster, K. J. (1983). *Commun. Statist. A*, **12**, 207–221.

[9] Mardia, K. V. and Kanazawa, M. (1983). *Commun. Statist. B*, **12**, 569–576.

[10] Mardia, K. V. and Zemroch, P. J. (1975). *Appl. Statist.*, **24**, 262–265.

[11] Schwager, S. J. and Margolin, B. H. (1982). *Ann. Statist.*, **10**, 943–954.

[12] Small, N. J. H. (1980). *Appl. Statist.*, **29**, 85–87.

(DEPARTURES FROM NORMALITY, TESTS FOR
MULTINORMAL DISTRIBUTION
MULTIVARIATE NORMALITY, TESTING FOR)

K. V. MARDIA

MARGINAL DISTRIBUTIONS *See* MARGINALIZATION

MARGINALIZATION

Marginalization is the process of deriving the marginal probability distribution of a random variable from a joint distribution involving the random variable and others. As a simple example, suppose X and Y are random variables that can take the value 0 or 1. Suppose that their joint probability function is given by the following table:

		X	
		0	1
Y	0	$\frac{1}{4}$	$\frac{1}{4}$
	1	$\frac{1}{3}$	$\frac{1}{6}$

so that, for example, $\Pr[X = 0 \text{ and } Y = 1] = \frac{1}{3}$. Then $\Pr[X = 0] = \Pr[X = 0 \text{ and } Y = 0] + \Pr[X = 0 \text{ and } Y = 1] = \frac{1}{4} + \frac{1}{3} = \frac{7}{12}$, and $\Pr[X = 1] = \frac{1}{4} + \frac{1}{6} = \frac{5}{12}$. These probabilities (7/12 and 5/12) define a probability distribution for X without regard to the value of Y; this is the marginal distribution of X. The probabilities correspond to the column sums of the table. Similarly, the marginal distribution of Y could be found from the row sums of the table. The obvious place to write these row and column sums is in the margin of the table; hence the term 'marginal distribution'.

		X		
		0	1	
Y	0	$\frac{1}{4}$	$\frac{1}{4}$	$\frac{1}{2}$
	1	$\frac{1}{3}$	$\frac{1}{6}$	$\frac{1}{2}$
		$\frac{7}{12}$	$\frac{5}{12}$	

Marginal distribution of Y.

Marginal distribution of X

If X and Y are continuous real-valued random variables with joint probability density function $f_{X,Y}(x, y)$, we can marginalize to produce the marginal probability density function of X by "integrating out" $Y: f_X(x) = \int_{-\infty}^{\infty} f_{X,Y}(x, y)\,dy$. Marginalization can be carried out in an analogous way if, for instance, X and Y are vector-valued random variables. The marginal distribution of X and of Y is uniquely defined by their joint distribution. The converse is not true; given marginal distributions for X and for Y, it is in general possible to find several different corresponding joint distributions.

It is sometimes useful to generalize the notation of marginalization slightly. If we think of the joint probability distribution of X and Y as a function on a suitable product σ-algebra, then "integrating out" Y amounts to restricting this function to the appropriate σ-algebra for X. In the example above, we can consider the joint distribution of X and Y as a distribution over the σ-algebra of all subsets of the sample space* $\{(0,0), (0,1), (1,0), (1,1)\}$ where $(0,0)$ means "$X = 0$ and $Y = 0$," etc. Calling this σ-algebra S, the joint distribution can be represented by a function, Π say, from S to the real numbers, where for example

$$\Pi(\{0,0\}) = \Pr[X = 0 \text{ and } Y = 0] = \tfrac{1}{4}.$$

In finding the marginal distribution for X, interest is confined to events like $[X = 0]$, which can be considered as ($[X = 0$ and $Y = 0]$ or $[X = 0$ and $Y = 1]$), that is, $\{(0,0), (0,1)\}$. The marginal distribution for X is thus defined by a probability distribution on the σ-algebra T containing the sets $\{(0,0), (0,1)\}$, $\{(1,0), (1,1)\}$ together with the whole sample space and the empty set. Now T is a sub-σ-algebra of S. If the marginal distribution of X is represented by a function $\Pi^{(T)}$ from T to the real numbers, then for example

$$\Pi^{(T)}(\{(0,0),(0,1)\}) = \Pr[X = 0] = 7/12$$
$$= \Pi(\{(0,0),(0,1\}).$$

In fact, if B is any set in the σ-algebra T, then $\Pi^{(T)}(B) = \Pi(B)$, since both sides represent the probability of the same event. The only difference between $\Pi^{(T)}$ and Π is that Π has a larger domain; thus, for example, $\Pi(\{(0,0)\} = \frac{1}{4}$ but $\Pi^{(T)}(\{(0,0)\})$ is not defined because the set $\{(0,0)\}$ is not in the σ-algebra T, though it is in S. Thus in general, if Π is a probability distribution on a σ-algebra S over some space and T is a sub-σ-algebra of S, then the marginal distribution $\Pi^{(T)}$ given by Π over T can be defined as the restriction of Π to T, so that $\Pi^{(T)}(B) = \Pi(B)$ for all sets B in T.

The process of marginalization is a relatively trivial matter in probability theory, but it has assumed a certain importance in the study of Statistical Inference*. In Bayesian inference*, the parameters of a statistical model are treated as random variables. Frequently a model involves parameters other than the parameter of interest. The initial outcome of the Bayesian analysis will be a joint posterior distribution for the parameter of interest *and* the so-called nuisance parameters*. The final inference is based on the posterior distribution of the parameter of interest, which can be found using marginalization. Fiducial inference* also involves probability distributions for parameters, and hence relies on marginalization in an analogous manner. In other approaches to statistical inference, though it is of course *used* in other ways, marginalization creates fewer theoretical problems and has therefore become less of an issue.

The idea of using marginalization to dispose of nuisance parameters lies behind the so-called marginalization paradoxes of Dawid et al. [2]. They consider problems where data X are to be observed, with a distribution depending on a parameter of interest θ and a nuisance parameter ϕ. Suppose it happens that the marginal posterior distribution of θ depends on the data X only through a statistic $t(X)$, and also that, given θ, $t(X)$ and ϕ are independent. Then it may seem reasonable, instead of involving ϕ and then integrating it out, to begin by finding a prior marginal distribution for θ, to work with $t(X)$ and ignore ϕ. A marginalization paradox arises when the resulting posterior distribution for θ differs from that found by marginalization from the joint posterior distribution of θ and ϕ. For a detailed discussion with an example *see* INVARIANT PRIOR DISTRIBUTIONS. Marginalization paradoxes cannot occur if all the densities involved are proper (i.e., integrate to one); the existence of these paradoxes has been used as an argument against using improper prior distributions (see, e.g., ref. 4). It has been argued that there is no good reason to suppose the two distributions for θ should be the same, so that there is no paradox [3]. On the other hand, such paradoxes cannot arise in at least one formulation of fiducial inference [1].

McConway [5] examined marginalization in the following context. Suppose a decision maker has asked a group of experts to assess subjective probability distributions over some space. In general the experts will differ, and the decision maker may wish to find a single distribution, the *aggregate* distribution over the space, as a function of the individual experts' distributions. Suppose the experts initially give joint distributions for X and Y and interest later focuses on X. An aggregate distribution for X could be formed in two different ways. First, the individual experts could each calculate marginal distributions for X, and these distributions could then be combined by the decision maker. Second, the decision maker could combine the experts' *joint* distributions for X and Y into an aggregate joint distribution, and then could calculate the marginal distribution for X defined by the aggregate joint distribution. Under some circumstances it would seem reasonable that the decision maker should end up with the same final distribution for X whether the first or second method is used. McConway showed that if all possible marginalizations are to commute with aggregation* in this way, then, under weak regularity conditions, a weighted average should be used to perform the aggregation.

References

[1] Dawid, A. P. and Stone, M. (1982). *Ann. Statist.*, **10**, 1054–1067.

[2] Dawid, A. P., Stone, M., and Zidek, J. V. (1973). *J. R. Statist. Soc.*, **B**, **35**, 189–233.

[3] Jaynes, E. T. (1980). Marginalization and prior probabilities (with comments by A. P. Dawid, M. Stone, and J. V. Zidek, and Reply). In *Bayesian Analysis in Econometrics and Statistics: Essays in Honor of Harold Jeffreys*, A. Zellner, ed. North Holland Publishing Co., Amsterdam, pp. 43–87.

[4] Lindley, D. V. (1978). *Scand. J. Statist.*, **5**, 1–26.

[5] McConway, K. J. (1981). *J. Amer. Statist. Ass.*, **76**, 410–414.

(BAYESIAN INFERENCE
SAMPLE SPACE)

K. J. McCONWAY

MARGINAL LIKELIHOOD *See* PSEUDO-LIKELIHOOD

MARGINAL SYMMETRY

Let t categorical responses $R_1, \ldots, R_t$, each taking one of the values $1, \ldots, r$, be measured on an Individual. (The reason for capital I will appear in the section "Matched Pairs and t-plets".) The hypothesis of *marginal symmetry* or, as it is often called, *marginal homogeneity*, says that $R_1, \ldots, R_t$ all have the same marginal distribution. The theory of testing this hypothesis can be dated from Stuart [11] who considered the case of $t = 2$.

The closest analogue in the longer-established continuous variable theory appears to be that of Hotelling's T^2-test* of the hypothesis that the expected values of t multivariate normal variables are equal. When $t = 2$ this reduces of course to Student's t-test* for the equality of two means using paired observations.

Example 1. Table 1 gives hypothetical data concerning the voting preferences in a sample of 1,811 voters one month before and one week before a mayoral election. There are three candidates $(1, 2, 3)$.

Table 1 Voting Preferences

		Week before 1	2	3	
Month before	1	319	17	72	408
	2	32	461	22	515
	3	15	35	838	888
		366	513	932	1811

Here the hypothesis of marginal symmetry maintains that there has been no change in the population distribution of voting preferences in the intervening three weeks.

Example 2. Table 2 gives actual data (unpublished) concerning activity ratings of 488 boys. Each child rated himself and was also rated by one of his parents and a teacher. The data were collected by the Tasmanian Department of Health and Professor I. C. Lewis of the University of Tasmania. Each rating is either 1 = inactive or 2 = active.

Table 2 Activity Ratings

Child 1

		Teacher 1	Teacher 2
Parent	1	9	22
	2	17	72

Child 2

		Teacher 1	Teacher 2
Parent	1	15	32
	2	35	286

Here the hypothesis of marginal symmetry holds that children have, on the average, the same perception as their parents and their teachers of the dividing line between "activity" and "inactivity."

In Example 1 we want to know if the marginal frequencies 408, 515, 888 differ significantly from 366, 513, 932. It is tempting to formulate this as a problem involving a 2×3 contingency table* but this is wrong because the same 1,811 people appear in both rows of the table. In Example 2 the inactive, active marginal frequencies are (120,368) for child, (78,410) for parent, and (76,412) for teacher. Again it would be wrong to test for differences between them by testing row $\times$ column independence in a 3×2 contingency table.

Let $\rho(j_1, \ldots, j_t)$ be the probability of the event $R_1 = j_1, \ldots, R_t = j_t$ and let $y(j_1, \ldots, j_t)$ be the frequency of this event in a sample of s Individuals. Let ρ_j^i and y_j^i be the marginal probability and marginal frequency, respectively, of the event $R_i = j$. Thus in Example 2, $t = 3$, $r = 2$ and if the three responses are considered in the order (child, parent, teacher), then, for instance, $y(1, 2, 2) = 72$ and $y_1^2 = 78$.

The hypothesis of marginal symmetry will be denoted by H_2 and is defined by

$$H_2 : \rho_j^i \text{ is constant with respect to } i.$$

In Example 1, H_2 is $\rho_j^1 = \rho_j^2$, $j = 1, 2, 3$ and, in Example 2, it is $\rho_j^1 = \rho_j^2 = \rho_j^3$, $j = 1, 2$.

In this entry we test H_2 when no assumptions are made about the probability function ρ. In the entry *quasi-symmetry**, H_2 is tested assuming that ρ satisfies the property H_1 of quasi-symmetry. The analogous problem in continuous variable theory appears to be that of testing equality of the means of t multivariate normal variables when their variance matrix is assumed to be symmetric with respect to permutations of the variables. The resulting test is an F-test* arising from a two-factor, mixed model, analysis of variance*. A more detailed discussion of marginal and quasi-symmetry, using the same notation as here, is given by Darroch [3].

THE WALD TEST OF H_2

We want to test H_2 against the alternative hypothesis H which places no restriction on the ρ_j^i. Consider Example 2; H_2 may be expressed as $\rho_1^1 - \rho_1^2 = 0$, $\rho_1^2 - \rho_1^3 = 0$ and the maximum likelihood* estimates of these two parameters under H are $s^{-1}e_1, s^{-1}e_2$, where

$$e_1 = y_1^1 - y_1^2, \qquad e_2 = y_1^2 - y_1^3.$$

The Wald test of H_2 against H calculates the estimated variance matrix $\tilde{\mathbf{\Sigma}}_{\mathbf{e}}$ say of $\mathbf{e}' = (e_1, e_2)$ and uses the fact that

$$\mathbf{e}'\tilde{\mathbf{\Sigma}}_{\mathbf{e}}^{-1}\mathbf{e} \mathrel{\dot{\div}} \chi_2^2 \qquad \text{on} \quad H_2,$$

where $\dot{\div}$ denotes approximately distributed as.

Now e_1 and e_2 are each linear combinations of the frequencies $y(\mathbf{j}), \mathbf{j} = (j_1, j_2, j_3)$. Using the standard multinomial formulae

$$\operatorname{var}(y(\mathbf{j})) = s\rho(\mathbf{j})(1 - \rho(\mathbf{j})),$$
$$\operatorname{cov}(y(\mathbf{j}), y(\mathbf{j}')) = -s\rho(\mathbf{j})\rho(\mathbf{j}'), \qquad \mathbf{j} \neq \mathbf{j}', \quad (1)$$

it is straightforward to prove that

$$\operatorname{var}(e_1) = s\big[\rho(1,2,1) + \rho(2,1,2) + \rho(1,2,2) + \rho(2,1,1)\big] - s\big(\rho_1^1 - \rho_1^2\big)^2. \quad (2)$$

The second term in (2) is zero under H_2 and smaller under H. It has been customary since Stuart [11] first constructed the Wald test for H_2 for $t = 2$, general r, to adjust the formula by omitting this term. The estimate $\tilde{\operatorname{var}}\,(e_1)$ of $\operatorname{var}(e_1)$ is obtained by replacing $s\rho(\mathbf{j})$ by $y(\mathbf{j})$. The resulting formula for $\tilde{\operatorname{var}}(e_1)$ can now be derived most simply by replacing (1) by the "estimated-adjusted" second moments

$$\tilde{\operatorname{var}}(y(\mathbf{j})) = y(\mathbf{j}), \qquad \tilde{\operatorname{cov}}(y(\mathbf{j}), y(\mathbf{j}')) = 0, \qquad \mathbf{j} \neq \mathbf{j}'. \quad (3)$$

The elements of $\tilde{\mathbf{\Sigma}}_{\mathbf{e}}$, evaluated for the data of example 2, are

$$\tilde{\operatorname{var}}(e_1) = y(1,2,1) + y(2,1,2) + y(1,2,2) + y(2,1,1) = 136,$$
$$\tilde{\operatorname{var}}(e_2) = y(1,1,2) + y(2,2,1) + y(2,1,2) + y(1,2,1) = 106,$$
$$\tilde{\operatorname{cov}}(e_1, e_2) = -y(1,2,1) - y(2,1,2) = -49.$$

Also $e_1 = 120 - 78 = 42$, $e_2 = 78 - 76 = 2$, and $\mathbf{e}'\tilde{\mathbf{\Sigma}}_{\mathbf{e}}^{-1}\mathbf{e} = 16.29$, which, on referring to χ_2^2 tables, is found to be very highly significant.

The corresponding test statistic for Example 1 can be constructed in a similar way. It is easy to see that neither test statistic depends on the "diagonal frequencies," namely $y(1,1)$, $y(2,2)$, $y(3,3)$ in Example 1 and $y(1,1,1)$, $y(2,2,2)$ in Example 2. An interesting curiosity, first noted by Mantel and Fleiss [9], is that there is a correspondence between the tests for 2^3 and 3^2 tables. To illustrate this the nondiagonal hypothetical frequencies of Example 1 were chosen to equal the nondiagonal frequencies of Example 2 and the result is that the value of the test statistic for Example 1 is also 16.29.

The simplest test of marginal symmetry occurs when $t = r = 2$. The test is named after McNemar* [10] and, for the 2^2 table

$y(1,1)$	$y(1,2)$
$y(2,1)$	$y(2,2)$

is based on

$$\frac{[y(1,2) - y(2,1)]^2}{y(1,2) + y(2,1)} \mathrel{\dot{\div}} \chi_1^2$$

under H_2; *see* MCNEMAR STATISTICS.

A GENERAL FORMULA

The construction of the test statistic via the e's is well suited to special values of t and r but is not suited to the general case because the e's are asymmetrically defined with respect to the t values of i and the r values of j.

Noting that H_2 can also be formulated as

$$H_2 : \rho_j^i - \rho_j^{\cdot} = 0,$$

where $\cdot$ denotes average, we define

$$d_j^i = y_j^i - y_j^{\cdot}.$$

Since

$$\sum_{i=1}^{t} d_j^i = \sum_{j=1}^{r} d_j^i = 0,$$

it is sufficient to work with $(t-1)(r-1)$ d's, say those for which $i = 1, \ldots, t-1$, and $j = 1, \ldots, r-1$. The test statistic can be constructed as $\mathbf{d}'\tilde{\boldsymbol{\Sigma}}_{\mathbf{d}}^{-1}\mathbf{d}$ using the second moment formulae (3). The elements of $\tilde{\boldsymbol{\Sigma}}_{\mathbf{d}}$ involve the univariate marginal frequencies y_j^i and also the bivariate marginal frequencies $y_{jj'}^{ii'}$, $i \neq i'$. The typical element of $\tilde{\boldsymbol{\Sigma}}_{\mathbf{d}}$ is

$$\tilde{\text{cov}}\left(d_j^i, d_{j'}^{i'}\right) = (1-\delta_{ii'})y_{jj'}^{ii'} - (1-t^{-1})\left(y_{jj'}^{i\cdot} + y_{jj'}^{\cdot i'} - y_{jj'}^{\cdot\cdot}\right) + \delta_{jj'}\left\{\delta_{ii'}y_j^i - t^{-1}\left(y_j^i + y_j^{i'} - y_j^{\cdot}\right)\right\}, \quad (4)$$

where δ is identified by $\delta_{ii'} = 1, 0$ according as $i = i'$, $i \neq i'$. The number of degrees of freedom equals the number of d's in $\mathbf{d}$, namely $(t-1)(r-1)$. This (adjusted) Wald test-statistic does not depend on the diagonal frequencies $y(j, \ldots, j)$ but is a function of the $r^t - r$ nondiagonal frequencies.

Other statistics for testing H_2 against H are the likelihood-ratio statistic, first discussed by Madansky [7] and the minimum discrimination information* statistic first discussed by Ireland et al. [5] for $t = 2$ and by Kullback [6] for general t (*see* INFORMATION, KULLBACK). These statistics cannot be expressed in closed form, so that analytical comparison with the Wald statistic is not possible. However, numerical evidence reported by Bishop et al. (ref. 1, p. 284) and by Gokhale and Kullback (ref. 4, p. 257) suggest that all three statistics are asymptotically close.

MATCHED PAIRS AND t-PLETS

Marginal symmetry is, as we have seen, a hypothesis associated with a t-dimensional Response measured on a sample of s Individuals. In Example 1 the Individual is a voter and in Example 2 a boy.

Another kind of example arises when the Individual is a matched t-plet of individuals, on each of which is measured a univariate response taking one of the values $1, \ldots, r$. Particularly important is the case $t = 2$ when the matched t-plet is a matched pair. Matched pairs* are very widely used in retrospective, case-control studies, when the two matched individuals are a person with a certain disease and one without it. The "response" is the level of an exposure factor of interest. There are very many papers on case-control studies in the epidemiology journals. The reader is also referred to the book by Breslow and Day [2]; *see also* EPIDEMIOLOGICAL STATISTICS.

Example 3. The hypothetical data in Table 3 concern 83 cases of people under 30 suffering from heart disease. Each was paired with a control according to sex, age, occupation, and place of residence. The response is level of cigarette smoking, classified as 1 = light, 2 = medium, 3 = heavy.

Table 3 Heart Disease and Smoking

		Control 1	2	3	
Case	1	13	4	6	23
	2	18	9	4	31
	3	19	5	5	29
		50	18	15	83

We find that $\mathbf{e}'\boldsymbol{\Sigma}_{\mathbf{e}}^{-1}\mathbf{e} = 15.52$, a highly significant value of χ_2^2.

By modelling the matching process we can see what the hypothesis of marginal symmetry means for the matched t-plets. It is sufficient to consider matched pairs; $\rho(j_1, j_2)$ is then the probability that, in a randomly selected pair, the case has response j_1 and the control has response j_2.

We imagine that the population under study is divided into a large number S of fine strata. A case and a control are said to be matched if they come from the same stratum. Let π_{h1j} be the probability that a case from stratum h has response j and π_{h2j} the corresponding probability for a control. The probability that a matched pair from stratum h exhibits Response (j_1, j_2) is

$$\rho_h(j_1, j_2) = \pi_{h1j_1}\pi_{h2j_2}.$$

Now suppose that a random sample of S strata is selected and then a matched pair is selected from each of these strata. This is not precisely what happens in practice because, invariably, the cases are obtained first, in this way selecting the strata. This consideration, however, does not materially affect our model. It follows that

$$\rho(j_1, j_2) = S^{-1} \sum_{h=1}^{S} \rho_h(j_1, j_2)$$
$$= S^{-1} \sum_{h=1}^{S} \pi_{h1j_1}\pi_{h2j_2}. \qquad (5)$$

From (5) the marginal probability ρ_j^1 is

$$\rho_j^1 = S^{-1} \sum_{h=1}^{S} \pi_{h1j} = \pi_{\cdot 1j},$$

where $\pi_{\cdot 1j}$ is the average probability that a case has response j, averaged over all strata. Similarly $\rho_j^2 = \pi_{\cdot 2j}$ and H_2 is thus equivalent to G_2, where $G_2 : \pi_{\cdot 1j} = \pi_{\cdot 2j}$. Thus marginal symmetry for matched pairs is equivalent to equality for case and control of the strata-average probability distributions of the response.

QUASI-SYMMETRY AND THE MANTEL–HAENSZEL TEST

As mentioned in the introduction, it is possible to test H_2 assuming H_1 : quasi-symmetry; a test statistic having closed form is given in the entry QUASI-SYMMETRY. An interesting feature of it is that, when applied to matched pairs of matched t-plets, it also can be interpreted as a Mantel–Haenszel* test statistic. This correspondence was reported as a numerical phenomenon by Mantel and Byar [8] and explored in detail by Darroch [3].

References

[1] Bishop, Y. M. M., Fienberg, S. E., and Holland, P. W. (1975). *Discrete Multivariate Analysis: Theory and Practice*. MIT Press, Cambridge, Massachusetts.

[2] Breslow, N. E. and Day, N. E. (1980). *Statistical Methods in Cancer Research*. IARC Scientific Publications No. 32, Lyons, France.

[3] Darroch, J. N. (1981). *International Statistical Review*, **49**, 285–307.

[4] Gokhale, D. V. and Kullback, S. (1978). *The Information in Contingency Tables*. Marcel Dekker, New York.

[5] Ireland, C. T., Ku, H. H., and Kullback, S. (1969). *J. Am. Stat. Assoc.*, **64**, 1323–1341.

[6] Kullback, S. (1971). *Ann. Math. Statist.*, **42**, 594–606.

[7] Madansky, A. (1963). *J. Am. Statist. Assoc.*, **58**, 97–119.

[8] Mantel, N. and Byar, D. P. (1978). *Commun. Statist. A*, **7**, 953–976.

[9] Mantel, N. and Fleiss, J. L. (1975). *Biometrics*, **31**, 727–729.

[10] McNemar, Q. (1947). *Psychometrika*, **12**, 153–157.

[11] Stuart, A. (1955). *Biometrika*, **42**, 412–416.

Bibliography

Haberman, S. (1979). *Analysis of Qualitative Data*, Vol. 2. Academic Press, New York.

(CONTINGENCY TABLES
MATCHED PAIRS
MCNEMAR STATISTICS
QUASI-SYMMETRY)

J. N. DARROCH

MARKETING, STATISTICS IN

INTRODUCTION

Statistical methods in marketing span almost every aspect of applied statistics that one can imagine. The study of statistics in marketing is essentially the study of marketing research methodology, since statistics is the cornerstone of this discipline. Marketing researchers take a very eclectic view of statistical methods. The past two decades have witnessed a burgeoning of activity in applied statistics, with particular emphasis on multivariate data analysis methods.

We first describe briefly the historical development and list key textbooks that dramatically changed marketing research from its earlier emphasis on data collection* to data analysis, statistical inference*, and interpretation. Following this we offer an overview of the major statistical methods employed by the problem areas they address. Given the extremely wide application of statistical methods in marketing, no review can do justice to them all. This review is not intended as a comprehensive coverage; the studies included are illustrative of the types of statistics used. There are many other excellent and creative uses of statistics in marketing not included here. The entry is organized, however, by the techniques used, not by the areas of application.

SOME EARLY CONTRIBUTIONS

While the origins of formal marketing research are still rather obscure, many students of the field credit Charles Coolidge Parlin, a former Wisconsin school principal, with being the first professional marketing researcher. Parlin began his research career in 1911, upon taking a job with the Curtis Publishing Company. His first task was to study the marketing problems associated with the agricultural implements industry. Following this, he studied wholesaling and retailing in the textile industry and developed a crude census of distribution. Soon other firms became aware of Parlin's efforts and commercial research departments were established in such companies as U.S. Rubber and Swift & Co. By the early 1920s over a score of firms in the New York area were employing people specializing in marketing research.

Many of the early marketing research activities were aimed at developing general statistics on distribution, culminating in 1929 with the implementation of the first Census of Distribution, the precursor to today's Census of Business (which covers wholesaling, retailing, and the service industries).

A few books and monographs on marketing research also appeared in the late 1920s and early 1930s; however, the first well-known text—*Marketing Research and Analysis*, by Lyndon O. Brown [3]—did not appear until 1937, and contained virtually no discussion of statistical methods. About a decade later Heidingsfeld and Blankenship [9] and Zeisel [11] came out with texts that did introduce descriptive statistical methods, such as tabular and graphical representation*. Still, up to this time little in the way of sampling* techniques and inferential statistics had appeared in marketing literature.

The decade of the 1940s was central in the development of marketing survey sampling methods and was largely spearheaded by Bureau of the Census* statisticians (Hauser and Hansen [22]; Hansen and Hurwitz [20]). Probably the greatest impact on the marketing research community, however, stemmed from the publication of Robert Ferber's *Statistical Techniques in Market Research* [6]. It was Ferber who pulled the sample survey* work together and coupled it with other material on descriptive statistics, hypothesis testing* and estimation*, and simple and multiple correlation*. The result was a lucid and comprehensive text on what was known at that time.

The decade of the 1950s witnessed a continued development in sampling methods, descriptive statistics, and simple correlation and regression* models. Meanwhile, the area of normative modeling in marketing was introduced to the research practitioner. This fledgling field began to come of age with the publication of Bass et al., *Mathematical Models and Methods in Marketing* [2], Frank, Kuehn, and Massy, *Quantitative Techniques in Marketing Analysis* [7], and Alderson and Green, *Planning and Problem Solving in Marketing* [1]. These books introduced the marketing researcher to then avant-garde techniques such as simulation*, Markov models (*see* MARKOV PROCESSES), game theory*, exponential smoothing*, inventory* control theory, multivariate techniques, statistical decision theory*, and mathematical programming*. Green and Tull's *Research for Marketing Decisions* [8] (analogous to Ferber's 1949 text) attempted to integrate the most recent developments in Bayesian methods (*see* BAYESIAN INFERENCE), attitude scaling, and multivariate analysis* into a working methodology for marketing research.

Since that time, statistical methods have continued to enhance the discipline, so much so that most of the statistical methods known to date have been applied to problems in marketing analysis.

The use of statistics in marketing research

outside of the U.S. has had a somewhat slower takeoff. The most notable efforts were in the U.K., which has an active market research society (with its own professional publication) and has had a number of important contributors to the discipline, such as Ehrenberg [4, 5]. More recently there have been significant statistical-based activities in other Western European countries, especially Belgium, Germany, France, the Netherlands and in Scandinavia.

The interested reader is referred to the major marketing journals which offer a continuous stream of applied statistical studies. These include the two publications of the American Marketing Association: The *Journal of Marketing* and the *Journal of Marketing Research* (*JMR*); an interdisciplinary publication, the *Journal of Consumer Research* (sponsored by 10 associations, including ASA); the *Journal of the Market Research Society* (U.K.); the TIMS/ORSA publication, *Marketing Science*; and the *International Journal of Research in Marketing*.

THE USES OF STATISTICS IN MARKETING RESEARCH

Statistical techniques are an integral part of modern marketing research, which in turn provides the input to many marketing decisions of both profit and nonprofit organizations. These decisions include the generation, evaluation, and selection of marketing strategies concerning the following:

- Market segments—groups of consumers (or industrial buyers) who respond similarly to the firm's marketing activities and offerings, and who share similar needs, benefits, and other identifiable characteristics;
- Product positioning and design—a product's unique characteristics and benefits which provide it with a differential advantage over its competitors and reasons for consumers to buy and use it;
- Price of each of the product(s), including promotional prices (e.g., coupons, trade promotions);
- The distribution system most effective in reaching consumers;
- The advertising and communication programs of the firm;
- The overall marketing program (the synergistic combination of the above components);
- Allocation of resources among the various products, market segments, and marketing activities;
- Scheduling the various marketing activities.

Marketing research procedures have focused on the generation and evaluation of creative options for each of these decisions. Statisti-

Table 1

Marketing Decisions	Sample Survey Methods	Multi-Variate Techniques	Forecasting Methods	Psychometric Methods	Experimentation	Probability and Normative Models
Market segmentation	√	√	√			
Positioning and product design	√	√	√	√	√	√
Pricing	√	√		√	√	
Distribution	√			√	√	
Advertising	√	√			√	
Marketing programs	√	√	√	√	√	√
Allocation of resources		√	√			
Scheduling of activities		√				

cal methods have played a major role in the evaluative phases. Some of the major classes of statistical techniques, aimed at helping executives make better marketing decisions are briefly summarized in Table 1.

SAMPLE SURVEY METHODS

The major source of primary data in marketing research is from sample surveys, employing personal, telephone, mail, or combination interviews. Since the 1940s statistical sampling procedures have been used both in selecting samples and in analyzing the survey results. In addition to simple random selection techniques, researchers have employed various kinds of stratified* and cluster sampling* procedures.

Researchers in marketing have been influenced by the classic books of Hansen, Hurwitz, and Madow [21] and Deming [15]. More recent texts by Sudman [31], Cochran [13], Jessen [23], Williams [34], and Namboodiri [28] also have presented material of interest.

Probably the most relevant of recent writings on sample survey methods appeared in the August 1977 issue of *JMR*, guest edited by Robert Ferber. In particular, a review article by Frankel and Frankel [16] dealt with a number of relatively new developments in sample survey methodology, including:

- Methods for minimizing total sampling error;
- Estimation of sampling error in complex surveys;
- Selecting telephone household samples, including random digit dialing;
- Methods for protecting respondent privacy.

Emphasis on the control of *total* sampling error (bias as well as sampling error) is warranted in marketing research, largely because of the frequent presence of various types of response bias. Frankel and Frankel describe a number of methods for optimal sampling in the presence of possible sources of bias; see also Sudman [31] for a Bayesian viewpoint.

Marketing researchers also are becoming aware of replicated sample designs in dealing with the estimation of standard errors in complex samples. In the design of complicated multistage samples for which simple standard error formulas are not applicable, replicated designs often are essential. A number of methods, including balanced repeated replication [25] and jackknifing* [32] have been proposed to select efficient sets of subsamples. Woodruff and Causey [35] have compared these (and other approaches) with regard to computer time, reliability, and validity.

A popular recent development in sample survey methods has been the trend toward random digit dialing in telephone interviewing. These methods have been developed largely in response to the growing proportion of households with unlisted numbers. The two most common methods used in marketing are: (1) directory listings, in which the last r digits ($r = 1, 2, 3,$ or 4) are replaced with random sequences [14, 17, 18, 29], and (2) the more commonly used random digit dialing with procedures which eliminate nonworking banks of numbers [12, 27].

Related to survey sampling procedures are developments in data collection methods. The historically popular personal interview has been losing ground to newer (and mostly cheaper) data collection* procedures involving telephone* interviews, mail, or combinations of mail and telephone. Increasing attention has been given to computerized data collection whereby the respondent interacts with a terminal [24, 26] and the use of two-way cable TV such as the QUBE system.

Developments in sampling and data collection procedures have been closely linked to the sampling of questionnaire items. Various experimental* designs such as orthogonal array* designs have been widely used in *conjoint analysis* type studies as a way to reduce the number of items presented to each respondent [59]. Sampling from a multivariate normal distribution has been proposed as an alternative [158], but most of the conjoint analysis type studies have relied on orthogonal main effects designs [140]. More recently, in response to the need to simplify the data collection task, Balanced Incomplete Block (BIB) designs* have been utilized, allowing for the study of a large number of factors and levels, while requiring each respondent to evaluate only few (typically less than 10) stimuli [137].

Another fairly recent application is the randomized response* technique [33], applied to sensitive questioning areas. Since Warner's publication other researchers [19] have extended the techniques beyond dichotomous questions to quantitative variables.

While the preceding topics by no means exhaust recent developments, they are indicative of the kinds of practical sample survey problems being addressed and for which solutions have been proposed. An extensive bibliography [30] on sample survey methods also appears in the same special issue of *JMR*.

MULTIVARIATE TECHNIQUES

Starting out with simple and multiple regression* models in the early 1950s, marketing researchers have gradually accumulated a full armamentarium of multivariate procedures. General sources appear in books by Aaker [39], Green and Carrol [59, 60], Sheth [86], Ehrenberg [5], and Hair et al. [63]. Most applications of these techniques appear in the various marketing journals listed earlier.

Dependence Structures

Marketing researchers have employed all of the following techniques for the analysis of dependence structures:

- multiple regression*;
- analysis of variance* (ANOVA) and analysis of covariance* (ANCOVA);
- discriminant analysis*—two-group and multiple-group;
- canonical correlation (*see* CANONICAL ANALYSIS), multivariate analysis of variance* (MANOVA) and covariance (MANCOVA);
- automatic interaction detection (AID)* and related procedures such as THAID, MAID, MCA, and CHAID;
- logit*, probit*, and tobit* models;
- covariance structure analysis.

In marketing—as in most applied areas—multiple regression is still the workhorse technique. Applications of regression models run the gamut from econometrics*-type applications to studies in the measurement of consumer attitudes [56, 96]. More methodologically based papers, dealing with discrete dependent variable regression and test reliability [77, 78], have been published in marketing journals.

ANOVA and ANCOVA have been used extensively in marketing research, primarily in analysis of data from field experiments [180]. Illustrative of these applications are studies by Curhan [46] and Kilbourne [69]. Articles also have appeared on ANOVA and ANCOVA that are more methodological in nature [57]. Virtually all classes of experimental designs—factorial*, repeated measures*, incomplete block designs*, and partial factorials of various types—have been utilized.

Discriminant analysis also has been applied to marketing problems; a priori selected groups include loyal customers versus switchers, buyers of different brands in a product class, good versus poor sales territories, and so on. Illustrative applications include studies by Ostlund [79], Massy [71], and Dillon et al. [49]. Excellent expository articles have also appeared [76]. Canonical correlation (*see* CANONICAL ANALYSIS) MANOVA, and MANCOVA have received less attention, primarily because of the difficulty in interpreting canonical variates. Early papers [37, 62, 82, 94] were primarily expository in nature.

On the other hand, model-free, data dredging methods, such as Automatic Interaction Detection* Multiple Classification Analysis [38], THAID [75], MAID [70] and CHAID [68, 80], have received quite a bit of attention. Automatic Interaction Detection, the prototype of methods of this type, takes an interval-scaled or dichotomous dependent variable and employs a binary splitting and search procedure that breaks down the total sample into more homogeneous subgroups whose between-groups sums of squares are large with respect to the dependent variable.

The popularity of such data dredging procedures is probably explained by the large data bases that marketing researchers are often faced with and by the paucity of substantive theory to guide the selection of appropriate predictor variables. Also, most of these techniques incorporate simple ways to summarize and display the output, such as a tree diagram showing the sequence of binary splits, dependent-variable means, and group sample sizes.

Logit models, in which the dependent variable is categorized, are becoming increasingly popular (see refs. 48, 58, and 83). Probit models (refs. 50 and 84) also have been applied. The papers by McFadden [72–74] are important contributions to quantal choice modeling. Essentially, logit models are being used to represent aspects of multiattribute choice in which the selection of some brands from a product class or some transportation mode from a set of alternatives are the types of behavior being modeled. The dependent variable is quantal (i.e., categorical) and the predictor variables can be either categorical, continuous, or mixtures thereof.

McFadden [72] extended the dichotomous outcome case to situations involving more than two outcomes—what has been called the multinomial logit*. Most applications of logit models in marketing assume that the logit is a linear (in the parameters) function of a set of predictor variables that denote various attributes of the choice objects (e.g., brands) and choosers of the choice objects. Parameters are typically estimated by some type of maximum likelihood procedure.

Perhaps the most recent methodology to be introduced into marketing is that of *covariance structure analysis* (also referred to as *structural equation models*, *causal modeling*, and *multivariate analysis with latent variables*). The primary methodological developments have come from psychometricians (e.g., Jöreskog [66]). An expository account from the viewpoint of a psychologist, appears in the review by Bentler [45]. Bagozzi [40] provides a comprehensive account of the methodology in applications to marketing and a special issue of *JMR* on the topic appeared in 1983.

A few applications of covariance structure modeling have recently appeared [41, 52, 53, 81]. The LISREL* computer program package [67] has been the methodology that most marketing researchers have applied.

Interdependence Structures

In the analysis of interdependence structures (where one is interested in the mutual relationships among a set of variables and/or objects), marketing researchers have applied two major sets of techniques, factor analysis and cluster analysis.

*Factor analysis** has a long tradition in marketing research, primarily due to the large number of variables on which buyer information usually is obtained. With the popularity of psychographic research (the study of consumers' life styles, attitudes, interests, activities, and personality measures), factor analysis has achieved a new round of popularity [91]. In a typical psychographic study consumers are shown a large list of statements (e.g., "Friends usually come to me for advice on what kinds of hair care products to buy.") and are asked to indicate on a seven-point scale how well each phrase describes them (from "not at all" to "very closely"). Factor analysis is then used to examine underlying groups of statements (i.e., factor loading patterns) and to compute factor scores of each respondent on the underlying dimensions of interest. The most commonly used version of factor analysis in marketing research is principal component analysis*, followed by varimax rotation of retained components to approximate "simple structure," in which each variable correlates "highly" with one and only one factor.

Early applications of factor analysis to marketing problems include studies by Ehrenberg [51] on the identification of segments who watch different types of TV programs, Sheth [85] on brand loyalty measures, Stoetzel [88] on liquor preferences, and Twedt [90] on advertising. Recently, marketers have become interested in various extensions of factor analytic methods, including three-mode* factor analysis [89], higher order factor analysis [95], confirmatory methods [65], and asymmetric factor analysis [64].

Cluster analysis also has received considerable attention in marketing, either in combination with factor analysis or on its own. Use of these techniques is a natural outgrowth of researchers' interest in market segmentation and product positioning; Wind [92, 93] reviews much of the material.

Early applications of cluster analysis include its use in grouping cities on the basis of common socioeconomic variables [61], studying the competitive structure of the computer market [54], and clustering to improve the efficiency of marketing experiments [47]. Recent methodological developments such as overlapping clusters have been adapted quickly for marketing research [39]; it is fair to say that clustering techniques currently play an important role in the delineation and description of market segments and product positioning.

In summary, virtually all multivariate techniques have received one or more applications to marketing problems, and many of these methods have been used in tandem—factor analysis followed by cluster analysis [61], factor analysis followed by regression or discriminant analysis [55], and many other combinations. As more applied researchers become familiar with these methods, the time lag between introduction of the tool and its use in practical marketing problems has been reduced dramatically.

FORECASTING METHODS

A major ingredient of any marketing strategy is a forecasting* model. This model provides management with a set of estimates (ranging, e.g., from the most optimistic to the most pessimistic) of market response (e.g., volume, share, profits, etc.) of each market segment in each of the relevant geographical areas in a defined time period, with defined environmental conditions, and under a set of alternative marketing programs.

A variety of forecasting models have been applied and developed by both academicians and business firms. (See, e.g., refs. 42, 97, 98, 101, and 113.) The various models used vary in terms of the purpose of the model, the types of products and services forecasted, the stochastic versus deterministic nature of the model, the dependent and independent variables, the required data, and the analytical procedures employed.

The focus of this section is not on those characteristics of the various forecasting models used in marketing (for such a review, see Wind [93], Chapt. 15), but rather on two related statistical issues; namely, the use of time series* analysis and the application of diffusion type models.

Estimation issues related to forecasting models (e.g., use of OLS or maximum likelihood estimation) have received considerable attention in marketing but with few exceptions, such as Schmittlein and Mahajan [114] most of the focus has been on straightforward applications of accepted estimation procedures.

TIME-SERIES ANALYSIS. The simplest and most commonly used forecasting models use time as the independent variable. Such models range from simple time series extrapolations through exponential smoothing*, regression, and Box–Jenkins* analysis. Straightforward applications of these procedures are widely used and discussed in the literature; see Green and Tull [8] and most marketing research and forecasting books. A number of expository articles have been offered for the newer techniques; see Helmer and Johansson [105] for an expository description of the Box–Jenkins approach. In addition there have been a number of creative applications of these procedures in tandem. Hanssens [104], e.g., combined Box–Jenkins analysis, simple time series modeling, and econometrics to model primary demand, competitive reaction, and feedback effects.

DIFFUSION TYPE MODELS (*see* DIFFUSION PROCESSES). The basic diffusion model can be presented as:

$$\frac{dN(t)}{dt} = a\big(\overline{N}(t) - N(t)\big) + bN(t)\big(\overline{N}(t) - N(t)\big),$$

where

$\frac{dN(t)}{dt}$ = rate of diffusion at time t (slope of $N(t)$);

$N(t)$ = cumulative number of adopters at time t;

$\overline{N}(t)$ = population of potential adopters (ceiling on the number of adopters at time t);

a = constant—coefficient of innovation;

b = constant—coefficient of imitation.

The various diffusion models vary in terms of their assumptions concerning: (a) the values of a and b; (b) whether $\overline{N}(t)$ (market potential) is constant or not; (c) the degree to which the diffusion model is only a function of time or also a function of the marketing strategies of the firm; (d) the relationship of the new brand to other products in the marketplace; and (e) whether the model takes into consideration the spatial (geographic) diffusion pattern.

Fourt and Woodlock [103], e.g., assumed that $b = 0$; i.e., the diffusion is a pure innovative effect model. Mansfield [110] and Fisher and Pry [107], on the other hand, assumed that $a = 0$, i.e., their models are pure imitation effect models. Bass [42] assumed positive values for both a and b and hence incorporated both innovativeness and imitation effects.

Diffusion models vary also with respect to the way they treat market potential. Most models (as in refs. 42, 103, and 107) assume that it is constant over time (either as the total population or a fraction of it), while others, such as Mahajan and Peterson [108] and Dodson and Muller [99], assume that the potential adopter population changes over time, i.e., $\overline{N}(t) = f(S(t))$, where $S(t)$ is a vector of all relevant exogenous and endogenous variables affecting $\overline{N}(t)$.

The pioneering diffusion models of Bass [42] and Fourt and Woodlock [103] do not include marketing strategy effects. This omission has been overcome in a number of extensions of the basic diffusion model. Robinson and Lakhani [112] incorporate marketing variables by representing b—the imitation coefficient—as a function of marketing strategy variables such as advertising, promotion, and price. In this case, b is a function of $M(t)$; $b = B(M(t))$, where $M(t)$ is a vector of marketing decision variables at time t. Horsky and Simon [106] incorporate marketing strategy variables such as advertising, not through their effect on b, but rather through their effect, a, on innovators. They assume that advertising serves as a source of information to innovators and hence affects the coefficient of innovation a.

Further extensions of the basic diffusion model were suggested by Peterson and Mahajan [111], who incorporated in the model the complementarity, substitutability, contingent and independent relations of the new brand with other brands in the market place. Furthermore, Mahajan and Peterson [109] have offered a limited spatial extension of the basic diffusion model, expressing a, b, and $\bar{N}$ as a function of distance.

Recently the basic structure of innovation diffusion models has been reexamined by Easingwood et al. [100] in terms of two mathematical properties: point of inflection and symmetry. They have shown that most of the diffusion models offer no or limited flexibility in terms of these two properties, thus restricting their ability to accommodate different diffusion patterns. Assuming that the word-of-mouth effect may vary over time, they have suggested a simple model, termed Non-Uniform Influence (NUI), which allows the diffusion curve to be symmetrical as well as nonsymmetrical with the point of inflection responding to the diffusion process.

Experience with the use of forecasting models in marketing has been encouraging. Of particular note is the progress in the new product forecasting area; many new such models have been developed, implemented, and validated; see Wind et al. [115].

PSYCHOMETRIC METHODS IN THE MEASUREMENT OF CONSUMER PERCEPTIONS AND PREFERENCES

Marketing researchers have benefited immeasurably from the introduction of new tools for analyzing consumer perceptions and preferences. These tools are also multivariate in nature and can be grouped into two broad classes:

- multidimensional scaling* (MDS);
- conjoint analysis and other choice models.

Review papers have been prepared on both MDS [135] and conjoint analysis applications [140]. Marketing-oriented applications of MDS have been described in the book by Green and Rao [138] and applications of MDS and conjoint analysis have appeared in the book by Green and Wind [141].

Multidimensional Scaling Techniques

Multidimensional scaling methods are concerned with the representation of individuals' judgments regarding the relative similarities of pairs of objects (e.g., brands of automobiles), as distance relations on points in a multidimensional space. Each point corresponds to a brand and interpoint distances among the brands correspond in certain ways to the judged similarities data.

In metric MDS, the distances are computed so as to best fit the numerical values of the dissimilarities, up to a constant of proportionality. In nonmetric MDS, the distances are found so that a monotonic transformation of the original dissimilarities data is best fitted to the distances in the sense of minimizing a (normalized) sum of squares for a given dimensionality. In the case of fallible data the basic idea is to achieve a reasonably good fit of the geometric model to the input data in as few dimensions as possible—usually two or three—so that visual contact with the data can be maintained.

Multidimensional scaling of similarities data has a long history, going back at least to the late 1930s (Young and Householder [176]) in the case of metric methods. Torgerson's work in metric MDS [167, 168] and Coombs' theoretical contributions in nonmetric procedures [130] were highly important precursors to the introduction of nonmetric algorithms by Shepard [163] and Kruskal [146, 147], followed by Guttman [142], Carroll [121], Lingoes [149], and Young [175].

Since these pioneering contributions to the MDS field, a large variety of computer programs have been prepared and distributed by the University of North Carolina and Bell Laboratories in Murray Hill, New Jersey. Moreover, the original scaling of two-way matrices of similarities data has been extended to three-way and higher-way data matrices (Carroll and Chang [124]; Takane et al. [166]; Carroll and Wish [125]).

In addition to the development of models

for dealing with the analysis of data on judged similarities, algorithms also have been prepared for analyzing preference data [121, 132], including unfolding models [130], and vector models [169]. Researchers have also linked MDS methods with cluster analysis [122].

Two excellent reviews on scaling (Cliff [129]; Carroll and Arabie [123]) are recommended. Cliff's review covers developments, prior to 1973, in both unidimensional and multidimensional scaling; Carroll and Arabie's review focuses on MDS methods since 1973. Interestingly, Carroll and Arabie's definition of MDS is broad in that they do not limit the term to spatial distance models for similarities data; their purview includes nonspatial (e.g., discrete geometric models, such as tree structures) and nondistance (e.g., scalar product or projection) models. They also provide a useful taxonomy for organizing the diverse methodological developments in this field.

Marketing researchers were among the first to apply MDS to substantive problems; see Volume II of *Multidimensional Scaling* (Shepard et al. [164]). Since then, numerous applications have appeared, primarily in the *Journal of Marketing Research*, the *Journal of Marketing*, and the *Journal of Consumer Research*. Today, MDS methodology is a commonly used technique in market segmentation research and product positioning.

Computer programs for carrying out the more basic types of MDS are routinely available both in university and commercial computer centers. Almost all recently published marketing research texts have at least a section on MDS methodology, and most marketing and product management texts use the output of MDS techniques to illustrate the concept of positioning. MDS has earned a place, primarily as a way to provide diagnostic information regarding consumers' perceptions of and preferences for a brand relative to its competitors. Moreover, the graphical aspects of the methodology—yielding perceptual and preference "maps" of competing products and services—has unquestionably contributed to its popularity in marketing.

Until recently MDS was geometric rather than statistical in nature; little was available in the form of statistical tests regarding appropriate dimensionality and the statistical stability of points in a particular configuration. Lately, however, Ramsay [160, 161] has developed a maximum likelihood procedure for MDS parameter estimation. Other researchers (Weinberg et al. [172] have used jackknifing methods to establish confidence regions; *see* MULTIDIMENSIONAL SCALING.

While marketing researchers have not adopted these recent developments as yet, there is every reason to believe that more statistically based MDS methods will gain favor in marketing, just as the precursor models have before them.

Conjoint Analysis

The second psychometric technique that has received considerable attention in marketing research is conjoint analysis. Not surprisingly, significant methodological developments have been made to this field by the same researchers (e.g., Kruskal [148], Young [174], Carroll [120]) who have contributed so importantly to MDS methodology. Since conjoint analysis was introduced to marketing [138], followed by the special case of tradeoff analysis [145], marketing researchers have enthusiastically applied these techniques to over 1,000 studies in industry and nonprofit organizations [126].

The basic model of conjoint analysis is monotonic analysis of variance, in which a respondent expresses preferences for a set of product profile descriptions by ranking them from most to least liked. In some cases real products, synthesized according to partial factorial designs*, may be evaluated (instead of hypothetical product or service descriptions). In either case, the basic idea is to find part-worth functions implied by the respondent's overall ranking and the factorial structure imposed on the design of the stimuli. The procedure can be likened to an analysis of variance* problem in which the attribute levels are dummy coded and the researcher attempts to find a monotonic function of the input data for which a set of (normalized) residuals between the function and the fitted model are minimized.

The seminal paper is that of Luce and Tukey [157], dealing with *conjoint measurement*—the specification of conditions under which there exist measurement scales for both the response and control variables, given the order of the joint effects of the control variables and a prespecified composition rule. Marketers, however, have been less interested in the axiomatica and hypothesis testing aspects than in the scaling procedure, itself *assuming* that a particular composition rule is applicable. Hence Green and Srinivasan [140] employ the term *conjoint analysis* to distinguish the scaling intention from the original motivation. Given its emphasis on the scaling aspects, conjoint analysis closely resembles other techniques proposed for modeling subjective judgments [117, 131, 144].

As described by Green and Srinivasan, the conjoint model embraces the special cases of the vector model and the ideal point model. Letting $p = 1, 2, \ldots, t$ denote the set of t attributes, y_{jp} denote the level of the pth attribute for the jth stimulus, and assuming that each y_{jp} is inherently a continuous variable, the *vector* model can be written as

$$s_j = \sum_{p=1}^{t} w_p y_{jp},$$

where s_j is the preference for the jth stimulus and the w_p's are the respondent's weights for the t attributes.

The *ideal point* model posits that preference s_j is negatively related to the (weighted) squared distance d_j^2 of the location y_{jp} of the jth stimulus from the individual's ideal point x_p, where

$$d_j^2 = \sum_{p=1}^{t} w_p (y_{jp} - x_p)^2.$$

Thus, stimuli closer to the ideal point (with smaller d_j^2) will be more preferred (i.e., will receive larger s_j values) than those more distant from the ideal.

The preceding models both are special cases of the *part-worth* model:

$$s_j = \sum_{p=1}^{t} f_p(y_{jp}),$$

where f_p is the function denoting the part-worths of different levels of y_{jp} for the pth attribute. In practice, however, $f_p(y_{jp})$ is estimated only for a small set of levels of y_{jp}; in the case of inherently continuous variables, the part-worths for intermediate y_{jp} are obtained by interpolation.

Conjoint methods are used to obtain each individual's utility function. Through the use of highly fractionated designs [136] part-worths can be estimated for a tiny fraction of all the combinations that would be implied by the full cartesian product of all attribute levels employed in designing the stimulus profiles. The two most commonly used techniques for obtaining the part-worths are OLS and the LINMAP algorithm (Shocker and Srinivasan, [165]). In a variety of empirical comparisons with competing algorithms, LINMAP has performed quite well.

Once the utility functions are obtained—one for each individual—they are entered into a consumer choice simulator in which new product ideas are tested to see what share of choices each would receive if introduced as a competitor to existing products in the marketplace. It is the choice simulation step that has contributed to the widespread popularity of the methodology.

Prompted by the apparent success of the methodology in choice simulation studies, marketing researchers are now developing extensions to provide *optimal* product designs, rather than simply the best of a relatively few profiles that may be tested in the simulator. Contributions by Shocker and Srinivasan [165], Urban [170], Albers and Brockhoff [116], Gavish et al [133], Hauser and Simmie [143], Zufryden [177], Pessemier [159], and Green et al. [134] illustrate efforts in this direction.

From a statistical standpoint, the last approach utilizes response surface* methods (Box et al. [119]) to find a function that relates the criterion variable of interest—cash flow, sales, market share, etc.—to the product design variables. It is this function (which models the choice simulator) that is then optimized.

As a relatively inexpensive way to estimate consumer preferences for alternative product formulations and marketing strategy variables (e.g., different price levels, promotional messages, etc.), conjoint analysis already has demonstrated its value to marketing decision makers over a wide variety of product and service classes (Cattin and Wittink [127]).

Other Choice Models

Other choice models have been used to assess consumer choices (for example, see Urban and Hauser [171]). Of special interest are some procedures used to assess management's own choice processes. The application of choice rules to management decision making goes back to early applications of Bayesian analysis [134]. Interest in Bayesian techniques has declined, even though they are frequently mentioned in texts and occasionally applied to marketing studies (see, e.g., Blattberg and Sen [118]). Of considerable interest are decision calculus type models and the recent application of the analytic hierarchy process.

Decision calculus models were first proposed by Little [150] who with his colleagues has developed a large battery of models covering most marketing mix areas (e.g., MEDIAC for media selection, Little and Lodish [154]; CALLPLAN for allocation of salesmen's time, Lodish [156]) as well as a model which relates brand sales and profit to the entire marketing program, competitive activities, and environmental conditions—BRANDAID (Little [151, 152]). These models are based on management's subjective judgments and more recently have been linked to marketing decision support systems—a coordinated collection of data, models, analytical tools, and computing power that help managers make better decisions (Little [153]). Some basic premises and approaches of these models have been recently under attack (Chakravarti et al. [128]). Yet management's experience with these models has been favorable and encouraging (Little and Lodish [155]).

The *Analytic Hierarchy Process* or AHP (Saaty [162]) uses management's subjective judgments (both for generation of alternative courses of action and their evaluation) as well as other market based data and as the central feature of the modeling approach. The AHP has been used in marketing applications as the modeling framework and methodology (Wind and Saaty [173]). Lacking statistical inferential procedures, this approach relies on sensitivity analysis and simulation.

Marketing researchers long have been interested in "normative choice models," yet their wide scale implementation had to wait for the incorporation of models with better measurement procedures (e.g., more reliable and valid data). Given the significant advances in the measurement area, one can thus expect a major revival of interest.

EXPERIMENTATION

Lab and field experiments have long been used to establish consumers' responses to alternative marketing strategies. Experiments have involved all individual marketing strategy variables (e.g., varying prices, different advertising messages, various product formulations, etc.) as well as overall marketing programs. Recent years have witnessed the increased use of adaptive experimentation, especially in direct mail marketing. Single factor experiments still dominate, but there is a steady increase in the use of factorial designs*, randomized blocks*, Latin* and Graeco-Latin squares*, and switch-over designs in both lab and field marketing experiments. Analytically, ANOVA and ANCOVA are the major techniques used, with occcasional applications of MANOVA and MANCOVA.

LAB EXPERIMENTS. Lab experiments are especially popular in testing consumers' reactions to alternative strategies of a single marketing variable. Copy testing procedures are available, for example, to evaluate alternative advertising messages. Similarly, product taste tests are often conducted as lab experiments comparing a number of alternative product formulations. These types of experiments usually employ a number of evaluative measures (dependent variables) such as intention to buy, attitudes toward the brand (overall liking as well as evaluation on specific attributes), and to the advertisement (in the case of advertising experiments). A recent development has been the *simulated test market*. In-home use tests and central location tests are the primary experimental procedures used to generate data for pre-test market simulators. A number of research firms offer such services on a regular basis, either as a substitute for a test market or as a precursor to one.

A prototypical test is aimed at assessing a new product's:

- trial rate;
- repeat purchase pattern, rate, and volume;
- source of business for the new brands (e.g., cannibalization);
- market segments most attracted to the new product;
- perceived product positioning;
- major strengths and weaknesses of the marketing plan—in particular, the effect of the advertising, sampling, and price on the product's usage and share.

To provide input to these and similar areas of investigation, typical experiments involve a before–after type design with one interview prior to using the product (but after exposure to the advertising of the brand and its major competitors) and one or more interviews following product usage.

Data from such experiments are among the major inputs to new product forecasting models aimed at predicting both trial and repeat purchase, and providing diagnostic insights into consumers' acceptance of the product. These can include analysis and evaluation of brand switching patterns, identification of the key demographic, psychographic and purchase/usage patterns which discriminate between buyers and nonbuyers, satisfied vs. dissatisfied users, etc., as well as assessment of the perceived positioning of the brands by various segments (such as benefit or usage segments).

Experience with simulated test market procedures has suggested extremely high predictive accuracy. Yankelovich et al. [224], for example, reported on a validation of over 160 cases with 92% success—the LTM prediction closely matched results of the test market.

FIELD EXPERIMENTS, not unlike lab experiments, have included both the study of single variables and the examination of programs. Two major types of field experiments are minitest markets and test market designs. In their simplest form, *minitests* are an extension of a simulated test market. Charlton et al. [181] report on a U.K. "minitest market" based on a panel of households which were visited weekly. After exposure to a color brochure that displayed a set of available brands, respondents were asked to place orders for brands from a variety of product categories which immediately were filled from a mobile van. The procedure allows the assessment of the level of repeat buying, the cumulative penetration level (percentage of households who buy the brand at least once in a given time period), and average buying frequency (the average number of times the buyers buy the brand in the same time period).

The U.K. minitest market, which was validated by Ehrenberg [183], is used primarily for the evaluation of new products and measuring the effect, under controlled conditions, of alternative marketing activities. Although similar facilities are not common in the U.S., a number of minitest markets are currently in operation in the U.S.

One of the better known and most comprehensive minitest market procedures is offered by AdTel. This system utilizes a split cable TV system to test alternative copy, spending level, media scheduling, and other TV advertising plans. It includes controlled store placement services for securing "instant distribution" of new products, and a 2,000-family ongoing weekly consumer diary panel within each city to assess trial, frequency, and amount of repeat purchase, brand switching, and cannibalization. AdTel minitest markets have been utilized in a large number of cases involving both packaged and nonpackaged consumer goods and services. The reported experience with the system is most encouraging.

Still the most commonly used field experiments are *test markets*, often a mandatory stage in the new product development process of many firms. They serve both as a vehicle to assess the market performance of a new product under alternative marketing strategies, and as a pilot operation.

A special type of field experiment is *adaptive experimentation* (Little [185]). The idea is to consider the entire market (whether a region or the entire country) and enter it with multiple strategies and simultaneously test them. Implementing an adaptive experimentation approach can be viewed as a large-scale test market operation that encompasses all the target market. Direct mail operations are especially suitable for such an approach and, in fact, this approach has been applied in many such cases as well as in the management of the advertising strategy of a number of firms (see, for example, Ackoff and Emshoff [178] and the commentary on it by Allaire [179]).

NEW DEVELOPMENTS. Three new technological developments are of special interest to the design of market experiments:

1. The availability of scanning data. The tremendous growth in scanning data offers new opportunities for store and market experiments. Alternative price, coupons, shelf space, point of purchase promotion, advertising, and other marketing variables now can be evaluated faster and more easily.
2. The development of two-way cable TV systems such as the QUBE system, which has been in operation for a number of years in Columbus, Ohio. Such systems facilitate more sensitive experimentation with alternative strategies and almost instantaneous response.
3. The proliferation of home computers and the development of computer data collection procedures which allow for varying the experimental stimuli presented to respondents as a function of earlier responses.

For an example of a procedure for the determination of an optimal advertising budget, see Dhalla [182].

PROBABILITY MODELS

Since the mid-fifties, considerable attention has been given to stochastic brand choice models, reviewed in the now classic book by Massy et al. [212] and in recent review articles [188, 213]. Stochastic models of buying behavior have encompassed both brand choice (e.g., [193, 198, and 205]) and purchase incidence models [103, 191, and 211], as related to interconsumer as well as intraconsumer situations.

Interconsumer type models have included both models which assume that consumers are similar in their choice probabilities (e.g., [197, 205, and 208]) and models which allow for consumer heterogeneity (e.g., [193, 198, 201, 206, and 210]). More recently, probability mixture models have been employed to model consumer responses in taste test type experiments [214] and more general choice situations [219].

Intraconsumer type models include both stationary models (assumed constant choice probability over a sequence of choices) and nonstationary choice probabilities. These latter models (and their counterparts of the interconsumer models; e.g., [218]) are of the greatest interest to marketing professionals since they relate brand purchase probabilities to factors such as purchase feedback, time effects (as a surrogate for external factors), explicit consideration of external factors, and, particularly, the marketing activities of the firm (advertising, price, promotions, etc.).

The basic stochastic choice models—the zero-order (Frank [193]), Markov (Lipstein [208]) and linear learning models (Kuehn [205])—have been widely used and adequately covered in the Massy, Montgomery and Morrison book as well as in other review articles. Of greater interest are the efforts to use stochastic models to forecast the sales of new products, to incorporate marketing variables in the stochastic choice models, to represent individual or segment behavior, and to suggest laws of market behavior.

Stochastic new product sales models go back to the pioneering work of Fourt and Woodlock [103], Parfitt and Collins [216], Eskin [192], and Massy [211]. The models are based on consumer panel data and distinguish between trial and repeat purchase probabilities. These models, and especially the simpler ones, have been widely accepted by practitioners but lately have been replaced with more complete models that incorporate marketing variables.

Incorporating marketing variables in stochastic choice models dates back to the early work of Telser [222] who used a Markov transition matrix in which the transition probabilities depended on advertising. His work did not consider population heterogeneity and other modifications of recent stochastic brand choice models. More recently, Blattberg and Jeuland [189] have incorporated advertising in a stochastic brand choice model, assuming that behavior is zero order but that the probability of buying depends on advertising. Pricing and promotional effects have been introduced to a number of stochastic brand choice models. Kuehn and Rohloff [206], using a linear learning model, examined the effect of a promotional deal on the probability of purchase. Similarly, Lilien [207] used a modified linear learning model to evaluate the effects of price on the probability of buying. An interesting addition to this literature is the finding of Shoemaker and Shoaf [221] that the probability of buying a brand after buying it on deal was lower than the probability of buying it after buying it with no deal. The effects of price and deals were studied by Bass [187], MacLachlan [209] and others. Jeuland [199] has incorporated distribution—brand availability—as an explicit factor in a brand switching matrix, and some pioneering efforts are being done in incorporating multiple marketing factors in stochastic brand choice markets [202, 203, 223].

The third area of interest is the use of stochastic models (1) to represent *individual* level behavior in the determination of *aggregate* sales-advertising relationships (see Horsky [196], and (2) to represent *segment* behavior in classifying individuals into various segments (see Blattberg and Sen [118]).

The fourth area of interest is *the use of stochastic brand choice models for developing laws of market behavior*. The three most notable efforts are that of Ehrenberg, with his NBD model and theory, the Bass and Herniter entropy-based models, and the Hendry system.

The Negative Binomial Distribution* (NBD) was proposed by Ehrenberg [183] as a model to estimate the long run penetration and usage rate of a brand. Assuming a Poisson purchase rate for a given period and a heterogeneous population that follows a gamma distribution*, the NBD distribution is derived and offers "simple and generalizable patterns of market behavior which suggest a stationary situation." Ehrenberg has creatively used a stochastic model (NBD) to examine empirical regularities in the marketplace and suggest simple rules of behavior. Recent efforts have also linked this and related product purchase models to some brand choice models listed above [200, 220]. The models can be used to forecast future market behavior based on past performance [194, 215] and thus provide a baseline for investigating the effects of marketing programs. The major weakness of this approach however, is, its failure to explicitly model the effect of marketing strategies on the brand's penetration and buying rate.

Using the entropy* concept Herniter [195] suggests that brand share information contains all the information needed to predict brand switching. Due to the complexity of solving the entropy equation, Herniter's approach has not been applied to a market with more than three brands. Bass [186] simplified the approach by assuming that there are k loyal segments and that all switching is due to one segment—the stochastic preference segment. The Bass and Herniter models suggest that consumers switch brands at a maximum level given their segment membership.

The Hendry system (Butler [190], Kalwani and Morrison [204], and Rubinson et al. [217]) is a widely used but incompletely documented stochastic brand choice model. It encompasses a heterogeneous multinomial process that assumes a zero order purchase behavior, a constant probability of purchase over time, and a Dirichlet distribution* for a heterogeneous population; the marginal distribution for a given brand's probability of being purchased is then beta binomial. Brand switching among directly competing brands is proportional to the brand shares.

References

GENERAL

[1] Alderson, W. and Green, P. (1964). *Planning and Problem Solving in Marketing*. Richard D. Irwin, Homewood, Illinois.

[2] Bass, F. M., et al., eds. (1961). *Mathematical Models and Methods in Marketing*. Richard D. Irwin, Homewood, Illinois.

[3] Brown, L. O. (1937). *Marketing Research and Analysis*. Ronald Press, New York.

[4] Ehrenberg, A. S. C. (1972). *Repeat Buying*. North-Holland, Amsterdam.

[5] Ehrenberg, A. S. C. (1975). *Data Reduction*. Wiley, New York.

[6] Ferber, R. (1949). *Statistical Techniques in Market Research*. McGraw-Hill, New York.

[7] Frank, R. E., Kuehn, A. A., and Massy, W. F., eds. (1962). *Quantitative Techniques in Marketing Analysis*. Richard D. Irwin, Homeward, Illinois.

[8] Green, P. E. and Tull, D. S. (1966). *Research Marketing Decisions*. Prentice-Hall, Englewood Cliffs, New Jersey.

[9] Heidingsfeld, M. S. and Blankenship, A. B. (1947). *Marketing and Marketing Analysis*. Henry Holt, New York.

[10] Lockley, L. C. (1974). In *Handbook of Marketing Research*, Robert Ferber (ed.). McGraw-Hill, New York, Chapt. 1, pp. 1–15.

[11] Zeisel, H. (1947). *Say It With Figures*. Harper & Brothers, New York. (This book is now in its 5th edition.)

SAMPLE SURVEYS

[12] Bortner, B. and Assael, H. (1967). *Continuous Tracking Studies Via WATS Lines and Personal Interviewing*. (Paper presented at the Annual Conference of the Advertising Research Foundation, New York.)

[13] Cochran, W. G. (1977). *Sampling Techniques*, 3rd ed. Wiley, New York.

[14] Cooper, S. L. (1964). *J. Market. Res.*, **1**, 45–52.

[15] Deming, W. E. (1950). *Some Theory of Sampling*. Wiley, New York.

[16] Frankel, M. R. and Frankel, L. R. (1977). *J. Market. Res.*, **14**, 280–293.

[17] Glasser, G. J. and Metzger, G. D. (1972). *J. Market. Res.*, **9**, 59–64.

[18] Glasser, G. J. and Metzger, G. D. (1975). *J. Market. Res.*, **12**, 359–361.

[19] Greenberg, B. G., Abernathy, J. R., and Horvitz, D. G. (1969). *Proc. Soc. Statist. Sec.*, American Statistical Association, pp. 40–43.

[20] Hansen, M. H. and Hurwitz, W. N. (1942). *J. Amer. Statist. Ass.*, **37**, 89–94.

[21] Hansen, M. H., Hurwitz, W. N., and Madow, W. G. (1953). *Sample Survey Methods and Theory*, Vols. I and II. Wiley, New York.

[22] Hauser, P. M. and Hansen, M. H. (1944). *J. Market.* **9**, 26–31.

[23] Jessen, R. J. (1978). *Statistical Survey Techniques*. Wiley, New York.

[24] Johnson, R. M. (1980). In *Market Measurement and Analysis*, D. B. Montgomery and D. R. Wittink, eds. Marketing Science Institute, Cambridge, Mass.

[25] McCarthy, P. J. (1969). *Rev. Internat. Statist. Inst.*, **37**, 239–264.

[26] Myers, J. (1976). In *Moving Ahead with Attitude Research*, Wind and Greenberg (eds.). AMA, Chicago.

[27] Nielsen, A. C. Co. (1976). *Total Telephone Frame*. Company brochure, Chicago.

[28] Namboodiri, N. K. (1978). *Survey Sampling and Measurement*. Academic Press, New York.

[29] Rich, C. L. (1977). *J. Market. Res.*, **14**, 300–305.

[30] Spaeth, M. A. (1977). *J. Market. Res.*, **14**, 403–409.

[31] Sudman, S. (1976). *Applied Sampling*. Academic Press, New York.

[32] Tukey, J. W. (1958). *Ann. Math. Statist.*, **29**, 614.

[33] Warner, S. L. (1965). *J. Amer. Statist. Ass.*, **60**, 63–69.

[34] Williams, W. (1978). *A Sampler on Sampling*. Wiley, New York.

[35] Woodruff, R. S. and Caersey, B. D. (1976). *J. Amer. Statist. Ass.*, **71**, 315–321.

MULTIVARIATE TECHNIQUES

[36] Aaker, D. A. (1981). *Multivariate Analysis in Marketing*, 2nd ed. Wadsworth, Belmont, California.

[37] Alpert, M. I. and Peterson, R. A. (1972). *J. Market. Res.*, **9**, 187–192.

[38] Andrews, F. M., Morgan, J. N., and Sonquist, J. A. (1971). *Multiple Classification Analysis*. University of Michigan Press, Ann Arbor, Mich.

[39] Arabie, P., Carroll, J. D., DeSarbo, W., and Wind, Y. (1981). *J. Market. Res.*, **18**, 310–317.

[40] Bagozzi, R. P. (1980). *Causal Models in Marketing*. Wiley, New York.

[41] Bagozzi, R. P. (1981). *J. Market. Res.*, **18**, 375–381.

[42] Bass, F. M. (1969). *J. Market. Res.*, **6**, 291–300.

[43] Bass, F. M. and Parsons, L. J. (1969). *Appl. Economics*, **1**, 103–124.

[44] Bass, F. M. and Wittink, D. R. (1975). *J. Market. Res.*, **12**, 414–425.

[45] Bentler, P. M. (1980). *Ann. Rev. Psych.*, **31**, 419–456.

[46] Curhan, R. C. (1974). *J. Market. Res.*, **11**, 280–294.

[47] Day, G. S. and Heeler, R. M. (1971). *J. Market. Res.*, **8**, 340–347.

[48] Desarbo, W. S. and Hildebrand, D. K. (1980). *J. Market.*, **44**, 40–51.

[49] Dillon, W., Goldstein, M., and Schiffman, L. G. (1978). *J. Market. Res.* **15**, 103–112.

[50] Doyle, P. (1977). *J. Business Res.*, **5**, 235–248.

[51] Ehrenberg, A. S. C. (1968). *J. Advertising Res.*, **8**, 55–70.

[52] Fornell, C. and Larcker, D. F. (1981). *J. Market. Res.*, **18**, 39–50.

[53] Fornell, C. and Larcker, D. F. (1981). *J. Market. Res.*, **18**, 382–388.

[54] Frank, R. E. and Green, P. E. (1968). *J. Market. Res.*, **5**, 83–94.

[55] Frank, R. E., Massy, W. F., and Wind, Y. (1972). *Market Segmentation*. Prentice-Hall, Englewood Cliffs, New Jersey.

[56] Ginter, J. L. (1974). *J. Market. Res.*, **11**, 30–40.

[57] Green, P. E. (1973). *Measurement of Judgmental Responses to Multiattribute Marketing Stimuli*. American Society for Testing and Materials, Philadelphia.

[58] Green, P. E., Carmone, F. J., and Wachspress, D. M. (1977). *J. Market. Res.*, **14**, 52–59.

[59] Green, P. E. and Carroll, J. D. (1976). *Mathematical Tools for Applied Multivariate Analysis*. Academic Press, New York.

[60] Green, P. E. and Carroll, J. D. (1978). *Analyzing Multivariate Data*. Dryden Press, Hinsdale, Illinois.

[61] Green, P. E., Frank, R. W., and Robinson, P. J. (1967). *Manag. Sci.*, **13**, 387–400.

[62] Green, P. E., Halbert, M. H., and Robinson, P. J. (1966). *J. Market. Res.*, **3**, 32–39.

[63] Hair, J. F., Anderson, R. E., Tatham, R. L., and Grablowsky, B. J. (1979). *Multivariate Data Analysis with Readings*. Petroleum Publishing Co., Tulsa, Oklahoma.

[64] Harshman, R., Green, P. E., Wind, Y., and Lundy, M. J. (1982). *Market. Sci.*, **1**, 205–242.

[65] Jöreskog, K. G. (1970). *Psychometrika*, **32**, 443–482.

[66] Jöreskog, K. G. (1970). *Biometrika*, **57**, 239–251.

[67] Jöreskog, K. G. and Sörbom, D. (1978). *LISREL: Analysis of Linear Structural Relationships by the Method of Maximum Likelihood*. National Educational Resources, Chicago, Illinois.

[68] Kass, G. V. (1976). *Significance Testing and Some Extensions of Automatic Interaction Detection*. Unpublished doctoral dissertation, University of Witwatersrand, Johannesburg, South Africa.

[69] Kilbourne, W. E. (1974). *J. Market. Res.*, **11**, 453–455.

[70] MacLachlan, D. L. and Johansson, J. K. (1981). *J. Market.*, **45**, 74–84.

[71] Massy, W. F. (1965). *J. Advertising Res.*, **5**, 39–48.

[72] McFadden, D. (1973). In *Frontiers of Econometrics*, P. Zarembka, ed. Academic Press, New York, pp. 105–141.

[73] McFadden, D. (1976). *Ann. Economic and Soc. Measurement*, **5**, 363–390.

[74] McFadden, D. (1980). *J. Business*, **53**, 513–530.

[75] Morgan, J. N. and Messenger, R. C. (1973). *THAID: A Sequential Analysis Program for the Analysis of Nominal Scale Dependent Variables*. University of Michigan, Ann Arbor, Mich.

[76] Morrison, D. G. (1969). *J. Market. Res.*, **6**, 156–163.

[77] Morrison, D. G. (1972). *J. Market. Res.*, **9**, 338–340.

[78] Morrison, D. G. (1973). *J. Consumer Res.*, **10**, 91–93.

[79] Ostlund, L. E. (1974). *J. Consumer Res.*, **11**, 23–29.

[80] Perreault, W. D. and Barksdale, H. C. Jr. (1980). *J. Market. Res.*, **16**, 503–515.

[81] Phillips, L. W. (1981). *J. Market. Res.*, **18**, 395–415.

[82] Pruden, H. O. and Peterson, R. (1971). *J. Market. Res.*, **8**, 501–504.

[83] Punj, G. N. and Staelin, R. (1978). *J. Market. Res.*, **15**, 588–598.

[84] Rao, V. R. and Winter, F. W. (1978). *J. Market. Res.*, **15**, 361–368.

[85] Sheth, J. N. (1968). *J. Market. Res.*, **4**, 395–404.

[86] Sheth, J. N., ed. (1977). *Multivariate Methods for Market and Survey Research*. AMA, Chicago.

[87] Sonquist, J. A., Baker, E. L., and Morgan, J. N. (1971). *Searching for Structure*. University of Michigan, Ann Arbor, Mich.

[88] Stoetzel, J. (1960). *J. Advertising Res.*, **1**, 7–11.

[89] Tucker, L. R. (1966). *Psychometrika*, **31**, 279–311.

[90] Twedt, S. W. (1952). *J. Appl. Psychol.*, **36**, 207–215.

[91] Wells, W. D. (1974). *Lifestyle and Psychographics.* AMA, Chicago.

[92] Wind, Y. (1978). *J. Market. Res.*, **15**, 317–337.

[93] Wind, Y. (1982). *Product Policy: Concepts, Methods, and Strategy*. Addison-Wesley, Reading, Mass.

[94] Wind, Y. and Denny, J. (1974). *J. Market. Res.*, **11**, 136–142.

[95] Wind, Y., Green, P. E., and Jain, A. K. (1973). *J. Market Res. Soc.*, **15**, 224–232.

[96] Winter, F. W. (1973). *J. Market. Res.*, **10**, 130–140.

FORECASTING METHODS

[97] Brown, D. A., Buck, S. F., and Pyatt, F. G. (1965). *J. Market. Res.*, **11**, 229–231.

[98] Claycamp, H. J. and Liddy, L. E. (1969). *J. Market. Res.*, **6**, 414–420.

[99] Dodson, J. A. and Muller, E. (1978). *Manag. Sci.*, **15**, 1568–1578.

[100] Easingwood, C., Mahajan, V., and Muller, E. (1982). A Non-uniform Influence Innovation Diffusion Model of New Product Acceptance. *Working Paper*, Wharton School, University of Pennsylvania, Philadelphia.

[101] Eskin, G. (1973). *J. Market. Res.*, **10**, 115–129.

[102] Fisher, J. C. and Pry, R. H. (1971). *Techn. Forecasting and Social Change*, **2**, 75–88.

[103] Fourt, L. A. and Woodlock, J. W. (1960). *J. Market.*, **25**, 31–38.

[104] Hanssens, D. M., (1980). *J. Market. Res.*, **17**, 470–485.

[105] Helmer, R. M. and Johansson, J. K. (1978). *J. Market. Res.*, **14**, 227–229.

[106] Horsky, D. and Simon, L. S. (1981). Advertising and the Diffusion of New Products. *Working Paper*, Graduate School of Management, University of Rochester, Rochester, New York.

[107] Mahajan, V. and Muller, E. (1979). *J. Market.*, **43**, 55–58.

[108] Mahajan, V. and Peterson, R. A. (1978). *Manag. Sci.*, **15**, 1589–1597.

[109] Mahajan, V. and Peterson, R. A. (1979). *Techn. Forecasting and Social Change*. **14**, 127–146.

[110] Mansfield, E. (1961). *Econometrica*, **29**, 741–766.

[111] Peterson, R. A. and Mahajan, V. (1978). In *Research in Marketing*, J. Sheth, ed. JAI Press, Greenwich, Connecticut.

[112] Robinson, B. and Lakhani, C. (1975). *Manag. Sci.*, **21**, 1113–1132.

[113] Ryans, A. F. (1974). *J. Market. Res.*, **11**, 434–443.

[114] Schmittlein, D. and Mahajan, V. (1982). *Market. Sci.*, **1**, 57–78.

[115] Wind, Y., Mahajan, V., and Cardozo, R., eds. (1981). *New Product Forecasting: Models and Applications*. Lexington Books, Lexington, Mass.

PSYCHOMETRIC METHODS IN CONSUMER PERCEPTIONS AND PREFERENCE MEASUREMENT

[116] Albers, S. and Brockhoff, K. (1977). *European J. Operat. Res.*, **1**, 230–238.

[117] Anderson, N. H. (1970). *Psychol. Rev.*, **77**, 153–170.

[118] Blattberg, R. C. and Sen, S. (1975). *Manag. Sci.*, **21**, 682–696.

[119] Box, G. E. P., Hunter, W. G., and Hunter, J. S. (1978). *Statistics for Experimenters*. Wiley, New York.

[120] Carroll, J. D. (1969). Categorical Conjoint Measurement. Meeting of Mathematical Psychology, Ann Arbor, Michigan.

[121] Carroll, J. D. (1972). In *Multidimensional Scaling*, Vol. I, R. N. Shepard, A. K. Romney, and

J. B. Nerlove, eds. Seminar Press, New York, pp. 105–157.

[122] Carroll, J. D. (1976). *Psychometrika*, **41**, 439–463.

[123] Carroll, J. D. and Arabie, P. (1980). In *Annu. Rev. Psychol.*, M. R. Rosenzweig and L. W. Porters, eds. Annual Reviews, Palo Alto, California.

[124] Carroll, J. D. and Chang, J. J. (1970). *Psychometrika*, **35**, 283–319.

[125] Carroll, J. D. and Wish, M. (1974). In *Handbook of Perception*, Vol. 2, E. C. Carterette and M. P. Friedman, eds. Academic Press, New York.

[126] Cattin, P. and Wittink, D. R. (1981). *J. Market. Res.*, **18**, 101–106.

[127] Cattin, P. and Wittink, D. R. (1982). *J. Market.*, **46**, 44–53.

[128] Chakravarti, D., Mitchell, A., and Staelin, R. (1981). *J. Market.*, **45**, 13–23.

[129] Cliff, N. (1973). In *Annu. Rev. Psychol.*, P. H. Mussen and M. R. Rosenzweig, eds. Annual Reviews, Palo Alto, California.

[130] Coombs, C. H. (1964). *A Theory of Data*. Wiley, New York.

[131] Dawes, R. M. and Corrigan, B. (1974). *Psychol. Bull.*, **81**, 95–106.

[132] DeSarbo, W. S. (1978). Three-Way Unfolding and Situational Dependence in Consumer Preference Analysis. Doctoral dissertation, Wharton School, University of Pennsylvania.

[133] Gavish, B., Horsky, D., and Srikanth, K. (1981). *Manag. Sci.*, **29**, 1277–1297.

[134] Green, P. E. (1963). *Business Horizons*, **5** (Fall), 101–109.

[135] Green, P. E. (1975). *J. Market.*, **38**, 24–31.

[136] Green, P. E., Carroll, J. D., and Carmone, F. J. (1976). *J. Business Res.*, **4**, 281–295.

[137] Green, P. E., Carroll, J. D., and Goldberg, S. (1981). *J. Market.*, **45**, 17–37.

[138] Green, P. E. and Rao, V. R. (1972). *Applied Multidimensional Scaling: A Comparison of Approaches and Algorithms*. Holt, Rinehart, and Winston, New York.

[139] Green, P. E. and Rao, V. R. (1971). *J. Market. Res.*, **8**, 355–363.

[140] Green, P. E. and Srinivasan, V. (1978). *J. Consumer Res.*, **5**, 103–123.

[141] Green, P. E. and Wind, Y. (1973). *Multiattribute Decisions in Marketing*. Dryden, Hinsdale, Illinois.

[142] Guttman, L. (1968). *Psychometrika*, **33**, 469–506.

[143] Hauser, J. R. and Simmie, P. (1981). *Manag. Sci.*, **27**, 38–56.

[144] Hoffman, P. J., Slovic, P., and Rorer, L. G. (1968). *Psych. Bull.*, **69**, 338–349.

[145] Johnson, R. M. (1974). *J. Market. Res.*, **11**, 121–127.

[146] Kruskal, J. B. (1964). *Psychometrika*, **29**, 1–27.

[147] Kruskal, J. B. (1964). *Psychometrika*, **29**, 115–129.

[148] Kruskal, J. B. (1965). *J. R. Statist. Soc. B*, **27**, 251–263.

[149] Lingoes, J. C. (1972). In *Multidimensional Scaling*, Vol. 1, R. N. Shepard, A. K. Romney, and S. B. Nerlove, eds. Seminar Press, New York, pp. 52–68.

[150] Little, J. D. C. (1970). *Manag. Sci.* **16**, 466–485.

[151] Little, J. D. C. (1975). *Operat. Res.*, **23**, 628–655.

[152] Little, J. D. C. (1975). *Operat. Res.*, **23**, 656–673.

[153] Little, J. D. C. (1979). *Operat. Res.*, **27**, 629–667.

[154] Little, J. D. C. and Lodish, L. M. (1969). *Operat. Res.*, **17**, 1–35.

[155] Little, J. D. C. and Lodish, L. M. (1981). *J. Market.*, **45**, 24–29.

[156] Lodish, L. M. (1971). *Manag. Sci.*, **18**, Part II, 25–40.

[157] Luce, R. D. and Tukey, J. W. (1964). *J. Math. Psychol.*, **1**, 1–27.

[158] Parker, B. R. and Srinivasan, V. (1976). *Operat. Res.*, **24**, 991–1025.

[159] Pessemier, E. A. (1982). *Product Management: Strategy and Organization*, 2nd ed. Wiley, New York.

[160] Ramsay, J .O. (1977). *Psychometrika*, **42**, 241–266.

[161] Ramsay, J .O. (1978). *Psychometrika*, **43**, 145–160.

[162] Saaty, T. L. (1977). *J. Math. Psychol.*, **15**, 234–281.

[163] Shepard, R. N. (1962). *Psychometrika*, **27**, 125–140; 219–246.

[164] Shepard, R. N., Romney, A. K., and Nerlove, S. B. (1972). *Multidimensional Scaling*, Vol. II. Seminar Press, New York.

[165] Shocker, A. D. and Srinivasan, V. (1977). *J. Market. Res.*, **15**, 101–103.

[166] Takane, Y., Young, F. W., and deLeeuw, J. (1977). *Psychometrika*, **42**, 7–67.

[167] Torgenson, W. S. (1952). *Psychometrika*, **17**, 401–419.

[168] Torgenson, W. S. (1958). *Theory and Methods of Scaling*. Wiley, New York.

[169] Tucker, L. R. (1960). In *Psychological Scaling: Theory and Applications*, H. Gulliksen and Samuel Messick, eds. Wiley, New York.

[170] Urban, G. L. (1975). *Manag. Sci.*, **21**, 858–871.

[171] Urban, G. L. and Hauser, J. R. (1980). *Design and Marketing of New Products*. Prentice-Hall, Englewood Cliffs, New Jersey.

[172] Weinberg, S. L., Carroll, J. D., and Cohen, H. S. (1981). Estimating Confidence Regions for INDSCAL-Derived Group Stimulus Points Using Jackknifing. *Working Paper*, Bell Laboratories, Murray Hill, New Jersey.

[173] Wind, Y. and Saaty, T. L. (1980). *Manag. Sci.*, **26**, 641–658.

[174] Young, F. W. (1969). Polynomial Conjoint Analysis of Similarities: Definitions for a Special Algorithm. *Research Paper #76*, Psychometric Laboratory, University of North Carolina, Chapel Hill, North Carolina.

[175] Young, F. W. (1972). In *Multidimensional Scaling*, Vol. I, R. N. Shepard, A. K. Romney, and S. B. Nerlove, eds. Seminar Press, New York, pp. 69–104.

[176] Young, G. and Householder, A. S. (1938). *Psychometrika*, **3**, 19–22.

[177] Zufryden, F. S. (1979). In *Analytic Approaches to Product and Market Planning*, A. D. Shocker, ed. Marketing Science Institute, Cambridge, Mass., pp. 100–114.

EXPERIMENTATION

[178] Ackoff, R. L. and Emshoff, J. R. (1976). *Sloan Manag. Rev.*, **16** (Winter 1975), 1–15 and (Spring 1976), 1–15.

[179] Allaire, Y. (1975). *Sloan Manag. Rev.*, **16**, 91–94.

[180] Banks, S. (1965). *Experimentation in Marketing*. McGraw-Hill, New York.

[181] Charlton, P., Ehrenberg, A. S. C., and Pymont, B. (1972). *J. Market. Res. Soc.*, **14**, 171–183.

[182] Dhalla, N. K. (1977). *J. Advertising Res.*, **17**, 11–17.

[183] Ehrenberg, A. S. C. (1971). *J. Advertising Res.*, **11**, 3–10.

[184] Green, P. E., Carroll, J. D., and Carmone, F. J. (1978). *Res. Market.*, **1**, 99–122.

[185] Little, J. D. C. and Lodish, L. M. (1969). *Operat. Res.*, **17**, 1–35.

PROBABILITY MODELS

[186] Bass, F. M. (1974). *J. Market. Res.*, **11**, 1–20.

[187] Bass, F. M. (1980). *J. Business*, **53**, 551–567.

[188] Blattberg, R. C. (1981). In *Marketing Decision Models*, Schultz and Zoltners, eds. North-Holland, New York.

[189] Blattberg, R. C. and Jeuland, A. P. (1981). *Manag. Sci.*, **27**, 988–1005.

[190] Butler, D. H. (1976). In *Speaking of Hendry*, The Hendry Corp.

[191] Ehrenberg, A. S. C. (1959). *Appl. Statist.*, **8**, 26–41.

[192] Eskin, G. (1974). Causal Structures in Dynamic Trial-Repeat Forecasting Models. Talk presented at the 57th Internat. Marketing Conf., Montreal.

[193] Frank, R. E. (1962). In *Quantitative Techniques in Marketing Analysis*, R. A. Frank, A. A. Kuehn, and W. F. Massy (eds.). Irwin, Homewood, Illinois.

[194] Goodhardt, G. J. and Ehrenberg, A. S. C. (1967). *J. Market. Res.*, **4**, 155–162.

[195] Herniter, J. D. (1973). *J. Market. Res.*, **10**, 361–375.

[196] Horsky, D. (1977). *J. Market. Res.*, **14**, 10–21.

[197] Howard, R. A. (1963). *J. Advertising Res.*, **3**, 35–42.

[198] Howard, R. A. (1965). *J. Operat. Res. Soc. America*, **13**, 712–733.

[199] Jeuland, A. P. (1979). *Manag. Sci.*, **25**, 671–682.

[200] Jeuland, A. P., Bass, F. M., and Wright, G. P. (1980). *Operat. Res.*, **28**, 255–277.

[201] Jones, J. M. (1969). A Nonstationary Probability Diffusion Model of Consumer Brand Choice Behavior. *Working Paper No. 146*, Operations Research Division, Management Science Institute, University of California, Los Angeles, Calif.

[202] Jones, J. M. and Zufryden, F. S. (1980). *J. Market. Res.*, **17**, 323–334.

[203] Jones, J. M. and Zufryden, F. S. (1982). *J. Market.*, **46**, 36–46.

[204] Kalwani, M. U. and Morrison, D. G. (1977). *Manag. Sci.*, **23**, 467–477.

[205] Kuehn, A. A. (1962). In *Quantitative Techniques in Marketing Analysis*, R. E. Frank, A. A. Kuehn, and W. F. Massy, eds. Irwin, Homewood, Illinois.

[206] Kuehn, A. A. and Rohloff, A. C. (1967). In *Promotional Decisions Using Mathematical Models*, P. Robinson, ed., Allyn & Bacon, Boston, Mass.

[207] Lilien, G. L. (1974). *Manag. Sci.*, **20**, 536–539.

[208] Lipstein, F. (1965). *J. Market. Res.*, **2**, 259–265.

[209] MacLachlan, D. L. (1972). *J. Market. Res.*, **9**, 378–382.

[210] Massy, W. F. (1966). *J. Market. Res.*, **3**, 48–54.

[211] Massy, W. F. (1969). *J. Market. Res.*, **6**, 405–413.

[212] Massy, W. F., Montgomery, D. B., and Morrison, D. G. (1970). *Stochastic Models of Buying Behavior*, MIT Press, Cambridge, Mass.

[213] Montgomery, D. F. and Ryans, A. B. (1973). In *Consumer Behavior*, S. Ward and T. Robertson, eds. Prentice-Hall, Englewood Cliffs, New Jersey.

[214] Morrison, D. G. (1981). *J. Market.*, **45**, 111–119.

[215] Morrison, D. G. and Schmittlein, D. C. (1981). *Manag. Sci.*, **27**, 1006–1023.

[216] Parfitt, J. H. and Collins, B. J. K. (1968). *J. Market. Res.*, 131–145.

[217] Rubinson, J. R., Vanhonacker, W. R., and Bass, F. M. (1980). *Manag. Sci.*, **26**, 215–226.

[218] Sabavala, D. J. and Morrison, D. G. (1981). *Manag. Sci.*, **27**, 637–657.

[219] Schmittlein, D. C. (1982). Assessing Validity and Test–Retest Reliability for 'Pick *K* of *N*' Data. *Working Paper*, Wharton School, University of Pennsylvania, Philadelphia.

[220] Schmittlein, D. C., Bemmaor, A. C., and Morrison, D. G. (1982). The Beta Binomial Negative Binomial Distribution: Some New Results for a Combining of Two Old Models. *Working Paper*, Graduate School of Business, Columbia University, New York.

[221] Shoemaker, R. W. and Soaf, F. R. (1977). *J. Advertising Res.*, **17**, 47–53.

[222] Telser, L. G. (1962). *Rev. Econ. Statist.*, **44**, 300–324.

[223] Wildt, A. A. and McCann, J. M. (1980). *J. Market. Res.*, **17**, 335–340.

[224] Yankelovich, Skelly, and White (1981). In *New Product Forecasting*, Y. Wind, V. Mahajan, and R. Cardozo, eds. Lexington Books, Lexington, Massachusetts. (Article on LTM estimating procedures.)

(FORECASTING
MANAGEMENT SCIENCE, STATISTICS IN
MULTIDIMENSIONAL SCALING
MULTIVARIATE ANALYSIS
SURVEY SAMPLING)

PAUL E. GREEN
YORAM WIND

MARKOV, ANDREI ANDREEVICH

Born: June 2 (o.s.), 1856, in Ryazan, Russia.

Died: July 20, 1922, in Petrograd (formerly St. Petersburg, now Leningrad), U.S.S.R.

Contributed to: number theory, analysis, calculus of finite differences, probability theory, statistics.

Markov entered the fifth Petersburg gymnasium in 1866, and was a poor student in all but mathematics. He revealed even at this educational level a rather rebellious nature, which was to manifest itself in numerous clashes with the czarist regime and academic colleagues.

Entering Petersburg University in 1874 he attended classes given by P. L. Chebyshev*, A. N. Korkin, and E. I. Zolotarev. At the completion of his university studies at the physicomathematical faculty in 1878, he received a gold medal and was retained by the university to prepare for a career as an academic. His master's and doctoral theses were in areas other than probability, but with the departure of Chebyshev from the university in 1883 Markov taught the course in this subject, which he continued to do yearly.

Markov was first elected to the St. Petersburg Academy of Science in 1886 at the proposal of Chebyshev, attaining full membership in 1896. Retiring from the university in 1905, he continued to teach probability theory there, and seeking to find practical applications for this, his then prime interest, he participated from the beginning in deliberations on the running of the retirement fund of the Ministry of Justice. He also took a keen interest in the teaching of mathematics in high schools, coming into conflict yet again with his old academic adversary, P. A. Nekrasov, an earlier error of whose had stimulated Markov to give thought to the Weak Law of Large Numbers* (WLLN) for dependent random variables, and led indirectly to the probabilistic concept of Markov chains (*see* MARKOV PROCESSES).

Markov was a member of the St. Petersburg mathematical "school" founded by Chebyshev and, with Liapunov*, became the most eminent of Chebyshev's disciples in probability. Their papers placed probability theory on the level of an exact mathematical science. Markov's doctoral dissertation, "On Some Applications of Algebraic Continued Fractions," results from which were published in 1884 in several journals, already had implicit connections with probability theory in that he proved and generalized certain inequalities of Chebyshev [11] pertaining to the theory of moments, and in turn based on notions of Bienaymé*. The method of moments* had been the tool used by Chebyshev in his development of the proof of a Central Limit Theorem* for not necessarily identically (but independently) distributed summands. Unlike Liapunov's later approach to this problem, by characteristic functions*, Markov's continued to be focused on the method of moments; it is often said that he was Chebyshev's closest student and best expositor of his teacher's ideas.

The stream of Markov's work in probability was initially, in fact, motivated by Chebyshev's [12] treatment of the Central Limit problem. In letters [2] to A. V. Vasiliev he notes that the significance of Chebyshev's result is obscured by the complexity and insufficient rigor of proofs, a situation he desires to correct. Chebyshev had asserted the Central Limit Theorem held for random variables $U_1, U_2, \dots,$ each described by a density, if (i) $EU_i = 0$, each i; and (ii) $|E\{U_i^k\}| \leqslant C$ for all i and positive integer $k \geqslant 2$ where C is a constant independent of i and k. In the letters Markov states that a further condition needs to be added to make the theorem correct, and suggests this can be taken to be: (iii) $(\sum_{i=1}^n \operatorname{Var} U_i)/n$ is uniformly bounded from zero in n. In the very last part of his actual correction [3] of Chebyshev's theorem, he replaces (iii) by the stronger (iiia): EU_n^2 is bounded from zero as $n \to \infty$. Liapunov's Theorem (see LIAPUNOV) published in 1901 differs not only in its approach (by characteristic functions) but in its much greater level of generality. This evidently led Markov to wonder whether the method of moments might not be suitably adapted to give the same result, and he finally achieved this in 1913 in the third edition of *The Calculus of Probabilities* [5]. The central idea is that of truncation* of random variables, much used in modern probability theory: For a random variable U_k, put $U_k' = U_k$ if $|U_k| < N$ and $U_k' = 0$ if $|U_k| \geqslant N$. The random variables $U_k', k = 1, 2, \dots,$ being confined to a finite interval, are easily tractable. The effects of the truncation can be eliminated by judicious manipulation as $N \to \infty$.

Markov's oustanding contribution to probability theory was the introduction of the concept of a Markov chain as one model for the study of dependent random variables, which is at the heart of a vast amount of research in the theory of stochastic processes*. Markov chains first appear in his writings in 1906 [4], in a paper which is primarily concerned with extending the WLLN to sums of dependent random variables. The motivation for this was twofold: First, to show that Chebyshev's approach to this problem, via the celebrated Bienaymé–Chebyshev moment Inequality (*see* CHEBYSHEV INEQUALITY) but only for independent random variables, could be taken further. The second motivation is less well known but the more relevant. P. A. Nekrasov in 1902 had noticed that "pairwise independence" (i.e., random variables uncorrelated pairwise) would yield the WLLN by Chebyshev's reasoning, but claimed erroneously, on mystical grounds, that this was not only sufficient but necessary for the WLLN to hold [7, 9]. Markov's deductions (to the contrary) are based on the inequality in the form

$$\Pr\{|S_n - ES_n|/n \geqslant \epsilon\} \leqslant \frac{\operatorname{Var} S_n}{(n\epsilon)^2} = \frac{1}{\epsilon^2}\left\{\frac{\sum_{i=1}^n \operatorname{Var} X_i}{n^2} + \frac{2}{n^2}\sum_{i<j}\operatorname{Cov}(X_i, X_j)\right\},$$

which motivated further work on the WLLN by Chuprov* and Slutsky* in a similar vein. Although Markov's interest in sequences of random variables forming a Markov chain was largely confined to investigating the WLLN and the Central Limit Theorem, the paper [4] nevertheless also contains an elegant treatment of the crucial averaging effect of a finite row-stochastic matrix $\mathbf{P}$ (applied to a vector $\mathbf{w}$ to produce a vector $\mathbf{z}$: $\mathbf{z} = \mathbf{P}\mathbf{w}$) at the heart of the ergodicity theory of finite inhomogeneous Markov chains [10], which Markov also uses to establish ergodicity for a finite homogeneous chain (*see* ERGODIC THEOREMS).

Markov's statistical work in part consisted of modelling the alternation of vowels and consonants in several Russian literary works by a two-state Markov chain, and his work in dispersion theory, which significantly influenced Chuprov [7]. An account of most aspects of his work may be found in his textbook [5], which was outstanding in its time, and served to influence figures such as S. N. Bernstein*, V. I. Romanovsky*, and Jerzy Neyman*—and through him the evolution of mathematical statistics. Markov was also interested, as regards statistics, again through the influence of Chebyshev, in the classical Linear Model (*see* GENERAL LINEAR MODEL) which he treated in his textbook in editions beginning in 1900, allowing different variances for the independently distributed residuals. The misnamed Gauss–Markov Theorem* appears to emanate from this source through the agency of Neyman [8]; there is little originality in Markov's treatment of the topic. The Markov inequality*, $\Pr\{X \geqslant \epsilon\} \leqslant (EX)/\epsilon$ for a random variable $X \geqslant 0$ and $\epsilon > 0$, first occurs in the third edition (1913) of the textbook.

References

[1] Maistrov, L. E. (1974). *Probability Theory. A Historical Sketch*. Academic Press, New York. (Translated and edited from the Russian-language work of 1967 by S. Kotz. Contains an extensive account of the work of Chebyshev, Markov, and Liapunov.)

[2] Markov, A. A. (1898). *Izv. Fiz.-Matem. Obsch. Kazan Univ.*, (2 ser.) **8**, 110–128. (In Russian. Also in ref. 6, pp. 231–251.)

[3] Markov, A. A. (1898). *Izv. Akad. Nauk S.-P.B.* (Ser 5.) **9**, 435–446. (Also in ref. 6, pp. 253–269.)

[4] Markov, A. A. (1906). *Izv. Fiz.-Matem. Obsch. Kazan Univ.*, (2 ser.) **15**, 135–156. (In Russian. Also in ref 6, pp. 339–361.)

[5] Markov, A. A. (1924). *Ischislenie Veroiatnostei* (4th, posthumous, ed.). Gosizdat, Moscow. (Markov's textbook: *The Calculus of Probabilities*.)

[6] Markov, A. A. (1951). *Izbrannie Trudy*. ANSSSR, Leningrad. [Selected works on number theory and the theory of probability. Pp. 319–338 contain Markov's proof, taken from the 3rd edition of ref. 5, of Liapunov's Theorem. There is a biography (with two photographs) by Markov's son, and commentaries by Yu. V. Linnik, N. A. Sapogov, and V. N. Timofeev, as well as a complete bibliography.]

[7] Ondar, Kh. O. (Ed.) (1981). *Correspondence Between Markov and Chuprov*. Springer-Verlag, New York. (An English translation by Charles and Margaret Stein, with an Introduction by Jerzy Neyman, of: *O Teorii Veroiatnostei i Matematicheskoi Statistike*. Nauka, Moscow.)

[8] Seal, H. L. (1967). *Biometrika* **54**, 1–24.

[9] Seneta, E. (1979). *Austral. J. Statist.*, **21**, 209–220. (Sect. 5 outlines less well-known aspects of work on the WLLN and Central Limit problem in prerevolutionary Russia.)

[10] Seneta, E. (1981). *Non-negative Matrices and Markov Chains*. (2nd ed.) Springer-Verlag, New York, pp. 80–91.

[11] Tchébichef, P. L. (1874). *Liouville's J. Math. Pures Appl.*, (2) **19**, 157–160. [Also in his *Oeuvres*. A. Markov and N. Sonin (Eds.) 2 Vols. Chelsea House, New York.]

[12] Tchébichef, P. L. (1887). Supplement to: *Zapiski Imp. Akad. Nauk* (S.P.-B.), **55**, No. 6. [Also in *Acta Math.*, **14** (1890–1891), pp. 305–315, and in his *Oeuvres*—see ref. 11.]

[13] Uspensky, J. V. (1937). *Introduction to Mathematical Probability*. McGraw-Hill, New York. (Contains an English-language discussion of less well-known contributions by Markov.)

(CHEBYSHEV, PAFNUTY LVOVICH
DISPERSION THEORY
LAWS OF LARGE NUMBERS
LIMIT THEOREM, CENTRAL
LIAPUNOV, ALEXANDER MIKHAILOVICH
MARKOV INEQUALITY
MARKOV PROCESSES)

E. SENETA

MARKOV DECISION PROCESSES

A FRAMEWORK FOR MARKOV DECISION PROCESSES

A Markov decision process is a Markov process* in which the transition probabilities take the form p_{ij}^k, dependent on action k taken. If the problem has a discrete set of *states*, S, and for each $i \in S$, a feasible set of *actions* $K(i)$ exists, p_{ij}^k is the probability that if the system is in state i, and takes action $k \in K(i)$, the next state will be j. In addition, given $i \in S$ and $k \in K(i)$, there exists an *immediate reward* r_i^k. A *decision rule* is a function from S to $\bigcup_{i \in S} K(i)$, i.e., $k = \delta(i)$ specifies which action $k \in K(i)$ to take for each state $i \in S$. Δ is the set of all decision rules. We ignore any history-dependent rules since the process is Markov, at least for scalar objective problems. A *policy* π is an infinite sequence of decision rules $(\delta_1, \delta_2, \dots, \delta_n, \dots)$, $\delta_n \in \Delta, \forall n$. Π is the set of all policies. We ignore randomized policies, at least for scalar problems. For each $\pi \in \Pi$, there will be a *reward vector*, $\mathbf{v}^\pi$, which depends on the measure of performance used.

When evaluating the performance of a decision rule over a time horizon, account must be taken of the possible different levels of importance of unit returns at different points in time. One way of doing this is to discount a unit return in interval n by a factor ρ^{n-1} to give an equivalent return at time 0, with $\rho < 1$. For such situations $\mathbf{v}^\pi$ will be given by

$$\mathbf{v}^\pi = \lim_{N \to \infty} \sum_{n=1}^{N} \rho^{n-1} \left(\prod_{t=0}^{n-1} \mathbf{P}(\delta_t) \right) \mathbf{r}(\delta_n),$$

where $\mathbf{P}(\delta_{t+1})$ is the transition matrix, $\mathbf{P}(\delta_0)$ is the identity matrix, and $\mathbf{r}(\delta_{t+1})$ the reward vector given δ_{t+1}.

If discounting is not used, then we may be interested in the average return per unit time. In this case the reward vector is given by

$$\lim_{N \to \infty} \sum_{n=1}^{N} \left(\prod_{t=0}^{n-1} \mathbf{P}(\delta_t) \right) \mathbf{r}(\delta_n)/N.$$

$\mathbf{P}$ and $\mathbf{r}$ may also depend upon t, but we ignore this for purposes of simplicity.

Note that $\mathbf{v}^\pi$ and $\mathbf{g}^\pi$ are vectors, each of whose components give the appropriate level of performance over the infinite time horizon for each state in which the process commences.

For many problems, even if infinite valued, $\mathbf{v}^\pi$ and $\mathbf{g}^\pi$ will exist, e.g., S and $K(i)$ finite $\forall i \in S$. If decisions are taken at the beginning of unit time intervals, $\mathbf{v}^\pi$ and $\mathbf{g}^\pi$ will give, respectively, the infinite horizon expected discounted return and the infinite horizon expected average return per unit time. If the process involves varying intervals between decisions (e.g., semi-Markov* decision processes), modifications may be made. It is also possible to consider total expected rewards, even with $\rho = 1$.

Let us illustrate these ideas with a simple example. We will consider the average return case only. The data in Table 1 describe a stock control problem in which the state i is the stock level at the beginning of a period when the decision k to be taken determines how much extra stock we might purchase, restricting $i + k$ to a maximal value of 4.

A typical decision rule δ might be as follows:

$$\delta(0) = 0, \quad \delta(1) = 0, \quad \delta(2) = 0, \quad \delta(3) = 0, \quad \delta(4) = 0.$$

For this decision rule the immediate reward vector is given by

$$\mathbf{r}(\delta) = (5.0, 5.5, 6.5, 7.0, 6.0).$$

Table 1

		j					
i	k	0	1	2	3	4	
0	0	1.0	0.0	0.0	0.0	0.0	5.0
	1	1.0	0.0	0.0	0.0	0.0	2.5
	2	0.7	0.3	0.0	0.0	0.0	3.5
	3	0.3	0.4	0.3	0.0	0.0	4.0
	4	0.0	0.3	0.4	0.3	0.0	3.0
1	0	1.0	0.0	0.0	0.0	0.0	5.5
	1	0.7	0.3	0.0	0.0	0.0	3.5
	2	0.3	0.4	0.3	0.0	0.0	4.0
	3	0.0	0.3	0.4	0.3	0.0	3.0
2	0	0.7	0.3	0.0	0.0	0.0	6.5
	1	0.3	0.4	0.3	0.0	0.0	4.0
	2	0.0	0.3	0.4	0.3	0.0	3.0
3	0	0.3	0.4	0.3	0.0	0.0	7.0
	1	0.0	0.3	0.4	0.3	0.0	3.0
4	0	0.0	0.3	0.4	0.3	0.0	6.0
				p_{ij}^k			r_{ij}^k

$r_i(\delta)$ is the immediate reward in state i if we use decision rule δ, i.e., $r_i(\delta) = r_i^{\delta(i)}$.

The transition matrix $\mathbf{P}(\delta)$ is as follows:

		j				
		0	1	2	3	4
	0	1.0	0.0	0.0	0.0	0.0
	1	1.0	0.0	0.0	0.0	0.0
i	2	0.7	0.3	0.0	0.0	0.0
	3	0.3	0.4	0.3	0.0	0.0
	4	0.0	0.3	0.4	0.3	0.0

If policy π is to use decision rule δ repeatedly, then

$$\mathbf{g}^\pi = (5.0, 5.0, 5.0, 5.0, 5.0)$$

i.e., the return per unit time is equal to 5 independently of the commencing state of the process. For this problem it so happens that the specified π is also optimal.

The objective in Markov decision processes is to find optimal policies in accordance with the definitions of optimality used. For the two examples given, π^* is optimal for $\mathbf{v}^\pi$ if $\mathbf{v}^{\pi^*} \geqslant \mathbf{v}^\pi$, $\forall \pi \in \Pi$, and for $\mathbf{g}^\pi$ if $\mathbf{g}^{\pi^*} \geqslant \mathbf{g}^\pi$, $\forall \pi \in \Pi$. Note that these are vector inequalities—i.e., π^* is a best policy $\forall i \in S$.

The key questions are: Do optimal policies exist, and do optimal policies exist which are stationary, i.e., for which $\delta_n = \delta$, $\forall n$ and some $\delta \in \Delta$? In general, the answer to these questions is no; for many problems the answer is yes, e.g., when S and $K(i), \forall i \in S$, are finite. In such cases it is known that the solutions satisfy the *optimality equations* in vector form which, for the discounted problem, take the form

$$\mathbf{v} = \mathbf{H}\mathbf{v} = \max_{\delta \in \Delta} \mathbf{H}^\delta \mathbf{v},$$

$$[\mathbf{H}^\delta \mathbf{v}]_i = r_i^{\delta(i)} + \rho \sum_{j \in S} p_{ij}^{\delta(i)} v_j .$$

For the average return case, the optimality equation depends on the structure of the transition matrices. If each transition matrix, $\mathbf{P}(\delta)$, $\delta \in \Delta$, has a single ergodic set (there may be transient states) then, for the optimal $\mathbf{g}$, each component g_i has the same value $g, \forall i \in S$, and is the unique solution to the equation

$$\mathbf{w} + g\mathbf{e} = \mathbf{H}\mathbf{w}$$

where $\mathbf{e} = (1, 1, \ldots 1)' \in R^m$ (if S has m states), $\mathbf{w}$ is a bias term, and g is now the common scalar component of the vector $\mathbf{g}$. For situations where more than one ergodic set can arise, modifications exist.

The simplest case is the one in which S and $K(i), \forall i \in S$, are finite. However, much work has been done for cases involving infinite sets and unbounded rewards, with the same objectives, broadly speaking, e.g., the use of Borel sets and bounded Baire functions.

There are many more facets of Markov decision processes in addition to the fundamental questions so far raised. A central one is the question of algorithms for solving the problems, specifically when the appropriate optimality equations are satisfied.

ALGORITHMS FOR MARKOV DECISION PROCESSES

The natural one for the discounted problem is the *successive approximations method*,

$$\mathbf{v}(n) = \mathbf{H}\mathbf{v}(n-1), \quad n \geqslant 1, \quad \mathbf{v}(0) \text{ arbitrary.}$$

Convergence to the requisite solution is assured in many cases and the δ which gives $\mathbf{H}\mathbf{v} = \mathbf{H}^\delta \mathbf{v}$ is optimal. Since in practice the computations have to be terminated after a finite number of steps, it is important to know how to use the finite iteration result to get approximately optimal solutions. The simplest result is

$$\mathbf{v}(n) - \sigma_n \mathbf{e} \leqslant \mathbf{v} \leqslant \mathbf{v}(n) + \sigma_n \mathbf{e},$$

$$\sigma_n = \rho^n \max_{i,k \in K(i)} |r_i^k| / (1 - \rho).$$

This gives bounds on $\mathbf{v}$, but we need to determine an approximately optimal δ. It may be shown that if $\mathbf{u} \in R^m$, with $\|\mathbf{u} - \mathbf{v}\| \leqslant \epsilon$ ($\|\ \|$ is the maximum norm), then $\mathbf{H}^\delta \mathbf{u} = \mathbf{H}\mathbf{u}$ implies that $\|\mathbf{v} - \mathbf{v}^\delta\| \leqslant 2\rho\epsilon/(1-\rho)$, where $\mathbf{v}^{\delta}$ represents $\mathbf{v}^{\pi}$, $\pi = (\delta, \delta, \ldots, \delta, \ldots)$. Thus δ gives a value close to $\mathbf{v}$ if $2\rho\epsilon/(1-\rho)$ is small.

Many other results exist, some better, and some results also exist for the average cost case. Problems with infinite state sets present special problems. One procedure for handling these is as follows:

$$\mathbf{v}(n) = \mathbf{H}(n)\mathbf{v}(n),$$

where $\mathbf{H}(n)$ is $\mathbf{H}$ restricted to the first n states, and $\mathbf{v}(n) \in R^n$.

An alternative to successive approximation is the *policy space method*, which for the discounted problem takes the form

$$\mathbf{v}(n) = \mathbf{H}^{\delta(n)}\mathbf{v}(n),$$

$$\mathbf{H}^{\delta(n)}\mathbf{v}(n-1) = \mathbf{H}\mathbf{v}(n-1),$$

$$\mathbf{v}(0) \text{ arbitrary.}$$

Other computational procedures involve the use of computational data already produced to eliminate actions from further consideration. Thus if $\mathbf{u}$, $\mathbf{l}$, are temporal upper and lower bounds on $\mathbf{v}$, and if, in the discounted problem,

$$r_i^k + \rho \sum_{j \in S} p_{ij}^k u_j < [\mathbf{H}\mathbf{l}]_i,$$

then k is nonoptimal for i and may be eliminated.

Alternative approaches make use of the equivalence of the Markov decision problem to a linear programming* problem. For the discounted case this is to minimize $\sum_{i \in S} \lambda_i v_i$ subject to

$$v_i \geqslant r_i^k + \rho \sum_{j \in S} p_{ij}^k v_j,$$

$\forall i, k \in K(i)$, where $\lambda_i > 0, \forall i$. This may be dualized and the dual variables x_i^k interpreted in terms of the probability of taking actions k and being in state i.

ADDITIONAL FEATURES OF MARKOV DECISION PROCESSES

With the considerable problems involved in solving large-scale Markov decision processes, attention has been given to finding structural properties of optimal solutions. The basic idea is that an ordering of the states may be reflected in an ordering of the optimal actions, e.g., in queueing* problems, larger queue sizes may require faster service rates.

Finally, other features of Markov decision processes which have been examined include:

(i) partially observable states, where information about states results in a probability distribution of the states and the use of Bayesian procedures;

(ii) multi-objective Markov decision processes in which $\mathbf{r}_i^k \in R^q$, $\forall i \in S$, $k \in K(i)$, and hence, for example for the discounted problem, $\mathbf{v} \in R^{m \times q}$;

(iii) Markov games in which several decision makers can make decisions and maximization is now replaced by max. min. operators.

No papers have been cited but the following texts are useful for their treatment of Markov decision processes from various points of view and for the wealth of references they inevitably include. Howard's book [3] accelerated the interest in Markov decision processes and is the best text for the beginner. The books by Derman [1], Mine and Osaki [5], and White [8] are intermediate textbooks dealing with finite state sets with slightly different emphasis. The book by Van Nunen [7] is useful for its emphasis on contraction processes fundamental to the area. Martin's book [4] deals with Bayesian adaptation of transition matrices. Gröenewegen [2] deals essentially with Markov games. Van der Wal's text [6] is much more advanced and deals with some of the trickier problems which arise when we move away from finite state set, finite action set problems. The remaining books are conference proceedings containing a considerable variety of material and references.

References

[1] Derman, C. (1970). *Finite State Markovian Decision Processes*. Academic Press, New York.

[2] Gröenewegen, L. P. J. (1981). *Characterisation of Optimal Strategies in Dynamic Games*. Mathematical Centre Tracts **90**, Mathematisch Centrum, Amsterdam.

[3] Howard, R. A. (1960). *Dynamic Programming and Markov Processes*. Wiley, New York.

[4] Martin, J. J. (1967). *Bayesian Decision Problems and Markov Chains*. Wiley, New York.

[5] Mine, H. and Osaki, S. (1970). *Markovian Decision Processes*. Elsevier, New York.

[6] Van der Wal, J. (1981). *Stochastic Dynamic Programming*. Mathematical Centre Tracts **139**, Mathematisch Centrum, Amsterdam.

[7] Van Nunen, J. A. E. E. (1976). *Contracting Markov Decision Processes*. Mathematical Centre Tracts **71**, Mathematisch Centrum, Amsterdam.

[8] White, D. J. (1978). *Finite Dynamic Programming*. Wiley, New York.

Bibliography

Hartley, R., Thomas, L. C., and White, D. J. (1980). *Recent Developments in Markov Decision Processes*. Academic Press, New York.

Puterman, M. (1978). *Dynamic Programming and Its Applications*. Academic Press, New York.

Tijms, H. C. and Wessels, J. (1977). *Markov Decision Theory*. Mathematical Centre Tracts **93**, Mathematisch Centrum, Amsterdam.

(DECISION THEORY
LINEAR PROGRAMMING
MARKOV PROCESSES)

D. J. WHITE

MARKOVIAN DEPENDENCE

A stochastic process $\{X_n : n = 1, 2, \dots\}$ is a *Markov process** if the conditional distribution of X_n given the entire past depends only on the last observation; for each x,

$$P[X_n \leqslant x \mid X_1, \dots, X_{n-1}] = P[X_n \leqslant x \mid X_{n-1}]$$

(this holding with probability one). If the X_n are independent then $\{X_n\}$ is Markov. As another example, consider a Markov process $\{X_n\}$ where each X_n takes the values 0 or 1, and assume $p_{ij} = P[X_n = j \mid X_{n-1} = i]$ and $p = P[X_n = 1]$ are the same for all n. It is easy to verify that (p_{11}, p) is the parameter for this model and that the X_n are independent if and only if $p_{11} = p$. Considering this model, David [3] proposes a test for the hypothesis of independence against the alternative of positive dependence*. Positive dependence means that the X_n's tend to remain in the same state, $p_{11} > p$ (and then $p_{00} < 1 - p$). The test statistic involves the number of runs*. Recall the notion of runs of ones and zeros; if, for example, in a sample of size 8 we observe 00101010 we say there are four runs of zeros, and if we observe 00011110 we say there are two runs of zeros. Given the number of ones in the sample, positive dependence suggests a clustering of ones and zeros, so that the number of runs is small. David's test for independence against positive dependence rejects the hypothesis if the number of runs, conditioned on the number of ones, is small. Lehmann [9] proves that the runs test introduced by David coincides with the most powerful similar test* at its natural significance levels. Klotz [8] gives an estimate of (p_{11}, p) related to the maximum likelihood estimate* and proves its asymptotic efficiency.*.

The runs* test and its generalizations (see Goodman [5] and Barton and David [1]) have a natural role in tests for Markovian dependence. Swed and Eisenhart [12] have tabled the exact distribution of the number of runs of ones and zeros. Tables for runs involving more than two types of elements are given in Shaughnessy [11]. Approximations to the distribution of the number of runs can be made using the hypergeometric distribution* or the normal distribution*.

Markovian dependence for integer-valued processes is often studied with a higher-order Markov dependence, defined as follows. Fix an integer $k \geqslant 1$ and assume the integer-valued process $\{X_n\}$ satisfies for all $n \geqslant k + 1$,

$$P[X_n = i \mid X_1 = j_1, \dots, X_{n-1} = j_{n-1}] = P[X_n = i \mid X_{n-k} = j_{n-k}, \dots, X_{n-1} = j_{n-1}],$$

with both sides not depending on n (the $i, j_1, \dots, j_{n-1}$ run through the possible values of the X_n). We call the right side of the equality the *stationary transition function* and say that the *stationary dependence extends k time units*. Processes satisfying this assumption are *higher-order Markov chains with stationary transitions*, or *Markov chains of order k*. (Higher-order Markovian dependence can be defined for processes other than chains.) It is convenient to call a process of independent identically distributed X_n a *Markov chain of order 0*.

A Markov chain of order $k \geqslant 2$ can be identified with a Markov chain of order 1, as follows: If $\{X_n\}$ is the chain of order k, let $Y_n = (X_n, X_{n-1}, \dots, X_{n-k+1}), n \geqslant k$. Then $\{Y_n\}$ is a Markov chain of order 1. Terminology about first-order chains thus carries over to higher-order chains and some problems about higher-order dependence reduce to problems of first-order dependence. Research has focused on determining the transition functions and the order k, on the basis of a sample $(x_1, \dots, x_N)$ taken from the chain.

Consider estimating the order of a higher-order Markov chain. Assume the following:

(i) each X_n takes only a finite number s of values;

(ii) the unknown order of the chain is less than a known integer m;

(iii) for $k < m$, if $\{X_n\}$ is a chain of order k and if $\{Y_n\}$ is the corresponding first-order chain, then $\{Y_n\}$ is irreducible and aperiodic.

For each $k = 0, \dots, m-1$, let $\lambda(k, m)$ be the likelihood ratio* for order k against order m. Letting

$$B(k) = -2 \ln \lambda(k, m) - (s^m - s^k)(s-1)\ln N,$$

$$B = \min(B(k) : k = 0, \dots, m-1),$$

Katz [7] proves that B is a consistent estimate of the order of the chain.

Hypothesis testing of the order of Markovian dependence involves sufficient statistics*. To obtain a sufficient statistic assume that $\{X_n\}$ is a chain of order 1 with s states, and let $(x_1, \dots, x_N), N \geqslant 2$, be a sample. With each pair of integers $1 \leqslant i, j \leqslant s$, let $f_N(i, j) =$ the number of integers $m, 1 \leqslant m \leqslant N-1$, for which $x_m = i$ and $x_{m+1} = j$. Then the vector with the $s^2 + 1$ components, $(x_1, f_N(i, j) : 1 \leqslant i, j \leqslant s)$, is a sufficient statistic. By identifying a chain of order $k > 1$ with a chain of order 1 we obtain a sufficient statistic for higher order chains. Whittle [14] obtains a formula for the distribution of the sufficient statistic (see also Billingsley [2]).

To obtain asymptotic tests of hypotheses about the order of a chain, Goodman [6] derives exact expressions for the conditional distribution of a sufficient statistic for a chain of order k, given a sufficient statistic for a chain of order $h < k$. These sufficient statistics are the higher-order analogues of the sufficient statistic for a chain of order 1. Goodman proves that the conditional distribution is asymptotically equivalent to a product of probabilities obtained from contingency tables* (the contingency tables have a natural interpretation) and obtains tests of the hypotheses of the order of the chain. Denny and Wright [4] prove that when x_1 is fixed the above sufficient statistics are complete (*see* COMPLETENESS), and then find a general form for the class of all admissible tests of the hypothesis of order h against order $k > h$ (*see* ADMISSIBILITY). These tests also involve contingency tables. On the other hand, if $\{X_n\}$ is stationary, so that x_1 is not fixed, and if the number of states exceeds two, then there does not exist a complete sufficient statistic for the chain [15]. Chi-square tests* for the order of a chain are discussed in Billingsley [2].

Research workers in the physical and social sciences have been interested in the order of Markov chains. Tong [13] describes some estimates in meteorological problems, his estimates depending on a cost function related to Kullback–Leibler information*. Tong does not claim consistency* of his estimates, and indeed, Katz [7] proves that Tong's estimates of the order of a chain are not consistent. Niederhoffer and Osborne [10] test the hypothesis that common stock transaction data is statistically independent, transaction data being the record of all trades in the stock. The hypothesis of independence is rejected in favor of Markov orders 1 and 2, at small critical values. The test statistic uses contingency tables with a chi-square approximation and is approximately equal to the test statistic for an admissible test for independence [4].

References

[1] Barton, D. E. and David, F. N. (1957). *Biometrika*, **44**, 168–177.

[2] Billingsley, P. (1961). *Ann. Math. Statist.*, **32**, 12–40. (Survey article with extensive bibliography.)

[3] David, F. N. (1947). *Biometrika*, **34**, 335–339.

[4] Denny, J. L. and Wright, A. L. (1978). *Zeit. Wahrsch. verw. Gebiete*, **43**, 331–338.

[5] Goodman, L. A. (1958). *Biometrika*, **45**, 181–197.

[6] Goodman, L. A. (1958). *Ann. Math. Statist.*, **29**, 476–490.

[7] Katz, R. W. (1981). *Technometrics*, **23**, 243–249.

[8] Klotz, J. (1973). *Ann. Statist.*, **1**, 373–379.

[9] Lehmann, E. L. (1959). *Testing Statistical Hypotheses*. Wiley, New York, pp. 155–156.

[10] Niederhoffer, V. and Osborne, M. F. M. (1966). *J. Amer. Statist. Ass.*, **61**, 897–916.

[11] Shaughnessy, P. W. (1981). *J. Amer. Statist. Ass.*, **76**, 732–736.

[12] Swed, F. and Eisenhart, C. (1943). *Ann. Math. Statist.*, **14**, 66–87.

[13] Tong, H. (1975). *J. Appl. Prob.*, **12**, 488–497.

[14] Whittle, P. (1955). *J. R. Statist. Soc. B*, **17**, 235–242.

[15] Wright, A. L. (1980). *Ann. Inst. Statist. Math.*, **32**, 95–97.

(ADMISSIBILITY
ASYMPTOTIC EFFICIENCY
CHI-SQUARE TESTS
CONSISTENT
CONTINGENCY TABLES
KULLBACK INFORMATION
MARKOV PROCESSES
MAXIMUM LIKELIHOOD ESTIMATION
RUNS
SIGNIFICANCE LEVEL
SUFFICIENT STATISTICS)

JOHN L. DENNY

MARKOV INEQUALITY

This result first appeared in the 1913 edition of Markov's book *The Calculus of Probabilities* [3].

Let Y be a non-negative random variable, so that $\Pr(Y \geqslant 0) = 1$ and suppose that $E(Y) = \mu, < \infty$, where E denotes expected value*. Suppose further that $E(Y^r) = \mu_r' < \infty$ for some integer $r > 1$; and let X be any random variable such that $E|X| = \nu < \infty$.

The Markov inequality takes several forms:

$$\Pr(Y \geqslant A\mu) \leqslant 1/A, \qquad A > 0; \quad (1a)$$

$$\Pr(Y \geqslant a) \leqslant \mu/a, \qquad a > 0; \quad (1b)$$

$$\Pr(Y \geqslant A\mu_r') \leqslant 1/A^r, \qquad A > 0; \quad (2a)$$

$$\Pr(Y \geqslant a) \leqslant \mu_r'/a^r, \qquad a > 0; \quad (2b)$$

$$\Pr(|X| \geqslant a) \leqslant \nu/a; \quad (3)$$

$$\Pr(Y - \mu \geqslant h) \leqslant \mu/(\mu + h), \qquad h > 0. \quad (4)$$

All of the preceding results can be expressed using complementary events. For example, (1b) is equivalent to

$$\Pr(Y < a) \leqslant 1 - (\mu/a).$$

The Markov inequality cannot be improved upon, in the sense that equality holds in (4), for example, if $Y = 0$ with probability $h/(\mu + h)$, and $Y = \mu + h$ with probability $\mu/(\mu + h)$. In 1874 Chebyshev* had proved essentially a one-sided inequality [1]

$$\Pr(X - E(X) \geqslant h) \leqslant \sigma^2/(\sigma^2 + h^2), \quad h > 0, \quad (5)$$

where X, however, is *any* random variable with finite variance σ^2.

If $Y_1, Y_2, \ldots, Y_n$ are non-negative independent random variables such that $\overline{Y} = n^{-1}\sum_{i=1}^n Y_i$ and $E(\overline{Y}) = \mu$, then (4) holds in the sense that

$$\Pr(\overline{Y} - \mu \geqslant h) \leqslant \mu/(\mu + h), \qquad h > 0; \quad (6)$$

this holds with equality if $Y_1 = 0$ with probability $h/(\mu + h)$, $Y_1 = n(\mu + h)$ with probability $\mu/(\mu + h)$, and $Y_2 = \ldots = Y_n = 0$ identically. Hoeffding [2] improved these results when $Y_1, \ldots, Y_n$ are bounded above as well as below; *see also* PROBABILITY IN-

EQUALITIES FOR SUMS OF BOUNDED RANDOM VARIABLES.

Chebyshev's inequality* follows from (2a) if $Y = |X - E(X)|$ and $r = 2$; Bernstein's inequality* in the form

$$\Pr(S \geqslant b) \leqslant e^{-bt}M(t), \qquad b > 0, \quad (7)$$

where $M(t)$ is the moment generating function* of S, follows from (2b) if $Y = e^S, r = t$, and $a = e^b$.

References

[1] Chebyshev, P. L. (1874). *J. Math. Pures Appliquées, Ser. 2*, **19**, 157–160.

[2] Hoeffding, W. (1963). *J. Amer. Statist. Ass.*, **58**, 13–30.

[3] Markov, A. A. (1913). *Ischislenie Veroiatnostei*, 3rd ed. Gosizdat, Moscow. (This is Markov's textbook, *The Calculus of Probabilities*.)

(BERNSTEIN'S INEQUALITY
CHEBYSHEV'S INEQUALITY
MARKOV, ANDREI ANDREEVICH
PROBABILITY INEQUALITIES FOR SUMS OF BOUNDED RANDOM VARIABLES)

CAMPBELL B. READ

MARKOV PROCESSES

A family or a collection of random variables indexed by a parameter is known as a stochastic process (or "random" or "chance" process). While defining stochastic processes* to represent random phenomena, time and space are the common index parameters used. Let S and T be the state space and the parameter space of the process respectively. Let $\{X(t),\ t \in T\}$ be the stochastic process defined on a probability space in the usual manner, taking values in the state space S. It is a *Markov process* under the following conditions.

Consider a finite (or countably infinite) set of points $(t_0, t_1, \dots t_n, t)$, $t_0 < t_1 < t_2, \dots < t_n < t$ and $t, t_r \in T$ $(r = 0, 1, 2, \dots n)$. The process $\{X(t),\ t \in T\}$ is a Markov process if for all sets of points $(t_0, t_1, \dots t_n, t)$ in T

$$P[X(t) \leqslant x \mid X(t_n) = x_n, X(t_{n-1}) = x_{n-1}, \dots, X(t_0) = x_0] = P[X(t) \leqslant x \mid X(t_n) = x_n]. \qquad (1)$$

This means that the conditional distribution of $X(t)$ for given values of $X(t_0), X(t_1), \dots X(t_n)$ depends only on $X(t_n)$, the most recent known value of the process.

Depending on whether S and T are continuous or discrete we may identify four distinct types of processes (see Note at the end of CHUNG PROCESSES). Some authors (e.g., Kolmogorov, Feller; also see Bailey [5], Cox and Miller [21], and Prabhu [75]) call a Markov process with a discrete parameter space a *Markov chain* and some others (e.g., Doob, Chung; also see Bartlett [7], Freedman [39], Karlin and Taylor [53], and Parzen [74]) use that term for a Markov process with discrete state space. Here we use the first convention and associate the term "chain" with a discrete parameter space. (It is interesting to note that BIRTH AND DEATH PROCESSES is in the Doob–Chung tradition and CHUNG PROCESSES is in the Kolmogorov–Feller tradition.)

While speaking of Markov chains we shall use the notation $\{X_n,\ n \in T\}$ for the process. In this case we may also define Markov chains of the kth order ($k \geqslant 1$) if they satisfy the condition

$$P[X_{n+k} = j_{n+k} \mid X_{n+k-1} = j_{n+k-1}, X_{n+k-2} = j_{n+k-2}, \dots, X_0 = j_0] = P[X_{n+k} = j_{n+k} \mid X_{n+k-1} = j_{n+k-1}, X_{n+k-2} = j_{n+k-2}, \dots, X_n = j_n]. \qquad (2)$$

When both the state and parameter spaces are continuous let

$$F(x_0, x; t_0, t_0 + t) = P[X(t_0 + t) \leqslant x \mid X(t_0) = x_0]. \qquad (3)$$

As a result of definition (1) of the Markov

process we have

$$F(x_0, x; t_0, t_0 + t) = \int_{y \in S} F(y, x; \tau, t_0 + t) dF(x_0, y; t_0, \tau), \quad (4)$$

where $t_0 < \tau < t_0 + t$. When the state space and parameter space are discrete let

$$P_{ij}^{(m,m+n)} = P(X_{m+n} = j \mid X_m = i). \quad (5)$$

Now equation (4) takes the form

$$P_{ij}^{(m,m+n)} = \sum_{k \in S} P_{ik}^{(m,r)} P_{kj}^{(r,m+n)} \quad (m < r < m+n). \quad (6)$$

Similar relations can be written for the other two cases. Equations (4) and (6), the *Chapman–Kolmogorov equations**, are basic to the study of Markov processes. Equations (3) and (5) define transition probability distributions. When they do not depend on the initial parameter value [i.e., t_0 in (3) and m in (5)] and depend only on the difference between initial and final parameter values, the processes are *time homogeneous* and transition distributions are said to be *stationary* (*see* STATIONARY PROCESSES).

Historically, the Markov process bears the name of A. A. Markov* (1856–1922) who in 1907 laid the foundation for the theory of finite Markov chains. Markov chains with countably infinite states were first studied by A. Kolmogorov in 1936 [65]. Markov chains gained their immense popularity as models for diverse natural random phenomena after the publication of the first edition of Feller [35] in 1950, which unified the theory for finite as well as infinite chains using the concept of recurrent events (*see* RENEWAL THEORY). Another milestone in the study of finite Markov chains was reached with Kemeny and Snell [57], published in 1960, which introduced and used extensively the concept of the fundamental matrix.

The foundations for the theory of Markov processes with a continuous parameter space were laid by Kolmogorov in 1931 [64], who considered denumerable processes. His results were extended to the general case by Feller in 1936 [33] and 1940 [34]. As against the analytical approach taken by these researchers, a treatment based on sample functions was given by Doeblin in 1939 [26] and Doob in 1942 [27] and 1945 [28] (see also Doob [29]). Theoretical investigations on both the discrete and continuous parameter Markov processes saw another jump with the publication of the monograph in 1960 by Chung [18], who continued and consolidated the work of previous researchers. For recent advances in the theory, see Dynkin and Yushkevich [31], Freedman [39], Gihman and Skorohod [41], Karlin and Taylor [54], Kelly [56], Kemeny, Snell and Knapp [58], Lévy [68], and Spitzer [80].

In the next four sections the four distinct types of Markov processes and some of their properties are described. These sections briefly describe and give references for the related topics, viz. nonhomogeneous processes, limit theorems, inference problems, and Markov decision processes*. Some remarks are made on applications of Markov processes in the final section.

MARKOV CHAINS WITH DISCRETE STATES

Let $\{X_n, n = 0, 1, 2, \ldots\}$ be a Markov chain with state space S. Without loss of generality we shall restrict S to non-negative real integers. We assume that the Markov chain is homogeneous on the parameter space and define

$$P_{ij}^{(1)} = P(X_{m+n} = j \mid X_m = i), \quad m = 0, 1, 2, \ldots.$$

$P_{ij}^{(1)}$ is the one-step transition probability and we write $P_{ij}^{(1)} \equiv P_{ij}$; let $\mathbf{P}$ be the matrix with P_{ij} as elements. Applying the Chapman–Kolmogorov equations recursively the n-step transition probabilities $P_{ij}^{(n)}$ are given by the elements of $\mathbf{P}^n$, the nth power of the *transition probability matrix* $\mathbf{P}$.

In analyzing systems modeled as Markov chains, classification of states based on the transition characteristics of Markov chains is the first major step. State j is *accessible* from state i if j can be reached from i in a finite number of steps. If two states i and j are accessible to each other they are said to *communicate*. This is an equivalence relation exhibiting properties of reflexivity, symmetry, and transitivity. Thus we can consider sets of states which communicate with each other as forming an *equivalence class*. States belonging to different equivalence classes do not communicate with each other even though one-way accessibility is possible. If a Markov chain has all its states belonging to one equivalence class it is *irreducible*.

Based on the inherent nature of the Markov chain with respect to different states we also have the following properties: A state i is *recurrent* if and only if, starting from state i, eventual return of the Markov chain to this state is certain. Otherwise it is *transient*. Let

$$N_{ij} = \min\{n \mid X_n = j, \text{ given } X_0 = i\} \quad \text{and}$$

$$f_{ij}^{(n)} = P(N_{ij} = n) \tag{7}$$

Also, let $f_{ii}^{*} = \sum_{n=1}^{\infty} f_{ii}^{(n)} = P(N_{ii} < \infty)$ and $\mu_{ii} = E[N_{ii}]$. Thus for a recurrent state i we have $f_{ii}^{*} = 1$ and for a transient state i, $f_{ii}^{*} < 1$. A recurrent state i can be further classified as *null-recurrent* if $\mu_{ii} = \infty$ and *positive-recurrent* if $\mu_{ii} < \infty$. State i is recurrent or transient according as

$$\sum_{n=1}^{\infty} P_{ii}^{(n)} = \infty \quad \text{or} \quad < \infty. \tag{8}$$

Using the communication relation it is shown that recurrence and transience are class properties. Thus, when equivalence classes are identified in a Markov chain model of a real system, classification of one state in a class defines the property of the class. A special positive recurrent state is an *absorbing state*. A state i is absorbing if $P_{ii} = 1$; clearly it belongs to a class of its own.

Some Markov chains may exhibit periodicity in their behavior. The *period* of a state i is defined as the greatest common divisor of all integers $n \geqslant 1$ for which $p_{ii}^{(n)} > 0$. Periodicity is also a class property and when the period is 1, the state or the class is *aperiodic*. (An aperiodic irreducible positive recurrent chain is known as *ergodic*.)

The state and class properties defined above lead directly to the following results which have proved to be extremely useful in the analysis of Markov models.

1. If state i has a period $d_i(\geqslant 1)$, then there exists an integer n such that for all integers $n \geqslant N$, $P_{ii}^{(nd_i)} > 0$.
2. Let P be the transition probability matrix (TPM) of an irreducible aperiodic finite Markov chain. Then there exists an N such that for all $n \geqslant N$ the n-step transition probability matrix P^n has no zero elements.
3. *Canonical representation* of the transition probability matrix. Consider the equivalence classes in a Markov chain as belonging to a hierarchy of classes, recurrent classes at the top and transient classes below them. Arrange transient classes such that accessibility is only upwards. Rearranging the states in this manner, the transition probability matrix (TPM) can be arranged to have the following structure:

$$\mathbf{P} = \tag{9}$$

$$\begin{bmatrix} \mathbf{P}_1 & & & & & & \\ \mathbf{0} & \mathbf{P}_2 & & & & & \\ \vdots & \vdots & & & & & \\ \mathbf{0} & \mathbf{0} & \cdots & \mathbf{P}_k & & & \\ \mathbf{R}_{k+1,1} & \mathbf{R}_{k+1,2} & \cdots & \mathbf{R}_{k+1,k} & \mathbf{0}_{k+1} & & \\ \vdots & \vdots & & \vdots & \vdots & & \\ \mathbf{R}_{n1} & \mathbf{R}_{n2} & \cdots & \mathbf{R}_{nk} & \mathbf{R}_{n,k+1} & \cdots & \mathbf{0}_n \end{bmatrix}$$

where submatrix $\mathbf{P}_i$ is the TPM of the ith ($i = 1, 2, \ldots k$) recurrent class, submatrix $\mathbf{Q}_i$ is the TPM of the ith ($i = k + 1, \ldots, n$) transient class, and the submatrix $\mathbf{R}_{ij}$ has transition probabilities from the ith ($i = k + 1, \ldots n$) transient class to the jth ($j = 1, 2, \ldots i - 1$) recurrent or transient class for its elements. This structure clarifies the interrelationships of different equivalence classes and shows that in the study of Markov chains, all we need is the analysis of

Markov chains with one transient and one or more recurrent classes (even absorbing states will do), and Markov chains with a single equivalence class.

4. In a finite Markov chain, as the number of steps tends to infinity, the probability that the process is in a transient state tends to zero irrespective of the state at which the process starts. When the initial state is also transient the convergence is geometric.
5. In a finite Markov chain not all states can be transient.

When the Markov chain is irreducible and finite, several other properties based on their eigenvalues* can be deduced as well (Takács [81], Cox and Miller [21, Chap. 3], and Karlin and Taylor [54, Chap. 10]; see also books on matrix theory, e.g., Gantmacher [40] and Karlin [52]). We list three such properties below.

1. Suppose the transition probability matrix $\mathbf{P}$ of an irreducible Markov chain with m states $\{1, 2, \ldots, m\}$ admits distinct eigenvalues λ_i $(i = 1, 2, \ldots, m)$. Let $\mathbf{X}_1, \mathbf{X}_2, \ldots, \mathbf{X}_m$ be the m linearly independent right eigenvectors (column) belonging to these eigenvalues. Let $\mathbf{Q}$ be the nonsingular matrix $(\mathbf{X}_1, \mathbf{X}_2, \ldots, \mathbf{X}_m)$. Noting that $\mathbf{P}\mathbf{X}_j = \lambda_j \mathbf{X}_j$ we get $\mathbf{Q}^{-1}\mathbf{P}\mathbf{Q} = \Lambda$ where Λ is the diagonal matrix with $(\lambda_1, \lambda_2, \ldots, \lambda_m)$ as diagonal elements. Using this relation we get $\mathbf{P}^n = \mathbf{Q}\Lambda^n\mathbf{Q}^{-1}$, which can be used to determine $\mathbf{P}^n$ $(n = 2, 3, \ldots)$. Where the transition probability matrix does not admit distinct eigenvalues one can derive similar results using the Jordan canonical form.
2. It is easily seen that 1 is always an eigenvalue of the transition probability matrix. Using the Perron–Frobenius theorem of matrix theory we find that for an irreducible finite Markov chain all other eigenvalues are less than or equal to 1 in modulus. Therefore, the rate of approach of $\mathbf{P}^n$ to its limiting form is determined by the eigenvalue of the largest modulus less than unity.
3. If the Markov chain is periodic, the periodicity of the chain is given by the number of eigenvalues of unit modulus.

In the classical analysis of Markov models characteristics of interest are: first passage times, recurrence times, and state distributions in finite time as well as in the long run (as time $\to \infty$). Extensive literature exists in all these aspects and it is not possible to review all these results in a short survey. Appropriate references are some of the books cited at various places in this article. Nevertheless, two aspects are worth special mention.

(i) **First passage times.** Consider a finite Markov chain with one transient class and one or more recurrent classes. Canonical representation of the TPM of this Markov chain has the form

$$\mathbf{P} = \begin{bmatrix} \mathbf{P}_1 & \mathbf{0} \\ \mathbf{R}_1 & \mathbf{Q} \end{bmatrix}, \qquad (10)$$

where $\mathbf{P}_1$ has transition probabilities among recurrent states for its elements, $\mathbf{Q}$ is a substochastic matrix (at least one row sum < 1) with transition probability only among transient states as its elements, and $\mathbf{R}_1$ has probabilities of one-step transition from transient states to the recurrent states. The matrix $\mathbf{H} = (\mathbf{I} - \mathbf{Q})^{-1}$ is the *fundamental matrix* and its elements give the mean number of steps needed for first passage from a transient state to a recurrent state (Kemeny and Snell [57], Bhat [11]). Several other system characteristics can be expressed as functions of $\mathbf{H}$ or its elements. For instance, let $\mathbf{F}^{(n)}$ be the matrix of first passage probabilities in n transitions from transient states to recurrent states. Then noting that

$$\mathbf{P}^n = \begin{bmatrix} \mathbf{P}_1^n & \mathbf{0} \\ \mathbf{R}_n & \mathbf{Q}^n \end{bmatrix},$$

$$\mathbf{R}_n = \mathbf{R}_{n-1}\mathbf{P}_1 + \mathbf{Q}^{n-1}\mathbf{R}_1, \quad (11)$$

we get $\mathbf{F}^{(n)} = \mathbf{Q}^{n-1}\mathbf{R}_1$ and $\mathbf{F}^* = \sum_{n=1}^{\infty}\mathbf{F}^{(n)} = \mathbf{H}\mathbf{R}_1$.

When the state space is countably infinite more analytical techniques are needed for the analysis. Recurrence relations between general transition probabilities and first passage probabilities play a significant role in providing additional insights into the behavior of Markov chains. One such relation is given as [see eq. 7]

$$P_{ij}^{(n)} = \delta_{ij}^{(0n)} + \sum_{k=0}^{n} f_{ij}^{(k)} P_{jj}^{(n-k)}, \quad n \geqslant 0, \tag{12}$$

where $\delta_{ij}^{(0n)} = 1$ when $n = 0$ and $j = i$ and 0 otherwise; also, $P_{ii}^{(0)} = 1$ and $f_{ii}^{(0)} = 0$.

First passage probabilities can be extended to define taboo probabilities that include events avoiding a state or a set of states.

(ii) Limit results. $\lim_{n\to\infty}\mathbf{P}^n$ can be thought of as the limiting distribution of the Markov chain. Depending on the structure of $\mathbf{P}$ and the nature of states, this limit may or may not exist. For instance, in a periodic chain (with period > 1) we can speak only of a stationary distribution which is obtained by taking time averages. But in ergodic chains (aperiodic, irreducible, and positive recurrent) we get limiting distributions (these are ensemble averages). The following two results are significant for ergodic chains.

a.

$$\lim_{n\to\infty} P_{ii}^{(n)} = 1/\mu_i \quad \text{and}$$
$$\lim_{n\to\infty} P_{ji}^{(n)} = \lim_{n\to\infty} P_{ii}^{(n)}. \tag{13}$$

This result is the solution of equation (12) (the discrete renewal equation) as $n \to \infty$.

b. When the Markov chain is positive recurrent $\lim_{n\to\infty} P_{ij}^{(n)} = \pi_j > 0$. The limiting probabilities $\{\pi_j\}$ are uniquely determined by the set of equations

$$\sum_{j=0}^{\infty} \pi_j = 1 \quad \text{and} \quad \pi_j = \sum_{i=0}^{\infty} \pi_i P_{ij}. \tag{14}$$

Since recurrence is a basic condition for the existence of the limiting distribution there are several theorems establishing criteria for recurrence. For details, see books such as Karlin and Taylor [53, Chap. 3] and [54, Chap. 11].

In the present context an overview of Markov chains is not complete without mentioning *stopping times*. The problem can be defined as follows. Let $\{X_n, n \geqslant 0\}$ be a discrete state Markov chain and with every outcome j of $\{X_n\}$ associate a reward $r(j)$. The Markov chain is observed sequentially and after each observation a decision has to be made whether to stop or continue observing. This decision is made based on the reward associated with each step. Let T be the stopping time. The event $\{T = n\}$ signifies that the observed value j_n of X_n is such that for the first time the reward $r(j_n)$ can be considered to be acceptable. The sequence of observations $X_0 = j_0, X_1 = j_1, \ldots, X_n = j_n$, with which a decision to stop is made after the nth observation, is called a *stopping sequence*, say s_n. The expected reward corresponding to the stopping time T can then be given as

$$E[r(X_T)|X_0 = i] = \sum_{n=0}^{\infty}\sum_{s_n} r(j_n)\Pr\{X_0 = j_0, \ldots, X_n = j_n | X_0 = i\}, \tag{15}$$

where the second sum is over all stopping sequences $s_n = \{j_0, j_1, \ldots j_n\}$ for T. Let the optimal reward vector be

$$v(i) = \sup\{E[r(X_T)\,|\,X_0 = i]\}.$$

Assuming $r(j)$ to be a bounded nonnegative sequence with $0 \leqslant r(j) \leqslant \sup r(j) < \infty$, the optimal reward vector is the smallest vector $v(i)$ for which

$$\sum_j P_{ij} v(j) \leqslant v(i) \tag{16}$$

and $v(i) \geqslant r(i)$ for all i. A nonnegative vector $v(i)$ satisfying (16) is a *right super regular vector* and the problem of stopping times can be represented now in terms of the properties of super regular vectors for Markov chains. (One may also define subregular as well as left vectors related to Markov chains in a similar fashion.) For a detailed discussion readers are referred to Kemeny et al. [58], Çinlar [19, Chap. 7], and Karlin and Taylor [54, Chap. 11].

There are some classical examples of Markov chains in the literature. Random walks* of various kinds (Feller [35]), branching processes* in discrete time (Harris [46], Karlin and Taylor [53]), and queue length processes in queueing theory* (Gross and Harris [42], Prabhu [76], Kleinrock [62]) are some of them. In each of these areas (and other areas of applications) Markov chains present special problems that call for special techniques of analysis.

MARKOV CHAINS WITH A CONTINUOUS INFINITY OF STATES

Consider a Markov chain $\{X_n,\ n = 0, 1, 2, \dots\}$ for which the state space S is the real line $(-\infty, \infty)$. Assuming that the chain is time homogeneous, we define the *transition distribution function* as

$$F_n(x, y) = P[X_{m+n} \leqslant y \mid X_m = x]. \quad (17)$$

In this case, the Chapman–Kolmogorov relation takes the form

$$F_{m+n}(x, y) = \int_{-\infty}^{\infty} d_z F_m(x, z) F_n(z, y). \quad (18)$$

For a general discussion of these Markov chains, see books such as Doob [29] and Feller [36]. Here we shall concentrate on a special class of random walk problems treated under fluctuation theory of sums of independent and identically distributed random variables. Specific references for this topic are Prabhu [75, Chap. 6], Feller, [36, Chap. 12], and Karlin and Taylor [54, Chap. 17]. Further references are Spitzer [80] and Takács [82].

Let $\{X_n,\ n = 1, 2 \dots\}$ be a sequence of independent and identically distributed random variables assuming values in $(-\infty, \infty)$. Also let $S_0 = 0$, $S_n = X_1 + X_2 + \cdots + X_n$ $(n \geqslant 1)$. The partial sum sequence $\{S_n,\ n = 0, 1, 2, \dots\}$ is a Markov chain. Define a sequence $\{N_k,\ k = 0, 1, 2, \dots\}$ of random variables as follows:

$$\begin{aligned} &N_0 \equiv 0; \qquad N_1 = \min\{n \mid S_n > 0\},\\ &N_2 = \min\{n > N_1 \mid S_n - S_{N_1} > 0\}, \dots,\\ &N_k = \min\{n > N_{k-1} \mid S_n - S_{N_{k-1}} > 0\},\\ &\qquad\qquad (k \geqslant 1). \quad (19)\end{aligned}$$

The random variables $\{N_k\}$ are the *ladder indices** of $\{S_n\}$. The partial sum values $\{S_{N_k}\}$ at the indices are *ladder heights*. Together, (N_k, S_{N_k}) are *ladder points*. As defined in (19), the ladder heights are in ascending order. In a symmetric fashion, one may define descending indices and heights as well. Also (19) uses strict inequalities in the definition; consequently, one may identify them as strict ladder indices and define weak indices using $\geqslant$ in the definition.

Combinatorial methods provide a simple setting for the derivation of properties of ladder variables. Some of the basic results are given below. Define the variables

$$T_k = N_k - N_{k-1}, \qquad H_k = S_{N_k} - S_{N_{k-1}}, \quad (20)$$

and

$$G_n(x) = P(T_1 = n, H_1 \leqslant x). \quad (21)$$

We have

(i) The pairs (T_k, H_k), $k = 1, 2, 3, \dots$ are mutually independent and have the common distribution given by (21).

(ii) The ladder points (N_k, S_{N_k}) form a two-dimensional renewal process*.

(iii) Let $F_n(x) = P(S_n \leqslant x)$. Then the transform (the probability generating function* of the Laplace–Stieltjes transform) of the joint distribution $G_n(x)$ is given by

$$\begin{aligned} &E(z^{T_1} e^{-\theta H_1})\\ &\quad = 1 - \exp\left[-\sum_{1}^{\infty} \frac{z^n}{n} \int_{0+}^{\infty} e^{-\theta x} dF_n(x)\right],\\ &\qquad\qquad (|z| < 1, \theta > 0). \quad (22)\end{aligned}$$

SIMPLE MARKOV PROCESSES

We shall identify a Markov process with discrete state space, continuous parameter space and stationary transition probabilities as a *simple Markov process*. For a general discussion of some of the structural properties of the process *see* CHUNG PROCESSES.

Let $\{X(t), t \geqslant 0\}$ be a discrete state, continuous parameter stochastic process, homogeneous in its parameter space and defined on a probability space in the usual manner. Define its transition probability as

$$P_{ij}(t) = P[X(t) = j \mid X(0) = i], \qquad i, j \in S. \tag{23}$$

The probabilities $P_{ij}(t)$ have the following properties for $t > 0$:

(i) $P_{ij}(t) \geqslant 0$,
(ii) $\sum_{j \in s} P_{ij}(t) = 1$,
(iii) $P_{ij}(t+s) = \sum_{k \in S} P_{ik}(t) P_{kj}(s)$, $s > 0$,
(iv) $P_{ij}(t)$ is continuous, and
(v) $\lim_{t \to 0} P_{ij}(t) = 1$ if $i = j$ and $= 0$ if $i \neq j$.

These properties also imply that $P_{ij}(t)$ have right-hand derivatives for every $t \geqslant 0$. It is also known that

$$\lim_{t \to 0} \left(\frac{1 - P_{ii}(t)}{t} \right) = P'_{ii}(0) = \lambda_{ii} \tag{24}$$

exists but may be infinite; and

$$\lim_{t \to 0} \left(\frac{P_{ij}(t)}{t} \right) = P'_{ij}(0) = \lambda_{ij}, \qquad i \neq j, \tag{25}$$

exists and is finite. Because of (24) and (25), when t is small, say Δt, as a linear approximation we may write

$$P_{ij}(\Delta t) = \lambda_{ij} \Delta t + o(\Delta t);$$

$$P_{ii}(\Delta t) = 1 - \lambda_{ii} \Delta t + o(\Delta t), \tag{26}$$

where $o(\Delta t)$ is such that $o(\Delta t)/\Delta t \to 0$ as $\Delta t \to 0$.

Setting $s = \Delta t$ in property (iii) above (which is the Chapman–Kolmogorov relation) yields the forward Kolmogorov equation

$$P'_{ij}(t) = -\lambda_{jj} P_{ij}(t) + \sum_{k \neq j} \lambda_{kj} P_{ik}(t) \tag{27}$$

and setting $t = \Delta s$ yields the backward Kolmogorov equation

$$P'_{ij}(t) = -\lambda_{ii} P_{ij}(t) + \sum_{k \neq i} \lambda_{ik} P_{kj}(t). \tag{28}$$

Of these equations the backward equation is considered to be more fundamental because in its derivation it is not necessary to assume that passage to the limit in (25) is uniform with respect to i. But in practice use of the forward equation is simpler and therefore more predominant, and because of the types of models considered, a unique solution for both forward and backward equations is obtained. Using $\mathbf{A}$ to denote the matrix of coefficients, and $\mathbf{P}(t)$ and $\mathbf{P}'(t)$ to denote matrices with $P_{ij}(t)$ and $P'_{ij}(t)$ as elements respectively, we may write (27) and (28) as

$$\mathbf{P}'(t) = \mathbf{P}(t)\mathbf{A} \quad \text{and} \quad \mathbf{P}'(t) = \mathbf{A}\mathbf{P}(t) \tag{29}$$

with the initial condition $\mathbf{P}(0) = \mathbf{I}$. A formal solution to this set of equations is obtained as

$$\mathbf{P}(t) = e^{\mathbf{A}t} \equiv \mathbf{I} + \sum_{n=1}^{\infty} \frac{\mathbf{A}^n t^n}{n!}. \tag{30}$$

When $\mathbf{A}$ is a finite matrix, the series in (30) is convergent and is the unique solution. Further when it is diagonalizable the solution can be expressed in terms of its eigenvalues. When $\mathbf{A}$ is of infinite dimensions however, special properties of $\mathbf{A}$ need to be considered.

As $t \to \infty$, when the limit exists, it is independent of parameter t and hence $P'_n(t) \to 0$. Denoting $\lim_{t \to 0} P_n(t) = p_n$, from (27) and (29) we get

$$0 = -\lambda_{jj} p_j + \sum_{k \neq j} \lambda_{kj} p_k \tag{31}$$

or $\mathbf{pA} = \mathbf{0}$, where $\mathbf{p} = (p_0, p_1, p_2, \ldots)$. If the Markov process is irreducible (all states communicate) the limiting distribution $\mathbf{p}$ exists as defined above. The limits $\{p_n, n \in S\}$ are such that they either vanish identically, or are all positive and form a probability distribution, i.e., $p_n > 0$ for all $n \in s$, $\sum_{n \in S} p_n = 1$. A Markov process with positive limiting probabilities is identified as being positive recurrent. Then the limiting distribution is also stationary (i.e., if $P_n(t) = p_n$, $P_n(t + \tau) = p_n$ for $\tau > 0$).

The following form of equation (31) is intuitively appealing and can be used in writing down such steady state equations without going through the Kolmogorov equations:

$$\lambda_{jj}p_j = \sum_{k \neq j} p_k \lambda_{kj} . \tag{32}$$

From (24) and (25) we see that λ_{jj} is the infinitesimal rate of transition out of state j and λ_{kj} is the infinitesimal rate into state j from state k. Associating these rates with corresponding probabilities of being in respective states, equation (32), in fact, represents the balance between transitions into and from state j (outflow = inflow, in the sense of fluids). Thus, when once the model is identified as a Markov process these balance equations can be readily written down when the limiting distribution exists.

A large class of simple Markov processes used as models of natural phenomena have the property of allowing only nearest neighbor transitions in (25) (i.e., $i-1$, or $i+1$ from i). These are *birth and death processes**.

A type of simple Markov process finding extensive use in population modeling and computer and communication systems is a *Markov population process*. It is defined on a network of population centers among which the transitions occur (see Whittle [86, 87], Kingman [61], and Bhat [11]).

MARKOV PROCESSES WITH CONTINUOUS STATE AND PARAMETER SPACES

This is the most general class of Markov processes and to derive meaningful results imposition of some structure becomes necessary. We shall identify three different classes: (i) purely discontinuous processes in which changes of state occur only by jumps (*see* JUMP PROCESSES); (ii) diffusion processes* in which changes of state occur continually; and (iii) general discontinuous processes in which changes of state can occur by jumps (of more than one kind) as well as otherwise.

The simple Markov process considered in the last section is a special case of the purely discontinuous processes. The denumerable state process was investigated thoroughly by Kolmogorov [64] using differential equations which now have become the lynch pin of analysis for similar processes. Feller [33, 34] extended Kolmogorov's results to the nondenumerable case and proved existence and uniqueness theorems necessary for the analysis. The Kolmogorov–Feller approach is purely analytical and is based on what we have identified earlier as Chapman–Kolmogorov relations for the process (see remarks leading to equations (27) and (28)).

Let $\{X(t), t > 0\}$ be a purely discontinuous Markov process for which the state space S is the real line $(-\infty, \infty)$. Let the transition distribution function be defined as in equation (3). When $X(t) = x$, let $\lambda(x, t)$ be the jump rate, with the probability of jump during $(t, t + \Delta t]$ being given by $\lambda(x, t)\Delta t + o(\Delta t)$. Let $h(x, y, t)$ be the conditional distribution of the magnitude of a jump when $X(t) = x$. With this structure the basic Kolmogorov–Feller equations for the process are given as

$$\begin{aligned} \frac{\partial}{\partial t} & F(x_0, x; t_0, t) \\ &= -\int_{-\infty}^{x} \lambda(y, t)\, d_y F(x_0, y; t_0, t) \\ &\quad + \int_{-\infty}^{\infty} \lambda(y, t) h(y, x, t)\, d_y F(x_0, y; t_0, t), \end{aligned} \tag{33}$$

which is the forward equation and

$$\begin{aligned} \frac{\partial}{\partial t_0} & F(x_0, x; t_0, t) \\ &= \lambda(x_0, t_0) F(x_0, x; t_0, t) - \lambda(x_0, t_0) \\ &\quad \times \int_{-\infty}^{\infty} h(x_0, y, t_0)\, d_y F(y, x; t_0, t), \end{aligned} \tag{34}$$

which is the backward equation. (Compare these equations with (27) and (28).) For a comprehensive discussion see Prabhu [75].

There have been alternative approaches to the analysis of Markov processes in this class. The treatment based on sample function properties has been given by Doeblin [26] and Doob [27, 28]. Also see Doob [29] for an extended treatment of this approach.

Another approach has been through the theory of semigroups as described in Loéve [69], Dynkin [31] and Feller [36].

Diffusion processes* are widely used in applications. The process is characterized by the condition

$$\lim_{\Delta t \to 0} \frac{1}{\Delta t} \int_{|y-x|>\delta} d_y F(x, y; t, t+\Delta t) = 0, \tag{35}$$

which states that only small changes of state occur during small intervals of time. It is also assumed that the infinitesimal mean and variance, defined in (36) and (37) below, also exist.

$$\lim_{\Delta t \to 0} \frac{1}{\Delta t} \int_{|y-x|\leqslant\delta} (y-x)\, d_y F(x, y; t, t+\Delta t) = \mu(x,t), \tag{36}$$

$$\lim_{\Delta t \to 0} \frac{1}{\Delta t} \int_{|y-x|\leqslant\delta} (y-x)^2\, d_y F(x, y; t, t+\Delta t) = \sigma^2(x,t). \tag{37}$$

See the article DIFFUSION PROCESSES, also Bharucha-Reid [10], Prabhu [75], Ito and McKean [51], Levy [68], and Karlin and Taylor [54].

For an investigation of the general discontinuous process with more than one type of change, see Moyal [73].

NONHOMOGENEOUS MARKOV PROCESSES

In the last four sections we have assumed that the processes are homogeneous in the parameter space. When this assumption does not hold we have nonhomogeneous (also called nonstationary) Markov processes and the techniques needed for their analysis are more elaborate. For instance, consider a finite Markov chain $\{X_n, n = 0, 1, 2 \dots\}$ with discrete states $\{1, 2, \dots, m\}$. Let $\mathbf{P}^{(1)}, \mathbf{P}^{(2)}, \mathbf{P}^{(3)}, \dots$ be the one-step transition probability matrices for this chain for parameter values $0, 1, 2, \dots$ respectively (note that the TPM is dependent on the parameter value). Let $\mathbf{P}^{(k,n)}$ be the transition probability matrix with element $P_{ij}^{(k,n)}$ given by

$$P_{ij}^{(k,n)} = P(X_n = j \mid X_k = i), \quad i, j = 1, 2, \dots, m. \tag{37}$$

Using the Chapman–Kolmogorov relations, one gets

$$\mathbf{P}^{(k,n)} = \mathbf{P}^{(k+1)} \cdot \mathbf{P}^{(k+2)} \dots \mathbf{P}^{(n)}. \tag{38}$$

Thus, much of the simplification achieved in the homogeneous case will be lost in the analysis of nonhomogeneous systems. Nevertheless, nonhomogeneous systems exhibit several limiting properties, much in the same way as homogeneous systems. These will be discussed in the next section.

LIMIT THEOREMS

For aperiodic, irreducible, and positive recurrent Markov chains we have identified two limit results in equation (13). Markov chains with these limiting properties are called *ergodic chains*. Questions naturally arise whether there are other chains that exhibit similar ergodic properties or other limiting results such as the law of large numbers* and the central limit theorem*.

The fact that homogeneous Markov chains obey the law of large numbers and the central limit theorem was one of the first few results established by Markov* in 1907 [70]. For more general forms of these results readers are referred to Doob [29, Sect. 5.6], Feller [36, Chap. 7], and Chung [18, Chap. I.16], who also establishes the law of the iterated logarithm*. The central limit theorem for nonhomogeneous Markov chains has been investigated by several researchers starting with Markov [71]; for a review and additional results see Dobrushin [24, 25].

The ergodic behavior of Markov chains is given through ergodic theorems* which establish conditions under which the limits of processes exist. In the case of homogeneous Markov chains see Chung [18, Chap. I.15] and Feller [36, Sect. VIII.7]. For nonhomogeneous Markov chains, a good discussion can be found in Isaacson and Madsen [50], who also list relevant references (also see

Hajnal [43, 44]). Two types of ergodic behavior can be defined.

Let $\mathbf{a}^{(0)}$ be the initial distribution of states (called the starting vector) of a Markov chain. Define $\mathbf{a}^{(k,n)} = \mathbf{a}^{(0)}\mathbf{P}^{(k+1)} \cdot \mathbf{P}^{(k+2)} \dots \mathbf{P}^{(n)}$, where $\mathbf{P}^{(i)}$ is the one-step transition probability for the ith transition. Similarly define $\mathbf{b}^{(k,n)}$ for the starting vector $\mathbf{b}^{(0)}$. Also define the norm of a vector $\mathbf{a} = (a_1, a_2, \dots)$ as $\|\mathbf{a}\| = \sum_{i=1}^{\infty}|a_i|$ and the norm of a square matrix $\mathbf{A}$ with elements a_{ij} $(i, j = 1, 2, \dots)$ as $\|\mathbf{A}\| = \sup_i \sum_{j=1}^{\infty}|a_{ij}|$. Now two types of ergodic behavior of Markov chains can be defined as given below.

1. A nonhomogeneous Markov chain is *weakly ergodic* if for all k and starting vectors $\mathbf{a}^{(0)}$ and $\mathbf{b}^{(0)}$

$$\lim_{n\to\infty} \sup_{\mathbf{a}^{(0)},\mathbf{b}^{(0)}} \|\mathbf{a}^{(k,n)} - \mathbf{b}^{(k,n)}\| = 0. \tag{39}$$

2. A nonhomogeneous Markov chain is *strongly ergodic* if there exists a vector $\boldsymbol{\pi} = (\pi_1, \pi_2, \dots)$ with $\|\boldsymbol{\pi}\| = 1$ and $\pi_i \geqslant 0$ for $i = 1, 2, 3, \dots$ such that for all k

$$\lim_{n\to\infty} \sup_{\mathbf{a}^{(0)}} \|\mathbf{a}^{(k,n)} - \boldsymbol{\pi}\| = 0 \tag{40}$$

where $\mathbf{a}^{(0)}$ is a starting vector.

The ergodic theorems* give necessary and sufficient conditions under which Markov chains are weakly or strongly ergodic. When the Markov chain is homogeneous in the parameter space, it turns out that if the chain is weakly ergodic, then it is also strongly ergodic as per above definitions.

INFERENCE FOR MARKOV PROCESSES

Considerable work has been done on inference problems on Markov processes. For an extensive bibliography readers are referred to Anderson and Goodman [4], Billingsley [12, 13] and Basava and Prakasa Rao [8]. Billingsley [13] gives a comprehensive treatment of the basic topics of estimation* and hypothesis testing* in Markov processes. Basava and Prakasa Rao [8] treat inference and stochastic processes* in general with specific discussion of Markov processes in two chapters and give an extensive bibliography.

Suppose a Markov chain has been observed until n transitions have taken place. Let n_{ij} be the number of transitions of the type $i \to j$. Thus we have $\sum_i \sum_j n_{ij} = n$. Let $P_{ij} (i, j \in S)$ be the one step transition probabilities of the Markov chain. The likelihood function based on sample observations can be given as

$$L = C \prod_{i,j} P_{ij}^{n_{ij}} \tag{41}$$

where C does not depend on P_{ij} and the product is taken over all i and j for which $n_{ij} > 0$. In the above expression we have ignored the contribution of the initial state of the process. Using (41) to obtain the maximum likelihood* estimate in the usual manner, one gets the estimate $\hat{P}_{ij}$ of P_{ij} as

$$\hat{P}_{ij} = n_{ij}/n_i, \qquad n_i = \sum_j n_{ij}. \tag{42}$$

Using this estimator one can develop a chi-square* (χ^2) statistic

$$\sum_i \sum_j n_i \frac{\left(\hat{P}_{ij} - P_{ij}^0\right)^2}{P_{ij}^0} \tag{43}$$

or a likelihood ratio* statistic Λ with

$$-2\ln\Lambda = 2\sum_i \sum_j n_{ij} \ln \frac{n_{ij}}{n_i P_{ij}^0} \tag{44}$$

for testing the null hypothesis $\mathbf{P} = \mathbf{P}^0$ (matrix $\mathbf{P}$ has P_{ij}, $i, j \in S$ as elements). Similar test statistics are easily derived for the order as well as homogeneity of Markov chains.

When the state space of the Markov chain is finite the asymptotic covariance matrix for the maximum likelihood estimators of transition probabilities has been derived by Handa [45].

If P_{ij} are known functions of a parameter $\boldsymbol{\theta} = (\theta_1, \theta_2, \dots, \theta_k)$, then estimators for $\boldsymbol{\theta}$ can be determined by solving

$$\sum_i \sum_j \left[n_{ij} \frac{\partial P_{ij}(\theta)}{\partial \theta_r} \cdot \frac{1}{P_{ij}(\theta)} \right] = 0, \qquad r = 1, 2, \dots k. \tag{45}$$

Suppose a simple Markov process is observed for a length of time t. Let t_i be the amount of time the process is observed in state i. With infinitesimal rates λ_{ii} and λ_{ij} defined in (24) and (25), a constructive development of the process can be accomplished by using the property that the process resides in a state i for an exponential length of time with mean $1/\lambda_{ii}$ and changes its state to j with probability $\lambda_{ij}/\lambda_{ii}$. Let n_{ij} be the number of state changes $i \to j$. Again ignoring the contribution of the initial state, the likelihood function can be written down as

$$L = C\Big(\prod_{i,j} \lambda_{ij}^{n_{ij}}\Big)\Big(\prod_{i} e^{-\lambda_{ii}t_i}\Big) \qquad (46)$$

where C does not depend on λ_{ij} $(i, j \in S)$. Likelihood ratio statistics for tests of hypotheses follow in the usual manner. (Also see Albert [2] for the asymptotic behavior of the maximum likelihood estimate and Keiding [55] for the estimation of parameters in a birth and death process with linear rates under conditions of extinction and nonextinction.)

In addition to the general references on inference on Markov processes given above, some special problems considered in the literature should be noted. Christiansen [17] has discussed the problem of estimating transition probabilities in a continuous parameter space Markov process using the maximum likelihood method. In the context of the measurement of competing risks*, Aalen [1] gives a nonparametric estimation procedure for the probability of absorption in a Markov process with a single transient state and several absorbing states. This problem has been extended by Fleming [37] to a nonhomogeneous Markov process with several transient states.

Construction of maximum likelihood estimators and likelihood ratio statistics is made by assuming an underlying model for the process. Where this assumption is no longer true, Foutz and Srivastava [38] derive asymptotic properties of estimators and test statistics in the context of a Markov process.

In many social phenomena, one may have only information on the number of individuals in each state at specified epochs of time, rather than the detailed information on transitions needed for using the maximum likelihood procedure. Then a least squares or a minimum χ^2 procedure may be appropriate. For a discussion of these methods the readers are referred to Lee et al. [67].

MARKOV DECISION PROCESSES

The influence of operations research* on the development of the theory of Markov processes is clearly seen in the growth of the topic known as *Markov decision processes** (also called *Markovian decision processes* and *controlled Markov processes*). It can also be considered as the meeting point between statistical decision theory* initiated by Wald [84, 85] and the general area of optimization developed under applied mathematics and operations research. A Markov decision problem aims at determining optimal policies for decision making in a Markovian setting. The stopping time problem introduced earlier is one of them. More specifically one of the simpler versions of the general problem can be formulated as follows.

Consider a finite state Markov chain (discrete parameter and discrete state) with state space S. Suppose with every state of the Markov chain we associate a decision to be chosen out of a set D. Let ${}^kP_{ij}$ be the probability of one step transition $i \to j$ $(i, j \in S)$ under decision $k \in D$. Also associate a reward ${}^kR_{ij}$ (or cost) with decision k and transition $i \to j$. Let ${}^kV_i^{(n)}$ be the expected total earnings in n future transitions if decision k is made when the process is in state i. For the optimal decision $k = 0$, we have

$${}^0V_i^{(n)} = \max_{k \in D} \sum_{j \in S} {}^kP_{ij}\big[{}^kR_{ij} + {}^0V_j^{(n-1)}\big]$$

$$n = 1, 2, \ldots, i \in S.$$

This recursive relation gives an iterative procedure to determine optimum decisions $d_i^{(n)} \in D$ for $i \in S$ and $n = 1, 2, \ldots$. This is a standard technique in dynamic programming* and it has been shown (Bellman [9],

Chap. XI) that this iteration process will converge on the best alternative for each state as $n \to \infty$. Because the iteration is based on the value of the policy, the procedure is called the *value iteration method*. An alternative procedure uses policy iteration and the formal basis for these procedures was laid by Howard [47] and Blackwell [14, 15]. Since then the theory of Markov decision processes has expanded at an increasing rate and various problems have been tackled. The following books may be cited as good references: Denardo [22], Derman [23], Dynkin and Yushkevich [32], Howard [48], and Ross [78].

The application of Bayesian decision theory to finite Markov chains with uncertain transition probabilities has been discussed extensively in Martin [72]. His results are obtained under the assumption that the prior distribution function of the matrix of transition probabilities belongs to a family of distributions closed under consecutive sampling. The results include expected steady state probabilities and related quantities for control processes as well as the distribution theory needed for the Bayesian analysis of Markov chains.

APPLICATIONS

Of all stochastic processes, without any doubt, Markov processes are the ones used most in applications. There are two major reasons for this. In simpler forms, the analyses and the results are simple enough to be meaningful even for applied researchers lacking in mathematical sophistication. Secondly, the first order dependence of a Markov process goes quite far in representing real phenomena. Thus Markov processes have found a variety of applications as mathematical models in disciplines such as biology, physics, chemistry, astronomy, operations research, and computer science (see Feller [35], Bharucha-Reid [10], and Bhat [11]). In specific areas the following additional references are in order: Archaeology*—Clarke [20]; biological sciences—Chiang [16], Iosifescu and Tăutu [49]; computer science—Kleinrock [63], Allen [3], and Trivedi [83]; geology*—Krumbein and Dacey [66] and Schwarzacher [79]; reliability*—Rau [76] and Trivedi [83]; and sociology*—Bartholomew [6].

Many times direct applications of the theory become impossible. Then modifications and extensions of the processes take place. For instance, even when the general model is non-Markovian, embedded Markov chains can be identified in the process and the analysis simplified (*see* EMBEDDED PROCESSES and Kendall [59, 60]). Markov renewal processes (also known as *semi-Markov processes**) define renewal probabilities in a Markovian setting. In order to represent dependencies spanning more than two time periods, higher-order Markov chains have been formulated. When the state space becomes too large and unmanageable, lumping of states into a smaller number has been tried. In geological applications it makes sense to reverse the process and talk about reversed processes. These are but a few examples where applications have spurred the growth of the theory of Markov processes.

References

[1] Aalen, O. (1978) *Ann. Statist.*, **6**, 534–546.

[2] Albert, A. (1962). *Ann. Math. Statist.*, **38**, 727–753.

[3] Allen, A. O. (1978). *Probability, Statistics and Queueing Theory*. Academic Press, New York.

[4] Anderson, T. W. and Goodman, L. A. (1959). *Ann. Math. Statist.*, **28**, 89–110.

[5] Bailey, N. T. J. (1964). *The Elements of Stochastic Processes*, Wiley, New York. (Covers mostly Markov processes.)

[6] Bartholomew D. J. (1973). *Stochastic Models for Social Processes* (2nd ed.) Wiley, New York.

[7] Bartlett, M. S. (1960). *An Introduction to Stochastic Processes*. Cambridge University Press, Cambridge. (One of the classics among books on stochastic processes.)

[8] Basava, I. V. and Prakasa Rao, B. L. S. (1980). *Statistical Inference for Stochastic Processes*. Academic Press, New York.

[9] Bellman, R. (1957). *Dynamic Programming*. Princeton University Press, Princeton, N.J.

[10] Bharucha-Reid, A. T., (1960). *Elements of the Theory of Markov Processes and Their Applications*. McGraw-Hill, New York.

[11] Bhat, U. N. (1984). *Elements of Applied Stochastic Processes* 2nd ed. Wiley, New York.

[12] Billingsley, P. (1961). *Ann. Math. Statist.*, **32**, 12–40.

[13] Billingsley, P. (1961). *Statistical Inference for Markov Processes*. University of Chicago Press, Chicago.

[14] Blackwell, D. (1962). *Ann. Math. Statist.*, **33**, 719–726.

[15] Blackwell, D. (1965). *Ann. Math. Statist.*, **36**, 226–235.

[16] Chiang, C. L. (1968). *Introduction to Stochastic Processes in Biostatistics*. Wiley, New York.

[17] Christiansen, H. D. (1978). *Scand. Actuar. J.*, 129–140.

[18] Chung, K. L. (1960). *Markov Chains*. Springer-Verlag, New York. (1967: 2nd ed.)

[19] Çinlar, E. (1975). *Introduction to Stochastic Processes*. Prentice Hall, Englewood Cliffs, N.J. (Covers only Markov and renewal processes.

[20] Clarke, D. L. (1972). *Models in Archaeology*, D. L. Clarke, ed. Methuen, London, pp. 1–60.

[21] Cox, D. R. and Miller, H. D. (1965). *The Theory of Stochastic Processes*, Wiley, New York. (Chaps. 1–5 are on Markov processes.)

[22] Denardo, E. V. (1982). *Dynamic Programming: Theory and Applications*. Prentice-Hall, Englewood Cliffs, N.J.

[23] Derman, C. (1970). *Finite State Markovian Decision Processes*. Academic Press, New York. (One of the best books on the subject even though the scope is rather limited.)

[24] Dobrushin, R. L. (1956). *Theory Prob. Appl.*, **1**, 65–80.

[25] Dobrushin, R. L. (1956). *Theory Prob. Appl.*, **1**, 329–383.

[26] Doeblin, W. (1939). *Skand. Aktuar.*, **22**, 211–222.

[27] Doob, J. L. (1942). *Trans. Amer. Math. Soc.*, **52**, 37–64.

[28] Doob, J. L. (1945). *Trans. Amer. Math. Soc.*, **58**, 455–473.

[29] Doob, J. L. (1953). *Stochastic Processes*. Wiley, New York. (A classic among books on the subject.)

[30] Dynkin, E. B. (1961). *Markov Processes* (2 volumes). Springer-Verlag, Berlin. (Trans. from Russian.)

[31] Dynkin, E. B. and Yushkevich, A. A. (1969). *Markov Processes: Theorems and Problems*. Plenum Press, New York. (Russian edition, 1967.)

[32] Dynkin, E. B. and Yushkevich, A. A. (1979). *Controlled Markov Processes*. Springer-Verlag, Berlin. (Russian edition, 1975. Coverage of theory and applications is to the liking of a probabilist.)

[33] Feller, W. (1936). *Math. Ann.*, **113**, 113–160.

[34] Feller, W. (1940). *Trans. Amer. Math. Soc.*, **48**, 488–515. (Errata ibid. (1945) **58**, p. 474.)

[35] Feller, W. (1950). *An Introduction to Probability Theory and its Applications, Vol. 1*. (1st ed.); 1968, 3rd ed. Wiley, New York. (A classic book which laid the foundation for the area of applied probability.)

[36] Feller, W. (1966). *An Introduction to Probability Theory and its Applications, Vol. II*. Wiley, New York.

[37] Fleming, T. R. (1978). *Ann. Statist.*, **6**, 1057–1070.

[38] Foutz, R. V. and Srivastava, R. C. (1979). *Adv. Appl. Prob.*, **11**, 737–749.

[39] Freedman, D. (1971). *Markov Chains*. Holden-Day, San Francisco, Ca. (The first of three Volumes. Covers advanced topics.)

[40] Gantmacher, F. R. (1960). *The Theory of Matrices, Vols. I and II*. Chelsea House, New York.

[41] Gihman, I. I. and Skorohod, A. V. (1975). *The Theory of Stochastic Processes II*. Springer-Verlag, Berlin. (Russian edition, 1973.)

[42] Gross, D. and Harris C. (1974). *Fundamentals of Queueing Theory*. Wiley, New York. (2nd ed. expected in 1985.)

[43] Hajnal, J. (1956). *Proc. Camb. Phil. Soc.*, **52**, 67–77.

[44] Hajnal, J. (1958). *Proc. Camb. Phil. Soc.*, **54**, 233–246.

[45] Handa, B. R. (1972). *Biometrika*, **59**, 407–414.

[46] Harris, T. (1963). *The Theory of Branching Processes*. Springer-Verlag, Berlin.

[47] Howard, R. A. (1960). *Dynamic Programming and Markov Processes*. Wiley, New York.

[48] Howard, R. A. (1971). *Dynamic Probabilistic Systems*, Vols. 1 and 2. Wiley, New York.

[49] Iosifescu, M. and Tăutu, P. (1973). *Stochastic Processes and Applications in Biology and Medicine, Vols. 1 and 2*. Springer-Verlag, Berlin. (Provides an extensive coverage of applications.)

[50] Isaacson, D. L. and Madsen, R. W. (1976). *Markov Chains*. Wiley, New York.

[51] Ito, K. and McKean, H. P., Jr. (1965). *Diffusion Processes and Their Sample Paths*. Springer-Verlag, Berlin.

[52] Karlin, S. (1968). *Total Positivity*. Stanford University Press, Stanford, Calif.

[53] Karlin, S. and Taylor, H. M. (1975). *A First Course in Stochastic Processes*. (2nd Ed.) Academic Press, New York. (First Edition, 1966. Several chapters on Markov processes have been moved to ref. 54 during revision.)

[54] Karlin, S. and Taylor, H. M. (1981). *A Second Course in Stochastic Processes*. Academic Press, New York.

[55] Keiding, N. (1975). *Ann. Statist.*, **3**, 363–372.

[56] Kelly, F. P. (1979). *Reversibility and Stochastic Networks*. Wiley, New York.

[57] Kemeny, J. G. and Snell, J. N. (1960). *Finite Markov Chains*. Van Nostrand, Princeton, N.J.

[58] Kemeny, J. G. Snell, J. N., and Knapp, A. W. (1966). *Denumerable Markov Chains*. Van Nostrand, Princeton, N.J.

[59] Kendall, D. G. (1951). *J. R. Statist. Soc. B.*, **13**, 151–185.

[60] Kendall, D. G. (1953). *Ann. Math. Statist.*, **24**, 338–354.

[61] Kingman, J. F. C. (1969). *J. Appl. Prob.*, **6**, 1–18.

[62] Kleinrock, L. (1975). *Queueing Systems (Vol. 1), Theory*. Wiley, New York.

[63] Kleinrock, L. (1976). *Queueing Systems (Vol. 2), Computer Applications*. Wiley, New York.

[64] Kolmogorov, A. (1931). *Math. Ann.*, **104**, 415–458.

[65] Kolmogorov, A. (1936). *Math. Sbornik. N.S.*, **1**, 607–610.

[66] Krumbein, W. C. and Dacey, M. F. (1969). *J. Int. Assoc. Math., Geology 1*, **1**, 79–96.

[67] Lee, T. C., Judge, G. G. and Zellner, A. (1970). *Estimating Parameters of the Markov Probability Model from Aggregate Time Series Data*. North-Holland, Amsterdam.

[68] Lévy, P. (1965). *Processes Stochastiques et Mouvement Brownien*, 2nd Ed. Gauthier-Villars, Paris.

[69] Loéve, M. (1963). *Probability Theory* (3rd ed.) Van-Nostrand, Princeton, N.J.

[70] Markov, A. A. (1907). (English translation) Appendix B, *Dynamic Probabilistic Systems*, Vol. 1 (R. A. Howard, ed.), pp. 552–576.

[71] Markov, A. A. (1910). *Zap. Akad. Nauk. Fiz. Mat. Otdel.*, VIII Ser. 25:3. *A. A. Markov Collected Works*, Izd. Akad. Nauk, USSR, 1951, pp. 465–509.

[72] Martin, J. J. (1967). *Bayesian Decision Problems and Markov Chains*. Wiley, New York.

[73] Moyal, J. E. (1957). *Acta Math.*, **98**, 221–264.

[74] Parzen, E. (1962). *Stochastic Processes*. Holden-Day, San Francisco, Calif.

[75] Prabhu, N. U. (1965). *Stochastic Processes*. The Macmillan Company, New York.

[76] Prabhu, N. U. (1965). *Queues and Inventories*. Wiley, New York.

[77] Rau, J. G. (1970). *Optimization and Probability in Systems Engineering*. Van Nostrand-Reinhold, New York.

[78] Ross, S. M. (1970). *Applied Probability Models with Optimization Applications*. Holden-Day, San Francisco. (Chapter 6 gives a good introduction to Markov decision processes.)

[79] Schwarzacher, W. (1969). *J. Int. Assoc. Math. Geology*, **1**, 17–39.

[80] Spitzer, F. (1964). *Principles of Random Walk*. Van Nostrand, Princeton, N.J.

[81] Takács, L. (1960). *Stochastic Processes*. Wiley, New York. (Methuen.)

[82] Takács, L. (1967). *Combinatorial Methods in the Theory of Stochastic Processes*. Wiley, New York.

[83] Trivedi, K. S. (1982). *Probability and Statistics with Reliability, Queueing and Computer Science Applications*. Prentice-Hall, Englewood Cliffs, N.J.

[84] Wald, A. (1947). *Sequential Analysis*. Wiley, New York.

[85] Wald, A. (1950). *Statistical Decision Functions*. Wiley, New York.

[86] Whittle, P. (1967). *Bull. Int. Stat. Inst.*, **42**, 642–646.

[87] Whittle, P. (1968). *J. Appl. Prob.*, **5**, 567–571.

Bibliography

The following books provide additional references on Markov processes.

Heyman, D. P. and Sobel, M. J. (1982). *Stochastic Models in Operations Research, Vol. I*. McGraw-Hill, New York. (For Markov processes and applications.)

Heyman, D. P. and Sobel, M. J. (1983). *Stochastic Models in Operations Research, Vol. II*. McGraw-Hill, New York. (For Markov decision processes.)

Kannan, D. (1979). *An Introduction to Stochastic Processes*. North Holland, Amsterdam.

Medhi, J. (1982). *Stochastic Processes*. Wiley Eastern, New Delhi, India.

Prabhu, N. U. (1980). *Stochastic Storage Processes*. Springer-Verlag, New York. (For Markov processes in queues, insurance risk, and dams.)

Ross, S. M. (1983). *Stochastic Processes*. Wiley, New York.

Ross, S. M. (1983). *Introduction to Stochastic Dynamic Programming*. Academic Press, New York. (For Markov decision processes.)

Whittle, P. (1982). *Optimization Over Time: Dynamic Programming and Stochastic Control*. Wiley, New York. (For Markov decision processes.)

(BIRTH-AND-DEATH PROCESSES
BRANCHING PROCESSES
CHAPMAN–KOLMOGOROV EQUATIONS
CHUNG PROCESSES
DIFFUSION PROCESSES
EMBEDDED PROCESSES
ERGODIC THEOREMS
GALTON–WATSON PROCESS
IMMIGRATION–EMIGRATION PROCESSES
JUMP PROCESSES
MARKOV DECISION PROCESSES
MARKOVIAN DEPENDENCE

QUEUEING THEORY
RANDOM WALKS
RENEWAL THEORY
SEMI-MARKOV PROCESSES
STATIONARY PROCESSES
STOCHASTIC PROCESSES)

U. NARAYAN BHAT

MARKOV RANDOM FIELDS

In dealing with a simple stochastic process* $X(t)$, where t ranges over some subset of the real line R^1, the notion of Markovian dependence* is reasonably straightforward to define. Essentially, we require that given the value of the process at some time t^* the (conditional) distributions of $X(s)$ and $X(t)$ will always be independent when $s < t^* < t$. In any attempt to define an analogous property for random fields* (i.e., stochastic processes $X(\mathbf{t})$ whose "time" parameter lies in some N-dimensional Euclidean space R^N) one runs into immediate and serious difficulties in trying to determine how to carry over to this setting the ideas of "past" and "future" which are so inherently important in the one-dimensional case. To do this successfully it is necessary to break up the study of Markov random fields (MRFs) into two quite distinct cases; one in which the parameter $\mathbf{t}$ is allowed to vary in a continuous fashion over the whole of R^N, and one in which it varies only over some lattice subset of Euclidean space. Both the theory, and the examples to which the theory can be applied, are quite distinct in each of these cases, and so they are treated independently below.

LATTICE INDEXED MRFs

There are two essentially different types of lattices that can be used to index MRFs, regular and irregular. Regular lattices in the plane commonly occur under experimental conditions in plant ecology, and include discrete-valued fields such as presence or absence of infection in an array of plants (e.g., Cochran [11] and Freeman [17]), and continuous-valued fields such as individual plant yields (e.g., Mead [20–22]). Irregular lattices in the plane often arise as the result of sampling natural populations, such as tree diameters in a forest (e.g., Brown [10]). They also have arisen, in a more theoretical setting, in connection with log-linear interaction models for contingency tables* (Darroch et al. [12]) and, in a very applied setting, in connection with interaction in social networks (Kinderman and Snell [19]).

For the sake of brevity we consider only one example of a lattice-indexed MRF, that being the case in which the lattice is the simple, regular, integer lattice in the plane. Thus we can write our process simply as $X(i, j) = X_{i,j}$, where i and j are integers. In the most famous example of this field, the sites represent the positions of atoms in some two-dimensional array, and the variables $X(i, j)$ the spins of electrons at these sites. If we now assume that each electron can spin in one of only two directions, that neighboring electrons want to spin in the same direction, and that each electron looks only at its four "nearest neighbors" (defined below) in deciding on which direction it will choose for its spin, we have just defined a peasant's version of the so-called *Ising model* of statistical mechanics (*see* GIBBS DISTRIBUTIONS; LATTICE SYSTEMS).

There are two main approaches to the precise mathematical formulation of such a system, which stem from two nonequivalent definitions of a "nearest neighbor" system. There is a so-called joint probability approach due to Whittle [25], in which it is required that the joint probability of the variates should be of the product form

$$\prod_{i,j} Q_{i,j}(x_{ij}\,; x_{i-1,j}\,, x_{i+1,j}\,, x_{i,j+1}\,, x_{i,j-1}).$$

Bartlett [2–4], however, noted that as a starting point what we would most like to be able to do is to write the conditional probability distribution of $X(i, j)$, given all the other values of the MRF, simply as a function of its values at the four sites nearest to (i, j); namely, $x_{i-1,j}$, $x_{i+1,j}$, $x_{i,j-1}$ and $x_{i,j+1}$. (Such a formulation would describe a "nearest neighbor" MRF. The extension to "second nearest neighbor" fields, and beyond, should be clear.) It is no easy matter to show that these demands are mathematically reasonable. However, under the assumption that $X(i, j)$ can take only a finite number of values, including zero, Hammersley and Clifford (see Besag [5], for an easy proof, and HAMMERSLEY–CLIFFORD THEOREM) showed not only that they were reasonable, but also managed to determine the exact form of the joint density of the $X(i, j)$ on a finite lattice as

$$K \cdot \exp\left\{ \sum_i \sum_j x_{ij} G_{ij}(x_{ij}) + \sum_i \sum_j \sum_k \sum_l x_{ij} x_{kl} G_{ijkl}(x_{ij} x_{kl}) \right\}.$$

Here the G are completely arbitrary functions, subject to the sole requirement that G_{ijkl} be nonzero only if the points (i, j) and (k, l) are neighbors in the sense described above. Generally they will be simple functions depending on only one or two parameters. The normalizing constant K is usually a complicated function of the G's, and is known as the *partition function*.

The Hammersley–Clifford theorem can be extended to more general situations. For example, if in the above the G's depend only on their subscripts and not on their arguments, and the x_{ij} vary over the whole of R^1, the resulting density belongs to the *autonormal* model. In this case, the partition function is simply the square root of the determinant of the appropriate covariance matrix. A little thought shows that this will make the partition function a very complex function of the G_{ij}.

Because of the complexity of the partition function, the statistical analysis of lattice indexed MRFs, in terms of parameter estimation and goodness of fit testing, is generally a rather difficult exercise. An early approach to parameter estimation was based on *coding schemes*. As an example, consider the above nearest neighbor model on a 3×9 lattice, with the points of the lattice "coded" either as crosses or circles, as depicted below.

○ × ○ × ○ × ○ × ○
× ○ × ○ × ○ × ○ ×
○ × ○ × ○ × ○ × ○

Given the X_{ij} indexed by the crosses, the X_{ij} indexed by the circles form a set of independent random variables, and standard maximum likelihood techniques can be employed, and similarly for the X_{ij} indexed by the crosses. Although the two sets of random variables are not themselves independent, the two sets of analyses can be combined in some reasonable way. In general, dependence structures more complicated than the nearest neighbor structure can be handled by this method with more sophisticated coding. More recently, maximum likelihood* type techniques have been developed for parameter estimation, and their efficiency calculated in special cases. All estimation procedures of this type have difficulties coping with border effects, which can be quite sizeable if the lattice is small. There are two quite distinct ways of getting around this difficulty. The first involves estimation conditional on fixed border effects (Gleeson and McGilchrist [18]). The second uses the theoretical physicist's trick of mapping the lattice onto a torus, thereby identifying sides and edges, so that there are no longer any borders (e.g., Besag [6, 8] and Besag and Moran [9]).

When classical methods of statistical analysis become too cumbersome mathematically to be useful, the statistician often turns to Monte Carlo* analysis. For MRFs this is generally very time consuming and, indeed, technically difficult, since the lack of ordering of "time" makes it difficult to know where to "start" the simulation*. Pickard [23, 24], however, describes a large class of MRFs whose joint distributions can be described in such a way as to make for easy simulation.

Most of what has been written above can be generalized, at least in principle, to MRFs defined on lattices in R^3 and higher dimensional spaces. In R^3, for example, each point on the integer lattice has six nearest neighbors to be considered in defining an appropriate nearest neighbor model. Higher dimensions are, however, significantly more complex mathematically, and far fewer exact results are available.

An excellent review of the statistical aspects of lattice indexed MRFs can be found in the Royal Statistical Society discussion paper of Besag [5] and in Besag [7]. The former paper, which also discusses the historical development of these MRFs, is required reading for anyone wishing to know more on this topic.

CONTINUOUS PARAMETER MRFs

Now let the parameter $\mathbf{t}$ of $X(\mathbf{t}) = X_{\mathbf{t}}$ vary in a continuous fashion over the whole of R^N. There is no totally satisfactory way of defining a Markov property for such processes, but a number of definitions do exist. All are based on the notion of a smooth surface ∂D in R^N that separates R^N into a bounded part D^- (the "interior" of ∂D) and an unbounded part, D^+ (the "exterior"). A simple definition of the Markov property is to demand that events that depend on $X_{\mathbf{t}}$ for $\mathbf{t} \in D^-$ and events that depend on $X_{\mathbf{t}}$ for $\mathbf{t} \in D^+$ are conditionally independent, given all the values of $X_{\mathbf{t}}$ for $\mathbf{t} \in \partial D$. (This definition, like those which follow, needs to be more precisely formulated in terms of σ-fields generated on ∂D, D^- and D^+, but we shall confine ourselves to a more heuristic setting. For a more rigorous treatment, and a wider treatment of continuous parameter MRFs in general, see the appendix in Adler [1].) As appealing as the above definition might be, it turns out to be excessively restrictive. For example, the only stationary Gaussian random fields satisfying this condition are either degenerate or deterministic (in the sense that the values of the field on one surface determine its value throughout the whole of R^N).

To find a more workable definition one has to condition on events in D^- and D^+, not by the information one has from ∂D, but rather the information contained in an infinitesimally small annulus around ∂D. A definition of MRFs based on this idea produces a rich theory with many examples and some surprising results. For example, the isotropic, or Lévy, Brownian motion* (the zero mean Gaussian field on R^N with covariance function $E\{X_{\mathbf{t}}X_{\mathbf{s}}\} = \frac{1}{2}\{|\mathbf{t}| + |\mathbf{s}| - |\mathbf{t} - \mathbf{s}|\}$) is, according to this definition, Markov in odd dimensions and non-Markov in even dimensions.

The theory of continuous parameter MRFs is at its most beautiful and useful when working with *generalized fields*, i.e., fields that are indexed not by points in R^N, but by classes of functions on R^N. As an example of this, imagine a fluid as a set of particles at points $\mathbf{t}_1, \mathbf{t}_2, \ldots$. The true fluid density is given by the random field $X(\mathbf{t}) = \sum_i \delta(\mathbf{t} - \mathbf{t}_i)$, where δ is the Dirac delta function*. We cannot generally observe $X(\mathbf{t})$, but rather blurred, locally averaged versions of it of the form $X(f) = \sum_i \int f(\mathbf{t} - \mathbf{t}_i) X(t_i)\, d\mathbf{t}$, where f is an averaging function. Varying f is like varying the resolution of the microscope or other instrument with which we measure density, or like looking at different regions. The important thing here is that we have replaced the time-indexed field $X(\mathbf{t})$ by a function-indexed (generalized) field $X(f)$. Such generalized fields are extremely important in theoretical physics, where they appear in both quantum field theory and statistical mechanics. In the latter they arise, for example, as continuum limits of the Ising model.

Generalized MRFs are the subject of vigorous current investigation by probabilists, much of it associated with the name of Dobrushin (see, in particular, refs. 13–16, and the references therein). However, since even the notation required to begin describing recent results in this area is long and involved, it must be left to interested readers themselves to find out more about this fascinating area, using perhaps refs. 15 and 16 following as a starting point.

References

[1] Adler, R. J. (1981). *The Geometry of Random Fields*. Wiley, Chichester. (The appendix deals with MRFs.)

[2] Bartlett, M. S. (1955). *An Introduction to Stochastic Processes*. Cambridge University Press, Cambridge, England.

[3] Bartlett, M. S. (1967). *J. R. Statist. Soc. A*, **130**, 457–477.

[4] Bartlett, M. S. (1968). *J. R. Statist. Soc. A*, **131**, 579–580.

[5] Besag, J. E. (1974). *J. R. Statist. Soc. B*, **35**, 192–236. (A well-written survey article with 12 pages of enlightening discussion.)

[6] Besag, J. E. (1975). *The Statistician*, **24**, 179–195.

[7] Besag, J. E. (1977). *Conf. on Information Theory, Prague*, **A**, 47–56.

[8] Besag, J. E. (1977). *Biometrika*, **64**, 616–618.

[9] Besag, J. E. and Moran, P. A. P. (1975). *Biometrika*, **62**, 555–562.

[10] Brown, G. S. (1965). *N.Z. For. Serv. Res.*, Note **38**, 1–11.

[11] Cochran, W. G. (1936). *J. R. Statist. Soc. Supp.*, **3**, 49–67.

[12] Darroch, J. N., Lauritzen, S. L., and Speed, T. P. (1980). *Ann. Statist.* **8**, 522–539.

[13] Dobrushin, R. L. (1979). *Ann. Prob.*, **7**, 1–28.

[14] Dobrushin, R. L. (1979). *Zeit. Wahrscheinlichkeitsth. verwand. Geb.*, **49**, 275–293.

[15] Dobrushin, R. L. and Sinai, Ya. G. (1980). *Multicomponent Random Systems*. Marcel Dekker, New York. (This is an important volume containing 21 papers by Russian authors on various aspects of MRFs.)

[16] Dynkin, E. B. (1980). *Bull. Amer. Math. Soc.*, **3**, 975–999.

[17] Freeman, G. H. (1953). *Biometrika*, **40**, 287–305.

[18] Gleeson, A. G. and McGilchrist, C. A. (1980). *Austral. J. Statist.*, **22**, 197–206.

[19] Kinderman, R. P. and Snell, J. L. (1980). *J. Math. Sociology*, **7**, 1–13.

[20] Mead, R. (1966). *Ann. Bot.*, **30**, 301–309.

[21] Mead, R. (1967). *Biometrics*, **23**, 189–205.

[22] Mead, R. (1968). *J. Ecol.*, **56**, 35–45.

[23] Pickard, D. K. (1977). *J. Appl. Prob.*, **14**, 717–731.

[24] Pickard, D. K. (1980). *Adv. Appl. Prob.*, **12**, 655–671.

[25] Whittle, P. (1963). *Bull. Internat. Statist. Inst.*, **40**, 974–994.

(GIBBS DISTRIBUTIONS
HAMMERSLEY–CLIFFORD THEOREM
LATTICE SYSTEMS
MARKOVIAN DEPENDENCE
RANDOM FIELDS)

ROBERT J. ADLER

MARRIAGE

Demographers study marriage mainly, indeed almost exclusively, as an element in fertility* measurement. If the birthrate changes, it could be the result of an alteration either in the proportion (or the age at) marrying or in intrinsic fertility. That deferment of marriage can have an important effect on fertility is shown by the population history of Ireland since the potato famine in the 1840s and by the introduction in China in recent years of a ban on early marriages as a means of birth control. In most countries, marriages are registered and information about marital status is obtained at censuses; the resulting statistics provide the basis for the calculation of rates* and proportions classified by age. For a general commentary on the significance and use of such tools, see DEMOGRAPHY.

Table 1 gives an idea of the distribution of marriages by age, for men and women, and of the relative ages of the brides and bridegrooms. It relates to the numbers of marriages in England and Wales in the year 1967.

Marriage also has an influence on other elements in population studies, for example mortality and migration*. Those who migrate to other countries tend to be single, but a wedding often leads to a change of address within a country. Sick and disabled people may be prevented by illness from marrying, and thus the unmarried include a higher proportion of people likely to die sooner than do the married. Such effects are of relatively minor importance but can be sig-

Table 1 Marriages in England and Wales, 1967 (numbers in thousands)

Age of wife	Age of husband					
	Under 20	20–24	25–29	30–34	35 and over	Total
Under 20	25	66	9	2	1	103
20–24	7	125	47	10	2	191
25–29		8	15	8	7	38
30–34		1	3	4	7	15
35 and over			2	2	35	39
Total	32	200	76	26	52	386

nificant in certain circumstances, e.g., in the construction of life tables* for use in estimates of the cost of social security and pensions, where the benefits differ according to marital status.

Historical studies have shown that in most earlier societies marriage was virtually universal at puberty, and this custom still continues in many countries today. For reasons probably connected with economic development, however, late marriage and abstention from marriage have emerged as a normal feature in Western and other countries. It is not surprising, therefore, that marriage is well worthy of study for its contribution to wider fields of knowledge. Psychologists take an interest in the relative ages of husbands and wives, and sociologists study the effect of marriage on women's status generally; the first of these two topics involves an analysis of the way in which variations in the relative numbers of nonmarried men and women influence the ages at which couples marry. The second includes consideration of the occupational and social-class distribution of single and married women.

While one man and one woman formally united form the general rule, in some areas (e.g., West Africa) polygamy is common; in others (e.g., the West Indies) there is a prevalence of "consensual" marriage, that is, couples living together for long periods and recognized by their fellow citizens as married, although they have not undergone any religious or civil ceremony together and may not ultimately do so. The study of these practices and their effect on fertility (which surprisingly tends to be depressive rather than heightening) is of much importance to sociologists and demographers. Customs also vary widely in regard to divorce and as to what should happen after the death of one spouse. Even within a country, remarriage rates vary, inter alia, according to the cause of the cessation of the previous marriage, the time elapsed since it ended, and the sex of the person in question. Although mortality has declined, and so reduced the chances of a marriage ending prematurely, divorce has increased in many countries; the study of remarriage thus remains very significant.

Bibliography

Cox, P. R. (1970). *J. Biosocial Sci.*, Suppl. **2**.

Cox, P. R. (1970). *J. Biosocial Sci.*, **2**, 111–121.

(These two papers show how the relative ages of partners at marriage vary according to place and time, and discuss the probable causes.)

Hajnal, J. (1955). European Marriage Patterns in Perspective. In *Population in History*. Aldine, Chicago. (An important paper and an exceptional one in reading matter which is otherwise very scattered. See also, however, the relevant parts of any standard demographic textbook, for example those in the Bibliography to DEMOGRAPHY.)

(DEMOGRAPHY
FERTILITY
RATES
VITAL STATISTICS)

PETER R. COX

MARSAGLIA'S TABLE METHOD

Marsaglia's table method [4] is a clever, fast, general-purpose method of generating discrete random variables over finite sample spaces (*see* GENERATION OF RANDOM VARIABLES). It is based upon representing the probabilities of the discrete probability density $f(\cdot)$ from which random variables are desired as a finite-length decimal (or other base) representation:

$$f(i) = d_{1i}B^{-1} + d_{2i}B^{-2} \dots + d_{ri}B^{-r},$$
$$i = 1,2, \dots, m, \quad m < \infty. \quad (1)$$

Here B denotes the base, r the length of the base B representation, and $d_{ji}, j = 1,2, \dots, r, i = 1,2, \dots, m$, nonnegative integers.

Using (1), $f(i)$ can be written as a mixture of discrete distributions $q_j(\cdot)$,

$$f(i) = \sum_{j=1}^{r} p(j)q_j(i) \qquad i = 1,2, \dots, m, \quad (2)$$

where $p(j) = n_j B^{-j}$ are the mixture weights and $q_j(i) = d_{ji}/n_j$ are the mixture densities, where $n_j = \sum_{i=1}^{m} d_{ji}$. (It is easy to show that $0 \leqslant p(j) \leqslant 1$, $\sum_{j=1}^{r} p(j) = 1$ and $\sum_{i=1}^{m} q_j(i) = 1$, $j = 1, \dots, r$.) This easily can be seen by multiplying and dividing the jth term of (1) by n_j and regrouping within each factor:

$$f(i) = \left(\frac{d_{1i}}{n_1}\right)(n_1 B^{-1}) + \left(\frac{d_{2i}}{n_2}\right)(n_2 B^{-2}) + \dots + \left(\frac{d_{ri}}{n_r}\right)(n_r B^{-r}).$$

Marsaglia's table algorithm uses the mixture representation (2) to generate realizations of random variables distributed as $f(\cdot)$ as follows:

For each random variable X desired from $f(\cdot)$ perform the following steps:

1. [Select at random a place in the desired representation.]
   Generate $J \sim p(\cdot)$.
2. [Select an outcome based only on the Jth digit in the base B expansions of $f(i), i = 1,2, \dots, m$.]
   Generate $X \sim q_J(\cdot)$.

As an example, consider a discrete distribution $f(\cdot)$ with $m = 3$ mass points:

$$f(1) = .1694$$
$$f(2) = .5135$$
$$f(3) = .3171.$$

Using base $B = 10$ and number of digits $r = 4$, $f(\cdot)$ is written as a mixture (2) with:

$$n_1 = 9, \quad p(1) = .9,$$
$$q_1(1) = \tfrac{1}{9}, \quad q_1(2) = \tfrac{5}{9}, \quad q_1(3) = \tfrac{3}{9},$$
$$n_2 = 8, \quad p(2) = .08,$$
$$q_3(1) = \tfrac{6}{8}, \quad q_2(2) = \tfrac{1}{8}, \quad q_2(3) = \tfrac{1}{8},$$
$$n_3 = 19, \quad p(3) = .019,$$
$$q_3(1) = \tfrac{9}{19}, \quad q_3(2) = \tfrac{3}{19}, \quad q_3(3) = \tfrac{7}{19},$$
$$n_4 = 10, \quad p(4) = .001,$$
$$q_4(1) = \tfrac{4}{10}, \quad q_4(2) = \tfrac{5}{10}, \quad q_4(3) = \tfrac{1}{10}.$$

Obviously the efficiency of the methods used for steps 1 and 2 will determine the performance of this algorithm. Despite the apparent additional complexity that *two* discrete random variables (J and X) are generated, the mixture representation (2) based on the base B expansion is quite efficient.

Step 2 is efficiently accomplished by using "urn sampling," which is very fast, from the distribution $q_J(\cdot)$. Since all the probabilities $q_J(i) = d_{Ji}/n_J$, $i = 1,2, \dots, m$, have a common denominator n_J, an array $T_J(\cdot)$ can be created with d_{J1} words set to "1", d_{J2} words to "2", . . . , d_{Jr} words to "r". We generate $I \sim U(1,2, \dots, n_J)$, and then $X \leftarrow T_J(I)$ will be a random variable from $q_J(\cdot)$. The array T is the "urn" and the uniform integer I selects a "ball" (mass point) at random from the urn.

Step 1 is less efficiently accomplished. But in the usual case where r, the number of digits, is chosen to be small, the random variable J in step 1 can be generated by using the inverse cumulative transformation linear-search method, that is, $J \leftarrow \min(j: U \leqslant F(j))$, where $U \sim$uniform $(0, 1)$ and $F(j) \equiv \sum_{i=1}^{j} p(i)$.

Marsaglia [4] implements these steps in a clever and efficient manner by using several devices. First he reduces the number of uniform random variables to one by obtaining "I" used above in step 2 from the uniform $(0, 1)$ used in step 1 by conditioning on J. Second, the arrays $T_J(\cdot)$ are stored one following the other in a one-dimensional array T. Finally, he uses integer arithmetic whenever possible to minimize the use of the more time-consuming floating point operations. Given below is the resulting table algorithm as it might be implemented in a high-order computer language:

M1. Generate $U \sim U(0, 1)$ and set $J \leftarrow 0$.

M2. $J \leftarrow J + 1, U \leftarrow U \times B, I \leftarrow \lceil U \rceil$ ($\lceil\ \rceil$ is the greatest integer function).

M3. If $I > N_J$, then go to M2.

M4. $X \leftarrow T(I - L_J)$, return X.

The arrays N, T, and L are computed in a setup program which is executed once prior to the use of the generation algorithm. These tables are defined as follows:

(1) $N_J = \sum_{K=1}^{J} n_K B^{J-K}$,

(2) $L_J = N_J - \sum_{K=1}^{J} n_K$, and

(3) T is defined as given above.

The looping and table look-ups for the N and L arrays can be eliminated by repeating code for each J and using constants for N_J and L_J. This adds to the length of the program but will increase the speed considerably.

For the $f(\cdot)$ given in our example,

$$N_1 = 9, \quad N_2 = 98, \quad N_3 = 999,$$

$$L_1 = 0, \quad L_2 = 90, \quad L_3 = 963,$$

and the T array has length $\sum_{K=1}^{4} n_K = 46$ and components

$$1, \overbrace{2, \ldots, 2}^{5}, 3, 3, 3, \overbrace{1, \ldots, 1}^{6}, 2, 3,$$

$$\overbrace{1, \ldots, 1}^{9}, 2, 2, 2, \overbrace{3, \ldots, 3}^{7},$$

$$\overbrace{1, \ldots, 1}^{4}, \overbrace{2, \ldots, 2}^{5}, 3.$$

Both the speed and the storage requirements of this method are a complex function of the number of digits r, the base B used to represent $f(\cdot)$, the distribution of the total probability mass among the r digits and the computer and language used. In practice, 4 digits, base 10 or 16, has been suggested [5]. For densities with probabilities requiring more digits than the chosen r, as is often the case, Marsaglia's table method is not exact. In this case, Marsaglia suggests using a truncated or rounded version of $f(\cdot)$.

For over 20 years Marsaglia's table method was the method of choice for generating large numbers of discrete random variables. Recently other methods have been developed, notably Walker's method* [7, 3] and the indexed-search method [1], that are exact and competitive in efficiency with the table method. For a comparison of the methods, see refs. 2 and 6.

References

[1] Atkinson, A. C. (1979a). *Appl. Statist.*, **28**, 29–35.

[2] Atkinson, A. C. (1979b). *Appl. Statist.*, **28**, 260–263.

[3] Kronmal, R. A. and Peterson, Jr., A. V. (1979). *Amer. Statist.*, **33**, 214–218.

[4] Marsaglia, G. (1963). *Commun. ACM*, **6**, 37–38.

[5] Norman, J. E. and Cannon, L. E. (1972). *J. Statist. Comput. Simul.*, **1**, 331–348.

[6] Peterson, Jr., A. V. and Kronmal, R. A. (1983). *Appl. Statist.*, **32**, 276–286.

[7] Walker, A. J. (1977). *ACM Trans. Math. Software*, **3**, 253–256.

(GENERATION OF RANDOM VARIABLES
MIXTURE METHOD
WALKER'S METHOD)

RICHARD A. KRONMAL
ARTHUR PETERSON

MARSHALL–EDGEWORTH–BOWLEY INDEX

The Marshall–Edgeworth–Bowley (MEB) Index is a price index number* which measures the relative change in the general price level between two periods, or its reciprocal, the purchasing power of money.

DEFINITION OF THE INDEX

Let the set of commodities under consideration be $A = \{a_1, \ldots, a_n\}$, observed in the base period o and the current period 1. The prices of a_i in these two periods are p_{io} and p_{i1} and the corresponding quantities are q_{io} and q_{i1}. The MEB Price Index in period 1 relative to period o, P_{o1}, is:

$$p_{o1} = \frac{\sum_i p_{i1}(q_{io} + q_{i1})}{\sum_i p_{io}(q_{io} + q_{i1})}. \tag{1}$$

This index was originally developed by Marshall [9] and advocated by Edgeworth [5]. Hence it is sometimes named after either or both of these two authors. Bowley [2] proved that this index provides a close approximation to the economic cost of living index, hence its name in this article.

PROPERTIES OF THE MEB INDEX

Most price indexes commonly used in practice are weighted averages of price ratios with the weights based on either q_{io} or q_{i1} or some combination of both q_{io} and q_{i1}. Indexes weighted by q_{io} or q_{i1}, like those of Laspeyres and Paasche (*see* PAASCHE–LASPEYRES INDEX NUMBERS), are usually simpler to interpret than those weighted by a combination of both q_{io} and q_{i1}, like the MEB index, the Walsh index (*see* INDEX NUMBERS), and Fisher's ideal index*. However, indexes that combine q_{io} and q_{i1} as weights are preferable in certain situations. For example, when the distribution of commodities changes drastically between the two periods under comparison, then both q_{io} and q_{i1} should enter the weighting system, and the MEB index which combines q_{io} and q_{i1} in a simple manner is particularly suitable. But the MEB index, unlike the Walsh index, is not suitable for multiperiod comparisons, and both these indexes fail the factor reversal test, unlike the Fisher ideal index.

Algebraically, the MEB index can be expressed in terms of Laspeyres and Paasche indexes as follows:

$$p_{o1} = \left(\frac{\sum_i p_{i1}q_{io}}{\sum_i p_{io}q_{io}}\right)\left[\frac{1 + \dfrac{\sum_i p_{i1}q_{i1}}{\sum_i p_{i1}q_{io}}}{1 + \dfrac{\sum_i p_{io}q_{i1}}{\sum_i p_{io}q_{io}}}\right]. \tag{2}$$

This equation is the product of two terms. The first is a Laspeyres price index and the second is the ratio of (a) one plus a Paasche quantity index to (b) one plus a Laspeyres quantity index.

THE BOWLEY APPROXIMATION

One advantage of the MEB index is that it is approximately equal to a version of the true economic cost of living index (Bowley [2]). In the base period the consumer chooses the quantities q_{io} that maximize his ordinal utility function $u(q_{1o}, \ldots, q_{no})$ subject to his budget constraint $y_o = \sum_i p_{io}q_{io}$. In the current period the consumer faces the generally different prices p_{i1} and the problem is to determine the optimal quantities $\tilde{q}_{i1}$ that minimize $\tilde{y}_1 = \sum_i p_{i1}\tilde{q}_{i1}$ subject to the constraint

$$u(\tilde{q}_{11}, \ldots, \tilde{q}_{n1}) = u(q_{1o}, \ldots, q_{no}).$$

The true economic cost of living index is given by

$$\tilde{P}_{o1} = \frac{\tilde{y}_1}{y_o} = \frac{\sum_i p_{i1}\tilde{q}_{i1}}{\sum_i p_{io}q_{io}}. \tag{3}$$

The Bowley approximation of $\tilde{P}_{o1}$ in (3) by P_{o1} in (1) is based on approximating $u(\cdot)$ by a Taylor expansion in the neighborhood of some point in the quantity space. The details of the derivation are in Bowley [2, pp. 223–226].

EXTENSIONS AND APPLICATIONS

Frisch [7, p. 28] modified the Bowley approximation and obtained an expression giving a closer approximation to $\tilde{P}_{o1}$ than P_{o1}. Instead of indexes based on Taylor series approximations of an arbitrary utility function, recent developments of the economic theory of index numbers provide various exact cost of living indexes derived from particular classes of quadratic utility functions. See refs. 1 and 4 for examples.

The average of the quantities q_{io} and q_{i1} used as weights in calculating P_{o1} is usually replaced in practice by the average of the expenditures $p_{io}q_{io}$ and $p_{i1}q_{i1}$ as, for example, in the Swedish cost of living index described in Hofsten [8].

The MEB index was originally designed as a price index between two time periods. However, a quantity index corresponding to P_{o1} may be defined by analogy as:

$$Q_{o1} = \frac{\sum_i q_{i1}(p_{io} + p_{i1})}{\sum_i q_{io}(p_{io} + p_{i1})}. \tag{4}$$

This is the quantity index to match P_{o1} suggested, for example, in Crowe [3, p. 63], and was applied to U.S. data by Fabricant [6, p. 358]. However, given the price index P_{o1}, an implicit quantity index $\tilde{Q}_{o1}$ may be defined using Fisher's weak factor reversal test:

$$\tilde{Q}_{o1} = \left(\frac{\sum_i p_{i1}q_{i1}}{\sum_i p_{io}q_{io}}\right) \bigg/ P_{o1}. \tag{5}$$

This can be interpreted as the real income index corresponding to the cost of living index $\tilde{P}_{o1}$.

References

[1] Balk, B. M. (1981). *Econometrica*, **49**, 1553–1558. (Discusses economic index numbers based on quadratic utility functions; intermediate, few references.)

[2] Bowley, A. R. (1928). *Econ. J.*, **38**, 216–237. (Detailed derivation of the Bowley approximation and numerical examples.)

[3] Crowe, W. R. (1965). *Index Numbers*. Macdonald and Evans, London. (Elementary.)

[4] Diewert, W. E. (1981). In *Essays in the Theory of Consumer Behaviour*, A. S. Deaton, ed. Cambridge University Press, London, pp. 163–208. (Excellent survey; intermediate, extensive references.)

[5] Edgeworth, F. W. (1925). *Papers Relating to Political Economy*, Vol. 1. Macmillan, London. (Papers H and I on index numbers.)

[6] Fabricant, S. (1940). *The Output of Manufacturing Industries, 1899–1937*. National Bureau of Economic Research, New York.

[7] Frisch, R. (1936). *Econometrica*, **4**, 1–38. (Survey article, contains alternative derivation and extension of the Bowley approximation, plus references.)

[8] Hofsten, E. von (1952). *Price Indexes and Quality Change*. Allen and Unwin, London. (Elementary, contains application of the Marshall–Edgeworth–Bowley index to Swedish data, plus references.)

[9] Marshall, A. (1887). *Contemporary Rev.*, **51**, 355–375.

(FISHER'S IDEAL INDEX NUMBER
INDEX NUMBERS
PAASCHE–LASPEYRES INDEX NUMBERS)

NURI T. JAZAIRI

MARTINGALES

The term *martingale* comes from the Provençal name of the French community Martigues. It has a long history in a gambling context, where originally it meant a system for recouping losses by doubling the stake after each loss. The modern mathematical concept of a martingale still may be described in terms of a gambling system, as follows. Suppose a gambler plays a sequence of games according to a strategy which incorporates information about the results of previous games. The games might be said to be "fair" if the expected size of the gambler's bank after the nth game, given the results of all previous games, is unchanged from the size after the $(n-1)$th game. That is, the gambler's average gain or loss after each game, given the previous history of the games, is zero. A martingale is just a mathematical formalization of this concept of a fair game. Its practical importance derives from the fact that many naturally occurring phenomena can be modeled by a sequence of fair games, or closely approximated by such a sequence.

Let $\{S_n, n \geq 1\}$ be a sequence of random variables with finite means. The sequence is called a *martingale* if

$$E[S_{n+1} \mid S_1, S_2, \ldots, S_n] = S_n, \qquad n \geq 1. \quad (1)$$

(The left-hand side equals the mean of S_{n+1}, given the values of $S_1, S_2, \ldots, S_n$.) This implies that

$$E[S_{n+1} \mid S_1, S_2, \ldots, S_m] = S_m$$

for all $m \leq n$. In a slightly more general context, we call a sequence $\{(S_n, \mathscr{F}_n), n \geq 1\}$ of random variables S_n and σ-fields $\mathscr{F}_n$, a martingale if the σ-fields are increasing, if S_n is measurable in $\mathscr{F}_n$ for $n \geq 1$, and if

$$E[S_{n+1} \mid \mathscr{F}_n] = S_n, \qquad n \geq 1. \quad (2)$$

Note that if $\{(S_n, \mathscr{F}_n), n \geq 1\}$ is a martingale in the sense of the second definition, then $\{S_n, n \geq 1\}$ is a martingale in the sense of the first. For the sake of simplicity we shall use the first definition in the discussion below.

The martingale property ensures that $E(S_n)$ does not depend on n. Since martingales are preserved under translation (that is, $\{S_n\}$ is a martingale if and only if $\{S_n - c\}$ is a martingale, for each constant c) then there is no essential loss of generality in assuming that a martingale has zero mean. This condition is often imposed in martingale limit theorems.

If we write $X_1 = S_1$ and $X_n = S_n - S_{n-1}$, $n \geq 2$, then the constraints (1) and (2) may be written as

$$E[X_{n+1} \mid X_1, X_2, \ldots, X_n] = 0 \quad \text{and}$$
$$E[X_{n+1} \mid \mathscr{F}_n] = 0, \qquad n \geq 1,$$

respectively. The variables X_n are known as *martingale differences*. If the constraint (1) is changed to

$$E[S_{n+1} \mid S_1, S_2, \ldots, S_n] \geq S_n, \qquad n \geq 1,$$

we say that $\{S_n\}$ is a *submartingale*, while if the direction of this inequality is reversed, we call the sequence a *supermartingale*. Chung [8, p. 319] introduced the term *smartingale* to describe both these processes. In the gambling example of the opening paragraph, S_n represents the size of the gambler's bank after the nth game, and so a submartingale corresponds to a sequence of gambling games which tend to run in favor of the gambler. The sequence $\{S_n\}$ is called a *reverse martingale* (or *backwards martingale*) if

$$E[S_n \mid S_{n+1}, S_{n+2}, \ldots] = S_{n+1}, \qquad n \geq 1.$$

For the reverse martingale, the role of the past is replaced by that of the future.

If $\{S_n\}$ is a martingale and Q is a convex function such that $Q(S_n)$ has finite mean for each n, then $\{Q(S_n)\}$ is a submartingale. In particular, if $p \geq 1$ and $E[|S_n|^p] < \infty$ for all n, then $\{|S_n|^p\}$ is a submartingale. Any submartingale can be broken up into a martingale and an increasing process, using Doob's decomposition theorem [8, Theorem 9.3.2, p. 321]. This states that a submartingale $\{S_n\}$ may be written in the form $S_n = M_n + A_n$, where $\{M_n\}$ is a martingale and $\{A_n\}$ is an increasing sequence of nonnegative random variables. If we stipulate that A_n be measurable in the σ-field generated by S_1, $S_2, \ldots, S_{n-1}$, then this decomposition is uniquely determined almost surely by the relationship

$$A_{n+1} - A_n = E[S_{n+1} \mid S_1, S_2, \ldots, S_n] - S_n, \qquad n \geq 1. \quad (3)$$

EARLY HISTORY

Work on the mathematical theory of martingales by Bernstein and Lévy predated the use of the word martingale in the mathematical literature. This early interest was due to the fact that martingales represent a generalization of the concept of sums of independent random variables. For example, if $S_n = \sum_1^n X_j$ where the X_j's are independent with zero means, then condition (1) holds. This relationship suggests that the vast paraphernalia of limit theory for sums of independent random variables, such as laws of large numbers*, central limit theorems*, and laws of the iterated logarithm*, might have a generalization to martingales. Indeed, the first moves in this direction were made by Bernstein [3] and Lévy [15], who derived martingale central limit theorems.

Doob [9, Chap. VII] was responsible for establishing martingale theory as a major tool in its own right, rather than just an extension of the theory of sums of independent random variables. Contemporary martingale theory may be said to date from Doob's martingale convergence theorem, which plays the role of a strong law of large numbers.

Theorem 1. If $\{S_n, n \geqslant 1\}$ is a submartingale and if $E[|S_n|]$ is bounded in n, then S_n converges almost surely as $n \to \infty$ to a random variable S_∞ with $E[|S_\infty|] < \infty$.

This fundamental result has numerous and wide-ranging applications. For example, it provides a very elementary proof of the rather deep probabilistic result that a sum of independent random variables which converges in distribution also converges almost surely; see Chung [8, p. 347]. As a second example, let Z_n denote the number of individuals in the nth generation of a simple supercritical Bienaymé–Galton–Watson process with finite mean m, and set $S_n = m^{-n}Z_n$, $n \geqslant 1$ (*see* BRANCHING PROCESSES*). Then $\{S_n\}$ is a martingale, and $E[|S_n|] = E[S_n] = 1$ for each n. Therefore the conditions of Theorem 1 are satisfied and so there exists a limit random variable W such that $m^{-n}Z_n \to W$ almost surely. The limit is nondegenerate if and only if $E[Z_1 \log Z_1] < \infty$.

The subsequent development of martingale theory has been based in a large part on limit theory, and has tended to parallel the earlier evolution of limit theory for sums of independent random variables. Thus, many applications of martingale theory have been to "large sample" problems in probability and statistics.

LIMIT THEORY

The principal laws of large numbers for sums of independent random variables may be written as *three series theorems*. In both the weak and strong cases these have direct analogues for martingales. They are stated as Theorems 2.13 and 2.16 of Hall and Heyde [13, pp. 29 and 33]. The major difference between the martingale results and those for sums of independent variables is that in the martingale case, the series conditions are sufficient but no longer necessary for convergence.

The history of the martingale central limit theorem* and its invariance principle* is outlined in Hall and Heyde [13, pp. 51–52]. Brown [5] proved a version of Lindeberg's theorem for martingales. The conditions of Brown's theorem were slightly weakened by McLeish [17], and put into a form sometimes easier to check in applications. We state here a central limit theorem for convergence to mixtures of normal laws. A more general version is presented as Theorem 3.2 of Hall and Heyde [13, p. 58].

Theorem 2. Let $\{S_n\}$ be a martingale with $E[S_n] = 0$ and $E[S_n^2] < \infty$ for each n, and let $X_1 = S_1$ and $X_n = S_n - S_{n-1}$, $n \geqslant 2$, denote the martingale differences. Suppose there exist positive constants c_n, $n \geqslant 1$, and an almost surely finite random variable η^2, such that

$$c_n^{-1} \max_{j \leqslant n} |X_j| \to 0$$

and

$$c_n^{-2} \sum_1^n X_j^2 \to \eta^2 \qquad \text{in probability,}$$

and

$$c_n^{-2} E\left[\max_{j \leqslant n} X_j^2\right] \text{ is bounded in } n. \tag{4}$$

Then $c_n^{-1}S_n$ has a limiting distribution with characteristic function*

$$\psi(t) = E\{\exp(-\eta^2 t^2/2)\}, \qquad -\infty < t < \infty.$$

The constants c_n would typically be taken equal to the standard deviation of S_n, $c_n = (E[S_n^2])^{1/2}$. In that case, condition (4) would hold trivially. The most familiar version of Theorem 2 is the one in which

$\eta^2 = 1$, for then the limit is standard normal. A standard normal limit may be obtained even when η^2 is not constant, by norming with a random variable rather than the constant c_n. Under the conditions of Theorem 2,

$$\left(\sum_1^n X_j^2\right)^{-1/2} S_n \to N(0, 1) \text{ in distribution,} \tag{5}$$

provided $P(\eta^2 > 0) = 1$; see Theorem 3.3 of Hall and Heyde [13, p. 64].

Early central limit theorems for martingales imposed conditions on the *conditional variance*,

$$V_n^2 = \sum_{j=1}^n E\left[X_j^2 \mid X_1, X_2, \ldots, X_{j-1}\right],$$

rather than on the sum of squares, $U_n^2 = \sum_1^n X_j^2$. If $c_n^{-2} V_n^2 \to \eta^2$ in probability, where η^2 is an almost surely finite random variable, then it is necessarily also true that $c_n^{-2} U_n^2 \to \eta^2$ in probability. Thus, norming by the random variable U_n, as in (5), is virtually equivalent to norming by V_n.

A convenient way of introducing the conditional variance is via Doob's decomposition of a submartingale. If we replace the zero mean martingale $\{S_n\}$ by the submartingale $\{S_n^2\}$, we may deduce from (3) that the increasing process in Doob's decomposition of S_n^2 is $A_n = V_n^2$.

The central limit theorem may be interpreted as a weak description of the rate of convergence in a law of large numbers. Strong rates of convergence are provided by laws of the iterated logarithm. Martingale laws may be stated using either a constant norming as in Theorem 2, or a random norming as in the result (5). See Section 4.4 of Hall and Heyde [13, p. 115] for examples of these laws, for invariance principles* in the law of the iterated logarithm, and for references to earlier work. We present here only a special case, to illustrate the form of more general laws with random norming.

Theorem 3. Let $\{S_n\}$ be a martingale whose differences X_n, $n \geqslant 1$, are uniformly bounded, and which satisfies $E[S_n] = 0$ for each n. Then with V_n denoting the conditional variance, we have

$$\limsup_{n\to\infty} (2V_n \log\log V_n)^{-1/2} S_n = 1,$$

$$\liminf_{n\to\infty} (2V_n \log\log V_n)^{-1/2} S_n = -1$$

almost surely on the set $[V_n \to \infty]$.

INEQUALITIES

Inequalities are the main tools for the derivation of limit theorems. Thus, given that limit theory for martingales so closely parallels that for sums of independent random variables, it is to be expected that many inequalities for sums of independent variables have martingale analogues. Several have variants for submartingales. Perhaps the most basic result is a version of *Kolmogorov's inequality**: If $\{S_n\}$ is a martingale then

$$\lambda \Pr\left[\max_{j \leqslant n} |S_j| > \lambda\right] \leqslant E\left[|S_n| I\left(\max_{j\leqslant n} |S_j| > \lambda\right)\right] \leqslant \lambda^{1-p} E\left[|S_n|^p\right]$$

for all $\lambda > 0$, $1 \leqslant p < \infty$, and $n \geqslant 1$. Here $I(E)$ denotes the indicator function of the event E. Doob's inequality compares the p-norm of a martingale with the p-norm of its supremum: If $\{S_n\}$ is a martingale then

$$\|S_n\|_p \leqslant \left\|\max_{j\leqslant n} S_j\right\|_p \leqslant q\|S_n\|_p$$

for all $1 < p < \infty$ and $n \geqslant 1$, where $p^{-1} + q^{-1} = 1$ and $\|Z\|_p \equiv (E[|Z|^p])^{1/p}$. Doob's *upcrossing inequality** [9, p. 316] is a basic tool in the proof of the martingale convergence theorem (Theorem 1 above), although Doob's original proof followed a different route. The upcrossing inequality may also be employed to prove a useful variant of Kolmogorov's inequality: If $\{S_n\}$ is a zero mean martingale then [5]

$$\lambda \Pr\left[\max_{j\leqslant n} |S_j| > 2\lambda\right] \leqslant E\left[|S_n| I(|S_n| > \lambda)\right]$$

for all $\lambda > 0$.

The square function inequalities developed by Burkholder and others form a major and comparatively recent addition to the armory of martingale tools. They imply a certain relationship between the behavior of a martingale and that of the sum of squares of its differences, which had been noticed earlier for sums of independent random variables. Let $\{S_n\}$ be a martingale with differences X_n. *Burkholder's inequality* declares that for any $1 < p < \infty$ there exist constants $C_2 > C_1 > 0$, depending only on p, such that

$$C_1 E\left[\left|\sum_1^n X_j^2\right|^{p/2}\right] \leqslant E[|S_n|^p] \leqslant C_2 E\left[\left|\sum_1^n X_j^2\right|^{p/2}\right]$$

for all n. *Rosenthal's inequality* states that if $2 \leqslant p < \infty$,

$$D_1\left(E[V_n^p] + \sum_1^n E[|X_j|^p]\right) \leqslant E|S_n|^p \leqslant D_2\left(E[V_n^p] + \sum_1^n E[|X_j|^p]\right)$$

for constants $D_2 > D_1 > 0$ depending only on p, where $V_n = (V_n^2)^{1/2}$ is the square root of the conditional variance. It is also true that for any $0 < p < \infty$, there exists a constant C depending only on p such that

$$E\left[\max_{j \leqslant n}|S_j|^p\right] \leqslant C\left(E[V_n^p] + E\left[\max_{j \leqslant n}|X_j|^p\right]\right)$$

for all n.

Martingale inequalities are discussed in detail by Burkholder [6] and Garsia [12]. The more elementary inequalities are proved in most graduate texts on probability theory, such as Billingsley [4, pp. 414–416], Chung [8, Section 9.4], and Loève [16, pp. 57 and 201]. The inequalities described above are all derived in Hall and Heyde [13, Chapter 2].

SKOROKHOD REPRESENTATION*

Proofs of the central limit theorem or law of the iterated logarithm for martingales usually take one of two routes. Either they mimic the proofs in the independence case, or they make use of the Skorokhod representation for martingales. An advantage of the Skorokhod approach is that it places a martingale in the context of a Gaussian process*, which permits relatively short proofs of several martingale results.

Let $\{S_n\}$ be a zero mean, square integrable martingale with differences X_n. On a rich enough probability space there exists a standard Wiener process* W and a sequence of nonnegative variables $\{\tau_n\}$, such that $S_n = W(\sum_1^n \tau_j)$ almost surely for $n \geqslant 1$. The relationship between $\{S_n\}$ and $\{\tau_n\}$ is described by several moment inequalities. In order to prove that $c_n^{-1}S_n$ is asymptotically normally distributed, it suffices to show that $c_n^{-2}\sum_1^n \tau_j \to 1$ in probability, which may be accomplished quite easily under regularity conditions on the martingale.

For more detailed accounts of the representation, see Strassen [24], Hall and Heyde [13, Appen. I], and the references therein.

CONTINUOUS PARAMETER MARTINGALES

Continuous parameter martingales are defined in the same way as the discrete parameter processes considered above. Thus, $\{S_t, t \geqslant 0\}$ is a martingale if $E[|S_t|] < \infty$ for each t and

$$E[S_t \mid S_u, \text{all } u \leqslant s] = S_s$$

for all $s \leqslant t$. More generally, if $\{\mathscr{F}_t, t \geqslant 0\}$ is an increasing sequence of σ-fields, we call $\{(S_t, \mathscr{F}_t), t \geqslant 0\}$ a martingale if $E[|S_t|] < \infty$ for each t, if S_t is $\mathscr{F}_t$-measurable for each t, and if

$$E[S_t \mid \mathscr{F}_s] = S_s$$

for all $s \leqslant t$. Submartingales and supermartingales are defined similarly. As examples, observe that if $W(t)$, $t \geqslant 0$, is a standard Wiener process, then $S_t \equiv W(t)$ and $S_t \equiv [W(t)]^2 - t$, $t \geqslant 0$, are martingales.

Several inequalities for discrete parameter martingales have continuous parameter versions. In particular, if $\{S_t\}$ is a martingale then

$$\lambda P\left(\sup_{s \leqslant t}|S_s| > \lambda\right) < E\left[|S_t| I\left(\sup_{s \leqslant t}|S_s| > \lambda\right)\right]$$

for all $t > 0$ and $\lambda > 0$. Meyer [19] has provided an analogue of Doob's decomposition theorem in the continuous parameter case. Kunita and Watanabe [14] have proved an embedding theorem for continuous parameter martingales with almost surely continuous sample paths.

The conditional variance of a zero mean, square integrable martingale $\{(S_t, \mathscr{F}_t)\}$ may be defined formally by

$$V_t^2 = \int_0^t E\left[(dS_s)^2 \mid \mathscr{F}_s\right],$$

where dS_s denotes the infinitesimal martingale difference. If the process S_t is almost surely right continuous, and if the σ-fields $\mathscr{F}_t$ are right continuous (meaning that $\mathscr{F}_t = \bigcap_{s>t} \mathscr{F}_s$ for $t \geqslant 0$), then V_t^2 is well defined, and is itself a right continuous, increasing process, with V_t^2 measurable in $\mathscr{F}_t$. It follows from the decomposition theorem that V_t^2 is uniquely determined almost surely by the relation

$$E\left[(S_t - S_s)^2 \mid \mathscr{F}_s\right] = E\left[(V_t^2 - V_s^2) \mid \mathscr{F}_s\right], \quad 0 \leqslant s \leqslant t.$$

The conditional variance plays an important role in Kunita and Watanabe's [14] embedding theorem.

Doob [9, Sec. 11 of Chap. VII], Meyer [20, Chap. IV] and Loève [16, Sec. 39] provide detailed introductions to the theory of continuous parameter martingales.

APPLICATIONS AND GENERALIZATIONS

Martingale theory may be applied in a very diverse range of situations, from pure mathematics to the theory of statistics. The examples listed below are by no means exhaustive.

Conditional expectations play a key role in the definition of a martingale. In several graduate texts these expectations are introduced via the Radon–Nikodym theorem* (*see* CONDITIONAL PROBABILITY AND EXPECTATION). However, it is possible to define a martingale without using this theorem, and in that case, martingale methods may be used to construct a new proof of the Radon–Nikodym theorem for finite measures; see Meyer [20, p. 153] and Bauer [2, p. 366]. Meyer [20, Chap. VIII] supplies several other illustrations of the use of martingale theory to derive mathematical results, such as the lifting theorem.

Since martingales represent a generalization of the concept of sums of independent random variables, martingale theory can be used to prove many results for sums of independent variables. See Doob [8, pp. 334–340] and Bauer [2, pp. 362–365]. Philipp and Stout [23] have used martingale methods to derive a very wide range of almost sure invariance principles for weakly dependent random variables, including mixing sequences, Markov sequences, and continuous parameter stochastic processes. Doob [9, pp. 388–390] has illustrated the application of continuous parameter martingale theory. Hall and Heyde [13, Chap. 5] have described the use of martingale methods in the theory of stationary processes*.

The area in which martingale theory has had perhaps its greatest recent impact is the theory of inference*. For example, let $X_1, X_2, \ldots$ be consecutive observations from a stochastic process* whose distribution depends on a parameter θ, and let $L_n(\theta)$ denote the likelihood function associated with the sample $X_1, X_2, \ldots, X_n$. (*See* MAXIMUM LIKELIHOOD ESTIMATION.) Under very mild regularity conditions on the likelihood, the sequence

$$\{d \log L_n(\theta)/d\theta, n \geqslant 1\}$$

is a zero mean martingale. In this case the conditional variance $V_n^2 = V_n^2(\theta)$ plays the role of a generalized form of Fisher information*, and often is denoted by $I_n(\theta)$. Martingale theory can be used to prove that under appropriate conditions, the likelihood equation

$$d \log L_n(\theta)/d\theta = 0$$

has a root $\hat{\theta}$ which is strongly consistent for θ, and which satisfies a central limit theorem with minimum asymptotic variance. See for example Basawa and Rao [1] and Hall and Heyde [13, Chap. 6]. The likelihood ratio statistic in the case of an independent sample is a positive martingale under the null hypothesis, and has mean equal to unity. It follows from Theorem 1 that such a martingale converges almost surely to a finite limit. The limit is zero under the null hypothesis. Similar methods can be used to prove that the limit is infinite under the alternative hypothesis.

Martingale theory plays an important role in areas of sequential analysis* (*see also* OPTIONAL SAMPLING). Indeed, Chow et al.'s account [7] of the theory of optimal stopping rules* is built upon a martingale foundation. These authors use martingale methods to prove Wald's equation* (see their Theorem 2.3, p. 23). Other applications of martingale theory include genetics*, stochastic approximation* and adaptive control of linear systems; see Hall and Heyde [13, Chapter 7].

There are several definitions of "asymptotic" or "weak" martingales; in particular, there are quasi-martingales (Fisk [11]), martingales in the limit (Mucci [21]), and amarts (Edgar and Sucheston [10]). Mixingales (McLeish [18]) are useful from the point of view of applications, and satisfy a version of the martingale convergence theorem.

References

[1] Basawa, I. V. and Prakasa Rao, B. L. S. (1980). *Statistical Inference for Stochastic Processes*. Academic Press, New York. (Martingale methods are used in various places to solve problems in inference.)

[2] Bauer, H. (1981). *Probability Theory and Elements of Measure Theory*. 2nd English ed. Academic Press, New York. (Chapter 11 contains a graduate level introduction to martingales.)

[3] Bernstein, S. (1927). *Math. Ann.*, **85**, 1–59. (Contains the first martingale central limit theorem.)

[4] Billingsley, P. (1979). *Probability and Measure*. Wiley, New York. (Section 35 contains a graduate-level introduction to martingales.)

[5] Brown, B. M. (1971). *Ann. Math. Statist.*, **42**, 59–66. (Contains the basic martingale central limit theorem.)

[6] Burkholder, D. L. (1973). *Ann. Prob.*, **1**, 19–42.

[7] Chow, Y. S., Robbins, H., and Siegmund, D. (1971). *Great Expectations: The Theory of Optimal Stopping*. Houghton Mifflin, Boston. (A theoretical introduction to optimal stopping and other aspects of sequential analysis.)

[8] Chung, K. L. (1974). *A Course in Probability Theory*. Academic Press, New York. (Chapter 9 contains a graduate-level introduction to martingales.)

[9] Doob, J. L. (1953). *Stochastic Processes*. Wiley, New York. (Chapter VII is the classic introduction to martingale theory, although somewhat outdated now.)

[10] Edgar, G. A. and Sucheston, L. (1976). *J. Multivariate Anal.* **6**, 572–591. (Introduces amarts.)

[11] Fisk, D. L. (1969). *Trans. Amer. Math. Soc.*, **120**, 369–388. (Describes quasi-martingales.)

[12] Garsia, A. M. (1973). *Martingale Inequalities: Seminar Notes on Recent Progress*. Benjamin Cummings, Reading, Mass.

[13] Hall, P. and Heyde, C. C. (1980). *Martingale Limit Theory and its Application*. Academic Press, New York. (Graduate- and research-level introduction to martingale theory.)

[14] Kunita, H. and Watanabe, S. (1967). *Nagoya Math. J.*, **30**, 209–245.

[15] Lévy, P. (1935). *J. Math. Pures Appl.*, **14**, 347–402. (Contains an early martingale central limit theorem.)

[16] Loève, M. (1978). *Probability Theory II*, 4th ed. Springer, New York. (Chapter IX contains a graduate-level introduction to martingales.)

[17] McLeish, D. L. (1974). *Ann. Prob.*, **2**, 620–628.

[18] McLeish, D. L. (1975). *Ann. Prob.*, **3**, 829–839. (Introduces mixingales.)

[19] Meyer, P. A. (1962). *Illinois J. Math.*, **6**, 193–205.

[20] Meyer, P. A. (1966). *Probability and Potentials*. Blaisdell, Waltham, Mass. (2nd edition coauthored with C. Dellachourie (1978), North-Holland, Amsterdam. Part B contains a mathematical account of martingale theory at graduate and research level.)

[21] Mucci, A. G. (1973). *Pacific J. Math.*, **48**, 197–202. (Introduces martingales in the limit.)

[22] Neveu, J. (1975). *Discrete Parameter Martingales*. North-Holland, Amsterdam. (Gives a mathematical account of martingale theory at graduate and research level.)

[23] Philipp, W. and Stout, W. F. (1975). *Mem. Amer. Math. Soc.*, **161**.

[24] Strassen, V. (1967). *Proc. 5th Berkeley Symp. Math. Statist. Prob.*, **2**, 315–343. (Provides an introduction to strong invariance principles and the Skorokhod embedding for martingales.)

Bibliography

Khmaladze, E. V. (1982). *Uspekhi Mat. Nauk*, **37**, 193–212. (In Russian. A martingale approch to goodness-of-fit and tests of homogeneity is presented.)

(BRANCHING PROCESSES
INVARIANCE PRINCIPLES AND
FUNCTIONAL LIMIT THEOREMS

KOLMOGROV'S INEQUALITY
LAWS OF THE ITERATED LOGARITHM
LAWS OF LARGE NUMBERS
MAXIMUM LIKELIHOOD ESTIMATION
OPTIMAL STOPPING RULES
SKOROKHOD REPRESENTATION
UPCROSSING INEQUALITY)

PETER HALL

MASKING AND SWAMPING

As the term masking is most commonly used, it arises when a sample contains multiple outliers* but on analysis by a particular outlier detection method, some or all of the outliers appear to be inlying. The converse problem of swamping arises when the method of analysis wrongly suggests that a good data point is outlying.

The earliest well-known example of masking was given by Pearson and Chandra Sekar [3]. Let $X_1, \dots, X_n$ have mean $\bar{X}$ and variance S^2, and consider the use of the studentized residuals $(X_i - \bar{X})/S$ for outlier detection. Pearson and Chandra Sekar showed by example that if n is sufficiently small, as X_n and X_{n-1} tend to infinity, the largest studentized residual may tend to a constant below the rejection level. Thus paradoxically as the two outliers become more outlying, the probability of identifying either of them as a significant outlier using the maximum absolute studentized residual goes to zero. This basic framework also illustrates swamping: fix X_{n-1} and increase X_n until $\bar{X} = X_{n-1}$. Then using the maximum absolute studentized residual, all of the $n-2$ good observations appear more outlying than X_{n-1}.

In this example the reason for the masking is that the outliers inflate S by an amount more than compensating for the matching increase in $\max(X_i - \bar{X})$. Another much less easily diagnosed problem of masking and swamping can arise in regression, where an additional complication is the leverage* of the predictors; i.e., the ability of data points with extreme values of the predictors to lever the regression line over toward themselves. See for example Belsley, et al. [2]. To illustrate these points, consider a set of (x_i, Y_i) pairs, $(20, 20)$, $(10, \Delta)$, $(-8, 0)$, and seven (x_i, Y_i) pairs that are independent $N(0, 1)$. The first two points are outliers (if $\Delta \neq 0$); the rest are good. The residuals* of the first three points obtained in a simulation with $\Delta = 12$ were 1.30, 1.90, and 5.56; using the known $\sigma = 1$, the studentized residuals were 2.25, 2.16, and 6.60. This shows both outliers being masked, with the $(-8, 0)$ inlier being swamped. With $\Delta = 0$, so that the second point is actually inlying, the residuals are 5.67, -7.45, and 4.95, and the studentized residuals are 10.58, -8.44, and 5.87. Here the outlier has been unmasked, but the second and third, inlying points, remain swamped.

As the preceding discussion suggests, masking and swamping are deficiencies not of the sample, but of the particular outlier detection method applied to it. For example, while with $\Delta = 12$ the second observation has a larger residual than the first, studentizing to correct for their different variances shows the first to be in fact the more extreme. In the Pearson–Chandra Sekar case, masking and swamping are easily avoided by replacing $\bar{X}$ and S with robust measures of location and scale or by removing the K most aberrant points and then successively testing them for reinclusion [5]. This method, however, will fail in the regression example by swamping the third, good observation pair. In principle the robust estimate remedy applies to the regression, but in practice there may be severe difficulties in finding consistent robust estimators when some points have high leverage.

For some further discussion of the avoidance of masking and swamping, see the discussion following Beckman and Cook [1] and OUTLIERS.

References

[1] Beckman, R. J. and Cook, R. D. (1983). *Technometrics*, **25**, 119–163.

[2] Belsley, D. A., Kuh, E., and Welsch, R. E. (1980). *Regression Diagnostics*. Wiley, New York.

[3] Pearson, E. S. and Chandra Sekar, C. (1936). *Biometrika*, **28**, 308–320.

[4] Rosner, B. (1975). *Technometrics*, **17**, 221–227.
[5] Rosner, B. (1983). *Technometrics*, **25**, 165–172.

(OUTLIERS)

DOUGLAS M. HAWKINS

MASON'S RULE *See* FLOWGRAPH ANALYSIS

MASTER SAMPLES

The term *master sample* is used to describe a sample which is selected in such a way that it may be subsampled on future occasions. In those instances where sampling from the master sample involves a further refinement of sampling units, the master sample is often described as a master sample frame.

Master samples offer a number of advantages over one-time or ad hoc samples when there are to be repeated and ongoing surveys of the same or related populations. The two most obvious advantages are lower cost and decreased lead time. Lower cost results because some portion of the cost of sample selection may be distributed among a number of studies. In addition, since some portion of the work required to produce the final sample takes place at the time the master sample is selected, the lead time required for final sample preparation is typically less than would be required for an ad hoc sample.

Master samples may range in complexity from designs based exclusively on simple random element selection to designs with a number of stages, involving a number of levels of stratification and clustering; *see* CLUSTER SAMPLING; MULTIPHASE SAMPLING; STRATIFIED MULTISTAGE SAMPLING.

Since most populations are dynamic in composition, master samples which are to be used over a number of years often make use of sampling units, which are relatively permanent in their definition, in which the definition of population elements is implicit until a specific sample is actually selected. For example, a master sample of households might initially consist of a sample of geographic areas (e.g., blocks). Selection of a specific sample from the master sample would begin by the selection of a subsample of these geographic areas. Within the areas selected for a specific sample an explicit listing of housing units would be constructed. These lists of housing units might define the final sample or might be further subsampled. At the time of actual data collection*, the households which exist within the selected units would comprise the final household sample. *See also* AREA SAMPLING.

If a geographic area were selected in which a housing unit list already existed from a prior use of the master sample, the listing might either be brought up to date in its entirety or in conjunction with a prespecified set of linking rules.

Because of their permanency, geographically defined area units are commonly used in master samples of households, farms, and business establishments. Typically the selection of small geographic areas is a multstage process beginning with the selection of conties or county groups (or equivalent units in other countries) and proceeding in subsequent stages of area sampling until the block level. When data collection is carried out by local interviewers, the use of small geographically defined selection units which are nested at a county or county group level permits the use of the same interviewers on successive surveys. This typically results in lower overall recruiting and training cost, and allows the development of an experienced interviewing staff. *See* SURVEY SAMPLING.

In addition to the advantages, which are usually measured in time and cost, master samples often allow designs which permit more efficient estimation than is possible with ad hoc samples.

When one of the purposes of an ongoing survey program is the measurement of change, the overlap in primary and possibly later units of sample selection resulting from the use of a master sample often results in positive correlation* in the successive sample estimates over time. This positive correlation, which does not require an overlap among sample individuals, decreases the standard error of estimates of change.

In addition, the structure associated with a master sample often allows for the use of more efficient multiphase estimates. In this case certain information collected from the specific survey as well as prior surveys is viewed as the data obtained from the first phase. Additional data obtained from the specific survey, but not the prior surveys, is viewed as information obtained in the second phase.

Survey organizations that use master samples typically reselect these samples on a cycle which coincides with the release of census* data. In the United States this has meant that master samples of households are reselected every ten years while master samples of farms and business establishments are reselected on either a five- or ten-year cycle.

Procedures have been developed by Keyfitz [1] and Kish and Scott [2] which maximize the retention of primary sampling units when multistage master samples are reselected. This maximization of overlap among primary selections in successive master sample cycles limits the number of areas where new interviewers must be recruited and trained. It also results in the retention of some of the existing positive correlation of survey estimates across the period of master sample changeover.

In the simplest case, where a single primary unit is selected within each stratum and the compositions of the strata are the same at times 1 and 2, the procedure developed by Keyfitz specifies the following rules:

Let p_{i1} and p_{i2} denote the probability of selection for unit i at times 1 and 2 respectively; if unit k was selected at time 1, it is retained at time 2 if its probability has not decreased (i.e., if $p_{k2} \geqslant p_{k1}$).

If unit k was selected at time 1, and its probability at time 2 is lower than its time 1 probability, it is eliminated with probability p_{k2}/p_{k1}. If unit k is eliminated, a selection is made among units remaining in the stratum that have increased in probability from time 1 to time 2. This selection is carried out with probability proportional to the size of the respective time 2 to time 1 increases.

Kish and Scott [2] provide extensions to this procedure when the definition of primary sampling units and/or the basic stratification changes from time 1 to time 2.

References

[1] Keyfitz, N. (1951). *J. Amer. Statist. Ass.*, **46**, 183–201.

[2] Kish, L. and Scott, A. (1971). *J. Amer. Statist. Ass.*, **66**, 461–470.

(AREA SAMPLING
CLUSTER SAMPLING
MULTIPHASE SAMPLING
STRATIFIED MULTISTAGE SAMPLING
SURVEY SAMPLING)

MARTIN R. FRANKEL

MATCHED PAIRS

ORIGINS OF DATA

Data in the form of matched pairs are frequently encountered in statistical practice, generally for one or the other of two reasons. First, some data naturally *occur* in this form: intelligence, height, and social class of husbands and wives are examples; before and after measurements of subjects in a study of learning are also examples. Second, an investigator may *arrange* to acquire data in this form, sometimes because it is the most feasible way to collect the data, sometimes with the intent to improve experimental precision [3]. If a judge at one sitting can compare two, but not three or more, complex protocols for some difficult-to-assess quality ("comprehension manifested by the subject" or "sympathy displayed"), then each sitting furnishes a pair of observations, for reasons rooted in feasibility.

Increased precision of comparison is a very common reason for acquiring data in pairs [1, 2]. Typically, a group of experimental units (animals, persons, test specimens), all receiving the *same treatment*, will produce different measured outcomes because of differences among the experimental units in qualities that affect the outcome measurement. This overlay of variation obscures the comparison of treatment effects. Deliberately assigning the two treatments to members of closely similar pairs (like twins) ensures that these "other" influential variables are constrained to have similar values within every pair, largely freeing the treatment comparison from extraneous influences. Consider comparing the durability of two different composition soles for children's shoes. The level of activity of the child, the kinds of surfaces (asphalt, concrete, gravel, grass) on which the child plays, the contact with water (mud puddles), etc. all would influence the life of the sole, and might vary greatly among different children. But let each child wear composition A on one shoe and composition B on the other; now all these influential extraneous variables will be very similar *within pairs*, for each child. Comparison of the durability of A and B will be much more precise than it would be with some children wearing A and others wearing B.

INFERENTIAL TASKS AND LIMITATIONS

If the observations are binary (say 0 or 1), then a pair of observations must be one of (0, 0), (0, 1), (1, 0), or (1, 1), where the first element is the outcome for the A member of the pair, and the second element that for the B member. The task of testing equality of treatment is done by applying McNemar's test for correlated proportions to the data from M pairs (*see* MCNEMAR'S STATISTICS).

If the observations x_i and y_i in the ith pair are real numbers, then their differences, $d_i = y_i - x_i$ $(i = 1, \ldots, M)$ may be analyzed by the matched pairs t-test*; confidence intervals for $\Delta = E(d_i)$ are based on that test. Alternatively the d_i may be treated by Wilcoxon's signed rank test*; confidence intervals for a translation parameter can be based on that test.

Yet another alternative is to use only the signs of the d_i; this is the sign test*, and it, too, has associated confidence intervals for a translation parameter. No one of these three approaches is uniformly better than the others. For example, if the data are normally distributed, t is best; if they are Cauchy distributed, t is worst; if they are logistically distributed, the signed rank test is best, and if double-exponential, the sign test is best.

Two kinds of possible limitation on inference from matched pairs data must not be lost sight of. First, it may be that knowledge about pairs leaves some doubts about inference to nonpairs; thus, conclusions from a comparative study of attitude-formation in identical twins might leave questions about whether processes of attitude-formation in only children are qualitatively similar to those in identical twins, whose social interactions are quite special. (A comparative study of physiological responses to two diets, done in identical twins, might be much less subject to such reservations.) Another possible difficulty may arise where two treatments are assigned, in sequence, to each subject; there is the prospect that the response to the earlier treatment may influence the response to the later one. If this happens, then A before B is not the same treatment as A after B—and inference is difficult. Such interference may occur not only through intertemporal effects, but sometimes through other mechanisms. Comparison of two ointments applied to matching sites on the same organism can provide a sensitive pairwise comparison—unless either treatment has systemic (as well as local) effects, in which case one or both sites will be exhibiting response to *both* treatments, and inference again is difficult.

RELATED PROCEDURES

The matched-pairs t-test is the special case of randomized blocks* where there are only two treatments in each block. The sign test is the special case of Friedman's analysis of variance by ranks where there are only two treatments in each block (*see* FRIEDMAN'S CHI-SQUARE TEST). McNemar's test for two correlated proportions is the special case of Cochran's Q where there are two (rather than $k > 2$) binary responses in each block (*see* COCHRAN'S Q-STATISTIC).

EXTENSIONS TO MORE COMPLEX PROBLEMS

Paired observations can be used for comparing $k > 2$ treatments, and a considerable body of methods exists.

The Method of Paired Comparisons* compares k treatments by acquiring data in pairs, where the result from the mth pair $(m = 1, \ldots M)$ is that treatment i is better (or worse) than treatment j; this can be expressed symbolically as $d_{ijm} = +1$ or -1. These binary data can enable comparison of the treatments; it is sufficient that all $\frac{1}{2}k(k-1)$ pairs be tested once; more replication gives more precise tests and estimates.

Incomplete randomized blocks designs compare real-valued data from k treatments in blocks of size r, with r smaller than k; the limiting case is $r = 2$. Sometimes the precision obtainable in blocks of size 2 is great enough to make this the preferred design. Sometimes blocks of size 2 are feasible, and larger blocks are not. Comparison of 3 shoe sole compositions could be done as follows; one complete replicate would have three subjects, (A, B), (A, C), (B, C). If there were five treatments, then $\frac{1}{2}(5)(4) = 10$ subjects would suffice for a complete replicate. Of course, more replication increases precision.

References

[1] Cox, D. R. (1958). *Planning of Experiments*. Wiley, New York, Sec. 3.2.

[2] Fisher, R. A. (1951). *The Design of Experiments*, 6th ed. Hafner, New York, Sects. 16, 17.

[3] Pratt, J. W. and Gibbons, J. D. (1981). *Concepts of Nonparametric Theory*. Springer-Verlag, New York, pp. 104–106.

Bibliography

Lehmann, E. L. (1975). *Nonparametrics: Statistical Methods Based on Ranks*. Holden-Day, San Francisco. (The sign test, Wilcoxon signed rank test, and the matched-pairs t-test are thoroughly discussed and compared in Chaps. 3 and 4.)

(MATCHED PAIRS *t*-TESTS
PAIRED COMPARISONS
RANDOMIZED BLOCKS)

LINCOLN E. MOSES

MATCHED PAIRS *t*-TESTS

The matched pairs t-test is appropriate for comparing two treatment means where the data have been acquired as *pairs of observations*, each pair having one observation with treatment 1 and the other with treatment 2 (*see* MATCHED PAIRS). Examples of pairs include littermates, left and right sides of an experimental animal, two plants grown in the same pot, early and late measurements on the same person, and two shoes on the same schoolboy. In each such case, similarity between observations in the same pair is likely; this circumstance renders the usual two-sample t-tests* inapplicable. The paired t-test is appropriate.

THE TEST

The analysis follows the pattern used in the (fictitious) example below. We write x_{ij} to denote the observed value of X_j $(j = 1, 2)$ in pair i $(i = 1, \ldots, M)$. The comparison between treatments is captured in the ith pair as the difference $d_i = x_{i2} - x_{i1}$. Each d_i is an estimate of $\Delta = E\{X_2 - X_1\}$. Table 1 shows the data, displayed in a form convenient for computation. The average d and the vari-

Table 1 Data Displayed for Computing Matched Pairs t

	Treatment		
Pair i	1	2	$d = x_{i2} - x_{i1}$
1	114	122	8
2	103	108	5
3	95	89	−6
4	111	112	1
5	86	85	−1
6	100	97	−3
7	104	106	2
8	91	105	14
9	79	88	9
10	112	115	3
11	63	70	7
12	94	100	6

ance S_d^2 of the d_i are computed, as is the estimated standard error of $\bar{d}$, s.e.($\bar{d}$),

$$\bar{d} = M^{-1}\sum d_i,$$

$$S_d^2 = (M-1)^{-1}\sum (d_i - \bar{d})^2,$$

$$\text{s.e.}(\bar{d}) = M^{-1/2}S_d.$$

The t-statistic, with $M - 1$ degrees of freedom, is

$$t = \bar{d}/\text{s.e.}(\bar{d}) = (S_d)^{-1}\sqrt{M}\,\bar{d}.$$

A two-sided α-level test of the hypothesis that $\Delta = 0$ is to reject if

$$(S_d)^{-1}\sqrt{M}\,|d| > t_{M-1,\alpha/2},$$

where the upper $\alpha/2$-point of the t-distribution* with $M - 1$ degrees of freedom is represented as $t_{M-1,\alpha/2}$.

For the example,

$$M = 12, \qquad \sum d_i = 45, \qquad \sum d_i^2 = 511,$$

whence $\bar{d}$ is 3.75, S_d is 5.58, s.e.($\bar{d}$) is 1.61, and $t = 3.75/1.61 = 2.33$. This value of t leads to rejecting the hypothesis that $\Delta = 0$ at $\alpha = .05$ but not at $\alpha = .01$ because this value, 2.33, exceeds 2.20 but does not reach 3.11, the values, respectively, of $t_{11,.025}$ and $t_{11,.005}$. A confidence interval* (confidence coefficient $1 - \alpha$) is formed in the usual way as $\bar{d} - t_{M-1,\alpha/2}\text{s.e.}(\bar{d}) \leqslant \Delta \leqslant \bar{d} + t_{M-1,\alpha/2}$ s.e.($\bar{d}$). Thus the 99% confidence interval for Δ in the example is

$$3.75 - 3.11(1.61) \leqslant \Delta \leqslant 3.75 + 3.11(1.61),$$

$$-1.26 \leqslant \Delta \leqslant 8.76.$$

ALTERNATE FORMS OF THE TEST

Two alternate representations of the matched pairs t-test afford additional insight into its character. If the reader were to plot the data in the example as 12 points, each having x_1 as its horizontal coordinate and x_2 as its vertical coordinate, he or she would see a strong pattern of association*, evidencing the correlation* between the x_1 and x_2 values in the 12 pairs. With the correlation coefficient r between x_1 and x_2 defined in the usual way, the following is an algebraic identity:

$$S_d^2 = S_{x_1}^2 - 2rS_{x_1}S_{x_2} + S_{x_2}^2.$$

The t-test* statistic already given can therefore be rendered in the form

$$t = \bar{d}/\sqrt{(S_{x_1}^2 - 2rS_{x_1}S_{x_2} + S_{x_2}^2)/M}.$$

Inspection shows that if $r > 0$ the denominator is *smaller* than $\sqrt{(S_{x_1}^2 + S_{x_2}^2)/M}$, which would be the appropriate standard error in the absence of matching. This consideration shows that incorrectly applying the unmatched formula incurs the penality of using *too large* a standard error if $r > 0$, the usual situation. (If there were a negative correlation within pairs, use of the unpaired analysis would now err in using too small a standard error.) In the example,

$$r = .9289, \qquad S_{x_1}^2 = 222.0, \qquad S_{x_2}^2 = 216.02,$$

$$\sqrt{S_{x_1}^2 - 2rS_{x_1}S_{x_2} + S_{x_2}^2} = 5.58,$$

agreeing with S_d as earlier computed; this illustrates the algebraic identity that was claimed.

The second alternate representation of the matched pairs t-test shows it as a special case of the randomized block design*, where two treatments are compared in M blocks (of size two). In particualr, the F-test* of equality of treatments is exactly the square of our t-statistic. That F-statistic may be written as

$$F = \frac{M\sum_{j=1}^{2}\left(\bar{x}_{\cdot j} - \bar{\bar{x}}\right)^2}{\sum_{i=1}^{M}\sum_{j=1}^{2}\left(x_{ij} - \bar{x}_{i\cdot} - \bar{x}_{\cdot j} + \bar{\bar{x}}\right)^2/(M-1)}.$$

Substituting $\bar{\bar{x}} = (\bar{x}_{\cdot 1} + \bar{x}_{\cdot 2})/2$ in the numerator quickly brings it to the form

$$(M/2)(\bar{x}_{\cdot 2} - \bar{x}_{\cdot 1})^2 = (M/2)(\bar{d})^2.$$

Slightly more elaborate algebra brings

$$\sum_{i=1}^{M}\sum_{j=1}^{2}\left(x_{ij} - \bar{x}_{i\cdot} - \bar{x}_{\cdot j} + \bar{\bar{x}}\right)^2$$

to the form

$$\frac{1}{2}\sum_{i=1}^{M}\left(d_i - \bar{d}\right)^2.$$

Rewriting F with numerator and denominator thus expressed in terms of d_i immediately shows that F is indeed the square of our t-statistic. Thus, the matched pairs t-test is the simplest case of randomized blocks; equivalently, the randomized blocks design and analysis for $t > 2$ treatments is a direct generalization of the matched pairs t-test.

Because the upper α percentage points of the $F_{1,M-1}$ distribution are exactly the squares of the two-sided α percentage points of t with $M - 1$ degrees of freedom, analysis by randomized blocks or by two-sided matched t-test must always agree.

Application of the randomized blocks analysis to the $2M$ observations of the example produces the analysis of variance in Table 2, from which

$$F_{1,11} = \frac{84.375}{15.557} = 5.42.$$

Table 2 Randomized Blocks; Analysis of Variance

Source	d.f.	S.S.	M.S.
Treatments	1	84.375	84.375
Blocks	11	4647.125	422.466
Error	11	171.125	15.557
Total		4902.625	

The value 2.33 of t obtained earlier is the square root of 5.42, and we have an illustration of the algebraic identity between matched pairs t and F for the randomized blocks analysis of the same data.

DISPLAY OF THE RESULTS

A common error in reporting paired data is to show $\bar{x}_1$, $\bar{x}_2$, $S_{x_1}^2$, $S_{x_2}^2$, and (perhaps) the correctly computed t-statistic, but to omit S_d^2 (or r). Only with knowledge of one of these (from which the other follows) in addition to $\bar{x}_1$, $\bar{x}_2$, $S_{x_1}^2$, and $S_{x_2}^2$ is the reader able to confirm the analysis. It is recommended that at least the following be reported:

$$\bar{x}_1, \quad \bar{x}_2, \quad \bar{d}, \quad S_d, \quad M.$$

Together these statistics permit appraisal of the evidence concerning a treatment difference.

APPLICABILITY; ROLE OF RANDOMIZATION

The t-statistic has, on the null hypothesis, *exactly* Student's t-distribution if the data are M independent observations on (X_1, X_2) where X_1 and X_2 have a bivariate normal distribution with any covariance matrix and $E(X_1) = E(X_2)$. Also, it has *exactly* Student's t-distribution if an analysis of variance* model with normal errors holds:

$$x_{i1} = \mu + \tau_1 + \rho_i + e_{1i},$$
$$x_{i2} = \mu + \tau_2 + \rho_i + e_{2i}$$

with e_{1i} and e_{2i} i.i.d. $N(0, \sigma^2)$ and $\tau_1 = \tau_2$.

The real usefulness of the test stems from its approximate (rather than exact) validity under much wider circumstances. First, if the data are independent, bivariate observations from any distribution (with finite second moments), then the central limit theorem* applies to $\bar{d}$, and S_d is a consistent estimator of σ_d; in sufficiently large samples the statistic t will be nearly standard normal, by a familiar argument [1, p. 388].

Second, and more important, is the justification of the test as a trustworthy approxi-

Table 3 Partial Enumeration of Possible Samples in Decreasing Order of $\sum d$

$\sum d$	Negative Sum	1	1	2	3	3	5	6	6	7	8	9	14	Number of Samples
65	0													1
63	1	—												2
			—											
61	2	—	—											2
				—										
59	3	—		—										4
			—	—										
					—									
						—								
57	4	—				—								5
		—			—									
			—			—								
			—		—									
		—	—	—										

mation to the exact permutation distribution of $\bar{d}$, a seminal idea due to Fisher [2, Sec. 21]. If the choice of which unit in the ith pair is to receive treatment 1 be made by the flip of a fair coin, and these choices are independently made from one pair to another (i.e., M coin flips, one for each pair), then under the null hypothesis the exact distribution of $\bar{d}$ can be calculated. Here is the reasoning. The value d_i that we see in the ith pair would have been $-d_i$ had the coin fallen the other way, for then the two experimental units (and their observed values) would have been exactly interchanged under the hypothesis that the treatments are identical in effect. (Notice the strong use of the null hypothesis that the two treatments are entirely equal in effect in each pair.) Continuing with this argument we see that all together there were 2^M possible outcomes for the experiment, having the form

$$\pm d_1, \pm d_2, \pm d_3, \ldots, \pm d_M .$$

We happen to have one of these 2^M possibilities as our data. For each of these 2^M outcomes there is a value of $\bar{d}$. All these possible outcomes can be listed in order of their value of $\bar{d}$. Each of the outcomes is equally likely with probability 2^{-M}, because the coin was actually used and induced these probabilities. If the *observed* value of $\bar{d}$ is one of the $\alpha \cdot 2^M$ most extremely large, or small, in the list—i.e., in the "permutation distribution"—then the hypothesis can be rejected at level α. Fisher proposed this idea and went further, indicating that the significance level found by making the list was obliged to agree closely with the significance level found by referring the t-statistic to Student's distribution. For the example one can list the permutation distribution of $\bar{d}$ by listing possible samples in order of the sum of negatively signed d's, beginning with 0. This is most easily done by first arranging the $|d_i|$ in increasing order and then tabulating configurations of $\pm$ in order of the resulting negative sum, as shown in Table 3.

The 14 possible outcomes with smallest negative sum are shown in Table 3, each identifying the $|d_i|$ carrying negative signs. The listing carried this far shows, for example, that of the $2^{12} = 4096$ equally likely samples, $1 + 2 + 2 + 4 + 5 = 14$ result in values of $\bar{d}$ as great as $57/12 = 4.75$. Extension of this list shows 95 of them produce $\bar{d}$ as great as 3.75, our obtained value. Then the two-sided significance level that our data attain is, by this permutation test*, $2 \times 95/4096 = .0464$. The t-statistic (with 11 degrees of freedom) had the value 2.33. A detailed table of the distribution [3] gives (using interpolation) an attained two-sided significance of .0400. The agreement of .0400 with .0464 is close, even though M is only 12.

The important fact is that *if the treatments are assigned randomly and independently within pairs*, then the exact permutation distribution is well approximated by the matched pairs t-test, provided M is large enough and none of the $|d_i|$ exert a preponderant influence. Observe that no assumptions at all about the distributions of X_1 and X_2 are needed; only a condition on the sample values is involved.

As with all asymptotic results it is hard to give crisp formulations of "large enough" sample size or just where the threshold of "preponderant influence" may lie. Fisher's example had $M = 15$ and $\max|d_i|^2/\sum d_i^2 = 3.22/M$, with a very close correspondence between the P-value* found by enumeration and that found by using the t table. Our example has $M = 12$ and $\max|d_i|^2/\sum d_i^2 = 196/511 = 4.6/M$, with less close agreement between t table and enumeration. These examples support such a rule of thumb as: If $M \geqslant 12$ and $\max|d_i|/\sum d_i^2$ is not much greater than $3/M$, then trust the t table to give a reliable indication of the exact permutation test's significance.

References

[1] Cramer, H. (1946). *Mathematical Methods of Statistics*. Princeton University Press, Princeton, N.J.

[2] Fisher, R. A. (1966). *The Design of Experiments*. Oliver and Boyd, Edinburgh.

[3] Pearson, E. and Hartley, H. O. (1966). *Biometrika Tables for Statisticians*, Vol. 1. Cambridge University Press, Cambridge, England.

Bibliography

Pratt, J. W. and Gibbons, J. D. (1981). *Concepts of Nonparametric Theory*. Springer-Verlag, New York. (This book gives a thorough treatment of permutation tests for matched pairs.)

(BLOCKS, RANDOMIZED COMPLETE
MATCHED PAIRS
PERMUTATION TESTS
RANDOMIZATION TESTS
t-TESTS)

LINCOLN E. MOSES

MATCHING DISTANCE *See* CLASSIFICATION

THE MATCHING PROBLEM

The matching problem, also known as the "problème de rencontre," has been discussed in the probability literature since the early eighteenth century. In the matching problem, n cards (numbered $1, 2, \ldots, n$) are randomly laid out onto n fixed positions, also marked $1, 2, \ldots, n$, with each position receiving one card. A *match* occurs at the ith position if the card designated i is placed there. Such an occurrence is denoted by event E_i, $i = 1, \ldots, n$. Events of interest are that *at least* r and *exactly* r matches occur, $r = 0, 1, \ldots, n$.

The problem can be framed in several entertaining scenarios, such as when a capricious checkroom attendant randomly returns coats to patrons; dinner guests seat themselves before noticing the place cards; and tipsy soldiers return to their barracks, randomly choosing a bed on which to sleep. Other variations can be found in Feller [9] and Parzen [19]. A modification is to randomly draw, *with replacement*, n cards from a deck of n cards, designated $1, 2, \ldots, n$. A match occurs if the number on the card coincides with the number of the draw.

If $E_1, E_2, \ldots, E_n$ are n random events, the probability that *exactly* r of them occurs is

$$P_{[r]} = \sum_{j=r}^{n} (-1)^{j-r} \binom{j}{j-r} S_j, \qquad (1)$$

where

$$S_j = \sum_{1 \leqslant \alpha_1 < \cdots < \alpha_j \leqslant n} \cdots \sum P\left[\bigcap_{h=1}^{j} E_{\alpha_h}\right]$$

and $S_0 = 1$. The probability that *at least* r of them occur is

$$P_r = \sum_{i=r}^{n} P_{[i]} = \sum_{j=r}^{n} (-1)^{j-r}\binom{j-1}{j-r} S_j. \quad (2)$$

(*see* INCLUSION-EXCLUSION METHOD and BOOLE'S INEQUALITY.)

For the matching problem,

$$P(E_i) = (n-1)!/n! = 1/n,$$

$$S_1 = nP(E_i) = 1;$$

$$P(E_i \cap E_j) = (n-2)!/n! = 1/\{n(n-1)\};$$

$$S_2 = [n(n-1)/2]P(E_i \cap E_j) = 1/2!;$$

in general,

$$P(E_{\alpha_1} \cap E_{\alpha_2} \cap \cdots \cap E_{\alpha_r}) = (n-r)!/n!,$$

$$S_r = \binom{n}{r} P(E_{\alpha_1} \cap E_{\alpha_2} \cap \cdots \cap E_{\alpha_r}) = 1/r!.$$

Thus

$$P_{[r]} = \frac{1}{r!}\sum_{i=0}^{n-r} \frac{(-1)^i}{i!}, \quad (3)$$

$$P_{[0]} = 1 - 1 + \frac{1}{2!} - \frac{1}{3!} + \cdots + (-1)^n\left(\frac{1}{n!}\right) = \sum_{j=0}^{n} (-1)^j/j!. \quad (4)$$

Note that $P_{[n-1]} = 0$, since $(n-1)$ matches cannot occur without n matches occurring. Since $e^{-1} = \sum_{j=0}^{\infty}(-1)^j/j!$, the probability of exactly r matches approaches $e^{-1}/r!$. Thus, the probability of no matches is approximately e^{-1} for larger n, while the probability of at least one match approaches $1 - e^{-1}$ as $n \to \infty$. This is demonstrated empirically in Table 1. Note that P_1 fluctuates, increasing for n odd and decreasing for n even [7].

Some of the aforementioned results and extensions can be derived by the use of indicator random variables. Define $X_i = 1$ if card i is in position i; otherwise, $X_i = 0$. For instance, if the observed sequence is (3, 2, 1), then $X_1 = 0$, $X_2 = 1$, and $X_3 = 0$. Since $X_i = 1$ is equivalent to the occurrence of event E_i, $P(X_i = 1) = 1/n$. The number of matches X is given by $X = X_1 + \cdots + X_n$; hence the expected number of matches $E(X)$ equals 1 for any n. Furthermore, the probability generating function* of X is given by $\phi_X(t) = \sum_{j=0}^{n}(t-1)^j/j!$, and $E[X^{(r)}] = 1$ for $r = 1, \ldots, n$, and 0 for $r > n$ [8], where $E[X^{(r)}]$ denotes the rth descending factorial moment*.

Table 1 Probabilities of at Least One Match (P_1) and Exactly r Matches ($P_{[r]}$) for n Cards

n	P_1	$P_{[r]}$
1	1.00000	0.3679
2	0.50000	0.1839
3	0.66667	0.0613
4	0.62500	0.0153
5	0.63333	0.0030
6	0.63194	0.0006
7	0.63214	0.0001
⋮	⋮	⋮
∞	$1 - e^{-1} = 0.63212$	0.0000

Pierre Rémond de Montmort* [15] treated the matching problem in 1713 for 13 cards. Abraham de Moivre* [14] generalized Montmort's results. L. Euler, J. Lambert*, and P. Laplace* also displayed an early interest in the problem [5, 22].

Chapman's interest [3] in the matching problem was sparked by its use in psychological studies. For instance, in order to ascertain whether certain aspects of personality are conveyed in handwriting, subjects attempt to match character sketches of a group of people with handwriting specimens. The actual number of matchings is compared with what is expected by chance to see if there is a significant difference. P. E. Vernon [23, 24] has written extensively on the use of the matching problem in the field of psychology*. In addition to finding the probabilities of exactly r and at least r correct matchings for a single chance arrangement between two sets of n elements each, Chapman also investigated the mean number of correct matchings if m independent chance arrangements of the two series are

Table 2 Probability that at Least Two People out of *n* Have the Same Birthday

n	10	20	22	23	24	30	40	50	60
$P(A)$	.129	.411	.476	.507	.538	.706	.891	.970	.994

made. In 1935, Chapman [4] generalized his earlier work by allowing for two sets of elements of unequal length. Huntington [11, 12] modified the simple matching problem by considering decks of s of each of t kinds of cards, denoted s^t, for $s = t = 3, 4, 5$. Sterne [21] obtained a solution for $t = 5$. Olds [18] developed a moment generating function to find the first four moments for the distribution of numbers of correct matchings. Battin [1] developed a method for obtaining the moments of the distribution of hits for matchings between three or more decks of cards of arbitrary composition. Barton [2] published a comprehensive treatment of the general matching problem in 1958 and made some important extensions. David and Barton [6] derive some of the aforementioned results in their book.

The well-known "birthday problem" asks for the probability that two or more people out of n have the same birthday (event A). It is assumed that all 365 days of the year (ignoring February 29) are equally likely for each person's birthday. Under these assumptions, it follows that

$$P(\bar{A}) = 365(364)\ldots(365 - n + 1)/(365)^n,$$

and $P(A)$ is easily calculated for various values of n (see Table 2). It is somewhat surprising that $P(A) > .5$ for as few as 23 people. This effect is discussed by Mosteller [16].

The absence of the uniformity assumption (all 365 days of the year are equally likely for each person's birthday) can distort $P(A)$. In fact, if days in one month are more likely to occur, then $P(A)$ for this case is larger than if all days are equally likely [17]. The effect of leap years and seasonal trends was treated by Rust [20]. Glick [10] proposes that the hijacking of planes to Cuba can be viewed as an updated version of the birthday problem. Although the birthday problem is actually an occupancy problem* rather than a matching problem, the number of matches (number of dates which correspond to at least two people for the group of n people) and the number of matched people (the number of people out of the group of n for whom at least one other person in the group has the same birthday) was studied by Meilijson et al. [13].

References

[1] Battin, I. L. (1942). *Ann. Math. Statist.*, **13**, 294–305.

[2] Barton, D. E. (1958). *J. R. Statist. Soc.* B, **20**, 73–92.

[3] Chapman, D. W. (1934). *Amer. J. Psychol.*, **46**, 287–298.

[4] Chapman, D. W. (1935). *Ann. Math. Statist.*, **6**, 85–95.

[5] David, F. N. (1962). *Games, Gods and Gambling*. Hafner, New York.

[6] David, F. N. and Barton, D. E. (1962). *Combinatorial Chance*. Hafner, New York.

[7] Dudewicz, E. J. (1976). *Introduction to Statistics and Probability*. American Sciences Press, Columbus, Ohio.

[8] Dwass, M. (1970). *Probability and Statistics: An Undergraduate Course*. W. A. Benjamin, New York.

[9] Feller, W. (1968). *An Introduction to Probability Theory and Its Applications*, Vol. 1, 3rd ed. Wiley, New York.

[10] Glick, N. (1970). *Amer. Statist.*, **24**, 41–44.

[11] Huntington, E. V. (1937). *J. Parapsych.*, **4**, 292–294.

[12] Huntington, E. V. (1937). *Science*, **86**, 499–500.

[13] Meilijson, I., Newborn, M. R., Tenenbein, A., and Yechiali, U. (1982). *Commun. Statist. B*, **11**, 361–370.

[14] de Moivre, A. (1718). *The Doctrine of Chances*. 1st ed. London.

[15] Montmort, P. R. (1713). *Essay d'Analyse sur les Jeux de Hazard*, 2nd ed. Paris. (Reprint: Chelsea House, New York, 1981.)

[16] Mosteller, F. (1962). *Math. Teacher*, **55**, 322–325.

[17] Munford, A. G. (1977). *Amer. Statist.*, **31**, 119.

[18] Olds, E. G. (1938). *Bull. Amer. Math. Soc.*, **44**, 407–413.

[19] Parzen, E. (1960). *Modern Probability Theory and Its Applications*. Wiley, New York.

[20] Rust, P. F. (1976). *Amer. Statist.*, **30**, 197–198.

[21] Sterne, T. E. (1937). *Science*, **86**, 500–501.

[22] Todhunter, I. (1865). *A History of the Mathematical Theory of Probability from the Time of Pascal to That of Laplace*. Cambridge University Press, Cambridge, England. (Reprint: Chelsea House, New York, 1949.)

[23] Vernon, P. E. (1936). *J. Ed. Psychol.*, **27**, 1–17.

[24] Vernon, P. E. (1936). *Psychol. Bull.*, **33**, 149–177.

(BOOLE'S INEQUALITY
COMBINATORICS
MONTMORT, PIERRE RÉMOND DE
OCCUPANCY PROBLEMS
URN MODELS)

FRANK B. ALT

MATCHING TEST *See* TAKACS' GOODNESS-OF-FIT DISTRIBUTION

MATCH STATISTICS *See* TAKACS' GOODNESS-OF-FIT DISTRIBUTION

MATHEMATICAL AND STATISTICAL PSYCHOLOGY, BRITISH JOURNAL OF

The *British Journal of Mathematical and Statistical Psychology* was founded by the British Psychological Society with the title *British Journal of Psychology (Statistical Section)* in 1947 under the joint editorship of Cyril Burt and Godfrey Thomson. It changed its title to *The British Journal of Statistical Psychology* in 1953, taking its current broader title in 1966.

On Thomson's death in 1954, Burt became sole editor until 1957, when J. W. Whitfield was appointed as joint editor with Burt. They were followed by R. J. Audley (1963–1968), A. R. Jonckheere (1969–1974), P. Levy (1975–1980), and C. Leach (1981–). The current editorial address is: The Editor, *British Journal of Mathematical and Statistical Psychology*, Department of Psychology, The University, Newcastle upon Tyne, NE1 7RU, U.K.

The journal publishes articles relating to any areas of psychology that have a greater mathematical or statistical or other formal aspect to their argument than is usually acceptable to other journals. Articles that have a clear reference to substantive issues in psychology are preferred. Recent issues have included a wide range of topics: fundamental work in the analysis of covariance* structures and factor analysis*; applications of measurement theory* to the combination of marks in examinations; mathematical models of choice and discrimination; techniques for analyzing binary and paired comparisons* data; reviews of work on generalizability theory, discriminant analysis*, and the early history of the use of ANOVA* in psychology*. The journal also publishes book reviews.

The journal is international, with most submissions coming from outside the U.K. All submissions are treated equally on their scientific merits within their area. The contents of a recent issue (Vol. 36, Part 2, 1983) included:

"Efficient estimation for a multiple matrix sample design" by H. Goldstein & A. N. James

"Genetic and environmental transmission in the Colorado Adoption Project: Path analysis" by D. W. Fulker & J. C. DeFries

"Latent structure analysis of repeated classifications with dichotomous data" by C. M. Dayton and G. B. Macready

"Moment approximations as an alternative to the F test in analysis of variance" by K. J. Berry & P. W. Mielke

"Rater agreement for complex assessments" by L. J. Hubert & R. G. Golledge

"A scale invariant model for three-mode factor analysis" by S. Y. Lee & W. K. Fong

"A solution to the St. Petersburg paradox" by M. Treisman

"Interval scale measurement of attitudes" by R. Westermann

"Decision making with unreliable probabilities" by P. Gardenfors & N. E. Sahlin

"Specific inference in ANOVA" by H. Rouanet & B. Lecoutre

"Serially correlated errors in some single-subject designs" by J. P. N. Phillips

"Correlation coefficients from 2 × 2 tables" by R. G. Chambers

Book reviews.

The journal is one of a group of seven psychology journals published by the British Psychological Society, which has recently also started publishing books, including *Recent Statistics for Psychologists* edited by A. D. Lovie and *Distribution-free Methods for Non-parametric Problems: A Classified and Selected Bibliography* by B. R. Singer. About the same time as the Society moved into book publishing, it took over responsibility for publishing its own journals. In the space of a few years, the Society has become a respected publisher in its own right.

C. LEACH

MATHEMATICAL BIOSCIENCES

The first volume of *Mathematical Biosciences* was published in 1967. Then as now, this journal published "carefully selected mathematical papers of both research and expository type devoted to the formulation, analysis and numerical solution of mathematical models in the biosciences."

The editor of the journal from 1967 until 1975 was Richard Bellman, and since 1976 (Volume 28) has been John A. Jacquez, to whom manuscripts for publication should be submitted at: The Department of Physiology, University of Michigan, 7712 Medical Sciences II, Ann Arbor, Michigan 48109. The publisher is Elsevier Science, New York.

MATHEMATICAL FUNCTIONS, APPROXIMATIONS TO

Statisticians often require the numerical calculation of values of cumulative distribution functions (CDFs), densities, expected values, and other functions. Where the function has no concise closed form or cannot be directly computed, a good approximation will be sought.

Some approaches are based on statistical reasoning. Under certain conditions the CDF F_k of the standardized sum of k independent and identically distributed random variables, for example, can be represented by the leading term of an Edgeworth expansion; see (3) of ASYMPTOTIC EXPANSIONS, and CORNISH–FISHER/EDGEWORTH EXPANSIONS. These approximate F_k by the CDF of a standard normal variable and an infinite series involving Chebyshev–Hermite polynomials*. Other CDFs and expected values can be evaluated by quadrature, covered in the article NUMERICAL INTEGRATION; *see also* n-DIMENSIONAL QUADRATURE for approximations of multivariable integrals. Finally, the method of statistical differentials* (or delta method) expresses expected values of functions $g(X)$ of a random variable X by expanding $g(X)$ as a Taylor series* about $E(X)$ and then taking expected values (*see* STATISTICAL DIFFERENTIALS, METHODS OF).

The *constructive theory of functions* is that branch of mathematics that studies the approximate representation of arbitrary functions in the simplest manner possible. Let $f(x)$ be the function of interest, and suppose that it is approximated by $p(x)$. Assume that $f \in \mathscr{C}[a,b]$, the space of functions continuous for $a \leqslant x \leqslant b$, or that $f \in \mathscr{C}_c$, the space of functions continuous on the real line and periodic of period c. For some norm $\|\cdot\|$, we seek to choose p from some class of approximants to minimize

$$\|f(x) - p(x)\|.$$

The following norms are of interest:

$$\|f(x) - p(x)\| = \max_{a \leqslant x \leqslant b} |f(x) - p(x)|, \quad (1)$$

the *uniform norm* or *Chebyshev norm*.

$$\|f(x) - p(x)\| = \left[\int_a^b |f(x) - p(x)|^q \, dx \right]^{1/q}, \qquad q > 0, \quad (2)$$

where $q = 2$ defines Euclidean distance, leading to a *least-squares approximation*, and $q = 1$ leads to a *least-first-power approximation*.

$$\|f(x) - p(x)\| = \max_{x \in [a,b]} \{|f(x) - p(x)|/w(x)\} \quad (3)$$

for some weight function $w(x)$. If $w(x) = f(x)$, this defines the *relative error norm*.

The functions p used to approximate f are generally restricted to be polynomials of degree no greater than n, say, to be rational functions, or to be splines* (linked piecewise polynomials). If $f \in \mathscr{C}[a,b]$, the polynomials considered are algebraic (of the form $\sum a_i x^i$); if $f \in \mathscr{C}_c$, trigonometric polynomials of the form $\sum(a_r \sin rx + b_r \cos rx)$ are considered, and $c = 2\pi$.

In this article attention is in the main confined to functions in $\mathscr{C}[a,b]$, using the preceding norms. Many functions can be expressed as infinite expansions:

(a) *Series expansions* that include *power series* $\sum a_j x^j$, derived as Taylor series, or in other ways [7]; but expansions in series of other functions may be more convenient, such as orthogonal polynomials, exponential functions, Bessel functions*, and various hypergeometric functions (*see* CONFLUENT HYPERGEOMETRIC FUNCTIONS). Good references for these are Erdelyi [4] and Luke [9]. Expansions in series of trigonometric functions apply to representation of functions in $\mathscr{C}_{2\pi}$, and involve *Fourier series*; see Tolstov [15]. Chapter 6 of the book *Computer Approximations* by Hart et al. [6] contains detailed information for computer subroutine writing and for subroutine users on several functions treated as expansions of the preceding types; these functions include square and cube roots, exponential and hyperbolic functions, logarithmic functions, trigonometric and inverse trigonometric functions, gamma and related functions, the standard normal CDF and the related error function (*see* NORMAL DISTRIBUTION), and Bessel functions. What is sought in series expansions is a form which converges rapidly and uniformly over ranges of interest, with a controlled error that can be measured accurately.

(b) *Continued fraction expansions* of the form

$$p_0(x) + \cfrac{p_1(x)}{q_1(x) + \cfrac{p_2(x)}{q_2(x) + \cfrac{p_3(x)}{q_3(x) + \cdots}}}$$

written notationally as

$$p_0(x) + \frac{p_1(x)}{q_1(x)+}\,\frac{p_2(x)}{q_2(x)+}\,\frac{p_3(x)}{q_3(x)+\ldots}.$$

For properties, see Khovanskii [8] and Blanch [2]. Continued fraction expansions of Mills' ratio* going back to Laplace*, for example, are given in Patel and Read [12, Sect. 3.4].

UNIFORM APPROXIMATION

For functions in $\mathscr{C}[a,b]$, let $\|g\| = \max_{a \leqslant x \leqslant b}|f(x)|$. The *Weierstrass approximation theorem* states: If f is continuous on $\mathscr{C}[a,b]$, then for every $\epsilon > 0$, there exists a polynomial p such that

$$\|f(x) - p(x)\| < \epsilon. \quad (4)$$

Bernstein polynomials* provide approximating functions which satisfy the uniform norm to any degree of required accuracy; see BERNSTEIN POLYNOMIALS and ref. 13, Section 1.1.1, for details. They can be improved upon, however, if restrictions are placed on the degree of $p(x)$.

The problem of best approximation of functions in $\mathscr{C}[a,b]$ by a polynomial of degree not greater than n was solved by Borel and Chebyshev* [Natanson, ref. 11, Chapter II]. If $\mathscr{P}_n$ denotes this class of polynomials, and if

$$E_n = \inf_{p \in \mathscr{P}_n} \|p(x) - f(x)\|, \quad (5)$$

then there exists a unique polynomial p_n in $\mathscr{P}_n$ such that $\|p_n(x) - f(x)\| = E_n$. Further, there exists an *alternating set* of $n + 2$ points $x'_1, \ldots, x'_{n+2}$, where

$$a \leqslant x'_1 < x'_2 < \cdots < x'_{n+2} \leqslant b,$$

such that $|p_n(x'_i) - f(x'_i)| = E_n$, $i = 1, \ldots, n+2$, and such that the deviations

$p_n(x'_i) - f(x'_i)$ alternate in sign as i takes successive integer values. An upper bound to E_n is given (ref. 13, Sect. 1.1) by

$$E_n \leqslant 6 \sup|f(x_1) - f(x_2)|, \qquad (6)$$

where the supremum is over all x_1 and x_2 in $[a,b]$ such that $|x_1 - x_2| \leqslant (b-a)/(2n)$.

If $f(x) = x^{n+1}$, the best approximating polynomial in $\mathscr{P}_n$ is given on $[-1,1]$ by

$$x^{n+1} - p_n(x) = 2^{-n}T_{n+1}(x), \qquad (7)$$

where $T_k(x)$ is the *Chebyshev polynomial* of degree k, i.e.,

$$\begin{aligned} T_k(x) &= \cos(k \arccos x). \\ &= x^k - \binom{k}{2}x^{k-2}(1-x^2) \\ &\quad + \binom{k}{4}x^{k-4}(1-x^2)^2 - \cdots. \qquad (8) \end{aligned}$$

Note that in (8), the coefficient of x^k is 2^{k-1}. For properties of these orthogonal polynomials, see Todd [14, Chap. 5] and Natanson [11, Sect. II.4]. In general, however, it is not a simple matter to derive best approximating polynomials to arbitrary functions in $\mathscr{C}[a,b]$. In the absence of exact solutions, methods have been developed for 'approximating' p_n.

(a) *Interpolation*. Here the interval $[a,b]$ is replaced by a finite set of m points in $[a,b]$, and the approximation problem is solved on the point set. If approximating polynomials belong to $\mathscr{P}_n$, then $m \geqslant n+2$, since a polynomial of degree n can be fitted to pass through all m points otherwise. *See* INTERPOLATION and ref. 13 (Sect. 1.3) for discussion and further references.

(b) *Chebyshev polynomial expansion* on $[-1,1]$. This method is representative of expansions using orthogonal polynomials which converge rapidly for a wide variety of functions. If $f \in \mathscr{C}[a,b]$ and

$$f(x) = g(\tfrac{1}{2}(b-a)x + \tfrac{1}{2}(b+a)),$$

then $g \in \mathscr{C}[-1,1]$. The expansion

$$g(y) = \frac{1}{2}a_0 + \sum_{r=1}^{\infty} a_r T_r(y) = \sum_{r=0}^{\infty}{}' a_r T_r(y), \qquad (9)$$

say, holds for many functions with no power series expansion $\sum b_r y^r$. Where

$$g(y) = \sum_{r=0}^{\infty} b_r y^r,$$

however, fewer terms in the Chebyshev series will be required than in the power series to achieve the same degree of accuracy in the uniform norm. Frequently $a_r \to 0$ rapidly, and if $g(y)$ is approximated by the truncated sum $\sum'^{n}_{r=0} a_r T_r(y)$, the first deleted term $a_{n+1}T_{n+1}(y)$ can be used to approximate the norm $\| g(y) - \sum'^{n}_{r=0} a_r T_r(y) \|$ over $[-1,1]$; see Hart et al. [6, Sect. 3.3]. The coefficients a_r satisfy

$$\begin{aligned} a_r &= \frac{2}{\pi}\int_{-1}^{1} g(y)T_r(y)(1-y^2)^{-1/2}\,dy \\ &= \frac{2}{\pi}\int_0^{\pi} g(\cos\theta)\cos r\theta\,d\theta, \qquad (10) \end{aligned}$$

but if these are not easily computed, they can be approximated by

$$\begin{aligned} a'_r &= \frac{2}{n+1}\sum_{k=0}^{n} T_r(\bar{y}_k)g(\bar{y}_k), \\ \bar{y}_k &= \cos\left(\frac{2k+1}{n+1}\,\frac{\pi}{2}\right) \qquad (11) \end{aligned}$$

or other algorithms; see Sect. 6.3 of ref. 5. The advantage in using Chebyshev over other polynomial expansions derives from properties associated with (7).

LEAST-SQUARES APPROXIMATIONS

Generalizing (2), we seek to approximate $f \in \mathscr{C}[a,b]$ by a polynomial p so that

$$\|f(x) - p(x)\| = \left[\int_a^b \{f(x) - p(x)\}^2 w(x)\,dx\right]^{1/2} \qquad (12)$$

is minimized for all p in $\mathscr{P}_n$, where $w(x)$ is a weight function positive almost everywhere in $[a,b]$. For $f \in \mathscr{C}[-1,1]$ a unique least-squares approximating polynomial p_n in $\mathscr{P}_n$ exists, and a necessary and sufficient condition for p_n to fill this role is that

$$\int_{-1}^{1} \{f(x) - p_n(x)\}\,p(x)w(x)\,dx = 0 \qquad (13)$$

for all polynomials p in $\mathscr{P}_n$ [13, Sect. 2.1]. This leads to *orthogonal polynomials* as solutions to the least-squares approximation problem. If $\{q_0, q_1, \ldots, q_n\}$ is a set of orthogonal polynomials with respect to $w(x)$, where the degree of q_i is i $(i = 1, \ldots, n)$, i.e.,

$$\int_{-1}^{1} q_i(x) q_j(x) w(x)\, dx = \begin{cases} \|q_i\|^2, & i = j, \\ 0, & i \neq j, \end{cases} \tag{14}$$

then the least-squares approximation to $f(x)$ in $\mathscr{P}_n$ is p_n, where

$$p_n(x) = a_0 q_0(x) + a_1 q_1(x) + \cdots + a_n q_n(x),$$

$$a_i = \frac{1}{\|q_i\|} \int_{-1}^{1} q_i(x) f(x) w(x)\, dx, \tag{15}$$

$i = 1, \ldots, n$. Orthogonal expansions built up in this way have the advantage that for the best approximating polynomial in $\mathscr{P}_{n+1}$,

$$p_{n+1}(x) = p_n(x) + (\text{constant}) x^{n+1}, \tag{16}$$

and the only new coefficient is that of x^{n+1}.

Weight functions $w(x) = (1 - x)^\alpha (1 + x)^\beta$ $(\alpha > -1, \beta > -1)$ lead to *Jacobi polynomials** [13, Sect. 2.2]. In particular, $\alpha = \beta = 0$ or $w(x) \equiv 1$ leads to *Legendre polynomials*, and $\alpha = \beta = -\frac{1}{2}$ leads to solutions proportional to Chebyshev polynomials, with $w(x) = (1 - x^2)^{-1/2}$;

$$p_n(x) = \frac{1}{2} b_0 + \sum_{r=1}^{m} b_r T_r(x),$$
$$b_r = \frac{2}{\pi} \int_{-1}^{1} f(x) T_r(x) (1 - x^2)^{-1/2}\, dx, \tag{17}$$

$r = 0, 1, \ldots, n$ [5, Sect. 5.4].

Since integrals like those in (15) and (17) may not be easily evaluated, the interval $[-1, 1]$ can be replaced by a point set as earlier and a solution to the problem is attempted on the point set. Suitable families of orthogonal polynomials are available (with integrals such as (14) replaced by summations over the point set) to yield good approximations to $p_n(x)$ [13, Sect. 2.3]; *see also* INTERPOLATION.

An important general question is that of convergence of least-squares approximations. The property

$$\lim_{n \to \infty} \|f(x) - p_n(x)\| = 0 \tag{18}$$

does not necessarily hold for all $f \in \mathscr{C}[-1, 1]$, and if it does hold, it does not necessarily follow that $p_n(x) \to f(x)$ pointwise as $n \to \infty$. See Handscomb [5, Chap. 5] for a discussion. An important case is the weight function giving rise to Chebyshev polynomial expansions. If

$$\int_{-1}^{1} \{f(x)\}^2 (1 - x^2)^{-1/2}\, dx < \infty,$$

then (18) holds, as does pointwise convergence as $n \to \infty$. The properties associated with (7) assure a superior rate of convergence for the Chebyshev series.

OTHER APPROACHES

An approximating function $p(x)$ may be sought from the class $\mathscr{R}_n^m$ of *rational functions* $r_n^m(x) = p_m(x)/p_n(x)$, where $p_m \in \mathscr{P}_m$, $p_n \in \mathscr{P}_n$, and $r_n^m(x)$ is irreducible, i.e., reduced to its lowest terms. A treatment based on the relative error norm (3) has been discussed by Curtis in ref. 5, Chap. 16, and on the uniform norm (1) by Rivlin [13, Chap. 5]. In both cases unique best approximations exist on an alternating set of points in $[a, b]$, but the theory has not been developed far enough in general to place bounds on the amount of error, and nothing like the Chebyshev expansions can be called upon. However, convergence of errors to zero should in principle be faster than with polynomial approximations, which form a subset of rational functions. An algorithm due to Maehly [10] is discussed by Curtis [5, Sect. 16.3]; see also Hart et al. [6, Sect. 3.4].

If $f \in \mathscr{C}[a, b]$, then there is a unique *least-first-power* approximation (see (2) for $q = 1$) p_n to f among all polynomials in $\mathscr{P}_n$; see Chap. 3 and Sect. 4.3 of Rivlin [13] for a discussion of least-first-power approximations including interpolation and computational aspects; in this case, the interpolation problem is one in linear programming*.

Rivlin [13, Sect. 4.4] also discusses approximation by *cubic splines* (*see* GRADUATION and SPLINE FUNCTIONS), i.e., by continuous functions which in every interval $[x_i, x_{i+1}]$ of a set of points or *nodes*

$$a = x_0 < x_1 < \cdots < x_n = b$$

agrees with a polynomial of degree no greater than 3. If f is known to be differentiable, one can require the splines to be 'smooth' in the sense of having unique derivatives at each node. See also de Boor [3].

The references and bibliography list books dealing with approximations, many of them written in the 1960s. For up to date research in the subject, a number of journals may be consulted. The *Journal of Approximation Theory* is devoted to basic theoretical aspects of approximation theory, but publishes significant applications as well as papers relating to numerical work. Approximation algorithms also appear in *Mathematics in Computation*, a journal which covers a wider field than that covered in this article; it publishes both theoretical articles and applications. Approximations of functions of special interest to statisticians are frequently published in the *Journal of Statistical Computation and Simulation*, founded in 1972. For a wealth of detail on approximating functions and expansions, see Abramowitz and Stegun [1]. There are some further books, referenced in Rivlin [13].

References

[1] Abramowitz, M. and Stegun, I. A. (eds.) (1964). *Handbook of Mathematical Functions*. National Bureau of Standards, Washington, D.C.

[2] Blanch, G. (1964). *SIAM Review*, **6**, 383–421. (A computer approach to continued fractions.)

[3] de Boor, C. (1978). *A Practical Guide to Splines*. Springer-Verlag, New York.

[4] Erdelyi, A. (1954). *Transcendental Functions*. McGraw-Hill, New York.

[5] Handscomb, D. C. (ed.) (1966). *Methods of Numerical Approximation*. Pergamon Press, Oxford, England.

[6] Hart, J. F., Cheney, E. W., Lawson, C. L., Maehly, H. J., Mesztenyi, C. K., Rice, J. R., Thacher, H. G., and Witzgall, C. (1968). *Computer Approximations*. Wiley, New York.

[7] Hirschman, I. I. (1962). *Infinite Series*. Holt, Rinehart and Winston, New York.

[8] Khovanskii, A. N. (1963). *The Application of Continued Fractions and their Generalization to Problems in Approximation Theory*. Noordhoff, Groningen. (English translation.)

[9] Luke, Y. L. (1975). *Mathematical Functions and Their Approximations*. Academic Press, New York.

[10] Maehly, H. J. (1963). *J. Ass. Comp. Machinery*, **10**, 257–277.

[11] Natanson, I. P. (1964). *Constructive Function Theory*, Volume I. Ungar, New York. (A mathematical treatment based on the uniform norm. Volume II is based on the norms in (2).)

[12] Patel, J. K. and Read, C. B. (1982). *Handbook of the Normal Distribution*. Dekker, New York.

[13] Rivlin, T. J. (1981). *An Introduction to the Approximation of Functions*. Dover, New York. (A republication of the 1969 work published by Blaisdell, Waltham, Mass.)

[14] Todd, J. (1963). *Introduction to the Constructive Theory of Functions*. Academic Press, New York.

[15] Tolstov, G. P. (1962). *Fourier Series* (translated by R. A. Silverman). Prentice-Hall, Englewood Cliffs, N.J.

(ASYMPTOTIC EXPANSIONS
BERNSTEIN POLYNOMIALS
CORNISH–FISHER/EDGEWORTH EXPANSIONS
GRADUATION
INTERPOLATION
JACOBI POLYNOMIALS
n-DIMENSIONAL QUADRATURE
NUMERICAL ANALYSIS
NUMERICAL INTEGRATION
PADÉ APPROACH
SPLINE FUNCTIONS
STATISTICAL DIFFERENTIALS
WHITTAKER–HENDERSON GRADUATION)

CAMPBELL B. READ

MATHEMATICAL MORPHOLOGY *See* CLASSIFICATION; DISCRIMINANT ANALYSIS

MATHEMATICAL PROGRAMMING

Mathematical programming is the study of the problem of finding an extreme value of a function of several variables when there are restrictions on the choice of variables. It is part of the general subject known as *optimization theory* which is concerned with the problem

$$\underset{\mathbf{x} \in B}{\text{minimize}}\, f(\mathbf{x})$$

where B is a specified subset of a vector space $\mathbf{X}$ and f is a real-valued functional on $\mathbf{X}$. Mathematical programming is commonly understood to refer to the case where $\mathbf{X}$ is a finite-dimensional space, thus distinguishing it from the related topics of the *calculus of variations* and *optimal control*.

Historically, the formal theory of optimization can be traced to the eighteenth century investigation of certain geometry problems such as the isoperimetric problem of maximizing the area enclosed within a planar region of fixed perimeter. The problems were formulated and studied by such eminent mathematicians as Euler, Bernoulli*, Jacobi, and Lagrange. Indeed, the oldest theoretical results in mathematical programming are the well-known *Lagrange multiplier** rules for the minimization of a differentiable function subject to side conditions expressed as equations in the variables. From its inception until World War II, the subject of optimization* received its primary impetus from geometrical problems. This development reached its culmination in the 1930s under the influence of the "Chicago School" of the calculus of variations.

The discovery of important applications of optimization theory in military and engineering problems during World War II coupled with contemporary advances in the science of computation revitalized the subject and produced a flourish of research, not only in the theory of optimization but also in the methods of computing solutions. That resurgence has continued apace, fueled by the accelerating development in computer technology. The increase in computational efficiency has made possible the solution of larger and more complex optimization problems which in turn has expanded the horizons for applications of the theory.

Mathematical programming involves three related areas of study: theory, computation, and application. The theory involves the characterization and classification of the solution (or solutions) as they depend on the structure, properties, and parameters of the function f and the set B. The computational aspects deal with the design and implementation of algorithms for obtaining solutions. Finally, application refers to the formulation of the optimization model from the actual "real world" problem.

Mathematical programming, perhaps because of its use by operational research groups during the war, is traditionally considered a branch of study in operations research*. However, the theory uses (and contributes to) methodologies of mathematics, probability, statistics, and computer science. Likewise, the algorithmic considerations of the subject cannot be divorced from the study of such subjects as combinatorics*, numerical analysis*, and computational complexity.

Applications of mathematical programming can be found in many areas of the engineering* and social sciences, especially in economics, planning, and management science*. In these applications a choice of variables (often called decision variables) typically represents a *policy* (an investment scheme, a transportation route, a production schedule) and a policy is sought which will impart an optimal value to some objective measure (income, delivery time, profit). Examples of constraints which can be imposed on the choice of policy are technological restrictions, resource limits, and governmental regulations. A collection of mathematical programming applications can be found in ref. 3.

Another important application, particularly to statisticians, is in the maximization of a likelihood* function to obtain estimates of the parameters of a distribution (*see* MAXIMUM LIKELIHOOD ESTIMATION). The variables (parameters) are often restricted in some way. They may be required to take on only values within specified intervals or there may be algebraic relations between them. In some cases the parameters are required to be integers. A study of these and other applications of mathematical programming to statistics is given in ref. 1. An example of the use of mathematical programming in regression analysis is given at the end of this article.

In the following sections a general formulation of the mathematical programming problem will be given and a classification attempted of the most important special cases, many of which have separate entries in this encyclopedia. For references to these articles and other books and research journals see the end of this article.

FORMULATION

The following formulation encompasses a wide range of standard mathematical programming problems.

$$(\text{MP})\begin{cases} \underset{\mathbf{x}}{\text{minimize}} & f(\mathbf{x},\mathbf{z}) \\ \text{subject to} & g_i(\mathbf{x},\mathbf{z}) \leqslant 0, \quad i \in I, \\ & h_j(\mathbf{x},\mathbf{z}) = 0, \quad j \in J, \\ & \mathbf{x} \in C. \end{cases}$$

The f, g_i, and h_j are real-valued continuous functions defined on $\mathbb{R}^n \times \mathbb{R}^m$, the n-dimensional vector $\mathbf{x}$ represents the decision variables, and the m-vector $\mathbf{z}$ is a vector of parameters. f is the *objective function* and the g_i and h_j are the *constraint functions*. I and J are (usually finite) index sets and C is a subset of $\mathbb{R}^n$ which cannot conveniently be functionally described or, for theoretic or computational reasons, is not so written.

The parameter vector $\mathbf{z}$ represents random data, i.e., variables occurring in the problem formulation which are not under the control of the optimizer. For example, in an economic application $\mathbf{z}$ might represent a demand vector for a set of goods. Since a production policy might have to be chosen before the demand is known, the parameter vector can introduce a degree of uncertainty into the problem MP. As will be seen, the assumptions which are made about the parameter vector are crucial in determining the way in which MP is solved.

For a given value of $\mathbf{z}$, a decision vector $\mathbf{x} \in C$ which satisfies the functional constraints will be called *$\mathbf{z}$-feasible*. A vector $\mathbf{x}^*(\mathbf{z})$ is an *optimal solution* of MP for a given $\mathbf{z}$, if $x^*(\mathbf{z})$ is $\mathbf{z}$-feasible and $f(\mathbf{x}^*(\mathbf{z}),\mathbf{z}) \leqslant f(\mathbf{x},\mathbf{z})$ for all other vectors $\mathbf{x}$ which are $\mathbf{z}$-feasible. The goal is to gain as much information as possible about $\mathbf{x}^*(\mathbf{z})$ and its dependence on $\mathbf{z}$. The extent to which success in this effort can be achieved depends upon the properties of the functions f, g_i, and h_j, the structure of C, and the assumptions concerning the parameter vector $\mathbf{z}$.

CLASSIFICATION

From the point of view of the model the most realistic assumption is that the values of $\mathbf{z}$ are observations of a random variable whose distribution (or at least some of its properties) are known. Thus for each $\mathbf{x}$, $f(\mathbf{x}, \cdot)$, $g_i(\mathbf{x}, \cdot)$, and $h_j(\mathbf{x}, \cdot)$ can be considered random variables. With this assumption MP is not a well-defined problem; appropriate interpretations for the constraints and the minimization process must be specified. The study of the problem under these conditions is called *stochastic programming*. Some of the common approaches to stochastic programming are discussed below.

If $\mathbf{z}$ is taken to have a fixed value, e.g., the mean value of its distribution, then the problem MP becomes one of *deterministic programming*; MP is well defined and, under reasonable assumptions, can theoretically be solved to obtain $\mathbf{x}^*(\mathbf{z})$. Theoretical or computational means then can be used to study the behavior of $\mathbf{x}^*(\mathbf{z})$ as $\mathbf{z}$ varies in value, if this information is desired. This post-optimality analysis is called *perturbation* or *stability theory* (see refs. 7, 8, and 9).

Because there is no ambiguity in their meaning and also because they are easier to analyze, deterministic programs are far more prevalent in the literature and serve more often as models for applications than stochastic programs. However, the assumption of fixed data limits the effectiveness of the deterministic model in many applications.

Deterministic Programming

A critical factor in the type of methodology used in solving a deterministic program is whether or not the set C restricts some or all of the decision variables to take on only values from a discrete set. If so, then MP falls into the category of *integer programming**. These problems arise in a variety of ways. In capital budgeting problems, a decision variable may represent the number of new plants to be built. The concept of a fractional plant is meaningless and the large cost of a single plant makes the rounding of a noninteger solution potentially nonoptimal or even nonfeasible. Therefore, one must require the variable to be a nonnegative *integer*. In other problems a binary variable may be used to model logical constraints of the either–or type. Basically the methodology of integer programs is combinatorial in nature. For a thorough description and references *see* INTEGER PROGRAMMING; *see also* COMBINATORICS.

Another major division of the deterministic program is into linear and nonlinear programs. The special case of MP where the functions f, g_i, and h_j are all affine and C has the form $\{\mathbf{x}: \alpha_j \leqslant x_j \leqslant \beta_j, j = 1, \ldots n\}$ is called a *linear program*, undoubtedly the most widely used version of MP. The *simplex algorithm*, based on the pivoting theory of linear algebra, provides an efficient method for solving this problem even for a very large number of variables. Moreover, there exists a *duality theory* which yields a very satisfactory analysis of the effects of changes in the value of the parameter $\mathbf{z}$. For a complete discussion and references *see* LINEAR PROGRAMMING.

The problems in which one or more of the objective and constraint functions are nonlinear are called, not surprisingly, *nonlinear programs*. Relevant entries are NONLINEAR PROGRAMMING and LAGRANGE MULTIPLIERS. Generally the objective and constraint functions are assumed to be differentiable so that the methods of multidimensional calculus can be used in the theory and computation. The subject is greatly complicated by the fact that the nonlinearities allow the existence of *local solutions* which are not optimal.

There are several subcategories of nonlinear programming important in applications and theory. The simplest example is the convex *quadratic program* which has the form

$$(\mathrm{QP})\begin{cases} \underset{\mathbf{x}}{\text{minimize}} & \left(\tfrac{1}{2}\mathbf{x}^T\mathbf{Q}\mathbf{x} + \mathbf{q}^T\mathbf{x}\right) \\ \text{subject to} & \mathbf{A}\mathbf{x} \leqslant \mathbf{a}, \quad \mathbf{B}\mathbf{x} = \mathbf{b}, \quad \mathbf{x} \geqslant \mathbf{0}, \end{cases}$$

where $\mathbf{A}$ and $\mathbf{B}$ are matrices of appropriate dimensions, $\mathbf{q}$ is a given n-vector, and $\mathbf{Q}$ is a symmetric, positive definite matrix. The notation $\mathbf{Ax} \leqslant \mathbf{a}$ means that $(\mathbf{Ax})_i \leqslant a_i$ for each i. In addition to its importance as a model, QP is used as a benchmark for testing theories and computational schemes for more general nonlinear problems. An analysis of this special case is included in Avriel [2].

A useful generalization of QP is the *convex program* in which f and the g_i are convex functions, the h_j are affine, and C is taken to be $\mathbb{R}^n$. For this problem every local solution is optimal, which greatly simplifies the analysis. Moreover, the techniques of convex analysis can be applied in the characterization and computation of solutions. This is especially helpful in *nondifferentiable programming* where the f and g_i may not be differentiable everywhere, e.g., $f(\mathbf{x}) = \max\{\mathbf{s}_k^T\mathbf{x} : k \in K\}$. In the convex nondifferentiable problem, the subgradients of the convex function can be used in place of the derivatives. Convexity and its importance in mathematical programming are exposited in Rockafellar [13]; *see also* GEOMETRY IN STATISTICS: CONVEXITY. For research articles on nondifferentiable programming, see [4] and [12].

Another special case of MP of some interest is *geometric programming*. Here the objective and (inequality) constraint functions have the form

$$\sum_{k=1}^{m} c_k \prod_{j=1}^{n} (x_j)^{a_{kj}}.$$

Problems of this special form find wide applicability, especially in engineering. They have been extensively studied; see, for example, the book by Duffin et al. [6].

Stochastic Programming

A categorization of stochastic programs analogous to that of deterministic programs can be provided. However, the inherent difficulties have limited most of the research to date to the linear stochastic program and the less complex nonlinear problems. For illustrative purposes the discussion will center on the simple linear problem with one constraint:

$$(\mathrm{SP})\begin{cases} \underset{\mathbf{x}}{\text{minimize}} & \mathbf{c}(\mathbf{Z})^T\mathbf{x} \\ \text{subject to} & \mathbf{a}(\mathbf{Z})^T\mathbf{x} \leqslant \mathbf{b}(\mathbf{Z}), \quad \mathbf{x} \geqslant \mathbf{0}, \end{cases}$$

where $\mathbf{Z}$ is a random variable on an appropriate probability space. The components of the vectors $\mathbf{c}$, $\mathbf{a}$, and $\mathbf{b}$ can likewise be thought of as random variables.

One approach to SP, called the *distribution problem*, is to attempt to find the distribution (or its characteristics) of the random variable $\mathbf{x}^*(\mathbf{Z})$, a difficult problem whose solution requires the application of perturbation theory from deterministic programming. Few general results are available. Moreover, even if the distribution of $\mathbf{x}^*(\mathbf{Z})$ were known it would not necessarily be of practical value; the value of $\mathbf{Z}$ would have to be observed before the optimal policy could be chosen. In many if not most applications the policy is chosen before the value of $\mathbf{Z}$ is observed. For this reason alternative interpretations of SP usually are adopted.

In *chance-constrained programming* the constraints must be satisfied with some given probability and the expected value of the objective function is minimized subject to that condition. In this formulation SP becomes

$$\begin{cases} \underset{\mathbf{x}}{\text{minimize}} & E\left[\mathbf{c}(\mathbf{Z})^T\mathbf{x}\right] \\ \text{subject to} & P\left[\mathbf{a}(\mathbf{Z})^T\mathbf{x} \leqslant \mathbf{b}(\mathbf{Z})\right] \geqslant \alpha, \quad \mathbf{x} \geqslant \mathbf{0}, \end{cases}$$

where $\alpha \in [0, 1]$ is fixed; this problem is now a deterministic mathematical program. Research on chance-constrained programming has centered on determining which classes of distributions lead to reasonable deterministic problems. Unfortunately, the resulting problems are almost always nonlinear and nondifferentiable and sometimes nonconvex.

A second formulation is called a *stochastic program with recourse*. If a given policy $\mathbf{x}$ is chosen prior to the determination of $\mathbf{Z}$, the actual observation of $\mathbf{Z}$ may cause $\mathbf{x}$ to be infeasible; this generally means that the optimizer must pay a penalty for having violated the constraints. For example, the producer must pay extra holding costs if production exceeds demand. In programs with recourse a penalty function for constraint violation is formulated and an optimal policy is defined as one which minimizes the expected value of the sum of the objective function and the penalty cost. An example is:

$$\underset{\mathbf{x} \geqslant \mathbf{0}}{\text{minimize}} \quad E\left[\mathbf{c}(\mathbf{Z})^T\mathbf{x} + p(\mathbf{x}, \mathbf{Z})\right], \quad \text{where}$$

$$p(\mathbf{x}, \mathbf{Z}) = \begin{cases} \underset{\mathbf{y}}{\text{minimum}}\ \mathbf{d}(\mathbf{Z})^T\mathbf{y} \text{ subject to} \\ \mathbf{B}(\mathbf{Z})\mathbf{y} = \left[\mathbf{a}(\mathbf{Z})^T\mathbf{x} - \mathbf{b}(\mathbf{Z})\right]_+, \\ \qquad \mathbf{y} \geqslant \mathbf{0}. \end{cases} \quad (1)$$

The vector $\mathbf{d}$ and matrix $\mathbf{B}$ are appropriately chosen to express the penalty concept; $[\mathbf{w}]_+$ represents the vector whose components are $\hat{w}_j = \max[0, w_j]$. Stochastic programs are also called *two-stage stochastic programs*. Under fairly reasonable assumptions the problem (1) is a deterministic convex program.

There are other interpretations of stochastic programs, including a multistage generalization of the recourse program. For a description of these and details of chance-constrained and two-stage programming see the text by Kolbin [11] and the conference proceedings edited by Dempster [5]. Kall [10] provides an elementary overview of stochastic programming and Stancu-Minasian and Wets [14] provide an excellent bibliography of articles appearing prior to 1976.

AN EXAMPLE

An important application of mathematical programming to statistics occurs in regression* analysis. For instance, suppose that it is desired to find coefficients a_i, $i = 1, \dots, n$, which yield the best fit to the linear system

$$\sum_{i=1}^{n} X_{ji} a_i = Y_j, \qquad j = 1, \dots, k.$$

Here the X_{ji}, $j = 1, \dots, k$, $i = 1, \dots, n$, represent the values of the independent variables X_i, $i = 1, \dots, n$, which produce the observations Y_j, $j = 1, \dots, k$. Rather than use the most common measure of goodness-of-fit, the least-squares* error, the object will be to choose the a_i so as to minimize the sum of the absolute deviations, i.e., to minimize

$$\sum_{j=1}^{k} \left| Y_j - \sum_{i=1}^{n} X_{ji} a_i \right|.$$

In many statistical applications this is considered a more appropriate measure of error than the sum of the squares of the deviations. Letting

$$\epsilon_j = Y_j - \sum_{i=1}^{n} X_{ji} a_i$$

and decomposing ϵ_j into nonnegative parts

$$\epsilon_j = \epsilon_j^+ - \epsilon_j^-, \qquad \epsilon_j^+ \geq 0, \qquad \epsilon_j^- \geq 0,$$

so that $|\epsilon_j| = \epsilon_j^+ + \epsilon_j^-$, permits the regression problem to be formulated as the linear program

$$(\text{LP})\begin{cases} \text{minimize} & \sum_{j=1}^{k} (\epsilon_j^+ + \epsilon_j^-) \\ \text{subject to} & \sum_{i=1}^{n} X_{ji} a_i + \epsilon_j^+ - \epsilon_j^- = Y_j, \\ & \qquad j = 1, \dots, k, \\ & \epsilon_j^+ \geq 0, \quad j = 1, \dots, k, \\ & \epsilon_j^- \geq 0, \quad j = 1, \dots, k, \\ & a_i \text{ unrestricted.} \end{cases}$$

This is a very simple example of a mathematical programming problem. It can be solved by the simplex algorithm in a straightforward manner (*see* LINEAR PROGRAMMING).

A more difficult problem results if the additional restriction that only $r < n$ of the independent variables can be used in the regression is placed on the model. In this setting a choice of the best r variables to use for regression is to be determined. In order to formulate this problem as a mathematical program, the binary variables δ_i, $i = 1, \dots, n$ are introduced, where

$$\delta_i = \begin{cases} 0 & \text{if variable } X_i \text{ is not used}, \\ 1 & \text{if variable } X_i \text{ is used}. \end{cases}$$

Now the program can be written

$$(\text{IP})\begin{cases} \text{minimize} & \left(\sum_{j=1}^{k} (\epsilon_j^+ + \epsilon_j^-) \right) \\ \text{subject to} & \sum_{i=1}^{n} X_{ji} a_i + \epsilon_j^+ - \epsilon_j^- = Y_j, \\ & \qquad j = 1, \dots, k, \\ & a_i - K\delta_i \leq 0, \quad i = 1, \dots, n, \quad (2) \\ & -a_i - K\delta_i \leq 0, \quad i = 1, \dots, n, \quad (3) \\ & \sum_{i=1}^{n} \delta_i = r, \quad (4) \\ & \epsilon_j^+ \geq 0, \quad j = 1, \dots, k, \\ & \epsilon_j^- \geq 0, \quad j = 1, \dots, k, \\ & \delta_i \in \{0, 1\}, \quad i = 1, \dots, n, \\ & a_i \text{ unrestricted}, \quad i = 1, \dots, n. \end{cases}$$

K is a constant chosen sufficiently large to ensure that the constraints (2) and (3) are not binding when $\delta_i = 1$. Constraints (2), (3), and (4) guarantee that no more than r of the a_i will be used; i.e., at least $n - r$ of the a_i will be zero.

This latter mathematical program is an *integer* program or, more precisely, a *mixed integer program*. It contains both continuous and discrete variables. A solution can be found by solving the $\binom{n}{r}$ linear programs which correspond to the possible choices of the r variables. This, however, can be so time consuming that it is impractical for even moderately sized r and n. There are algorithms available (e.g., the branch-and-bound* algorithm) which, in practice, solve integer programs more rapidly (*see* INTEGER PROGRAMMING).

If it is desired to fit the data using a nonlinear function $f(a_1, \ldots, a_n; X_1, \ldots, X_l)$ then the problem becomes a nonlinear program in which the linear constraints

$$\sum_{i=1}^{n} X_{ji}a_i + \epsilon_j^+ - \epsilon_j^- = Y_j$$

in (LP) are replaced by the nonlinear equations

$$f(a_1, \ldots, a_n; X_{j1}, \ldots, X_{jl}) + \epsilon_j^+ - \epsilon_j^- = Y_j,$$

where the $a_1, \ldots, a_n$ are parameters that determine the function. For example,

$$f(a_1, a_2, a_3, a_4; X_1, X_2) = a_1\exp(a_2X_1) + a_3\exp(a_4X_2).$$

In this case the computation of the optimal values of the parameters is performed by approximation algorithms; exact optimal values cannot in general be obtained (*see* NONLINEAR PROGRAMMING).

References

[1] Arthanari, T. S. and Dodge, Y. (1981). *Mathematical Programming in Statistics*. Wiley, New York. (Emphasizes linear programming applications.)

[2] Avriel, M. (1976). *Nonlinear Programming: Analysis and Methods*. Prentice-Hall, Englewood Cliffs, N.J. (The most complete reference currently available on nonlinear programming.)

[3] Balinski, M. L. and Lemarechal, C., eds. (1978). *Mathematical Programming Study* **9**. North-Holland, Amsterdam.

[4] Balinski, M. L. and Wolfe, P., eds. (1975). *Mathematical Programming Study* **3**. North-Holland, Amsterdam. (Research articles.)

[5] Dempster, M. A. H., ed. (1980). *Stochastic Programming*. Academic Press, London. (Research articles.)

[6] Duffin, R. J., Peterson, E. L., and Zener, C. (1967). *Geometric Programming—Theory and Applications*. Wiley, New York. (An introductory text.)

[7] Fiacco, A. V., (1983). *Introduction to Sensitivity and Stability Analysis in Nonlinear Programming*. Academic Press, New York. (The best reference text available on the perturbation of nonlinear programs.)

[8] Fiacco, A. V., ed. (1982). *Mathematical Programming with Data Perturbations*. Marcel Dekker, New York. (Research articles.)

[9] Gal, T. (1979). *Postoptimal Analysis, Parametric Programming, and Related Topics*. McGraw-Hill, New York.

[10] Kall, P. (1982). *European Journal of Operations Research*, **10**, 125–130. (A good short survey article on stochastic programming.)

[11] Kolbin, V. V. (1977). *Stochastic Programming*. Reidel, Dordrecht, Holland.

[12] Lemarechal, C. and Mifflin, R., eds. (1978). *Nonsmooth Optimization*. Pergamon Press, Oxford, England. (Research articles.)

[13] Rockafellar, R. T. (1970). *Convex Analysis*. Princeton University Press, Princeton, N.J.

[14] Stancu-Minasian, I. M. and Wets, M. J. (1976). *Operat. Res.*, **24**, 1078–1119. (A bibliography.)

(COMBINATORICS
INTEGER PROGRAMMING
LAGRANGE MULTIPLIERS
LINEAR PROGRAMMING
MAXIMUM LIKELIHOOD ESTIMATION
NONLINEAR PROGRAMMING)

JON TOLLE

MATHEMATICAL THEORY OF POPULATION

Statistical data about the sizes and structures of populations have been gathered since the dawn of history. The Sumerians, five thousand years ago, for example, found it necessary for taxation purposes, and the Romans somewhat later needed to do so to enforce military conscription. In the sixteenth, seventeenth, and eighteenth centuries, the Christian church compiled a mass of population data in the form of parish registers, and much of this material is still available. Today, central statistical offices compile enormous amounts of demographic data annually (*see* DEMOGRAPHY).

In earlier times, population data were usually gathered for taxation, military, or ecclesiastical purposes. The Englishman John Graunt (1620–1674) appears to have been the first to study population data from the purely scientific point of view and to publish an account of his work, although others soon were to follow his lead. Graunt's most notable achievement was undoubtedly the life table* [16], but his other work concerning the population of London was also of a high standard, and indicates a keen mind (Sutherland, [39]).

Mathematical models of population growth appeared toward the end of the eighteenth century with the work of Euler [13] and Malthus* [28]. A little later, Verhulst [44] studied the exponential and logistic curves as models of population growth. The mathematicians Bienaymé* [4], Galton* and Watson [15], and others became embroiled in the problem of the extinction of family names (*see* GALTON–WATSON PROCESS) and their mathematical models form the basis of what is now known as the Theory of Branching Processes*, which finds wide application in a number of areas.

The paper that gave the greatest impetus to the mathematical theory of population and influenced all subsequent developments in mathematical demography was that of Sharpe and Lotka [38].

SHARPE AND LOTKA MODEL

Sharpe and Lotka considered the male population only and assumed that the growth and development of the female population would be such as to justify the assumption of constant age-specific fertility rates and mortality rates for the males (i.e., rates which depended on the age x of the male, but *not* on the epoch t). The roles of the two sexes in the model may be reversed, and present-day demographers usually apply it to the female sex because of its shorter reproductive lifespan and the fact that exnuptial births are more readily attributable to the mother.

The theory leads to a renewal integral equation (*see* RENEWAL THEORY) and the prediction that the population will grow exponentially asymptotically at rate r (the *intrinsic growth rate*) and adopt a *stable age distribution* with density proportional to

$$ {}_xp_0^f e^{-rx}, \tag{1}$$

where x is the age under consideration, and ${}_xp_0^f$ is the proportion of females surviving from birth to age x [20, 33]. The male component of the population also adopts a stable age distribution, with density proportional to

$$ {}_xp_0^m e^{-rx}, \tag{2}$$

where ${}_xp_0^m$ is the proportion of males surviving from birth to age x. A systematic presentation of time-continuous stable population theory in modern probabilistic dress can be found in Hoem and Keiding [17].

The stable population is a consequence of the constant mortality and fertility rates assumed in the theory. In practice, of course, mortality and fertility tend to change with time. Nevertheless, the stable model provides a reasonably accurate representation of quite a wide range of populations and permits the derivation of some important and useful results.

MODEL LIFE TABLES AND STABLE POPULATION THEORY

Model life tables have been developed by various authors [5, 11, 42, 43] for use in countries with limited or nonexistent data on mortality and survival. Knowing only a few details about the mortality of the population (e.g., the approximate expectation of life at birth, the approximate mortality rate during the first year of life, etc.) or guessing such parameters in the light of knowledge of mortality levels in similar populations, it is possible to choose a suitable model life table with these characteristics and hence estimate any other required life table function, e.g., the probability of surviving from age x to age y (*see* LIFE TABLES and Coale and Demeny [11]).

With the aid of a model life table and having only limited knowledge of the rate of growth of a population, stable population theory can be used to make a wide range of inferences about the structure and composition of that population (e.g., number of children under 5 years of age, number of persons in the productive age range 18–65, sex ratio of adults, etc.).

MOMENTUM OF POPULATION GROWTH

Using stable population theory, Keyfitz [21] has shown that if, in a presently high fertility stable population with intrinsic growth rate r, the birth rate $\lambda(x)$ is suddenly reduced to the lower level $\lambda(x)/R_0$ corresponding to replacement (but no growth) in the long term, the population will continue to grow for some time before becoming stationary, and the ratio of the final population size to its size at the time of the fertility transition is

$$\frac{b\mathring{e}_0}{r\kappa_1}\left(\frac{R_0 - 1}{R_0}\right). \tag{3}$$

In this formula, b is the birth rate in the population prior to the change, and κ_1 is the average age of childbearing; $\mathring{e}_0$ is the expectation of life at birth, and R_0 the *net reproduction rate* before the transition (the average number of daughters a newborn female will have during her lifetime).

Although natural populations are not strictly stable, (3) provides accurate results for both developed and less-developed populations. The effects of gradual reductions in fertility* to replacement level have also been studied [8, 9, 41].

DISCRETE MODELS

Deterministic models of population growth exist in two forms: those using a continuous time variable and a continuous age scale (following Sharpe and Lotka, above), and those using a discrete time variable and a discrete age scale. Both types have their advantages, but the discrete formulation is the closer to actuarial practice and is preferable when the age-specific birth and death rates are to be given on the basis of empirical data, rather than as analytical formulae.

The discrete analogue of the Sharpe and Lotka theory was developed independently by several different authors about the same time: H. Bernardelli [3]; E. G. Lewis [26], and P. H. Leslie [25]. Leslie's work is the most detailed and best known. He considered the female population only and assumed that the growth and development of the male component of the population would be such as to justify the assumption of constant age-specific fertility rates and mortality rates for the females.

Leslie listed the numbers of females in the various age groups at time t in a column vector $\mathbf{n}_t$, and showed that the numbers in the various age groups at successive points of time obeyed the vector recurrence equation

$$\mathbf{n}_{t+1} = \mathbf{A}\mathbf{n}_t, \tag{4}$$

where $\mathbf{A}$ is a square matrix with a particularly simple form. The solution to the recurrence equation (4) indicates that the population will grow exponentially asymptotically at rate $\lambda - 1$ per time unit, where λ is the dominant eigenvalue of the matrix $\mathbf{A}$, and adopt a stable age distribution proportional to the dominant right eigenvector $\mathbf{x}$ of matrix $\mathbf{A}$ [20, 33].

WEAK ERGODICITY

The stable age distribution predicted by both the continuous and discrete population models does not depend on the initial age structure of the population. In other words, under the strong conditions of constant fertility and mortality rates the population tends to forget its initial age distribution with the passage of time. A population subject to changing regimes of fertility and mortality also tends to forget its initial age structure with the passage of time and adopt a structure related to the fertility and mortality it has experienced [27]. This property is referred to as *weak ergodicity* because of the weaker assumptions made concerning fertility and mortality [20, 33].

STOCHASTIC MODELS

In recent years, population mathematicians have turned their attention more and more to stochastic models. The natural stochastic analogue of the Leslie deterministic model turns out to be a branching process [31, 33]. The branching process approach does not, however, explain all the variability observed in actual populations, much of which is due to environmental factors, and various alternative models have been proposed to overcome this deficiency. Time series* methods also have been tried. As yet, however, there is no widely accepted practical stochastic model.

A full account of the various stochastic models which have been developed for population analysis is given in STOCHASTIC DEMOGRAPHY.

MIGRATION

In the absence of migration*, the models of Sharpe and Lotka and of Leslie provide reasonably adequate deterministic descriptions of the growth and development of most, if not all, human populations. The fertility and survival rates in the models vary of course from population to population.

The structure of migration is different in different populations. For this reason, no single mathematical model has been developed which provides an adequate description in all cases. Some populations, for example, are subject to substantial emigration. If the number of emigrants of a particular age in a particular year can be assumed to be approximately proportional to the number of persons in the age group at that time, the effect of emigration on the population may be accounted for by simply reducing the survival rates in the Sharpe and Lotka or Leslie model. This simple model, however, would not be satisfactory for studying another population with substantial immigration but subject to a strict quota. A possible approach in this case would be to introduce a vector of immigrants **b** into the Leslie model and adopt the recurrence equation.

$$\mathbf{n}_{t+1} = \mathbf{A}\mathbf{n}_t + \mathbf{b}. \tag{5}$$

Stochastic versions of these models are also available [33].

THE TWO-SEX PROBLEM

All the above deterministic and stochastic models involve the female sex alone. The growth of the male component of the population is assumed to be consistent with the assumptions of constant fertility and survival rates for the females. There is no mathematical reason why the roles of the two sexes should not be interchanged. The only problem is that different results are obtained, and these are sometimes quite contradictory [23]. This *two-sex problem* has been discussed sporadically over more than half a century since the pioneering work of Karmel [18], D. G. Kendall [19], and Pollard [30] and many ingenious methods have been proposed for dealing with it and achieving consistency [35].

The search for a practical realistic two-sex model continues to present an intellectual challenge to mathematicians (see, e.g., Asmussen [1], Mitra [29]). Fortunately, most of the problems regularly encountered by demographers are handled adequately by the traditional Sharpe and Lotka, and Leslie methods.

POPULATION PROJECTIONS

Government and private planning in many areas depends heavily upon estimates of future population. Water storage facilities and electricity-generating systems, for example, need long-term planning, and as the demand for these utilities depends upon the size and structure of the relevant population, population projections* are required well into the future. Projections also are needed to estimate future demand for kindergartens, schools, hospitals, playing fields, senior citizen facilities, old age pensions, etc.

The mathematical model underlying most national population projections is Leslie's matrix model, or a variation of it (e.g., the immigration model represented by (5)). To distinguish it from a simple projection of total numbers it is often referred to as the "component method" [12]. Trends in fertility, mortality, and migration are readily taken into account in the projection process [2]. Age-specific fertility rates also can be adjusted to take account of perceived trends in average completed family size and average age at marriage*, even though these factors are not included explicitly in the Leslie model.

The model also can be made more complicated by subdividing women according to their *parity* (the number of children a woman already has) as well as age. The recurrence matrix **A** becomes much larger and parity progression rates are required, but the mathematical theory remains substantially unaltered.

Subdivision by marital status is sometimes attempted in population projection programs, with different fertility and survival rates for married and unmarried women. Problems of internal consistency arise (e.g., ensuring that the number of married females of a certain age is no greater than the total number of females of that age and that there is an adequate supply of males—the two-sex problem), and divorce must also be taken into account.

Population projections often turn out to be very inaccurate, and the question naturally arises: "Can a stochastic population model be used to obtain measures of reliability in population projections in the form of variances and covariances?" The lack of reliability is caused mainly by

(a) the use of an incorrect model (e.g., omitting parity);

(b) an incorrect assessment of future trends in transition probabilities;

(c) stochastic variation in the numbers of females surviving and reproducing; and

(d) random fluctuations in the underlying probabilities (due to variations in climatic conditions, etc.).

Existing stochastic models make allowance for (c) and sometimes (d). They make no allowance for the dominant sources (a) and (b) (*see also* STOCHASTIC DEMOGRAPHY).

MULTIREGIONAL MODELS

The continuous-time model of Sharpe and Lotka and the discrete matrix methods of Bernardelli, Lewis, and Leslie have been generalized by A. Rogers [37] to a population experiencing different mortality and fertility in different regions and migration between regions. Rogers shows how the one-dimensional single region theory can be extended in vector form to deal with the multiregional population, and suggests methods for estimating basic multiregional demographic measures from incomplete data. Braun [6] presents a modern probability-theoretic approach to multiregional populations.

Individual Behavior

Although macro models such as those outlined above usually are employed for population projection purposes, models of individual behavior are essential for certain purposes, e.g., modeling reproduction and studying the effects of contraception and the nonfecund period following pregnancy [20, pp. 390–398; 22, pp. 303–320].

Models of individual behavior also are required to study the effects (if any) of couples in a population following various rules in their decision to have further children. As an example, consider the case of a population in which each couple desires a son and will stop having children as soon as a son is obtained. What will the average family size be? Will the sex ratio in the community be affected? What will be the average proportion of boys in a family?

Let us assume that at each birth for each fecund couple, the child is a son with probability p and a daughter with probability $q = 1 - p$. A proportion p of couples will achieve their goal with one child, qp with two children, q^2p with three children, and so on. The average family size will be

$$p + 2qp + 3q^2p + \cdots = p/(1-q)^2$$
$$= 1/p, = 1.96$$

if $p = 0.51$. The ratio of boy births to total births will be

$$\frac{p + qp + q^2p + \cdots}{p + 2qp + 3q^2p + \cdots} = p.$$

In other words, the desire to have boy babies has no effect on the sex ratio of children born, despite a popular belief to the contrary.

With this stopping rule, however, the average proportion of boys in a family is

$$p + \tfrac{1}{2}qp + \tfrac{1}{3}q^2p + \tfrac{1}{4}q^3p + \cdots = -(p/q)\ln p$$

or about 0.70 if $p = 0.51$, which is markedly different from p.

PRACTICAL APPLICATION OF POPULATION MODELS

There are many difficult and challenging problems for the model builder in population studies. Some of them are outlined in this article. Others can be found in the references, particularly Keyfitz [20, 22], Pollard [33], and Brillinger [7].

It is essential that the model builder keep both feet on the ground, however. The purpose for formulating a particular model must be kept clearly in mind. It is pointless developing an extremely complex model to answer a simple question when the validity of some of the assumptions in the model is open to dispute and the data are unreliable. The level of complexity of a model must be determined by the accuracy required, the available data, and what is known about the underlying social processes.

References

[1] Asmussen, S. (1980). *Ann. Prob.*, **8**, 727–744.

[2] Benjamin, B. (1968). *Demographic Analysis*. George Allen and Unwin, London, pp. 112–122.

[3] Bernardelli, H. (1941). *J. Burma Res. Soc.*, **31**, 1–18.

[4] Bienaymé, I. J. (1845). *Soc. Philomath. Paris Extraits*, Series 5, 37–39.

[5] Brass, W., ed. (1971). *Biological Aspects of Demography*. Taylor and Francis, London, pp. 357–358.

[6] Braun, H. (1978). *Scand. Actuarial J.*, 185–203.

[7] Brillinger, D. B. (1981). *Canad. J. Statist.*, **9**, 173–194.

[8] Cerone, P. and Keane, A. (1978). *Demography*, **15**, 131–134.

[9] Cerone, P. and Keane, A. (1978). *Demography*, **15**, 135–137.

[10] Coale, A. J. (1957). *Cold Spring Harbor Symposia on Quantitative Biology*, K. B. Warren, ed. **22**, 83–89.

[11] Coale, A. J. and Demeny, P. (1966). *Regional Model Life Tables and Stable Populations*. Princeton University Press, Princeton, N.J. (Useful introduction on use of model life tables.)

[12] Cox, P. R. (1970). *Demography*, 4th ed. Cambridge University Press, Cambridge, England, pp. 234–272.

[13] Euler, L. (1767). *Histoire de l'Académie Royale des Sciences et Belles Lettres, Année 1760.* Preussische Akademie der Wissenschaften zu Berlin, pp. 144–164.

[14] Galton, F. (1873). *Educational Times* (1 April), p. 17.

[15] Galton, F. and Watson, H. W. (1873). *Educational Times* (1 August), p. 115.

[16] Graunt, J. (1662). *Natural and Political Observations Mentioned in a Following Index, and Made upon the Bills of Mortality, with Reference to the Government, Religion, Trade, Growth, Air, Diseases, and the Several Changes of the Said City.* John Martyn, London.

[17] Hoem, J. M. and Keiding, N. (1976). *Scand. Actuarial J.*, 150–175.

[18] Karmel, P. H. (1947). *Population Studies*, **1**, 249–274. (Useful exposition on the two-sex problem.)

[19] Kendall, D. G. (1949). *J. R. Statist. Soc. B*, **11**, 230–264. (Important contribution to stochastic population theory and the two-sex problem.)

[20] Keyfitz, N. (1968). *Introduction to the Mathematics of Population*. Addison-Wesley, Reading, Mass., pp. 271–292. (Useful general reference on the mathematical theory of population.)

[21] Keyfitz, N. (1971). *Demography*, **8**, 71–80.

[22] Keyfitz, N. (1977). *Applied Mathematical Demography*. Wiley-Interscience, New York. (Useful

general reference on the mathematical theory of population.)

[23] Kuczynski, R. R. (1932). *Fertility and Reproduction*. Falcon Press, New York, pp. 36–38.

[24] Ledermann, S. (1969). *Nouvelles Tables—Types de Mortalité*, Institut National d'Études Démographiques.

[25] Leslie, P. H. (1945). *Biometrika*, **33**, 183–212. (Important original paper on the matrix analysis of population growth.)

[26] Lewis, E. G. (1942). *Sankhyā*, **6**, 93–96.

[27] Lopez, A. (1961). *Problems in Stable Population Theory*. Office of Population Research, Princeton, N.J., pp. 47–62.

[28] Malthus, T. R. (1798). *An Essay on the Principle of Population*, 1st ed., printed for J. Johnson in St. Paul's Churchyard, London, p. 21.

[29] Mitra, S. (1978). *Demography*, **15**, 541–548.

[30] Pollard, A. H. (1948). *J. Inst. Actuaries*, **74**, 288–318.

[31] Pollard, J. H. (1966). *Biometrika*, **53**, 397–415.

[32] Pollard, J. H. (1968). *Biometrika*, **55**, 589–590.

[33] Pollard, J. H. (1973). *Mathematical Models for the Growth of Human Populations*. Cambridge University Press, Cambridge, England. (Useful general reference on the mathematical theory of population.)

[34] Pollard, J. H. (1975). *Austral. J. Statist.*, **17**, 63–76.

[35] Pollard, J. H. (1977). *Proc. Internat. Population. Conf. (Mexico)*, **I**, 291–309. (Useful survey of the two-sex literature.)

[36] Rhodes, E. C. (1940). *J. R. Statist. Soc.*, **103**, 61–89, 218–245, 362–387. (Useful introduction to the Sharpe and Lotka theory of population.)

[37] Rogers, A. (1975). *Introduction to Multiregional Mathematical Demography*. Wiley, New York.

[38] Sharpe, F. R. and Lotka, A. J. (1911). *Philos. Mag. 6th Series*, **21**, 435–438.

[39] Sutherland, I. (1963). *J. R. Statist. Soc. A*, **126**, 537–556.

[40] Todhunter, I. (1865). *A History of the Mathematical Theory of Probability*. Reprinted 1965, Chelsea House, New York, pp. 240–241.

[41] Tognetti, K. P. (1976). *Demography*, **13**, 507–512.

[42] United Nations (1955). *Age and Sex Patterns of Mortality. Model Life Tables for Underdeveloped Countries.* UNO.

[43] United Nations (1965). *The Concept of a Stable Population. Applications to the Study of Populations of Countries with Incomplete Demographic Statistics*. UNO, New York.

[44] Verhulst, P. F. (1847). *Nouveaux Mémoires de l'Académie Royale des Sciences et Belles Lettres de Bruxelles*, **20**, 1–32.

[45] Watson, H. W. and Galton, F. (1874). *J. Anthropol. Inst. Great Britain and Ireland*, **4**, 138–144.

[46] Yellin, J. and Samuelson, P. A. (1974). *Proc. Nat. Acad. Sci.* (USA), **2**, 2813–2817. (More advanced treatment of certain two-sex models.)

(BIRTH-AND-DEATH PROCESSES
BRANCHING PROCESSES
DEMOGRAPHY
FERTILITY
GALTON–WATSON PROCESS
HUMAN GENETICS, STATISTICS IN
LIFE TABLES
MARRIAGE
MIGRATION
POPULATION PROJECTIONS
POPULATION PYRAMID
RENEWAL THEORY
STOCHASTIC DEMOGRAPHY
VITAL STATISTICS)

J. H. POLLARD

MATHEMATICS OF COMPUTATION

This journal was first published in 1943 by the National Research Council as *Mathematical Tables and Other Aids to Computation*, "a quarterly journal edited on behalf of the Committee on Mathematical Tables and Other Aids to Computation by Raymond Clare Archibald and Derrick Henry Lehmer." In a description of the goals and format appearing on pp. 1–2 of the first issue, the journal is described as "a clearing-house for information."

By 1959 it had become clear that the scope of the journal needed to be enlarged. This was due to the developing technology of computers and concomitant advances in numerical methods. An announcement at the end of Volume 13 stated that the journal would change its name with " . . . increased emphasis on modern advances in the theory and application of computational methods." The philosophy of the journal today remains as it was summarized by Harry Polachek on pp. 1–2 of Volume 14 (1960): "With this entire field of modern mathematics, embracing (1) advances in numerical analysis, (2) the application of numerical methods and high-speed calculator devices, (3) the computation of mathematical tables, (4) the development of new mathematical disciplines related to computation, (5) the theory of high-speed calculating devices and other aids to computation, we have associated the title, *Mathematics of Computation*."

From 1962 to 1965 the journal was published for the National Academy of Sciences–National Research Council by the American Mathematical Society (AMS), but since July 1965 the announcement has listed the AMS as sole publisher. The current managing editor is James H. Bramble.

One policy of the journal, announced in each issue, is noteworthy: "The editorial office of the journal maintains a repository of Unpublished Mathematical Tables (UMT). When a table is deposited in the UMT repository, a brief summary of its contents is published in the section *Reviews and Descriptions of Tables and Books*. Upon request, the chairman of the editorial committee will supply copies of any table for a nominal cost per page. All tables and correspondence concerning the UMT should be sent to James H. Bramble, Chairman, Department of Mathematics, White Hall, Cornell University, Ithaca, NY 14853." Manuscripts for publication should be sent to the managing editor at the same address.

MATHEMATISCHE OPERATIONSFORSCHUNG UND STATISTIK, SERIES STATISTICS

The journal *Mathematische Operationsforschung und Statistik* was founded at the former Institute of Applied Mathematics and Mechanics of the Academy of Sciences of the German Democratic Republic (GDR) in 1970. Since then it has been published by the Akademie-Verlag, Berlin. The founding editor was O. Bunke (Humboldt University, Berlin), who served until 1977, when the journal, having grown from 331 pages in Vol. 1 (1970) to 977 pages in Vol. 7 (1976), was divided into two series, Statistics and Optimization. Since 1977 Bunke has been editor-in-chief of the *Series Statistics*.

The international editorial board is composed of J. Anděl (Prague), H. Bandemer (Freiberg), R. Bartoszyński (Warsaw), H. Bunke (Berlin), O. Bunke (Berlin), V. Dupač (Prague), V. V. Fedorov (Moscow), B. W. Gnedenko (Moscow), J. Jurečkova (Prague), W. Klonecki (Wrocław), E. Lyttkens (Uppsala), G. M. Mania (Tbilissi), P. H. Müller (Dresden), B. I. Penkov (Sofia), M. L. Puri (Bloomington, Indiana), C. R. Rao (New Delhi). P. Révész (Budapest), J. Roy (Calcutta), K. Sarkadi (Budapest), J.-L. Soler (Grenoble), I. Văduva (Bukarest), I. Vincze (Budapest), W. Winkler (Dresden), and H. Wold (Göteborg).

At the end of the 1960s the GDR and other socialist countries still needed a special journal for mathematically oriented papers in operations research* and statistics. The foundation of the journal was intended to enable the necessary scientific exchange of research results and to inform scientists working in applied fields about new theoretical developments, procedures and algorithms, and interesting experiences in case studies.

The editorial policy of *Series Statistics* has not changed since its foundation and is the following: The journal publishes theoretical and applied papers related to the different fields of statistics like regression* and analysis of variance*, design of experiments*, foundations of statistical inference*, statistical decision theory*, testing hypotheses, parameter estimation, nonparametric methods, sequential procedures*, time series* and statistical problems for stochastic processes*, statistical data analysis, etc. Submitted papers are expected to be interesting and to provide new contributions to theoretical problems at a good mathematical level or a stimulating presentation of ideas in foundations or applications. All papers should have a clear introduction and include practical or theoretical results and connections to the relevant literature. The trend to highest generality should not obscure the essential ideas. A special section is devoted to survey papers on theory and methods in interesting fields of statistics. Furthermore, the journal publishes proceedings of conferences, book reviews, and announcements on related topics.

The scope of the journal is international and both subscribers to the journal and submitters of papers are from various countries. The percentage of papers from the GDR fluctuates around 35%. Though the majority of the submitted papers are in English, they may also be published in French, German, or Russian. It is required that they have not been published previously. All submissions are treated equally on their scientific merits. They are refereed, normally by two or three referees. Usually the published papers are already revised versions.

A selection from the contents of Vol. 13 (1982) gives an insight into the range of the journal:

"Fitting Models in Time Series Analysis" by J. Anděl

"Effects of the Presence of a Harmonic Term on the Spectral Factorisation Procedure" by R. J. Bhansali and S. H. Sarhan

"Iterative Algorithms for Non-orthogonal Analysis of Variance" by J. J. Daudin

"Statistical Estimation of Nonlinear Regression Functions" by W. Grossmann

"Simple Construction of Least Favourable Pairs of Distributions and of Robust Tests for Prokhorov-Neighbourhoods" by R. Hafner

"Tables of the Cumulative Non-Central Chi-Square Distribution—Part 1" by G. E. Haynam, Z. Govindarajulu, F. C. Leone, and P. Siefert

"Sufficiency, Generalized Likelihood Function and Exponential Family" by K. Hoffmann

"Simultaneous M-estimator of the Common Location and the Scale-ratio in the Two-sample Problem" by J. Jurečková and P. K. Sen

"On the Maximal Deviation of the Kernel Regression Function Estimate" by H. Liero

"Test of χ^2 in the Generalized Linear Model" by J. R. Mathieu

"A Remark on the Estimation of the Rose of Directions of Fibre Processes" by J. Ohser

"Non-linear Models" by G. J. S. Ross

"Testing Hypotheses in Nonlinear Regression (with Discussions)" by W. H. Schmidt

"Asymptotic Theory of Some Tests for Constancy of Regression Relationships Over Times" by P. K. Sen

"D-Optimality of Complete Latin Squares" by E. Sonnemann

"Sequential Statistical Procedures for Processes of the Exponential Class with Independent Increments" by W. Winkler, J. Franz and I. Küchler

The current editorial address is: Prof. O. Bunke, Editor-in-Chief, Statistics, I. Math., AdW der DDR, DDR—1086 Berlin, Mohrenstraße 39, German Democratic Republic.

OLAF BUNKE

MATHISEN'S MEDIAN TEST

Let $X_1, \ldots, X_{2n+1}$ and $Y_1, \ldots, Y_m$ be independent samples from continuous distributions with distribution functions $F(x)$ and $G(y)$, respectively. Let $\operatorname{med} X$ denote the median of the X sample. For testing $H_0: F = G$ vs. $H_A: F \neq G$, Mathisen [4] proposed the simple statistic

$$M = \sum_{i=1}^{m} I(Y_i < \operatorname{med} X)$$

where $I(\cdot)$ is the indicator function. This differs from the Brown–Mood statistic S which counts the number of Y observations which are less than the combined sample median; *see* BROWN–MOOD MEDIAN TEST.

Under the null hypothesis, Mathisen derived the probability function for M as

$$P(M = k) = \frac{(2n+1)!\,m!\,(n+k)!\,(n+m-k)!}{n!\,n!\,k!\,(m-k)!\,(2n+1+m)!},$$

$k = 0, 1, \ldots, m$, with $EM = m/2$ and $\operatorname{var} M = m(m+2n+2)/\{4(2n+3)\}$. Hence, exact critical values for the test can be tabulated. Furthermore, $(M - EM)/(\operatorname{var} M)^{1/2}$ has an approximate standard normal distribution for large m and n, and this can be used to approximate the critical values of the test. Bowker [1] studied the consistency of the test based on M. He showed that the test is not consistent if F and G are identical in a neighborhood of their medians.

Gastwirth [2] studied the use of Mathisen's statistic for curtailed sampling in life testing* contexts (*see* MODIFIED OR CURTAILED SAMPLING PLANS). In comparison with S, Gastwirth found that the test based on M typically reaches a decision before the test based on S.

In contrast to S, Mathisen's statistic is not a linear rank statistic; however, it has properties similar to S. For example, the Pitman efficiency* of M relative to S is one. Using Bahadur efficiency*, Killeen et al. [3] showed that Mathisen's test should use the median of the smaller sample. They investigated the Bahadur efficiency of M relative to S for normal, logistic*, and double exponential* shift alternatives. The Brown–Mood test is more efficient for equal sample sizes and Mathisen's test only becomes more efficient after the larger sample size is at least three times the smaller sample size.

References

[1] Bowker, A. H. (1944). *Ann. Math. Statist.*, **15**, 98–101.

[2] Gastwirth, J. L. (1968). *J. Amer. Statist. Ass.*, **63**, 692–706.

[3] Killeen, T. J., Hettmansperger, T. P., and Sievers, G. L. (1972). *Ann. Math. Statist.*, **43**, 181–192.

[4] Mathisen, H. C. (1943). *Ann. Math. Statist.*, **14**, 188–194.

(BROWN–MOOD MEDIAN TEST
DISTRIBUTION-FREE METHODS
RANK TESTS)

THOMAS P. HETTMANSPERGER

MATRICES, COMPOUND

Let **A** be an $n \times n$ matrix. Consider all possible minors of **A** of order p, $1 \leqslant p \leqslant n$,

$$\mathbf{A}\begin{pmatrix} i_1, i_2, \ldots, i_p \\ j_1, j_2, \ldots, j_p \end{pmatrix} = \mathbf{A}\begin{pmatrix} \mathbf{i} \\ \mathbf{j} \end{pmatrix}$$

such that

$$1 \leqslant \begin{matrix} i_1 < i_2 < \cdots < i_p \\ j_1 < j_2 < \cdots < j_p \end{matrix} \leqslant n.$$

(This is the matrix formed by deleting all rows and columns of **A** except those in **i** and **j** respectively.) There are $\binom{n}{p}$ selections of *ordered* p indices (**i** and **j**). The matrix $\mathbf{A}_{[p]}$ with the elements $\tilde{a}_{\tilde{i}\tilde{j}} = \mathbf{A}\binom{\mathbf{i}}{\mathbf{j}}$ is called the pth component of **A**. Conventionally, the rows and columns of $\mathbf{A}_{[p]}$ are arranged according to lexicographical order of **i** and **j** respectively. The operation of compounding a matrix is used in the study of Pólya frequency functions and sequences.

Bibliography

Karlin, S. (1968). *Total Positivity*, Vol. 1. Stanford University Press, Stanford, Calif.

(PÓLYA TYPE 2 FREQUENCY
DISTRIBUTIONS)

MATRICES, RANDOM *See* RANDOM MATRICES

MATRIC-*t* DISTRIBUTION

Since 1908, when W. S. Gosset* first introduced the Student t-distribution*, a proliferation of related distributions has developed.

Extension of these to multivariate situations can take a number of forms, with the result that, as Johnson and Kotz [12] stated, there is no unique "multivariate-t (central or noncentral) distribution." They review many possibilities, among them the Ando and Kaufman [1] version, namely the distribution of the random vector

$$\mathbf{T} : p \times l = \sqrt{n}\,\mathbf{A}^{-1/2}\mathbf{X} \tag{1}$$

where $\mathbf{A} : p \times p \sim W(\boldsymbol{\Sigma}; n)$ (*see* WISHART DISTRIBUTION), $n \geqslant p$, independently of $\mathbf{X} : p \times 1 \sim N(\mathbf{0}, \mathbf{I}_p)$ (*see* MULTINORMAL DISTRIBUTION), and $\mathbf{A}^{-1/2}$ is the symmetric square root of $\mathbf{A}^{-1}$. The density function of $\mathbf{T}$ is

$$f_{\mathbf{T}}(\mathbf{t}) = (n\pi)^{-p/2}\Gamma(\tfrac{1}{2}(n+1)) \times \{\Gamma(\tfrac{1}{2}(n-p+1))\}^{-1}|\boldsymbol{\Sigma}|^{1/2} \cdot \{1 + n^{-1}\mathbf{t}'\boldsymbol{\Sigma}\mathbf{t}\}^{-(n+1)/2}, \qquad \mathbf{t} \in R^p. \tag{2}$$

The synthetic representation (1) lends itself in a very natural generalization to a matrix case. Once again there is also no uniquely defined matric-t variate. We will mention a few different approaches, but will give preference to one specific form which "generalizes" the matric-t variate and its distribution.

The right of existence of a matric-t variate was first underlined in the early 1960s when Kshirsagar [18] proved that the unconditional distribution of the usual estimate of the parameter matrix of regression coefficients actually follows a matric-t distribution. Later Olkin and Rubin [25] and Geisser [8] proved that density functions equivalent to that of a matric-t variate arose in Bayesian analysis. It was Dickey [6], however, who reconciled these results in a comprehensive discussion of some properties of the matric-t distribution.

Dickey generalized the vector version (1), where the elements of the vector $\mathbf{X} : p \times 1$ are independent, to the matrix case, where the rows of the matrix $\mathbf{X} : p \times m$ are independent and $\mathbf{X} : p \times m$ is distributed $N(\mathbf{0}, \mathbf{I}_p \otimes \boldsymbol{\Sigma}_2)$. The density function of this matric-t variate, $\mathbf{T} : p \times m = \sqrt{n}\,\mathbf{A}^{-1/2}\mathbf{X}$, which we present below, may appear to differ slightly from the original definition. This apparent discrepancy can be accounted for by the more convenient choice of degrees of freedom as well as the introduction of an arbitrary multiplicative constant in the definition of $\mathbf{T}$. This constant is usually chosen as the square root of the degrees of freedom of the Wishart distribution.

$$f_{\mathbf{T}}(\mathbf{t}) = \frac{\Gamma_p(\tfrac{1}{2}(m+n))}{(n\pi)^{mp/2}\Gamma_p(\tfrac{1}{2}n)} \cdot \frac{|\boldsymbol{\Sigma}|^{m/2}}{|\boldsymbol{\Sigma}_2|^{p/2}} \times |\mathbf{I}_p + n^{-1}\boldsymbol{\Sigma}\mathbf{t}\boldsymbol{\Sigma}_2^{-1}\mathbf{t}'|^{-(m+n)/2}, \qquad \mathbf{t} \in R^{pm}. \tag{3}$$

Marx [20] developed Dickey's generalization even further by considering the case where both the rows and columns of the normally distributed matrix $\mathbf{X}$ are dependent and defined the *central matric-t distribution* as that of the matrix variate

$$\mathbf{T} : p \times m = \sqrt{n}\,\boldsymbol{\Sigma}_1^{1/2}\mathbf{A}^{-1/2}\boldsymbol{\Sigma}_1^{-1/2}\mathbf{X} + \boldsymbol{\mu}, \tag{4}$$

where $\mathbf{A} : p \times p \sim W(\boldsymbol{\Sigma}; n)$, $n \geqslant p$, independently of $\mathbf{X} : p \times m \sim N(\mathbf{0}, \boldsymbol{\Sigma}_1 \otimes \boldsymbol{\Sigma}_2)$ and $\boldsymbol{\mu} : p \times m$ is a constant matrix. If $\mathbf{A}^{-1/2}$ denotes the symmetric square root of $\mathbf{A}^{-1}$, the density function of $\mathbf{T}$ is given by

$$f_{\mathbf{T}}(\mathbf{t}) = (n\pi)^{-mp/2}\Gamma_p(\tfrac{1}{2}(m+n))\{\Gamma_p(\tfrac{1}{2}n)\}^{-1} \times |\boldsymbol{\Sigma}_1^{1/2}\boldsymbol{\Sigma}^{-1}\boldsymbol{\Sigma}_1^{1/2}|^{-m/2}|\boldsymbol{\Sigma}_2|^{-p/2} \cdot |\mathbf{I}_p + n^{-1}\boldsymbol{\Sigma}_1^{-1/2}\boldsymbol{\Sigma}\boldsymbol{\Sigma}_1^{-1/2} \times (\mathbf{t}-\boldsymbol{\mu})\boldsymbol{\Sigma}_2^{-1}(\mathbf{t}-\boldsymbol{\mu})'|^{-(m+n)/2}, \qquad \mathbf{t} \in R^{pm}. \tag{5}$$

We will denote this distribution by $\mathbf{T} \sim T_{p,m}(n; \boldsymbol{\mu}, \boldsymbol{\Sigma}_1^{1/2}\boldsymbol{\Sigma}^{-1}\boldsymbol{\Sigma}_1^{1/2} \otimes \boldsymbol{\Sigma}_2)$, where the parameters refer respectively to the degrees of freedom, the expected value, and the variance of $\mathbf{T}$;

$$\mathbf{var}(\operatorname{vec}\mathbf{T}') = n(n-p-1)^{-1}\boldsymbol{\Sigma}_1^{1/2}\boldsymbol{\Sigma}^{-1}\boldsymbol{\Sigma}_1^{1/2} \otimes \boldsymbol{\Sigma}_2$$

(see Marx [20], Kaufman [16]). Here $\operatorname{vec}\mathbf{T} : p \times m$ is the column-stacked form of $\mathbf{T} : p \times m$.

By virtue of its generality, definition (4) of the matric-t variate seems to be useful in most situations. This is particularly apparent in that the same parameters may evolve from different underlying assumptions; given $\mathbf{T} \sim T_{p,m}(n;\ \boldsymbol{\mu}, \boldsymbol{\Sigma} \otimes \boldsymbol{\Sigma}_2)$, the parameter $\boldsymbol{\Sigma}$ may evolve from the fact that $\mathbf{A} \sim W(\boldsymbol{\Sigma}^{-1}; n)$ independently of $\mathbf{X} \sim N(\mathbf{0}, \mathbf{I}_p \otimes \boldsymbol{\Sigma}_2)$, or from the fact that $\mathbf{X}$ may be distributed $N(\mathbf{0}, \boldsymbol{\Sigma} \otimes \boldsymbol{\Sigma}_2)$ independently of $\mathbf{A} \sim W(\mathbf{I}_p; n)$. Matric-*t* distributions also have been approached in ways differing from those presented above, with the main contributor being Tan [28–30]. Tan [29] defined the *restricted matric-t distribution* as that of the matrix $\mathbf{T}: p \times m = \mathbf{Z}\mathbf{S}^{-1/2}$, where $\mathbf{S}: m \times m \sim W(\boldsymbol{\Sigma}_2, n)$, $n \geqslant m$, $\mathbf{S}^{-1/2}$ is the symmetric square root of $\mathbf{S}^{-1}$, while $\mathbf{Z}: p \times m$, independent of $\mathbf{S}$, has a singular normal distribution with expected value $\mathbf{0}: p \times m$, and covariance matrix $\boldsymbol{\Sigma}_1 \otimes \boldsymbol{\Sigma}_2$, subject to conditions (of rank r) $\mathbf{BZ} = \mathbf{0}$. A complex analogue of this distribution is discussed by Tan [28]. Tan [30] also considered the distribution of $\mathbf{T}: p \times m = \mathbf{Z}\mathbf{S}_1^{-1}$, where $\mathbf{Z}: p \times m \sim N(\mathbf{0}, \boldsymbol{\Sigma}_1 \otimes \boldsymbol{\Sigma}_2)$ independently of $\mathbf{S} = \mathbf{S}_1'\mathbf{S}_1 : m \times m \sim W(\boldsymbol{\Sigma}_1, n)$. This distribution of $\mathbf{T}$ depends on how $\mathbf{S}$ is partitioned. He considered the case where $\mathbf{S}_1$ has positive diagonal elements and is either lower triangular or upper triangular. The distributions of $\mathbf{T}$ corresponding to these two decompositions are respectively called *lower* and *upper disguised matric-t distribution*.

The importance and applicability of the matric-*t* distribution is accentuated as it appears more and more often in statistical methodology, for instance in a Bayesian context as described by Dickey [6], Press [26], Box and Tiao [2] (with an updated section based on Dickey [6]) and Nel et al. [24].

DISTRIBUTIONS OF TRANSFORMATIONS AND SUBMATRICES

Let $\mathbf{T}: p \times m \sim T_{p,m}(n;\ \boldsymbol{\mu}, \boldsymbol{\Psi} \otimes \boldsymbol{\Phi})$. If $\mathbf{B}: p \times p$ and $\mathbf{C}: m \times m$ are constant nonsingular matrices, then

$$\mathbf{Z} = \mathbf{BTC} \sim T_{p,m}(n; \mathbf{B}\boldsymbol{\mu}\mathbf{C}, \mathbf{B}\boldsymbol{\Psi}\mathbf{B}' \otimes \mathbf{C}'\boldsymbol{\Phi}\mathbf{C}), \tag{6}$$

[20, 2]. On the other hand, if $\mathbf{B}: b \times p$ is of rank b and $\mathbf{C}: m \times C$ is of rank c, then

$$\mathbf{Y} = \mathbf{TC} \sim T_{p,c}(n;\ \boldsymbol{\mu}\mathbf{C}, \boldsymbol{\Psi} \otimes \mathbf{C}'\boldsymbol{\Phi}\mathbf{C}), \tag{7}$$

and

$$\mathbf{W} = \mathbf{BT} \sim T_{b,m}(n - p + m; \mathbf{B}\boldsymbol{\mu}, \mathbf{B}\boldsymbol{\Psi}\mathbf{B}' \otimes \boldsymbol{\Phi}), \tag{8}$$

[22], with the density function of $\mathbf{W}$ given by

$$\begin{aligned} f_{\mathbf{W}}(\mathbf{w}) &= (n\pi)^{-bm/2}\Gamma_b(\tfrac{1}{2}(m + n - p + b)) \\ &\quad \times \left\{\Gamma_b(\tfrac{1}{2}(n - p + m))\right\}^{-1} \\ &\quad \times |\mathbf{B}\boldsymbol{\Psi}\mathbf{B}'|^{-m/2}|\boldsymbol{\Phi}|^{-b/2} \\ &\quad \times |\mathbf{I}_b + n^{-1}(\mathbf{B}\boldsymbol{\Psi}\mathbf{B}')^{-1}(\mathbf{w} - \mathbf{B}\boldsymbol{\mu})\boldsymbol{\Phi}^{-1} \\ &\quad \cdot (\mathbf{w} - \mathbf{B}\boldsymbol{\mu})'|^{-(m+n-p+b)/2}, \\ &\qquad \mathbf{w} \in R^{bm}. \end{aligned} \tag{9}$$

Equations (7) and (8), with appropriate choices of $\mathbf{C}$ and $\mathbf{B}$ respectively, facilitate the derivation of the distributions of submatrices of a matric-*t* variate (see Marx and Nel [22]). These distributions are also obtainable with a method described by Dickey [6]; see also Box and Tiao [2].

Conditional distributions of submatrices of a matric-*t* variate as well as a matric-*t* variate, expressed as a product of multivariate-*t* distributions, are well exposed by Box and Tiao [2]. Marx [20] updated these results if $\mathbf{T}$ is defined as in (4).

QUADRATIC FORMS

A result on a quadratic form of a matrix normal variate by Crowther [3], which was later extended by De Waal [5], plays an important role in the derivation of the distribution of quadratic forms of a matric-*t* variate.

If $\mathbf{T}: p \times m \sim T_{p,m}(n;\ \boldsymbol{\mu}, \boldsymbol{\Sigma}_1^{1/2}\boldsymbol{\Sigma}^{-1}\boldsymbol{\Sigma}_1^{1/2} \otimes \boldsymbol{\Sigma}_2)$, $m \leqslant p$, then the density function of $\mathbf{R} = \mathbf{T}'\mathbf{T}$ is given in Marx [21]. It is quite complicated and cumbersome; however, if

$\boldsymbol{\mu} : p \times m = \mathbf{0}$, it reduces to

$$
\begin{aligned}
f_{\mathbf{R}}(\mathbf{r}) &= \Gamma_p\left(\tfrac{1}{2}(m+n)\right)\left\{\Gamma_m\left(\tfrac{1}{2}p\right)\Gamma_p\left(\tfrac{1}{2}n\right)\right\}^{-1} \\
&\quad \times |n\boldsymbol{\Sigma}_1^{1/2}\boldsymbol{\Sigma}^{-1}\boldsymbol{\Sigma}_1^{1/2}|^{-m/2}|\boldsymbol{\Sigma}_2|^{-p/2} \\
&\quad \cdot |\mathbf{r}|^{(p-m-1)/2}{}_1F_0\left(\tfrac{1}{2}(m+n);\right. \\
&\qquad \left. n^{-1}\boldsymbol{\Sigma}_1^{-1/2}\boldsymbol{\Sigma}\boldsymbol{\Sigma}_1^{-1/2}, -\boldsymbol{\Sigma}_2^{-1}\mathbf{r}\right), \\
&\qquad \mathbf{r} > \mathbf{0}, \qquad (10)
\end{aligned}
$$

where ${}_1F_0(\cdot\,;\,\cdot,\cdot)$ is the hypergeometric function with double matrix argument [11, eq. 13, p. 477].

The joint density function of the characteristic roots of $\mathbf{R}$, the distribution function of the largest characteristic root of $\mathbf{R}$ and the density function of $\operatorname{tr}\mathbf{R}$ was derived by Marx [21]. Khatri [17] and Olkin and Rubin [25] also investigated the distribution of the quadratic form $\mathbf{TT}'$ for some special cases.

ASYMPTOTIC RESULTS

Analogous to the univariate case, a matric-*t* distribution ($\mathbf{T} \sim T_{p,m}(n;\, \boldsymbol{\mu}, \boldsymbol{\Psi} \otimes \boldsymbol{\Phi})$) tends to a matrix normal distribution if the degrees of freedom are increased, i.e., $\mathbf{T}$ is approximately distributed $N(\boldsymbol{\mu}, \boldsymbol{\Psi} \otimes \boldsymbol{\Phi})$. By utilizing the Kullback–Leibler divergence (Kullback [19]) Marx [20] (see also Johnson and Geisser [13]) proved that the adjustment of the covariance structure of the normal density resulted in a better limiting distribution of $\mathbf{T}$. As a rule we can state that a good approximation of the distribution of $\mathbf{T}$, $\mathbf{T} \sim T_{p,m}(n;\, \boldsymbol{\mu}, \boldsymbol{\Psi} \otimes \boldsymbol{\Phi})$ is that of a variate $\mathbf{X}$, where $\mathbf{X} : p \times m \sim N(\boldsymbol{\mu},\, \mathbf{var}(\mathbf{T}) = n(n-p-1)^{-1}\boldsymbol{\Psi} \otimes \boldsymbol{\Phi})$. The applicability of this result is illustrated in Nel et al. [24]. By using a Cornish–Fisher* type expansion of a Wishart matrix [23], the asymptotic distribution of the matric-*t* variate was derived by Marx [20]. The asymptotic distributions of extensive linear combinations of independent matric-*t* variates of the quadratic forms of $\mathbf{T}$ were also derived.

THE NONCENTRAL MATRIC-*t* VARIATE

If we define

$$
\begin{aligned}
\mathbf{T} &: p \times m \\
&= \sqrt{n}\,\boldsymbol{\Sigma}_1^{1/2}\mathbf{A}^{-1/2}\boldsymbol{\Sigma}_1^{-1/2}\mathbf{X} + \boldsymbol{\mu}_1, \qquad (11)
\end{aligned}
$$

where $\mathbf{X} : p \times m \sim N(\boldsymbol{\mu}_2, \boldsymbol{\Sigma}_1 \otimes \boldsymbol{\Sigma}_2)$ independent of $\mathbf{A} : p \times p \sim W(\boldsymbol{\Sigma}; n, \boldsymbol{\Omega})$, then $\mathbf{T}$ has a *doubly noncentral matric-t* distribution. If the matrix $\mathbf{A}$ in (11) is central Wishart instead of noncentral, $\mathbf{T}$ has an *upper noncentral matric-t* distribution. Marx [20] investigated the asymptotic distributions of these two cases. If the mean, $\boldsymbol{\mu}_2$, of the normal matrix is equal to zero, then $\mathbf{T}$ follows a *lower noncentral matric-t* distribution with density function

$$
\begin{aligned}
f_{\mathbf{T}}(\mathbf{t}) &= \Gamma_p\left(\tfrac{1}{2}(m+n)\right)\left\{\Gamma_p\left(\tfrac{1}{2}n\right)(n\pi)^{pm/2}\right\}^{-1} \\
&\quad \times \operatorname{etr}\left(-\tfrac{1}{2}\boldsymbol{\Omega}\right)|\boldsymbol{\Sigma}_1^{1/2}\boldsymbol{\Sigma}^{-1}\boldsymbol{\Sigma}_1^{1/2}|^{-m/2} \\
&\quad \times |\boldsymbol{\Sigma}_2|^{-p/2}|\mathbf{g}(\mathbf{t})|^{-(m+n)/2} \\
&\quad \cdot {}_1F_1\left(\tfrac{1}{2}(m+n); \tfrac{1}{2}n; \tfrac{1}{2}\boldsymbol{\Omega}(\mathbf{g}(\mathbf{t}))^{-1}\right), \\
&\qquad \mathbf{t} \in R^{pm},
\end{aligned}
$$

where [14, 15, 20]

$$
\mathbf{g}(\mathbf{t}) = \mathbf{I}_p + n^{-1}\boldsymbol{\Sigma}_1^{-1/2}(\mathbf{t} - \boldsymbol{\mu}_1)\boldsymbol{\Sigma}_2^{-1}(\mathbf{t} - \boldsymbol{\mu}_1)'\boldsymbol{\Sigma}_1^{-1/2}\boldsymbol{\Sigma}.
$$

Juritz [14] (see also Juritz and Troskie [15]) emphasizes the importance of this distribution by applying it to regression analysis. She proved that the distribution of a submatrix of a lower noncentral matric-*t* variate is also distributed as a lower noncentral matric-*t*. The distribution of a quadratic form $\mathbf{TT}'$ for the special case where $\boldsymbol{\Sigma}_1 = \boldsymbol{\Sigma} = \mathbf{I}_p$, $\boldsymbol{\Sigma}_2 = \mathbf{I}_m$, and $\boldsymbol{\mu}_1 = \mathbf{0}$, is also given in Juritz [14]. Marx [20] derived the distribution of $\mathbf{T}'\mathbf{T}$ for the case where $\boldsymbol{\mu}_1 = \mathbf{0}$.

References

[1] Ando, A. and Kaufman, G. M. (1965). *J. Amer. Statist. Ass.*, **60**, 347–358.

[2] Box, G. E. P. and Tiao, G. C. (1973). *Bayesian Inference in Statistical Analysis*. Addison-Wesley, Reading, Mass. (The only textbook with an extensive section on a matric-*t* distribution.)

[3] Crowther, N. A. S. (1975). *S. Afr. Statist. J.*, **9**, 27–36.

[4] Dawid, A. P. (1981). *Biometrika*, **68**, 265–274. (Discussion of sphericity properties of the matric-t distribution.)

[5] De Waal, D. J. (1979). *S. Afr. Statist. J.*, **13**, 103–122.

[6] Dickey, J. M. (1967). *Ann. Math. Statist.*, **38**, 511–518.

[7] Dickey, J. M. (1976). *J. Multivariate Anal.*, **6**, 343–346.

[8] Geisser, S. (1965). *Ann. Math. Statist.*, **36**, 151–159. (Geisser refers to a matric-t distribution as a determinantal distribution in his publications.)

[9] Hayakawa, T. (1972). *Ann. Inst. Statist. Math.*, **24**, 205–230.

[10] Hayakawa, T. (1982). On the Distribution of a Quadratic Form of Matric-t Variate. *Paper presented at the Pacific area Statistical Conference*, Tokyo, Japan, December 15–17, 1982. (Discussion of the quadratic form $\mathbf{TBT}'$ where the Wishart matrix has a noncentral distribution.)

[11] James, A. T. (1964). *Ann. Math. Statist.*, **35**, 475–501.

[12] Johnson, N. L. and Kotz, S. (1972). *Distributions in Statistics. Continuous Multivariate Distributions.* Wiley, New York.

[13] Johnson, W. and Geisser, S. (1983). *J. Amer. Statist. Ass.*, **78**, 137–144. (They derived a normal approximation to the multivariate-t distribution.)

[14] Juritz, J. M. (1973). Aspects of Non-central Multivariate t-Distributions. Ph.D. dissertation, University of Cape Town, South Africa.

[15] Juritz, J. M. and Troskie, C. G. (1976). *S. Afr. Statist. J.*, **10**, 1–7.

[16] Kaufman, G. M. (1967). Some Bayesian Moment Formulae. *Discussion Paper No. 6710*, Center for Operations Research & Econometrics, Katholieke Universiteit, Leuven.

[17] Khatri, C. G. (1966). *Ann. Math. Statist.*, **37**, 468–479.

[18] Kshirsager, A. M. (1960). *Proc. Camb. Phil. Soc.*, **57**, 80–85.

[19] Kullback, S. (1959). *Information Theory and Statistics.* Wiley, New York.

[20] Marx, D. G. (1981). Aspects of the Matric-t Distribution. Ph.D. dissertation, University of the Orange Free State, South Africa.

[21] Marx, D. G. (1983). *Ann. Inst. Statist. Math.*, **35**, 347–353.

[22] Marx, D. G. and Nel, D. G. (1982). A Note on the Linear Combinations of Matric-, Vector-, and Scalar-t Variates. *Tech. Rep. No. 85*, Dept. of Mathematical Statistics, University of the Orange Free State, Bloemfontein, South Africa.

[23] Nel, D. G. and Groenwald, P. C. N. (1979). On a Fisher–Cornish Type Expansion of Wishart Matrices. *Tech. Rep. No. 47*, Dept. of Mathematical Statistics, University of the Orange Free State, Bloemfontein, South Africa.

[24] Nel, D. G., De Waal, D. J., and Marx, D. G. (1985). To appear in *Commun. Statist. Theor. Meth.* (Uses matric-t distributions for predictive and detective purposes in the multivariate linear regression model.)

[25] Olkin, I. and Rubin, H. (1964). *Ann. Math. Statist.*, **35**, 261–269.

[26] Press, S. J. (1972). *Applied Multivariate Analysis.* Holt, Rinehart and Winston, New York. (An application of the matric-t distribution to multivariate regression.)

[27] Rinco, S. (1974). *Proc. Sympos. Statist. and Related Topics*, Carleton University, Ottawa, Canada.

[28] Tan, W. Y. (1968). *Tamkang J.*, **7**, 263–302.

[29] Tan, W. Y. (1969). The Restricted Matric-t Distribution and Its Applications in Deriving Posterior Distributions of Parameters in Multivariate Regression Analysis. *Tech. Rep. No. 205*, Dept. of Statistics, University of Wisconsin, Madison, Madison.

[30] Tan, W. Y. (1973). *Canad. J. Statist.*, **1**, 181–199.

(MATRIX-VALUED DISTRIBUTIONS
MULTINORMAL DISTRIBUTION
MULTIVARIATE DISTRIBUTIONS
t-DISTRIBUTION
WISHART DISTRIBUTION)

D. G. Marx
D. G. Nel

MATRIX BETA DISTRIBUTION *See* MATRIX VARIATE BETA DISTRIBUTION

MATRIX DERIVATIVES: APPLICATIONS IN STATISTICS

Rules for writing down matrix derivatives, i.e., for obtaining a number of partial derivatives simultaneously, are useful in solving optimization* problems, calculating Jacobians*, etc. Their uses in statistical problems were demonstrated in early papers by Dwyer and MacPhail [1] and Rao [7]. Since then systematic methods were developed by Schönemann [10], Tracy and Dwyer [11], Neudecker [6], MacRae [3], and others. For a complete bibliography on matrix derivatives, see recent books by Graham [2] and Rogers [9], and a review paper by Nel [5].

Notations. For a scalar function f of a matrix variable $\mathbf{X} = (x_{ij}) : m \times n$ we define its *matrix derivative* by

$$\frac{\partial f}{\partial \mathbf{X}} = \left(\frac{\partial f}{\partial x_{ij}} \right) : (m \times n). \tag{1}$$

When $n = 1$, $\mathbf{X}$ is a column vector $\mathbf{x}$ and the corresponding vector $\partial f/\partial \mathbf{x}$ is called the *vector derivative*. For a matrix function $\mathbf{F} = (f_{ij}) : p \times q$ of a matrix variable $\mathbf{X} : m \times n$, the matrix derivative is defined by the partitioned matrix (with pq partitions)

$$\frac{\partial \mathbf{F}}{\partial \mathbf{X}} : pm \times qn = \left(\frac{\partial f_{ij}}{\partial \mathbf{X}} \right),$$

$$i = 1, \ldots, p; \quad j = 1, \ldots, q, \tag{2}$$

where $\partial f_{ij}/\partial \mathbf{X}$ is the (i, j)th partition of order $m \times n$.

Note that if f is a differentiable scalar function of $\mathbf{X} : m \times n$, then

$$\lim_{t \to 0} \frac{f(\mathbf{X} + t\mathbf{Y}) - f(\mathbf{X})}{t} = \operatorname{tr}\left(\mathbf{Y}' \frac{\partial f}{\partial \mathbf{X}} \right) \quad \text{for arbitrary } \mathbf{Y}. \tag{3}$$

Thus to determine $\partial f/\partial \mathbf{X}$, we need to evaluate the limit on the left-hand side of (3), express it in the form of the right-hand side, and choose the factor $\partial f/\partial \mathbf{X}$ which is the desired matrix derivative. For the derivative with a transposed matrix, it is seen that $\partial f/\partial \mathbf{X}' = (\partial f/\partial \mathbf{X})'$ and when $n = m$ and $\mathbf{X}_s$ is symmetric,

$$\frac{\partial f}{\partial \mathbf{X}_s} = \left\{ \frac{\partial f(\mathbf{Y})}{\partial \mathbf{Y}} + \frac{\partial f(\mathbf{Y})}{\partial \mathbf{Y}'} - \operatorname{diag} \frac{\partial f}{\partial \mathbf{Y}} \right\} \Bigg|_{\mathbf{Y} = \mathbf{X}_s}, \tag{4}$$

where in $\partial f/\partial \mathbf{Y}$, all the elements of $\mathbf{Y}$ are treated as independent.

Using (3), if f and g are scalar functions of $\mathbf{X}$, then

$$\begin{aligned} \frac{\partial fg}{\partial \mathbf{X}} &= f \frac{\partial g}{\partial \mathbf{X}} + g \frac{\partial f}{\partial \mathbf{X}}, \\ \frac{\partial (f/g)}{\partial \mathbf{X}} &= \frac{1}{g} \frac{\partial f}{\partial \mathbf{X}} - \frac{f}{g^2} \frac{\partial g}{\partial \mathbf{X}}. \end{aligned} \tag{5}$$

If f is a scalar function of $\mathbf{H} = (h_{ij}) : p \times q$ and each h_{ij} is a function of $\mathbf{X} : m \times n$, then

$$\frac{\partial f[\mathbf{H}(\mathbf{X})]}{\partial \mathbf{X}} = \sum \sum \frac{\partial f}{\partial h_{ij}} \frac{\partial h_{ij}}{\partial \mathbf{X}}, \tag{6}$$

where the sum on the right side is built up with the elements of the matrices $\partial f/\partial \mathbf{H} = (\partial f/\partial h_{ij})$ and $\partial \mathbf{H}/\partial \mathbf{X} = (\partial h_{ij}/\partial \mathbf{X})$. The expression (6) may be symbolically written as $\langle \partial f/\partial \mathbf{H}, \partial \mathbf{H}/\partial \mathbf{X} \rangle$. The formulae (3–6) are used in determining the derivatives in the following sections.

VECTOR DERIVATIVES
(Scalar Function)

Consider $f(\mathbf{x}) = \mathbf{x}'\mathbf{A}\mathbf{x}$, a scalar function of an m-vector $\mathbf{x}$, where $\mathbf{A} : m \times m$. Since

$$[(\mathbf{x} + t\mathbf{y})'\mathbf{A}(\mathbf{x} + t\mathbf{y}) - \mathbf{x}'\mathbf{A}\mathbf{x}]/t \to \mathbf{y}'(\mathbf{A} + \mathbf{A}')\mathbf{x}$$

as $t \to 0$, we deduce that $\partial(\mathbf{x}'\mathbf{A}\mathbf{x})/\partial \mathbf{x} = (\mathbf{A} + \mathbf{A}')\mathbf{x}$. The results exhibited in Table 1 are easily established.

Applications. (1) To minimize $\mathbf{x}'\mathbf{A}\mathbf{x}$ subject to $\mathbf{B}\mathbf{x} = \mathbf{p}$, where $\mathbf{A}$ is nonnegative definite (n.n.d.) and $\mathbf{x}$ is a vector, consider $L(\mathbf{x}, \boldsymbol{\lambda}) = \mathbf{x}'\mathbf{A}\mathbf{x} + 2\boldsymbol{\lambda}'(\mathbf{B}\mathbf{x} - \mathbf{p})$ where $\boldsymbol{\lambda}$ is the vector

Table 1 Vector Derivatives ($\mathbf{x} : p \times 1$)

f	$\partial f/\partial x$
$\mathbf{a}'\mathbf{x}$ ($\mathbf{a}$ is constant)	$\mathbf{a}$
$\mathbf{x}'\mathbf{x}$	$2\mathbf{x}$
$\mathbf{x}'\mathbf{A}\mathbf{x}$, $\mathbf{A} : p \times p$	$(\mathbf{A} + \mathbf{A}')\mathbf{x}$
$\mathbf{x}'\mathbf{A}\mathbf{x}$, $(\mathbf{A} = \mathbf{A}')$	$2\mathbf{A}\mathbf{x}$

Table 2 Matrix Derivatives (1) (X : $n \times n$)

f	$\partial f/\partial \mathbf{X}$	$\partial f/\partial \mathbf{X}_s$
$\lvert\mathbf{X}\rvert$	$\lvert\mathbf{X}\rvert(\mathbf{X}^{-1})'$	$\lvert\mathbf{X}\rvert\{(2\mathbf{X}^{-1})' - \operatorname{diag}\mathbf{X}^{-1}\}$
$\log\lvert\mathbf{X}\rvert$	$\frac{1}{\lvert\mathbf{X}\rvert}\frac{\partial\lvert\mathbf{X}\rvert}{\partial\mathbf{X}} = (\mathbf{X}^{-1})'$	$(2\mathbf{X}^{-1})' - \operatorname{diag}\mathbf{X}^{-1}$
$\lvert\mathbf{X}\rvert^r$	$r\lvert\mathbf{X}\rvert^{r-1}\frac{\partial\lvert\mathbf{X}\rvert}{\partial\mathbf{X}} = r\lvert\mathbf{X}\rvert^r(\mathbf{X}^{-1})'$	$r\lvert\mathbf{X}\rvert^r\{(2\mathbf{X}^{-1})' - \operatorname{diag}\mathbf{X}^{-1}\}$

of Lagrangian multipliers*. Using the results of Table 1, the equations for $\mathbf{x}$ and $\boldsymbol{\lambda}$ are

$$\frac{\partial L(\mathbf{x},\boldsymbol{\lambda})}{\partial\mathbf{x}} = 2\mathbf{A}\mathbf{x} + 2\mathbf{B}'\boldsymbol{\lambda} = \mathbf{0},$$

$$\frac{\partial L(\mathbf{x},\boldsymbol{\lambda})}{\partial\boldsymbol{\lambda}} = \mathbf{B}\mathbf{x} - \mathbf{p} = \mathbf{0},$$

which are linear in $\mathbf{x}$ and $\boldsymbol{\lambda}$.

The above optimization problem arises in the minimum variance unbiased estimation* of a linear function $\mathbf{p}'\boldsymbol{\theta}$ where $\boldsymbol{\theta}$ is the vector of unknown parameters in a linear model $\mathbf{Y} = \mathbf{B}'\boldsymbol{\theta} + \boldsymbol{\epsilon}$, where $\boldsymbol{\epsilon}$ has expectation zero and variance–covariance matrix $\mathbf{A}$. It is seen that if $\mathbf{x}'\mathbf{y}$ is an unbiased estimator of $\mathbf{p}'\boldsymbol{\theta}$ then $\mathbf{B}\mathbf{x} = \mathbf{p}$ and the variance of $\mathbf{x}'\mathbf{y}$ is $\mathbf{x}'\mathbf{A}\mathbf{x}$. Then the problem is that of minimizing $\mathbf{x}'\mathbf{A}\mathbf{x}$ subject to $\mathbf{B}\mathbf{X} = \mathbf{p}$.

(2) To find the stationary values of $\mathbf{x}'\mathbf{A}\mathbf{x}/\mathbf{x}'\mathbf{B}\mathbf{x}$ where $\mathbf{B}$ is p.d. Applying (5) with $f = \mathbf{x}'\mathbf{A}\mathbf{x}$ and $g = \mathbf{x}'\mathbf{B}\mathbf{x}$, and the formulae of Table 1

$$\frac{\partial(\mathbf{x}'\mathbf{A}\mathbf{x}/\mathbf{x}'\mathbf{B}\mathbf{x})}{\partial\mathbf{x}} = \frac{2}{\mathbf{x}'\mathbf{B}\mathbf{x}}\mathbf{A}\mathbf{x} - \frac{2\mathbf{x}'\mathbf{A}\mathbf{x}}{(\mathbf{x}'\mathbf{B}\mathbf{x})^2}\mathbf{B}\mathbf{x} = \mathbf{0}$$

$$\Rightarrow (\mathbf{A} - \lambda\mathbf{B})\mathbf{x} = \mathbf{0}, \qquad \lambda = \mathbf{x}'\mathbf{A}\mathbf{x}/\mathbf{x}'\mathbf{B}\mathbf{x},$$

which leads to the determinantal equation $\lvert\mathbf{A} - \lambda\mathbf{B}\rvert = 0$.

The problem of maximizing the ratio of two quadratic forms* occurs in the generalization of univariate analysis of variance* to the multivariate case by considering a linear function of the variables, which leads to tests based on roots of determinantal equations of the type $\lvert\mathbf{A} - \lambda\mathbf{B}\rvert = 0$.

MATRIX DERIVATIVES (Scalar Function)

Consider the scalar function $f(\mathbf{X}) = \lvert\mathbf{X}\rvert$, $\mathbf{X}: m \times m$. Expanding up to first order terms in t,

$$\lvert\mathbf{X} + t\mathbf{Y}\rvert - \lvert\mathbf{X}\rvert = t\sum\sum y_{ij}X^{ij} = \operatorname{tr}(\mathbf{Y}'\mathbf{X}^c),$$

where $\mathbf{X}^c$ is the matrix of cofactors of $\mathbf{X}$. Hence taking the limit as in (2), we have

$$\partial\lvert\mathbf{X}\rvert/\partial\mathbf{X} = \mathbf{X}^c = \lvert\mathbf{X}\rvert(\mathbf{X}^{-1})' \qquad \text{if} \quad \lvert\mathbf{X}\rvert \neq 0,$$

$$\partial\lvert\mathbf{X}\rvert/\partial\mathbf{X}_s = 2\mathbf{X}^c - \operatorname{diag}\mathbf{X}^c = \lvert\mathbf{X}\rvert\{(2\mathbf{X}^{-1})' - \operatorname{diag}\mathbf{X}^{-1}\}.$$

The results of Table 2 are similarly derived.

The results of Table 3 are derived by a direct application of (2). See Rao [8]).

Let $\mathbf{U}: p \times q$ and $\mathbf{V}: q \times p$ be two matrix functions of $\mathbf{X}: m \times n$. Applying (4) to each term $U_{ij}(\mathbf{X})V_{ji}(\mathbf{X})$ in $\operatorname{tr}(\mathbf{U}(\mathbf{X})\mathbf{V}(\mathbf{X}))$,

$$\frac{\partial\operatorname{tr}(\mathbf{U}(\mathbf{X})\mathbf{V}(\mathbf{X}))}{\partial\mathbf{X}} = \left.\frac{\partial\operatorname{tr}(\mathbf{U}(\mathbf{X})\mathbf{V}(\mathbf{Y}))}{\partial\mathbf{X}}\right|_{\mathbf{Y}=\mathbf{X}} + \left.\frac{\partial\operatorname{tr}(\mathbf{U}(\mathbf{Y})\mathbf{V}(\mathbf{X}))}{\partial\mathbf{X}}\right|_{\mathbf{Y}=\mathbf{X}}, \tag{7}$$

which is the product rule of Schönemann [10]. Also

$$\frac{\partial f(\mathbf{U}(\mathbf{X}),\mathbf{V}(\mathbf{X}))}{\partial\mathbf{X}} = \left\langle\frac{\partial f(\mathbf{U},\mathbf{V})}{\partial\mathbf{U}}, \frac{\partial\mathbf{U}(\mathbf{X})}{\partial\mathbf{X}}\right\rangle + \left\langle\frac{\partial f(\mathbf{U},\mathbf{V})}{\partial\mathbf{V}}, \frac{\partial\mathbf{V}(\mathbf{X})}{\partial\mathbf{X}}\right\rangle. \tag{8}$$

Table 3 Matrix Derivatives (2) ($\mathbf{X}: m \times n$)

f	$\partial f/\partial \mathbf{X}$	$\partial f/\partial \mathbf{X}_s$ (when $m = n$)
$\operatorname{tr}(\mathbf{AX}), (\mathbf{A} : n \times m)$	$\mathbf{A}'$	$2\mathbf{A}' - \operatorname{diag} \mathbf{A}$
$\operatorname{tr}(\mathbf{A}'\mathbf{X}), (\mathbf{A} : n \times m)$	$\mathbf{A}$	$2\mathbf{A} - \operatorname{diag} \mathbf{A}$
$\operatorname{tr}(\mathbf{X}'\mathbf{AXB}), \begin{Bmatrix} \mathbf{A} : m \times n \\ \mathbf{B} : n \times n \end{Bmatrix}$	$\mathbf{A}'\mathbf{XB}' + \mathbf{AXB}$	$\mathbf{A}'\mathbf{XB}' + \mathbf{AXB} + \mathbf{BX}'\mathbf{A} + \mathbf{B}'\mathbf{X}'\mathbf{A}'$ $- \operatorname{diag}(\mathbf{A}'\mathbf{XB}' + \mathbf{AXB})$
$\operatorname{tr}(\mathbf{XAXB}), \begin{Bmatrix} \mathbf{A} : n \times m \\ \mathbf{B} : n \times m \end{Bmatrix}$	$\mathbf{B}'\mathbf{X}'\mathbf{A}' + \mathbf{A}'\mathbf{X}'\mathbf{B}'$	$\mathbf{B}'\mathbf{X}'\mathbf{A}' + \mathbf{A}'\mathbf{X}'\mathbf{B}' + \mathbf{AXB} + \mathbf{BXA}$ $- \operatorname{diag}(\mathbf{B}'\mathbf{X}'\mathbf{A}' + \mathbf{A}'\mathbf{X}'\mathbf{B}')$
$\operatorname{tr}(\mathbf{XX}')$	$2\mathbf{X}$	$2(\mathbf{X} + \mathbf{X}' - \operatorname{diag} \mathbf{X})$
$\operatorname{tr} \mathbf{X}^n$	$n\mathbf{X}^{n-1}$	$n(\mathbf{X}^{n-1} + (\mathbf{X}')^{n-1} - \operatorname{diag} \mathbf{X}^{n-1})$
$\mathbf{y}'\mathbf{Xz}, \begin{Bmatrix} \mathbf{y} : m \times 1 \\ \mathbf{z} : n \times 1 \end{Bmatrix}$	$\mathbf{yz}'$	$\mathbf{yz}' + \mathbf{zy}'$

Applying (7) or (8),

$$\frac{\partial \operatorname{tr}(\mathbf{U}^{-1})}{\partial \mathbf{X}} = -\left.\frac{\partial \operatorname{tr}\big(\mathbf{U}^{-2}(\mathbf{Y})\mathbf{U}(\mathbf{X})\big)}{\partial \mathbf{X}}\right|_{\mathbf{Y}=\mathbf{X}}, \tag{9}$$

$$\frac{\partial \operatorname{tr}(\mathbf{U}^{-1}\mathbf{A})}{\partial \mathbf{X}} = -\left.\frac{\partial \operatorname{tr}\big(\mathbf{U}^{-1}(\mathbf{Y})\mathbf{A}\mathbf{U}^{-1}(\mathbf{Y})\mathbf{U}(\mathbf{X})\big)}{\partial \mathbf{X}}\right|_{\mathbf{Y}=\mathbf{X}}, \tag{10}$$

$$\frac{\partial \operatorname{tr}(\mathbf{AX}^{-1}\mathbf{B})}{\partial \mathbf{X}} = -(\mathbf{X}^{-1}\mathbf{BAX}^{-1})', \tag{11}$$

where in (10) and (11) **A** and **B** are constant matrices. Further if $\mathbf{U} : p \times p$, then using (6),

$$\frac{\partial|\mathbf{U}|}{\partial \mathbf{X}} = |\mathbf{U}(\mathbf{X})| \left.\frac{\partial \operatorname{tr}\big(\mathbf{U}(\mathbf{Y})^{-1}\mathbf{U}(\mathbf{X})\big)}{\partial \mathbf{X}}\right|_{\mathbf{Y}=\mathbf{X}} \tag{12}$$

and in particular, since $\partial(\operatorname{tr} \mathbf{AX})/\partial \mathbf{X} = \mathbf{A}'$ from Table 3,

$$\frac{\partial|\mathbf{AX}|}{\partial \mathbf{X}} = |\mathbf{AX}| \left.\frac{\partial \operatorname{tr}\big((\mathbf{AY})^{-1}\mathbf{AX}\big)}{\partial \mathbf{X}}\right|_{\mathbf{Y}=\mathbf{X}} = |\mathbf{AX}|\big((\mathbf{AX})^{-1}\mathbf{A}\big)'. \tag{13}$$

Applications. (1) Maximum likelihood estimates of parameters of a multivariate normal distribution (*see* MULTINORMAL DISTRIBUTION). The likelihood of the parameters given the sample mean vector $\overline{\mathbf{X}}$ and the covariance matrix **S** based on n observations is

$$L = c - \frac{n}{2}\log|\boldsymbol{\Sigma}| - \frac{n}{2}\operatorname{tr}\Big(\boldsymbol{\Sigma}^{-1}\big[\mathbf{S} + (\overline{\mathbf{X}} - \boldsymbol{\mu})(\overline{\mathbf{X}} - \boldsymbol{\mu})'\big]\Big).$$

Ignoring the symmetry of $\boldsymbol{\Sigma}$ and using results of Table 3 and (11),

$$\frac{\partial L}{\partial \boldsymbol{\mu}} = \boldsymbol{\Sigma}^{-1}(\overline{\mathbf{X}} - \boldsymbol{\mu}) = \mathbf{0} \Rightarrow \overline{\mathbf{X}} = \boldsymbol{\mu}$$

$$\frac{\partial L}{\partial \boldsymbol{\Sigma}} = -(\boldsymbol{\Sigma}^{-1})' + \Big(\boldsymbol{\Sigma}^{-1}\big[\mathbf{S} + (\overline{\mathbf{X}} - \boldsymbol{\mu})(\overline{\mathbf{X}} - \boldsymbol{\mu})'\big]\boldsymbol{\Sigma}^{-1}\Big)' = \mathbf{0} \Rightarrow (\boldsymbol{\Sigma}^{-1})'\mathbf{S} = \mathbf{I} \quad \text{or} \quad \boldsymbol{\Sigma} = \mathbf{S}$$

so that the optimum is attained at $\boldsymbol{\mu} = \overline{\mathbf{X}}$ and $\boldsymbol{\Sigma} = \mathbf{S}$ which is symmetric.

(2) Given a matrix **A** with the singular value decomposition $\mathbf{PDQ}'$, to find an orthogonal matrix **X** (i.e., $\mathbf{X}'\mathbf{X} = \mathbf{I}$) such that $\operatorname{tr}((\mathbf{A} - \mathbf{X})'(\mathbf{A} - \mathbf{X}))$ is a minimum. Consider

$$\begin{aligned} F(\mathbf{X}, \boldsymbol{\Lambda}) &= \operatorname{tr}((\mathbf{A} - \mathbf{X})'(\mathbf{A} - \mathbf{X})) + \operatorname{tr}(\boldsymbol{\Lambda}'(\mathbf{X}'\mathbf{X} - \mathbf{I})) \\ &= \operatorname{tr}(\mathbf{A}'\mathbf{A} - \mathbf{X}'\mathbf{A} - \mathbf{A}'\mathbf{X} + \mathbf{I}) + \operatorname{tr}(\boldsymbol{\Lambda}'(\mathbf{X}'\mathbf{X})) - \operatorname{tr} \boldsymbol{\Lambda}, \end{aligned}$$

where $\boldsymbol{\Lambda}$ is the matrix of Lagrangian multipliers. Using the results of Table 3

$$\frac{\partial F}{\partial \mathbf{X}} = 2\mathbf{A} + \mathbf{X}(\Lambda + \Lambda') = \mathbf{0},$$

$$\frac{\partial F}{\partial \Lambda} = \mathbf{X}'\mathbf{X} - \mathbf{I} = \mathbf{0},$$

from which it follows that $2^{-1}(\Lambda + \Lambda') = \mathbf{X}'\mathbf{A} = \mathbf{A}'\mathbf{X}$ and that $\mathbf{X} = \mathbf{PQ}'$ is a solution providing the minimum.

Vec OPERATOR

For a given $\mathbf{A} : p \times q$, we denote by Vec $\mathbf{A}$ or simply $\bar{\mathbf{A}}$ the vector obtained by writing the columns of $\mathbf{A}$ one below the other starting with the first. If $\mathbf{y}$ is an r-vector and $\mathbf{x}$ is an s-vector, then we define

$$\frac{\partial \mathbf{y}}{\partial \mathbf{x}'} = \left(\frac{\partial y_i}{\partial x_j} \right) : r \times s$$

consistent with the definition (2). The results given in Table 4 can be derived from first principles. Details can be found in Tracy and Dwyer [11], Neudecker [6], McDonald and Swaminathan [4], and Nel [5].

The vec operation is useful in finding Jacobians of transformations $J(\mathbf{Y} : m \times n \to \mathbf{X} : m \times n) = |\partial \bar{y}/\partial \bar{\mathbf{X}}'|_+$, where the suffix + indicates positive value of the determinant. Let $\mathbf{Y} = \mathbf{AXB}$ where $\mathbf{A} : m \times m$ and $\mathbf{B} : n \times n$ are fixed nonsingular matrices. Note that the transformation could be written $\mathbf{Y} = \mathbf{AZ}$, $\mathbf{Z} = \mathbf{XB}$ and

$$J(\mathbf{Y} \to \mathbf{X}) = J(\mathbf{Y} \to \mathbf{Z})J(\mathbf{Z} \to \mathbf{X}),$$

$$\left| \frac{\partial \bar{\mathbf{Y}}}{\partial \bar{\mathbf{X}}'} \right|_+ = \left| \frac{\partial \bar{\mathbf{Y}}}{\partial \bar{\mathbf{Z}}'} \right|_+ \left| \frac{\partial \bar{\mathbf{Z}}}{\partial \bar{\mathbf{X}}'} \right|_+ .$$

Applying the formulae in lines 1 and 2 of Table 4 to the factors on the right-hand side,

$$\left| \frac{\partial \bar{\mathbf{Y}}}{\partial \bar{\mathbf{X}}'} \right|_+ = |\mathbf{I}_n \otimes \mathbf{A}|_+ |\mathbf{B}' \otimes \mathbf{I}_m|_+$$

$$= |\mathbf{A}|_+^n |\mathbf{B}_+^m ,$$

which is the desired Jacobian.

Table 4 Vec Derivatives ($\mathbf{X} : m \times n$; $\otimes$ = Kronecker product; $\mathbf{A} * \mathbf{B}$ = Hadamard product)

Matrix	Vec Derivative	
$\mathbf{AX}$, $\mathbf{A} : p \times m$	$\mathbf{I}_n \otimes \mathbf{A}$	
$\mathbf{XB}$	$\mathbf{B}' \otimes \mathbf{I}_m$	
$\mathbf{AXB}$, $\mathbf{A} : p \times m$, $\mathbf{B} : n \times r$	$\mathbf{B}' \otimes \mathbf{A}$	
$\mathbf{AX'B}$, $\mathbf{A} : p \times n$, $\mathbf{B} : m \times r$	$(\mathbf{A} \otimes \mathbf{B}')\mathbf{P}$	(1)
$\mathbf{U(X)V(X)}$, $\mathbf{U} : p \times q$, $\mathbf{V} : q \times r$	$(\mathbf{V} \otimes \mathbf{I}_p)' \dfrac{\partial \bar{\mathbf{U}}}{\partial \bar{\mathbf{X}}'} + (\mathbf{I}_r \otimes \mathbf{U}) \dfrac{\partial \bar{\mathbf{V}}}{\partial \bar{\mathbf{X}}'}$	
$\mathbf{X'AX}$, $\mathbf{A} : m \times m$	$(\mathbf{X'A'} \otimes \mathbf{I})\mathbf{P} + (\mathbf{I} \otimes \mathbf{X'A})$	
$\mathbf{AX}^{-1}\mathbf{B}$, $m = n$	$-(\mathbf{X}^{-1}\mathbf{B})' \otimes (\mathbf{AX}^{-1})$	
$z(\mathbf{X})\mathbf{U(X)}$, $z : 1 \times 1$, $\mathbf{U} : p \times q$	$\bar{\mathbf{U}} \dfrac{\partial z}{\partial \bar{\mathbf{X}}'} + z \dfrac{\partial \bar{\mathbf{U}}}{\partial \bar{\mathbf{X}}'}$	
$[\mathbf{U(X)}]^{-1}$, $\mathbf{U} : p \times p$	$-([\mathbf{U}^{-1}]' \otimes \mathbf{U}^{-1}) \dfrac{\partial \bar{\Omega}}{\partial \bar{\mathbf{X}}'}$	
$\mathbf{Z[Y(X)]}$, $\mathbf{Z} : r \times s$, $\mathbf{Y} : p \times q$	$\dfrac{\partial \bar{\mathbf{Z}}}{\partial \bar{\mathbf{Y}}'} \dfrac{\partial \bar{\mathbf{Y}}}{\partial \bar{\mathbf{X}}'}$	
$\mathbf{Z(X)} * \mathbf{Y(X)}$, $\mathbf{Z} : p \times q$, $\mathbf{Y} : p \times q$	$\mathbf{D(Z)} \dfrac{\partial \bar{\mathbf{Y}}}{\partial \bar{\mathbf{X}}'} + \mathbf{D(Y)} \dfrac{\partial \bar{\mathbf{Z}}}{\partial \bar{\mathbf{X}}'}$	(2)
$\mathbf{Z(X)} * \mathbf{B}$, $\mathbf{B}$ (constant)	$\mathbf{D(B)} \dfrac{\partial \bar{\mathbf{Z}}}{\partial \bar{\mathbf{X}}'}$	

(1) $\mathbf{P}$ is the permutation matrix which transforms the vector $\bar{\mathbf{X}}$ to $\bar{\mathbf{X}}'$.
(2) $\mathbf{D(Z)} = \text{diag}(z_{11}, \ldots, z_{1q}, \ldots, z_{p1}, \ldots, z_{pq})$ where $\mathbf{Z} = (z_{ij})$.

References

[1] Dwyer, P. S. and MacPhail, M. S. (1948). *Ann. Math. Statist.*, **19**, 517–534.

[2] Graham, A. (1981). *Kronecker Products and Matrix Calculus with Applications*. Ellis Horwood and Wiley, New York.

[3] MacRae, E. C. (1974). *Ann. Statist.*, **2**, 337–346.

[4] McDonald, R. P. and Swaminathan, H. (1973). *General Systems*, **18**, 37–54.

[5] Nel, D. G. (1980). *S. Afr. Statist. J.*, **14**, 137–193.

[6] Neudecker, H. (1969). *J. Amer. Statist. Ass.*, **64**, 953–963.

[7] Rao, C. R. (1952). *Advanced Statistical Methods in Biometric Research*. Wiley, New York.

[8] Rao, C. R. (1973). *Linear Statistical Inference and Its Applications*, 2nd ed. Wiley, New York.

[9] Rogers, G. S. (1980). *Matrix Derivatives*. Dekker, New York.

[10] Schönemann, P. H. (1965). *Research Memo 27*, Psychometric Laboratory, University of North Carolina, Chapel Hill, North Carolina.

[11] Tracy, D. S. and Dwyer, P. S. (1969). *J. Amer. Statist. Ass.*, **64**, 1576–1594.

(GENERAL LINEAR MODEL
LINEAR ALGEBRA, COMPUTATIONAL
QUADRATIC FORMS)

C. RADHAKRISHNA RAO

MATRIX, ILL-CONDITIONED

Although the term *ill-conditioned* was used informally before 1948 to refer to systems of linear equations whose solutions are likely to be calculated inaccurately, the notion was first quantified by Alan Turing [4], who introduced the *condition number*

$$\kappa(\mathbf{A}) = \|\mathbf{A}\|\,\|\mathbf{A}^{-1}\| \tag{1}$$

(here $\|\cdot\|$ is a norm satisfying $\|\mathbf{Ax}\| \leqslant \|\mathbf{A}\|\,\|\mathbf{x}\|$). The *raison d'être* for the condition number is the following result. Let $\mathbf{A}$ be nonsingular and consider the system

$$\mathbf{Ax} = \mathbf{b}. \tag{2}$$

For any matrix $\mathbf{E}$, if

$$\kappa(\mathbf{A})\frac{\|\mathbf{E}\|}{\|\mathbf{A}\|} < 1,$$

then $\mathbf{A} + \mathbf{E}$ is nonsingular, and the solution of the system

$$(\mathbf{A} + \mathbf{E})\tilde{\mathbf{x}} = \mathbf{b} \tag{3}$$

satisfies

$$\frac{\|\tilde{\mathbf{x}} - \mathbf{x}\|}{\|\tilde{\mathbf{x}}\|} \leqslant \kappa(\mathbf{A})\frac{\|\mathbf{E}\|}{\|\mathbf{A}\|}. \tag{4}$$

If the quantity $\|\mathbf{E}\|/\|\mathbf{A}\|$ is interpreted as a relative error in $\mathbf{A}$ and $\|\tilde{\mathbf{x}} - \mathbf{x}\|/\|\mathbf{x}\|$ as a relative error in $\mathbf{x}$, then (4) says that $\kappa(\mathbf{A})$ is a magnification factor bounding the relative error in the solution of (3) in terms of the relative error in $\mathbf{A}$. Systems whose matrices have large condition numbers are said to be *ill-conditioned*, and by extension the same terminology is applied to the matrix of the system itself.

The bound (4) applies to any perturbation, wherever it comes from, and for a statistician its most useful function may be to bound the effects of errors from a variety of real-life sources (e.g., observation errors). However, the condition number is most closely associated with errors arising in the numerical solution of (2). This comes about as follows.

If a "good" algorithm (such as Gaussian elimination with partial pivoting) is used to solve (2) in t-digit, decimal arithmetic, the computed solution $\bar{\mathbf{x}}$ will satisfy (3), where $\|\mathbf{E}\|$ is of order 10^{-t}. Thus if $\kappa(\mathbf{A}) = 10^k$, the relative error in the computed solution will be about 10^{k-t}. Since the magnitude of the common logarithm of a relative error can be loosely identified with the number of accurate digits, the foregoing may be summarized in the folk theorem: *If* $\kappa(\mathbf{A}) = 10^k$, *one can expect to lose k decimal digits in computing the solution of* (1). For more on this and further references see LINEAR ALGEBRA, COMPUTATIONAL.

The notion of ill-conditioning and condition number have been extended to rectangular matrices $\mathbf{X}$ in connection with the solution of the least-squares* problem

$$\text{minimize } \|\mathbf{y} - \mathbf{Xb}\|^2, \tag{5}$$

where $\|\cdot\|$ now denotes the usual Euclidean

norm. If $\tilde{\mathbf{b}}$ is the solution of the perturbed problem

$$\text{minimize } \|\mathbf{y} - (\mathbf{X} + \mathbf{E})\tilde{\mathbf{b}}\|^2,$$

then with

$$\kappa(\mathbf{X}) = \|\mathbf{X}\| \, \|(\mathbf{X}^T\mathbf{X})^{-1}\mathbf{X}^t\| \tag{6}$$

we have (asymptotically as $\|E\| \to 0$)

$$\frac{\|\tilde{\mathbf{b}} - \mathbf{b}\|}{\|\mathbf{b}\|} \lesssim \left[\kappa(\mathbf{X}) + \kappa^2(\mathbf{X}) \frac{\|\mathbf{y} - \mathbf{Xb}\|}{\|\mathbf{Xb}\|} \right] \frac{\|\mathbf{E}\|}{\|\mathbf{X}\|}. \tag{7}$$

Thus $\kappa(\mathbf{X})$ controls the accuracy of least-squares solutions. Note that (6) and (7) reduce to (1) and (4) when $\mathbf{X}$ is square. For a derivation of (7) see ref. 3.

From its definition it would appear that to calculate $\kappa(\mathbf{A})$ one must undertake the expensive and (usually) unnecessary computation of $\mathbf{A}^{-1}$. Fortunately there are techniques that reliably estimate $\kappa(\mathbf{A})$ without requiring $\mathbf{A}^{-1}$ [1, 2].

The notion of ill-conditioning and its embodiment in the condition number are useful in a variety of applications. However, two caveats are in order. In the first place, the notion of ill-conditioning is tied to the problem being solved—in this article the solution of linear systems and least-squares problems. It is quite possible for a matrix to be ill-conditioned with respect to one problem and well-conditioned with respect to another. For example, if $\mathbf{A}$ is symmetric, its eigenvalues will be well-conditioned no matter how large $\kappa(\mathbf{A})$ is.

Second, the condition number is not invariant under row and column scaling; indeed by reducing the size of a row or column the condition number can be made arbitrarily large, a situation sometimes called artificial ill-conditioning. Just how to scale a matrix so that its condition number is meaningful is not well understood, although it is clear that some knowledge of the error matrix $\mathbf{E}$ is required (see the introduction to [2]). Thus a large condition number must not be taken blindly as a token of disaster but instead should be regarded as a warning that the problem needs closer inspection.

Bibliographical note. Ill-conditioning and the condition number are now treated in elementary numerical analysis textbooks as a matter of course. J. H. Wilkinson gives a detailed discussion in his book *Rounding Errors in Algebraic Processes* [5], which in spite of its age remains one of the best introductions to the subject.

References

[1] Cline, A. K., Moler, C. B., Stewart, G. W., and Wilkinson, J. H. (1979). *SIAM J. Num. Anal.*, **16**, 368–375.

[2] Dongarra, J. J., Bunch, J. R., Moler, C. B., and Stewart, G. W. (1979). *LINPACK Users' Guide*, SIAM, Philadelphia.

[3] Stewart, G. W. (1977). *SIAM Rev.*, **19**, 634–666.

[4] Turing, A. M. (1948). *Quart. J. Mech. Appl. Math.*, **1**, 287–308.

[5] Wilkinson, J. H. (1963), *Rounding Errors in Algebraic Processes*. Prentice-Hall, Englewood Cliffs, N.J.

(LINEAR ALGEBRA, COMPUTATIONAL)

G. W. STEWART

MATRIX NORMAL DISTRIBUTION
See MATRIX-VALUED DISTRIBUTIONS

MATRIX-VALUED DISTRIBUTIONS

A set of p variables measured at q time units can be represented in a $p \times q$ matrix variable and the joint distribution of the set of pq random variables is referred to as a *matrix-valued distribution*. This example of a random matrix may lead to the introduction of a matrix-valued normal distribution which is one of the important matrix-valued distributions in statistics (see, e.g., Roy [43], p. 18). We shall say more about it in the next section. Many other matrix-valued distributions have been introduced, such as the Wishart, inverse Wishart, t, F, beta, and Dirichlet distributions*. (See Dawid [6], Olkin and Rubin [39], Perlman [40], and Press [41].) We will pay special attention to the normal and some quadratic forms* of matrix-normal variables. Consider a matrix variable $\mathbf{X} = (X_{ij})$, $i = 1, \ldots, p$; $j = 1, \ldots, q$. The expected value of $\mathbf{X}$ is defined as $E[\mathbf{X}] = (E[X_{ij}])$ and the covariance matrix of $\mathbf{X}$ is the $pq \times pq$ matrix

$$\mathbf{var}(\mathbf{X}) = E\left[\operatorname{vec}(\mathbf{X} - E[\mathbf{X}])\operatorname{vec}'(\mathbf{X} - E[\mathbf{X}])\right], \tag{1}$$

where $\operatorname{vec}\mathbf{X}$ is a $pq \times 1$ vector containing the row vectors of $\mathbf{X}$ successively written as column vectors to form one column vector. $\operatorname{vec}'\mathbf{X}$ denotes the transpose of $\operatorname{vec}\mathbf{X}$. This notation was introduced by Koopmans et. al. [31]. According to (1) it follows that if $\mathbf{var}(\mathbf{X}) = \boldsymbol{\Sigma} \otimes \boldsymbol{\Psi}$, where $\boldsymbol{\Sigma} = (\sigma_{ij})$, $i, j = 1, \ldots, p$ and $\boldsymbol{\Psi} = (\psi_{ij})$, $i, j = 1, \ldots q$ are positive definite symmetric (p.d.s.) matrices, then the covariance matrix beween the ith and jth rows of $\mathbf{X}$ is $\sigma_{ij}\mathbf{R}$. The covariance matrix between the ith and jth columns of $\mathbf{X}$ is $\psi_{ij}\boldsymbol{\Sigma}$.

If $\mathbf{X}(p \times p)$ is a symmetric matrix, however, then the vector containing only the different elements of $\mathbf{X}$ can be obtained through the transition matrix $\mathbf{K}_p$ [3, 35] with typical element

$$(\mathbf{K}_p)_{ij,gh} = \tfrac{1}{2}(\delta_{ig}\delta_{jh} + \delta_{ih}\delta_{jg}),$$

$$i \leqslant p, \quad j \leqslant p, \quad g \leqslant h \leqslant p, \tag{2}$$

$$\delta_{ij} = \begin{cases} 0, & i \neq j, \\ 1, & i = j. \end{cases}$$

The vector is denoted by Nel [35] as $\operatorname{vec} p\mathbf{X}$ of order $\frac{1}{2}p(p+1) \times 1$ and is equal to

$$\operatorname{vec} p\mathbf{X} = \mathbf{K}_p' \operatorname{vec}\mathbf{X}. \tag{3}$$

The covariance matrix of the different elements of $\operatorname{vec} p\mathbf{X}$, denoted by $\mathbf{var}(\mathbf{X} = \mathbf{X}')$ and of order $\frac{1}{2}p(p+1) \times \frac{1}{2}p(p+1)$, is

$$\mathbf{var}(\mathbf{X} = \mathbf{X}') = \mathbf{K}_p' \mathbf{var}(\mathbf{X}) \mathbf{K}_p. \tag{4}$$

If, for instance, $\mathbf{var}(\mathbf{X}) = \boldsymbol{\Sigma} \otimes \boldsymbol{\Psi}$, then a typical element of $\mathbf{var}(\mathbf{X} = \mathbf{X}')$ will be [36]

$$\operatorname{cov}(X_{ij}, X_{kl}) = \tfrac{1}{4}(\sigma_{ik}\psi_{jl} + \sigma_{jk}\psi_{il} + \sigma_{il}\psi_{jk} + \sigma_{jl}\psi_{ik}). \tag{5}$$

The characteristic function* of $\mathbf{X}(p \times q)$ with density function $f_{\mathbf{X}}(\mathbf{x})$ is defined as

$$\phi_{\mathbf{X}}(\mathbf{T}) = E\left[\exp(i \operatorname{tr}\mathbf{T}'\mathbf{X})\right] \tag{6}$$

for $\mathbf{T}(p \times q)$ real and $i = \sqrt{-1}$; tr denotes the trace of the matrix. Instead of using the characteristic function, the Laplace transform (*see* INTEGRAL TRANSFORMS)

$$g_{\mathbf{X}}(\mathbf{Z}) = E\left[\exp(-\operatorname{tr}\mathbf{Z}'\mathbf{X})\right] \tag{7}$$

can be used. $\mathbf{Z} = \mathbf{T}_R + i\mathbf{T}_C$ is a complex $p \times q$ matrix with $R(\mathbf{Z}) = \mathbf{T}_R$ and $\mathbf{T}_C$ real. It is assumed that the integral in (7) converges in the half-plane $R(\mathbf{Z}) = \mathbf{T}_R > \mathbf{T}_0$ for some positive definite $\mathbf{T}_0$. $g_{\mathbf{X}}(\mathbf{Z})$ is then an analytic function of $\mathbf{Z}$ in the half-plane. (See Constantine [4] and Herz [24] for more details.)

THE MATRIX NORMAL DISTRIBUTION

Suppose $\mathbf{X}(p \times q)$ real is normally distributed with mean $E(\mathbf{X}) = \boldsymbol{\mu}$ and covariance matrix $\boldsymbol{\Sigma} \otimes \boldsymbol{\Psi}$, $\boldsymbol{\Sigma}(p \times p)$ and $\boldsymbol{\Psi}(q \times q)$ p.s.d.; then $\mathbf{X}$ has density function given by

$$f_{\mathbf{X}}(\mathbf{x}) = (2\pi)^{-pq/2}|\boldsymbol{\Sigma}|^{-q/2}|\boldsymbol{\Psi}|^{-p/2} \times \exp\left(-\tfrac{1}{2}\operatorname{tr}\boldsymbol{\Sigma}^{-1}(\mathbf{x} - \boldsymbol{\mu})\boldsymbol{\Psi}^{-1}(\mathbf{x} - \boldsymbol{\mu})'\right),$$

$$-\infty < \mathbf{x} < \infty. \tag{8}$$

This can easily be shown from the fact that $\operatorname{vec}\mathbf{X}$ is distributed multivariate normal with mean $\operatorname{vec}\boldsymbol{\mu}$ and covariance matrix $\boldsymbol{\Sigma} \otimes \boldsymbol{\Psi}$ (*see* MULTINORMAL DISTRIBUTION). Note that $|\boldsymbol{\Sigma} \otimes \mathbf{R}| = |\boldsymbol{\Sigma}|^q|\boldsymbol{\Psi}|^p$, and

$$\operatorname{vec}'(\mathbf{x} - \boldsymbol{\mu})(\boldsymbol{\Sigma} \otimes \boldsymbol{\Psi})^{-1}\operatorname{vec}(\mathbf{x} - \boldsymbol{\mu}) = \operatorname{tr}\left(\boldsymbol{\Sigma}^{-1}(\mathbf{x} - \boldsymbol{\mu})\boldsymbol{\Psi}^{-1}(\mathbf{x} - \boldsymbol{\mu})'\right).$$

Equation (8) is the *matrix normal density function* and we say that $\mathbf{X}(p \times q)$ is distributed $N(\boldsymbol{\mu}, \boldsymbol{\Sigma} \otimes \boldsymbol{\Psi})$. The Laplace transform of $f_{\mathbf{X}}(\mathbf{x})$ is given by

$$g_{\mathbf{X}}(\mathbf{Z}) = \exp(-\operatorname{tr}\mathbf{Z}'\boldsymbol{\mu} + \boldsymbol{\Sigma}\mathbf{Z}\boldsymbol{\Psi}\mathbf{Z}'). \tag{9}$$

Using (9) it can be shown that for any $\mathbf{D}(r \times p)$ of rank $r \leqslant p$ and $\mathbf{C}(q \times t)$ of rank $t \leqslant q$, $\mathbf{DXC}$ is distributed $N(\mathbf{D}\boldsymbol{\mu}\mathbf{C}, \mathbf{D}\boldsymbol{\Sigma}\mathbf{D}' \otimes \mathbf{C}'\boldsymbol{\Psi}\mathbf{C})$.

To obtain moments of $\mathbf{X}$ such as

$$E[\mathbf{XX}'] = \boldsymbol{\mu\mu}' + \boldsymbol{\Sigma}\operatorname{tr}\boldsymbol{\Psi},$$

and many more (see Nel [38]; Groenewald [20], pp. 70–75; van der Merwe [48]; Marx [32], pp. 43–48), the following expectations are useful, where we assume for simplicity that the mean $\boldsymbol{\mu} = 0$:

(i) $E[X_{ij}X_{kl}] = \operatorname{cov}(X_{ij}, X_{kl}) = \sigma_{ik}\psi_{jl}$

(ii) $E[X_{ij}X_{kl}X_{mn}X_{op}]$

$$= \operatorname{cov}(X_{ij}, X_{kl})\operatorname{cov}(X_{mn}, X_{op}) + \operatorname{cov}(X_{ij}, X_{mn})\operatorname{cov}(X_{kl}, X_{op}) + \operatorname{cov}(X_{ij}, X_{op})\operatorname{cov}(X_{kl}, X_{mn}).$$

These results follow, recalling that $\operatorname{vec}\mathbf{X}$ is distributed multivariate normal with mean $\operatorname{vec}\mathbf{0}$ and covariance matrix $\boldsymbol{\Sigma}\otimes\boldsymbol{\Psi}$. By picking out the appropriate elements from $\boldsymbol{\Sigma}\otimes\boldsymbol{\Psi}$, (i) and (ii) follow from Anderson [1, p. 39]. It also follows that the covariance matrix between the ith and jth rows of $\mathbf{X}$ is $\sigma_{ij}\boldsymbol{\Psi}$.

An example where the matrix normal distribution is applicable is the distribution of the estimator of the regression matrix $\boldsymbol{\beta} = \boldsymbol{\Sigma}_{12}\boldsymbol{\Sigma}_{22}^{-1}$. Assume that the vector

$$\mathbf{X}(p\times 1) = \begin{bmatrix} \mathbf{X}^{(1)} \\ \mathbf{X}^{(2)} \end{bmatrix} \begin{matrix} (q\times 1) \\ ((p-q)\times 1) \end{matrix}$$

is distributed multivariate normal with mean

$$\boldsymbol{\mu}(p\times 1) = \begin{bmatrix} \boldsymbol{\mu}^{(1)} \\ \boldsymbol{\mu}^{(2)} \end{bmatrix} \begin{matrix} (q\times 1) \\ ((p-q)\times 1) \end{matrix}$$

and covariance matrix

$$\boldsymbol{\Sigma}(p\times p) = \begin{bmatrix} \boldsymbol{\Sigma}_{11}(q\times q) & \boldsymbol{\Sigma}_{12} \\ \boldsymbol{\Sigma}_{21} & \boldsymbol{\Sigma}_{22} \end{bmatrix};$$

then it is well known that, conditional on $\mathbf{X}^{(2)} = \mathbf{x}^{(2)}$,

$$E[\mathbf{X}^{(1)} \mid \mathbf{X}^{(2)} = \mathbf{x}^{(2)}] = \boldsymbol{\mu}^{(1)} + \boldsymbol{\beta}(\mathbf{x}^{(2)} - \boldsymbol{\mu}^{(2)}),$$

where $\boldsymbol{\beta} = \boldsymbol{\Sigma}_{12}\boldsymbol{\Sigma}_{22}^{-1}$ $(q\times(p-q))$ is the regression matrix of $\mathbf{X}^{(1)}$ on $\mathbf{X}^{(2)}$. The maximum likelihood estimate of $\boldsymbol{\beta}$ is $\hat{\boldsymbol{\beta}} = \mathbf{A}_{12}\mathbf{A}_{22}^{-1}$, where

$$\mathbf{A}(p\times p) = \begin{bmatrix} \mathbf{A}_{11}(q\times q) & \mathbf{A}_{12} \\ \mathbf{A}_{21} & \mathbf{A}_{22} \end{bmatrix}$$

and $\mathbf{A}/N$ is the maximum likelihood estimate* of $\boldsymbol{\Sigma}$ based on a sample of N observations. Then conditional on $\mathbf{X}^{(2)} = \mathbf{x}^{(2)}$ or $\mathbf{A}_{22}$, $\hat{\boldsymbol{\beta}}$ has a matrix normal distribution $N(\boldsymbol{\beta}, \Sigma_{11.2}\otimes\mathbf{A}_{22}^{-1})$, where

$$\boldsymbol{\Sigma}_{11.2} = \boldsymbol{\Sigma}_{11} - \boldsymbol{\Sigma}_{12}\boldsymbol{\Sigma}_{22}^{-1}\boldsymbol{\Sigma}_{21}.$$

The unconditional distribution of $\hat{\boldsymbol{\beta}}$ is an interesting but complicated problem and arises in time series analysis. An asymptotic distribution for $\hat{\boldsymbol{\beta}}$, unconditionally, has been derived by Groenewald and de Waal [21].

THE SYMMETRIC MATRIX NORMAL DISTRIBUTION

The symmetric matrix normal distribution frequently appears in asymptotic distributions of symmetric matrices such as the Wishart distribution*.

If $\mathbf{X} = \mathbf{X}'(p\times p)$ is symmetric and matrix normally distributed, then $\operatorname{vec} p\mathbf{X}$ is multivariate normal distributed. Suppose the mean is $\operatorname{vec} p\boldsymbol{\mu}$ and the covariance matrix is $\mathbf{K}_p'(\boldsymbol{\Sigma}\otimes\boldsymbol{\Psi})\mathbf{K}_p$; then the density function of $\operatorname{vec} p\mathbf{X}$, which we will denote by $f_{\mathbf{X}=\mathbf{X}'}(\mathbf{x})$, is given for $-\infty < \mathbf{x} < \infty$ by

$$f_{\mathbf{X}=\mathbf{X}'}(\mathbf{x}) = (2\pi)^{-p(p+1)/4}|\mathbf{K}_p'(\boldsymbol{\Sigma}\otimes\boldsymbol{\Psi})\mathbf{K}_p|^{-1/2} \times \exp\left\{-\tfrac{1}{2}\left[\operatorname{vec} p'(\mathbf{x}-\boldsymbol{\mu}) \times (\mathbf{K}_p'(\boldsymbol{\Sigma}\otimes\boldsymbol{\Psi})\mathbf{K}_p)^{-1} \times \operatorname{vec} p(\mathbf{x}-\boldsymbol{\mu})\right]\right\}. \quad (10)$$

In the special case $\boldsymbol{\Psi} = 2\boldsymbol{\Sigma}$, (10) reduces to

$$f_{\mathbf{X}=\mathbf{X}'}(\mathbf{x}) = (2\pi)^{-p(p+1)/4}2^{-p/2}|\boldsymbol{\Sigma}|^{-(p+1)/2} \times \exp\left\{-\tfrac{1}{4}\operatorname{tr}\left[\boldsymbol{\Sigma}^{-1}(\mathbf{x}-\boldsymbol{\mu})\right]\right\}^2, \quad -\infty < \mathbf{x} < \infty, \quad (11)$$

which is a useful distribution in asymptotic theory. Since the symmetric normal distribution (10), especially (11), plays such an important role in asymptotic theory as indicated later in this section, we briefly consider the derivation of some properties such as the moments of $\mathbf{X}$. If $\mathbf{X}$ is distributed according to (10), then we say $\mathbf{X} = \mathbf{X}'(p \times p)$ is distributed $SN(\boldsymbol{\mu}, \mathbf{K}_p'(\boldsymbol{\Sigma} \otimes \boldsymbol{\Psi})\mathbf{K}_p)$. The moments of $\mathbf{X} = \mathbf{X}'$, distributed $SN(\boldsymbol{\mu}, \mathbf{K}_p'(\boldsymbol{\Sigma} \otimes \boldsymbol{\Psi})\mathbf{K}_p)$, can be obtained from the moments of the nonsymmetric variable $\mathbf{Y}(p \times p)$ distributed $N(\mathbf{0}, \boldsymbol{\Sigma} \otimes \boldsymbol{\Psi})$ by substituting $\frac{1}{2}(\mathbf{Y} + \mathbf{Y}')$ for $\mathbf{X} - \boldsymbol{\mu}$. For example, $E[\mathbf{XX}'] = E[\frac{1}{4}(\mathbf{Y} + \mathbf{Y}')(\mathbf{Y} + \mathbf{Y}')' + \boldsymbol{\mu}\boldsymbol{\mu}']$. For further details and expected values see Nel [33, 36, 38], Groenewald [20], Marx [32], Fujikoshi [18, 19], and Hayakawa and Kikuchi [23]. The Laplace transform of $\mathbf{X} = \mathbf{X}'(p \times p)$ distributed $SN(\boldsymbol{\mu}, \boldsymbol{\Sigma} \otimes \boldsymbol{\Sigma})$ is given by

$$g_{\mathbf{X}=\mathbf{X}'}(\mathbf{Z}) = \exp(-\operatorname{tr}\mathbf{Z} + \mathbf{Z\Sigma Z\Sigma}). \quad (12)$$

Applications of the symmetric matrix normal distribution can be found in the derivations of the asymptotic distributions of many test statistics. We shall consider an application to the Wishart matrix (*see* WISHART DISTRIBUTION). Let $\mathbf{A}(p \times p)$ be distributed Wishart with n degrees of freedom and covariance matrix $\boldsymbol{\Sigma}$; it follows from Anderson [1, Theorem 4.2.4] that $\mathbf{Z} = \lim(1 + n^{1/2})(\mathbf{A} - n\boldsymbol{\Sigma})$ is distributed $SN(\mathbf{0}, \mathbf{K}_p'(\boldsymbol{\Sigma} \otimes 2\boldsymbol{\Sigma})\mathbf{K}_p)$, a symmetric matrix normal distribution. This result implies a Fisher–Cornish type of expansion of the Wishart matrix, namely,

$$\mathbf{A}/n \simeq \boldsymbol{\Sigma} + n^{-1/2}\mathbf{Z}, \quad (13)$$

and has widely been applied in deriving asymptotic distributions of statistics where Wishart matrices are involved (see for instance Fujikoshi [13–18]). Nel and Groenewald [37] consider the expansion (13) further and have shown that $\mathbf{A}/n$ can be better approximated by

$$\begin{aligned}\mathbf{A}/n \simeq\ & \boldsymbol{\Sigma} + n^{-1/2}\mathbf{Z} \\ & + \tfrac{1}{3}n^{-1}(\mathbf{Z\Sigma}^{-1}\mathbf{Z} - (p+1)\boldsymbol{\Sigma}) \\ & + \frac{1}{36}n^{-3/2}(\mathbf{Z\Sigma}^{-1}\mathbf{Z\Sigma}^{-1}\mathbf{Z} - 3\boldsymbol{\Sigma}\operatorname{tr}(\boldsymbol{\Sigma}^{-1}\mathbf{Z}) \\ & - (4p+7)\mathbf{Z}),\end{aligned} \quad (14)$$

where $\mathbf{Z}$ is distributed $SN(\mathbf{0}, \mathbf{K}_p'(\boldsymbol{\Sigma} \otimes 2\boldsymbol{\Sigma})\mathbf{K}_p)$. This result coincides in the special case with that for the chi-square variable derived by Fisher and Cornish [12]. The expansions (13) or (14) are useful in deriving asymptotic distributions for matrix variables such as $\mathbf{S}_H\mathbf{S}_E^{-1}$ where $\mathbf{S}_H$ and $\mathbf{S}_E$ are independent Wishart variables. Other types of matrix variables are of the t-type (*see* MATRIC t-DISTRIBUTION) such as

$$\mathbf{T} = \mathbf{A}^{-1/2}\mathbf{X} + \boldsymbol{\mu}, \quad (15)$$

where $\mathbf{A}(p \times p)$ is distributed Wishart with n degrees of freedom and covariance matrix $\boldsymbol{\Sigma}$ independent of $\mathbf{X}(p \times q)$ distributed $N(\mathbf{0}, \boldsymbol{\Sigma} \otimes \boldsymbol{\Psi})$ (for exact and asymptotic distributions of $\mathbf{T}$ see Marx [32]). If, however, $\mathbf{A}(p \times p)$ is distributed noncentral Wishart with n degrees of freedom, covariance matrix $\boldsymbol{\Sigma}$ and noncentrality parameter $\boldsymbol{\Omega}$, then two limiting cases received attention, namely

$$\operatorname{plim}\mathbf{A}/n = \begin{cases}\boldsymbol{\Sigma} & \text{if } \boldsymbol{\Omega} = O(1) \\ \boldsymbol{\Sigma}(\mathbf{I} + \boldsymbol{\theta}) & \text{if } \boldsymbol{\Omega} = O(n) = n\boldsymbol{\theta}, \text{ say.}\end{cases}$$

It has been shown by Fujikoshi [12] (see also de Waal [8]) that the limiting distribution of $(1/n^{1/2})(\mathbf{A} - n\boldsymbol{\Sigma}(\mathbf{I} + \boldsymbol{\theta}))$ is $SN(\mathbf{0}, \mathbf{K}_p'(\boldsymbol{\Sigma}(\mathbf{I} + 2\boldsymbol{\theta}) \otimes 2\boldsymbol{\Sigma})\mathbf{K}_p)$ if $\boldsymbol{\Omega} = O(n) = n\boldsymbol{\theta}$. A typical element of the covariance matrix can be found from (5). If $\boldsymbol{\Omega} = O(1)$ then it can be shown (de Waal [8]) that the limiting distribution of $(1/n^{1/2})(\mathbf{A} - n\boldsymbol{\Sigma})$ is $SN(\mathbf{0}, \mathbf{K}_p'(\boldsymbol{\Sigma} \otimes 2\boldsymbol{\Sigma})\mathbf{K}_p)$. Steyn and Roux [46] showed by the method of moments* and ignoring terms of order higher than $O(n^{-1})$, that $\mathbf{A}$ is approximately distributed as a central Wishart with n degrees of freedom and covariance matrix $\boldsymbol{\Sigma}(\mathbf{I} + \boldsymbol{\Omega}/n)$. Nel and Groenewald [37] also looked at asymptotic expansions* such as (14) for the noncentral Wishart matrix.

DISTRIBUTIONS OF QUADRATIC FORMS OF NORMAL MATRICES

Let $\mathbf{X}(p \times q)$ be distributed $N(\boldsymbol{\mu}, \boldsymbol{\Sigma} \otimes \boldsymbol{\Psi})$; then the distribution of the quadratic form* $\mathbf{S} = \mathbf{X}'\mathbf{X}$, $q \leqslant p$, is of interest. Note that $\mathbf{S}$ is a symmetric random matrix. If $q > p$, the distribution of $\mathbf{XX}'$ can be obtained from that of $\mathbf{S}$ since $\mathbf{X}'$ is distributed $N(\boldsymbol{\mu}', \boldsymbol{\Psi} \otimes \boldsymbol{\Sigma})$. Authors such as Johnson and Kotz [26, 27], Crowther [5], Khatri and

Mardia [29, 30], Khatri [28], and de Waal [10, 11] have made contributions to the distribution of $\mathbf{S}$. The density of $\mathbf{S}$ is given in Crowther [5] and de Waal [10], using an extension of the Hayakawa polynomial [22]. Interesting special cases can be derived such as the Wishart distribution, the distribution of the sum of weighted noncentral chi-square variables, etc. [11]. If $\mathbf{\Sigma} = \mathbf{I}_p$, then $\mathbf{S}$ has a noncentral Wishart distribution with p degrees of freedom, covariance matrix $\mathbf{\Psi}$, and noncentrality parameter $\mathbf{\Omega} = \mathbf{\Psi}^{-1}\boldsymbol{\mu}\boldsymbol{\mu}'$. If $\boldsymbol{\mu} = \mathbf{0}$, the density of $\mathbf{S}$ becomes

$$\left(2^{pq/2}\Gamma_q\left(\tfrac{1}{2}p\right)\right)^{-1}|\mathbf{\Sigma}|^{-q/2}|\mathbf{\Psi}|^{-p/2}|\mathbf{s}|^{(p-q-1)/2} \times \exp\left(-\tfrac{1}{2}\lambda \operatorname{tr}\mathbf{\Psi}^{-1}\mathbf{s}\right) \times {}_0F_0\left(\mathbf{\Sigma}^{-1} - \lambda\mathbf{I}_p, \tfrac{1}{2}\mathbf{\Psi}^{-1}\mathbf{s}\right), \quad \mathbf{s} > \mathbf{0}, \quad (16)$$

where ${}_0F_0(\cdot,\cdot)$ is a hypergeometric function of double matrix argument [25].

MATRIX BETA DISTRIBUTIONS

Another type of quadratic form in normal matrices is $\mathbf{F} = \mathbf{X}'\mathbf{A}^{-1}\mathbf{X}$, where $\mathbf{X}(p \times q)$ is distributed $N(\boldsymbol{\mu}, \mathbf{\Sigma} \otimes \mathbf{I}_q)$, $q < p$, and $\mathbf{A}(p \times p)$ is independently distributed Wishart with $n \geqslant p$ degrees of freedom and covariance matrix $\mathbf{\Sigma}$. The density of $\mathbf{F}$ is given by James [25] as

$$\frac{\Gamma_q\left(\frac{1}{2}(n+q)\right)}{\Gamma_q\left(\frac{1}{2}(n+q-p)\right)\Gamma_q\left(\frac{1}{2}p\right)} \cdot \frac{|\mathbf{f}|^{(p-q-1)/2}}{|\mathbf{I}_q + \mathbf{f}|^{(n+q)/2}} \times {}_1F_1\left(\tfrac{1}{2}(n+q); \tfrac{1}{2}p; \tfrac{1}{2}\mathbf{\Omega}\mathbf{f}(\mathbf{I}_q + \mathbf{f})^{-1}\right) \times \exp\left(-\tfrac{1}{2}\operatorname{tr}\mathbf{\Omega}\right), \qquad \mathbf{f} > \mathbf{0}, \quad (17)$$

where ${}_1F_1$ is the confluent hypergeometric function* of matrix argument given in Constantine [4].

The density is a *noncentral matrix F distribution* with p and $n + q - p$ degrees of freedom and noncentrality parameter $\mathbf{\Omega} = \boldsymbol{\mu}'\mathbf{\Sigma}^{-1}\boldsymbol{\mu}$. According to Johnson and Kotz [27], it is also referred to as a *noncentral inverted multivariate beta distribution*. Various properties of (17) have been derived (de Waal [7, 8, 9]; Saw [44]; Shah and Khatri [45]; and others), for instance [9],

$$E\operatorname{tr}_j\mathbf{F} = \left(1/(n+j-p-2)^{(j)}\right) \times \sum_{i=1}^{j}\binom{q-i}{j-i}(p-i)^{(j-i)}\operatorname{tr}_i\mathbf{\Omega}, \quad (18)$$

where $\operatorname{tr}_j\mathbf{F}$ is the jth elementary symmetric function of $\mathbf{F}$ and $(q)^{(j)} = q(q-1)\dots(q-j+1)$. Special cases such as $E[\operatorname{tr}\mathbf{F}] = E[\operatorname{tr}_1\mathbf{F}]$ and $E[|\mathbf{F}|] = E[\operatorname{tr}_p\mathbf{F}]$ follow.

The density (17) is derived from the density of $\mathbf{L} = \mathbf{X}'(\mathbf{A} + \mathbf{X}\mathbf{X}')^{-1}\mathbf{X}$. The density of $\mathbf{L}$ is given by

$$\frac{\Gamma_q\left(\frac{1}{2}(n+q)\right)}{\Gamma_q\left(\frac{1}{2}(n+q-p)\right)\Gamma_q\left(\frac{1}{2}p\right)}|\boldsymbol{l}|^{(p-q-1)/2} \times |\mathbf{I}_q - \boldsymbol{l}|^{(n-p-1)/2}\exp\left(-\tfrac{1}{2}\operatorname{tr}\mathbf{\Omega}\right) \times {}_1F_1\left(\tfrac{1}{2}(n+q); \tfrac{1}{2}p; \tfrac{1}{2}\mathbf{\Omega}\boldsymbol{l}\right), \quad \mathbf{I}_q - \boldsymbol{l} > \mathbf{0}, \quad \boldsymbol{l} > \mathbf{0}. \quad (19)$$

The density (19) is of the noncentral multivariate beta type with p and $n - p + q$ degrees of freedom and noncentrality parameter $\mathbf{\Omega}$. The central multivariate beta distributions follow as special cases of the noncentral distributions if $\mathbf{\Omega} = \mathbf{0}$.

Other forms of (17) and (19) can be obtained. The density of $\mathbf{V} = \mathbf{B}^{1/2}\mathbf{A}^{-1}\mathbf{B}^{1/2}$, for instance, where $\mathbf{B} = \mathbf{X}\mathbf{X}'$ $(q \geqslant p)$ has a noncentral Wishart distribution with q degrees of freedom, covariance matrix $\mathbf{\Sigma}$, and noncentrality parameter $\mathbf{\Omega} = \mathbf{\Sigma}^{-1}\boldsymbol{\mu}\boldsymbol{\mu}'$, is a *noncentral inverted beta* with q and n degrees of freedom. $\mathbf{B}^{1/2}$ is the symmetric square root of $\mathbf{B}$ (see de Waal [8]). Other forms of matrix-variable beta distributions are defined and their properties are considered. See for instance Olkin and Rubin [39], Perlman [40], and Roux and Ratnaparkhi [42].

These forms are of importance mainly because they appear in various test statistics. For instance, the test statistic for testing the equality of the means of several multivariate normal populations with the same unknown covariance matrix can be written as

(i) $\text{tr}\,\mathbf{V}$ (or $\text{tr}\,\mathbf{F}$), which is Hotelling's generalized T_0^2 statistic (*see* HOTELLING'S T^2; HOTELLING'S TRACE);

(ii) $|\mathbf{L}|$ (or $|\mathbf{A}|/|\mathbf{A}+\mathbf{B}|$), which is the likelihood ratio statistic (*see* LAMBDA CRITERION, WILKS'S);

(iii) $\text{tr}\,\mathbf{L}$ (or $\text{tr}\,A(\mathbf{A}+\mathbf{B})^{-1}$), which is Pillai's trace* statistic; and

(iv) max characteristic root $\mathbf{L}$ (or $\mathbf{A}(\mathbf{A}+\mathbf{B})^{-1}$) which is Roy's largest root statistic. Various properties of $\mathbf{F}$, $\mathbf{V}$, and $\mathbf{L}$ can be found in de Waal [8].

A matrix variable that is also distributed multivariate beta and that is of interest in multivariate analysis* is the *generalized correlation matrix* $\mathbf{R} = \mathbf{A}_{11}^{-1/2}\mathbf{A}_{12}\mathbf{A}_{22}^{-1}\mathbf{A}_{21}\mathbf{A}_{11}^{-1/2}$, where

$$\mathbf{A}(p\times p) = \begin{bmatrix} \mathbf{A}_{11}(q\times q) & \mathbf{A}_{12} \\ \mathbf{A}_{21} & \mathbf{A}_{22} \end{bmatrix}$$

is distributed Wishart with n degrees of freedom and covariance matrix

$$\boldsymbol{\Sigma}(p\times p) = \begin{bmatrix} \boldsymbol{\Sigma}_{11}(q\times q) & \boldsymbol{\Sigma}_{12} \\ \boldsymbol{\Sigma}_{21} & \boldsymbol{\Sigma}_{22} \end{bmatrix}.$$

If the population generalized correlation matrix $\mathbf{P} = \boldsymbol{\Sigma}_{11}^{-1/2}\boldsymbol{\Sigma}_{12}\boldsymbol{\Sigma}_{22}^{-1}\boldsymbol{\Sigma}_{21}\boldsymbol{\Sigma}_{11}^{-1/2}$ is the zero matrix, then $\mathbf{R}$ is distributed multivariate beta with $n-p+q$ and $p-q$ degrees of freedom. If $\mathbf{P}\neq\mathbf{0}$, then $\mathbf{R}$ follows a noncentral multivariate beta type distribution which is not in an explicit form (Troskie [47]). Some properties of $\mathbf{R}$ can be found in de Waal [8].

The multivariate test statistic associated with tests of the equality of the means and covariances of several multivariate normal populations can be written in terms of matrix Dirichlet variables (*see* DIRICHLET DISTRIBUTION). It is therefore of interest to consider the joint density of $\mathbf{V}_j = \mathbf{B}^{1/2}\mathbf{A}^{-1}\mathbf{B}^{1/2}$; $j=1,\ldots,m$, where $\mathbf{A}_j(p\times p)$ are independently distributed Wishart with n_j degrees of freedom and covariance matrix $\boldsymbol{\Sigma}$ for $j=1,\ldots,m$, independently of $\mathbf{B}(p\times p)$ distributed as a noncentral Wishart with q degrees of freedom, covariance matrix $\boldsymbol{\Sigma}$, and noncentrality parameter $\boldsymbol{\Omega}$. The joint density is given by

$$\frac{\Gamma_p(\frac{1}{2}(n+q))}{\Gamma_p(\frac{1}{2}q)\prod_{j=1}^m\Gamma_p(\frac{1}{2}n_j)}\cdot\frac{\prod_{j=1}^m|\mathbf{v}_j|^{(n_j-p-1)/2}}{|\mathbf{I}_p+\sum_{j=1}^m\mathbf{v}_j|^{(n+q)/2}}$$
$$\times\,{}_1F_1\left(\frac{1}{2}(n+q);\frac{1}{2}q;\frac{1}{2}\boldsymbol{\Omega}\left(\mathbf{I}_p+\sum_{j=1}^m\mathbf{v}_j\right)^{-1}\right)$$
$$\times\exp\left(-\frac{1}{2}\text{tr}\,\boldsymbol{\Omega}\right),\quad \mathbf{v}_j>\mathbf{0};\ j=1,\ldots,m;$$
$$n=\sum_{j=1}^m n_j. \tag{20}$$

The density (20) is the *noncentral inverted multivariate Dirichlet* density, according to Johnson and Kotz [27]. The joint density of $\mathbf{L}_j = (\sum_{i=1}^m\mathbf{A}_i+\mathbf{B})^{-1/2}\mathbf{A}_j(\sum_{i=1}^m\mathbf{A}_i+\mathbf{B})^{-1/2}$, $j=1,\ldots,m$, is the *noncentral multivariate Dirichlet* density (de Waal [8]). If $\boldsymbol{\Omega}=\mathbf{0}$, the central distributions follow as special cases.

THE VON MISES–FISHER MATRIX DISTRIBUTION

Another class of matrix-valued distributions derived from the matrix normal is that of the von Mises–Fisher type distributions. If $\mathbf{X}(p\times q)$, $q\leqslant p$, is distributed $N(\boldsymbol{\mu},\boldsymbol{\Sigma}\otimes\boldsymbol{\Psi})$ with $\boldsymbol{\Sigma}(p\times p)$ and $\boldsymbol{\Psi}(q\times q)$, p.s.d., then $\mathbf{X}$ conditional on $\mathbf{X}'\mathbf{X}=\mathbf{S}$, is distributed as a generalization of the von Mises–Fisher matrix distribution [11] with density given by

$$C\exp\left(-\tfrac{1}{2}\text{tr}\boldsymbol{\Sigma}^{-1}\mathbf{x}\boldsymbol{\Psi}^{-1}\mathbf{x}'+\text{tr}\,\boldsymbol{\Sigma}^{-1}\mathbf{x}\boldsymbol{\Psi}^{-1}\boldsymbol{\mu}'\right),$$
$$\mathbf{x}'\mathbf{x}=\mathbf{s},\quad -\infty<\mathbf{x}<\infty, \tag{21}$$

where

$$C=\left(2^q\pi^{pq/2}/\Gamma_q(\tfrac{1}{2}p)\right)$$
$$\times\sum_{k=1}^{\infty}\sum_{K}\left((\tfrac{1}{2}p)_K k!\right)^{-1}$$
$$\times P_K\left(-\sqrt{\tfrac{1}{2}}\,\boldsymbol{\Psi}^{-1/2}\boldsymbol{\mu}'\boldsymbol{\Sigma}^{-1/2},\boldsymbol{\Sigma}^{-1},\tfrac{1}{2}\boldsymbol{\Psi}^{-1}\right),$$

and where P_K is the Hayakawa polynomial; *see also* DIRECTIONAL DISTRIBUTIONS. If $\boldsymbol{\Sigma}=\mathbf{I}_p$, $\boldsymbol{\Psi}=\mathbf{I}_q$, and $\mathbf{s}=\mathbf{I}_q$, then (21) reduces to the von Mises–Fisher matrix distribution [30] with density proportional to $\text{etr}(\mathbf{x}\boldsymbol{\mu}')$. Further generalizations have been obtained by Bingham [2] and de Waal [10, 11].

References

[1] Anderson, T. W. (1958). *Introduction to Multivariate Statistical Analysis*. Wiley, New York.

[2] Bingham, C. (1974). *Ann. Statist.*, **2**, 1201–1225.

[3] Browne, M. W. (1974). *S. Afr. Statist. J.*, **8**, 1–24.

[4] Constantine, A. G. (1963). *Ann. Math. Statist.*, **34**, 1270–1285. (Basic paper on noncentral distributions and applications of hypergeometric functions of matrix argument.)

[5] Crowther, N. A. S. (1975). *S. Afr. Statist. J.*, **9**, 27–36. (Extends the Hayakawa polynomial.)

[6] Dawid, A. P. (1981). *Biometrika*, **68**, 265–274. (Basic paper on matrix-valued distributions.)

[7] de Waal, D. J. (1972). *Ann. Math. Statist.*, **43**, 344–347.

[8] de Waal, D. J. (1974). Parametric multivariate analysis. Part 2. *Monograph*, Department of Statistics, University of North Carolina, Chapel Hill. (Applications of zonal polynomials are given and properties of some matrix-valued distributions.)

[9] de Waal, D. J. (1978). *S. Afr. Statist. J.*, **12**, 75–82.

[10] de Waal, D. J. (1979). *S. Afr. Statist. J.*, **13**, 103–122. (General theorem on quadratic forms of matrix normal variables is given.)

[11] de Waal, D. J. (1983). Quadratic forms and manifold normal distributions. *Contributions to Statistics: Essays in Honor of N. L. Johnson.* P. K. Sen, ed. North-Holland, Amsterdam.

[12] Fisher, R. A. and Cornish, E. A. (1960). *Technometrics*, **2**, 209–255. (Basic paper on asymptotic expansion of a chi-square variable.)

[13] Fujikoshi, Y. (1968). *J. Sc. Hiroshima Univ.* Ser. A-1, **34**, 73–144. (Asymptotic expansion of the distribution of the generalized variance in the noncentral case.)

[14] Fujikoshi, Y. (1973). *Ann. Inst. Statist. Math.*, **25**, 423–437. (Asymptotic formulas for the distribution of three statistics for multivariate linear hypotheses.)

[15] Fujikoshi, Y. (1974). *J. Multivariate Anal.*, **4**, 327–340. (The likelihood ratio tests for the dimensionality of regression coefficients.)

[16] Fujikoshi, Y. (1975). *Multivariate Analysis IV*, P. R. Krishnaiah, ed. Academic Press, New York. (Asymptotic expansions for the distributions of some multivariate tests.)

[17] Fujikoshi, Y. (1975). *Ann. Inst. Statist. Math.*, **27**, 99–108. (Asymptotic formulas for the non-null distributions of three statistics for multivariate linear hypotheses.)

[18] Fujikoshi, Y. (1977). *J. Multivariate Anal.*, **7**, 386–396. (Asymptotic expansions of the distributions of the latent roots in MANOVA and canonical correlations.)

[19] Fujikoshi, Y. (1978). *J. Multivariate Anal.*, **8**, 63–72. (Asymptotic expansions of the distributions of some functions of the latent roots of matrices in three situations.)

[20] Groenewald, P. C. N. (1978). On Asymptotic Distributions of Certain Regression Matrices with an Applicatoin to Multiple Time Series Analysis. Unpublished Ph.D. thesis, University of the Orange Free State, Bloemfontein, South Africa.

[21] Groenewald, P. C. N. and de Waal, D. J. (1979). *S. Afr. Statist. J.*, **13**, 15–28. (The asymptotic distribution of the regression matrix is derived.)

[22] Hayakawa, T. (1972). *Ann. Inst. Statist. Math.*, **24**, 205–230. (A distribution of a quadratic form of matrix normal variables are considered.)

[23] Hayakawa, T. and Kikuchi, Y. (1979). *S. Afr. Statist. J.*, **13**, 71–82. (Properties of the symmetric matrix normal distribution are considered.)

[24] Herz, E. S. (1964). *Ann. Math.*, **61**, 474–523. (Basic paper on hypergeometric functions of matrix argument.)

[25] James, A. T. (1964). *Ann. Math. Statist.*, **35**, 475–501. (Basic paper on applications of zonal polynomials.)

[26] Johnson, N. L. and Kotz, S. (1970). *Distributions in Statistics: Continuous Univariate Distributions*, Vol. 2. Wiley, New York.

[27] Johnson, N. L. and Kotz, S. (1972). *Distributions in Statistics: Continuous Multivariate Distributions*, Wiley, New York.

[28] Khatri, C. G. (1977). *S. Afr. Statist. J.*, **11**, 167–180. (A distribution of a quadratic form in normal vectors.)

[29] Khatri, C. G. and Mardia, K. V. (1975). The von Mises–Fisher Matrix Distributions. *Research Report No. 1*, Department of Statistics, University of Leeds, Leeds, England.

[30] Khatri, C. G. and Mardia, K. V. (1977). *J. R. Statist. Soc. B*, **39**, 95–106. (A matrix distribution in orientation statistics.)

[31] Koopmans, T. C. et al. (1950). *Statistical Inference in Dynamic Economic Models*. Cowles Commission for Research in Economics. *Monograph no. 10.* Wiley, New York.

[32] Marx, D. G. (1981). Aspects of the Matric-t Distribution. Unpublished Ph.D. thesis, University of the Orange Free State, Bloemfontein, South Africa. (Covers derivations of exact distributions, asymptotic distributions, and quadratic forms.)

[33] Nel, D. G. (1978). *S. Afr. Statist. J.*, **12**, 145–159. (Contains results on the symmetric matrix normal.)

[34] Nel, D. G. (1979). On Patterned Matrix Normal Distributions. *Tech. Report No.* 48, Dept. of Math. Statist., University of the Orange Free State, Bloemfontein, South Africa.

[35] Nel, D. G. (1980). *S. Afr. Statist. J.*, **14**, 137–193. (Basic paper on matrix differentiation.)

[36] Nel, D. G. (1981). On Patterned Matrices and Their Applications in Multivariate Statistics. *Tech. Report No.* 67, Dept. of Math. Statist., University of the Orange Free State, Bloemfontein, South Africa.

[37] Nel, D. G. and Groenewald, P. C. N. (1979). On a Fisher–Cornish Type Expansion of Wishart Matrices. *Tech. Report No.* 47, Dept. of Math. Statist., University of the Orange Free State, Bloemfontein, South Africa.

[38] Nel, H. M. (1977). On Distributions and Moments Associated with Matrix Normal Distributions. *Tech. Report No.* 24, Dept. of Math. Statist., University of the Orange Free State, Bloemfontein, South Africa.

[39] Olkin, I. and Rubin, H. (1964). *Ann. Inst. Statist. Math.*, **35**, 261–269. (Contains results on multivariate beta distributions.)

[40] Perlman, M. D. (1977). *Sánkya*, **39A**, 290–298. (An invariance property of the matrix-variate *F* distribution is discussed.)

[41] Press, S. J. (1972). *Applied Multivariate Analysis*. Holt, Rinehart and Winston, New York, pp. 109–112.

[42] Roux, J. J. J. and Ratnaparkhi, M. V. (1981). In *Statistical Distributions in Scientific Work*, Vol. 4 (B. A. Baldessari, C. Taillie, and G. P. Patil, eds.). D. Reidel, Dordrecht, pp. 375–378. (Discussion of a characterization of the Wishart distribution using matrix-variate beta distributions.)

[43] Roy, S. N. (1957). *Some Aspects of Multivariate Analysis*. Wiley, New York, p. 18.

[44] Saw, J. G. (1973). *Ann. Statist.*, **1**, 580–582.

[45] Shah, B. H. and Khatri, C. G. (1974). *Ann. Statist.*, **2**, 833–836.

[46] Steyn, H. S. and Roux, J. J. J. (1982). *S. Afr. Statist. J.*, **6**, 165–173. (Considers an approximation for the noncentral Wishart distribution.)

[47] Troskie, C. G. (1969). *S. Afr. Statist. J.*, **13**, 109–121. (Basic paper on the generalized correlation matrix.)

[48] Van der Merwe, C. A. (1980). Expectations of the Traces of Functions of a Multivariate Normal Variable. *Tech. Report No.* 56, Dept. of Math. Statist., University of the Orange Free State, Bloemfontein, South Africa.

(DIRICHLET DISTRIBUTION
INVERTED BETA DISTRIBUTION
INVERTED DIRICHLET DISTRIBUTIONS
INVERTED WISHART DISTRIBUTIONS
LATENT ROOT DISTRIBUTIONS
MATRIC t-DISTRIBUTION
MULTINORMAL DISTRIBUTION
MULTIVARIATE DISTRIBUTIONS
QUADRATIC FORMS
WISHART DISTRIBUTIONS
ZONAL POLYNOMIALS)

D. J. DE WAAL

MATRIX-VARIATE BETA DISTRIBUTION

A symmetric positive definite $m \times m$ matrix $\mathbf{U}$ is said to have a *matrix-variate beta distribution* with parameters $n_1/2$ and $n_2/2$ if its density function is given by

$$\frac{\Gamma_m\left[\frac{1}{2}(n_1+n_2)\right]}{\Gamma_m(\frac{1}{2}n_1)\Gamma_m(\frac{1}{2}n_2)}(\det \mathbf{U})^{(n_1-m-1)/2} \times \det(\mathbf{I}_m - \mathbf{U})^{(n_2-m-1)/2},$$

where $\mathbf{I}_m - \mathbf{U}$ is a positive definite matrix, $\mathbf{I}_m$is an identity matrix, and $\Gamma_m(a)$ is the $m \times m$ "multivariate gamma function" defined by

$$\Gamma_m(a) = \int_{\mathbf{A}>\mathbf{0}} \operatorname{etr}(-\mathbf{A})|\mathbf{A}|^{a-(m+1)/2}\,d\mathbf{A}.$$

The matrix beta distribution is a generalization of the univariate beta distribution*, analogous to the Wishart* distribution's generalization of the chi-squared distribution. A discussion of this distribution and its application to the theory of multivariate analysis is given in Johnson and Kotz [1] and in more detail in Muirhead [2].

References

[1] Johnson, N. L. and Kotz, S. (1972) *Continuous Multivariate Distributions*. Wiley, New York.

[2] Muirhead, R. J. (1982). *Aspects of Multivariate Statistical Theory*. Wiley, New York.

(BETA DISTRIBUTION
MATRIX-VALUED DISTRIBUTIONS
WISHART DISTRIBUTION)

MATUSITA'S DISTANCE

INTRODUCTION

Let F_1 and F_2 be two distribution functions admitting probability densities f_1 and f_2, respectively, with respect to some measure μ. Matusita's distance between F_1 and F_2 is defined for $r \geqslant 1$ by

$$\|F_1 - F_2\|_r = \left| \int \left(f_1^{1/r}(x) - f_2^{1/r}(x)\right)^r d\mu(x) \right|^{1/r}, \quad (1)$$

Note that if

$$\rho(F_1, F_2) = \int (f_1(x) f_2(x))^{1/2} d\mu(x),$$

then $\|F_1 - F_2\|_2^2 = 2(1 - \rho(F_1, F_2))$. Here and elsewhere, integrals will be taken over the common support of f_1 and f_2. The distance defined in (1) is also known as the *Hellinger distance**; see Beran [4] and Rao [15].

The duality between $\|F_1 - F_2\|_2$ and $\rho(F_1, F_2)$, known as the *affinity* between F_1 and F_2, is one of the most important aspects in the applicability of the distance (1) in statistical inference*. The dual notion of affinity between F_1 and F_2 can be extended to measure the closeness between individual members of a finite family of distributions, $F_1, \ldots, F_m$, all admitting densities $f_1, \ldots, f_m$, respectively, with respect to some measure μ. This can be done as follows:

$$\rho_m(F_1, \ldots, F_m) = \int f_1^{r_1}(x) \ldots f_m^{r_m}(x) \, d\mu(x), \quad (2)$$

where $r_i \geqslant 0$, $i = 1, \ldots, m$ and $\sum_{i=1}^m r_i = 1$. When $r_i = 1/m$, $i = 1, \ldots, m$, we obtain the notion defined and studied by Matusita [12, 13]. The extension in (2) was first proposed by Toussaint [16]. The Matusita distance defined in (1) is both of mathematical and statistical interest.

MATHEMATICAL PROPERTIES

First, note that with $r_i = 1/m$

$$\begin{aligned} 0 &\leqslant \rho_m^m(F_1, \ldots, F_m) \\ &\leqslant \rho_{m-1}^{m-1}(F_{i_1}, \ldots, F_{i_{m-1}}) \\ &\leqslant \ldots \leqslant \rho_2^2(F_{i_l}, F_{i_p}) \leqslant 1, \end{aligned} \quad (3)$$

where $i_1, \ldots, i_{m-1}$ is a permutation subset of $1, \ldots, m$ and $\{i_l, i_p\} \subset \cdots \subset \{i_1, \ldots, i_{m-1}\} \subset \{1, \ldots, m\}$ [12]. Also, note that $\rho_m(F_1, \ldots, F_m) = 1$ whenever $F_1 = \cdots = F_m$ and that

$$\begin{aligned} 1 - (m-1)\delta &\leqslant \rho_m(F_1, \ldots, F_m) \\ &\leqslant \min \rho^{2/m}(F_i, F_j), \end{aligned} \quad (4)$$

where $\delta > 0$ is such that $\|F_i - F_j\|_r \leqslant \delta$ for all i, $i = 1, \ldots, m$.

A more refined inequality than (4) is given by Toussaint [16]:

$$\begin{aligned} 1 - m^{-2} &\textstyle\sum_{i<j} J(F_i, F_j) \\ &\leqslant \rho_m(F_1, \ldots, F_m) \\ &\leqslant \left[\frac{2}{m(m-1)} \right]^{1/2} \textstyle\sum_{i<j} \rho_2(F_i, F_j), \end{aligned} \quad (5)$$

$$J(F_1, F_2) = \int [f_1(x) - f_2(x)] \ln \left[\frac{f_1(x)}{f_2(x)} \right] d\mu(x).$$

On the other hand we have that

$$\|F_1 - F_2\|_{r-1}^{r-1} \geqslant \|F_1 - F_2\|_r^r, \quad r \geqslant 1, \quad (6)$$

and also that $\|F_1 - F_2\| \leqslant r \|F_1 - F_2\|_r$.

If $\{F_{in}\}$, $i = 1, \ldots, m$ are m sequences of distributions, then $\rho_m(F_{1n}, \ldots, F_{mn}) \to \rho(F_{10}, \ldots, F_{m0})$ provided that $f_{in}(x) \to f_{i0}(x)$, a.e., as $n \to \infty$. Next, if we partition the support of $F_1, \ldots, F_m$ into subsets $E_1, E_2, \ldots,$ then

$$\rho_m(F_1, \ldots, F_m) \leqslant \textstyle\sum_i \left\{ \prod_{j=1}^m \int_{E_i} f_j(x) \, d\mu(x) \right\}^{1/m}, \quad (7)$$

and also

$$\rho_m(F_1, \ldots, F_m) = \inf \textstyle\sum_i \left\{ \prod_{j=1}^m \int_{E_i} f_j(x) \, d\mu(x) \right\}^{1/m}, \quad (8)$$

where the inf is taken over all partitions $\{E_i\}$. Other properties of ρ when transformations of variables are sought are in Matusita [13].

Kirmani [5, 6] proved that $\rho_2(F_1, F_2)$ and also $\|F_1 - F_2\|_2$ have limiting properties similar to those of the Kullback–Leibler information number $J(F_1, F_2)$, while Kirmani [7] gave more bounds for $\rho_2(F_1, F_2)$.

STATISTICAL APPLICATIONS AND PROPERTIES

When μ is a finite counting measure, many inferential problems have been treated using Matusita's measure of distance or equivalently its dual affinity measure. Matusita [8] used $\|F_1 - F_2\|_2$ in the one sample goodness-of-fit* problem, showed the size of the test based on $\|F_0 - F_n\|_2$, and that its limiting distribution under the null is chi-square and under the alternative is normal; see also Ahmad and VanBelle [3]. He also discussed the two-sample problem and showed that in this case the null distribution is only approximately a weighted sum of chi-squares. Similar results also were obtained by Rao [15]. In the decision problem, Matusita and Akaike [14] and Matusita [9, 10] used similar ideas in various statistical problems such as independence, invariance*, two-sample goodness-of-fit, classification and pattern recognition*, and interval estimation. Formulation of the statistical decision problem in terms of distance functions in general terminology is presented in Matusita [11].

Other statistical applications include proving that a limit of a sequence of sufficient statistics is also sufficient (see Matusita [13]) and also its use as a measure of discrimination.

When μ is Lebesgue measure, the use of $\rho(F_1, F_2)$ in statistical inference is much scarcer. In parametric estimation, Beran [4] shows that an estimate of a vector of parameters $\boldsymbol{\theta}$ of $f_{\boldsymbol{\theta}}$ which minimizes $\|F_{\boldsymbol{\theta}} - F_n\|_2$ always exists (under certain conditions on the parameter space θ) and is unique. This estimate is robust against moderate perturbations. Ahmad [2] has proven, however, that if one is to estimate $\rho(F_1, F_2)$ using suitable density estimators $\hat{f}_1$ and $\hat{f}_2$, then one can find conditions under which $\hat{\rho}(F_1, F_2) = \int[\hat{f}_1(x)\hat{f}_2(x)]^{1/2}\,dx$ is consistent in the mean and is a strongly consistent estimate of $\rho(F_1, F_2)$.

RELATED DISTANCES

While in the case of two distributions, there is a complete duality between the distance measure and the affinity measure of Matusita, this is not clear when there are more than two distributions. The affinity between several distributions was investigated by Matusita [12, 13], but the distance between two sets of distributions is still open. The usage of the affinity or distance in statistical inference needs further investigation, in particular as a method for providing robust estimation* and as a vehicle to do hypothesis testing*.

An affinity measure that has proved a particularly successful alternative to $\rho(F_1, F_2)$ is

$$\lambda(F_1, F_2) = \frac{2\int f_1(x) f_2(x)\,d\mu(x)}{\int f_1^2(x)\,d\mu(x) + \int f_2^2(x)\,d\mu(x)}. \tag{9}$$

Note that $\lambda(F_1, F_2)$ assumes that f_i, $i = 1, 2$, are square integrable.

In the case when μ is the counting measure, $\lambda(F_1, F_2)$ enjoys some very attractive properties and is useful in many inference problems, while in the absolutely continuous case, it is more suitable than $\rho(F_1, F_2)$ for statistical inference when F_1, F_2 admit square integrable densities; see Ahmad [1], where the two-sample and one-sample goodness-of-fit problems are discussed as well as problems of independence and symmetry.

References

[1] Ahmad, I. (1980). *Ann. Inst. Statist. Math.*, **32**, 223–240.

[2] Ahmad, I. (1980). *Ann. Inst. Statist. Math.*, **32**, 241–245.

[3] Ahmad, I. and VanBelle, G. (1974). *In Reliability and Biometry, Statistical Analysis of Lifetesting*, F. Proschan and R. J. Serfling, eds. *SIAM*, Philadelphia, pp. 651–668.

[4] Beran, R. (1977). *Ann. Statist.*, **5**, 445–463.

[5] Kirmani, S. N. (1968). *J. Indian Statist. Ass.*, **6**, 89, 98.

[6] Kirmani, S. N. (1971). *Ann. Inst. Statist. Math.*, **23**, 157–162.

[7] Kirmani, S. N. (1979). *Ann. Inst. Statist. Math.*, **31**, 289–291.

[8] Matusita, K. (1955). *Ann. Math. Statist.*, **26**, 631–640.

[9] Matusita, K. (1956). *Ann. Inst. Statist. Math.*, **8**, 67–77.

[10] Matusita, K. (1961). *Bull. Inter. Statist. Inst.*, **38**, 241–244.

[11] Matusita, K. (1964). *Ann. Inst. Statist. Math.*, **16**, 305–315.

[12] Matusita, K. (1967). *Ann. Inst. Statist. Math.*, **19**, 181–192.

[13] Matusita, K. (1971). *Ann. Inst. Statist. Math.*, **23**, 137–155.

[14] Matusita, K. and Akaike, H. (1956). *Ann. Inst. Statist. Math.*, **7**, 67–90.

[15] Rao, C. R. (1963). *Sankhyā*, **25**, 189–206.

[16] Toussaint, G. T. (1974). *Ann. Inst. Statist. Math.*, **26**, 389–394.

(HELLINGER DISTANCE
MEASURES OF SIMILARITY,
DISSIMILARITY AND DISTANCE)

IBRAHIM A. AHMAD

MAURICE MODEL *See* COLTON'S MODEL

MAVERICK *See* OUTLIERS

MAXIMAL INVARIANT STATISTICS *See* INVARIANCE CONCEPTS IN STATISTICS

MAXIMIN TESTS *See* MINIMAX TESTS

MAXIMUM ALLOWABLE PERCENT DEFECTIVE (MAPD)

The proportion of defectives in a single sampling plan corresponding to the inflection point of the OC curve. It was introduced as a quality standard by Mayer [1]. See also Soundararajan [2] for a discussion of single sample attribute plans indexed by MAPD.

References

[1] Mayer, P. L. (1967). *Ann. Inst. Statist. Math.*, **19**, 537–542.

[2] Soundararajan, V. (1975). *J. Quality Tech.*, **7**, No. 4, 173–177.

(ACCEPTANCE SAMPLING)

MAXIMUM AUTOCORRELATIONS FOR MOVING AVERAGES PROCESSES *See* MOVING AVERAGES

MAXIMUM ENTROPY PRINCIPLE, CLASSICAL APPROACH

In statistical inference problems, it is sometimes necessary to estimate individual state probabilities of a system when all that is known are the probabilities (or frequencies) of certain aggregate states. For instance, one may be given the marginal probabilities in a two-way contingency table* and one is asked to estimate the individual table entries. If no other information were available, the classical statistician would assume that the two factors are statistically independent, and assign to each table entry the product of its two marginals. This technique raises two questions:

1. What theoretical justification is there for making the independence assumption?
2. While the solution based on independence assumptions was perfectly clear in this example, it is not so in more complicated examples, where even the formulation of the independence assumptions may be ambiguous.

Both of these problems are resolved by the *maximum entropy principle*, which is based on *information theory**. According to the latter, the amount of information (relative to the value of a random variable) contained in a probability distribution with frequency (or density) function $p(\ )$ is $I(p) = E[\log p]$, where $E[\ ]$ is the expectation. $I(p)$ is also known as the *negative entropy* of the distribution.

The maximum entropy principle can now be stated as follows:

> Of all probability distributions which satisfy the constraints imposed by the known aggregate probabilities, choose the one that has the maximum entropy or, equivalently, contains the least information.

Theoretically, the principle is justified by noting that if *all* the information that we have concerning the unknown distribution is contained in the known aggregates, then we are only entitled to select the least informative distribution that is consistent with this knowledge. Adopting any other distribution would be tantamount to assuming that we know more about the random variable than we actually do. In the contingency table problem, let p_{ij} be the probability of the (i, j) entry, and let $p_{i\cdot}$ and $p_{\cdot j}$ be the marginal probabilities of the ith row and jth column, respectively. Then $I(p) = \sum_{i,j} p_{ij} \log p_{ij}$. We must find the p_{ij} which minimizes $I(p)$ subject to the constraints $p_{i\cdot} = \sum_j p_{ij}$ (for each i) and $p_{\cdot j} = \sum_i p_{ij}$ (for each j). One easily verifies that in this case the solution is $p_{ij} = p_{i\cdot} p_{\cdot j}$, just as one would obtain from the independence assumptions. Indeed, it can be shown [1] that for many cases in which the independence assumptions are obvious, the maximum entropy and independence based solutions coincide. In more complicated problems, where the independence assumptions become intractable, maximum entropy still leads to a solution in a straightforward manner.

Suppose we wish to determine the discrete probabilities $p_i = \Pr[X = x_i]$, given a set of known aggregate probabilities $q_j = \Pr[X \in Q_j)$, where Q_j is some subset of the possible values of the random variable X. Then the maximum entropy solution can be formulated as finding $p_1, p_2, \ldots, p_n$, such that $I(p) = \sum_{i=1}^{n} p_i \log p_i$ is minimum, and such that $\sum_{x_i \in Q_j} p_i = q_j$ $(j = 1, 2, \ldots, m)$. By using standard Lagrangian methods, one is led to the solution

$$p_i = \prod_{(j \mid x_i \in Q_j)} u_j \qquad (i = 1, 2, \ldots, n)$$

where the u_j $(j = 1, 2, \ldots, m)$ are constants satisfying

$$\sum_{(i \mid x_i \in Q_j)} \ \prod_{(k \mid x_i \in Q_k)} u_k = q_j\,.$$

In many cases explicit solutions to these equations can be found (see ref. 2 for a fairly complicated example involving computer performance analysis). Where this is impossible, numeric methods such as Newton–Raphson* can be used [3]. The maximum entropy solution exists and is unique, and the Newton–Raphson method is bound to converge to it. Note that in the above formulas, the entire system should usually be regarded as an aggregate state with known probability 1.

The procedure for obtaining a solution may be expressed in words as follows: Associate an unknown u_j with each known aggregate. Express each state probability p_i as the product of the u_j belonging to all the aggregates containing state i. Substitute these expressions in the equations defining the q_j, yielding a set of m (= number of aggregates) equations in the m unknown u_j. After solving those, substitute back into the expressions for the p_i to evaluate the latter.

We illustrate with a simple example. Jack and Jill work in an office with several coworkers. The probabilities q_1 that there is somebody in the office, q_2 that Jack is in the office, and q_3 that Jill is in the office are known. What is the probability that both Jack and Jill are in the office? In the absence of any additional information (say, about the relations between Jack and Jill), we start out by distinguishing between the following states of our system:

1. Somebody in the office, but neither Jack nor Jill.
2. Jack, but not Jill, in the office.
3. Jill, but not Jack, in the office.

4. Both Jack and Jill in the office.
5. Nobody in the office.

The probability of state 5 is known ($= 1 - q_1$), and need not be considered further. This also takes care of the entire system aggregate. The remaining probabilities satisfy

$$p_1 + p_2 + p_3 + p_4 = q_1,$$
$$p_2 + p_4 = q_2,$$
$$p_3 + p_4 = q_3.$$

Associating the unknowns u_1, u_2, and u_3 with the three aggregates, we have

$$p_1 = u_1,$$
$$p_2 = u_1 u_2,$$
$$p_3 = u_1 u_3,$$
$$p_4 = u_1 u_2 u_3.$$

The last equation, for example, follows because p_4 is present in the equations for q_1, q_2, and q_3. Substituting these expressions in the previous equations and factoring, we obtain

$$u_1(1 + u_2)(1 + u_3) = q_1,$$
$$u_1 u_2(1 + u_3) = q_2,$$
$$u_1 u_3(1 + u_2) = q_3,$$

so that

$$p_4 = u_1 u_2 u_3 = q_2 q_3 / q_1,$$

which is the required probability.

References:

[1] Bard, Y. (1980). *IBM J. Res. Develop.*, **24**, 563–569.

[2] Bard, Y. (1980). *Comm. ACM*, **23**, 564–572.

[3] Agmon, N., Alhassid, Y., and Levine, R. D. (1979). An Algorithm for Determining the Lagrange Parameters in the Maximum Entropy Formalism. In *The Maximum Entropy Formalism*, R. D. Levine and M. Tribus, eds., pp. 207–210. MIT Press, Cambridge, Mass.

(ENTROPY
INFORMATION THEORY AND CODING THEORY
MAXIMUM ENTROPY SPECTRAL ANALYSIS)

Y. BARD

MAXIMUM ENTROPY SPECTRAL ANALYSIS

In a variety of applications, it is desired to characterize a zero-mean weakly stationary time series* $\{x_n\}$ through its associated spectral density representation. This representation is particularly useful in those situations where the time series is thought either to contain hidden periodicities or to be narrowband in nature. The spectral density is formally given by

$$S_x(\omega) = \sum_{k=-\infty}^{\infty} r_x(k) e^{-j\omega k}, \qquad (1)$$

and is seen to be equal to the Fourier transform (*see* INTEGRAL TRANSFORMS) of the time series' covariance sequence

$$r_x(k) = E\{x_{n+k} x_n^*\}, \quad k = 0, \pm 1, \pm 2, \ldots, \qquad (2)$$

in which E denotes the expected value operator. The spectral density and the covariance sequence thereby constitute a Fourier transform pair.

Clearly, the determination of $S_x(\omega)$ requires a complete knowledge of the covariance sequence. In most practical applications, however, one has available only the first few terms of the covariance sequence or, more typically, only a finite length observation of the time series upon which to estimate the spectral density. A number of different procedures have been proposed for obtaining spectral density estimates. Undoubtedly, the procedure generally referred to as the *maximum entropy method* (MEM) has received the most attention. An excellent treatment of this and other spectral density estimation methods (and a rather thorough bibliography) is to be found in Haykin [4] and Childers [3].

MAXIMUM ENTROPY METHOD: KNOWN COVARIANCE ELEMENTS

In this case, it will be assumed that the time series is a bandlimited Gaussian process* whose first $q + 1$ covariance elements

$$r_x(0), r_x(1), \ldots, r_x(q) \qquad (3)$$

are given. The premise of the MEM, as originally suggested by Burg [2], was to extend the covariance elements (i.e., assign values to $r_x(n)$ for $n > q$) so that the resultant covariance characterization would correspond to the most "random" time series which would be consistent with the a priori information (3). Using concepts drawn from information theory*, this randomness* is measured by the time series' entropy*, which for a Gaussian bandlimited process is proportional to

$$\int_{-W}^{W} \ln[\hat{S}_x(\omega)]\, d\omega, \qquad (4)$$

where $[-W, W]$ designates the bandlimit interval.

It is then desired to find a nonnegative-valued spectral density estimate $\hat{S}_x(\omega)$ which will maximize this entropy measure of randomness over the set of nonnegative functions whose inverse Fourier transforms satisfy constraints (3). The solution to this constrained optimization problem is given by

$$\hat{S}_x(\omega) = \frac{P_q}{2W|1 + \sum_{k=1}^{q} a_k e^{-jk\omega}|^2}, \qquad (5a)$$

where the coefficients $\{a_k\}$ satisfy the system of linear equations

$$\sum_{k=1}^{q} r_x(n-k)a_k = -r_x(n), \quad n = 1, 2, \ldots, q, \qquad (5b)$$

while the nonnegative scalar P_q is specified by

$$P_q = \sum_{k=0}^{q} r_x^*(k)a_k. \qquad (5c)$$

Interestingly, the set of coefficients $\{a_k\}$ satisfying (5b) also constitutes the set of coefficients for the optimum one-step predictor, as well as the optimal autoregressive spectral estimate [4].

An important computational consideration in implementing the MEM relates to the method used for generating the required modeling coefficients $\{a_k\}$. The Levinson algorithm provides an efficient method for recursively performing this task [6]. If $a_k^{(m)}$ for $k = 1, 2, \ldots, m$ denotes the set of coefficients which satisfy the optimal mth order relationship (5b), the Levinson algorithm enables one to recursively obtain the $(m+1)$st order MEM coefficients according to

$$a_k^{(m+1)} = \begin{cases} a_k^{(m)} + \rho_{m+1} a_{m+1-k}^{(m)*}, & k = 0, 1, \ldots, m, \\ \rho_{m+1}, & k = m+1, \end{cases} \qquad (6a)$$

where $a_0^{(m)} \equiv 1$ and ρ_{m+1} is the reflection coefficient specified by

$$\rho_{m+1} = -\frac{1}{P_m} \sum_{k=0}^{m} r_x(m+1-k)a_k^{(m)}, \quad m = 0, 1, \ldots, q-1, \qquad (6b)$$

while the prediction error variance is given by

$$P_{m+1} = [1 - |\rho_{m+1}|^2] P_m, \quad m = 0, 1, \ldots, q-1, \qquad (6c)$$

in which the initial prediction error variance is $P_0 = r_x(0)$. Using relationships (6), it is then possible to recursively generate the optimal qth order MEM spectral estimate from the given covariance values (3) commencing at $m = 0$.

MAXIMUM ENTROPY METHOD: FINITE LENGTH OBSERVATIONS

In most applications, one does not have available the underlying covariance knowledge to implement the above MEM estimate. Typically, the only information concerning the time series is available in the form of a finite set of observations as given by

$$x_1, x_2, \ldots, x_N \qquad (7)$$

Using this finite data set, Burg developed a procedure for generating the required spectral estimate which utilized the Levinson recursion formulas (6). This first entailed generating the forward and backward prediction error elements

$$f_n^{(m)} = x_n + \sum_{k=1}^{m} a_k^{(m)} x_{n-k},$$
$$m = 0, 1, \ldots, q-1, \quad \text{(8a)}$$

$$b_n^{(m)} = x_n + \sum_{k=1}^{m} a_k^{(m)*} x_{n+k}$$
$$m = 0, 1, \ldots, q-1, \quad \text{(8b)}$$

respectively, where $f_n^{(0)} = b_n^{(0)} = x_n$. The $(m+1)$st reflection coefficient is next calculated according to

$$\rho_{m+1} = \frac{-2\sum_{k=1}^{N-m-1} f_{k+m+1}^{(m)} b_k^{(m)*}}{\sum_{k=1}^{N-m-1} \left\{ |f_{k+m+1}^{(m)}|^2 + |b_k^{(m)}|^2 \right\}}$$
$$m = 0, 1, \ldots, q-1. \quad \text{(8c)}$$

One then updates the order of the spectral density estimate commencing at $m = 0$, where

$$a_1^{(1)} = \rho_1 \quad \text{and} \quad P_0 = \frac{1}{N} \sum_{k=1}^{N} |x_k|^2. \quad \text{(8d)}$$

Finally, relationships (8), (6a), and (6c) are recursively used to generate the higher order spectral estimates (5a) for $m = 1, 2, \ldots, q-1$. Burg has shown that this procedure always results in minimum phase prediction error filters.

References

[1] Akaike, H. (1974). *IEEE Trans. Autum. Control*, **AC-19**, 716–723. (Methods for determining the order of the linear predictor operator used in the MEM are given.)

[2] Burg, J. P. (1967). Maximum Entropy Spectral Analysis. Paper presented at the 37th Ann. Intern. Meeting Soc. of Explor. Geophys., Oklahoma City, Oklahoma. (The first published treatment of the maximum entropy method by its originator.)

[3] Childers, D. G., ed. (1978). *Modern Spectral Analysis*. IEEE Press, New York. (A collection of original key papers which give a historical portrait of spectral analysis' evolvement.)

[4] Haykin, S., ed. (1979). *Nonlinear Methods of Spectral Analysis*. Springer-Verlag, Berlin. (An excellent treatment of maximum entropy and other methods of spectral estimation written by recognized researchers in this field.)

[5] Levine, R. D. and M. Tribus, eds. (1979). *The Maximum Entropy Formalism*. MIT Press, Cambridge, Mass. (Specifically, see E. T. Jaynes, "Where do we stand on maximum entropy?", pp. 15–118, for nonspectral estimate applications of the maximum entropy principle.)

[6] Levinson, N. (1947). *J. Math. Phys*., **25**, 261–278.

[7] Parzen, E. (1969). Multiple Time Series Modeling. In *Multivariate Analysis-II*, P. R. Krishnaiah, ed., Academic Press, New York. (A least-squares autoregressive modeling procedure equivalent to the MEM is presented.)

[8] Van Den Bos, A. (1971). *IEEE Trans. Inform. Theory*, **IT-17**, 493–494. (Equivalency of the maximum entropy, least-squares error linear prediction, and autoregression is established.)

(ENTROPY
SPECTRAL ANALYSIS
TIME SERIES)

JAMES A. CADZOW

MAXIMUM LIKELIHOOD ESTIMATION

Maximum likelihood is by far the most popular general method of estimation. Its widespread acceptance is seen on the one hand in the very large body of research dealing with its theoretical properties, and on the other in the almost unlimited list of applications.

To give a reasonably general definition of maximum likelihood estimates, let $\mathbf{X} = (X_1, \ldots, X_n)$ be a random vector of observations whose joint distribution is described by a density $f_n(\mathbf{x} \mid \mathbf{\Theta})$ over the n-dimensional Euclidean space R^n. The unknown parameter vector $\mathbf{\Theta}$ is contained in the parameter space $\Omega \subset R^s$. For fixed $\mathbf{x}$ define the likelihood* function of $\mathbf{x}$ as $L(\mathbf{\Theta}) = L_{\mathbf{x}}(\mathbf{\Theta}) = f_n(\mathbf{x} \mid \mathbf{\Theta})$ considered as a function of $\mathbf{\Theta} \in \Omega$.

Definition 1. Any $\hat{\mathbf{\Theta}} = \hat{\mathbf{\Theta}}(\mathbf{x}) \in \Omega$ which maximizes $L(\mathbf{\Theta})$ over Ω is called a *maximum likelihood estimate* (MLE) of the unknown true parameter $\mathbf{\Theta}$.

Often it is computationally advantageous to derive MLEs by maximizing $\log L(\mathbf{\Theta})$ in place of $L(\mathbf{\Theta})$.

Example 1. Let X be the number of successes in n independent Bernoulli trials with success probability $p \in [0, 1]$; then

$$L_x(p) = f(x \mid p) = P(X = x \mid p)$$
$$= \binom{n}{x} p^x (1-p)^{n-x} \quad x = 0, 1, \ldots, n.$$

Solving

$$\frac{\partial}{\partial p} \log L_x(p) = x/p - (n-x)/(1-p) = 0$$

for p, one finds that $\log L_x(p)$ and hence $L_x(p)$ has a maximum at

$$\hat{p} = \hat{p}(x) = x/n.$$

This example illustrates the considerable intuitive appeal of the MLE as that value of p for which the probability of the observed value x is the largest.

It should be pointed out that MLEs do not always exist, as illustrated in the following natural mixture example; see Kiefer and Wolfowitz [32].

Example 2. Let $X_1, \ldots, X_n$ be independent and identically distributed (i.i.d.) with density

$$f(x \mid \mu, \nu, \sigma, \tau, p)$$
$$= \frac{p}{\sqrt{2\pi}\,\sigma} \exp\left[-\frac{1}{2}\left(\frac{x-\mu}{\sigma}\right)^2\right] + \frac{1-p}{\sqrt{2\pi}\,\tau} \exp\left[-\frac{1}{2}\left(\frac{x-\nu}{\tau}\right)^2\right],$$

where $0 \leqslant p \leqslant 1$, $\mu, \nu \in R$, and $\sigma, \tau > 0$.

The likelihood function of the observed sample $x_1, \ldots x_n$, although finite for any permissible choice of the five parameters, approaches infinity as, for example, $\mu = x_1$, $p > 0$ and $\sigma \to 0$. Thus the MLEs of the five unknown parameters do not exist.

Further, if an MLE exists, it is not necessarily unique as is illustrated in the following example.

Example 3. Let $X_1, \ldots, X_n$ be i.i.d. with density $f(x \mid \alpha) = \frac{1}{2} \exp(-|x - \alpha|)$. Maximizing $f_n(x_1, \ldots, x_n \mid \alpha)$ is equivalent to minimizing $\sum |x_i - \alpha|$ over α. For $n = 2m$ one finds that any $\hat{\alpha} \in [x_{(m)}, x_{(m+1)}]$ serves as MLE of α, where $x_{(i)}$ is the ith order statistic of the sample.

The method of maximum likelihood estimation is generally credited to Fisher* [17–20], although its roots date back as far as Lambert*, Daniel Bernoulli*, and Lagrange in the eighteenth century; see Edwards [12] for an historical account. Fisher introduces the method in [17] as an alternative to the method of moments* and the method of least squares*. The former method Fisher criticizes for its arbitrariness in the choice of moment equations and the latter for not being invariant under scale changes in the variables. The term *likelihood** as distinguished from (*inverse*) *probability* appears for the first time in [18]. Introducing the measure of information named after him (*see* FISHER INFORMATION) Fisher [18–20] offers several proofs for the efficiency of MLEs, namely that the asymptotic variance of asymptotically normal estimates cannot fall below the reciprocal of the information contained in the sample and, furthermore, that the MLE achieves this lower bound. Fisher's proofs, obscured by the fact that assumptions are not always clearly stated, cannot be considered completely rigorous by today's standards and should be understood in the context of his time. To some extent his work on maximum likelihood estimation was anticipated by Edgeworth [11], whose contributions are discussed by Savage [51] and Pratt [45]. However, it was Fisher's insight and advocacy that led to the prominence of maximum likelihood estimation as we know it today.

For a discussion and an extension of Definition 1 to richer (nonparametric) statistical models which preclude a model description through densities (i.e., likelihoods will be missing), see Scholz [52]. At times the primary concern is the estimation of some function g of $\boldsymbol{\Theta}$. It is then customary to treat $g(\hat{\boldsymbol{\Theta}})$ as an "MLE" of $g(\boldsymbol{\Theta})$, although strictly speaking, Definition 1 only justifies this when g is a one-to-one function. For arguments toward a general justification of

$g(\hat{\mathbf{\Theta}})$ as MLE of $g(\mathbf{\Theta})$, see Zehna [58] and Berk [7].

CONSISTENCY

Much of maximum likelihood theory deals with the large sample (asymptotic) properties of MLEs; i.e., with the case in which it is assumed that $X_1, \ldots, X_n$ are independent and identically distributed with density $f(\cdot \mid \mathbf{\Theta})$ (i.e., $X_1, \ldots, X_n$ i.i.d. $\sim f(\cdot \mid \mathbf{\Theta})$). The joint density of $\mathbf{X} = (X_1, \ldots, X_n)$ is then $f_n(\mathbf{x} \mid \mathbf{\Theta}) = \prod_{i=1}^n f(x_i \mid \mathbf{\Theta})$. It further is assumed that the distributions P_θ corresponding to $f(\cdot \mid \mathbf{\Theta})$ are identifiable, i.e., $\mathbf{\Theta} \neq \mathbf{\Theta}'$, and $\mathbf{\Theta}, \mathbf{\Theta}' \in \Omega$ implies $P_{\mathbf{\Theta}} \neq P_{\mathbf{\Theta}'}$. For future reference we state the following assumptions:

A0: $X_1, \ldots, X_n$ i.i.d. $f(\cdot \mid \mathbf{\Theta})$ $\mathbf{\Theta} \in \Omega$;
A1: the distributions $P_{\mathbf{\Theta}}$, $\mathbf{\Theta} \in \Omega$, are identifiable.

The following simple result further supports the intuitive appeal of the MLE; see Bahadur [3]:

Theorem 1. Under **A0** and **A1**

$$P_{\mathbf{\Theta}'}[f_n(\mathbf{X} \mid \mathbf{\Theta}') > f_n(\mathbf{X} \mid \mathbf{\Theta})] \to 1$$

as $n \to \infty$ for any $\mathbf{\Theta}, \mathbf{\Theta}' \in \Omega$ with $\mathbf{\Theta} \neq \mathbf{\Theta}'$. If, in addition, Ω is finite, then the MLE $\hat{\mathbf{\Theta}}_n$ exists and is consistent.

The content of Theorem 1 is a cornerstone in Wald's [56] consistency proof of the MLE for the general case. Wald assumes that Ω is compact, which by a familiar compactness argument reduces the problem to the case in which Ω contains only finitely many elements. Aside from the compactness assumption on Ω, which often is not satisfied in practice, Wald's uniform integrability conditions (imposed on $\log f(\cdot \mid \mathbf{\Theta})$) often are not satisfied in typical examples.

Many improvements in Wald's approach toward MLE consistency were made by later researchers. For a discussion and further references, see Perlman [42]. Instead of Wald's theorem or any of its refinements, we present another theorem, due to Rao [47], which shows under what simple conditions MLE consistency may be established in a certain specific situation.

Theorem 2. Let **A0** and **A1** be satisfied and let $f(\cdot \mid \mathbf{\Theta})$ describe a multinomial experiment with cell probabilities $\boldsymbol{\pi}(\mathbf{\Theta}) = (\pi_1(\mathbf{\Theta}), \ldots, \pi_k(\mathbf{\Theta}))$. If the map $\mathbf{\Theta} \to \boldsymbol{\pi}(\mathbf{\Theta})$, $\mathbf{\Theta} \in \Omega$, has a continuous inverse (the inverse existing because of A1), then the MLE $\hat{\mathbf{\Theta}}_n$, if it exists, is a consistent estimator of $\mathbf{\Theta}$.

For a counterexample to Theorem 2 when the inverse continuity assumption is not satisfied see Kraft and LeCam [33].

A completely different approach toward proving consistency of MLEs was given by Cramér [9]. His proof is based on a Taylor expansion of $\log L(\mathbf{\Theta})$ and thus, in contrast to Wald's proof, assumes a certain amount of smoothness in $f(\cdot \mid \mathbf{\Theta})$ as a function of $\mathbf{\Theta}$. Cramér gave the consistency proof only for $\Omega \subset R$. Presented here are his conditions generalized to the multiparameter case, $\Omega \subset R^s$:

C1: The distributions $P_{\mathbf{\Theta}}$ have common support for all $\mathbf{\Theta} \in \Omega$; i.e., $\{x : f(x \mid \mathbf{\Theta}) > 0\}$ does not change with $\mathbf{\Theta} \in \Omega$.

C2: There exists an open subset ω of Ω containing the true parameter point $\mathbf{\Theta}_0$ such that for almost all x the density $f(x \mid \mathbf{\Theta})$ admits all third derivatives

$$\frac{\partial^3}{\partial\Theta_i \partial\Theta_j \partial\Theta_k} f(x \mid \mathbf{\Theta}) \qquad \text{for all} \quad \mathbf{\Theta} \in \omega.$$

C3:

$$E_{\mathbf{\Theta}}\left[\frac{\partial}{\partial\Theta_j} \log f(X \mid \mathbf{\Theta})\right] = 0,$$

$j = 1, \ldots, s$ and

$$I_{jk}(\mathbf{\Theta}) := E_{\mathbf{\Theta}}\left[\left(\frac{\partial}{\partial\mathbf{\Theta}_j} \log f(X \mid \mathbf{\Theta})\right) \times \left(\frac{\partial}{\partial\mathbf{\Theta}_k} \log f(X \mid \mathbf{\Theta})\right)\right]$$

$$= -E_{\mathbf{\Theta}}\left[\frac{\partial^2}{\partial\Theta_j \partial\Theta_k} \log f(X \mid \mathbf{\Theta})\right]$$

exist and are finite for $j, k = 1, \ldots, s$ and all $\Theta \in \omega$.

C4: The Fisher information matrix $\mathbf{I}(\Theta) = (I_{jk}(\Theta))_{j,k=1,\ldots,s}$ is positive definite for all $\Theta \in \omega$.

C5: There exist functions $M_{ijk}(x)$ independent of Θ such that for all $i, j, k, = 1, \ldots, s$,

$$\left| \frac{\partial^3}{\partial\theta_i \partial\theta_j \partial\theta_k} \log f(x \mid \Theta) \right| \leqslant M_{ijk}(x) \quad \text{for all} \quad \Theta \in \omega,$$

where

$$E_{\Theta_0}(M_{ijk}(X)) < \infty.$$

We can now state Cramér's consistency theorem.

Theorem 3. Assume **A0**, **A1**, and **C1**–**C5**. Then with probability tending to one as $n \to \infty$ there exist solutions $\tilde{\Theta}_n = \tilde{\Theta}_n(X_1, \ldots, X_n)$ of the likelihood equations

$$\frac{\partial}{\partial\Theta_j} \log f_n(\mathbf{X} \mid \Theta) = \frac{\partial}{\partial\Theta_j} \sum_{i=1}^{n} \log f(X_i \mid \Theta) = 0, \quad j = 1, \ldots, s,$$

such that $\tilde{\Theta}_n$ converges to Θ_0 in probability; i.e., $\tilde{\Theta}_n$ is consistent.

For a proof see Lehmann [37, Sect. 6.4]. The theorem needs several comments for clarification:

(a) If the likelihood function $L(\Theta)$ attains its maximum at an interior point of Ω then the MLE is a solution to the likelihood equation. If in addition the likelihood equations only have one root, then Theorem 3 proves the consistency of the MLE($\hat{\Theta}_n = \tilde{\Theta}_n$).

(b) Theorem 3 does not state how to identify the consistent root among possibly many roots of the likelihood equations. One could take the root $\tilde{\Theta}_n$ which is closest to Θ_0, but then $\tilde{\Theta}_n$ is no longer an estimator since its construction assumes knowledge of the unknown value of Θ_0. This problem may be overcome by taking that root which is closest to a (known) consistent estimator of Θ_0. The utility of this approach becomes clear in the section on efficiency.

(c) The MLE does not necessarily coincide with the consistent root guaranteed by Theorem 3. Kraft and LeCam [33] give an example in which Cramér's conditions are satisfied, the MLE exists, is unique, and satifies the likelihood equations, yet is not consistent.

In view of these comments, it is advantageous to establish the uniqueness of the likelihood equation roots whenever possible. For example, if $f(x \mid \Theta)$ is of nondegenerate multiparameter exponential family type, then $\log L(\Theta)$ is strictly concave. Thus the likelihood equations have at most one solution. Sufficient conditions for the existence of such a solution may be found in Barndorff-Nielsen [4]. In a more general context Mäkeläinen et al. [38] give sufficient conditions for the existence and uniqueness of roots of the likelihood equations.

EFFICIENCY

The main reason for presenting Cramér's consistency theorem and not Wald's is the following theorem which specifically addresses consistent roots of the likelihood equations and not necessarily MLEs.

Theorem 4. Assume **A0**, **A1**, and **C1** – **C5**. If $\tilde{\Theta}_n$ is a consistent sequence of roots of the likelihood equations, then as $n \to \infty$,

$$\sqrt{n}\,(\tilde{\Theta}_n - \Theta_0) \xrightarrow{L} N_s(\mathbf{0}, \mathbf{I}(\Theta_0)^{-1});$$

i.e., in large samples the distribution of $\tilde{\Theta}_n$ is approximately s-variate normal with mean Θ_0 and covariance matrix $\mathbf{I}(\Theta_0)^{-1}/n$. This theorem is due to Cramér [9], who gave a proof for $s = 1$. A proof for $s \geqslant 1$ may be found in Lehmann [37].

Because of the form, $\mathbf{I}(\Theta)^{-1}$, of the asymptotic covariance matrix for $\sqrt{n}\,(\tilde{\Theta}_n - \Theta)$, one generally regards $\tilde{\Theta}_n$ as an efficient estimator for $\mathbf{\Theta}$. The reasons for this are now discussed. Under regularity conditions (weaker than those of Theorem 4) the Cramér–Rao* lower bound (CRLB) states that

$$\operatorname{var}(T_{jn}) \geqslant \left(\mathbf{I}(\mathbf{\Theta})^{-1}\right)_{jj}/n$$

for any unbiased estimator T_{jn} of Θ_j which is based on n observations. Here $(\mathbf{I}(\mathbf{\Theta})^{-1})_{jj}$ refers to the jth diagonal element of $\mathbf{I}(\mathbf{\Theta})^{-1}$. Note, however, that the CRLB refers to the actual variance of an (unbiased) estimator and not to the asymptotic variance of such estimator. The relationship between these two variance concepts is clarified by the following inequality. If as $n \to \infty$

$$\sqrt{n}\,(T_{jn} - \Theta_j) \xrightarrow{L} N(0, v_j(\mathbf{\Theta})), \qquad (1)$$

then

$$\lim_{n\to\infty} \left[n \operatorname{var}(T_{jn}) \right] \geqslant v_j(\mathbf{\Theta}),$$

where equality need not hold; see Lehmann [37].

Thus for unbiased estimators T_{jn} which are asymptotically normal, i.e., satisfy (1), and for which

$$\lim_{n\to\infty} \left[n \operatorname{var}(T_{jn}) \right] = v_j(\mathbf{\Theta}),$$

the CRLB implies that $v_j(\mathbf{\Theta}) \geqslant (\mathbf{I}(\mathbf{\Theta})^{-1})_{jj}$. It was therefore thought that $(\mathbf{I}(\mathbf{\Theta})^{-1})_{jj}$ is a lower bound for the asymptotic variance of any asymptotically normal unbiased estimate of Θ_j. Since the estimators $\tilde{\Theta}_{jn}$ of Theorem 4 have asymptotic variances equal to this lower bound, they were called *efficient estimators*. In fact, Lehmann [36] refers to $\tilde{\mathbf{\Theta}}_n$ as an *efficient likelihood estimator* (ELE) in contrast to the MLE, although the two often will coincide. For a discussion of the usage of ELE versus MLE refer to his paper. As it turns out, $(\mathbf{I}(\mathbf{\Theta})^{-1})_{jj}$ will not serve as a true lower bound on the asymptotic variance of asymptotically normal estimators of Θ_j unless one places some restrictions on the behavior of such estimators. Without such restrictions Hodges (see LeCam [35]) was able to construct so called *superefficient estimators* (*see* HODGES SUPEREFFICIENCY for a simple example). It was shown by LeCam [35] (see also Bahadur [2]) that the set of superefficiency points must have Lebesgue measure zero for any particular sequence of estimators. LeCam (see also Hájek [28]) further showed that falling below the lower bound at a value $\mathbf{\Theta}_0$ entails certain unpleasant properties for the mean squared error* (or other risk functions) of such superefficient estimators in the vicinity of $\mathbf{\Theta}_0$. Thus it appears not advisable to use superefficient estimators.

Unpleasant as such superefficient estimators are, their existence led to a reassessment of large sample properties of estimators. In particular, a case was made to require a certain amount of regularity not only in the distributions but also in the estimators. For example, the simple requirement that the asymptotic variance $v_j(\mathbf{\Theta})$ in (1) be a continuous function in $\mathbf{\Theta}$ would preclude such estimator from being superefficient since, as remarked above, such phenomenon may occur only on sets of Lebesgue measure zero. For estimators satisfying (1) with continuous $v_j(\mathbf{\Theta})$ one thus has $v_j(\mathbf{\Theta}) \geqslant (\mathbf{I}(\mathbf{\Theta})^{-1})_{jj}$. Rao [49] requires the weak convergence in (1) to be uniform on compact subsets of Ω, which under mild assumption on $f(\cdot\,|\,\mathbf{\Theta})$ implies the continuity of the asymptotic variance $v_j(\mathbf{\Theta})$.

Hájek [27] proved a very general theorem which gives a succinct description of the asymptotic distribution of regular estimators. His theorem will be described in a somewhat less general form below. Let $\mathbf{\Theta}(n) = \mathbf{\Theta}_0 + \mathbf{h}/\sqrt{n}$, $\mathbf{h} \in R^s$ and denote by $\xrightarrow{L_{\mathbf{\Theta}(n)}}$ convergence in law when $\mathbf{\Theta}(n)$ is the true parameter.

Definition 2. An estimator sequence $\{\mathbf{T}_n\}$ is called *regular* if for all $\mathbf{h} \in R^s$

$$\sqrt{n}\,(\mathbf{T}_n - \mathbf{\Theta}(n)) \xrightarrow{L_{\mathbf{\Theta}(n)}} \mathbf{T}$$

as $n \to \infty$, where the distribution of the random vector $\mathbf{T}$ is independent of $\mathbf{h}$.

The regularity conditions on $f(\cdot\,|\,\mathbf{\Theta})$ are formulated as follows.

Definition 3. $f(\cdot \mid \Theta)$ is called *locally asymptotically normal* (LAN) if for all $\mathbf{h} \in R^s$

$$\log[f_n(\mathbf{X} \mid \Theta(n))/f_n(\mathbf{X} \mid \Theta_0)] = \mathbf{h}'\Delta_n(\Theta_0) - \tfrac{1}{2}\mathbf{h}'\mathbf{I}(\Theta_0)\mathbf{h} + Z_n(\mathbf{h}, \Theta_0)$$

with

$$\Delta_n(\Theta_0) \xrightarrow{L_{\Theta_0}} N_s(\mathbf{0}, \mathbf{I}(\Theta_0))$$

and

$$Z_n(\mathbf{h}, \Theta_0) \xrightarrow{P_{\Theta_0}} 0 \quad \text{as} \quad n \to \infty.$$

Comment. Under the conditions **A0**, **A1**, and **C1–C5**, one may show that $f(\cdot \mid \Theta)$ is LAN and in that case

$$\Delta_n'(\Theta_0) = \left(\frac{\partial}{\partial \Theta_1} \log f_n(\mathbf{X} \mid \Theta), \ldots, \frac{\partial}{\partial \Theta_s} \log f_n(\mathbf{X} \mid \Theta)\right) / \sqrt{n}.$$

Theorem 5 (Hájek). If $\{\mathbf{T}_n\}$ is regular and $f(\cdot \mid \Theta)$ is LAN then $\mathbf{T} = \mathbf{Y} + \mathbf{W}$, where $\mathbf{Y} \sim N_s(\mathbf{0}, \mathbf{I}(\Theta_0)^{-1})$ and $\mathbf{W}$ is a random vector independent of $\mathbf{Y}$. The distribution of $\mathbf{W}$ is determined by the estimator sequence.

Comment. Estimators for which $P_{\Theta_0}(\mathbf{W} = \mathbf{0}) = 1$ are most concentrated around Θ_0 (see Hájek [27]) and may thus be considered efficient among regular estimators. Note that this optimality claim is possible because competing estimators are required to be regular; on the other hand, it is no longer required that the asymptotic distribution of the estimator be normal.

As remarked in the comments to Theorem 3 the ELE $\tilde{\Theta}_n$ of Theorem 4 may be chosen by taking that root of the likelihood equations which is closest to some known consistent estimate of Θ. The latter estimate need not be efficient. In this context we note that the consistent sequence of roots generated by Theorem 3 is essentially unique. For a more precise statement of this result see Huzurbazar [31] and Perlman [43]. Roots of the likelihood equations may be found by one of various iterative procedures offered in several statistical computer packages; *see* ITERATED MAXIMUM LIKELIHOOD ESTIMATES. Alternatively on may just take a one-step iteration estimator in place of a root. Such one step estimators use as starting point $\sqrt{n}$-consistent estimators (not necessarily efficient) and are efficient. A precise statement is given in Theorem 6. First note the following definition.

Definition 4. An estimator $\mathbf{T}_n$ is $\sqrt{n}$*-consistent* for estimating the true Θ_0 if for every $\epsilon > 0$ there is a K_ϵ and an N_ϵ such that

$$P_{\Theta_0}\left(\|\sqrt{n}(\mathbf{T}_n - \Theta_0)\| \leqslant K_\epsilon\right) \geqslant 1 - \epsilon$$

for all $n \geqslant N_\epsilon$, where $\|\cdot\|$ denotes the Euclidean norm in R^s.

Theorem 6. Suppose the assumptions of Theorem 4 hold and that Θ_n^* is a $\sqrt{n}$-consistent estimator of Θ. Let $\boldsymbol{\delta}_n' = (\delta_{1n}, \ldots, \delta_{sn})$ be the solution of the linear equations

$$\sum_{k=1}^{s} (\delta_{kn} - \Theta_{kn}^*) R_{jk}''(\Theta_n^*) = -R_j'(\Theta_n^*),$$

$j = 1, \ldots, s$, where

$$R_j'(\Theta) = \frac{\partial}{\partial \Theta_j} \log L(\Theta),$$

and

$$R_{jk}''(\Theta) = \frac{\partial^2}{\partial \Theta_j \partial \Theta_k} \log L(\Theta).$$

Then

$$\sqrt{n}(\boldsymbol{\delta}_n - \Theta_0) \xrightarrow{L} N_s\left(\mathbf{0}, \mathbf{I}(\Theta_0)^{-1}\right)$$

as $n \to \infty$, i.e., $\boldsymbol{\delta}_n$ is asymptotically efficient.

For a proof of Theorem 6 see Lehmann [37].

The application of Theorem 6 is particularly useful in that it does not require the solution of the likelihood equations. The two estimators $\tilde{\Theta}_n$ (Theorem 4) and $\boldsymbol{\delta}_n$ (Theorem 6) are asymptotically efficient. Among other estimators that share this property are the

Bayes estimators. Because of this multitude of asymptotically efficient estimators one has tried to discriminate between them by considering higher order terms in the asymptotic analysis. Several measures of second-order efficiency* have been examined (*see* EFFICIENCY, SECOND ORDER) and it appears that with some qualifications MLEs are second-order efficient, provided the MLE is first corrected for bias of order $1/n$. Without such bias correction one seems to be faced with a problem similar to the nonexistence of estimators with uniformly smallest mean squared error in the finite sample case (*see* ESTIMATION, POINT). For investigations along these lines see Rao [48], Ghosh and Subramanyam [23], Efron [13], Pfanzagl and Wefelmeyer [44] and Ghosh and Sinha [22]. For a lively discussion of the issues involved see also Berkson [8].

Other types of optimality results for MLEs, such as the local asymptotic admissibility* and minimax* property, were developed by LeCam and Hájek. For an exposition of this work and some perspective on the work of others, see Hájek [28]. For a simplified introduction to these results, see also Lehmann [37].

The following example (see Lehmann [37]) illustrates a different kind of behavior of the MLE when the regularity condition **C1** is not satisfied.

Example 4. Let $X_1, \ldots, X_n$ be i.i.d. $\sim U(0, \Theta)$ (uniform on $(0, \Theta)$). Then the MLE is $\hat{\Theta}_n = \max(X_1, \ldots, X_n)$ and its large sample behavior is described by

$$n(\Theta - \hat{\Theta}_n) \xrightarrow{L} \Theta \cdot E$$

as $n \to \infty$, where E is an exponential random variable with mean 1. Note that the normalizing factor is n and not $\sqrt{n}$. Also the asymptotic distribution ΘE is not normal and is not centered at zero. Considering instead $\delta_n = (n+1)\hat{\Theta}_n/n$ one finds as $n \to \infty$

$$n(\Theta - \delta_n) \xrightarrow{L} \Theta(E - 1).$$

Further

$$E\left[n(\hat{\Theta}_n - \Theta)\right]^2 \to 2\Theta^2$$

and

$$E\left[n(\delta_n - \Theta)\right]^2 \to \Theta^2.$$

Hence the MLE, although consistent, may no longer be asymptotically optimal.

For a different approach which covers regular as well as nonregular problems, see Weiss and Wolfowitz [57] on maximum probability estimators*.

MLEs IN MORE GENERAL MODELS

The results concerning MLEs or ELEs discussed so far all assumed **A0**. In many statistical applications the sampling structure gives rise to independent observations which are not identically distributed. For example one may be sampling several different populations or with each observation one or more covariates may be recorded. For the former situation Theorems 3 and 4 are easily extended, see Lehmann [37]. In the case of independent but not identically distributed observations, results along the lines of Theorem 3 and 4 were given by Hoadley [29] and Nordberg [40]. Nordberg deals specifically with exponential family* models presenting the binary logit model and the log-linear Poisson model as examples. See also Haberman [26], who treats maximum likelihood theory for diverse parametric structures in multinomial and Poisson counting data models.

The maximum likelihood theory for incomplete data* from an exponential family is treated in Sundberg [55]. Incomplete data models include situations with grouped*, censored*, or missing data and finite mixtures. See Sundberg [55] and Dempster et al. [10] for more examples. The latter authors present the EM algorithm for iterative computation of the MLE from incomplete data.

Relaxing the independence assumption of the observations opens the way to stochastic process applications. Statistical inference problems concerning stochastic processes* only recently have been treated with vigor. The maximum likelihood approach to parameter estimation here plays a prominent role. Consistency and asymptotic normality* of MLEs (or ELEs) may again be established by using appropriate martingale* limit theorems. Care has to be taken to define the concept of likelihood function when the observations consist of a continuous time stochastic process. As a start into the growing literature on this subject see Feigin [16], Basawa and Prakasa Rao [5, 6].

Another assumption made so far is that the parameter space Ω has dimension s, where s remains fixed as the sample size n increases. For the problems one encounters as s grows with n or when Ω is so rich that it cannot be embedded into any finite dimensional Euclidean space, see Kiefer and Wolfowitz [32] and Grenander [25] respectively. Huber [30] examines the behavior of MLEs derived from a particular parametric model when the sampled distribution is different from this parametric model. These results are useful in studying the robustness properties of MLEs (*see* ROBUST ESTIMATION).

MISCELLANEOUS REMARKS

Not much can be said about the small sample properties of MLEs. If the MLE exists it is generally a function of the minimal sufficient statistic*, namely the likelihood function $L(\cdot)$. However, the MLE by itself is not necessarily sufficient. Thus in small samples some information may be lost by considering the MLE by itself. Fisher [19] proposed the use of ancillary statistics* to recover this information loss, by viewing the distribution of the MLE conditional on the ancillary statistics. For a recent debate on this subject, see Efron and Hinkley [15], who make an argument for using the observed and not the expected Fisher information* in assessing the accuracy of MLEs. Note, however, the comments to their paper. See also Sprott [53, 54], who suggests using parameter transformations to achieve better results in small samples when appealing to large sample maximum likelihood theory. Efron [14], in discussing the relationship between maximum likelihood and decision theory*, highlights the role of maximum likelihood as a summary principle in contrast to its role as an estimation principle.

Although MLEs are asymptotically unbiased in the regular case, this will not generally be the case in finite samples. It is not necessarily clear whether the removal of bias from an MLE will result in a better estimator, as the following example shows (*see also* UNBIASEDNESS).

Example 5. Let $X_1, \dots, X_n$ be i.i.d. $N(\mu, \sigma^2)$; then the MLE of σ^2 is

$$\hat{\sigma}^2 = \sum_{i=1}^{n} \left(X_i - \overline{X}\right)^2 / n,$$

which has mean value $\sigma^2(n-1)/n$, i.e., $\hat{\sigma}^2$ is biased. Taking instead $\tilde{\sigma}^2 = \hat{\sigma}^2 n/(n-1)$ we have an unbiased estimator of σ^2 which has uniformly smallest variance among all unbiased estimators; however,

$$E(\hat{\sigma}^2 - \sigma^2)^2 < E(\tilde{\sigma}^2 - \sigma^2)^2$$

for all σ^2.

The question of bias removal should therefore be decided on a case by case basis with consideration of the estimation objective. See, however, the argument for bias correction of order $1/n$ in the context of second-order efficiency of MLEs as given in Rao [48] and Ghosh and Subramanyam [23].

Gong and Samaniego [24] consider the concepts of pseudo maximum likelihood estimation which consists of replacing all nuisance parameters* in a multiparameter model by suitable estimates and then solving a reduced system of likelihood equations for the remaining structural parameters. They present consistency and asymptotic normality results and illustrate them by example.

AN EXAMPLE

As an illustration of some of the issues and problems encountered, consider the Weibull* model. Aside from offering much flexibility in its shape, there are theoretical extreme value type arguments (Galambos [21]), recommending the Weibull distribution as an appropriate model for the breaking strength of certain materials.

Let $X_1, \ldots, X_n$ be i.i.d. $W(t, \alpha, \beta)$, the Weibull distribution with density

$$f(x \mid t, \alpha, \beta) = \frac{\beta}{\alpha}\left(\frac{x-t}{\alpha}\right)^{\beta-1} \times \exp\left(-\left(\frac{x-t}{\alpha}\right)^{\beta}\right),$$

$$x > t, \quad t \in R, \quad \alpha, \beta > 0.$$

When the threshold parameter t is known, say $t = 0$ (otherwise subtract t from X_i) the likelihood equations for the scale and shape parameters α and β have a unique solution. In fact, the likelihood equations may be rewritten

$$\hat{\alpha} = \left(\frac{1}{n}\sum_{i=1}^{n} x_i^{\hat{\beta}}\right)^{1/\hat{\beta}},$$

$$\sum_{i=1}^{n} x_i^{\hat{\beta}} \log x_i \left(\sum_{i=1}^{n} x_i^{\hat{\beta}}\right)^{-1} - \hat{\beta} - \frac{1}{n}\sum_{i=1}^{n}\log x_i = 0,$$

i.e., $\hat{\alpha}$ is given explicitly in terms of $\hat{\beta}$, which in turn can be found from the second equation by an iterative numerical procedure such as Newton's method. The regularity conditions of Theorems 3 and 4 are satisfied. Thus we conclude that

$$\sqrt{n}\left(\begin{pmatrix}\hat{\alpha}\\ \hat{\beta}\end{pmatrix} - \begin{pmatrix}\alpha\\ \beta\end{pmatrix}\right) \xrightarrow{L} N_2(\mathbf{0}, \mathbf{I}^{-1}(\alpha, \beta)),$$

where

$$\mathbf{I}^{-1}(\alpha, \beta) = \begin{bmatrix} 1.109\left(\frac{\alpha}{\beta}\right)^2 & .257\alpha \\ .257\alpha & .608\beta^2 \end{bmatrix}.$$

From these asymptotic results it follows that, as $n \to \infty$,

$$\sqrt{n}\,(\hat{\beta} - \beta)/\beta \xrightarrow{L} N(0, .608),$$

$$\sqrt{n}\;\hat{\beta}\log(\hat{\alpha}/\alpha) \xrightarrow{L} N(0, 1.109),$$

from which large sample confidence intervals* for β and α may be obtained. However, the large sample approximations are good only for very large samples. Even for $n = 100$ the approximations leave much to be desired. For small to medium sample sizes a large collection of tables is available (see Bain [1]) to facilitate various types of inference. These tables are based on extensive Monte Carlo* investigations. For example, instead of appealing to the asymptotic $N(0, .608)$ distribution of the pivot $\sqrt{n}\,(\hat{\beta} - \beta)/\beta$, the distribution of this pivot was simulated for various sample sizes and the percentage points of the simulated distributions were tabulated. Another approach to finite sample size inference is offered by Lawless [34]. His method is based on the conditional distribution of the MLEs given certain ancillary statistics. It turns out that this conditional distribution is analytically manageable; however, computer programs are ultimately required for the implementation of this method.

Returning to the three-parameter Weibull problem, so that the threshold is also unknown, we find that the likelihood function tends to infinity as $t \to T = \min(X_1, \ldots, X_n)$ and $\beta < 1$, i.e. the MLEs of t, α, and β do not exist. It has been suggested that the parameter β be restricted a priori to $\beta \geqslant 1$ so that the likelihood function remains bounded. In that case the MLEs will always exist, but with positive probability will not be a solution to the likelihood equations. It is not clear what the large sample properties of the MLEs are in this case.

Appealing to Theorems 3 and 4 one may attempt to find efficient roots of the likelihood equations or appeal to Theorem 6, since $\sqrt{n}$-consistent estimates are easily found (e.g., method of moments, method of quantiles*). However, the stated regularity conditions, notably **C1**, are not satisfied. In addition to that the likelihood equations will, with positive probability, have no solution at all and if they have a solution they have at least two, one yielding a saddle point and one yielding a local maximum of the likelihood function (Rockette et al. [50]). A further problem in the three-parameter Weibull model concerns identifiability for large shape parameters β. Namely, if

$X \sim W(t, \alpha, \beta)$ then uniformly in α and t, as $\beta \to \infty$

$$\frac{\beta}{\alpha}(X - t - \alpha) \xrightarrow{L} E_0$$

where E_0 is a standard extreme value* random variable with distribution function $F(y) = 1 - \exp(-\exp(y))$. Hence for large β,

$$X \overset{L}{\simeq} t + \alpha + \frac{\alpha}{\beta} E_0 = u + bE_0$$

and it is clear that (t, α, β) can not be recovered from u and b. This phenomenon is similar to the one experienced for the generalized gamma distribution*; see Prentice [46]. In the Weibull problem the identifiability problem may be remedied by proper reparametrization. For example, instead of (t, α, β) one can easily use three quantiles. However, because of the one-to-one relationship between these two sets of parameters the abovementioned problems concerning maximum likelihood and likelihood equation estimators still persist.

It is conceivable that the conditions of Theorem 6 may be weakened to accommodate the three-parameter Weibull distribution. Alternatively one may use the approach of Gong and Samaniego [24] described above, in that one estimates the threshold t by other means. Mann and Fertig [39] offer one such estimate; *see* MANN–FERTIG STATISTIC. Neither of these two approaches seems to have been explored rigorously so far. For an extensive account of maximum likelihood estimation as well as other methods for complete and censored samples from a Weibull distribution see Bain [1] and Lawless [34]. Both authors treat maximum likelihood estimation in the context of many other models.

For the very reason that applications employing the method of maximum likelihood are so numerous no attempt is made here to list them beyond the few references and examples given above. For a guide to the literature see the survey article on MLEs by Norden [41]. Also see Lehmann [37] for a rich selection of interesting examples and for a more thorough treatment.

Acknowledgments

In writing this article I benefited greatly from pertinent chapters of Erich Lehmann's book *Theory of Point Estimation* [37]. I would like to thank him for this privilege and for his numerous helpful comments.

I would also like to thank Jon Wellner for the use of his notes on maximum likelihood estimation and Michael Perlman and Paul Sampson for helpful comments. This work was supported in part by The Boeing Company.

References

[1] Bain, L. J. (1978). *Statistical Analysis of Reliability and Life-Testing Models*. Dekker, New York.

[2] Bahadur, R. R. (1964). *Ann. Math. Statist.*, **35**, 1545–1552.

[3] Bahadur, R. R. (1971). *Some Limit Theorems in Statistics*, SIAM, Philadelphia.

[4] Barndorff-Nielsen, O. (1978). *Information and Exponential Families in Statistical Theory*. Wiley, New York.

[5] Basawa, I. V. and Prakasa Rao, B. L. S. (1980). *Stoch. Proc. Appl.*, **10**, 221–254.

[6] Basawa, I. V. and Prakasa Rao, B. L. S. (1980). *Statistical Inference for Stochastic Processes*. Academic Press, London.

[7] Berk, R. H. (1967). *Math. Rev.*, **33**, No. 1922.

[8] Berkson, J. (1980). *Ann. Statist.*, **8**, 457–469.

[9] Cramér, H. (1946). *Mathematical Methods of Statistics*. Princeton University Press, Princeton, N.J.

[10] Dempster, A. P., Laird, N. M., and Rubin, D. B. (1977). *J. R.. Statist. Soc. B*, **39**, 1–22.

[11] Edgeworth, F. Y. (1908/09). *J. R. Statist. Soc.*, **71**, 381–397, 499–512, and *J. R. Statist. Soc.*, **72**, 81–90.

[12] Edwards, A. W. F. (1974). *Internat. Statist. Rev.*, **42**, 4–15.

[13] Efron, B. (1975). *Ann. Statist.*, **3**, 1189–1242.

[14] Efron, B. (1982). *Ann. Statist.*, **10**, 340–356.

[15] Efron, B. and Hinkley, D. V. (1978). *Biometrika*, **65**, 457–487.

[16] Feigin, P. D. (1976). *Adv. Appl. Prob.*, **8**, 712–736.

[17] Fisher, R. A. (1912). *Messenger of Mathematics*, **41**, 155–160.

[18] Fisher, R. A. (1922). *Philos. Trans. R. Soc. London A*, **222**, 309–368.

[19] Fisher, R. A. (1925). *Proc. Camb. Phil. Soc.*, **22**, 700–725.

[20] Fisher, R. A. (1935). *J. R. Statist. Soc.*, **98**, 39–54.

[21] Galambos, J. (1978). *The Asymptotic Theory of Extreme Order Statistics*. Wiley, New York.

[22] Ghosh, J. K. and Sinha, B. K. (1981). *Ann. Statist.*, **9**, 1334–1338.

[23] Ghosh, J. K. and Subramanyam, K. (1974). *Sankhya (A)*, **36**, 325–358.

[24] Gong, G. and Samaniego, F. J. (1981). *Ann. Statist.*, **9**, 861–869.

[25] Grenander, U. V. (1980). *Abstract Inference*. Wiley, New York.

[26] Haberman, S. J. (1974). *The Analysis of Frequency Data*. The University of Chicago Press, Chicago.

[27] Hájek, J. (1970). *Zeit. Wahrscheinlichkeitsth. Verw. Geb.*, **14**, 323–330.

[28] Hájek, J. (1972). *Proc. Sixth Berkeley Symp. Math. Statist. Prob.*, **1**, 175–194.

[29] Hoadley, B. (1971). *Ann. Math. Statist.*, **42**, 1977–1991.

[30] Huber, P. J. (1967). *Proc. Fifth Berkeley Symp. Math. Statist. Prob.*, **1**, 221–233.

[31] Huzurbazar, V. S. (1948). *Ann. Eugen.*, **14**, 185–200.

[32] Kiefer, J. and Wolfowitz, J. (1956). *Ann. Math. Statist.*, **27**, 887–906.

[33] Kraft, C. and LeCam, L. (1956). *Ann. Math. Statist.*, **27**, 1174–1177.

[34] Lawless, J. F. (1982). *Statistical Models and Methods for Lifetime Data*. Wiley, New York.

[35] LeCam, L. (1953). *Univ. of Calif. Publ. Statist.*, **1**, 277–330.

[36] Lehmann, E. L. (1980). *Amer. Statist.*, **34**, 233–235.

[37] Lehmann, E. L. (1983). *Theory of Point Estimation*. Wiley, New York, Chaps. 5, 6.

[38] Mäkeläinen, T., Schmidt, K., and Styan, G. (1981). *Ann. Statist.*, **9**, 758–767.

[39] Mann, N. R. and Fertig, K. W. (1975). *Technometrics*, **17**, 237–245.

[40] Nordberg, L (1980). *Scand. J. Statist.*, **7**, 27–32.

[41] Norden, R. H. (1972/73). *Internat. Statist. Rev.*, **40**, 329–354; **41**, 39–58.

[42] Perlman, M. (1972). *Proc. Sixth Berkeley Symp. Math. Statist. Prob.*, **1**, 263–281.

[43] Perlman, M. D. (1983). In *Recent Advances in Statistics: Papers in Honor of Herman Chernoff on his 60th Birthday*, M. H. Rizvi, J. S. Rustagi, and D. Siegmund, eds. Academic Press, New York, pp. 339–370.

[44] Pfanzagl, J. and Wefelmeyer, W. (1978/79). *J. Multivariate Anal.*, **8**, 1–29; **9**, 179–182.

[45] Pratt, J. W. (1976), *Ann. Statist.*, **4**, 501–514.

[46] Prentice, R. L. (1973), *Biometrika*, **60**, 279–288.

[47] Rao, C. R. (1957), *Sankhyā*, **18**, 139–148.

[48] Rao, C. R. (1961). *Proc. Fourth Berkeley Symp. Math. Statist. Prob.*, **1**, 531–546.

[49] Rao, C. R. (1963). *Sankhyā*, **25**, 189–206.

[50] Rockette, H., Antle, C., and Klimko, L. (1974). *J. Amer. Statist. Ass.*, **69**, 246–249.

[51] Savage, L. J. (1976). *Ann. Statist.*, **4**, 441–500.

[52] Scholz, F.-W. (1980). *Canad. J. Statist.*, **8**, 193–203.

[53] Sprott, D. A. (1973). *Biometrika*, **60**, 457–465.

[54] Sprott, D. A. (1980). *Biometrika*, **67**, 515–523.

[55] Sundberg, R. (1974). *Scand. J. Statist.*, **1**, 49–58.

[56] Wald, A. (1949). *Ann. Math. Statist.* **20**, 595–601.

[57] Weiss, L. and Wolfowitz, J. (1974). *Maximum Probability Estimators and Related Topics*. Springer-Verlag, New York. (Lect. Notes in Math., No. 424.)

[58] Zehna, P. W. (1966). *Ann. Math. Statist.*, **37**, 744.

Bibliography

Akahira, M. and Takeuchi, K. (1981). *Asymptotic Efficiency of Statistical Estimators: Concepts of Higher Order Asymptotic Efficiency*. Springer-Verlag, New York. Lecture Notes in Statistics 7. (Technical Monograph on higher order efficiency with an approach different from the references cited in the text.)

Barndorff-Nielsen, O. (1983). *Biometrika*, **70**, 343–365. (Discusses a simple approximation formula for the conditional density of the maximum likelihood estimator given a maximal ancillary statistic. The formula is generally accurate (in relative error) to order $O(n^{-1})$ or even $O(n^{-3/2})$, and for many important models, including arbitrary transformation models, it is in fact, exact. The level of the paper is quite mathematical. With its many references it should serve as a good entry point into an area of research of much current interest although its roots date back to R. A. Fisher.)

Fienberg, S. E. and Hinkley, D. V. (eds.) (1980). *R. A. Fisher: An Appreciation*. Springer-Verlag, New York. (Lecture Notes in Statistics 1. A collection of articles by different authors highlighting Fisher's contributions in statistics.)

Ibragimov, I. A. and Has'minskii (1981). *Statistical Estimation, Asymptotic Theory*. Springer-Verlag, New York. (Technical monograph on the asymptotic behavior of estimators (MLEs and Bayes estimators) for regular as well as irregular problems, i.i.d. and non-i.i.d. cases.)

LeCam, L. (1970). *Ann. Math. Statist.*, **41**, 802–828. (Highly mathematical, weakens Cramér's third-order differentiability conditions to first-order differentiability in quadratic mean.)

LeCam, L. (1979). *Maximum Likelihood, an Introduction.* Lecture Notes No. 18, Statistics Branch, Department of Mathematics, University of Maryland. (A readable and humorous account of the pitfalls of MLEs illustrated by numerous examples.)

McCullagh, P. and Nelder, J. A. (1983). *Generalized Linear Models*. Chapman and Hall, London. (This monograph deals with a class of statistical models that generalizes classical linear models in two ways: (i) the response variable may be of exponential family type (not just normal), and (ii) a monotonic smooth transform of the mean response is a linear function in the predictor variables. The parameters are estimated by the method of maximum likelihood through an iterated weight least-squares algorithm. The monograph emphasizes applications over theoretical concerns and represents a rich illustration of the versatility of maximum likelihood methods.)

Rao, C. R. (1962). *Sankhyā A*, **24**, 73–101. (A readable survey and discussion of problems encountered in maximum likelihood estimation.)

(ANCILLARY STATISTICS
ASYMPTOTIC NORMALITY
CRAMÉR–RAO LOWER BOUND
EFFICIENCY, SECOND ORDER
EFFICIENT SCORE
ESTIMATION, POINT
FISHER INFORMATION
FULL INFORMATION ESTIMATORS
GENERALIZED MAXIMUM LIKELIHOOD ESTIMATION
HODGES SUPEREFFICIENCY
INFERENCE, STATISTICAL
ITERATED MAXIMUM LIKELIHOOD ESTIMATES
LARGE-SAMPLE THEORY
LIKELIHOOD
LIKELIHOOD PRINCIPLE
LIKELIHOOD RATIO TESTS
M-ESTIMATORS
METHOD OF SIEVES
MINIMUM CHI-SQUARE
PENALIZED MAXIMUM LIKELIHOOD ESTIMATOR
PSEUDO LIKELIHOOD
UNBIASEDNESS)

F. W. SCHOLZ

MAXIMUM-MODULUS TEST

This is essentially a test of the hypothesis (H_0) that each of k independent normal variables $X_1, X_2, \ldots, X_k$ with known standard deviations $\sigma_1, \sigma_2, \ldots, \sigma_k$ has expected value zero. The test statistic is

$$T = \max_{1 \leqslant j \leqslant k} \left(\sigma_j^{-1} |X_j| \right).$$

If H_0 is valid,

$$\Pr\{T \leqslant t_0\} = \{1 - 2\Phi(t_0)\}^k$$

where $\Phi(t_0) = (\sqrt{2\pi})^{-1} \int_{-\infty}^{t_0} \exp(-\frac{1}{2}u^2)\,du$.

A test at significance level* ϵ is obtained with the critical region* $(T > t_0)$, where

$$\{1 - 2\Phi(t_0)\}^k = 1 - \epsilon,$$

that is,

$$\Phi(t_0) = \tfrac{1}{2}\left[1 - (1 - \epsilon)^{-1/k}\right].$$

In applications we usually have $\sigma_1 = \sigma_2 = \cdots = \sigma_k$, while $X_1, X_2, \ldots, X_k$ are actually linear functions of directly observed random variables and H_0 is defined by hypotheses on the latter. Thus if Y_j is distributed normally $N(\eta_j, \sigma^2)$ and $Y_1, Y_2, \ldots, Y_{k+1}$ are mutually independent, the hypothesis $\eta_1 = \eta_2 = \cdots = \eta_{k+1}$ is equivalent to the hypothesis H_0 on the k random variables

$$X_1 = (Y_1 - Y_2)/\sqrt{1 \times 2}$$

$$X_2 = (Y_1 + Y_2 - 2Y_3)/\sqrt{2 \times 3}$$

$$\vdots$$

$$X_k = (Y_1 + Y_2 + \cdots Y_k - kY_{k+1})/\sqrt{k(k+1)}$$

with $\sigma_1 = \sigma_2 = \cdots = \sigma_k = \sigma$ (*see* HELMERT, FRIEDRICH ROBERT).

If σ is not known, and it is replaced by an estimator independent of the X's and distributed as $\sigma\sqrt{(\chi_\nu^2/\nu)}$ the resulting test statistic leads to the Studentized maximum modulus test*.

(STUDENTIZED MAXIMUM MODULUS)

MAXIMUM PROBABILITY ESTIMATION

HISTORICAL NOTE

The most widely used general method of estimation is the method of maximum likelihood.* Maximum likelihood estimators have desirable large-sample properties in a large number of problems, and there have been various attempts to explain this (*see* LARGE SAMPLE THEORY). One such attempt is described in the article GENERALIZED MAXIMUM LIKELIHOOD ESTIMATION, where it is shown that under mild conditions, when estimating a single unknown parameter, a generalization of the maximum likelihood estimator gives the maximum asymptotic probability of being within a given distance of the true parameter value. When estimating a vector of unknown parameters, if the generalized maximum likelihood estimators are asymptotically independent, a similar result holds; the vector of generalized maximum likelihood estimators gives the maximum asymptotic probability of being within a rectangular parallelepiped centered at the true parameter vector. But there are many important cases where the generalized maximum likelihood estimators are not asymptotically independent. It is to deal with such cases, as well as with more general regions than rectangular parallelepipeds, that the theory of maximum probability estimation was developed by Weiss and Wolfowitz [7, 9–11].

INTRODUCTION

The theory of maximum probability estimation is predominantly asymptotic, so we introduce an index n, with positive integral values, and examine what happens as n increases.

For each n, let $\mathbf{X}(n)$ denote the random vector to be observed. $\mathbf{X}(n)$ does not necessarily contain n components, nor are the components necessarily independent or identically distributed. The joint distribution of the components of $\mathbf{X}(n)$ depends on m unknown parameters $\theta_1, \ldots, \theta_m$; $f_n(\mathbf{x}(n); \theta_1, \ldots, \theta_m)$ denotes the joint probability density function of the components if $\mathbf{X}(n)$ is continuous, and denotes $P(\mathbf{X}(n) = \mathbf{x}(n) \mid \theta_1, \ldots, \theta_m)$ if $\mathbf{X}(n)$ is discrete, where $\mathbf{x}(n)$ is a vector of the same dimension as $\mathbf{X}(n)$, and $P(E \mid \theta_1, \ldots, \theta_m)$ denotes the probability of the event E when the parameters are $\theta_1, \ldots, \theta_m$. The set Ω of all possible values of the vector $(\theta_1, \ldots, \theta_m)$ is an open subset of m-dimensional space.

Let R be a fixed bounded measurable subset of m-dimensional space, and let $K_1(n), \ldots, K_m(n)$ be given positive values. For any given values $D_1, \ldots, D_m$, denote by $R_n^*(D_1, \ldots, D_m)$ the set of all vectors $(\theta_1, \ldots, \theta_m)$ in Ω such that the m-dimensional point $(K_1(n)(D_1 - \theta_1), \ldots, K_m(n)(D_m - \theta_m))$ is in R. Denote the integral

$$\int_{R_n^*(D_1, \ldots, D_m)} \cdots \int f_n(\mathbf{X}(n); \theta_1, \ldots, \theta_m)\, d\theta_1 \ldots d\theta_m$$

by $I_n(D_1, \ldots, D_m)$. Denote the values of $D_1, \ldots, D_m$ which maximize $I_n(D_1, \ldots, D_m)$ by $\bar{\theta}_1(n), \ldots, \bar{\theta}_m(n)$, which we call *maximum probability estimators of* $\theta_1, \ldots, \theta_m$ *with respect to* R. (It would be more precise to call them "maximum probability estimators of $\theta_1, \ldots, \theta_m$ with respect to R and $K_1(n), \ldots, K_m(n)$," but in practice this is not necessary, as will be demonstrated.)

ASYMPTOTIC PROPERTIES OF MAXIMUM PROBABILITY ESTIMATORS

In practice, as n gets larger, it means that we can estimate $\theta_1, \ldots, \theta_m$ more precisely, and $K_1(n), \ldots, K_m(n)$ are chosen so that the m random variables

$$K_1(n)\big(\bar{\theta}_1(n) - \theta_1\big), \ldots, K_m(n)\big(\bar{\theta}_m(n) - \theta_m\big)$$

have a joint limiting distribution. This means that $K_i(n)$ approaches infinity as n increases, for $i = 1, \ldots, m$. Now suppose that $T_1(n), \ldots, T_m(n)$ are estimators of $\theta_1, \ldots, \theta_m$, based on $\mathbf{X}(n)$. Under the condition that for large n,

$$P\big[(K_1(n)(T_1(n) - \theta_1), \ldots, K_m(n)(T_m(n) - \theta_m)) \text{ in } R \mid \theta_1, \ldots, \theta_m\big]$$

is a continuous function of $\theta_1, \ldots, \theta_m$, it was shown in Weiss and Wolfowitz [7] that there exists a nonnegative quantity $\Delta_n(\theta_1, \ldots, \theta_m)$, with $\lim_{n\to\infty} \Delta_n(\theta_1, \ldots, \theta_m) = 0$, such that

$$\begin{aligned} P[(K_1(n)(T_1(n) - \theta_1), \ldots, & \\ K_m(n)(T_m(n) - \theta_m)) \text{ in } R \mid \theta_1, \ldots, \theta_m] & \\ \leqslant P\Big[\big(K_1(n)\big(\bar{\theta}_1(n) - \theta_1\big), \ldots, & \\ K_m(n)\big(\bar{\theta}_m(n) - \theta_m\big)\big) \text{ in } R \mid \theta_1, \ldots, \theta_m\Big] & \\ + \Delta_n(\theta_1, \ldots, \theta_m). & \end{aligned}$$

This explains the phrase "maximum probability estimators with respect to R" applied to $\bar{\theta}_1(n), \ldots, \bar{\theta}_m(n)$; among a wide class of estimators $\{T_1(n), \ldots, T_m(n)\}$, $\bar{\theta}_1(n), \ldots, \bar{\theta}_m(n)$ achieve the highest asymptotic probability that the m-dimensional point $(K_1(n)(T_1(n) - \theta_1), \ldots, K_m(n)(T_m(n) - \theta_m))$ will be in R. In practice, R consists of points in a neighborhood of the m-dimensional origin $(0, 0, \ldots, 0)$, so $\bar{\theta}_i(n)$ has a high probability of being close to θ_i, since $K_i(n)$ approaches infinity as n increases.

$K_i(n)$ is chosen so that the vector random variable $(K_1(n)(\bar{\theta}_1(n) - \theta_1), \ldots, K_m(n)$ $\bar{\theta}_m(n) - \theta_m))$ has a limiting distribution as n increases, but the choice is not unique: we could use $c_i K_i(n)$ instead of $K_i(n)$, where c_i is a fixed positive value. But using $c_i K_i(n)$ in place of $K_i(n)$ can be interpreted as using $K_i(n)$ but modifying the region R. Thus, suppose $m = 1$, $R = (-1, 1)$, and $K_1(n) = \sqrt{n}$. Then we are trying to maximize the probability that our estimator falls in the interval $(\theta_1 - 1/\sqrt{n}, \theta_1 + 1/\sqrt{n})$. If we use $K_1(n) = 2\sqrt{n}$, we are trying to maximize the probability that our estimator falls in the interval $(\theta_1 - 1/(2\sqrt{n}), \theta_1 + 1/(2\sqrt{n}))$, which can be considered as using $K_1(n) = \sqrt{n}$ and $R = (-\frac{1}{2}, \frac{1}{2})$. This explains why we speak of "maximum probability estimators with respect to R" rather than "maximum probability estimators with respect to R and $K_1(n), \ldots, K_m(n)$."

A CLASS OF REGULAR PROBLEMS

In this section we discuss a large class of estimation problems which satisfy some simple regularity conditions. Suppose that for each vector $\boldsymbol{\theta} = (\theta_1, \ldots, \theta_m)$ in the interior of Ω,

$$\frac{\partial^2}{\partial\theta_i\,\partial\theta_j} \log_e f_n(X(n); \theta_1, \ldots, \theta_m)$$

$$\equiv D_n(i, j, \boldsymbol{\theta}),$$

say, exists for $i, j = 1, \ldots, m$, and that there are nonrandom positive values $M_1(n), \ldots, M_m(n)$ with $\lim_{n\to\infty} M_i(n) = \infty$ and $\lim_{n\to\infty}(M_i(n)/K_i(n)) = 0$ for $i = 1, \ldots, m$, such that

(a) $-D_n(i, j, \boldsymbol{\theta})/[K_i(n)K_j(n)]$ converges stochastically as n increases to a nonrandom quantity $B_{ij}(\theta)$ for $i, j = 1, \ldots, m$. $B_{ij}(\boldsymbol{\theta})$ is a continuous function of $\boldsymbol{\theta}$, and the $m \times m$ matrix $\mathbf{B}(\boldsymbol{\theta})$ with $B_{ij}(\boldsymbol{\theta})$ in row i and column j is positive definite.

(b) $M_i(n)M_j(n)| - D_n(i, j, \boldsymbol{\theta})/[K_i(n)K_j(n)] - B_{ij}(\boldsymbol{\theta})|$ converges stochastically to zero as n increases, for $i, j = 1, \ldots, m$.

(c) Roughly speaking, the convergence in **(b)** is uniform for all vectors $\boldsymbol{\theta}$ whose ith coordinates are no further apart than $M_i(n)/K_i(n)$ for $i = 1, \ldots, m$. (The exact condition is given in Weiss [5].)

Under these three assumptions, it was shown in Weiss [4, 5] that if the maximum likelihood estimators $\hat{\theta}_1(n), \ldots, \hat{\theta}_m(n)$ of $\theta_1, \ldots, \theta_m$ exist, they have the following asymptotic properties:

(a) The asymptotic joint distribution of the vector $(K_1(n)(\hat{\theta}_1(n) - \theta_1), \ldots, K_m(n)$ $(\hat{\theta}_m(n) - \theta_m))$ is normal, with zero means and covariance matrix $(\mathbf{B}(\boldsymbol{\theta}))^{-1}$.

(b) Suppose R is an arbitrary convex region symmetric about the m-dimensional origin, and $\bar{\theta}_1(n), \ldots, \bar{\theta}_m(n)$ are maximum probability estimators with respect to R. Then for every vector $\boldsymbol{\theta}$ in Ω,

$$\lim_{n\to\infty} \Big| P\big[\big(K_1(n)(\hat{\theta}_1(n) - \theta_1), \ldots, K_m(n)(\hat{\theta}_m(n) - \theta_m)\big) \text{ in } R \mid \boldsymbol{\theta}\big] - P\Big[\Big(K_1(n)\big(\bar{\theta}_1(n) - \theta_1\big), \ldots, K_m(n)\big(\bar{\theta}_m(n) - \theta_m\big)\Big) \text{ in } R \mid \boldsymbol{\theta}\Big]\Big| = 0.$$

Considering the asymptotic properties of maximum probability estimators, this is a very strong asymptotic optimality property of maximum likelihood estimators.

Many estimation problems satisfy the regularity conditions of this section, including the following. Suppose $\mathbf{X}(n) = (X_1, \ldots, X_n)$, where $X_1, \ldots, X_n$ are independent and identically distributed, so that $f_n(\mathbf{X}(n); \boldsymbol{\theta}) = \prod_{i=1}^{n} f_{\mathbf{X}}(X_i; \boldsymbol{\theta})$, where $f_{\mathbf{X}}(x; \boldsymbol{\theta})$ is the marginal probability density function of X_1 if X_1 is continuous, or $f_{\mathbf{X}}(x; \boldsymbol{\theta}) = P[X_1 = x \mid \boldsymbol{\theta}]$ if X_1 is discrete. If $(\partial^2/\partial\theta_i\,\partial\theta_j)\log_e f_{\mathbf{X}}(X_1; \boldsymbol{\theta})$ exists, has an expected value which is a continuous function of $\theta_1, \ldots, \theta_m$, and a finite variance for $i, j = 1, \ldots, m$, then all the regularity conditions are satisfied and we can take $K_i(n) = \sqrt{n}$ and $M_i(n) = n^{(1/6)-\delta}$ for some δ in the open interval $(0, \frac{1}{6})$. A very large number of common statistical problems are of this type.

Many problems in which $\mathbf{X}(n)$ does not consist of independent and identically distributed components also satisfy the regularity conditions of this section. Weiss and Wolfowitz [11] and Roussas [3] discussed maximum probability estimation when $\mathbf{X}(n)$ consists of the observed states in a Markov process*. These cases satisfied the regularity conditions.

GENERALIZATIONS AND MODIFICATIONS

The following generalizations and modifications of the definition of maximum probability estimators are described more fully in Weiss and Wolfowitz [11].

The region R can be made to depend on $\theta_1, \ldots, \theta_m$, and so would be written as $R(\theta_1, \ldots, \theta_m)$. Then $R_n^*(D_1, \ldots, D_m)$ would be defined as the set of all vectors $(\theta_1, \ldots, \theta_m)$ in Ω such that the m-dimensional point $(K_1(n)(D_1 - \theta_1), \ldots, K_m(n)(D_m - \theta_m))$ is in $R(\theta_1, \ldots, \theta_m)$. $I_n(D_1, \ldots, D_m)$ and $\bar{\theta}_1(n), \ldots, \bar{\theta}_m(n)$ then are defined as above. As an example of when this may be desirable, suppose $m = 1$ and θ_1 is an unknown positive scale parameter, so $\Omega = (0, \infty)$. If we define $R(\theta_1)$ as the interval $(\theta_1 - c\theta_1, \theta_1 + c\theta_1)$ for a fixed c in the open interval $(0, 1)$, then

$$P\left[K_1(n)\left(\bar{\theta}_1(n) - \theta_1\right) \text{ in } R(\theta_1) \mid \theta_1\right]$$
$$= P\left[\theta_1 - c\theta_1/K_1(n) < \bar{\theta}_1(n) < \theta_1 + c\theta_1/K_1(n) \mid \theta_1\right].$$

$\bar{\theta}_1(n)$ will be invariant under a change of scale, so this latter probability will be independent of θ_1, which might be considered an advantage.

The region R can be unbounded. Thus, suppose $m = 2$, but we only want to estimate θ_1. Then we could define the region R in two-dimensional space as $\{(w_1, w_2) : -c < w_1 < c, -\infty < w_2 < \infty\}$.

We can modify the definition of $\bar{\theta}_1(n), \ldots, \bar{\theta}_m(n)$ to be values such that

$$I_n\left(\bar{\theta}_1(n), \ldots, \bar{\theta}_m(n)\right) \geqq (1 - L_n) \sup_{D_1, \ldots, D_m} I_n(D_1, \ldots, D_m),$$

where L_n is a nonrandom positive quantity and $\lim_{n\to\infty} L_n = 0$. Such estimators have the same asymptotic properties as the maximum probability estimators defined above. This modified definition is sometimes easier to use. Also, there are cases where many different estimators share the asymptotic properties, and in order to find them all this modified definition may be needed; see for example Weiss and Wolfowitz [6, 8].

Suppose $\theta_1^*(n), \ldots, \theta_m^*(n)$ are consistent estimators of $\theta_1, \ldots, \theta_m$. We can modify the definition of $\bar{\theta}_1(n), \ldots, \bar{\theta}_m(n)$ by defining the region $R_n^*(D_1, \ldots, D_m)$ as the set of all vectors $(\theta_1, \ldots, \theta_m)$ in the intersection of Ω and a neighborhood $N_n(\theta_1^*(n), \ldots, \theta_m^*(n))$ of $(\theta_1^*(n), \ldots, \theta_m^*(n))$ such that the m-dimensional point $(K_1(n)(D_1 - \theta_1), \ldots, K_m(n)(D_m - \theta_m))$ is in R. That is, we force the point $(\bar{\theta}_1(n), \ldots, \bar{\theta}_m(n))$ to be close to the point $(\theta_1^*(n), \ldots, \theta_m^*(n))$. The neighborhood shrinks as n increases, but slowly enough so that the probability approaches one that it contains the true parameter point $(\theta_1, \ldots, \theta_m)$. This modification is necessary in certain cases where $f_n(X(n); \theta_1, \ldots, \theta_m)$ becomes infinite for values of $\theta_1, \ldots, \theta_m$

which are not the true parameter values; the restriction on $R_n^*(D_1, \ldots, D_m)$ avoids these troublesome values. (See the example in the next section for an illustration.)

Finally, in decision-theoretic terms maximum probability estimators are constructed with a two-value loss function in mind: The loss is -1 if the vector of estimates $(\bar{\theta}_1(n), \ldots, \bar{\theta}_m(n))$ is such that

$$\Big(K_1(n)\big(\bar{\theta}_1(n) - \theta_1\big), \ldots, K_m(n)\big(\bar{\theta}_m(n) - \theta_m\big)\Big)$$

is in R, and is zero otherwise. The techniques described above can be used to derive estimators which are asymptotically optimal with respect to more general loss functions.

A REGULAR CASE WHERE MAXIMUM LIKELIHOOD ESTIMATORS DO NOT EXIST

Suppose $\mathbf{X}(n) = (X_1, \ldots, X_n)$, where $X_1, \ldots, X_n$ are independent and identically distributed, with common probability density function

$$\frac{1}{2\sqrt{2\pi}} \exp\left[-\frac{1}{2}(x-\theta_1)^2\right] + \frac{1}{2\theta_2\sqrt{2\pi}} \exp\left[-\frac{1}{2}\left(\frac{x-\theta_1}{\theta_2}\right)^2\right],$$

where Ω consists of all (θ_1, θ_2) with $-\infty < \theta_1 < \infty$ and $\theta_2 > 0$. This case satisfies the regularity conditions given above, but maximum likelihood estimators do not exist, because $f_n(\mathbf{X}(n); X_1, \theta_2)$ approaches infinity as θ_2 approaches zero from above. But by restricting $R_n^*(D_1, D_2)$ to be in a neighborhood of consistent estimators of θ_1, θ_2, estimators can be constructed which are asymptotically maximum probability estimators simultaneously with respect to all bounded convex R symmetric about $(0, 0)$. Details are given in Weiss [4]. This example of the nonexistence of maximum likelihood estimators was discovered by Kiefer and Wolfowitz [2].

NONREGULAR CASES

The regularity conditions can fail because $D_n(i, j, \boldsymbol{\theta})$ fails to exist. A simple example of this with $m = 1$ is $\mathbf{X}(n) = (X_1, \ldots, X_n)$, where $X_1, \ldots, X_n$ are independent and identically distributed, the common probability density function being $e^{-(x-\theta_1)}$ if $x > \theta_1$, zero if $x < \theta_1$. Here Ω consists of all θ_1 with $-\infty < \theta_1 < \infty$. We take R as the open interval $(-r, r)$, for fixed positive r, and $K_1(n) = n$; then

$$\bar{\theta}_1(n) = \min(X_1, \ldots, X_n) - r/n.$$

Asymptotically $P[n(\bar{\theta}_1(n) - \theta_1) \leqslant y \mid \theta_1]$ is equal to zero if $y < -r$, and is equal to $1 - e^{-(r+y)}$ if $y > -r$. Several other examples of the computation of maximum probability estimators where $D_n(i, j, \boldsymbol{\theta})$ fails to exist are given in Weiss and Wolfowitz [7, 9–11].

The regularity conditions can also fail even when $D_n(i, j, \boldsymbol{\theta})$ exists, because no values $K_1(n), \ldots, K_m(n)$ exist that will make $-D_n(i, j, \boldsymbol{\theta})/[K_i(n)K_j(n)]$ converge stochastically to quantities $B_{ij}(\boldsymbol{\theta})$ such that the matrix $\mathbf{B}(\boldsymbol{\theta})$ is positive definite. An example with $m = 1$ is: $\mathbf{X}(n)$ consists of one element, which has probability density function $\sqrt{n}/[\pi + \pi n(x - \theta_1)^2]$. Ω consists of all θ_1 with $-\infty < \theta_1 < \infty$. We take R as the open interval $(-r, r)$ for fixed positive r. Here $D_n(1, 1, \theta)$ exists, but there is no sequence $\{K_1(n)\}$ of nonrandom quantities such that $-D_n(1, 1, \theta)/K_1^2(n)$ converges stochastically as n increases to a positive constant, which would be required for the one-by-one matrix $B(\theta)$ to be positive definite. In this case, $\bar{\theta}(n) = X(n)$, and the probability density function of $Y_n = \sqrt{n}\,(\bar{\theta}(n) - \theta)$ is then $1/(\pi + \pi y^2)$.

Other examples where $D_n(i, j, \boldsymbol{\theta})$ exists, but the regularity conditions fail, are described by Basawa and Prakasa Rao [1].

SMALL SAMPLE PROPERTIES OF MAXIMUM PROBABILITY ESTIMATORS

The optimal properties of maximum probability estimators described above are asymptotic, as n approaches infinity. But how good are the estimators for fixed n? The type of argument used in the following example can often be used to show that for fixed n maximum probability estimators are arbitrarily close to admissible estimators (*see* ADMISSIBILITY).

Suppose we observe the vector $(X_1, \ldots, X_m)$, where $X_1, \ldots, X_m$ are mutually independent, each with a normal distribution with standard deviation 1, and $E\{X_i\} = \theta_i$, with $\theta_1, \ldots, \theta_m$ unknown. Ω consists of m-dimensional space. Set $K_i(n) = 1$ for $i = 1, \ldots, m$, and define R as

$$\{(w_1, \ldots, w_m) : |w_i| \leqq r, i = 1, \ldots, m\},$$

where r is fixed and positive. Then $\bar{\theta}_i(n) = X_i$ for $i = 1, \ldots, m$.

To investigate properties of $\bar{\theta}_i(n)$, we set up the following statistical decision problem. Based on $(X_1, \ldots, X_m)$, we have to choose a vector $(D_1, \ldots, D_m)$. The loss is -1 if $|D_i - \theta_i| \leqq r$ for $i = 1, \ldots, m$, and the loss is 0 otherwise. We construct a Bayes decision rule (*see* BAYESIAN INFERENCE; DECISION THEORY) using the a priori distribution which makes $\theta_1, \ldots, \theta_m$ mutually independent, each with probability density function $(1/2v)e^{-|\theta|/v}$ for $-\infty < \theta < \infty$, where v is positive. A simple calculation shows that the Bayes decision rule chooses

$$D_i = X_i - v^{-1} \quad \text{if} \quad X_i > r + v^{-1},$$

$$D_i = X_i + v^{-1} \quad \text{if} \quad X_i < -(r + v^{-1}),$$

$$D_i = \frac{rX_i}{r + v^{-1}}$$

$$\text{if} \quad -(r + v^{-1}) < X_i < r + v^{-1}.$$

Call this choice $D_i(\beta, v)$. Then all cases, $|D_i(\beta, v) - \bar{\theta}_i(n)| \leqq v^{-1}$; this difference can be made arbitrarily close to 0 by taking v large enough. From elementary decision theory, the decision rule which chooses $D_i(\beta, v) = D_i$ is admissible among all decision rules whose expected losses are continuous functions of θ. Thus the decision rule which chooses $D_i = \bar{\theta}_i(n)$ is very close to an admissible decision rule.

References

[1] Basawa, I. V. and Prakasa Rao, B. L. S. (1980). *Stoch. Processes Appl.*, **10**, 221–254. (Discusses estimation of parameters of nonergodic stochastic processes.)

[2] Kiefer, J. and Wolfowitz, J. (1956). *Ann. Math. Statist.*, **27**, 887–906. (Contains an example where the maximum likelihood estimator does not exist.)

[3] Roussas, G. G. (1977). *Ann. Inst. Statist. Math.* **29**, 203–219. (Discusses maximum probability estimators of parameters of Markov processes.)

[4] Weiss, L. (1971). *J. Amer. Statist. Ass.*, **66**, 345–350. (Demonstrates that maximum likelihood estimators and maximum probability estimators are asymptotically equivalent in some nonstandard cases.)

[5] Weiss, L. (1973). *J. Amer. Statist. Ass.*, **68**, 428–430. (Strengthens the results of the preceding reference.)

[6] Weiss, L. and Wolfowitz, J. (1966). *Theory Prob. Appl.*, **11**, 58–81. (Introduces generalized maximum likelihood estimation, the precursor of maximum probability estimation. This paper also appears (in English) in the Russian journal *Teor. Veroyatn. ee Primen.*, **11**, 68–93.)

[7] Weiss, L. and Wolfowitz, J. (1967). *Ann. Inst. Statist. Math.*, **19**, 193–206. (Introduces maximum probability estimation.)

[8] Weiss, L. and Wolfowitz, J. (1968). *Theory Prob. Appl.*, **13**, 622–627. (Discusses estimation in a particular case involving certain complications. This paper also appears (in English) in the Russian journal *Teor. Veroyatn. ee Primen.*, **13**, 657–662.)

[9] Weiss, L. and Wolfowitz, J. (1969). *Proc., Internat. Symp. Prob. Inform. Theory, McMaster University April 4 and 5, 1968*, M. Behara, ed. Springer-Verlag, New York. (Lecture Notes in Mathematics 89, pp. 232–256. Develops the theory of maximum likelihood estimation with a general loss function.)

[10] Weiss, L. and Wolfowitz, J. (1970). *Ann. Inst. Statist. Math.*, **22**, 225–244. (Demonstrates that maximum probability estimators are asymptotically sufficient in a variety of cases.)

[11] Weiss, L. and Wolfowitz, J. (1974). *Maximum Probability Estimators and Related Topics*. Springer-Verlag, New York. (Lecture Notes in Mathematics 424. A largely self-contained exposition of the theory of maximum probability estimation.)

Bibliography

Grossmann, W. (1979). *Metrika*, **26**, 129–138. (Gives conditions under which maximum probability estimators are consistent.)

Kuss, U. (1972). *Zeit. Wahrscheinlichkeitsth. verw. Geb.*, **24**, 123–133. (Gives conditions under which maximum likelihood estimators and maximum probability estimators are asymptotically equivalent.)

Kuss, U. (1980). *Statist. Hefte*, **21**, 2–13. (Shows that for certain utility functions, maximum probability estimators asymptotically maximize the expected utility.)

Wegner, H. (1976). *Ann. Inst. Statist. Math.*, **28**, 343–347. (Gives conditions under which maximum probability estimators exist.)

(GENERALIZED MAXIMUM LIKELIHOOD ESTIMATION
LARGE-SAMPLE THEORY
MAXIMUM LIKELIHOOD ESTIMATION)

LIONEL WEISS

MAXIMUM PRODUCT OF SPACINGS (MPS) ESTIMATION

MPS estimation is a technique introduced by Cheng and Amin [1] for estimation of parameters ($\boldsymbol{\theta}$) of continuous distributions. If $X_1, X_2, \ldots, X_n$ are mutually independent random variables with common probability density function (PDF) $f_X(x \mid \boldsymbol{\theta})$ then the likelihood function is

$$L(X_1, \ldots, X_n) = \prod_{i=1}^{n} f_X(X_i \mid \boldsymbol{\theta})$$

and maximum likelihood* estimators (MLEs) $\hat{\boldsymbol{\theta}}$ are obtained by maximizing L with respect to $\boldsymbol{\theta}$.

MPS estimators $\boldsymbol{\theta}^*$, on the other hand, are obtained by maximizing

$$M = \prod_{i=1}^{n+1} \left[\int_{X'_{i-1}}^{X'_i} f_X(x \mid \boldsymbol{\theta})\, dx \right],$$

where $X'_1 \leqslant X'_2 \leqslant \cdots \leqslant X'_n$ are the order statistics* corresponding to $X_1, X_2, \ldots, X_n$, $X'_0 = -\infty$ and $X'_{n+1} = \infty$. This method of estimation is especially suggested for use in cases where the MLEs do not solve the likelihood* equations ($\partial L / \partial \boldsymbol{\theta} = \mathbf{0}$), because maximizing values occur at the edge(s) of regions of variation. In particular, it is to be expected that MPS estimators will be useful when there is a parameter (or two) defining boundaries of the range of variation of X.

Note that M must lie between 0 and 1, unlike L which can be unboundedly great.

The following examples are given in ref. 1:

(a) If $f_X(x \mid \theta) = \theta^{-1}$, $k\theta \leqslant x \leqslant (k+1)\theta$; $k > 0$, the MLE of θ is

$$\hat{\theta} = \max(X_1, \ldots, X_n);$$

the MPS estimator is

$$\theta^* = A - \left\{ A^2 - \frac{n+1}{n-1} \frac{T_n}{k(k+1)} \right\}^{1/2},$$

where

$$T_n = \min(X_1, \ldots, X_n)\max(X_1, \ldots, X_n),$$
$$A = nV_n / \{2(n-1)k(k+1)\},$$
$$V_n = (k+1)\min(X_1, \ldots, X_n) - k\max(X_1, \ldots, X_n).$$

Both are asymptotically unbiased, but for large k,

$$\operatorname{Var}(\theta^*)/\operatorname{Var}(\hat{\theta}) = \tfrac{1}{2}.$$

(b)

$$f_X(x \mid \boldsymbol{\theta}) = \beta\gamma^{-\beta}(x-\alpha)^{\beta-1} \times \exp\left[-\{(x-\alpha)/\beta\}^{\beta} \right],$$

($x > \alpha$), so that $\boldsymbol{\theta} = (\alpha, \beta, \gamma)$, a Weibull distribution*; then

(i) if $\beta > 2$ there is in probability an MPS estimator $\boldsymbol{\theta}^*$ with the same asymptotic properties as the MLE;

(ii) if $0 < \beta < 2$ there is in probability an MPS estimator $\boldsymbol{\theta}^*$ with

$$\alpha^* - \alpha = O_p(n^{-1/\beta})$$

and (β^*, γ^*) have the same asymptotic properties as the corresponding MLEs *with α known.*

For the MLE $\hat{\theta}$, however, property (ii) holds only for $1 < \beta < 2$. For $\beta < 1$ there is no consistent MLE.

Reference

[1] Cheng, R. C. H. and Amin, N. A. K. (1983). *J. R. Statist. Soc. B*, **45**, 394–403.

(ESTIMATION, POINT
LIKELIHOOD
MAXIMUM LIKELIHOOD ESTIMATION
SPACINGS)

MAXVAR *See* GENERALIZED CANONICAL VARIABLES

MAXWELL–BOLTZMANN STATISTICS *See* FERMI–DIRAC STATISTICS

MAXWELL DISTRIBUTION

A continuous random variable X has the chi-distribution* (χ-distribution) with parameter ν (> 0)—called degrees of freedom*—if its density function is

$$f_X(x) = \begin{cases} \dfrac{1}{2^{(\nu/2)-1}\Gamma(\nu/2)} x^{\nu-1} e^{-x^2/2}, & x > 0, \\ 0, & x < 0. \end{cases}$$

If $X_1, X_2, \ldots, X_n$ are independent random variables each with the standard normal distribution, then $X = (\sum X_i^2)^{1/2}$ is a χ-distributed variable with n degrees of freedom. (*See* CHI-DISTRIBUTION for more details.) For $n = 3$, the χ-distribution apart from a scale factor describes the distribution of velocities of gas molecules and is called the *Maxwell distribution*. (*See also* MAXWELL, JAMES CLERK.)

(CHI-DISTRIBUTION)

MAXWELL, JAMES CLERK

Born: June 13, 1831, in Edinburgh, Scotland.

Died: November 5, 1879, in Cambridge, England.

Contributed to: mathematical physics (including statistical mechanics).

James Clerk Maxwell (1831–1879) was a native Scot, educated at Edinburgh Academy and at the Universities of Edinburgh and Cambridge. He held senior professorial positions in natural philosophy and experimental physics at Aberdeen University, King's College in London, and at the University of Cambridge, where he supervised the planning and early administration of the Cavendish Laboratory.

Max Planck [9] asserted that among Maxwell's many contributions to diverse areas in mathematical physics his work materially influenced two main areas, those dealing with the physics of continuous media such as his electromagnetic theory, and with the physics of particles. The latter includes Maxwell's study of the kinetic theory of gases [5, 6], which initiated a new period in physics by describing physical processes in terms of a statistical function rather than a mechanical or deterministic one; *see* PHYSICS, EARLY HISTORY OF STATISTICS IN.

DYNAMICAL SYSTEMS

Maxwell became acquainted with probabilistic arguments while in Edinburgh, under his friend and teacher James Forbes, and through a review by Sir John Herschel of Quetelet's* treatise on probability in the *Edinburgh Review* in 1850; see Everitt [3]. He was also aware that Rudolf Clausius [2] had used statistical arguments in 1859 to show that, under the assumption that all molecules have equal velocity, the collisions of any given particle occur in what we would describe as a Poisson process*.

In his first paper on the dynamical theory of gases [5] in 1860, Maxwell used geometry to determine the distributions $f(x)$, $f(y)$, and $f(z)$ of mutually orthogonal components of velocity. If N particles start from the origin together, then the number of particles in a "box" of volume $dx\,dy\,dz$ after a large number of collisions has occurred is

$$Nf(x)f(y)f(z)\,dx\,dy\,dz;$$

the independence of the densities $f(x)$, $f(y)$, $f(z)$ follows from the orthogonality of the coordinate axes. "But the directions of the coordinates are perfectly arbitrary, and therefore . . .

$$f(x)f(y)f(z) = \phi(x^2 + y^2 + z^2)."$$

This functional equation solves to give the normal law in the form

$$f(x) = \frac{1}{\alpha\sqrt{\pi}} e^{-x^2/\alpha^2}. \qquad (1)$$

"It appears from this proposition," Maxwell wrote, "that the velocities are distributed among the particles according to the same law as the errors are distributed among the observations in the theory of the method of least squares," a reference to Gauss's contributions to statistics (*see* GAUSS, C. F.).

The velocity $v = \sqrt{(x^2 + y^2 + z^2)}$ has the *Maxwell distribution** with density

$$g(v) = \frac{4}{\alpha^3\sqrt{\pi}} v^2 e^{-v^2/\alpha^2}, \qquad v > 0, \quad (2)$$

with mean velocity $2\alpha/\sqrt{\pi}$ and variance $((3/2) - (4/\pi))\alpha^2$. It would then have been easy to show that the molecular kinetic energy has a chi-square distribution* with three degrees of freedom*. In 1867 Maxwell [6] derived similar results when the collisions of the particles are incorporated; the above distributions are then stationary. His approach attracted the keen attention of Ludwig Boltzmann*, who built upon it the foundations of statistical mechanics*.

THERMODYNAMICS AND MAXWELL'S DEMON

Maxwell had a penetrating mind, which, as his friend P. G. Tait wrote, "could never bear to pass by any phenomenon without satisfying itself of at least its general nature and causes" [10]. Having resolved the crux of the phenomenon, he could then leave others to work out the details. Nowhere does this mark of his character appear more forcefully than in his insistence that the Second Law of Thermodynamics is a statistical one; at the same time other, such as Clausius and Boltzmann, were trying to explain it as a mechanical one, to the effect that heat can never pass from a colder to a warmer body without some change (in the form of external work) occurring at the same time. Challenging the universality of this assertion, Maxwell [7] considered gas in a compartmentalized container, with a diaphragm between a section A with hotter gas and a Section B with colder gas; he postulated "a finite being who knows the paths and velocities of all the molecules by simple inspection but who can do no work except open and close a hole in the diaphragm by means of a slide without mass." Labeled 'Maxwell's demon' by William Thomson (see ref. 4 for a full discussion), this being would allow molecules from A with sufficiently low velocities to pass into B, and would allow molecules from B with sufficiently high velocities to pass into A, carrying out these procedures alternately. The statistical distribution (2) guarantees the availability of such molecules; the gas in A would become hotter and that in B colder, no work would be done, and the second law would be violated. "Only we can't, not being clever enough" [7].

It was again to be Ludwig Boltzmann who would give explicit form to Maxwell's insight, developing in his H-theorem the relationship between entropy* and probability [1]. Maxwell returned to his theme in 1878, while reviewing the second edition of Tait's *Sketch of Thermodynamics* [8].

The law is continually violated, he says,

> ". . . in any sufficiently small group of molecules belonging to a real body. As the number of molecules in the group is increased, the deviations from the mean of the whole become smaller and less frequent; and when the number is increased till the group includes a sensible portion of the body, the probability of a measurable variation from the mean occurring in a finite number of years becomes so small that it may be regarded as practically an impossibility.
>
> This calculation belongs of course to molecular theory and not to pure thermodynamics, but it shows that we have reason for

> believing the truth to the second law to be of the nature of a strong probability, which, though it falls short of certainty by less than any assignable quantity, is not an absolute certainty."

Maxwell stood apart from the controversy between Tait and the German school over credit and priorities in thermodynamics, but quantitative data in support of the statistical molecular theory which he and Boltzmann had initiated did not appear until after 1900. Klein [4] gives a very readable detailed narrative and discussion, and includes an extensive bibliography.

References

[1] Boltzmann, L. E. (1877). *Wiss. Abh.*, **2**, 164–223.

[2] Clausius, R. (1859). *Philos. Mag., 4th Ser.*, **17**, 81–91.

[3] Everitt, C. W. F. (1975). *James Clerk Maxwell, Physicist and Natural Philosopher*. Scribners, New York.

[4] Klein, M. J. (1970). *Amer. Scientist*, **58**, 84–97.

[5] Maxwell, J. C. (1860). *Philos. Mag., 4th Ser.*, **19**, 19–32; *ibid.*, **20**, 21–37.

[6] Maxwell, J. C. (1866). *Philos. Mag., 4th Ser.*, **32**, 390–393.

[7] Maxwell, J. C. (1867). Letter to P. G. Tait. (See Knott, C. B. (1911), *Life and Scientific Work of Peter Guthrie Tait*, Cambridge University Press, Cambridge, England, pp. 213–214.)

[8] Maxwell, J. C. (1878). *Nature*, **17**, 278–280.

[9] Planck, M. (1931). In *James Clerk Maxwell: A Commemoration Volume, 1831–1931.* Cambridge University Press, Cambridge, England.

[10] Tait, P. G. (1880). *Nature*, **21**, 321.

Bibliography

Campbell, L. and Garnett, W. (1883). *Life of James Clerk Maxwell*. MacMillan, London.

Domb, C. (1980/81). *Notes and Records R. S. London*, **35**, 67–103. (An account of Maxwell's life and work in London and Aberdeen, 1860–1871.)

(BOLTZMANN, LUDWIG EDWARD
LAWS OF ERROR II
PHYSICS, EARLY HISTORY OF STATISTICS IN
STOCHASTIC MECHANICS)

CAMPBELL B. READ

McCALL *T*-SCORES *See* NORMALIZED *T*-SCORES

McCARTHY–BIRNBAUM CONJECTURE

Let $D_n^+ = \sup_{-\infty<x<\infty}(F_n(x) - F(x))$ denote the Kolmogorov–Smirnov statistic (*see* KOLMOGOROV–SMIRNOV STATISTICS for the definitions of $F(x)$ and $F_n(x)$). Define for $x \geqslant 0$

$$P_n^+(x) = \Pr\left[D_n^+(x) \leqslant x/\sqrt{n}\right],$$
$$P_n^+(x) = 0, \qquad x < 0.$$

(The random variable $\sqrt{n}\,D_n^+$ has been investigated by Smirnov [5], Birnbaum and Tingey [2], and Dvoretzky et al. [3], among others.)

It was conjectured by Birnbaum and McCarthy [1] that

$$P_n^+(x) \geqslant 1 - e^{-2x^2}$$

for all $x \geqslant 0$ and all positive integers n. Recently T. Matsunawa [4] gave a proof for the validity of this conjecture.

References

[1] Birnbaum, Z. W. and McCarthy, R. C. (1958). *Ann. Math. Statist.*, **29**, 558–562.

[2] Birnbaum, Z. W. and Tingey, F. H. (1951). *Ann. Math. Statist.*, **22**, 592–596.

[3] Dvoretzky, A., Kiefer, J. and Wolfowitz, J. (1956). *Ann. Math. Statist.*, **27**, 642–669.

[4] Matsunawa, T. (1982). Verification of the Birnbaum–McCarthy Conjecture. In *Proc. Pacific Area Statistical Conference*, December 15–17, 1982, Tokyo, Japan. (Short-form papers), pp. 315–318, Institute of Statistical Mathematics, Tokyo, Japan.

[5] Smirnov, N. V. (1944). *Uspehi Matem. Nauk*, **10**, 179–206 (in Russian).

(KOLMOGOROV–SMIRNOV STATISTICS
KOLMOGOROV–SMIRNOV TYPE TESTS OF FIT)

McINTYRE–TAKAHASHI–WAKIMOTO SELECTIVE PROCEDURE *See* MILLER–GRIFFITHS PROCEDURE

McKAY'S APPROXIMATION

An approximation to the distribution of the sample coefficient of variation* from a normal distribution* $N(\mu, \beta^2\mu^2)$. McKay [2] derived an approximate probability density function for

$$V = S/\overline{X},$$

where

$$S = \left[\sum_{i=1}^{n} (X_i - \overline{X})^2 / n \right]^{1/2};$$

$$\overline{X} = \frac{\sum_{i=1}^{n} X_i}{n};$$

and $X_1, X_2, \dots, X_n$ are independently distributed $N(\mu, \beta^2\mu^2)$. McKay showed that for small $\beta(\leqslant 0.3)$ and assuming that $\Pr\{V < 0 | \beta\} \doteqdot 0$, $nV^2(1+\beta^2)/(1+V^2)$ has approximately a chi-square distribution* with $(n-1)$ degrees of freedom. Iglewicz [1] compared this approximation with several others.

References

[1] Iglewicz, B. (1967). Some Properties of the Sample Coefficient of Variation. Ph.D. Thesis, Virginia Polytechnic Institute, Blacksburg, Va.

[2] McKay, A. (1932). *J. Roy. Statist. Soc.*, **96**, 695–698.

(COEFFICIENT OF VARIATION
NONCENTRAL t-DISTRIBUTION)

McNEMAR STATISTIC

When comparing two proportions based on independent samples an asymptotic *chi-square test** has long been available and can be found in nearly every elementary statistical text. Often, however, in order to increase the precision of comparison, members from one sample are matched with members of the other sample on variables associated with the response being studied. Besides pairwise observations from two distinct samples, matched samples also occur when individuals are given two different treatments, asked two different questions, or the same question at two different time periods (*see* MATCHED PAIRS).

McNemar [13] developed a test statistic for comparing two matched samples when the response is a dichotomy. The data for comparing two matched proportions may be placed in a 2^2 contingency table formed by responses $i_j = 1$ (success) or 2 (failure) for $j = 1$ or 2 representing the two treatments. For such a situation let n_{i_1,i_2} be the number of observations having response (i_1, i_2). Thus n_{11} is the number of matched pairs where the response to both treatments is "success," n_{12} is the number of matched pairs where the response to the first treatment is a success and the second treatment is a failure, etc. Using the * notation for summation over a starred subscript $n_{1*} = n_{11} + n_{12}$ is the number of matched pairs where the response to the first treatment is a success, etc., and $n = n_{11} + n_{12} + n_{21} + n_{22}$ is the number of matched pairs in the experiment.

Likewise, let $\pi_{i_1i_2}$ represent the probability of an observation (matched pair) falling in cell (i_1, i_2). Then $\pi_1 = \pi_{1*} = \pi_{11} + \pi_{12}$ is the probability of success for the first treatment and $\pi_2 = \pi_{*1} = \pi_{11} + \pi_{12}$ is the probability of success for the second treatment. The null hypothesis may be stated as:

$$H_0 : \pi_1 = \pi_2 .$$

For testing H_0 the estimate of the difference of $\pi_1 - \pi_2$ is $p_1 - p_2$ where $p_1 = n_{1*}/n$ and $p_2 = n_{*1}/n$. This estimate reduces to $(n_{12} - n_{21})/n$, a difference in discordant pairs, and is not dependent on n_{11} or n_{22}, the number of concordant pairs. Thus under the null hypothesis, which reduces to $\pi_{12} = \pi_{21}$, the total number of discordant pairs should be evenly divided between the (1, 2) and (2, 1) cells. The expected number of observations in each of these cells is $(n_{12} + n_{21})/2$ and the *chi-square** (χ^2) statistic [(observed-expected)^2/expected] for the two relevant cells reduces to

$$\chi^2 = \frac{(n_{12} - n_{21})^2}{n_{12} + n_{21}}$$

which has an asymptotic $\chi^2(1)$ distribution. A continuity correction* which changes the numerator from $(n_{12} - n_{21})^2$ to $(|n_{12} - n_{21}| - 1)^2$ was given by Edwards [8]. Stuart [20] developed a statistic equivalent to McNemar's by replacing the ordinary binomial variance for independent sampling by the unbiased within-stratum estimate where each matched pair may be looked at as a stratum. Mosteller [14] pointed out that an exact null distribution may be derived for testing H_0 by noting that the $n_{12} + n_{21}$ discordant pairs may be treated as binomial trials with probability of being assigned to cell (1,2) or (2,1) being $\frac{1}{2}$. Hence critical values of $n_{12} + n_{21}$ are obtained from binomial tables. Schork and Williams [19] gave sample size tables based on the exact power function for the matched pairs design with binary response.

Cox [7] also looked at the problem of comparing two matched proportions. He assumed a *logistic* model* and demonstrated the McNemar test to be optimal for testing the difference between two treatments. Gart [10] suggested the order of the treatments making up the matched pairs may have an effect. In particular he considered a crossover design (*see* CHANGEOVER DESIGNS) for a subject receiving each of two drugs at two different time periods. Using an extension of the logistic model of Cox, Gart derived both an exact and an asymptotic test which are optimal for the order effect situation. His test for both order and drug effect reduces to the product of binomial variables. (This is not the same as in GART TEST.) Nam [15] demonstrated that if an order effect is not present, McNemar's test is indeed most powerful but if an order effect exists McNemar's test is biased and Gart's should be preferred. Altham [1] considered a *Bayesian* analysis* for the matched pairs problem with a dichtomous response and compared it with the previously mentioned work of Cox. Roy [18] also offered a $C(\alpha)$ test which is locally asymptotically optimal for the crossover design. (*see* OPTIMAL $C(\alpha)$ TESTS).

An application of the McNemar statistic is seen in the following hypothetical example. Suppose we want to compare two treatments, chemotherapy and surgery, for a particular type of cancer with regard to three-year survival rate. In order to utilize smaller sample sizes a matched pairs design is chosen. Possible participants are matched on stage of cancer, age of patient, and the patient's mental outlook. These variables are thought to most influence the binomial outcome of survival after three years of treatment. Within a matched pair, patients are randomly assigned to receive one of the two treatments. If there are 100 matched pairs the data may be put in the following table:

		Chemotherapy		
		Alive	Dead	
Surgery	Alive	5	20	25
	Dead	15	60	75
		20	80	100

From the table it is seen that 25 of the 100 patients treated with surgery are alive after three years while the chemotherapy three year survivial rate is 20/100. However, by using the McNemar statistic,

$$\chi^2 = \frac{(20 - 15)^2}{20 + 15} = \frac{25}{35} = 0.71,$$

we see that the difference is not statistically significant.

McNemar's statistic has been generalized to various situations by different authors. For the situation of comparing R matched controls per case with a dichotomous response there are available the *Miettinen statistic** and the *Mantel–Haenszel statistic**. The latter statistic also is applicable if the number of controls per case varies between matching groups. Cochran [6] generalized McNemar's statistic for comparing more than two matched samples with a dichotomous response (*see* COCHRAN'S Q-STATISTIC). Bhapkar [2] and Bennett [3,4] also offer a statistic for comparing c ($\geqslant 2$) matched samples with a dichotomous response. However, they assume a multinomial* model and test for marginal homogeneity in a 2^c contingency table.

Hamdan et al. [11] extended McNemar's original problem from one population to several independent populations which need to be tested simultaneously. They considered four situations each with different assumptions. Bowker [5] developed a test for symmetry (*see* BOWKER'S TEST FOR SYMMETRY*) in an m^2 contingency table. This situation occurs when comparing two matched samples, where an observation within a sample may take on m possible outcomes instead of two. Read [16, 17] did further work along these lines to develop a test for *proportional symmetry*. (*See also* QUASI-SYMMETRY.)

McNemar's statistic also may be generalized to testing marginal homogeneity for the above m^2 contingency table. Mantel and Fleiss [12] have demonstrated the equivalence of the generalized McNemar tests for 2^3 and 3^2 tables.

For further readings dealing with McNemar's statistic the previously mentioned articles by McNemar or Mosteller are recommended, or the appropriate section of the book by Fleiss [9].

References

[1] Altham, P. M. E. (1971). *Biometrika*, **58**, 561–576.

[2] Bhapkar, V. P. (1970). In *Random Counts in Scientific Work*, Vol. 2, G. P. Patil, ed. Pennsylvania State University Press, University Park, pp. 255–267.

[3] Bennett, B. M. (1967). *J. R. Statist. Soc. B*, **29**, 408–472.

[4] Bennett, B. M. (1968). *J. R. Statist. Soc. B*, **30**, 368–370.

[5] Bowker, A. H. (1948). *J. Amer. Statist. Ass.*, **43**, 572–574.

[6] Cochran, W. G. (1950). *Biometrika*, **37**, 256–266.

[7] Cox, D. R. (1958). *Biometrika*, **45**, 562–565.

[8] Edwards, A. L. (1948). *Psychometrika*, **13**, 185–187.

[9] Fleiss, J. L. (1973). *Statistical Methods for Rates and Proportions*. Wiley, New York, pp. 83–87.

[10] Gart, J. J. (1969). *Biometrika*, **56**, 75–80.

[11] Hamdan, M. A., Pirie, W. R. and Arnold, J. C. (1975). *Psychometrika*, **40**, 153–162.

[12] Mantel, N. and Fleiss, J. L. (1975). *Biometrics*, **31**, 727–729.

[13] McNemar, O. (1947). *Psychometrika*, **12**, 153–157.

[14] Mosteller, F. (1952). *Biometrics*, **8**, 220–226.

[15] Nam, J. (1971). *Biometrics*, **27**, 945–959.

[16] Read, C. B. (1977). *Commun. Statist. A*, **6**, 553–562.

[17] Read, C. B. (1978). *Psychometrika*, **43**, 409–420.

[18] Roy, R. M. (1976). *Commun. Statist. A*, **5**, 545–563.

[19] Schork, M. A. and Williams, G. W. (1980). *Commun. Statist. B*, **9**, 349–357.

[20] Stuart, A. (1975). *Brit. J. Statist. Psychol.*, **10**, 29–32.

(BOWKER'S TEST FOR SYMMETRY
CHI-SQUARE TESTS
COCHRAN'S *Q*-STATISTIC
MANTEL–HAENSZEL STATISTIC
MATCHED PAIRS
MIETTINEN'S STATISTIC
QUASI-SYMMETRY)

GRANT W. SOMES

McNEMAR'S TEST *See* McNEMAR STATISTIC

m-DEPENDENCE

A sequence of random variables $\{X_n\}$ is said to be m-dependent if there exists a positive integer r such that any subsequence $\{X_{n_j}, j \geqslant 1\}$ of $\{X_n\}$ with $n_j + m < n_{j+1}$ for every $j \geqslant 1$ and $n_1 \geqslant r$ is a sequence of *independent* random variables.

(DEPENDENCE, CONCEPTS OF)

MEAN *See* ARITHMETIC MEAN; EXPECTED VALUE; GEOMETRIC MEAN; HARMONIC MEAN; MEAN, MEDIAN, MODE, AND SKEWNESS

MEAN, ARITHMETIC *See* ARITHMETIC MEAN

MEAN, ESTIMATION OF *See* LOCATION PARAMETER, ESTIMATION OF

MEAN, GEOMETRIC *See* GEOMETRIC MEAN

MEAN, HARMONIC *See* HARMONIC MEAN

MEAN, MEDIAN, MODE, AND SKEWNESS

The *mean* μ of a random variable X with distribution function $F(x)$ is defined as $\int x\, dF(x)$, provided this exists; it is the first moment of X about the origin, and is also known as the expected value*, expectation, or average value of X.

A *median* m of X satisfies $\Pr(X \geqq m) \geqq \frac{1}{2}$ and $\Pr(X \leqq m) \geqq \frac{1}{2}$, so that for $F(x)$ right continuous, $F(m) \geqq \frac{1}{2}$ and $F(m-) \leqq \frac{1}{2}$, and the line $y = \frac{1}{2}$ intersects the graph of $F(x)$ at $x = m$. If $F(x)$ is not strictly increasing, m may not be unique and we speak of a median class or a weak median [13]; for a strictly increasing $F(x)$, m is the unique solution of $F(m) = \frac{1}{2}$. The median is the 50% quantile* of the distribution.

A *mode* may be described empirically as a locally most frequently occurring value of X. Let X have a probability density function $f(x)$ or probability function $\phi(x)$; then M is a mode of the distribution if $f(x)$ or $\phi(x)$ has a local maximum at M. The abscissae of local minima are called anti-modes (and a density with one anti-mode is called U-shaped). It is often of importance to know if a distribution is unimodal, bimodal, or multimodal. The Pearson system* was set up in 1895 to generalize the normal or Gaussian distribution to a family of unimodal densities; another unimodal family is formed by distributions with monotonic hazard rates*. The unimodal property gives rise to other properties, for example, a number of Tchebycheff* inequalities [19], and certain moment inequalities which follow from the moment problem applied to unimodal distributions [12].

Multivariate analogues of mean and mode are readily defined but are not so straightforward for the median. One approach [8] is provided by recognizing that the univariate mean and median minimize respectively the expected squared deviation and the expected absolute deviation from a point. This characterization of the median provides an alternative multivariate median to the vector of the univariate medians of the components of the vector random variable.

The sample mean (*see* ARITHMETIC MEAN), median, and mode may be defined as above with respect to the sample distribution function. As above, the sample median and mode may not be unique. For an even-sized sample with no ties, the sample median is usually defined to be the average of the two middle sample observations. The sample mean and the sample median minimize respectively the sum of the squares and the sum of the absolute values of the deviations around a point; they are the equally weighted least-squares estimate and least absolute deviations estimate, respectively, of the population values [3, 23].

MEASURES OF LOCATION

The parameters mean, median, and mode are measures of the center of the distribution and so may be interpreted as location parameters. In a symmetric distribution, the mean and median coincide at a mode or anti-mode, giving the center of symmetry which is thus the natural location parameter of F, and the estimation of location in this case has received much attention; see, for example, refs. 1, 10, 11, and 24 for surveys of work on this and other cases with a given location parameter. The sample mean and median provide two such estimators, each satisfying a different criterion as given above.

From the mid-sixties there has been much work on robust statistical procedures (*see* ROBUST ESTIMATION) which aim to minimize the effect of departures from any prior assumptions about the underlying distribution of the observed random variables [11]. For example, the sample mean is highly sensitive to outliers* and hence also to sampling from heavy-tailed distributions*. Van Zwet [26] illustrates his measure of kurtosis* by showing that the relative efficiency* of the sample median to the sample mean in estimating the center of symmetry, increases with increasing kurtosis. Considerations of robustness have led to many alternative estimators of location; in ref. 1, 58 estimators of the center of symmetry are surveyed. There are three main classes of such estimators. The L-estimators are linear combinations of order statistics and include trimmed and Winsorized means [25] (*see* L-STATISTICS). The R-estimators are derived from rank tests* for shift and include the Hodges–Lehmann* statistic. The M-estimators* [9] generalize maximum likelihood* estimators.

Assumption breakdowns considered, other than the much-treated departure from normality, include unequal scale [3] and asymmetry [1]. As interest in robust estimation of location increased, the question of location measures for asymmetric distributions also received attention in a general way [2, 4, 18], with authors maintaining that it is possible to speak about location for asymmetric distributions, and with the problem of location measure very much influenced by the need to estimate location [9, 10]. A basic theoretical approach considered in refs. 2, 4, and 18 is to define a partial ordering of distributions according to location, and to consider functionals as location measures if they preserve this ordering, are linear, and odd. The ordering considered is the stochastic ordering [17], although Oja [18] also presents other possibilities and the resulting measures include the classic measures. Bickel and Lehmann [2] consider the location measures among the quantities estimated by the three classes of location estimators, giving the symmetrically weighted quantile average (special cases of which are the symmetrically trimmed means and the *generalized median*

$$m_F(u) = \tfrac{1}{2}\left[F^{-1}(u) + F^{-1}(1-u)\right]),$$

the R-estimator quantities, and a small subclass of the M-estimator quantities. On the basis of parameter robustness, and availability and efficiency of estimation, measures from the first class are recommended. Doksum's approach [4] is to approximate F as closely as possible above and below by symmetric distributions, and to obtain a location interval which contains the values of all the location measures that satisfy the order-preserving approach.

In the case of mean versus median, one approach [23] defines two coefficients comparing the mean and median with respect to the variations they minimize, and aims to relate these coefficients to skewness* and kurtosis* in both the sampling and population situations.

MEAN, MEDIAN, MODE, AND SKEWNESS

For an asymmetric distribution, comparisons of measures of location may be used as measures of skewness; the first of these to be considered was the relationship among the mean, median, and mode. About the turn of the century, Pearson* [20] used $(\mu - M)/\sigma$ as a measure of skewness, and found empirically that for his Type III distributions (gamma*),

$$M - \mu \simeq 3(m - \mu).$$

In 1917 with limited success Doodson [5] considered establishing this relationship for a class of densities with small Pearson skewness, and for the Pearson family. Haldane [7] shows that for certain distributions close to normal in the sense of having small cumulants, Pearson's relationship holds approximately. The conditions apply to Pearson Type III and some sampling distributions*; counterexamples include other Pearson distributions.

Although it has been recognized that Pearson's empirical relationship does not generally hold, it still is often stated that μ, m, and M occur either in this or the reverse order. A sufficient condition for this to hold is that $1 - F(m + x) - F(m - x)$ be of one sign in $x > 0$, nonnegative, for example, giving $\mu > m$ ($> M$, for a unimodal distribution) [27, 15]. A stronger condition is that $f(m + x) - f(m - x)$ change sign once in $x > 0$, negative to positive, for example, giving $\mu > m > M$ [27, 15, 22, 6]. It is interesting that theorems foreshadowing some of these results were introduced in 1897 by Fechner and in 1915 by Timerding (see ref. 22). Fechner shows that $\mu < m < M$ for an asymmetric density defined using two normal densities with different variances. Timerding considers the special case of a differentiable unimodal density falling to zero at the boundaries of its interval support, and his conditions for $\mu < m < M$ actually imply that $f(m+x) - f(m-x)$ has one zero, which is a change of sign from positive to negative values. It is only recently that attention has again focused on the mean-median-mode inequality, with authors independently arriving at similar conclusions.

The other classical measure of skewness is the third central moment μ_3; a sufficient condition for $\mu_3 > 0$ is that $1 - F(\mu + x) - F(\mu - x)$ or, more strongly, $f(\mu + x) - f(\mu - x)$, change sign once from negative to positive values [15, 22]. These results are essentially due to the variation-diminishing properties of the totally positive (TP) kernel x^r [14], and because x^r is strictly TP, $f(\mu + x) - f(\mu - x)$ and $f(m+x) - f(m-x)$ *must* change sign at least once to ensure that

$$\int x\{f(\mu + x) - f(\mu - x)\}\,dx = 0$$

and

$$\int \{f(m+x) - f(m-x)\}\,dx = 0.$$

This last condition plus the use of the indicator function, another TP kernel, facilitates checking the one change of sign condition, which is obeyed by a number of well-known distributions including most of the Pearson family, the inverse Gaussian*, and the lognormal* [15].

Sometimes the change of sign occurs because of a discontinuity at a support boundary of $f(x)$, for example the exponential distribution*, and so truncated distributions* may provide counterexamples. An example of a distribution family for which one change of sign of $1 - F(\zeta + x) - F(\zeta - x)$, $\zeta = \mu$ or m, does not hold for all members is the asymmetric lambda* family [21] defined by its percentile* function

$$F^{-1}(p) = \lambda_1 + \left[p^{\lambda_3} - (1-p)^{\lambda_4}\right]/\lambda_2.$$

This family can be shown to provide examples of unimodal asymmetric distributions with $\mu_3 = 0$, or with $\mu = m$, or with $\mu - m$ of opposite sign to μ_3; examples occur with the density either continuous or discontinuous at its support boundaries [16].

The relationship between the mean, medians, and modes for discrete distributions is of course a more difficult problem, as illustrated by the binomial distribution* [13] for which it is shown that if median and mode differ, the mean lies in between.

MEASURES OF SKEWNESS

Van Zwet [26] comments that the ordering of distributions by μ_3 is not necessarily interpretable, that the class of distributions being compared is too large. He proposes a partial ordering according to skewness of distributions based on convex transformations and that a distribution is *skewed to the right* if $F^{-1}(1 - F(x))$ is convex in x. This implies [27] that the generalized median, $m_F(\mu)$, is $\geqq m$ for $0 < \mu < 1$, which is equivalent to

$$1 - F(m+x) - F(m-x) \geqq 0.$$

Oja [18], in unifying and developing the various orderings of distributions according to location, scale, skewness, and kurtosis, considers skewness measures that preserve the van Zwet ordering (μ_3/σ^3 does, and $(\mu - m)/\sigma$ does not), and extends the concept to two weaker orderings. A different approach to the problem [4] proposes a symmetry function

$$\theta_F(x) = \tfrac{1}{2}\{x + F^{-1}(1 - F(x))\},$$

such that $X - 2\theta_F(X)$ has the same distribution as $-X$. For each x, $\theta_F(x)$ satisfies the location axioms, with the location interval as its range. The function $\theta_F(x) - m$ is proposed as a skewness measure, with F said to be *skewed to the right* if $\theta_F(x)$ attains its minimum at m, and *strongly skewed to the right* if $\theta_F(x)$ is U-shaped; other shapes also may be interpreted. It appears that orderings and measures of skewness are more complex questions than those of location. Not only does the concept of skewness depend to some extent on the conception of location, but it also seems to be more difficult to formalize.

References

[1] Andrews, D. F., Bickel, P. J., Hampel, F. R., Huber, P. J., Rogers, W. H., and Tukey, J. W. (1972). *Robust Estimates of Location: Surveys and Advances*. Princeton University Press, Princeton, N.J. (A user-oriented extensive survey of the properties of 58 location estimators.)

[2] Bickel, P. J. and Lehmann, E. L. (1975). *Ann. Statist.*, **3**, 1038–1069. (One of three papers on measures of location, scale and kurtosis; discusses some distributions illustrating some of the location measures and their properties.)

[3] Cressie, N. (1980). *Statist. Neerl.*, **34**, 19–32.

[4] Doksum, K. A. (1975). *Scand. J. Statist.* **2**, 11–22. (Includes estimates and examples of his symmetry function.)

[5] Doodson, A. T. (1917). *Biometrika*, **11**, 425–429.

[6] Groeneveld, R. A. and Meeden, G. (1977). *Amer. Statist.*, **31**, 120–121.

[7] Haldane, J. B. S. (1942). *Biometrika*, **32**, 294–299.

[8] Haldane, J. B. S. (1948). *Biometrika*, **35**, 414–415.

[9] Huber, P. J. (1964). *Ann. Math. Statist.*, **35**, 73–101.

[10] Huber, P. J. (1972). *Ann. Math. Statist.*, **43**, 1041–1067.

[11] Huber, P. J. (1981). *Robust Statistics*. Wiley, New York. (A theoretical overview with the applied statistician in mind.)

[12] Johnson, N. L. and Rogers, C. A. (1951). *Ann. Math. Statist.*, **22**, 433–439.

[13] Kaas, R. and Buhrman, J. M. (1980). *Statist. Neerl.*, **34**, 13–18.

[14] Karlin, S. (1968). *Total Positivity*, Vol. I. Stanford University Press, Stanford, Calif., pp. 20–21.

[15] MacGillivray, H. L. (1981). *Austral. J. Statist.*, **23**, 247–250.

[16] MacGillivray, H. L. (1982). *Commun. Statist. A*, **11**, 2239–2248.

[17] Mann, H. B. and Whitney, D. R. (1947). *Ann. Math. Statist.*, **18**, 50–60.

[18] Oja, H. (1981). *Scand. J. Statist.*, **8**, 154–168. (Proposes a theoretical structure linking the concepts of location, scale, skewness and kurtosis.)

[19] Patel, J. K., Kapadia, C. H., and Owen, D. B. (1976). *Handbook of Statistical Distributions*. Dekker, New York, pp. 51–55, 104–105.

[20] Pearson, K. (1895). *Philos. Trans. R. Soc. London*, **186**, 343.

[21] Ramberg, J. S., Tadikamalla, P. R., Dudewicz, E. J., and Mykytka, E. F. (1979). *Technometrics*, **21**, 201–214.

[22] Runnenburg, J. T. (1978). *Statist. Neerl.*, **32**, 73–79. (This paper gives some of the history of the mean-median-mode inequality.)

[23] Stavig, G. and Gibbons, J. D. (1977). *Intern. Statist. Rev.*, **45**, 63–70.

[24] Stigler, S. M. (1973). *J. Amer. Statist. Ass.*, **68**, 872–879.

[25] Tukey, J. W. (1960). In *Contributions to Probability and Statistics*, Vol. I, I. Olkin et al., eds. Stanford University Press, Stanford, Calif., pp. 448–485.

[26] Zwet, W. R. van (1964). *Statist. Neerl*, **18**, 433–441. (This contains interesting comments on skewness and examples.)

[27] Zwet, W. R. van (1979). *Statist. Neerl*, **33**, 1–5.

Bibliography

Landers, D. and Rogge, L. (1983). *Statist. Decis.*, **1**, 269–284. (The authors investigate consistent estimation of the *natural median* m' of a random variable X. $m' = \lim_{s\downarrow 1} m_s$, where

$$(E|X - m_s|^s)^{1/s} = \min_c (E|X - c|^s)^{1/s}.$$

The natural median is the usual median m of X when m is unique, but unlike m, m' always has a unique value.)

(ARITHMETIC MEAN
KURTOSIS
L-STATISTICS
MIDRANGES
QUANTILES
SKEWNESS, MEASURES OF
TRIMMING AND WINSORIZATION
TOTAL POSITIVITY
UNIMODALITY)

H. L. MacGillivray

MEAN, TRIMMED *See* *L*-STATISTICS; TRIMMING AND WINSORIZATION

MEAN, WINSORIZED *See* *L*-STATISTICS; TRIMMING AND WINSORIZATION

MEAN CONCENTRATION FUNCTION

Given a one-dimensional cumulative distribution function $F(x)$ the function

$$C_F(l) = \frac{1}{2l}\int_{-\infty}^{\infty}\{F(x+l) - F(x-l)\}^2 dx$$

is called the *mean concentration function* and, along with Lévy's concentration function*, is used to study properties of sums of independent random variables.

(LÉVY CONCENTRATION FUNCTION)

MEAN DEVIATION

The mean deviation d of a data set $x_1, \ldots, x_n$ is given by

$$d = \left(\sum_{i=1}^{n} |x_i - \bar{x}|\right) \Big/ n,$$

where $\bar{x}$ is the arithmetic mean* $\sum x_i/n$. The *population mean deviation* is ν_1, where

$$\nu_1 = E|X - \mu|, \qquad \mu = E(X),$$

X being a random variable. Then ν_1 is a measure of dispersion of the distribution of X.

POPULATION MEAN DEVIATION

Exact and approximate expressions for ν_1 in the binomial distribution* appear in Johnson [6], and for ν_1 in the hypergeometric, negative binomial, Poisson, logarithmic series, and geometric distributions in Ramasubban [13].

Karl Pearson [12] (see also [9]) established for normal and Pearson Type III (gamma) distributions (*see* FREQUENCY CURVES, SYSTEMS OF) the relationship

$$\nu_1 = 2(\text{variance}) \times (\text{probability density function at } \mu),$$

where μ is the mean of the distribution. Note that for a normal distribution with mean μ and variance σ^2, $\nu_1 = (\sqrt{2/\pi})\sigma$, and for the Laplace distribution* or exponential distribution* with scale factor a, $\nu_1 = a$. Another result along these lines was attributed to N. L. Johnson [13] and explored further by Kamat in two short papers [7, 8]; whenever the mean μ of a hypergeometric*, binomial, negative binomial* or geometric distribution* is an integer value, and for any Poisson distribution* [2],

$$\nu_1 = 2(\text{variance}) \times (\text{probability mass function at } \mu).$$

SAMPLE MEAN DEVIATION

In a random sample, d is analogously a measure of the dispersion of the sample. Due to mathematical difficulties in studying its properties, d has seen less use than the sample variance or standard deviation either in theory or in practice. Young [15] derived the first two moments of d in data observed from a symmetric multinomial distribution and set up a procedure using d to test the hypothesis of equal cell probabilities. But most attention has been given to the case of normally distributed data; we suppose the mean and variance to be μ and σ^2, respectively.

Since

$$E(d) = \sigma\sqrt{\left(1 - \frac{1}{n}\right)\frac{2}{\pi}} = \sqrt{1 - \frac{1}{n}}\,\nu_1,$$

$c_1 d$ is an unbiased estimator of the standard deviation σ, where $c_1 = \sqrt{n\pi/\{2(n-1)\}}$. R. A. Fisher* [3] showed, however, that the asymptotic efficiency* of d with respect to the unbiased estimator $c_2 s$ of σ, where s is the sample standard deviation, i.e., $s = \sqrt{\sum(x_i - \bar{x})^2/(n-1)}$, and

$$c_2 = \sqrt{\frac{2}{n-1}}\,\frac{\Gamma(n/2)}{\Gamma\{(n-2)/2\}},$$

is only $1/(\pi - 2)$ or 88%.

The variance of d [10, Sect. 5.6] is

$$\operatorname{var}(d) = \frac{2(n-1)\sigma^2}{n^2\pi}\left[\frac{\pi}{2} + \sqrt{n(n-2)} - n + \arcsin\left(\frac{1}{n-1}\right)\right].$$

The exact distribution was obtained by Godwin [5], but is cumbersome to use. The two volumes of *Biometrika Tables for Statisticians* [11] give some upper and lower percentiles and values of the cumulative distribution function, respectively, for $n \leqslant 10$. See Patel and Read [10, Sect. 5.6] for some large-sample approximations, higher moments, and further references.

GEARY'S RATIO

The studentized sample mean deviation, known as Geary's ratio or Geary's statistic, is $a = \sqrt{n/(n-1)}\, d/s$. Geary [4] determined approximate percentiles of a for $n \geqslant 10$, and derived moments of a in order to obtain those of d, since a and s are independent (one proof is an application of Basu's theorem*). He obtained asymptotic expansions of moments of both a and d in powers of $(n-1)^{-1}$, but there are some errors in these; see refs. 14 and 1; in the latter, percentiles are given for $3 \leqslant n \leqslant 9$, and the expansion of the fourth moment of a is extended. (See also ref. 10.)

For the first two moments,

$$E(a) = \sqrt{\frac{n-1}{\pi}}\ \frac{\Gamma\{(n-1)/2\}}{\Gamma(n/2)},$$

$$E(a^2) = \frac{1}{n} + \frac{2}{n\pi}\left[\sqrt{n(n-2)} + \arcsin\left(\frac{1}{n-1}\right)\right].$$

Geary's statistic has been used to test departures from normality.*

References

[1] Bowman, K. O., Lam, H. K., and Shenton, L. R. (1980). *Reports of Statist. Application Res., Union of Japanese Scientists and Engineers*, **27**, 1–15.

[2] Crow, E. L. (1958). *Biometrika*, **45**, 556–559.

[3] Fisher, R. A. (1920). *Mthly Notes R. Astronomical Soc.*, **80**, 758–769.

[4] Geary, R. C. (1936). *Biometrika*, **28**, 295–305.

[5] Godwin, H. J. (1943–1946). *Biometrika*, **33**, 254–255.

[6] Johnson, N. L. (1957). *Biometrika*, **44**, 532–533.

[7] Kamat, A. R. (1965). *Biometrika*, **52**, 288–289.

[8] Kamat, A. R. (1966). *Biometrika*, **53**, 285–287.

[9] Kamat, A. R. (1967). *Biometrika*, **54**, 333.

[10] Patel, J. K. and Read, C. B. (1982). *Handbook of the Normal Distribution*. Dekker, New York.

[11] Pearson, E. S. and Hartley, H. O. (1966, 1972). *Biometrika Tables for Statisticians*, Vols. 1 and 2. Cambridge University Press, London. (See Vol. 1, Table 21 and Vol. 2, Table 8.)

[12] Pearson, K. (1924). *Biometrika*, **16**, 198–200.

[13] Ramasubban, T. A. (1958). *Biometrika*, **45**, 549–556.

[14] Shenton, L. R., Bowman, K. O., and Lam, H. K. (1979). *Biometrika*, **66**, 400–401.

[15] Young, D. H. (1967). *Biometrika*, **54**, 312–314.

(DEPARTURES FROM NORMALITY, TESTING FOR)

CAMPBELL B. READ

MEAN DIFFERENCE *See* GINI'S MEAN DIFFERENCE

MEAN HALF-SQUARE SUCCESSIVE DIFFERENCE *See* SUCCESSIVE DIFFERENCES

MEAN OF FINITE POPULATIONS *See* SEQUENTIAL ESTIMATION OF MEAN IN FINITE POPULATIONS

MEAN RECIPROCAL VALUES

Given the random variable X, a quantity of the type $E(1/X)$, which is the expectation of the reciprocal of the random variable, crops up at times as an ingredient in the evaluation of certain quantities (e.g., variance, average life) in statistical theory. We call $E(1/X)$ the *mean reciprocal value* of X in contradistinction to the mean value or expectation $E(X)$ of X. It is worth mentioning that the reciprocal of $E(1/X)$ is the harmonic mean* of X. For the mean reciprocal value to be finite, it is mathematically necessary to make zero a value of the random variable having probability zero.

We could think in general of the mean

reciprocal values of powers of X denoted by $E(X^{-r})$ for $r > 0$ and designated as inverse or negative moments of X (also known as moments of negative order). But what we really encounter in certain applications is the mean reciprocal value of X.

The formula for $E(1/X)$ is simple in the case of some discrete distributions, as for example the geometric distribution*. However, the expression for $E(1/X)$ is generally not as simple as that for $E(X)$. When X is a positive random variable, then

$$E(1/X) \geqslant \frac{1}{E(X)}.$$

Gurland [10, 11] has established the following more general inequalities for a positive random variable:

$$E(1/X) \geqslant \frac{E(X^{\alpha-1})}{E(X^{\alpha})}; \qquad (\alpha > 0),$$

$$E(1/X^{\beta}) \geqslant (E(1/X))^{\beta}$$

$$\geqslant \left(\frac{1}{E(X)}\right)^{\beta} \geqslant \frac{1}{E(X^{\beta})};$$

$$\beta \geqslant 0, \quad \beta \leqslant -1,$$

where all the expectations appearing in the inequalities are assumed to exist.

The mean reciprocal value of an inverse gamma distribution* (employed in Bayesian analysis) is obtainable as the mean value of a gamma distribution*, and vice-versa; for X has an inverse gamma (gamma) distribution if $1/X$ has a gamma (inverse gamma) distribution.

GENERAL DEVELOPMENT

The first well-known work in the direction of evaluating mean reciprocal values of positive binomial and hypergeometric variates was due to Stephan [20], who underscored and illustrated the need of such values in the context of certain sampling problems. Grab and Savage [9] have tabulated $E(1/X)$ for positive binomial and Poisson variates; see also refs. 6–8, 22.

Evaluation of certain mean reciprocal values (spelt out below) in survey sampling* has engaged the attention of many a statistician over the last two decades or so [1, 2, 5, 12, 13, 16, 19, and 21]. A component of the type $E(1/X)$ is involved in some problems in life-testing* [3, 14] and marketing* [15].

SOME RESULTS AND APPLICATIONS

1. *Poststratified random sampling*: Let a population of size N be divided into k strata, and let n_i (> 0) be the number of units falling in the ith stratum of size $N_i (i = 1, 2, \ldots, k)$ if a sample of size n is drawn from the population. Then (see Stephan [20]) we can write to terms of order n^{-2},

$$E(1/n_i) = \frac{N}{nN_i}\left(1 + \frac{N - N_i}{nN_i}\right),$$

$$i = 1, 2, \ldots, k,$$

which is required to arrive at the average variance of the usual estimator in poststratified random sampling. For details, see Cochran [4, p. 135].

2. *Simple random sampling with replacement*: Let η be the number of distinct units in a sample of size n drawn from a population of size N according to simple random sampling with replacement. Then we have

$$E(1/\eta) = \frac{1}{N} \sum_{j=1}^{N} (j/N)^{n-1},$$

which has been worked out by several authors through a variety of methods [2, 12, 13, 16, 19, and 21]. The above mean reciprocal value is required to find the variance of an estimator defined as the average of the η distinct units.

3. *Inverse simple random sampling with replacement*: In the case of simple random sampling with replacement, we may continue sampling from a population of size N until a prespecified number n of dis-

tinct units is drawn. For this scheme, called inverse simple random sampling with replacement, the variance of the mean $\bar{y}_D$, say, based on all the D draws needed to obtain the n distinct units, involves certain mean reciprocal values of functions of D (see Lanke [13]). Dahiya et al. [5] and Agrawal [1] have evaluated the desired mean reciprocal values, i.e., if $h(n, N) = (-1)^{n-1}\binom{N-1}{n-1}$ and $g(n, j) = (-1)^j j^{-1}\binom{n-1}{j}$,

$$E(1/D) = h(n, N)\left[1 - N\sum_{j=1}^{n-1} g(n, j)\log(1 - j/N)\right],$$

$$E(1/(D+1)) = h(n, N)\left[\frac{1}{2} + N\sum_{j=1}^{n-1}\frac{1}{j} - N^2\sum_{j=1}^{n-1}\frac{1}{j}g(n, j)\log\left(1 - \frac{j}{N}\right)\right],$$

$$E(1/(D+1)^2) = h(n, N)\left[\frac{1}{4} + N\sum_{j=1}^{n-1}\sum_{r=1}^{\infty} g(n, j)\left(\frac{j}{n}\right)^r\left(\frac{1}{r+1}\right)^2\right].$$

4. *Interpenetrating subsampling* (replicated sampling)*: If we consider a sampling scheme in which a sample is chosen in the form of k subsamples of equal size, say m, drawn independently according to simple random sampling without replacement, and if η^* is the number of distinct units in the sample so drawn, then

$$E(1/\eta^*) = \binom{N}{m}^{-k}\sum_{r=0}^{N-m}\frac{\binom{N-r}{m}^k}{N-r},$$

where N is the population size. Pathak [17] and Agrawal [2] have evaluated the above mean reciprocal value which is needed to find the variance of an estimator defined as the average of the η^* distinct units.

5. *Two-phase ratio and regression estimators*: Rao [18, 19] defines certain two-phase ratio and regression estimators, the variances of which involve a certain mean reciprocal value. Rao [19] and Agrawal [2] have offered methods to evaluate the same.

6. *Problems in life-testing*: Mendenhall and Lehman [14] discussed the mean reciprocal values in the context of life-testing. Bartholomew [3] considered a problem in life-testing with a view to estimating the average life of an equipment, the solution of which depends on a certain mean reciprocal value.

7. *Dollar averaging*: Morris [15] investigated a purchasing policy under fluctuating prices called "dollar averaging" in which the amount purchased is a function of price. He obtained results which involve a certain mean reciprocal value.

References

[1] Agrawal, M. C. (1981). Statistical Research Report No. 1981-9, Institute of Mathematical Statistics, University of Umea, Sweden. (To appear in *Metron*.)

[2] Agrawal, M. C. (1982). *Math. Operationsforschung Statistik, Ser. Statist.*, **13**, 191–197.

[3] Bartholomew, D. J. (1957). *J. Amer. Statist. Ass.*, **52**, 350–355.

[4] Cochran, W. G. (1977). *Sampling Techniques*, 3rd ed. Wiley, New York.

[5] Dahiya, R. C., Korwar, R. M., and Lew, R. A. (1977). *Sankhyā C*, **39**, 162–169.

[6] Govindarajulu, Z. (1962). *J. Amer. Statist. Ass.*, **57**, 906–913.

[7] Govindarajulu, Z. (1963). *J. Amer. Statist. Ass.*, **58**, 468–473.

[8] Govindarajulu, Z. (1964). *Sankhyā B*, **26**, 217–236.

[9] Grab, E. L. and Savage, J. R. (1954). *J. Amer. Statist. Ass.*, **49**, 169–177.

[10] Gurland, J. (1967). *Amer. Statist.*, **21**(2), 24–25.

[11] Gurland, J. (1968). *Amer. Statist.*, **22**(2), 26–27.

[12] Korwar, R. M. and Serfling, R. J. (1970). *Ann. Math. Statist.*, **41**, 2132–2134.

[13] Lanke, J. (1975). Some Contributions to the Theory of Survey Sampling. University of Lund, Sweden. (Unpublished manuscript.)

[14] Mendenhall, W. and Lehmann, E. H. (1960). *Technometrics*, **2**, 233–239.

[15] Morris, W. T. (1959). *Manag. Sci.*, **5**, 154–169.

[16] Pathak, P. K. (1961). *Sankhyā A*, **23**, 415–420.

[17] Pathak, P. K. (1964). *Ann. Math. Statist.*, **35**, 795–808.

[18] Rao, P. S. R. S. (1972). *Sankhyā A*, **34**, 473–476.

[19] Rao, P. S. R. S. (1975). *J. Amer. Statist. Ass.*, **70**, 839–845.

[20] Stephan, F. F. (1945). *Ann. Math. Statist.*, **16**, 50–61.
[21] Thionet, P. (1967). *Rev. Statist. Appl.*, **15**, 35–46.
[22] Tiku, M. L. (1964). *J. Amer. Statist. Ass.*, **59**, 1220–1224.

(EXPECTED VALUE)

M. C. AGRAWAL

MEAN RESIDUAL LIFE FUNCTION (MRLF)

Let X be a nonnegative random variable (often describing the life of a component). The conditional expectation

$$E(X - t \mid X > t)$$

is the mean residual life function. It is called the *expectation of life at t* in survival analysis*. It is used in reliability theory* and survival analysis* and has been studied extensively (see, e.g., Hollander and Proschan [1], Kotz and Shanbhag [2], and the bibliography therein).

References

[1] Hollander, M. and Proschan, F. (1975). *Biometrika*, **62**, 585–593.
[2] Kotz, S. and Shanbhag, D. N. (1980). *Adv. Appl. Prob.*, **12**, 903–921.

(HAZARD RATE AND OTHER CLASSIFICATIONS OF DISTRIBUTIONS
RELIABILITY THEORY
SURVIVAL ANALYSIS)

MEAN SLIPPAGE PROBLEMS

In a collection of populations, *mean slippage* occurs when one or several of the population means differ from the common mean of the remaining populations. Any populations whose means deviate from this common mean are said to have slipped. A *slippage test* is a rule for determining whether slippage has occurred and identifying which populations, if any, have slipped. The study of slippage problems centers on the search for rules that perform these tasks well.

The mean slippage framework encompasses a wide range of situations. The populations may be normal* or nonnormal (e.g., gamma*), or a nonparametric* approach may be followed. The model of interest may be either a single slippage, in which the number of slipped populations is known to be at most one, or multiple slippage, in which there is the possibility of several slipped populations; in the latter case, the slipped means may be equal or unequal among themselves. A control group, known not to have slipped, may be either present or absent. The direction of mean slippage may be known, as when any slippage that occurs must be positive, or unknown, as when either positive or negative slippage is possible. Sample sizes from the populations may be equal or unequal. Observations may be univariate or multivariate. This list, though not exhaustive, illustrates the diversity of mean slippage problems; fortunately, many problems can be treated by a common approach.

Mean slippage was first studied by Mosteller [12]. He considered equal-sized random samples from n continuous populations, with the null hypothesis that all n populations are identical and the alternative that one among them has slipped to the right, the rest remaining identical. His rule is to find the sample containing the largest observation, determine how many observations in this sample exceed all observations in all other samples, and reject the null hypothesis when this number is sufficiently large. Another rule, due to Doornbos and Prins [5] and likely to be more powerful than Mosteller's rule [9], is to reject the null hypothesis if the greatest sample rank sum is sufficiently large, where the ith sample rank sum is the sum of the overall ranks of the observations from the ith sample.

To illustrate these methods, consider the data in Table 1 on seven varieties of guayule, with five observations of rubber yield per variety.

Each observation represents the rubber yield in grams obtained from two plants

Table 1 Rubber Yield from Seven Varieties of Guayule

Variety	Observations					Mean
1	12.15	8.20	8.94	12.27	7.32	9.776
2	12.19	4.09	8.86	8.26	7.44	8.168
3	10.54	11.71	13.90	4.96	8.51	9.924
4	7.18	9.29	5.32	5.67	5.94	6.680
5	11.82	9.88	12.62	2.88	7.34	8.908
6	14.33	9.80	12.89	13.72	17.55	13.658
7	8.21	9.08	9.90	6.62	9.10	8.582

randomly selected in a plot. To test for slippage of one population to the right with Mosteller's rule, observe that Variety 6 has the largest observation, 17.55. The number r of observations of Variety 6 that exceed all observations in other samples (17.55, 14.33) is two. For seven samples of size 5, $\Pr[r \geqslant 2] = .118$ when all populations are identical [1, Sect. 5.1.1], so the null hypothesis is not rejected at the $\alpha = .05$ level. (Mosteller's rule would reject the null hypothesis if the 13.90 observed for Variety 3 were 13.70, resulting in $r = 3$, because $\Pr[r \geqslant 3] = .0107$ under the null hypothesis.) Variety 6 has the greatest sample rank sum of $34 + 21 + 31 + 32 + 35 = 153$. This exceeds the $\alpha = .01$ critical value of 150 [9, Appendix 7], so the rule of Doornbos and Prins rejects the null hypothesis at this level. (The data are from Federer [6, p. 122], but each observation x has been replaced here by $20 - x$, e.g., 7.85 by 12.15, 11.80 by 8.20, for consistency with the discussion above of slippage to the right. With obvious adjustments, these methods could be used to examine the original data for slippage to the left.)

For extensions, modifications, and competitors of these nonparametric rules, see Barnett and Lewis [1], David [3], and Hawkins [9]. Multiple slippage, unequal sample sizes, and other cases are treated. Hashemi-Parast and Young [8] dealt with distribution-free* procedures based on sample linear rank statistics, in particular on exponential scores. Neave [13] discussed several quick, simple tests based on extreme observations; Joshi and Sathe [10] proposed another such test.

The earliest work on a parametric mean slippage model was by Paulson [14]. He took a multiple decision approach to the single slippage problem. The mutually independent $N(\mu_i, \sigma^2)$ random variables X_{ij} $(i = 1, \dots, n;\ j = 1, \dots, m)$ form n random samples of equal size m. Let D_0 be the decision that all of the means are equal, and D_i $(i = 1, \dots, n)$ the decision that population i has slipped to the right, that is, $\mu_i = \mu + \delta$ where μ is the mean of each of the $n - 1$ populations other than the ith and $\delta > 0$. Statistical rules that choose optimally, in some sense, among these $n + 1$ decisions are desired. Three reasonable restrictions on a rule are (i) when all means are equal, making D_0 correct, the probability of selecting D_0 is $1 - \alpha$; (ii) the rule is invariant under the transformation $y = ax + b$ of the observations, where $a > 0$ and b are constants; and (iii) the rule is symmetric, that is, the probability of selecting D_i when population i has slipped is the same for every i. Under these restrictions, the probability of making the correct decision when one population has slipped to the right is maximized by the rule: compute

$$T = m(\bar{x}_M - \bar{x}) \Big/ \left[\sum_{i=1}^{n} \sum_{j=1}^{m} (x_{ij} - \bar{x})^2 \right]^{1/2},$$

where $\bar{x}$ is the mean of all mn observations, and the sample mean $\bar{x}_M$ from population M is the largest of the n sample means; if $T \leqslant c_\alpha$, select D_0; if $T > c_\alpha$, select D_M, where the constant c_α is chosen to make $P[T > c_\alpha] = \alpha$ when all means are equal.

Returning to the guayule data, it is routine to calculate $T = 5(13.658 - 9.385)/(352.57)^{1/2} = 1.138$. This is greater than the $\alpha = .01$ critical value of 1.01 [1, Sect. 5.3.1], so Paulson's rule rejects the null hypothesis at this level, concluding that Variety 6 $(= M)$ has slipped.

This rule has been modified for use with slippage in an unspecified direction, additional external information about the variance σ^2, known variance, and unequal sample sizes. Rules for nonnormal populations with gamma, Poisson, binomial, and other distributions are discussed in Doornbos [4]. For details on these and related matters, see refs. 1 and 3.

A Bayesian treatment of slippage problems was given by Karlin and Truax [11]. For the case of a single slippage, they derived optimal rules under very general conditions by characterizing the class of Bayes rules (*see* BAYESIAN INFERENCE) within the set of all rules that obey certain natural restrictions of invariance* and symmetry, and then showing that Bayes procedures are uniformly most powerful*. Many special cases were examined in detail, including nonparametric situations, Paulson's model, multivariate observations, and the presence of a control group.

There is a close connection between slippage and outliers*, since a slippage problem with one observation from each population can be formulated as an outlier problem, with each slipped population corresponding to an outlier. For example, let $X_1, \ldots, X_n$ be independent normal observations with variance σ^2, one of which has mean $\mu + \delta$, where $\delta \neq 0$, and the remaining $n - 1$ of which have common mean μ. This is a mean slippage model with one slipped population. The outlier literature refers to this as Model A, and to the observation from the slipped population as an outlier caused by mean slippage. Thus outlier results for Model A apply immediately to mean slippage problems with normal populations and samples of equal size. Schwager and Margolin [15] treated a problem of this type with an unknown number of outliers.

Under the *multiple slippage* model, several population means may deviate from the common mean of the rest. For example, if there are n populations and the distribution of population i is $N(\mu_i, \sigma^2)$, $n - 2$ of the means μ_i having the common value μ, and the remaining two means the values $\mu + \delta_1$ and $\mu + \delta_2$, where $\delta_1 > 0$ and $\delta_2 > 0$, then two populations have slipped, possibly by differing amounts. When the number k of slipped populations is fixed, the multiple decision approach has the null hypothesis of no slippage and $\binom{n}{k}$ slippage alternatives that some unknown set of k populations differ from the remaining $n - k$. Butler [2] and Singh [16] treated this situation, also addressed in the outlier literature as Model A with multiple outliers.

References

[1] Barnett, V. and Lewis, T. (1978). *Outliers in Statistical Data*. Wiley, New York. (An excellent review of outlier methodology. Includes extensive bibliography and tables. Chapter 5 deals with slippage.)

[2] Butler, R. W. (1981). *Ann. Statist.*, **9**, 960–973.

[3] David, H. A. (1981). *Order Statistics*, 2nd ed. Wiley, New York. (Section 8.3 is on slippage problems.)

[4] Doornbos, R. (1966). *Slippage Tests*. Math. Centre Tracts No. 15, Mathematisch Centrum, Amsterdam.

[5] Doornbos, R. and Prins, H. J. (1958). *Indag. Math.*, **20**, 38–55, 438–447.

[6] Federer, W. T. (1955). *Experimental Design*. Macmillan, New York.

[7] Ferguson, T. S. (1967). *Mathematical Statistics: A Decision Theoretic Approach*. Academic Press, New York, Sect. 6.3.

[8] Hashemi-Parast, S. M. and Young, D. H. (1979). *J. Statist. Comp. Simul.*, **8**, 237–251.

[9] Hawkins, D. M. (1980). *Identification of Outliers*. Chapman and Hall, New York. (This monograph discusses slippage and contains a useful bibliography and tables.)

[10] Joshi, S. and Sathe, Y. S. (1981). *J. Statist. Plan. Infer.*, **5**, 93–98.

[11] Karlin, S. and Truax, D. R. (1960). *Ann. Math. Statist.*, **31**, 296–324. (An important paper; deals with Bayes rules for a variety of single slippage situations.)

[12] Mosteller, F. (1948). *Ann. Math. Statist.*, **19**, 58–65.

[13] Neave, H. R. (1979). *J. Quality Tech.*, **11**, 66–79.

[14] Paulson, E. (1952). *Ann. Math. Statist.*, **23**, 610–616.

[15] Schwager, S. J. and Margolin, B. H. (1982). *Ann. Statist.*, **10**, 943–954.

[16] Singh, A. K. (1978). *Canad. J. Statist.*, **6**, 201–218.

(DECISION THEORY
OUTLIERS
VARIANCE SLIPPAGE PROBLEMS)

STEVEN J. SCHWAGER

MEAN SQUARED ERROR

Let $\hat{\theta} = \hat{\theta}(X_1, \ldots, X_n)$ be an estimator of a parameter θ, where $X_1, \ldots, X_n$ is a random sample of size n from a distribution indexed by θ, where $\theta \in \Theta$, a parameter space. The *squared-error loss* in estimating θ by $\hat{\theta}$ is $(\hat{\theta} - \theta)^2$, and the mean square error (MSE) is its expected value* $E[(\hat{\theta} - \theta)^2]$, when this exists.

If we set $\hat{\theta} \equiv \theta_0$ for some θ_0 in Θ, then the MSE is zero at θ_0 and hence no (uniformly) minimum mean square error estimator of θ exists. It is therefore customary to seek a minimum MSE estimator under some restriction like unbiasedness* (i.e., $E(\hat{\theta}) = \theta$), for which

$$\mathrm{MSE}(\hat{\theta}) = \mathrm{Var}(\hat{\theta});$$

see ESTIMATION, POINT. In general, however,

$$\mathrm{MSE}(\hat{\theta}) = \mathrm{Var}(\hat{\theta}) + \left[E(\hat{\theta}) - \theta\right]^2, \quad (1)$$

where $E(\hat{\theta}) - \theta$ is the *bias*. The property $\lim_{n\to\infty}\mathrm{MSE}(\hat{\theta}) = 0$ is *mean square error consistency*, and implies that both the variance and the bias of θ tend to zero as $n \to \infty$.

Example. In a sample from a normal population with unknown mean and variance σ^2, that minimum MSE estimator of σ^2 which is a multiple of the sample variance is $(n+1)^{-1}\sum_{i=1}^{n}(X_i - \bar{X})^2$, where $\bar{X} = \sum X_i/n$. This improves upon the sample variance $(n-1)^{-1}\sum_{i=1}(X_i - \bar{X})^2$, which is the minimum variance unbiased* estimator of σ^2.

In most textbooks, the MSE is preferred over the *mean absolute error* (MAE) $E|\hat{\theta} - \theta|$ as a criterion for evaluating estimator performance on the grounds that the former is mathematically convenient while the latter is not. Bickel and Doksum [1, Sect. 4.1] point out, however, that if $\hat{\theta}$ is approximately normal with mean θ, as happens if $\hat{\theta}$ has asymptotic normality* and if n is large, then

$$E|\hat{\theta} - \theta| \simeq \sqrt{2/\pi}\,\sqrt{\mathrm{MSE}(\hat{\theta})}, \quad (2)$$

so that a minimum MSE estimator is then approximately a minimum MAE estimator also.

REGRESSION

The performance of a predictor of a variable Y which depends on a vector $(X_1, \ldots, X_k)$ to be observed is often evaluated by least squares*, but the 1960s and more particularly the 1970s saw a shift of emphasis, to biased regression*. In general $E(Y|X_1, \ldots, X_k)$ is the unique minimum MSE predictor of Y, but may depend on unknown regression coefficients. Denote the multiple regression model by

$$\mathbf{Y} = \mathbf{X}\boldsymbol{\beta} + \boldsymbol{\epsilon}, \quad (3)$$

where $\mathbf{Y}$ is $n \times l$ (response variables), $\mathbf{X}$ is $n \times p$ (known constants), $\boldsymbol{\beta}$ is $p \times l$ (unknown regression coefficients), and $\boldsymbol{\epsilon} \sim N(\mathbf{0}, \sigma^2\mathbf{I}_n)$ (normally distributed errors). One purpose in introducing biased regression estimators $\hat{\boldsymbol{\beta}}$ is to achieve some reduction in MSE; *see* JAMES–STEIN ESTIMATORS; RIDGE REGRESSION, and a review in ref. 3. Performance criteria include:

(a) *Total MSE,*

$$E\left[(\hat{\boldsymbol{\beta}} - \boldsymbol{\beta})'(\hat{\boldsymbol{\beta}} - \boldsymbol{\beta})\right];$$

(b) *Generalized MSE,*

$$E\left[(\hat{\boldsymbol{\beta}} - \boldsymbol{\beta})'\mathbf{A}(\hat{\boldsymbol{\beta}} - \boldsymbol{\beta})\right],$$

in which the $p \times p$ nonnegative definite matrix $\mathbf{A}$ is chosen to weight certain regression coefficients (or combinations thereof) more than others [5];

(c) *Integrated MSE,*

$$\int \cdots \int_R E[\{\hat{Y}(\mathbf{x}) - E(Y \mid \mathbf{X} = \mathbf{x})\}^2] W(\mathbf{x})\, d\mathbf{x}$$

where $\hat{Y}(\mathbf{x}) = \mathbf{x}'\hat{\boldsymbol{\beta}}$ predicts Y when $\mathbf{X} = \mathbf{x}$, $W(\mathbf{x})$ is a normalized nonnegative function which weights some $\mathbf{x}$-vector values more than others, and R is the region of $\mathbf{x}$-values of interest [2].

Hess and Gunst [4] give some background to the use of these criteria and prove that if (3) is correct and if $\mathbf{A} = \int \cdots \int_R \mathbf{x}\mathbf{x}' W(\mathbf{x})\, d\mathbf{x}$, then

$$(\text{Integrated MSE}) = (\text{Generalized MSE}) + \sigma^2/n.$$

References

[1] Bickel, P. J. and Doksum, K. A. (1977). *Mathematical Statistics: Basic Ideas and Selected Topics*. Holden-Day, San Francisco, Calif.

[2] Box, G. E. P. and Draper, N. R. (1959). *J. Amer. Statist. Ass.*, **54**, 622–654.

[3] Draper, N. R. and van Nostrand, R. C. (1979). *Technometrics*, **21**, 451–466.

[4] Hess, J. L. and Gunst, R. F. (1980). *Comm. Statist. A*, **9**, 321–326.

[5] Theobald, C. M. (1974). *J. R. Statist. Soc., B* **36**, 103–106.

(ESTIMATION, POINT
JAMES–STEIN ESTIMATORS
MINIMUM VARIANCE UNBIASED ESTIMATORS
REGRESSION, LINEAR
RIDGE REGRESSION
UNBIASEDNESS)

CAMPBELL B. READ

MEAN SQUARE SUCCESSIVE DIFFERENCE *See* SUCCESSIVE DIFFERENCES

MEAN SUCCESSIVE DIFFERENCE *See* SUCCESSIVE DIFFERENCES

MEAN TIME TO FAILURE (MTTF)

This is another name for expectation of life, or expected lifetime. It issued more particularly in reliability theory* and is very often the mean θ of an exponential distribution* given by the density function

$$f(y) = \frac{1}{\theta} e^{-y/\theta}, \qquad y > 0, \quad \theta > 0,$$

although this is by no means always the case. It is also called MTBF (mean time between failures). (In practice MTBF is calculated via system availability* or unavailability* and the failure frequency).

(EXPONENTIAL DISTRIBUTION
RELIABILITY THEORY)

MEAN UNBIASED ESTIMATOR *See* ESTIMATION; LOCATION SCALE PARAMETER

MEAN-VARIANCE ANALYSES

Mean-variance analyses, used principally in financial economics, involve preference-choice models among risky alternatives over future returns that focus exclusively on expected returns and variances as the bases of preference, and presume that greater expected return and smaller variance are desirable. In particular, if $\succ$ denotes a partial order for preferences of an individual or firm on a set $\{F, G, H, \ldots\}$ of probability distribution functions that characterize risky alternatives (read $F \succ G$ as "F is preferred to G"), and if $\mu_F = \int x\, dF(x)$ and $\sigma_F^2 = \int (x - \mu_F)^2\, dF(x)$, the mean-variance models assume

(i) If $(\mu_G, \sigma_G) = (\mu_H, \sigma_H)$, then $F \succ G \Rightarrow F \succ H$, and $G \succ F \Rightarrow H \succ F$;

(ii) If $\sigma_F \leqslant \sigma_G$ and $\mu_F > \mu_G$, then $F \succ G$;

(iii) If $\mu_F \geqslant \mu_G$ and $\sigma_F < \sigma_G$, then $F \succ G$.

Three specific mean-variance models are the dominance model, the tradeoff model, and the lexicographic model. The *dominance model* defines $\succ$ by

$F \succ G$ if and only if the hypothesis of (ii) or (iii) holds, and refers to the set $\{F : G \succ F \text{ for no } G\}$ of undominated distributions as the efficient $(E - V, \mu - \sigma)$ frontier. Dominance analyses identify efficient alternatives on the frontier as a step toward a final choice.

A *tradeoff model* assumes that there is a utility* function U defined on (μ, σ) pairs that increases in μ, decreases in σ, and completely orders preferences in the sense that

$$F \succ G \quad \text{if and only if} \quad U(\mu_F, \sigma_F) > U(\mu_G, \sigma_G). \tag{1}$$

Tradeoffs between μ and σ are identified by (μ, σ)-sets over which U is constant. Tradeoff analyses seek to determine feasible F that maximize $U(\mu_F, \sigma_F)$.

The *lexicographic model* takes

$$F \succ G \quad \text{if and only if} \quad \mu_F > \mu_G \quad \text{or} \quad (\mu_F = \mu_G, \sigma_F < \sigma_G),$$

so that expected return is the dominant criterion. Although this model is much less frequent in the literature than the others, it is a logical consequence of (i)–(iii) when distributions that can have arbitrarily large or small returns are included [8, 9].

The mean-variance approach was popularized by Markowitz [18, 19], Tobin [28, 29], Sharpe [25, 26], and Lintner [15, 16] in financial economics, especially with regard to portfolio selection (*see* PORTFOLIO ANALYSIS) and capital budgeting. Its inclusion of variance as a rough measure of riskiness (*see* RISK MEASUREMENT, FOUNDATIONS OF) is viewed as an improvement over the traditional criterion of maximizing expected return. When multiple investment opportunities can be mixed in various proportions to construct risky alternatives, their covariance matrix for returns forms an integral part of the analysis.

Mean-variance models are sometimes portrayed as approximations of expected utility models (*see* UTILITY THEORY and DECISION THEORY) that represent $\succ$ by

$$F \succ G \quad \text{if and only if} \quad \int u(x)\,dF(x) > \int u(x)\,dG(x), \tag{2}$$

where u is an increasing function on returns. In general (1) and (2) can characterize the same preference relation only if either u is quadratic with x bounded or the distributions are in a two-parameter family [2, 3, 6, 11]. Without such restrictions, the mean-variance approach may still provide a good approximation to expected utility [23, 24, 30] in certain circumstances. Prime among these is that u be concave, in which case it is said to be *risk averse*.

Comparisons have also been made between the mean-variance dominance order, defined by (ii) and (iii), and the second-degree stochastic dominance order [31] $>_2$ defined by $F >_2 G$ if

$$\int u(x)\,dF(x) > \int^{u}(x)\,dG(x)$$

for all strictly increasing and concave utility functions u. The interface between these approaches is explicated in Fishburn [9]; Dybvig and Ross [5] characterize the efficient frontier based on $>_2$ and compare it to the $\mu - \sigma$ frontier.

Apart from direct comparisons to expected utility and stochastic dominance, there have been two main criticisms of the mean-variance approach. The first is that it does not go far enough, since higher moments may also affect preferences. Arguments for inclusion of skewness* have been made by several authors [1, 12, 13, 30].

The second main criticism is that, insofar as variance is used to measure the riskiness of distributions, it is the wrong measure. As Markowitz [19] and others observe, variance is a symmetric parameter that does not differentiate between large and small returns, whereas risk is commonly associated with possible losses or shortfalls below an acceptable target return. Consequently, other measures of risk have been proposed. The resultant collection of mean-risk models includes dominance, tradeoff, and lexicographic forms. An example of the last of these is a safety-first model [22] in which the probability of ruin is dominant.

Other specific risk measures include weighted losses [4], loss probability [19, 20], and below-mean or below-target semivariance [10, 17, 19, 21]. A fuller discussion of these measures and related parameters is given in [14]. General classes of risk measures are discussed by Stone [27] and Fishburn [7], the latter of which makes a detailed analysis—including congruence with expected utility and stochastic dominance—of the two-parameter risk measure

$$F_\alpha(t) = \int_{-\infty}^{t} (t-x)^\alpha \, dF(x),$$

where t is the target return and $\alpha \geqslant 0$ is a risk parameter. The values 0, 1, and 2 for α characterize, respectively, below-target probability, below-target expectation, and below-target semivariance.

References

[1] Alderfer, C. P. and Bierman, H. (1970). *J. Business*, **43**, 341–353. (Includes experiments to test moments' salience.)

[2] Bawa, V. S. (1975). *J. Finan. Econ.*, **2**, 95–121.

[3] Chipman, J. S. (1973). *Rev. Econ. Studies*, **40**, 167–190. (Mathematical analysis of moments versus expected utilities.)

[4] Domar, E. V. and Musgrave, R. A. (1944). *Q. J. Econ.*, **58**, 389–422. (Risk analysis in economics.)

[5] Dybvig, P. H. and Ross, S. A. (1982). *Econometrica*, **50**, 1525–1546. (Second-degree efficient sets.)

[6] Feldstein, M. S. (1969). *Rev. Econ. Studies*, **36**, 5–12.

[7] Fishburn, P. C. (1977). *Amer. Econ. Rev.*, **67**, 116–126.

[8] Fishburn, P. C. (1979). *Theory and Decision*, **10**, 99–111. (Logical implications of mean-variance assumptions.)

[9] Fishburn, P. C. (1980). *Res. Finance*, **2**, 69–97.

[10] Hogan, W. W. and Warren, J. M. (1974). *J. Finan. Quant. Anal.*, **9**, 1–11. (Capital markets and semivariance.)

[11] Klevorick, A. (1973). *Rev. Econ. Studies*, **40**, 293–296.

[12] Kraus, A. and Litzenberger, R. H. (1976). *J. Finance*, **31**, 1085–1100.

[13] Levy, H. (1969). *J. Finance*, **24**, 715–719.

[14] Libby, R. and Fishburn, P. C. (1977). *J. Acct. Res.*, **15**, 272–292. (Survey of models with risk in business decisions.)

[15] Lintner, J. (1965). *Rev. Econ. Stat.*, **47**, 13–37.

[16] Lintner, J. (1965). *J. Finance*, **20**, 587–615. (Value of portfolio diversification in the mean-variance setting.)

[17] Mao, J. C. T. (1970). *J. Finan. Quant. Anal.*, **4**, 657–675.

[18] Markowitz, H. (1952). *J. Finance*, **7**, 77–91. (Major stimulus for mean-variance development.)

[19] Markowitz, H. (1959). *Portfolio Selection*. Wiley, New York. (A classic in financial economics.)

[20] Payne, J. W. (1975). *J. Exp. Psych.: Human Percep., Performance*, **104**, 86–94. (Experiments on perceived risk.)

[21] Porter, R. B. (1974). *Amer. Econ. Rev.*, **64**, 200–204.

[22] Roy, A. D. (1952). *Econometrica*, **20**, 431–449.

[23] Samuelson, P. A. (1970). *Rev. Econ. Studies*, **37**, 537–542. (Approximation theory for portfolio analysis.)

[24] Samuelson, P. A. and Merton, R. C. (1974). *J. Finance*, **29**, 27–40.

[25] Sharpe, W. F. (1963). *Manag. Sci.*, **9**, 277–293.

[26] Sharpe, W. F. (1964). *J. Finance*, **19**, 425–442. (Capital asset pricing model.)

[27] Stone, B. K. (1973). *J. Finance*, **28**, 675–685.

[28] Tobin, J. (1958). *Rev. Econ. Studies*, **25**, 65–85.

[29] Tobin, J. (1965). In *The Theory of Interest Rates*, F. H. Hahn and F. P. R. Brechling, eds. Macmillan, London, pp. 3–51. (Extensive discussion of the portfolio selection problem.)

[30] Tsiang, S. C. (1972). *Amer. Econ. Rev.*, **62**, 354–371.

[31] Whitmore, G. A. and Findlay, M. C., eds. (1978). *Stochastic Dominance*. Heath, Lexington, Mass. (Broad introduction to stochastic dominance and its applications. R. C. Burgess compares mean-variance and mean-semivariance models.)

(DECISION THEORY
FINANCE, STATISTICS IN
RISK MEASUREMENT, FOUNDATIONS OF
UTILITY THEORY)

PETER C. FISHBURN

MEASUREMENT ERROR

Errors of measurement are the differences between observed values recorded under "identical" conditions and some fixed true value. The observed values are recorded, realized. The "true value," the *quaesitum*, is a concept that has attracted much discussion [1]. Here we take the true value to be a constant η, and further assume that the measurement process provides observations in a state of statistical control. More formally, we assume the measurement process provides errors $\epsilon_1, \epsilon_2, \ldots$ that are independent identically distributed random variables with a fixed probability density function having mean $E(\epsilon) = 0$ and finite variance σ^2. The observations are $y_i = \eta + \epsilon_i$, $i = 1, 2, \ldots, n$. We now define the true value η as the limit of the average $\bar{y} = \sum_i y_i/n$ as the number of observations $n \to \infty$; then

$$\Pr\{|\bar{y} - \eta| > \delta\} \to 0.$$

To demonstrate that a measurement process is in statistical control it is common to employ Shewhart charts, in particular the $\bar{y}$ and R charts (*see* CONTROL CHARTS), and to inspect historically recorded observations for evidences of nonrandomness. Good practice also suggests that whenever possible the metric chosen for the measurements should provide a symmetrically distributed error distribution.

Consider a continuous quantitative measurement $y_o = \eta + \epsilon$ recorded without bias*. The value actually recorded, y, will be a discretized approximation to y_o. When all values of y_o falling within the cell boundaries $y \pm \delta/2$ are recorded as y the measurements are said to be *rounded*. The measurements are said to be *truncated* when all values of y_o falling within the cell boundaries $y, y + \delta$ are recorded as y. The random quantification error d is assumed to have a uniform distribution with $E(d) = 0$ and $V(d) = \sigma^2/12$. Given n independently recorded measurements $y_1, y_2, \ldots, y_n$, and statistics $\bar{y} = \sum_i y_i/n$ and $s^2 = \sum_i (y_i - \bar{y})^2/(n-1)$ then:

$$E(\bar{y}) = \eta \qquad \text{for rounded data,}$$

$$E(\bar{y}) = \eta - \delta \qquad \text{for truncated data,}$$

$$E(s^2) = \sigma^2 + \delta^2/12.$$

These approximations hold well for values of $\delta < \sigma/3$. Exact series expressions for $E(y)$ and $E(s^2)$ are given in [2].

Repeated measurements* recorded under identical conditions will not be equal. This intrinsic variability is sometimes identified as the *repeatability variance* of the measurement process. Estimating the repeatability variance is not easy since, once an assignable cause of variability is proposed, it is presumably possible to design an experimental program to eliminate its contribution from the estimate. The *reproducibility variance* of a measurement process is taken to reflect all random contributions to measurement as, e.g., contributions due to different conditions, instruments, operators, samples, days, laboratories, and environments.

In practice, definitions of repeatability variance and reproducibility variance depend on local decisions as to which components of variance to include, or exclude, in the estimate. For most measurement processes, repeatability variance is taken to include all those natural sources of variability occurring within a laboratory, for example, due to operators and samples. Reproducibility variance commonly considers components external to the laboratory, in particular the variance between laboratories and test materials.

Experimental designs useful for estimating repeatability and reproducibility variance can be hierarchical or nested (e.g., determinations made within samples, on samples within batches, and on batches within days) or composed of cross-classification designs (e.g., laboratories by materials by days) with possible additional hierarchical classifications* within the cells. The estimation of components of variance* requires model II analysis of variance [3–5], or mixed model I and II analyses [6, 7] (*see also* FIXED-, RANDOM-, AND MIXED-EFFECTS MODELS). Bayesian* estimation procedures also have been developed [8].

In its simplest form a hierarchical data array consists of k laboratories each making n measurements on a single item. The model for the observations y_{ij}, $i = 1, 2, \ldots, k$; $j = 1, 2, \ldots, n$ is

$$y_{ij} = \eta + \lambda_i + \epsilon_{ij},$$

where the λ_i are independent normally distributed random effects attributable to laboratories with $E(\lambda_i) = 0$ and $V(\lambda_i) = \sigma_R^2$; and where the ϵ_{ij} also are normally and independently distributed with $E(\epsilon_{ij}) = 0$ and $V(\epsilon_{ij}) = \sigma_r^2$. The λ_i are independent of the ϵ_{ij}. The

Table 1 Model II Analysis of Variance

Source	Sum of Squares	Degrees of Freedom	Mean Squares	Expected Mean Squares
Between	$n\sum_i(\bar{y}_i - \bar{y})^2 = S_b$	$k-1$	$S_b/(k-1)$	$\sigma_w^2 + n\sigma_b^2$
Within	$\sum_i\sum_j(y_{ij} - \bar{y}_i)^2 = S_w$	$k(n-1)$	$S_w/\{k(n-1)\}$	σ_w^2
Total	$\sum_i\sum_j(y_{ij} - \bar{y})^2$	$nk-1$		

associated analysis of variance* Model II table thus would be as in Table 1. The repeatability variance σ_r^2 (equivalently, the within variance σ_w^2) is estimated by

$$\hat{\sigma}_w^2 = \hat{\sigma}_r^2 = S_w/\{k(n-1)\}$$

and the reproducibility variance σ_R^2 by

$$\sigma_R^2 = [S_b/(k-1) - S_w/k(n-1)]/n.$$

The laboratory analyst may be more interested in repeatability and reproducibility *intervals*. The repeatability interval I_r is defined to be the width of the $100(1-\alpha)\%$ confidence interval* (commonly 95%) appropriate to the difference between two single observations recorded within a laboratory, that is, approximately $2\sqrt{2}\,\hat{\sigma}_w$. The reproducibility interval I_R usually is defined as the width of the $100(1-\alpha)\%$ confidence interval appropriate to the difference between two observations, one observation from each of two laboratories. Thus, very approximately,

$$I_R = 2\sqrt{2}\,[\hat{\sigma}_R^2 + \hat{\sigma}_r^2]^{1/2}.$$

Experimental designs and analyses, called "round robin" trials, often are used to test the performance of different measurement protocols, or to provide a check upon the performance of collaborating laboratories. Many practical difficulties occur in the statistical analyses of such programs: missing or aberrant observations, unbalanced numbers of observations, nonhomogeneous variance, cross classification interactions and nonnormality [9, 10].

Many responses, particularly those in the physical and chemical sciences, are rarely measured directly. Surrogate, easily measured responses are used instead, e.g., electrical resistance or color in place of chemical concentrations. A calibration curve is required therefore to relate the surrogate response to the actual desired response. The construction of calibration* curves, particularly when both the surrogate and true responses are subject to errors, remains an unresolved problem in applied statistics. The surrogate response also must be specific to the desired response, and care must be taken to ensure that the instrument probe or the structure of the sample does not interfere with the recording of the measurement. Clearly, the very act of measurement can itself contribute to both bias and variance. Finally, all measurements should be traceable to national standards* [11].

References

[1] Eisenhart, C. (1963). *J. Res. Nat. Bur. Stand.*, **67C**, 161–187.

[2] Schwartz, L. M. (1980). *Anal. Chem.*, **52**, 1141–1147.

[3] Cochran, W. G. and Cox, G. M. (1957). *Experimental Design*. Wiley, New York.

[4] Anderson, R. L. (1975). In *Statistical Design and Linear Models*, J. N. Srivastava, ed. North-Holland, Amsterdam, Chap. 1.

[5] Snedecor, G. W. and Cochran, W. G. (1967). *Statistical Methods*, 6th ed. Iowa State University Press, Ames, Iowa.

[6] Scheffé, H. (1959). *The Analysis of Variance*. Wiley, New York.

[7] Searle, S. R. (1971). *Linear Models*. Wiley, New York.

[8] Box, G. E. P. and Tiao, G. (1973). *Bayesian Inference in Statistical Analysis*. Addison-Wesley, Reading, MA.

[9] Mandel, J. (1977). *ASTM Stand. News*, **5/3**.

[10] Mandel, J. (1978). *ASTM Stand. News*, **6/12**.

[11] Hunter, J. S. (1980). *Science*, **210**, 869–874.

(CALIBRATION
ERROR ANALYSIS

GRUBBS' ESTIMATORS
PRECISION)

J. S. HUNTER

MEASUREMENT STRUCTURES AND STATISTICS

MEASUREMENT STRUCTURES AND STATISTICS

Consider a simple two-group design. There are two independent random samples, one from a continuous population A, another from a continuous population B. It is desired to test the null hypothesis that the population means are identical against the alternative that they are different. Assume that the populations are normal in distribution and have common variance; then a two-sample t-test is easily constructed to evaluate the hypothesis, and it is of interest to focus on certain statistical issues related to measurements of the variable of interest. A *measurement* is simply a numerical assignment to some element, usually nonnumerical. Measurements convey certain information about the element and its relationship to other elements. In a certain sense, some measurements are richer in the information conveyed than others. Those that convey less information present certain statistical difficulties.

Suppose that the random variable of interest is height. Measurements of height or weight are *ratio scale measurements* because measurements of an object in two different metrics are related to each other by an invariant ratio. For instance, if x is measurement in centimeters, and y is measurement in inches, then $x/y = 2.54$. Similarly, a change from one ratio measurement scale to another is performed by a transformation of the form $y = \alpha x$, $\alpha > 0$. In measurement theory such transformations are *permissible transformations*.

Although positive, height measurements usually are assumed to be roughly normal in distribution. It is obvious that the value of the t-test statistic is invariant regardless of a ratio scale's unit of height measurement, because multiplication of the observed responses by a positive constant does not change the t-test.

Alternatively, suppose the observations were measurements in degrees Fahrenheit (°F). Conversion to Celsius degrees (°C) is obtained by a permissible transformation of the form $y = \alpha x + \beta$, $\alpha > 0$. Scales of this form are interval scales because while ratios are not invariant, ratios of differences of intervals are invariant. For example, 136°F $= 57\frac{7}{9}$°C; 68°F = 20°C; 34°F $= 1\frac{1}{9}$°C; 17°F $= -8\frac{1}{3}$°C. Clearly 68°F/20°C ≠ 34°F/$1\frac{1}{9}$°C but

$$\frac{136°\text{F} - 68°\text{F}}{34°\text{F} - 17°\text{F}} = \frac{57\frac{7}{9}°\text{C} - 20°\text{C}}{1\frac{1}{9}°\text{C} - (-8\frac{1}{3}°\text{C})}.$$

As with ratio scales, a permissible transformation from one interval scale to another leaves the two-sample t-test invariant.

If measurements were always made on ratio or interval scales, as may be the case in some areas of physical science, there would be a few problems. But most measurements of interest in the behavioral sciences and education are poorly understood; it often is not clear what such scales are, in fact, measuring. Furthermore, the construction of such scales frequently uses arbitrary conventions. Thus there is no obvious reason why certain scale values (measurements) receive one numerical value rather than another. Ordinary classroom tests, opinion poll attitude measurements, and measurements of psychiatric disturbance are examples. This does not mean that such measurements are useless; but claims regarding their properties often tend to be inflated, and upon close scrutiny typically are statements of belief without conceptual foundation. Often the claim is made that the scales have interval scale properties. A familiar example is IQ test scores which are claimed to be interval scale measurements (cf. Thomas [22]), yet there has never appeared any conceptual justification for this belief. What would constitute adequate conceptual justification is discussed below. For the moment, regard IQ scores as *ordinal scale measurements*, which means that numerical assignments simply order elements (*see* ORDINAL DATA). For instance, if children a, b, and c have received IQ scores such that $\mathrm{IQ}(a) = 110$, $\mathrm{IQ}(b) = 117$, and $\mathrm{IQ}(c) = 89$, then the measurements simply order the children b, a, c in order of decreasing intelligence (note the fuzziness in meaning of the term intelligence when contrasted with the terms length or mass). A permissible scale transformation for an ordinal scale is any order preserving transformation, i.e., $f: \mathrm{IQ} \rightarrow \mathrm{IQ}'$ where f is order preserving. Thus $\mathrm{IQ}'(a) = -33$, $\mathrm{IQ}'(b) = 0$, and $\mathrm{IQ}'(c) = -100$ would do as well if IQ measurements were regarded as ordinal scale measurements.

Suppose that the measurements of concern are ordinal, that normality holds, and that the null hypothesis regarding the equality of population means is rejected by the t-test*. There is immediately a problem; with ordinal scale measurements any order preserving permissible scale transformation of the scores would do as well. It is clear that by a suitable monotonic transformation —perhaps simply changing a value or two— the value of the t statistic may be changed, because the t-statistic is not invariant under monotonic (order preserving) transformations. Thus the statistical decision could be changed at will, simply by appropriate monotonic transformation, at least for many data sets. (It might be argued that a monotonic transformation will change the distribution of the scores, thus invalidating the use of the t-test. But with sample size sufficiently large, t is largely distribution-free.) Clearly this is an unhealthy state of affairs; it has troubled social scientists for years.

The classification of measurements into three basic scale types, ratio, interval, and ordinal, is due to S. S. Stevens in 1946 [20] and 1951 [21]. Stevens also proposed that scale types dictated the appropriate statistics. Those deemed appropriate were termed *permissible statistics*, viewed as those which obeyed certain (loosely defined) invariance rules. The sample mean, sample variance, and Pearson correlation* coefficients were regarded as permissible statistics for interval and ratio scales but not for ordinal scales. (There is a certain inconsistency in Stevens' meaning of invariance here. While the correlation coefficient remains numerically identical when a permissible interval scale transformation is applied to either or both marginal variables, clearly the mean and standard deviation do not remain invariant; and of course only the mean changes when the observations are incremented by an additive constant, not the variance.) The geometric mean* and coefficient of variation* were regarded as permissible for ratio scale measurements only. For ordinal measurements the sample median and rank correlation were regarded as permissible.

Since there is a nested relation among the three scale types in terms of their permissible transformations, i.e., a ratio scale implies an interval scale which implies an ordinal scale, permissible statistics of the more general ordinal scale were regarded as appropriate for the special cases.

Stevens was not a statistician, and he was not concerned with the problem of inference. But a natural extension of his viewpoint is to cast statistical inference into a scale-theoretical orientation. The idea was not long in coming, and over the years there has been a steady stream of applied statistics texts with this viewpoint. Two early books were Senders (1958) [18] and Siegel (1956) [19]. Siegel's book, *Nonparametric Statistics for the Behavioral Sciences*, has remained in the applied literature as a widely cited text.

Siegel's general approach has considerable appeal. In the two-sample location problem, for example, it is perfectly reasonable to demand that the value of the test statistic and thus its probability value remain invariant with permissible transformations of the data. For ordinal data, rank-based tests such as those of the Wilcoxon type have this essential feature, which classical normal theory t-tests lack. Thus at least for some problems, the nonparametric approach skirts a troublesome measurement problem, and simultaneously it allows for weaker distribution and sampling assumptions. Furthermore, even if the normal theory t-test is regarded as appropriate, nonparametric tests retain good power, are only slightly less efficient than the normal theory test when the normal model holds, and are more powerful for some nonnormal models (cf. Lehmann [13]). A measurement-theoretical view of this same result would be that even if the scale structure were interval, little statistical information is lost when the observations are replaced by a monotonic transformation, the ranks. This result seems surprising. *See also* DISTRIBUTION-FREE METHODS.

It is interesting that in the behavioral sciences the use of nonparametric methods often is viewed largely as a solution to measurement concerns, not necessarily statistical concerns. This was a main thrust of Siegel's (1956) book. Statisticians, however (Siegel was a psychologist), typically do not mention this motivation for using nonparametric methods.

From a practical viewpoint the nonparametric approach in earlier years lacked alternatives to complex analysis of variance* designs. The situation has not improved much. Procedures in the literature often are difficult to compute and are not available in statistical computer packages, and little is known about the behavior of the tests in small samples [5]. However, from a measurement viewpoint, the approach of Conover and Iman [7] has considerable appeal. Their suggestion is simple; merely replace the observations with their ranks and compute t or F statistics in the usual way, based on these rankings. However, from a statistical viewpoint the approach has shortcomings. For example, in complex models it is not clear how the rankings should be done. The most serious difficulty is that their suggestion lacks an adequate conceptual foundation. Thus in some cases it may not be clear what hypotheses are tested or how the resulting estimates are to be viewed. The hypotheses of interest in conventional normal theory settings may not, unfortunately, readily translate into corresponding hypotheses on the ranks [9].

It has been argued that measurement issues should be viewed independently of statistical issues. Lord's [14] light-hearted classic paper "On the Statistical Treatment of Football Numbers" makes the critical point that in terms of scale structure, "The numbers don't remember where they came from" (p. 751), suggesting that statistical issues are one matter, measurement issues another. One point of the controversy focused on the belief, clearly stated by Siegel, that underlying the t-test was an assumption of interval scale measurement [19, pp. 11 and 31]. Whatever the apparent source of this wrong belief, which was not Stevens' position, it dies hard. Even after periodic reminders that such statements such as Siegel's are wrong [10, 11] books with wrong statements continue to appear.

The measurement issue became intertwined with another issue, the appropriate role of nonparametric versus parametric methods in behavioral data analysis. Analysis of variance* models had become firmly entrenched in psychology* by the mid-1950s, not just as a procedure for evaluating experimental effects but as a general conceptual framework for viewing certain psychological phenomena, particularly learning and motivational phenomena. Since there were no nonparametric alternatives to complex experimental designs, there was little to argue about when more complicated experiments were of interest. Still, the conceptual defenses for employing parametric procedures were, from a measurement viewpoint, weak. For instance, in a lengthy defense, Anderson noted, "Although rank order* tests do possess some logical advantage over parametric tests when only permissible transformations are considered, this advantage is, in the writer's opinion, very slight" [4, p. 313]. The defense had become largely a belief statement.

In the 1950s and 1960s there was a substantial number of typically brief papers or comments focusing on the general measurement–statistics issue. Several were stimulated by Lord's (1953) paper [14]. Other than being able to divide writers into what Burke

[6] characterized as the measurement-directed or measurement-independent positions, there was little conceptual progress beyond noting that statistical procedures do not depend on measurement scale assumptions.

However, some progress was made by Adams et al. [3]. Their paper is the most thoroughgoing attempt to deal with fundamental issues. In part, the paper may be viewed as a rigorous extension and development of Stevens' ideas [21]. The paper is largely definitional in character; although there are few general results, they have clarified certain difficulties. For instance, Stevens' vague definition of invariance* is now precisely defined by distinguishing between different types of invariance. Their main conclusion is that scale types do not dictate which statistics may be computed, but, depending on the use to be made of the statistics, some statistics may not be empirically meaningful.

Suppose it is desired to order two-sample means based on numerical data, $x_1, \ldots, x_n$ and $y_1, \ldots, y_m$. The mean would be regarded as meaningful when

$$\begin{aligned} n^{-1}\Sigma x_i > m^{-1}\Sigma y_j \qquad \text{iff} \\ n^{-1}\Sigma g(x_i) > m^{-1}\Sigma g(y_j) \end{aligned} \tag{1}$$

holds for all scale-permissible transformations g. Otherwise, the mean would not be regarded as meaningful. If $g(w) = \alpha w + \beta$, $\alpha > 0$, (1) holds for ratio and interval scales. If g is simply order preserving, (1) will not hold and thus the mean is not regarded as meaningful for this statement. On the other hand, a condition like (1) defined for medians would be meaningful for ordinal scales. More recent considerations of scales and statistics have focused on similar or identical definitions of meaningfulness [16, 17].

As Adams et al. [3] note, definitions of meaningfulness like (1) are critically dependent on being able to specify a particular measurement structure's class of permissible transformations. To do so requires that the structure be well understood. As noted above, most measurement structures in common use in the behavioral sciences are not well understood. Thus their class of permissible transformations is not clear. While it was assumed that test scores such as IQ scores were ordinal, even this point is debatable. Thus definitions tied to permissible transformations appear to be restricted in usefulness to a fairly narrow class of well-understood measurement structures, possibly excluding most of the common measurement procedures used in the behavioral sciences.

Since the early 1960s there has been considerable research in representational measurement theory. This axiomatic approach might be regarded as having two main goals: first, to put certain commonly used measurement structures on a firm conceptual foundation, and second, to develop new measurement procedures particularly for the behavioral sciences. In each case there is an attempt to specify certain structural conditions (axioms) which often are viewed as empirically testable conditions. Once these conditions have been specified, a mapping into the reals is sought that numerically represents the empirical structural conditions. To illustrate, consider the construction of an ordinal scale. (The following is after Krantz et al. [12].) Let the pair (A, R) denote a finite set A of typically nonnumerical elements and R a binary order relation on A. A might be a collection of persons or of beers. R might be viewed as, for example, a preference relation if the elements were beers. Then if the conditions (i) either aRb or bRa (connectedness) and (ii) if aRb and bRc then aRc (transitivity) hold for all elements of A, then there exists a map from (A, R) into $(\mathrm{Re}, \geqslant)$ such that

$$aRb \qquad \text{iff} \quad f(a) \geqslant f(b), \tag{2}$$

which states the representation theorem for ordinal measurement. Note that f *is* the measurement. Furthermore, as is evident from the representation desired, there are other functions that will serve as well. Suppose a function f' also satisfies (2); then to state how f and f' are functionally related proves the uniqueness of the representation. In this case $h(f) = f'$ where h is an order-preserving real-valued function of a real variable and is the permissible transformation for Stevens [21]. Very light conditions are required for ordinal measurement, and of course relatively weak measurement results. For ratio scale measurement, considerably more machinery is required, and a structure of the form $(A, R, *)$ is the starting point, where * is a binary operation, $(A, *)$ a group, and (A, R) satisfies certain order conditions as in ordinal measurement. A map from $(A, R, *)$ into (Re, $>$, $+$) is sought which has the important feature that $f(a*b) = f(a) + f(b)$, i.e., the representation is additive. Familiar ratio scale uniqueness results. Of course different types of measurement may require very different axiom systems and in each case it is desirable for the axioms to be empirically testable. For example, it might not be that preferences for beers are transitive. If they are not, no ordinal scale for beers can be constructed. In practice, however, certain axioms in some systems have no empirical "bite" or are difficult to evaluate in some settings.

In the physicist's laboratory concatenation operations, identified as binary operations, are in evidence. For instance, identical length rods can be concatenated (joined) end-to-end to match long lengths of objects being measured. The behavioral analog of this binary operation is typically missing, which is a main reason why strong scales, such as ratio scales, are not in evidence in the behavioral sciences.

There are many different measurement procedures with different axiom systems that lead to interval scales, such as the scale for temperature measurement. Binary operations are not required for such scales. To justify the widely held viewpoint that test scores like IQ scores are interval scale measurements would require the specification of the axioms that lead to interval scale measurement. It has never been done. However, one of the most interesting new measurement structures in psychology that leads to interval scales is *conjoint measurement* (cf. Krantz et al. [12] or Roberts [17]). In spirit, simple conjoint measurement is like two-factor analysis of variance. Its starting point is with order relations on pairs of a cross-product set. Suppose that one variable is temperature, the other is humidity, and the judgment rendered is whether a particular combination of temperature and humidity seems hotter than some other combination. From these order relations (note order relations only), interval scales of measurement of each variable are obtained if all the axioms are satisfied.

In practice, however, not all the axioms of conjoint measurement are satisfied, and the measurement scale that results is between an ordinal and interval scale, what Coombs [8] has called an *ordered-metric scale*. To illustrate, suppose that the elements are beers, and suppose $aRbRcRd$, where R means "preferred over." Let $[a, b]$ denote the psychological difference in taste between the two beers, a and b. It might be that $[a, b]$ is larger than $[c, d]$, which is larger than $[b, c]$. The scale that results has ordinal properties, but with an ordering on pairwise psychological differences as well. There are many examples of such structures in psychology, so the idea makes sense. But the uniqueness properties of such a representation are difficult to specify. Note, furthermore, that Stevens' [21] classification system breaks down in such cases, and of course definitions of meaningful such as (1) are difficult to apply. It is interesting, however, that many different axiomatizations do lead to one of Stevens' three types of measurement scales, ordinal, interval, and ratio. Why there seem to be such few general scale types is of concern to Narens [15].

Abelson and Tukey [1, 2] have provided numerical assignment rules useful for ordinal and certain other scale structures, including the ordered-metric case, which fall between interval and ordinal scales. For ordinal scales, while any order-preserving assignment is as good as any other from the standpoint of representational uniqueness, their rationale narrows sharply the class of possible assignments. Consider the ordinal case. View the problem in the spirit they propose, as a game against Nature, with Nature possessing the "true" assignments. We select an order-preserving assignment. Nature now selects a corresponding assignment, but selected so that the squared Pearson correlation coefficient (r^2) between the corresponding pairs of the two sequences is as small as possible. Abelson and Tukey's solution is to select the sequence which maximizes the minimum r^2; thus their maximin sequence bounds r^2 from below, for any sequence Nature happens to select. An easily remembered approximation to the ordinal maximin sequence is the linear 2–4 rule. Suppose k elements need assignments. Write down an equally spaced sequence of k numbers with sum zero. Quadruple the end values, and double those one unit in from each end. For instance, $-20, -6, -1, 1, 6, 20$ is a linear 2–4 sequence if $k = 6$. The sequence can be linearly transformed if nonnegative values are desired. Unfortunately, the Abelson and Tukey solution is not well known and thus is rarely employed.

It may be that the measurement-statistical issues will be satisfactorily resolved only when broader measurement issues are clearly resolved. Unfortunately, formal representational measurement gives little understanding to many "everyday" measurements in the behavioral sciences. The approach also suffers because in most formalized measurement structures there is no allowance made for measurement error*. Until these difficulties are resolved, when measurement is of concern (and it should be wherever behavior is of concern), from a statistical viewpoint it appears that nonparametric rank test procedures [13] advocated by Siegel [19] still offer the best possibility of avoiding certain measurement-statistical dilemmas.

References

[1] Abelson, R. P. and Tukey, J. W. (1959). *Amer. Statist. Ass.: Proc. Soc. Sci. Sect.*, pp. 226–230.

[2] Abelson, R. P. and Tukey, J. W. (1963). *Ann. Math. Statist.*, **34**, 1347–1369.

[3] Adams, E. W., Fagot, R. F., and Robinson, R. E. (1965). *Psychometrika*, **30**, 99–127.

[4] Anderson, N. H. (1961). *Psychol. Bull.*, **58**, 305–316.

[5] Aubuchon, J. C. and Hettmansperger, T. P. (1982). *On the Use of Rank Tests and Estimates in the Linear Model. Tech. Rep. 41*. Dept. of Statistics, Pennsylvania State University, University Park, Philadelphia.

[6] Burke, C. J. (1963). In *Theories in Contemporary Psychology*, M. H. Marx, ed. Macmillan, New York.

[7] Conover, W. J. and Iman, R. L. (1981). *Amer. Statist.*, **35**, 124–129.

[8] Coombs, C. H. (1964). *Theory of Data*. Wiley, New York.

[9] Fligner, M. A. (1981). *Amer. Statist.*, **35**, 131–132.

[10] Gaito, J. (1960). *Psychol. Rev.*, **67**, 277–278.

[11] Gaito, J. (1980). *Psychol. Bull.*, **87**, 564–567.

[12] Krantz, D. H., Luce, R. D., Suppes, P., and Tversky, A. (1971). *Foundations of Measurement*, Vol. 1. Academic Press, New York.

[13] Lehmann, E. L. (1975). *Nonparametrics: Statistical Methods Based on Ranks*. Holden-Day, San Francisco.

[14] Lord, F. M. (1953). *Amer. Psychol.*, **8**, 750–751.

[15] Narens, L. (1981). *J. Math. Psychol.*, **24**, 249–275.

[16] Pfanzagl, J. (1968). *Theory of Measurement*. Wiley, New York.

[17] Roberts, F. S. (1979). *Measurement Theory with Applications to Decision-Making, Utility, and the Social Sciences.* (*Encyclopedia of Mathematics and Its Applications, Vol. 7.*) Addison-Wesley, Reading, MA.

[18] Senders, V. L. (1958). *Measurement and Statistics: A Basic Text Emphasizing Behavioral Science Applications*. Oxford, New York.

[19] Siegel, S. (1956). *Nonparametric Statistics for the Behavioral Sciences*. McGraw-Hill, New York.

[20] Stevens, S. S. (1946). *Science*, **103**, 677–680.

[21] Stevens, S. S. (1951). In *Handbook of Experimental Psychology*, S. S. Stevens, ed. Wiley, New York.

[22] Thomas, H. (1982). *Psychol. Bull.*, **91**, 198–202.

(DISTRIBUTION-FREE METHODS
MEASUREMENT THEORY
ORDINAL DATA
PSYCHOLOGICAL TESTING THEORY
PSYCHOLOGY, STATISTICS IN
RANK PROCEDURES IN EXPERIMENTAL DESIGN
RANKING PROCEDURES
SOCIOLOGY, STATISTICS IN)

HOBEN THOMAS

MEASUREMENT THEORY

Measurement is the assignment of numbers (or vectors) to objects or events in such a way that specified relations among the numbers represent certain empirical relations among the objects. Measurement theory, sometimes called axiomatic measurement theory or the foundations of measurement, is concerned with the nonnumerical, qualitative conditions that an empirical relational system must meet in order for its elements to be measured in an internally consistent manner. Qualitative axioms about the empirical system are sought that allow representation, uniqueness, and meaningfulness theorems to be proved. *Representation theorems* establish the existence of a measurement scale, i.e., the mapping from the ordered empirical set with its zero or more empirical operations into the positive real numbers. *Uniqueness theorems* establish the measurement scale characteristics in terms of the transformation that relates two equally acceptable numerical mappings. *Meaningfulness theorems* specify the kinds of empirical conclusions that may be drawn on the basis of measurement, but that are independent of the particular measurement scale employed. The meaningfulness issue is the most controversial one. *See* MEASUREMENT STRUCTURES AND STATISTICS for a discussion relating meaningful statements to statistical tests, see Luce [10] for a discussion relating meaningful statements to dimensionally invariant laws, and see Narens [14] for more general coverage of the issues.

Modern approaches to the foundations of measurement date back to the work of Helmholtz [4] and Hölder [5], who provided an axiomatic analysis of extensive attributes such as mass and length. By now measurement theory is reasonably well understood for many variables of classical physics. It is less well understood for most variables of relativistic physics and quantum mechanics, as well as for those of the behavioral and social sciences [12]. Indeed, much of the most recent work in measurement theory has been aimed at clarifying measurement structures in the behavioral sciences. Seminal research in this regard includes the successful axiomatization of expected utility theory by von Neumann and Morgenstern [19] (*see* DECISION THEORY* and UTILITY THEORY*), as well as the work of Suppes and Zinnes [17] and of Pfanzagl [15]. A thorough treatment with many major results is provided by Krantz et al. [8]. A recent relatively simple text emphasizing behavioral measurement is available [16], as are more up-to-date reviews [11, 12].

It is useful to distinguish certain classes of measurement structures. *Extensive structures* with one operation are involved in the measurement of variables such as mass or length. These structures were among the earliest studied [4, 5] and are among the best understood (see Krantz et al. [8]). As a specific example let $a = \{A, B, \ldots\}$ be a set of weights that are empirically ordered by means of a pan balance, and let $\succsim$ denote this ordering. Further, define an empirical concatenation o by the operation "placing two weights in the same pan." Then $\langle a, \succsim, o\rangle$ is an extensive structure with one operation. If certain qualitative axioms (e.g., the set provided by ref. 8) hold for $\langle a, \succsim, o\rangle$ then

1. The structure has an additive representation in the positive real numbers, and
2. Pairs of such representations are related through multiplication by a positive constant, i.e., the measurement is on a ratio scale.

Extensive structures with two operations have also been studied [8, 12], but little work has been done on the more general case involving n operations.

It has been claimed (e.g., by Campbell [1]) that ratio scale measurement and the weaker interval scale measurement (in which no empirical zero is defined) are impossible in the behavioral and social sciences because empirical concatenation corresponding to addition cannot be operationalized. An important insight achieved by measurement theory is that this view is wrong. Other procedures that are operationally definable in the behavioral sciences can be substituted for empirical addition, and the resulting structures, which are generalizations of extensive structures, do yield ratio or interval scales. Examples of these procedures include *bisection* (in which the "subject" selects a stimulus that is halfway between two others) and *difference judgment* (in which the "subject" states which of two pairs of stimuli differs more in the attribute of interest). Another interesting structure, involving the empirical ordering "at least as likely as," relates to the foundations of probability*.

Conjoint structures differ from extensive ones in that an empirical ordering is obtained on a Cartesian product, say $A_1 \times A_2$, and the concatenation is derived indirectly from tradeoffs among the factors. Conjoint structures appear throughout the behavioral and physical sciences, e.g., gambles that differ in terms of probability and outcome can be ordered by preference, and objects that differ in mass and volume can be ordered by density. Primary attention to date has been directed thus far to two- and three-dimensional conjoint structures $\langle A_1 \times A_2, \succsim \rangle$ and $\langle A_1 \times A_2 \times A_3, \succsim \rangle$, respectively, that yield additive representations over the separate factors. The representations form an interval scale in that they are unique up to multiplication by a common positive constant and addition by separate constants. Other polynomial representations have been studied as well; see Krantz et al. [8].

An important problem, particularly for the behavioral sciences, that is beginning to receive investigation concerns the real-valued representation of nonadditive relational systems. Although the numerical structures are understood, the axiomatizations for specific representations remain to be worked out [13].

The axioms about empirical relational systems are of three types [11]. There are *first-order necessary axioms*, which follow from the desired representation. Much effort is devoted to formulating necessary axioms that are empirically interesting, because their experimental validation establishes them as qualitative, scientific laws [6, 7]. This has been of particular importance in behavioral measurement, where the underlying structures are still relatively poorly understood. There are *second-order necessary axioms*, which indicate how the numerical mapping is to be accomplished. These axioms frequently provide insight as to how the actual scaling should be performed. Finally, there are *structural axioms*, which limit the kinds of empirical structures that will satisfy the representation. These are needed both to avoid trivial structures and for mathematical convenience, but they are rarely of empirical interest.

Procedures have been developed for empirically evaluating conjoint structure axioms, particularly with regard to measurement in psychology (e.g., [9]), and some conjoint measurement experiments have been carried out (e.g., [2, 18, 20]) with encouraging results. The volume of experimental work will undoubtedly increase as suitable statistics are developed for testing the axioms; work along these lines is proceeding [3].

References

[1] Campbell, N. R. (1920). *Physics: The Elements*. Cambridge University Press, New York. Reprinted (1957) as *Foundations of Science: The Philosophy and Theory of Measurement*. Dover, New York.

[2] Coombs, C. H. and Huang, L. C. (1970). *J. Math. Psychol.*, **7**, 317–338.

[3] Falmagne, J. C. (1976). *Psychol. Rev.*, **83**, 65–79.

[4] von Helmholtz, H. (1887). Zahlen und Messen erkenntnis-theoretisch Betracht. *Philosophische Aufsätze Eduard Zeller gewidmet*, Leipzig. Reprinted (1895) *Gesamm. Abhandl.*, **3**, 356–391. English transl. (1930) in C. L. Bryan, *Counting and Measuring*. Van Nostrand, Princeton, NJ.

[5] Hölder, O. (1901). *Berichte über die Verhandlungen der Königlich Sächsischen Gesellschaft der Wissenschaften zu Leipzig. Mathematisch-physische Klasse*, **53**, 1–64.

[6] Krantz, D. (1972). *Science*, **175**, 1427–1435.

[7] Krantz, D. (1974). In *Contemporary Developments in Mathematical Psychology, Vol. 2*, D. H. Krantz, R. C. Atkinson, R. D. Luce, and P. Suppes, eds. Freeman, San Francisco, pp. 160–199.

[8] Krantz, D. H., Luce, R. D., Suppes, P., and Tversky, A. (1971). *Foundations of Measurement, Vol. 1*. Academic Press, New York.
[9] Krantz, D. H. and Tversky, A. (1971). *Psychol. Rev.*, **78**, 151–169.
[10] Luce, R. D. (1978). *Philos. Sci.*, **45**, 1–16.
[11] Luce, R. D. (1978). In *Foundations and Applications of Decision Theory, Vol. 1*, C. A. Hooker, J. J. Leach, and E. F. McClennen, eds. D. Reidel, Dordrecht, Holland.
[12] Luce, R. D. and Narens, L. (1981). *SIAM-AMS Proc.*, **13**, 213–235.
[13] Luce, R. D. and Narens, L. (in press). *J. Math. Psychol.*
[14] Narens, L. (1981). *Theory and Decision*, **13**, 1–70.
[15] Pfanzagl, J. (1968, 1971). *Theory of Measurement*. Wiley, New York.
[16] Roberts, F. S. (1979). *Measurement Theory*. Addison-Wesley, Reading, MA.
[17] Suppes, P. and Zinnes, J. L. (1963). In *Handbook of Mathematical Psychology, Vol. 1*, R. D. Luce, R. R. Bush, and E. Galanter, eds. Wiley, New York, pp. 1–76.
[18] Tversky, A. (1967). *J. Math Psychol.*, **4**, 175–201.
[19] von Neumann, J. and Morgenstern, O. (1944, 1947, 1953). *Theory of Games and Economic Behavior*. Princeton University Press, Princeton, NJ.
[20] Wallsten, T. S. (1976). *J. Math. Psychol.*, **14**, 144–185.

(FOUNDATIONS OF PROBABILITY
MEASUREMENT STRUCTURES AND STATISTICS
ORDINAL DATA
PSYCHOLOGICAL TESTING THEORY)

THOMAS S. WALLSTEN

MEASURES OF AGREEMENT

Measures of agreement are special cases of measures of association* or of correlation* that are designed to be sensitive not merely to deviations from independence, but specifically to deviations indicating agreement. These are measures most commonly used to assess reliability (*see* GROUP TESTING) or reproducibility of observations. In this context, it is usually assured that one has better than chance agreement. Consequently, the statistical problems of interest revolve not around the issue of testing the null hypothesis of independence but around estimation of the population measure of agreement.

Typically, one samples n subjects and has m observations on each, say $X_{i1}, X_{i2}, \ldots, X_{im}$ $(i = 1, 2, \ldots, n)$, where the marginal distributions of the observations are the same for all $j = 1, 2, \ldots, m$. Typically, a measure of agreement is zero when all ratings are independent and identically distributed, and 1.0 if $\Pr\{X_{ij} = X_{ij'}\} = 1$ for all i and $j \neq j'$.

Controversies as to the *validity* of certain measures of agreement can be generated if the assumption of equal marginal distributions is violated [1]. This assumption, however, imposes no major limitation. One need only randomly assign the m observations for each subject to positions $1, 2, \ldots, m$ prior to analysis.

Suppose that

$$X_{ij} = \mu + \xi_i + \epsilon_{ij},$$

where $\xi_i \sim N(0, \sigma_\xi^2)$, $\epsilon_{ij} \sim N(0, \sigma_\epsilon^2)$, $\rho = \sigma_\xi^2/(\sigma_\xi^2 + \sigma_\epsilon^2)$, and the "true" values of ξ_i and "errors" ϵ_{ij} are independent for all i and j. For this type of interval or ratio-level data, the intraclass correlation coefficient* is one such measure of agreement. This measure is most readily computed by applying a one-way analysis of variance* with each subject constituting a group. The intraclass correlation coefficient then is

$$r_I = \frac{F - 1}{F + m - 1},$$

where F is the F-statistic* to test for subject differences.

Since

$$F \sim \frac{(m-1)\rho + 1}{1 - \rho} F_{n-1, n(m-1)},$$

nonnull tests and confidence intervals for the parameter ρ can be structured on this basis [2].

A nonparametric analogue of the same procedure is based on the use of coefficient of concordance*. In this case the n observations for each value of j are rank ordered $1, 2, \ldots, n$, with ties given the average of ranks that would have been assigned had there been no ties. One may then calculate the intraclass correlation coefficient based on the ranks r_S. This statistic, r_S, is the average Spearman rank correlation coefficient* between pairs of ratings and is related to the coefficient of concordance W by the relationship

$$r_S = \frac{mW - 1}{m - 1}.$$

The distribution of r_S is approximately of the same form as that of r_I[3].

These two measures, r_I and r_S, are designed to measure agreement for measurements taken on ordinal, interval, or ratio scales. For measurements taken on the nominal scale, the kappa coefficient* performs the same function [4]; its nonnull distribution is not known theoretically, and jackknife* or bootstrap* methods are suggested for estimation and testing of this measure [5].

To illustrate these methods we use scores on a test of memory for 11 subjects ($n = 11$) each tested three times ($m = 3$). The three scores per subject are listed in random time order in Table 1. The intraclass correlation coefficient was found to be $r_I = 0.59$. The rank orders for each rating appear as superscripts in the table. The average Spearman coefficient based on these data was $r_S = 0.45$.

Finally, one may dichotomize (or classify) the scores in innumerable ways. For illustration we defined a "positive" test as a score of 50 or above, a "negative" test as a score below 50. The kappa coefficient for this dichotomization was $k = 0.05$.

This example illustrates an important point in evaluating magnitudes of measures of agreement. The measure of agreement reflects the nature of the population sampled (i.e., σ_ξ^2), the accuracy of the observation (i.e., σ_ϵ^2), and the nature of the observation itself (interval vs. ordinal vs. nominal). Consequently, poor measures of agreement are obtained if the observation is insensitive to the variations inherent in the population either because of an intrinsic scaling problem or because of inaccuracy of measurement.

Further, there are many more measures of agreement than these common ones proposed in the literature, because there are many ways of conceptualizing what constitutes agreement and how to measure disagreement. To see this, let $D(X_{ij}, X_{rs})$ be any metric reflecting agreement between two observations X_{ij} and X_{rs} (even if the X's are multivariate) with $D(X_{ij}, X_{rs}) = 1$ if and only if $X_{ij} \equiv X_{rs}$. If D_w is the mean of D's between all $n\binom{m}{2}$ pairs of observations within subjects, and D_t the mean of D's between all $\binom{nm}{2}$ pairs of observations, then a measure of agreement is

$$\frac{D_w - D_t}{1 - D_t}.$$

For example, when X_{ij} is a rank order vector, one might propose that $D(X_{ij}, X_{rs})$ be a rank correlation coefficient between such vectors. Such a measure has been proposed as an extension of kappa [5] for multiple choices of categories, or as a measure of intergroup rank concordance [6].

As a result, the magnitude of a measure of agreement is determined not only by the nature of the population, the accuracy of the observation, and the nature of the observation itself, but, finally, by the metric of agreement between observations selected as the basis of the measure of agreement. *See also* MEASURES OF SIMILARITY, DISSIMILARITY, AND DISTANCE.

Table 1

	Rating		
Subject	1	2	3
1	44^{4}	48^{7}	54^{5}
2	57^{8}	45^{6}	$40^{2.5}$
3	$46^{5.5}$	32^{3}	58^{7}
4	66^{9}	50^{8}	55^{6}
5	$46^{5.5}$	43^{4}	50^{4}
6	72^{11}	83^{11}	70^{11}
7	35^{3}	28^{2}	$64^{9.5}$
8	67^{10}	66^{10}	63^{8}
9	26^{2}	44^{5}	$64^{9.5}$
10	16^{1}	21^{1}	19^{1}
11	48^{7}	69^{9}	$40^{2.5}$

References

[1] Bartko, J. J. (1976). *Psychol. Bull.*, **83**, 762–765.
[2] Bartko, J. J. (1966). *Psychol. Rep.*, **19**, 3–11.
[3] Fleiss, J. L. and Cuzick, J. (1979). *Appl. Psychol. Meas.*, **3**, 537–542.
[4] Kraemer, H. C. (1981). *Biometrika*, **68**, 641–646.
[5] Kraemer, H. C. (1980). *Biometrics*, **36**, 207–216.
[6] Kraemer, H. C. (1976). *J. Amer. Statist. Ass.*, **71**, 608–613.

(ASSOCIATION, MEASURES OF
COEFFICIENT OF CONCORDANCE (*W*)
INTRACLASS CORRELATION COEFFICIENT
KAPPA COEFFICIENT
MEASURES OF SIMILARITY, DISSIMILARITY, AND DISTANCE
PSYCHOLOGICAL TESTING THEORY
SPEARMAN RANK CORRELATION COEFFICIENT
VARIANCE COMPONENTS)

HELENA CHMURA KRAEMER

MEASURES OF ASSOCIATION *See* ASSOCIATION, MEASURES OF

MEASURES OF DISPERSION *See* SPREAD, MEASURES OF

MEASURES OF INFORMATION

The concept of information is used in many fields such as statistics, probability theory, communication theory, ecology, biology, manpower planning*, psychometry, economics, law, systems, and computer science. Related with the concept are those of entropy*, disorder, uncertainty*, unexpectedness, affinity, intrinsic accuracy, weight of evidence, divergence, diversity*, and dispersion. These concepts have been quantified as measures of information and are used in a technical sense in a variety of contexts. The term information is also used in a generic fashion, meaning knowledge or data. The primary application of this interpretation is in information processing and retrieval, which constitute a portion of computer science.

There is a proliferation of concepts and measures, sometimes almost synonymous and other times apparently very different. All cases, however, have a basic underlying connecting factor. Technically, information means the amount by which uncertainty for an unknown entity is reduced as a result of the outcome of an experiment.

The discussion here is limited to *statistical information*. *See also* DIVERSITY INDICES, ENTROPY, INFORMATION THEORY AND CODING THEORY, and the bibliography at the end of this article.

There are several measures of varied applicability and generality that seem to play a mysterious role by appearing in connection with various important results in statistics. Analogous is the role of entropy in communication theory*. The foremost example is the Cramér-Rao lower bound* on the variance of estimators and the asymptotic variance of maximum likelihood estimators. Information is used as a criterion of estimation*, to construct tests of hypotheses, to prove theoretical results, and so on.

The treatment of information theory follows two patterns: axiomatic and operational. The axiomatic approach requires laying out intuitively acceptable axioms and constructing numerical quantities (functions, measures) that satisfy the axioms. The operational significance comes later in results and applications. The operational approach establishes the applicability of an even ad hoc measure from the properties it satisfies. This is the dominating approach in statistical information theory and will be followed in the present discussion.

STATISTICAL INFORMATION

The background is a set of data X and an unknown density f which may depend on a parameter θ or not:

1. Before the experiment we have "zero information" about f or θ. (This is the

non-Bayesian point of view; the ideas have been carried over to the Bayesian case.)

2. Making observations on a random variable X with distribution f or $f(x,\theta)$ reduces the uncertainty regarding f or θ by an extent equal to the information on f or θ contained in the data X.
3. Statistical observations contain a fixed amount of information. Two statistics are equally informative if they are identically distributed.
4. If there is a unique observation for every value of f or θ, the data X contain maximum information about f or θ. If the true f or θ can be obtained from the data without error, the data contain maximum information about f or θ.
5. Random variables (or statistics generated by the data) having the same distribution for all f or θ contain no information about f or θ. These are ancillary statistics*.
6. Two independent and identically distributed observations contain twice as much information as a single one. (This is a controversial proposition; some authors [3] do not accept its validity.)
7. The information contained in the data cannot increase by data condensation.
8. A sufficient statistic* contains all the information about f or θ contained in the sample.

The *intrinsic accuracy* of a distribution is the sensitivity of the random variable with respect to the parameter(s). Fisher* identified this type of sensitivity with statistical information. It may be judged by the extent to which the distribution of the variable is altered by changes in the value(s) of the parameter(s).

Variance and information are related to some extent. This is illustrated by the extreme case of $f(x,\theta)$ having zero variance. Then a single observation from $f(x,\theta)$ possesses maximum information. Information, however, is a broader concept than variance.

One kind of statistical information is *discriminatory*; it is that provided by the data for discriminating in favor of a certain hypothesis against another. It is natural to require that discriminatory information has the same properties as general information (*see* INFORMATION, KULLBACK).

*Uncertainty** is related not only to an unknown density f or parameter θ but also to the outcome of a random experiment, the possible results of which follow a specific probability distribution. Then we have the uncertainty or *entropy* of a distribution, closely related with entropy or disorder in thermodynamics and leading to information as developed in communication theory*; *see* ENTROPY, INFORMATION THEORY, AND CODING THEORY.

An experiment can be identified with a random variable or data X and the notions *information contained in experiment* X and *information contained in a random variable* X are identical. The usual approach is Bayesian; θ is treated as a random variable. In contrast with **1** above, before the experiment we have some knowledge (information, uncertainty) about θ expressed by a prior distribution ξ on θ. After performing experiment X our knowledge about θ is modified in the posterior distribution* of θ. The reduction in uncertainty between the prior and posterior distributions is the information provided by the experiment. This is identical with **2** above. Such information has been used to compare pairs of experiments [17, 18], in sequential* experimentation [9], and so on.

EXISTING MEASURES

There are several measures of information; each enjoys certain axiomatic, heuristic, or operational properties. A convenient way to differentiate among them is to classify them as parametric, nonparametric, and entropy-type measures.

The *parametric measures* refer to parametric families $\{f(x,\theta), \theta \in \Theta\}$ of distributions; they measure the amount of information supplied by the data about the unknown θ

and are functions of θ. The main parametric measures of information are those of:

Fisher [12]

$$I_X^F(\theta) = \begin{cases} E_\theta\left\{\left[(\partial/\partial\theta)\ln f(X,\theta)\right]^2\right\}, \\ \qquad\qquad \theta \text{ univariate}, \\ \|E_\theta\left\{\left[(\partial/\partial\theta_i)\ln f(X,\theta)\right]\right. \\ \qquad \left.\left[(\partial/\partial\theta_j)\ln f(X,\theta)\right]\right\}\|_{k\times k}, \\ \qquad\qquad \theta\ k\text{-variate}, \end{cases}$$

where $\| \quad \|_{k\times k}$ denotes a $k \times k$ matrix (*see* FISHER INFORMATION), and, for θ univariate, those of Vajda [24], Mathai [19], and Boekee [5].

Vajda

$$I_X^{\mathrm{V}}(\theta) = E_\theta|(\partial/\partial\theta)\ln f(X,\theta)|^\alpha, \qquad \alpha \geqslant 1,$$

Mathai

$$I_X^{\mathrm{Mat}}(\theta) = \left[E_\theta|(\partial/\partial\theta)\ln f(X,\theta)|^\alpha\right]^{1/\alpha}, \qquad \alpha \geqslant 1,$$

Boekee

$$I_X^{\mathrm{Bo}}(\theta) = \left[E_\theta|(\partial/\partial\theta)\ln f(X,\theta)|^{s/(s-1)}\right]^{s-1}, \qquad 1 < s < \infty.$$

The Vajda measure is also known as *Fisher's information of order* $\alpha \geqslant 1$, and is a lim inf product of Vajda's nonparametric measure of information given below. It was introduced as a more general measure of the sensitivity (intrinsic accuracy) with respect to θ of a parametric distribution, by means of a wide class of measures of divergence between two probability distributions. The measures of Mathai and Boekee are related with $I_X^{\mathrm{V}}(\theta)$ as follows:

$$I_X^{\mathrm{Mat}}(\theta) = \left[I_X^{\mathrm{V}}(\theta)\right]^{1/\alpha} \quad \text{and}$$

$$I_X^{\mathrm{Bo}}(\theta) = \left[I_X^{\mathrm{V}}(\theta)\right]^{1/(\alpha-1)}.$$

If θ is k-variate, Fisher's information matrix is the only parametric measure of information available. One-dimensional parametric measures of information for k-variate θ have been given by Papaioannou and Kempthorne [21] and Ferentinos and Papaioannou [10] as by-products of Fisher's information matrix. These are:

$$I_X^*(\theta) = \mathrm{tr}\left[I_X^F(\theta)\right],$$

$$D_X(\theta) = \det\left[I_X^F(\theta)\right],$$

$$\lambda_X^i(\theta) = \lambda_i\left[I_X^F(\theta)\right],$$

where λ_i is the ith largest eigenvalue of $I_X^F(\theta)$. Other parametric measures derived from the nonparametric have been given by Kagan [15], Aggarwal [1], Boekee [6], and Ferentinos and Papaioannou [10].

Nonparametric measures of information express the amount of information in the data for discriminating in favor of a distribution f_1 against another f_2, or measure the distance or affinity between f_1 and f_2. Frequently they are related to or are equivalent to measures of divergence between two probability distributions. The main nonparametric measures are those of Bhattacharyya [4], Kullback and Leibler [16], Rényi [22], Kagan [15], Csiszár [8], Matusita [20], and Vajda [24].

Bhattacharyya

$$I_X^B(f_1, f_2) = -\ln\int\sqrt{f_1}\sqrt{f_2}\,d\mu,$$

the negative natural logarithm of the "affinity" $\rho = \int\sqrt{f_1}\sqrt{f_2}\,d\mu$ between the distributions f_1 and f_2; ρ is in turn related to the Hellinger distance* $\int(\sqrt{f_1} - \sqrt{f_2})^2\,d\mu$ between f_1 and f_2.

Kullback–Leibler

$$I_X^{\mathrm{KL}}(f_1, f_2) = \int f_1 \ln(f_1/f_2)\,d\mu$$

(*see* INFORMATION, KULLBACK), also known as the *expected weight of evidence*, *cross-entropy* [14], *discrimination information*, *directed divergence*, and *gain of information* in replacing distribution f_2 by distribution f_1. If f_1 is the density of $X = (U, V)$ and f_2 is the product of the marginal densities of U and V, then I_X^{KL} becomes the mutual or relative information in coding theory or the mean information in U about V or in V about U.

Rényi

$$I_X^R(f_1, f_2) = (\alpha - 1)^{-1} \ln \int f_1^{\alpha} f_2^{1-\alpha} \, d\mu,$$
$$\alpha > 0, \quad \alpha \neq 1,$$

also known as the *information of order* α when f_2 is replaced by f_1; it is a generalization of the Bhattacharyya measure. For $\alpha = 2$,

$$I_X^R(f_1, f_2) = 2 I_X^B(f_1, f_2).$$

In the discrete case the Rényi measure is obtained from a system of postulates analogous to those of entropy and is an alternative to the Kullback-Leibler information. As a matter of fact,

$$\lim I_X^R(f_1, f_2) \to I_X^{KL}(f_1, f_2) \quad \text{as} \quad \alpha \to 1.$$

Kagan

$$I_X^{Ka}(f_1, f_2) = \int \left[1 - (f_1/f_2)\right]^2 f_2 \, d\mu,$$

also known as the χ^2 *divergence* of the probability distributions f_1, f_2; it appears in Chapman–Robbins type inequalities.

Csiszár

$$I_X^C(f_1, f_2) = \int \phi(f_1/f_2) f_2 \, d\mu,$$

where ϕ is a real valued-convex function on $[0, \infty)$ with $\phi(\mu) \to 0$ as $\mu \to 0$, $0\phi(0/0) = 0$, $0\phi(u/0) = u\phi_\infty$, $\phi_\infty = \lim_{u \to \infty}[\phi(u)/u]$. This is a measure of divergence between f_1 and f_2, sometimes called *f-divergence*. It is a general measure of information in that special choices of ϕ lead to other nonparametric measures, for example, the Kullback–Leibler measure if $\phi(u) = u \ln u$, the Kagan measure if $\phi(u) = (1 - u)^2$, the Matusita* measure if $\phi(u) = (1 - \sqrt{u})^2$, and the Vajda measure if $\phi(u) = |1 - u|^\alpha$, $\alpha \geqslant 1$. The same measure was independently introduced by Ali and Silvey [2].

Matusita

$$I_X^M(f_1, f_2) = \left[\int \left(\sqrt{f_1} - \sqrt{f_2}\right)^2 d\mu\right]^{1/2},$$

a measure of the distance between f_1 and f_2. Moreover, $I_X^M(f_1, f_2) = [2(1 - \rho)]^{1/2}$, where ρ is the affinity between f_1 and f_2. Along with affinity it has found applications in statistical decision theory*.

Vajda

$$I_X^V(f_1 f_2) = \int |1 - (f_1/f_2)|^\alpha f_2 \, d\mu,$$

an extension of Kagan's measure known as the $\chi - \alpha$ *divergence* of f_1 and f_2.

Entropy-type measures give the amount of information contained in a distribution. The classical measures of this type are due to Shannon [23] and Renyi [22]; *see* ENTROPY.

The previous measures have been used in a Bayesian context to define measures of information contained in an experiment X. Let $\xi(\theta)$, $p(x)$, and $\xi(\theta \mid x)$ be the prior, marginal, and posterior distributions and $p(x \mid \theta)$ be the density of X given θ. Then Lindley [17] defines the information in experiment X to be

$$I_X^L = -\int \xi(\theta) \ln \xi(\theta) \, d\theta - \left\{\int \left[-\int \xi(\theta \mid x) \ln \xi(\theta \mid x) \, d\theta\right] p(x) \, dx\right\}.$$

This is equal to the Kullback–Leibler measure with $f_1 = p(x, \theta)$ and $f_2 = p(x)\xi(\theta)$, where $p(x, \theta)$ is the joint density for X and θ. Mallows [18], taking $W(\theta, \xi(\cdot))$ to be a numerical measure of the experimenter's knowledge about θ, uses the measure

$$I_X^{Mal}(\theta) = \int p(x \mid \theta)\{W(\theta, \xi(\theta \mid x)) - W(\theta, \xi(\theta))\} \, dx.$$

Information in experiments can be measured by either the parametric measures with $f(x, \theta) = p(x \mid \theta)$ or the nonparametric measures with $f_1 = \xi(\theta)$, $f_2 = \xi(\theta \mid x)$ or $f_1 = p(x \mid \theta)$, $f_2 = p(x)$ or $f_1 = p_1(x)$, $f_2 = p_2(x)$ in the case of the two priors ξ_1, ξ_2.

The measures listed above are not defined for all cases and all distributions; not all measures satisfy all properties and there is no measure satisfying all of them. Sometimes the validity of a property is achieved by imposing conditions on the distribution(s). These are regularity conditions and families of distributions satisfying them are *regular families*. Fisher's measure, for instance, is applicable only to regular families. The parametric measures are suitable for estimation theory, the nonparametric for testing hypotheses, and the entropy-type measures for coding theory.

PROPERTIES

Let I_X be any measure of (statistical) information contained in the data X, and $T(X)$ a statistic. The desirable properties of I_X and the behavior of some of the preceding measures of information with respect to these properties are as follows. Regularity conditions [10] are required for the parametric measures.

1. **Nonnegativity:** $I_X \geqslant 0$. Equality occurs if and only if the distribution is independent of θ or the family of distributions is a singleton. Nonnegativity is satisfied for all measures except Csiszár's and the continuous versions of Shannon's and Rényi's entropies. For example, if f is uniform over the interval $(0, 1/2)$, Shannon's entropy with natural logarithms is $-\ln 2$. Csiszár's measure satisfies the more general inequality $I_X^{\mathrm{C}} \geqslant \phi(1)$.
2. **Additivity-Subadditivity.**

   Weak additivity $I_{X,Y} = I_X + I_Y$ if X, Y are independent
   Strong additivity $I_{X,Y} = I_X + I_{Y|X}$, $I_{Y|X} = E_X(I_{Y|X=x})$
   Subadditivity $I_{X,Y} \leqslant I_X + I_Y$, with equality if X is independent of Y

   Strong additivity is expected to imply weak additivity. Weak additivity is satisfied for all measures except those of Vajda (and therefore Mathai and Boekee), Matusita, and Kagan. Strong additivity is satisfied for I^{F}, I^{KL}, and the Shannon entropy but not for $I^{\mathrm{V}}(I^{\mathrm{Mat}}$, $I^{\mathrm{Bo}})$, I^{R} (and therefore I^{B}), and I^{M}. Subadditivity is not satisfied for any measure except Shannon's entropy. The validity of this property for Csiszár's measure depends on the choice of ϕ.
3. **Conditional Inequality:** $I_{Y|X} \leqslant I_X$. If any two of the properties (strong additivity, subadditivity, conditional inequality) hold, then the third is also satisfied. Conditional inequality is satisfied by Shannon's entropy but not by the measures of Fisher and Kullback–Leibler.
4. **Maximal Information:** $I_{T(X)} \leqslant I_X$. This is satisfied for all measures except the Shannon and Rényi entropies in the continuous case.
5. **Invariance under Sufficient Transformations:** $I_X = I_{T(X)}$ if and only if $T(X)$ is sufficient. This is satisfied for all measures except the Shannon and Rényi entropies.
6. **Convexity.** Let $\alpha_1 \geqslant 0$, $\alpha_2 \geqslant 0$, $\alpha_1 + \alpha_2 = 1$ and f_1, f_2 be densities; then

   $$I_X(\alpha_1 f_z + \alpha_2 f_2) \leqslant \alpha_1 I_X(f_1) + \alpha_2 I_X(f_2).$$

   Besides its mathematical interest, convexity has obvious applications in mixtures of probability distributions, in Cramér–Rao type lower bounds, the probability of error in communication theory, and so on. Recently it led to the development of measures of divergence based on entropy functions [7]. Convexity is satisfied for all measures except the Rényi (information gain) measure when $\alpha > 1$, the Matusita measure, and the Shannon and Rényi entropies.
7. **Loss of Information.** Let G be the set of all partitions g of R^k and I_g the measure of information on g; then

   $$\sup_{g \in G} I_g = I_X \quad \text{or} \quad I_g \to I_X$$

   as the refinement $\lambda(g)$ of $g \to 0$. This means that loss of information due to a grouping or discretization g can be made arbitrarily small by properly selecting g. It is satisfied for all measures except for the Shannon and Rényi entropies in the continuous case.

8. **Sufficiency in Experiments.** If experiments $\mathscr{E}_X$ and $\mathscr{E}_Y$ have the same parameter space Θ and $\mathscr{E}_X$ is sufficient for $\mathscr{E}_Y$ according to Blackwell's definition, then $I_X \geqslant I_Y$.

   Experiment $\mathscr{E}_X$ is *sufficient* for experiment $\mathscr{E}_Y$ in the Blackwell sense if there exists a stochastic transformation of X to a random variable $Z(X)$ such that, for each $\theta \in \Theta$, the random variables $Z(X)$ and Y have identical distributions. In other words $\mathscr{E}_X$ is sufficient for $\mathscr{E}_Y$ if, regardless of θ, a value on X and a random mechanism makes it possible to generate a value on Y. An elementary example is the following. Let X and Y be normally distributed $N(\theta, 1)$ and $N(\theta, 4)$ respectively. An observation on X is more informative than an observation on Y because of smaller variance. Also $\mathscr{E}_X$ is sufficient for $\mathscr{E}_Y$ because if $Z(X) = X + U$, where U is normally distributed $N(0,3)$ independent of X, then $Z(X)$ has the same distribution as Y. Sufficiency in experiments is satisfied for the measures $I^{\mathrm{F}}, I^{\mathrm{V}}, I^{\mathrm{Mat}}, I^{\mathrm{Bo}}$ all with one prior and $f(x,\theta) = p(x \mid \theta)$, for the measures $I^{\mathrm{KL}}, I^{\mathrm{C}}$ with one prior, $f_1 = p(x \mid \theta)$, $f_2 = p(x)$, and for the measures $I^{\mathrm{KL}}, I^{\mathrm{C}}$, $I^{\mathrm{B}}, I^{\mathrm{R}}$ with $0 < \alpha < 1$, $I^{\mathrm{Ka}}, I^{\mathrm{M}}$ all with two priors, and $f_1 = p_1(x)$ and $f_2 = p_2(x)$ [11, 13, 17].

9. **Appearance in Cramér–Rao Inequalities.** The parametric measures given above appear in inequalities of Cramér–Rao* type provided that a few regularity conditions are satisfied. If, for instance, $g(\theta)$ is a parametric function of θ, then for Fisher's measure

$$E\left[T(x) - g(\theta)\right]^2 \geqslant \left[g'(\theta)\right]^2 / I_X^{\mathrm{F}}(\theta);$$

   see FISHER INFORMATION.

10. **Invariance under Parametric Transformations.** The measures I^{F}, I^{V}, I^{Mat}, and I^{Bo} are not invariant under reparametrizations.

11. **Nuisance Parameter Inequality: $I_X(\theta_1, \theta_2) \leqslant I_X(\theta_1)$.** The information decreases as the number of parameters increases.

12. **Order Preserving Property.** Let I^1 be a measure of information which is accepted as standard and I^2 any other measure. Then

$$I^1_{X_1} \leqslant I^1_{X_2} \rightarrow I^2_{X_1} \leqslant I^2_{X_2}.$$

   This is satisfied if I^1 and I^2 are the Shannon and Rényi (with $\alpha > 1$) entropies respectively.

13. **Asymptotic Behavior.** The sequence of random variables $\{X_n\}$ converges in some sense to the random variable X as $n \rightarrow \infty$ if and only if $I_{X_n} \rightarrow I_X$. Under some conditions this is satisfied for the Kullback–Leibler and Rényi measures [22].

Other areas of statistical applications for measures of information are updating subjective probability, hypothesis formulation and tests in contingency tables*, model identification in time series* and dynamic systems, tests of independence and goodness of fit*, theory of ancillary statistics, randomized response* models in sample surveys, pattern recognition*, and so on.

References

[1] Aggarwal, J. (1974). In *Théories de l'Information*, J. Kampé de Feriet, ed. Springer, Berlin, pp. 11–117.

[2] Ali, S. M. and Silvey, S. D. (1966). *J. R. Statist. Soc. B*, **28**, 131–142.

[3] Basu, D. (1958). *Sankhyā*, **20**, 223–226.

[4] Bhattacharyya, A. (1943). *Calcutta Math. Soc. Bull.*, **35**, 99–109.

[5] Boekee, D. (1977). In *Topics in Information Theory*, I, Csiszár and P. Elias, eds. North-Holland, Amsterdam, pp. 113–123.

[6] Boekee, D. (1979). *Trans. 8th Prague Conf. Inf. Theory, Statist. Decision Functions and Random Processes*. Prague, 1978, pp. 55–66.

[7] Burbea, J. and Rao, C. R. (1982). *IEEE Trans. Inf. Theory*, **IT-28**, 489–495.

[8] Csiszár, I. (1963). *Publ. Math. Inst. Hungar. Acad. Sci.*, **8**, 85–108.

[9] DeGroot, M. (1962). *Ann. Math. Statist.*, **33**, 404–419.

[10] Ferentinos, K. and Papaioannou, T. (1981). *Inf. Control*, **51**, 193–208.

[11] Ferentinos, K. and Papaioannou, T. (1982). *J. Statist. Planning and Inference*, **6**, 309–317.

[12] Fisher, R. A. (1925). *Proc. Cambridge Philos. Soc.*, **22**, 700–725.

[13] Goel, P. and DeGroot, M. (1979). *Ann. Statist.*, **7**, 1066–1077.

[14] Good, I. J. (1950). *Probability and the Weighing of Evidence*. Griffin, London.

[15] Kagan, A. M. (1963). *Soviet Math. Dokl.*, **4**, 991–993.

[16] Kullback, S. and Leibler, A. (1951). *Ann. Math. Statist.*, **22**, 79–86.

[17] Lindley, D. V. (1956). *Ann. Math. Statist.*, **27**, 986–1005.

[18] Mallows, C. (1959). *J. R. Statist. Soc. B*, **21**, 67–72.

[19] Mathai, A. M. (1967). *Metron*, **26**, 1–12.

[20] Matusita, K. (1967). *Ann. Inst. Statist. Math.*, **19**, 181–192.

[21] Papaioannou, T. and Kempthorne, O. (1971). On Statistical Information Theory and Related Measures of Information. *Tech. Rep. No. ARL 71-0059*. Aerospace Research Laboratories, Wright–Patterson A.F.B., Ohio.

[22] Rényi, A. (1961). *Proc. 4th Berkeley Symp. Math. Stat. Prob.*, Vol. 1. University of California Press, Berkeley, pp. 547–561.

[23] Shannon, C. E. (1948). *Bell Syst. Tech. J.*, **27**, 379–423, 623–656.

[24] Vajda, I. (1973). *Trans. 6th Prague Conf. Inf. Theory, Statist. Decision Functions and Random Processes*. Prague, 1971, pp. 873–886.

Bibliography

Aczél, J. and Daróczy, Z. (1975). *On Measures of Information and Their Characterizations*. Academic Press, New York. (A highly technical treatment of the Shannon and Rényi entropies, their generalizations, and their properties, all discrete.)

Csiszár, I. (1977). *Trans. 7th Prague Conf. Inf. Theory, Statist. Decision Functions and Random Processes*. Prague, 1974, pp. 73–86. (Excellent review paper.)

Guiasu, S. (1977). *Information Theory with Applications*. McGraw-Hill, New York. (A general theoretical treatment of entropy, transmission of information, and algebraic coding with applications to statistical inference, classification theory, pattern-recognition, and game theory.)

Kendall, M. G. (1973). *Int. Stat. Rev.*, **41**, 59–68. (An excellent account of the developments in entropy, communication theory, and their relation to statistical information.)

Kullback, S. (1959). *Information Theory and Statistics*. Wiley, New York. (Classical reference on statistical discrimination information at the beginning graduate level.)

Mathai, A. M. and Rathie, P. N. (1976). In *Essays in Probability and Statistics*, S. Ikeda et al., eds. Shinko Tsusho, Tokyo. (An excellent review paper.)

Rényi, A. (1970). *Probability Theory*. North-Holland, Amsterdam. (The last chapter contains an excellent introduction to the basic notions of information theory and the Shannon and Rényi measures devoid of coding theory, mostly discrete and with statistical interpretations and applications to limit theorems.)

(ALGORITHMIC INFORMATION THEORY
DIVERSITY INDICES
ENTROPY
FISHER INFORMATION
INFORMATION, KULLBACK
INFORMATION THEORY AND CODING THEORY
MEASURES OF SIMILARITY, DISSIMILARITY, AND DISTANCE
MINIMUM DISCRIMINATION INFORMATION
STATISTICAL INFERENCE
SUFFICIENCY)

T. PAPAIOANNOU

MEASURES OF LOCATION *See* MEAN, MEDIAN, MODE, AND SKEWNESS

MEASURES OF SIMILARITY, DISSIMILARITY, AND DISTANCE

Suppose that $\mathbf{x}_i = (x_{i1}, x_{i2}, \ldots, x_{ip})$ is the row vector of observations on p variables associated with a unit labeled i. When the same variables relate to a set of n units, the similarity between units i and j is defined as $s_{ij} = f(\mathbf{x}_i, \mathbf{x}_j)$, some function of the observed values. Many functions have been proposed, depending partly on the types of variable concerned (e.g., quantitative, nominal, qualitative, metristic, dichotomous, categorical, ordinal) and partly on the type of unit (sampling unit or population). Before discussing particular instances of similarity coefficients some general comments can be made.

Similarity is usually regarded as a symmetric relationship requiring $s_{ij} = s_{ji}$. Most similarity coefficients are nonnegative and bounded by unity, $0 \leqslant s_{ij} \leqslant 1$, some of a correlational nature satisfy $-1 \leqslant s_{ij} \leqslant 1$, and a few are unbounded. Associated with every similarity bounded by zero and unity is a dissimilarity $d_{ij} = 1 - s_{ij}$, which is symmetric and nonnegative. The degree of similarity between two units increases with s_{ij} and decreases with increasing d_{ij}. It is natural for a unit to have maximal similarity with itself so that $s_{ii} = 1$ and $d_{ii} = 0$. More generally, with many coefficients $s_{ij} = 1$ iff $\mathbf{x}_i = \mathbf{x}_j$; it follows that if $s_{ij} = 1$, then $s_{ik} = s_{jk}$ for all units k. This seems to be a fundamental property of the notion of similarity that allows the n units to be represented as the n nodes of a graph linked by edges with lengths d_{ij} and with coincident nodes for identical sets of units (*see* GRAPH THEORY). Other, more structured, representations are of interest. Some dissimilarity coefficients have the metric property that

$$d_{ij} + d_{ik} \geqslant d_{jk}$$

for all units i, j, k; this implies that if $s_{ij} = 1$, then $s_{ik} = s_{jk}$, though not necessarily that $\mathbf{x}_i = \mathbf{x}_j$. When the metric inequality is satisfied among all triplets (d_{ij}, d_{ik}, d_{jk}) it must also be satisfied among all triplets $(d_{ij}^{1/r}, d_{ik}^{1/r}, d_{jk}^{1/r})$, $r \geqslant 1$. Another standard result of this kind is that if the values d_{ij} satisfy the metric inequality, then so do the values $d_{ij}/(d_{ij} + c^2)$, where c is any constant.

Compliance with the metric inequality is necessary but not sufficient for the n units to have a representation as n points in a metric space (usually Euclidean) such that for all pairs i, j the distance between the ith and jth points is d_{ij}. When this can be done the points are said to be *embeddable*. Embedding in a Euclidean space normally requires $n - 1$ dimensions so that dimensionality increases with the number of units being represented. Some coefficients are inherently Euclidean and require the same number of dimensions (usually p) however many units are represented. Suppose $\mathbf{D}$ is the $n \times n$ symmetric matrix with elements $-\frac{1}{2} d_{ij}^2$, $\mathbf{1}$ is an n-vector of units, and $\mathbf{t}$ is any n vector such that $\mathbf{1}'\mathbf{t} = 1$ and $\mathbf{Dt} \neq 0$; then Euclidean embedding with distances d_{ij} is possible iff the matrix $(\mathbf{I} - \mathbf{1t}')\mathbf{D}(\mathbf{I} - \mathbf{t1}')$ is positive semidefinite (p.s.d.). It follows that a Euclidean representation can be found with distances $d_{ij}^{1/2}$ if $\mathbf{S}$ (the similarity matrix containing values s_{ij}) is positive semi-definite (p.s.d). Coefficients that are nonmetric, metric, and Euclidean embeddable may differ only by near-linear monotonic transformations of d_{ij}. Indeed a Euclidean embedding can always be found with distances $(d_{ij}^2 + c)^{1/2}$, where c is any constant greater than some minimal value, and a Euclidean embedding can always be found with distances $d_{ij} + b$, where b is any constant greater than some calculable minimal value. Thus too much should not be made of these properties, though they have interest for the many methods that operate on similarity coefficients and assume explicitly or implicitly that there is an underlying Euclidean representation (*see* MULTIDIMENSIONAL SCALING; HIERARCHICAL CLUSTER ANALYSIS; CLASSIFICATION; HIERARCHICAL CLASSIFICATION; GRAPH THEORY).

SIMILARITY BETWEEN PAIRS OF SAMPLES

The simplest similarity coefficients relate to dichotomous variables where each variable has two values only. The values may represent qualities of equal standing such as black/white or rough/smooth, or they may represent presence or absence of some character. The difference between these two uses is a fundamental one but in the following discussion the term *character* is used for both, the meaning being interpreted from context. With $p(> 0)$ dichotomous variables and two units i, j we can form the usual 2×2 table with entries a_{ij} (the number of characters common to both units), b_{ij} (the number of characters present in the ith unit and absent in the jth), c_{ij} (the number of characters absent in the ith unit and present in the jth), and d_{ij} (the number of characters absent from both units). Thus $a_{ij} + b_{ij} + c_{ij} + d_{ij} = p$ for all pairs i, j. Many similarity coefficients have been proposed that combine the quantities a_{ij}, b_{ij}, c_{ij}, and d_{ij}. Table 1 lists some coefficients, where for clarity the suffices i, j are omitted. Further coefficients, discussion, and detailed references may be found in Anderberg [1], Legendre and Le-

Table 1 Similarity Coefficients for Dichotomous Variables

Variable	Similarity Coefficient		Range	Metric	S p.s.d.
S_1	$\frac{a}{b+c}$	(Kulczynski)	$0, \infty$	Dissimilarity undefined	Yes
S_2	$\frac{a}{a+b+c+d}$	(Russell and Rao)	0, 1	Yes	Yes
S_3	$\frac{a}{a+b+c}$	(Jaccard)	0, 1	Yes	Yes
S_4	$\frac{a+d}{a+b+c+d}$	(simple matching)	0, 1	Yes	Yes
S_5	$\frac{a}{a+2(b+c)}$	(Anderberg)	0, 1	Yes	Yes
S_6	$\frac{a+d}{a+2(b+c)+d}$	(Rogers and Tanimoto)	0, 1	Yes	Yes
S_7	$\frac{a}{a+\frac{1}{2}(b+c)}$	(Sørensen, Dice, and Czekanowski)	0, 1	No	Yes
S_8	$\frac{a+d}{a+\frac{1}{2}(b+c)+d}$	(Sneath and Sokal)	0, 1	No	No
S_9	$\frac{a-(b+c)+d}{a+b+c+d}$	(Hamman)	$-1, 1$	Yes	Yes
S_{10}	$\frac{1}{2}\left(\frac{a}{a+b}+\frac{a}{a+c}\right)$	(Kulczynski)	0, 1	No	No
S_{11}	$\frac{1}{4}\left(\frac{a}{a+b}+\frac{a}{a+c}+\frac{d}{c+d}+\frac{d}{b+d}\right)$	(Anderberg)	0, 1	No	No
S_{12}	$\frac{a}{\sqrt{\{(a+b)(a+c)\}}}$	(Ochiai)	0, 1	No	Yes
S_{13}	$\frac{ad}{\sqrt{\{(a+b)(a+c)(d+b)(d+c)\}}}$		0, 1	No	Yes
S_{14}	$\frac{ad-bc}{\sqrt{\{(a+b)(a+c)(d+b)(d+c)\}}}$	(Pearson's ϕ)	$-1, 1$	No	Yes
S_{15}	$\frac{ad-bc}{ad+bc}$	(Yule)	$-1, 1$	No	No

gendre [6], Sneath and Sokal [8], and Spath [9].

In all coefficients, b and c may be interchanged—as is necessary for a symmetric coefficient. In those coefficients that refer to both a and d, these too may be interchanged. When absence of a character in both units is deemed to convey no information, then d should not occur in the coefficient (see S_1, S_3, S_5, S_7, S_{10}, and S_{12}). Writing $S_1 = x$, we have

$$x = S_1 = \frac{S_3}{1 - S_3} = \frac{2S_5}{1 - S_5} = \frac{S_7}{2(1 - S_7)},$$

showing that these coefficients are simply related by monotonic functions and hence are monotonic functions of each other. Similarly defining $y = (a + d)/(b + c)$, itself proposed as an unbounded similarity coefficient, we have

$$y = \frac{S_4}{1 - S_4} = \frac{2S_6}{1 - S_6} = \frac{S_8}{2(1 - S_8)} = \frac{1 + S_9}{1 - S_9},$$

so that these coefficients too are monotonic functions of each other. The equation

Table 2 Scores and Weights

Character i	Character j	$s_k(x_{ik}, x_{jk})$	$w_k(x_{ik}, x_{jk})$	$s_k(x_{ik}, x_{jk})$	$w_k(x_{ik}, x_{jk})$
Present	Present	1	1	1	1
Present	Absent	0	2	0	2
Absent	Present	0	2	0	2
Absent	Absent	1	1	Immaterial	0

$S_2 = x/(1+y)$ links the two sets of coefficients. The coefficients S_1, S_3, S_5, S_7, S_{10}, S_{11}, S_{12}, S_{13}, S_{14}, and S_{15} can be undefined for zero settings of a, b, c, d, in various rarely occurring combinations. These coefficients are then conventionally assigned some appropriate value, usually zero.

Besides dealing with presence/absence of characters, the coefficients described above deal with qualitative variables at two levels. Multilevel qualitative variables may be treated by a simple scheme allocating a score s_{ijk} (usually zero or unity) when comparing units i and j on the kth variable. The simplest rule is to score unity when the kth variable has the same form for both units (e.g., both are black), otherwise score zero. This score is averaged over all variables to give the coefficient

$$S_{ij} = \sum_{k=1}^{p} s_{ijk}/p.$$

This scheme is preferable to the one sometimes used where each form is treated as a separate dichotomous variable with values like black/not black, white/not white, and so on, which introduces spurious weights into the coefficient.

More generally the score s_{ijk} may be a function of the actual values x_{ik}, x_{jk} leading to

$$S_{ij} = \sum_{k=1}^{p} s_k(x_{ik}, x_{jk})/p,$$

a form indicating that different functions s_k may be chosen for each variable.

Weights w_k, perhaps reflecting the a priori importance of a variable (say for identification) or its reliability, may be associated with the kth variable. More generally the weight may be a function $w_k(x_{ik}, x_{jk})$ of the values concerned, leading to a very general coefficient

$$S_{ij} = \frac{\sum_{k=1}^{p} w_k(x_{ik}, x_{jk}) s_k(x_{ik}, x_{jk})}{\sum_{k=1}^{p} w_k(x_{ik}, x_{jk})}.$$

The weights provide one way of ignoring "comparisons between absences" of characters as described above. For example, consider the scores and weights in Table 2, in which the first choice of settings defines the coefficient S_6 and the second set defines the coefficient S_5; similarly, the coefficients S_1–S_9 are all subsumed in the general formula. Occasionally characters can be ranked into primary and secondary in such a way that for each primary character there is at least one secondary character. The weight for the kth primary character can then be chosen to be the similarity between its associated secondary characters. Clearly this process may be generalized to any depth of nesting of primary, secondary, tertiary, etc., characters. Another simple use of weights is to deal with missing values by assigning zero weight when either (or both) of x_{ik} and x_{jk} is missing and unit weight otherwise. Matrices that are p.s.d. without missing values may lose this property when missing values are treated in this way; metric coefficients may become nonmetric.

When variables are measured quantitatively it is natural to measure distance or dissimilarity D_{ij} between pairs of units. How such quantities can be converted into similarity measures and combined with information on qualitative and dichotomous variables to give an overall measure of similarity is shown below. Table 3 lists some commonly occurring *distance*/dissimilarity coef-

Table 3 Distance/Dissimilarity Coefficients for Quantitative Variables

		Metric
$D_1^2 = \frac{1}{p} \sum_{k=1}^{p} (x_{ik} - x_{jk})^2$		Yes
$D_2^2 = \frac{1}{p} \sum_{k=1}^{p} (x_{ik} - x_{jk})^2 / r_k^2$	(taxonomic distance)	Yes
$D_3 = \frac{1}{p} \sum_{k=1}^{p} \lvert x_{ik} - x_{jk} \rvert / r_k$		Yes
$D_4^t = \frac{1}{p} \sum_{k=1}^{p} \lvert x_{ik} - x_{jk} \rvert^t / r_k^t$	(Minkowski)	Yes, $t \geqslant 1$
$D_5^2 = \frac{1}{p} \sum_{k=1}^{p} \frac{(x_{ik} - x_{jk})^2}{(x_{ik} + x_{jk})^2}$	(divergence)	Yes[a]
$D_6 = \frac{1}{p} \sum_{k=1}^{p} \frac{\lvert x_{ik} - x_{jk} \rvert}{\lvert x_{ik} + x_{jk} \rvert}$		Yes[a]
$D_7 = \sum_{k=1}^{p} \frac{\lvert x_{ik} - x_{jk} \rvert}{\lvert x_{ik} \rvert + \lvert x_{jk} \rvert}$	(Canberra metric)	Yes
$D_8 = \frac{\sum_{k=1}^{p} \lvert x_{ik} - x_{jk} \rvert}{\sum_{k=1}^{p} (x_{ik} + x_{jk})}$	(Bray and Curtis)	No
$D_9 = \frac{\sum_{k=1}^{p} \lvert x_{ik} - x_{jk} \rvert}{\sum_{k=1}^{p} \max(x_{ik}, x_{jk})}$	(Soergel)	Yes[a]
$D_{10} = \frac{1}{p} \sum_{k=1}^{p} \left(1 - \frac{\min(x_{ik}, x_{jk})}{\max(x_{ik} x_{jk})}\right)$	(Ware and Hedges)	Yes[a]

[a]Not necessarily metric for negative data.

ficients for quantitative variables. Further details may be found in the same references as those given for Table 1.

In Table 3 the subscripts i, j should be associated with each definition (e.g., $D_{1,i,j}$) but for clarity and for consistency with the similarities defined in Table 1, they are dropped. The quantities r_k are normalizers, introduced to eliminate the effects of different scales of measurement. The only logical constraint on choosing r_k is that it should be in the same measurement units as the variable X_k that it normalizes. The usual choices are that r_k is the standard deviation of X_k in the total sample of size n, or that r_k is the range of X_k in the sample. Occasionally r_k may be the supposed range of X_k in the population and hence greater than the sample range. Another form of normalization, suitable for ratio scales, is to log transform each value x_{ik} before using the formulae. When r_k is chosen as the sample range then unity is the maximum value of D_2, D_3, and D_4, as is desirable for a dissimilarity. The coefficients from D_5 on are self-normalizing; for positive observations D_6 and D_7 are identical. Except for D_8, all coefficients are metric for positive observations, but D_5, D_6, D_9

and D_{10} may be nonmetric when negative values (perhaps the result of data transformation) enter the formulae. The form of D_7 is such that differences between large values get less weight than the same differences between small values. D_4 is the well-known *Minkowski metric* with the special cases $D_2(t = 2)$ and $D_3(t = 1)$. When the variables are binary, D_3 often is termed *Hamming distance* and is the dissimilarity equivalent of S_4. Similarly, D_9 for binary variables is the dissimilarity equivalent of S_3.

Qualitative and quantitative variables may be combined in the general coefficient discussed earlier by setting $w_k(x_{jk}, x_{jk}) = 1$ (or any other weight thought appropriate). The process is illustrated for D_4 where we set

$$s_k(x_{ik}, x_{jk}) = 1 - |x_{ik} - x_{jk}|^t / r_k^t .$$

For this to give nonnegative scores, r_k must be properly chosen; usually the sample range or greater will suffice. Further, except when $t \neq 1$, this gives a similarity corresponding to D^t rather than to D itself. When $t = 1$ and sample range is the normalizer the resulting similarity matrix $\mathbf{S}$ is p.s.d.; when sample standard deviation is the normalizer then $\mathbf{S}$ is not necessarily p.s.d. A similar method of scoring may be used to convert the other coefficients of Table 3 to similarities. In particular the similarity associated with D_{10} becomes

$$S_{ij} = \frac{1}{p} \sum_{k=1}^{p} \frac{\min(x_{ik}, x_{jk})}{\max(x_{ik}, x_{jk})},$$

which is an element of a p.s.d. similarity matrix.

Table 3 contains no coefficients of correlational type but they are often used. Although algebraically identical to sample correlations between pairs of variables, the calculation of correlational similarities is performed on pairs of units. Two forms of normalization are involved, one explicit and one implicit. Explicit normalization by dividing each variable by r_k, as above, usually is needed to make the variables comparable, possibly but not necessarily associated with replacing each x_{ik} by its deviation from the mean to give $y_{ik} = (x_{ik} - x_{.k})/r_k$. The simplest thing is then to calculate an uncentered correlation ρ_{ij} from

$$\rho_{ij}^2 = \frac{\left(\sum_{k=1}^{p} y_{ik} y_{jk} \right)^2}{\sum_{k=1}^{p} y_{ik}^2 \sum_{k=1}^{p} y_{jk}^2};$$

ρ_{ij} is the cosine of the angle subtended by the ith and jth units at the mean unit (or null unit when deviations from the mean are ignored). Frequently, however, a centered correlation is used, which carries with it implicit forms of normalization. Now y_{ik} is replaced by $z_{ik} = y_{ik} - y_{i.}$ but the justification for summing across variables to obtain $y_{i.}$ is unclear, and $\sum z_{ik}^2$ has no ready interpretation, even though the values y_{ik} may be standardized to nondimensional forms.

SIMILARITY BETWEEN POPULATIONS

The preceding concerns the evaluation of similarity between a pair of units. Nothing has been said as to whether the units are to be regarded as samples from some larger population of units or whether the units themselves represent larger groups or complete populations. In practice the same coefficients often are used in both situations but their use when units are samples raises many difficult problems. It is sometimes plausible to envisage populations described by qualitative variables that vary only between and not within populations. The similarity coefficients of Table 1 may then be interpreted as measuring similarity between populations. Quantitative variables always vary within populations, though the ratio of within to between population variation may be small. The mean values $\bar{x}_{ik}$ $(k = 1, 2, \ldots, p)$ for the ith population may be used to represent the entire population, and these values may replace x_{ik} in Table 3 to give between-population distances/dissimilarities for quantitative variables. D_2 is then known as the *coefficient of racial likeness* (CRL) and often is used in a form adjusted for unequal sample sizes and their resulting biases. To allow for differences in variances within populations, different standardizations may be adopted for each population. These adjustments usually destroy the metric properties of D_2. The population version of D_3 is known as the *mean character difference* or as *Czekanowski distance*.

The distribution of character values in a population is naturally described by a multivariate probability distribution function $f(\mathbf{x})$. Assessing similarity between two populations then becomes a problem of comparing two probability functions $f_1(\mathbf{x})$ and $f_2(\mathbf{x})$. This may be done in several ways, but the general notion is that of measuring overlap; the greater the overlap the more similar are two populations. One way of measuring overlap is to set up a discriminant rule R for assigning a given sample $\mathbf{x}$ either to $f_1(\mathbf{x})$ or to $f_2(\mathbf{x})$ (*see* DISCRIMINANT ANALYSIS). If $\mathbf{x}$ is from $f_1(\mathbf{x})$ there will be some probability $\alpha(1,2)$ that the rule R assigns $\mathbf{x}$ to $f_2(\mathbf{x})$. Similarly, there is a probability $\alpha(2,1)$ that R will assign incorrectly a sample from $f_2(\mathbf{x})$ to $f_1(\mathbf{x})$. The quantities $\alpha(1,2)$ and $\alpha(2,1)$ are termed *errors of classification*. The problem of discrimination is to choose R to minimize some function of $\alpha(1,2)$ and $\alpha(2,1)$, perhaps involving costs of making incorrect decisions and the a priori probabilities of the occurrences of samples from the two populations, but we shall ignore these quantities. The most simple rule is to choose R so that $\alpha = \frac{1}{2}(\alpha(1,2) + \alpha(2,1))$ is minimized. Unlike discriminant analysis we are not so much concerned with R itself but more with the values of α. Ideally we would like to have $\alpha = 0$ so that there is no overlap and the two populations cannot be confused; this corresponds to an extreme distance between the populations. At the other end of the scale when $\alpha = 1$ there is complete overlap and the populations are inseparable, corresponding to zero distance. Clearly $\delta = 1 - \alpha$ is one measure of dissimilarity appropriate to such circumstances; δ satisfies the metric inequality. Although α has a very general form and is potentially widely applicable, it seems to have had little use in practice. One reason for this is undoubtedly the difficulty of specifying with any confidence the functional forms of $f_1(\mathbf{x})$ and $f_2(\mathbf{x})$, especially when the variables of the vector $\mathbf{x}$ have mixed modes. When multinormal with $f_1(\mathbf{x}) = N(\boldsymbol{\mu}_1, \boldsymbol{\Sigma})$ and $f_2(\mathbf{x}) = N(\boldsymbol{\mu}_2, \boldsymbol{\Sigma})$ we have

$$1 - \frac{1}{2}\alpha^2 = \frac{1}{\sqrt{2\pi}} \int_{-\infty}^{D/2} e^{-x^2/2}\, dx$$

where $D^2 = (\boldsymbol{\mu}_1 - \boldsymbol{\mu}_2)'\boldsymbol{\Sigma}^{-1}(\boldsymbol{\mu}_1 - \boldsymbol{\mu}_2)$, the square of Mahalanobis distance, itself a Euclidean metric. The population values, $\boldsymbol{\mu}_1$, $\boldsymbol{\mu}_2$, and $\boldsymbol{\Sigma}$ are usually unknown and therefore have to be replaced by sample values $\mathbf{m}_1, \mathbf{m}_2$, and $\mathbf{S}$ (here a sample dispersion matrix, not a similarity matrix). The corresponding estimate of D then has a distribution whose mathematical properties can be studied, together with the estimates of $\alpha(1,2)$ and $\alpha(2,1)$. When $\mathbf{S}$ is taken to be the unit matrix $\mathbf{I}$, Mahalanobis D^{2*} becomes the population version of D_1^2. When $\mathbf{S}$ is the diagonal matrix of sample variances then D^2 becomes D_2^2 with $r_k = S_{kk}$, i.e., D becomes CRL. Like CRL, Mahalanobis distance often is used in a form adjusted for unequal sample sizes, a form in which it can lose its metric properties.

Writing $d_k = x_{ik} - x_{jk}$, where variables may be measured in standardized units and $\bar{d} = \sum_{k=1}^{p} d_k / p$, an average of this difference over all variables, we have

$$D_1^2 = D_2^2 = \text{CRL} = \frac{1}{p} \sum_{k=1}^{p} d_k^2$$
$$= \frac{1}{p} \sum_{k=1}^{p} \left(d_k - \bar{d}\right)^2 + \bar{d}^2.$$

The component $\bar{d}^2$ is an average size difference; it is usually denoted by C_Q^2 and is referred to as Penrose distance. The remaining component $(1/p)\sum_{k=1}^{p}(d_k - \bar{d})^2$ may be regarded as a measure of shape difference and is denoted by $(p-1)C_Z^2/p$ after Zarapkin; thus

$$C_Z^2 = \frac{1}{p-1} \sum_{k=1}^{p} \left(d_k - \bar{d}\right)^2 \quad \text{and}$$

$$\text{CRL} = \frac{p-1}{p} C_Z^2 + C_Q^2.$$

A similar decomposition into size and shape components can be made for Mahalanobis

D^2. The study of size and shape has a considerable literature, e.g., Bookstein [2].

Another general approach to evaluating interpopulation dissimilarity that has been proposed is to consider each population as a point in a Riemannian space with the population parameters as coordinates. Thus for a multinormal population the coordinates are $(\boldsymbol{\mu}, \boldsymbol{\Sigma})$ and the space has $p + \frac{1}{2}p(p+1)$ dimensions. Similar populations will map into neighboring points in this space and, provided a suitable metric can be defined, the distance between populations is merely a matter of integrating along the geodesic connecting the points representing the two populations. Rao [7] suggested using a metric with elements

$$g_{kl} = E\left(\frac{1}{f} \frac{\partial f}{\partial \theta_k} \frac{1}{f} \frac{\partial f}{\partial \theta_l} \right), \quad k, l = 1, 2, \ldots, q$$

and $\theta_1, \theta_2 \ldots, \theta_q$ are the parameters of $f(\mathbf{x}, \boldsymbol{\theta})$ and where expectation is taken with respect to $f(\mathbf{x})$. For multinormal distributions with equal dispersion matrices $\boldsymbol{\Sigma}$, this again leads to Mahalanobis D^2.

For discrete multivariate distributions with t classes and associated probabilities $\pi_1, \pi_2, \ldots, \pi_t$ we have

$$\pi_1 + \pi_2 + \cdots + \pi_t = 1,$$

so that the population may be represented as a point in this plane and distance between populations i and j evaluated from $\sum_{k=1}^{t} (\pi_{ik} - \pi_{jk})^2$ similarly to D_1^2. Alternatively we note that the point $(\sqrt{\pi_1}, \sqrt{\pi_2}, \ldots, \sqrt{\pi_t})$ lies on a unit sphere, so that distance may be evaluated as the angle (equivalent to arc length) given by

$$\cos^{-1}\left(\sqrt{\pi_{i1}\pi_{j1}} + \sqrt{\pi_{i2}\pi_{j2}} + \cdots + \sqrt{\pi_{it}\pi_{jt}} \right),$$

a form due to Bhattacharyya that has proved quite popular. Various maplike projections of the points on the unit sphere onto the plane $\Sigma\pi_i = 1$ have also been used, but the differences between the resulting variant forms of distance tend to be slight.

For continuous distributions $f_1(\mathbf{x}), f_2(\mathbf{x})$ the angular form may be generalized to give

$$\cos^{-1} \int \sqrt{f_1(\mathbf{x}) f_2(\mathbf{x})}\, d\mathbf{x},$$

which can be shown to be related to Rao's geodesic approach.

A more recent measure of population difference proposed by Sibson [5] is based on the concept of information gain, and is measured by

$$I(1,2) = \int_{-\infty}^{\infty} \log_2 \left(\frac{f_1(\mathbf{x})}{f_2(\mathbf{x})} \right) f_1(\mathbf{x})\, d\mathbf{x}.$$

This may be thought of as the average information for rejecting population 2 in favor of population 1 when $\mathbf{x}$ belongs to population 1. There is a clear relationship with the probability of misclassification $\alpha(1,2)$, discussed above, and as with discriminant functions the likelihood ratio $f_1(\mathbf{x})/f_2(\mathbf{x})$ plays a prominent role. Usually $I(1,2) \neq I(2,1)$, so these two measures have to be combined in some symmetric way; Sibson defines symmetric information gain by

$$J(1,2) = \tfrac{1}{2}\left[I(1,2) + I(2,1) \right].$$

When the populations do not overlap this measure is undefined and is replaced by *information radius*, which measures information gain relative to the mixture distribution of the two populations $g(x) = \frac{1}{2}[f_1(\mathbf{x}) + f_2(\mathbf{x})]$, to give

$$K(1,2) = \frac{1}{4} \int_{-\infty}^{\infty} \left[f_1(\mathbf{x}) \log_2 f_1(\mathbf{x}) + f_2(\mathbf{x}) \log_2 f_2(\mathbf{x}) - g(\mathbf{x}) \log_2 g(\mathbf{x}) \right] g(\mathbf{x})\, d\mathbf{x}.$$

This may be generalized to incorporate weights associated with the various populations (perhaps a priori probabilities) and also to give a measure of association* between more than two populations. For two multinormal distributions with unequal disperson matrices $\boldsymbol{\Sigma}_i, \boldsymbol{\Sigma}_j$, information radius is given by

$$N_{ij} = \log\left[\frac{\det\left(\frac{1}{2}(\boldsymbol{\Sigma}_i + \boldsymbol{\Sigma}_j)\right)}{\sqrt{\{(\det \boldsymbol{\Sigma}_i)(\det \boldsymbol{\Sigma}_j)\}}} \right] + \frac{1}{2} \log_2 \left\{ 1 + \frac{1}{4} D_{ij}^2 \right\},$$

where

$$D_{ij}^2 = (\boldsymbol{\mu}_i - \boldsymbol{\mu}_j)' \left[\tfrac{1}{2}(\boldsymbol{\Sigma}_i + \boldsymbol{\Sigma}_j) \right]^{-1} (\boldsymbol{\mu}_i - \boldsymbol{\mu}_j).$$

When $\mathbf{\Sigma}_i = \mathbf{\Sigma}_j$ then D_{ij}^2 is exactly Mahanalobis' D^2 and the first term in the expression for N_{ij} vanishes. Thus N_{ij} can be considered a symmetric generalization of Mahalanobis' D^2 for two normal populations with different dispersions.

This article has been concerned with definitions of dissimilarity rather than with applications. One application, that of comparing two different classifications of the same n samples, sometimes uses the coefficients described above in a very direct manner. Suppose the n units have been classified into k classes; then for each of the $p = \frac{1}{2}n(n-1)$ sample pairs we may score unity if they are in the same class, zero otherwise. This gives a row vector $\mathbf{x}_1$ of p dichotomous variables. A second row vector $\mathbf{x}_2$ may be derived from a second classification. Provided the two classifications represented by $\mathbf{x}_1$ and $\mathbf{x}_2$ have the same number of classes, they may be compared by any of the coefficients given in Table 1. Gower [4] discusses the many methods of comparing classifications, detailed expositions of which are given by other authors [4]. In particular, Milligan describes the above approach. A hierarchical classification (*see* HIERARCHICAL CLUSTER ANALYSIS) may be represented by a tree which when cut at an appropriate level will generate k exclusive classes. Thus the above method may be used to compare trees. Fowlkes and Mallows [3] have used this approach with a coefficient B_k (which turns out to be the same as Ochiai's coefficient S_{12} of Table 1), and are especially interested in the interpretation of the plot of B_k against k.

References

[1] Anderberg, M. R. (1973). *Cluster Analysis for Applications*. Academic Press, New York.

[2] Bookstein, F. L. (1979). *The Measurement of Biological Shape and Shape Change*. Lecture Notes in Biomathematics, 24. Springer-Verlag, Berlin-Heidelberg-New York.

[3] Fowlkes, E. B. and Mallows, C. L. (1983). *J. Amer. Statist. Ass.*, **78**, 553–569 (with discussion).

[4] Gower, J. C. (1983). In *Numerical Taxonomy*, J. Felsenstein, ed. Springer-Verlag, New York.

[5] Jardine N. and Sibson R. (1971). *Mathematical Taxonomy*. Wiley, New York.

[6] Legendre L. and Legendre P. (1982). *Numerical Ecology*. Elsevier, Amsterdam.

[7] Rao, C. R. (1949). *Sankhyā*, **9**, 246–248.

[8] Sneath, P. H. and Sokal, R. R. (1973). *Numerical Taxonomy*. Freeman, San Francisco.

[9] Spath H. (1980). *Cluster Analysis Algorithms*, Trans. Ursula Bull. Ellis Horwood (Halsted/Wiley), Chichester, England.

(CLASSIFICATION
DENDRITES
DISCRIMINANT ANALYSIS
J-DIVERGENCES AND RELATED CONCEPTS
MAHALANOBIS D^2
MULTIDIMENSIONAL SCALING
PROBABILITY SPACES, METRICS AND DISTANCES ON)

J. C. GOWER

MEASURES OF UNIFORMITY

To analyze issues of equity such as the fairness of the allocation of seats (representatives) in a legislature or the assessed value of homes in a county one requires a measure of how far a set of observations is from the ideal of perfect equality (e.g., each district has exactly the same population or each house is assessed at the same fraction of its market value). Different measures are used in different applications, in part because of the historical development of the subject matter area and in part because of special characteristics of the data such as nonnormality. In measuring tax assessment uniformity the data consist of assessed values and sales prices for individual homes, while in apportioning a legislature, one must divide the population into a fixed number of districts, each of which should contain the same number of people. If there are geographical constraints on the allocation as in the division of the 435 seats in the U.S. Congress to the states, then absolute equality is impossible to attain and mathematical procedures for minimizing the disparity have been investigated [2]. Evaluating the degree of integration (uniformity) of pupils in a school system is further complicated by the fact that reasonable measures are affected by both the number of schools and the minority fraction of the entire population of students.

There are two main types of measures of uniformity, those related to measures of inequality and those developed for measuring integration. A description of several important indices of each type and their use is given in the next two sections.

MEASURES OF UNIFORMITY RELATED TO MEASURES OF RELATIVE INEQUALITY

A large family of measures of uniformity is related to measures of relative inequality such as the coefficient of variation*, the ratio of the standard deviation to the mean, and the *Gini index* (*see* INCOME INEQUALITY MEASURES). They compare the spread among observations to a central value. Although the coefficient of variation is a standard measure for approximately normal data, applications involving nonnormal data [5, 9] such as real estate assessments use the median, a more robust measure of the center than is the mean. The measure used in that application is the *coefficient of dispersion* (CD), which is the ratio of the average absolute deviation from the median of the data to the median. Mathematically, let x_i, $i = 1, \ldots, n$ be n observations, m their median; then the CD is defined as

$$\mathrm{CD} = \sum_{i=1}^{n} |x_i - m|/(nm). \tag{1}$$

In the tax assessment application, x_i is the ratio of assessed value (V_i) to the sales price (S_i) of the ith property. By expressing (1) as

$$\mathrm{CD} = n^{-1} \sum_{i=1}^{n} \left| \frac{V_i}{S_i} - m \right| \Big/ m$$

$$= n^{-1} \sum_{i=1}^{n} |m^{-1}V_i - S_i|/S_i, \tag{2}$$

the CD can be interpreted as a measure of the accuracy of assessments as forecasts of sales prices, where the sales price is predicted by $m^{-1}V_i$. The right side of (2) is the average relative error of the forecast to the actual price, a standard measure of forecast accuracy [3]. The assessment literature [5, 11, 12] indicates that assessors consider a CD of .20 as acceptable, but mathematical considerations [9] indicate that a CD of .10 or less is needed to ensure that only a small proportion of homes would be overassessed or underassessed. For 1976, the U.S. Bureau of the Census* reported [23] that only 42.3% of the areas surveyed had a CD less than .20, so this topic continues to be of interest to legal and government bodies [11, 17].

Courts have used the Lorenz curve* and measures related to it to measure the disparity of legislative districts in one-person-one-vote cases. If n districts are ordered according to their population P_i from smallest to largest, the percentage of the total population in the smallest i districts is

$$L_i = 100\left(\sum_{j=1}^{i} P_j \Big/ \sum_{j=1}^{n} P_j \right).$$

The Lorenz curve is a plot of the values $(i/n, L_i)$, $i = 1, \ldots, n$. Courts often have considered the *minimum* population percentage needed to control the legislature, essentially $L(0.5)$, as a criterion in examining equality of representation. Two related criteria that courts and legal commentators have used are the ratio of the population $P_{(n)}$ of the largest district to that of the smallest district ($P_{(1)}$), called the *population ratio* or *citizen population variance*, and the *total variation measure*. The second measure is the sum of $(P_{(n)} - \bar{P})/\bar{P}$ and $[\bar{P} - P_{(1)}]/\bar{P}$ where $\bar{P}$ is the average of the P_i, which respectively measure the largest degree of underrepresentation and overrepresentation. The larger of the two ratios, the *maximum deviation*, has also been used. The U.S. Supreme Court has not set a maximum allowable tolerance from perfect population equality because it has stated that any deviation from equality has to be justified by legitimate considerations. The court has required a stricter adherence to population equality for congressional districts, rejecting a 5.97 total variation in *Kirkpatrick v. Priesler*, 394 U.S. 526 (1969), than for state and county legislatures [allowing a total variation of 16.4% in *Mehan v. Howell*, 410 U.S. 315 (1973)].

Political scientists have used the Gini index, derived from the Lorenz curve, as a measure [1]. It can be shown that if the maximum deviation is $y\%$, then the Gini index must be less than $y/200$, i.e., if the maximum deviation is less than 10%, the Gini index is less than .05. Several of these measures have been used in the opinion and in the many articles concerning the landmark *Baker v. Carr* [369 U.S. 186 (1962)] case in which the U.S. Supreme Court found the apportionment of the legislature of the state of Tennessee in violation of the equal protection clause. The state had not been reapportioned since 1900 and the suit first reached the court in 1960. In 1900 the 33 districts ranged in population from 9466 to 19,992 and the Gini index equaled .112, while the minimum population percentage for control, the fraction of the population residing in the smallest 17 districts, was 43.4%. By 1960 the Gini index had risen to .305 and the minimum population percentage for control had decreased to 29.3%.

Theil's coefficient [20, 21] is a measure of relative inequality, based on information theory*, which has been used to measure the uniformity of expenditures per pupil [19]. For data from a random variable Z with a continuous distribution and mean μ, Theil's measure is defined by

$$E\left[(Z/\mu)\log(Z/\mu)\right] = \left\{E\left[Z(\log Z)\right] - \mu\log\mu\right\}/\mu. \quad (3)$$

If divided by $\log\mu$ this can be put in the general form

$$\frac{E\left[h(Z)\right] - h(\mu)}{h(\mu)}, \quad (4)$$

where h is a convex function (h is $\log z$ for Theil's index and is z^2 for the coefficient of variation). These measures [10, 20] allow for a decomposition of the total inequality or nonuniformity into between- and within-group components. In a study by the National Center for Educational Statistics [19], the database consisted of district level data, so the formula used was

$$\sum(z_i/\mu)\log(z_i/\mu), \quad (5)$$

where $\mu = (\Sigma P_i z_i)/\Sigma P_i$, P_i is the school population of the ith district and z_i the resource (expenditures per pupil, teachers per 100 pupils). The data available for use in this application consists of school expenditures and teacher/pupil ratios for each district. Thus the calculation of Theil's measure or any other measure of uniformity (e.g., the coefficient of variation [9]) implicitly assumes that the distribution of resources in each district is equal for all students, so that the true index is higher than the calculated one. Nevertheless, the fact that Theil's measure can easily be decomposed into within- and between-state components makes it a more appropriate index than the Gini index, whose decomposition is very involved.

INDICES OF SEGREGATION

The *dissimilarity index D* has been the most commonly used measure [7, 18] of the difference between the actual distribution of a minority racial group in the census tracts of a city from their percentage of the total population of the city. If there are M_i minority persons and W_i whites in the ith census tract and $M = \sum_{i=1}^{k} M_i$ minorities and $W = \sum_{i=1}^{k} W_i$ whites in the city, then

$$D = \frac{1}{2}\sum_{i=1}^{k}\left|\frac{M_i}{M} - \frac{W_i}{W}\right|, \quad (6)$$

where k is the number of census tracts. When the same percentage of each racial group (minority and white) is contained in each tract, $D = 0$. The index can be interpreted as the proportion of the minority population that would have to change census tracts to make $D = 0$ [6, 7], assuming that the minority residents are *not* replaced by majority ones. The index D can also be interpreted as the maximum vertical distance between the "segregation curve" and the line of equality, where the segregation curve plots the cumulative proportion of whites versus the cumulative proportion of minorities, where the respective cumulatives are calculated after arranging the census tracts in order of increasing nonwhite proportions. This approach was used by Alker [1] to analyze the segregation of blacks in schools. The dissimilarity index D is also expressible as

$$D = \frac{1}{2PQ} \sum_{i=1}^{k} \frac{T_i}{T} |P_i - P|, \qquad (7)$$

where $T_i = M_i + W_i$, $T = \sum_{i=1}^{k} T_i$, $P_i = M_i/T_i$, $P = M/T$, and $Q = 1 - P$, which is a weighted average of the absolute differences between the minority proportions P_i for subareas (census tracts, schools, etc.) and their proportion P in the entire area. A related index considered by the U.S. Bureau of the Census [18] is

$$H = \frac{1}{TPQ} \sum_{i=1}^{k} \frac{T_i}{T} (P_i - P)^2. \qquad (8)$$

Since the values of both indices, D and H, on a given set of data depend on the size of the units (subareas), recent studies [6, 22, 25] suggest that one use the difference from their expected values under a random distribution as a measure. Letting D_0 and H_0 denote these expected values, the standardized indices [25] are

$$D_s = \frac{D - D_0}{1 - D_0}, \quad \text{ranging from } \frac{-D_0}{1 - D_0} \text{ to } 1$$

and

$$H = \frac{H - H_0}{1 - H_0}, \quad \text{ranging from } \frac{-H_0}{1 - H_0} \text{ to } 1.$$

An alternative standardization of D using its approximate normal distribution under the assumption of random assignment note above has been suggested by Cortese et al. [6]. This has the advantage of incorporating the size variations of the subareas into the calculation.

Another measure of the lack of uniformity or relative segregation was proposed by Theil [21] and measures the difference between the subarea P_i's and overall P by the information measure:

$$I_i = P_i \log(P_i/P) + Q_i \log(Q_i/Q), \quad \text{where } Q_i = 1 - P_i, \qquad (9)$$

and weights the individual indices by their proportion of the overall population. The measure is standardized by taking its ratio to the information value of the most equal assignment.

All the segregation measures discussed thus far measure the closeness of the minority proportions in all subareas to their proportion in the total area. Another index, due to Coleman, measures the difference between the cross-racial experience (contact) of a group and what would exist under perfect integration (all $P_i = P$). The cross-racial contact for a majority (white) individual in the ith subarea is measured by the black fraction, P_i, in that location. The overall measure, the average percent black for a white, defined by

$$P_{b/w} = \sum_{i=1}^{k} \frac{W_i}{W} P_i, \qquad (10)$$

is the average black fraction of "neighbors" a white has. The standardized version

$$S = \frac{P - P_{b/w}}{P} = 1 - \frac{P_{b/w}}{P} \qquad (11)$$

equals 0 when $P_{b/w} = P$ (perfect integration) and 1.0 when all locations are completely segregated. The U.S. Commission on Civil Rights [20] considers a school system to have a high (medium, low) degree of segregation when S is greater than 0.5 (between 0.2 and 0.5, below 0.2, respectively). This index has the desirable property that its expected value under random assignment [4, 22] is independent of P and rapidly approaches zero for subareas of moderate size [4]. Moreover, it can be regarded as the between-location proportion of the overall sum of squares in an analysis of variance* where $X_{1j} = 1$ if the jth person in the ith location is minority (and 0 otherwise). This interpretation yields another formula for S:

$$S = \frac{\sum T_i P_i^2 - TP^2}{TP(1 - P)}. \qquad (12)$$

This shows that S mathematically is the same as H [see (8)]. Because of its various meanings and desirable statistical characteristics, the index S or related diversity measures [13–16] developed to summarize the diversity of species and gene frequencies in populations probably will be used more frequently in future research.

References

[1] Alker, H. R., Jr. (1965). *Mathematics and Politics*. Macmillan, New York.

[2] Balinski, M. L. and Young, H. P. (1982). *Fair Representation*. Yale University Press, New Haven, Connecticut.

[3] Basi, B. A., Carey, K. J., and Twark, R. D. (1976). *Accounting Rev.*, 244–254.

[4] Becker, H. J. (1978). *Amer. Statist. Ass. Proc.: Soc. Statist.*, pp. 349–353.

[5] Behrens, J. O. (1977). *J. Educ. Finance*, **3**, 158–164.

[6] Cortese, C. F., Falk, F., and Cohen, J. K. (1976). *Amer. Sociol. Rev.*, **41**, 630–637.

[7] Duncan, O. D. and Duncan, B. (1955). *Amer. Sociol. Rev.*, **20**, 210–217.

[8] Friedman, L. S. and Wiseman, M. (1978). *Harvard Educ. Rev.*, **48**, 193–226.

[9] Gastwirth, J. L. (1982). *J. Statist. Plan. Inf.*, **6**, 1–12.

[10] Gastwirth, J. L. (1975). *J. Econometrics*, **3**, 61–70.

[11] Noto, N. A. (1978). *Business Rev. Federal Reserve Bank of Phil.* May–June, 13–23.

[12] Oldman, D. and Aaron, H. (1965). *National Tax J.*, **18**, 36–49.

[13] Patil, G. P. and Taillie, C. (1982). *J. Amer. Statist. Ass.*, **77**, 548–567.

[14] Pielou, E. C. (1975). *Ecological Diversity*. Wiley, New York.

[15] Rao, C. R. (1982). *Theoret. Pop. Biol.*, **21**, 24–43.

[16] Rao, C. R. (1982). *Sankhyā*, **44A**, 1–22.

[17] Robertson, J. L. (1977). *Mississippi Law J.*, **48**, 201–207.

[18] Taueber, K. E. and Taueber, A. F. (1965). *Negroes in Cities*. Aldine, Chicago.

[19] The Condition of Education (1980). *Statistical Report of the National Center for Education Statistics*.

[20] Theil, H. (1967). *Economics and Information Theory*. North-Holland, Amsterdam.

[21] Theil, H. (1972). *Statistical Decomposition Analysis*. North-Holland, Amsterdam.

[22] U.S. Bureau of the Census (1971). Proposal for Developmental Work on Measures for Residential Similarity (unpublished memorandum).

[23] U.S. Bureau of the Census (1979). *Taxable Property Values and Assessment Sales Ratio*. (1977 Census of Governments, Vol. 2.)

[24] U.S. Civil Rights Commission (1977). *Reviewing a Decade of School Desegregation*.

[25] Winship, C. (1977). *Social Forces*, **55**, 1058–1066.

(DIVERSITY, INDICES OF
INCOME INEQUALITY MEASURES
INDEX NUMBERS
LORENZ CURVE
MEASURES OF SIMILARITY, DISSIMILARITY, AND DISTANCE)

JOSEPH L. GASTWIRTH

MEASURE THEORY IN PROBABILITY AND STATISTICS

While much of probability and statistics can be developed and understood with no more than the tools provided by calculus, large portions remain inaccessible without a proper understanding of measure theory. Fortunately, in several important instances where measure theory is necessary for the development of a result, after the measure-theoretic "dust" has settled, the final result can be understood and applied without a knowledge of measure theory. Thus for the user of probability and statistics, the need for measure theory is indeed limited to certain special areas only. However, for the researcher in the areas of probability and statistics, measure-theoretic probability and its ramifications in mathematical statistics provide tools and a clarity of view which are indispensable, as evidenced by the fact that some form of measure theory is required, or strongly encouraged, by all major Ph.D. programs in mathematical statistics.

This article begins with some motivation, makes some brief historical comments, lists representative texts, summarizes the basic results of measure theory relevant to probability and statistics, and concludes with some additional examples of usages of measure theory in probability and statistics.

MOTIVATION

There is naturally some reluctance among teachers and textbook writers to develop probability and, particularly, statistics from a measure-theoretic perspective; many students are not prepared for it, and many of the basic concepts can be introduced and developed more simply, particularly in special cases, unhampered by measurability questions. While this approach has considerable appeal, there remain quite good reasons for using measure theory in the development of probability and mathematical statistics:

1. Any rigorous development of probability or statistics requires a well-defined notion of a random variable, and, for this, the notion of a sample space* must be clearly understood. When the sample space is finite or countable, this can be done without measure theory. But when the sample space is even as simple as a real interval, measure theory is required (*see* PROBABILITY THEORY).
2. An infinite sequence $X_1, X_2, \ldots$ of (nondegenerate) independent random variables requires an uncountable sample space and cannot be studied rigorously without measure theory. Thus to state the strong law of large numbers* carefully, even in its simplest nontrivial setting, for independent and identically distributed Bernoulli random variables, measure theory is needed; there is no simpler countable-sample-space analog of the strong law to be described with ordinary calculus.
3. Various results from analysis are required to justify the interchange of limits. These are most easily summarized within the context of measure theory: the monotone and the dominated convergence theorems, Fatou's lemma, and Fubini's theorem.
4. Measure theory is required to answer important basic questions of the following type, although the answers are usually "clear" on intuitive grounds:
   (a) Is a function of one or more random variables a random variable?
   (b) When X and Y are independent random variables, is a function of X independent of a function of Y?
   (c) If the random variable X depends on (is a function of) the random variables Y and Z, and if Z is jointly independent of X and Y, does X depend only on Y (is X a function of Y only)?
5. The "transformation theorem" enables one to express the expected value of a function g of the random variables $X_1, \ldots, X_n$ in terms of their joint distribution function $H(x_1, \ldots, x_n)$ by

$$Eg(X_1, \ldots, X_n) = \int_{-\infty}^{\infty} \cdots \int_{-\infty}^{\infty} g(x_1, \ldots, x_n) \times dH(x_1, \ldots, x_n).$$

   The proof of this very useful result requires measure theory (even when the integral above is expressible as a Riemann integral).
6. Various writers of statistical textbooks, notably Lehmann [26], have found it convenient, for the sake of unity, to use measure theory in a modest way by working with ("dominated") families of density functions defined with respect to some fixed ("dominating") measure μ. When μ is Lebesgue measure, a density function becomes the usual "probability density function." When μ is "counting measure," the density function becomes a "probability function." And, of course, there are other possibilities for μ.
7. It is widely recognized that the notions of conditional probability and expectation* depend on measure theory, notably the Radon–Nikodym theorem*, whenever one is conditioning on an event that has probability zero. The fol-

lowing paradox illustrates the point (see Billingsley [2], p. 392). The task is to evaluate the regression $E(R^2 | X = Y)$, where X and Y are independent standard (unit) normal random variables and R, Θ are the polar coordinates of X, Y. Since R and Θ are independent, by one line of argument

$$E(R^2 | X = Y) = E\left(R^2 | \Theta = \frac{\pi}{4} \text{ or } \frac{5\pi}{4}\right) = E(R^2) = E(X^2 + Y^2) = 1 + 1 = 2.$$

It is also easily seen that $Z = (X + Y)/\sqrt{2}$ and $W = (X - Y)/\sqrt{2}$ are independent standard normal random variables and that $Z^2 + W^2 = X^2 + Y^2 = R^2$. Thus by another line of reasoning

$$E(R^2 | X = Y) = E(Z^2 + W^2 | W = 0) = E(Z^2) = 1.$$

It appears that $2 = 1$! The problem is that no clear meaning can be attached to "$E(R^2 | X - Y = 0)$"; one may make sense of "$E(R^2 | X - Y = v)$" for *almost every* v, but not for *every* v, nor for any particular v. This difficulty is completely resolved, as far as mathematics allows, by defining conditional expectations in terms of Radon–Nikodym derivatives. It should be emphasized that there would be no difficulties of this sort if all of the random variables were discrete.

8. Because of the difficulties referred to in point **7**, (except for the "discrete case") Fisher's notion of sufficiency requires measure theory to be fully understood. The highly appealing "factorization theorem," by which one finds sufficient statistics*, can be used without knowing anything about measure theory, but its proper understanding requires measure theory. (See Halmos and Savage [22].)

HISTORICAL COMMENTS AND TEXTS

The foundations of modern probability were laid by Kolmogorov [25], who first developed a coherent and general framework based on measure theory. Elements of measure theory are required for the construction of a sample (probability) space*, to introduce and develop the properties of random variables, distribution functions, mathematical expectation, and important tools such as the characteristic function*, and to define and elaborate the notions of convergence*, independence, and conditioning. Perhaps the clearest distinction between what is achievable with calculus only and what requires measure theory is contained in Feller's two volumes [10, 11]. The first is restricted to discrete sample spaces, and the second describes probability theory in general using measure theory.

There are several excellent texts which develop the general theory of probability based on measure theory such as Loève [27], Neveu [28], Feller's second volume [11], Breiman [4], Chung [6], Chow and Teicher [5], and Billingsley [2], whose special feature "is the alternation of probability and measure, probability motivating measure theory and measure theory generating further probability."

As statistics is broadly based on probability, most of the impact of measure theory in statistics is evidenced via probability. The first text to take advantage of measure-theoretic probability fully in laying the mathematical foundations of statistics was written by Cramér [7]. Virtually every area of statistics has seen some use of measure theory. Some of the areas of greatest impact are hypothesis testing*, estimation* and prediction, sequential analysis*, nonparametric* statistics, decision-theoretic inference and asymptotic inference. Typical texts, taken from a variety of statistical areas, which use (at least some) measure theory are Cramér [7], Fraser [14], Lehmann [26], Rao [31], Ferguson [12], Hájek and Šidák [20], Ghosh [15], Hannan [23], Puri and Sen [30], Serfling [34], and Grenander [19].

Measure theory also plays an important role in the study of stochastic processes* when the parameter space or state space or both are not discrete. The first and most influential text in this area is Doob [9]. Several aspects of stochastic processes are covered in some of the standard probability texts mentioned above: Loève [27], Neveu [28], Breiman [4], Feller [11]; some typical broad texts devoted to stochastic processes are Cramér and Leadbetter [8], Gihman and Skorohod [16, 17, 18], and Wentzell [35]. The influence of measure theory is even more conspicuous in monographs on more specialized topics in stochastic processes such as Markov* and diffusion processes*, martingales*, and stochastic calculus. While these are topics that traditionally have been of greater interest to probabilists, numerous applications of stochastic processes* are currently being made by statisticians.

BASIC INGREDIENTS OF MEASURE THEORY

The first notion of measure theory is that of (a σ-field and of) measurability. A very useful way of thinking of measurability is provided by Littlewood's three principles for functions of a real variable (see Royden [32]): "Every measurable set is nearly a finite union of intervals; every measurable function is nearly continuous; every convergent sequence of measurable functions is nearly uniformly convergent." The word "nearly" in these statements has a specific measure-theoretic meaning which enables one to pass from such well understood concepts as intervals and continuous functions to the more elusive measurable sets and measurable functions. Analogous principles are also true for general measurable spaces (rather than the real line considered in the above formulation). Here is a list of the basic results of measure theory which are relevant to probability and statistics. The adjective "measurable" is omitted throughout for simplicity.

1. **Extension Theorem.** This shows that a measure originally defined on a semiring of sets can be extended to a measure on the σ-field generated by the semiring (and that the extension is unique and σ-finite if the measure is σ-finite on the semiring).
2. **Integration and Limit Theorems.** The monotone convergence theorem and the dominated convergence theorem permit the interchange of integration and pointwise limit under appropriate conditions, justifying the formula $\lim_n \int f_n \, d\mu = \int \lim_n f_n \, d\mu$; Fatou's lemma provides a useful inequality between the limit infimum of a sequence of integrals and the integral of the limit infimum of the sequence of integrands: $\int \liminf_n f_n \, d\mu \leqslant \liminf_n \int f_n \, d\mu$.
3. **Transformation of Integrals (Change of Variables).** The transformation theorem enables us to transform integrals with respect to the measure induced by a transformation: if T is a transformation from X to Y, μ a measure on X, and μT^{-1} the measure induced on Y by
$$\mu T^{-1}(E) = \mu\{T^{-1}(E)\}$$
$$= \mu\{x \in X : T(x) \in E\}$$
then
$$\int_X f(T(x)) \, d\mu(x) = \int_Y f(y) \, d\mu T^{-1}(y).$$
4. **Inequalities and L_p Spaces.** Minkowski's inequality* shows that for $p \geqslant 1$,
$$\left(\int |f+g|^p \, d\mu\right)^{1/p} \leqslant \left(\int |f|^p \, d\mu\right)^{1/p} + \left(\int |g|^p \, d\mu\right)^{1/p},$$
and Hölder's inequality* shows that when $p, q > 1$ and $1/p + 1/q = 1$, then $\int |fg| \, d\mu \leqslant (\int |f|^p \, d\mu)^{1/p} \cdot (\int |g|^q \, d\mu)^{1/q}$. When $p \geqslant 1$, the space L_p of all functions satisfying $\int |f|^p \, d\mu < \infty$ is a Banach space with norm $\|f\|_p = (\int |f|^p \, d\mu)^{1/p}$.
5. **Absolute Continuity* and Singularity: Lebesgue Decomposition and Radon–Nikodym Theorem.** Fix a (σ-finite) measure μ on a space. Then every other (σ-finite) measure ν on the same space can be decomposed into two measures: $\nu = \nu_a + \nu_s$, of which one is absolutely continuous with respect to μ (i.e., sets of

small μ measure have small ν_a measure) and the other is singular to μ (i.e., ν_s and μ "live" on disjoint sets). The Radon–Nikodym theorem* shows that integration with respect to ν_a (or with respect to ν, if ν is absolutely continuous with respect to μ) can be expressed in terms of integration with respect to μ via the Radon–Nikodym derivative f of ν_a with respect to μ (which is uniquely determined by μ and ν): $\int g\, d\nu_a = \int gf\, d\mu$, i.e., $d\nu_a = f\, d\mu$.

6. **Iterative Evaluation of Multiple Integrals, Fubini's Theorem.** The double integral of a function of two variables can be evaluated iteratively in any order, i.e., the order of repeated integrals can be inverted:

$$\int_Y \left\{ \int_X f(x, y)\, d\mu(x) \right\} d\nu(y)$$

$$= \int_X \left\{ \int_Y f(x, y)\, d\nu(y) \right\} d\mu(x)$$

$$\left(= \int_{X \times Y} f(x, y)\, d(\mu \times \nu)(x, y) \right),$$

provided f is nonnegative or product integrable ($\int\int |f(x, y)|\, d\mu(x)\, d\nu(y) < \infty$). This statement includes as a special case the general form of integration by parts; a similar iterative property holds for n-tuple integrals.

Results (**2**), (**3**), and (**6**) comprise the operational calculus of integration, while results (**1**), (**4**), and (**5**) describe the basic structural properties of measure and integration. Among the many excellent texts covering the body of measure theory, we mention the classical book of Halmos [21] and the real analysis texts of Royden [32] and Rudin [33]. Some of the standard texts on probability develop the necessary measure theory, such as Loève [27], Neveu [28], Chow and Teicher [5], and Billingsley [2].

EXAMPLES

The following examples illustrate a variety of uses of the notions and results of measure theory in probability and statistics.

Construction of Sample Spaces

In the case of experiments with an uncountably infinite number of outcomes the sample space is constructed by first defining the probabilities of a simple class of events (which usually form a semiring) and then appealing to the extension theorem (**1**) above to obtain a probability on the σ-field of events they generate. Most texts on calculus-based probability discuss such examples and appeal to measure theory for the construction of the sample space.

Classification of Distribution Functions

As a consequence of the Lebesgue decomposition (5) above, we find there are three types of univariate distribution functions: discrete, absolutely continuous, and continuous singular, and every distribution function is a convex combination of distribution functions of these three types. Use of measure theory enables us to unify results and expressions using general distribution functions rather than treating the discrete and absolutely continuous cases separately. Even though continuous singular distribution functions have rather pathological behavior and are rarely discussed, they nevertheless occur in certain situations; for an interesting example in gambling see Billingsley [3]. In a multivariate setting, continuous singular distributions arise more frequently and quite naturally, e.g., when there is a linear dependence among the random variables.

Series of Independent Random Variables

The study of the properties of series of independent random variables, such as (almost sure) convergence and the zero-one law of tail events, requires measure theory. Series of independent random variables are used very effectively to provide more complex models of random processes with dependence, e.g., via moving averages*. They also have integral analogs, where integrals, with respect to a random measure with independent increments, can be used to model a wide variety of random processes, including stationary processes*.

Conditioning

The notions of conditional probability* and expectation are defined (in the general nondiscrete case), by appealing to the Radon–Nikodym* theorem (**5**), as Radon–Nikodym derivatives (with respect to the basic probability measure) of an appropriately defined measure (possibly a signed measure in the case of conditional expectation) and their properties are developed using the machinery of measure theory. Thus measure theory helps unify the mathematically straightforward discrete case and the general case. Once the general notions and properties of conditioning are understood, their application in other areas of probability and statistics may not require further explicit use of measure theory. Conditioning is used heavily in connection with martingales*, sequential* inference, prediction and filtering, and Markovian models.

Statistical Decision Theory*

Randomized decision rules are probability measures on the action space and thus the general framework of decision theory is cast in the language of measure theory.

Likewise, proper prior distributions* are probability measures on the state space. An interesting statistical example for which it is impossible to avoid a discussion of measure is provided by the "Dirichlet (process) prior": The objective is to define a prior, in a nonparametric setting, on the family of all distributions F on the real line. Associated with the Dirichlet prior is an arbitrary finite (nonnull) measure α on the real line. Briefly, for each partition of the real line into intervals $I_1, \ldots, I_k$, the vector of probabilities $(F(I_1), \ldots, F(I_k))$ is assigned, by the prior, a Dirichlet distribution* with associated parameters $(\alpha(I_1), \ldots, \alpha(I_k))$. Using (**1**) above, Ferguson [13] is able to show that a genuine prior (random) measure is thereby determined. Moreover, using this approach, he finds reasonable explicit Bayesian* procedures for a variety of nonparametric problems.

Contiguity

A sequence of alternatives to the null hypothesis is *contiguous* (to the null hypothesis) if two appropriately defined sequences of probabilities are asymptotically absolutely continuous as the sample size tends to infinity. See CONTIGUITY for a more precise statement and for general comments.

Neyman–Pearson Test and Likelihood Ratio

The Neyman–Pearson fundamental lemma*, which determines the most powerful test of a specified size for two simple hypotheses, is expressed in terms of a likelihood ratio*, i.e., the ratio of the "densities" associated with the two hypotheses with respect to an appropriate references measure (*see* HYPOTHESIS TESTING). This likelihood ratio is in fact the Radon–Nikodym derivative* of the probability measures corresponding to the two hypotheses, and the Neyman–Pearson Lemma is intimately related to the Lebesgue decomposition of the two probabilities. In most cases, such as when a fixed number of observations are taken, the densities and consequently the likelihood ratio are easily computed. However, there are other cases, such as when a random or infinite number of observations are taken, where the relevant densities are elusive (e.g., there is no natural reference measure) and the likelihood ratio is best computed as the Radon–Nikodym derivative of the two probability measures. For instance, when the two hypotheses concern the distribution of a time series* observed over a time interval (e.g., detection of a signal in noise), the probabilities associated with the two hypotheses are the distributions of two time series, and determining their Lebesgue decomposition and Radon–Nikodym derivative is a very complex problem. For a few examples see Grenander [19, Chapters 3 and 5] and for a related discussion see the "Signal Detection" section of COMMUNICATION THEORY, STATISTICAL.

Consider also the following example: Suppose $X_1, X_2, \ldots$ are independent exponentially distributed random variables with common means θ_0^{-1} and θ_1^{-1} under the hypotheses H_0 and H_1, respectively ($\theta_1 > \theta_0 > 0$). If one observes $X_1, \ldots, X_n$ only, the relevant densities for $\theta = \theta_0, \theta_1$ are

$$f_\theta(x_1, \ldots, x_n) = \theta^n \exp\left(-\theta \sum_1^n x_i\right),$$

from which the likelihood ratio is easily constructed. Suppose instead that $X_1, X_2, \ldots$ are lifetimes which are observed sequentially (one after the other) for T units of time. Effectively, what is "observed" are the values of the minimum of T and $X_1 + \cdots + X_n$ for $n = 1, 2, \ldots$. What are the relevant densities? The task here is aided by the fact that $N = \max(n \geqslant 0 : X_1 + \cdots + X_n \leqslant T)$ is a sufficient statistic for θ, and that N has a Poisson distribution with mean θT. The likelihood ratio associated with N is

$$L(N, T) = \left(\frac{\theta_1}{\theta_0}\right)^N \exp(-(\theta_1 - \theta_0)T),$$

which is also the Radon–Nikodym derivative pertinent to the Neyman–Pearson theory. Both testing situations, as well as many others, can be accommodated within one general framework requiring measure theory: Suppose $X_1, X_2, \ldots$ are observed sequentially in time for some possibly random amount of time T, where the choice of T is allowed to depend on what is seen. (T must be a "stopping time.") Then the Radon–Nikodym derivative required for the Neyman–Pearson theory is $L(N, T)$, where N has the same definition as before. (For the first testing situation $N = n$ and $T = X_1 + \cdots + X_n$.)

Complete Statistics

A statistic S is said to be *complete* for a family of distributions indexed by θ if the only real-valued functions of $S, h(S)$, which have expectation zero for every θ, are those for which $h(S) = 0$ with probability one for every θ. For a detailed discussion see COMPLETENESS. The point here is that this useful statistical notion requires for its comprehension at least an intuitive understanding of the meaning of "probability zero" in measure theory. (The definition does not require h to be identically zero, or necessarily zero almost everywhere.)

Consistent Estimation and the Singularity of Probability Measures

There is a close relationship between the existence of (strongly) consistent estimators and the singularity of related probability measures.

A sequence of estimators $\hat{\theta}_n = \hat{\theta}_n(X_1, \ldots, X_n)$ of a real parameter θ is *consistent* if it tends to θ with probability one under the probability P_θ determined by θ. This means that $P_\theta\{\hat{\theta}_n \to \theta\} = 1$ for each value of θ, and it shows that the distributions of the infinite sequence $X_1, X_2, \ldots$ of potential observations under P_θ and $P_{\theta'}$ are singular when $\theta \neq \theta'$. (When the observations are real-valued and independent for θ and θ', these distributions are infinite product measures on the space of all sequences.) Thus a consistent (sequence of) estimator(s) of the parameter θ exists only if the infinite sequence distributions determined by θ are singular, and this may be used as a test prior to seeking consistent estimators. (The test for θ and θ' determines, as well, whether one can distinguish between the simple hypotheses θ and θ', to any required level of precision, with a sufficiently large sample size.) It is well known that the infinite sequence distributions are singular whenever (for each n) $X_1, \ldots, X_n$ is a random sample and the distributions of X_1 are distinct for different values of θ. A simple (necessary and sufficient) test for singularity is provided by Kakutani's theorem whenever the sequence $X_1, X_2, \ldots$ consists of independent random variables under P_θ and $P_{\theta'}$. The case of dependent samples is only slightly more involved. Analogous considerations apply to time series observed over an interval $[0, T]$ as $T \to \infty$.

Relevant references are Doob [9, Section III.1], Grenander [19, Section 3.3] and Neveu [29, Section 3.1].

Asymptotic Statistical Theory via Weak Convergence

The central limit theorem* asserts that the distribution of the nth partial sum of independent and identically distributed random variables with finite variance, when centered at its mean and divided by its standard deviation, converges in law to the standard normal distribution. This has led to a variety of important statistical applications such as the determination of the critical values of test statistics for many common statistical tests. Additional applications to probability and statistics require an approximation of the distribution of the maximum of the first n partial sums, and of the distributions of various other functionals of these partial sums. A now standard way of obtaining such approximations is to embed the sequence of partial sums into a sequence of stochastic processes and to study the convergence of the distributions of these processes, i.e., the weak convergence of probability measures on function spaces, usually metric spaces of continuous functions. The variety of possible applications is enormous. An elegant application of weak convergence theory is to the derivation of the asymptotic distribution of the Kolmogorov–Smirnov* test statistics. For more details see Billingsley [1] and Serfling [34].

Robustness*

The notion of robust statistics involves measuring distances between distributions of samples and of statistics. When independent samples are used, as is the case throughout Huber's book [24], it suffices to work with probabilities on the real line. However, when dependent samples are used, as in a time series setting, large sample robustness would require working with probability measures on infinite dimensional spaces (of sequences or functions). The case of dependent samples is only now beginning to be an object of research activity.

Stochastic Processes

The specification of a stochastic process $\{X(t), t \in T\}$ by means of a consistent family of its finite dimensional distributions [i.e., of the distributions of the random vectors $X(t_1), \ldots, X(t_n)$] is achieved by Kolmogorov's extension theorem, a result like **(1)** above on the space of all functions on the parameter set T. Thus the definition and study of stochastic processes are generally intimately related with the study of probability measures on function spaces (for which there is now a vast literature).

References

[1] Billingsley, P. (1968). *Convergence of Probability Measures*. Wiley, New York.

[2] Billingsley, P. (1979). *Probability and Measure*. Wiley, New York.

[3] Billingsley, P. (1983). *Amer. Scientist*, **71**, 392–397.

[4] Breiman, L. (1968). *Probability*. Addison-Wesley, Reading, MA.

[5] Chow, Y. S. and Teicher, H. (1978). *Probability Theory*. Springer-Verlag, New York.

[6] Chung, K. L. (1974). *A Course in Probability Theory*, 2nd ed. Academic Press, New York.

[7] Cramér, H. (1946). *Mathematical Methods of Statistics*. Princeton University Press, Princeton, NJ.

[8] Cramér, H. and Leadbetter, M. R. (1965). *Stationary and Related Stochastic Processes*. Wiley, New York.

[9] Doob, J. L. (1953). *Stochastic Processes*. Wiley, New York.

[10] Feller, W. (1968). *An Introduction to Probability Theory and Its Application*, Vol. 1, 3rd ed. Wiley, New York.

[11] Feller, W. (1971). *An Introduction to Probability Theory and Its Application*, Vol. 2. Wiley, New York.

[12] Ferguson, T. S. (1967). *Mathematical Statistics: A Decision-Theoretic Approach*. Academic Press, New York.

[13] Ferguson, T. S. (1973). *Ann. Statist.*, **1**, 209–230.

[14] Fraser, D. A. S. (1957). *Nonparametric Methods in Statistics*. Wiley, New York.

[15] Ghosh, B. K. (1970). *Sequential Tests of Statistical Hypotheses*. Addison-Wesley, Reading, MA.

[16] Gihman, I. I. and Skorohod, A. V. (1974). *The Theory of Stochastic Processes*, Vol. 1. Springer-Verlag, New York.

[17] Gihman, I. I. and Skorohod, A. V. (1975). *The Theory of Stochastic Processes*, Vol. 2. Springer-Verlag, New York.

[18] Gihman, I. I. and Skorohod, A. V. (1979). *The Theory of Stochastic Processes*, Vol. 3. Springer-Verlag, New York.

[19] Grenander, V. (1981). *Abstract Inference*. Wiley, New York.

[20] Hájek, J. and Šidák, Z. (1967). *Theory of Rank Tests*. Academic Press, New York.

[21] Halmos, P. R. (1950). *Measure Theory*. Van Nostrand, Princeton, NJ.

[22] Halmos, P. R. and Savage, L. J. (1949). *Ann. Math. Statist.*, **20**, 225–241.

[23] Hannan, E. J. (1970). *Multiple Time Series*. Wiley, New York.

[24] Huber, P. J. (1981). *Robust Statistics*. Wiley, New York.

[25] Kolmogorov, A. N. (1933). *Grundbegriffe der Wahrscheinlichkeitsrechnung*. Springer-Verlag, Berlin; and (1956). *Foundations of the Theory of Probability*. Chelsea, New York.

[26] Lehmann, E. L. (1959). *Testing Statistical Hypotheses*. Wiley, New York.

[27] Loève, M. (1977). *Probability Theory*, Vols. 1 and 2, 2nd ed. Springer-Verlag, New York.

[28] Neveu, J. (1965). *Mathematical Foundations of the Calculus of Probability*. Holden-Day, San Francisco.

[29] Neveu, J. (1975). *Discrete-Parameter Martingales*. North-Holland, Amsterdam.

[30] Puri, M. L. and Sen, P. K. (1971). *Nonparametric Methods in Multivariable Analysis*. Wiley, New York.

[31] Rao, C. R. (1973). *Linear Statistical Inference and Its Applications*, 2nd ed. Wiley, New York.

[32] Royden, H. L. (1968). *Real Analysis*, 2nd ed. Macmillan, New York.

[33] Rudin, W. (1966). *Real and Complex Analysis*. McGraw-Hill, New York.

[34] Serfling, R. J. (1980). *Approximation Theorems of Mathematical Statistics*. Wiley, New York.

[35] Wentzell, A. D. (1981). *A Course in the Theory of Stochastic Processes*. McGraw-Hill, New York.

(ABSOLUTE CONTINUITY
ASYMPTOTIC NORMALITY
CHARACTERISTIC FUNCTIONS
CONDITIONAL PROBABILITY AND EXPECTATION
CONVERGENCE OF SEQUENCES OF RANDOM VARIABLES
ESTIMATION, POINT
HYPOTHESIS TESTING
INFERENCE, STATISTICAL—I
INVARIANCE CONCEPTS IN STATISTICS
LAWS OF LARGE NUMBERS
LIMIT THEOREM, CENTRAL
MULTIDIMENSIONAL CENTRAL LIMIT THEOREM
PROBABILITY THEORY
PRIOR DISTRIBUTIONS
RADON–NIKODYM THEOREM
SAMPLE SPACES
STATISTICAL FUNCTIONALS
STOCHASTIC PROCESSES
SUFFICIENT STATISTICS)

S. CAMBANIS
G. SIMONS

MEDIAN *See* MEAN, MEDIAN, MODE, AND SKEWNESS

MEDIAN EFFECTIVE DOSE *See* BIOASSAY, STATISTICAL METHODS IN; QUANTAL RESPONSE ANALYSIS

MEDIAN ESTIMATES AND SIGN TESTS

The median of a sample of numbers is the "middle number," that is, a number which exceeds half the sample and is exceeded by the other half. As a measure of centrality, it is very old, predating even least squares* procedures. Its companion statistical tests are *sign tests*; and throughout statistics, median estimates and sign tests* are enjoying renewed popularity because they have strong *robustness* properties and because the electronic computer has eliminated earlier computational difficulties.

Since the sample median minimizes the sum of absolute distances to other observations, median estimates in other statistical settings usually are defined as minimizing the sum of absolute residuals and are then called *least-absolute deviations* (LAD) *esti-*

mates. In the following sections, median estimates and sign tests as well as their main properties are summarized for the one- and k-sample problems, regression*, simple ANOVA*, and other statistical designs.

ROBUSTNESS PROPERTIES

Robustness is the property possessed by some statistical procedures of being resistant to the breakdown of basic assumptions (*see* ROBUST ESTIMATION). Qualitatively, the most robust of all are medianlike procedures. This fact is well understood for robustness against outliers* in the one-sample problem, through Hampel's [8] robustness measures *break down point* and *influence curve* (*see* INFLUENCE FUNCTIONS). What is more generally true is that medianlike procedures are the most robust against (1) outliers, in other statistical analyses, (2) incorrectly chosen weights in regression [11], and (3) unequal error variances in one-way ANOVA with unequal sample sizes [4].

Not all medianlike procedures are robust against all assumption breakdown. For instance, in simple linear regression*, LAD estimates are not robust against outliers at a large design point; but another median estimate—the unweighted Brown-Mood estimate—is; see Kildea [9] and the regression section that follows.

THE ONE-SAMPLE PROBLEM

M, the median of the ordered sample $X_1' < X_2' < \cdots < X_n'$, is the value of θ that minimizes $\sum|X_i' - \theta|$. It is X_{r+1}' when $n = 2r + 1$, odd, and any value between X_r' and X_{r+1}' when $n = 2r$, even, although the usual custom is to take $M = \frac{1}{2}\{X_r' + X_{r+1}'\}$.

If the X_i' are originally drawn independently from a population with probability density function (PDF) f, median θ, and variance σ^2, then for large n, the distribution of M is approximately normal with mean θ and variance $\{4nf^2(\theta)\}^{-1}$. The asymptotic efficiency* e of the sample median to the sample mean is therefore $e = 4\sigma^2 f^2(\theta)$. For normal f, this is $2/\pi = 0.637$, but it is higher in sampling from other distributions. For instance, Laplace f yields $e = 2$, logistic f yields $e = \pi^2/8 = 1.234$, and long-tailed f like the Cauchy yield $e = +\infty$. Objections to using M because of low normal efficiency are unrealistic. Simple normal mixtures as in Tukey [12], of form indistinguishable from normality to the naked eye, make the efficiency of M relative to sample mean about unity. For normal variance mixtures, $e > 1$ if $E(\sigma^2)\{E(\sigma^{-1})\}^2 > \frac{1}{2}\pi$, where E denotes expectation over the distribution of variance in the mixture.

The estimating equation for $\theta = M$ is $S(\theta) = \sum \operatorname{sgn}(X_i' - \theta) = 0$ (where "$= 0$" may mean "changes sign") and companion tests to the sample median are *sign tests*, which to test $\theta = \theta_0$ use $S(\theta_0)$ as test statistic. The exact null distribution of $\frac{1}{2}(S + n)$ is binomial $(n, \frac{1}{2})$, and for n moderate or large, the approximate null distribution of S is $N(0, n)$.

REGRESSION

The general linear regression* model is $\mathbf{Y} = \mathbf{X}\boldsymbol{\theta} + \boldsymbol{\epsilon}$, where $\mathbf{Y}$ is an $(n \times 1)$ observed vector, $\mathbf{X}$ is a known $(n \times p)$ design matrix, $\boldsymbol{\theta}$ is an unknown $(p \times 1)$ parameter vector, and $\boldsymbol{\epsilon}$ is an $(n \times 1)$ error vector. The LAD estimate $\hat{\boldsymbol{\theta}}$ of $\boldsymbol{\theta}$ minimizes $\sum_i |Y_i - \mathbf{X}\boldsymbol{\theta})_i|$. To compute $\hat{\boldsymbol{\theta}}$, linear programming* can be used; $\hat{\boldsymbol{\theta}}$ is the optimal solution to the problem:

$$\begin{aligned}
&\text{Choose } \boldsymbol{\theta}, u_1, \ldots, u_n \text{ to minimize}\\
&u_1 + \cdots + u_n,\\
&\text{subject to } u_i \geqslant Y_i - (\mathbf{X}\boldsymbol{\theta})_i \text{ and}\\
&u_i \geqslant -Y_i + (\mathbf{X}\boldsymbol{\theta})_i, \qquad i = 1, \ldots, n.
\end{aligned}$$

Gentle [6] and articles in the same issue discuss LAD computation.

Asymptotic properties of $\hat{\boldsymbol{\theta}}$ correspond to those of the sample median. If the components of $\boldsymbol{\epsilon}$ are independent and identically distributed with PDF f, symmetric about zero, then for large n the approximate distribution of $\hat{\boldsymbol{\theta}}$ is multivariate normal, with mean $\boldsymbol{\theta}$ and covariance matrix $(\mathbf{X}^T\mathbf{X})^{-1}\{4f^2(0)\}^{-1}$; see Koenker and Bassett [10], Adichie [1], and Brown [2]. Therefore, efficiency comparisons between $\hat{\boldsymbol{\theta}}$ and least squares are the same as those between M and sample mean in the one-sample problem.

Special results hold in simple linear regression, when $p = 2$, $\boldsymbol{\theta}^T = (\alpha, \beta)$ and

$$\mathbf{X}^T = \begin{pmatrix} 1 & 1 & \cdots & 1 \\ x_1 & x_2 & \cdots & x_n \end{pmatrix}.$$

LAD estimates can be computed graphically [2] and for a median line alternative to LAD, namely the unweighted Brown–Mood estimate [9], a very simple graphic method exists: Plot the points (X_i, Y_i), and consider the two groups $x_i < \bar{x}$ and $x_i > \bar{x}$ separately. Find the chord connecting one point from each group which leaves equal numbers of positive and negative residuals in each group.

Signlike tests of slope β use $S(\beta) = \sum_i x_i \operatorname{sgn}\{y_i - m(\beta)\}$ as test statistic, where $m(\beta) =$ median of $\{y_i - \beta x_i, i = 1, \ldots, n\}$. S has an exact null permutation distribution*, which is approximately

$$N\left(0, 2r(n-1)^{-1}\sum(x_i - \bar{x})^2\right),$$

for either $n = 2r$ or $n = 2r + 1$; see Brown and Maritz [5]. Confidence intervals are available as the set of slopes β not rejected by such tests.

THE k-SAMPLE PROBLEM, OR ONE-WAY ANOVA

The two-sample problem tests or estimates location shift between two samples $y_{11}, \ldots, y_{1m}$ and $y_{21}, \ldots, y_{2n}$. The usual model is $y_{i1} = \alpha + \epsilon_{i1}$ and $y_{2j} = \alpha + \beta + \epsilon_{2j}$, where all ϵ are independent identically distributed errors. This form is a special case of simple linear regression with values of design variable x being either 0 (first sample) or 1 (second sample). The location shift β corresponds to slope in simple linear regression.

The LAD estimate of β is simply $\hat{\beta} =$ median of $\{y_{1i}\}$ − median of $\{y_{2j}\}$, and signlike tests of $\beta = \beta_0$ employ

$$S(\beta_0) = \textstyle\sum_j \operatorname{sgn}\{y_{2j} - \beta_0 - m(\beta_0)\},$$

where $m(\beta_0)$ is the median of all $y_{1i}, y_{2j} - \beta_0$ combined. The exact null permutation distribution of S is related to the hypergeometric*, and its normal approximation is $N(0, 2rmn(m+n)^{-1}(m+n-1)^{-1})$, where $m + n = 2r$ or $2r + 1$.

Confidence intervals* related to these tests have end points of the form $y_{2j} - y_{1i}$, since it is only at these β_0 values that $S(\beta_0)$ changes.

In the k-sample problem, or one-way ANOVA, median estimates are individual sample medians. The sign test analogue for between-sample testing scores all observations by their signs after subtracting the combined median. If the ith sample has n_i observations and sum of scores S_i, and $N = \sum_1^k n_i$, then $(N-k)(2r)^{-1}\sum_1^k S_i^2/n_i$ is the test statistic, $N = 2r$ or $2r + 1$, with approximate χ_{k-1}^2 null distribution. The form of this test is analogous to the Kruskal–Wallis test*, with scores ± 1 or 0 instead of ranks.

OTHER MEDIANLIKE PROCEDURES

Sign tests in two-way ANOVA are available but require a balanced design (for the procedure to be simple) and a reasonable amount of replication (for chi-squared approximations to be valid). In a balanced block design*, complete or incomplete, with m blocks, t treatments, and n plots per block, $n = 2r$ or $2r + 1$, score observations by signs after extracting within-block medians, and let $S_1, \ldots, S_t$ be treatment-score totals. Then $(t-1)(2mr)^{-1}\sum S_i^2 \sim \chi_{t-1}^2$ approximately, under a null hypothesis of no treatment difference. The permutation theory derivation of this result is as for the Kruskal–Wallis test*.

Sign tests in higher-order ANOVA such as Latin squares* are as yet unavailable because of the lack of a unique way of scoring observations within a two-way layout. The best-known median estimate in two-way tables is Tukey's [13, p. 366] median polish*, which removes medians successively from rows, columns, rows, etc. It is a valuable data-analytic tool but highly nonunique.

Another median estimate is as a location measure for multivariate spatial data, defined by Gower [7] to minimize the sum of distances to observations. Its efficiency properties are better than those of the univariate sample median; its sign-test analogues have an angular aspect and are called *angle tests* (see Brown, [3]). For computational aspects, see Gower [7].

Because of their strong robustness properties, medianlike procedures can sometimes be used to handle a small number of missing observations (*see* INCOMPLETE DATA), where the missing elements are replaced by arbitrary values, which are varied to yield bounds on test significance levels.

References

[1] Adichie, J. N. (1967). *Ann. Math. Statist.*, **38**, 894–904.

[2] Brown, B. M. (1980). *Austral. J. Statist.*, **22**, 154–165.

[3] Brown, B. M. (1983). *J. R. Statist. Soc. B*, **45**, 25–30.

[4] Brown, B. M. (1982). *Austral. J. Statist.*, **24**, 283–295.

[5] Brown, B. M. and Maritz, J. S. (1982). *Austral. J. Statist.*, **24**, 318–331.

[6] Gentle, J. E. (1977). *Commun. Statist. B*, **6**, 313–328.

[7] Gower, J. C., (1974). *Appl. Statist.*, **23**, 466–470.

[8] Hampel, F. R. (1974). *J. Amer. Statist. Ass.*, **69**, 383-393.

[9] Kildea, D. G. (1981). *Ann. Statist.*, **9**, 438–442.

[10] Koenker, R. and Bassett, G. (1978). *Econometrica*, **46**, 33–50.

[11] Sen, P. K. (1968). *Ann. Math. Statist.*, **39**, 1724–1730.

[12] Tukey, J. W. (1960). In *Contributions to Probability and Statistics*, I. Olkin, ed. Stanford University Press, Stanford, CA.

[13] Tukey, J. W. (1977). *Exploratory Data Analysis*. Addison-Wesley, Reading, MA.

(DISTRIBUTION-FREE METHODS
MEDIAN ESTIMATION, INVERSE
ROBUST ESTIMATION
SIGN TESTS)

B. M. BROWN

MEDIAN ESTIMATION, INVERSE

The traditional measures of location of a univariate distribution F are its mean μ, median m, and mode M. The median, or 50% point, is defined by the inverse cumulative distribution function $F^{-1}(u) = \inf(x : F(x) \geqslant u)$ for $0 < u < 1$, as $m = F^{-1}(0.5)$. In symmetric populations all three measures coincide, thereby defining a unique, natural measure of central tendency. In positively skewed populations the relation $M < m < \mu$ generally holds. The reverse relation generally is valid for negatively skewed populations; see Van Zwet [16] and Kaas and Buhrman [7] for precise conditions under which these relations hold in continuous and discrete populations respectively. *See also* MEAN, MEDIAN MODE, AND SKEWNESS.

Robust estimates of location, designed to estimate the median of symmetric populations, give rise to yet other distinct *measures of* location in the asymmetric case. Thus the Hodges–Lehmann estimate*, an R-statistic, gives rise to the pseudo-median, or the median of the distribution of $(X_1 + X_2)/2$ where X_1 and X_2 are independent random variables from distribution F [6]. Tukey's trimean* $\frac{1}{4}(F^{-1}(\frac{1}{4}) + 2F^{-1}(\frac{1}{2}) + F^{-1}(\frac{3}{4}))$ derives from an L-statistic*, a linear combination of order statistics*. The α-trimmed means* $(1 - 2\alpha)^{-1}\int_{F^{-1}(\alpha)}^{F^{-1}(1-\alpha)} x\, dF(x)$ are also L-measures of location. Andrews' sine M-estimate* gives rise to an M-measure of location [1]; see Serfling [14] for details.

Bickel and Lehmann [2] extended Hampel's [4] concept of robustness to compare the robustness of different measures of location. Informally, a measure of location $m_{l1}(F)$ is robust if small changes in F produce small changes in the measure, i.e., the measure is continuous in F. The location measure $m_{l2}(F)$ is more robust than $m_{l1}(F)$ if it is continuous at all distributions F at which $m_{l1}(F)$ is continuous. The median m was shown by Bickel and Lehmann to be the most robust among the L- and M-measures of location. Their choice among robust measures rested on the accuracy with which the measures could be estimated. When asymmetric populations are admitted (essentially) the only consistent estimate of a location measure of the form $ml = g(f)$ is $T_{ml} = g(F_n)$. Here F_n denotes the empirical cumulative distribution function (CDF). Using the ratio of asymptotic variances of the estimates as a measure of relative efficiency* of the corresponding measures, they showed the following:

1. The relative efficiency of the population median to the population mean may be zero; however, it is bounded away from zero for unimodal populations (*see* UNIMODALITY).
2. If the distribution F has heavier tails than G, then the relative efficiency of the median to the mean is larger for F than for G (*see* HEAVY-TAILED DISTRIBUTIONS).
3. Although the relative efficiency of the median to the mean in normal populations is only .637, it is considerably higher for scale-contaminated normal distributions and other heavy-tailed distributions like the double exponential and Cauchy. See Kubat [9] for an extensive collection of numeric relative efficiencies in contamination models.
4. The 5% trimmed means* are less robust but more efficient than the population median.

The median emerges therefore as a highly robust, fairly efficient, and easily interpretable measure of location of an arbitrary population.

In the following sections median estimates derived from the sample quantile function $F_n^{-1}(u) = \inf(x: F_n(x) \geqslant u)$ for $0 < u < 1$ are discussed along with estimates of their variance and related confidence intervals. The natural extension of median estimates to multisample, regression, and other designs are the LAD estimates; *see* MEDIAN ESTIMATES AND SIGN TESTS* for further discussion. Extensions of median estimates that are defined by the quantile function of more complex samples arise naturally in median estimation from stratified* and cluster samples*. Such estimates as well as estimates of their variances and their properties are summarized in the last section.

MEDIAN ESTIMATION IN THE ONE-SAMPLE CASE

Numerous estimates of the median m of a symmetric population based on a random sample $X_1, \ldots, X_n$, or the corresponding ordered sample $X_{(1)}, \ldots, X_{(n)}$, are available. Andrews et al. [1], in their celebrated Princeton study, compared the robustness and efficiency properties of 68 estimates in a large variety of symmetric populations. They derived small sample distributions and relative efficiencies of the estimates and related confidence intervals, using Monte Carlo methods*. Summaries of their results are given in ref. 1, Chap. 7. The sample median $T_m = F_n^{-1}(0.5)$ if $n = 2k + 1$ or $T_m = (X_{(k)} + X_{(k+1)})/2$ if $n = 2k$ did not emerge as the most highly efficient of the lot. Nonetheless, their results clearly demonstrate the sample median to be a robust and reasonable efficient estimate of location in medium-sized samples from contaminated or heavy-tailed symmetric distributions. Most of the estimates included in their study do not estimate the median consistently in asymmetric populations and will not be discussed here. All the known consistent competitors of the sample median are quantile* estimates derived from the sample quantile function F_n^{-1}.

Mosteller's [11] quasimedians

$$Q_n = (X_{[n/2]-m+1} + X_{[n/2]+m+1})/2$$

for suitable $1 \leqslant m \leqslant [n/2]$ were investigated by Reiss [13]. Harrell and Davis (HD) [5] and Kaigh and Lachenbruch (KL) [8] proposed "robustified" median estimates using bootstrapping* and jackknifing*, respectively. The HD estimate is defined by

$$T_{\text{HD}} = \int_0^1 F_n^{-1}(u) h_n(u)\, du,$$

or equivalently by

$$T_{\text{HD}} = \sum_{i=1}^{n} W_{ni} X_{(i)},$$

a linear combination of order statistics with

weights

$$W_{ni} = \int_{(i-1)/n}^{i/n} h_n(u)\,du.$$

The functions $h_n(u)$ are the symmetric beta $((n+1)/2, (n+1)/2)$ densities. This estimate is obtained by bootstrapping the first moment of the sample median. The KL estimate is obtained by averaging the subsample medians of all subsamples of size k for a suitable $1 \leqslant k \leqslant n$.

In terms of ease of computation, the sample median and quasimedian are obtained easily from the sorted sample. The KL estimate requires automated computation in most cases but avoids the numerical integration* essential to computing the HD estimator.

When the distribution F possesses a density f, positive and continous at the median m, $\sqrt{n}\,(T_m - m)$ is asymptotically normal $(0, \sigma^2)$ with $\sigma^2 = 1/(4f^2(m))$. Far stronger asymptotic results, including a Berry–Esseen* bound and a law of iterated logarithms*, are available for the sample median; see Serfling [14, Sect. 2.5.1] for precise statements and references. Pfanzagl [12] obtained a strong efficiency result for sample quantiles. Informally it may be stated as follows: under certain regularity conditions on F, for all equivariant and asymptotically median-unbiased estimates T_n of the median m,

$$\begin{aligned}\limsup P\left(m - \sqrt{n}\,t' \leqslant T_n \leqslant m + \sqrt{n}\,t''\right) \\ \leqslant \limsup P\left(m - \sqrt{n}\,t' \leqslant T_m \right. \\ \left. \leqslant m + \sqrt{n}\,t''\right) \\ = \Phi(t''/\sigma) - \Phi(t'/\sigma)\end{aligned}$$

where Φ is the standard normal CDF and $t', t'' \geqslant 0$ are arbitrary. The estimate T_n is equivariant if $T_n(X_1 + a, \dots, X_n + a) = T_n(X_1, \dots, X_n) + a$ (*see* EQUIVARIANT ESTIMATORS). It is median unbiased if $P(T_n \leqslant m) \geqslant 1/2$ and $P(T_n \geqslant m) \geqslant 1/2$ (*see* MEDIAN UNBIASED ESTIMATORS).

Both the HD and KL estimates satisfy Pfanzagl conditions and hence the probability that they fall in any given interval about the median cannot exceed that of the sample median for large samples. However, the HD estimate is asymptotically equivalent to the sample median in attaining the bound in the equation above, provided the subsample size $k \to \infty$ and k/n remains bounded away from zero as $n \to \infty$. Reiss [13] proved that with an appropriate choice of $m \to \infty$ (of order determined by the smoothness of f), the Pfanzagl coverage probabilities of the quasimedian exceed those of the sample median asymptotically. No finite sample results are available.

Under mild conditions on the tails of the distribution F, and under the conditions stated above for the subsample size of the KL estimate, both the HD and KL estimates are asymptotically normal with the same mean and variance as the sample median [5, 8]. Thus all three estimates are asymptotically equivalent. However, both the HD estimate and the KL estimate were shown to be superior to the sample median in moderate samples from a variety of symmetric and asymmetric populations. Their finite sample comparisons were based on relative mean square error* obtained by simulation* or exact computations.

VARIANCE ESTIMATION AND CONFIDENCE INTERVALS

Maritz and Jarrett [10] proposed a small sample bootstrap estimate for the variance of the sample median. The jackknife estimate of the variance is known to be inconsistent in this case. Harrell and Davis investigated the jackknife estimate of the variance of their estimate. Both variance estimates appear adequate for moderate sample sizes. It appears that no variance estimates have been investigated for the quasimedians and the KL estimates.

Nonparametric confidence intervals* for the median are based on order statistics and binomial probability tables [14, Sect. 2.6.1]. Approximate intervals may be obtained from the sample median and T_{HD} using the estimates of their variance given above. Fixed-width confidence intervals* for the median are discussed by Swanepoel and Lombard [15].

MEDIAN ESTIMATION IN STRATIFIED SAMPLES FROM FINITE POPULATIONS

An approximate confidence interval for the median m, computed from a stratified sample, was proposed by Woodruff [17]. Gross [3] used a similar procedure to derive a weighted median estimate from a combined sample quantile function. If F_{ni} denote the empirical CDFs of the independent samples of size n_i from stratum i of size N_i, the quantile function F_n^{-1} is the inverse of the weighted empirical CDF $F_n = \sum_{i=1}^{k} p_i F_{ni}$ with $p_i = N_i/N$. The weighted median is defined by $T_m = F_n^{-1}(0.5)$. It may be expressed in terms of the order statistics $Y_{(r)}$, $r = 1, \ldots, \sum_{i=1}^{k} n_i$ of the combined sample as follows: if $Y_{(r)}$ originated in the ith stratum, assign it a weight $w_i = p_i/n_i$. Proceed to accumulate the weights of $Y_{(1)}, Y_{(2)}, \ldots$ until 0.5 is first crossed, say at $Y_{(j)}$. Then $T_m = Y_{(j)}$. As in simple random samples the customary version of the weighted sample median is $T_m = (Y_{(j)} + Y_{(j-1)})/2$. If $n_i, N_i \to \infty$ so that $n_i/N_i \to f_i$ with $f_i > 0$ for all strata $1 \leqslant i \leqslant k$, and if the population CDF approaches a limiting distribution with density f with $f(m) > 0$, then $\sqrt{N}(T_m - m)$ is asymptotically normal $(0, \sigma^2)$ with

$$\sigma^2 = \{f(m)\}^{-2} \sum_{i=1}^{k} \left[(p_i/f_i)(1 - f_i) \times F_i(m)(1 - F_i(m)) \right].$$

Here $F_i(m)$ denotes the asymptotic CDF of the ith stratum at the population median m. The optimal sampling fractions f_i are proportional to $N_i(F_i(m)(1 - F_i(m)))^{1/2}$. Thus the larger allocation should be made to the stratum whose median is close to the population median m.

A modified version of the bootstrap for finite populations was introduced by Gross [3], who used it to obtain a Maritz–Jarrett-type variance estimate for the variance of the sample median. Simulation studies with a variety of skewed and heavy-tailed distributions proved its usefulness in small samples. Unfortunately its computation requires extensive tables of hypergeometric* probabilities. An HD-type estimate using the quantile function F_n^{-1} is possible here as well and it is likely to enjoy similar properties relative to the weighted median.

A weighted median estimate from stratified samples of clusters, when no subsampling is done within clusters, is available. Clustering does not alter the weighting introduced by stratification. For a precise description of the estimate and its variance estimates, see Gross [3].

References

[1] Andrews, D. F., Bickel, P. J., Hampel, F. R., Huber, P. J., Rogers, W.H., and Tukey, J. W. (1972). *Robust Estimates of Location, Survey and Advances*. Princeton University Press, Princeton, NJ.

[2] Bickel, P. J. and Lehmann, E. L. (1975). *Ann. Statist.*, **3**, 1045–1069.

[3] Gross, S. T. (1980). *Proc. ASA Survey Res.*, 181–184.

[4] Hampel, F. R. (1971). *Ann. Math. Statist.*, **42**, 1887–1896.

[5] Harrell, F. E. and Davis, C. E. (1982). *Biometrika*, **69**, 635–640.

[6] Hodges, J. L. and Lehmann, E. L. (1963). *Ann. Math. Statist.*, **34**, 598–611.

[7] Kaas, R. and Buhrman, J. M. (1980). *Statist. Neerlandica*, **34**, 13–18.

[8] Kaigh, W. D. and Lachenbruch, P. A. (1982). *Commun. Statist. A*, **11**, 2217–2238.

[9] Kubat, P. (1979). *Statist. Neerlandica*, **33**, 191–196.

[10] Maritz, J. S. and Jarrett, R. G. (1978). *J. Amer. Statist. Ass.*, **73**, 194–196.

[11] Mosteller, F. (1946). *Ann. Math. Statist.*, **17**, 377–408.

[12] Pfanzagl, J. (1975). In *Statistical Methods in Biometry*, W. J. Ziegler, ed. Birkhauser Verlag, Basel, pp. 111–126.

[13] Reiss, R. D. (1980). *Ann. Math. Statist.*, **8**, 87–105.

[14] Serfling, R. J. (1980). *Approximation Theorems of Mathematical Statistics*. Wiley, New York.

[15] Swanepoel, J. W. H. and Lombard, F. (1978). *Commun. Statist. A*, **7**, 829–835.

[16] Van Zwet, W. R. (1979). *Statist. Neerlandica*, **33**, 1–5.
[17] Woodruff, R. S. (1952). *J. Amer. Statist. Ass.*, **47**, 635–646.

(DISTRIBUTION-FREE METHODS
MEAN, MEDIAN, MODE, AND SKEWNESS
MEDIAN ESTIMATES AND SIGN TESTS
NONPARAMETRIC CONFIDENCE INTERVALS
ROBUST ESTIMATION
ROBUSTIFICATION)

SHULAMITH T. GROSS

MEDIAN LETHAL DOSE *See* BIOASSAY, STATISTICAL METHODS IN; QUANTAL RESPONSE ANALYSIS

MEDIAN POLISH

For a two-way table (a "matrix") Tukey [2, Chaps. 10 and 11] suggests a procedure for fitting an additive structure, representing the table as closely as possible by **A** + **B**, where **A** is a vector of row constants and **B** is a vector of column constants. The procedure, called median polish, involves repeated calculations of row and column medians.

At step 1 the median of each row is found and subtracted from every entry in the row. At step 2 the median of each resulting column is found and subtracted from each entry in the column. At steps 3 and 4, the procedure is repeated on rows and columns respectively, and so on. If at any step after the first every one of a set of medians (whether of rows or of columns) becomes zero, the procedure stops. The matrix of residuals has zero median for every row and for every column.

Anscombe [1] discusses this procedure and its execution using the programming language APL.

References

[1] Anscombe, F. (1981). *Computing in Statistical Science Through APL*. Springer-Verlag, New York.
[2] Tukey, J. W. (1977). *Exploratory Data Analysis*. Addison-Wesley, Reading, MA.

(EXPLORATORY DATA ANALYSIS)

MEDIAN TEST FOR TWO POPULATIONS *See* BROWN–MOOD MEDIAN TEST

MEDIAN UNBIASED ESTIMATORS

An estimator $\hat{\theta}$ of a parameter θ is median unbiased if

$$\Pr_\theta(\hat{\theta} \leqslant \theta) \geqslant \tfrac{1}{2}, \qquad \Pr_\theta(\hat{\theta} \geqslant \theta) \geqslant \tfrac{1}{2}. \quad (1)$$

Here $\hat{\theta}$ is based on an underlying distribution depending on θ, which is unknown but belongs to a parameter set Ω. If the distribution of $\hat{\theta}$ is continuous, then the criterion (1) becomes

$$\Pr_\theta(\hat{\theta} \leqslant \theta) = \tfrac{1}{2}, \qquad \Pr_\theta(\hat{\theta} \geqslant \theta) = \tfrac{1}{2}.$$

If the underlying distribution is symmetric about θ and if (with suitable regularity conditions) $\hat{\theta}$ is the (unique) maximum likelihood* estimator of θ based on a random sample, then (see Pfanzagl [4]) $\hat{\theta}$ is also median unbiased. The first of the two following examples is included to illustrate this property.

Example (i). Let $\overline{X}$ be the sample mean of n independent observations from a normal population. Then $\overline{X}$ is a median unbiased estimator (MUE) of the population mean [3, pp. 175–176].

Example (ii). Let x_p be the 100pth percentile of a normal population with unknown mean and unknown standard deviation. An MUE of x_p in a sample of n independent observations is $\overline{X} + \alpha S$, where $\overline{X}$ is the sample mean, S the sample standard deviation, and α a constant depending on n and p. For details, see Dyer et al. [2], where values of α are tabulated and where it is shown that this estimator is closer to x_p than any other estimator of the form $\overline{X} + kS$ (*see* CLOSENESS OF ESTIMATORS).

OPTIMUM PROPERTIES

Lehmann [3, pp. 80, 83] showed that some families of distributions with monotone likelihood ratio* (MLR) admit MUEs which are optimum in the sense that among all MUEs they minimize the expected loss for any loss function that is zero at the true parameter value θ^* and that is nondecreasing as θ moves away from θ^* in either direction. If $\hat{\theta}$ is the optimum MUE and $\hat{\hat{\theta}}$ is any other MUE, then in particular,

$$\Pr_\theta\{-\Delta \leqslant \theta - \hat{\hat{\theta}} \leqslant \Delta\}$$

is maximized by $\hat{\hat{\theta}} = \hat{\theta}$ for any choice of nonnegative constant Δ. The conditions are that the family of probability density functions $p_\theta(x)$ for θ in Ω have MLR in T, say, and that the distribution function F_θ of T is continuous with region of positive density independent of θ. If $T = t$ is observed, then the optimum MUE is determined from the equation

$$F_{\hat{\theta}}(t) = \Pr(T \leqslant t \mid \theta = \hat{\theta}) = \tfrac{1}{2}.$$

By generalizing (1) to randomized MUEs, Pfanzagl [5] extended Lehmann's result to include some discrete families, such as binomial* and Poisson* populations.

Certain median-unbiased estimators have some remarkable properties analogous to those of unbiased estimators, with convex loss functions, which are functions of complete sufficient statistics. Brown et al. [1, Corollary 4.1] and Pfanzagl [9] have obtained results analogous to the Rao–Blackwell theorem* and the Lehmann–Scheffé theorem*, respectively. In the former, let $\hat{\theta}(x)$ be an MUE of θ not based only on a sufficient statistic $T = T(X_1, \ldots, X_n)$, where $X_1, \ldots, X_n$ is a random sample from a member of the exponential family indexed by θ and having a strict MLR. Then an MUE $\hat{\hat{\theta}}(T)$ can be constructed, such that $\hat{\hat{\theta}}$ is a function of T and $\hat{\hat{\theta}}$ is strictly better than $\hat{\theta}$. Pfanzagl's result [9] applies to certain exponential families* with sample joint density

$$C(\theta, \boldsymbol{\eta})h(\mathbf{x})\exp\left[a(\theta)T(\mathbf{x}) + \sum_{i=1}^{p} a_i(\theta, \boldsymbol{\eta})S_i(\mathbf{x})\right],$$

where $\boldsymbol{\eta}$ are nuisance parameters*, a is increasing and continuous in θ for each sample size, and the loss function is monotone, i.e., nondecreasing as the estimate moves away from the true θ-value in either direction. Then there exists an MUE of θ of minimal risk in the class of MUEs of θ which is a function of a complete sufficient statistic. The existence of an MUE of minimal risk for any monotone function of θ is also implied.

ASYMPTOTIC PROPERTIES

It is a common practice to compare asymptotically normal sequences of estimators by comparing their asymptotic variances (*see* EFFICIENCY, ASYMPTOTIC RELATIVE). The latter can be considered as measures of "asymptotic concentration." An alternative measure for a more general sequence $\{\hat{\theta}_n\}$ of estimators of θ is

$$\lim_{n\to\infty} \Pr_\theta\left(\theta - t/\sqrt{n} < \hat{\theta}_n < \theta + t'/\sqrt{n} \mid \theta\right) \tag{2}$$

for positive constants t and t'. (*See also* LARGE-SAMPLE THEORY.)

Definition. A sequence $\{\hat{\theta}_n\}$ of estimators is *asymptotically optimal* in a class $\mathscr{C}$ if

$$\lim_{n\to\infty} \Pr_\theta\{\theta - t/\sqrt{n} < \hat{\theta}_n < \theta + t'/\sqrt{n}\}$$
$$\geqslant \lim_{n\to\infty} \Pr_\theta\{\theta - t/\sqrt{n} < \hat{\hat{\theta}}_n < \theta + t'/\sqrt{n}\} \tag{3}$$

for any other sequence $\{\hat{\hat{\theta}}_n\}$ of estimators in $\mathscr{C}$, and for all $t, t' > 0$ and all θ in Ω.

The limit (2) exists for asymptotically normal sequences of estimators, but also for a broader class of estimators. Pfanzagl [4] gives an example of a sequence of MUEs $\{\hat{\theta}_n\}$ for which the distributions of $\sqrt{n}(\hat{\theta}_n - \theta)$ do not converge. He also gives, roughly speaking, an upper bound to (2) for sequences of MUEs; this is the *maximal asymptotic concentration*. It is attained by certain sequences of maximum likelihood estimators, but he shows elsewhere [6]:

1. Any sequence $\{\hat{\theta}'_n\}$ of estimators of θ which is asymptotically normal is also *asymptotically median unbiased* (AMU) in the sense that

$$\lim_{n\to\infty}\Pr(\hat{\theta}_n \leqslant \theta) = \lim_{n\to\infty}\Pr(\hat{\theta}'_n \geqslant \theta) = \tfrac{1}{2}. \tag{4}$$

2. Under certain conditions an asymptotically efficient sequence of estimators can be adjusted to give a sequence of asymptotically optimal MUEs having the same asymptotic behavior, and in which the adjustment amount $\to 0$ as $n \to \infty$.

Pfanzagl [5] shows that MUEs with maximal asymptotic concentration exist for all exponential families having certain regularity conditions. See also Pfanzagl [7] and Strasser [10]; in general, asymptotically optimal sequences need not exist, but if they do, there exist infinitely many of them.

A final result applies to use of a sample quantile $\hat{q}_n$ as an estimate of the corresponding population quantile q. The sequence $\{\hat{q}_n\}$ is AMU. Pfanzagl [8] showed that when the shape of the underlying distribution is unknown, no location-equivariant and asymptotically uniformly median unbiased estimator of q is asymptotically more concentrated about q than $\{\hat{q}_n\}$. This result does not hold if the shape of the distribution is known or if q is the median and the distribution is known to be symmetric.

References

[1] Brown, L. D., Cohen, A., and Strawderman, W. E. (1976). *Ann. Statist.*, **4**, 712–722.

[2] Dyer, D. D., Keating, J. P., and Hensley, O. L. (1977). *Commun. Statist. B*, **6**, 269–283.

[3] Lehmann, E. L. (1959). *Testing Statistical Hypotheses*. Wiley, New York.

[4] Pfanzagl, J. (1970). *Ann. Math. Statist.*, **41**, 1500–1509.

[5] Pfanzagl, J. (1970). *Metrika*, **17**, 30–39.

[6] Pfanzagl, J. (1971). *Metrika*, **18**, 154–173.

[7] Pfanzagl, J. (1975). In *Statistical Inference and Related Topics*, M. L. Puri, ed. Proc. Summer Research Inst. on Statistical Inference and Stochastic Processes.

[8] Pfanzagl, J. (1975). In *Statistical Methods in Biometry*, W. J. Ziegler, ed. Birhauser Verlag, Basel, West Germany, pp. 111–126.

[9] Pfanzagl, J. (1979). *Ann. Statist.*, **7**, 187–193.

[10] Strasser, H. (1978). *Ann. Statist.*, **6**, 867–881.

(ASYMPTOTIC NORMALITY
EQUIVARIANT ESTIMATORS
ESTIMATION, POINT
LARGE-SAMPLE THEORY
MAXIMUM LIKELIHOOD ESTIMATION
MINIMUM VARIANCE UNBIASED ESTIMATORS)

CAMPBELL B. READ

MEDICAL DIAGNOSIS, STATISTICS IN

Medical diagnosis is one area in which the application of statistical analysis has made a rather enigmatic impact. The associated problems can be abstracted neatly as well-developed statistical ones, and real-life applications are reported in ever increasing numbers, but acceptance of statistical methods has not yet been widespread for reasons that will be delineated below.

THE DIAGNOSIS PROBLEM

Formally, it is assumed that patients belong each to one of d disease classes,

$\Pi_1, \ldots, \Pi_d$, say, but it is not known to which. The objective is to use all information available from "past experience" and from data gathered from a particular patient to make some inference or decision about the unknown class membership of that patient. The data from a patient usually take the form of records of the presence or absence of various symptoms or the results of relevant clinical tests or measurements. These variables may be grouped together under the term *indicants*. In a very simple case, with $d = 2$, Π_1 represents patients with appendicitis, for which the appropriate treatment is surgery, and Π_2 subsumes patients suffering from abdominal pain resulting from causes which require less drastic treatment. Indicants are used from which, statistically or otherwise, the members of Π_1 can be identified without having to operate on everyone.

In some applications, the $\{\Pi_i\}$ may represent *prognostic* categories [13].

Even at this level of generality there may be practical difficulties. Is the term "disease" appropriate for a particular condition? Can all possible disease classes be identified? Is the presence of a symptom well defined? (Often patients are asked to describe the degree of some pain or other, which is likely to lead to very subjective, nonstandardized responses.) Is even a particular clinical test properly standardized?

Such very real difficulties beset both clinical and statistical diagnosis. They have helped to maintain antipathy toward statistical methods which depend on well-defined structure and have promoted some exploration of "fuzzy" methods; see [14] and FUZZY SET THEORY.

RELEVANT STATISTICAL METHODS

Discriminant Analysis*

If the indicants measured on a given patient are represented by a $k \times 1$ vector $\mathbf{x}$, then of direct interest are the conditional probabilities* $\{p(\Pi_i | \mathbf{x}),\ i = 1, \ldots, d\}$, or ratios such as $p(\Pi_1 | \mathbf{x})/p(\Pi_2 | \mathbf{x})$, sometimes called *odds-ratios**. On the basis of these, a patient may be assigned to a particular disease class (that corresponding to the largest of the $\{p(\Pi_i | \mathbf{x})\}$, for instance) and treated accordingly. If the consequences of the various misclassifications are not equally costly, then some modification of the above based on decision theory* can be used, provided relative costs can be assessed—another point of dispute in the medical context.

Statistical modeling of the $\{p(\Pi_i | \mathbf{x})\}$, or of some valid discriminant rule, is required, with subsequent estimation using "past experience," often in the form of data sets from patients already identified as belonging to the various disease classes. Modeling may be of the $\{p(\Pi_i | \mathbf{x})\}$ or their ratios directly, using logistic* models, for instance. Alternatively, modeling may be based on the components of the right-hand side of the following expression of Bayes' theorem*:

$$p(\Pi_i | \mathbf{x}) \propto p(\Pi_i, \mathbf{x}) = p_i f_i(\mathbf{x}).$$

Here $f_i(\cdot)$ denotes the sampling density of $\mathbf{x}$ within the ith class and $\{p_i\}$ the prior probabilities or incidence rates of the disease classes. This latter formulation is by far the more prevalent in the medical literature, although Dawid [2] argues that, particularly for medical problems, direct modeling of the $\{p(\Pi_i | \mathbf{x})\}$ is more sensible in that the $\{p(\Pi_i | \mathbf{x})\}$ are (1) less influenced by any bias incurred by the method of selecting patients; (2) often more stable, over time, for instance, than the joint probabilities $\{p(\Pi_i, x)\}$.

The Bayes theorem* approach can be illustrated by the analysis in Tables 1 and 2. The two "disease categories" indicate pres-

Table 1 $\{f_i(\mathbf{x})\}$ **for Presence or Absence of Disease;** x **= Diastolic Blood Pressure**

	Low $x = 1$	Moderate $x = 2$	High $x = 3$	p_i
Presence Π_1	0.2	0.4	0.4	0.3
Absence Π_2	0.7	0.2	0.1	0.7

Table 2 $\{p(\Pi_i \mid \mathbf{x})\}$ **and Odds Ratios for the Model of Table 1**

	$x = 1$	$x = 2$	$x = 3$
Π_1	6/55	6/13	12/19
Π_2	49/55	7/13	7/19
Odds ratios ($\Pi_1 : \Pi_2$)	6/49	6/7	12/7

ence (Π_1) or absence (Π_2) of a certain disease the indicant is diastolic blood pressure, coded low ($x = 1$), moderate ($x = 2$), or high ($x = 3$). Table 1 gives the $\{f_i(\mathbf{x})\}$ and the $\{p_i\}$, with the resulting $\{p(\Pi_i \mid \mathbf{x})\}$ and odds ratios $\{p(\Pi_1 \mid \mathbf{x})/p(\Pi_2 \mid \mathbf{x})\}$ in Table 2.

The information in Table 1 can be used to illustrate the use of blood pressure in a *diagnostic screening test*. One sensible diagnostic screening test is derived by specifying a cutoff value for blood pressure above which we decide to treat a patient as diseased (*positive* diagnosis) and below which we treat as unaffected (*negative*). Table 3 shows the proportions of positives classified correctly as positives and falsely as negatives, and the corresponding proportions of negatives, for two such screening procedures. In Table 3A a +, that is, Π_1, is predicted if $x = 2$ or 3, and in Table 3B only if $x = 3$.

The proportion of actual positives correctly predicted as such is called the *sensitivity* of the test; the proportion of actual negatives correctly predicted is the *specificity*. Note that the sensitivity is higher in Table 3A than in Table 3B (0.8 vs. 0.4) but that the corresponding specificities show, in compensation, an opposite trend (0.7 vs. 0.9). Which of the tables represents the more acceptable compromise depends critically on the relative costs of the two types of error; *see* HYPOTHESIS TESTING* *and* DECISION THEORY*.

Table 3 Proportions of True/False Positives and Negatives for Two Screening Procedures[a]

		A. Predicted		
		− ($x = 1$)	+ ($x = 2, 3$)	Total
True	+ (Π_1)	0.2	0.8	1
	− (Π_2)	0.7	0.3	1

		B. Predicted		
		− ($x = 1, 2$)	+ ($x = 3$)	Total
True	+	0.6	0.4	1
	−	0.9	0.1	1

[a] Based on the model in Table 1.

In the general structure all the special methods and considerations of discriminant analysis* are relevant. This is reflected in the medical literature, at least some of which is right up to date with developments in discriminant analysis and pattern recognition*. It is largely the result of direct involvement in medical diagnosis of appropriate experts in the methodology [8].

Two points have particular importance to the treatment of the medical problems.

THE USE OF INDEPENDENCE MODELS. Many early applications of Bayes' theorem incorporated independence models for the class-conditional densities $\{f_i(\mathbf{x})\}$; that is, it was assumed that

$$f_i(\mathbf{x}) \equiv \prod_{j=1}^{k} f_{ij}(x_j) \equiv f_i^{I}(\mathbf{x}),$$

say, where the $\{f_{ij}(\cdot)\}$ are all univariate densities. This occurred mainly because indicants often are categorical and, in particular, binary (symptom present or absent). The independence model is estimated easily and the only elementary alternative model, the multinomial*, which for large k has many cells, cannot usually be well estimated. (A typical data set might initially contain about 300 patients in 6 disease classes with up to 50 binary indicants.) Such was the popularity of the independence model that some medical writers still seem to be under the misapprehension that the "Bayes" method has as one of its precepts that the indicants be independent, given the disease class [5]! Some empirical studies have shown that the independence model can perform quite well in diagnostic terms even when it is wrong. Indeed, odds ratios can be calculated correctly using the independence model [4] even when the more general assumption is made that

$$f_i(\mathbf{x}) = c(\mathbf{x})f_i{}^I(\mathbf{x}), \qquad \text{for each } \mathbf{x} \text{ and } i.$$

Recently, models intermediate between the independence and multinomial models have also been used [13]. General multivariate normal assumptions have also been made, leading to linear and quadratic discriminant functions.

VARIABLE SELECTION AND CREATED INDICANTS. In medical data there are often very many indicants, as exemplified earlier. In contrast, seldom are more than a very few variables required to extract the best out of a data set from the point of view of discriminant analysis.

In the head-injury study reported by Titterington et al. [13], for instance, over 200 variables could be reduced to six and yet still provide a good discrimination procedure. Variable selection is therefore a vital part of the statistical exercise, which may then be a stepwise discriminant analysis, for instance. Reduction of dimensionality also allows us to explore beyond the independence models toward more highly parameterized and realistic structures. Variable selection has an appealing parallel in clinical practice, where the experienced diagnostician often is distinguished by asking a few pertinent questions instead of accumulating a mass of possibly irrelevant information [3]. Another means of reducing dimensionality is to create new indicants, each as a function of several original variables. This might be to cope with missing data (provided at least one of the original variables is available, the created indicant may be evaluated) or to produce a set of indicants for which the conditional independence assumption is more reasonable.

Diagnostic Keys and Sequential Testing

In an ideal situation there would be enough information in any patient's **x** to identify the disease class uniquely, whatever **x** is. In particular, with data about k binary symptoms, the sample space* of 2^k possible **x**'s could be partitioned into d subsets corresponding to the various disease classes. If k is large, it is advantageous to streamline the testing procedure into a sequential pattern in which the symptoms are investigated one by one, the order depending on the results of previous tests. Such a rule is called a *diagnostic key* [9] and it can be represented diagrammatically by a binary tree. An optimal key would minimize, say, the expected number of tests required to complete a diagnosis. Construction of such a key is, combinatorially, a difficult problem, for large k and d.

In medical problems perfect discrimination is seldom possible and a further requirement of a good key is that, when testing is complete, the discrimination achieved is as good as possible [12]. If $\{\theta_i\}$ denote probabilities of the disease classes for a given patient after testing, a good key should lead to the $\{\theta_i\}$ being close to the correct degenerate probability measure as frequently as possible. This closeness can be measured by expected entropy*. Furthermore, expected change in entropy can be used as a criterion for sequential selection of tests and can be played off against the cost of a test to decide when to end the testing procedure (*see* SEQUENTIAL ANALYSIS).

APPLICATION AND ACCEPTANCE IN PRACTICE

Performance of statistical techniques often is assessed by simulation*, so that properties can be discovered or checked under known conditions. In medical diagnosis, because of the variety of data sets that arise, empirical comparison of various methods on the same real data sets is also of great importance. Some comparisons (e.g., [1, 13]) have shown that the choice of technique is not as important as, for instance, variable selection and data preparation. Refinement of technique is far less of a priority to Croft [1] than the following:

1. Improved standardization of tests and symptom definition.

2. Generation of large data sets from which rules can be derived that can also be reliably transferred for use in other establishments.
3. Increased acceptance by the medical profession.

Doctors have been wary of the anonymous power of computers, which conflicts with the fact that clinical diagnosis follows a variety of patterns, depending on the individual doctor, who blends hard data with subjective experience, opinions, and even emotion; see Mitchell [6]. The decision trees associated with diagnostic keys and sequential testing are far more acceptable attempts to model this logical-emotional process than the probabilistic perfection of Bayes' theorem and discriminant analysis. Gradually, however, the latter methods are finding more favor. For instance, the use of Bayes' theorem can produce more accurate diagnoses in practice than those of clinicians; see Stern et al. [10]. Taylor et al. [11] show that clinicians often do not choose the most efficient tests and do not then process the results appropriately, according to Bayes' theorem, in particular being very conservative about the resulting probabilities or odds ratios. They are also liable to ignore the prior probabilities of disease classes in their mental assessments. If we add to this the difficulty of interpreting multivariate data intuitively, particularly the correlations* therein, the advantages of computer-assisted diagnosis seem great.

A final problem is that a procedure, once developed, seldom lives up to expectations. Error rates claimed in the development stage often are not met in later application, although this may be because the program has been tested on the training set, which gives a falsely optimistic indication of performance. Variable selection may also lead to bias in that the "best subset" as judged by one data set may well not be the best for another. This happens even in artificially generated situations in which the data sets have identical statistical properties [7].

References

[1] Croft, D. J. (1972). *Comput. Biomed. Res.*, **5**, 351–367.

[2] Dawid, A. P. (1976). *Biometrics*, **32**, 647–648.

[3] De Dombal, F. T. (1978). *Meth. Inform. Med.*, **17**, 28–35.

[4] Hilden, J. (1982). Personal communication.

[5] Kember, N. F. (1982). *An Introduction to Computer Applications in Medicine*. Arnold, London, Chap. 10.

[6] Mitchell, J. H. (1970). *Internat. J. Biomed. Comput.*, **1**, 157–166.

[7] Murray, G. D. (1977). *Appl. Statist.*, **26**, 246–250.

[8] Patrick, E. A., Stelmack, F. P., and Shen, L. Y. L. (1974). *IEEE Trans. Syst. Man. Cybern.*, **SMC-4**, 1–16. (Includes well-annotated references to applications.)

[9] Payne, R. W. and Preece, D. A. (1980). *J. R. Statist. Soc. A.*, **143**, 253–292. (Good reference list.)

[10] Stern, R. B., Knill-Jones, R. P., and Williams, R. (1975). *Brit. Med. J.*, **2**, 659–662.

[11] Taylor, T. R., Aitchison, J., and McGirr, E. M. (1971). *Brit. Med. J.*, **3**, 35–40.

[12] Teather, D. (1974). *J. R. Statist. Soc. A.*, **137**, 231–244.

[13] Titterington, D. M., Murray, G. D., Murray, L. S., Spiegelhalter, D. J., Skene, A. M., Habbema, J. D. F., and Gelpke, G. J. (1981). *J. R. Statist. Soc. A*, **144**, 145–175. (Includes discussion.)

[14] Wechsler, H. (1976). *Internat. J. Biomed. Comput.*, **7**, 191–203.

Bibliographical Remarks

The literature on discriminant analysis is highly relevant, as is that on decision theory. Direct references to medical diagnosis can be obtained by browsing through several specialized journals; sporadic references in many of the most well-known medical periodicals also are useful. Recent volumes of the following are recommended: *Computers in Biology and Medicine*, *Computers and Biomedical Research*, *International Journal of Biomedical Computing*, and *Methods of Information in Medicine*. Also useful are the following:

Lusted, L. B. (1968). *Introduction to Medical Decision Making*. Thomas, Springfield, IL. (Bayes-based basic text.)

Miller, M. C., Westphal, M. C., Reigart, J. P., and Barner, C. (1977). University Microfilms International, Ann Arbor, MI. (Bibliography.)

Rogers, W., Ryack, B., and Moeller, G. (1979). *Internat. J. Biomed. Comput.*, **10**, 267–289. (Review, with 58 annotated empirical studies.)

Wagner, G., Tautu, P., and Woller, U. (1978). *Meth. Inform. Med.*, **17**, 55–74. (Bibliography of 827 references.)

Wardle, A. and Wardle, L. (1978). *Meth. Inform. Med.*, **17**, 15–28. (Another good review and reference list.)

(BIOSTATISTICS
CLASSIFICATION
DECISION THEORY
DISCRIMINANT ANALYSIS
ODDS RATIO ESTIMATORS)

D. M. Titterington

MEDICINE, STATISTICS IN *See* STATISTICS IN MEDICINE

MEDICO-ACTUARIAL INVESTIGATIONS

Medico-actuarial studies originated in the United States in the 1890s from concerted efforts to improve the underwriting of life insurance risks [15]. The mortality investigations undertaken were aimed to isolate and measure the effects of selected risk factors, such as occupational hazards, medical conditions, and build. The underlying hypothesis was that each of the factors (or certain combinations of factors) influencing mortality could be regarded as an independent variable and the total mortality risk could be treated as a linear compound of a number of independent elements.

The first comprehensive study, known as the *Specialized Mortality Investigation*, was carried out by the Actuarial Society of America and published in 1903 [1]. It covered the experience of 34 life insurance companies over a 30-year period and focused attention on the mortality in 35 selected occupations, 32 common medical conditions, and several other factors affecting mortality. It was followed in 1912 by the *Medico-Actuarial Mortality Investigation* [2], sponsored jointly by the Actuarial Society of America and the Association of Life Insurance Company Medical Directors, which included a much wider variety of occupations, medical conditions, and other factors, among them abuse of alcohol. This study laid the broad lines on which such investigations have been conducted since.

The assessment of the long-term risk in life insurance was seen as requiring analysis of mortality by sex, age at issue, and duration since issue of insurance for policies issued under similar underwriting rules. Cohorts of policyholders were followed over long periods of time. Attention was focused on the mortality in the years following issue of insurance in order to trace the effects on mortality of the selection exercised by insurance companies through medical examinations and other screening for insurance, as well as effects of antiselection by applicants for insurance who withheld information relevant to their health. The extent of class selection, that is, the reflection of the underlying mortality in the segments of the population from which the insured lives were drawn, was brought out in the mortality experienced among the insured lives under study after many years since issue of insurance had elapsed. Most important, however, the patterns of the mortality experienced over longer periods of time indicated the incidence of the extra mortality by duration, which permitted classifying the long-term risk as one of decreasing extra mortality, relatively level extra mortality, or increasing extra mortality.

Analyses of mortality by cause shed light on the causes mainly responsible for excess mortality and also on the causes of death which could be controlled to some degree by screening applicants for life insurance. Successive medico-actuarial investigations permitted some conclusions regarding the trends in mortality associated with different factors affecting mortality, notably occupational hazards, build, blood pressure, various medical conditions, and changing circumstances.

METHODOLOGY

In selecting a particular cohort of policyholders for study, because of interest in some particular factor influencing mortality, it was the practice in medico-actuarial investigations to exclude individuals who also presented other kinds of risks. Specifically, all individuals were excluded from the study if they were also subject to any other kind of risk that would have precluded issuing insurance at standard premium rates. Consequently, such extra mortality as was found in the study could properly be associated with the factor of interest rather than with the combined effects of this factor and other elements at risk.

The findings of medico-actuarial investigations have been customarily expressed as ratios of actual deaths in the cohort of policyholders under study to the expected deaths, which were calculated on the basis of contemporaneous death rates among otherwise similar life insurance risks accepted at standard premium rates. Such mortality ratios usually were computed by sex, age groups at issue of the insurance, duration since issue of the insurance, and causes of death. The calculation of expected deaths involves accurate estimates of the exposed to risk. Because of the varying forms of records, individual and grouped, different tabulating procedures have been employed, as a rule considering deaths within a unit age interval as having occurred at the end of that interval.

Ratios of actual to expected mortality provide very sensitive measures of mortality and therefore may fluctuate widely in finer subdivisions of the experience level. They have the merit, however, of revealing even small departures from expected mortality in broad groupings. In some circumstances the patterns of excess mortality are more clearly perceived from the extra deaths per 1000 than from corresponding mortality ratios; this often is the case during the period immediately following surgery for cancer. It is important to keep in mind that mortality ratios generally decrease with age, so that the mortality ratios for all ages combined can be materially affected by the age composition of a population.

Proportions surviving a specified period of time, even though they provide absolute measures of longevity in a population, have rarely been used in medico-actuarial investigations, because they are relatively insensitive to appreciable changes in mortality; small differences in proportions surviving may be difficult to assess. Relative proportions surviving have been used occasionally in medico-actuarial studies where very high mortality rates occur, as among cancer patients.

In all mortality comparisons, but particularly in comparisons of mortality ratios and relative proportions surviving, it is the suitability of the basis for calculating expected deaths that makes the figures meaningful. If such a basis is regarded as a fixed yardstick, then the reliability of comparisons based on small numbers of deaths can be tested by determining whether an observed deviation from this basis is or is not significant in probability terms; if it is significant, then what are the limits in probability terms within which the "true" value of the observed deviation can be expected to lie [14]?

In medico-actuarial mortality investigations the numbers of deaths in most classifications have usually been quite large. The mortality ratios shown for such classifications have therefore been taken as reasonably reliable estimates of the "true" values of the mortality ratios in the underlying population. As a rule of thumb, when the number of policies terminated by death was 35 or greater and some doubt attached to the significance of the mortality ratio, confidence limits were calculated at the 95% confidence level on the assumption of a normal distribution; when the number of policies terminated by death was less than 35, confidence limits have been calculated on the assumption of a Poisson distribution.

INTERPRETING THE FINDINGS

Medico-actuarial investigations have been based on the experience among men and women insured under ordinary life insurance policies. These insured lives have been drawn predominantly from the middle-class and better-off segments of the population and have passed the screening for life insurance which results in the rejection of about 2% of all applicants and the charging of extra premiums to about 6% of all applicants. Initially at least more than 9 out of 10 persons are accepted for life insurance at standard premium rates and those issued insurance at standard premium rates are in ostensibly good health. In recent years the death rates of men aged 25 or older insured under standard ordinary policies have ranged from 25 to 35% of the corresponding population death rates in the first two policy years, from 40 to 50% of the corresponding population death rates at policy year durations 3–5, and from 55 to 75% of the corresponding population death rates after 15 or more years have elapsed since issue of insurance. The corresponding figures for women insured under standard ordinary policies have been similar to those of male insured lives at ages over 50, but were closer to population death rates at the younger ages.

Inasmuch as the underwriting rules determine which applicants are accepted for standard insurance and which for insurance at extra premium rates, the mortality experience in medico-actuarial investigations has occasionally been affected to a significant degree by changes in underwriting practices to more lenient or stricter criteria.

The mortality findings in medico-actuarial investigations relate to the status of an individual at time of issue of the insurance. The experience therefore reflects not only the effects of some individuals becoming poorer risks with the passage of time, but also of some individuals becoming better risks (e.g., leaving employment in a hazardous occupation or benefiting from medical or surgical treatment) and withdrawing from the experience. Where the effects of employment in hazardous occupations or of certain physical impairments are deferred, it is essential that the study cover a sufficiently long period of time for the deferred mortality to become manifest. This is particularly important in the case of overweight and hypertension [13].

The results of medico-actuarial investigations have been relatively free from bias arising from failure to trace the experience among those withdrawing. Considerable evidence has been accumulated to show that insured lives who cease paying premiums and thus automatically remove themselves from observation are as a group subject to somewhat lower mortality [12].

It should also be kept in mind that the mortality ratios shown in medico-actuarial studies were computed on the basis of the number of policies (or amounts of insurance) and not on the basis of lives. In classifications involving small numbers of policies terminated by death it has been necessary to look into the data to determine whether the results had been affected by the death of a single individual with several policies (or with large amounts of insurance). This has usually been noted in the descriptive text.

The data in medico-actuarial investigations are very accurate with respect to reported ages and remarkably complete in the follow-up*. The information obtained on applications for life insurance with respect to past medical histories requires some qualification. The great majority of applicants for life insurance admit some physical impairment or medical history; if the impairment or history appears to be significant, attention is focused on it in the medical examination for insurance and a statement from the attending physician may be obtained. The medical examination on modest amounts of life insurance is not as comprehensive as a diagnostic examination in clinical practice, where the physician is in position to study a patient for a longer period of time, more intensively, and with the patient's full cooperation. Applicants for insurance not infrequently forget or try to conceal unfavorable aspects of their personal or family medical histories, particularly with respect to questionable habits. Even when reasonably complete details are elicited, there are usually practical limits on the extent to which it is feasible to check up on indefinite statements and vague diagnoses reported on applications for life insurance. Only on applications for large amounts of insurance would two or more medical examinations by different physicians be called for and intensive effort made to clarify obscure findings. Broadly speaking, the medical findings on life insurance medical examinations stand up very well, but the medical impairments studied in medico-actuarial mortality investigations often represent less differentiated conditions which cannot be characterized as precisely as is sometimes possible in clinical studies [13].

On the other hand, it has proved feasible on applications for life insurance to obtain fuller details of occupation and avocation (even approximate exposure to occupational hazards) than has been possible in many epidemiological* studies.

FINDINGS OF MEDICO-ACTUARIAL INVESTIGATIONS

The *Medico-Actuarial Mortality Investigation* of 1912 covered the period from 1885 to 1909 [2]. It produced tables of average weights for men and women by age and height which remained in general use as a weight standard until 1960. The mortality experienced according to variations in build indicated some extra mortality among underweights at ages under 35, associated with materially greater risk of tuberculosis and pneumonia, but at ages 35 and older the lowest mortality was found among those 5 to 10 pounds underweight. Overweight was found to be associated with increased death rates from heart disease, diabetes, and cerebral hemorrhage. The investigation also included 76 groups of medical impairments, 68 occupations, four categories of women studied according to marital status, and insured blacks and North American Indians.

The *Occupational Study 1926* dealt with some 200 occupations or groups of occupations, separately for those where occupational accidents were the dominant element of extra risk and those where nonoccupational accidents, pneumonia, cirrhosis of the liver, cancer, or other causes, were suspect as responsible for the extra risk [3].

The *Medical Impairment Study 1929* [4], which covered the period from 1909 to 1928, and its 1931 Supplement[11] broadly confirmed the findings of the Medico-Actuarial Investigation as to average weights by age and height at ages 25 and older and the effects on mortality of departures from average weight. The study centered on 122 groups of medical impairments, including a number of combinations of two impairments treated as a single element of risk. The more important findings related to the extra mortality on heart murmurs, elevated blood pressure, and albumin and sugar in the urine. The findings on elevated blood pressure indicated clearly that systolic blood pressures in excess of 140 mm were associated with significant extra mortality, which at the time was contrary to medical opinion [10].

Smaller investigations of the mortality among insured lives with various medical impairments followed, published under the titles *Impairment Study 1936* [6] and *Impairment Study 1938* [19]. Together they included 42 groups of medical impairments, among them persons with a history of cancer, gastric and duodenal ulcers, gall bladder disease, and kidney stone, including surgery for these conditions.

In 1938 an extensive investigation was also undertaken of the mortality according to variations in systolic and diastolic pressure. This study, known as the *Blood Pressure Study 1938*, covered the period from 1925 to 1938 [8, 9]. It confirmed earlier findings that diastolic pressures in excess of 90 mm as well as systolic blood pressures in excess of 140 mm were associated with at least 25% extra mortality, and it brought out clearly that various minor impairments accompanying elevated blood pressure, notably overweight, increased the risk appreciably.

The *Occupational Study 1937* covered numerous occupations over the period 1925 to 1936 [7]. It developed the extra mortality among those employed in the manufacturing, distribution, and serving of alcoholic beverages. It also indicated some decline since the early 1920s in the accidental death rates in many occupations.

The *Impairment Study 1951*, which covered the period 1935 to 1950, reviewed the mortality experience for 132 medical impairment classifications on policies issued during the years 1935 through 1949 [16]. It showed lower extra mortality than that found in the Medical Impairment Study 1929 for medical impairments due to infections, for conditions treated surgically, for diseases of the respiratory system, and for some women's diseases. Because of the inclusion in the study of smaller groups of lives with specific impairments, greater use was made of confidence intervals based on the Poisson distribution.

The *Build and Blood Pressure Study 1959* covered the experience on about 4,500,000 policies over the period 1935 to 1953 [17]. It focused on changes in the mortality experienced among underweights and overweights, elevated blood pressures, and combinations of overweight and hypertension with other impairments. New tables of average weights for men and women by age and height were developed, which showed that men had gained weight while women had reduced their average weights since the 1920s. Moderate underweights showed very favorable mortality, while marked overweights recorded somewhat higher relative mortality. The mortality on slight, moderate, and marked elevations in blood pressure registered distinctly higher mortality than found in earlier investigations.

The *Occupational Study 1967* covered the period from 1954 to 1964 and was limited to occupations believed to involve some extra mortality risks [18]. Only the following occupations, on which there was substantial experience, recorded a crude death rate in excess of 1.5 per 1000:

Lumberjacks
Mining operators
Explosive workers
Construction crane workers
Shipbuilding operators
Structural iron workers
Railroad trainmen and switchmen
Taxi drivers
Marine officers and crew
Guards and watchmen
Marshals and detectives
Sanitation workers
Porters
Elevator operators
Persons selling, delivering, or serving alcoholic beverage

The mortality in most occupations decreased from that reported in the *Occupational Study 1937*. Significant reductions occurred among mining officials and foremen, workers in metal industry, telecommunication linemen, longshoremen, firefighters, police officers, window cleaners, hotelkeepers, saloonkeepers and bartenders, and most laborers. Relative mortality increased for lumberjacks, railroad trainmen and switchmen, truck drivers, marine crew and guards, and watchmen.

The *Build Study 1979* [21] and the *Blood Pressure Study 1979* [22] each covered about 4,250,000 policies over the period 1954 to 1971. They showed that the average weights for men had continued to increase, as did the average weights for women under 30; women 30 and older registered decreases in average weights as compared with the *Build and Blood Pressure Study 1959*. The excess mortality among overweights was found to be substantially the same as in the earlier study, but somewhat higher mortality was recorded among moderate overweights. Nevertheless, the optimum weights (those associated with the lowest mortality) were again found to be in the range of 5 to 10% below average weight, even though the average weights for men had increased significantly.

The excess mortality on elevated blood pressures was found to be distinctly lower than in the *Build and Blood Pressure Study 1959*. A cohort of 24,000 men who had been treated for hypertension exhibited virtually normal mortality among those whose blood pressures had been reduced to below 140 systolic and 90 diastolic after treatment. The study adduced the most convincing evidence thus far available that recent treatment for hypertension was highly effective for many years. In progress at this time is another medico-actuarial investigation of the mortality among insured lives with a wide variety of medical impairments, covering the period from 1955 through 1974.

References

References 1, 3 to 10, and 16 to 21 are original reports.

[1] Actuarial Society of America (1903). *Specialized Mortality Investigation*. New York.

[2] Actuarial Society of America and Association of Life Insurance Medical Directors of America (1912–1914). *Medico-Actuarial Mortality Investigation*, 5 vols. New York. (Original reports; some of these cover basic design of studies.)

[3] Actuarial Society of America and Association of Life Insurance Medical Directors of America (1926). *Occupational Study 1926*. New York.

[4] Actuarial Society of America and Association of Life Insurance Medical Directors of America (1929). *Medical Impairment Study 1929*. New York.

[5] Actuarial Society of America and Association of Life Insurance Medical Directors of America (1932). *Supplement to Medical Impairment Study 1929*. New York.

[6] Actuarial Society of America and Association of Life Insurance Medical Directors of America (1936). *Impairment Study 1936*. New York.

[7] Actuarial Society of America and Association of Life Insurance Medical Directors of America (1937). *Occupational Study 1937*. New York.

[8] Actuarial Society of America and Association of Life Insurance Medical Directors of America (1939). *Blood Pressure Study 1939*. New York.

[9] Actuarial Society of America and Association of Life Insurance Medical Directors of America (1940). *Supplement to Blood Pressure Study 1939*. New York.

[10] Association of Life Insurance Medical Directors of America and the Actuarial Society of America (1925). *Blood Pressure Study*. New York.

[11] Batten, R. W. (1978). *Mortality Table Construction*. Prentice-Hall, Englewood Cliffs, N.J.

[12] Benjamin, B. and Haycocks, H. W. (1970). *The Analysis of Mortality and Other Actuarial Statistics*. Cambridge University Press, London.

[13] Lew, E. A. (1954). *Amer. J. Publ. Health*, **44**, 641–654. (Practical considerations in interpretation of studies.)

[14] Lew, E. A. (1976). In *Medical Risks*. R. B. Singer and L. Levinson, eds. Lexington Books, Lexington, Mass., Chap. 3. (Limitations of studies.)

[15] Lew, E. A. (1977). *J. Inst. Actuaries*, **104**, 221–226. (A history of medico-actuarial studies.)

[16] Society of Actuaries (1954). *Impairment Study 1951*. New York. (A highly readable original report.)

[17] Society of Actuaries (1960). *Build and Blood Pressure Study 1959*. Chicago. (A highly readable original report.)

[18] Society of Actuaries (1967). *Occupational Study 1967*. New York.

[19] Society of Actuaries and Association of Life Insurance Medical Directors of America (1938). *Impairment Study 1938*. New York.

[20] Society of Actuaries and Association of Life Insurance Medical Directors of America (1980). *Blood Pressure Study 1979*. Chicago. (Very readable.)

[21] Society of Actuaries and Association of Life Insurance Medical Directors of America (1980). *Build Study 1979*. Chicago. (Highly readable.)

(ACTUARIAL STATISTICS
CLINICAL TRIALS
EPIDEMIOLOGICAL STATISTICS
FOLLOW UP
LIFE TABLES
STANDARDIZED MORTALITY RATIO
VITAL STATISTICS)

EDWARD A. LEW

MEIXNER POLYNOMIALS

The nth-order Meixner polynomial is defined as

$$M_n(x;\,\beta,\gamma) = \sum_{k=0}^{n} \binom{n}{k}(-1)^k \binom{x}{k} \times k!\,(x+\beta)^{[n-k]}\gamma^{-k},$$

$n = 1, 2, \ldots, \quad 0 < \gamma < 1$, and $\beta > 0$. They are orthogonal with respect to a discrete measure with jumps at $k = 0, 1, 2, \ldots$ of size $(1-\gamma)^n[\beta^{[k]}/k!]\gamma^k$, where $\beta^{[k]} = \beta(\beta+1)\cdots(\beta+k-1)$. This measure can be viewed as a discrete analog of the gamma measure on which Laguerre polynomials are based (*see* LAGUERRE SERIES) and is related to the negative binomial distribution*.

A useful property of these polynomials is the generating function relation

$$\sum_{n=0}^{\infty} \frac{M_n(x;\beta,\gamma)}{n!} t^n = \left(1 - \frac{t}{\gamma}\right)^x (1-t)^{-x-\beta}.$$

These polynomials are used in the theory of total positivity*, particularly in connection with Polya* frequency functions and sequences, and also in statistical distribution theory.

Bibliography

Eagelson, G. K. (1964). *Ann. Math. Statist.*, **35**, 1208–1215.

Erdélyi, A., Magnus, W., Oberhettinger, F., and Tricomi, F. (1953). *Higher Transcendental Functions*, Vol. II. McGraw-Hill, New York, p. 225.

Karlin, S. (1968). *Total Positivity*, Vol. I. Stanford University Press, Stanford, CA, pp. 447–448.

(KRAWTCHOUK POLYNOMIALS
LAGUERRE SERIES
ORTHOGONAL EXPANSIONS)

MELLIN TRANSFORM *See* INTEGRAL TRANSFORMS

MEMORYLESS PROPERTY *See* CHARACTERIZATION OF DISTRIBUTIONS

MENON ESTIMATORS

These are estimators for the parameters θ, c of the Weibull distribution* with density function

$$f(t) = \frac{ct^{c-1}}{\theta}\exp(-t/\theta)^c, \qquad t \geqslant 0, \quad \theta > 0, \quad k > 0.$$

Menon's estimators for $d = 1/c$ and $\alpha = \log\theta$ based on n independent values $T_1, T_2, \ldots, T_n$ are

$$d = \left[(6/\pi^2)s_Y^2\right]^{1/2} \quad \text{and} \quad \alpha = \overline{Y} + 0.5772,$$

where $Y_i = \log T_i$ $i = 1, \ldots, n$, and $\overline{Y}$ and s_Y are the "usual" sample mean and standard deviation of $Y_1, \ldots, Y_n$. The number 0.5772 is Euler's constant truncated at the fourth significant figure. The Menon estimators of c and θ are

$$\hat{c} = 1/\hat{\alpha}, \qquad \hat{\theta} = \exp(\hat{\alpha}),$$

respectively.

Although these estimators are computationally attractive, recent investigations of Engeman and Keefe [1] indicate both the generalized least squares* and the maximum likelihood* estimator of Weibull parameters are substantially more efficient.

Reference

[1] Engeman, R. M. and Keefe, T. J. (1982). *Commun. Statist. A*, **11**, 2181–2193.

Bibliography

Menon, M. V. (1963). *Technometrics*, **5**, 175–182.

(MAXIMUM LIKELIHOOD ESTIMATION
WEIBULL DISTRIBUTION)

MERRIMAN, MANSFIELD

Born: March 27, 1848, in Connecticut.
Died: June 7, 1925, in New York.
Contributed to: estimation, especially least squares; engineering.

Mansfield Merriman was the author of the most successful textbook on statistical methods published in America in the nineteenth century. Merriman was born on March 27, 1848, the son of a Connecticut farmer. He studied mathematics and surveying in district schools and was appointed a county surveyor in 1867, before going on to a course of study in engineering at Yale University's Sheffield scientific school. He earned a Ph.D. in 1876 with a thesis on the method of least squares*, the earliest American doctorate on a statistical topic.

Merriman's dissertation was remarkable in two respects. The first was its extensive historical review, which he published in 1877 as "A list of writings relating to the method of least squares, with historical and critical notes" [1]. The "list" was in fact a nearly exhaustive bibliography of 408 titles published between 1722 and 1876, and it remains to this day an invaluable resource for historians of statistics. The second distinguishing feature of Merriman's thesis was that it led to his earliest statistics textbook, *Elements of the Method of Least Squares*, which he published in London, also in 1877 [2]. In 1884, he rewrote and expanded this work as *A Text-Book on the Method of Least Squares* [3]. This latter version was published by John Wiley & Sons, the earliest of their long series of statistics texts. Merriman's book was not the first handbook of its type published in America (Chauvenet's *Treatise on the Method of Least Squares* was issued in 1868), but it was the most successful. It was even adopted as a standard text in England.

In the 1900 paper that introduced chi-square* Karl Pearson* took Merriman to task for presenting as normal a set of data that did not pass Pearson's test [4, 5]. Pearson's criticism was somewhat unfair, however, since the specific data set he analyzed was not the one Merriman presented as being in "very satisfactory" agreement with the normal distribution, but a motivating illustration, whose imperfect nature was pointed out by Merriman himself.

Merriman taught civil engineering and astronomy at Yale during 1877/1878, before accepting the professorship of civil engineering at Lehigh University. Aside from consulting duties (including summer surveying for the U.S. Coast and Geodetic Survey from 1880 to 1885), he remained at Lehigh until 1907. During and after this period he wrote and edited a seemingly endless stream of texts and handbooks on engineering topics. By the time he died in New York on June 7, 1925, 340,000 copies of his works had been published. He is said to have been one of the greatest engineering teachers of his day, and although his statistics texts do not display unusual depth, they are uncommonly lucid.

References

[1] Merriman, M. (1877). *Trans. Conn. Acad. Arts Sciences*, **4**, 151–232. (Reprinted in ref. 6.)

[2] Merriman, M. (1877). *Elements of the Method of Least Squares*. Macmillan, London.

[3] Merriman, M. (1884). *A Text-Book on the Method of Least Squares*. Wiley, New York. (8th ed., 1907.)

[4] Pearson, K. (1900). *Philos. Mag., 5th Ser.*, **50**, 157–175. (Reprinted in *Karl Pearson's Early Statistical Papers*. Cambridge University Press, Cambridge, 1956.)

[5] Pearson, K. (1901). *Philos. Mag., 6th Ser.*, **1**, 670–671. (Comments further on Merriman's data, correcting the calculation Pearson had given earlier.)

[6] Stigler, S. M., ed. (1980). *American Contributions to Mathematical Statistics in the Nineteenth Century*. 2 vols. Arno Press, New York. (Includes photographic reprints of Merriman's 1877 bibliography as well as two of Merriman's papers on the history of statistics.)

Bibliography

The National Cyclopedia of American Biography (1933). Vol. 23, pp. 70–71. James T. White, New York. (Biographical article.)

(LEAST SQUARES)

STEPHEN M. STIGLER

MERRINGTON–PEARSON APPROXIMATION

This is an approximation to the distribution of noncentral chi-square* obtained by fitting a Pearson Type III (gamma*) distribution with the same first three moments (*see* PEARSON'S DISTRIBUTIONS). The approximation can be described by saying

> (noncentral χ^2 with ν degrees of freedom and noncentrality parameter λ) is approximately distributed as $(a\chi_f^2 + b)$ with
>
> $$a = (\nu + 3\lambda)/(\nu + 2\lambda),$$
>
> $$b = -\lambda^2/(\nu + 3\lambda),$$
>
> $$f = \nu + \lambda^2(3\nu + 8\lambda)(\nu + 3\lambda)^{-2}.$$

This approximation is good for the upper tails of the distribution, but not for the lower tails, where the effect of the negative $[-\lambda^2/(\nu + 3\lambda)]$ start is important.

Bibliography

Pearson, E. S. (1959). *Biometrika*, **46**, 364.

(NONCENTRAL CHI-SQUARE DISTRIBUTION)

MESOKURTIC CURVE

A frequency curve with a zero coefficient of kurtosis

$$\gamma_2 = (\mu_4/\mu_2^2) - 3$$

(e.g., the family of normal, or Gaussian, curves).

(KURTOSIS
LEPTOKURTIC CURVE
PLATYKURTIC CURVE)

MESSY DATA

Every experimenter has a different idea as to what constitutes messy data. The less informed an experimenter is about statistical methodology, the more messy the data seem to be. Many think messy data are found because of sloppy experimental techniques, resulting from lack of control over the experiment; others believe that data are messy because they fail to satisfy the required assumptions for particular statistical methods. Thus data that are not normally distributed may be considered messy by some experimenters but not by others. Still others believe messy data result from unequal numbers of observations from the different populations under study. Thus the definition depends on the situation at hand. Here data are defined to be messy *if a proper analysis requires more than the standard statistical methods* available in most subject matter methods textbooks.

Nonstandard analyses frequently are needed because assumptions necessary to apply standard techniques are violated, sample sizes are not equal, the data are from a distribution different from the one for which the standard analysis was developed, variances are unequal, censoring* has occurred, outliers* are present, a mixture* of distributions is present, etc.

In years past, much attention was given to attempts to make messy data "nice" so that they could be analyzed by standard techniques and practices. As a result of modern computing software, analysis of messy data is being reconsidered, and new theories are continually being proposed. The existence of computers and computing software allows analyses of messy data to be practical and feasible, whereas, in earlier times, such analyses could only be imagined.

This entry is concerned with messy data which result from designed experiments where the assumptions required for standard statistical analyses are not satisfied. First a general hypothesis testing* procedure that utilizes matrix notation is reviewed. This procedure then is generalized so that data failing to satisfy the usual homogeneity* of variance assumption can be analyzed. In the next section the procedure is applied to two-way treatment structure designs where there are unequal numbers of observations per treatment combination. Finally, a few other methods available for analyzing nonstandard data from designed experiments are reviewed. Numerical examples illustrating each of the techniques can be found in Milliken and Johnson [11].

A GENERAL HYPOTHESIS TESTING PROCEDURE

The one-way* analysis of variance* model is no doubt the most useful model that occurs in the field of statistics. Many experimental situations are special cases of this model. Situations which appear to require more complicated models can often be analyzed by using one way models.

Suppose that a sample of N experimental units is selected completely at random from a population of experimental units and that an experimenter wishes to compare the effects of t treatments.

Let y_{ij} denote the response of the (i, j)th experimental unit to the ith treatment. The usual assumptions are that the y_{ij}'s can be modeled by

$$y_{ij} = \mu_i + \epsilon_{ij},$$

$$i = 1, 2, \dots, t; \quad j = 1, 2, \dots, n_i, \quad (1)$$

where $\epsilon_{ij} \sim N(0, \sigma^2)$. Let $N = \sum n_i$.

For this situation, the best estimate of each μ_i is $\hat{\mu}_i = (1/n_i)\sum_j y_{ij}$, and the best estimate of σ^2 is

$$\hat{\sigma}^2 = \frac{1}{N - t} \sum_{i=1}^{t} \sum_{j=1}^{n_i} (y_{ij} - \hat{\mu}_i)^2.$$

Let $\boldsymbol{\mu} = [\mu_1 \mu_2 \dots \mu_t]'$ be the vector of population means and let $\hat{\boldsymbol{\mu}} = [\hat{\mu}_1 \hat{\mu}_2 \dots \hat{\mu}_t]'$. Consider testing a general hypothesis, H_0: $\mathbf{C}\boldsymbol{\mu} = \mathbf{a}$, where $\mathbf{C}$ is a $k \times t$ matrix of known constants which has rank k. A test statistic [7] for testing $H_0: \mathbf{C}\boldsymbol{\mu} = \mathbf{a}$ versus $H_a: \mathbf{C}\boldsymbol{\mu} \neq \mathbf{a}$ is given by

$$F = \frac{SS_{H_0}/k}{\hat{\sigma}^2},$$

$$SS_{H_0} = (\mathbf{C}\hat{\boldsymbol{\mu}} - \mathbf{a})'[\mathbf{CDC}']^{-1}(\mathbf{C}\hat{\boldsymbol{\mu}} - \mathbf{a}),$$

$$\mathbf{D} = (1/\sigma^2)\text{cov}(\hat{\boldsymbol{\mu}})$$

$$= \text{DIAG}(1/n_1, 1/n_2, \dots, 1/n_t).$$

H_0 is rejected at the $\alpha \cdot 100\%$ significance level if $F > F_{\alpha,k,N-t}$ (*see* F-TESTS) and accepted otherwise. When

$$\mathbf{C} = \begin{bmatrix} 1 & -1 & 0 & \cdots & 0 \\ 1 & 0 & -1 & \cdots & 0 \\ \vdots & \vdots & \vdots & \ddots & \vdots \\ 1 & 0 & 0 & \cdots & -1 \end{bmatrix},$$

the above F-statistic tests $H_0: \mu_1 = \mu_2 = \dots = \mu_t$, the one-way analysis of variance hypothesis.

A basic understanding of the above test procedure allows experimenters to deal with many messy experimental design situations. In the next section this procedure is generalized to the nonhomogeneous variances case. In the following section it is applied to two-way unbalanced data sets (*see* INCOMPLETE DATA) from which generalizations to more complicated treatment structures should be apparent.

UNEQUAL EXPERIMENTAL ERROR VARIANCES

Suppose now that $y_{ij} = \mu_i + \epsilon_{ij}$, where $\epsilon_{ij} \sim$ independent $N(0, \sigma_i^2)$. The best estimates of the parameters in this model are

$$\hat{\mu}_i = \sum_{j=1}^{n_i} y_{ij}/n_i,$$

$$\hat{\sigma}_i^2 = \left[\sum_{j=1}^{n_i} (Y_{ij} - \hat{\mu}_i)^2\right] \Big/ (n_i - 1),$$

$$i = 1, 2, \dots, t.$$

Then

$$\hat{\mu}_i \sim \text{independent } N(0, \sigma_i^2/n_i),$$

$$(n_i - 1)\hat{\sigma}_i^2/\sigma_i^2 \sim \text{independent } \chi^2(n_i - 1),$$

and all $\hat{\mu}_i$'s and $\hat{\sigma}_i^2$'s are independently distributed.

Suppose $H_0: \sigma_1^2 = \sigma_2^2 = \dots = \sigma_t^2$ has been tested and rejected (*see* BARTLETT'S TEST; HARTLEY'S F-MAX TEST). Now consider testing $H_0: \mathbf{c}'\boldsymbol{\mu} = a$ versus $H_a: \mathbf{c}'\boldsymbol{\mu} \neq a$ where $\mathbf{c}$ is a $t \times 1$ vector of known constants. The best estimate of $\mathbf{c}'\boldsymbol{\mu}$ is $\mathbf{c}'\hat{\boldsymbol{\mu}}$, and

$$\mathbf{c}'\hat{\boldsymbol{\mu}} \sim N(\mathbf{c}'\boldsymbol{\mu}, \mathbf{c}'\mathbf{V}\mathbf{c})$$

where $\mathbf{V} = \text{DIAG}(\sigma_1^2/n_1, \sigma_2^2/n_2, \ldots, \sigma_t^2/n_t)$. Hence

$$Z = \frac{\mathbf{c}'\hat{\boldsymbol{\mu}} - a}{\sqrt{\mathbf{c}'\mathbf{V}\mathbf{c}}} \sim N(0,1)$$

when H_0 is true. This fact can be used for an approximate test of H_0 whenever those sample sizes corresponding to nonzero c_i's are large. This is accomplished by substituting $\hat{\sigma}_i^2$ for σ_i^2 in the expression for Z and rejecting H_0 if $|Z| > Z_{\alpha/2}$. In other cases, an intuitively appealing statistic for testing H_0 is given by

$$U = (\mathbf{c}'\hat{\boldsymbol{\mu}} - a)/\sqrt{\mathbf{c}'\hat{\mathbf{V}}\mathbf{c}}\ ,$$

$$\hat{\mathbf{V}} = \text{DIAG}(\hat{\sigma}_1^2/n_1, \hat{\sigma}_2^2/n_2, \ldots, \hat{\sigma}_t^2/n_t).$$

The distribution of U is not known but can be approximated by Satterthwaite's* method [7, p. 642]. Note that u can be written as

$$U = Z/\sqrt{\mathbf{c}'\hat{\mathbf{V}}\mathbf{c}/\mathbf{c}'\mathbf{V}\mathbf{c}}\ . \qquad (2)$$

Since the numerator and denominator in (2) are independent and since the numerator has a standard normal distribution, U would be approximately distributed as $t(\nu)$, if one could find ν so that $W = \nu \cdot \mathbf{c}'\hat{\mathbf{V}}\mathbf{c}/\mathbf{c}'\mathbf{V}\mathbf{c}$ is approximately chi-square distributed with ν degrees of freedom. Satterthwaite's method determines ν so that the first two moments of W are equal to the first two moments of a chi-square distribution, $\chi^2(\nu)$. This gives

$$\nu = \left[\sum_i c_i^2\sigma_i^2/n_i\right]^2 \Big/ \sum_i \left[c_i^4\sigma_i^4/\left(n_i^2(n_i - 1)\right)\right].$$

Since ν depends on unknown parameters, it cannot be determined exactly. Usually it is estimated by

$$\hat{\nu} = \left[\sum_i c_i^2\hat{\sigma}_i^2/n_i\right]^2 \Big/ \sum_i \left[c_i^4\hat{\sigma}_i^4/\left(n_i^2(n_i - 1)\right)\right].$$

Summarizing, one rejects $H_0 : \mathbf{c}'\boldsymbol{\mu} = a$ if

$$|t_c| = |\mathbf{c}'\hat{\boldsymbol{\mu}} - a|/\sqrt{\mathbf{c}'\hat{\mathbf{V}}\mathbf{c}} > t_{\alpha/2,\hat{\nu}}\ .$$

One unfortunate aspect of this test is that the degrees of freedom, $\hat{\nu}$, must be reestimated for every different value of $\mathbf{c}$.

No general procedure now exists for testing the matrix hypothesis $\mathbf{C}\boldsymbol{\mu} = \mathbf{a}$. However, several approximate tests exist for testing $\mu_1 = \mu_2 = \cdots = \mu_t$; the first was suggested by Box [3], another was proposed by Li (1964). The usual F-test for equal means is very robust, and one should only need to resort to one of these two tests when the population variances are extremely unequal.

TWO-WAY TREATMENT STRUCTURES

In this section the hypothesis testing procedure reviewed earlier is applied to two-way treatment structures. It is assumed that there are two sets of treatments $T_1, T_2, \ldots, T_t$ and $B_1, B_2, \ldots, B_b$. Each one of the T treatments is combined with each one of the B treatments and applied to an experimental unit. Thus a total of bt populations are being sampled.

Let μ_{ij} denote the response expected when treatments T_i and B_j are used together on a randomly selected experimental unit, $i = 1, 2, \ldots, t$; $j = 1, 2, \ldots, b$. Since bt different populations are being sampled, there are $bt - 1$ degrees of freedom in the sum of squares for the hypothesis of equal treatment means; i.e., $\mu_{11} = \mu_{12} = \cdots = \mu_{tb}$. All results pertaining to the one-way treatment structure apply to two-way treatment structures if one considers the bt treatment combinations as bt different populations.

A model for this case, called a *cell means model*, is

$$y_{ijk} = \mu_{ij} + \epsilon_{ijk}\ , \qquad i = 1, 2, \ldots, t;$$
$$j = 1, 2, \ldots, b; \quad k = 1, 2, \ldots, n_{ij}\ ,$$

where it is assumed that $\epsilon_{ijk} \sim$ i.i.d.$N(0, \sigma^2)$.

When treatments are applied in a two-way manner, experimenters are often interested in how the T treatments affect the response and how the B treatments affect the response. To provide adequate answers to these two questions, it must first be determined whether these two sets of treatments interact.

Each of the above questions, as well as many others that may be of interest, can be answered using the procedure described in

"A General Hypothesis Testing Procedure" above. It is important to note that it does not matter whether all the n_{ij} are equal provided that $n_{ij} > 0$ for all i and j. If some of the n_{ij} are equal to zero, then hypotheses about the corresponding μ_{ij}'s cannot be tested unless restrictions are placed on the μ_{ij}. One restriction often used is that the treatment effects do not interact. Such a restriction should not be used without strong evidence that the treatment effects do, indeed, not interact. The methods described earlier can be applied to these messy data situations as well as balanced data cases.

The hypothesis of no interaction* can be written as

$$H_{01}: \mu_{ij} - \mu_{i'j} - \mu_{ij'} + \mu_{i'j'} = 0$$

for all i, i', j, and j'. For illustration purposes, suppose $t = 3$, $b = 4$, and let

$$\boldsymbol{\mu} = [\mu_{11}\ \mu_{12}\ \mu_{13}\ \mu_{14}\ \mu_{21}\ \mu_{22}\ \mu_{23}\ \mu_{24}\ \mu_{31}\ \mu_{32}\ \mu_{33}\ \mu_{34}]'.$$

Then one matrix $\mathbf{C}$ for testing H_{01} is given by

$$\mathbf{C}_1 = \begin{bmatrix} 1 & -1 & 0 & 0 & -1 & 1 & 0 & 0 & 0 & 0 & 0 & 0 \\ 1 & -1 & 0 & 0 & 0 & 0 & 0 & 0 & -1 & 1 & 0 & 0 \\ 1 & 0 & -1 & 0 & -1 & 0 & 1 & 0 & 0 & 0 & 0 & 0 \\ 1 & 0 & -1 & 0 & 0 & 0 & 0 & 0 & -1 & 0 & 1 & 0 \\ 1 & 0 & 0 & -1 & -1 & 0 & 0 & 1 & 0 & 0 & 0 & 0 \\ 1 & 0 & 0 & -1 & 0 & 0 & 0 & 0 & -1 & 0 & 0 & 1 \end{bmatrix}.$$

The hypotheses of equal T effects and equal B effects can be written, respectively, as

$$H_{02}: \bar{\mu}_{1\cdot} = \bar{\mu}_{2\cdot} = \bar{\mu}_{3\cdot},$$

$$H_{03}: \bar{\mu}_{\cdot 1} = \bar{\mu}_{\cdot 2} = \cdots = \bar{\mu}_{\cdot 4}.$$

When $t = 3$ and $b = 4$, corresponding hypothesis matrices can be given by

$$\mathbf{C}_2 = \begin{bmatrix} 1 & 1 & 1 & 1 & -1 & -1 & -1 & -1 & 0 & 0 & 0 & 0 \\ 1 & 1 & 1 & 1 & 0 & 0 & 0 & 0 & -1 & -1 & -1 & -1 \end{bmatrix},$$

$$\mathbf{C}_3 = \begin{bmatrix} 1 & -1 & 0 & 0 & 1 & -1 & 0 & 0 & 1 & -1 & 0 & 0 \\ 1 & 0 & -1 & 0 & 1 & 0 & -1 & 0 & 1 & 0 & -1 & 0 \\ 1 & 0 & 0 & -1 & 1 & 0 & 0 & -1 & 1 & 0 & 0 & -1 \end{bmatrix}.$$

There are other sums of squares which are often obtained for unbalanced treatment structures, usually a result of sequential model fitting procedures. In most unbalanced data cases the hypotheses tested by such procedures are not meaningful and hence such tests cannot be recommended. See, for example, Speed and Hocking [16] and Urquhart and Weeks [17].

To conclude this section some remarks are made about some of the available statistical software* packages. SAS GLM, SPSS MANOVA, SPSS ANOVA, BMD-P2V, and BMD-P4V all have options that test hypotheses generated by matrices when there are no missing treatment combinations. When there are missing treatment combinations the user should *always* develop specific hypotheses. SAS, SPSS, and BMD allow this ([6, pp. 406–408], [12, pp. 72–75], and [13, p. 146], for example).

To summarize, a user faced with unbalanced data situations should use a cell means model as above, and be prepared to generate his or her own hypothesis and the corresponding appropriate hypothesis tests using the procedure described earlier.

OTHER NONSTANDARD METHODS

Milliken and Johnson [11] give many additional nonstandard data analysis techniques. Some topics they discuss are: two-way and higher order treatment structures when there is only one observation per treatment combination, multiple comparison* techniques for unbalanced treatment designs, split-plot* and repeated measures* designs with unbalanced and missing data, random effects models, mixed models, and covariance models. They also provide more details and examples on the topics discussed here.

Bancroft [1] provides many techniques for analyzing unbalanced experiments. Graybill [7], Searle [14], and Seber [15] provide the theoretical background necessary for a complete understanding of many messy data techniques. Conover, et al. [4] give up-to-date comparisons of many of the tests for homogeneity of variances. Velleman and Hoaglin [18] discuss graphical techniques that can be used to help make sense of messy data (*see* GRAPHICAL REPRESENTATION

OF DATA). Gnanadesikan [6] gives many procedures that can be used with multivariate data. Gunst and Mason [8] and Cook and Weisberg [5] give many techniques useful for regression* type data, as do Belsley, et al. [2].

References

[1] Bancroft, T. A. (1968). *Topics in Intermediate Statistical Methods*. Iowa State University Press, Ames, Iowa.

[2] Belsley, D. A., Kuh, E., and Welsch, R. E. (1980). *Regression Diagnostics*. Wiley, New York.

[3] Box, G. E. P. (1954). *Ann. Math. Statist.*, **25**, 290–302.

[4] Conover, W. J., Johnson, M. E., and Johnson, M. M. (1981). *Technometrics*, **23**, 351–361.

[5] Cook, R. D. and Weisberg, S. (1982). *Influence and Residuals in Regression*. Chapman & Hall, London.

[6] Dixon, W. J., Brown, M. B., Engelman, L., Frane, J., Hill, M., Jennrick, R., and Toporek, J., eds. (1981). *BMDP Biomedical Computer Programs*. University of California Press, Berkeley, CA.

[6] Gnanadesikan, R. (1977). *Methods for Statistical Data Analysis of Multivariate Observations*. Wiley, New York.

[7] Graybill, F. A. (1976). *Theory and Application of the Linear Model*. Duxbury Press, North Scituate, MA.

[8] Gunst, R. F. and Mason, R. L. (1980). *Regression Analysis and Its Application*. Dekker, New York.

[9] Hull, C. H. and Nie, N. H. (1981). *SPSS Update 7–9: New Procedures and Facilities for Releases 7–9*. McGraw-Hill, New York.

[10] Li, C. C. (1964). *Introduction to Experimental Statistics*. McGraw-Hill, New York.

[11] Milliken, G. A. and Johnson, D. E. (1984). *Analysis of Messy Data, Vol. 1: Designed Experiments*. Lifetime Learning Publications, Belmont, CA.

[12] Nie, N. H., Hull, C. H., Jenkins, J., Steinbrenner, K., and Best, D. H. (1975). *Statistical Package for the Social Sciences*, 2nd ed. McGraw-Hill, New York.

[13] SAS Institute Inc. (1982). *SAS User's Guide: Statistics, 1982 Edition*. SAS Institute Inc., Cary, NC.

[14] Searle, S. R. (1971). *Linear Models*. Wiley, New York.

[15] Seber, G. A. F. (1977). *Linear Regression Analysis*. Wiley, New York.

[16] Speed, F. M. and Hocking, R. R. (1976). *Amer. Statist.*, **30**, 30–33.

[17] Urquhart, N. S. and Weeks, D. L. (1978). *Biometrics*, **34**, 696–705.

[18] Velleman, P. F. and Hoaglin, D. C. (1981). *Applications, Basics, and Computing of Exploratory Data Analysis*. Duxbury Press, Boston.

(ANALYSIS OF VARIANCE
GENERAL LINEAR MODEL
INCOMPLETE DATA
ONE-WAY CLASSIFICATION)

D. E. JOHNSON
G. A. MILLIKEN

M-ESTIMATORS

The term *M*-estimator was introduced by Huber [20] to denote estimators of *maximum* likelihood type. Maximum likelihood* is perhaps the most important and widely used estimation method. Although there are examples where the maximum likelihood estimate (MLE) is not even consistent, for most parametric models of practical interest the MLE is, in fact, an asymptotically efficient (minimum asymptotic variance) estimator. Efficiency here, of course, means only when all the assumptions of the statistical model actually hold. In applications, we can normally expect that the assumptions are at most only approximately true. It is now becoming widely accepted that an estimator should not only have high efficiency at the ideal model, but it also should be well behaved when the data have a distribution only close to the model. Such an estimator is said to be *distributionally robust*.

The statistical models most widely employed in practice, as well as most frequently studied theoretically, are those based on the normal distribution. For a normal population, the MLEs of the mean and variance are simply the sample mean and variance. Unfortunately, these are typically not distributionally robust. This point was made especially clear by Tukey [37]. The difficulty here is that both sample mean and variance can be changed greatly in value by the addition of an outlier* (extreme value) to the data. If one modifies the normal distribution by moving a small amount of probability from the center to the tails, then this creates a distribution very close to the normal yet much more prone to outliers. Under this new distribution, the normal model MLE will have a greatly inflated variance, and if the probability is not transferred symmetrically to the tails it may be highly biased as well. Thus the nonrobustness of the MLE is due to its extreme sensitivity to outliers.

Huber [20] showed that one can modify the MLE slightly to reduce its sensitivity to outliers and yet still retain high efficiency at the model. He was particularly concerned with the symmetric location problem, that is, estimation of the center of a symmetric distribution. He modified the MLE (the sample mean) to minimize the maximum asymptotic variance, where the maximum is taken over all symmetric distributions which are close, in a certain natural sense, to the normal distribution. This estimator, now called Huber's M-estimator, apparently was the first robust estimator proposed not ad hoc but rather to maximize some measure of robustness*. Huber's M-estimator turned out to be similar to a trimmed mean (*see* TRIMMING AND WINSORIZATION). An α-trimmed mean is simply the mean of the sample after all observations below the αth quantile and above the $(1-\alpha)$th quantile have been removed. The advantage of the M-estimate over the trimmed mean is that, as seen below, M-estimators can be defined easily for more general estimation problems.

Since Huber's pioneering paper, definitions and measures of robustness have been introduced, and robust estimators have been found for many statistical problems. Among all robust estimators M-estimators seem to be the most flexible (*see* ROBUST ESTIMATION). They can be defined whenever there is an ideal parametric model, so they can handle multidimensional parameters, nonidentical distributions, and dependent observations. M-estimators for scale parameters and for the simultaneous estimation of location and scale parameters were already discussed by Huber [20]. M-estimation for linear statistical models, now relatively well understood, offers an appealing alternative to classical least-squares* estimation. M-estimators have also been studied in the contexts of covariance and correlation* estimation and time series*, as well as other estimation problems. Hogg [19] gives a good introduction to robust estimation.

DEFINITION, INFLUENCE CURVE, ASYMPTOTIC THEORY

We begin with the simplest case, where $X_1, \ldots, X_n$ are independent and identically distributed (i.i.d.) according to a distribution F with density f. The ideal assumed model is that f belongs to a parametric family of densities $\{f_\theta : \theta \in \Theta\}$ where Θ is a subset of the real line. The MLE maximizes

$$\sum_{i=1}^{n} \log f_{\hat{\theta}}(X_i).$$

The basic idea behind M-estimation is to replace $\log f_\theta(x)$ by another function, say $\rho_\theta(X_i)$, which is less sensitive to outliers. For example, if $X_1, \ldots, X_n$ are i.i.d. $N(\theta, 1)$, that is, normal with mean θ and variance 1, then $\log f_\theta(X) = -(X-\theta)^2/2$ plus a constant. Huber's M-estimator [20] replaces $(x-\theta)^2/2$ by

$$\rho(x-\theta) = \begin{cases} (x-\theta)^2/2 & \text{if } |x-\theta| \leqslant k \\ k|x-\theta| - k^2/2 & \text{if } |x-\theta| > k, \end{cases}$$

where k is a constant, typically between 1 and 2, which is discussed later. The M-estimator is the minimizer of

$$\sum_{i=1}^{n} \rho(X_i - \hat{\theta}).$$

Because ρ grows only linearly as $x \to \infty$, the M-estimator is less sensitive to outliers than the MLE. Let $\psi(x) = (d/dx)\rho(x)$. Then

$$\psi(x) = \begin{cases} -k, & x < -k \\ x, & -k \leqslant x \leqslant k \\ k, & x > k. \end{cases}$$

The M-estimator solves

$$\sum_{i=1}^{n} \psi(X_i - \hat{\theta}) = 0, \tag{1}$$

which implies that $\hat{\theta}$ is the weighted average,

$$\hat{\theta} = \left(\sum_{i=1}^{n} X_i w_i \right) \Big/ \left(\sum_{i=1}^{n} w_i \right),$$

where $w_i = |\psi(X_i - \hat{\theta})/(X_i - \hat{\theta})|$ are the weights. Now, w_i is 1 if $|X_i - \hat{\theta}| \leqslant k$ and w_i decreases to 0 as $|X_i - \hat{\theta}|$ increases to infinity, so one sees that outliers are downweighted.

For general parametric problems, *M*-estimators are defined by generalizing (1). One chooses a function $\psi_\theta(x)$ and solves

$$\sum_{i=1}^{n} \psi_\theta(X_i) = 0.$$

Usually $\psi_\theta(x)$ is chosen so that

$$\int \psi_\theta(x) f_\theta(x)\, dx = 0 \qquad \text{for all } \theta, \qquad (2)$$

a condition called *Fisher consistency*, which typically implies consistency, i.e., that $\hat{\theta} \to \theta$ in probability under F_θ as $n \to \infty$.

For robustness, $\psi_\theta(x)$ should also be chosen so that $\hat{\theta}$ is relatively insensitive to outliers. Toward this end, an extremely important tool is Hampel's *influence function** or *influence curve* (IC) [15, 17]. The basic idea is that the IC is a function $\mathrm{IC}_\theta(x)$ so that

$$(\hat{\theta} - \theta) = n^{-1} \sum_{i=1}^{n} \mathrm{IC}_\theta(X_i) + R_n,$$

where R_n is a remainder which is negligible compared to the sum. Intuitively, $n^{-1}\mathrm{IC}_\theta(x)$ measures the effect of replacing one randomly chosen X_i by an outlier with value x.

The gross error sensitivity is defined to be

$$\gamma^* = \sup_x |\mathrm{IC}(x)|.$$

If $\gamma^* < \infty$, then a single outlier has at most a finite influence and the estimator is said to be *infinitesimally robust*. This definition, along with other definitions of robustness, was introduced by Hampel [15–17]. Also see Huber [24].

Huber [24, p. 45] shows that for an *M*-estimator

$$\mathrm{IC}(x) = \frac{\psi_{\theta(F)}(x)}{-\dfrac{d}{d\theta} \displaystyle\int \psi_{\theta(F)}(x)\, dF(x)}.$$

Under suitable assumptions (Huber [21, 24]), $\hat{\theta} \to \theta(F)$ almost surely and $n^{1/2}(\hat{\theta} - \theta(F))$ converges to the $N(0, \sigma^2(F, \psi))$ distribution, where the asymptotic variance is given by

$$\sigma^2(F, \psi) = \int \mathrm{IC}^2(x)\, dF(x)$$

$$= \frac{\displaystyle\int \psi^2_{\theta(F)}(x)\, dF(x)}{\left[\dfrac{d}{d\theta} \displaystyle\int \psi_{\theta(F)}(x)\, dF(x) \right]^2}.$$

Hampel proved a very significant result which enables one to find efficient and robust estimators for quite general univariate parametric estimation problems [15]. Let $l_\theta(x) = (d/d\theta) \log f_\theta(x)$, and fix θ_0 in Θ. For positive a and b, let

$$l_{\theta,a,b}(x) = \begin{cases} -b, & l_\theta(x) - a < -b \\ l_\theta(x) - a, & |l_\theta(x) - a| \leqslant b \\ b, & l_\theta(x) - a > b. \end{cases}$$

Hampel showed that for each fixed b there exists $a(b)$ such that with $l_{\theta_0,b}(x) = l_{\theta_0,a,b}(x)$, one has Fisher consistency, that

$$\int l_{\theta_0,b}(x)\, dF_{\theta_0}(x) = 0.$$

Moreover, since $|l_{\theta_0,b}(x)| \leqslant b$, its gross error sensitivity γ^* is bounded by

$$k = \frac{b}{\left| \left[(d/d\theta) \int l_{\theta,b}(x)\, dF_{\theta_0}(x) \right]_{\theta=\theta_0} \right|}.$$

Hampel showed further that among all ψ_θ which are Fisher consistent and have γ^* less than or equal to k, $l_{\theta_0,b}$ minimizes $\sigma^2(F_{\theta_0}, \psi_{\theta_0})$. In summary, then, Hampel found an estimator which, among all Fisher consistent *M*-estimators satisfying the robustness constraint given by $\gamma^* \leqslant k$, is most efficient in the sense of minimizing the asymptotic variance.

LOCATION PARAMETER ESTIMATION

Suppose that $X_1, \ldots, X_n$ are observed and $X_i = \theta + \epsilon_i$, where ϵ_i is distributed according to F, which is symmetric about 0. The ideal model will be $F = \Phi$, the standard normal distribution. According to Hampel's results just discussed, we can minimize the asymptotic variance $\sigma^2(\Phi, \psi)$, subject to a bound on γ^*, by solving (1). Huber [20], whose work of course predates Hampel's, showed that this estimator also minimizes

$$\max_{G \in N_\epsilon} \sigma^2(\psi, G),$$

where N_ϵ is the symmetric contamination neighborhood of Φ defined by

$$N_\epsilon = \{G : G = (1 - \epsilon)\Phi + \epsilon H, \quad H \text{ a symmetric distribution}\}.$$

An element G in N_ϵ has a simple interpretation. If most of the data have distribution Φ, but a small fraction ϵ have distribution H, which is symmetric but otherwise arbitrary, then the sample has distribution G. The elements with distribution H can be considered contaminants. The value of k depends on ϵ. Of course, in practice we may have no particular value of ϵ in mind, and k is chosen by other considerations. The probability under the assumed model that X_i is downweighted ($w_i < 1$) is $2\Phi(-k)$. As a rule of thumb, $1.0 \leqslant k \leqslant 2.0$ can be recommended, with $k = 1.5$ being a good choice. It is probably better that k be too small than too large. As $k \to 0$, the estimate converges to the sample median, while as $k \to \infty$ it converges to the sample mean. With $k = 1.5$, the estimator is robust like the median, but unlike the median it is reasonably efficient at the normal model. A natural question is "How efficient is an *M*-estimate if the assumed model actually is correct?" Table 1 of ref. 21 shows that under the normal distribution, the asymptotic variance of $\hat{\theta}$ is only 1.107, 1.037, and 1.010 times greater than that of the sample mean for $k = 1.0$, 1.5, and 2.0, respectively.

If the ideal model is the normal distribution with standard deviation σ then one should use $\psi_k((x - \theta)/\sigma)$ with the same k as above. If σ is unknown, then one can use $\psi_k((x - \theta)/\hat{\sigma})$ where $\hat{\sigma}$ is a robust scale estimator. There are two scale estimators which are generally recommended. (The sample standard deviation is highly nonrobust and is not recommended.)

The first is the median absolute deviation (MAD). Let m be the median of $X_1, \dots, X_n$. Then the MAD is the median of $|X_1 - m|, \dots, |X_n - m|$. The second, called Huber's proposal 2 [20], solves simultaneously

$$\sum_{i=1}^{n} \psi_k((X_i - \hat{\theta})/\hat{\sigma}) = 0,$$

$$\sum_{i=1}^{n} \chi((X_i - \hat{\theta})/\hat{\sigma}) = 0,$$

for a suitable function χ, for example,

$$\chi(x) = \psi_k^2(x) - \int \psi_k^2(x)\, d\Phi(x).$$

Besides Huber's ψ_k, ψ's which equal 0 for all large x have been recommended since these ignore gross outliers, rather than merely bounding their influence. Such ψ are called *descending* or *redescending* (*see* R-DESCENDING *M*-ESTIMATORS). With a descending ψ there usually will be multiple solutions to

$$\sum_{i=1}^{n} \psi((X_i - \hat{\theta})/\hat{\sigma}) = 0$$

and one must take care that the "right" solution is used.

If Φ is contaminated with an asymmetric distribution, a difficult conceptual problem arises: What parameter are we to estimate? For partial answers, see Collins [10] and Bickel and Lehmann [5]. Huber [24, Sect. 4.2] shows that the maximum asymptotic bias is minimized by the median, which is the *M*-estimate for $\psi(x) = \operatorname{sign} x$. Here we note only that if F is close to F_{θ_0}, then a robust estimator will consistently estimate a parameter θ close to θ_0, but nonrobust estimates will not in general share this property.

Andrews et al. [2] report on a large-scale simulation* study of robust location estimates, including many *M*-estimators. Gross [14] uses asymptotic theory to develop confidence interval methodology, which he studies by Monte Carlo methods*. In general, his methods work quite respectably, even for small sample sizes ($n = 10$). Stigler [35] applies various robust estimators to real data sets.

MULTIVARIATE *M*-ESTIMATORS

If $\boldsymbol{\theta}$ is k-dimensional, then an M-estimator is still a solution to

$$\sum_{i=1}^{n} \psi_{\theta}(\mathbf{X}_i) = 0,$$

but now the range space of ψ is k-dimensional. The dimension of $\mathbf{X}_i$, which plays no essential role here, need not be k. The IC is

$$\mathbf{IC}(\mathbf{x}) = \left[-(d/d\boldsymbol{\theta}) \int \psi_{\theta}(\mathbf{x})\, dF(\mathbf{x}) \right]^{-1} \psi_{\theta}(\mathbf{x}) \tag{3}$$

where $(d/d\boldsymbol{\theta})$ denotes the differential. (If $\mathbf{F}(\boldsymbol{\theta}) = (F_1(\boldsymbol{\theta}), \dots, F_I(\boldsymbol{\theta}))$ is an I-dimensional function of the J-dimensional parameter $\boldsymbol{\theta} = (\theta_1, \dots, \theta_J)'$, then $(d/d\boldsymbol{\theta})\mathbf{F}(\boldsymbol{\theta})$ is an $I \times J$ matrix with (i, j)th element equal to $(\partial/\partial\theta_j)F_i(\boldsymbol{\theta})$.) Under suitable conditions [21] again $\hat{\boldsymbol{\theta}} \to \hat{\boldsymbol{\theta}}(F)$ in probability and $n^{1/2}(\hat{\boldsymbol{\theta}} - \boldsymbol{\theta}(F))$ is asymptotically normal with mean $\mathbf{0}$ and variance-covariance matrix

$$\boldsymbol{\Sigma}(F, \psi) = \int [\mathbf{IC}(\mathbf{x})][\mathbf{IC}(\mathbf{x})]'\, dF(\mathbf{x}).$$

MULTIVARIATE LOCATION AND SCALE (COVARIANCE AND CORRELATION)

Suppose that $\mathbf{X}_1, \dots, \mathbf{X}_n$ are i.i.d. k-dimensional vectors with density

$$|\det \mathbf{V}|^{1/2} f\big((\mathbf{x} - \boldsymbol{\mu})'\mathbf{V}^{-1}(\mathbf{x} - \boldsymbol{\mu})\big),$$

where $f(\|\mathbf{x}\|)$ is a k-dimensional, spherically symmetric density. Here $\|\cdot\|$ is the Euclidean norm. Then, provided f has finite second moments, $\mathbf{V}$ is a scalar multiple of the covariance matrix. Maronna [28] proposed estimating $\boldsymbol{\mu}$ and $\mathbf{V}$ by solving

$$\sum_{i=1}^{n} \mu_1(d_i)(\mathbf{X}_i - \hat{\boldsymbol{\mu}}) = 0,$$

$$\frac{1}{n} \sum_{i=1}^{n} \mu_2(d_i^2)(\mathbf{X}_i - \hat{\boldsymbol{\mu}})(\mathbf{X}_i - \hat{\boldsymbol{\mu}})' = \hat{\mathbf{V}},$$

where $d_i = (\mathbf{X}_i - \hat{\boldsymbol{\mu}})'\hat{\mathbf{V}}^{-1}(\mathbf{X}_i - \hat{\boldsymbol{\mu}})$ and μ_1 and μ_2 are nonnegative functions. If $\mu_1(x) \equiv \mu_2(x) \equiv 1$, then we obtain the sample mean and covariance matrix. If $\mathbf{x}\mu_1(\mathbf{x})$ and $\mathbf{x}\mu_2(\mathbf{x})$ are bounded, the estimates will be robust, but $\hat{\boldsymbol{\mu}}$ and $\hat{\mathbf{V}}$ can in general be found only iteratively. As in the univariate case, the estimators are weighted sample moments with data-dependent weights. Many multivariate methods, for example principal components and discriminant analysis* (*see* COMPONENT ANALYSIS), can be made robust by replacing the sample covariance matrix by a robust $\hat{\mathbf{V}}$.

See Huber [23, 24] and Devlin et al. [11, 12] for further work on robust covariance and correlation estimation.

The investigation of robust correlation is still in the preliminary stages, but given the fundamental importance of covariance and correlation matrices to multivariate analysis and the nonrobustness of the traditional estimators, the area should become one of major theoretical and practical interest.

LINEAR MODELS

Suppose that $y_i = \mathbf{x}_i'\boldsymbol{\beta} + \epsilon_i$, where $\mathbf{x}_i$ is a known p-dimensional vector of independent variables, $\boldsymbol{\beta}$ is an unknown vector of regression coefficients, and $\epsilon_1, \dots, \epsilon_n$ are i.i.d. with symmetric distribution F. To estimate $\boldsymbol{\beta}$ one can generalize the location estimator by solving, for one of the ψ functions used for the location problem,

$$\sum_{i=1}^{n} \psi\big((y_i - \mathbf{x}_i'\hat{\boldsymbol{\beta}})/\hat{\sigma}\big)\mathbf{x}_i = 0.$$

Here $\hat{\sigma}$ is a simultaneous estimate of scale. Let $r_i = y_i - \mathbf{x}_i'\hat{\boldsymbol{\beta}}$ be the ith residual*. Then two frequently recommended scale estimators are the MAD of $r_1, \dots, r_n$ and Huber's proposal 2, which solves

$$\sum_{i=1}^{m} \chi(r_i/\hat{\sigma}) = 0.$$

The estimates $\hat{\boldsymbol{\beta}}$ and $\hat{\sigma}$ must be computed iteratively, for example, by iteratively reweighted least-squares. See Huber [24, Sect. 7.8] for details and further references on computational aspects.

If we regard $\mathbf{x}_1, \dots, \mathbf{x}_n$ as i.i.d. with a distribution independent of $\boldsymbol{\beta}$, then the IC can be calculated from (3) and is

$$\mathrm{IC}(\mathbf{x}, y) = \sigma \mathbf{S}^{-1}\mathbf{x}\psi((y - \mathbf{x}'\boldsymbol{\beta})/\sigma),$$

where $\mathbf{S} = E[\mathbf{x}_1\mathbf{x}_1^t\dot{\psi}(\epsilon_1)]$, $\dot{\psi}(x) = (d/dx)\psi(x)$. If $\mathbf{x}_1, \ldots, \mathbf{x}_n$ are not i.i.d. (they may not even be random), then we can still define the influence curve for y at $\mathbf{x}_i$. The result is simply $\mathrm{IC}(\mathbf{x}_i, y)$. By the usual theory, $n^{1/2}(\hat{\boldsymbol{\beta}} - \boldsymbol{\beta})$ is asymptotically normal with mean $\mathbf{0}$ and variance-covariance matrix

$$\boldsymbol{\Sigma}(\psi, F) = \sigma \mathbf{S}^{-1}\frac{E\psi^2(\epsilon_1/\sigma)}{\left(E\dot{\psi}(\epsilon_1/\sigma)\right)^2}.$$

See Yohai and Maronna [38] and Maronna and Yohai [29].

The confidence regions and tests associated with least-squares theory have analogs based on *M*-estimators. See Schrader and Hettmansperger [33] and Huber [24, Chap. 7].

Calculation of the estimates is only a small part of least-squares software packages. Fortunately, once software is developed for calculating the robust estimates, existing least-squares software can be employed, for example, to construct robust analysis of variance* tables, and confidence intervals for the coefficients. Let K be the bias correction factor of order (p/n) discussed by Huber [22; 24, Sects. 7.8 and 7.10). The ith pseudo-observation [3] is

$$y_i^* = \mathbf{x}_i^t\hat{\boldsymbol{\beta}} + \lambda\psi(r_i/\hat{\sigma}),$$

where $\lambda = K\hat{\sigma}(n^{-1}\sum_{i=1}^{n}\dot{\psi}(r_i/\hat{\sigma}))$. If p/n is small, then one might replace K by 1, though Schrader and Hettmansperger [33] found K to be "absolutely essential" in their simulation studies. The pseudo-observations permit one to adapt least-squares software to robust regression. When one puts $\{(y_i^*, \mathbf{x}_i)\}$ into a linear models, least-squares package, then the calculated least-squares estimator is the robust estimator $\hat{\boldsymbol{\beta}}$ (which is already known), the estimated covariance matrix of the regression coefficient estimates is the consistent estimate of $\boldsymbol{\Sigma}(\psi, F)$ given by Huber [24, eq. 7.10.2], and all resulting tests and confidence regions are valid asymptotically. Also, stepwise variable selection routines can be used with the pseudo-observations, provided that $\hat{\boldsymbol{\beta}}$ is computed for the model which includes all variables under consideration.

When F is asymmetric but the model includes an intercept, then the intercept parameter is not well defined, but the slope parameters are well defined and consistently estimated. There are difficulties when estimating the variances of the estimators [6].

There are several published applications of robust linear models, including Andrews [1], Carroll [7], and Eddy and Kadane [13]. The first two papers show how well *M*-estimators can find anomalous data points, which are difficult, although not impossible, to locate with least-squares. Regression *M*-estimators probably will never replace least-squares, but they are now becoming well accepted as a way of obtaining another, possibly very informative, view of data.

Krasker [26] and Krasker and Welsch [27] proposed regression *M*-estimators that limit the influence of outlying values of **x**. These so-called bounded-influence estimates have been studied by Huber [25] from a minimax* viewpoint.

OTHER TOPICS

Besides the estimation problems mentioned above, *M*-estimators have been proposed for time series* [30, 31], estimation of a power transformation parameter [8, 4], estimation in heteroscedastic linear models [9], quantal bioassay* [32], and radioimmunoassay [36].

For adaptive *M*-estimators, see Hogg's review [18].

References

[1] Andrews, D. F. (1974). *Technometrics*, **16**, 523–531. (Discusses one application in detail.)

[2] Andrews, D. F., Bickel, P. J., Hampel, F. R., Huber, P. J., Rogers, W. H., and Tukey, J. W. (1972). *Robust Estimates of Location: Survey and Advances*. Princeton University Press, Princeton, NJ.

[3] Bickel, P. J. (1976). *Scand. J. Statist.*, **3**, 145–168.

[4] Bickel, P. J. and Doksum, K. A. (1981). *J. Amer. Statist. Ass.*, **76**, 296–311.

[5] Bickel, P. J. and Lehmann, E. L. (1975). *Ann. Statist.*, **3**, 1045–1069.

[6] Carroll, R. J. (1979). *J. Amer. Statist. Ass.*, **74**, 674–679.

[7] Carroll, R. J. (1980). *Appl. Statist.*, **29**, 246–251. (Good article on applications.)

[8] Carroll, R. J. (1980). *J. R. Statist. Soc. B*, **42**, 71–78.

[9] Carroll, R. J. and Ruppert, D. (1982). *Ann. Statist.*, **10**, 429–441.

[10] Collins, J. R. (1976). *Ann. Statist.*, **4**, 68–85.

[11] Devlin, S. J., Gnanadesikan, R., and Kettenring, J. R. (1975). *Biometrika*, **62**, 531–545.

[12] Devlin, S. J., Gnanadesikan, R., and Kettenring, J. R. (1981). *J. Amer. Statist. Ass.*, **76**, 354–363.

[13] Eddy, W. F. and Kadane, J. B. (1982). *J. Amer. Statist. Ass.*, **77**, 262–269.

[14] Gross, A. M. (1976). *J. Amer. Statist. Ass.*, **71**, 409–416.

[15] Hampel, F. R. (1968). *Contributions to the Theory of Robust Estimation.* Ph.D. Dissertation, University of California, Berkeley, Calif. (This and refs. 16 and 17 are an important part of the foundation of robustness theory.)

[16] Hampel, F. R. (1971). *Ann. Math. Statist.*, **42**, 1887–1896.

[17] Hampel, F. R. (1974). *J. Amer. Statist. Ass.*, **62**, 1179–1186.

[18] Hogg, R. V. (1974). *J. Amer. Statist. Ass.*, **69**, 909–927.

[19] Hogg, R. V. (1979). In *Robustness in Statistics*, R. L. Launer and G. N. Wilkinson, eds. Academic Press, New York.

[20] Huber, P. J. (1964). *Ann. Math. Statist.*, **35**, 73–101. (The first and perhaps the single most important paper on M-estimators. Still well worth reading, though the mathematical level is high.)

[21] Huber, P. J. (1967). *Proc. Fifth Berkeley Symp. Math. Statist. Prob.*, **1**, 221–233. (General asymptotic theory.)

[22] Huber, P. J. (1973). *Ann. Statist.*, **5**, 799–821.

[23] Huber, P. J. (1977). In *Statistical Decision Theory and Related Topics*, Vol. II; S. S. Gupta and D. S. Moore, eds. Academic Press, New York.

[24] Huber, P. J. (1981). *Robust Statistics*. Wiley, New York. (Detailed survey of the theory of robust statistics. Mathematical level varies, but much requires a mathematically sophisticated reader.)

[25] Huber, P. J. (1983). *J. Amer. Statist. Ass.*, **78**, 66–72. (Followed by discussion.)

[26] Krasker, W. S. (1980). *Econometrics*, **48**, 1333–1346.

[27] Krasker, W. S. and Welsch, R. E. (1982). *J. Amer. Statist. Ass.*, **77**, 595–604.

[28] Maronna, R. A. and Yohai, V. J. (1976). *Ann. Statist.*, **4**, 51–67.

[29] Maronna, R. A. (1981). *Zeit. Wahrscheinlichkeitsth. verw. Gebiete*, **58**, 7–20.

[30] Martin, R. D. (1979). In *Robustness in Statistics*, R. L. Launer and G. L. Wilkinson, eds. Academic Press, New York.

[31] Martin, R. D. (1980). In *Directions in Time Series*, D. R. Brillinger and G. C. Tiao, eds. Institute of Mathematical Statistics, Hayward, Calif.

[32] Miller, R. G. and Halpern, J. W. (1980). *Biometrika*, **67**, 103–110.

[33] Schrader, R. M. and Hettmansperger, T. P. (1980). *Biometrika*, **67**, 93–101.

[34] Serfling, R. J. (1980). *Approximation Theorems of Mathematical Statistics*. Wiley, New York.

[35] Stigler, S. M. (1977). *Ann Statist.*, **5**, 1055–1098.

[36] Tiede, J. J. and Pagano, M. (1979). *Biometrics*, **35**, 567–574.

[37] Tukey, J. W. (1960). In *Contributions to Probability and Statistics*, I. Olkin, ed. Stanford University Press, Stanford, CA.

[38] Yohai, V. J. and Maronna, R. A. (1979). *Ann. Statist.* **7**, 258–268.

(INFLUENCE FUNCTIONS
MAXIMUM LIKELIHOOD ESTIMATION
OUTLIERS
R-DESCENDING M-ESTIMATORS
ROBUST ESTIMATION
ROBUSTIFICATION AND ROBUST SUBSTITUTES)

D. RUPPERT

METAMATHEMATICS OF PROBABILITY

This term usually deals with formalized theories for probability and more generally the study of relationships between logic and probability.

The metamathematical structure of probability theory comes out of the necessity to study the *formal* linguistic structure which goes into the description of the concepts of probability. The method is a glueing of probability and *logic* methods, intended to give us a better understanding of the foundations of probability* and statistics; it may be a new starting point for incorporating theories such as those of fuzzy sets*, probabilistic metric spaces, Boolean and Heyting-valued models, nonstandard analysis, topos theory, and so on. As a result we may also have a general theory of uncertainty* not necessarily probabilistic in nature.

We review the more mathematical aspects of the subject. The reader who is interested in foundational problems and different axiomatic theories for probability should see ref. 7. The article is divided into three sections: Assigning Probabilities to Logical Formulas, Hyperfinite Model Theory, and Final Remarks.

ASSIGNING PROBABILITIES TO LOGICAL FORMULAS

Historically this idea goes back to the Cambridge philosopher W. E. Johnson, who influenced both Keynes and Jeffreys in their well-known treatises on probability [10]. Also, one can find similar ideas in Koopman's comparative logical probability [7, pp. 183–186].

Carnap was the first to study probabilities on *formalized* sentences [7, pp 187–204]. His study, however, was restricted to a portion of a formalized language and to symmetric probabilities. It was Gaifman [8] who first used the full power of a first-order language and arbitrary probabilities. His work was extended considerably by Scott and Krauss [19] and others [4–6], most recently Gaifman and Snir [9].

Let $\mathscr{L}$ be a first-order language and let C be the set of its constants with $C \subseteq T$. A probability model for $\mathscr{L}(T)$ is an ordered pair (T, μ) where $\mu: \mathscr{S}_0 \to [0, 1]$ is a probability on the quantifier-free sentences $\mathscr{S}_0(T)$ such that, for $\phi, \psi \in \mathscr{S}_0(T)$,

(i) $\mu(\phi) = 1$ if ϕ is provable in $\mathscr{L}$ and $\mu(\phi) = \mu(\psi)$ if ϕ and ψ are logically equivalent in $\mathscr{L}$,

(ii) $\mu(\phi \vee \psi) + \mu(\phi \wedge \psi) = \mu(\phi) + \mu(\psi)$,

(iii) $\mu(\neg \phi) = 1 - \mu(\varphi)$.

Gaifman [8] first proved the following interesting extension property.

Theorem. Let (T, μ) be a probability model. Then there exists a unique probability μ^* on the set $\mathscr{S}(T)$, of all closed sentences of $\mathscr{L}(T)$, which extends μ and satisfies the Gaifman condition:

$$(G)\ \mu^*(v\phi(v)) = \sup_{F \in T^{(\omega)}} \mu^*\Big(\bigvee_{t \in F} \phi(t)\Big),$$

where $T^{(\omega)}$ is the set of all finite subsets of T. It follows that Doob's separability condition is a special case of the (G) condition.

Another important aspect of assigning probabilities to logical formulas is that it is a special case of assigning truth values in a complete Boolean algebra $\mathscr{B}$. More preciesly, if we take as the Boolean algebra $\mathscr{B}$ the quotient algebra of the σ-field $\mathscr{A}$ of a probability space $(\Omega, \mathscr{A}, P)$, modulo the σ-ideal of sets of probability zero, then assigning probabilities to logical formulas is the same as assigning truth values in $\mathscr{B}$. This gives us a connection to Boolean-valued models [1, 23] and finally to more general Heyting-valued models [13] and topos theory. These structures are generalizations toward covering uncertainties which are not necessarily probabilistic in nature.

HYPERFINITE MODEL THEORY

Hyperfinite model theory was introduced by Keisler [16] and is aimed at models that arise in probability and applied mathematics, in contrast with classical model theory which is more appropriate for problems in algebra and set theory. Hyperfinite models are appropriate for the study of the limiting behavior of finite models through the use of finite combinatorial arguments in infinite settings.

Keisler's work [16], along with Hoover's [14, 15], constitutes a novel approach to the metamathematics of probability. It is based on hyperfinite probability [21], which in turn is based on Robinson's infinitesimal analysis.

A hyperfinite model is a structure $\mathscr{A} = \langle A, \mu, a_i, S_j \rangle$, $i \in I, j \in J$, where $\langle A, \mu \rangle$ is a hyperfinite probability space, $a_j \in A$, and each S_j is an internal relation on A. The logics $L_{\omega P}, L_{\omega_1 P}$ involved are similar to the ordinary logics but instead of the quantifiers $\exists x$ and $\forall x$, one uses the probability quantifiers $(Px > r)$, $(Px \geqslant r)$, where $(Px > r)$ means that the set $\{x, \phi(x)\}$ has probability greater than r in the Loeb measure involved.

An interpretation of a hyperfinite model can be given in terms of the so-called opinion poll model [16]. We can think of A as a very large set of people, S_j as a statement about the tuple $\mathbf{y}$, and $S_j(a, \mathbf{b})$ is true iff the person a agrees with statement S_j about the n-tuple of people $\mathbf{b}$. Thus

$$(Px > \tfrac{1}{2})(Py > \tfrac{1}{2})S(x, y)$$

means that a majority of x's agree with S about a majority of y's. The sentence

$$(Py > \tfrac{1}{2})(Px > \tfrac{1}{2})S(x, y)$$

means that for a majority of y's, a majority of x's agree with S about y. Neither of the foregoing sentences implies the other.

In addition to the above logics, an internal logic $\int \mathscr{L}$ is introduced with hyperfinite truth values and three quantifiers (max), (min), and $\int \ldots dx$. This internal logic is needed to get deeper model-theoretic results about the probability logic $L\omega p$, but it is also of interest in its own right. An interesting discovery is the connection between apparently unrelated theorems from model theory and probability theory. Downward Löwenheim-Skolem theorems for $\int \mathscr{L}$ are closely related to the weak law of large numbers*, whereas the elementary chain theorem is related to the strong law of large numbers.

In the opinion-poll model an elementary submodel $\mathscr{B}$ of $\mathscr{A}$ is a sample which is good in the sense that, for any statement S_j and any n-tuple $\mathbf{b} \in A^n$, the percentage of $a \in B$ who agree with S_j about $\mathbf{b}$ is infinitely close to the percentage of $a \in A$ who agree with S_j about $\mathbf{b}$. In a similar way one can interpret the Löwenheim-Skolem theorem.

FINAL REMARKS

In addition to the above two mainstreams of metamathematics of probability there are interesting connections between logic and probability.

Hyperfinite Stochastic Analysis

This is based on Robinson's infinitesimal analysis and is a formal way of reducing continuous problems to discrete ones. For example, Brownian motion* is treated as a hyperfinite random walk*, stochastic differential equations* as stochastic difference equations, stochastic integrals* as nonstandard Stieltjes integrals, etc. For details, extensions and further references, see ref. 21.

Statistical Model Theory

This idea starts with Suppes [22], who studies models of (statistical) theories, models of experiments, and models of data. Then we have the book of Hájek and Harvánek [11], and Harvánek's work [12] toward a model theory of statistical theories. These works can be considered a starting point for a metamathematical structure of statistics.

From a Bayesian viewpoint, Gaifman and Snir [9] study a language $\mathscr{L}$ which has also an empirical part L_0, consisting of finitely many so-called empirical predicates and/or empirical function symbols.

Recursion Theory and Randomness

The basic issue here is what it means for a sequence to be random. R. von Mises* based his axiomatic treatment of probability on the notion of "kollektiv," a notion of randomness* for a sequence; see ref. 7, pp. 92–100. Church [2] was the first to apply recursion theory to von Mises' concept of a kollektiv. The names of Ville, Kolmogorov, Martin-Löf, Knuth, Schnorr, and others [17] are connected with the efforts to develop a consistent theory of random sequences, with the help of recursion theory. The most recent developments of the notion of randomness can be found in Gaifman and Snir [9]. For details and exact references that are not mentioned here, see refs. 7, 9, 12, and 17.

Fuzzy Sets, Heyting-Valued Models, and Topos Theory

Examining the work of Higgs [13] and other developments of topos theory one can easily convince oneself that Heyting-valued models and topos theory constitute a formalization of the theory of fuzzy sets. The logic involved is of course intuitionistic. On the other hand some of the work on elementary topoi was conducted much in the spirit of nonstandard analysis. The elementary topoi is the appropriate frame to do nonstandard mathematics. The development of nonstandard models (not only Robinson's infinitesimal models [1, 13, 21, 23]), has shown that they contain the concept of "randomness" or more generally the concept of "uncertainty" as a basic element. For example, in Takeuti [23, Chap. 2] the nonstandard reals are random variables, and in Solovay [20] the concept of a "random real" is introduced for the adoption of Cohen's forcing into measure-theoretic problems. Furthermore, Boolean algebras and probabilities on them represent the qualitative and quantitative aspects of "randomness" respectively, whereas Heyting algebras represent more general aspects of uncertainty, not necessarily probabilistic in nature, e.g., fuzzy sets [3].

It is believed that fuzzy sets* [3] and probabilistic metric spaces [18] can be expressed formally using Heyting-valued and Boolean-valued models.

References

[1] Bell, J. L. (1977). *Boolean-Valued Models*. Clarendon, Oxford, England. (A standard textbook on the subject. Presupposes acquaintance with mathematical logic and axiomatic set theory.)

[2] Church, A. (1940). *Bull. Amer. Math. Soc.*, **46**, 130–135.

[3] Dubois, D. and Prade, H. (1980). *Fuzzy Sets and Systems: Theory and Applications*. Academic Press, New York.

[4] Eisele, K. Th. (1976). *Booleschwertige Modelle mit Wahrscheinlichkeitsmasses und Darstellungen Stochastischer Prozesse*. Ph.D. Thesis, University of Heidelberg, Germany.

[5] Fenstad, J. E. (1967). In *Sets, Models and Recursion Theory*, J. N. Crossley, ed. North-Holland, Amsterdam, pp. 156–172.

[6] Fenstad, J. E. (1968). *Synthèse*, **18**, 1–23.

[7] Fine, T. (1973). *Theories of Probability*. Academic Press, New York. (Examines the foundations of probability. Perhaps the only textbook of this sort.)

[8] Gaifman, H. (1964). *Israel J. Math.*, **2**, 1–18. (Original work; deals with finitary formulas.)

[9] Gaifman, H. and Snir, M. (1982). *J. Symbolic Logic*, **47**, 495–548. (A recent landmark paper which introduces new concepts; results from a Bayesian viewpoint and summarizes and extends the works in refs. 4, 5, 16, and 14.)

[10] Good, I. J. (1965). *The Estimation of Probabilities*. MIT Press, Cambridge, MA.

[11] Hájek, P. and Havránek, T. (1978). *Mechanizing Hypothesis Formation: Mathematical Foundations for a General Theory*. Universitexte, Springer-Verlag, Heidelberg.

[12] Havránek, T. (1977). *Synthèse*, **36**, 441–458. (Technical paper, worth reading by those interested in model theory of statistical theories. Also, there are further references on this subject.)

[13] Higgs, D. (1973). *A Category Approach to Boolean-Valued Set Theory*. University of Waterloo, preprint. (Excellent technical paper for Heyting-valued models; can be considered as a formalization of the theory of fuzzy sets.)

[14] Hoover, D. N. (1978). *Ann. Math. Logic*, **14**, 287–313. (Technical paper on probability logic extending Keisler's work in ref. 11.)

[15] Hoover, D. N. (1982). *J. Symbolic Logic*, **47**, 605–624. (Recent extensions of refs. 11 and 13.)

[16] Keisler, H. J. (1977). In *Logic Colloquium 76*. North-Holland, Amsterdam, pp. 5–110. (Source paper; develops hyperfinite model theory, based on hyperfinite probabilities; see ref. 16.)

[17] Schnorr, C. P. (1971). *Zufälligkeit und Wahrscheinlichkeit*. Lecture Notes in Math., No. 218, Springer, Berlin. (Studies random sequences using recursion theory.)

[18] Schweizer, B. and Sklar, A. (1983). *Probabilistic Metric Spaces*. North-Holland, Amsterdam. (The only introductory text to this very interesting subject.)

[19] Scott, D. and Krauss, P. (1966). In *Aspects of Inductive Logic*, J. Hintikka and P. Suppes, eds. North-Holland, Amsterdam, pp. 219–264. (Extends ref. 7 to infinitary languages.)

[20] Solovay R. M. (1970). *Ann. Math.*, **92**, 1–59. (Source paper; adopts forcing to measure theoretic problems.)

[21] Stroyan, K. D. and Bayod, J. M. (1983). *Foundations of Infinitesimal Stochastic Analysis*. North-Holland, Amsterdam. (A graduate introduction to

hyperfinite probability and infinitesimal stochastic analysis.)

[22] Suppes, P. (1962). In *Logic, Methodology and Philosophy of Science; Proceedings of the 1960 International Congress*, E. Nagel, P. Suppes, and A. Tarski, eds. Stanford University Press, Stanford, CA, pp. 252–261.

[23] Takeuti, G. (1978). *Two Applications of Logic to Mathematics*. Iwanami Shoten, Tokyo; Princeton University Press, Princeton, NJ. (A technical book that develops Boolean analysis.)

(FOUNDATIONS OF PROBABILITY
FUZZY SET THEORY
RANDOMNESS)

CONSTANTIN A. DROSSOS

METAMETER

This term is used to denote a *transformed value*, as opposed to one that is directly observed. Usually the transformed value is used because a model is more conveniently expressed in terms of it. In quantal response analysis*, for example, the actual dose (x) of a drug is often replaced by the logarithm ($\log x$) of this quantity (termed the "dosage" or dose metameter) because it is believed that the tolerance distribution* is lognormal*, so that it is convenient to use $\log X$ rather than X.

Sometimes the term *natural transform* (or something similar) is used—especially with regard to the logarithm of the hazard rate. This is unfortunate, because the word "natural" may be felt to imply some inherent respectability, rather than mere convenience, for the transform used (and the model on which it is based). Similar remarks apply to the use of the term "natural conjugate"*.

(TRANSFORMATION)

METASTATISTICS *See* NONEXPERIMENTAL INFERENCE

METEOROLOGY, STATISTICS IN

Meteorology, the science of the atmosphere, is concerned with the study of the Earth's weather. This includes such general phenomena as solar radiation and its effect on the Earth's surface, water in the atmosphere and its precipitation, the global circulation of the atmosphere, the formation of air fronts, cyclones, hurricanes and typhoons, and the evolution of weather systems. In their analysis of these, meteorologists rely on daily records of temperature, atmospheric pressure, humidity and precipitation, wind speed and direction, concentration of carbon dioxide, and various other quantitative measurements of weather conditions. These are collected by weather stations, ships, aircraft, buoys, radiosondes, radar, and most recently by satellites. The information is used to produce synoptic charts and maps which describe the weather system over a particular region such as Western Europe, the continental United States, Australia, Japan, or some sections of them. The numerical data provide a basis for mathematical models used to describe the Earth's weather, and these in turn lead to the predictive methods of weather forecasting. For a general introduction to the subject, the reader is referred to Barry and Chorley [1], Cole [9], Dutton [17], Linacre and Hobbs [33], Neiburger et al. [38], and Riehl [43].

Statistics enters into many areas of meteorological studies; three representative fields of statistical application are (1) the reduction of meteorological data, (2) the construction of meteorological models and their use in weather prediction, and (3) experiments on artificial weather modification*. We consider each of these in turn.

THE REDUCTION OF METEOROLOGICAL DATA

The earliest observations of temperature and pressure relied on the thermometer (invented by Galileo at the end of the sixteenth century) and the barometer (invented by Toricelli in the middle of the seventeenth century). By the eighteenth century, these instruments had become sufficiently standardized to allow data recorded in different parts of the world to be compared. Weather maps began to appear in the nineteenth century, the observations for them being collected by mail. It was only when the telegraph became current in the twentieth century that weather maps could be prepared daily. Since the 1960s, satellite pictures and observations have provided more detailed and accurate information on a variety of meteorological phenomena including cloud formations, vertical temperature profiles of the atmosphere, and even the salinity of coastal waters. A useful monthly compilation of meteorological data is to be found in the United Nations publication *Climatic Data for the World*.

Meteorological rainfall records usually consist of daily precipitation measured in millimeters (or 0.01 inch); at some stations the rate at which rain has fallen is also recorded. Table 1 is representative of daily rainfall observations (in mm) for the month of June 1970 in a rainy temperate coastal city. Observed data of this kind may be aggregated to give weekly, monthly, and annual rainfall totals. If $X_{1j}^{(6)}, \ldots, X_{30,j}^{(6)}$ are daily rainfalls for the 30 days of June in the year j, one can obtain the total rainfall for June as $Z_j^{(6)} = \sum_{i=1}^{30} X_{ij}^{(6)}$, and the mean daily rainfall as $Z_j^{(6)}/30$. For the June 1970 data above, this is 9.7 mm; sometimes median rather than mean rainfall data are quoted. The mean monthly rainfall for June over the 10 years $j = 1970, \ldots, 1979$ would be given by $\sum_{j=1970}^{1979} Z_j^{(6)}/10$; standard deviations are also often computed to provide some measure of variability. In meteorological literature, monthly rainfall graphs are plotted, indicating the amount of rainfall for each month computed over a period of, say, 10–100 years, together with percentiles as in Fig. 1. Annual average rainfall is often used to draw isohyets on maps, these being lines of equal precipitation. In some cases, the frequency of days with rain is computed; for June 1970 data above, for example, there were 20 days of rain.

Table 1 Daily Precipitation $X_{i,1970}^{(6)}$ for June 1970 (mm)

Day i	$X_{i,1970}^{(6)}$	Day i	$X_{i,1970}^{(6)}$	Day i	$X_{i,1970}^{(6)}$
1	0	11	4	21	36
2	0	12	10	22	20
3	30	13	28	23	9
4	18	14	5	24	7
5	0	15	0	25	3
6	5	16	0	26	0
7	24	17	0	27	0
8	35	18	0	28	0
9	12	19	10	29	5
10	6	20	17	30	6

Other standard measurements recorded at weather stations are minimum and maximum daily temperatures, humidity, evaporation, pressure, cloud type and cover, and wind speed and direction, both at the Earth's surface and in the upper atmosphere. Daily temperatures are recorded (in degrees Celsius or Fahrenheit) and monthly or yearly averages for temperature, similar to those for rainfall, can be computed for them at different locations. Extremes of temperature observed in the air at the Earth's surface are −88.3°C or −126.9°F in Antarctica, and 54°C or 129.2°F in Death Valley, California. To discuss trends in rainfall, temperature, or other meteorological observations, the data may be smoothed, for example, when 10-year moving averages* for rainfall are graphed.

Sometimes a parameter combining several of the elements used to specify weather conditions is measured. Bean and Dutton [2] used the radio refractivity N as a climatic index; this is a function

$$N = K_1 \frac{P_d}{T} + K_2 \frac{e}{T} + K_3 \frac{e}{T^2} + K_4 \frac{P_c}{T}$$

of the atmospheric pressure P_d of dry air, the absolute temperature T, the partial pressure e of water vapor, and the partial pressure P_c of CO_2, where $K_1, \ldots, K_4$ are constants. They calculated the mean $\overline{N}$ of k refractivity observations (adjusted for height) obtained from 18 radiosonde stations throughout the world for the five-year period 1952–1957 and derived its standard deviation $s(\overline{N}) =$

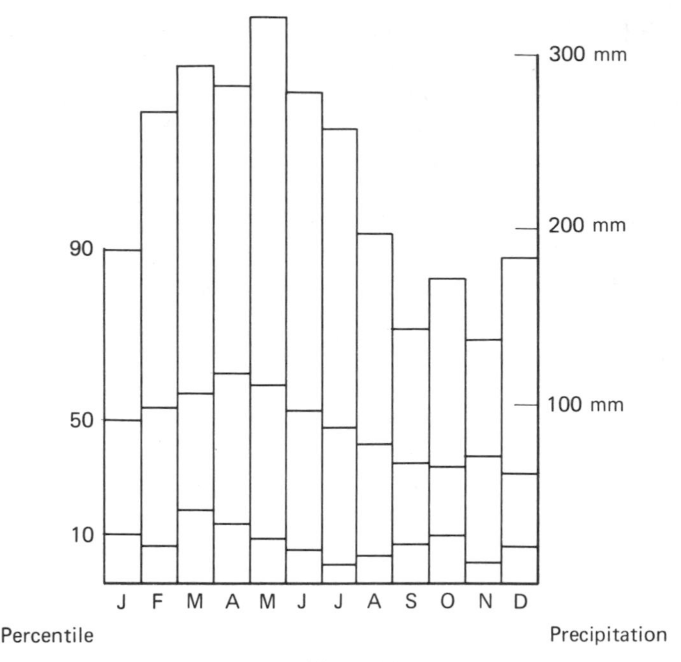

Figure 1 Monthly rainfall graph.

$s(N)/\sqrt{k}$. Charts of the diurnal, seasonal, and vertical variation of N for each of the 18 stations, together with estimates of error, have also been produced.

Several statistical studies of meteorological data have been carried out by meteorologists. For example, Lamb [31] used elementary statistical methods to study weather patterns; these involve graphical diagrams of rainfall decade averages as percentages of the 1900–1939 average, or numbers of days in the year with westerly type winds in the British Isles. Winstanley [51] attempted to draw conclusions about future rainfall trends up to the year 2030 in the Mediterranean and Middle East, the Sahel, and N.W. India on the basis of an analysis of past rainfall records. His general conclusions have been questioned, however, by Tanaka et al. [48] and Bunting et al. [7]. Mitchell [35] has been concerned with fluctuations in global mean temperature of the Earth and has argued that there has been a distinct warming of the Earth from about 1880, followed by a cooling since about 1940. Wahl and Bryson [50] have studied changes in surface temperature of the Atlantic. Despite attempts by Julian [29], Thom and Thom [49], and others, to popularize the use of statistical methods among meteorologists, such methods have penetrated the meteorological literature slowly; for example, errors in observed or sampled data are not always taken into account.

Statisticians have also analyzed rainfall and temperature data and developed stochastic models for them. The modeling of rainfall occurrence has been considered among many others by Gabriel and Neumann [21], who used a Markov process* of order 1 to describe the occurrence of dry and wet days in Tel-Aviv, and Green [25], who assumed that the sequence of these dry and wet spells formed an alternating renewal sequence. Trends, periodicities, and secular variations of rainfall in India have been studied by Jagannathan and Parthasarathy [27], and Parthasarathy and Dhar [40] while Smith and Schreiber [47] have used point process* methods on thunderstorm rainfall. Buishand [6] has made an extensive analysis of rainfall in the Netherlands, fitted distributions to monthly and annual rainfall totals, and discussed the homogeneity of rainfall data. He has also considered topics such as the serial correlation* of monthly totals and the relation between rain intensity and the length of wet spells.

In Australia, Cornish [10, 12] studied the secular variation of rainfall using precipitation records for over 100 years at Adelaide. His regression analysis indicated a regular oscillation of period 23 years and amplitude 30 days respectively in the duration and incidence of winter rains; there also appeared to be a superimposed long-term trend which meant that spring rains occurred about three weeks later in 1950 than a century earlier. Further studies were carried out by Cornish and Stenhouse [14] on interstation correlation of rainfall, and by Cornish and Evans [13] on daily temperatures at Adelaide; some of the variability in the temperature was also attributed to secular changes. Gani [22] has indicated that the claim for increased variability of rainfall, at least in Eastern and Southern New South Wales and Northern Victoria over the years 1915 to 1974, cannot be substantiated. There is, in addition, some statistical literature on the relation of weather to agricultural yields, such as Cornish's study [11] of the influence of rainfall on wheat yield in South Australia.

These examples are only a small selection from the statistical literature on meteorological data. Such studies help to summarize extensive records accurately and to indicate relationships and trends, but they fall short of providing comprehensive scientific explanations of the phenomena considered.

METEOROLOGICAL MODELS AND WEATHER PREDICTION

Meteorological phenomena have a physical basis, and their main outlines can be modeled mathematically. The most comprehensive model dealing with motion in the Earth's entire atmosphere, and the associated distributions of pressure, temperature, and humidity, is known as the *general circulation model*. It consists of sets of partial differential equations (with time as the independent variable) relating the atmospheric pressure, density, water concentration and temperature with various flux densities such as the vertical eddy flux density of momentum, the vertical radiative heat flux density, and the vertical eddy flux density of water substance, subject to appropriate boundary conditions. Because these equations are nonlinear, they are difficult to solve analytically, though they are more tractable in their simpler linearized versions.

With the development of high-speed electronic computers, it has become possible to obtain numerical solutions to the nonlinear partial differential equations. The values of the relevant variables are taken on a lattice of points in space, and the differential equations are approximated by difference equations which are solved numerically. A numerical model of atmospheric regions or of the entire atmosphere can thus be obtained. Limited local versions of this model are at the basis of regional weather forecasting.

To achieve rapid numerical weather forecasts, the model must be restricted in size and greatly simplified; the simpler the model, the shorter its predictive horizon. The simplest possible regional model is the barotropic model, which uses only one lattice point at the 500-mb level along each vertical. In this, the set of governing equations for each particle of the system can be very rapidly approximated by

$$f + \zeta = \text{constant in time},$$

where f is the Coriolis parameter (twice the angular velocity of the earth around the local vertical) and

$$\zeta = \mathbf{k} \cdot \operatorname{curl} \mathbf{v}_\zeta,$$

where $\mathbf{v}_\zeta$ is that component of the horizontal velocity field which carries all the vorticity but none of the divergence. The barotropic model provides useful predictions for up to three days ahead.

By increasing the number of lattice points on the vertical, one can obtain more accurate models with longer predictive horizons of one to two weeks. Computers not only solve the equations of the model; they also store and process observations from weather stations, radiosondes, and satellites, plot weather maps based on their predictions, and convert their results into information for the layperson (such as the likelihood of rain). Murphy and Winkler [37] have discussed the reliability of weather predictions of this kind by forecasters, while Dawid [15] has given a Bayesian interpretation of the effectiveness of their calibration*. Further improvements in the storage capacity and speed of computers should lead to improved weather forecasting. For details, the reader may refer to Hasselman [26], Frankignoul and Hasselman [20], Lemke [32], Gauntlett [23], Dutton [18] and Bengtsson et al. [3].

Statistical aspects of numerical weather prediction have been studied by Jones [28], Klein [30], and Ghil et al. [24], among others. Ghil et al. [24] pointed out that the initial data fed into the system of partial differential equations are incomplete and inaccurate; they suggested the application of estimation theory to previous observations on the system as a corrective. Their work applies a discrete Kalman*–Bucy filter to the finite difference* equations approximating the partial differential equations, and they use a sequential estimation* procedure to improve predictions based on the input data.

Simpler statistical methods have been used to predict short-term trends; for example, local short-term rainfall predictions have relied on the use of discriminant analysis*. Statistical correction of predictions, where the prognosis error fields are correlated with the prognoses, has proved successful in contexts where the input data for the initial conditions are inaccurate because of the coarseness of the lattice. Feller and Schemm [19], among others, used statistical methods to correct numerical prediction equations.

Statistical procedures are also used to check meteorological observations for spatial and temporal consistency in what is referred to as objective analysis. Here the values of atmospheric variables on a regular lattice are derived at a given time, when the actual observations are irregularly distributed in both space and time. The problem, basically one of multiple linear regression*, has been studied by Petersen [41], Schlatter [44], Schlatter et al. [45], and Bergman [4]. Finally, statistical methods are important in the verification of the fit of mathematical models to the real weather patterns; see Klein [30] and Miller [34].

ARTIFICIAL WEATHER MODIFICATION

Several imaginative suggestions have been made to modify the Earth's weather on a large scale by schemes such as the introduction of ice crystals into the atmosphere to decrease radiation on the Earth's surface or the pumping of warmer Atlantic water into the Arctic to eliminate ice. These have not been followed up; among the largest systematic weather modification experiments so far carried out are those on hurricanes.

On the smaller scale, trees have long been used as windbreaks to protect crops from wind, while orchard heaters have helped to reduce damage from frost. The aspect of weather modification that has possibly been of greatest statistical interest has been the stimulation of rainfall. For more detailed accounts, the reader should consult Byers [8], Breuer [5], and WEATHER MODIFICATION, STATISTICS IN.

In 1946 Vincent Schaefer found that dry ice pellets dropped in a supercooled cloud could initiate precipitation, while Bernard Vonnegut noted that silver iodide particles would do the same. There have since been many experiments on the effects of cloud seeding. An airplane above or in a cloud may be used to introduce dry ice pellets or silver iodide particles into the cloud; alternatively, generators on the ground can release silver iodide smoke into the cloud (see Dennis [16]).

To evaluate the effectiveness of such cloud seeding experiments is a statistically complex problem, since the variability of the rainfall is large compared with the potential effects of seeding. Mostly, two experimental designs have been used; the first involves a comparison of seeded target areas with control areas which have closely correlated amounts of rainfall (the target-control method), and the second uses the randomization* of seeding occasions on the same area over a group of periods considered to be equally suitable for the success of seeding (the randomization method). A third method is the crossover design (*see* CHANGEOVER DESIGNS), in which one of two targets is chosen for seeding at random on equally suitable occasions.

In the United States a randomized experiment, project Whitetop, carried out in Missouri in 1960–1964, in which silver iodide was seeded from an airplane, resulted in smaller amounts of precipitation on seeded than unseeded days. In Israel, however, similar experiments in 1961–1966 indicated an increase of 18% in precipitation when winter storm clouds were seeded. In Tasmania, target-control experiments on cloud seeding carried out by CSIRO in the 1960s resulted in some apparent rainfall increase but were not entirely conclusive. Clearly, further research into cloud physics is necessary to determine under what conditions it is necessary to seed clouds in order to increase precipitation.

The Precipitation Enhancement Project (PEP) currently being carried out under the auspices of the World Meteorological Organization and the Precipitation Augmentation for Crops Experiment (PACE) started in the midwestern United States by the National Oceanic and Atmospheric Administration should help to decide the value of cloud-seeding procedures. Both projects include a heavy component of statistical analysis (see [52]). The PEP is currently considering the seeding of a site in the Duero basin in Spain, roughly between February and June, as Shaw [46] has outlined. Statisticians, closely involved with this project from its inception, are assisting in the analysis of past data and the preparation of the experimental design, which they have recommended to be of the fixed target-control type.

Similar experiments have been carried out to test methods of reducing damage from hail; these have included the explosion of rockets and artillery shells in clouds, as well as more standard seedings with dry ice and silver iodide. Large amounts of silver iodide have also been released from airplanes into the eye wall of hurricanes in an effort to reduce the speed of their winds. Although some reduction of these speeds was achieved, it lay within the normal range of variability in hurricanes. The value of such seeding in both cases is not fully decided, and further experiments are planned.

Statistics enters into many aspects of meteorology, as the 1969 Symposium on the Applications of Statistical Methods in Meteorology held in Leningrad [53], and the 1978 volume of Pittock et al. [42] indicate. While some meteorologists such as Mitchell [36] and Panofsky and Brier [39] recommend the use of statistical methods in their field, there remains a great need for statistical analysis and the application as well as development of statistical techniques in meteorology. This brief account is designed not only to inform the reader but also to point out to statisticians the opportunities available to them in meteorological research.

References

[1] Barry, R. G. and Chorley, R. J. (1976). *Atmosphere, Weather and Climate*, 3rd ed. Methuen, London.

[2] Bean, B. R. and Dutton, E. J. (1966). *Radio Meteorology*. National Bureau of Standards Monograph 92, Washington, DC.

[3] Bengtsson, L., Ghil, M., and Källén, E., eds. (1981). *Dynamic Meteorology: Data Assimilation Methods*. Springer-Verlag, New York.

[4] Bergman, K. H. (1979). *Mon. Wea. Rev.*, **107**, 1423–1444.

[5] Breuer, G. (1979). *Weather Modification—Prospects and Problems*. Cambridge University Press, Cambridge, England.

[6] Buishand, T. A. (1977). *Stochastic Modelling of Daily Rainfall Sequences*. Veenman and Zonen, Wageningen.

[7] Bunting, A. H., Dennett, M. D., Elston, J., and Milford, J. R. (1975). *Nature*, **253**, 622–623.

[8] Byers, H. R. (1974). In *Weather and Climate Modification*, W. N. Hess, ed. Wiley, New York, pp. 3–44.

[9] Cole, F. (1980). *Introduction to Meteorology*, 3rd ed. Wiley, New York.

[10] Cornish, E. A. (1936). *Q. J. R. Met. Soc.*, **62**, 481–498.

[11] Cornish, E. A. (1950). *Aust. J. Sci. Res., Series B, Biol. Sci.*, **3**, 178–218.

[12] Cornish, E. A. (1954). *Aust. J. Phys.*, **7**, 334–346.

[13] Cornish, E. A. and Evans, M. J. (1964). *An Analysis of Daily Temperatures at Adelaide, South Australia.* CSIRO Div. Math. Statist. Tech. Paper No. 17.

[14] Cornish, E. A. and Stenhouse, N. S. (1958). *Inter-Station Correlations of Monthly Rainfall in South Australia*. CSIRO Div. Math. Statist. Tech. Paper No. 5.

[15] Dawid, A. P. (1982). *J. Amer. Statist. Ass.*, **77**, 605–613.

[16] Dennis, A. (1980). *Weather Modification by Cloud Seeding*. Academic Press, New York.

[17] Dutton, J. A. (1976). *Ceaseless Wind: An Introduction to the Theory of Atmospheric Motion*. McGraw-Hill, New York.

[18] Dutton, J. A. (1982). *SIAM Review*, **24**, 1–33.

[19] Feller, A. J. and Schemm, C. E. (1977). *Mon. Wea. Rev.*, **105**, 37–56.

[20] Frankignoul, C. and Hasselman, K. (1977). *Tellus*, **29**, 289–305.

[21] Gabriel, K. R., and Neumann, J. (1962). *Q. J. R. Met. Soc.*, **88**, 90–95.

[22] Gani, J. (1975). *Search*, **6**, 504–508.

[23] Gauntlett, D. J. (1975). *The Application of Numerical Models to the Problems of Meteorological Analysis and Prognosis over the Southern Hemisphere*. Meteorological Study No. 28, Dept. of Science, Bureau of Meteorology. Australian Government Publishing Service, Canberra.

[24] Ghil, M., Cohn, S., Tavantzis, J., Bube, K., and Isaacson, E. (1981). In *Dynamic Meteorology: Data Assimilation Methods*, L. Bengtsson, M. Ghil, and E. Källén, eds. Springer-Verlag, New York.

[25] Green, J. K. (1964). *J. R. Statist. Soc. B*, **26**, 345–353.

[26] Hasselman, K. (1976). *Tellus*, **28**, 473–485.

[27] Jagannathan, P. and Parthasarathy, B. (1973). *Mon. Wea. Rev.*, **101**, 371–375.

[28] Jones, R. H. (1965). *J. Atmos. Sci.*, **22**, 658–663.

[29] Julian, P. R. (1970). *Mon. Wea. Rev.*, **98**, 142–153.

[30] Klein, W. H. (1982). *Bull. Amer. Met. Soc.*, **63**(2), 170–177.

[31] Lamb, H. H. (1974). *Ecologist*, **4**, 10–15.

[32] Lemke, P. (1977). *Tellus*, **29**, 385–392.

[33] Linacre, E. and Hobbs, J. (1977). *The Australian Climatic Environment*. Wiley, Brisbane, Queensland, Australia.

[34] Miller, A. (1982). *CSIRO Div. Math. Statist. Newsletter*, No. 84, 1–6.

[35] Mitchell, J. M. (1963). In *Changes of Climate*, Arid Zone Res. 20. UNESCO, Paris, pp. 161–181.

[36] Mitchell, J. M., ed. (1966). *Climatic Change*. Tech. Note No. 9. World Meteorological Organization, Geneva.

[37] Murphy, A. H. and Winkler, R. L. (1977). *J. R. Statist. Soc. C*, **26**, 41–47.

[38] Neiburger, M., Edinger, J. G., and Bonner, W. D. (1982). *Understanding Our Atmospheric Environment*, 2nd ed. Freeman, San Francisco, Calif.

[39] Panofsky, H. A., and Brier, G. W. (1963). *Some Applications of Statistics to Meteorology*. Pennsylvania State University, University Park.

[40] Parthasarathy, B. and Dhar, O. N. (1974). *Q. J. R. Met. Soc.*, **100**, 245–257.

[41] Petersen, D. P. (1973). *J. Appl. Met.*, **12**, 1093–1101.

[42] Pittock, A. B., Frakes, L. A., Jenssen, D., Petersen, J. A., and Zillman, J. (1978). *Climate Change and Variability: A Southern Perspective.* Cambridge University Press, Cambridge, England.

[43] Riehl, H. (1978). *Introduction to the Atmosphere*, 3rd ed. McGraw-Hill, New York.

[44] Schlatter, T. W. (1975). *Mon. Wea. Rev.*, **103**, 246–257.

[45] Schlatter, T. W., Branstator, G. W., and Thiel, L. G. (1976). *Mon. Wea. Rev.*, **104**, 765–783.

[46] Shaw, D. E. (1978). In *Weather Modification Programme, PEP Design Document*. Report No. 9. World Meteorological Organization, Geneva, pp. 65–83.

[47] Smith, R. E. and Schreiber, H. A. (1973, 1974). *Water Resources Res.*, **9**, 871–884; **10**, 418–423.

[48] Tanaka, M., Weare, B. C., Navato, A. R., and Newell, R. E. (1975). *Nature*, **255**, 201–203.

[49] Thom, H. C. S. and Thom, M. D. (1972). *Mon. Wea. Rev.*, **100**, 503–508.

[50] Wahl, E. N. and Bryson, R. A. (1975). *Nature*, **254**, 45–46.

[51] Winstanley, D. (1973). *Nature*, **245**, 190–194.

[52] World Meteorological Organization (1980). *Climate Research Programme and the Global Atmo-*

spheric Research Programme. World Meteorological Organization, Geneva.

[53] Yudin, M. I., et al. (1971). *Applications of Statistical Methods in Meteorology* (in Russian). Gidrometeorologicheskoye Izdatelsvo, Leningrad.

(GEOSTATISTICS
HYDROLOGY, STOCHASTIC
WEATHER MODIFICATION,
STATISTICS IN)

J. GANI

METHOD OF AVERAGES

This method is an alternative to the method of least squares* for determining the coefficients in a linear or nonlinear regression* equation of a dependent variable on one or more independent variables. The simplest case is that of estimating the linear relation $y = \alpha + \beta x$, given a set of points (x_i, y_i), $i = 1, 2, \ldots, n$, where the values x_i are exact and the values y_i are subject to error, the error in the ith measurement being denoted by ϵ_i. As in the method of least squares, we have the observational equations

$$y_i = \alpha + \beta x_i + \epsilon_i = a + bx_i + e_i, \quad i = 1, \ldots, n, \quad (1)$$

where a, b, and e_i are estimates of α, β, and ϵ_i, respectively. In the method of averages the observational equations, (1) in the simplest case or analogous equations in other cases, are divided into as many subsets as there are coefficients to be determined, the division being made according to the values of (one of) the independent variable(s), those having the largest values of this variable being grouped together, then the next largest in another group, etc. Then the equations in each group are added together. The resulting equations, whose number is equal to the number of coefficients to be determined, are then solved simultaneously. The fitted equation is satisfied exactly by the coordinates of the centroid of each subset.

This method, one of the earliest to be proposed, was developed independently by Euler [1] and Mayer [5]. Lambert* [3] also used the method of averages. A drawback is that when there are two or more independent variables, the results depend on which independent variable is used in dividing the observational equations into subsets and are therefore arbitrary and subjective. Nevertheless, because of its simplicity, the method of averages has been suggested by authors of some twentieth-century textbooks, including Whittaker and Robinson [8] and Scarborough [6], as a viable alternative to the method of least squares. Wald [7] proposed its use in fitting a straight line when both variables are subject to error, and Linnik [4] devoted a chapter in his book to this procedure.

The method of averages is applicable (1) when the error distribution is unknown and the experimenter is unwilling to make the distributional assumptions underlying one of the other methods or (2) as a quick approximation even when another method is applicable. Further information on the applicability of this and other alternatives to the method of least squares was given by Harter [2].

Example. Let it be required to find the straight line which best fits the 24 points:

$$(1, 2.3394),\ (2, -0.4190),\ (3, 3.9360),$$
$$(4, 5.7104),\ (5, 4.9803),\ (6, 5.0956),$$
$$(7, 5.6350),\ (8, 5.7270),\ (9, 6.6854),$$
$$(10, 6.5562),\ (11, 8.2599),\ (12, 7.7398),$$
$$(13, 8.4688),\ (14, 9.1321),\ (15, 9.7863),$$
$$(16, 9.8558),\ (17, 10.3997),\ (18, 11.2892),$$
$$(19, 11.4872),\ (20, 11.8735),\ (21, 11.0908),$$
$$(22, 13.5829),\ (23, 13.6063),\ (24, 13.7849).$$

These points (x_i, y_i) yield 24 observational equations of the form (1). Summing the first 12 of these equations yields $62.2460 = 12a + 78b$ and summing the other 12 yields $134.3575 = 12a + 222b$, the terms involving e_i being omitted since it is assumed that within each group of 12 equations, $\Sigma e_i = 0$. Simultaneous solution of these two equations **yields** $a = 1.932$, $b = 0.5008$, so that the best fitting straight line by the method of averages is $y = 1.932 + 0.5008x$. The method of least squares yields the line $y = 1.860 + 0.5065x$, while the method of least absolute values* yields the line $y = 2.21 + 0.486x$, which also results from using Harter's adaptive robust method*.

References

[1] Euler, L. (1749). *Pièce qui a Remporté le Prix de l'Académie Royale des Sciences en 1748, sur les Inégalités du Mouvement de Saturn et de Jupiter*. Martin, Coignard et Guerin, Paris. Reprint, *Leonhardi Euleri Opera Omnia* **II 25** (*Commentationes Astronomicae* **I**). Orell Fussli, Turici, 1960, pp. 45–157.

[2] Harter, H. L. (1974–1976). *Int. Statist. Rev.*, **42** (1974), 147–174, 235–264, 282; **43** (1975), 1–44, 125–190, 269–278; **44** (1976), 113–159. (A comprehensive survey of the literature on the method of least squares and various alternatives, including the method of averages.)

[3] Lambert, J. H. (1765). *Beyträge zum Gebrauche der Mathematik und deren Anwendung*, **1**, 428–488.

[4] Linnik, Yu. V. (1958). *Method of Least Squares and Principles of the Theory of Observations*. Fizmatgiz, Moscow. (in Russian). English translation by Regina C. Elandt (edited by N. L. Johnson), Pergamon Press, New York, 1961.

[4] Mayer, J. T. (1750). *Kosmographische Nachrichten und Sammlungen*, **1**, 52–183.

[5] Scarborough, J. B. (1930). *Numerical Mathematical Analysis*. Johns Hopkins Press, Baltimore. 4th ed., 1958.

[6] Wald, A. (1940). *Ann. Math. Statist.*, **11**, 284–300.

[7] Whittaker, E. T. and Robinson, G. (1924). *The Calculus of Observations*. Blackie & Son, London-Glasgow. 4th ed., 1944.

(HARTER'S ADAPTIVE ROBUST METHOD
LEAST SQUARES
LINEAR REGRESSION
METHOD OF GROUP AVERAGES
METHOD OF LEAST ABSOLUTE VALUES
METHOD OF LEAST *p*TH POWERS
MINIMAX METHOD
NONLINEAR REGRESSION)

H. LEON HARTER

METHOD OF ELEMENTS *See* COMMONALITY ANALYSIS; NEWTON-SPURRELL METHOD

METHOD OF GROUP AVERAGES

This is a modification of the method of averages* which allows the possibility of omitting the equation(s) obtained by summing the observational equations in one or more groups (subsets) (*see* (1) *in* METHOD OF AVERAGES). For example, a straight line may be fitted by dividing the observed points into three groups and connecting the centroids of the first and third groups by a straight line, ignoring the second group. This method was proposed by Eddington [2] and later by Nair and Shrivastava [6], who showed that it gives better results than the method of averages. Nair and Banerjee [5] found that this superiority extends to the case in which both variables are subject to error (see Wald [7]). Eddington considered only the case of fitting a straight line when the first and third groups have the same number of observations and showed that the best estimates are obtained when all three groups have the same number of observations. Nair and Shrivastava considered the fitting of both straight lines and parabolas, with various numbers of observations in the groups. In the case of fitting a parabola, they found that the best estimates are obtained by dividing the observations into five equal groups and then rejecting the second and fourth groups.

In the linear case, Bartlett [1] proposed a further modification in which the observational points are divided into three groups containing, as nearly as possible, equal numbers of points, and taking as the result the line through the centroid of the middle group with slope equal to that of the line joining the centroids of the two extreme groups. He showed that his method has, in general, greater efficiency than that of Wald (method of averages), and stated that it is theoretically preferable to the method of Nair and Shrivastava. Madansky [4] studied the efficiency of various forms of the method of group averages, including those of Nair and Shrivastava and of Bartlett, for various proportions of the observations included in the several groups. Further results on this method and its applicability were given by Harter [3].

References

[1] Bartlett, M. S. (1949). *Biometrics*, **5**, 207–212.

[2] Eddington, A. S. (1933). *Proc. Phys. Soc. (London)*, **45**, 271–282; discussion, 282–287.

[3] Harter, H. L. (1974–1976). *Int. Statist. Rev.*, **42** (1974), 147–174, 235–264, 282; **43** (1975), 1–44, 125–190, 269–278; **44** (1975), 113–159. (A comprehensive survey of the literature on the method of least squares and various alternatives, including the method of group averages.)

[4] Madansky, A. (1959). *J. Amer. Statist. Ass.*, **54**, 173–205.

[5] Nair, K. R. and Banerjee, K. S. (1943). *Sankhyā*, **6**, 331.

[6] Nair, K. R. and Shrivastava, M. P. (1942). *Sankhyā*, **6**, 121–132.

[7] Wald, A. (1940). *Ann. Math. Statist.*, **11**, 284–300.

(LINEAR REGRESSION
METHOD OF AVERAGES
NONLINEAR REGRESSION)

H. LEON HARTER

METHOD OF LEAST ABSOLUTE VALUES

In this method (also known as the *method of least absolute deviations* or L_1-*estimation*), the values of the unknown coefficients in a regression equation are determined so that the sum of the absolute values of the deviations from regression is a minimum. This alternative to the method of least squares* has the advantage of being less sensitive to outlying observations which may result from gross errors (*see* OUTLIERS). The least absolute values estimate of central value is the median, since it is the value from which the sum of the absolute deviations is less than from any other value, and the corresponding estimate of dispersion is a constant multiple of the average deviation from the median. These estimates of central value and dispersion and the least absolute values estimates of regression coefficients are maximum likelihood* estimates when the error distribution is double exponential (*see* LAPLACE DISTRIBUTION).

Boscovich* [7] proposed two criteria for determining the best fitting straight line $y = a + bx$ to three or more points: (1) the sums of the positive and negative residuals (in the y-direction) shall be numerically equal; (2) the sum of the absolute values of the residuals shall be a minimum. His first criterion requires that the line pass through the centroid $(\bar{x}, \bar{y})$ of the observations, and his second criterion is then applied subject to the restriction imposed by the first. The result is a hybrid method which uses a least squares estimate (the mean) of central value and a least absolute values estimate of the slope. Boscovich [8] gave a geometric method of solving the equations resulting from the criteria stated in his earlier paper. Laplace* [26] developed an analytic procedure based on Boscovich's criteria. He called the method based on these criteria the "method of situation". Laplace [27] summarized his earlier results. Later authors dropped Boscovich's first criterion and applied the second without restriction.

Fourier [17] considered the problem of fitting a linear equation in n variables to a set of m observed points $(m > n)$ so as to minimize the average absolute deviation. He formulated this problem as what would now be called a linear programming* problem, as he had done for the minimax* problem, and stated that it can be solved analogously (method of descent).

Edgeworth [12, 13] gave a procedure for unrestricted minimization of the sum of the absolute values of the residuals. Turner [33] noted the advantages Edgeworth claimed for his method: (1) it is considerably less laborious than the method of least squares; (2) in the presence of discordant observations, it is theoretically better. Turner stated that (1) is very doubtful and (2) is somewhat counterbalanced by the failure to give a *unique* solution. Edgeworth [14] restated the method given in his earlier papers. He disposed of Turner's second criticism by proposing adoption of the *middle* of the indeterminate tract as the best point (just as the median of an even number of values is defined as the value midway between the two middle ones). He then endeavored (somewhat less successfully) to answer Turner's first criticism. Edgeworth [15] restated the rationale of the method he had proposed much earlier and amplified the directions for its application given by Turner.

Rhodes [28] gave a simpler method than that of Edgeworth for fitting a curve (a parabola in his example) by the method of minimum deviations (least absolute values). Singleton [29] pointed out that the method of Rhodes, which he presented without proof, is iterative and recursive. Singleton used geometric methods and terminology to develop proofs for various procedures and to reduce the labor by eliminating the recursive feature. Harris [20] gave a simple explanation, with a numerical example, of a procedure, essentially that of Edgeworth [15] and Rhodes. He pointed out the relation between this problem and linear programming. The use of linear programming in fitting by the method of least absolute values has been discussed by various authors, including Charnes et al. [9], Bejar [5, 6], Karst [23], Wagner [35], Fisher [16], Davies [11], Usow [34], Crocker [10], Zorilescu [36], Kiountouzis [24, 25], Appa and Smith [1], Barrodale and Roberts [4], Spyropoulos et al. [32], and Sposito and Smith [31]. Davies, Usow, and Barrodale and Roberts gave practical computational algorithms. Armstrong et al. [3] gave a revised simplex algorithm.

Glahe and Hunt [19] compared the results of least absolute values estimation with least squares estimation of regression in four major sampling experiments. Their general conclusion was that the L_1 norm estimators should prove equal to or superior to the L_2 norm estimators for models using a structure similar to the overidentified one specified for their study, with randomly distributed error terms and very small sample sizes. Spyropoulos et al. [32] gave a sufficient (though not necessary) condition for uniqueness of the least absolute values regression. Appa and Smith [1] established necessary conditions for fitting a linear model to a set of points by use of the L_1-criterion (least absolute values). Sposito and Smith [31] derived a sufficient condition and an additional necessary condition to determine an optimal plane using this criterion. Harter [22] gave examples of situations where the regression line is unique and where it has 2, 3, or 4 limiting positions, and proposed compromise solutions when the regression is not unique. Gentle et al. [18] and Sposito et al. [30] enumerated various useful properties of least absolute values (L_1) estimators. Armstrong and Frome [2] discussed least-absolute-value estimators for one-way* and two-way tables.

The method of least absolute values is optimal (maximum likelihood*) when the error distribution is double exponential. It is nearly optimal for other leptokurtic (long-tailed) distributions. Further information about the applicability of this and other alternatives to the method of least squares was given by Harter [21].

References

[1] Appa, G. and Smith, C. (1973). *Math. Program.*, **5**, 73–87.

[2] Armstrong, R. D. and Frome, E. L. (1979). *Nav. Res. Logist. Q.*, **26**, 79–96.

[3] Armstrong, R. D., Frome, E. L., and Kung, D. S. (1979). *Commun. Statist. B*, **8**, 175–190.

[4] Barrodale, I. and Roberts, F. D. K. (1973). *SIAM J. Numer. Anal.*, **10**, 839–848.

[5] Bejar, J. (1956). *Trabajos Estadíst.*, **7**, 141–158.

[6] Bejar, J. (1957). *Trabajos Estadíst.*, **8**, 157–173.

[7] Boscovich, R. J. (1757). *Bononiensi Scientiarum et Artum Instituto atque Academia Commentarii*, **4**, 353–396.

[8] Boscovich, R. J. (1760). *Philosophiae Recentioris, a Benedicto Stay . . .*, **2**, 406–426. Romae.

[9] Charnes, A., Cooper, W. W., and Ferguson, R. O. (1955). *Manag. Sci.*, **1**, 138–151.

[10] Crocker, D. C. (1969). *AIIE Trans.*, **1**, 112–126.

[11] Davies, M. (1967). *J. R. Statist. Soc. B*, **29**, 101–109.

[12] Edgeworth, F. Y. (1887). *Hermathena*, **6**, 279–285.

[13] Edgeworth, F. Y. (1887). *Philos. Mag., 5th Ser.*, **24**, 222–223.

[14] Edgeworth, F. Y. (1888). *Philos. Mag., 5th Ser.*, **25**, 184–191.

[15] Edgeworth, F. Y. (1923). *Philos. Mag., 6th Ser.*, **46**, 1074–1088.

[16] Fisher, W. D. (1961). *J. Amer. Statist. Ass.* **56**, 359–362.

[17] Fourier, J. B. J. (1823, 1824). *Hist. Acad. Sci. Paris*, 1823, 29 ff.; 1824, 47–55.

[18] Gentle, J. E., Kennedy, W. J., and Sposito, V. A. (1977). *Commun. Statist. A*, **6**, 839–846.

[19] Glahe, F. R. and Hunt, J. G. (1970). *Econometrica*, **38**, 742–753.

[20] Harris, T. E. (1950). *Amer. Statist.*, **4**(1), 14–15.

[21] Harter, H. L. (1974–1976). *Int. Statist. Rev.*, **42** (1974), 147–174, 235–264, 282; **43** (1975), 1–44, 125–190, 269–278; **44** (1976), 113–159. (A comprehensive survey of the literature on the method of least squares and various alternatives, including the method of least absolute values).

[22] Harter, H. L. (1977). *Commun. Statist. A*, **6**, 829–838.

[23] Karst, O. J. (1958). *J. Amer. Statist. Ass.*, **53**, 118–132.

[24] Kiountouzis, E. A. (1971). *Bull. Soc. Math. Grèce* (N.S.), **12**, 191–206.

[25] Kiountouzis, E. A. (1973). *Appl. Statist.*, **22**, 69–73.

[26] Laplace, P. S. (1793). *Mém. Acad. R. Sci. Paris Année 1789*, 1–87.

[27] Laplace, P. S. (1799). *Traité de Mécanique Céleste*, Vol. 2 J.B.M. Duprat, Paris.

[28] Rhodes, E. C. (1930). *Philos. Mag., 7th Ser.*, **9**, 974–992.

[29] Singleton, R. R. (1940). *Ann. Math. Statist.*, **11**, 301–310.

[30] Sposito, V. A., Kennedy, W. J., and Gentle, J. E. (1980). *Commun. Statist. A*, **9**, 1309–1315.

[31] Sposito, V. A. and Smith, W. C. (1976). *Appl. Statist.*, **25**, 154–157.

[32] Spyropoulos, K., Kiountouzis, E., and Young, A. (1973). *Comput. J.*, **16**, 180–186.

[33] Turner, H. H. (1887). *Philos. Mag., 5th Ser.*, **24**, 466–470.

[34] Usow, K. H. (1967). *SIAM J. Numer. Anal.*, **4**, 233–244.

[35] Wagner, H. M. (1959). *J. Amer. Statist. Ass.*, **54**, 206–212.

[36] Zorilescu, D. (1970). *Stud. Cerc. Mat.*, **22**, 209–212.

(HARTER'S ADAPTIVE ROBUST METHOD
LEAST SQUARES
LINEAR REGRESSION
MAXIMUM LIKELIHOOD ESTIMATION
METHOD OF AVERAGES
METHOD OF LEAST pTH POWERS
MINIMAX METHOD
NONLINEAR REGRESSION)

H. LEON HARTER

METHOD OF LEAST pTH POWERS

This method of fitting a linear or nonlinear regression* equation to a set of observed points involves minimizing the sum of the pth powers of the absolute values of the residuals. Since this is equivalent to minimizing the L_p norm (the pth root of the sum of the pth powers of the absolute values of the residuals), this type of regression is also called L_p regression. Special cases include the method of least squares* ($p = 2$), the method of least absolute values* ($p = 1$), and the minimax method* ($p \to \infty$).

Fechner [6] discussed power means, which he defined as values such that the sums of powers of deviations are minimal when taken from them, and probability laws under which such power means are valid averages. Jackson [13], given a set of m simultaneous equations in n unknowns ($m > n$), studied the question of determining values for the unknowns so that these equations would be approximately solved, in the sense that the sum of the pth powers of the absolute values of the residual errors is a minimum. He showed that there is at least one solution for $p > 0$ and a unique solution for $p > 1$.

Bruen [3] considered various methods of combining observations based on the concept of power means, as defined by Fechner. The pth order power mean of a set of observations x_i ($i = 1, 2, \ldots, n$), is that value x which makes the sum $\Sigma|x_i - x|^p$ a minimum. It is well known that the median is the first-order power mean, the arithmetic mean* is the second-order power mean, and the midrange* is the limiting value of the pth order power mean as $p \to \infty$. Not so well known is the fact, which Bruen attributed to R. M. Foster (see Rietz [20, p. 7]), that the mode is the limiting value of the pth order power mean as $p \to 0$. Bruen generalized the concept of the power mean from the case of direct observations to that of indirect observations or of implicit functional observations, for which it leads to the method of least power sums of the absolute values of the deviations. Corresponding to mode, median, mean, and midrange one has then the methods of least number (least sum of zero powers), least sum of first powers, least sum of squares, and least maximum (least sum of infinite powers) of the absolute deviations, respectively. Bruen discussed the choice of method, pointing out that the choice depends on the presumed error distribution—the mode in one variable or the modal point in two or more variables for a spike distribution (single isolated value), the median or median loci for a symmetric exponential (first Laplacean) distribution, the mean or mean loci for a normal (Gaussian or second Laplacean) distribution, and the midrange or midpoint of least range for a uniform (rectangular) distribution.

Computational methods for least pth power (L_p) regression, based on linear programming*, have been discussed by numerous authors, including Goldstein et al [11], Kelley [15], Stiefel [23–25], Lawson [17], Davis [4], Rice [18], Rice and White [19], Barrodale and Young [1], Gentleman [10], Usow [26], Kiountouzis [16], Forsythe [7], Kahng [14], Sielken and Hartley [22], Watson [27], Boggs [2], and Ekblom [5]. Goldstein et al. applied the method of descent (see Fourier [8, 9]). Kelley remarked that except for the case $p = 2$, which has been adequately solved by classical means, the case $p = \infty$ is probably the most interesting. He applied the simplex method to a dual formulation of the problem of fitting an equation involving n terms whose coefficients are to be determined to a set of m ($> n$) points and constructed an algorithm in which n need not be specified in advance but can be chosen to meet a preassigned tolerance. Stiefel developed the exchange algorithm, which is the dual of the simplex algorithm and hence yields the same results but is computationally simpler. Lawson solved the problem for vector-valued approximations which are no longer linear programs, so that the exchange absolute is no longer applicable. Rice and White reported theoretical and experimental results of applying several of the L_p norms ($1 \leqslant p \leqslant \infty$) to the problem of determining one or two parameters (mean or regression parameters) from data subject to several symmetric error distributions for various sample sizes. They concluded that the L_p norm one should use depends on the distribution of the errors, no single norm being good (or even mediocre) in all situations. In the presence of wild points, however, they found that the L_1 norm is markedly superior among the L_p norms, $1 \leqslant p \leqslant \infty$. Gentleman considered robust estimation* of multivariate location by minimizing pth power deviations, with emphasis on values of p between 1 and 2. Usow presented a general method for computing best linear discrete approximations, based on the characteristics of the solution set (not on linear programming, which he considered unwieldy), and gave results of experience in its use on an IBM 7094 after coding it in FORTRAN IV. His algorithm converges in a finite number of steps, but considerably more slowly for L_1 than for L_∞ or L_2 approximation. Kiountouzis discussed numerical methods that have been proposed for minimizing the sum of the pth powers, favoring an algorithm based on linear programming. Forsythe proposed the use of L_p estimators of straight line regression coefficients ($1 < p < 2$), which have been shown to be more robust than least squares estimators, and showed that a reasonably fast and widely available computer subroutine [IBM Scientific Subroutine Package (1968)] is available to solve the problem. Kahng presented a new algorithm (a modification of the Newton-Raphson method*) with quadratic convergence for the best L_p approximation of a continuous function over a discrete set or a finite interval for $2 < p < \infty$. He also presented methods to accelerate the convergence of an extension by Rice and

Usow of Lawson's algorithm as well as that of the new method, and gave a numerical example. Boggs developed an algorithm for finding the best solution to an overdetermined system of linear equations. Letting $x(p)$ be the point which minimizes the residual of a linear system in the L_p norm, he derived a differential equation describing $x(p)$, from which he devised an iterative scheme, for which he gave a convergence analysis and presented numerical results. Ekblom discussed algorithms for L_p methods. In a Monte Carlo* experiment, he compared the statistical "goodness" of the different methods when applied to regression problems. He concluded that in a large family of distributions an L_p method with a p value around 1.25 is a good choice, while for error densities with very long tails, and perhaps also for strongly asymmetric ones, an L_p method with $p \leqslant 1$ is to be preferred.

Shier and Witzgall [21] explored a relation between various norm approximation problems (arising from fitting linear models to data) and corresponding statistical measures (norm statistics), and established that for any optimal solution to an approximation problem defined with respect to a norm, the resulting residuals have zero as their norm statistic.

The method of least pth powers is optimal (maximum likelihood*) when the error distribution is double exponential (*see* LAPLACE DISTRIBUTION) ($p = 1$), normal ($p = 2$), or uniform ($p \to \infty$). It is optimal, or nearly so, for many other distributions for appropriate values of p ($1 < p < 2$ gives excellent results in many cases). Further information on its applicability was given by Harter [12].

References

[1] Barrodale, I. and Young, A. (1966). *Numer. Math.*, **8**, 295–306.

[2] Boggs, P. T. (1974). *Math. Comp.*, **28**, 203–217.

[3] Bruen, C. (1938). *Metron*, **13**(2), 61–140.

[4] Davis, P. J. (1963). *Interpolation and Approximation*. Blaisdell, Waltham, MA.

[5] Ekblom, H. (1974). *BIT*, **14**, 22–32.

[6] Fechner, G. Th. (1874). *Abh. K. Sachs. Gesell. Wiss. Leipzig, (Math-Phys. Cl.* **11**), **18**, 1–76.

[7] Forsythe, A. B. (1972). *Technometrics*, **14**, 159–166.

[8] Fourier, J. B. J. (1823–1824). *Hist. Acad. Sci. Paris*, 1823, 29ff.; 1824, 47–55.

[9] Fourier, J. B. J. (1831). *Analyse des Équations Déterminées, Première Partie*. Didot Frères, Paris.

[10] Gentleman, W. M. (1966). Robust Estimation of Multivariate Location by Minimizing pth Power Deviations. Ph.D. dissertation, Princeton University. (University Microfilms, Ann Arbor, MI.)

[11] Goldstein, A. A., Levine, N., and Hereshoff, J. B. (1957). *J. Ass. Comput. Mach.*, **4**, 341–347.

[12] Harter, H. L. (1974–1976). *Int. Statist. Rev.*, **42** (1974), 147–174, 235–264, 282; **43** (1975), 1–44, 125–190, 269–278; **44** (1976), 113–159. (A comprehensive survey of the literature on the method of least squares ($p = 2$) and various alternatives, including the method of least pth powers for various values of $p \neq 2$.)

[13] Jackson, D. (1924). *Ann. Math., Ser. 2*, **25**, 185–192.

[14] Kahng, S. W. (1972). *Math. Comp.*, **26**, 505–508.

[15] Kelley, J. E., Jr. (1958). *J. Soc. Indust. Appl. Math.*, **6**, 15–22.

[16] Kiountouzis, E. A. (1971). *Bull. Soc. Math. Grèce* (N.S.), **12**, 191–206.

[17] Lawson, C. L. (1961). Contributions to the Theory of Linear Least Maximum Approximation. Ph.D. dissertation, University of California, Los Angeles, Calif.

[18] Rice, J. R. (1964). *The Approximation of Functions*, Vol. I: *Linear Theory*. Addison-Wesley, Reading, MA.

[19] Rice, J. R. and White, J. S. (1964). *SIAM Review*, **6**, 243–256.

[20] Rietz, H. L., ed. (1924). *Handbook of Mathematical Statistics*. Houghton Mifflin, Boston; Riverside Press, Cambridge.

[21] Shier, D. R. and Witzgall, C. J. (1978). *J. Res. Nat. Bur. Stand.*, **83**, 71–74.

[22] Sielken, R. L., Jr. and Hartley, H. O. (1973). *J. Amer. Statist. Ass.*, **68**, 639–641.

[23] Stiefel, E. L. (1959). In *On Numerical Approximation*, R. E. Langer, ed. University of Wisconsin Press, Madison, WI, pp. 217–232.

[24] Stiefel, E. L. (1959). *Numer. Math.*, **1**, 1–28.

[25] Stiefel, E. L. (1960). *Numer. Math.*, **2**, 1–17.

[26] Usow, K. H. (1967). *SIAM J. Numer. Anal.*, **4**, 233–244.

[27] Watson, G. A. (1973). *Math. Comp.*, **27**, 607–620.

(LEAST SQUARES
LINEAR PROGRAMMING
LINEAR REGRESSION
MEAN, MEDIAN, MODE, AND SKEWNESS

METHOD OF AVERAGES
METHOD OF LEAST ABSOLUTE VALUES
MINIMAX METHOD
NONLINEAR REGRESSION)

H. LEON HARTER

METHOD OF MOMENTS

The method of moments is a method of estimating parameters based on equating population and sample values of certain moments of a distribution. For example, in the univariate case the h parameters $\theta_1, \theta_2, \ldots, \theta_h$ of the density $f(x; \boldsymbol{\theta})$ have moment estimators $\{\boldsymbol{\theta}^*\}$ given by

$$\mu'_r(\boldsymbol{\theta}^*) = m'_r, \qquad r = 1, 2, \ldots, h, \quad (1)$$

where m'_r is the rth noncentral sample moment, and

$$\mu'_r(\boldsymbol{\theta}) = \int_a^b x^r f(x; \boldsymbol{\theta})\, dx,$$
$$r = 0, 1, \ldots k; \quad k \geqslant h \quad (2)$$

is assumed to exist. If $k > h$, then the values of r used in (1) need not be consecutive, and indeed this might be necessary to avoid inconsistencies.

Solutions of (1) may be invalid or not exist for the vector $(m'_1, m'_2, \ldots, m'_h)$; for $\theta_1, \theta_2, \ldots, \theta_h$ may be subject to constraints. For example, the parameters may refer to the mixing proportions when $f(x; \boldsymbol{\theta})$ is a mixture* of component densities, or they may refer to variances of individual normal components. Karl Pearson* [21] studied the two-component normal mixture model

$$P(x; p_1, p_2, \sigma_1, \sigma_2, \lambda_1, \lambda_2) = \sum_{r=1}^{2} p_i g_i(x),$$

where

$$g_i(x) \equiv \frac{1}{\sigma_i \sqrt{(2\pi)}} \exp\left\{ - \frac{(x - \lambda_i)^2}{2\sigma_i^2} \right\}$$

and here solutions to the moment method abort if the value obtained for σ_1^2 (or σ_2^2) is negative; again a mixing proportion could be negative, at least in theory, without sacrificing the nonnegativity condition implied in the probability density.

This problem involves five moments, but in most applications, no more than four moments are used. The use of higher sample moments is viewed with some suspicion because of the high variability likely to be encountered. With the Pearson system* (*see* APPROXIMATIONS TO DISTRIBUTIONS) as a model, Type IV requires four moments, Type III (with unknown origin) involves three moments, and Type I (with known range limits) involves two moments. In the case where (1) is used with $h = 4$, there is no guarantee if a Type I curve is involved that the endpoints will be those desired. An early account of the Pearson system and the moment approach is the treatment by Elderton [8], later revised as Elderton and Johnson [9].

With the Johnson system* the skewness* and kurtosis* are first made equivalent for model and data. Next, the equivalence of location and origin is used to complete a solution.

Multivariate data can also be the basis for estimation by moment methods—for example, meteorological measurements of temperature, wind velocity, pressure, etc.

The method of moments has a long history, involves an enormous amount of literature, has been through periods of severe turmoil associated with its sampling properties compared to other estimation procedures, yet survives as an effective tool, easily implemented and of wide generality.

In the sequel we use m_r and μ_r for sample and population central moments, respectively.

Two Illustrations

THE NEGATIVE BINOMIAL. If the negative binomial* probability function is the model, then one parametric form is

$$P(x; p, k) = \binom{x + k - 1}{x} p^x (1 + p)^{-x-k},$$
$$x = 0, 1, \ldots; \quad p; k > 0, \quad (2a)$$

with

$$\mu_1'(p,k) = kp, \qquad \mu_2(p,k) = kp(1+p)$$

leading to the moment estimators

$$p^* = \frac{m_2}{m_1'} - 1, \qquad k^* = \frac{m_1'^2}{m_2 - m_1'} \qquad (2b)$$

These are explicit solutions, but they abort if $m_2 < m_1'$.

THE TWO-PARAMETER WEIBULL. The two-parameter Weibull model has the density

$$f(t;b,c) = \frac{ct^{c-1}}{b^c}\exp\left\{-\left(\frac{t}{b}\right)^c\right\},$$

$$t > 0; \quad c,b > 0 \quad (3a)$$

with moment estimators (b^*, c^*) given by

$$\begin{cases} \Gamma(1+2/c^*)/\Gamma^2(1+1/c^*) = m_2'/m_1'^2, \\ b^* = m_1'/\Gamma(1+1/c^*). \end{cases} \qquad (3b)$$

These are implicit solutions.

We now describe some properties of the estimators.

VARIANCES OF THE ESTIMATORS

Maximum likelihood estimation* focuses attention on estimators whose variance is as small as possible for large samples. Indeed, Fisher introduced the ratio of variances as an efficiency index, but the variances were first-order terms in the sample size n—often called asymptotic variances. Thus for an estimator t of a parameter τ, we consider the expansion

$$\operatorname{Var} t \sim \tau_1/n + \tau_2/n^2 + \cdots, \qquad (4)$$

where $\tau_1, \tau_2, \ldots,$ are functions of the parameters involved in the model. However, whereas first-order terms such as τ_1 can in general be found, higher order terms soon get out of reach, and frequently an inordinate amount of labor using pencil and paper tactics can be involved to reach n^{-2} and n^{-3} terms [25]. It should be remembered that the topic attracted widespread interest in the first half of the twentieth century, when mathematical methods were prevalent. The situation now is very different due to the impact of digital computers.

Another aspect of the properties of estimators of importance in the present context concerns asymptotic normality*—a property holding in most cases. For example, for a function $f(m_r, m_s)$ of sample central moments, Cramér [4, pp. 353–366] shows that $f(\cdot,\cdot)$ is asymptotically normal with mean $f(\mu_r, \mu_s)$ and variance

$$\left(\frac{\overline{\partial f}}{\partial m_r}\right)^2 \mu_2(m_r) + 2\frac{\overline{\partial f}}{\partial m_r}\frac{\overline{\partial f}}{\partial m_s}\mu_{11}(m_r, m_s) + \left(\frac{\overline{\partial f}}{\partial m_s}\right)^2 \mu_2(m_s), \qquad (5)$$

where

(i) $\displaystyle \frac{\overline{\partial f}}{\partial m_t} = \left.\frac{\partial f}{\partial m_t}\right|_{m_t = \mu_t},$

(ii) $\mu_2(m_r) \equiv$ Variance of m_r to order n^{-1}, and similarly $\mu_{1,1}$ is the first-order term in $E\{(m_r - E(m_r))(m_s - E(m_s))\}$,

provided that in a neighborhood of $(m_r = \mu_r, m_s = \mu_s)$ the function $f(\cdot,\cdot)$ is continuous and has continuous derivatives of the first and second order with respect to m_r, m_s.

Since the normal distribution is determined by its mean and variance, an estimator with smallest variance is to be preferred against others, and if one exists, it is called an "efficient estimator." Unfortunately, the search for efficiency* very often overlooked its asymptotic nature, so the era "of best estimators provided n is sufficiently large" was born. One cannot quibble with this concept except that all too frequently the largeness aspect becomes a matter of pure guesswork, or at best a figure (such as 100) which sounds safe enough. This cavalier attitude to asymptotics continues to this day in spite of the insistence on clear definitions for many other aspects of statistical theory in modern studies.

The controversy between Karl Pearson* and R. A. Fisher* a half-century or so ago centered on the so-called inefficiency of the method of moments. Both were well versed in mathematics and could not (one surmises) be unaware of usages of asymptotic series, especially by theoretical astronomers; again Borel [1] at this time was giving the substance of his epoch-making studies on divergent series. So from a present-day vantage point, the heated arguments ignored a fundamental aspect: What information does an asymptotic supply?

MOMENTS OF SAMPLE MOMENTS

It is well known that noncentral sample moments are unbiased ($Em'_r = \mu'_r$), whereas central moments are biased; thus $Em_2 = (1 - 1/n)\mu_2$. The difficulty resides in the fact that central sample moments are nonlinear functions of noncentral sample moments. For the latter, from independence,

$$Ee^{\alpha(m'_r - \mu'_r)} = \{Ee^{\alpha(x^r - \mu'_r)/n}\}^n$$
$$= 1 + \frac{\alpha^2}{2!}A_2 + \frac{\alpha^3}{3!}A_3 + \cdots \quad (6)$$

where $A_i = E(m'_r - \mu'_r)^i$. For example

$$A_2 = E(x^r - \mu'_r)^2/n$$
$$= (\mu'_{2r} - \mu'^2_r)/n.$$

Higher moments are most easily found using the relation (differentiating (6) with respect to α):

$$\left(\frac{\alpha A_2}{1!} + \frac{\alpha^2}{2!}A_3 + \cdots\right)$$
$$\times\left(1 + \frac{\alpha^2 a_2}{n^2} + \frac{\alpha^3 a_3}{n^3} + \cdots\right)$$
$$= \left(\frac{\alpha a_2}{1!n} + \frac{\alpha^2 a_3}{2!n^2} + \cdots\right)$$
$$\times\left(1 + \frac{\alpha^2}{2!}A_2 + \frac{\alpha^3}{3!}A_3 + \cdots\right) \quad (7)$$

and equating coefficients of powers of α; thus

$$A_2 = a_2/n, \qquad A_3 = a_3/n^2,$$

and in general

$$\sum_{s=0}^{r-1}\binom{r-1}{s}\frac{a_s A_{r-s}}{n^s} = \sum_{s=0}^{r-1}\binom{r-1}{s}\frac{A_s a_{r-s}}{n^{r-s-1}}, \quad (8)$$

whereas $a_s = E(x^r - \mu'_r)^s$. This formulation is suitable for digital implementation. Equation (8) can be used to derive the recursive expression

$$A_s^{(k)} = \sum_{t=1}^{s}\binom{s}{t}\left(a_{t+1}A_{s-t}^{(k-t)} - a_t A_{s-t+1}^{(k-t)}\right),$$
$$A_s^{(k)} = 0 \quad \text{for } k < 0 \text{ and } k \geqq s, \quad (9)$$

where $A_s^{(k)}$ is the coefficient of n^{-k} in A_s. The fundamental entities are now $a_s = E(x^r - \mu'_r)^s$, which can be evaluated as univariate moments or as a single Stieltjes integral (using quadrature or summation).

The bivariate analogue of (6) is readily set up (for the general case, see ref. 24). We define

$$A_{r,s} = E\left[(m'_R - \mu'_R)^r(m'_S - \mu'_S)^s\right],$$
$$a_{r,s} = E\left[(x^R - \mu'_R)^r(x^S - \mu'_S)^s\right]. \quad (10)$$

The recursion for the coefficient of n^{-k} in $A_{s,t}$ is now defined by the two systems

$$A_{s+1,t}^{(k)} = \sum_{\lambda=0}^{s}\sum_{\mu=0}^{t}\binom{s}{\lambda}\binom{t}{\mu}$$
$$\times a_{s+1-\lambda,t-\mu}A_{\lambda,\mu}^{(k+\lambda+\mu-s-t)}$$
$$-(1-\delta_{\lambda,\mu})\sum_{\lambda=0}^{s}\sum_{\mu=0}^{t}\binom{s}{\lambda}\binom{t}{\mu}$$
$$\times a_{\lambda,\mu}A_{s+1-\lambda,t-\mu}^{(k-\lambda-\mu)}, \quad (11a)$$

$$A_{s,t+1}^{(k)} = \sum_{\lambda=0}^{s}\sum_{\mu=0}^{t}\binom{s}{\lambda}\binom{t}{\mu}$$
$$\times a_{s-\lambda,t+1-\mu}A_{\lambda,\mu}$$
$$-(1-\delta_{\lambda,\mu})\sum_{\lambda=0}^{s}\sum_{\mu=0}^{t}\binom{s}{\lambda}\binom{t}{\mu}$$
$$\times a_{\lambda,\mu}A_{s-y,t+1-\mu}^{(k-\lambda-\mu)}, \quad (11b)$$

where δ is the Kronecker function, and

(i) $\left[\frac{s+t+1}{2}\right] \leqq k \leqq s+t$;

(ii) $A_{s,t}^{(k)} = 0$ for $k < 0$ or $k \geqq s+t$;

(iii) $[x]$ is the integer part of x;

(iv) $A_{0,0} = a_{0,0} = 1$; $A_{10} = A_{01} = a_{10} = a_{01} = 0$.

From these, successive values of $A_{s,t}$ can be

evaluated, and the digital implementation, while not quite straightforward, cannot be regarded as a major programming task. Here the fundamental entities are univariate expectations which can be assessed in various ways depending on the background information. For example,

$$E(x - \mu_1')^r (x^2 - \mu_2')^s$$

$$= Ey^r (y^2 + 2\mu_1' y - \mu_2)^s$$

$$= \sum_{0 \leqq l+m \leqq s}\sum \frac{s!\, p^m q^{s-l-m}}{l!\, m!\, (s-l-m)!} \mu_{r+2l+m},$$

$$p = 2\mu_1', \quad q = -\mu_2,$$

and this would be convenient to use if there is a closed form for μ_r (the rth central moment) or if there exists a recursive scheme for these moments. This recursive approach and its generalization is one solution to the problem of moments of sample moments. For a central moment the formula

$$m_r = \sum_{s=0}^{r} (-1)^{r-s} \binom{r}{s} m_s' m_1'^{r-s} \qquad (12)$$

also carries over to population moments.

Moments of a statistic $t = g(m_1', m_2', \ldots, m_h')$ are now found from the multivariate Taylor expansion* (written in summatory form)

$$t = \tau + \epsilon_r \bar{t}_r + \frac{\epsilon_r \epsilon_s}{2!} \bar{t}_{r,s} + \cdots,$$

where

$$\tau = g(\mu_1', \mu_2', \ldots, \mu_h') \quad \text{and}$$

$$\bar{t}_r = \left. \frac{\partial t}{\partial m_r'} \right|_{\mathbf{m}' = \boldsymbol{\mu}'}$$

with obvious extensions to higher dimensions. Further details are given in Shenton et al. [26].

Illustrations

THE NEGATIVE BINOMIAL. In (2b) the moment estimator k^* for the index k is a function of m_1' and m_2', namely,

$$k^* = m_1'^2 / (m_2' - m_1' - m_1'^2).$$

Sheehan [23] evaluated terms up to n^{-6} in the bias and variance. For example,

(i) $k = 0.5, p = 1.0, n = 850;$ (13)
$\operatorname{Var} k^* \sim .00706 + .00044 + .000167 - .00005 + .00005 - .00006$
showing the successive n^{-1} through n^{-6} terms (The % error in using the n^{-1} term is 7.2);

(ii) $k = 5, p = 0.1, n = 6100;$
$\operatorname{Var} k^* \sim 1.190 + 0.417 + 0.153 + 0.065 + 0.032 + 0.018$ (Here the % error in using the n^{-1} term is 36.6.);

(iii) $k = 25, p = 0.2, n = 1600;$
$\operatorname{Var} k^* \sim 29.25 + 10.30 + 3.74 + 1.56 + 0.75 + 0.42$ (The % error in using the n^{-1} term is 36.5.).

Comment. The sample size required to make the series approximants acceptable is noteworthy.

THE WEIBULL TWO-PARAMETER MODEL. Equation (3b) has to be inverted and expanded (using Faà di Bruno's formula* [10] for the derivatives of a function of a function) to get c^*, the moment estimator of the shape parameter c as a function of m_1' and m_2' [2]. We find for $c = 1$ that

$$\operatorname{Var} c^* \sim 1/n - 1.162\text{E}01/n^2 + 1.047\text{E}03/n^3$$
$$- 1.1868\text{E}05/n^4$$
$$+ \cdots - 3.685\text{E}25/n^{12}. \qquad (14)$$

Summatory techniques are used on these seemingly divergent series to provide the comparisons in Table 1.

Comment. It will not go unnoticed that for $c = 1$, the *error increases with sample size* (note also the remarkable agreement for samples of $n = 15$). The results for $c = 2.5$ show one aspect of the problem of deciding how large a sample must be for the first asymptotic to a moment to be reliable.

Table 1[a]

c	n	Extended Series Values of $\sigma(c^*)$	Asymptotic First Order $\sigma(c^*)$	% Error
1	15	0.2589	0.2582	0.3
	20	0.2188	0.2236	2.2
	25	0.1936	0.2000	3.3
	30	0.1758	0.1826	3.9
	50	0.1358	0.1414	4.1
	75	0.1113	0.1154	3.7
2.5	15	0.6245	0.5067	18.9
	20	0.5105	0.4388	14.0
	25	0.4419	0.3925	11.2

[a](% error = 100|difference|/ext. ser. value).

EARLY WORK

Tchouproff's [28, 29] 50-page papers in *Biometrika* included (in modern terminology) (1) formulas to convert noncentral to central moments (sample and population), (2) moments of m_1' up to the eighth; (3) $E\{(m_r' - \mu_r')(m_s' - \mu_s')(m_t' - \mu_t')\}$; (4) Em_r, $r = 2, 3, \ldots, 6$; (5) Em_2^r, $r = 1, 2, 3, 4$, ; (6) $\mu_r(m_2)$, $r = 2, 3, 4$ ($\mu_4(m_2)$ involves terms to order n^{-7} and some 38 coefficients).

Thiele's [30] semi-invariants (or cumulants*) held out some hope that moment expressions might be simplified (e.g., the n^{-2} term in $E\prod_{s=1}^{4}(m_{\lambda_s} - \mu_{\lambda_s})$ consists of some 60 terms), for the cumulants κ corresponding to moments μ are defined by

$$1 + \mu_1'\alpha + \mu_2'\alpha^2/2! + \cdots = \exp\left\{k_1\alpha + k_2\frac{\alpha^2}{2!}\cdots\right\}$$

and for normal densities (univariate or otherwise) cumulants higher than the second are zero. Thiele pointed out that $\kappa_2, \kappa_3, \ldots,$ are origin free, and that $\{\kappa_r/\kappa_2^{r/2}\}$, $r \geqslant 2$ are origin and scale free. According to Hald [13], Thiele may have seen some association with cumulants in the Gram-Charlier series*

$$g(y) = \phi(y) - \frac{a_3}{3!}\phi^{(3)}(y) + \frac{a_4}{4!}\phi^{(4)}(y) + \cdots, \tag{15}$$

where $a_3 = \kappa_3/\kappa_2^{3/2}$, $a_4 = \kappa_4/\kappa_2^2$, and the derivatives involve $y^3 - 3y$ and $y^4 - 6y^2 + 3$.

Church [3] gave the first four moments of m_2 (noncentral and central) using Tchouproff's works. "Sophister" [27] considered the distribution of m_2 from a gamma* population ($y = y_0 x^7 \exp(-x)$) for samples of 5 and 20, fitting a Pearson Type VI curve using four moments.

There was doubtless some common ground in the studies of Thiele and Fisher [11]. The notion of unbiased moment estimators (k-statistics), developed by Fisher in their precise form and their sampling properties certainly took the subject a considerable step forward at a time when digital computers were scarcely imagined (*see* FISHER'S k-STATISTICS). If the cumulants under normality were mostly zero, then the central limit theorem* would hold hope for the same property for statistics, at least asymptotically. So there is a three-stage process:

1. Express the moment statistic in question in terms of k-statistics.
2. Expand the function by Taylor series (if necessary) in terms of incremental quantities such as $k_1 - \kappa_1, k_2 - \kappa_2, \ldots,$ with expectation zero. Using Fisher's patterns, derive the cumulants of expressions such as $E(k_1 - \kappa_1)^{r_1}(k_2 - \kappa_2)^{r_2}$.
3. Convert the cumulants to moments.

The main references are Fisher [11, 12], Wishart [31–34], Dressel [7], Kendall [15–18], Kendall and Stuart [19], Schaeffer and Dwyer [22], and James [14]. There are many applications (rarely going beyond terms of order n^{-3}); see, for example, Pearson [20], David [5], and David and Johnson [6].

CONCLUDING REMARKS

If there is sufficient need, moments of functions of sampling moments can be carried out to terms of order n^{-k} $(4 \leqslant k < 30)$ using a digital computer. If the population sampled has limited range (Type I Pearson, for example) the available set of coefficients may not show strongly divergent tendencies; otherwise, summation techniques may be necessary.

There is practically no point in endless studies of asymptotic efficiency unless there is some indicator of how large n must be for a substantial region of the parameter space. Guesswork has its limitations.

Moment methods are, in the latter part of the twentieth century, still a powerful tool in the interpretation of chance data—especially since there is no longer a predilection for normality. In some complicated cases they may present the only possibility.

References

[1] Borel, É. (1920). *Leçons sur les Séries Divergentes*. Gauthier-Villars, Paris. (This book along with its second edition (1927) summarizes lectures given at l'Ecole Normale, 1899–1900. It sets out a fascinating history of the subject. One chapter clearly describes the bridge supplied by Stieltjes' continued fraction connecting series to definite integrals, a concept of basic importance. An English translation (Critchfield and Vakar) appeared in 1975).

[2] Bowman, K. O. and Shenton, L. R. (1982). *Proc. Comp. Sci. and Statist: 14th Symposium on the Interface*. Springer-Verlag, New York.

[3] Church, A. E. R. (1925). *Biometrika*, **17**, 79–83.

[4] Cramér, H. (1946). *Mathematical Methods of Statistics*. Princeton University Press, Princeton, NJ. (A foundation classic, mathematically oriented.)

[5] David, F. N. (1949). *Biometrika*, **36**, 383–393.

[6] David, F. N. and Johnson, N. L. (1951). *Biometrika*, **38**, 43–57.

[7] Dressel, P. L. (1940). *Ann. Math. Statist.*, **11**, 33–57.

[8] Elderton, W. P. (1960). *Frequency Curves and Correlation*, Cambridge University Press, Cambridge, England.

[9] Elderton, W. P. and Johnson, N. L. (1969). *Systems of Frequency Curves*. Cambridge University Press, Cambridge, England. (A classic treatise and guide to fitting the Pearson systems with comments on other approximating systems.)

[10] Faá di Bruno, F. (1876). *Théorie des Formes Binaires*. Librairie Breno, Turin. (Considers the basic formula for derivatives of functions of functions. Cumulant-moment identities provide examples of Bruno's formula.)

[11] Fisher, R. A. (1928). *Proc. Lond. Math. Soc.*, **30**, 199–238.

[12] Fisher, R. A. (1930). *Proc. R. Soc. A*, **130**, 16–28.

[13] Hald, A. (1981). *Int. Statist. Rev.*, **49**, 1–20.

[14] James, G. S. (1958). *Sankhȳa*, **20**, 1–30.

[15] Kendall, M. G. (1940). *Ann. Eugen.* (Lond.), **10**, 106–111.

[16] Kendall, M. G. (1940). *Ann. Eugen.* (Lond.), **10**, 215–222.

[17] Kendall, M. G. (1940). *Ann. Eugen.* (Lond.), **10**, 392–402.

[18] Kendall, M. G. (1942). *Ann. Eugen.* (Lond.), **11**, 300.

[19] Kendall, M. G. and Stuart, A. (1977). *The Advanced Theory of Statistics*, Vol. 1, 4th ed. Charles Griffin & Co., London; Macmillan, New York. (Contains much basic material on univariate and multivariate moments and cumulants, with applications.)

[20] Pearson, E. S. (1930). *Biometrika*, **22**, 239–249.

[21] Pearson, K. (1894). *Phil. Trans. R. Soc. Lond. A.*, **185**, 1–40.

[22] Schaeffer, E. and Dwyer, P. S. (1963). *J. Amer. Statist. Ass.*, **58**, 120–151.

[23] Sheehan, D. M. (1967). The Computational Approach to Sampling Moments. Ph.D. dissertation, Virginia Polytechnic Institute, Blacksburg, VA.

[24] Shenton, L. R. and Bowman, K. O. (1975). *Int. Statist. Rev.*, **43**, 317–334.

[25] Shenton, L. R. and Bowman, K. O. (1977). *Maximum Likelihood Estimation in Small Samples*, Monograph 38. Macmillan, New York. (Shows unexpected complexity of series for moments of m.l.e.'s in the general case; even with modern computers, third-order contributions are generally beyond reach, and whether series ultimately diverge is not known. Gunner Kulldorff had tackled a similar problem for grouped samples—see *Contributions to the Theory of Estimation From Grouped and Partially Grouped Samples*, Wiley; Almquist and Wicksell, 1961.)

[26] Shenton, L. R., Bowman, K. O., and Sheehan, D. (1971). *J. R. Statist. Soc. B*, **33**, 444–457.

[27] Sophister (1928). *Biometrika*, **20**, 389–423.
[28] Tchouproff, A. A. (1918). *Biometrika*, **12**, 140–169.
[29] Tchouproff, A. A. (1919). *Biometrika*, **12**, 185–210.
[30] Thiele, T. N. (1903). *Theory of Observations*. C. and E. Layton, London. (Some of the basic concepts relating to cumulants (semi-invariants) are due to Thiele—see recent comments by Hald [15].)
[31] Wishart, J. (1929). *Proc. Lond. Math. Soc., Ser. 2*, **29**, 309–321.
[32] Wishart, J. (1929). *Proc. R. Soc. Edinburgh*, **49**, 78–90.
[33] Wishart, J. (1930). *Biometrika*, **22**, 224–238.
[34] Wishart, J. (1933). *Biometrika*, **25**, 52–60.

Acknowledgment

Research sponsored by the Applied Mathematical Sciences Research Program, Office of Energy Research, U.S. Department of Energy under contract W-7405-eng-26 with the Union Carbide Corporation. (K.O.B.)

(ASYMPTOTIC NORMALITY
EFFICIENCY
ESTIMATION, POINT
FISHER'S k-STATISTICS)

K. O. BOWMAN
L. R. SHENTON

METHOD OF SIEVES

Grenander's method of sieves is a general technique through which parametric approaches to *estimation* can be applied to nonparametric problems. Typically, classical approaches such as maximum likelihood* and least squares* fail to produce consistent estimators when applied to nonparametric (infinite dimensional) problems. Thus, for example, the unconstrained maximum likelihood estimator for a density function is not consistent (not even well defined) in the nonparametric case (see Examples 1 and 2 below), and direct application of least squares similarly fails for the nonparametric estimation of a regression function (see Examples 3 and 4 below). Speaking loosely, it might be said that in each case the parameter space (a space of functions) is too large. Grenander [11] suggests the following remedy: perform the optimization* (maximization of the likelihood, minimization of the sum of squared errors, etc.) within a subset of the parameter space, choosing increasingly dense subsets with increasing sample sizes. He calls this sequence of subsets from which the estimator is drawn a *sieve*, and the resulting estimation procedure is his method of sieves. It leads to consistent nonparametric estimators, with different sieves giving rise to different estimators.

The details and versatility of the method are best illustrated by examples; other applications can be found in Grenander [11], wherein the method was first introduced, and in some of the other references.

Example 1. Histogram. Let $x_1, \dots, x_n$ be an independent and identically distributed (i.i.d.) sample from an absolutely continuous distribution with unknown probability density function (p.d.f.) $\alpha_0(x)$. The maximum likelihood estimator for α_0 maximizes the likelihood function

$$\prod_{i=1}^{n} \alpha(x_i). \tag{1}$$

But the maximum of (1) is not achieved within any of the natural parameter spaces for the nonparametric problem (e.g., the collection of all nonnegative functions with area 1). Thus unmodified maximum likelihood is not consistent for nonparametric density estimation.

A sieve is a sequence of subsets of the parameter space indexed by sample size. For each $\lambda > 0$ let us define

$$S_\lambda = \left\{ \alpha : \alpha \text{ is a p.d.f. which is constant on } \left[\frac{k-1}{\lambda}, \frac{k}{\lambda} \right), k = 0, \pm 1, \pm 2, \dots \right\},$$

and allow $\lambda = \lambda_n$ to grow with sample size. $\{S_{\lambda_n}\}$ constitutes a sieve, and the associated (maximum likelihood) method of sieves estimator solves the problem:

$$\text{maximize} \prod_{i=1}^{n} \alpha(x_i) \qquad \text{subject to} \quad \alpha \in S_{\lambda_n}.$$

The well-known solution is the function

$$\hat{\alpha}(x) = \frac{\lambda}{n} \# \left\{ x_i : \frac{k-1}{\lambda_n} \leqslant x_i < \frac{k}{\lambda_n} \right\}$$

$$\text{for} \quad x \in [(\frac{k-1}{\lambda_n}, \frac{k}{\lambda_n})],$$

i.e., the histogram* with bin width λ_n^{-1}. If $\lambda_n \uparrow \infty$ sufficiently slowly, then $\hat{\alpha}$ is consistent, e.g., in the sense that $\int |\hat{\alpha}(x) - \alpha_0(x)| \, dx \to 0$ a.s.

Example 2. Convolution Sieve for Nonparametric Density Estimation*. For the same problem, a different and more interesting sieve is the convolution sieve:

$$S_{\lambda_n} = \left\{ \alpha : \alpha(x) = \int \frac{\lambda_n}{\sqrt{2\pi}} \exp\left[-\frac{\lambda_n^2}{2}(x-y)^2 \right] F(dy), \right.$$

$$\left. F \text{ an arbitrary c.d.f.} \right\},$$

where λ_n is a nonnegative sequence increasing to infinity. The method of sieves estimator $\hat{\alpha}$ maximizes (1) within the sieve S_{λ_n}. It can be shown [10] that $\hat{\alpha}$ has the form

$$\hat{\alpha}(x) = \sum_{i=1}^{n} p_i \frac{\lambda_n}{\sqrt{2\pi}} \exp\left\{ -\frac{\lambda_n^2}{2}(x - y_i)^2 \right\}$$

for some $y_1, \ldots, y_n$ and $p_1, \ldots, p_n$ satisfying $p_i \geqslant 0$, $1 \leqslant i \leqslant n$, $\sum_{i=1}^{n} p_i = 1$. It can also be shown that $\{y_1, \ldots, y_n\} \neq \{x_1, \ldots, x_n\}$ (with probability 1). Thus the convolution sieve defines an estimator closely related to, but distinct from, the Parzen-Rosenblatt Gaussian kernel estimator. Observe that the latter is in the sieve S_{λ_n}: take F to be the empirical distribution function. But the maximum of the likelihood is achieved by using a different distribution. As with the Parzen-Rosenblatt estimator, if $\lambda_n \uparrow \infty$ sufficiently slowly (i.e., the "window width" is decreased sufficiently slowly), then the estimator is consistent. For details see Geman and Hwang [9], and for an interesting discussion of this and related estimators from a different point of view, see Blum and Walter [2].

Example 3. Splines* for Nonparametric Regression. Let X and Y be random variables and let $(x_1, y_1), \ldots, (x_n, y_n)$ be an i.i.d. sample from the bivariate distribution of (X, Y). The least squares estimator of the regression function $E(Y \mid X = x)$ minimizes

$$\sum_{i=1}^{n} \{ y_i - \alpha(x_i) \}^2. \qquad (2)$$

Observe that the minimum is zero and is achieved by any function that passes through all of the points of observation, $(x_1, y_1), \ldots, (x_n, y_n)$. Excepting some very special cases, this set does not in any useful sense converge to the true regression.

For any nonnegative sequence $\lambda_n \uparrow \infty$ define a sieve $\{S_{\lambda_n}\}$ as follows:

$$S_{\lambda_n} = \left\{ \alpha : \alpha \text{ absolutely continuous}, \int \left| \frac{d}{dx} \alpha(x) \right|^2 dx \leqslant \lambda_n \right\}.$$

The least squares method of sieves estimator, $\hat{\alpha}$, for the regression function is the function in S_{λ_n} minimizing (2). The unique minimum is a first-degree polynomial smoothing spline, i.e., $\hat{\alpha}$ is continuous and piecewise linear with discontinuities in $d\hat{\alpha}/dx$ at $x_1, \ldots, x_n$ (see ref. 15). It is possible to show that if λ_n increases sufficiently slowly, then the estimator is strongly consistent for $E(Y|X = x)$ in a suitable metric (details are in ref. 8).

Example 4. Dirichlet Kernel for Nonparametric Regression. Recall the nonparametric regression problem discussed in the previous example. Let us here take x, the "independent" variable, to be deterministic. We then think of the distribution of Y as being an unknown function of x, $F_x(\cdot)$. For this example, we assume $x \in [0, 1]$. The problem is then to estimate

$$\alpha_0(x) = E_x[Y] \equiv \int_{-\infty}^{\infty} y F_x(dy), \quad x \in [0, 1],$$

from independent observations $y_1, \ldots, y_n$, where $y_i \sim F_{x_i}$, and $x_1, \ldots, x_n$ is a deterministic, so-called design, sequence. For example, assume that the design sequence for fixed n is equally spaced on the interval [0, 1] with

$$x_i = \frac{i}{n}, \qquad i = 1, 2, \ldots, n.$$

As with the previous example, unconstrained minimization of the sum of squares of errors, (2), does not produce a useful estimator. Introduce the Fourier sieve

$$S_m = \left\{ \alpha(x) : \alpha(x) = \sum_{k=-m}^{m} a_k e^{2\pi i k x} \right\};$$

S_m is particularly tractable and makes for a good illustration of the method in this setting. The sieve size is governed by the parameter m, which is allowed to increase to infinity with n. If we restrict m_n so that $m_n \leqslant n$ for all n, then $\hat{\alpha}$ is uniquely defined by requiring that it minimize (2) subject to $\alpha \in S_{m_n}$. A simple calculation gives the explicit form:

$$\hat{\alpha}(x) = \frac{1}{n} \sum_{i=1}^{n} y_i D_{m_n}(x - x_i)$$

where D_m is the Dirichlet kernel

$$D_m(x) = \frac{\sin \pi(2m+1)x}{\sin \pi x}.$$

Kernel estimators for nonparametric regression have been widely studied, although from a somewhat different point of view. See refs. 1, 4, 6, 16, and 17 for some recent examples. It is not difficult to exploit this simple form for $\hat{\alpha}$. Depending on the rate at which $m_n \uparrow \infty$, and depending on assumptions about α_0, consistency, rates of convergence, and asymptotic distribution can be established [8].

What makes this example particularly tractable is that the estimator is based on a sieve that consists of increasing subspaces of a Hilbert space. Nguyen and Pham [14] used sieves of this type to estimate the drift function of a repeatedly observed nonstationary diffusion.

Example 5. Nonparametric Estimation of the Drift Function of a Diffusion. From an observation of a sample path of a diffusion process* one can construct consistent estimators for the diffusion drift. If the form of the drift function is known up to a finite collection of parameters, then it is possible to use maximum likelihood and obtain consistent and asymptotically normal estimators (see Brown and Hewitt [3], Feigin [5], Lee and Kozin [12], and Lipster and Shiryayev [13]). But unconstrained maximum likelihood fails in the nonparametric case.

More precisely, let us consider a diffusion process x_t defined by

$$dx_t = \alpha_0(x_t)\,dt + \sigma\,dw_t, \qquad x_0 = x_0,$$

with w_t a standard (one-dimensional) Brownian motion* and x_0 a constant. α_0 and σ are assumed to be unknown; we wish to estimate α_0 from an observation of a sample path of x_t. It is well known that the distribution of x_s, $s \in [0, t]$, is absolutely continuous with respect to the distribution of σw_s, $s \in [0, t]$ (assuming some mild regularity condition on α_0). A likelihood function for the process x_s, $s \in [0, t]$ is the Radon-Nikodym derivative:

$$\exp\left\{ \int_0^t \alpha_0(x_s)\,dx_s - \frac{1}{2} \int_0^t \alpha_0(x_s)^2\,ds \right\}. \quad (3)$$

The maximum likelihood estimator for α_0 maximizes (3) over a suitable parameter space, most appropriately the space of uniformly Lipschitz continuous functions. But the maximum of the likelihood is not attained, either in this or in any other of the usual function spaces. In a manner analogous to the previous examples, a sieve S_t can be introduced (here indexed by time) and an estimator $\hat{\alpha}$ defined to maximize (3) subject to $\alpha \in S_t$. Provided that the sieve growth is sufficiently slow with respect to t, this method of sieves estimator can be shown to be consistent: $\hat{\alpha} \to \alpha_0$, in a suitable norm, as $t \to \infty$. Details are in Geman [7].

References

[1] Ahmad, I. A. and Lin, P. E. (1976). *Bull. Math. Statist.*, **17**, 63–75.

[2] Blum, J. and Walter, G. (1982). *Technical Report Series of the Intercollege Division of Statistics*, No. 43, University of California, Davis, Calif.

[3] Brown, B. M. and Hewitt, J. I. (1975). *J. Appl. Prob.*, **12**, 228–238.

[4] Devroye, L. P. and Wagner, T. J. (1980). *Ann. Statist.* **8**, 231–239.

[5] Feigin, P. D. (1976). *Adv. Appl. Prob.*, **8**, 712–736.

[6] Gasser, T. and Muller, H. G. (1979). *Smoothing Techniques for Curve Estimation*. T. Gasser and Rosenblatt, eds. Lecture Notes in Mathematics, Springer-Verlag, New York.

[7] Geman, S. (1980). *Colloquia Mathematica Societatis Janos Bolyai*, Vol. *32*. Nonparametric Statistical Inference. North-Holland, Budapest.

[8] Geman, S. (1981). *Reports on Pattern Analysis*, No. 99, Division of Applied Mathematics, Brown University, Providence, RI.

[9] Geman, S. and Hwang, C. R. (1982). *Ann. Statist.*, **10**, 401–414.

[10] Geman, S. and McClure, D. E. (1982). *Proceedings of the NASA Workshop on Density Estimation and Function Smoothing*. L. F. Guseman, Jr., ed. Texas A & M University, College Station, TX, 38–47.

[11] Grenander, U. (1981). *Abstract Inference*. Wiley, New York.

[12] Lee, T. S. and Kozin, F. (1977). *J. Appl. Prob.*, **14**, 527–537.

[13] Lipster, R. S. and Shiryayev, A. N. (1978). *Statistics of Random Processes* Vol II: *Applications*. Springer-Verlag, New York, Chap. 17.

[14] Nguyen, H. T. and Pham, T. D. (1982). *SIAM J. Control and Optimization*.

[15] Schoenberg, I. J. (1964). *Proc. Nat. Acad. Sci.*, **52**, 947–950.

[16] Schuster, E. and Yakowitz, S. (1979). *Ann. Statist.*, **7**, 139–149.

[17] Spiegelman, C. and Sacks, J. (1980). *Ann. Statist.*, **8**, 240–246.

(DENSITY ESTIMATION
ESTIMATION, POINT
KERNEL ESTIMATORS
LEAST SQUARES
MAXIMUM LIKELIHOOD ESTIMATION)

STUART GEMAN

METRIC NUMBER THEORY *See* PROBABILISTIC NUMBER THEORY

METRICS AND DISTANCES ON PROBABILITY SPACES *See* PROBABILITY SPACES, METRICS AND DISTANCES ON

METRICS, IDEAL

This concept was introduced by Zolotarev [1], who discussed applications to mathematical statistics in some detail [2], and later presented further developments [3]. The notion is useful in problems of approximating distributions of random variables obtained from independent random variables by successive application of addition, multiplication, taking maxima, or some other "group operations."

References

[1] Zolotarev, V. M. (1976). *Mat. Sbornik*, **101** (**144**), No. 3 (11), 416–454 (in Russian).

[2] Zolotarev, V. M. (1979). *Austr. J. Statist.*, **21**, 193–208.

[3] Zolotarev, V. M. (1981). In *Stability Problems for Stochastic Models, Proc. Sem. Moscow Inst. System Studies*, 1981, pp. 30–39.

(APPROXIMATIONS TO DISTRIBUTIONS)

METRIKA

The journal *Metrika* bears the subtitle *International Journal for Theoretical and Applied Statistics*. It appears quarterly, starting with volume 1 in 1958. In the course of time the number of pages has increased up to nearly 300 per volume (= 4 fasc.). There are no auxiliary publications.

Research papers and, very rarely, survey papers are published. As expressed in the title, published articles belong to the field of mathematical statistics (see Fig. 1). During the starting years this concept was understood in a wider sense, but now, because of the large number of submitted manuscripts, only articles on statistics in a narrower sense are accepted, i.e., only those on statistical methods and mathematical statistics. Great importance is attached to applicability of proposed and investigated methods.

Articles written in German or in English are acceptable. Far more than half the papers are submitted in English. Besides the actual articles each volume also contains book reviews. Whereas formerly there was a large number of brief reviews, future issues will review fewer books in greater detail.

M. J. Ahsan and S. U. Khan. Optimum allocation in multivariate stratified random sampling with overhead cost
M. A. Beg. Optimal tests and estimators for truncated exponential families
S. K. Bhattacharya and G. S. Pandey. Applications of an inequality involving conditional expectations
K. F. Cheng. Contributions to nonparametric generalized failure rate function estimation
P. Findeisen. Die Charakterisierung der Normalverteilung nach Gauß
P. Findeisen. Charakterisierung der zweiseitigen Exponentialverteilung
J. Fischer. Nonparametric maximum likelihood estimation of a probability density via mathematical programming
P. Grant. Secretary problems with inspection costs as a game
N. Henze. Verhalten des Chi-Quadrat-Tests für Prüfung der Gleichwahrscheinlichkeit der Lotto-Zahlen bei Nichtgültigkeit der Hypothese
D. Kalin. Zum Problem des zweiarmigen Bernoulli-Banditen mit einer bekannten Erfolgswahrscheinlichkeit und unendlich vielen Spielen
E. Kremer. Local comparison of linear rank tests, in the Bahadur sense
P. Mukhopadhyay. Optimum strategies for estimating the variance of a finite population under a superpopulation model
J. Panaretos. An extension of the damage model
S. Schach. An elementary method for the statistical analysis of growth curves
W. Sendler. On functionals of order statistics
S. Sengupta. On interpenetrating samples of unequal sizes
W. Stadje. Some remarks on the average sample number of sequential and non-sequential tests
I. Strauss. On the admissibility of estimators for the finite population variance
M. T. Subrahmanya. The combined inverse binomial sampling procedure
K.-H. Waldmann. On two-state quality control under Markovian deterioration
P. G. Wong and S. P. Wong. A curtailed test for the shape parameter of the Weibull distribution
E. Zinzius. Unigünstige Verteilungen bei einparametrigen Parameterschätzproblemen

Figure 1 Contents page from *Metrika*, Vol. 29 (1982).

Metrika came into being when two journals—*Mitteilungsblatt für mathematische Statistik* and *Statistische Vierteljahresschrift*—merged into one. The former was edited by O. Anderson, H. Kellerer, H. Münzner, and K. Stange; there appeared nine volumes in the years 1949 to 1957. The latter journal appeared in ten volumes from 1948 to 1957 and was edited by W. Winkler.

The founders of *Metrika* are O. Anderson, H. Kellerer, A. Linder, H. Münzner, S. Sagoroff, and W. Winkler. Anderson, Kellerer, Linder and Winkler worked subsequently as editors and Kellerer and Sagoroff as managing editors. Since 1973 (volume 20), there are two editors, F. Ferschl and W. Uhlmann. The current editorial adressess are Prof. Dr. F. Ferschl, Seminar für Angewandte Stochastik, Universität München, Akademiestraße 1/IV, D 8000 München 40, West Germany, and Prof. Dr. W. Uhlmann, Institut für Angewandte Mathematik und Statistik, Universität Würzburg, Sanderring 2, D 8700 Würzburg, West Germany.

In the future Prof. Dr. O. Krafft, Technische Hochschule Aachen, Institut für Statistik und Wirtschaftsmathematik, Wüllnerstraße 3, D 5100 Aachen, will take over the function of Prof. Ferschl.

For many years P. Sint was responsible for the book reviews; in 1981 H. Heyer took over this task. The address is Prof. Dr. H. Heyer, Universität Tübingen, Mathematisches Institut, Auf der Morgenstelle 10, D 7400 Tübingen, West Germany.

Metrika is independent of any professional society. Its publishers are Physica-Verlag Ges.m.b.H., Seilerstätte 18, A-1010 Vienna, Austria. Papers intended for publication in *Metrika* are to be sent to one of the editors. Authors are requested to observe the "instructions to authors" to be found in almost every issue. This is the procedure: the respective editor sends the paper to one of the referees and on the basis of the referee's report he will decide whether the paper is to be accepted or not, if necessary after some revision. The referee remains anonymous to the author. A list of the experts acting as referees in any one year is published together with the index that appears in the last issue of that particular year.

WERNER UHLMANN

METRON, INTERNATIONAL JOURNAL OF STATISTICS

Metron was founded in Italy by Corrado Gini in 1920. Its international character is attested to by the many non-Italian authors who contributed to it, including F. Y. Edgeworth, E. Czuber, R. A. Fisher, E. Slutsky, A. A. Tchouprow, M. Fréchet, R. Frisch, A. J. Lotka, Ch. Jordan, V. Romanowsky, R. C. Geary, S. S. Wilks, M. Greenwood, "Student," E. J. Gumbel, S. Bernstein, F. Bernstein, G. Darmois, A. L. Bowley, N. Georgescu-Roegen, A. A. Konüs, and many others. The program presented by Gini in *Metron*'s first volume could still be valid today; he said

> It is hoped that METRON may be a bond of union between statistical workers in different branches, perhaps at length an organ of scientific coordination. With this object METRON will be catholic; its pages will be open to those who employ no methods beyond the scope of ordinary cultivated men as well to those who delight in the most refined and subtle developments of mathematical science . . . . Between these extremes are insensible gradations and both orders of inquiry interest science in general and statistical science in particular. It is hoped that both will find in METRON an appropriate treatment.

Today with the enormous diffuseness of mathematics there has been a kind of flattening in scientific output. The few basic ideas that have marked the main stages in the growth of statistical methods, have been, in the last decennia, objects of detailed mathematical exercises by a growing army of scholars who remind us of Poincaré's

E. SLUTSKY. Ueber stochastische Asymptotem und Grenzwerte
R. A. FISHER. Application of "Student" 's distribution
"STUDENT". New tables for testing the significance of observations
R. A. FISHER. Expansion of "Student"'s integral in powers of n^{-1}
M. BOLDRINI. Capacità contributiva e gravame fiscale di alcuni stati
G. H. KNIBBS. The growth of human populations and the laws of their increase

Figure 1 Contents page from *Metron*, Vol. 5, No. 3 (1925).

C. BENEDETTI. Carlo Emilio Bonferroni (1892–1960)
A. M. KSHIRSAGAR and B. MCKEE. A unified theory of missing plots in experimental designs
L. DE CAROLIS, G. GRECO, M. IANNARELLI, L. LIONETTI, and P. ZECCHI. Analisi statistica tramite computer della fase di depolarizzazione del tracciato elettrocardiografico
B. BALDESSARRI and F. GALLO. Linear structural dependence
A. CRESCIMANNI. Di una particolare classe di "poligonali sghembe"
O. CUCCONI and G. DIANA. On exact sequential test for continous random variables
NAVATNA RATHIE. On exact distribution of Pearson and Wilks likelihood test statistic
K. GUPTA and A. K. RATHIE. Distribution of the likelihood ratio criterion for the problem of k samples
E. BALLATORI. Sui tests statistici per il confronto tra due frequenze in tabelle 2×2
P. DADDI. On the moments of the sample partial autocorrelation function
V. AMATO. Il tensore di Ricci e la correlazione statistica
M. MONTINARO. Considerazioni sul campionamento a due stadi

Figure 2 Contents page from *Metron*, Vol. 40, No. 3–4 (1982).

"microbes of science" and who have added little to the "real stream" of those basic ideas. *Metron*, like similar statistical journals, has had to face an invasion of such mass output. *Metron* has five languages admitted for publication: Italian, French, English, German, and Spanish. Although Italian, French, and German are less widespread than others, the journal seeks to offer the opportunity to the scholars of Italy, France, and Germany—to whom science owes much—to express their ideas in the most genuine form. *Metron* is published in two double issues a year, which make up one volume. From 1920 to 1962 the journal was the personal property of Gini; since 1962 it has belonged to the University of Roma. Gini continued his direction of *Metron* until 1965, the year of his death, after which V. Castellano succeeded him as editor.

In 1982 the direction of *Metron* passed to C. Benedetti. In Volume 28, Numbers 1–4 (1970), which marked the fiftieth anniversary of the journal's founding, Castellano and Benedetti tried to evaluate the actual state of statistics in the light of a new program (Castellano) and in light of the past and present situation (Benedetti). After a brief preface by Benedetti in Volume 39, Numbers 1–2 (1981), Castellano, in his article "Leave-Taking," explains the reasons that led him to leave the direction of *Metron* and also analyzes in a penetrating yet passionate manner the statistician's situation in the worldwide crisis of today. Figures 1 and 2 are tables of contents of an early and a recent volume.

CARLO BENEDETTI

MICKEY'S UNBIASED RATIO AND REGRESSION ESTIMATORS

Mickey [5] defines a class of ratio and regression type estimators which are unbiased for random sampling without replacement from a finite population (see FINITE POPULATIONS, SAMPLING FROM). The class of estimators is based on the fact that the estimator

$$\bar{t} = \bar{y} - \sum_{i=1}^{p} a_i(\bar{x}_i - \mu_{x_i})$$

is an unbiased estimator of μ_y for any choice of constants $a_1, \ldots, a_p$, where

$$\mu_y = \frac{1}{N}\sum_{\beta=1}^{N} y_\beta, \qquad \mu_{x_i} = \frac{1}{N}\sum_{\beta=1}^{N} x_{i\beta}, \qquad i = 1, \ldots, p,$$

and $\bar{y}$ and $\bar{x}_i$ are sample means.

This is used as follows: for any chosen α $(\alpha < n)$ of the sample elements, the remaining $n - \alpha$ elements constitute a random sample of size $n - \alpha$ from the finite population of $N - \alpha$ elements that results from excluding the selected α elements from the population. Therefore, by choosing the a_i as functions of the α selected elements and applying estimator $\bar{t}$ above for the sample of size $n - \alpha$ yields an unbiased estimator for the derived population. This leads to an unbiased estimator for the original population, because the relation between the population characteristics for the two populations is determined by α, N, and the α selected elements.

Formally, let Z_α denote the ordered set of observations on the first α sample elements, $1 \leqslant \alpha < n$, and let $a_i(Z_\alpha)$ denote functions of these observations where the sample elements are drawn one at a time, so that the sample is an ordered set by order of draw. Let $\bar{y}(\alpha)$, $\bar{x}_i(\alpha)$, $i = 1, \ldots, p$, denote the means of the indicated observations on the first α sample elements. Then the estimator $\bar{t}_M$ can be written as

$$\bar{t}_M = \bar{y} - \sum_{i=1}^{p} a_i(Z_\alpha)[\bar{x}_i - \mu_{x_i}] - \frac{\alpha(N-n)}{(n-\alpha)N} \times \left\{ \bar{y}(\alpha) - \bar{y} - \sum_{i=1}^{p} a_i(Z_\alpha)[\bar{x}_i(\alpha) - \bar{x}_i] \right\},$$

where $\bar{y} - \sum_{i=1}^{p} a_i(Z_\alpha)[\bar{x}_i - \mu_{x_i}]$ is a biased estimator of μ_y and the remainder is an estimate of the bias.

A general class of unbiased estimators can be constructed by including all estimators of the form $\bar{t}_M$ applied to any permutation of the ordering of sample elements and weighted averages of such estimators. Whether or not an estimator of the class defined above is ratio (invariant to scale changes in the x variable only) or regression (invariant with respect to location and scale changes in the x variable) depends on the properties of the coefficient function $a_i(Z_\alpha)$.

A special case of the Mickey estimator is the unbiased ratio estimator* of Hartley and Ross [3]:

$$\bar{t}_{\mathrm{HR}} = \mu_x \bar{r} + \frac{(N-1)n}{N(n-1)}(\bar{y} - \bar{r}\bar{x})$$

where $r_i = y_i/x_i$ and $\bar{r}$ = sample mean of the r_i.

The variance estimator is readily obtained only for a subclass of the estimators. If $t(\alpha, n)$ is the estimator $\bar{t}_M$ based on a sample of size n, let $0 < \alpha_1 < \alpha_2 < \cdots < \alpha_{k+1} = n$ where the α_i are integers, and consider the k estimators

$$t(\alpha_1, \alpha_2), t(\alpha_2, \alpha_3), \ldots, t(\alpha_k, n)$$

where the estimator $t(\alpha_j, \alpha_{j+1})$ is based on the first α_{j+1} sample elements. Then

$$\bar{t}_1 = \frac{1}{k} \sum_{j=1}^{k} t(\alpha_j, \alpha_{j+1})$$

is an unbiased estimator of μ_y, because it is a linear combination of unbiased estimators. An unbiased nonnegative estimator of the variance of $\bar{t}_1$ is

$$s^2(\bar{t}_1) = \frac{1}{k(k-1)} \sum_{j=1}^{k} \left[t(\alpha_j, \alpha_{j+1}) - \bar{t}_1 \right]^2$$

because the $t(\alpha_j, \alpha_{j+1})$, $j = 1, \ldots, k$, are uncorrelated.

These ideas underlying construction of unbiased estimators for random sampling are extended to unequal probability sampling. Some new unbiased estimators are obtained as well as unbiased estimators of the variance of estimators. These estimators are approximately as efficient as the Horvitz-Thompson estimator*.

Several authors examine the efficiency of Mickey's estimator in random sampling, using the special case $\bar{t}_{M1} = \bar{r}_g + (g/\bar{X})(\bar{y} - \bar{r}_g\bar{x})$, where the sample n is divided into g groups, each of size p, where $n = pg$, $\bar{r}_g = \sum r_j/g$, and where r_j is the classical ratio estimator computed from the sample after omitting the jth group. Rao [6] compares $\bar{t}_{\mathrm{M1}}$ analytically with $\bar{t}_{\mathrm{HR}}$, and the approximately unbiased Quenouille, Tin, and Pascual ratio estimators. He assumes two linear models: (1) $y_i = \alpha + \beta x_i + \mu_i$; $x_i \sim N(1, \sigma^2)$, $E(\mu_i \mid x_i) = 0$, $V(\mu_i \mid x_i) = n\delta$, where δ is constant of order $1/n$; and (2) the same except that $x \sim$ gamma (h). He determines that $\bar{t}_{\mathrm{M1}}$ is considerably more efficient than $\bar{t}_{\mathrm{HR}}$ but that the other estimators are preferable in efficiency unless freedom of bias is important; *see* PASCUAL ESTIMATOR; QUENOUILLE ESTIMATOR; TIN ESTIMATOR.

Rao and Beegle [7] and Hutchinson [4] compare several ratio estimators including $\bar{t}_{\mathrm{M1}}$, $\bar{t}_{\mathrm{HR}}$, and Quenouille's, Beale's, and Tin's ratio estimators. Again $\bar{t}_{\mathrm{M1}}$ was more efficient than $\bar{t}_{\mathrm{HR}}$ but less efficient than other estimators considered.

Ek [1] compares $\bar{t}_{\mathrm{M1}}$, $\bar{t}_{\mathrm{HR}}$, the ratio-of-means, the mean-of-ratios, the Horvitz-Thompson, and some simple linear and polynomial regression estimators on three forest populations where specific linear and nonlinear relations often can be assumed. Even for samples of size 4, the bias of the ratio estimators generally was not important. In general, $\bar{t}_{\mathrm{M1}}$ was not one of the better estimators for any sample size studied.

DeGraft-Johnson and Sedransk [2] compare a two-phase counterpart of Mickey's estimator with other two-phase estimators. They find that the Mickey estimator is typically less efficient than those of Beale or Tin.

In summary, Mickey's class of unbiased ratio and regression estimators is an interesting contribution to the literature. Generally, the bias of ratio estimators is rarely important in practice. In those cases where it could be, e.g., very small sample sizes, the estimator $\bar{t}_{\mathrm{M1}}$ should be preferred to $\bar{t}_{\mathrm{HR}}$ for efficiency.

References

[1] Ek, A. R. (1971). *Forest Science*, **17**, 2–13.

[2] deGraft-Johnson, K. T. and Sedransk, J. (1974). *Ann. Inst. Statist. Math.*, **26**, 339–350.

[3] Hartley, H. O. and Ross, A. (1954). *Nature*, **174**, 270–271.

[4] Hutchinson, M. C. (1971). *Biometrika*, **58**, 313–321.

[5] Mickey, M. R. (1959). *J. Amer. Statist. Ass.*, **54**, 594–612.

[6] Rao, J. N. K. (1967). *Biometrika*, **54**, 321–324.

[7] Rao, J. N. K. and Beegle, L. D. (1967). *Sankhyā B*, **29**, 47–56.

(FINITE POPULATIONS, SAMPLING FROM
HORVITZ-THOMPSON ESTIMATOR
PASCUAL ESTIMATOR
QUENOUILLE ESTIMATOR
RATIO ESTIMATORS
TIN ESTIMATOR)

H. T. SCHREUDER

MID-MEAN

This is the average of all observations between (and including) the quartiles*. It serves as a measure of location. More precisely, let $Y_1 \leqslant Y_2 \leqslant \cdots \leqslant Y_n$ represent the ordered data. Define the sample inverse cumulative distribution function F_n^{-1} at point $(i - 0.5)/n$ to be Y_i, $i = 2, \ldots, n - 1$, with $Y_1 = 0$, $Y_n = 1$. At other points, F_n^{-1} is defined by linear interpolation. Now define

$$I(\alpha, \beta) = \frac{1}{\beta - \alpha} \int_\alpha^\beta F_n^{-1}(t)\, dt \qquad \text{for} \quad 0 \leqslant \alpha < \beta \leqslant 1.$$

The mid-mean is then defined as $I(\frac{1}{4}, \frac{3}{4})$. See Tukey [2] and Cleveland and Kleiner [1] for more details.

References

[1] Cleveland, W. S. and Kleiner, B. (1975). *Technometrics*, **17**, 447–454.

[2] Tukey, J. W. (1977). *Exploratory Data Analysis*. Addison-Wesley, Reading, MA.

(EXPLORATORY DATA ANALYSIS
SEMI-MID-MEAN
TRIMMED MEAN)

MIDRANGES

The midrange of a set of data points is the average of the largest and the smallest. If

$$X_{(1)} \leqslant X_{(2)} \leqslant \cdots \leqslant X_{(n)}$$

represent order statistics* in a random sample from a population with cumulative distribution function $F(\cdot)$, then the midrange is a random variable M, defined by

$$M = \{X_{(1)} + X_{(n)}\}/2.$$

If F has probability density function (PDF) f, then M has PDF

$$f_M(x) = 2n(n-1) \times \int_{-\infty}^{x} \{F(2x - y) - F(y)\}^{n-2} \times f(y) f(2x - y)\, dy.$$

ESTIMATION

The main use of the sample midrange M is to estimate the center of a symmetric population; less frequently it can be used to estimate the midrange of a population having a finite range, and it was once proposed as a measure of central tendency in subgroups for control charts* in statistical quality control* [2]. Here we confine attention to estimation in symmetric populations.

If the underlying population is symmetrical, the sample midrange may be more efficient as an estimator of the common mean/median than the arithmetic mean* $\overline{X}$. This holds for uniform populations, for certain "double Weibull" distributions, and for some truncated normal populations. A key factor appears to be the value of the kurtosis* α_4 [4, 6], where

$$\alpha_4 = \mu_4 / \mu_2^2,$$

and μ_r is the rth central moment of the population.

Table 1 Efficiency of the Sample Midrange Relative to the Sample Mean for Some Symmetric Distributions[a]

		Relative Efficiency				
Distribution	Kurtosis	$n = 3$	$n = 5$	$n = 10$	$n = 30$	$n = 100$
Uniform	1.80	1.11	1.40	1.67	5.12	18.77
Triangular	2.40			0.82	0.70	0.97
Two-point	1.00	1.33	3.20	51.20		
Truncated normal						
$c = 1$	1.97			1.66	2.54	8.08
$c = 1.50$	2.13			1.37	1.58	3.34
$c = 2.00$	2.36			0.95	1.01	0.94
$c = 2.50$	2.64			0.64	0.54	0.40
Double Weibull						
$a = 2$ (normal)	3.00			0.61	0.23	0.14
$a = 4$	2.18			0.91	0.69	0.27
$a = 6$	2.02			1.34	1.43	0.76
$a = 8$	1.94			1.36	2.23	1.81
$a = 20$	1.86			2.33	2.74	5.12

[a]Sources: Table 1 of ref. 4 and Table 4 of ref. 6.

The preceding assertion may appear surprising, considering that in normal populations, the relative efficiency of M to $\bar{X}$ decreases from 0.920 to 0.350 as n increases from 3 to 20, and decreases thereafter to zero as n becomes infinitely large. Relative efficiency is defined in this context as

$$(\text{Variance of } \bar{X})/(\text{Variance of } M).$$

Simulation results by Hand and Sposito [4] and exact results by Rider [6] indicate that if $\alpha_4 < 2$, M is more efficient than $\bar{X}$ for samples of size 10 or more. Table 1 presents a few of their results; the truncated normal cutoff points c (*see* TRUNCATION) are determined by their PDFs

$$f_c(x) = \begin{cases} \exp(-x^2/2) \Big/ \int_{-c}^{c} \exp(-y^2/2)\,dy, \\ \qquad\qquad -c < x < c, \\ 0, \qquad\qquad |x| \geqslant c, \end{cases}$$

and the double Weibull distributions have PDFs

$$f_a(x) \propto \exp\{-|x|^a\}, \qquad |x| < \infty.$$

The discrete two-point distribution has probability mass 1/2 at $x = \pm 1$ and corresponds to the minimum value of α_4 at 1. Rider points out that for a population uniform over $c \leqslant x \leqslant d$, M is the "best possible estimate of the center" $(c + d)/2$ (*see* RECTANGULAR (UNIFORM) DISTRIBUTIONS).

*r*th MIDRANGE

The rth midrange of the sample is M_r, where

$$M_r = \tfrac{1}{2}\{X_{(r)} + X_{(n-r+1)}\};$$

for $r \geqslant 2$, M_r has been considered as an alternative estimator to M, which is subject to the effect of outliers*. Most of the work, done with normally distributed populations, has been directed at finding the value of r which gives the smallest variance for any sample size n. As a rule of thumb, $r = [n/3]$ comes close to this, where $[x]$ denotes the smallest integer less than or equal to x; see ref. 5 for tables of the cumulative distribution function and the standard deviation for values of $n \leqslant 21$ and of $r \leqslant 4$; ref. 1 includes a table of relative efficiencies and variances of the optimum choice of M_r.

Sen [7] considered the same problem for a wide class of distributions. He showed that the sample median cannot have asymptotically (as $n \to \infty$) the smallest variance among rth midranges $(1 \leqslant r \leqslant [(n+1)/2])$ for any parent distribution such that the PDF f and its first two derivatives are continuous in some interval containing the population median as an interior point and such that $E(|X^\delta) < \infty$ for some $\delta > 0$, X having the parent distribution of interest. However, for an underlying Laplace distribution*, which violates the first condition at its point of symmetry, he shows that the sample median does have such an asymptotic smallest variance. See also Gumbel [3] for further asymptotic properties of rth midranges.

References

[1] Dixon, W. J. (1957). *Ann. Math. Statist.*, **28**, 806–809.

[2] Ferrell, D. B. (1953). *Industrial Quality Control*, **9**, 30–32.

[3] Gumbel, E. J. (1944). *Ann. Math. Statist.*, **15**, 414–422.

[4] Hand, M. L. and Sposito, V. A. (1979). *Proc. Statist. Comp. Section, Amer. Statist. Ass.*, 332–335.

[5] Leslie, R. T. and Culpin, D. (1970). *Technometrics*, **12**, 311–325.

[6] Rider, P. R. (1957). *J. Amer. Statist. Ass.*, **52**, 537–542.

[7] Sen, P. K. (1961). *J. R. Statist. Soc. B*, **23**, 453–459.

(ARITHMETIC MEAN
MEAN, MEDIAN, MODE, AND SKEWNESS
ORDER STATISTICS)

CAMPBELL B. READ

MIDSPREAD

A term used in exploratory data analysis*, denoting the distance between the lower quartile (or hinge) and the upper quartile of a distribution of a variable quantity; almost synonymous with interquartile range*, it is sometimes called the H-spread.

(FIVE-NUMBER SUMMARIES)

MIDZUNO SAMPLING DESIGN *See* RATIO ESTIMATORS

MIETTINEN'S STATISTIC

The technique of matched samples often is employed in the hope of securing more accurate comparisons by eliminating possible sources of variation. The purpose is to control factors known or suspected of being related to the outcome variable.

Here we are concerned with the response variable being a dichotomy of "success" or "failure." Consider comparing two matched samples, but instead of one observation for each matched sample there are now R (say) observations in one sample and one observation in the other. This will be referred to as an *R-to-one matching situation.* A case-control study design where, to gain efficiency, each case is matched with R controls would be an example of an R-to-one matching situation.

Miettinen [5] developed a statistic for this situation which assumes the number of successes for a given matched $(R+1)$-tuple to be fixed. His statistic tests the null hypothesis of no difference between the proportions of success in the two treatments assuming a fixed number of successes for each matched set of observations. Under the null hypothesis Miettinen's statistic is asymptotically chi-square with one degree of freedom, or $\chi^2(1)$. Pike and Marrow [6] independently developed the same statistic.

To derive Miettinen's statistic assume we have n $(R+1)$-tuples and let $X_{1j} = 1$ if the outcome from treatment 1 in the jth matched set of observations is a success and 0 otherwise for $j = 1, 2, \ldots, n$. Likewise let X_{2j} be the number of successes for treatment 2 from the jth matched set of observations, so $X_{2j} = 0, 1, 2, \ldots, R$. The test statistic given by Miettinen is

$$T = \frac{\left[\sum_{j=1}^{n} (RX_{1j} - X_{2j})\right]^2}{\sum_{j=1}^{n} \{X_j(1 + R - X_j)\}},$$

where $X_j = X_{1j} + X_{2j}$ for $j = 1, 2, \ldots, n$.

An application of Miettinen's statistic is seen in the following hypothetical situation.

Table 1 Frequency of Successes for 40 Smokers with Three Matched Controls

		Number of Successes in Control Group				
		3	2	1	0	Totals
Smoking	S	4	5	4	1	14
Group	F	7	11	5	3	26
Totals		11	16	9	4	40

Suppose we want to study the relationship of cigarette smoking and baby size independent of mother's race, prepregnancy weight or weight gain. For each smoker or propositus we will match on the previously mentioned factors three nonsmokers or controls. The outcome of success will be assigned to a low-risk baby or one which has a body weight greater than or equal to 2500 grams, the alternative outcome being failure. We want to test the null hypothesis that the proportion of success is the same in both smoking and nonsmoking mothers. If we have 40 matched samples of four individuals (three control and one propositus) the data may be presented as in Table 1.

From this table $T = 8.98$, which is significant at the 0.005 level. This hypothetical example implies that smoking mothers have significantly fewer low-risk babies than nonsmoking mothers.

If $R = 1$, T is the statistic given by McNemar [4] for the 1–1 matching situation (*see* MCNEMAR STATISTIC). If $R = c - 1$, T is also the statistic given by Cochran [2] for partitioning his Q statistic into components comparing one treatment with the remaining $c - 1$ treatments. Cochran's Q statistic* tests the null hypothesis of no difference between the c treatments conditional on the number of successes for a given matched c-tuple. Under the above null hypothesis Cochran's Q statistic is asymptotically $\chi^2(c - 1)$. Bhapkar [1] offers an unconditional Wald [8] statistic for testing equality of c matched treatments; this is asymptotically $\chi^2(c - 1)$ also under its respective null hypothesis.

Ury [7] studied the efficiency of Miettinen's statistic with multiple controls to McNemar's statistic utilizing one-to-one matching. He found the Pitman efficiency* of Miettinen's statistic with R controls per propositus to McNemar's statistic to be

$$2R/(R + 1).$$

He also showed the efficiency of $R + k$ controls relative to R controls to be

$$(R + k)(R + 1)/\{R(R + k + 1)\}.$$

For further reading, I suggest Miettinen's original article [5], which encompasses cost considerations in choosing a design. Ury [7] also gives a fine review in his article dealing with the efficiency of multiple controls per case. Mantel and Haenszel [3] have a very fine paper on controlling factors in the analysis of categorical data. Their statistic can handle the situation of multiple matched observations per case where the number of matched observations may vary. In fact, as pointed out by Miettinen, his statistic is a special case of the Mantel-Haenszel statistic*.

References

[1] Bhapkar, V. P. (1970). In *Random Counts in Scientific Work*, Vol. 2, G. P. Patil, ed. Pennsylvania State University Press, University Park, PA.

[2] Cochran, W. G. (1950). *Biometrika*, **37**, 256–266.

[3] Mantel, N. and Haenszel, W. (1959). *J. Nat. Cancer Inst.*, **22**, 719–748.

[4] McNemar, Q. (1947). *Psychometrika*, **12**, 153–157.

[5] Miettinen, O. S. (1969). *Biometrics*, **25**, 339–355.

[6] Pike, M. C. and Marrow, R. H. (1970). *Brit. J. Prev. Soc. Med.*, **24**, 42–44.

[7] Ury, H. K. (1975). *Biometrics*, **31**, 643–649.

[8] Wald, A. (1943). *Trans. Amer. Math. Soc.*, **54**, 426–482.

(CHI-SQUARE TESTS
COCHRAN'S Q-STATISTIC
MANTEL-HAENSZEL STATISTIC

MATCHED PAIRS
McNEMAR STATISTIC)

GRANT W. SOMES

MIGRATION

Changes of address of residence occur frequently among most populations, but many of these changes are of very short term, for example, holidays or visits to relatives. "Migration" usually means a more permanent move: people going for longer periods to find employment, to render military service, to find a more suitable house or district, to escape political or religious persecution, or for a host of other possible reasons. Statistics of these longer-term population movements are collected for a number of purposes—for example, the improvement of population assessments and forecasts, or as a measure of social and economic change, or to ascertain the need for, or effectiveness of, governmental control over such movements. Permanent changes of address within a country or administrative region are often called "internal" migration and those which cross international boundaries are called "external." The significance and analytical treatment of external migration are generally different from those for internal movements. "Immigration" is the word used to describe movement into a country, and "emigration" signifies the corresponding outward movement.

ASSEMBLY OF DATA

How successful the collection of migration statistics can be depends on the circumstances. Some international borders are not policed effectively enough to enable any accurate figures to be obtained at the time when the movements occur; examples are the border between Mexico and the United States and that between Eire and Northern Ireland. Similarly, few governments attempt to get information about internal movements as they happen. At ports and airports, however, and at major roads, where they cross a border, officials ask travelers to state the nature of their journey; where this is an intended migration, forms have to be filled in giving such personal details as sex, age, occupation and places of start and end of journey. These data cannot be very searching unless the journey is to be extensively interrupted.

If the results of two successive population censuses are compared, and if allowance is made for recorded births and deaths in the period between them, an estimate of the amount of migration in that period can be arrived at by deduction; this can be done for a country as a whole or for any part of it. As, however, neither censuses nor registrations can ever be wholly accurate, any errors must fall into the migration estimate, and their disruptive effect upon that estimate can be severe where—as is often the case—natural population movement by birth and death is large relative to migration. Nevertheless, there is another useful way in which measures of spatial movements can be obtained: questions can be asked at censuses about length of stay at present address and about places of residence in the past—for instance 12 months ago, or 10 years earlier. Birthplace is also often included as an item to be recorded on the census* form, as is nationality.

NATURE OF MIGRATION

The aims of statistical analysis are materially influenced by what has been found about migration from past studies. These emphasize a number of important points. First, the amount of movement is restricted by the limitations of the available transport—the longer the journey the more severe the curb; among large populations, only a small proportion can migrate in a year. Second, economic pressure or opportunity is a powerful cause of change of address; for instance, where population increase in a region exceeds the economic growth there, there is likely to be an overspill into an area of better prospects. Third, migration can be very selective as regards occupation and skill at work; demand for certain types of expert can cause a "brain drain" from some places. Fourth, changes in time in the magnitude and direction of migratory flows are much more subject to fluctuation than are the numbers of births or deaths. Finally, for any place there is nearly always both inward and outward movement occurring simultaneously and it is often not gross changes but net balance which is the most important feature.

Migrants, both internally and externally, are likely to be young adults; people aged between 15 and 35 are the most mobile, and the rate of movement is lower for children and for people in middle life and old age. An illustrative age-pattern is shown in Fig. 1, which is based on general experience and does not relate to any particular situation; some data referring to actual internal movement (into cities of the United States in the 1930s) are given in Table 1, showing how rates of migration per thousand population vary by age. On the whole, men are more likely than women to change their address. Race can also have an important effect on mobility; this arises because of both outward pressure in the place of origin and a warmer welcome, or better chance of assimilation, in some destinations than in others. Healthy people have a much better chance of a journey than the unfit, not only because the latter are less likely to want or to be able to travel, but also because, internationally, receiving countries impose the condition that only those who can satisfy their medical tests can be permanently admitted.

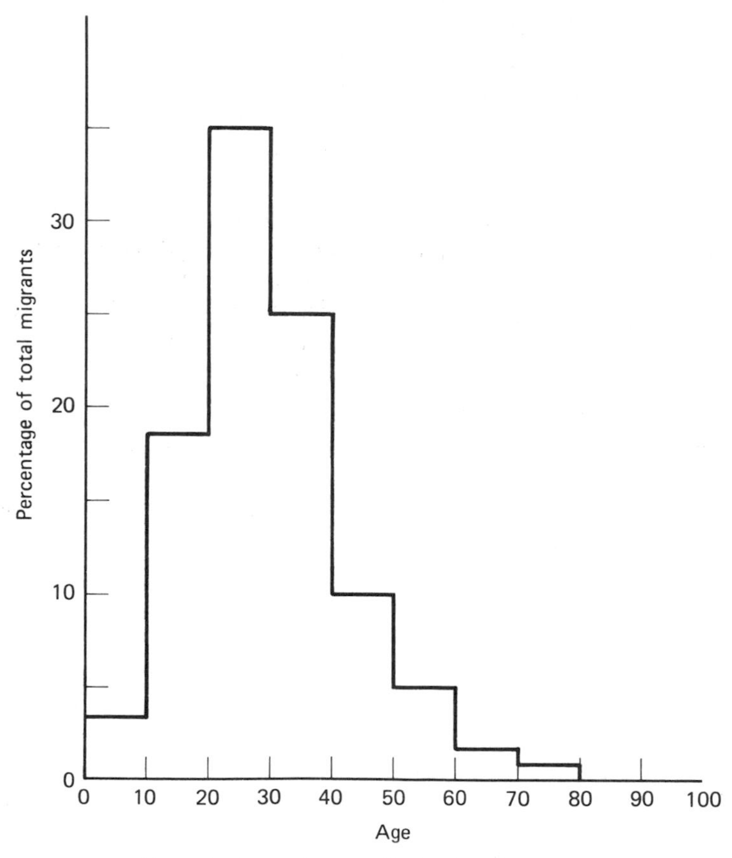

Figure 1 Illustrative age-distribution of migrants.

Table 1 Indexes of Migration Rates (per 1000 population) by Age; USA, 1930–1939

Age Group	Index of Migration Rate (all ages = 10)
14–19	9
20–24	14
25–34	15
35–44	10
45–54	7
55 and over	5

STATISTICAL ANALYSIS

Migration data often are incomplete, and they tend to act as a pointer rather than as a precise measure. Their form differs from place to place, and occasionally from time to time. There may well be different types of statistics covering the same period in the same area, which it is difficult to reconcile, e.g., census* statistics of past movements and direct measures of current transfers. The machinery of collection of international migration statistics may differ in some countries between immigrants and emigrants. One country's record of exits to another country may not agree with the second country's statistics of migrants said to have been received from the first. In these circumstances, methods of presentation and analysis of the data can follow no routine pattern but must be decided ad hoc. Elaborate measures such as standardization and life tables* are seldom used; a simple statement of numbers or of crude rates* or proportions is probably as much as can be justified. But the best denominator in any fraction of this kind is not necessarily the total population; for example, net migration can be expressed as a ratio to the total flow, inward and outward combined, or it may be related to the ceding population as well as the receiving population. It is most important to look for essentials, some of which have been found to be:

1. Transfers from rural areas to the towns.
2. Relationship between distance and extent of migration.
3. Effects of fluctuations in trade upon the incentive to move.
4. Influence of wars and particular political developments.

EFFECT OF MIGRATION ON POPULATION

A material influx of migrants can have an appreciable effect on the experience of a population as regards fertility*, marriage*, and even mobility. This can arise not merely from a changed age distribution but also from modification in the rates at individual ages. How marriage rates are affected will depend on the balance of the sexes among immigrants and on the extent to which they are assimilated, e.g., blacks being perhaps less likely to be accepted as marriage partners in a predominantly white country. Fertility will depend on the habits in the ceding country, at least in the early years, and upon the extent of marriage. As to mortality, the beneficial effect of selection for good health may be undone because of adverse climatic conditions or inability to find suitable work or housing. It may prove difficult, however, to measure these migration effects on the experience of the population, unless special statistics can be collected which bear on them. Political pressure may be brought against the collection of such statistics, for fear that if immigrants are separately distinguished in inquiries they may be adversely affected socially.

Emigration could, *pari passu*, have a selective effect on the population experience of the ceding country. It may remove those who are specially healthy, ambitious, or philoprogenitive, or members of a particular race. Moreover, where high population pressure causes an outward flow of people, it may represent another type of association between fertility and migration.

For all these reasons, in population projection* work it is customary to treat migrants, particularly immigrants, where they are likely to be significant in number, as a separate group subject to a mortality or fertility experience different from the remainder of the population.

Bibliography

Matras, J. *Populations and Societies*. Prentice-Hall, New York, 1973. (Contains some valuable sociological analyses of migration.)

Methods of Measuring Internal Migration. United Nations, New York, 1970. (A manual designed as an aid to demographers generally, with special reference to the developing countries.)

Selected Studies of Migration since World War II. Milbank Memorial Fund, New York, 1958. (These studies illustrate some of the techniques used in migration analysis.)

Thomas, B. (1973). *Migration and Economic Growth: A Study of Great Britain and the Atlantic Economy*, 2nd. ed. Cambridge University Press, Cambridge, England. (1st ed. 1954. A classic study of variations according to economic circumstances in the volume of migration to the United States.)

(CENSUS
DEMOGRAPHY
FERTILITY
MARRIAGE
POPULATION PROJECTIONS)

P. R. Cox

MILITARY STANDARDS FOR FIXED-LENGTH LIFE TESTS

Military standards have as their primary purpose the provision of a standard set of acceptance testing procedures. These procedures provide each specification writer, inspector, or tester with a limited range of plans through which design and manufacturing information is communicated. The commonly used military standard, MIL-STD-781C, entitled "Reliability Design Qualification and Production Acceptance Tests: Exponential Distribution," is discussed in MILITARY STANDARDS FOR SEQUENTIAL LIFE TESTING. There, our discussion emphasizes the sequential testing aspects of the standard. Here we shall focus attention on the fixed-length tests of this standard. MIL-STD-781C is applicable only when the batch (lot) of items has life lengths that are assumed to have an exponential distribution*. The quantitative requirement is expressed in terms of the mean time between failures (MTBF), or equivalent expressions, such as mean cycles between failures. The effect of this standard is generally to cause an increase in the quality of the product as measured by the specified tests, by motivating the manufacturer or producer to search for those causes and factors related to better quality (cf. Neathammer et al. [8]).

In a fixed-length test, several items are simultaneously tested for a specified fixed length of time, and the test is terminated with a decision either to accept or to reject the lot. The decision for acceptance or rejection is based on the number of failures observed during the test. The statistical basis underlying the fixed-length test portion of MIL-STD-781C is given in Epstein and Sobel [2]. Before giving an outline of the statistical basis of such tests, it is necessary to say a few words about the general scheme of the standard.

A BRIEF DESCRIPTION OF THE STANDARD

Military Standard 781C consists of 10 environmental test levels and 30 test plans, allowing a total of 300 choices for test levels and test plans for every specified MTBF. The 10 test levels are based on temperature, vibration, temperature cycling, and equipment on/off cycling. The user must select those conditions most appropriate to the expected conditions of usage of the equipment. The 30 test plans consist of 16 fixed-length tests, labeled Test Plan X through Test Plan XXV, and the details of these plans are discussed here. The remaining test plans are either sequential plans or fixed-length plans having a special character.

Suppose that the random life length X of an item can be described by the exponential density function $g_X(x;\theta) = \theta^{-1}\exp(-x/\theta)$, $x > 0$, where θ is the mean life, assumed unknown. Making a decision to accept or reject a lot reduces to testing the simple hypothesis $H_0 : \theta = \theta_0$ against the simple alternative hypothesis $H_1 : \theta = \theta_1$ (where $\theta_1 < \theta_0$); θ_0 is the specified MTBF, and θ_1 the minimum acceptable MTBF. The type I and type II errors equal the preassigned values α and β, respectively. The ratio θ_0/θ_1 is denoted by the letter d, and is the discrimination ratio; α and β are the producer's and consumer's risks, respectively (*see* ACCEPTANCE SAMPLING).

In all fixed-length plans, the risks are equal and are either 10, 20, or 30%, with discrimination ratios of 1.25, 1.5, 2.0, and 3.0. The test duration time is in multiples of θ_0 and gives the minimum accumulated test time on all samples necessary before an acceptance decision can be made. With each of the 16 test plans referenced before, MIL-STD-781C gives the risks, the discrimination ratio, the test duration time, and the "reject" and "accept" numbers. If the observed number of failures in the specified duration of test time is equal to or greater (less) than the reject (accept) number, the lot is rejected (accepted). Another feature of MIL-STD-781C is the notion of "switching rules" for reduced and tightened testing. These allow a reduction in testing when the reliability expressed as MTBF exceeds requirements and impose tightened testing on inferior products. Reduced testing is possible upon the acceptance of eight consecutive production lots, and the achievement of an observed MTBF at least $1\frac{1}{2}$ times the specified MTBF. Reduced testing continues until a lot is rejected or until production is interrupted for 60 days or more. Tightened testing is imposed when any two lots in any sequence of five lots give rise to a reject decision. The switching rules keep the decision risks the same but increase (decrease) the discrimination ratio when switching from normal to reduced (tightened) testing.

THE STATISTICAL BASIS OF FIXED-LENGTH TESTS

To simplify the calculations, suppose that θ_0, the specified MTBF, is set equal to 1, so that θ_1, the minimum acceptable MTBF, equals $1/d$, where d is the discrimination ratio.

Since the items under test are assumed to have exponentially distributed life lengths, the number of failures in an interval of length t will have a Poisson distribution*, with parameter $\lambda = \theta^{-1}$, where θ is the mean time to failure. That is, the probability of observing x failures in time $[0, t)$ is

$$P(x;\theta,t) = \frac{e^{-(t/\theta)}(t/\theta)^x}{x!}, \qquad x = 0, 1, 2, \ldots .$$

The probability of observing r or fewer failures in time $[0, t)$ is given by the cumulative distribution function

$$P(r;\theta,t) = \sum_{x=0}^{r} \frac{e^{-(t/\theta)}(t/\theta)^x}{x!}.$$

For the given hypotheses H_0 and H_1 we can, for specified values of α, β, and d, determine a test time t and an acceptable number of failures r by finding the smallest values r and t such that

$$\sum_{x=0}^{r} \frac{e^{-t}t^x}{x!} \geqslant 1 - \alpha \quad \text{and}$$

$$\sum_{x=0}^{r} \frac{e^{-td}(td)^x}{x!} \leqslant \beta.$$

Techniques that facilitate the computations involved in solving the preceding inequalities are given in Mann et al. [4, p. 316].

Barlow and Proschan [1] have studied the consequences of using the fixed-length tests discussed here when the underlying distribution of life lengths is not exponential. These tests favor the producer (consumer) when the underlying distribution of life lengths has an increasing (decreasing) failure rate.

Harter [3] is also relevant to the material discussed here. Other military standards that consider fixed-length tests are MIL-STD-471 for maintainability demonstration and MIL-

STD-690B, in which the parameter of interest is the failure rate θ^{-1}.

References

[1] Barlow, R. E. and Proschan, F. (1967). *J. Amer. Statist. Ass.*, **62**, 548–560.

[2] Epstein, B. and Sobel, M. (1953). *J. Amer. Statist. Ass.*, **48**, 486–502.

[3] Harter, H. L. (1978). *J. Qual. Technol.* **10**, 164–169.

[4] Mann, N. R., Schafer, R. E., and Singpurwalla, N. D. (1974). *Methods for Statistical Analysis of Reliability and Life Data*, Wiley, New York.

[5] Military Standard 471 (1966). "Maintainability demonstration." Department of Defense, Washington, D.C., 15 February.

[6] Military Standard 690B (1968). "Failure rate sampling plans and procedures." Department of Defense, Washington, D.C., 17 April.

[7] Military Standard 781C (1977). "Reliability design qualification and production acceptance tests: Exponential distribution." Department of Defense, Washington, D.C., AMSC No. 22333, 20 October.

[8] Neathammer, R. D., Pabst, W. R., Jr., and Wiggington, C. G. (1969). *J. Qual. Technol.*, **1**, 91–100.

(ACCEPTANCE SAMPLING
LIFE TESTING
MILITARY STANDARDS FOR
SEQUENTIAL LIFE TESTING
QUALITY CONTROL, STATISTICAL
RELIABILITY, STATISTICAL METHODS IN
SAMPLING PLANS)

Acknowledgment

Supported by the Army Research Office under Grant DAAG 29-80-C-0067, and Office of Naval Research Contract N00014-77-C-0263, Project NR 042-372, with The George Washington University, Washington, D.C.

NOZER D. SINGPURWALLA

MILITARY STANDARDS FOR SEQUENTIAL LIFE TESTING

The U.S. Department of Defense issues a series of documents called military standards (abbreviated MIL-STD) that specify procedures and rules for undertaking several activities of interest to the department. One such activity, a major one, is the procurement of goods and services from industry and contractors. Several standards, each appropriate for various circumstances of this activity, have been issued over the past several years. Although the MIL-STDs were developed for use within the Department of Defense, they have proved to be so successful that they have also been adopted by other agencies of the U.S. government, some foreign governments, and private industry. An example of a commonly used military standard is MIL-STD-781C, entitled "Reliability Design Qualification and Production Acceptance Tests: Exponential Distribution." This standard describes the various testing environments and specifies rules for accepting or rejecting a batch (lot) of items based on their life length characteristics when these are assumed to have an exponential distribution*. The statistical basis underlying the sequential life testing part of MIL-STD-781C was given by Epstein and Sobel [2] and is an adaptation of Wald's [8] sequential procedures for the exponential distribution.

In sequential life testing, the items are subjected to a life test one after another until a decision to accept or reject a lot is made. This is in contrast to a fixed sample-size life test, in which items are subjected to a life test* simultaneously, and the test terminates when either some or all of the items under test fail. Thus in sequential life tests, the total number of items tested is random. The key advantage of sequential life testing is the possibility of saving on the number of items that are tested. This happens because of an early acceptance (rejection) of the lot whenever the items tested reveal highly desirable (undersirable) life characteristics.

In MIL-STD-781C, since the items tested are assumed to have an exponential distribution of life lengths, the simultaneous testing of a group of items, as opposed to testing one item at a time sequentially, is allowed. Group testing is undertaken to save on test time and is theoretically made possible by the lack of memory property of the exponential distribution. Another feature of MIL-STD-781C is that known as truncation*.

Truncation implies the termination of the sequential test by a premature decision to accept or reject the lot. Truncation sets an upper limit on the time that a sequential life test will take to complete, and its consequences are an increase in the risks of incorrect decisions. *See also* SEQUENTIAL ANALYSIS.

SEQUENTIAL LIFE TESTS IN THE EXPONENTIAL CASE

Suppose that the random life length X of an item can be described by the exponential density function $g_X(x;\theta) = \theta^{-1}\exp(-x/\theta)$, where $\theta > 0$ is the mean life, assumed to be unknown. Making a decision either to accept or reject a lot of items whose life lengths have the density function $g_X(x;\theta)$ reduces to testing the simple hypothesis $H_0 : \theta = \theta_0$ against the simple alternative hypothesis $H_1 : \theta = \theta_1$ (where $\theta_1 < \theta_0$). The type I and type II errors equal the preassigned values α and β, respectively.

In MIL-STD-781C, θ_0 is known as the *upper test mean*, θ_1 the *lower test mean*, and the ratio θ_0/θ_1 the *discrimination ratio*, denoted by the letter d. The error probabilities α and β are known as the producer's and consumer's risks, respectively (*see* ACCEPTANCE SAMPLING). The criteria for accepting or rejecting the lot are given for values of α and β between 0.01 and 0.20, and d is between 1.5 and 3.0.

The test is conducted by placing $n \geqslant 1$ items on a life test and continuously monitoring the value of $V(t)$, the total time on test, where

$$V(t) = \begin{cases} nt, & \text{if failed items are replaced by fresh ones,} \\ \sum_{i=1}^{r} X_{(i)} + (n-r)(t - X_{(r)}), & \text{otherwise;} \end{cases}$$

$X_{(1)} \leqslant X_{(2)} \leqslant \cdots \leqslant X_{(r)}$ denote the r-ordered times to failure in $[0, t)$. Information about θ is continuously available through $V(t)$, and if it happens that no decision is made by the time all the n items have failed, then n (or even less than n) more items are put on the test, and $V(t)$ continues to be monitored until a decision is reached.

The decision rule calls for a continuation of the life test when

$$B < (\theta_0/\theta_1)^r \exp[-((1/\theta_1) - (1/\theta_0))V(t)] < A, \qquad (1)$$

where $B = \beta/(1-\alpha)$ and $A = (1-\beta)/\alpha$. Testing is stopped with a decision to accept H_0 when the left-hand side of the inequality (1) is violated, and a decision to accept H_1 when the right-hand side of inequality (1) is violated (*see* SEQUENTIAL PROBABILITY RATIO TEST).

A Plot of the Sample Path of $V(t)$

A visual display of the sample path of $V(t)$ vs. r, the observed number of failures in $[0, t)$, is instructive. To do this, it is convenient to write (1) as

$$-h_1 + rs < V(t) < h_0 + rs, \qquad (2)$$

where for $\tilde{\theta} = \{(1/\theta_1) - (1/\theta_0)\}$, $h_0 = -\log B/\tilde{\theta}$, $h_1 = \log A/\tilde{\theta}$, and $s = \log d/\tilde{\theta}$. A plot of the boundaries $-h_1 + rs$, $h_0 + rs$, and the regions for acceptance, rejection, and continuance of the sequential life test is shown in Fig. 1; also shown are two possible sample paths for $V(t)$, one in solid lines and the other in dashed lines. The former shows an acceptance of H_0 after observing three failures, but before observing the fourth failure. The latter shows an acceptance of H_1 (rejection of H_0) upon the occurrence of the eighth failure. An important feature of MIL-STD-781C, and one that is obvious from an inspection of Fig. 1, is that whereas it is possible to accept H_0 without observing a single failure, the rejection of H_0 calls for observing at least r^* failures, where $r^* =$ largest integer contained in $(h_1/s) + 1$. In Fig. 1, $r^* = 5$. This is reasonable, since long life lengths are supportive of H_0, whereas small life lengths are supportive of H_1, and the rejection of a lot should be undertaken only when a sufficient number of failures have been observed.

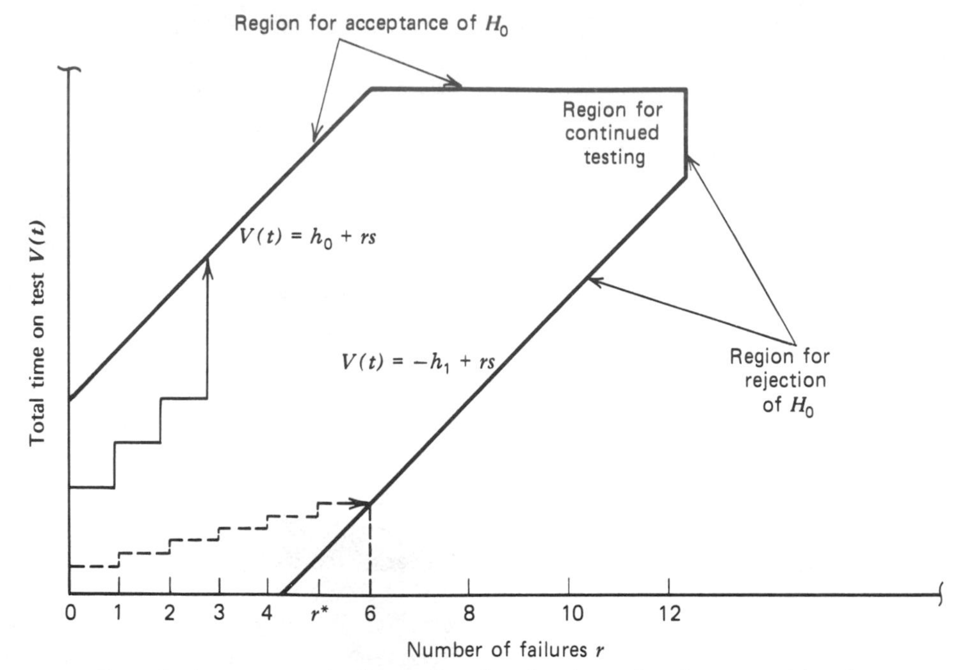

Figure 1 Acceptance, rejection, and continued testing regions for sequential testing.

In Fig. 1, toward the right-hand end of the lines $h_0 + rs$ and $-h_1 + rs$, respectively, a horizontal and a vertical line are shown. These lines help define the boundary of the region for continued testing and are based on the nature of the truncation rule, which is desired for the sequential life test. Thus, according to Fig. 1, the test must terminate at or before the occurrence of the twelfth failure. Note that whenever a sequential life test terminates by reason of $V(t)$ reaching either the horizontal or the vertical truncation line, the risks α and β will no longer have their specified values.

The Operating Characteristic Curve and the Expected Number of Failures

Let $L(\theta)$ denote the probability that the sequential life test will terminate with an acceptance of H_0, when θ is the true value of the mean life. A plot of $L(\theta)$ versus θ is the operating characteristic (OC) curve of the life test. Given α, β, and d, the desirability or lack of desirability of a sequential test is judged by the nature of its OC curve and also the expected number of failures to arrive at a decision. The ideal OC curve for any test of the hypotheses given before is one for which $L(\theta) = 0$ for $\theta < \theta_0$ and $L(\theta) = 1$ for $\theta \geqslant \theta_0$. However, even an infinite amount of testing will not help achieve this ideal, and so one must focus attention on procedures whose OC curves come close to the ideal curve. The closer the OC curve is to the ideal one, the better the underlying sequential life test. In MIL-STD-781C, a series of OC curves for various choices of α, β, and d are given.

An expression for obtaining $L(\theta)$ for different values of θ has been given by Epstein and Sobel [2]. This expression, besides being an approximation, is difficult to explain without the introduction of additional notation. However, the following points on the operating characteristic curve are easy to specify:

$L(\theta_1) = \beta$, $L(\theta_0) = 1 - \alpha$,

$L(s) = \log A / (\log A - \log B)$,

$L(0) = 0$, $L(\infty) = 1$.

Let $E_\theta(r)$ denote the expected number of failures to arrive at a decision when θ is the true value of the mean life; $E_\theta(r)$ is known as the average sample number*, and [2]

$$E_\theta(r) \simeq \begin{cases} \dfrac{h_1 - L(\theta)(h_0 + h_1)}{(s - \theta)}, & \theta \neq s, \\ h_0 h_1 / s^2, & \theta = s. \end{cases}$$

It can be easily verified that $E_\theta(r)$ attains its maximum when $\theta = s$ and that $E_{\theta_0}(r) \leqslant E_{\theta_1}(r)$; note that $\theta_1 < s < \theta_0$. Thus the sequential life test described here calls for a maximum amount of testing when $\theta = s$, and the amount of testing when θ is close to θ_0 is less than that required when θ is close to θ_1. An explanation for the latter result is that the rejection of H_0 can only occur at the time of a failure, whereas its acceptance can occur between failure times, even before the first failure is observed.

Some Recent Results

Even though MIL-STD-781C clearly specifies that the underlying procedures are appropriate when the life distribution is exponential, there does remain the possibility that these procedures can be misused. Harter and Moore [3] show the consequences of using the sequential part of MIL-STD-781C when the underlying life distribution is Weibull. In particular, if the shape parameter of the Weibull distribution* is greater than 1, then both the producers' and consumers' risks are lowered, whereas if the shape parameter is less than 1, both these risks increase. Montagne and Singpurwalla [5] generalize the results of Harter and Moore when the underlying distribution of life lengths has a monotone failure rate (*see* HAZARD RATE CLASSIFICATION OF DISTRIBUTIONS).

Sequential life tests for situations in which there is some prior knowledge about the scale parameter of the exponential distribution have been considered by Schafer and Singpurwalla [6]. Sequential life test procedures for other life distributions such as the Weibull, or the gamma*, are not available.

Once a sequential life test is terminated, with a decision being made as to whether to accept or to reject the batch, the observed life lengths can be used to estimate the parameters of the life distribution in question. Methods for doing this have been discussed by Bryant and Schmee [1] and by Siegmund [7].

References

[1] Bryant, C. and Schmee, J. (1979). *Technometrics*, **21**, 33–38.

[2] Epstein, B. and Sobel, M. (1955). *Ann. Math. Statist.*, **26**, 82–93.

[3] Harter, H. L. and Moore, A. H. (1976). *IEEE Trans. Reliab.*, **R-25**, 100–104.

[4] Military Standard 781C, "Reliability Design Qualification and Production Acceptance Tests: Exponential Distribution," U.S. Department of Defense, Washington, D.C., AMSC No. 22333, 21 October 1977.

[5] Montagne, E. R. and Singpurwalla, N. D. (1982). "On the Robustness of the Exponential Sequential Life Testing Procedures when the Distribution has Monotone Failure Rate." Technical Paper IRRA 82/2, Institute for Reliability and Risk Analysis, George Washington University, Washington, D.C.

[6] Schafer, R. E. and Singpurwalla, N. D. (1970). *Nav. Res. Logist. Q.*, **17**, 55–67.

[7] Siegmund, D. (1979). *Nav. Res. Logist. Q.*, **26**, 57–67.

[8] Wald, A. (1947). *Sequential Analysis*, Wiley, New York.

(ACCEPTANCE SAMPLING
LIFE TESTING
MILITARY STANDARDS FOR FIXED-LENGTH LIFE TESTS
QUALITY CONTROL, STATISTICAL
RELIABILITY, STATISTICAL METHODS IN
SAMPLING PLANS
SEQUENTIAL ANALYSIS)

Acknowledgment

Supported by the Army Research Office, Grant DAAG 39-80-C-0067, and the Office of Naval Research, Contract N00014-77-C-0263, Project NR 042-372, with The George Washington University, Washington, D.C.

NOZER D. SINGPURWALLA

MILITARY STATISTICS

Statistics and probability have a long history in the military. While not always warmly received in that setting, these disciplines have persisted because of their usefulness and strong proponents who could visualize and promote their applications. Much of what we will discuss relates to the military in the United States of America. Other nations (e.g., Britain) have employed statistics in the armed services. Some of us are familiar with Florence Nightingale* and her efforts in the mid-nineteenth century to improve the sanitary services of the British army and the use of statistics by her as an important component in this drive [7]. This was echoed in the United States at roughly the same time in such publications as the six-volume *Medical and Surgical History of the War of the Rebellion* reporting on the Civil War and a two-volume report published in 1875 *Medical Statistics of the Provost Marshal General's Bureau* [9] that contained the results of the anthropometric examination of more than a million men. In this way army statisticians were responsible for exceedingly valuable data on Civil War recruits. There is an interesting table, along similar lines, of height and chest measurements of 5,732 Scottish militia men compiled by an army contractor and listed in 1817 [16].

In World War I, the army Alpha examination supplied statisticians with data on human mental abilities. Since then tests of this sort have become commonplace in all branches of the armed services for purposes of classification of recruits. Records derived from World War II and its aftermath included attitudinal, demographic, and personality variables on recruits leading to many studies to provide policies in personnel selection for military occupational specialties. The advent of the All Volunteer Service in the United States in the early 1970s increased the use of such statistical information in recruitment and reenlistment procedures.

For what might be considered more traditional programs in the military, one can examine Daniel Bernoulli's work in 1777 [3], in which he introduced the example of the firing of a marksman in discussing the estimation of the most probable value for a set of different observations. About this time, cryptanalysis was already being employed [8]. Subsequent to these efforts, there was widespread military use of statistics and probability in the nineteenth century in land survey work performed by army engineers. A classic American text in least squares* was written by Wright [15] in the late nineteenth century based on his experiences with the U.S. Army Corps of Engineers.

The Civil War was a testing ground for many innovations in ordnance material. However, this material had not gone through development cycles and congressional criticism was leveled at the army afterwards for lack of controls in testing and for ignoring operational conditions in testing. The notion of probable error* in measuring firing accuracy also received attention after the Civil War. In 1885, in a handbook published for the use of cadets at the Naval Academy [9], Army Captain James M. Ingalls included an appendix that discussed the theory of least squares, examination of residuals*, outliers*, and the probability of at least one hit from several shots.

In the decade after World War I, the Ballistic Research Laboratories at the U.S. Army Aberdeen Proving Ground became a focal point for statistical work on ordnance problems. The sample range* as the basis for an estimate of variance was investigated and its probability distribution developed. This was followed in the 1930s by surveillance testing of ammunition and the introduction of concepts of quality control* and quality assurance in ammunition surveillance. These concepts originated with and were developed by Bell Telephone Laboratories engineers for Western Electric production. Prominent among this group of investigators were Walter A. Shewhart, Harold F. Dodge, and Harry G. Romig. Colonel Leslie E. Simon of Army Ordnance along with people at Bell Laboratories set the stage for standard acceptance sampling* procedures for the military in connection with their huge procurement activities in World War II. Before the United States entered that war, Colonel Simon and his associates in Army Ordnance had developed procedures for sampling stockpiles of ammunition that had been in storage for some years to assess the reliability of these reserves. It should also be mentioned here that Colonel Simon helped secure support for the development by John Mauchly and J. Presper Eckert of the first operational electronic computer, the ENIAC, at the Aberdeen Proving Ground.

The coming of World War II led to the institutionalization of statistics and probability to aid military programs that served as a model for post-World War II developments. In 1940, before the United States entered the war, the National Defense Research Committee (NDRC) was established to provide assistance to the military services. This was part of the Office of Scientific Research and Development (OSRD), an agency of the executive office of the president, headed by Vannevar Bush. The NDRC concerned itself with submarine warfare, radar, electronic countermeasures, explosives, rocketry, and other programs.

By 1942, demands for analytical studies increased rapidly and so in the fall of that year a new unit, the Applied Mathematics Panel (AMP), was incorporated into the NDRC. The AMP established contracts to exploit the talents of mathematicians with eleven universities, including Princeton, Columbia, and the University of California (Berkeley) in which statistical and probabilistic work in military problems were featured. Prominent leaders at each were Jerzy Neyman* at Berkeley, Samuel S. Wilks* at Princeton, Harold Hotelling* and Abraham Wald* at Columbia. Their colleagues included a host of younger people who were to have intellectual and leadership roles in statistics and probability for the next forty years. Estimation and distribution theory in bombing problems, operations analysis of submarine patterns and convoy size, sampling inspection procedures for procurement and acquisition, testing procedures for new ordnance, hit probabilities for weapon systems, firing tactics in aerial warfare, estimating vulnerability of aircraft to flak damage, detection methods, and other subjects occupied these groups during World War II. The uses of statistics and probability in coding and decoding went on in great secrecy, and results are not as available as those in the other areas just described. In short, a wide variety of statistical and probabilistic questions motivated by World War II problems led to new developments in theory and methods.

One of the widespread uses of statistics in the military during World War II and after arose in the subject of acceptance sampling. The Statistical Research Group (SRG) at Columbia University under an AMP contract had extended the pioneering work of Shewhart, Dodge, and Romig at Bell Labs on acceptance sampling to include multiple sampling* plans and sequential sampling* plans. Previously, single and double sampling* plans for product acceptance had been developed and employed by Army Ordnance where items received attribute inspection (defective–nondefective). Other results along these lines awaited postwar developments. Abraham Wald's development of sequential analysis* in the war years had a great intellectual impact on statistics, but in the postwar years it did not receive the practical acclaim in acceptance sampling, the subject that motivated it, that one might have expected for it. Administrative and management reasons were cited for not employing it. Nevertheless, it did have a fair amount of success at the end of World War II when several thousand sequential inspection plans were operating at war plants. *See also* SAMPLING PLANS.

The establishment of the Office of Naval Research (ONR) in 1946 reflected the concern of some naval officers and former leaders in the uses of science in the war effort to continue the successful relationships established between the military and scientists and engineers. Mina S. Rees, an AMP employee during World War II, became the first head of mathematical research at ONR in 1946 and within a few years established a statistics branch to handle the ongoing statistics and probability program under Herbert Solomon, who had joined ONR earlier and had previous experience in the SRG group at Columbia. This served as a model for counterparts in the army and air force that were to come about several years later. Under the ONR statistics program, contracts with a number of universities were developed to cover a wide spectrum of statistical subjects in the belief that these basic research efforts would be useful to the military and their contractors as well as to the expansion of the field itself.

In the early days of this program, support to Will Feller led to the first of his two volumes unifying results on probability theory scattered throughout the literature. Work on the Kolmogorov–Smirnov* approach for goodness of fit* by Z. W. Birnbaum provided small-sample-size tables for this test. Small-sample studies of various kinds were conducted by Samuel S. Wilks. Harold Hotelling continued his research in multivariate analysis* of work, initially considered in connection with a wartime problem of air testing sample bombsights, that developed his T_0^2 statistic (*see* HOTELLING'S T^2). In addition, there was work on distributions of quadratic forms* and various aspects of correlation coefficients of importance in personnel selection. Jerzy Neyman had been occupied with statistics in astronomy* and problems of medical diagnosis*. A visit to Neyman by engineers from the Naval Ordnance Test Station led to Best Asymptotically Normal estimates of the variance of a bivariate normal distribution and estimates of a rocket's rotational velocity. Abraham Wald, until his untimely death in 1950, continued his work on sequential analysis and statistical decision theory* under military (ONR) sponsorship.

Work on acceptance sampling plans developed during World War II continued at Stanford under Albert H. Bowker and M. A. Girshick. The SRG at Columbia had developed and published a catalogue of acceptance sampling plans including single, double, and multiple sampling plans, that was soon used in procurement. These plans were for attribute inspection and after the unification of the Department of Defense early in 1949 appeared as JAN (Joint Army–Navy) Standard 105. A revision MIL-STD-105A superseded JAN-STD-105 in 1950. Several revisions occurred in later years, MIL-STD-105D emerging in 1963. The international designation is ABC-STD-105 (American, British, Canadian).

Sampling plans on a variables basis (item characteristic measured on a continuous scale) to match the attribute plans developed during the war were completed. Matching occurs by equating operating characteristic curves. These now appear as Defense Department Handbook Military Standard 414 (MIL-STD-414).

Continuous sampling plans had been developed by Harold F. Dodge and his colleagues at Bell Telephone Laboratories. These allowed for sampling when product was good and 100% inspection when product was defective. All defective products found by inspection were replaced with good product. This still permits some defectives in the process, and the inspection scheme is measured by average outgoing quality* (AOQ). Dodge et al. proposed several minor variations of this inspection scheme. In the early 1950s, Gerald J. Lieberman and Herbert Solomon proposed multilevel continuous sampling plans (MLP) in which inspection varies over two or more levels of sampling fractions up to 100% inspection and the different sampling fractions depend on the quality deduced by the inspection process. Several minor variations of the MLP were then developed. These plans now appear in Defense Department Handbook MIL-STD-1235 (ORD). This was published in 1962 and combined, in a single document, previously published handbooks 106 and 107 that appeared in 1958 and 1959 containing, respectively, the Lieberman–Solomon multilevel plans and the Dodge plans.

As the mathematics division began to mature at ONR, it became more structured in 1949 and 1950. In addition to the contract research program in mathematics (which originally included statistics), programs were developed in computers, in mechanics, and in logistics. The latter area began just before and was developed at least partly in response to the Korean War and represents an area with substantial statistical content initiated and developed by the military. Fred Rigby was the first director of the Logistics Branch within the mathematics division. Working with C. B. Tompkins, at George Washington University in Washington, D.C., he developed a very large logistics research program that became the centerpiece of ONR's logistics program. Tompkins became the first director of the George Washington University (GWU) program. The Korean War afforded an opportunity to investigate empirically the logistics operations of the navy and a plan known as the Korean Data Collection Plan was formulated. Huge amounts of data were collected which stimulated not only the need for improved modeling of logistics operations, but also clarified the need for and role of computers in logistics. George Washington University acquired one of the early large computers (developed by Engineering Research Associates of Minneapolis) and began work that anticipated much of the modern work on data base management. The Logistics Computer was a prototype for the UNIVAC File Computer that served the airlines in the early days of reservations by computer. Rigby's program in logistics was not limited to the GWU project, but in fact supported a variety of other statisticians and mathematicians. Notable among these were J. Wolfowitz and his associates at Columbia University, who worked on inventory control models. Much of the early work in queueing* and inventory theory* was motivated by the logistics requirements of the military. Just after World War II, the air force program, Scientific Computation of Optimal Programs (SCOOP) directed by Marshall Wood was another military logistics activity dealing with large data bases for planning and control purposes. It was within this program that George Dantzig developed and promoted linear programming* methods that were motivated by earlier work on Neyman–Pearson* theory in his dissertation.

The GWU project in logistics continues on to the present time but has shifted emphasis from time to time. With the advent of nuclear submarines in the late 1950s, emphasis shifted to inventory* problems aboard a nuclear vessel. Admiral William Rayborn was head of the Special Projects Office in the Pentagon that had cognizance of the Polaris-class submarine in the late 1950s. He and later his deputy, Admiral Levering Smith, worked with the GWU project on the development of allowance lists for stores to be carried aboard the submarine. Such an approach was motivated by the volume-limited storage capacities aboard nuclear submarines that, for the first time, would be away from ports for very long times. This presented new requirements for stores that were addressed at GWU in the late 1950s. The work on allowance lists benefited from previous GWU efforts on the Korean Data Collection Plan.

In the mid to late 1960s, attention turned to standardizing the navy data collection* system. By this time the Statistics and the Logistics Branches at ONR had been consolidated and Robert Lundegard, Marvin Denicoff, and Captain Fred Bellar began an effort to standardize data collection in the navy to reduce redundant collection of information. Known formally as the Standard Navy Maintainance and Material Management Data Collection System, it was known less formally as the navy's 3M system. Attention was turned again as in the Korean War era to the problems of collecting and analyzing very large data sets with a strong operational emphasis. That system has been superseded, but at the time it reflected the first systematic attempt to improve the quality of statistical reporting navy-wide. In recent years, the GWU logistics project has focused on readiness issues, and the research content has migrated away from more traditional statistics.

The scientific/mathematical approach to logistics in the early 1950s was revolutionary. Since an established forum did not exist, the *Naval Research Logistics Quarterly** was established in Fred Rigby's logistics branch under the editorship of Jack Laderman in 1953. Laderman was succeeded by Alan J. Hoffman in 1955; M. E. Rose in 1956; Commanders M. Rosenberg, H. Jones, and H. D. Moore in 1957, 1959, and 1960, respectively, Jack Laderman again in 1961; and finally by Seymour Selig in 1963. The quarterly continues to be published, with Herbert Solomon assuming editorship in 1983. In the early issues, papers tended to be somewhat discursive while in more recent years, the quarterly has become a much more mathematically oriented operations research journal. Of course, the traditional emphasis on logistics has persisted throughout its long history.

While the navy with its office of Naval Research was the first service to establish a contract research program in statistics, the other services were also involved in exploiting statistics. Frank Grubbs at the Aberdeen Proving Ground was one of the first statisticians working in-house on projectile dispersion and targeting problems (*see* TARGET COVERAGE). Army support of the Statistical Research Group at Columbia University was mentioned earlier. The early exploitation of statistical techniques by the army was mainly in connection with ordnance acceptance sampling and testing. In 1950, the Army Office of Ordnance Research was established at Duke University under Professor W. Gergen. The scope of the office was broadened in 1952 and it officially became the Army Research Office—Durham (ARO–D) at that time. While the statistics component of ARO has only been highlighted recently as a separate program, statistics has always been a part of the mathematics program. Sam Wilks at Princeton had been an army contractor since the Second World War, and with ARO support began the annual Army Design of Experiments Conference in 1955. The first of these conferences was held at the Bureau of Standards, and this series of conferences continues to the present time.

The air force also had an early research program located at Wright–Patterson Air Force Base (then Wright Field) in about 1950. This later became the *Air Force Office of Scientific Research* (AFOSR) in 1952. Like the army, the air force has not had a separately identified statistics program until relatively recent times.

However, an important interagency group was formed in 1951. Called the Joint Services Advisory Group (JSAG), the group had two representatives from each of the services acting in concert to identify mathematical and statistical problems common to the services. Its initiation was motivated by Korean War concerns. In the beginning major university groups were formed under the JSAG, at the University of Chicago under W. A. Wallis, at Stanford University under A. Bowker and M. Girshick, and at Princeton under S. Wilks and J. Tukey. They were joined later by a group at North Carolina State University under J. Cell.

From the founding of ONR in 1946 through the postwar Korean War era to the launch of the Soviet satellite SPUTNIK in 1957, we have a particularly noteworthy era in the history of military statistics. It was characterized by close interaction of the statistical community with the military. Many important results were discovered, and many programs initiated. Not only were ONR, ARO, and AFOSR founded, but the National Science Foundation (NSF) was founded in 1951 on the model of ONR and a number of its senior leaders, including the first director, came directly from ONR. The JSAG gained great visibility for statistics in the military and, conversely, drew the attention of the academic statisticians to problems in the military. Concomitantly, there was a Joint Services Program in Quality Control that developed and codified much of the work in acceptance sampling which became the basis of many of the military standards still in use as of this writing. The service contract research programs in statistics were firmly established in this era. In particular, the ONR program was established as a separate component of the mathematics division. Following Solomon's tenure as head of statistics from 1949 to 1952, Ed Paulson became director through 1953. He was followed by Eugene Lukacs until 1956, who was then succeeded by Dorothy Gilford in 1956.

The launch of the Soviet satellite in 1957 had very strong repercussions on the political and technical leadership in the United States. Long convinced of their own technological superiority, U.S. political leadership quickly mobilized programs and funds to close the imagined gap between Soviet and U.S. science.

Funding at the Department of Defense (DOD) agencies as well as the civilian agencies (notably NSF) grew rapidly. Of particular concern was the possible lack of an adequate base of scientific personnel in the United States. The National Science Foundation began its fellowship program in the early 1960s while its defense counterpart, called National Defense Education Act (NDEA) fellowships also began in the early 1960s. Within the DOD, funding for large-scale interdisciplinary programs became available under the name of THEMIS in the late 1960s. The THEMIS program as well as most of the other programs just mentioned were attempts to build the science base within the United States. As a result of this shift in emphasis, the close coupling between the academic statistics community and the military enjoyed in the postwar Korean War era eroded.

This decoupling in the post-Sputnik era was due largely to the increased availability of funding and the consequent seller's market for academic statisticians. Particularly notable was the development of Bayesian inference* methods through the 1960s with strong impulse given by L. J. Savage in his 1954 book on subjective probability. This work was supported initially by ONR while Savage was at the University of Chicago. The seminal papers, in probability density and spectral density estimations*, were published in this era, notably papers by Rosenblatt [12], Parzen [10], and Watson and Leadbetter [14], all sponsored by ONR. The foundations of the theory of inference under order* restrictions were laid by many people during this era and the well-known monograph of Barlow et al. [2] resulted from this work. The theory of second-order stochastic processes was codified by Cramér and Leadbetter [6] under navy sponsorship while the ubiquitous fast Fourier transform was derived by Tukey and Cooley [5] under army sponsorship. Throughout the 1960s, also primarily under army sponsorship, the autoregressive–moving average models* in time series* were developed, resulting in the text by Box and Jenkins [4]. Advances in density estimation were translated into advances in clustering and classification* methodology. The inauguration of support by the National Institutes of Health for statistics research encouraged rapid growth of the biostatistics community. The area of nonparametric statistics was extensively developed in this era, in particular, the theory of rank tests, which was developed in large part by E. Lehmann under ONR sponsorship. The development of reliability theory*, particularly nonparametric life testing and work on lower bounds for coherent systems, received considerable attention in this era.

This flourishing era tapered off in 1968 and 1969 for several reasons. The achievement of the lunar landing in general completed the national commitment to building up the science base in the United States. Supply of scientists in general and mathematicians in particular caught up with demand. Although statisticians were not as oversubscribed as mathematicians, the general perception was that the United States was once again the clear technological leader. The separation of academic community from the military was highlighted by the Mansfield Amendment in 1969, which basically required the DOD funding agencies to fund only research directly relevant to the mission of the agency. This amendment had little effect on the statistics program in the DOD agencies, but caused major perturbations in other disciplines.

The Vietnam War also had a chilling effect on the academic–military relationship.

The culmination of a declining relationship occurred with the U.S. invasion of Cambodia in October 1970, the subsequent shooting incident at Kent State University, and the resultant widespread student strikes. The Arab oil embargo in the winter of 1973 initiated a recession that completed a sequence of events that reduced the level of DOD support for academic research to an unprecedented low.

The post-Vietnam War era, say 1972 to the present time, has been an era of substantial invigoration of the academic–military relationship. In the AFOSR, the statistics program became much more visible as a separate program under the leadership of Dr. David Osteye in 1974 and then experienced dramatic growth under Dr. Ismail Shimi, who became director of the program at AFOSR in 1975. The air force, in particular, has been a strong proponent and sponsor of research in reliability theory, notably in the project at Florida State University under Frank Proschan, in stochastic process theory at Northwestern University under E. Çinlar, and much more recently in signal processing at the University of North Carolina at Chapel Hill.

The ARO also highlighted the statistics program as a separate program in 1976 under the direction of Robert Launer. The ARO has been particularly active in developing exploratory data* techniques by J. Tukey and his associates and in developing much of the modern work in robustness*. The landmark book by Andrews et al. [1] was sponsored by ARO and set the tone for subsequent development in the area of robustness through the later 1970s.

In the post-Sputnik era, Robert Lundegard succeeded Dorothy Gilford as the statistics program director at ONR. By 1969, Lundegard had built the program, then called the Mathematical Statistics and Logistics Branch, to sizable proportions. In 1969, he became director of the mathematics division at ONR and subsequently reorganized the division. A new branch in Operations Research was carved out of the old Mathematical Statistics and Logistics Branch with the old branch being redesignated as the Statistics and Probability Branch. Bruce McDonald succeeded Lundegard as branch head, followed by Edward Wegman in 1978.

The program under McDonald was organized into two major pieces: the early work in inspection sampling grew into a major subprogram in quality control and reliability while more general statistical interests were manifested in a major subprogram in modeling and estimation. In 1969, McDonald inaugurated work in statistical signal analysis that grew into a major program by 1972. This work was closely coupled with the Naval Electronic Systems Command and represented a reinitiation of the close coupling between academic and military. This subprogram began as an effort to address naval ocean surveillance. The work by Parzen [11] originating and developing the so-called high-resolution AR spectral estimates was funded by this program and has found very widespread usage in ocean surveillance.

Beginning in 1979, the Office of Naval Research has used a planning process called POM (for program objectives memoranda). The POM process is a method of developing five-year defense plans in which program directors of ONR compete for money for their discipline areas based on proposals to high-level management to complete more focused research projects. The statistics and probability program at ONR has been singularly successful in these competitions with a sixfold increase in funds between 1978 and 1982. In addition to the aforementioned subprograms of modeling and estimation (now called mathematical statistics), quality control and reliability (now called quality assurance), and statistical signal analysis, the Statistics and Probability Branch also has subprograms in logistics, computational statistics, simulation* methodology, statistical target tracking, remote sensing data analysis, statistical aspects of electronic warfare, non-Gaussian signal processing, and stochastic modeling in the neurosciences. All three of the DOD research funding agencies have fostered a resurgence in the relationship between the various military commands and the academic statistics community.

Perhaps the salient feature of the military–academic statistics interaction is the wide support for training of statisticians by the defense agencies both under the NDEA program in the 1960s and also by the agencies themselves as part of research contracts. (Research support in statistics by the DOD agencies was more than 2.4 times the NSF support in 1982). The training made possible by the DOD has resulted in a generation of computer-literate statisticians that has had an undeniably potent influence on statistics in the late 1970s and early 1980s. Notable among the advances facilitated by the new computer literacy are those in exploratory data analysis, time-series analysis, robustness, jackknife*, bootstrap*, principal components (*see* COMPONENTS ANALYSIS) and, in general, the increasingly nonparametric interactive approach to statistical analysis.

References

[1] Andrews, D. F., Bickel, P. J., Hampel, F. R., Huber, P. J., Rogers, W. H., and Tukey, J. W. (1972). *Robust Estimates of Location: Survey and Advances*. Princeton University Press, Princeton, NJ.

[2] Barlow, R. E., Bartholomew, D. J., Bremner, J. M., and Brunk, H. D. (1972). *Statistical Inference under Order Restrictions*. Wiley, New York.

[3] Bernoulli, D. (1777). *Acta Acad. Sci. Imp. Petropolitanae* Pt. I, 3–23. (English transl.: *Biometrika*, **48**, 3–13 (1961).)

[4] Box, G. E. P. and Jenkins, G. M. (1970). *Time Series Analysis Forecasting and Control*. Holden-Day, San Francisco.

[5] Cooley, J. and Tukey, J. W. (1965). *Math. Comput.*, **19**, 297–301.

[6] Cramér, H. and Leadbetter, M. R. (1967). *Stationary and Related Stochastic Processes*. Wiley, New York.

[7] Diamond, M. and Stone, M. (1981). *J. R. Statist. Soc. A*, **144**, 66–79.

[8] Friedman, W. F. (1928). "Report on the History of the Use of Codes and Code Languages." Washington, DC.

[9] Maloney, C. J. (1962). *Amer. Statist.*, **16** (3), 13–17.

[10] Parzen, E. (1962). *Ann. Math. Statist.*, **33**, 1065–1076.

[11] Parzen, E. (1969). In *Multivariate Analysis*, Vol. II, P. R. Krishnaiah, ed. Academic Press, New York.

[12] Rosenblatt, M. (1956). *Ann. Math. Statist.*, **27**, 832–837.

[13] Savage, L. J. (1954). *The Foundations of Statistics*. Wiley, New York.

[14] Watson, G. S. and Leadbetter, M. R. (1963). *Ann. Math. Stat.*, **34**, 480–491.

[15] Wright, T. W. (1884). *Treatise on the Adjustment of Observations*. D. Van Nostrand, Princeton, NJ.

[16] *Edinburgh Med. Surgical J.* (1817), **13**, p. 260.

(ACCEPTANCE SAMPLING
MILITARY STANDARDS FOR FIXED-LENGTH LIFE TESTS
MILITARY STANDARDS FOR SEQUENTIAL LIFE TESTING
NAVAL RESEARCH LOGISTICS QUARTERLY
QUALITY CONTROL, STATISTICAL
TARGET COVERAGE)

HERBERT SOLOMON
EDWARD J. WEGMAN

MILLER–GRIFFITHS PROCEDURE

The Miller–Griffiths (M-G) selective procedure for estimating the population mean μ in ranked set sampling* (useful when the sample elements can easily be ordered by inspection, but exact measurements are costly) was suggested by A. J. Miller and D. A. Griffiths of CSIRO (Commonwealth Scientific and Industrial Research Organization, Australia) in 1979.

Assume that the population possesses a finite mean μ and finite variance. Let $n = 2l$ sets of size m be selected independently from the population and the elements in each set be ordered by the order of magnitude of the characteristic to be estimated. Let $I_1, I_2, \ldots I_l$ be independent random variables such that

$$\Pr[I_i = j] = P_{ij},$$
$$i = 1, 2, \ldots, l; \quad j = 1, \ldots, m,$$

with $\sum_{j=1}^{m} P_{ij} = 1$ for each i. $\{P_{ij}\}$ is called the *selective probability matrix*. If $I_1 = i_1$, then select the i_1th smallest element from the first set and the $(m + 1 - i_1)$th smallest element from the nth set and measure them (and so on). In general, if $(I_1, I_2, \ldots, I_l) = (i_1, i_2, \ldots, i_l)$, we have

$$\begin{array}{ll} X_{1,i_1} & X_{n,m+1-i_1}; \\ X_{2,i_2} & X_{n-1,m+1-i_2}; \\ \vdots & \vdots \\ X_{l,i_l} & X_{l+1,m+1-i_l}, \qquad n = 2l, \end{array}$$

where $X_{\alpha,i}$ is the ith smallest order statistic* of a random sample of size m selected from the αth set, $\alpha = 1, \ldots, n$. The M-G estimator of μ is given by the arithmetic mean* of measured values:

$$\bar{X} = \frac{1}{n} \sum_{j=1}^{l} \left[X_{j,i_j} + X_{n+1-j,m+1-i_j} \right].$$

If $m = n$ and $P_{ii} = 1$, and $P_{ij} = 0$ for $i \neq j$, $i = 1, \ldots, l$; $j = 1, \ldots, m$, this procedure reduces to the McIntyre–Takahashi–Wakimoto selective procedure [1,2].

The worst estimator in the M-G procedure is the one obtained by simple random selection of an element from each set; this estimator is still more efficient than the sample mean based on n observations.

Detailed discussion of the M-G procedure including suggestions for optimal choices of $\{P_{ij}\}$ is given by Yanagawa and Chen (1980).

References

[1] McIntyre, G. A. (1952). *Aust. J. Agric. Res.*, **3**, 985–990.

[2] Takahashi, K. and Wakimoto, K. (1968). *Ann. Inst. Statist. Math.*, **20** 1–31.

[3] Yanagawa, T. and Chen S-H. (1980). *J. Statist. Plan. Infer.*, **4**, 33–44.

(RANKED SET SAMPLING)

MILLER'S PARADOX

The following fallacious argument has been given a surprising amount of attention in the philosophical literature, where it has been called *Miller's paradox of information*.

Let p and q denote a coin's objective probabilities for heads and tails, respectively. Assume that the coin never lands on edge, so that $q = 1 - p$. A Bayesian statistician's subjective probability* for the coin landing heads on a particular toss, conditional on the event $p = \frac{1}{2}$, is $\frac{1}{2}$. In symbols,

$$P(\text{heads} \mid p = \tfrac{1}{2}) = \tfrac{1}{2}. \tag{1}$$

Substituting for $p = \frac{1}{2}$ the equivalent condition $p = q$, we obtain

$$P(\text{heads} \mid p = q) = \tfrac{1}{2}. \tag{2}$$

But if $p = q$, then $\frac{1}{2} = q$, so we can substitute q for $\frac{1}{2}$ in (2), obtaining

$$P(\text{heads} \mid p = q) = q, \tag{3}$$

which looks like a generalization of (1). Combining (1) and (3), we obtain the condition

$$q = P(\text{heads} \mid p = q) = P(\text{heads} \mid p = \tfrac{1}{2}) = \tfrac{1}{2}.$$

In other words, the objective probability q of tails is necessarily equal to $\frac{1}{2}$.

The fallacy in the argument is obviously in the step from (2) to (3). The condition $p = q$ in (2) is only a condition defining the conditional probability on the left-hand side. It is not an assumption that can be used for purposes other than the computation of that conditional probability.

Further Reading

For a review of the philosophical literature on this fallacy, see R. C. Jeffrey's 1970 paper in the *Journal of Symbolic Logic* (**35**, 125–127). A 1979 article by Colin Howson and Graham Oddie in *The British Journal for the Philosophy of Science* (**30**, 253–261) translates the fallacy into symbolic logic in such a way that it is seen to depend on the illegitimate substitution of a bound term for a free variable.

(FALLACIES, STATISTICAL
LOGIC OF STATISTICAL REASONING)

GLENN SHAFER

MILLIKEN'S ESTIMABILITY CRITERION

Throughout this entry, $\mathbf{A}'$, $C(\mathbf{A})$, $N(\mathbf{A})$, $r(\mathbf{A})$, and $\mathrm{tr}(\mathbf{A})$ will denote the transpose, column space, null space, rank, and trace, respectively, of a real matrix $\mathbf{A}$. Moreover, $\mathbf{A}^-$ will stand for a generalized inverse* of $\mathbf{A}$, i.e., for any solution to the equation $\mathbf{A}\mathbf{A}^-\mathbf{A} = \mathbf{A}$, while $\mathbf{I}$ will denote the identity matrix of order understood by the context.

Consider a general linear model*

$$\{\mathbf{Y}, \mathbf{X}\boldsymbol{\beta}, \sigma^2\mathbf{V}\}, \tag{1}$$

with no assumption made on the ranks of the matrices $\mathbf{X}$ and $\mathbf{V}$. The basic estimability* criterion asserts that linear parametric functions $\mathbf{L}\boldsymbol{\beta}$, specified by a given matrix $\mathbf{L}$, are estimable in the model (1) if and only if

$$C(\mathbf{L}') \subset C(\mathbf{X}'). \tag{2}$$

Several alternative formulations of the inclusion (2) have been developed. Among them are the two equalities derived by Milliken [8] in terms of the ranks and traces of certain matrices. Taking into account generalizations established in Alalouf [1] and Baksalary and Kala [6], the label Milliken's estimability criterion (MEC) may be attached to the following theorem.

Theorem 1. The parametric functions $\mathbf{L}\boldsymbol{\beta}$ are estimable in the model $\{\mathbf{Y}, \mathbf{X}\boldsymbol{\beta}, \sigma^2\mathbf{V}\}$ if and only if

$$r\left[\mathbf{X}(\mathbf{I} - \mathbf{L}^-\mathbf{L})\right] = r(\mathbf{X}) - r(\mathbf{L}), \tag{3}$$

or, alternatively, if and only if

$$\mathrm{tr}\left\{\left[\mathbf{X}(\mathbf{I} - \mathbf{L}^-\mathbf{L})\right]^-\mathbf{X}(\mathbf{I} - \mathbf{L}^-\mathbf{L})\right\} = \mathrm{tr}(\mathbf{X}^-\mathbf{X}) - \mathrm{tr}(\mathbf{L}^-\mathbf{L}). \tag{4}$$

The equivalence between (2) and (3) follows by observing that (2) holds if and only if $r\begin{pmatrix}\mathbf{X}\\ \mathbf{L}\end{pmatrix} = r(\mathbf{X})$ and by reexpressing $r\begin{pmatrix}\mathbf{X}\\ \mathbf{L}\end{pmatrix}$ according to the equality

$$r\begin{pmatrix}\mathbf{A}_1\\ \mathbf{A}_2\end{pmatrix} = r(\mathbf{A}_2) + r\left[\mathbf{A}_1(\mathbf{I} - \mathbf{A}_2^-\mathbf{A}_2)\right],$$

which is valid for any matrices $\mathbf{A}_1$ and $\mathbf{A}_2$ with the same number of columns (cf. Marsaglia and Styan [7, p. 274]). The equivalence between (3) and (4) is an immediate consequence of the fact that for any matrix $\mathbf{A}$, $r(\mathbf{A}) = \mathrm{tr}(\mathbf{A}^-\mathbf{A})$. On the other hand, (3) can equivalently be expressed in the form

$$r(\mathbf{X}\mathbf{L}_0) = r(\mathbf{X}) - r(\mathbf{L}), \tag{5}$$

where $\mathbf{L}_0$ is any matrix such that $C(\mathbf{L}_0) = N(\mathbf{L})$. This variant of the rank part of MEC may be traced back to Roy and Roy [10], who obtained (5) as a necessary and sufficient condition for testability* of the linear hypothesis $\mathbf{L}\boldsymbol{\beta} = \mathbf{0}$ (cf. Alalouf and Styan [2, p. 194] and Srivastava and Khatri [12, p. 137]).

Milliken [8] pointed out the attractiveness of his criterion from the computational standpoint and, actually following Pyle [9], provided an algorithm for calculating orthogonal projectors of the type $\mathbf{A}^+\mathbf{A}$, where $\mathbf{A}^+$ is the Moore–Penrose inverse of a given matrix $\mathbf{A}$. Some gain in numerical efficiency is achievable when taking advantage of the possibility of using any generalized inverse $\mathbf{A}^-$ in place of $\mathbf{A}^+$ as revealed in Theorem 1. An algorithm for calculating projectors of the type $\mathbf{A}^-\mathbf{A}$ is given in Baksalary et al. [4]. It should be noted, however, that there is not a common agreement as to considering MEC to be the computationally most convenient estimability criterion (cf. Alalouf and Styan [2, p. 195]).

For a restricted linear model, where $\boldsymbol{\beta}$ is subject to some consistent linear restrictions $\mathbf{R}\boldsymbol{\beta} = \mathbf{r}$, the basic estimability condition is the inclusion

$$C(\mathbf{L}') \subset C(\mathbf{X}' \,\vdots\, \mathbf{R}').$$

Consequently, the rank part of MEC takes the form of the following theorem (cf.

Alalouf and Styan [3, p. 195] and Baksalary and Kala [6, p. 640]).

Theorem 2. The parametric functions $\mathbf{L}\boldsymbol{\beta}$ are estimable in the model $\{\mathbf{Y}, \mathbf{X}\boldsymbol{\beta} \mid \mathbf{R}\boldsymbol{\beta} = \mathbf{r}, \sigma^2\mathbf{V}\}$ if and only if

$$r\left[\begin{pmatrix}\mathbf{X}\\ \mathbf{R}\end{pmatrix}(\mathbf{I} - \mathbf{L}^-\mathbf{L})\right] = r\begin{pmatrix}\mathbf{X}\\ \mathbf{R}\end{pmatrix} - r(\mathbf{L}),$$

or, alternatively, if and only if

$$r\left[\begin{pmatrix}\mathbf{X}\\ \mathbf{R}\end{pmatrix}(\mathbf{I} - \mathbf{L}^-\mathbf{L})\right] = r\left[\mathbf{X}(\mathbf{I} - \mathbf{R}^-\mathbf{R})\right] + r(\mathbf{R}) - r(\mathbf{L}).$$

If $\boldsymbol{\beta}$ in the model (1) is partitioned as $\boldsymbol{\beta} = (\boldsymbol{\beta}_1' \vdots \boldsymbol{\beta}_2')'$, where $\boldsymbol{\beta}_2$ comprises nuisance parameters*, then only the functions of the form $\mathbf{L}_1\boldsymbol{\beta}_1$ are of interest (*see* GENERAL LINEAR MODEL). In this case, an analogue to (2) is (cf. Seely and Birkes [11, p. 401])

$$C(\mathbf{L}_1') \subset C\{[(\mathbf{I} - \mathbf{X}_2\mathbf{X}_2^-)\mathbf{X}_1]'\}. \quad (6)$$

Using (6), two rank conditions can be established according to the spirit of MEC.

Theorem 3. The parametric functions $\mathbf{L}_1\boldsymbol{\beta}_1$ are estimable in the model $\{\mathbf{Y}, \mathbf{X}_1\boldsymbol{\beta}_1 + \mathbf{X}_2\boldsymbol{\beta}_2, \sigma^2\mathbf{V}\}$ if and only if

$$r\left[(\mathbf{I} - \mathbf{X}_2\mathbf{X}_2^-)\mathbf{X}_1(\mathbf{I} - \mathbf{L}_1^-\mathbf{L}_1)\right] = r\left[(\mathbf{I} - \mathbf{X}_2\mathbf{X}_2^-)\mathbf{X}_1\right] - r(\mathbf{L}_1),$$

or, alternatively, if and only if

$$r\left[(\mathbf{I} - \mathbf{X}_2\mathbf{X}_2^-)\mathbf{X}_1(\mathbf{I} - \mathbf{L}_1^-\mathbf{L}_1)\right] = r(\mathbf{X}_1 \vdots \mathbf{X}_2) - r(\mathbf{X}_2) - r(\mathbf{L}_1).$$

In multivariate models, more general parametric functions **LBM**, specified by given matrices **L** and **M**, are considered. For a usual multivariate linear model $\{\mathbf{Y}, \mathbf{XB}, \boldsymbol{\Sigma} \otimes \mathbf{V}\}$ no additional condition over that in (2) is required for **LBM** to be estimable, thus preserving the validity of MEC as expressed in Theorem 1. This is not the case, however, for a growth curve* model $\{\mathbf{Y}, \mathbf{XBT}, \boldsymbol{\Sigma} \otimes \mathbf{V}\}$, with the matrix **T** being deficient in rank. Then the basic estimability criterion consists of two inclusions:

$$C(\mathbf{L}') \subset C(\mathbf{X}') \quad \text{and} \quad C(\mathbf{M}) \subset C(\mathbf{T}).$$

Consequently, the rank part of MEC has the form of the following result (cf. Baksalary and Kala [5]).

Theorem 4. The parametric functions **LBM** are estimable in the model $\{\mathbf{Y}, \mathbf{XBT}, \boldsymbol{\Sigma} \otimes \mathbf{V}\}$ if and only if

$$r\left[\mathbf{X}(\mathbf{I} - \mathbf{L}^-\mathbf{L})\right] = r(\mathbf{X}) - r(\mathbf{L})$$

and

$$r\left[(\mathbf{I} - \mathbf{MM}^-)\mathbf{T}\right] = r(\mathbf{T}) - r(\mathbf{M}).$$

References

[1] Alalouf, I. S. (1975). "Estimability and testability in linear models." Ph.D. dissertation. McGill University, Montreal, Canada.

[2] Alalouf, I. S. and Styan, G. P. H. (1979). *Ann. Statist.*, **7**, 194–200.

[3] Alalouf, I. S. and Styan, G. P. H. (1979). *Math. Operationsforsch. Statist. Ser. Statist.*, **10**, 189–201.

[4] Baksalary, J. K., Dobek, A., and Kala, R. (1976). *Ž. Vyčisl. Mat. Mat. Fiz.*, **16**, 1038–1040.

[5] Baksalary, J. K. and Kala, R. (1976). *Math. Operationsforsch. Statist.*, **7**, 5–9.

[6] Baksalary, J. K. and Kala, R. (1976). *Ann. Statist.*, **4**, 639–641.

[7] Marsaglia, G. and Styan, G. P. H. (1974). *Linear and Multilinear Algebra*, **2**, 269–292.

[8] Milliken, G. A. (1971). *Ann. Math. Statist.*, **42**, 1588–1594.

[9] Pyle, L. D. (1964). *J. Ass. Comput. Mach.*, **11**, 422–428.

[10] Roy, S. N. and Roy, J. (1959). *Ann. Math. Statist.*, **30**, 577–581.

[11] Seely, J. and Birkes, D. (1980). *Ann. Statist.*, **8**, 399–406.

[12] Srivastava, M. S. and Khatri, C. G. (1979). In *An Introduction to Multivariate Statistics*. North-Holland, Amsterdam, Section 5.3.

(ESTIMABILITY
GENERAL LINEAR MODEL
GROWTH CURVES)

JERZY K. BAKSALARY

MILLS' RATIO

In 1926, B. H. Camp wrote in *Biometrika** [4] that "the area A of the tail of a frequency curve could be found approximately by the formula $A = \phi(\cdot)R_x$", where $\phi(\cdot)$ is the density function at x, and noted that this had appeared in approximating binomial and geometric probabilities in terms of the standard normal distribution*.

For a standard normal distribution with cumulative distribution function $\Phi(\cdot)$ and density function $\phi(\cdot)$, this becomes

$$1 - \Phi(x) = \phi(x)R_x,$$

where R_x is *Mills' ratio*, the reciprocal of the hazard rate* or failure rate. Since some expansions for $\Phi(x)$ involve its derivatives $\phi(x), \phi'(x)$, and so on, and since

$$\frac{d^m}{dx^m}\phi(x) = (-1)^m H_m(x)\phi(x), \qquad m = 1, 2, \ldots,$$

where $H_m(x)$ is a Chebyshev–Hermite polynomial*, one would expect that approximations to R_x frequently correspond to approximations to $\Phi(x)$ and vice versa, and such is indeed the case.

Approximations for Mills' ratio are largely derived from expansions and inequality bounds. As examples of the latter,

$$\frac{1}{x + x^{-1}} < R_x < \frac{1}{x},$$

$$\frac{2}{x + \sqrt{x^2 + 4}} < R_x < \frac{4}{3x + \sqrt{x^2 + 8}},$$

the second [1, 9] being sharper than the first [2]. For $x > 2$, it turns out that $R_x \simeq 4[3x + \sqrt{x^2 + 8}\,]^{-1} - 2x^{-7}$. Among many approximations, a good one is that of Patry and Keller [7]:

$$R_x = \frac{t(x)}{xt(x) + \sqrt{2}}, \qquad x \geqslant 0,$$

$$t(x) = \sqrt{\pi} + x(2 - a(x)/b(x)),$$

$$a(x) = 0.858{,}407{,}657 + x\left[0.307{,}818{,}193 + x\{0.063{,}832{,}389{,}1 - (0.000{,}182{,}405{,}075)x\}\right],$$

$$b(x) = 1 + x\left[0.650{,}974{,}265 + x\{0.229{,}485{,}819 + (0.034{,}030{,}182{,}3)x\}\right].$$

If this is used as an approximation to $\Phi(x) = 1 - \phi(x)R_x$, the error is less than 12.5×10^{-9} for the range $0 \leqslant x \leqslant 6.38$. Detailed listings of expressions and bounds for R_x, along with sources for tables, appear in Johnson and Kotz [3, Section 7.1] and in Patel and Read [6, Chapter 3]. Values of R_x to five decimal places appear in Mills [4] and in Owen [5], for example.

As a further application of Mills' ratio, let Y_x be a standard normal variable truncated on the left at x. Then the mean and variance of Y_x are, respectively, $[R_x]^{-1}$ and $1 + x[R_x]^{-1} - [R_x]^{-2}$.

A multivariate generalization of Mills' ratio is given by

$$R(\mathbf{a}, \mathbf{\Sigma}) = F(\mathbf{a})/f(\mathbf{a}),$$

where $f(\mathbf{a})$ is the joint PDF at $\mathbf{a}$ of a $p \times 1$ random vector $\mathbf{X}$ with a multivariate normal distribution* having mean $\mathbf{0}$ and variance–covariance matrix $\mathbf{\Sigma}$, and where $F(\mathbf{a}) = \Pr(\mathbf{X} \geqslant \mathbf{a})$; in this expression $\mathbf{X} \geqslant \mathbf{a}$ means that $X_i \geqslant a_i$ for all i; $i = 1, \ldots, p$; $\mathbf{X}' = (X_1, \ldots, X_p)$ and $\mathbf{a}' = (a_1, \ldots, a_p)$ is a vector of constants. Savage [10] obtained upper and lower bounds, Ruben [8] gives an asymptotic expansion, and Steck [11] has three approximations to $R(\mathbf{a}, \mathbf{\Sigma})$.

References

[1] Birnbaum, Z. W. (1942). *Ann. Math. Statist.*, **13**, 245–246.

[2] Gordon, R. D. (1941). *Ann. Math. Statist.*, **12**, 364–366.

[3] Johnson, N. L. and Kotz, S. (1970). *Distributions in Statistics. Continuous Univariate Distributions, 2.* Wiley, New York, Section 7.1.

[4] Mills, J. F. (1926). *Biometrika*, **18**, 395–400. (The introduction was written by B. H. Camp, who directed the project of compiling tables of R_x. The computation was done by J. F. Mills after whom the ratio R_x was named.)

[5] Owen, D. B. (1962). *Handbook of Statistical Tables*. Addison-Wesley, Reading, MA.

[6] Patel, J. K. and Read, C. B. (1982). *Handbook of the Normal Distribution*. Dekker, New York, Chapter 3.

[7] Patry, J. and Keller, J. (1964). *Numerische Math.*, **6**, 89–97. (The original version of the approximation is given in German in terms of the error function.)

[8] Ruben, H. (1964). *J. Res. Nat. Bur. Stand. B*, **68**, 3–11.
[9] Sampford, M. R. (1953). *Ann. Math. Statist.*, **24**, 130–132.
[10] Savage, I. R. (1962). *J. Res. Nat. Bur. Stand. B*, **66**, 93–96.
[11] Steck, G. P. (1979). *Ann. Prob.*, **7**, 547–551.

(APPROXIMATIONS TO DISTRIBUTIONS
NORMAL DISTRIBUTION)

CAMPBELL B. READ

MINIMAL CUT SET *See* COHERENT STRUCTURES

MINIMAL PATH SET *See* COHERENT STRUCTURES

MINIMAL SUFFICIENT STATISTIC *See* SUFFICIENT STATISTICS

MINIMAX DECISION RULES

ZERO-SUM, TWO-PERSON GAMES

Minimax decision rules were introduced originally in conjunction with the mathematical *theory of games* (*see* GAME THEORY) as developed by John von Neumann in the 1920s. [See, e.g., von Neumann and Morgenstern (1947). Although von Neumann's development was carried out independently, he was anticipated to some extent by E. Borel.] Consider a game, or decision* problem, involving two players in which player 1 must choose a decision a from some given finite set A of available decisions and simultaneously player 2 must choose a decision b from some available finite set B. Suppose that a payoff function $L(a,b)$ is defined which specifies, for each possible pair of decisions (a,b) that the players might choose, the amount in monetary or other appropriate units that player 1 must pay player 2. [A negative value of $L(a,b)$ for some pair (a,b) indicates that player 2 must pay the amount $-L(a,b)$ to player 1.]

A game of this type, in which one player gains exactly the amount that the other player loses, is called a zero-sum two-person game. Under these conditions, player 1 can evaluate, for each decision $a \in A$, the value

$$\max_{b \in B} L(a,b)$$

of the maximum loss that he or she could suffer. A decision $a^* \in A$ for which this maximum is as small as possible is called a *minimax decision*. In other words, a minimax decision a^* satisfies the relation

$$\max_{b \in B} L(a^*,b) = \min_{a \in A} \max_{b \in B} L(a,b) = M^0, \quad \text{say.} \tag{1}$$

Similarly, a *maximin decision* for player 2 is a decision $b^* \in B$ that satisfies the relation

$$\min_{a \in A} L(a,b^*) = \max_{b \in B} \min_{a \in A} L(a,b) = M_0, \quad \text{say.} \tag{2}$$

Thus by choosing a minimax decision player 1 can guarantee a loss not exceeding M^0, and by choosing a maximin decision player 2 can guarantee that his or her gain will be at least M_0. It follows that $M_0 \leqslant M^0$.

It should be emphasized that although this discussion has been presented in terms of the loss to player 1 and the gain to player 2, the analysis is completely symmetric for both players. Minimizing the maximum possible loss that a player can suffer is precisely the same as maximizing the minimum possible gain that the player can achieve.

To illustrate these concepts, consider a zero-sum two-player game in which the amount that player 1 must pay player 2 is given by Table 1. In this example, the set A contains three decisions and the set B contains four decisions. It can be seen from Table 1 that if player 1 chooses decision a_1,

Table 1

	b_1	b_2	b_3	b_4
a_1	3	0	2	1
a_2	0	1	4	7
a_3	4	2	3	1

a_2, or a_3, his or her maximum possible loss is 3, 7, or 4, respectively. Therefore, the minimax decision is $a^* = a_1$ and $M^0 = 3$. Similarly, if player 2 chooses decision b_1, b_2, b_3, or b_4, his or her minimum possible gain is 0, 0, 2, or 1, respectively. Therefore, the maximin decision is $b^* = b_3$ and $M_0 = 2$.

The situation described in this example is not in equilibrium, as is reflected by the fact that $M_0 \neq M^0$. If player 2 is aware of the reasoning used by player 1, then player 2 knows that player 1 will choose decision a_1. But if player 1 is going to choose a_1, then player 2 can maximize his or her own gain by choosing decision b_1 rather than b_3. However, if player 1 knows that player 2 is going to choose b_1, then player 1 will choose a_2 and keep his or her loss at 0. In turn, however, the choice of a_2 by player 1 would lead player 2 to switch to b_4, etc. These considerations lead to the following concepts.

Randomized Decisions

Rather than directly choosing one of the finite number of available decisions in A, player 1 can leave his or her choice to chance, assigning probabilities $\alpha(a)$ to the elements of A and then, by means of an auxiliary randomization*, choosing a decision from A in accordance with the probability distribution α. Similarly, player 2 can choose a decision in accordance with some probability distribution β on the elements of B. In this context, a probability distribution α on the elements of A or a distribution β on the elements of B is a *randomized decision*. The original decisions in A and B, which correspond to degenerate distributions that place all their probability on a single element, are *pure decisions*.

Let $\mathscr{A}$ and $\mathscr{B}$ denote the spaces of all possible randomized decisions of players 1 and 2, respectively. For any given choices of α and β from $\mathscr{A}$ and $\mathscr{B}$, the expected payoff from player 1 to player 2 will be

$$R(\alpha, \beta) = \sum_{\alpha \in A} \sum_{\beta \in B} L(a,b)\alpha(a)\beta(b). \quad (3)$$

It is assumed that when the players use randomized strategies it is only the expected payoff $R(\alpha, \beta)$ that is relevant to them. In other words, it is assumed that the payoffs are expressed in units of *utility** for both players.

A minimax randomized decision α^* for player 1 and a maximin randomized decision β^* for player 2 can now be defined analogously to (1) and (2). Thus

$$\max_{\beta \in \mathscr{B}} R(\alpha^*, \beta) = \min_{\alpha \in \mathscr{A}} \max_{\beta \in \mathscr{B}} R(\alpha, \beta) = \mathscr{M}^0, \quad \text{say} \quad (4)$$

and

$$\min_{\alpha \in \mathscr{A}} R(\alpha, \beta^*) = \max_{\beta \in \mathscr{B}} \min_{\alpha \in \mathscr{A}} R(\alpha, \beta) = \mathscr{M}_0, \quad \text{say.} \quad (5)$$

The famous minimax theorem of von Neumann states that $\mathscr{M}_0 = \mathscr{M}^0 = R(\alpha^*, \beta^*)$. The common value $\mathscr{M}$ of $\mathscr{M}_0$ and $\mathscr{M}^0$ is called the *value of the game*. Randomized decisions α^* and β^* that satisfy (4) and (5) are called optimal. Since a player can only reduce his or her expected payoff by not using an optimal randomized decision when the opponent is using an optimal one, it is often argued that in games between two intelligent players both should use optimal randomized decisions.

As an illustration, consider again the game for which the payoffs from player 1 to player 2 are given by Table 1. It can be shown that the minimax randomized decision α^* for player 1 is to choose decision a_1 with probability 0.8 and decision a_2 with probability 0.2. The maximin randomized decision β^* for player 2 is to choose decision b_1 with probability 0.4 and decision b_3 with probability 0.6. The value of the game is $\mathscr{M}_0 = \mathscr{M}^0 = 2.4$.

Statistical Decision Problems

The concept of a minimax decision can be directly carried over to statistical decision problems (*see* DECISION THEORY). Consider a problem in which a decision maker (DM) must choose a decision a from some given set A, and the consequences of the decision depend on some parameter Θ whose unknown value θ lies in the parameter space Ω. As before, let $L(a,\theta)$ denote the loss to the DM in choosing decision a when the true value of Θ is θ. As in (1) or (4), one could again define a minimax decision or a minimax randomized decision for the DM.

Before choosing a decision from A, the DM will often have the opportunity of observing sample data x which provide information about θ. In this situation, the DM must choose a decision rule or decision function δ which specifies the decision in A that is to be chosen for each possible observed x. For each value $\theta \in \Omega$ and each decision rule δ, he or she can then calculate the expected loss or risk $R(\delta,\theta)$, and a minimax decision rule or minimax randomized decision rule can be determined for this risk function.

One basic difference between a statistical decision problem and a two-person game is that the value of θ typically represents an unknown state of the world rather than the intelligent choice of an opponent. For this reason, it has been suggested [see, e.g., Savage (1972)] that rather than working directly with the loss function $L(a,\theta)$, the DM should express his or her problem in terms of a modified loss function $L^*(a,\theta)$ defined as follows for $a \in A$ and $\theta \in \Omega$:

$$L^*(a,\theta) = L(a,\theta) - \min_{a' \in A} L(a',\theta) \quad (6)$$

The function $L^*(a,\theta)$ is sometimes called the *regret* function* because the DM's loss from choosing decision a when $\Theta = \theta$ is regarded as the difference between the DM's actual cost $L(a,\theta)$ and the minimum cost that he or she could have achieved. A minimax decision and a minimax randomized decision can then be defined with respect to the function L^*.

Minimax Estimation

Problems of estimation form an important special class of statistical decision problems. A common estimation problem is one in which an estimate a must be chosen for some real-valued parameter θ subject to the squared-error loss function $L(a,\theta) = (a-\theta)^2$. Suppose, for example, that a random sample $X_1, \dots, X_n$ is to be drawn from a normal distribution with an unknown mean θ and known variance σ^2, and let $\bar{X}$ denote the sample mean. Then the risk function of the estimator $\delta^* = \bar{X}$ is readily found to be $R(\delta^*,\theta) = \sigma^2/n$ for $-\infty < \theta < \infty$. In this example, it can be shown that there is no other estimator δ such that $R(\delta,\theta) \leqslant \sigma^2/n$ for all values of θ and $R(\delta,\theta) < \sigma^2/n$ for at least one value of θ; that is, δ^* is an admissible estimator (*see* ADMISSIBILITY). It follows that for any other estimator δ, $R(\delta,\theta) \geqslant \sigma^2/n$ for at least one value of θ. Hence δ^* is a minimax estimator of θ.

As another example, suppose that $X_1, \dots, X_k$ are independent observations and that, for $i = 1, \dots, k$, the observation X_i has a normal distribution with unknown mean θ_i and variance 1. Suppose that a vector of estimates $\mathbf{a} = (a_1, \dots, a_k)$ must be chosen for the vector of parameters $\boldsymbol{\theta} = (\theta_1, \dots, \theta_k)$ subject to the loss function

$$L(\mathbf{a},\boldsymbol{\theta}) = \sum_{i=1}^{k} (a_i - \theta_i)^2. \quad (7)$$

Let $\boldsymbol{\delta}^*$ denote the vector of estimators in which θ_i is estimated by X_i for $i = 1, \dots, k$. Then the risk function of $\boldsymbol{\delta}^*$ is $R(\boldsymbol{\delta}^*,\boldsymbol{\theta}) = k$ for all values of $\boldsymbol{\theta}$. It was shown in a landmark paper by Stein (1956) that although $\boldsymbol{\delta}^*$ is a minimax estimator of $\boldsymbol{\theta}$ for all values of k, it is an admissible estimator only for $k = 1$ and $k = 2$, not for $k \geqslant 3$. In particular, for $k \geqslant 3$, Stein constructed estimators $\boldsymbol{\delta}$ such that $R(\boldsymbol{\delta},\boldsymbol{\theta}) < k$ for all values of $\boldsymbol{\theta}$. However, it must be true that $\sup_{\boldsymbol{\theta}} R(\boldsymbol{\delta},\boldsymbol{\theta}) = k$ for any such estimator $\boldsymbol{\delta}$, since $\boldsymbol{\delta}^*$ is minimax.

Properties of Minimax Decision Rules

The concept and mathematical properties of minimax decision rules were introduced and extensively studied by A. Wald* as part of his creation and development of statistical decision theory in the 1940s. Thus minimax decision rules played an important role in the history of the modern theory of statistical decisions. However, they possess certain features that seem undesirable in any practical decision problem. Four of these features will now be described.

1. Minimax decision rules do not make use of any information regarding the relative likelihood* that different values of Θ are correct. In the search for a minimax decision rule, the only relevant feature of the risk function $R(\delta,\theta)$ is its maximum value, no matter how unlikely or improbable the value θ may be at which this maximum occurs.
2. The minimax decision rule is often a randomized rule. However, since the DM is trying to choose the best decision that he or she can from A in the face of uncertainty about θ, there would seem to be no reason for resorting to an auxiliary randomization to make that choice. There will usually be some decision in A that is felt to be best under the prevailing uncertainty. The use of randomization only introduces the possibility of the DM choosing a decision other than the one regarded as best.
3. The minimax decision rule with respect to the regret function L^* defined in (6) is sensitive to irrelevant extra decisions. For example, suppose that the parameter Θ can take only two values, that there are three decisions in the set A, with the loss function $L(a,\theta)$ in Table 2. In this example, $L^*(a,\theta)=L(a,\theta)$ and it can be shown that the minimax randomized decision rule is simply to choose the pure decision a_2.

   Now suppose that a fourth decision a_4 is made available such that $L(a,\theta_1) = -3$ and $L(a_4,\theta_2)=50$. Then $L^*(a,\theta)$ is given in Table 3, and it can be shown that the minimax randomized decision rule is now simply to choose the pure decision a_3. Thus introducing a new decision a_4 causes the DM to switch from a_2 to a_3. It is as if a person, after choosing among steak, fish, and chicken in a restaurant, ordered steak and then on being informed that duck is also available, changed the order to fish!
4. The minimax decision rule will depend on whether the loss function L or L^* is used. Each of these functions yields minimax decision rules with undesirable features, and neither is clearly better than the other.

Table 2

	θ_1	θ_2
a_1	3	0
a_2	1	1
a_3	0	3

Table 3

	θ_1	θ_2
a_1	6	0
a_2	4	1
a_3	3	3
a_4	0	50

Bayes Decision Rules

Because of the undesirable features just described, many statisticians feel that rather than attempting to find a minimax decision rule, it is preferable to assign a reasonable nonnegative weight function $\xi(\theta)$ to the values of Θ and to choose a decision rule δ for which the average risk

$$\sum_{\theta\in\Omega} R(\delta,\theta)\xi(\theta) \qquad (8)$$

is minimized. In many problems, the weight function $\xi(\theta)$ can be chosen to represent the DMs prior probabilities that the different values of Θ are correct. A decision rule δ for which (8) is a minimum is then called a Bayes decision rule.

Bayes decision rules do not suffer from the undesirable features that plague minimax decision rules. In particular, (a) the weight function $\xi(\theta)$ can be chosen to reflect the relative likelihoods of different values of θ; (b) a Bayes decision rule need never involve an auxiliary randomization; (c) a Bayes decision rule will not change when other decisions that the DM does not wish to choose are made available; and (d) the Bayes decision rules will be the same for L and L^*.

Group Decisions

Problems of group decision-making provide another area for the possible use of minimax decisions. Suppose that together a group of J persons must select a decision a from some set A. For $a \in A$ and $j = 1, \ldots, J$, let $L(a, j)$ denote the loss to person j if decision a is chosen. In this context, a minimax decision a^* would be one for which

$$\max_j L(a^*, j) = \min_a \max_j L(a, j). \quad (9)$$

In other words, a minimax decision is one for which the maximum loss among the persons in the group is as small as possible. A minimax randomized decision can be defined similarly.

Bibliography

Berger, J. O. (1980). *Statistical Decision Theory: Foundations, Concepts, and Methods*. Springer-Verlag, New York. (A graduate-level textbook that treats minimax decision functions and minimax estimators in a variety of statistical problems.)

Blackwell, D., and Girshick, M. A. (1954). *Theory of Games and Statistical Decisions*. Wiley, New York. (A graduate-level textbook with extensive discussions of minimax decision rules. Although the presentation is restricted to discrete distributions, the treatment is highly technical.)

Chernoff, H., and Moses, L. E. (1959). *Elementary Decision Theory*. Wiley, New York. (An elementary introduction to the concepts of, and approaches to, statistical decision theory.)

DeGroot, M. H. (1970). *Optimal Statistical Decisions*. McGraw-Hill, New York. (An intermediate-level text devoted to the study of Bayes decision rules.)

Jones, A. J. (1980). *Game Theory: Mathematical Models of Conflict*. Halsted Press, New York. (An intermediate-level exposition of some of the modern developments and applications of game theory.)

Luce, R. D., and Raiffa, H. (1957). *Games and Decisions*. Wiley, New York. (A famous and wide-ranging exposition of the theory and applications of decision-making.)

Savage, L. J. (1972). *The Foundations of Statistics*, 2nd ed. Dover, New York. (One of the original developments of the minimax and Bayesian approach to statistical inference and decisions. Although the book uses only elementary mathematics, it is thorough and rigorous.)

Stein, C. (1956). "Inadmissibility of the usual estimator for the mean of a multivariate normal distribution." *Proceedings of the Third Berkeley Symposium on Mathematical Statistics and Probability, Vol. 1*, University of California Press, Berkeley, pp. 197–206. (A landmark paper on inadmissible minimax estimators in some standard problems of statistical estimation.)

von Neumann, J. and Morgenstern, O. (1947). *Theory of Games and Economic Behavior*, 2nd ed. Princeton University Press, Princeton, NJ. (A path-breaking book by the originators of the theory of games and minimax decisions.)

Wald, Abraham (1950). *Statistical Decision Functions*. Wiley, New York. (The first book to present the mathematical basis of statistical decision theory and minimax decision rules, written by the man who developed most of the theory himself.)

(DECISION THEORY
GAME THEORY
MINIMAX ESTIMATION
REGRET)

M. H. DEGROOT

MINIMAX ESTIMATION

INTRODUCTION AND DEFINITIONS

The minimax criterion has enjoyed considerable popularity for estimation* in statistical decision theory*. In the usual decision-theoretic statistical model for estimation, one observes a random variable X with density (for convenience) $f(x \mid \theta)$ and desires to estimate θ based on the observation x using some decision function $\delta(x)$. There is assumed to be a loss involved if θ is incorrectly estimated by $\delta(x)$, usually denoted by $L(\theta, \delta(x))$. The overall performance of $\delta(x)$ is typically measured by its expected loss, or risk function

$$R(\theta,\delta) = E_\theta L(\theta,\delta(X)) = \int L(\theta,\delta(x))f(x \mid \theta)\,dx.$$

Example 1. Suppose $X = (1/n)\sum_{i=1}^{n} X_i$ is the sample mean from a normal distribution with mean θ and variance σ^2. Then X itself has a normal distribution with mean θ and variance σ^2/n. Suppose it is desired to estimate θ under squared error loss, i.e., $L(\theta,\delta) = (\theta - \delta)^2$. The usual estimator of θ is, of course, $\delta(x) = x$, and

$$R(\theta,\delta) = E_\theta(\theta - \delta(X))^2$$
$$= E_\theta(\theta - X)^2 = \sigma^2/n.$$

It will generally be the case that there is no estimator $\delta(x)$ that simultaneously minimizes $R(\theta,\delta)$ for all θ. Among the criteria that have been proposed for comparing risk functions is the *minimax criterion*, which states that a small value of

$$M(\delta) = \sup_\theta R(\theta,\delta)$$

is desired. That is, one looks at the maximum (or supremum) risk that could be incurred if δ were used and seeks a small value of this maximum risk. A minimax estimator, δ^M, is an estimator that minimizes this maximum risk, i.e., an estimator for which

$$M(\delta^M) = \min_\delta M(\delta).$$

Example 1 (continued). Consider estimators of the form $\delta_c(x) = cx$. A calculation gives

$$R(\theta,\delta_c) = E_\theta(\theta - \delta_c(X))^2 = E_\theta(\theta - cX)^2$$
$$= \theta^2(1-c)^2 + c^2\sigma^2/n.$$

Clearly

$$M(\delta_c) = \sup_\theta R(\theta,\delta_c) = \begin{cases} \infty & \text{if } c \neq 1 \\ \sigma^2/n & \text{if } c = 1. \end{cases}$$

Hence $\delta_1(x) = x$ is the best estimator in this class according to the minimax principle. Indeed, we will see later that δ_1 minimizes $M(\delta)$ among *all* estimators and thus is a minimax estimator of θ.

Minimax theory is quite closely related to game theory*. Indeed the statistical situation can be viewed as a two-person zero-sum game with the statistician as player 2, "nature" (which chooses θ) as player 1, and $R(\theta,\delta)$ as the payoff (to nature) if nature chooses θ and the statistician chooses δ. A minimax procedure is the strategy usually suggested for player 2 in such a game, assuming that player 1 is an intelligent opponent out to win—a point that will be discussed later. Minimax theory could apply in this fashion to any statistical decision problem; it is most often used in estimation problems, however.

As in game theory, it is sometimes necessary to allow randomized decision rules to achieve the minimum possible value of $M(\delta)$. A *randomized estimator* δ can be thought of as a probability distribution, depending only on the observed x, according to which an estimate of θ is chosen. In Example 1, for instance, a possible randomized estimator would be to estimate θ by x, $x - 1$, or $x + 1$, with probability $\frac{1}{3}$ each. When the loss is a convex function of δ (as is the loss in Example 1), it can be shown by Jensen's inequality* that any randomized estimator can be improved upon by a nonrandomized estimator.

DETERMINING A MINIMAX ESTIMATOR

No known method is guaranteed to produce a minimax estimator. The two most commonly used methods are the least favorable prior approach and the invariance approach.

Least Favorable Prior Approach

In Bayesian analysis (see BAYESIAN INFERENCE), one presumes that θ is itself a random quantity with some density (for convenience) $\pi(\theta)$, which is called a *prior* density (*see* PRIOR DISTRIBUTIONS). Of interest then is the *Bayes risk* of an estimator δ, defined by

$$r(\pi,\delta) = E^\pi R(\theta,\delta) = \int R(\theta,\delta)\pi(\theta)\,d\theta.$$

This is the overall average loss that will be incurred if δ is repeatedly used when θ varies according to π. An estimator δ^π which minimizes $r(\pi,\delta)$ is a *Bayes estimator* with respect to π (and L). The following theorem can often be used to find a minimax estimator.

Theorem 1. (a) If δ is a Bayes estimator with respect to π and $M(\delta) = r(\pi, \delta)$, then δ is a minimax estimator. (b) If δ_m, $m = 1, 2, \ldots$, are Bayes estimators with respect to prior densities π_m and δ is an estimator such that

$$M(\delta) = \lim_{m\to\infty} r(\pi_m, \delta_m),$$

then δ is a minimax estimator.

Example 2. Suppose X has a binomial distribution with parameters n and θ and that it is desired to estimate θ under squared-error loss. If π is chosen to be a beta density with parameters $\sqrt{n}/2$ and $\sqrt{n}/2$, then the Bayes rule is

$$\delta^{\pi}(x) = \frac{x + \sqrt{n}/2}{n + \sqrt{n}}.$$

Also,

$$\begin{aligned} R(\theta, \delta^{\pi}) &= E_{\theta}(\theta - \delta^{\pi}(X))^2 \\ &= n \Big/ \left[4\left(n + \sqrt{n}\right)^2 \right] = r(\pi, \delta^{\pi}). \end{aligned}$$

Clearly, $M(\delta^{\pi}) = r(\pi, \delta^{\pi})$ (since the risk equals this constant for all θ), so that, by Theorem 1(a), δ^{π} is a minimax estimator.

The prior π in Theorem 1(a) is the *least favorable prior* and can be interpreted as the (random) choice of θ that would make life most difficult for the statistician. (From the game-theoretic viewpoint, π would be the maximin strategy for player 1—nature.) Although this intuition is not very helpful in Example 2, there are many situations in which it is possible to guess what a least favorable prior might be and hence determine a minimax estimator. In Example 1, for instance, it seems reasonable that a least favorable prior density would be one giving equal weight to all possible values of θ, i.e., $\pi(\theta) = c$ for $-\infty < \theta < \infty$. Unfortunately, no choice of c can make this a proper density. It is at such a point that Theorem 1(b) comes into play in that the π_m can be chosen to approximate such an "improper" prior and still give the minimax result.

Example 1 (continued). Choose the π_m to be normal densities with mean zero and variance m. (As $m \to \infty$, these become nearly constant densities, approximating the intuitive least favorable prior.) The Bayes estimators are

$$\delta_{\pi_m}(x) = \frac{m}{m + \sigma^2/n} x$$

and

$$r(\pi_m, \delta_{\pi_m}) = \sigma^2 m/(\sigma^2 + mn).$$

Since $\lim_{m\to\infty} r(\pi_m \delta_{\pi_m}) = \sigma^2/n$ and $M(\delta_1) = \sigma^2/n$ (where, recall, $\delta_1(x) = x$ is the sample mean), δ_1 is a minimax estimator.

An interesting feature of both the minimax estimators in Examples 1 and 2 is that they have constant risk functions. Any estimator having a constant risk function is called an *equalizer rule*, and often the search for a minimax estimator can be successfully carried out among the equalizer rules.

INVARIANCE APPROACH

If a statistical decision problem is invariant under a group of transformations (*see* INVARIANCE CONCEPTS IN STATISTICS), and the group happens to be what is called amenable (see Bondar and Milnes [2]), then the best invariant estimator is minimax (under mild conditions). Since finding the best invariant estimator is often relatively straightforward, this provides a powerful tool for finding minimax estimators.

Example 3. Suppose $X = (X_1, \ldots, X_n)$, where the X_i are independent observations from a location density $f(x_i - \theta)$, $-\infty < x_i < \infty$ and $-\infty < \theta < \infty$. It is desired to estimate θ under a loss function of the form $L(\theta - \delta)$ [i.e., a function that depends only on $(\theta - \delta)$]. This problem is invariant under the additive group on R^1, since transforming the X_i, θ, and δ by adding a constant does not change the structure of the problem. This group is amenable, and hence the best invariant estimator, which turns out to be the value of δ that minimizes $\int L(\theta - \delta) \prod_{i=1}^{n} f(x_i - \theta)\, d\theta$, is minimax. If L is squared-error loss, the minimizing δ is given by the Pitman estimator*

$$\delta = \int \theta \prod_{i=1}^{n} f(x_i - \theta)\, d\theta \Big/ \int \prod_{i=1}^{n} f(x_i - \theta)\, d\theta.$$

The theorem relating invariance to minimaxity is the Hunt–Stein theorem*, and a general development of it can be found in Bondar and Milnes [2], along with earlier references. The theorem is based on a sophisticated application of Theorem 1(b).

OTHER APPROACHES

Since the statistical minimax problem can be posed as a problem in game theory, game–theoretic techniques may sometimes be useful in finding minimax rules, especially when the parameter space or the decision space is finite. Also, there are general theorems that establish the existence of minimax rules under weak conditions (see Bibliography).

Due to the extreme difficulty of finding minimax rules in many problems, a somewhat more tractable approach involving *asymptotic minimaxity* has been developed. This involves letting the sample size n go to infinity in a given situation and attempting to find a minimax estimator up to second-order terms of the asymptotic risk. For example, many estimation problems behave, asymptotically, like location problems, and Pitman's estimator is thus often asymptotically minimax. For an analysis of such situations and earlier references on asymptotic minimaxity, see Strasser [6].

MULTIPARAMETER MINIMAX ESTIMATION

The vector of sample means is minimax for estimating a p-variate normal mean under loss $\sum_{i=1}^{p}(\theta_i - \delta_i)^2$ (by either the reasoning in Example 1 or the reasoning in Example 3), yet if $p \geqslant 3$, it is inadmissible (*see* ADMISSIBILITY). That is, it can be improved on in terms of risk. (The improvement cannot, of course, be larger than any fixed $\epsilon > 0$ for *all* θ, because of minimaxity.) This is the so-called Stein effect*. (*See also* JAMES–STEIN ESTIMATORS.) Since any estimator that improves on the vector of sample means must itself be minimax, much of the literature on this subject goes under the name of minimax estimation. The effect is not limited to normal distributions and occurs in most multiparameter estimation problems. Examples in this area are Berger [1], Ghosh and Parsian [3], Hwang [4], and Judge and Bock [5], which also have good bibliographies.

DISCUSSION OF THE MINIMAX CRITERION

While the minimax criterion is certainly useful in game theory, there are several problems with its use in statistics. Clearly nature is not an intelligent opponent out to maximize loss, and pretending that this is the case seems somewhat artificial. Indeed, if the "worst" θ or least favorable prior are very implausible, then use of a minimax rule seems unwise. In Example 2, for instance, when $n = 10{,}000$ the least favorable prior density [the beta* (50, 50) density] is enormously concentrated about $\theta = \frac{1}{2}$. Hence unless θ is deemed to almost certainly be near $\frac{1}{2}$, the minimax rule is suspect. For Examples 1 and 3, in contrast, the least favorable (improper) prior is sensible, reducing concerns as to the adequacy of the minimax rule in these situations.

It is commonly claimed that a minimax rule δ^M is desirable if conservative behavior is sought. The difficulty with this claim is that the measure used to evaluate δ^M is $R(\theta, \delta^M)$, involving an average over all possible x. Since x will be known before making a conclusion, truly conservative behavior would involve trying to choose a particular estimate $\delta(x)$ to minimize $\sup_\theta L(\theta, \delta(x))$.

References

[1] Berger, J. (1982). *Ann. Statist.*, **10**, 81–92.

[2] Bondar, J. V. and Milnes, P. (1981). *Zeit. Wahrscheinlichkeitsth. verw. Geb.*, **57**, 103–128.

[3] Ghosh, M. and Parsian, A. (1982). *J. Multivariate Anal.*, **10**, 551–564.

[4] Hwang, J. T. (1982). *Ann. Statist.*, **10**, 857–867.

[5] Judge, G. and Bock, M. E. (1977). *Implications of Pre-Test and Stein Rule Estimators in Econometrics*. North-Holland, Amsterdam.

[6] Strasser, H. (1982). *Zeit. Wahrscheinlichkeitsth. verw. Geb.*, **60**, 223–247.

Bibliography

Berger, J. (1980). *Statistical Decision Theory: Foundations, Concepts, and Methods*. Springer-Verlag, New York. (Extensive discussion of techniques and applicability at a mixed level of difficulty.)

Blackwell, D. and Girshick, M. A. (1954). *Theory of Games and Statistical Decisions*. Wiley, New York. (An extremely thorough mathematical investigation of the discrete case.)

Ferguson, T. S. (1967). *Mathematical Statistics: A Decision-Theoretic Approach*. Academic Press, New York. (Concentrates on development of theory, but at a readable level.)

Wald, A. (1950). *Statistical Decision Theory*. McGraw-Hill, New York. (Advanced mathematically.)

(DECISION THEORY
ESTIMATION, POINT
GAME THEORY
HUNT–STEIN THEOREM
INVARIANCE CONCEPTS IN STATISTICS
MINIMAX DECISION RULES)

JAMES BERGER

MINIMAX METHOD

This method is an alternative to the method of least squares* for determining the coefficients in a linear or nonlinear regression* equation. In this method, the values of the unknown coefficients in the *regression equation* are determined so as to minimize the maximum deviation from regression. The minimax estimate of central value is the midrange*, since it is the value from which the maximum deviation (the semirange) is less than from any other value, and the corresponding estimate of dispersion is a constant multiple of the semirange. These estimates of central value and dispersion and the minimax estimates of regression coefficients are maximum likelihood* estimates when the error distribution is uniform (rectangular*).

This method has a long history, having been proposed more than half a century before the method of least squares. Euler [3] used the minimax method and Lambert* [13] stated the minimax principle, but confessed that he did not know how to use it in a general and straightforward manner. Laplace* [14] gave a procedure for using the minimax method to determine the values of a and b in the linear relation $y = a + bx$ from three or more noncollinear points. In a later paper Laplace [15] gave a simpler procedure and pointed out that when the absolute value of the largest residual is minimized, there are actually three residuals*, two with one sign and one with the other, that have this same absolute value. Laplace [16] summarized his earlier results.

Fourier [4] considered the problem of fitting a linear equation in n variables to a set of m observed points ($m > n$) so as to minimize the maximum absolute deviation and gave a geometric solution for the case $n = 2$. He formulated the problem as what would now be called a linear programming* problem, i.e., minimization of an objective function (the largest absolute diviation) subject to constraints in the form of linear inequalities. The method he used, which has come to be known as *Fourier's method of descent*, is also given in his posthumous book [5].

Poncelet [19] considered a particular case of the problem of approximation of a continuous function on an interval by a linear expression. Chebyshev [2] developed a general procedure for the solution of such problems. Kirchberger [12], apparently unaware of the still earlier work of Laplace and Fourier, stated that the method of approximation that is best in the sense that the maximum error is as small as possible was first proposed by Poncelet and was systematically worked out by Chebyshev* (*see* MATHEMATICAL FUNCTIONS, APPROXIMATIONS TO). He discussed approximations in two cases: a continuous function over an interval and over a finite set of points, with emphasis on the latter. Among other results, he showed that the best approximation to a set of $m(> n)$ points by a function containing n

constants to be fitted is identical to the best approximation to some subset of $n+1$ points. This was not, however, a new result. Mansion [17] pointed out that Gauss [6] criticized the minimax method precisely on the ground that it uses for the final calculation of the unknowns only a number of equations equal to the number of unknowns, the other equations being used only to decide the choice that one should make.

Goedseels [7] advocated use of the minimax method (which he called the *method of minimum approximation*) when the limits of the errors are not known. He recommended a modification, the *most approximative method**, otherwise. Further results on the minimax method were given by Goedseels [8, 9], de la Vallée Poussin [25], Tits [24], Mansion, [18], and Haar [10]. Haar stated the condition (that the determinant of the coefficients of each subset of n of the m equations be of rank n) that has come to be known as the Haar condition, though priority is believed to belong to de la Vallée Poussin. He showed that this condition guarantees uniqueness of the minimax solution of m ($>n$) linear equations in n unknowns and hence of the hyperplane in Euclidean n-space from which the maximum deviation of m ($>n$) points is a minimum. These results were summarized by de la Vallée Poussin [26].

Use of the theory of linear programming to solve the minimax regression problem has been discussed by a number of authors. Rice [20] has pointed out that Stiefel [23] showed that the method of descent (Zuhovickiĭ [27]) and the exchange method (Stiefel [22]), which is an ascent method, are duals of one another, so that either can be used, but that the exchange method is computationally more efficient. Bartels and Golub [1] presented an implementation and a computer algorithm (in ALGOL 60) for Stiefel's exchange method. Schryer [21] developed two modifications (improvements) of the Bartels–Golub algorithm.

The minimax method is optimal (maximum likelihood) when the error distribution is uniform (rectangular), e.g., when the data have been rounded from exact values, as in the following example. It is nearly optimal for other platykurtic (short-tailed) error distributions. Further information about the applicability of this and other alternatives to the method of least squares was given by Harter [11].

Example. Find the minimax regression line for the following points:

$$(0, 0.03), (1, 0.53), (2, 1.02), (3, 1.52),$$
$$(4, 2.02), (5, 2.51), (6, 3.01), (7, 3.51),$$
$$(8, 4.01), (9, 4.50), (10, 5.00).$$

As has been known since the time of Laplace, the minimax line for $m > 3$ points is the minimax line for some subset of three points. In this example, it can be shown that the three points which determine the minimax regression line are $(1, 0.53)$, $(5, 2.51)$, and $(8, 4.01)$. Specifically, the minimax line is the line parallel to the line joining $(1, 0.53)$ and $(8, 4.01)$ such that the (positive) vertical distance of those two points from the line is numerically equal to the (negative) vertical distance of $(5, 2.51)$ from the line. The required line satisfying these conditions is

$$y = 0.0286 + 0.49714x,$$

points $(1, 0.53)$ and $(8, 4.01)$ being 0.0043 units above the line and point $(5, 2.51)$ the same distance below it.

References

[1] Bartels, R. H. and Golub, G. H. (1968). *Commun. ACM*, **11**, 401–406, 428–430.

[2] Chebyshev, P. L. [Tchébychef, P. L.] (1854). *Mémoires Présentés à l'Académie Impériale des Sciences de St. Pétersbourg par Divers Savants*, **7**, 539–568. Reprinted in *Oeuvres de P. L. Tchébychef*, Vol. I (1899), A. Markoff and N. Sonin, eds., pp. 111–143. Imprimerie de l'Académie Impériale des Sciences, St. Pétersbourg.

[3] Euler, L. (1749). *Pièce qui a Remporté le Prix de l'Académie Royale des Sciences en 1748, sur les Inégalités du Mouvement de Saturn et de Jupiter.* Martin, Coignard et Guerin, Paris. Reprint, *Leonhardi Euleri Opera Omnia* **II 25** (*Commentationes Astronomicae* **I**), 45–157. Orell Fussli, Turici, 1960.

[4] Fourier, J. B. J. (1823–24). *Hist. Acad. Sci. Paris*, 1823, 29ff.; 1824, 47–55.

[5] Fourier, J. B. J. (1831). *Analyse des Équations Déterminées*, Part 1. Didot Frères, Paris.

[6] Gauss, C. F. (1809). *Theoria Motus Corporum Coelestium in Sectionibus Conicis Solem Ambientium.* Frid. Perthes et I. H. Besser, Hamburg.

[7] Goedseels, P. J. E. (1909). *Théorie des Erreurs d'Observation*, 3rd ed. Charles Peeters, Louvain; Gauthier-Villars, Paris.

[8] Goedseels, P. J. E. (1910). *Ann. Soc. Sci. Bruxelles*, **34**(2), 257–287.

[9] Goedseels, P. J. E. (1911). *Ann. Soc. Sci. Bruxelles*, **35**(1), 351–368.

[10] Haar, A. (1918). *Math. Ann.*, **78**, 294–311.

[11] Harter, H. L. (1974–76). *Int. Statist. Rev.*, **42** (1974), 147–174, 235–264, 282; **43** (1975), 1–44, 125–190, 269–278; **44** (1976), 113–159. (A comprehensive survey of the literature on the method of least squares and various alternatives, including the minimax method.)

[12] Kirchberger, P. (1903). *Math. Ann.*, **57**, 509–540.

[13] Lambert, J. H. (1765). *Beyträge zum Gebrauche der Mathematik und deren Anwendung*, **1**, 1–313.

[14] Laplace, P. S. (1786). *Mém. Acad. R. Sci. Paris*, 1783, 17–46.

[15] Laplace, P. S. (1793). *Mém. Acad. R. Sci. Paris*, 1789, 1–87.

[16] Laplace, P. S. (1799). *Traité de mécanique céleste*, Vol. 2. J. B. M. Duprat, Paris.

[17] Mansion, P. (1906). *Ann. Soc. Sci. Bruxelles*, **30**(1), 169–174.

[18] Mansion, P. (1913). *Ann. Soc. Sci. Bruxelles*, **37**(2), 107–117.

[19] Poncelet, J. V. (1835). *J. Reine Angew. Math*, **13**, 277–291.

[20] Rice, J. R. (1964). *Linear Theory* Vol I of *The Approximation of Functions*. Addison-Wesley, Reading, MA.

[21] Schryer, N. L. (1969). *Commun. ACM*, **12**, 326.

[22] Stiefel, E. L. (1959). *Numer. Math.*, **1**, 1–28.

[23] Stiefel, E. L. (1960). *Numer. Math.*, **2**, 1–17.

[24] Tits, L. (1912). *Ann. Soc. Sci. Bruxelles*, **36**(2), 253–263.

[25] de la Vallée Poussin, Ch. J. (1911). *Ann. Soc. Sci. Bruxelles*, **35**(2), 1–16.

[26] de la Vallée Poussin, Ch. J. (1919). *Leçons sur l'Approximation des Fonctions d'une Variable Réelle*. Gauthier-Villars, Paris.

[27] Zuhovickiĭ, S. I. (1951). *Dokl. Akad. Nauk SSSR* (N.S.), **79**, 561–564.

(HARTER'S ADAPTIVE ROBUST METHOD
LEAST SQUARES
LINEAR PROGRAMMING
LINEAR REGRESSION
MATHEMATICAL FUNCTIONS, APPROXIMATIONS TO
MAXIMUM LIKELIHOOD ESTIMATION
METHOD OF LEAST *p*TH POWERS
MOST APPROXIMATIVE METHOD
NONLINEAR REGRESSION)

H. LEON HARTER

MINIMAX TESTS

The basic ideas of statistical testing are presented in the entry HYPOTHESIS TESTING. As described there, when both the null hypothesis* H_0 and the alternative hypothesis* H_1 specify the distribution of the data completely, the best test is based on the ratio of the probability (or probability density) of the data under H_1 to the probability (or probability density) of the data under H_0 (*see* NEYMAN–PEARSON LEMMA). In many situations, however, the hypotheses are composite and there is no best test of H_0 against all the distributions specified by H_1. For example, suppose we have observed x heads in n independent tosses of a coin and we wish to test the null hypothesis that the probability p of heads for the coin is 0.5 against the alternative that p is either 0.25 or 0.75. Suppose we agree that falsely rejecting H_0 for a proportion α of the samples when H_0 is true would be tolerable. We might then want to require the proportion of samples leading to false acceptance of H_0 when H_1 is true to be as small as possible both when $p = 0.25$ and when $p = 0.75$. Unfortunately, no such test exists. The test that is best when $p = 0.25$ is not best when $p = 0.75$ and vice versa. Lacking a best test, we might take a conservative stance and ask that for a given α the maximum probability of false acceptance under H_1 be minimized. That is, we could ask for a minimax test that protects against the worst performance under the alternative.

Minimax tests do exist under general conditions (see Lehmann [5]), but they are often difficult to construct. One method that sometimes produces a minimax test is to determine the most difficult testing problem based on simple components of H_0 and H_1 and then to construct the best test for this worst case problem. In many cases, the worst case can be guessed by assuming that an intelligent adversary will be allowed to confuse the researcher by randomly choosing the components of H_0 and H_1 to be tested. In the preceding example, the adversary may choose the alternative to be either $p = 0.25$ with probability λ or $p = 0.75$ with probability $1 - \lambda$, where λ is chosen to make the testing problem as difficult as possible. The probability of x successes in n tosses of the coin under the adversary's alternative is then

$$\lambda(0.25)^x(0.75)^{n-x} + (1 - \lambda)(0.75)^x(0.25)^{n-x}.$$

The probability of the data under H_0 is 0.5^n. Since the distribution of the data is specified completely under the hypotheses set up by the adversary, the best test against the adversary rejects when the ratio

$$\{\lambda(0.25)^x(0.75)^{n-x} + (1 - \lambda)(0.75)^x(0.25)^{n-x}\}/0.5^n$$

is large, or, equivalently, when $\lambda 3^{(n/2)-x} + (1 - \lambda)3^{x-n/2}$ is large. Finally, the symmetry of the alternative distributions around $p = 0.5$ suggests that the adversary can make testing most difficult by choosing $\lambda = 0.5$. The best test for $\lambda = 0.5$, which is the minimax test, rejects when $|x - n/2|$ is large.

The worst case distributions are called *least favorable*. Sufficient conditions for least favorable distributions on H_0 and H_1 to exist and for the corresponding best test to be minimax are given in Lehmann [5]. Other methods for determining minimax tests are also given there. In Lehmann [5] and in some other references as well, minimax tests are called *maximin* because they have the property of maximizing the minimum probability of correctly deciding to reject H_0 when the alternative H_1 is true.

The study of minimax tests has led to some elegant mathematics. The minimax principle itself is often criticized, however, for being too pessimistic. (See, e.g., Berger [1] and Cox [2].) One criticism is that the worst case, which the minimax test protects against, may correspond only to a least favorable distribution that emphasizes implausible components of H_0 and H_1. In such cases it would be preferable to replace the least favorable distributions with other distributions on H_0 and H_1 that better represent the researcher's opinions about the components of H_0 and H_1 and to use the Bayes test that corresponds to these distributions (*see* BAYESIAN INFERENCE). On the other hand, the pessimism of the minimax test may be justifiable if the sample size is so large that even the minimax test, which protects against the worst case, has a small probability of false acceptance of H_0. In this situation, the minimax test provides inexpensive insurance against the worst case.

The minimax test may also be warranted if the distributions specified under one (or both) of the hypotheses are difficult to distinguish and the researcher cannot specify which, if any, of these distributions are implausible. For example, H_0 and H_1 might include all distributions within a small distance of two completely specified distributions. In the context discussed, an enlarged null hypothesis might specify that the probability of heads is $0.5 \pm \delta_0$ and the enlarged alternative hypothesis might specify that the probability of heads is $0.75 \pm \delta_1$. If δ_0 and δ_1 are small, the researcher may be unable to distinguish the components of the hypotheses and, therefore, may need to protect against the worst case. Minimax tests for hypotheses built of such neighborhoods of distributions are called robust* and are constructed in Huber [3, 4].

References

[1] Berger, J. O. (1980). *Statistical Decision Theory*. Springer-Verlag, New York.

[2] Cox, D. R. and Hinkley, D. V. (1974). *Theoretical Statistics*. Chapman & Hall, London.

[3] Huber, P. J. (1965). *Ann. Math. Statist.*, **36**, 1753–1758.

[4] Huber, P. J. (1968). *Zeit. Wahrscheinlichkeitsth. verw. Geb.*, **10**, 269–278.

[5] Lehmann, E. L. (1959). *Testing Statistical Hypotheses*. Wiley, New York.

(DECISION THEORY
HYPOTHESIS TESTING
MINIMAX DECISION RULES
MOST STRINGENT TESTS)

DIANE LAMBERT

MINIMUM CHI-SQUARE

Berkson [9, 10] has published opinions questioning the sovereignty of maximum likelihood estimation*. Moreover, in many circumstances the effective determination of maximum likelihood estimates (MLE) can be difficult. In 1945, J. Neyman* [21] presented for multinomial situations a class of estimates (originally termed best asymptotically normal, BAN, and now called regular best asymptotically normal or RBAN) based on the minimization of a special kind of distance function, that is, the alternatively defined chi-square goodness-of-fit* expressions (*see* LARGE-SAMPLE THEORY). This class of estimates was extended and methods of generating BAN estimates as roots of certain linear forms were obtained by Amemiya [1–3]; Barankin and Gurland [6]; Berkson [7–10]; Chiang [15]; Ferguson [17]; Taylor [23]; Zellner and Lee [25], etc. Simple methods of testing hypotheses also followed from the derivations of such BAN estimates (e.g., Bhapkar [12, 13]; Grizzle et al. [19]; Neyman [21]).

MINIMUM CHI-SQUARE METHODS OF GENERATING BAN ESTIMATES

The Pearson chi-square* test statistic,

$$\rho = \sum \frac{(\text{observed} - \text{expected})^2}{\text{expected}},$$

may be viewed as an expression of weighted sums of squares of the deviations between observed and expected values. Various minimum chi-square methods of generating BAN estimates are in fact weighted least-squares* methods with the weights chosen in a particular way. They vary in computational difficulty but all have the same asymptotic behavior. Here we use a multinomial case to illustrate the most widely used methods.

Suppose there are $n = \sum_{t=1}^{T} n_t$ independent observable random variables Y_{tv} ($t = 1, \dots, T$, $v = 1, \dots, n_t$), each capable of producing any $s + 1$ outcomes according to the probability function $p_{ti} = F_i(\mathbf{x}_t, \boldsymbol{\theta}) = F_{ti}$ for $i = 1, \dots, s + 1$, where $\mathbf{x}_t$ is a K-dimensional vector of known constants, $\boldsymbol{\theta}$ is a K-dimensional vector of unknown parameters in Ω, and $K \leqslant s$. The functions F_{ti} satisfy the identity $\sum_{i=1}^{s+1} F_{ti} = 1$ and are assumed continuous with respect to $\boldsymbol{\theta}$ as well as possessing continuous partial derivatives up to the second order. Let r_{ti} denote the proportion of the tth trials which result in the ith outcome.

Method 1: Pearson Chi-square

Choosing $\hat{\boldsymbol{\theta}}$ to minimize

$$\rho = \sum_{t=1}^{T} \sum_{i=1}^{s+1} \frac{(n_t r_{ti} - n_t p_{ti})^2}{n_t p_{ti}} \tag{1}$$

yields the minimum chi-square estimate of $\boldsymbol{\theta}$. It may be shown that (1) is algebraically equal to the quadratic form

$$\sum_{t=1}^{T} [\mathbf{r}_t - \mathbf{P}_t]' \boldsymbol{\Sigma}_t(\boldsymbol{\theta})^{-1} [\mathbf{r}_t - \mathbf{P}_t], \tag{2}$$

where $\mathbf{r}_t$ and $\mathbf{P}_t$ are $s \times 1$ vectors of $(r_{t1}, \dots, r_{ts})'$ and $E\mathbf{r}_t = \mathbf{P}_t(\boldsymbol{\theta})$, respectively, and $\boldsymbol{\Sigma}_t(\boldsymbol{\theta}) = E(\mathbf{r}_t - \mathbf{P}_t)(\mathbf{r}_t - \mathbf{P}_t)'$ is the $s \times s$ covariance matrix. For instance, consider the dichotomous case ($s + 1 = 2$): we have

$$\begin{aligned}\rho &= \sum_{t=1}^{T} n_t \left\{ \frac{(r_{t1} - p_{t1})^2}{p_{t1}} + \frac{[(1 - r_{t1}) - (1 - p_{t1})]^2}{(1 - p_{t1})} \right\} \\ &= \sum_{t=1}^{T} \left\{ \frac{n_t}{p_{t1}(1 - p_{t1})} (r_{t1} - p_{t1})^2 \right\}. \end{aligned} \tag{3}$$

The advantage of (2) lies in the fact that it also describes a method for estimating parameters of continuous distributions.

Method 2: Modified or Reduced Chi-square

Let $\mathbf{M}_t(\mathbf{r}_t)$ be a $s \times s$ symmetric positive definite matrix that depends only on $\mathbf{r}_t$ and $\mathbf{M}_t \to \boldsymbol{\Sigma}_t(\boldsymbol{\theta})^{-1}$ as $n_t \to \infty$. The estimates that minimize the quadratic form

$$\rho_A = \sum_{t=1}^{T} (\mathbf{r}_t - \mathbf{P}_t)' \mathbf{M}_t (\mathbf{r}_t - \mathbf{P}_t) \tag{4}$$

or

$$\rho_A = \sum_{t=1}^{T} n_t \sum_{i=1}^{s+1} \frac{(r_{ti} - p_{ti})^2}{r_{ti}} \tag{5}$$

are called the *minimum modified* or *reduced chi-square* estimates.

This method has the advantage that when F_{ti} are linear functions of $\boldsymbol{\theta}$, the BAN estimates are determined by a system of linear equations. For example, Zellner and Lee's [25] joint estimation procedure in the linear probability model is a generalization of this method.

Method 3: Substitution of P for θ

The problem of estimating $\boldsymbol{\theta}$ is equivalent to that of estimating all the probabilities p_{ti}. Eliminating $\boldsymbol{\theta}$ from the equations $p_{ti} = F_{ti}$, we have $L = Ts - K$ side conditions on the Ts independent p_{ti},

$$h_j(\mathbf{P}_1, \dots, \mathbf{P}_T) = h_j(\mathbf{P}) = 0, \qquad j = 1, \dots, L. \tag{6}$$

To obtain the BAN estimates of $\mathbf{P}$, Neyman [21] suggests minimizing (5) subject to (6) and $\sum_{i=1}^{s+1} p_{ti} = 1$ by the method of Lagrange multipliers*. When $h_j(\mathbf{P})$ is linear, the minimization problem reduces to finding the solution of a system of linear equations. When $h_j(\mathbf{P})$ is nonlinear, minimizing (5) subject to a linearized version of (6), (i.e., the first two terms of the Taylor series expansion about $\mathbf{r}_t$) again reduces to finding the solution of a system of linear equations. Neyman proves that this solution also yields BAN estimates of the p_{ti}'s.

If θ_k is a function of some p_{ti}, then a BAN estimate of θ_k is obtained by substituting the BAN estimates of p_{ti}'s into the expression.

Method 4: Transformed Chi-square

Let $\mathbf{g}(\mathbf{r})$ be any function from R^s to R^m ($m \leqslant s$) with the $m \times s$ continuous first partial derivative matrix $\dot{\mathbf{g}}(\mathbf{r}) = (\partial \mathbf{g}/\partial \mathbf{r}')$. The quadratic form

$$Q = \sum_{t=1}^{T} \left[\mathbf{g}(\mathbf{r}_t) - \mathbf{g}(\mathbf{P}_t)\right]' \left[\dot{\mathbf{g}}(\mathbf{P}_t)\boldsymbol{\Sigma}_t(\boldsymbol{\theta})\dot{\mathbf{g}}(\mathbf{P}_t)'\right]^{-1} \times \left[\mathbf{g}(\mathbf{r}_t) - \mathbf{g}(\mathbf{P}_t)\right] \tag{7}$$

or

$$Q_A = \sum_{t=1}^{T} \left[\mathbf{g}(\mathbf{r}_t) - \mathbf{g}(\mathbf{P}_t)\right]' \mathbf{M}_t \left[\mathbf{g}(\mathbf{r}_t) - \mathbf{g}(\mathbf{P}_t)\right], \tag{8}$$

where $\mathbf{M}_t \to [\dot{\mathbf{g}}(\mathbf{P}_t)\boldsymbol{\Sigma}_t(\boldsymbol{\theta})\dot{\mathbf{g}}(\mathbf{P}_t)']^{-1}$ in probability, is called transformed or modified transformed chi-square. Taylor [23] has proved that if

$$\frac{\partial^2 Q}{\partial r_{ti} \partial \theta_k} = c n_t \frac{p_{tik}}{p_{ti}}$$

and

$$\frac{\partial^2 Q}{\partial \theta_k \partial \theta_j} = -c \sum_{t=1}^{T} n_t \sum_{i=1}^{s+1} \frac{p_{tik} p_{tij}}{p_{ti}},$$

then the $\hat{\boldsymbol{\theta}}$ that minimize Q or Q_A are RBAN estimates, where $p_{tik} = \partial p_{ti}/\partial \theta_k$, and c is a constant.

Method 5: Constrained Chi-square

In many econometric* applications, in addition to the sample information there is prior information on $\boldsymbol{\theta}$ in the form of known constraints $\boldsymbol{\psi}(\boldsymbol{\theta}) = \mathbf{0}$. Let $\hat{\boldsymbol{\theta}}^*$ be a consistent and asymptotically efficient estimate of $\boldsymbol{\theta}$ when there are no constraints, and $\hat{\mathbf{R}}$ be its asymptotic covariance matrix $\mathbf{R}$ evaluated at $\boldsymbol{\theta} = \hat{\boldsymbol{\theta}}^*$. The estimate that minimizes the quadratic form

$$(\hat{\boldsymbol{\theta}}^* - \boldsymbol{\theta})' \hat{\mathbf{R}}^{-1} (\hat{\boldsymbol{\theta}}^* - \boldsymbol{\theta}), \tag{9}$$

subject to the constraint $\boldsymbol{\psi}(\boldsymbol{\theta}) = \mathbf{0}$ or the linearized version $\boldsymbol{\psi}(\hat{\boldsymbol{\theta}}^*) + \boldsymbol{\Psi}(\boldsymbol{\theta} - \hat{\boldsymbol{\theta}}^*) = \mathbf{0}$, is referred to by Malinvaud [20] as the minimum distance*estimator, where

$$\boldsymbol{\Psi} = \left(\frac{\partial \boldsymbol{\psi}}{\partial \boldsymbol{\theta}'}\right)_{\boldsymbol{\theta} = \hat{\boldsymbol{\theta}}^*}.$$

Its general principle and extension to simultaneous equations models are further discussed by Rothenberg [22].

The reason a large class of asymptotically equivalent methods of deriving BAN estimates are proposed is that the computation required to minimize the original Pearson chi-square statistic can be complicated even in the simplest case. Various procedures may, in a suitable manner, simplify the equations obtained by the original method and lead to solutions of linear equations. For instance, if an easy-to-calculate, efficient estimator $\hat{\boldsymbol{\theta}}^*$ for the unconstrained problem exists, the constrained minimum chi-square method of (9) is practical and is easily obtained as

$$\hat{\boldsymbol{\theta}} = \hat{\boldsymbol{\theta}}^* - \hat{\mathbf{R}}\boldsymbol{\Psi}'(\boldsymbol{\Psi}\hat{\mathbf{R}}\boldsymbol{\Psi}')^{-1}\psi(\hat{\boldsymbol{\theta}}^*). \quad (10)$$

In other cases transformed functions $\mathbf{g}(\mathbf{r})$ are linear functions of parameters $\boldsymbol{\theta}$. Hence, finding the value of $\boldsymbol{\theta}$ that minimizes the transformed chi-square is reduced to solving K linear equations in K unknowns. For instance, consider the dichotomous logit model (i.e., $s + 1 = 2$) with

$$\Pr(Y_{tv} = 1) = \left[1 + \exp(-\boldsymbol{\theta}'\mathbf{x}_t)\right]^{-1} = p_{t1}, \quad (11)$$

then

$$g(p_t) = \log \frac{p_{t1}}{1 - p_{t1}} = \boldsymbol{\theta}'\mathbf{x}_t\,.$$

Berkson's *minimum logit chi-square* estimator is defined by minimizing (8), which leads to the linear function

$$\hat{\boldsymbol{\theta}} = \left[\sum_{t=1}^{T} n_t r_{t1}(1 - r_{t1})\mathbf{x}_t\mathbf{x}_t'\right]^{-1} \times \left\{\sum_{t=1}^{T} n_t r_{t1}(1 - r_{t1})\left[\log \frac{r_{t1}}{1 - r_{t1}}\right]\mathbf{x}_t\right\}. \quad (12)$$

Alternatively, if

$$\Pr(Y_{tv} = 1) = \Phi(\boldsymbol{\theta}'\mathbf{x}_t), \quad (13)$$

where Φ is an integrated standard normal and $\Phi^{-1}(p_{t1}) = \boldsymbol{\theta}'\mathbf{x}$. Berkson's *minimum normit* chi-square estimator is that which minimizes

$$\sum_{t=1}^{T} \frac{n_t}{r_{t1}(1 - r_{t1})}\left[\frac{1}{\sqrt{2\pi}}\exp\left\{-\frac{1}{2}\Phi^{-1}(r_{t1})\right\}^2\right]^2 \times \left(\Phi^{-1}(r_{t1}) - \boldsymbol{\theta}'\mathbf{x}_t\right)^2. \quad (14)$$

For further discussion of the general principle and proof of minimum chi-square methods, see Amemiya [1, 2], Chiang [15], Ferguson [17], and Taylor [23], etc. For illustrative examples, see Amemiya [5], Berkson [9, 10], Bishop et al. [14] and Theil [24].

HYPOTHESIS TESTING

When the hypothesized model holds, the various quadratic forms discussed above using any of the BAN estimates of p_{ti}'s all converge in distribution to a chi-square variate. Hence one may use these statistics to test the goodness of fit of the model.

The chi-square statistic also allows us to test more complicated hypotheses (*see* CHI-SQUARE TESTS). Consider the hypothesis ascribing $\boldsymbol{\theta}$ to a parameter set ω that is a subset of Ω. Let $\hat{p}_{ti}(\omega)$ and $\hat{p}_{ti}(\Omega)$ be BAN estimates for p_{ti} under the hypothesis H_ω: $f_j(\boldsymbol{\theta}) = 0$, $j = 1, \ldots, m$, and $H_\Omega: \boldsymbol{\theta} \in \Omega$, respectively. Neyman [21] suggests testing H_ω using

$$D = \sum_{t=1}^{T} n_t \sum_{i=1}^{s+1} \frac{\left[r_{ti} - \hat{p}_{ti}(\omega)\right]^2}{\hat{p}_{ti}(\omega)} - \sum_{t=1}^{T} n_t \sum_{i=1}^{s+1} \frac{\left[r_{ti} - \hat{p}_{ti}(\Omega)\right]^2}{\hat{p}_{ti}(\Omega)} \quad (15)$$

or

$$D_A = \sum_{t=1}^{T} n_t \sum_{i=1}^{s+1} \frac{\left[r_{ti} - \hat{p}_{ti}(\omega)\right]^2}{r_{ti}} - \sum_{t=1}^{T} n_t \sum_{i=1}^{s+1} \frac{\left[r_{ti} - \hat{p}_{ti}(\Omega)\right]^2}{r_{ti}}. \quad (16)$$

Both (15) and (16) are asymptotically chi-square distributed with $(K - M)$ degrees of freedom under H_ω and are asymptotically equivalent to the likelihood ratio test* statistic. Exploiting this framework, Grizzle et al. [19] also propose a general procedure to test the goodness of fit* of the model and linear hypotheses about the parameters in terms of weighted least-squares analysis.

Moreover, the computation of chi-square statistics may be simplified further without having actually to compute the BAN estimates of the p_{ti}'s. Using Method 3, Bhapkar [12] has shown that minimizing (5) subject to (6) yields the value $\mathbf{a}'\mathbf{B}^{-1}\mathbf{a}$, where $\mathbf{a}' = (h_1(\mathbf{r}), \dots, h_L(\mathbf{r}))$, $\mathbf{B} = \mathbf{HVH}$,

$$\mathbf{H} = \left(\frac{\partial h_j}{\partial p_{ti}} \right)_{p_{ti} = r_{ti}}$$

is an $L \times Ts$ matrix, $\mathbf{V}$ is a block diagonal matrix having the $s \times s$ matrix $\mathbf{V}_t$ on the main diagonal, and $\mathbf{V}_t = (1/n_t)$ [diagonal $(r_{t1}, \dots, r_{ts}) - \mathbf{r}_t \mathbf{r}_t'$]. In other words, the minimum reduced chi-square is exactly equivalent to the weighted sum of squares of the unbiased estimates $h_j(\mathbf{P})$, with its variance–covariance matrix estimated by the sample variance–covariance matrix.

ADVANTAGES AND DISADVANTAGES

Although the MLE and the minimum chi-square estimates of the $\boldsymbol{\theta}$ have the same asymptotic covariance matrix (to the order n^{-1}, where n now stands for the average number of observations for each t), "statistics is an applied science and deals with finite samples" [11]. Berkson [9–11] has evaluated the exact mean square error* (MSE) of the minimum chi-square and the MLE for certain simple models and found that the MSE of the minimum chi-square estimator is smaller in all cases considered. Table 1 reproduces one of his experiment's results for the MLE, the minimum Pearson, and the minimum modified transformed (minimum logit) chi-square estimates of the logistic function with binomial variation of the dependent variable,

$$P_{t1} = \{1 + \exp[-(\theta_1 + \theta_2' x_t)]\}^{-1} \quad \text{and}$$

$$P_{t2} = 1 - P_{t1}.$$

Amemiya [4] further confirms Berkson's

Table 1 Comparison of Statistics of the Three Estimators for Various Positions of Three Equally Spaced Doses, 10 ($= n_t$) at Each Dose (x_t)[a]

True P_{t1} per Dose			Mean			Mean Square Error		
Low	Mid	High	MLE	Minimum Pearson chi-square	Minimum logit chi-square	MLE	Minimum Pearson chi-square	Minimum logit chi-square
				Estimate of θ_1 ($= 0$)				
0.3	0.5	0.7	0	0.002	0	0.187	0.179	0.154
0.391	0.6	0.778	− 0.006	− 0.013	− 0.020	0.230	0.218	0.206
0.5	0.7	0.845	− 0.021	− 0.011	− 0.013	0.430	0.412	0.394
0.632	0.8	0.903	− 0.026	0.037	0.084	1.103	0.972	0.689
				Estimate of θ_2 ($= 0.8743$)				
0.3	0.5	0.7	0.095	0.0624	0.048	0.322	0.280	0.271
0.391	0.6	0.778	0.100	0.0620	0.038	0.341	0.307	0.272
0.5	0.7	0.845	0.108	0.037	0.004	0.404	0.323	0.274
0.632	0.8	0.903	0.088	− 0.019	− 0.077	0.466	0.392	0.208

Source. Berkson [9, Table 2].

[a] Statistics of the MLE and minimum logit chi-square estimates are based on the total sampling population, those of minimum Pearson chi-square on a stratified random sample of 1,000 at each dosage arrangement. Samples not yielding finite estimates by maximum likelihood are omitted in calculating all statistics.

results by analytically and numerically evaluating the MSE of the MLE and the minimum logit chi-square estimator for the dichotomous logit regression model to the order of n^{-2}. It is only after the correction for the bias to the order of n^{-1} that the MLE is superior (see also Ghosh and Subramanyam [18], who proved a general theorem about the second-order efficiency* of the MLE in the exponential family). However, apart from the question whether the correction could be applied at all, the numerical evaluations conducted by Amemiya [4] show that the difference in the MSE between the minimum chi-square estimator and the bias-corrected MLE is never so large as to negate the computational advantage of the minimum chi-square estimator.

Despite the statistical attractiveness, the minimum chi-square method is probably less useful than the maximum likelihood method in analyzing survey data, and more suitable for laboratory settings. Application of it requires repeated observations for each value of the vector of explanatory variables. In survey data, most explanatory variables are continuous. The survey sample size has to be extremely large for the possible configurations of explanatory variables. Moreover, the maximum likelihood method can still be applied when some r_{ti} are zero, where the reduced or transformed chi-squares are not defined. Replacement of zero values by some positive value has been suggested; however, except in the dichotomous random variable cases, the effects have not been definitively investigated [9, 16, 19].

References

[1] Amemiya, T. (1974). *J. Amer. Statist. Ass.*, **69**, 940–944.

[2] Amemiya, T. (1976). *J. Amer. Statist. Ass.*, **71**, 347–351.

[3] Amemiya, T. (1977). *J. Econometrics*, **5**, 295–299.

[4] Amemiya, T. (1980). *Ann. Statist.*, **8**, 488–505.

[5] Amemiya, T. (1981). *J. Econ. Lit.*, **19**, 1483–1536.

[6] Barankin, E. W. and J. Gurland (1951). *Univ. Calif. Publ. Stat.*, **1**, 89–130.

[7] Berkson, J. (1944). *J. Amer. Statist. Ass.*, **39**, 357–365.

[8] Berkson, J. (1953). *J. Amer. Statist. Ass.*, **48**, 565–599.

[9] Berkson, J. (1955). *J. Amer. Statist. Ass.*, **50**, 130–136.

[10] Berkson, J. (1957). *Biometrika*, **44**, 411–435.

[11] Berkson, J. (1980). *Ann. Statist.*, **8**, 457–487.

[12] Bhapkar, V. P. (1961). *Ann. Math. Statist.*, **32**, 72–83.

[13] Bhapkar, V. P. (1966). *J. Amer. Statist. Ass.*, **61**, 228–235.

[14] Bishop, Y. M. M., Fienberg, S. E. and Holland, P. W., (1975). *Discrete Multivariate Analysis*. MIT Press, Cambridge, Mass.

[15] Chiang, C. L. (1956). *Ann. Math. Statist.*, **27**, 336–351.

[16] Cox, D. R. (1970). *Analysis of Binary Data*. Methuen, London.

[17] Ferguson, T. S. (1958). *Ann. Math. Statist.*, **29**, 1046–1062.

[18] Ghosh, J. K. and K. Subramanyam (1974). *Sankhyā A*, **36**, 325–358.

[19] Grizzle, J. E., Starmer, C. F., and Koch, G. G. (1969). *Biometrics*, **25**, 489–503.

[20] Malinvaud, E. (1970). *Statistical Methods of Econometrics*. North-Holland, Amsterdam.

[21] Neyman, J. (1949). In *Proc. Berkeley Symp. Math. Stat. Prob.*, 239–273.

[22] Rothenberg, T. (1973). *Efficient Estimation with A Priori Information*. Yale University Press, New Haven, Conn.

[23] Taylor, W. F. (1953). *Ann. Math. Statist.*, **24**, 85–92.

[24] Theil, H. (1970). *Amer. J. Sociol.*, **76**, 103–154.

[25] Zellner, A. and Lee, T. H. (1965). *Econometrica*, **33**, 382–394.

Acknowledgment

This work was supported in part by the Social Sciences and Humanities Research Council of Canada. The author also wishes to thank T. Amemiya, T. W. Anderson, A. Melino, and A. Yatchew for helpful comments.

(CHI-SQUARE TESTS
LARGE-SAMPLE THEORY
MAXIMUM LIKELIHOOD ESTIMATION
MINIMUM DISTANCE ESTIMATION)

CHENG HSIAO

MINIMUM-DESCRIPTION-LENGTH PRINCIPLE

ESTIMATION AND CODING

The MDL principle (MDL for minimum description length) [4, 5], provides a unified approach to statistical estimation*, and it allows the estimation of parameters along with their number without separate hypothesis testing*. The approach may be viewed as a coding theoretic formulation of earlier philosphically attractive but vague attempts to incorporate the elusive concepts of simplicity of a model and its prior knowledge in the estimation process. For example, see Kemeny's fascinating paper [2].

The subject matter falls somewhere in the void between traditional *information theory* and statistics, with a philosophical flavor from the algorithmic notion of information, (*see* INFORMATION THEORY AND CODING THEORY, ALGORITHMIC INFORMATION THEORY, and Kolmogorov [3]) and it would be misleading to guide the reader to the standard texts on communication theory* for prerequisites. However, we can warmly recommend the classical paper by Shannon [8] and the elementary textbook by Abramson [1] as sources for the basic coding theoretic notions needed. Slightly more advanced coding techniques and concepts are given in Rissanen and Langdon [7] and Rissanen [5, 6].

We begin by a review of the basic philosophy with the aim of convincing the reader that coding and estimation are closely related. This exposition is meant for those without any knowledge of coding. Just as in the maximum likelihood* (ML) technique, we select a parametric class of probability functions $P_{\boldsymbol{\theta}}(\mathbf{x})$, each assigning a probability to any possible observed sequence $\mathbf{x} = (x_1, \ldots, x_n)$. The parameter is a vector $\boldsymbol{\theta} = (\theta_1, \ldots, \theta_k)$, $k = 0, 1, \ldots,$ to be estimated along with the number of its components, and we also require $P_{\boldsymbol{\theta}}(\mathbf{x})$ to satisfy the usual compatibility conditions for a random process. It seems more natural to work with probabilities rather than densities, because each observed number x_i is always written in some finite precision, say, with q fractional binary digits. Often, for example in time series*, the observations consist of both an "input" sequence $\mathbf{y}$ and an "output" sequence $\mathbf{x}$. Then the appropriate probability function is the conditional one $P_{\boldsymbol{\theta}}(\mathbf{x} \mid \mathbf{y})$. However, the situation is in essence as before, and we will discuss the basic case for simplicity.

By just listing the numbers x_i sequentially in their binary notation, we see that the entire sequence x can be written down with something in excess of nq bits (= binary digits), the excess depending on the size of the integral parts of the numbers and whatever it takes to supply the necessary commas. But such a trivial "coding" or description of the observed sequence does not take into account the possible correlations that exist between the numbers x_i nor the relative frequency with which each observation occurs. If such dependencies were taken advantage of, we might be able to reduce the total number of binary digits in the description of $\mathbf{x}$. The very shortest description would result if all the statistical regular features were utilized, which clearly is possible only if we use the "true" data-generating probability $P_{\boldsymbol{\theta}^0}(\mathbf{x})$ in the code design. This in broad terms is the rationale behind the MDL principle.

What makes our approach more than a vague philosophical speculation is that we can form an excellent idea of the least number of bits that we have to spend to encode a sequence $\mathbf{x}$. But we first describe the most important properties for us of codes. We can take a code to be a one-to-one function C, which maps each sequence $\mathbf{x}$ of the same length n to a binary sequence of $L(\mathbf{x})$ symbols long such that the code length function satisfies the important *Kraft inequality* [1],

$$\sum_{\mathbf{x}} 2^{-L(\mathbf{x})} \leqslant 1. \tag{1}$$

This inequality necessarily holds if we require the code strings to have the so-called prefix property [i.e., no code string $C(\mathbf{x})$ is a prefix of another $C(\mathbf{x}')$]. In order to understand the meaning of the prefix property, consider a code string $C(\mathbf{x})$ immediately followed by other binary symbols y_i thus: $s = C(\mathbf{x})y_1 y_2 \ldots$. If we imagine the code strings to be listed in a table, we can readily see that the prefix property is equivalent with our ability to read off the code string $C(\mathbf{x})$ as an initial portion of s. We are not allowed to use a comma, which would amount to a third symbol. The prefix property with its associated inequality then has the important connotation that the code string $C(\mathbf{x})$ is a self-contained description of $\mathbf{x}$, and particularly important, it includes its own length information. It is no accident that with such descriptions of objects the code lengths just about define a probability distribution. In fact, whenever such a code is also efficient in any reasonable sense, say, having the minimum mean length, then equality in (1) will hold and $2^{-L(\mathbf{x})}$ defines a probability.

We next interpret the ML estimation criterion in coding theoretic terms. If we pick just any "model" $P_{\boldsymbol{\theta}}(\mathbf{x})$ in the class and assign a binary code sequence to $\mathbf{x}$ of $-\log P_{\boldsymbol{\theta}}(\mathbf{x})$ symbols long (all logarithms here are binary), or rather this quantity rounded to an integral upper bound, then the mean code length over all data sequences $\mathbf{x}$ of length n, $-\sum P_{\boldsymbol{\theta}^0}(\mathbf{x}) \log P_{\boldsymbol{\theta}}(\mathbf{x})$, where $\boldsymbol{\theta}^0$ denotes the "true" parameter, cannot be smaller than the entropy* $-\sum P_{\boldsymbol{\theta}^0}(\mathbf{x}) \log P_{\boldsymbol{\theta}^0}(\mathbf{x})$. Moreover, the equality is achieved only with $\boldsymbol{\theta} = \boldsymbol{\theta}^0$. Hence if somebody told us that the observed sequences turn up with probability $P_{\boldsymbol{\theta}}(\mathbf{x})$, $\boldsymbol{\theta}$ regarded as fixed, then all we need to do to encode the sequences efficiently is to pick for each string $\mathbf{x}$ its code $C(\mathbf{x})$ as a distinct binary sequence of about $-\log P_{\boldsymbol{\theta}}(\mathbf{x})$ symbols long. This can be done, and we are justified in calling $-\log P_{\boldsymbol{\theta}}(\mathbf{x})$ the *ideal code length* for $\mathbf{x}$ under the circumstances. If instead of a single distribution we are given a parametric family of them, and we wish to design the best code, we clearly ought to pick $\boldsymbol{\theta}$ so as to minimize the ideal length $-\log P_{\boldsymbol{\theta}}(\mathbf{x})$. The result is the familiar ML estimator—albeit in a new interpretation. We conclude that code design and estimation are indeed closely related.

We can immediately see from the new interpretation of estimation the main shortcoming of the ML estimation criterion and related techniques. If somebody would indeed construct a code for $\mathbf{x}$ as a binary string with the ideal code length $L(\mathbf{x}) = -\log P_{\boldsymbol{\theta}^*(\mathbf{x})}(\mathbf{x})$ relative to some estimated parameter $\boldsymbol{\theta}^*(\mathbf{x})$, then another person, when presented with the code string, would have no means of decoding $\mathbf{x}$ out of the string, unless he or she was told which parameter vector $\boldsymbol{\theta}^*(\mathbf{x})$ the encoding person had used. In other words, the code string of about $-\log P_{\boldsymbol{\theta}^*(\mathbf{x})}(\mathbf{x})$ binary symbols long is not a complete description of $\mathbf{x}$. Indeed, the inequality (1) fails. What is missing is a code for the parameter $\boldsymbol{\theta}^*(\mathbf{x})$ as a preamble in the total code string.

An efficient coding of the parameters is quite different from the coding of the random observations $\mathbf{x}$. This is because they may be deterministic objects without any readily given probability distribution to restrain their values. The issues confronting us are really the same as in the bitter Bayesian versus non-Bayesian dispute, involving the essence of prior knowledge, and we can hardly expect to get away with quite as rudimentary coding theoretic notions as have taken us this far. Nevertheless, with or without prior distributions, it makes sense to contemplate the coding of the parameters and the associated code length, the real question remaining of how many digits we have to spend on the code. First, note the following simple observation. Regardless of how we design a code for the parameters, the decoder must be able to tell which initial portion of the total code string includes the code of the parameters. One can do that only if the code length for the parameters satisfies the inequality

$$\sum 2^{-L(\boldsymbol{\theta})} \leqslant 1, \tag{2}$$

where $\boldsymbol{\theta}$ runs through all possible values of the estimates $\boldsymbol{\theta}^*(\mathbf{x})$.

PRIOR KNOWLEDGE AND CODE LENGTH OF PARAMETERS

There are two sources of information in estimation problems. The first consists of observed data $\mathbf{x}$, and the second, called prior information, consists of everything else, based on some earlier observations no longer available to us. In one of the two main cases to be studied further, the case in which the number of observations exceeds the number of parameters, a part of the prior information is used to define a parametric probability function $P_{\boldsymbol{\theta}}(\mathbf{x})$ on the set of all possible data sequences. The rest of the prior information concerns the parameters, the precise nature of which has been a matter of considerable controversy. However, for best estimates, the prior information about the parameters, even when it is lacking, must be taken into account.

The vague notion of prior information becomes tangible if we imagine a contest where the objective is to encode the parameters with as efficient a code as possible using all the available knowledge—except for the observed data. We make the notion of the available prior knowledge precise by defining it to be a set Γ of "test" distributions on the parameter values. These act as constraints such that to a small amount of prior knowledge there corresponds a large set Γ and vice versa. For example, in the classical case where the parameters are known to range over a finite set, the set Γ should consist of all distributions in the given finite set. Having defined Γ it makes sense to ask for a code length function $L(\boldsymbol{\theta})$ satisfying (2), which minimizes the ratio of the mean length $E_P L(\boldsymbol{\theta})$ to the entropy H_P for the worst case (maximizing) P in Γ. The result is a code length function $L^*(\boldsymbol{\theta})$ that satisfies (2) with equality, and $Q(\boldsymbol{\theta}) = 2^{-L^*(\boldsymbol{\theta})}$ defines a prior distribution for the parameters. This, then, is the meaning of the prior probabilities, and they are defined whether or not the unknown parameter is an individual object. If the parameter admits a single frequency distribution, then $Q(\boldsymbol{\theta})$ will agree with that, and if the parameter is known to range over a finite set of M values, then $Q(\boldsymbol{\theta}) = 1/M$, so that our formalism at the very least does not violate one's intuition in these commonly accepted solutions.

Of particular interest is the case where the parameters are integers about which nothing else is known; our formalism produces an equivalence class of minimax code lengths from which a particular member is singled out, i.e., the length given by

$$L^0(n) = c + \log n + \log\log n + \cdots, \quad (3)$$

where the sum includes all the positive iterates and c is a constant. The other optimum lengths deviate only slightly from (3). The associated probability $Q^0(n) = 2^{-L^0(n)}$ is seen to turn Jeffreys' improper prior to a universal proper prior for the integers.

The coding of vector parameters, again without additional knowledge, is done by first truncating the vector and then converting the result to an integer, which in turn is encoded with the length (3). The truncation involves an error, which in optimum truncation behaves like $1/\sqrt{n}$ for each component, so that the optimum code length for k parameters is $\frac{1}{2}k\log n$, where n is the number of observations. Therefore, the total ideal code length for the data $\mathbf{x}$ with use of a k-component parameter vector $\boldsymbol{\theta}$ is given to terms of order $O(\log n)$ by

$$L(\mathbf{x}) = -\log P_{\boldsymbol{\theta}}(\mathbf{x}) + \tfrac{1}{2}k\log n. \quad (4)$$

This is the MDL criterion, derived in Rissanen [4, 5] under the assumption that there is no prior knowledge about $\boldsymbol{\theta}$. The minimizing parameters $\theta^*(\mathbf{x})$ and $k^*(\mathbf{x})$ are the MDL estimates; the minimized length to within terms of order $O(\log n)$, i.e.,

$$L^*(\mathbf{x}) = -\log P_{\theta^*(\mathbf{x})}(\mathbf{x}) + \tfrac{1}{2}k^*(\mathbf{x})\log n, \quad (5)$$

satisfies (1) for all n.

In the special case where the prior information fixes the number of parameters, this criterion degenerates to the ML estimation criterion. But in other cases, even when we known nothing about the parameters, the prior information term plays a crucial role. If it is ignored, as with the ML estimation criterion, the estimation results in a complete failure: the best model is the one with as many parameters as there are observations.

OPTIMALITY OF MDL ESTIMATORS

The natural question is whether the minimized length $L^*(\mathbf{x})$ is indeed the shortest achievable among all conceivable codes, if all we know is that some process in the parametric class considered has generated the sequence. After all, we have only examined one rather particular way of doing the coding, based on a prior on the integers that is universal only in the sense we have defined it. To get an answer we must first sharpen the definition of the class of the allowed codes. Call a code *regular*, if its length $L(\mathbf{x})$ satisfies (1) for all n, and if $L(\mathbf{x}z) \geqslant L(\mathbf{x})$, where $\mathbf{x}z$ denotes the sequence obtained by tagging one more value to the end of $\mathbf{x}$; in fact all known codes are regular. The interest in regular codes stems from the simple fact that if (1) holds with equality for all n, then $2^{-L(\mathbf{x})}$ defines a random process, and, conversely, any random process defining probability function $P(\mathbf{x})$ immediately defines a regular code by $L(\mathbf{x}) = -\log P(\mathbf{x})$. (Here we conveniently drop the quite irrelevant requirement that a code must have an integer-valued length function.) Hence a regular code with (1) holding with equality is a coding-theoretic equivalent for a random process.

The length (5) is optimum among all regular codes in the sense of the following theorem, proved in Rissanen [6], and we define it to be the *information* in the data $\mathbf{x} = \mathbf{x}^n = x_1 \ldots x_n$, relative to the considered class of processes. This information measure is a combination of Shannon's probabilistic information about the random observations and Kolmogorov's algorithmic or combinatorial information, measuring the complexity of the parameters that are usually nonrandom (*see also* MEASURES OF INFORMATION).

Theorem 1. Let the central limit theorem* hold for the ML-estimates $\boldsymbol{\theta}^*(\mathbf{x}^n)$ of each $\boldsymbol{\theta}$ in the interior of a compact k-dimensional set Ω^k with a nonempty interior.

(a) If $L(\mathbf{x}^n)$ is a length function of a regular code, then for all k,

$$n^{-1}E_{\boldsymbol{\theta}}L(\mathbf{x}^n) \geqslant n^{-1}H_n(\boldsymbol{\theta}) + \tfrac{1}{2}(k/n)\log n + r(n), \quad (6)$$

for all points $\boldsymbol{\theta}$ except in a set of Lebesgue measure zero. Here, $r(n)n/\log n \to 0$ and $H_n(\boldsymbol{\theta})$ denotes the entropy of strings of length n.

(b) The length (5) is *optimum* in that the inequality $\leqslant$ opposite to the one in (6) holds for every $\boldsymbol{\theta}$ in Ω, the union of Ω^k over k.

The regular codes given by parametrically defined probability functions are of special interest, because we can use their code length to measure the goodness of estimators. If $\boldsymbol{\theta}''(\mathbf{x})$ is an estimator, then $-\log P_{\boldsymbol{\theta}''(\mathbf{x})}(\mathbf{x})$ does not satisfy (1) and hence does not define a regular code. However, if we add the term $\frac{1}{2}k''(\mathbf{x})\log n$, where $k''(\mathbf{x})$ denotes the number of components in $\boldsymbol{\theta}''(\mathbf{x})$ just as in (5), then the result does define a regular code. By comparing its length with the optimum (i.e., the information), we can assess the goodness of the estimator. There is also another basic way to get a regular code from an estimator. Form an estimate of the distribution for the possible values of the observation x_{t+1}, immediately following the sequence $\mathbf{x}^t = x_1 \ldots x_t$, as the conditional probability $P_{\boldsymbol{\theta}''(\mathbf{x}^t)}(x_{t+1} \mid \mathbf{x}^t)$, and design a code for these values with the associated ideal code length. This results in the code length for the string $\mathbf{x}$ of length n as

$$L(x) = -\sum_{t=0}^{n-1} \log P_{\boldsymbol{\theta}''(\mathbf{x}^t)}(x_{t+1} \mid \mathbf{x}^t). \quad (7)$$

Here, we set x^0 arbitrarily to some constant, say 0. This length is seen to be regular. Observe that there is no preamble in the code string to include the code of the parameters. None is needed, because the decoder is thought to know the rule for calculation of the estimated parameters. Nevertheless, one cannot avoid the second "penalty" term in the right-hand side of (6), which may be thought of as the inherent cost of the estimated model.

References

[1] Abramson, N. (1968). *Information Theory and Coding*. McGraw-Hill, New York.

[2] Kemeny, J. (1953). *Philos. Rev.*, **62**, 391–315.

[3] Kolmogorov, A. N. (1965). *Prob. Inf. Transm. (USSR)*, **1**, 4–7.

[4] Rissanen, J. (1978). *Automatica*, **14**, 465–471. (This paper uses only the most primitive notions of information theory.)

[5] Rissanen, J. (1983). *Ann. Statist.*, **11**, 416–431.

[6] Rissanen, J. (1984). *IEEE Trans. Inf. Theory*, **IT-30**, 629–636.

[7] Rissanen, J. and Langdon, G. G., Jr. (1981). *IEEE Trans. Inf. Theory*, **IT-27**, 12–23.

[8] Shannon, C. E. (1948). *Bell Syst. Tech. J.*, **27**, 379–423.

(ALGORITHMIC INFORMATION THEORY
ENTROPY
INFORMATION THEORY AND CODING THEORY
MAXIMUM LIKELIHOOD ESTIMATION
PRIOR DISTRIBUTIONS)

JORMA RISSANEN

MINIMUM DISCRIMINATION INFORMATION (MDI) ESTIMATION

This article is essentially a continuation of the article KULLBACK INFORMATION. The reader is urged to consider the latter article as prerequisite reading. References to numbered equations in KULLBACK INFORMATION will be indicated by KI followed by the equation number or numbers in parentheses.

If the expected values θ_i in KI(20, 23, 25) are not known, then, as an estimate $\hat{\theta}_i$ of θ_i, we take the observed values $c_i(\omega) = \hat{\theta}_i$. The related estimate of τ_i is $\hat{\tau}_i(\theta) = \tau_i(\hat{\theta})$ such that

$$c_i(\omega) = \hat{\theta}_i = (\partial/\partial\tau_i)\ln M(\tau_1, \tau_2, \ldots, \tau_r)|_{\tau_1=\hat{\tau}_1, \ldots, \tau_r=\hat{\tau}_r}, \quad i = 1, 2, \ldots, r, \tag{1}$$

or

$$c_i(\omega) = E_{p^*}(c_i(\omega))|_{\tau_1=\hat{\tau}_1, \ldots, \tau_r=\hat{\tau}_r}.$$

Since

$$(\partial/\partial\tau_i)\ln p^*(\omega) = c_i(\omega) - (\partial/\partial\tau_i)\ln M(\tau_1, \ldots, \tau_r), \tag{2}$$

it is seen that $\hat{\tau}_1, \ldots, \hat{\tau}_r$ are maximum likelihood* estimates of $\tau_1, \ldots, \tau_r$, as parameters of the MDI estimate $p^*(\omega)$ [1; 8, pp. 79–80; 10, p. 573], where $p^*(\omega)$ is given in KI(21) with $\tau_1 = \hat{\tau}_1, \ldots, \tau_r = \hat{\tau}_r$. A more extensive discussion is to be found in Kullback [9].

An important area of application of the foregoing is to categorical* or count data which includes contingency tables*. Here Ω is a space of n cells or categories and ω takes on values identifying the cells. For example, in a $4 \times 3 \times 2$ contingency table, ω takes on 24 values in lexicographic order (1, 1, 1), (1, 1, 2), (1, 2, 1), (1, 2, 2), (1, 3, 1), (1, 3, 2), (2, 1, 1), . . . , (2, 3, 2), . . . , (4, 3, 2). The lexicographic ordering is a form of numerical alphabetizing. Suppose that a total N of cell counts have been observed. Let us write $x(\omega) = Np(\omega)$ for the observed cell counts and define $x^*(\omega) = Np^*(\omega)$.

When the values of the moment constraints are derived from the observed data, we have a class of problems whose objectives are data fitting or smoothing or model building. We designate this class of problems *internal constraint problems*, ICP. When the values of the moment constraints are a consequence of certain hypotheses, we have a class of problems designated *external constraints problems*, ECP. For the ICP, the constraints KI(20) are

$$\sum_{\Omega} c_i(\omega)x^*(\omega) = \sum_{\Omega} c_i(\omega)x(\omega), \quad i = 0, 1, \ldots, r. \tag{3}$$

In particular, $\sum_{\Omega} x^*(\omega) = \sum_{\Omega} x(\omega) = N$. The goodness-of-fit* or MDI statistic for ICP is

$$2I(x : x^*) = 2\sum_{\Omega} x(\omega)\ln(x(\omega)/x^*(\omega)). \tag{4}$$

The MDI statistic in (4) is asymptotically distributed as chi-square with $n - r - 1$ degrees of freedom. In the case of ICP the distribution $\pi(\omega)$ is usually taken so that $N\pi(\omega) = N/n$. As was seen from (2), in the ICP case the results of the MDI estimation procedure are the same as the maximum-likelihood estimates, and the MDI statistic in (4) is the log-likelihood ratio statistic. This fact is not true for the ECP case although the MDI estimates are BAN.

Suppose we have two nested ICP models M_a and M_b where every moment constraint in M_a is explicitly or implicitly contained in the set of moment constraints of M_b. Let $x_a^*(\omega)$ be the MDI estimate corresponding to the set of $r_a + 1$ moment constraints in M_a and let $x_b^*(\omega)$ be the MDI estimate corresponding to the set of $r_b + 1$ moment constraints in M_b. Then, the Pythagorean type property KI(26) is the analysis of information relation

$$2I(x : x_a^*) = 2I(x_b^* : x_a^*) + 2I(x : x_b^*), \tag{5}$$

$$(n - r_a - 1) = (r_b - r_a) + (n - r_b - 1), \quad r_b > r_a, \text{ d.f.}$$

The relation in (5) is an additive analysis of $2I(x : x_a^*)$ into components which are MDI statistics, with an additive relation for the associated degrees of freedom of their asymptotic chi-square distributions.

For the ECP case the moment constraints are

$$\sum_{\Omega} c_i(\omega)x^*(\omega) = N\theta_i, \qquad i = 0, 1, \ldots, r,$$

and the MDI statistic to test the hypothesis is

$$2I(x^* : x) = 2\sum_{\Omega} x^*(\omega)\ln(x^*(\omega)/x(\omega)). \tag{6}$$

The MDI statistic in (6) is asymptotically distributed as chi-square with r degrees of freedom. For the cases of ECP the distribution $\pi(\omega)$ is usually taken so that $x(\omega) = N\pi(\omega)$. In ECP if

$$\sum_{\Omega} c_{bi}(\omega)x_b^*(\omega) = N\theta_{bi}, \quad i = 0, 1, 2, \ldots, r_b$$

implies

$$\sum_{\Omega} c_{ai}(\omega)x_a^*(\omega) = N\theta_{ai}, \quad i = 0, 1, 2, \ldots, r_a,$$

with $r_b > r_a$, then the analysis of information relation is

$$2I(x_b^* : x) = 2I(x_b^* : x_a^*) + 2I(x_a^* : x), \quad r_b = (r_b - r_a) \quad + r_a, \text{ d.f.} \tag{7}$$

The relation in (7) is an additive analysis of $2I(x_b^* : x)$ into components which are MDI statistics with an additive relation for the associated degrees of freedom of their asymptotic chi-square distributions.

The statistical literature contains a variety of quadratic expressions that have been proposed as chi-square statistics and the basis for a minimum-chi-square* type of estimation [3]. The principle of MDI estimation provides a unified approach. Quadratic approximations to the MDI statistics either in terms of the moment parameters or the dual set of exponential or natural parameters are described in Gokhale and Kullback [6] and Kullback [9]. These approximations can be shown to be equivalent to many of the quadratic chi-square statistics that have been proposed. In refs 6 and 9 are given examples of quadratic approximations to MDI statistics; these are Pearson's chi-square*, Neyman's modified chi-square, minimum modified chi-square, Wald-type statistics, minimum logit chi-square (*see also* MINIMUM CHI-SQUARE) as presented in a number of papers [2–5, 7, 11–15].

For further discussion, computer algorithms, and many applications to real multidimensional contingency tables* see Gokhale and Kullback [6].

References

[1] Barton, D. E. (1956). *Biometrika*, **43**, 200–202.

[2] Berkson, J. (1972). *Biometrics*, **28**, 443–468.

[3] Berkson, J. (1980). *Ann. Statist.*, **8**, 457–469 and 485–487.

[4] Bhapkar, V. P. and Koch, G. G. (1968). *Technometrics*, **10**, 107–23.

[5] Fisher, R. A. (1950). *Biometrics*, **6**, 17–24.

[6] Gokhale, D. V. and Kullback, S. (1978). *The Information in Contingency Tables*, Dekker, New

York. (Presentation of MDI estimation at an intermediate level emphasizing methodology in the analysis of categorical or count data; contains many practical examples. Extensive bibliography.)

[7] Ireland, C. T., Ku, H. H., and Kullback, S. (1969). *J. Amer. Statist. Ass.*, **64**, 1323–1341.

[8] Khinchin, A. I. (1949). *Mathematical Foundations of Statistical Mechanics*. Dover, New York.

[9] Kullback, S. (1959). *Information Theory and Statistics*. Wiley, New York (Dover, New York, 1968; Peter Smith Publisher, Magnolia, MA., 1978). (First five chapters contain a measure-theoretic presentation of theory. Chapters 6–13 consider in particular applications to classification and hypothesis testing at an intermediate level. Contains many examples, problems, an extensive bibliography, tables, and a glossary.)

[10] Kupperman, M. (1958). *Ann. Math. Statist.*, **29**, 571–574.

[11] Neyman, J. (1929). *XVIII Session de l'Institut International de Statistique*, Varsovie, Poland, pp. 1–48.

[12] Pearson, K. (1911). *Biometrika*, **8**, 250–253.

[13] Stuart, A. (1958). *Biometrika*, **42**, 412–416.

[14] Wald, A. (1943). *Trans. Amer. Math. Soc.*, **54**, 426–482.

[15] Woolf, B. (1955). *Ann. Hum. Genet.*, **19**, 251–253.

(CHI-SQUARE TESTS
CONTINGENCY TABLES
FISHER INFORMATION
KULLBACK INFORMATION
LIKELIHOOD RATIO
MAXIMUM-LIKELIHOOD ESTIMATION
MULTIDIMENSIONAL CONTINGENCY TABLES
PARTITION OF CHI-SQUARE)

S. KULLBACK

MINIMUM DISTANCE ESTIMATION

Minimum distance (MD) is a method of parameter estimation designed to clearly reflect the scientific modeler's desire to construct a model reproducing the probabilistic structure of the real-world phenomenon under study. Although Smith [12] mentioned the method in 1916, the pioneering theoretical work was done by Wolfowitz in a series of papers culminating in 1957 [13]. Wolfowitz's motivation was a desire to provide consistent parameter estimators for problems where other methods had not proved successful.

The method is best explained by consideration of one of the simplest cases—repeated sampling from a distribution known to lie in some parametric set. Let $X_1, X_2, \ldots,$ be independent and identically distributed real-valued random variables with cumulative distribution function G, thought to be an element of $\Gamma = \{F_\theta,\ \theta \in \Theta\}$, a parametrized set of continuous distribution functions, and let G_n denote the usual empirical distribution function. Let $\delta(G_n F_\theta)$ be some measure of the "distance" between G_n and F_θ (possibly a metric), such as

$$\delta_K(G_n, F_\theta) = \sup_{-\infty < x < \infty} |G_n(x) - F_\theta(x)|$$

or

$$\delta_C(G_n, F_\theta) = \int_{-\infty}^{\infty} (G_n(x) - F_\theta(x))^2 w_\theta(x)\, dF_\theta(x),$$

the Kolmogorov and Cramér–von Mises discrepancies, respectively. The MD estimator of θ is chosen to be any value $\hat{\theta}_n$ in Θ such that

$$\delta(G_n, F_{\hat{\theta}_n}) = \inf_{\theta \in \Theta} \delta(G_n, F_\theta),$$

a value $\hat{\theta}_n$ minimizing the distance between G_n and F_θ. An extensive bibliography on MD estimation, classified by subject matter and choice of discrepancy for minimization, is given in Parr [7]. Discrepancies can be chosen to measure the distance between empirical and theoretical distribution functions, characteristic functions*, quantile functions, density functions, or other such quantities, depending on the particular application. In the following, we concentrate on discrepancies such as δ_K and δ_C between distribution functions, which is currently the most fully explored case, occasionally with more general references. For references and an introduction to minimum chi-square estimation, not covered here, *see* MINIMUM CHI-SQUARE.

Estimators minimizing discrepancies such as those mentioned are, under extremely modest regularity conditions, strongly consistent. Sets of sufficient conditions are given in Boos [4], Parr [8, 9], and Wolfowitz [13]. Estimators minimizing δ_C are typically also asymptotically normal. In the case of a location parameter, $G \in \Gamma$, and use of δ_C with $w_\theta(x) \equiv 1$, a sufficient condition for asymptotic normality* is that F_θ possess a uniformly continuous density. Other less stringent sets of conditions for asymptotic normality of minimum δ_C estimators can be found in refs. 4, 8, and 9. (However, those MD estimators based on distances involving sup-norms such as δ_K are typically not asymptotically normal (see Rao [11]). Parr and DeWet [9] and Boos [4] give methods for deriving fully efficient MD estimators for location and scale problems using appropriate choice of the function w_θ, the former paper showing how to obtain prespecified influence curves by appropriate choice of the weight function w_θ (*see* INFLUENCE FUNCTIONS).

Among the desirable features of MD estimators are (1) natural robustness* properties, (2) a concrete interpretation for the value to which the estimator converges even when the model is wrong (i.e., $G \notin \Gamma$), (3) ease of application to problems not involving symmetries or invariance* properties, (4) for the minimum δ_C estimator, simplicity of computation, and (5) extremely competitive small-sample behavior in the several situations thus far explored by the Monte Carlo method* in, for example ref. 8. (See also the simulation* results given later.)

Millar has demonstrated that, in an appropriate methematical framework, MD estimators based on δ_C possess excellent robustness properties against local deviations from the model [6]. In the correct model case, the influence curve of a minimum δ_C estimator is clearly bounded and continuous if

$$\int_{-\infty}^{\infty} \left| \frac{\partial F_\theta(x)}{\partial \theta} w_\theta(x) \right| dF_\theta(x) < \infty.$$

(Continuity and boundaries of the influence curve are good indicators that an estimator is robust against both gross outliers* and rounding errors*.) In fact, the influence curve (IC) reduces in the location problem to

$$\mathrm{IC}_{T,F_\theta}(c) = \frac{\int_{-\infty}^{\infty} \{F_\theta(x) - I(c \leqslant x)\} w_\theta(x) f_\theta^2(x)\,dx}{\int_{-\infty}^{\infty} w_\theta(x) f_\theta^3(x)\,dx}.$$

For the case of normality, in the location problem the minimum δ_C estimator with $w_\theta(x) \equiv 1$ has an asymptotic variance of $1.095\sigma^2$. That is, only a 9.5% "insurance premium" need be paid in order to achieve a considerable degree of robustness, as evidenced by the results already referenced above the simulations referred to.

Even in the case where the practitioner has picked the wrong model ($G \notin \Gamma$), the MD estimator possesses an interpretation as giving a suitable approximation to G_n. For instance, if we are using δ_K then $F_{\hat\theta_n}$ is the L^∞ projection of G_n into Γ. Similarly, if we use δ_C as our discrepancy with $w_\theta(x) = 1/f_\theta(x)$, $F_{\hat\theta_n}$ is an L^2 projection of G_n into Γ. Similar comments hold for F_{θ_0} and G, when $\theta_0 = \text{plim}_{n\to\infty}\theta_n$. Thus, under suitable regularity, the MD estimator converges to a point in the parameter space giving in some sense the best approximation to G. (Other methods of estimation, such as R-estimators, L-estimators, or M-estimators*, do not possess this property of minimizing a distance in probability units between the data and the statistical model.)

Construction of a robust MD estimator is typically a straightforward process: take a discrepancy measure δ not overly sensitive to small changes in probability assignments and minimize $\delta(G_n, F_\theta)$ to obtain the minimum distance estimator of θ. General construction of fully efficient MD-estimators seems problematic for nonlocation-or-scale problems.

Computation of MD estimators using the discrepancy δ_C with $w_\theta(x) \equiv 1$ is quite simple. Using the computational formula for the Cramér–von Mises statistic*, we observe that

$$\delta_C(G_n, F_\theta) = \frac{1}{12n} + \sum_{i=1}^{n} \left(\frac{2i-1}{2n} - F_\theta(X_{(i)}) \right)^2,$$

where $X_{(i)}$ denotes the ith-order statistic. Thus computation of the MD estimator becomes a nonlinear least-squares problem. (In the case of estimating the mixing proportions for an unknown mixture of known components, the problem is linear.) Thus a Marquardt-type algorithm can be used to obtain the estimator. Most simply, the data can be fed into a nonlinear regression* program with $(2i - 1)/(2n)$ playing the role of the dependent variable and $F_\theta(X_{(i)})$ the role of the model. Experience with location and scale models indicates rapid convergence and robustness to poor starting values. For problems of higher dimension, good starting values are much more important.

Parr and Schucany [8] report on part of a massive simulation study of MD estimators, primarily of minimum δ_C type, for the problem of estimation of the point of symmetry of a symmetric population based on samples of size 20. A large variety of symmetric long, medium, and short-tailed distributions were considered. Acronymns for the non-MD estimators, chosen as among the better M-estimators from the Princeton study, are as in Andrews et al. [1]. H15 and H20 are Huber's, using ψ-functions designed to produce estimators asymptotically equivalent to trimmed means. 17A and 22A are Hampel's, designed to have redescending influence curves and give absolutely no weight to extreme outliers. MD estimators have acronyms as follows: CNF [a minimum Cramér–von Mises estimator, with F_θ a normal cumulative distribution function, using a rescaled interquartile range* as the fixed (noniterative) scale estimator], CTF (the same as CNF, but F_θ is a t-distribution* with 4 degrees of freedom), CNS (the same as CNF but with scale estimated simultaneously with location by minimizing the Cramér–von Mises statistic), and KSN (a minimum Kolmogorov–Smirnov* estimator as discussed in Rao [11], with F_θ a normal cumulative distribution function). Table 1 gives some representative results. From this we observe that the minimum Cramér–von Mises estimators are quite competitive, except for the most extremely long-tailed situations, such as the Cauchy, where they are slightly inferior to the Hampel's 17A and 22A (the only estimators with redescending influence curves tablulated here).

To numerically illustrate the effects of model choice (the choice of Γ) on the obtained estimators, Table 2 gives values of the minimum Cramér–von Mises location estimates using Student t-models with various degrees of freedom for the Cushny and Peebles data on soporific drugs as discussed by Hampel [5]. The various t-distributions are chosen to give a good range of thickness of tails. Scale was estimated by minimizing δ_C simultaneously in location and scale. The data were:

$$0, 0.8, 1.0, 1.2, 1.3, 1.3, 1.4, 1.8, 2.4, 4.6.$$

Table 1 Monte Carlo Estimates of 20 times Variance Based on 1,000 Repetitions[a]

	Distribution					
Estimator	$N(0,1)$	$T(8)$	$T(4)$	$T(2)$	$T(1)$	Laplace
CNF	1.06	1.24	1.46	2.10	4.39	1.41
CTF	1.09	1.25	1.45	2.00	3.83	1.35
CNS	1.09	1.25	1.45	2.00	3.65	1.36
KSN	1.09	1.31	1.58	2.44	5.67	1.53
H15	1.04	1.24	1.51	2.33	5.78	1.54
H20	1.01	1.25	1.60	2.70	8.55	1.67
17A	1.11	1.28	1.47	1.96	3.10	1.39
22A	1.09	1.29	1.52	2.09	3.47	1.51

[a] The variance reduction methods of Andrews et al. [1] are used to reduce Monte Carlo variability.

Table 2

Estimation	Numerical Value
Minimum δc; $F_\theta = t(1)$	1.31
Minimum δc; $F_\theta = t(2)$	1.32
Minimum δc; $F_\theta = t(3)$	1.33
Minimum δc; $F_\theta = t(4)$	1.33
Minimum δc; $F_\theta = t(8)$	1.33
Minimum δc; $F_\theta = t(\infty)$	1.34
Mean	1.58
10% trimmed mean	1.40
20% trimmed mean	1.33
Median	1.30

All of the MD estimators in Table 2 can be seen to fall between the sample median and 20% trimmed mean. In any case, the effect of model choice does not seem to be substantial.

In developments related to MD estimation, the use of minimized goodness-of-fit statistics for testing goodness of fit* in the composite null case has been studied in Boos [4] and Pollard [10]. Although the asymptotic null distributions of the statistics depend on the family Γ if not also on the specific point in the parameter space, excellent and workable approximations not depending even on Γ are given in ref. 4 for the Anderson–Darling* statistic.

References

[1] Andrews, D. F., Bickel, P. J., Hampel, F. R., Huber, P. J., Rogers, W. H., and Tukey, J. W. (1972). *Robust Estimates of Location: Survey and Advances*. Princeton University Press, Princeton, NJ.

[2] Beran, R. (1977). *Ann. Statist.*, **5**, 445–463.

[3] Berkson, J. (1980). *Ann. Statist.*, **8**, 457–487.

[4] Boos, D. D. (1981). *J. Amer. Statist. Ass.* **76**, 663–670.

[5] Hampel, F. R. (1973). *Zeit. Wahrsheinlichkeitsth. verw. Geb.*, **27**, 87–104.

[6] Millar, P. W. (1981). *Zeit. Wahrsheinlichkeitsth. verw. Geb.*, **55**, 73–89.

[7] Parr, W. C. (1981). *Commun. Statist. A*, **10**, 1205–1224.

[8] Parr, W. C. and Schucany, W. R. (1980). *J. Amer. Statist. Ass.*, **75**, 616–624.

[9] Parr, W. C. and DeWet, T. (1981). *Commun. Statist. A*, **10**, 1149–1166.

[10] Pollard, D. (1980). *Metrika*, **27**, 43–70.

[11] Rao, P. V., Schuster, E. F., and Littell, R. C. (1975). *Ann. Statist.*, **3**, 862–873.

[12] Smith, (1916). *Biometrika*, **11**, 262–276.

[13] Wolfowitz, J. (1957). *Ann. Math. Statist.*, **28**, 75–88.

(CRAMÉR–VON MISES STATISTIC
GOODNESS OF FIT
KOLMOGOROV–SMIRNOV STATISTICS
M-ESTIMATORS
MINIMUM CHI-SQUARE
ROBUST ESTIMATION)

W. C. PARR

MINIMUM NORM QUADRATIC ESTIMATION *See* MINQE

MINIMUM RANK FACTOR ANALYSIS

The factor analysis* model assumes that the $p \times p$ population covariance* matrix $\mathbf{\Sigma}$ can be represented in the form

$$\mathbf{\Sigma} = \mathbf{\Lambda}\mathbf{\Lambda}' + \mathbf{\Psi}$$

where $\mathbf{\Lambda}$ is the $p \times r$ matrix of factor loadings and $\mathbf{\Psi}$ is the $p \times p$ diagonal matrix of the residual variances. This suggests that $\mathbf{\Sigma}$ is decomposable into two parts,

$$\mathbf{\Sigma} = (\mathbf{\Sigma} - \mathbf{\Psi}) + \mathbf{\Psi},$$

where $\mathbf{\Sigma} - \mathbf{\Psi}$ is a Gramian matrix of rank* r and $\mathbf{\Psi}$ is diagonal. (A matrix is Gramian if it is symmetric and nonnegative definite.) It is clear that without any additional assumptions there exist infinitely many such decompositions. In factor analysis* one usually tries to explain the data in terms of as few factors as possible. Since the rank of $\mathbf{\Sigma} - \mathbf{\Psi}$ is equal to the number r of common factors this approach leads to the concept of *minimum rank* solution. Thus the objective of the *minimum rank factor analysis* is to determine an appropriate diagonal matrix $\mathbf{\Psi}$ in order to obtain $\mathbf{\Sigma} - \mathbf{\Psi}$ of minimal rank. By an appropriate matrix $\mathbf{\Psi}$ we mean that choice of $\mathbf{\Psi}$ must be consistent with the factor analysis model. Eventually this indicates that the (reduced) matrix $\mathbf{\Sigma} - \mathbf{\Psi}$ as well as $\mathbf{\Psi}$ must be Gramian.

Given the matrix $\mathbf{\Sigma}$, two questions naturally arise:

1. To what extent can one reduce the rank of $\mathbf{\Sigma} - \mathbf{\Psi}$ by an appropriate choice of $\mathbf{\Psi}$?
2. Is a given minimum rank solution unique?

Example. Consider the following 3×3 covariance matrix

$$\mathbf{\Sigma} = \begin{bmatrix} 1 & 0.5 & 0.2 \\ 0.5 & 1 & 0.5 \\ 0.2 & 0.5 & 1 \end{bmatrix}$$

Unless $\mathbf{\Sigma}$ is diagonal, the zero-rank solution is impossible, and the question is whether we can substitute the diagonal elements of $\mathbf{\Sigma}$ by numbers x_1, x_2, x_3, say, to give a Gramian matrix of rank one. The obtained (reduced) matrix will be of rank one if and only if all its second-order minors are zero. In particular, considering the minor involving the first and second rows and the first and third columns, the equation

$$0.5x_1 - 0.5 \times 0.2 = 0$$

must hold.

This implies the solution

$$\psi_{11} = 1 - x_1 = 0.8.$$

Similarly $\psi_{22} = -0.25$ and $\psi_{33} = 0.8$. Here the one-rank factor analysis leads to an inappropriate solution where one or more diagonal elements of $\mathbf{\Psi}$ are negative (the so-called *Heywood case*).

It should be mentioned that there exist infinitely many appropriate two-rank solutions for $\mathbf{\Sigma}$.

RANK-REDUCIBILITY AND LEDERMANN'S BOUND

In the earlier days of factor analysis, it was believed that a relatively small rank could be attained for a covariance matrix of mental-test data by an appropriate choice of diagonal elements. In those days an extensive effort was made to find algebraic conditions ensuring reducibility to a given rank r. Charles Spearman, the founder of factor analysis, knew even then that in order for a $p \times p$ covariance matrix $\mathbf{\Sigma}$ to be reducible to rank one, all its second-order minors not including diagonal elements must vanish [8]. (See, e.g., Harman [3, Section 5.3] for an extension to $r = 2$.)

For $p = 3$, no relationships among the off-diagonal elements of $\mathbf{\Sigma}$ are necessary to attain rank one. In general, $p(p-3)/2$ equations must hold for the one-rank solubility. This indicates that for $p \geqslant 4$ the rank cannot be reduced to one unless $\mathbf{\Sigma}$ has an exact algebraic structure. Generally the rank cannot be reduced below the so-called Ledermann bound

$$\phi(p) = \frac{2p + 1 - (8p + 1)^{1/2}}{2}$$

for *almost every* symmetric matrix $\mathbf{S}$. More precisely: *Symmetric $p \times p$ matrices that can be reduced to rank r by modifying their diagonal elements form a set of Lebesgue measure* zero (in the vector space of symmetric matrices) if and only if $r < \phi(p)$* (Shapiro, [6]).

Thus the number $\phi(p)$ represents with probability one a *lower* bound to the minimal reduced rank of a *sample* covariance matrix $\mathbf{S}$. Some values of p and the corresponding lower bound k [k is the smallest integer greater than or equal to $\phi(p)$] are

p	3	4	5	6	7	9	10	15	21	55	91
k	1	2	3	3	4	6	6	10	15	45	78

It can be seen that the ratio k/p tends to one as p increases in its value.

It has been argued by Ledermann [5] and some other authors (e.g., Thurstone [9, p. 293], Harman [3, section 5.2]), that rank can always be reduced to the bound $\phi(p)$. However, Guttman [2] gave an example wherein it is impossible to reduce rank below $p - 1$; it deals with covariance matrices that are composed of zero elements everywhere apart from p elements on the main diagonal and $2(p - 1)$ elements immediately above and below the main diagonal. Such matrices are exceptional and unlikely to appear in practice. In general, the rank of $\mathbf{S}$ cannot be reduced below $p - 1$ if there exists a diagonal matrix $\mathbf{D} = \text{diag}(\pm 1, \pm 1, \ldots, \pm 1)$ such that all off-diagonal elements of the (scaled) matrix $\mathbf{DSD}$ are negative (see Shapiro [7], p. 247).

The preceding discussion shows that a reduced rank of the *sample* covariance matrix cannot be small relative to p. If an exact low-rank solution is impossible, one should think about an approximate solution. This leads to the statistical problem of parameter estimation. A detailed treatment (maximum likelihood approach) can be found in Lawley and Maxwell [4].

IDENTIFICATION* PROBLEM

We consider now the uniqueness of the solution. It will be assumed that at least one r-rank solution exists. The parameters $\mathbf{\Lambda}$ and $\mathbf{\Psi}$ are estimable only if this solution is unique, or, following common statistical terminology, if $\mathbf{\Lambda}$ and $\mathbf{\Psi}$ are *identified* (*see also* IDENTIFIABILITY). Since any solutions $\mathbf{\Lambda}, \mathbf{\Psi}$ can be replaced by $\mathbf{\Lambda T}, \mathbf{\Psi}$, where $\mathbf{T}$ is an $r \times r$ orthogonal matrix, some additional conditions must be put on $\mathbf{\Lambda}$. A detailed study of the identification problem has been given by Anderson and Rubin [1, Section 5]. They have shown that *sufficient conditions for identification of* $\mathbf{\Psi}$ *and* $\mathbf{\Lambda}$ *up to multiplication by an orthogonal matrix are that, if any row of* $\mathbf{\Lambda}$ *is deleted, there remain two disjoint submatrices of* $\mathbf{\Lambda}$ *of rank* r.

These sufficient conditions concern cases where $r < p/2$ while at least *local* identification holds for *almost every* $\mathbf{\Lambda}$ and $\mathbf{\Psi}$ if $r \leqslant \phi(p)$. (By local identification we mean that there does not exist another solution in a *neighborhood* of a given one.) There is a considerable gap between $p/2$ and $\phi(p)$. No conditions that are both necessary and sufficient have been found yet.

References

[1] Anderson, T. W. and Rubin, H. (1956). *Proc. 3rd Berkeley Symp. Math. Statist. Prob.*, Vol. 5., University of California Press, Berkeley, Calif., pp. 111–150. (Contains most complete, even up-to-date, discussion of the identification problem in factor analysis.)

[2] Guttman, L. (1958). *Psychometrika*, **23**, 297–308.

[3] Harman, H. H. (1976). *Modern Factor Analysis*, 3rd ed. University of Chicago Press, Chicago.

[4] Lawley, D. N. and Maxwell, A. E. (1971). *Factor Analysis as a Statistical Method*. Butterworth, London.

[5] Ledermann, W. (1937). *Psychometrika*, **2**, 85–93.

[6] Shapiro, A. (1982). *Psychometrika*, **47**, 187–199.

[7] Shapiro, A. (1982). *Psychometrika*, **47**, 243–264.

[8] Spearman, C. (1927). *The Abilities of Man*. Macmillan, New York.

[9] Thurstone, L. L. (1947). *Multiple Factor Analysis*. University of Chicago Press, Chicago.

(FACTOR ANALYSIS
IDENTIFIABILITY)

A. SHAPIRO

MINIMUM RISK EQUIVARIANT ESTIMATORS *See* EQUIVARIANT ESTIMATORS

MINIMUM SPANNING TREE *See* DENDRITES

MINIMUM TRACE FACTOR ANALYSIS

In the factor analysis* model the $p \times p$ population covariance* matrix $\mathbf{\Sigma}$ is supposed to be decomposable into two parts,

$$\mathbf{\Sigma} = (\mathbf{\Sigma} - \mathbf{\Psi}) + \mathbf{\Psi}, \tag{1}$$

where $\boldsymbol{\Psi}$ is a diagonal matrix representing the residual variances of the error part and $\boldsymbol{\Sigma} - \boldsymbol{\Psi}$ corresponds to the common (true-score) part. For a given $\boldsymbol{\Sigma}$ the specific decomposition (1) is not unique and in the so-called *minimum trace factor analysis* (MTFA) one selects $\boldsymbol{\Psi}$ in such a manner that the trace of the (reduced) matrix $\boldsymbol{\Sigma} - \boldsymbol{\Psi}$ is minimal while keeping $\boldsymbol{\Sigma} - \boldsymbol{\Psi}$ Gramian (*see* MINIMUM RANK FACTOR ANALYSIS).

The MTFA could lead to an inappropriate solution where one or more diagonal elements of $\boldsymbol{\Psi}$ are negative (the so-called *Heywood case*). Therefore, we consider the MTFA with further constraint on $\boldsymbol{\Psi}$ to be Gramian, and call it the *constrained minimum trace factor analysis* (CMTFA).

The concept of MTFA has been discussed in the early work of Ledermann [5] who showed that minimum-trace and minimum-rank solutions do not always coincide even in the case of reduced rank one. Bentler [1] rediscovered the MTFA in connection with an investigation of reliability in factor analysis. The *reliability coefficient* ρ for a specific decomposition (1) is defined as

$$\rho = \frac{\mathbf{1}'(\boldsymbol{\Sigma} - \boldsymbol{\Psi})\mathbf{1}}{\mathbf{1}'\boldsymbol{\Sigma}\mathbf{1}} = 1 - \frac{\operatorname{tr}\boldsymbol{\Psi}}{\mathbf{1}'\boldsymbol{\Sigma}\mathbf{1}},$$

where $\mathbf{1} = (1, \ldots, 1)'$ is a $p \times 1$ column vector of ones. As long as the decomposition (1) is not known, one may try to find the smallest value ρ_0 among all admissible values of ρ. Clearly, ρ_0 represents the *greatest lower bound* (g.l.b) to reliability. This leads to the CMTFA problem of maximizing $\operatorname{tr}\boldsymbol{\Psi}$ subject to $\boldsymbol{\Sigma} - \boldsymbol{\Psi}$ and $\boldsymbol{\Psi}$ being Gramian. Naturally

$$\rho_0 = 1 - \frac{\operatorname{tr}\boldsymbol{\Psi}_0}{\mathbf{1}'\boldsymbol{\Sigma}\mathbf{1}},$$

where $\boldsymbol{\Psi}_0$ is the CMTFA solution (cf. Bentler and Woodward [2] and Jackson and Agunwamba [4]).

COMPUTATIONAL ALGORITHM

The MTFA as well as the CMTFA solution is always unique (see Della Riccia and Shapiro [3] or Ten Berge et al. [9]). An effective and simple computational algorithm for MFTA has been proposed by Bentler [1] and further developed in refs. 2 and 9 to deal with the CMTFA problem.

Following Bentler's method, let $\mathbf{T}$ be a $p \times m$ matrix such that

$$\operatorname{diag}(\mathbf{T}\mathbf{T}') = \mathbf{I}_p, \tag{2}$$

where $\mathbf{I}_p$ is the $p \times p$ identity matrix. Then

$$\begin{aligned}\operatorname{tr}(\mathbf{T}'\boldsymbol{\Sigma}\mathbf{T}) &= \operatorname{tr}(\mathbf{T}'(\boldsymbol{\Sigma} - \boldsymbol{\Psi})\mathbf{T}) + \operatorname{tr}(\mathbf{T}'\boldsymbol{\Psi}\mathbf{T})\\ &\geqslant \operatorname{tr}(\mathbf{T}'\boldsymbol{\Psi}\mathbf{T}) = \operatorname{tr}(\boldsymbol{\Psi}\mathbf{T}\mathbf{T}') = \operatorname{tr}\boldsymbol{\Psi}.\end{aligned}$$

It follows that

$$\operatorname{tr}(\mathbf{T}'\boldsymbol{\Sigma}\mathbf{T}) \geqslant \operatorname{tr}\boldsymbol{\Psi} \tag{3}$$

whenever $\boldsymbol{\Sigma} - \boldsymbol{\Psi}$ is Gramian. Moreover, equality in (3) holds if and only if

$$(\boldsymbol{\Sigma} - \boldsymbol{\Psi})\mathbf{T} = \mathbf{0}. \tag{4}$$

There exists a matrix $\mathbf{T}$ satisfying (2) and (4) and the corresponding $\boldsymbol{\Psi}$ is the MTFA solution ([3], [9]). Therefore, the MTFA problem of maximizing $\operatorname{tr}\boldsymbol{\Psi}$ is equivalent to the problem:

$$\min_{\mathbf{T}} \operatorname{tr}(\mathbf{T}'\boldsymbol{\Sigma}\mathbf{T}), \tag{5}$$

where $\mathbf{T}$ must satisfy (2). In order to solve (5), Bentler proposed a Gauss–Seidel*-type procedure. Starting from an arbitrary $\mathbf{T}$ satisfying (2), the procedure iteratively solves problem (5) with respect to the ith row of $\mathbf{T}$, $i = 1, \ldots, p, 1, \ldots,$ while keeping the remaining rows of $\mathbf{T}$ fixed. The minimization problem for a chosen row of $\mathbf{T}$ can be easily solved explicitly. For the CMTFA problem a similar algorithm has been developed (see refs. 2 and 9), replacing (2) with

$$\operatorname{diag}(\mathbf{T}\mathbf{T}') \geqslant \mathbf{I}_p.$$

NUMERICAL EXAMPLE

Consider the example analyzed by Bentler and Woodward [2 p. 262]. The covariance matrix, based on a sample of size $n = 1416$, is presented in the left-hand part of Table 1. Bentler's algorithm simultaneously calculates the solution values of $\mathbf{T}$ and $\boldsymbol{\Psi}$ which are given in the right-hand part of Table 1. The obtained values of the diagonal ele-

Table 1

	Covariance Matrix				T		diag Ψ
	1	2	3	4			
1	94.7				1.000	0	47.58
2	87.3	212.0			− 0.996	0.092	36.99
3	63.9	138.7	160.5		0.262	− 0.965	42.77
4	58.4	128.2	109.8	115.4	0.395	0.919	13.28

ments of Ψ are all positive and therefore the MTFA and CMTFA solutions coincide.

SAMPLING CONSIDERATIONS

In practice, the population covariance matrix $\boldsymbol{\Sigma}$ is unknown. A consistent estimate $\hat{\rho}_0$ of ρ_0 can be obtained by the CMTFA method applied to the *sample* covariance matrix $\mathbf{S}$. Asymptotic properties of $\hat{\rho}_0$ have been studied by Shapiro [6]. It turns out that the asymptotic behavior of $\hat{\rho}_0$ is closely related to the uniqueness of the solution in problem (5). [Note that any solution $\mathbf{T}$ of (5) can be replaced by $\mathbf{TQ}$, where $\mathbf{Q}$ is an orthogonal matrix.] Suppose that a sample of size n is drawn from a normally distributed population. Then the following result holds: $n^{1/2}(\hat{\rho}_0 - \rho_0)$ *has an asymptotically normal distribution with zero mean and variance*

$$2(\mathbf{1}'\boldsymbol{\Sigma}\mathbf{1})^{-2}\Big[\operatorname{tr}(\mathbf{T}'\boldsymbol{\Psi}_0\mathbf{T}\mathbf{T}'\boldsymbol{\Psi}_0\mathbf{T}) - 2(1-\rho_0)(\mathbf{1}'\boldsymbol{\Psi}_0\mathbf{T}\mathbf{T}'\boldsymbol{\Psi}_0\mathbf{1}) + (\operatorname{tr}\boldsymbol{\Psi}_0)^2\Big] \qquad (6)$$

if and only if the solution $\mathbf{T}$ *of problem* (5) *is unique up to multiplication by an orthogonal matrix* [6].

Uniqueness of $\mathbf{T}$ can be ensured by simple regularity conditions (see Shapiro [6]). Unfortunately these conditions concern cases where the rank r of $\boldsymbol{\Sigma} - \boldsymbol{\Psi}_0$ is not less than the Ledermann bound $\phi(p)$,

$$\phi(p) = \frac{2p + 1 - (8p+1)^{1/2}}{2}.$$

(*see* MINIMUM RANK FACTOR ANALYSIS for a further discussion of the Ledermann bound.)

If r is less than $\phi(p)$, then the asymptotic distribution of $n^{1/2}(\hat{\rho}_0 - \rho_0)$ is very complicated and a general theory of such distributions is not yet available (cf. Shapiro [8]). We shall only remark that the sample estimate $\hat{\rho}_0$ becomes heavily biased when r is small relative to $\phi(p)$.

Example. Consider now the covariance matrix in Table 1 as a *sample* covariance matrix $\mathbf{S}$. The values of $\mathbf{T}$ and $\boldsymbol{\Psi}$ in the right part of Table 1 are considered as sample estimates $\hat{\mathbf{T}}$ and $\hat{\boldsymbol{\Psi}}$. The sample estimate $\hat{\rho}_0 = 0.920$. Substituting estimates $\mathbf{S}$, $\hat{\mathbf{T}}$ and $\hat{\boldsymbol{\Psi}}$ in place of their population counterparts in (6) gives a consistent estimate

$$\left[\widehat{\operatorname{var}}(\hat{\rho}_0)\right]^{1/2} = 0.00373$$

of the asymptotic standard deviation of $\hat{\rho}_0$.

WEIGHTED MINIMUM TRACE FACTOR ANALYSIS

One of the difficulties with the MTFA is that it is not scale-independent. Let $\mathbf{W}$ be a diagonal matrix representing a scale transformation and consider the new (scaled) covariance matrix

$$\boldsymbol{\Sigma}_1 = \mathbf{W}\boldsymbol{\Sigma}\mathbf{W}.$$

The corresponding reliability coefficient

$$\rho = \frac{\mathbf{1}'\mathbf{W}(\boldsymbol{\Sigma} - \boldsymbol{\Psi})\mathbf{W}\mathbf{1}}{\mathbf{1}'\mathbf{W}\boldsymbol{\Sigma}\mathbf{W}\mathbf{1}}$$

can be written as

$$\rho = \frac{\mathbf{w}'(\boldsymbol{\Sigma} - \boldsymbol{\Psi})\mathbf{w}}{\mathbf{w}'\boldsymbol{\Sigma}\mathbf{w}},$$

where $\mathbf{w}$ is the $p \times 1$ vector (of weights) composed from the diagonal elements of $\mathbf{W}$.

Suppose that for some choice of $\boldsymbol{\Psi}$, the Gramian matrix $\boldsymbol{\Sigma} - \boldsymbol{\Psi}$ is singular and $\mathbf{w}$ is chosen in such a way that

$$(\boldsymbol{\Sigma} - \boldsymbol{\Psi})\mathbf{w} = \mathbf{0}.$$

Then $\rho = 0$ and consequently the g.l.b. to reliability corresponding to $\boldsymbol{\Sigma}_1$ is zero! Although such a situation will rarely, if ever, occur in practice, some systematic procedure for weights selection must be developed. Shapiro [7] proposed to solve the scaling problem by maximization of the g.l.b. as a function of the weight vector $\mathbf{w}$. Then in the so-called *weighted minimum trace factor analysis* (WMTFA) the maximal g.l.b. ρ^* is defined by

$$\rho^* = \max_{\mathbf{w}} \min_{\boldsymbol{\Psi}} \frac{\mathbf{w}'(\boldsymbol{\Sigma} - \boldsymbol{\Psi})\mathbf{w}}{\mathbf{w}'\boldsymbol{\Sigma}\mathbf{w}},$$

subject to $\boldsymbol{\Sigma} - \boldsymbol{\Psi}$ being Gramian. Computational methods and the asymptotic theory of WMTFA are discussed in Shapiro [7].

References

[1] Bentler, P. M. (1972). *Soc. Sci. Res.*, **1**, 343–357.

[2] Bentler, P. M. and Woodward, J. A. (1980). *Psychometrika*, **45**, 279–267.

[3] Della Riccia, G. and Shapiro, A. (1982). *Psychometrika*, **47**, 443–448.

[4] Jackson, P. H. and Agunwamba, C. C. (1977). *Psychometrika*, **42**, 567–578.

[5] Ledermann, W. (1939). *Proc. R. Soc. Edinburgh*, **60**, 1–17.

[6] Shapiro, A. (1982). *Psychometrika*, **47**, 187–199.

[7] Shapiro, A. (1982). *Psychometrika*, **47**, 243–264.

[8] Shapiro, A. (1983). *S. Afr. Statist. J.*, **17**, 33–81. (Sections 3 and 5 discuss an asymptotic distribution theory of an extremal-value function of the sample covariance matrix. The MTFA is given as an example in Section 8.)

[9] Ten Berge, J. M. F., Snijders, T. A. B., and Zegers, F. E. (1981). *Psychometrika*, **46**, 201–213.

(FACTOR ANALYSIS
MINIMUM RANK FACTOR ANALYSIS)

A. SHAPIRO

MINIMUM UNLIKELIHOOD ESTIMATION *See* UNLIKELIHOOD

MINIMUM VARIANCE UNBIASED ESTIMATION

A statistic U is unbiased for estimating a parametric function $g(\theta)$ if $E_\theta U = g(\theta)$. If $\text{Var}_\theta U \leqslant \text{Var}_\theta U_1$ for all θ for any unbiased competitor U_1, then U is called a UMVU (uniformly minimum variance unbiased) estimator. The UMVU property can sometimes be established using the Cramér–Rao theory of lower bounds for variances and sometimes by the theory of complete sufficient statistics, as indicated below. Neither linear models and linear unbiased estimators (see GENERAL LINEAR MODEL) nor quadratic unbiased estimators* will be discussed here.

If $f(x;\theta)$ is the probability law of data x, then

$$I(\theta) = E\left[(\partial \log f/\partial\theta)^2\right]$$
$$= -E(\partial^2 \log f/\partial\theta^2)$$

is the Fisher information*. If U is any unbiased estimator of θ, then $\text{Var}_\theta U$ cannot be less than $1/I(\theta)$, the Cramér–Rao lower bound* [6, 15]. An estimator that achieves the bound is sometimes called efficient [6]. The bound is achieved if and only if U is a linear function of the score function $\partial \log f/\partial\theta$ (*see* EFFICIENT SCORE). Achievement of the bound implies that U is a UMVU estimator, but the converse is false (see Example 2).

Several generalizations are known. Rao [15] showed that if $\boldsymbol{\theta} = (\theta_1, \ldots, \theta_k)'$ is a vector parameter and $\mathbf{U} = (U_1, \ldots, U_k)'$ satisfies $E_\theta \mathbf{U} = \boldsymbol{\theta}$, then for any constant vector $\mathbf{c}$, $\text{Var}_{\boldsymbol{\theta}}\mathbf{c}'\mathbf{U} \geqslant \mathbf{c}'\mathbf{I}^{-1}\mathbf{c}$, where $\mathbf{I}$ is the Fisher information matrix whose elements are

$$I_{ij} = E_{\boldsymbol{\theta}}\left[(\partial \log f/\partial\theta_i)(\partial \log f/\partial\theta_j)\right]$$
$$= -E_{\boldsymbol{\theta}}\left(\partial^2 \log f/\partial\theta_i\partial\theta_j\right).$$

A second generalization involves improvements in the bounds. By considering the regression of U on $(1/f)d^jf/d\theta^j$ for $j = 1, \ldots, r$ one can obtain the rth bound in a nondecreasing sequence [2] (*see* BHATTACHARYYA BOUNDS). Implicit in the above are regularity conditions associated with differentiation with respect to $\boldsymbol{\theta}$ [6]. By considering differences rather than derivatives Chapman and Robbins [5] circumvented regularity conditions in establishing a lower bound

$$\sup_{\theta'}\left\{\frac{[g(\theta')-g(\theta)]^2}{\mathrm{Var}_\theta[f(X;\theta')/f(X;\theta)]}\right\}$$

for the variance of any unbiased estimator of $g(\theta)$.

The theory of complete sufficient statistics* is a powerful tool for establishing the UMVU property. If $U_0(x)$ is any unbiased estimator of $g(\theta)$, and $T(x)$ is a sufficient statistic, then by the Rao–Blackwell theorem* [3, 15], $U(t) = E_\theta((U_0(X) \mid T(x) = t)$ is also an unbiased estimator of $g(\theta)$ and $\mathrm{Var}_\theta U \leqslant \mathrm{Var}_\theta U_0$ for all θ.

Sometimes two different functions $U_1(t)$ and $U_2(t)$ of a sufficient statistic $T(x)$ are both unbiased estimators of $g(\theta)$. Typically neither is then UMVU (see Example 6), and their difference $H(t)$ is a (nontrivial) "unbiased estimator of zero," $E_\theta H(T) = 0$. Nonexistence of such an $H(t)$ is equivalent to completeness* of the family of distributions of $T(x)$. Then T is complete sufficient, and according to the Lehmann–Scheffé theorem* [10], any function of $T(x)$ is then a UMVU estimator of its expectation.

Evaluation of $E_\theta(U_0 \mid t)$ is sometimes simplified by using Basu's theorem* [7, 17].

Alternatives to the Rao–Blackwell method that do not require an initial unbiased estimator are the Laplace transform method [18, 19] (*see* INTEGRAL TRANSFORMS) and the expansion method [12].

UMVU estimation for finite populations has been studied, for example, by Hartley and Rao [8] and Särndal [16].

Maximum likelihood* estimators are not always unbiased or sufficient and may or may not be UMVU, but in some cases they are asymptotically equivalent to UMVU estimators [14].

Blom [4] has shown that if U_n is a UMVU estimator of $g(\theta)$ based on a sample of size n, then $m < n$ implies that $m\,\mathrm{Var}_\theta U_m \geqslant n\,\mathrm{Var}_\theta U_n$.

Example 1. Estimate $g(\theta) = \mathrm{Var}_\theta X = \theta(1-\theta)$ using a single observation X from a Bernoulli distribution* with $E_\theta X = \theta$. Here only linear functions of θ have unbiased estimators (are *estimable*), so there can be no UMVU estimator of $\theta(1-\theta)$. Thus in any problem we must restrict ourselves to estimable functions $g(\theta)$.

Example 2. Estimate $\theta = (E_\lambda X)^2 = \lambda^2$ using one Poisson observation X. Here $U = X^2 - X$ is unbiased but fails to achieve the Cramér–Rao bound. Nevertheless U is UMVU by the Rao–Blackwell–Lehmann–Scheffé theory.

Example 3. If $f(x;\theta) = a(\theta)b(x)e^{\theta t(x)}$, then the distributions are said to constitute an exponential family*. For a sample of size n, $T(x_1, \ldots, x_n) = \Sigma t(x_i)$ is a complete sufficient statistic [9, p. 132], and a UMVU estimator of $g(\theta) = nE_\theta t(X)$.

Example 4. If $X_1, \ldots, X_n$ are normal with mean zero and variance θ, then $U = n^{-1}\Sigma x_i^2$ is a UMVU estimator of θ, but $nU/(n+2)$ has smaller mean squared error*. Thus UMVU estimators need not be admissible (*see* ADMISSIBILITY) with respect to a squared error loss function.

Example 5. It is known that the vector of order statistics* is a complete sufficient statistic for the family of all continuous distributions. This implies that the sample cumulative distribution function (CDF) $F_n(x)$ is complete sufficient. For any x, $EF_n(x) = F(x)$, the population CDF, and thus $F_n(x)$ is a UMVU estimator of $F(x)$.

Example 6. In a Bernoulli process let S = success, F = failure, $\theta = P(\mathrm{S}) = 1 - P(\mathrm{F})$. A certain stopping rule ensures one of six realizations: FFF, FFS, FS, SF, SSF, SSS. If $(X_\mathrm{F}, X_\mathrm{S}) =$ (number of F's, number of S's), then $T = (X_\mathrm{F}, X_\mathrm{S})$ is a minimal sufficient statistic. An unbiased estimator of θ is given by $U(T) = U(X_\mathrm{F}, X_\mathrm{S})$, where $U(3,0) = U(2,1) = 0$, $U(1,1) = \frac{1}{2}$, $U(1,2) = U(0,3) = 1$. But if $H(T) = H(X_\mathrm{F}, X_\mathrm{S})$ is defined by $H(0,3) = H(3,0) = 0$, $H(2,1) = H(1,2) = 2$, $H(1,1) = -1$, then $E_\theta H = 0$, and $U_\lambda = U + \lambda H$ is unbiased for every λ. For any given θ, $\mathrm{Var}_\theta U_\lambda$ is minimized when $\lambda = (1 - 2\theta)/6$, and no estimator is UMVU. Thus in the absence of completeness, a function of a minimal sufficient statistic T may not be UMVU.

Example 7. Let $g(\theta) = P_\theta[X \leqslant c]$ for fixed c. To estimate $g(\theta)$ using a sample of size n, one can begin with an initial unbiased estimator depending on x_1 only, that is, $U_0(x_1, \ldots, x_n) = 1$ if $x_1 \leqslant c$ (0 otherwise). For applications of this idea, see Basu [1], Lieberman and Resnikoff [11], and Patil and Wani [13].

Example 8. Accidents per year follow a Poisson law with mean θ. Estimate the probability of no accidents in the next four years given that last year there were five accidents. Here $g(\theta) = e^{-4\theta}$, and $U(x)$ is unbiased if $\sum_0^\infty U(x) e^{-\theta}\theta^x/x! = e^{-4\theta}$. The only unbiased estimator $U(x) = (-3)^x$ is UMVU, and our estimate is $U(5) = -243$. This shows that the seemingly reasonable UMVU criterion can lead to a preposterous solution. The procedure tries too hard to be exactly unbiased, and it would be preferable to use the maximum likelihood estimate e^{-20}, accepting a small bias.

References

[1] Basu, A. P. (1964). *Technometrics*, **6**, 215–219.

[2] Bhattacharyya, A. (1946). *Sankhyā*, **8**, 1–14, 201–218, and 315–328.

[3] Blackwell, D. (1947). *Ann. Math. Statist.*, **18**, 105–110.

[4] Blom, G. (1978). *Biometrika*, **65**, 642–643.

[5] Chapman, D. G. and Robbins, H. (1951). *Ann. Math. Statist.*, **22**, 581–586.

[6] Cramér, H. (1946). *Mathematical Methods of Statistics*. Princeton University Press, Princeton, NJ.

[7] Eaton, M. L. and Morris, C. N. (1970). *Ann. Math. Statist.*, **5**, 1708–1716.

[8] Hartley, H. O. and Rao, J. N. K. (1968). *Biometrika*, **55**, 547–557.

[9] Lehmann, E. L. (1959). *Testing Statistical Hypotheses*. Wiley, New York.

[10] Lehmann, E. L. and Scheffé, H. (1950). *Sankhyā*. **10**, 305–340.

[11] Lieberman, G. J. and Resnikoff, G. J. (1955). *J. Amer. Statist. Ass.*, **50**, 457–516.

[12] Neyman, J. and Scott, E. L. (1960). *Ann. Math. Statist.*, **31**, 643–655.

[13] Patil, G. P. and Wani, J. K. (1966). *Ann. Inst. Statist. Math.*, **18**, 39–47.

[14] Portnoy, S. (1977). *Ann. Statist.*, **5**, 522–529.

[15] Rao, C. R. (1945). *Bull. Calcutta Math. Soc.*, **37**, 81–91.

[16] Särndal, C. E. (1976). *Ann. Statist.*, **4**, 993–997.

[17] Sathe, Y. S. and Varde, S. D. (1969). *Ann. Math. Statist.*, **40**, 710–714.

[18] Tate, R. F. (1959). *Ann. Math. Statist.*, **30**, 341–366.

[19] Washio, Y., Morimoto, H., and Ikeda, N. (1956). *Bull. Math. Statist.*, **6**, 69–93.

(BHATTACHARYYA BOUNDS
COMPLETENESS
CRAMÉR–RAO LOWER BOUND
ESTIMABILITY
ESTIMATION, POINT
FISHER INFORMATION
LEHMANN–SCHEFFÉ THEOREM
RAO–BLACKWELL THEOREM
SUFFICIENT STATISTICS
UNBIASEDNESS)

ROBERT J. BUEHLER

MINITAB™

Minitab is one of the general-purpose statistical computing systems that have been developed to make it easier to use computers in analyzing data. It is perhaps the easiest to learn among the widely available systems. Its strengths include its general ease of use, its interactive nature, and its availability on a wide range of computers.

FEATURES

Like most other computer software, Minitab is in a state of evolution. Its features as of 1982 include the following:

Descriptive statistics including a variety of plotting capabilities.

Multiway tables that can contain counts, percentages, and descriptive statistics on associated variables in a wide selection of layouts.

Multiple regression* analysis including residual analysis, diagnostics, and stepwise procedures.

One-way and two-way analysis of variance*.

Exploratory data analysis* including procedures developed by John Tukey.

Time-series* analysis including procedures developed by Box and Jenkins and robust smoothers.

Flexible data management including sorting, ranking, selecting, and eliminating subsets of the data, and joining data sets.

Convenient transformations as well as arithmetic and matrix operations.

Flexibly formatted output that can be conveniently tailored for visual display screens or line printers.

Compatible interactive and batch operation.

On-line HELP.

HOW MINITAB WORKS

Minitab maintains a worksheet in which data may be stored as either constants, columns, or matrices. Constants are denoted by K, columns by C or by a name and matrices by M. Most commands resemble English sentences, for example

```
PLOT 'INCOME' VS 'AGE'
```

There are about 150 commands such as READ, PRINT, SAVE, RETRIEVE, MEAN, TINTERVAL, TABLE, and REGRESS. Extra annotating text may appear on commands but is ignored by the processor.

Example 1. Retrieve a stored data file, make histograms, and print descriptive statistics on all 12 variables.

```
RETRIEVE DATA IN FILE 'TAXES'
HISTOGRAM C1–C12
DESCRIBE C1–C12
```

Example 2. Make a table that gives the mean and standard deviations of the variables height and weight when the individuals are categorized by sex and activity level.

```
TABLE BY 'SEX' AND 'ACTIVITY';
STATISTICS ON 'HEIGHT' AND 'WEIGHT';
LAYOUT 2 BY 0.
```

Here STATISTICS and LAYOUT are subcommands; such subcommands may be used to modify execution or formatting. STATISTICS requests the calculation of means and standard deviations while LAYOUT modifies the output such that both factors, SEX and ACTIVITY, will be used to define the rows of the table, and none will be used to define the columns.

HISTORY

The first version of Minitab was patterned after Omnitab, an important step in the evolution of modern statistical computing systems. About 1960, Joseph Hilsenrath, a thermodynamicist at the National Bureau of Standards*, saw the need for an omnibus table maker—an easy-to-use computer program that could manipulate columns of numbers, perform various arithmetic functions, and print the resulting tables. Hilsenrath was an early believer in the notion that *computers* should be taught to understand the user's language rather than the other way around. Thus Omnitab was designed to understand English commands and to operate on columns of numbers. All input was designed from the beginning to be free format. Users soon saw the advantages of this structure and asked for more capabilities, such as plotting and regression.

Brian Joiner, one of Minitab's developers, made extensive use of Omnitab from 1963 to 1971 while he worked at the National Bureau of Standards. When he went to Pennsylvania State University in 1971, he tried to use Omnitab in elementary statistics courses. But Omnitab proved to be inconvenient for this purpose, primarily because it was too large to run in the amount of memory allocated for student use.

After one academic term, Joiner considered carving out a subset of Omnitab that would be usable by students. Thomas A. Ryan, Jr., who had been at Pennsylvania State for several years, had developed several programs for teaching statistics and thus knew better the computing environment there and the type of programs the students would be able to use comfortably. They joined forces and produced the first version of Minitab, a program that looked externally similar to a subset of Omnitab, but was almost completely rewritten internally to provide better bookkeeping, as well as being smaller and more efficient for student-size problems. In 1974, Barbara F. Ryan, who had been consulting on Minitab, became a full partner in the development effort.

The early availability of Minitab on a wide range of computers came from several sources. Omnitab was rewritten in standard Fortran at the National Bureau of Standards because of early portability problems. In spite of this, Joiner experienced extreme difficulty in moving Omnitab and other programs from computer to computer, so he initiated steps to ensure Minitab's portability. Under Tom Ryan's leadership, Minitab followed methods developed by Roald Buhler to maintain machine-specific features along with an ANSI standard Fortran version.

In 1974, Joiner left Pennsylvania State to direct the Statistical Laboratory at the University of Wisconsin. Development of Minitab continued, centered at Pennsylvania State, with programming directed by Tom Ryan and design and documentation directed by Barbara Ryan. Joiner continued to provide input, especially on statistical aspects of the system and on data analysis problems encountered in the Lab.

IMPACT ON TEACHING

Although Minitab is now widely used for data analysis in industry and government, it continues to be popular in its original teaching environment. Minitab allows students to analyze more realistic data sets and thus gain a better perspective on how statistics should be done. Students are relieved of the need to carry out tedious calculations and thus can concentrate on the important concepts of statistics such as how to plot data, how to choose appropriate procedures, how to interpret results, and the pitfalls of blind mechanical calculations. Simulation* can be used to advantage to teach such concepts as the central limit theorem*.

MACHINE REQUIREMENTS

Minitab is written almost exclusively in a machine-compatible subset of ANSI Standard Fortran and will compile under Fortran IV, Fortran 77, or ASCII compilers. The source code consists of about 30,000 executable statements and 10,000 lines of comments. The code is highly modular and may be overlayed into areas as small as 56K bytes. A nonexecutable master source "program" is maintained along with a selector program. The selector is used to choose among a variety of versions of Minitab, all maintained simultaneously in the master source. This makes it very convenient to tailor the source code to a wide range of computing environments.

At this time, fully supported conversions of Minitab are available for the following computers: Burroughs Large Systems, CDC under NOS, Data General Eclipse, Harris, HP 3000, IBM under MVS and CMS, ICL 2900, LSI-11, NORD, PDP-11, Prime, Sperry Univac 1100, Terak, and VAX under VMS. Minitab also installs routinely on most other mainframe and minicomputers and on some large microcomputers.

Bibliography

The two major sources of information about Minitab are Ryan et al. (1976) and Ryan et al. (1982). The Tukey-style exploratory data analysis* features are based on the programs in Velleman and Hoaglin (1981).

References

Ryan, T. A., Jr., Joiner, B. L., and Ryan, B. F. (1976). *Minitab Student Handbook*. Duxbury Press, Boston, Mass.

Ryan, T. A., Jr., Joiner, B. L., and Ryan, B. F. (1982). *Minitab Reference Manual*. Pennsylvania State University, University Park, PA.

Velleman, P. F. and Hoaglin, D. C. (1981). *Applications, Basics and Computing of Exploratory Data Analysis*. Duxbury Press, Boston.

(COMPUTERS AND STATISTICS
STATISTICAL SOFTWARE)

THOMAS A. RYAN
BRIAN L. JOINER
BARBARA F. RYAN

MINKOWSKI'S INEQUALITY

Let X and Y be random variables such that $E(|X|^p) < \infty$ and $E(|Y|^p) < \infty$, $p \geqslant 1$, where E denotes expected value. Then [1]

$$\{E(|X+Y|^p)\}^{1/p} \leqslant \{E(|X|^p)\}^{1/p} + \{E(|Y|^p)\}^{1/p}.$$

Reference

[1] Chung, K. L. (1974). *A Course in Probability Theory*, 2nd ed. Academic Press, New York, Section 3.2.

(CAUCHY–SCHWARZ INEQUALITY
HÖLDER'S INEQUALITY)

MINQE

MIXED LINEAR MODELS

Consider the linear model

$$\mathbf{Y} = \mathbf{X}\boldsymbol{\beta} + \mathbf{U}_1\boldsymbol{\phi}_1 + \cdots + \mathbf{U}_p\boldsymbol{\phi}_p \qquad (1)$$

where $\mathbf{Y}$ is a vector of n observations, $\mathbf{X}(n \times m), \mathbf{U}_i(n \times n_i)$, $i = 1, \ldots, p$, are known matrices, $\boldsymbol{\beta}$ is an m-vector of unknown fixed parameters and $\boldsymbol{\phi}_i$ an n_i-vector of unobserved random variables, $i = 1, \ldots, p$, such that

$$E(\boldsymbol{\phi}_i) = 0, \qquad E(\boldsymbol{\phi}_i\boldsymbol{\phi}_j') = 0; \qquad i \neq j,$$
$$E(\boldsymbol{\phi}_i\boldsymbol{\phi}_i') = \sigma_i^2\mathbf{I}_{n_i}, \qquad (2)$$

where $\sigma_1^2, \ldots, \sigma_p^2$, the *variance components**, are unknown. The statistical problems associated with the model are: (a) estimation of $\boldsymbol{\beta}$, (b) estimation of σ_i^2 and (c) prediction of $\boldsymbol{\phi}_i$, the *structural variables*.

The earliest use of such models was in astronomy by Airy [1] (*see* STATISTICS IN ASTRONOMY), but the first systematic investigation of problems (a)–(c), was due to Fisher* [2]. Reference may be made to the contributions by Yates and Zacopancy and Cochran in survey sampling*, Yates and Rao in combining intra- and interblock estimates in design of experiments*, Fairfield Smith, Henderson, Panse, and Rao in the construction of selection indices in genetics*, and Brownlee in industrial applications (see Rao and Kleffe [15] for references).

Henderson [7] made a systematic study of the estimation of variance components through equations obtained by equating certain quadratic forms* in $\mathbf{Y}$ to their expected values. The entries in ANOVA* tables giving sums of squares due to various factors were natural candidates for the choice of quadratic forms. Except in the case of balanced designs (*see* BALANCE IN EXPERIMENTAL DESIGN), such estimators are not efficient, as was shown by Seely [17]. A completely different approach is the maximum likelihood (ML) method initiated by Hartley and J. N. K. Rao [4], assuming the normality of the distribution of the observed $\mathbf{Y}$. Patterson and Thompson [10] considered the marginal likelihood based on the maximal invariant of $\mathbf{Y}$ and obtained what are called marginal maximum likelihood (MML) estimators. Harville [6] reviewed the ML and MML methods and the computational algorithms associated with them.

Rao proposed a general method called MINQE in 1970, the scope of which has been considerably extended to cover a wide variety of situations by Kleffe (see Rao [13],

Rao and Kleffe [15], and Kleffe [8] for detailed references to papers by the authors and other principal contributors Chaubey, Drygas, Hartung, J. N. K. Rao, P. S. R. S. Rao, Pukelsheim, Sinha and Wienad).

MINQE THEORY

For developing the MINQE theory of estimation, we consider a linear model more general than (1):

$$\mathbf{Y} = \mathbf{X}\boldsymbol{\beta} + \mathbf{U}\boldsymbol{\phi}, \qquad E(\boldsymbol{\phi}\boldsymbol{\phi}') = \mathbf{F}_\theta = \theta_1\mathbf{F}_1 + \cdots + \theta_p\mathbf{F}_p, \tag{3}$$

where $\mathbf{F}_i$ are known matrices and θ_i (not necessarily nonnegative) are unknown parameters and $\mathbf{F}_\theta$ is p.d. (positive definite) in the admissible region of $\boldsymbol{\theta}$. Denoting

$$\mathbf{F}_\alpha = \alpha_1\mathbf{F}_1 + \cdots + \alpha_p\mathbf{F}_p,$$

where α_i are a priori values of θ_i and $\boldsymbol{\psi} = \mathbf{F}_\alpha^{-1/2}\boldsymbol{\phi}$, we may define a natural estimator of $\mathbf{f}'\boldsymbol{\theta} = f_1\theta_1 + \cdots f_p\theta_p$ in terms of $\boldsymbol{\psi}$, if observable, as

$$\hat{\gamma}_n = \boldsymbol{\psi}'\mathbf{F}_\alpha^{-1/2}(\Sigma\mu_i\mathbf{F}_i)\mathbf{F}_\alpha^{-1/2}\boldsymbol{\psi} = \boldsymbol{\psi}'\mathbf{N}\boldsymbol{\psi} \tag{4}$$

(say), where for unbiasedness the μ_i are chosen to satisfy the equations

$$\left(\operatorname{tr}\left(\mathbf{F}_i\mathbf{F}_\alpha^{-1}\mathbf{F}_1\mathbf{F}_\alpha^{-1}\right)\right)\mu_1 + \cdots + \left(\operatorname{tr}\left(\mathbf{F}_i\mathbf{F}_\alpha^{-1}\mathbf{F}_p\mathbf{F}_\alpha^{-1}\right)\right)\mu_p = f_i, \quad i = 1, \ldots, p.$$

A quadratic estimator $\hat{\gamma} = \mathbf{Y}'\mathbf{A}\mathbf{Y}$ of $\gamma = \mathbf{f}'\boldsymbol{\theta}$ can be written in terms of the parameters of the model (3)

$$\mathbf{Y}'\mathbf{A}\mathbf{Y} = \begin{pmatrix}\boldsymbol{\psi}\\ \boldsymbol{\beta}\end{pmatrix}'\begin{pmatrix}\mathbf{U}_*'\mathbf{A}\mathbf{U}_*' & \mathbf{A}\mathbf{X}\\ \mathbf{X}'\mathbf{A}\mathbf{U}_* & \mathbf{X}'\mathbf{A}\mathbf{X}\end{pmatrix}\begin{pmatrix}\boldsymbol{\psi}\\ \boldsymbol{\beta}\end{pmatrix}, \tag{5}$$

where $\mathbf{U}_* = \mathbf{U}\mathbf{F}_\alpha^{1/2}$, while a natural estimator is $\boldsymbol{\psi}'\mathbf{N}\boldsymbol{\psi}$ as defined in (4). The difference between $\mathbf{Y}'\mathbf{A}\mathbf{Y}$ and $\boldsymbol{\psi}'\mathbf{N}\boldsymbol{\psi}$ is

$$\begin{pmatrix}\boldsymbol{\psi}\\ \boldsymbol{\beta}\end{pmatrix}'\begin{pmatrix}\mathbf{U}_*'\mathbf{A}\mathbf{U}_* - \mathbf{N} & \mathbf{U}_*'\mathbf{A}\mathbf{X}\\ \mathbf{X}'\mathbf{A}\mathbf{U}_* & \mathbf{X}'\mathbf{A}\mathbf{X}\end{pmatrix}\begin{pmatrix}\boldsymbol{\psi}\\ \boldsymbol{\beta}\end{pmatrix}. \tag{6}$$

The MINQE is the one obtained by choosing $\mathbf{A}$ so as to minimize a suitable chosen norm of the matrix in (6),

$$\left\|\begin{matrix}\mathbf{D}_{11}\mathbf{D}_{12}\\ \mathbf{D}_{21}\mathbf{D}_{22}\end{matrix}\right\| = \left\|\begin{matrix}\mathbf{U}_*'\mathbf{A}\mathbf{U}_* - \mathbf{N} & \mathbf{U}_*'\mathbf{A}\mathbf{X}\\ \mathbf{X}'\mathbf{A}\mathbf{U}_* & \mathbf{X}'\mathbf{A}\mathbf{X}\end{matrix}\right\|. \tag{7}$$

We consider two kinds of norms, one a simple Euclidean norm,

$$\operatorname{tr}(\mathbf{D}_{11}\mathbf{D}_{11}) + 2\operatorname{tr}(\mathbf{D}_{12}\mathbf{D}_{21}) + \operatorname{tr}(\mathbf{D}_{22}\mathbf{D}_{22}), \tag{8}$$

and another a weighted Euclidean norm,

$$\operatorname{tr}(\mathbf{D}_{11}\mathbf{W}\mathbf{D}_{11}\mathbf{W}) + 2\operatorname{tr}(\mathbf{D}_{12}\mathbf{K}\mathbf{D}_{21}\mathbf{W}) + \operatorname{tr}(\mathbf{D}_{22}\mathbf{K}\mathbf{D}_{22}\mathbf{K}), \tag{9}$$

where $\mathbf{W}$ and $\mathbf{K}$ are n.n.d. (nonnegative definite) matrices. The norm (7) gives different weights to $\boldsymbol{\psi}$ and $\boldsymbol{\beta}$ in the quadratic form (5).

We impose other restrictions on $\mathbf{A}$ and indicate the MINQE so obtained by adding a symbol in parentheses. For example,

a. Unbiasedness*: MINQE(U).

b. Invariance* for translation in $\boldsymbol{\beta}$: MINQE(I).

c. Satisfies both **a** and **b**: MINQE(U, I).

d. Unbiasedness and nonnegativity: MINQE(U, NND).

The properties of the estimators strongly depend on the a priori value of $\boldsymbol{\theta}$, the norm chosen, and the restrictions imposed on $\mathbf{A}$. It is also possible to obtain a series of iterated MINQEs (IMINQE) by repeatedly solving the MINQE equations using the solution at any stage as a priori values.

MINQE(U, I)

For the class of invariant unbiased quadratic estimators $\mathbf{Y}'\mathbf{A}\mathbf{Y}$ of $\mathbf{f}'\boldsymbol{\theta}$ under the model (3), $\mathbf{A}$ belongs to the set

$$\mathscr{C}_{UI}^f = \{\mathbf{A} : \mathbf{A}\mathbf{X} = \mathbf{0}, \operatorname{tr}(\mathbf{A}\mathbf{V}_i) = f_i, \; i = 1, \ldots, p\},$$

where $\mathbf{V}_i = \mathbf{U}\mathbf{F}_i\mathbf{U}'$. Let $\boldsymbol{\alpha} = (\alpha_1, \ldots, \alpha_p)'$ be an a priori value of $\boldsymbol{\theta} = (\theta_1, \ldots, \theta_p)'$ and define

$$\mathbf{T} = \mathbf{V}_\alpha + \mathbf{X}\mathbf{X}', \quad \mathbf{V}_\alpha = \alpha_1\mathbf{V}_1 + \cdots + \alpha_p\mathbf{V}_p,$$

$$\mathbf{P}_T = \mathbf{X}(\mathbf{X}'\mathbf{T}^{-1}\mathbf{X})^-\mathbf{X}'\mathbf{T}^{-1}, \quad \mathbf{M}_T = (\mathbf{I} - \mathbf{P}_T),$$

where $\mathbf{G}^-$ is a g-inverse of $\mathbf{G}$ (*see* GENERALIZED INVERSES). Theorem 1 provides the MINQE(U, I).

Theorem 1. If $\mathscr{C}_{UI}^f$ is not empty, then under the Euclidean norm (8), the MINQE(U, I) of $\mathbf{f}'\boldsymbol{\theta}$ is $\mathbf{f}'\hat{\boldsymbol{\theta}}$, where $\hat{\boldsymbol{\theta}}$ is a solution of

$$[H_{UI}^{\alpha}]\boldsymbol{\theta} = \mathbf{h}_I^{\alpha}(\mathbf{Y}). \qquad (10)$$

In (10),

$$\mathbf{H}_{UI}^{\alpha} = (\operatorname{tr} \mathbf{A}_i \mathbf{V}_j), \qquad \mathbf{A}_i = \mathbf{T}^{-1}\mathbf{M}_T\mathbf{V}_i\mathbf{M}_T'\mathbf{T}^{-1},$$
$$\mathbf{h}_I^{\alpha}(\mathbf{Y}) = (\mathbf{Y}'\mathbf{A}_1\mathbf{Y}, \ldots, \mathbf{Y}'\mathbf{A}_p\mathbf{Y})'.$$

Notes

1. If $\mathbf{V}_\alpha$ is nonsingular, $\mathbf{T}$ can be replaced by $\mathbf{V}_\alpha$ in Theorem 1.
2. $\mathscr{C}_{UI}^f$ is not empty iff $\mathbf{f} \in \mathscr{S}(\mathbf{H}_M)$, i.e., $\mathbf{f}$ belongs to the linear manifold* generated by the columns of $\mathbf{H}_M$, where
$$\mathbf{H}_M = (h_{ij}), \qquad h_{ij} = \operatorname{tr}(\mathbf{M}\mathbf{V}_i\mathbf{M}\mathbf{V}_j),$$
$$\mathbf{M} = \mathbf{I} - \mathbf{X}(\mathbf{X}'\mathbf{X})^-\mathbf{X}'.$$
3. If $\mathbf{H}_{UI}^{\alpha}$ is nonsingular, then the equation (10) has a unique solution which is the MINQE(U, I) estimator of $\boldsymbol{\theta}$.
4. If $\boldsymbol{\alpha}$ is not known, we start with a first approximation of $\boldsymbol{\alpha}$ and compute the solution of (10), which may be chosen as the second approximation of $\boldsymbol{\alpha}$. We may repeat this process to find the third approximation, and so on. The limiting value of $\boldsymbol{\theta}$ satisfies the equation
$$[\mathbf{H}_{UI}^{\theta}]\boldsymbol{\theta} = \mathbf{h}_I^{\theta}(\mathbf{Y}) \qquad (11)$$
and is called the IMINQE(U, I), the iterated MINQE(U, I); Equation (11) is also the marginal maximum likelihood equation derived by Patterson and Thompson [10] under the assumption of normality.
5. If $\mathbf{Y}$ in the linear model (3) has a multivariate normal distribution* with mean $\mathbf{X}\boldsymbol{\beta}$ and dispersion matrix
$$\mathbf{V}_\theta = \mathbf{U}(\theta_1\mathbf{F}_1 + \cdots + \theta_p\mathbf{F}_p)\mathbf{U}',$$
then the MINQE(U, I) as determined in (10) is the locally minimum variance invariant unbiased estimator (LMVIUE) of $\mathbf{f}'\boldsymbol{\theta}$ at the point $\boldsymbol{\theta} = \boldsymbol{\alpha}$.

MINQE(U)

If the invariance condition is dropped, then it would be appropriate to consider estimates of the form

$$(\mathbf{Y} - \mathbf{X}\boldsymbol{\beta}_0)'\mathbf{A}(\mathbf{Y} - \mathbf{X}\boldsymbol{\beta}_0),$$

where $\boldsymbol{\beta}_0$ is an a priori value of $\boldsymbol{\beta}$. If this is unbiased for $\mathbf{f}'\boldsymbol{\theta}$, then $\mathbf{A}$ belongs to the class

$$\mathscr{C}_U^f = \{\mathbf{A} : \mathbf{X}'\mathbf{A}\mathbf{X} = \mathbf{0}, \operatorname{tr} \mathbf{A}\mathbf{V}_i = f_i,$$
$$i = 1, \ldots, p\}.$$

The following theorem provides the MINQE(U).

Theorem 2. If $\mathscr{C}_U^f$ is not empty, then under the Euclidean norm, the MINQE(U) of $\mathbf{f}'\boldsymbol{\theta}$ is $\mathbf{f}'\hat{\boldsymbol{\theta}}$, $\hat{\boldsymbol{\theta}}$ being a solution of

$$[\mathbf{H}_U^{\alpha}]\boldsymbol{\theta} = \mathbf{h}_U^{\alpha,\beta_0}, \qquad (12)$$

where

$$\mathbf{H}_U^{\alpha} = (\operatorname{tr} \mathbf{A}_i\mathbf{V}_j),$$
$$\mathbf{h}_u^{\alpha,\beta_0} = ((\mathbf{Y} - \mathbf{X}\boldsymbol{\beta}_0)'\mathbf{A}_1(\mathbf{Y} - \mathbf{X}\boldsymbol{\beta}_0), \ldots,$$
$$(\mathbf{Y} - \mathbf{X}\boldsymbol{\beta}_0)'\mathbf{A}_p(\mathbf{Y} - \mathbf{X}\boldsymbol{\beta}_0))',$$
$$\mathbf{A}_i = \mathbf{T}^{-1}(\mathbf{V}_i - \mathbf{P}_T\mathbf{V}_i\mathbf{P}_T')\mathbf{T}^{-1},$$
$$\mathbf{T} = \mathbf{V}_\alpha + \mathbf{X}\mathbf{X}'.$$

As in the case of MINQE(U, I), we can obtain IMINQE(U), the iterated MINQE(U) of $\boldsymbol{\theta}$, and the estimate of $\boldsymbol{\beta}$ as solutions of the equations

$$\mathbf{X}'\mathbf{V}^{-1}\mathbf{X}\boldsymbol{\beta} = \mathbf{X}'\mathbf{V}_\theta^{-1}\mathbf{Y},$$
$$[\mathbf{H}_U^{\theta}]\boldsymbol{\theta} = \mathbf{h}_U^{\theta,\beta}.$$

Notes

1. The class $\mathscr{C}_U^f$ is not empty iff $\mathbf{f} \in \mathscr{S}(\mathbf{H})$, where $\mathbf{H} = (h_{ij}) = (\operatorname{tr}(\mathbf{V}_i\mathbf{M}\mathbf{V}_j))$.
2. The estimator in (12) is obtained by minimizing the Euclidean norm
$$\|\mathbf{U}_*'\mathbf{A}\mathbf{U}_* - \mathbf{N}\|^2 + 2\|\mathbf{U}_*'\mathbf{A}\mathbf{X}\|^2. \qquad (14)$$
Focke and Dewess [3] gave different weights to the two terms in [14] and

obtained what is called r-MINQE(U). Let us choose $\mathbf{W} = \mathbf{I}$ and $\mathbf{K} = \mathbf{r}^2\mathbf{K}$ in (9) and minimize the weighted Euclidean norm. In such a case, the estimator of $\mathbf{f}'\boldsymbol{\theta}$ has the same form as in (12), with $\mathbf{T}$ replaced by $(\mathbf{V}_\alpha + r^2\mathbf{XKX}')$. The optimal $\mathbf{A}$ so obtained is denoted by $\mathbf{A}_r$.

3. Let $\mathbf{A}_\infty = \lim \mathbf{A}_r$ as $r \to \infty$. $\mathbf{A}_\infty$ exists iff $\mathbf{f} \in \mathscr{C}_U^f$; we call the corresponding estimator $\mathbf{Y}'\mathbf{A}_\infty\mathbf{Y}$, the ∞ − MINQE. Theorem 3, from Kleffe, provides the formula for ∞ − MINQE.

Theorem 3. Let $\mathscr{C}_U^f$ be not empty and

$$\mathbf{B} = \left(\mathrm{tr}(\mathbf{MV}_\alpha\mathbf{M})^+ \mathbf{V}_i(\mathbf{XKX}')_*^-\mathbf{V}_j\right),$$

$$(\mathbf{XKX}')_*^- = \mathbf{T}^{-1/2}\left(\mathbf{T}^{-1/2}\mathbf{XKX}'\mathbf{T}^{-1/2}\right)^+ \mathbf{T}^{-1/2},$$

where $\mathbf{G}^+$ denotes the Moore–Penrose inverse (*see* GENERALIZED INVERSES). Then the ∞ − MINQE(U) of $\mathbf{f}'\boldsymbol{\theta}$ is $\mathbf{Y}'\mathbf{A}_*\mathbf{Y}$, where

$$\begin{aligned} A^* = {} & (\mathbf{XKX}')_*^-\mathbf{V}_c(\mathbf{MV}_\alpha\mathbf{M})^+ \\ & + (\mathbf{MV}_\alpha\mathbf{M})^+ \mathbf{V}_c(\mathbf{XKX}')_*^- \\ & + (\mathbf{MV}_\alpha\mathbf{M})^+ V_b(\mathbf{MV}_\alpha\mathbf{M})^+, \end{aligned}$$

$$\mathbf{V}_c = \Sigma c_i\mathbf{V}_i, \qquad \mathbf{V}_b = \Sigma b_i\mathbf{V}_i,$$

and where $\mathbf{c} = (c_1, \ldots, c_p)'$, $\mathbf{b} = (b_1, \ldots, b_p)'$ satisfy the equations

$$\mathbf{H}_{UI}^\alpha\mathbf{b} + 2\mathbf{Bc} = \mathbf{f}, \qquad \mathbf{H}_{UI}^\alpha\mathbf{c} = 0.$$

Notes

4. The ∞ − MINQE(U) coincides with MINQE(U, I) if it exists.
5. If in the linear model (3), $\mathbf{Y}$ has the multivariate normal distribution defined in Note **5** of the preceding section, then the MINQE(U) estimator as determined in (12) is the same as the locally minimum variance unbiased estimator* (LMVUE) of $\mathbf{f}'\boldsymbol{\theta}$ at the parameter point $\boldsymbol{\beta} = \boldsymbol{\beta}_0$ and $\boldsymbol{\theta} = \boldsymbol{\alpha}$.

MINQE WITHOUT UNBIASEDNESS

Instead of unbiasedness, we may impose other conditions by restricting the symmetric matrix $\mathbf{A}$ to one of the following classes:

a. $\mathscr{C} = \{\mathbf{A}\}$.

b. $\mathscr{C}_{U'} = \{\mathbf{A} : \mathbf{X}'\mathbf{AX} = \mathbf{0}\}$ (i.e., bias is independent of $\boldsymbol{\beta}$).

c. $\mathscr{C}_I = \{\mathbf{A} : \mathbf{AX} = \mathbf{0}\}$, invariance*.

The MINQEs obtained subject to restrictions **a**, **b** and **c** are denoted by MINQE, MINQE(U') and MINQE(I), respectively.

Theorem 4. Let $\mathbf{W} = \Sigma\mu_i\mathbf{UF}_i\mathbf{U}'$, where μ_i are as defined in (4). Then under the Euclidean norm (8), the optimal matrices $\mathbf{A}_*$ providing MINQEs of $\mathbf{f}'\boldsymbol{\theta}$ are as follows:
MINQE:

$$\mathbf{A}_* = (\mathbf{V}_\alpha + \mathbf{XX}')^{-1}\mathbf{W}(\mathbf{V}_\alpha + \mathbf{XX}')^{-1}, \quad (15)$$

MINQE(U'):

$$(\mathbf{V}_\alpha + \mathbf{XX}')^{-1}(\mathbf{W} - \mathbf{P}_\alpha\mathbf{WP}_\alpha)(\mathbf{V}_\alpha + \mathbf{XX}')^{-1}, \quad (16)$$

$$\mathbf{P}_\alpha = \mathbf{X}(\mathbf{X}'\mathbf{V}_\alpha\mathbf{X})^-\mathbf{X}'\mathbf{V}_\alpha^{-1},$$

MINQE(I):

$$\mathbf{A}_* = \mathbf{V}_\alpha^{-1}(\mathbf{I} - \mathbf{P}_\alpha)\mathbf{W}(\mathbf{I} - \mathbf{P}_\alpha)\mathbf{V}_\alpha^{-1}. \quad (17)$$

Notes

1. The MINQE in (15) and the MINQE(I) in (17) are nonnegative if the natural estimator is nonnegative.
2. The MINQE(I) of $\mathbf{f}'\boldsymbol{\theta}$ in c can be written as $\mathbf{f}'\hat{\boldsymbol{\theta}}$, where $\hat{\boldsymbol{\theta}}$ is a solution of

$$[\mathbf{H}_I^\alpha]\boldsymbol{\theta} = \mathbf{h}_I^\alpha(\mathbf{Y}), \quad (18)$$

$$\mathbf{H}_I^\alpha = \mathrm{tr}\,\mathbf{V}_\alpha^{-1}\mathbf{V}_i\mathbf{V}_\alpha^{-1}\mathbf{V}_j,$$

with the ith element of $\mathbf{h}_I^\alpha(\mathbf{Y})$ as

$$\mathbf{Y}'\mathbf{V}_\alpha^{-1}(\mathbf{I} - \mathbf{P}_\alpha)\mathbf{UV}_i\mathbf{U}'(\mathbf{I} - \mathbf{P}'_\alpha)\mathbf{V}_\alpha^{-1}\mathbf{Y}.$$

The form of the solution (18) enables us to obtain IMINQE(I), the iterated MINQE(I), as the solution of the equation

$$[\mathbf{H}_I^\theta]\boldsymbol{\theta} = \mathbf{h}_I^\theta(\mathbf{Y}), \quad (19)$$

which in the special case $\mathbf{U} = \mathbf{I}$ reduces to the maximum likelihood estimate.

MINQE(NND)

The MINQEs obtained under the condition of unbiasedness and/or invariance need not be nonnegative. But if nonnegativity is desirable, then we have to impose the additional condition that the matrix $\mathbf{A}$ is NND.

First observe that, if a nonnegative unbiased estimator exists, it automatically satisfies the invariance condition. In such a case we need only consider the model

$$\mathbf{t} = \mathbf{Z}'\mathbf{Y}, \qquad E(\mathbf{t}) = \mathbf{0},$$
$$E(\mathbf{tt}') = \Sigma\theta_i\mathbf{Z}'\mathbf{V}_i\mathbf{Z} = \Sigma\theta_i\mathbf{B}_i,$$

where $\mathbf{Z} = \mathbf{X}^{\perp}$ (with rank s, say), and quadratic estimators of the type $\mathbf{t}'\mathbf{C}\mathbf{t}$, where $\mathbf{C}$ belongs to the class

$$\mathscr{C}_{UD}^{f} = \{\mathbf{C} : \mathbf{C} \geqslant 0, \operatorname{tr}\mathbf{CB}_i = f_i\}.$$

Theorem 5. $\mathscr{C}_{UD}^{f}$ is not empty iff $\mathbf{f} \in$ convex span $\{q(\mathbf{b}), \mathbf{b} \in R^s\}$, where $q(\mathbf{b}) = (\mathbf{b}'\mathbf{B}_1\mathbf{b}, \ldots, \mathbf{b}'\mathbf{B}_p\mathbf{b})'$.

Since $\mathscr{C}_{UD}^{f}$ is convex, it has a maximal member C_0 such that $\mathscr{S}(C_0) \supseteq \mathscr{S}(C)$ for all $C \in \mathscr{C}_{UD}^{f}$. With such a C_0, so far as the estimation of $\mathbf{f}'\boldsymbol{\theta}$ is concerned, we may further reduce the model (19) to

$$\mathbf{t}_f = \mathbf{Z}_f'\mathbf{t}, \qquad E(\mathbf{t}_f) = 0,$$
$$E(\mathbf{t}_f\mathbf{t}_f') = \Sigma\theta_i\mathbf{Z}_f'\mathbf{B}_i\mathbf{Z}_f, \tag{20}$$

where $\mathbf{Z}_f$ is any matrix (with rank $s_f \leqslant s$) such that $\mathscr{S}(\mathbf{Z}_f) = \mathscr{S}(C_0)$.

If $\mathscr{C}_{UD}^{f}$ is not empty, then the MINQE(U, NND) is obtained by solving the problem

$$\text{minimize } \operatorname{tr}(\mathbf{AV}_\alpha\mathbf{AV}_\alpha) \qquad \text{for} \quad \mathbf{A} \in \mathscr{C}_{UD}^{f}. \tag{21}$$

There is no closed-form solution to this problem, but the following theorem is helpful in recognizing a solution.

Theorem 6. Let $\mathscr{C}_{UD}^{f}$ be not empty and contain an element $\mathbf{A}_0$ with $\mathscr{S}(\mathbf{A}_0) = \mathscr{S}(\mathbf{MV}_\alpha\mathbf{M})$, which if necessary may be achieved in terms of the reduced model (20). Then

(a) Problem (21) admits a solution of the form

$$\mathbf{A}_* = (\mathbf{MV}_\alpha\mathbf{M})^+\mathbf{C}_*(\mathbf{MV}_\alpha\mathbf{M})^+.$$

(b) $\mathbf{A}_* = (\mathbf{MV}_\alpha\mathbf{M})^+\mathbf{C}_*(\mathbf{MV}_\alpha\mathbf{M})^+$ is a solution to (21) iff

$$\mathbf{A}_* \in \mathscr{C}_{UD}^{f},$$
$$\mathbf{A}_* + \Sigma b_i(\mathbf{MV}_\alpha\mathbf{M})^+\mathbf{V}_i(\mathbf{MV}_\alpha\mathbf{M})^+ \geqslant 0,$$
$$\operatorname{tr}\mathbf{A}_*(\mathbf{C}_* + \Sigma b_i\mathbf{V}_i) = 0$$
$$\text{for some } b_1, \ldots, b_p.$$

Another characterization of MINQE(U, NND) brings out its connection with MINQE(U, I). For any symmetric matrix $\mathbf{B}$, let $\mathbf{B}^0 = \mathbf{B} - \Sigma s_i\mathbf{V}_i$ where $\mathbf{s} = (s_1, \ldots, s_p)'$ is a solution of $\mathbf{H}_{UI}^{\alpha}\mathbf{s} = h(\mathbf{B})$ and where in the vector $\mathbf{h}(\mathbf{B})$, the ith element is $\operatorname{tr}[\mathbf{B}(\mathbf{MV}_\alpha\mathbf{M})^+\mathbf{V}_i(\mathbf{MV}_\alpha\mathbf{M})^+]$. Note that by construction MINQE(U, I) is $\mathbf{Y}'\mathbf{AY}$ with $\operatorname{tr}\hat{\mathbf{A}}\mathbf{B}^0 = 0$ for any given $\mathbf{B}$.

Theorem 7. Let $\mathscr{C}_{UD}^{f}$ be not empty and $\mathbf{Y}'\hat{\mathbf{A}}\mathbf{Y}$ be the MINQE(U, I). Then:

a. There are solutions to (21) of the form

$$\mathbf{A}_* = \hat{\mathbf{A}} + (\mathbf{MV}_\alpha\mathbf{M})^+\mathbf{B}^0{}_*(\mathbf{MV}_\alpha\mathbf{M})^+$$

for some $\mathbf{B}_* \geqslant 0$.

b. A sufficient condition that $\mathbf{A}_*$ in (a) provides the MINQE(U, NND) is $\mathbf{A}_* \geqslant 0$ and $\operatorname{tr}\mathbf{A}_*\mathbf{B}_* = 0$.

c. The optimality condition $\operatorname{tr}\mathbf{A}_*\mathbf{B}_* = 0$ is necessary if $\mathscr{C}_{UD}^{f}$ has a member $\mathbf{Y}'\mathbf{AY}$ with $\mathscr{S}(\mathbf{A}) = \mathscr{S}(\mathbf{MV}_\alpha\mathbf{M})$.

If $\mathscr{C}_{UD}^{f}$ is empty, we consider the class of nonnegative invariant quadratic estimators for which the bias is a minimum and minimize $\operatorname{tr}(\mathbf{AV}_\alpha\mathbf{AV}_\alpha)$ for $\mathbf{A}$ in this class. The details of the computations involved in obtaining MINQE(NND) with unbiasedness or minimum bias are contained in the papers by Hartung [5], Kleffe [8] and Pukelsheim [11].

There are other methods used in practice for obtaining nonnegative estimators of variance components. One is to use any of the methods described from the second section to this one and compute the estimator. If t is the value obtained, the estimator is chosen as t if $t \geqslant 0$ and ϵ if $t < 0$, where ϵ is some small positive value. The choice of ϵ depends on the purpose for which the variance components are estimated. Such a procedure was suggested by J. N. K. Rao and Subramaniam [16] for using estimated variance components as weights in pooling estimates of the common mean with different precisions.

Another method suggested by Rao [14] which involves some additional computations is as follows. Suppose the estimator is $\mathbf{Y}'\mathbf{A}_*\mathbf{Y}$ and the matrix $\mathbf{A}_*$ is not NND. Let

$$\mathbf{A}_* = \lambda_1 \mathbf{P}_1 \mathbf{P}_1' + \cdots + \lambda_n \mathbf{P}_n \mathbf{P}_n'$$

by the spectral decomposition of $\mathbf{A}_*$ and suppose that $\lambda_1, \ldots, \lambda_r$ are nonnegative while $\lambda_{r+1}, \ldots, \lambda_n$ are negative. Then the estimator $\mathbf{Y}'\mathbf{A}_*\mathbf{Y}$ is modified as

$$\mathbf{Y}'(\lambda_1 \mathbf{P}_1 \mathbf{P}_1' + \cdots + \lambda_r \mathbf{P}_r \mathbf{P}_r'),$$

or, more generally, as

$$\mathbf{Y}'(\lambda_1 \mathbf{P}_1 \mathbf{P}_1' + \cdots + \lambda_r \mathbf{P}_r \mathbf{P}_r' + \lambda_{r+1} \mathbf{P}_{r+1} \mathbf{P}_{r+1}' + \cdots + \lambda_m \mathbf{P}_m \mathbf{P}_m')\mathbf{Y}$$

by including the terms involving the negative eigenvalues one by one so long as the result is nonnegative. (Note that m is a function of $\mathbf{Y}$.)

SOME REMARKS

Early in this article MINQEs were obtained under different kinds of restrictions. In practice, the choice of restrictions depends on the nature of the a priori information available and also on the model for the observations. Consider the model with four observations

$$Y_1 = \beta_1 + \epsilon_1, \qquad Y_2 = \beta_1 + \epsilon_2,$$
$$Y_3 = \beta_2 + \epsilon_3, \qquad Y_4 = \beta_2 + \epsilon_4,$$

where $V(\epsilon_1) = V(\epsilon_3) = \sigma_1^2$ and $V(\epsilon_2) = V(\epsilon_4) = \sigma_2^2$ and all the ϵ_i are uncorrelated. The matrices $\mathbf{X}, \mathbf{V}_1$, and $\mathbf{V}_2$ are

$$\mathbf{X} = \begin{bmatrix} 1 & 0 \\ 1 & 0 \\ 0 & 1 \\ 0 & 1 \end{bmatrix}, \qquad \mathbf{V}_1 = \begin{bmatrix} 1 & 0 & 0 & 0 \\ 0 & 0 & 0 & 0 \\ 0 & 0 & 1 & 0 \\ 0 & 0 & 0 & 0 \end{bmatrix},$$

$$\mathbf{V}_2 = \begin{bmatrix} 0 & 0 & 0 & 0 \\ 0 & 1 & 0 & 0 \\ 0 & 0 & 0 & 0 \\ 0 & 0 & 0 & 1 \end{bmatrix}.$$

The matrices $\mathbf{H} = (\operatorname{tr} \mathbf{V}_i \mathbf{M} \mathbf{V}_j)$ of Theorem 2, Note 1 and $\mathbf{H}_M = (\operatorname{tr} \mathbf{M} \mathbf{V}_i \mathbf{M} \mathbf{V}_j)$ of Theorem 1, Note 2 are

$$\mathbf{H} = \begin{pmatrix} 1 & 0 \\ 0 & 1 \end{pmatrix} \quad \text{and} \quad \mathbf{H}_M = \begin{pmatrix} \frac{1}{2} & \frac{1}{2} \\ \frac{1}{2} & \frac{1}{2} \end{pmatrix}.$$

Since $\mathbf{H}$ is of full rank, we find from Theorem 2 Note 1 that σ_1^2 and σ_2^2 are each unbiasedly estimable. But $\mathbf{H}_M$ is of rank one, and the unit vectors do not belong to the space $\mathscr{S}(\mathbf{H}_M)$. Then from Theorem 1, Note 2 it follows that σ_1^2 and σ_2^2 are not estimable unbiasedly by invariant quadratic forms. Thus, in the preceding problem only the MINQE(U) is possible.

Consider the model $\mathbf{Y} = \mathbf{X}\boldsymbol{\beta} + \mathbf{X}\boldsymbol{\phi} + \boldsymbol{\epsilon}$, where $\boldsymbol{\beta}$ is a fixed parameter and $\boldsymbol{\phi}$ is a vector of random effects such that $E(\boldsymbol{\phi}) = \mathbf{0}$, $E(\boldsymbol{\epsilon}) = \mathbf{0}$ and $E(\boldsymbol{\phi}\boldsymbol{\phi}') = \sigma_2^2 \mathbf{I}_m$, $E(\boldsymbol{\phi}\boldsymbol{\epsilon}') = \mathbf{0}$ and $E(\boldsymbol{\epsilon}\boldsymbol{\epsilon}') = \sigma_1^2 \mathbf{I}_n$. If $\mathbf{Y}'\mathbf{A}\mathbf{Y}$ is unbiased for σ_2^2, then

$$\mathbf{X}'\mathbf{A}\mathbf{X} = \mathbf{0}, \qquad \operatorname{tr} \mathbf{A}\mathbf{X}\mathbf{X}' = 1, \qquad \operatorname{tr} \mathbf{A} = 0,$$

which are not consistent. Hence an unbiased estimator of σ_2^2 does not exist.

It is seen that the formulae for the various types of MINQEs involve the inversion of large matrices. But in problems that arise in practice, the matrices involved are of special types and their inverses can be built up from inverses of lower-order matrices. References to such methods are given in Kleffe and Seifert [9].

References

[1] Airy, G. B. (1961). *On the Algebraical and Numerical Theory of Errors of Observations and Combina-*

tion of Observations. Macmillan, Cambridge, England.
[2] Fisher, R. A. (1918). *Trans. R. Soc. Edinburgh*, **52**, 399–433.
[3] Focke, J. and Dewess, G. (1972). *Math. Operationsforsch. Statist.*, **3**, 129–143.
[4] Hartley, H. O. and Rao, J. N. K. (1967). *Biometrika*, **54**, 93–108.
[5] Hartung, J. (1981). *Ann. Statist.*, **9**, 278–292.
[6] Harville, D. A. (1977). *J. Amer. Statist. Ass.*, **72**, 320–340.
[7] Henderson, C. R. (1953). *Biometrics*, **9**, 226–252.
[8] Kleffe, J. (1980). *Math. Operationsforsch. Statist. Ser. Statist.*, **11**, 563–588.
[9] Kleffe, J. and Seifert, B. (1982). *Proc. Seventh Conf. Prob. Statist.*, Brasov, Roumania.
[10] Patterson, H. D. and Thompson, R. (1975). *Proc. Eighth Int. Biometric Conf.*, 197–207.
[11] Pukelsheim, F. (1981). *Ann. Statist.*, **9**, 293–299.
[12] Rao, C. R. (1973). *Linear Statistical Inference and its Applications*, 2nd ed. Wiley, New York.
[13] Rao, C. R. (1979). *Sankhyā B*, **41**, 138–153.
[14] Rao, C. R. (1984). In *W. G. Cochran's Impact on Statistics*, P. S. R. S. Rao and J. Sedransk, eds. Wiley, New York, pp. 191–202.
[15] Rao, C. R. and Kleffe, J. (1980). In *Handbook of Statistics*, Vol. 1, P. R. Krishnaiah, ed. Elsevier North Holland, New York, pp. 1–40.
[16] Rao, J. N. K. and Subramaniam, K. (1971). *Biometrics*, **27**, 971–990.
[17] Seely, J. (1975). *Biometrika*, **62**, 689–690.

(ESTIMATION, POINT
FIXED-, RANDOM-, AND MIXED-EFFECTS MODELS
GENERAL LINEAR MODEL
GEOMETRY IN STATISTICS
INVARIANCE PRINCIPLES IN STATISTICS
MINIMUM VARIANCE UNBIASED ESTIMATION
UNBIASEDNESS
VARIANCE COMPONENTS)

C. RADHAKRISHNA RAO
J. KLEFFE

MINRES

An abbreviation of minimum residual, with special reference to methodology used in factor analysis*.

MINRET METHOD *See* FACTOR ANALYSIS

MISCLASSIFICATION PROBABILITY *See* DISCRIMINANT ANALYSIS

MISES' DISTRIBUTION *See* DIRECTIONAL DISTRIBUTIONS

MISES, RICHARD VON *See* VON MISES, RICHARD

MISSING DATA *See* INCOMPLETE DATA

MISSING INFORMATION PRINCIPLE

Maximum likelihood estimation* with missing or otherwise incomplete data* is often facilitated by first posing the maximization problem for some apposite (and possibly hypothetical) "complete" data problem. The likelihood of the observed or incomplete data can be expressed as an integral of the complete data likelihood. This characterization of the observed data likelihood leads to an attractive representation for the likelihood equations and the information matrix based on the complete data, as well as a simple, iterative computing algorithm. This special feature of maximum likelihood with incomplete data is described in a general context in a number of key papers. Orchard and Woodbury [5] coined the phrase missing information principle (MIP) to characterize *maximum likelihood estimation with incomplete data*; Beale and Little [1] elucidated the principle and applied it to missing values in multivariate analysis*. Sundberg [6] derived an equivalent set of relationships specialized to the case where the completed data likelihood has an exponential family* form, and thus a set of sufficient statistics*. Dempster et al. [2] integrated previous work, broadened the definition of incomplete data to include several new examples, and coined the name E-M (expectation-maximization) for the algorithm. Their paper contains an extensive literature review of the subject. Laird [4] considered nonparametric applications with incomplete data, discussing the link between the MIP for parametric applications and the self-consistency* principle of Efron [3] for nonparametric estimation problems.

In its most general form, the incomplete data problem is stated in terms of two data vectors $\mathbf{x}$ and $\mathbf{y}$ and their corresponding sample spaces $\mathscr{X}$ and $\mathscr{Y}$. The observed data vector $\mathbf{y}$ defines a point in $\mathscr{Y}$. The completed data vector $\mathbf{x}$ defines a point in $\mathscr{X}$, but is not observed directly. Rather we assume a many-to-one mapping of $\mathbf{x}$ to the observed data $\mathbf{y}$. After observing $\mathbf{y}$, $\mathbf{x}$ is only known to lie in the subset of $\mathscr{X}$, say $\mathscr{X}(\mathbf{y})$ defined by the mapping $\mathbf{x} \to \mathbf{y}$. In many applications, $\mathbf{x}$ models physically observable data that is not observed due to particular circumstances, e.g., missing values, censoring*, or truncation*. In other problems, such as factor analysis and mixture problems, $\mathbf{x}$ may include parameters, latent variables, or other inherently unobservable quantities.

We let $\boldsymbol{\phi}$ denote the vector of unknown parameters to be estimated by maximum likelihood, $g(\mathbf{y} \mid \boldsymbol{\phi})$ the density of $\mathbf{y}$, and $f(\mathbf{x} \mid \boldsymbol{\phi})$ the density of $\mathbf{x}$. Then

$$g(\mathbf{y} \mid \boldsymbol{\phi}) = \int_{\mathscr{X}(\mathbf{y})} f(\mathbf{x} \mid \boldsymbol{\phi})\, d\mathbf{x}.$$

For a given $g(\mathbf{y} \mid \boldsymbol{\phi})$, $f(\mathbf{x} \mid \boldsymbol{\phi})$ is not unique, and its particular representation in a given problem can be a matter of convenience. We define $k(\mathbf{x} \mid \mathbf{y}, \boldsymbol{\phi})$ to be the conditional density of the complete data given the observed data. Thus

$$k(\mathbf{x} \mid \mathbf{y}, \boldsymbol{\phi}) = f(\mathbf{x} \mid \boldsymbol{\phi}) / g(\mathbf{y} \mid \boldsymbol{\phi}). \quad (1)$$

Taking logarithms of both sides of (1) and rearranging, we have

$$L(\boldsymbol{\phi}; \mathbf{x}) = L(\boldsymbol{\phi}; \mathbf{y}) + L(\boldsymbol{\phi}; \mathbf{x} \mid \mathbf{y}), \quad (2)$$

where $L(\boldsymbol{\phi}; \mathbf{x})$ and $L(\boldsymbol{\phi}; \mathbf{y})$ are the log-likelihoods of $\boldsymbol{\phi}$ associated with the complete and observed data vectors, respectively, and $L(\boldsymbol{\phi}; \mathbf{x} \mid \mathbf{y})$ is the log-likelihood based on the conditional density of $\mathbf{x}$ given $\mathbf{y}$. Let $\boldsymbol{\phi}_A$ denote any assumed value of $\boldsymbol{\phi}$ and take expectations of both sides of (2) with respect to $k(\mathbf{x} \mid \mathbf{y}, \boldsymbol{\phi}_A)$. Then

$$Q(\boldsymbol{\phi} \mid \boldsymbol{\phi}_A) = L(\boldsymbol{\phi}; \mathbf{y}) + H(\boldsymbol{\phi} \mid \boldsymbol{\phi}_A), \quad (3)$$

where

$$Q(\boldsymbol{\phi} \mid \boldsymbol{\phi}_A) = E\{L(\boldsymbol{\phi}; \mathbf{x}) \mid \mathbf{y}, \boldsymbol{\phi}_A\},$$
$$H(\boldsymbol{\phi} \mid \boldsymbol{\phi}_A) = E\{L(\boldsymbol{\phi}; \mathbf{x} \mid \mathbf{y}) \mid \mathbf{y}, \boldsymbol{\phi}_A\}.$$

Equation (3) leads directly to the MIP as follows.

The value of $\boldsymbol{\phi}$ which maximizes $H(\boldsymbol{\phi} \mid \boldsymbol{\phi}_A)$ is precisely $\boldsymbol{\phi}_A$. Let $\boldsymbol{\phi}^*$ denote the maximum likelihood estimate, i.e., the value of $\boldsymbol{\phi}$ which maximizes $L(\boldsymbol{\phi}; \mathbf{y})$. The value of $\boldsymbol{\phi}$ that maximizes $Q(\boldsymbol{\phi} \mid \boldsymbol{\phi}_A)$ is some function of $\boldsymbol{\phi}_A$, say

$$\boldsymbol{\phi} = M(\boldsymbol{\phi}_A).$$

If we now set $\boldsymbol{\phi}_A = \boldsymbol{\phi}^*$, the right-hand side of (3), i.e.,

$$L(\boldsymbol{\phi}; \mathbf{y}) + H(\boldsymbol{\phi} \mid \boldsymbol{\phi}^*),$$

is maximized when $\boldsymbol{\phi} = \boldsymbol{\phi}^*$. Thus $\boldsymbol{\phi}^*$ must also maximize $Q(\boldsymbol{\phi} \mid \boldsymbol{\phi}^*)$. As a result, we have Lemma 1.

Lemma 1. *The maximum likelihood estimate of $\boldsymbol{\phi}$ based on the observed data $\mathbf{y}$ must satisfy the fixed point equation*

$$\boldsymbol{\phi}^* = M(\boldsymbol{\phi}^*). \quad (4)$$

Lemma 1 not only characterizes the maximum likelihood estimate in incomplete data problems, but also suggests an iterative computing algorithm, taking $\boldsymbol{\phi}^{(p)}$ into $\boldsymbol{\phi}^{(p+1)}$ as

$$\boldsymbol{\phi}^{(p+1)} = M(\boldsymbol{\phi}^{(p)}).$$

In various forms, this has been called the E-M, the MIP, the successive substitutions,

Lemma 2. $L(\boldsymbol{\phi}^{(p+1)}; \mathbf{y}) \geqslant L(\boldsymbol{\phi}^{(p)}; \mathbf{y})$, *with equality if and only if*

$$Q(\boldsymbol{\phi}^{(p+1)} \mid \boldsymbol{\phi}^{(p)}) = Q(\boldsymbol{\phi}^{(p)} \mid \boldsymbol{\phi}^{(p)})$$

and

$$k(\mathbf{x} \mid \mathbf{y}, \boldsymbol{\phi}^{(p+1)}) = k(\mathbf{x} \mid \mathbf{y}, \boldsymbol{\phi}^{(p)}).$$

If, further, $Q(\phi \mid \phi_A)$, $L(\phi; \mathbf{y})$, and $H(\phi \mid \phi_A)$ are differentiable, and we can reverse the order of differentiation and integration, then differentiating both sides of (3) leads to the following lemmas.

Lemma 3. *Any solution* ϕ^* *to equation* (4) *defines a maximum or a stationary point of* $L(\phi; \mathbf{y})$.

Lemma 4.

$$I(\phi; \mathbf{x}) = I(\phi; \mathbf{y}) + E\{I(\phi; \mathbf{x} \mid \mathbf{y})\},$$

where $I(\phi; \mathbf{x})$, $I(\phi; \mathbf{y})$ *and* $I(\phi; \mathbf{x} \mid \mathbf{y})$ *are the Fisher information* matrices for* ϕ, *based on* $f(\mathbf{x} \mid \phi)$, $g(\mathbf{y} \mid \phi)$, *and* $k(\mathbf{x} \mid \mathbf{y}, \phi)$, *respectively, and* $E\{I(\phi; \mathbf{x} \mid \mathbf{y})\}$ *is expectation with respect to* $g(\mathbf{y} \mid \phi)$.

Orchard and Woodbury [5] define this last term to be the "lost information" resulting from the incompleteness in the data. Lemmas 1 and 4 describe the essence of the MIP. Orchard and Woodbury [5] further derive variance inequalities based on Lemma 4.

When the complete data density can be put in the exponential family* form, Lemmas 1 and 4, as well as the definition of the algorithm, take on a paricularly striking representation. This was noted in Sundberg [6] who attributes the original results to Martin-Löf. For this case, we assume that

$$f(\mathbf{x} \mid \phi) = b(\mathbf{x}) \exp\{\boldsymbol{\theta}^T \mathbf{t}(\mathbf{x})\} / a(\boldsymbol{\theta}), \quad (6)$$

where $\boldsymbol{\theta}$ is a k-dimensional one-to-one transformation of ϕ (into the natural parameter space), and $\mathbf{t}(\mathbf{x})$ is a k-dimensional vector of sufficient statistics, based on the complete data $\mathbf{x}$. Substituting (6) into (1), taking logs and differentiating yields Lemma 1′.

Lemma 1′. *The maximum likelihood estimate of* $\boldsymbol{\theta}$ *based on* $\mathbf{y}$ *satisfies*

$$E\{\mathbf{t}(\mathbf{x}) \mid \boldsymbol{\theta}^*\} = E\{\mathbf{t}(\mathbf{x}) \mid \mathbf{y}, \boldsymbol{\theta}^*\}.$$

The expectation on the left-hand side is with respect to $f(\mathbf{x} \mid \boldsymbol{\theta}^*)$ and that on the right-hand side is with respect to $k(\mathbf{x} \mid \mathbf{y}, \boldsymbol{\theta}^*)$. Differentiating (1) again and taking expectations leads to Lemma 4′.

Lemma 4′.

$$I(\boldsymbol{\theta}; \mathbf{y}) = \operatorname{var}(\mathbf{t}(\mathbf{x}) \mid \boldsymbol{\theta}) - E\{\operatorname{var}(\mathbf{t}(\mathbf{x}) \mid \mathbf{y}, \boldsymbol{\theta})\}.$$

To specify the algorithm in this case, we note that if we had observed $\mathbf{x}$, and thus $\mathbf{t}(\mathbf{x})$, the maximum likelihood estimate of $\boldsymbol{\theta}$ would satisfy

$$E\{\mathbf{t}(\mathbf{x}) \mid \hat{\boldsymbol{\theta}}\} = \mathbf{t}(\mathbf{x}).$$

Conversely, given some value for $\boldsymbol{\theta}$, we may estimate $\mathbf{t}(\mathbf{x})$ by setting it equal to its expected value, given $\mathbf{y}$ and θ. These facts lead to the two explicit steps of the E-M algorithm:

E-Step: Set

$$\mathbf{t}^{(p)} = E\{\mathbf{t}(\mathbf{x}) \mid \mathbf{y}, \phi^{(p)}\}.$$

M-Step: Define $\phi^{(p+1)}$ as the solution to

$$\mathbf{t}^{(p)} = E\{\mathbf{t}(\mathbf{x}) \mid \phi\}.$$

These two steps are equivalent to maximizing $Q(\boldsymbol{\theta} \mid \boldsymbol{\theta}^{(p)})$ as a function of $\boldsymbol{\theta}$, when $f(\mathbf{x} \mid \boldsymbol{\theta})$ has the form given in (6).

Lemmas 1 and 4 (or 1′ and 4′) are almost always useful in incomplete data estimation problems, since they characterize the MLE and its asymptotic variance. The usefulness of the E-M algorithm depends on the particular context. It has linear convergence and thus may require many more iterations than alternate computing algorithms such as Newton–Raphson* or Fisher–Scoring. In addition, its convergence is not guaranteed (the proof of convergence given in Dempster et al. [2] is incorrect). If the algorithm does converge, it is only guaranteed to be a stationary point of the likelihood. Offsetting these disadvantages is the fact that it always increases the likelihood. In addition, each iteration of the E and M steps often takes especially simple forms—thus numerous iterations are not expensive and programming the algorithm may be very easy.

Besides the usual application to missing, censored, truncated, and grouped-data* problems, the theory presented can be applied in mixture problems*, hyperparameter estimation, variance component* models, factor analysis*, latent variable analysis, and other special cases, such as estimating gene frequencies (*see* GENETICS, STATISTICS IN).

AN APPLICATION

The following simple example suggested by Milton Weinstein illustrates the application of the E-M algorithm. The problem arises in developing screening tests for a disease, when a standard for determining disease status is unavailable (*see* MEDICAL DIAGNOSIS, STATISTICS IN). Suppose that the observed data consist of a random sample of patients measured on two screening tests, each test giving a dichotomous result. Their true disease status is not known. If we assume that the two test results are conditionally independent given disease status, and further, that the false positive rates for each test are zero, then we can use the E-M algorithm to calculate maximum likelihood estimates of disease prevalence and the sensitivities of the two tests.

To implement the E-M algorithm, notice that if we could observe disease status for each individual, the estimation problem would be trivial. Our estimate of the disease prevalence π would be the observed proportion diseased. Because the false positive rates are zero, each test sensitivity would be estimated as the total number of positives on that test divided by the total number of diseased patients. These calculations form the basis of the M-step. Conversely, if we know the disease prevalence rate and each test's sensitivity, it is straightforward to predict disease status conditional on test outcome. This calculation forms the basis of the E-step.

More formally, the observed data can be arrayed as

		Test 2 +	Test 2 −
Test 1	+	y_{11}	y_{12}
	−	y_{21}	y_{22}

The complete data are $x_{11}, x_{12}, x_{21}, x_{221}, x_{222}$, where $y_{ij} = x_{ij}$ unless $(i, j) = (2, 2)$. Then $y_{22} = x_{221} + x_{222}$, where x_{221} is the number of diseased with negative outcomes on both tests and x_{222} is the number of nondiseased patients. Letting S_1 and S_2 denote the test sensitivities, the complete data likelihood is

$$L(\pi, S_1, S_2) = (S_1S_2)^{x_{11}}\{S_1(1-S_2)\}^{x_{12}}\{S_2(1-S_1)\}^{x_{21}} \times \{(1-S_1)(1-S_2)\}^{x_{221}}\pi^{N-x_{222}}(1-\pi)^{x_{222}},$$

where N is the sample size $(\sum_{ij} y_{ij})$. Rearranging terms, and letting N_D be the number of diseased patients $(N - x_{222})$, we see that

$$L(\pi, S_1S_2) = S_1^{x_{1+}}(1-S_1)^{N_D-x_{1+}}S_2^{x_{+1}} \times (1-S_2)^{N_D-x_{+1}}\pi^{N_D}(1-\pi)^{x_{222}}.$$

It now follows that x_{1+}, x_{+1}, and N_D are jointly sufficient for S_1, S_2, and π.

Table 1

(a) Observed Data

		Test 2 +	Test 2 −	
Test 1	+	20	5	25
	−	15	60	
		35		100

(b) Starting Values

$N_D^{(0)} = 40$
$\pi^{(0)} = 0.4$
$S_1^{(0)} = \frac{25}{40} = 0.625$
$S_2^{(0)} = \frac{35}{40} = 0.875$

(c) Iterated Values

Iteration	$\pi^{(p)}$	$S_1^{(p)}$	$S_2^{(p)}$
0	0.4	0.625	0.875
1	0.4182	0.5978	0.8370
2	0.4270	0.5855	0.8197
3	0.4317	0.5792	0.8107
4	0.4342	0.5757	0.8060
5	0.4356	0.5739	0.8034
6	0.4364	0.5728	0.8019
7	0.4371	0.5720	0.8008
⋮	⋮	⋮	⋮
∞	0.4375	0.5714	0.8000

The two steps of each iteration of the E-M are now easily obtained. Let $(\pi^{(p)}, S_1^{(p)}, S_2^{(p)})$ denote the estimated parameters at the end of the pth iteration. Then setting the complete data sufficient statistics equal to their expectations given $\mathbf{y}$ yields the E-step:

$$E(x_{1+} \mid \mathbf{y}, \pi^{(p)}, S_1^{(p)}, S_2^{(p)}) = y_{1+},$$

$$E(x_{+1} \mid \mathbf{y}, \pi^{(p)}, S_1^{(p)}, S_2^{(p)}) = y_{+1},$$

$$N_D^{(p+1)} = E(N_D \mid \mathbf{y}, \pi^{(p)}, S_1^{(p)}, S_2^{(p)})$$
$$= N - y_{22}(1 - \pi^{(p)})/g^{(p)},$$

where

$$g^{(p)} = (1 - S_1^{(p)})(1 - S_2^{(p)})\pi^{(p)} + (1 - \pi^{(p)}).$$

To get the M-step, we solve the equations defined by setting the complete data "sufficient statistics" equal to their unconditional expectations, substituting the results of the preceding E-step for the missing "sufficient statistics":

$$\pi^{(p+1)} = N_D^{(p+1)}/N,$$

$$S_1^{(p+1)} = y_{1+}/N_D^{(p+1)},$$

$$S_2^{(p+1)} = y_{+1}/N_D^{(p+1)}.$$

As with many simple applications of the E-M, the likelihood equations can be solved directly, so iterative calculations are unnecessary. If more than two tests are considered, or a different set of assumptions about test independence, false positive rates, etc., are applied, then exact solutions may be infeasible, but implementation of the E-M is a straightforward extension of the simple case considered here.

Table 1 illustrates the calculations for a set of numbers; note that

$$N_D^{(1)} = 100 - 60(0.6/[0.375 \times 0.125 \times 0.4 + 0.6]) = 41.82$$

$$\pi^{(1)} = 41.82/100, \quad S_1^{(1)} = 25/41.82,$$
$$S_2^{(1)} = 35/41.82.$$

References

[1] Beale, E. M. L. and Little, R. J. A. (1975). *J. R. Statist. Soc. B.*, **37**, 129–145. (Considers missing data in the multivariate problem.)

[2] Dempster, A. P., Laird, N. M., and Rubin, D. B. (1977). *J. R. Statist. Soc. B.*, **39**, 1–38. With discussion. (Introduces the general form of the E-M algorithm and gives many examples.)

[3] Efron, B. (1967). *Proc. 5th Berkeley Symp. Math. Statist. Prob.*, **4**, 831–853. (Discusses the two-sample problem with censored data and introduces the self-consistency algorithm.)

[4] Laird, N. M. (1978). *J. Amer. Statist. Ass.*, **73**, 805–811. (Considers nonparametric applications of the E-M algorithm.)

[5] Orchard, T. and Woodbury, M. A. (1972). *Proc. 6th Berkeley Symp. Math. Statist. Prob.*, **1**, 697–715. (Introduces the missing information principle and considers missing data in the multivariate normal.)

[6] Sundberg, R. (1974) *Scand. J. Statist.*, **1**, 49–58. (Gives the Sundberg formulas for exponential family data.)

(INCOMPLETE DATA
MAXIMUM LIKELIHOOD ESTIMATION
MEDICAL DIAGNOSIS, STATISTICS IN
SELF-CONSISTENCY ALGORITHM)

NAN LAIRD

MISSPECIFICATION, TESTS FOR

The problem of misspecification arises when the conditions necessary for the desired behavior of a statistic are violated. The desired behavior may be the unbiasedness*, consistency*, or efficiency* of an estimator for a parameter of interest, or it may be that a test statistic has a particular exact or asymptotic distribution. Tests for misspecification test the validity of the conditions ensuring the desired behavior of the statistic. Such tests are valuable in alerting an investigator to problems with the validity or interpreation of statistical results. In some cases, they also provide guidance as to how to remove the misspecification.

Tests for misspecification have been extensively studied in the framework of the linear regression* model and its extensions, so we shall illustrate the relevant concepts and techniques within this framework. The assumptions that we consider are the following, where $\overset{A}{\sim} N(\cdot, \cdot)$ denotes convergence to the normal distribution, i.e., asymptotic normality*:

A0. The data-generating process is $y_t = g(X_t) + \epsilon_t$, $t = 1, \ldots, n$, where $\{y_t\}$ and $\{X_t\}$ are observable stochastic processes*, $\{\epsilon_t\}$ is an unobservable stochastic process and $g : \mathbb{R}^k \to \mathbb{R}^1$ is a measurable function.

A1. $g(X_t) = X_t\beta_0$, where β_0 is an unknown finite $k \times 1$ vector of constants to be estimated.

A2. $n^{-1}\sum_{t=1}^{n} X_t'X_t \overset{\text{a.s.}}{\to} M_{xx}$, a finite nonsingular matrix.

A3. $E(X_t'\epsilon_t) = 0$ and $n^{-1}\sum_{t=1}^{n} X_t'\epsilon_t \overset{\text{a.s.}}{\to} 0$.

A4. $E(\epsilon_t^2) = \sigma_0^2 < \infty$ for all t, $E(\epsilon_t\epsilon_\tau) = 0$ for all $t \neq \tau$ and $n^{-1}\sum_{t=1}^{n} \epsilon_t^2 \overset{\text{a.s.}}{\to} \sigma_0^2$.

A5. $n^{-1/2}\sum_{t=1}^{n} X_t'\epsilon_t \overset{A}{\sim} N(0, V)$, $V = \sigma_0^2 M_{xx}$.

A6. $\epsilon_t \sim N(0, \sigma_0^2)$.

Using these conditions one can show that the ordinary least-squares* estimators

$$\hat{\beta}_n = (X'X)^{-1}X'y,$$

$$\hat{\sigma}_n^2 = (y - X\hat{\beta}_n)'(y - X\hat{\beta}_n)/(n - k)$$

exhibit the following behavior:

B1. Given **A0–A2**, $\hat{\beta}_n$ and $\hat{\sigma}_n^2$ exist almost surely (a.s.) for all n sufficiently large.

B2. Given **A0–A3**, $\hat{\beta}_n \overset{\text{a.s.}}{\to} \beta_0$.

B3. Given **A0–A4**, $\hat{\sigma}_n^2 \overset{\text{a.s.}}{\to} \sigma_0^2$ and $\hat{\sigma}_n^2(X'X/n)^{-1} \overset{\text{a.s.}}{\to} \sigma_0^2 M_{xx}^{-1}$.

B4. Given **A0–A5**, $\sqrt{n}\,(\hat{\beta}_n - \beta_0) \overset{A}{\sim} N(0, \sigma_0^2 M_{xx}^{-1})$.

B5. Given **A0–A6**, $\hat{\beta}_n$ is efficient asymptotically in that it has smallest asymptotic variance within the class of consistent uniformly asymptotically normal estimators.

Assumption **A0** is a definition, and is only testable once g has been specified, as it is in **A1**. Consequently, we treat **A0** as always valid and ignore it as a potential source of misspecification.

Table 1 Effects of Misspecification

Violated Assumption	Behavior				
	B1	B2	B3	B4	B5
A1		X^{a}	X	X	X
A2	X	X	X	X	X
A3		X	X	X	X
A4			X	X	X
A5				X	X
A6					X

[a] X = Behavior generally violated.

Table 1 shows the consequences of the failure of any single assumption. For example, when **A4** fails (e.g., as a result of heteroscedasticity* or serial correlation*) then **B3–B5** also fail, although **B1** and **B2** remain valid. Further, since inference is based on **B3** and **B4**, failure of either of these invalidates standard hypothesis testing* techniques.

Tests for misspecification of **A1–A6** typically involve specifying a more general form of the particular assumption in question and testing the appropriate statistical hypothesis using the implied more general analogues of **B1–B5**. A useful early discussion is given by Theil [22]. Additional early results are given by Ramsey [21].

To test **A1**, one may investigate whether relevant variables Z have been excluded by specifying $g(X_t, Z_t) = X_t\beta_0 + Z_t\gamma_0$ instead of $g(X_t) = X_t\beta_0$ and testing the hypothesis $\gamma_0 = 0$ using standard techniques (i.e., Wald*, Lagrange multiplier*, and likelihood ratio tests*). Alternatively, one can test the hypothesis that $g(X_t, Z_t) = X_t\beta_0$ against the alternative that $g(X_t, Z_t) = Z_t\gamma_0$ using a Cox test (Cox [4, 5], Pesaran and Deaton [19], Davidson and MacKinnon [6]).

Failure of **A2** is evidenced by singularity or near singularity of $X'X$, indicating that one or more of the regressors is a linear combination of the others. Detecting the source of the multicollinearity* can be accomplished by computing

$$R_j^2 = 1 - (1 - R^2)t_j^2/\{(n - k - 1)\hat{\beta}_{nj}^2\},$$

where R_j is the multiple correlation* coefficient of X_{tj} on all the other regressors, R^2 is the multiple correlation coefficient of the regression, t_j is the jth t-statistic and $\hat{\beta}_{nj}$ is the jth element of $\hat{\beta}_n$. This formula is due to Lemieux [17]. Regressors with high values for R_j^2 are relatively more responsible for the multicollinearity, and will tend to reduce the multicollinearity if deleted.

A test of **A3** could be based on $n^{-1}\sum_{t=1}^{n} X_t'\epsilon_t$ if this were observable. Of course, it is not; and by construction $n^{-1}\sum_{t=1}^{n} X_t'\hat{\epsilon}_t \equiv 0$, where $\hat{\epsilon}_t = y_t - X_t\hat{\beta}_n$. Given $\tilde{\beta}_n$ such that $\tilde{\beta}_n \overset{\text{a.s.}}{\to} \beta_0$ (e.g., let $\tilde{\beta}_n$ be a consistent instrumental variables* estimator) a test can be based on $n^{-1}\sum_{t=1}^{n} X_t'\tilde{\epsilon}_t$, where $\tilde{\epsilon}_t = y_t - X_t\tilde{\beta}_n$. This is equivalent to basing a test on $\hat{\beta}_n - \tilde{\beta}_n$. When **A3** holds, $\hat{\beta}_n - \tilde{\beta}_n \overset{\text{a.s.}}{\to} 0$, but generally not otherwise. This fact has been exploited by Wu [30], Hausman [15], White [26], and others, to obtain misspecification tests in various contexts. The statistic is computed as

$$n(\hat{\beta}_n - \tilde{\beta}_n)'[\widehat{\text{avar}}\ \tilde{\beta}_n - \widehat{\text{avar}}\ \hat{\beta}_n]^{-1}(\hat{\beta}_n - \tilde{\beta}_n),$$

where $\widehat{\text{avar}}$ is a consistent estimator of the asymptotic covariance matrix indicated; this statistic has the χ_k^2 distribution asymptotically.

There are numerous tests for potential ways in which **A4** can be violated. If $E(\epsilon_t^2) \neq \sigma_0^2$ for some t, one has heteroscedasticity; if $E(\epsilon_t\epsilon_\tau) \neq 0$ for some $t \neq \tau$, one has serial correlation*. Tests for heteroscedasticity* have been given by Goldfeld and Quandt [13], Glejser [10], Breusch and Pagan [1], Godfrey [11], and White [24], among others.

The most widely used test for serial correlation is provided by the Durbin–Watson test* statistic,

$$d = \sum_{t=2}^{n} (\hat{\epsilon}_t - \hat{\epsilon}_{t-1})^2 / \sum_{t=1}^{n} \hat{\epsilon}_t^2 .$$

The distribution of d is tabulated in Durbin and Watson [8]. A modified version, suitable for use when X_t contains lagged values of y_t is given by Durbin [7]. Other tests have been given by Breusch and Pagan [2], Godfrey [12], and Harvey and Phillips [14]. A general approach that is particularly useful in testing for violations of **A4** is the application of the Lagrange multiplier test* is discussed by Engle [9].

Assumption **A5** is violated if **A1**, **A3**, or **A4** are invalid, generally speaking. Direct tests of the hypothesis that $V = \sigma_0^2 M_{xx}$ have been given by White [25–27] and White and Domowitz [28]. Given a consistent estimator of V, say $\hat{V}_n$, these tests are based on $\hat{V}_n - \hat{\sigma}_n^2(X'X/n)$. Computation is particularly simple when the ϵ_t are independent: the test statistic is simply n times the R^2 from the regression of $\hat{\epsilon}_t^2$ on a constant and the cross-products of the regressors, $X_{it}X_{jt}$, $i, j = 1, \ldots, k$. This statistic has the $\chi^2_{k(k+1)/2}$ distribution asymptotically. Rejection of the null hypothesis (i.e., $V \neq \sigma_0^2 M_{xx}$) implies that **A1**, **A3**, and/or **A4** are violated, so that the previously discussed tests for misspecification may be helpful in isolating the difficulty. Acceptance of the null hypothesis lends support to the validity of **B1**–**B4** and suggests that standard hypothesis testing techniques are legitimate.

Departures from normality* causing the violation of **A6** (e.g., skewness or kurtosis*) can be tested for by applying standard tests for normality to the estimated residuals $\hat{\epsilon}_t$, as shown by Huang and Bolch [16], White and MacDonald [29], and Pierce and Kopecky [20]. Particularly convenient are the skewness and kurtosis measures

$$\sqrt{\hat{b}_1} = n^{-1} \sum_{t=1}^{n} \hat{\epsilon}_t^3 / (\tilde{\sigma}_n^2)^{3/2},$$

$$\hat{b}_2 = n^{-1} \sum_{t=1}^{n} \hat{\epsilon}_t^4 / (\tilde{\sigma}_n^2)^2,$$

where $\tilde{\sigma}_n^2 = n^{-1}\sum_{t=1}^{n} \hat{\epsilon}_t^2$. Critical values for these statistics are given by Pearson and Hartley [18, pp. 207–208].

Although the present discussion has been in terms of the linear model, similar results and techniques apply to nonlinear models, instrumental variables estimation* or general maximum likelihood estimation* (Burguete et al. [3] and White [26, 27]).

References

[1] Breusch, T. S. and Pagan, A. R. (1979). *Econometrica*, **47**, 1287–1294.

[2] Breusch, T. S. and Pagan, A. R. (1980). *Rev. Econ. Studies.*, **47**, 239–253.

[3] Burguete, J., Gallant, A., and Souza, G. (1982). *Econometric Rev.*, **1**, 151–190.

[4] Cox, D. R. (1961). In *Proc. 4th Berkeley Symp. Math. Stat. Probab.*, University of California Press, Berkeley, pp. 105–123.

[5] Cox, D. R. (1962). *J. R. Statist. Soc. B*, **24**, 406–424.

[6] Davidson, R. and MacKinnon, J. (1981). *Econometrica*, **49**, 781–793.

[7] Durbin, J. (1970). *Econometrica*, **38**, 410–421.

[8] Durbin, J. and Watson, G. S. (1951), *Biometrika*, **38**, 159–178.

[9] Engle, R. (1982). In *Handbook of Econometrics*, Vol. 2, Z. Griliches and M. Intrilligator, eds. North-Holland, Amsterdam.

[10] Glejser, H. (1969). *J. Amer. Statist. Ass.*, **64**, 316–323.

[11] Godfrey, L. G. (1978a). *J. Econometrics*, **8**, 227–236.

[12] Godfrey, L. G. (1978b). *Econometrica*, **46**, 1293–1302.

[13] Goldfeld, S. M., and Quandt, R. E. (1965). *J. Amer. Statist. Ass.* **60**, 539–559.

[14] Harvey, A. C. and Phillips, G. D. A. (1980). *Econometrica*, **48**, 747–759.

[15] Hausman, J. A. (1978). *Econometrica*, **46**, 1251–1272.

[16] Huang, C. J. and Bolch, B. W. (1974). *J. Amer. Statist. Ass.*, **69**, 330–335.

[17] Lemieux, P. (1978). *Amer. J. Pol. Sci.*, **22**, 183–186.

[18] Pearson, E. S. and Hartley, H. O. (1972). *Biometrika Tables for Statisticians*, Vol. 2. Cambridge University Press, Cambridge, England.

[19] Pesaran, M. H. and Deaton, A. (1978). *Econometrica*, **46**, 677–694.

[20] Pierce, D. A. and Kopecky, K. J. (1979). *Biometrika*, **66**, 1–5.

[21] Ramsey, J. B. (1969). *J. R. Statist. Soc. Ser. B*, **31**, 350–371.

[22] Theil, H. (1957). *Rev. Int. Statist. Inst.*, **25**, 41–51.

[23] White, H. (1980a). *Int. Econ. Rev.*, **21**, 149–170.

[24] White, H. (1980b). *Econometrica*, **48**, 817–838.

[25] White, H. (1981). *J. Amer. Statist. Assoc.*, **76**, 419–433.

[26] White, H. (1982a). *Econometrica*, **50**, 1–26.

[27] White, H. (1982b). *Econometrica*, **50**, 483–500.

[28] White, H. and Domowitz, I. (1984). *Econometrica*, **51**, 143–162.

[29] White, H. and MacDonald, G. (1980). *J. Amer. Statist. Ass.*, **75**, 16–27.

[30] Wu, D. (1973). *Econometrica*, **41**, 733–750.

(ECONOMETRICS
HETEROSCEDASTICITY
INSTRUMENTAL VARIABLES
ESTIMATION
LEAST SQUARES
LINEAR REGRESSION
MODEL CONSTRUCTION:
SELECTION OF DISTRIBUTIONS
MULTICOLLINEARITY
ROBUST ESTIMATION
SERIAL CORRELATION)

HALBERT WHITE

MITSCHERLICH'S LAW

Mitscherlich's law is represented by the curve

$$Y = A - B\rho^x,$$

$$A > 0, \quad B > 0, \quad 0 < \rho < 1.$$

This curve asymptotically approaches the maximum value A as $x \to \infty$. In agricultural applications Mitscherlich's law represents the relation between the yield Y of a crop (grown in pots) and the amount of fertilizer X added to the soil. The terms fixed order reaction curve or asymptotic regression are also used to denote this relation in various sciences.

Bibliography

Mitscherlich, E. A. (1909). *Landwirtsch. Jahrb.*, **38**, 537–.

Snedecor, G. W. and Cochran, W. G. (1967). *Statistical Methods*, 6th ed. Iowa University Press, Ames.

(LOGISTIC CURVE
NONLINEAR REGRESSION)

MIXED MODEL *See* FIXED-, RANDOM-, AND MIXED-EFFECTS MODELS; MODELS I, II, AND III

MIXED MOMENT *See* PRODUCT MOMENT

MIXED POISSON PROCESS

A mixed Poisson process (MPP) $\{N_t;\ t \geqslant 0\}$ is a birth process (*see* BIRTH-AND-DEATH PROCESSES) with state space $\mathscr{N}_0$, possessing a counting distribution $P_n(t) = \Pr(N_t = n)$ of the form

$$P_n(t) = \int_0^\infty e^{-\lambda t}(\lambda t)^n (n!)^{-1}\, dU(\lambda),$$

where $U(\lambda)$ is a distribution function with $U(0) = 0$. More generally, the MPP is obtained from the homogeneous Poisson process* (HPP) with parameter λ by assuming that the Poisson parameter (the mean rate of points occurring per time unit) is itself a random variable Λ concentrated on $(0, \infty)$ with distribution function $U(\lambda)$. U is called the mixing (weighting, compounding) distribution of the process. In a Bayesian or empirical Bayesian approach, U can be interpreted to be the a priori distribution of the Poisson parameter λ. Conditionally on $\Lambda = \lambda$, the MPP is a HPP; the same is true when Λ is concentrated in one point $\lambda > 0$. The MPP belongs to the class of doubly stochastic Poisson processes* [12, p. 31].

In insurance mathematics [7, pp. 63–71] and in the context of accident theory [15, p. 104], the MPP is used to model the "accident proneness" of the members of a collective of risks. The individual risk parameters λ are independent outcomes of a random experiment characterized by the distribution function $U(\lambda)$. U, the "structure function", describes the risk structure of the considered collective, i.e., the potential variation of the individual risk parameters. In reliability* theory, a similar interpretation is used [23, p. 287].

The counting distribution of the MPP is a mixed Poisson distribution; however, this property does not characterize the MPP. In fact [18, p. 123], there are point processes* with this property that are not even Markov processes*. The terminology is not unique in literature; some authors speak of a "compound Poisson process" [10, 20], some of "weighted Poisson process" [7, 12]. Other ways of defining the MPP are given in Bühlmann [7, pp. 65–71] and Grandell [12, p. 31].

PROBABILISTIC PROPERTIES

We have $E(N_t) = tE(\Lambda)$, $\mathrm{Var}(N_t) = t^2\mathrm{Var}(\Lambda) + tE(\Lambda)$, and in general, for the kth factorial moment [3, p. 2]

$$E[N_t(N_t - 1)\ldots(N_t - k + 1)] = t^k E[\Lambda^k].$$

The MPP is always "overdispersed" compared with the corresponding HPP with the same mean value; a similar property for $P_0(t)$ is given in Feller [10, p. 392]. The transition probability function of the MPP is given by

$$\begin{aligned} P(N(t) = n \mid N(s) = m) &\\ = \binom{n}{m}\left(\frac{s}{t}\right)^m &\left(1 - \frac{s}{t}\right)^{n-m} \frac{P_n(t)}{P_m(s)} \end{aligned}$$

and the transition density functions $q_{mn}(t)$ by

$$\begin{aligned} q_{m,m+1}(t) &= -P_0^{(m+1)}(t)/P_0^{(m)}(t) \\ &= -q_{mm}(t), \qquad m \geqslant 0, \\ q_{mn}(t) &= 0, \qquad n < m,\ n > m + 1. \end{aligned}$$

Conversely, if we have a completely monotonic function $H(t)$ [i.e., $(-t)^n H^{(n)}(t) \geqslant 0$ for all $n \in \mathscr{N}_0$, $t > 0$] with $\lim_{h \to 0} H(t) = 1$, then [20, p. 73] the birth process with the intensities

$$\begin{aligned} q_{m,m+1}(t) &= -q_{mm}(t) \\ &= -H^{(m+1)}(t)/H^{(m)}(t) \end{aligned}$$

is a MPP. This is a way of introducing the MPP without making explicit reference to a mixing distribution. For example, the Pólya process to be introduced in the following section can also be defined by the birth process with the intensities $q_{m,m+1}(t) = (m + a)/(t + b)$. Various interesting properties of the intensities of a MPP are given in Lundberg [20, pp. 76, 77].

The MPP [2, 22] is a process with stationary increments*; moreover, it is a stationary point process*, but not a stationary (homogeneous) Markov chain—with the exception of the HPP. The increments are *not* independent but are exchangeable [11, p. 29]. Let $0 = T_0 < T_1 < \cdots < T_n < \cdots$ denote the sequence of occurrence times and $\{W_i = T_{i+1} - T_i;\ i \geqslant 1\}$ the sequence of interoccurrence times. One has [4, p. 250] for the joint distribution of occurrence times

$$f_{T_1 \dots T_n}(t_1, \dots, t_n) = (-1)^n P_0^{(n)}(t_n) = n!\, P_n(t_n)/t_n^n, \quad 0 < t_1 < \cdots < t_n;$$

in particular one has

$$f_{T_n}(x) = nP_n(x)/x, \qquad x > 0.$$

The joint density of the interoccurrence times is given [4, p. 254; 22, p. 287] by

$$f_{W_1 \dots W_n}(x_1, \dots, x_n) = (-1)^n P_0^{(n)}(x_1 + \cdots + x_n), \quad x_i > 0;$$

the interoccurrence times are identically distributed with density

$$f_{W_n}(x) = -P_0'(x) = \frac{P_1(x)}{x}, \qquad x > 0;$$

they are *not* independent (but are exchangeable; *see* EXCHANGEABILITY). That is, the MPP is not a renewal process*—with the exception of the HPP.

Further probabilistic properties of the MPP are given in Haight [15, pp. 36–38, 80], Jung and Lundberg [18, p. 122–126], and McFadden [22]. Characterizations of the MPP that can be used for constructing goodness-of-fit* tests (see the last section) are

1. *Lundberg's "Binomial Criterion"* [20, p. 89]: Every birth process possessing the "inversely conditioned probability function"

$$P(N(s) = m \mid N(t) = n) = \binom{n}{m}\left(\frac{s}{t}\right)^m \left(1 - \frac{s}{t}\right)^{n-m} \quad (s < t;\ n \geqslant m)$$

must be a MPP.

2. Every birth process with

$$f_{T_1 \dots T_n \mid N(T) = n}(t_1, \dots, t_n) = n!/T^n$$

or with

$$f_{T_1 \dots T_n \mid T_{n+1} = T}(t_1, \dots, t_n) = n!/T^n, \quad 0 < t_1 < \cdots < t_n < T$$

must be a MPP [5]. This means that the corresponding distribution is the same as that of the order statistics* corresponding to n independent random variables uniformly distributed over $[0, T]$.

Other characterizations are given in Albrecht [2, p. 245], Gerber [11, p. 29], and McFadden [22, p. 90].

SPECIAL MIXED POISSON PROCESSES

Special types of MPP are obtained by considering special distributional types of the mixing distribution U. The most well-known is the Pólya process, obtained by assuming that U is a gamma-distribution*; the resulting counting distribution is the negative binomial*. The properties of the Pólya process are reviewed in Albrecht [1, pp. 186–189], Haight [15, pp. 39–40], Jung and Lundberg [18, p. 128], and Lundberg [20, pp. 96–100]. A very general class ("finite mixtures of Poisson processes") is obtained by taking U as a step function with finitely many steps; for details, see Albrecht [1, pp. 190–191]. A special case is the *double Poisson process*: U is a two-point distribution. For details, see Albrecht [1, p. 191], Haight [15, p. 41], and Rider [25]. In Johnson and Kotz [17], the binomial* (p. 187) and the hypergeometric distributions* (p. 188) are considered mixing distributions. A more general class (*discrete mixtures of Poisson processes*) is obtained by taking U as a step function with countable steps. The most important case is when U itself is a Poisson distribution, the resulting counting distribution being Neyman type-A* (cf. Albrecht [1, pp. 192–193]). Other discrete mixing distributions are the logarithmic series* ([17, pp. 187 and 188]) and the negative binomial ([17, p. 187]) distributions. In the class of continuous distributions, the rectangular* [1, p. 195; 6; 17, pp. 184–185], the truncated normal [1, pp. 195–196; 17, p. 185], the truncated gamma [1, pp. 196–197; 19] and the lognormal* distributions [17, pp. 185–186] have been considered. A very general class of MPP is obtained by assuming that U is a member of the Pearson family* of distributions. Expressions for the corresponding counting distributions have been developed by Philipson [24] (see also Albrecht [1, pp. 198–201] and Haight [15, pp. 41–42]).

STATISTICAL ANALYSIS

We begin with results for *parameter estimation*. One possibility is to estimate the parameters of the counting distribution corresponding to a certain type of mixing distribution. Here, especially for the negative binomial and the Neyman type-A distribution, well-developed estimation methods exist that are reviewed in Albrecht [1, pp. 368–389] and Johnson and Kotz [17, pp. 131–135, 222–226]. Further results for other mixing distributions are reviewed in Albrecht [1, pp. 389–393]. Hasselblad [16] develops an algorithm for maximum likelihood estimation of the parameters (unknown jump-points, unknown jump-heights of U) of an arbitrary finite mixture of Poisson processes. Further results in this connection are reviewed in Albrecht [1, pp. 393–404; 3, pp. 2–8]. Another possibility is to estimate the parameters of the mixing distribution U or U itself, based on observations of N_t. In the empirical Bayes' context, this means to estimate the a priori distribution (cf. Maritz [21, Chap. 2]). As mixtures of Poisson distributions are *identifiable* mixtures, i.e.,

$$\int F(x \mid y)\, dG(y) \equiv \int F(x \mid y)\, dG^*(y)$$

$$\Rightarrow G = G^*,$$

(cf. Maritz [21, pp. 30–31], this way of proceeding is sound. A very rich class of mixing distributions is the family of Pearson distributions; details of the estimation problem in this connection are given in Albrecht [1, pp. 404–414; 3, pp. 8–10]. Results for nonparametric estimation of U are reviewed in Albrecht [1, pp. 414–418] and are due to De Vylder [9, pp. 69–77] and Tucker [28]; the most general results have been obtained by Simar [27], who considers maximum likelihood estimation.

We now come to the problem of testing goodness of fit*. Two cases have to be distinguished. Either one tests whether the data are adequately described by a MPP, or one assumes this, and tests whether a certain parametric form of the counting distribution fits the data well. For the first problem, the results of Lundberg [20, pp. 147–155] and Albrecht [5] are available, based on the characterizations of the MPP mentioned in the first section. For the second problem, the usual goodness-of-fit tests can be applied; some details are reviewed in Albrecht [1, pp. 421–425]. A few more statistical results in connection with the MPP can be found in the literature. Grenander [14, pp. 78–81] develops a procedure for testing whether the MPP is a HPP, and De Oliveira [8] develops a procedure for testing whether the double Poisson process is a HPP. The problem of examining the hypothesis that the mixing distribution is stable in time is treated by Seal [26, p. 25], Jung and Lundberg [18, pp. 126–127], and Lundberg [20, pp. 142–144].

References

[1] Albrecht, P. (1981). *Dynamische Statistische Entscheidungsverfahren für Schadenzahlprozesse*. Verlag Versicherungswirtschaft, Karlsruhe, Germany. (Reviews in detail the probabilistic features of the homogeneous, inhomogeneous, and mixed Poisson process and the statistical methods available for examining these processes.)

[2] Albrecht, P. (1981). *Mitt. Ver. Schweiz. Versicherungsmath.*, 241–249.

[3] Albrecht, P. (1982). *Scand. Actuarial J.*, 1–14.

[4] Albrecht, P. (1982). *Blä. Dtsch. Ges. Versicherungsmath.*, **15**, 249–257. (Examines properties of the occurrence and interoccurrence times of the MPP.)

[5] Albrecht, P. (1982). *Insurance: Mathematics and Economics*, **1**, 27–33. (Considers the problem of testing the goodness of fit of a MPP.)

[6] Bhattacharya, S. K. and Holla, M. S. (1965). *J. Amer. Statist. Ass.*, **60**, 1060–1066.

[7] Bühlmann, H. (1970). *Mathematical Methods in Risk Theory*. Springer, Berlin. (Considers mixtures of processes to describe the "risk in the collective"; already the classical book on risk theory.)

[8] De Oliveira, J. T. (1965). In *Classical and Contagious Discrete Distributions*, G. P. Patil, ed. Calcutta, pp. 379–384.

[9] De Vylder, F. (1975). *Introduction aux Théories Actuarielles de Credibilité*. Office des Assureurs de Belgique, Louvain, Belgium.

[10] Feller, W. (1943). *Ann. Math. Statist.*, **14**, 389–400.

[11] Gerber, H. U. (1981). *An Introduction to Mathematical Risk Theory*. Hübner Foundation, University of Pennsylvania, Philadelphia.

[12] Grandell, J. (1976). "Doubly Stochastic Poisson Processes." *Lect. Notes Math.*, **529**. (Treats the MPP as a special case of the doubly stochastic Poisson process.)

[13] Greenwood, M. and Yule, G. U. (1920). *J. R. Statist. Soc.*, **83**, 255–279. (The classical work on accident proneness and the Pólya distribution.)

[14] Grenander, U. (1957). *Scand. Actuarial J.*, **40**, 71–84. (Develops various estimation and testing techniques in connection with the MPP.)

[15] Haight, F. A. (1967). *Handbook of the Poisson Distribution*. Wiley, New York. (Also gives an account of the MPP.)

[16] Hasselblad, V. (1969). *J. Amer. Statist. Ass.*, **64**, 1459–1471. (Treats maximum likelihood estimation for finite mixtures of MPP.)

[17] Johnson, N. L. and Kotz, S. (1969). *Distributions in Statistics: Discrete Distributions*. Wiley, New York. (Also reviews various mixtures of Poisson distributions.)

[18] Jung, J. and Lundberg, O. (1969). *Scand. Actuarial J.*, **52**, *Suppl.*, 118–131.

[19] Kemp, A. W. (1968). *Scand. Actuarial J.*, **51**, 198–203.

[20] Lundberg, O. (1940). *On Random Processes and their Application to Sickness and Accident Statistics*. Almqvist & Wiksells, Uppsala (rpt. 1964). (The classical work on the MPP. A must.)

[21] Maritz, J. S. (1970). *Empirical Bayes Methods*. Methuen, London.

[22] McFadden, J. A. (1965). *Sankhyā A*, **27**, 83–92. (Develops various probabilistic properties of the MPP.)

[23] McNolty, F. (1964). *Sankhyā A*, **26**, 287–292.

[24] Philipson, C. (1960). *Scand. Actuarial J.*, **43**, 136–162. (Develops expressions for the counting distribution of the MPP when the mixing distribution is a member of the Pearson family of distributions. Difficult to read.)

[25] Rider, P. (1961). *Bull. Inst. Int. Statist.*, **39**, 225–232.

[26] Seal, H. L. (1969). *Stochastic Theory of a Risk Business*. Wiley, New York.

[27] Simar, L. (1976). *Ann. Statist.*, **4**, 1200–1209. (Considers maximum likelihood estimation of the mixing distribution.)

[28] Tucker, H. G. (1963). *Theory of Probability and Its Applications*, **8**, 195–200.

(BIRTH-AND-DEATH PROCESSES
MIXTURE DISTRIBUTIONS
POISSON DISTRIBUTION
POISSON PROCESS
PROCESSES, DISCRETE)

PETER ALBRECHT

MIXING *See* ERGODIC THEOREMS

MIXTURE DISTRIBUTIONS

Consider taking a random sample of people living in New York and recording for each member of the sample their height. What might be a sensible model for the distribution of this variable in the population? First, our sample is made up of males and females, and it is well known that, on average, males are taller than females. Second, it would probably be reasonable to assume that for males and females taken separately height is normally distributed. Such considerations lead naturally to the following density function for height:

$$f(\text{height}) = pN(\mu_F, \sigma_F) + (1-p)N(\mu_M, \sigma_M), \quad (1)$$

where p is the proportion of females in the population, and (μ_F, σ_F) and (μ_M, σ_m) are, respectively, the mean and standard deviation of height for females and males. This is an example from a class of density functions known as *mixtures*, the general definition of which will be given in the next section. For this example, the major concern would be to use the sample of recorded heights to estimate the five parameters of the density function: p, μ_F, σ_F, μ_M, σ_M. (Of course, if we had been sensible enough to record the sex of each member of the sample, estimation of these quantities would have been straightforward; here this could have been done very simply, even allowing for the eccentricities of dress in New York! In other areas, however, sexing of species is more difficult and estimation of the parameters must be made from the unlabeled sample. Examples will be described in the section "Commonly Occurring Mixture Distributions.")

MIXTURE DENSITIES: DEFINITIONS

In (1) the density function of height has been expressed as the sum of two conditional density functions, the height density function of women and that of men. Such superpositions of simpler component densities are usually termed *mixture* or *compound* density functions and have the general form

$$f(x) = \int g(x;\theta) \cdot h(\theta)\, d\theta, \qquad (2)$$

where $g(x;\theta)$ is a conditional density function depending on the parameter θ, itself subject to chance variation described by the density function $h(\theta)$, the *mixing density*. (This definition is easily extended to include both densities dependent on multidimensional parameter vectors and multidimensional density functions; see Everitt and Hand [16] for details.)

Cases such as the density function in (1) arise from the general definition (2) when $h(\theta)$ is discrete and assigns positive probability to only a finite number of parameter values, $\theta_1, \ldots, \theta_c$. In such cases the integral is replaced by a sum to give a *finite mixture* density function of the form

$$f(x) = \sum_{i=1}^{c} g(x;\theta_i)h(\theta_i). \qquad (3)$$

The terms $h(\theta_i)$ are now the *mixing proportions* and are generally denoted by p_i, subject to the constraint $\sum_{i=1}^{c} p_i = 1$.

MIXTURE DENSITIES: APPLICATIONS

Mixture density functions are important in a wide range of applications. For example, in biology it is often required to measure certain characteristics in natural populations of a particular species; samples of individuals are taken from the natural habitat of the species and the characteristic under investigation is recorded for each individual in the sample. Now the distribution of many such characteristics may vary greatly with the age of the individuals, and age is frequently difficult to ascertain in samples from wild populations; consequently the biologist observing the population as a whole is dealing with a mixture distribution, the mixing being over a parameter depending on the unobservable variate age. Such a situation arises in particular in the investigation of the distribution of length of fish, discussed in detail by Cassie [6] and Bhattacharya [3]. The same problem arises in the not uncommon event that the biologist is unable to determine the sex of an individual.

In another general area where mixtures of distributions are important, the observations are times to failure of a sample of items. Often failure can occur for more than one reason (see, e.g., Davies [10]), and the failure time distribution for each reason can be adequately approximated by a simple density function such as the negative exponential. The overall failure distribution is then a mixture. Several attempts have been made to fit such mixtures to the failure times of electronic values. For example, Mendenhall and Hader [36] fit mixtures with exponential components and Kao [31] mixtures with Weibull components.

Clark et al. [8] illustrate an area in which mixture distributions are being applied more frequently—the study of disease distributions. The general question is: Is there more than one type of disease here? Their investigation involved patients with hypertension, and the main interest was in determining whether or not there were different types of hypertension. Part of the study involved assessing if a frequency curve of the form given by (1) provided an adequate fit for a sample of blood-pressure data. If so, it would provide evidence in favor of regarding hypertension as more than a single condition. On the whole, the results were inconclusive, although the mixture density of (1) did provide a reasonable fit for diastolic blood pressures.

COMMONLY OCCURRING MIXTURE DISTRIBUTIONS

Finite Mixtures

The frequency curve specified by (1), involving a mixture of two normal components, is perhaps the most commonly used mixture distribution. It was first studied by Karl Pearson* [38], who applied the method of moments* (see the next section) to the problem of estimating its five parameters. More recently mixtures of both univariate and multivariate normals have been considered by Hasselblad [23], Day [11], Wolfe [48], Duda and Hart [14], Hosmer [25], Quandt and Ramsey [39], and Fowlkes [17]. An example of the mixture density given by (1) with $p = 0.4$, $\mu_1 = 0.0$, $\sigma_1 = 1.0$, $\mu_2 = 2.0$, and $\sigma_2 = 0.5$ is given in Fig. 1. Similarly, Fig. 2 is an example of a three-component bivariate normal mixture (BVN) i.e., a density function of the form

$$f(\mathbf{x}) = \sum_{i=1}^{3} p_i \mathrm{BVN}(\mu_1^{(i)}, \mu_2^{(i)}, \sigma_1^{(i)}, \sigma_2^{(i)}, \rho^{(i)}), \tag{4}$$

where

$$\mathrm{BVN}(\mu_1, \mu_2, \sigma_1, \sigma_2, \rho) = \left(2\pi\sqrt{1-\rho^2}\,\sigma_1\sigma_2\right)^{-1} \times \exp\left\{-\frac{1}{2(1-\rho^2)}\left[\left(\frac{x_1-\mu_1}{\sigma_1}\right)^2 - 2\rho\left(\frac{x_1-\mu_1}{\sigma_1}\right)\left(\frac{x_2-\mu_2}{\sigma_2}\right) + \left(\frac{x_2-\mu_2}{\sigma_2}\right)^2\right]\right\}. \tag{5}$$

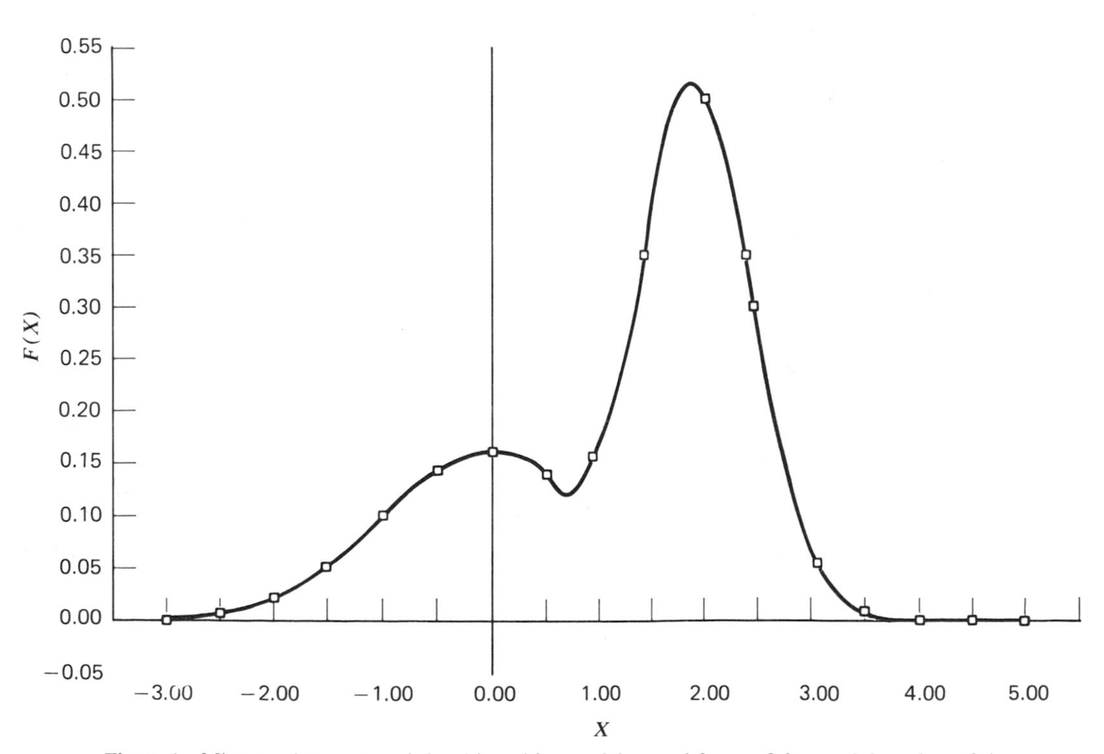

Figure 1 Mixture of two normal densities with $\mu_1 = 0.0$, $\sigma_1 = 1.0$, $\mu_2 = 2.0$, $\sigma_2 = 0.5$, and $p = 0.4$.

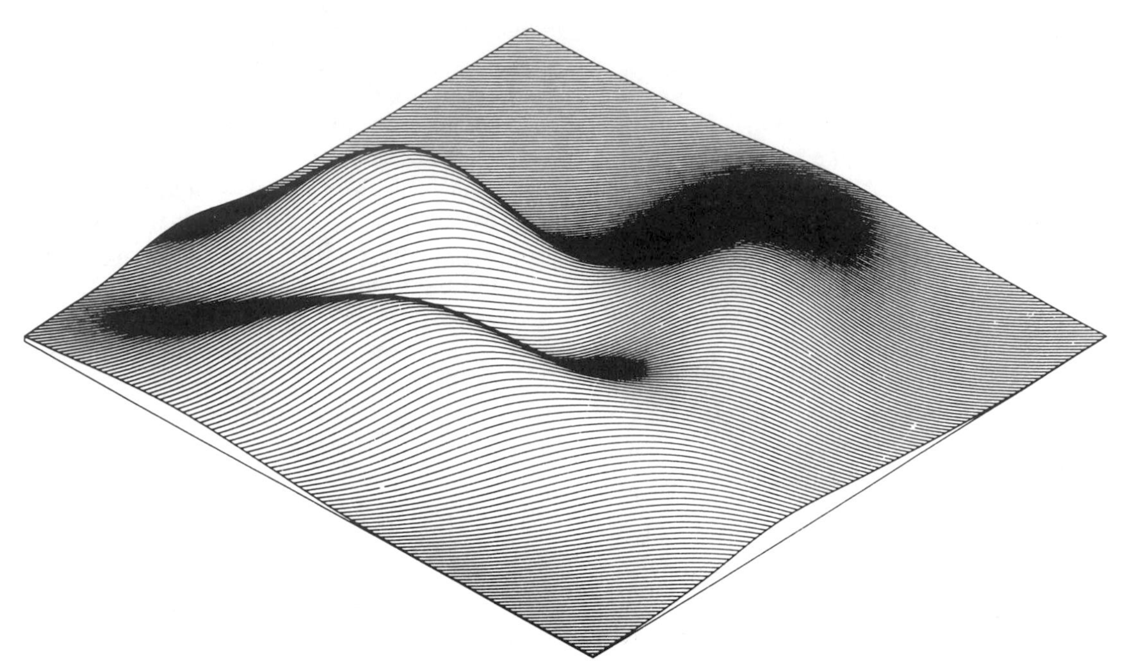

Figure 2 Mixture of three bivariate normal densities with parameter values as specified in the text.

In Fig. 2, the parameter values are as follows:

$$\mu_1^{(1)} = 1.0, \quad \mu_1^{(2)} = 3.0, \quad \mu_1^{(3)} = 5.0,$$
$$\sigma_1^{(1)} = 1.5, \quad \sigma_1^{(2)} = 2.0, \quad \sigma_1^{(3)} = 1.5,$$
$$\mu_2^{(1)} = 2.0, \quad \mu_2^{(2)} = 0.0, \quad \mu_2^{(3)} = 4.0,$$
$$\sigma_2^{(1)} = 1.0, \quad \sigma_2^{(2)} = 1.0, \quad \sigma_2^{(3)} = 2.0,$$
$$\rho^{(1)} = 0.5, \quad \rho^{(2)} = 0.6, \quad \rho^{(3)} = -0.5,$$
$$p_1 = 0.3, \quad p_2 = 0.3$$

A further interesting finite mixture distribution is the basis of *latent class analysis*. The basic idea (first suggested by Green [19]) is that the observed associations between a set of d dichotomous variables are generated by the presence of different "latent" classes within which the variables are independent. This model may be translated into mixture terms by supposing that the random vector $\mathbf{x}$ of dichotomous variables has a probability density function of the form

$$p(\mathbf{x}) = \sum_{j=1}^{c} p_j P_j(\mathbf{x}; \boldsymbol{\theta}_j), \tag{6}$$

where

$$P_j(\mathbf{x}; \boldsymbol{\theta}_j) = \prod_{l=1}^{d} \theta_{jl}^{x_l} (1 - \theta_{jl})^{1 - x_l}; \tag{7}$$

$x_1, \ldots, x_d$ are the elements of $\mathbf{x}$, taking values zero or one, and $\theta_{j1}, \ldots, \theta_{jd}$, are the elements of $\boldsymbol{\theta}_j$ which give the probability of variables in class j taking the value one; c is the number of latent classes.

Other examples of finite mixture distributions are given in Everitt and Hand [16].

Other Mixture Distributions

In a model of 'accident-proneness' suggested by Greenwood and Yule [20], the number of accidents for a particular individual is assumed to follow a Poisson distribution* with parameter λ, itself assumed to vary from individual to individual according to a gamma distribution* with parameters α and β. So we have

$$f(\text{number of accidents} \mid \lambda) = \frac{e^{-\lambda}\lambda^n}{n!}, \tag{8}$$

$$h(\lambda) = \beta^{\alpha}[\Gamma(\alpha)]^{-1}\lambda^{\alpha-1}\exp(-\lambda \mid \beta), \quad \lambda > 0, \quad \alpha > 0, \quad \beta > 0. \tag{9}$$

Consequently, the unconditional density function of the number of accidents is given by the mixture density

g(number of accidents)

$$= \int_0^\infty \frac{e^{-\lambda}\lambda^n}{n!} \beta^\alpha [\Gamma(\alpha)]^{-1} \lambda^{\alpha-1} e^{-\lambda/\beta} \, d\lambda \tag{10}$$

$$= \binom{\alpha+n-1}{\alpha-1} \left(\frac{\beta}{\beta+1}\right)^n \left(\frac{1}{\beta+1}\right)^a, \tag{11}$$

which is a negative binomial* density with parameters α and β.

The distribution of the number of items observed to be defective in samples from a finite population when detection of defectiveness is not certain is considered by Johnson et al. [30]. A sample of n items is chosen at random from a population of N items and examined for errors. Inspection is less than perfect, and the probability of detecting an error given that an item is defective is p. Assuming that X of the items in the population are defective and that no nondefective items are classified as defective, Johnson et al. consider the probability that z of the sampled items are pronounced defective. Since the probability of choosing y defective items is the hypergeometric* probability

$$p(y) = \binom{X}{y}\binom{N-X}{n-y} \bigg/ \binom{N}{n}, \tag{12}$$

$$\max(0, n-N+X) \leqslant y \leqslant \min(n, X),$$

and the probability of detecting z of these y items as defective is

$$p(z \mid y) = \binom{y}{z} p^z (1-p)^{y-z},$$

$$z = 0, 1, \ldots, y, \tag{13}$$

the unconditional distribution of z is given by

$$P(z) = \sum_{y \geqslant z} \binom{X}{y}\binom{N-X}{n-y}\binom{y}{z} p^z \times (1-p)^{y-z} \bigg/ \binom{N}{n}. \tag{14}$$

Extensions of this distribution are considered in Kotz and Johnson [32].

A further mixture distribution arises from an alternative model of the variation in the number of defective items found in routine sampling inspection. Here the number of defectives found in any inspection is assumed to follow a binomial distribution with probability p of a defective. The latter, however, is assumed to vary from inspection to inspection according to some density function. The most common assumption for the mixing density is a beta distribution, and this leads to the *Pólya–Eggenberger* density for the number of defective items [see (28) and (29)].

Mixtures where both g and h in (2) are continuous density functions are perhaps less important in practice than those already considered. Nevertheless, Johnson and Kotz [29] consider a variety of such mixtures, many of which lead to fascinating and complex density functions. For example, the Laplace distribution* has the form

$$g(x; \theta, \phi) = \tfrac{1}{2}\phi^{-1}\exp[-|x-\theta|/\phi]. \tag{15}$$

Allowing the parameter θ to be normally distributed with mean μ and variance σ^2 gives

$$f(x; \phi) = \int_{-\infty}^{\infty} \frac{1}{2}\phi^{-1}\exp[-|x-\theta|/\phi] \cdot \frac{1}{\sqrt{2\pi}\,\sigma}\exp\left[-\frac{1}{2}\left(\frac{\theta-\mu}{\sigma}\right)^2\right] d\theta. \tag{16}$$

By integrating from $-\infty$ to x and from x to ∞ and replacing $|x-\theta|$ by $(\theta - x)$ and $(x-\theta)$, (16) may be reduced to

$$f(x; \phi) = \frac{1}{2\phi}\left\{\exp\left(\frac{1}{2}\frac{\sigma^2}{\phi^2}\right)\right\} \times \left[\Phi\left(\frac{x-\mu}{\sigma} - \frac{\sigma}{\phi}\right)\exp\left(-\frac{x-\mu}{\sigma}\right) + \Phi\left(-\frac{x-\mu}{\sigma} - \frac{\sigma}{\phi}\right)\exp\left(\frac{x-\mu}{\sigma}\right)\right], \tag{17}$$

where

$$\Phi(x) = \int_{-\infty}^{x} \frac{1}{\sqrt{2\pi}} \exp\left(-\frac{1}{2}u^2\right) du.$$

Further examples arising when g and h are continuous are given by Romanowski [40], Shah [42], and Holla and Bhattacharya [24].

IDENTIFIABILITY

Of major concern in the study of mixture distributions is the estimation of their parameters. However, before estimation can be undertaken, we have to consider the identifiability* or unique characterization of the mixture. A mixture is identifiable if there exists a one-to-one correspondence between the mixing distribution and the resulting mixture. Mixtures that are not identifiable cannot be expressed uniquely as functions of component and mixing distributions. For example, the following two mixtures of univariate, uniform densities are identical.

$$f(x) = \tfrac{1}{3}U(-1,1) + \tfrac{2}{3}U(-2,2), \qquad (18)$$

$$f(x) = \tfrac{1}{2}U(-2,1) + \tfrac{1}{2}U(-1,2), \qquad (19)$$

where $U(a,b)$ is a uniform density with range (a,b).

Identifiability is crucial since it is not possible to estimate or test hypotheses about parameters of unidentifiable mixtures. See Teicher [46, 47] and Yakowitz and Spragins [50]; the latter authors present a useful theorem that helps to show which distributions yield identifiable finite mixtures. Other investigations are provided by Tallis [45], Blum and Susarla [5], and Chandra [7].

ESTIMATION

Many methods have been devised and used for estimating the parameters of mixture distributions ranging from the method of moments through formal maximum likelihood approaches to informal graphical techniques.

Method of Moments

If the mixture density under investigation is a function of s parameter values, estimation by the method of moments* involves writing any s population moments (usually the first s) as equations in the s unknown parameters, solving these to obtain expressions for the parameters as functions of the moments, and defining estimators by substituting sample moments for population moments in this solution. For example, consider the following mixture of two Poisson densities

$$f(x) = \frac{\alpha e^{-\lambda_1}\lambda_1^x}{x!} + (1-\alpha)\frac{e^{-\lambda_2}\lambda_2^x}{x!}, \quad (x = 0, 1, \ldots). \qquad (20)$$

The kth factorial moment* of x may be written as

$$\mu_{(k)} = \alpha\lambda_1^k + (1-\alpha)\lambda_2^k = \alpha(\lambda_1^k - \lambda_2^k) + \lambda_2^k. \qquad (21)$$

Corresponding sample moments are defined as

$$v_{(k)} = \sum_{x=k}^{R} x(x-1)\ldots\frac{(x-k+1)n_x}{n}, \qquad (22)$$

where R is the largest observed value of x, n_x is the sample frequency of x, and $n = \sum_{x=0}^{R} n_x$ is the total sample size. ($v_{(1)}$ is, of course, the sample mean $\bar{x}$.) Equating the corresponding sample and theoretical moments leads to the equations

$$\begin{aligned} \bar{x} - \lambda_2 &= \alpha(\lambda_1 - \lambda_2), \\ v_{(2)} - \lambda_2^2 &= \alpha(\lambda_1^2 - \lambda_2^2), \qquad (23) \\ v_{(3)} - \lambda_2^3 &= \alpha(\lambda_1^3 - \lambda_2^3). \end{aligned}$$

Cohen [9] shows that λ_1 and λ_2 are obtained as the roots of the quadratic equation

$$\lambda^2 - \theta\lambda + \Gamma = 0, \quad \text{where} \qquad (24)$$

$$\theta = (v_{(3)} - \bar{x}v_{(2)})/(v_{(2)} - \bar{x}^2) \quad \text{and} \qquad (25)$$

$$\Gamma = \bar{x}\theta - v_{(2)} \qquad (26)$$

Having determined λ_1 and λ_2 from equation (24), the mixing proportion α is found from

$$\alpha = (\bar{x} - \lambda_2)/(\lambda_1 - \lambda_2). \qquad (27)$$

A numerical example is given in Everitt and Hand [16, Chap. 4].

Skellam [41] considers estimation by the method of moments for the two parameters in the beta-binomial mixture described at the end of the preceding section. This density function has the form

$$P(x) = \int_0^1 \binom{n}{x} p^x (1-p)^{n-x} \frac{p^{\alpha-1}(1-p)^{\beta-1}}{B(\alpha, \beta)} dp \tag{28}$$

$$= \binom{n}{x} \frac{\alpha^{[x]}\beta^{[n-x]}}{(\alpha+\beta)^{[n]}}, \qquad x = 0, 1, 2, \ldots, n, \tag{29}$$

where $a^{[b]}$ denotes $a(a+1)\ldots(a+b-1)$.

The moment estimators for α and β are

$$\hat{\alpha} = \frac{m_{(1)}m_{(2)} - (n-1)m_{(1)}^2}{(n-1)m_{(1)}^2 - nm_{(2)}}, \tag{30}$$

$$\hat{\beta} = \hat{\alpha}\frac{n - m_{(1)}}{m_{(1)}}, \tag{31}$$

where $m_{(1)}$ and $m_{(2)}$ are the observed values of the first and second factorial moments about the origin.

Moment estimators were considered for mixture densities as early as 1894 by Karl Pearson, who applied the technique to a mixture of two normal densities [see (1)]. In this case, the solution involves finding the negative real roots of a ninth-degree polynomial. It is of no great difficulty with a present-day computer but this must have presented formidable problems in the latter part of the nineteenth century!

Maximum Likelihood Estimation

The method of maximum likelihood* has a number of desirable statistical properties. For example, under very general conditions the estimators obtained from the technique are consistent (they converge in probability to the true parameter values), and they are asymptotically normal. In the past, however, the method has not been considered particularly suitable for estimating the parameters of a mixture distribution primarily because of the computational difficulties involved. With modern computers this is no longer a serious problem, and maximum likelihood estimation for many mixtures is now fairly routine.

A detailed account is given in Everitt and Hand [16], who show that the likelihood equations for finite mixtures are weighted averages of the likelihood equations arising from each component density in the mixture separately. The weights are the posterior probabilities* of an observation arising from a particular component. Generally, the equations must be solved by some type of iterative procedure, the most useful being the E-M algorithm of Dempster et al. [12]. This algorithm has two steps, the first being to estimate the membership probabilities of each observation for each component, and the second being equivalent to c separate estimation problems with each observation contributing to the log-likelihood associated with the separate components, weighted by the estimated membership probability. These two steps are then repeated iteratively until some convergence criterion is satisfied; for more details, *see* MISSING INFORMATION PRINCIPLE.

This algorithm has proved successful in the maximum likelihood estimation of the parameters in mixtures with a variety of components, for example, multivariate normal (see Wolfe [48]), binomial and Poisson (see Everitt and Hand [16, Chap. 4]), and multivariate Bernoulli, i.e., latent class analysis (see Aitken et al. [1]).

Laird [34] considers the problem of estimating the mixing distribution in (2) without assuming a specific parametric form. Maximim likelihood methods lead to an estimate that is a step function with a finite number of steps under certain conditions. Such "nonparametric" maximum likelihood estimation is also considered by Simaz [44] and Jewell [26].

Error Function Methods

A number of suggested estimation techniques involve the minimization of some error function measuring the difference between some theoretical characteristic of the distribution and the corresponding observed value. For example, Quandt and Ramsey [39] consider the difference between the observed and theoretical moment generating function* as measured by the function

$$F = \sum_{j=1}^{m} \left\{ E(\exp \beta_j x) - n^{-1} \sum_{i=1}^{n} \exp(\beta_j x_i) \right\}^2, \tag{32}$$

where $x_1, \ldots, x_n$ are the observed values of the random variable x. For given values of $\beta_1, \ldots, \beta_m$, F is a function of the parameters of the distribution of x. Minimization of F by some suitable algorithm leads to the required estimates. The method essentially generalizes the method of moments described earlier. Rather than considering a handful of low-order moments, however, the moment generating function implicitly uses all the moments. Quandt and Ramsey give a numerical example of the method applied to the two-component normal mixture specified by equation (1). Results in Kumar et al. [33] suggest that the choice of $\beta_1, \ldots, \beta_m$ is critical. An essentially similar approach is the minimum chi-square* criterion discussed by Fryer and Robertson [18].

Other Methods

Before the advent of computers, many of the proposed procedures for estimating mixture density parameters were unattractive because of the computational load; the result was a number of graphical techniques designed for use without large numbers of calculations. Many of these involved some adaptation of probability plotting*. Examples are the techniques proposed by Cassie [6] and Harding [22]. Such graphical procedures might be useful in some circumstances, but to be effective the approach requires well-separated components and/or large sample sizes.

Medgyessi [35] describes a method of localizing the different components of a finite normal mixture using Fourier transformation methods; a suitable algorithm is provided by Gregor [21]. There appear to have been no attempts to compare this method with statistically more conventional techniques such as moments or maximum likelihood.

DETECTING MIXTURE DISTRIBUTIONS

The decision to fit a mixture distribution to a set of data should, in the majority of cases, be the result of a priori considerations of what might be a suitable model for the process under investigation. In some situations, however, data is collected for which there may be several alternative models, some of which would be supported if the distribution of the data suggested some form of mixture. In such cases, the investigators will be interested in any indications from the sample frequency distribution that they are dealing with a mixture.

Perhaps the most natural statistic to consider when assessing whether a mixture distribution might be appropriate is the number of modes of the sample histogram*, since a common property of mixtures is that under certain circumstances they are multimodal. This criterion, though helpful in some situations, can be misleading. For example, Murphy [37] gives a number of histograms arising from samples of size 50 generated from a *single* normal distribution. Many of the histograms show signs of bimodality and a few even indicate trimodality. Therefore, examination of the sample histogram is unlikely to be conclusive in detecting the presence of a mixture, and we are led to consider other possibilities.

For examining data for the presence of mixtures of normal distributions a variety of techniques based on probability plotting have been suggested. For detecting mixtures of normal distributions, the simplest type of probability plot which might be useful is that of the sample quantiles* against those of a standard normal curve. If a single normal distribution is appropriate for the data, such a plot should be approximately linear. In the case of a mixture, the curve will be to some degree S-shaped, the extent of the departure from linearity depending on the

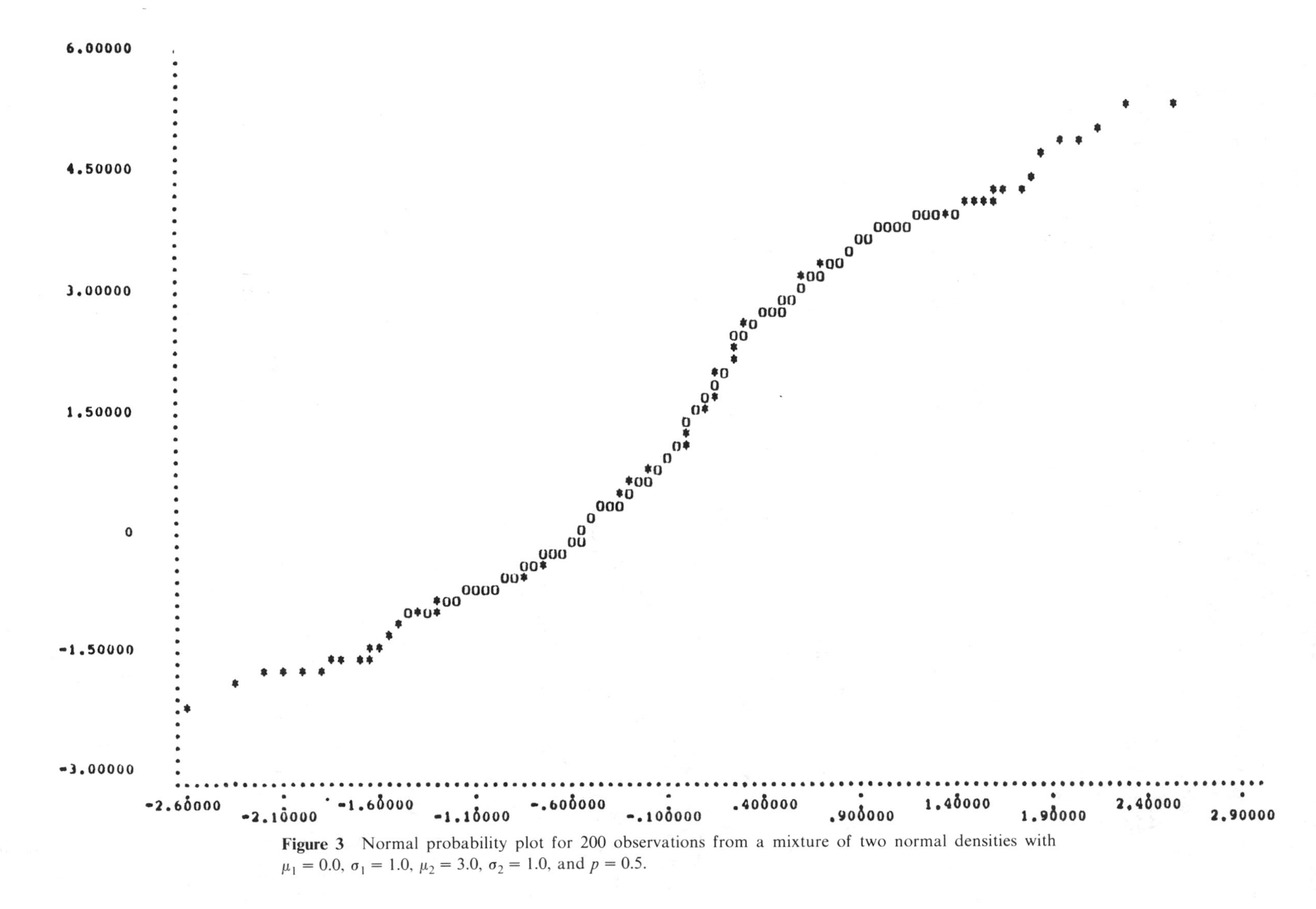

Figure 3 Normal probability plot for 200 observations from a mixture of two normal densities with $\mu_1 = 0.0$, $\sigma_1 = 1.0$, $\mu_2 = 3.0$, $\sigma_2 = 1.0$, and $p = 0.5$.

separation of the components in the mixed distribution. An example of such a plot for 200 observations from a two-component normal distribution [see (1)] appears in Fig. 3. The parameter values were $p = 0.5$, $\mu_1 = 0.0$, $\sigma_1 = 1.0$, $\mu_2 = 3.0$, $\sigma_2 = 1.0$. A further probability plotting technique for detecting normal mixtures, which it is claimed is more sensitive than the one just given, has been described by Fowlkes [17]. However, a number of examples in Everitt and Hand [16] indicate that any increase in sensitivity is rather small.

For finite mixtures, an obvious candidate for testing hypotheses about the number of components c is the *likelihood ratio test**. However, in the context of mixture distributions, there are some problems in determining the appropriate asymptotic null distribution of the test statistic (see Wolfe [49], Everitt [15], and Aitken et al. [1]).

A number of attempts have been made to produce specific tests for mixture distributions. For example, Johnson [27] describes two methods for testing whether an observed sample is consistent with being from a mixture (in unknown proportions) of two specified symmetrical populations. The first of these is based on the difference in two estimators of the unknown mixing proportion, the second on the proportion of sample values falling between the expected values of the two component distributions. Test statistics are derived and the approximate power of each test considered.

Shaked [43] derives a particularly simple test for assessing whether a set of observed data arises from a mixture distribution (see also Baker [2], De Oliveira [13], and Binder [4]).

References

[1] Aitken, M., Anderson, D., and Hinde, J. (1981). *J. R. Statist. Soc. A*, **144**, 419–448. (Describes a number of interesting applications of finite mixtures of multivariate Bernoulli distributions.)

[2] Baker, G. A. (1958). *J. Amer. Statist. Ass.*, **53**, 551–557.

[3] Bhattacharya, C. G. (1967). *Biometrics*, **23**, 115–135. (A graphical method for estimating the parameters in a finite mixture of normal distributions.)

[4] Binder, D. A. (1978). *Biometrika*, **65**, 31–38.

[5] Blum, J. R., and Susarla, V. (1977). *Ann. Prob.*, **5**, 200–209.

[6] Cassie, R. M. (1954). *Aust. J. Mar. Freshwater Res.*, **5**, 513–522.

[7] Chandra, S. (1977). *Scand. J. Statist.*, **4**, 105–112.

[8] Clark, V. A., Chapman, J. M., Coulson, A. H., and Hasselblad, V. (1968). *Johns Hopkins Med. J.*, **122**, 77–83.

[9] Cohen, A. C. (1963). *Proceedings of the International Symposium on Classical and Contagious Discrete Distributions*, G. P. Patil, ed. Pergamon Press, Montreal, Canada.

[10] Davis, D. J. (1952). *J. Amer. Statist. Ass.*, **47**, 113–150.

[11] Day, N. E. (1969). *Biometrika*, **56**, 463–474. (Describes a modified E-M algorithm for estimating the parameters of a mixture of two normal distributions.)

[12] Dempster, A. P., Laird, N. M., and Rubin, D. B. (1977). *J. R. Statist. Soc. B.*, **39**, 1–38. (Detailed account of the properties of the E-M algorithm.)

[13] De Oliveira, J. (1963). *Proceedings of the International Symposium on Classical and Contagious Discrete Distributions*, G. P. Patil, ed. Pergamon Press, Montreal, Canada, pp. 379–384.

[14] Duda, R. and Hart, P. (1973). *Pattern Classification and Scene Analysis*. Wiley, New York. (An excellent account of finite mixtures of normal distributions is given in Chapter 6.)

[15] Everitt, B. S. (1981). *Multiv. Behav. Res.*, **16**, 171–180.

[16] Everitt, B. S. and Hand, D. J. (1981). *Finite Mixture Distributions*, Chapman and Hall, London.

[17] Fowlkes, E. B. (1979). *J. Amer. Statist. Ass.*, **74**, 561–575. (Describes a graphical technique based on probability plotting for assessing whether a sample arises from a mixture of two normal components.)

[18] Fryer, J. G. and Robertson, C. A. (1972). *Biometrika*, **59**, 639–648.

[19] Green, B. F. (1951). *Psychometrika*, **16**, 151–166.

[20] Greenwood, M. and Yule, G. U. (1920). *J. R. Statist. Soc.*, **83**, 255–279.

[21] Gregor, J. (1969). *Biometrics*, **25**, 79–93.

[22] Harding, J. P. (1949). *J. Mar. Biol. Assoc. U.K.*, **28**, 141–153.

[23] Hasselblad, V. (1969). *Technometrics*, **14**, 973–976.

[24] Holla, M. S. and Bhattacharya, S. K. (1968). *Ann. Inst. Statist. Math.*, **20**, 331–336.

[25] Hosmer, D. W. (1973). *Biometrics*, **29**, 761–770.

[26] Jewell, N. P. (1982). *Ann. Statist.*, **10**, 479–484.

[27] Johnson, N. L. (1973). *Commun. Statist.*, **1**, 17–25.

[28] Johnson, N. L. and Kotz, S. (1970). *Continuous Univariate Distributions*—1, Wiley, New York.

[29] Johnson, N. L. and Kotz, S. (1970). *Continuous Univariate Distributions*, —2, Wiley, New York. (This and the previous reference contain a comprehensive account of continuous univariate distributions with a section in each chapter on mixtures.)

[30] Johnson, N. L., Kotz, S., and Sorkin, H. L. (1980). *Commun. Statist. A*, **9**, 917–922.

[31] Kao, J. H. K. (1959). *Technometrics*, **1**, 389–407.

[32] Kotz, S. and Johnson, N. L. (1982). *Commun. Statist. A*, **11**, 1997–2016.

[33] Kumar, K. D., Nicklin, E. H., and Paulson, A. S. (1979). *J. Amer. Statist. Ass.*, **74**, 52–55.

[34] Laird, N. (1978). *J. Amer. Statist. Ass.*, **74**, 52–55. (Describes the nonparametric maximum likelihood estimation of a mixing distribution.)

[35] Medgyessi, P. (1961). *Decomposition of Superpositions of Distributions*. Publishing house of the Hungarian Academy of Sciences, Budapest.

[36] Mendenhall, W. and Hader, R. J. (1958). *Biometrika*, **45**, 504–520.

[37] Murphy, E. A. (1964). *J. Chron. Dis.*, **17**, 301–324. (Contains a very clear discussion on the dangers of arguing for the existence of distinct groups on the basis of an observed bimodal frequency distribution.)

[38] Pearson, K. (1894). *Philos. Trans. A*, **185**, 71–110. (The original paper applying the method of moments to estimating the parameters in a mixture of two normal distributions.)

[39] Quandt, R. E. and Ramsey, J. B. (1978). *J. Amer. Statist. Ass.*, **73**, 730–738.

[40] Romanowski, M. (1969). *Metrologia*, **4**, 84–86.

[41] Skellam, J. G. (1948). *J. R. Statist. Soc. B*, **10**, 257–261.

[42] Shah, B. K. (1963). *J. M. S. University of Baroda, (Science Issue)*, **12**, 21–22.

[43] Shaked, M. (1980). *J. R. Statist. Soc. B*, **42**, 192–198.

[44] Simaz, L. (1976). *Ann. Statist.*, **4**, 1200–1209.

[45] Tallis, G. M. (1969). *J. Appl. Prob.*, **6**, 389–398.

[46] Teicher, H. (1961). *Ann. Math. Statist.*, **32**, 244–248.

[47] Teicher, H. (1963). *Ann. Math. Statist.*, **34**, 1265–1269.

[48] Wolfe, J. H. (1970). *Multivar. Behav. Res.*, **5**, 329–350. (Estimation of the parameters of a finite mixture of multivariate normal distributions using the E-M algorithm.)

[49] Wolfe, J. H. (1971). *Technical Bulletin STB 72-2*, Naval Personnel and Training Research Laboratory, San Diego, Calif.

[50] Yakowitz, S. J. and Spragins, J. D. (1968). *Ann. Math. Statist.*, **39**, 209–214.

(IDENTIFIABILITY
MAXIMUM LIKELIHOOD ESTIMATION
METHOD OF MOMENTS
MINIMUM CHI-SQUARE
MODEL CONSTRUCTION: SELECTION OF DISTRIBUTIONS
PROBABILITY PLOTTING)

B. S. EVERITT

MIXTURE EXPERIMENTS

In a mixture experiment two or more ingredients are mixed or blended together to form an end product. Measurements are taken on several blends of the ingredients in an attempt to find the blend that produces the "best" response. The measured response is assumed to be a function only of the proportions of the ingredients (components) present in the mixture and is not a function of the total amount of the mixture. For example, stainless steel is a mixture of iron, copper, nickel, and chromium, and the tensile strength of the steel depends only on proportions of each of the four components in the alloy. The topic of mixture experiments is quite distinct from the topic of mixtures of distributions.

In mixture experiments, since the controllable variables or components are nonnegative proportionate amounts of the mixture, when expressed as fractions of the mixture, they sum to unity. Clearly, if the number of components in the system is denoted by q and if the proportion of the ith component in the mixture is represented by x_i, then

$$x_i \geqslant 0, \qquad i = 1, 2, \dots, q \tag{1}$$

and

$$\sum_{i=1}^{q} x_i = x_1 + x_2 + \cdots + x_q = 1.0. \tag{2}$$

The experimental region or factor space of interest, which is defined by the values of x_i, is a regular $(q-1)$-dimensional simplex. Since the proportions sum to unity as shown in (2), the x_i are constrained variables and altering the proportion of one component in a mixture will cause a change in the proportion of at least one other component in the experimental region. For $q = 2$ components, the factor space is a straight line, for three components ($q = 3$) an equilateral triangle, and for four components the factor space is a tetrahedron.

The coordinate system for mixture proportions is that of a simplex. With three components, for example, the vertices of the triangle represent single-component mixtures and are denoted by $x_i = 1$, $x_j = x_k = 0$ for $i, j, k = 1, 2$, and 3, $i \neq j \neq k$. The interior points of the triangle represent mixtures where all of the component proportions are nonzero, that is, $x_1 > 0$, $x_2 > 0$, and $x_3 > 0$. The centroid of the triangle corresponds to the mixture with equal proportions $(\frac{1}{3}, \frac{1}{3}, \frac{1}{3})$ from each of the components.

In mixture problems, the experimental data is defined on a quantitative scale such as the yield or some physical characteristic of the product formed from the blend. The purpose of the experimental program is to model the blending surface with some form of mathematical equation so that:

> Predictions of the response for *any* mixture or combination of the ingredients can be made empirically, or;
>
> Some measure of the influence on the response of each component singly and in combination with the other components can be obtained.

In setting up a mathematical equation, it is assumed that there exists some functional relationship

$$\eta = \phi(x_1, x_2, \ldots, x_q) \tag{3}$$

that defines the dependence of the response η on the proportions $x_1, x_2, \ldots, x_q$ of the components. The function ϕ is a continuous function in the x_i and is represented usually by a first- or second-degree polynomial. On some occasions a third-degree equation or a reduced form of a cubic equation with certain terms omitted from the complete cubic equation may be used to represent the surface.

In an experimental program consisting of N trials, the observed value of the response in the uth trial, denoted by y_u, is assumed to vary about a mean of η_u with a common variance σ^2 for all $u = 1, 2, \ldots, N$. The observed value contains an additive experimental error ϵ_u,

$$y_u = \eta_u + \epsilon_u, \qquad 1 \leqslant u \leqslant N,$$

where the errors ϵ_u are assumed to be uncorrelated and identically distributed with zero mean and common variance σ^2.

Statistical methods are used to measure product characteristics and improve product performance. These methods include (a) choosing the design program that defines which blends to study, (b) selecting the type of model equation to be fitted to the resulting data, and (c) using the appropriate techniques in the analysis of such data. We shall discuss these methods for experiments where all combinations of the ingredients are possible as well as for experiments in which only certain combinations are feasible. Some of the expository papers written on methods for analyzing data from mixture experiments are by Cornell [4, 6], Gorman and Hinman [18], Hare [19], and Snee [31, 32, 34]. Mixture designs, models, and techniques used in the analysis of data are discussed in considerable detail in Cornell [7].

THE SIMPLEX-LATTICE DESIGNS

For investigating the response surface* over the entire simplex region, a natural choice for a design would be one with points that are positioned uniformly over the simplex factor space. Such a design is the $\{q, m\}$ simplex-lattice introduced by Scheffé [29].

A $\{q, m\}$ simplex-lattice design for q components consists of points defined by the following coordinate settings: the proportions assumed by each component take the

$m + 1$ equally spaced values from 0 to 1,

$$x_i = 0, \frac{1}{m}, \frac{2}{m}, \ldots, 1 \qquad (4)$$

and all possible combinations (mixtures) of the components are considered, using the proportions in (4) for each component.

For a $q = 3$ component system, suppose each component is to take the proportions $x_i = 0, \frac{1}{2}$, and 1, for $i = 1, 2$, and 3, which is the same as setting $m = 2$ in (4). Then the $\{3, 2\}$ simplex-lattice consists of the six points on the boundary of the triangular factor space,

$$(x_1, x_2, x_3) = (1,0,0), (0,1,0), (0,0,1), \\ (\tfrac{1}{2}, \tfrac{1}{2}, 0), (\tfrac{1}{2}, 0, \tfrac{1}{2}), (0, \tfrac{1}{2}, \tfrac{1}{2}).$$

The three vertices $(1,0,0)$, $(0,1,0)$, $(0,0,1)$ represent the individual component mixtures, while the points $(\frac{1}{2}, \frac{1}{2}, 0)$, $(\frac{1}{2}, 0, \frac{1}{2})$, and $(0, \frac{1}{2}, \frac{1}{2})$ represent the binary blends or two-component mixtures and are located at the midpoints of the three sides of the triangle.

Table 1 lists the number of points in a $\{q, m\}$ simplex-lattice for values of q and m from $3 \leqslant q \leqslant 10$, $1 \leqslant m \leqslant 4$. The number of points in a $\{q, m\}$ simplex-lattice is $\binom{q+m-1}{m} = (q + m - 1)!/[m!(q - 1)!]$.

An alternative arrangement to the $\{q, m\}$ simplex-lattice is the simplex-centroid design introduced by Scheffé [30]. In a q-component simplex-centroid design, the number of points is $2^q - 1$. The design points correspond to the q permutations of $(1, 0, 0, \ldots, 0)$, the $\binom{q}{2}$ permutations of $(\frac{1}{2}, \frac{1}{2}, 0, 0, \ldots, 0)$, the $\binom{q}{3}$ permutations of $(\frac{1}{3}, \frac{1}{3}, \frac{1}{3}, 0, \ldots, 0), \ldots,$ and the centroid point $(1/q, 1/q, \ldots, 1/q)$. A four-component simplex-centroid design consists of $2^4 - 1 = 15$ points.

Besides experimental regions, mixture experiments differ from the ordinary regression problems also in the form of the polynomial model to be fitted. Scheffé [29] introduced canonical polynomials for use with the simplex-lattices and simplex-centroid designs.

THE CANONICAL FORM OF MIXTURE POLYNOMIALS

The canonical form of the mixture polynomial is derived by applying the restriction $x_1 + x_2 + \cdots + x_q = 1$ to the terms in the standard polynomial and then simplifying. For example, with two components, x_1 and x_2, where $x_1 + x_2 = 1$, the first-degree polynomial is

$$\begin{aligned}\eta &= \beta_0 + \beta_1 x_1 + \beta_2 x_2 \\ &= (\beta_0 + \beta_1)x_1 + (\beta_0 + \beta_2)x_2 \\ &= \beta_1' x_1 + \beta_2' x_2,\end{aligned}$$

so that the constant term β_0 is removed from the model. For the second-degree polynomial in x_1 and x_2, we also replace x_1^2 by

Table 1 Number of Points in the $\{q, m\}$ Simplex-Lattice for $3 \leqslant q \leqslant 10$, $1 \leqslant m \leqslant 4$, Where the Number of Levels for Each Component Is $m + 1$[a]

Degree of Model	Number of Components, q							
m	3	4	5	6	7	8	9	10
1	3	4	5	6	7	8	9	10
2	6	10	15	21	28	36	45	55
3	10	20	35	56	84	120	165	220
4	15	35	70	126	210	330	495	715

[a] *Source.* Cornell [7].

$x_1(1 - x_2)$ and x_2^2 by $x_2(1 - x_1)$ to get

$$\begin{aligned}\eta &= \beta_0(x_1 + x_2) + \beta_1 x_1 + \beta_2 x_2 + \beta_{12}x_1x_2 \\ &\quad + \beta_{11}x_1(1 - x_2) + \beta_{22}x_2(1 - x_1) \\ &= (\beta_0 + \beta_1 + \beta_{11})x_1 \\ &\quad + (\beta_0 + \beta_2 + \beta_{22})x_2 \\ &\quad + (\beta_{12} - \beta_{11} - \beta_{22})x_1x_2 \\ &= \beta_1' x_1 + \beta_2' x_2 + \beta_{12}' x_1 x_2,\end{aligned}$$

so that the quadratic terms $\beta_{11}x_1^2$ and $\beta_{22}x_2^2$ are also removed from the model along with the constant term β_0. Thus the mixture models have fewer terms than the standard polynomials.

In general, the canonical forms of the mixture models (with the primes removed from the β_i's) are:

Linear:

$$\eta = \sum_{i=1}^{q} \beta_i x_i, \tag{5}$$

Quadratic:

$$\eta = \sum_{i=1}^{q} \beta_i x_i + \sum_{i<j}^{q}\sum \beta_{ij} x_i x_j, \tag{6}$$

Full cubic:

$$\begin{aligned}\eta &= \sum_{i=1}^{q} \beta_i x_i + \sum_{i<j}^{q}\sum \beta_{ij} x_i x_j \\ &\quad + \sum_{i<j}^{q}\sum \delta_{ij} x_i x_j (x_i - x_j) \\ &\quad + \sum_{i<j<k}^{q}\sum\sum \beta_{ijk} x_i x_j x_k,\end{aligned} \tag{7}$$

Special cubic:

$$\begin{aligned}\eta &= \sum_{i=1}^{q} \beta_i x_i + \sum_{i<j}^{q}\sum \beta_{ij} x_i x_j \\ &\quad + \sum_{i<j<k}^{q}\sum\sum \beta_{ijk} x_i x_j x_k,\end{aligned} \tag{8}$$

(The special cubic equation is a reduced form of third-degree polynomial that provides measures of ternary blends of the three components i, j, and k. It represents the lowest form of polynomial of degree higher than two and contains $q(q^2 + 5)/6$ terms while the number of terms in the full cubic model is $q(q + 1)(q + 2)/6$.)

The canonical form of the polynomial in q components that is to be fitted to data collected at the points of the simplex-centroid design is

$$\begin{aligned}\eta &= \sum_{i=1}^{q} \beta_i x_i + \sum_{i<j}^{q}\sum \beta_{ij} x_i x_j \\ &\quad + \sum_{i<j<k}^{q}\sum\sum \beta_{ijk} x_i x_j x_k \\ &\quad + \cdots + \beta_{12\ldots q} x_1 x_2 \ldots x_q\end{aligned} \tag{9}$$

The model of (9) contains $2^q - 1$ terms.

The terms in the canonical polynomial models have simple interpretations. In (6), for example, if $x_i = 1$ and therefore $x_j = 0$ for $j \neq i$, then $\eta = \beta_i$, that is, β_i is the expected response to the pure component i. If the blending is additive, the model is $\eta = \sum_{i=1}^{q} \beta_i x_i$, which is the equation of a planar surface. With a quadratic model, the second-degree terms describe quadratic departure of the response surface from a plane. Higher-order terms (such as $\beta_{123}x_1x_2x_3$) describe additional perturbations of the response surface beyond those described by first- and second-degree terms.

The estimates of the coefficients in the canonical polynomials are simple functions of the observed values of the response collected at the points of design. For instance, if the average of r_i responses to the pure component i ($x_i = 1, x_j = 0, j \neq i$) is denoted by $\bar{y}_i$ and the average of r_{ij} responses to the 50% : 50% binary mixture ($x_i = \frac{1}{2}$, $x_j = \frac{1}{2}$, $x_k = 0$ for all $i < j < k$) of components i and j is denoted by $\bar{y}_{ij}$, then the formulas for calculating the coefficient estimates b_i and b_{ij} in the second-degree model of (6), are

$$\begin{aligned} b_i &= \bar{y}_i, \qquad i = 1, 2, \ldots, q \\ b_{ij} &= 4\bar{y}_{ij} - 2(\bar{y}_i + \bar{y}_j), \\ &\qquad i, j = 2, \ldots, q, \quad i < j.\end{aligned} \tag{10}$$

Formulas for estimating the parameters in models up to the fourth degree are presented in Cornell [7, Chap. 2].

Since the estimates b_i and b_{ij} are linear

functions of random variables (the $\bar{y}_i$ and $\bar{y}_{ij}$), the estimate $\hat{y}(\mathbf{x})$ of the response at $\mathbf{x}$ is also a random variable. When the fitted model is of the same degree in the x_i's as the true surface, then the expectation of $\hat{y}(\mathbf{x})$ is $E[\hat{y}(\mathbf{x})] = \eta$.

A formula for the variance of the estimate $\hat{y}(\mathbf{x})$ can be written in terms of the variances of the averages $\bar{y}_i$ and $\bar{y}_{ij}$ at the lattice points, since

$$\hat{y}(\mathbf{x}) = \sum_{i=1}^{q} a_i \bar{y}_i + \sum\sum_{i<j}^{q} a_{ij} \bar{y}_{ij},$$

where $a_i = x_i(2x_i - 1)$ and $a_{ij} = 4x_i x_j$, $i, j = 1, 2, \ldots, q, i < j$, so that

$$\text{Var}[\hat{y}(\mathbf{x})] = \sigma^2 \left\{ \sum_{i=1}^{q} \frac{a_i^2}{r_i} + \sum\sum_{i<j}^{q} \frac{a_{ij}^2}{r_{ij}} \right\}. \tag{11}$$

When σ^2 is unknown, a sample measure s^2 from the replicated observations can be used to estimate σ^2.

AN EXAMPLE OF A THREE-COMPONENT FRUIT PUNCH EXPERIMENT

Watermelon (x_1), pineapple (x_2), and orange (x_3) juice concentrates were used as primary ingredients in a fruit punch. In addition to individual fruit juices, blends consisting of pairs of fruit juices were to be studied. The six blends of the three juice concentrates were evaluated for overall general acceptance by a panel of judges. The ingredient proportions and the adjusted acceptance values for three replications of each blend are listed in Table 2. The coefficient estimates in the fitted second-degree model are

$$b_1 = (4.3 + 4.7 + 4.8)/2 = 4.60,$$

$$b_2 = 19.0/3 = 6.33, \qquad b_3 = 21.3/3 = 7.10,$$

$$b_{12} = 4(18.2/3) - 2[(13.8/3) + (19.0/3)]$$
$$= 24.27 - 21.87 = 2.40,$$
$$b_{13} = 1.27,$$
$$b_{23} = -2.20.$$

The fitted second-degree model is

$$\begin{aligned}\hat{y}(\mathbf{x}) = {} & \underset{(0.13)}{4.60x_1} + \underset{(0.13)}{6.33x_2} + \underset{(0.13)}{7.10x_3} + \underset{(0.66)}{2.40x_1x_2} \\ & + \underset{(0.66)}{1.27x_1x_3} - \underset{(0.66)}{2.20x_2x_3}.\end{aligned} \tag{12}$$

The quantities in parentheses below the coefficient estimates are the corresponding estimated standard errors where the sample estimate of the variance of an observation is calculated using the replicated data from the six blends and is $s^2 = 0.054$.

The linear estimates (the b_i's) in the model of equation (12) represent the heights of the estimated surface at the vertices of the triangle and are the scores of the individual pure juice types. Since $b_3 > b_2 > b_1$, orange juice appears to be the most acceptable pure juice while watermelon juice is the least accept-

Table 2 General Acceptance Values of Fruit and Punch Blends

% Juice/100			
Watermelon (x_1)	Pineapple (x_2)	Orange (x_3)	General Acceptance Value (y)
1	0	0	4.3, 4.7, 4.8
0	1	0	6.2, 6.5, 6.3
$\frac{1}{2}$	$\frac{1}{2}$	0	6.3, 6.1, 5.8
0	0	1	7.0, 6.9, 7.4
$\frac{1}{2}$	0	$\frac{1}{2}$	6.1, 6.5, 5.9
0	$\frac{1}{2}$	$\frac{1}{2}$	6.2, 6.1, 6.2

able juice type when each is considered individually.

The cross-product coefficient estimates (the b_{ij}'s) represent the effects of blending pairs of juice types. In particular, there is evidence of the synergistic blending of watermelon (x_1) juice with pineapple (x_2) juice (meaning the watermelon-pineapple combination is more acceptable than would be expected by averaging the acceptability of each), since the coefficient $b_{12} = 2.40$ (relative to its standard error) is greater than zero when tested with a Student's t-statistic, assuming the errors are normally distributed. Watermelon juice (x_1) and orange (x_3) juice do not appear to blend synergistically, since $b_{13} = 1.27$, relative to its standard error, is not greater than zero. The sign of $b_{23}(-2.20)$ and the corresponding value of the t-statistic $(-2.20/0.66 = -3.33)$ lead us to infer that pineapple and orange juice are antagonistic when blended together only in terms of average overall acceptance. Thus other combinations of watermelon with pineapple or watermelon with orange should be tried as well as some three-component blends, in seeking to find a fruit punch of high acceptability.

Before predictions of the acceptability of other blends can be made, the accuracy of the prediction equation (12) should be established. This is accomplished by checking the fit or testing for "lack of fit" of the model. However, since the polynomials contain the same number of terms as there are points in the associated lattice designs (this is true also with (9) and the simplex-centroid design), there are no degrees of freedom with which to test "lack of fit" of the model without augmenting the design by including additional points. When additional points are included, the formulas of (10) no longer are valid for estimating the parameters. But if one has serious reason to question the adequacy of the model, particularly in some subregion of interest, additional points will not only allow for checking "lack of fit" but also improve the estimate of $\text{Var}[\hat{y}(\mathbf{x})]$ in the subregion of interest.

There are cases when only mixtures with none of the components absent can be considered. An example is the formulation of a bleach used to remove ink stains where the blending agents in the bleach are bromine, hypochlorite powder, and dilute HCl. In this case only mixtures in which all three agents are present can be used. Consequently, only interior points of the triangle are used, and these blends are called *complete mixtures*.

THE USE OF INDEPENDENT VARIABLES

An alternative to working with the q mixture components whose proportions are dependent is to work with a system of $q - 1$ mathematically independent variables. Reasons cited by several authors for preferring to work with independent variables are: a familiarity with standard factorial-type designs for fitting models that are expressed in the independent variables and the ease in interpreting the estimates of the parameters in the models.

Draper and Lawrence [12] present designs for three- and four-component mixture systems where the design settings are constructed in the metric of the variables w_i, $i = 1, 2$ and $i = 1, 2,$ and 3, respectively. (Becker [2] extends the methodology to q components mapped into $q - 1$ variables.) In the case of three components, the variables w_1 and w_2 are defined as $w_1 = \frac{1}{2}(x_3 - x_2)$ and $w_2 = (\sqrt{3}/6)(2x_1 - x_2 - x_3)$, so that for values of the $x_i \geqslant 0$, $i = 1, 2, 3$, where $x_1 + x_2 + x_3 = 1$, the corresponding values of w_1 and w_2 are defined in a two-dimensional coordinate system where the axes of w_1 and w_2 are at right angles. A second-degree polynomial in w_1 and w_2,

$$y = \beta_0 + \beta_1 w_1 + \beta_2 w_2 + \beta_{11} w_1^2 + \beta_{22} w_2^2 + \beta_{12} w_1 w_2 + \epsilon,$$

is then fitted to data collected at the points of any of several designs listed by Draper and Lawrence, who were the first to consider both the variance and the bias of the prediction equation for the purpose of optimizing mixture designs.

Thompson and Myers [37] suggest an alternative design strategy by defining an ellipsoidal region of interest in the simplex factor space where the region is centered about a point of main interest. This point, $\mathbf{x}_0 = (x_{01}, x_{02}, \ldots, x_{0q})'$, could be the current operating conditions or simply a convenient base point from which to construct a design for exploring the response surface in the neighborhood around $\mathbf{x}_0$. The ellipsoidal region of interest is

$$\sum_{i=1}^{q} \left(\frac{x_i - x_{0i}}{h_i} \right)^2 \leqslant 1,$$

where the x_{0i} and h_i are chosen by the experimenter. The point x_{0i} denotes the center of the interval of interest for the ith component and h_i is a constant that allows for the spread of the symmetric interval of interest for the ith component.

Thompson and Myers define the intermediate variables $v_i = (x_i - x_{0i})/h_i$, $1 \leqslant i \leqslant q$, and the ellipsoidal region becomes a unit spherical region in the v_i. From the intermediate variables, a transformation is performed to $q - 1$ independent variables w_i. This latter transformation to independent variables enables the easy use of standard response surface designs (they choose to use rotatable designs* in the w_i) as well as the use of the integrated mean square error of $\hat{y}(\mathbf{w})$ as a design criterion in which to choose optimal designs. A very similar development, in which the mixture components belong to several distinct categories (i.e., acid constituents, base constituents, etc.), was presented by Cornell and Good [8].

To this point, we have considered different methods for constructing designs and fitting models in mixture experiments where the constraints on the proportions are specified as in (1) and (2). Often, however, because of physical or economic limitations, the mixtures may be confined to a subregion in the composition space. Although it may be possible to employ the methodology of Thompson and Myers for defining a feasible region in which to work, in most cases, the complete simplex-lattice designs, the simplex-centroid designs and the symmetric simplex designs of Murty and Das [25] cannot be used. When one is interested in a component not so much from the standpoint of its percentage present but rather its relationship to still another variable, Kenworthy [21] shows how factorial arrangements can be used to analyze the response to ratios of components. For example, in glass production, it may be desirable to think of the ratio silica/soda instead of the actual percentage of each. With a third component, lime, the ratios soda/lime and silica/lime are of interest. Only complete mixtures are considered since each component proportion is required to be greater than zero.

CONSTRAINED MIXTURE EXPERIMENTS

In many experiments it is not possible to explore the total range, $0 \leqslant x_i \leqslant 1$, with all of the components. An example was the bleach used for removing ink stains where each of the ingredients was required to be present in the dye. Another example is the formulation of a rocket propellant (see Kurotori [22]), where useful blends required lower bounds to be placed on the components proportions in the form: binder (x_1), oxidizer (x_2), and fuel (x_3), so that

$$x_1 \geqslant 0.20, \qquad x_2 \geqslant 0.40, \qquad x_3 \geqslant 0.20. \tag{13}$$

The coordinates of several design blends in the restricted experimental region are listed in Table 3.

When one or more of the original components are constrained by lower bounds of the form $x_i \geqslant a_i$, $i = 1, 2, \ldots, q$, an alternative system of coordinates can be used to simplify the construction of designs and the fitting of model forms. Pseudocomponents x_i' are defined as

$$x_i' = \frac{x_i - a_i}{1 - L}, \qquad L = \sum_{i=1}^{q} a_i < 1.0.$$

In the x_i' a simplex-lattice arrangement can be set up and the design coordinates of the x_i' are then mapped back to provide settings in the original coordinates using

$$x_i = a_i + (1 - L)x_i'.$$

Table 3 Rocket Propellant Experimental Data

Constrained Component Proportions			Pseudocomponent Proportions			Elasticity
x_1	x_2	x_3	x'_1	x'_2	x'_3	(y)
0.40	0.40	0.20	1	0	0	2350
0.20	0.60	0.20	0	1	0	2450
0.20	0.40	0.40	0	0	1	2650
0.30	0.50	0.20	$\frac{1}{2}$	$\frac{1}{2}$	0	2400
0.30	0.40	0.30	$\frac{1}{2}$	0	$\frac{1}{2}$	2750
0.20	0.50	0.30	0	$\frac{1}{2}$	$\frac{1}{2}$	2950
0.27	0.46	0.27	$\frac{1}{3}$	$\frac{1}{3}$	$\frac{1}{3}$	3000
0.33	0.43	0.23	$\frac{2}{3}$	$\frac{1}{6}$	$\frac{1}{6}$	2690
0.23	0.53	0.23	$\frac{1}{6}$	$\frac{2}{3}$	$\frac{1}{6}$	2770
0.23	0.43	0.33	$\frac{1}{6}$	$\frac{1}{6}$	$\frac{2}{3}$	2980

A model in the pseudocomponents x'_1, x'_2, and x'_3, fitted to the 10 elasticity values listed in Table 3, is

$$\hat{y}(\mathbf{x}') = 2351x'_1 + 2446x'_2 + 2653x'_3 - 6x'_1x'_2 + 1008x'_1x'_3 + 1597x'_2x'_3 + 6141x'_1x'_2x'_3. \quad (14)$$

An equivalent model in the original constrained components is

$$\hat{y}(\mathbf{x}) = 50201x_1 + 16991x_2 + 35674x_3 - 153686x_1x_2 - 281852x_1x_3 - 113595x_2x_3 + 767647x_1x_2x_3. \quad (15)$$

The surface representations (14) and (15) are the same in terms of providing the estimated response surface for all blends in the region (13), although the pseudocomponent model is usually the preferred model to work with. For additional reading on problems encountered when fitting models in constrained regions, see Gorman [17].

When the feasible region for the mixture blends is defined by placing both upper and lower bounds on the component proportions of the form

$$0 \leqslant a_i \leqslant x_i \leqslant b_i \leqslant 1, \qquad 1 \leqslant i \leqslant q, \quad (16)$$

the resulting factor space is a convex polyhedron. For locating design points of the polyhedron, McLean and Anderson [24] developed an algorithm that computes the coordinates of the vertices of the polyhedron. The vertices of the region and convex combinations of some of the vertices are used as design points at which to observe the response for fitting a first- or second-degree polynomial.

Once the limits a_i and b_i in (16) are specified, the design is fixed so that there may be clusters of points in some areas of the factor space and only a few points in other areas. In this case, the precision of the estimate of the response, Var $\hat{y}(\mathbf{x})$, will be affected: poor precision will result in areas of sparse experimentation while there will be good precision in areas with clusters of points.

Snee and Marquardt [35] proposed an algorithm that selects a subset of the extreme vertices of a polyhedron when fitting the Scheffé first-degree model. The algorithm is referred to as XVERT; designs selected by XVERT produce model coefficient esimates with a smaller average variance than do designs selected by competitive algorithms. Snee [33] extended the design construction procedure to support the fit of a quadratic model.

Goel [15] suggests a modification of the McLean and Anderson algorithm and of XVERT in a new algorithm called UNIEXP, which is designed to provide a more uniform coverage of the constrained region. Goel lists five features of UNIEXP for generating design points and claims it is slightly superior to the other two algorithms.

Saxena and Nigam [28] show how to obtain a uniform coverage of a highly constrained factor space for the purpose of fitting the Scheffé second-degree model. The procedure is to express the x_i as linear functions of a set of design components z_i, set up a design in the z_i, and then transform back to the x_i. The appeal of Saxena and Nigam's procedure lies in its freedom from dependence on the computer for generating designs, even with extremely complicated factor spaces. The relative ease in applying a linear transformation from the x_i's to the z_i's and vice versa is the major advantage stated by the authors for preferring their method of generating designs in constrained regions.

ALTERNATIVE MODELS

Response surface methodology is based on the assumption that the response can be approximated within the range of the data by a low-order polynomial. Quenouille [27] criticized the Scheffé polynomial for not being able to account correctly for components that are inert or that have additive effects. To counter this difficulty, Becker [1] introduced the following models, which are homogeneous of degree one.

$$\textbf{H1}:\quad E(y) = \sum_{i=1}^{q} \alpha_i x_i + \sum\sum_{i<j}^{q} \alpha_{ij} \min(x_i, x_j) + \sum\sum\sum_{i<j<l}^{q} \alpha_{ijl} \min(x_i, x_j, x_l);$$

$$\textbf{H2}:\quad E(y) = \sum_{i=1}^{q} \alpha_i x_i + \sum\sum_{i<j}^{q} \alpha_{ij} \frac{x_i x_j}{(x_i + x_j)} + \sum\sum\sum_{i<j<l}^{q} \alpha_{ijl} \frac{x_i x_j x_l}{(x_i + x_j + x_l)^2};$$

$$\textbf{H3}:\quad E(y) = \sum_{i=1}^{q} \alpha_i x_i + \sum\sum_{i<j}^{q} \alpha_{ij} (x_i x_j)^{1/2} + \sum\sum\sum_{i<j<l}^{q} \alpha_{ijl} (x_i x_j x_l)^{1/3}.$$

Snee [31] discusses briefly the types of curvature covered by **H1**, **H2**, and **H3** and shows a plot of the curvature of each model over the range $0 \leqslant x_i \leqslant 1$. Contour plots of estimated surfaces of mite population numbers using the models **H1**, **H2**, and **H3**, over a three component triangle are drawn in Cornell [7, p. 231]; other papers that discuss the modeling of the additive property of one or more of the components are by Becker [1, 3] and by Cornell and Gorman [9].

In an attempt to obtain model parameter estimates with an interpretation closer to that of the parameters in the ordinary polynomial, Cox [11] defined the first- and second-degree polynomials, respectively, as

$$E(y) = \beta_0 + \sum_{i=1}^{q} \beta_i x_i, \tag{17}$$

$$E(y) = \beta_0 + \sum_{i=1}^{q} \beta_i x_i + \sum_{i=1}^{q} \sum_{j=1}^{q} \beta_{ij} x_i x_j, \quad \beta_{ij} = \beta_{ji}, \tag{18}$$

where a single constraint is imposed on the parameters in (17) and $q + 1$ constraints are imposed on the parameters in (18). The effect of component i on the response is defined by selecting a reference mixture $\mathbf{s} = (s_1, s_2, \ldots, s_q)'$ and some point $\mathbf{x}$ other than $\mathbf{s}$ where at $\mathbf{x}$, $x_i = s_i + \Delta_i$ and $x_j = s_j(1 - x_i)/(1 - s_i)$, $j \neq i$, and calculating the weighted difference between the observed response values $y(\mathbf{x})$ and $y(\mathbf{s})$ at $\mathbf{x}$ and $\mathbf{s}$, respectively, in the manner of

$$b_i = \frac{1 - s_i}{\Delta_i} [y(\mathbf{x}) - y(\mathbf{s})]. \tag{19}$$

The formula of (19) is an estimate of the difference between what the observed heights of the surface would be at the vertex $x_i = 1$ and what it is at $\mathbf{s}$. A similar definition of the effect of component i is given by Snee and Marquardt [36] as

$$E_i = b_i - (q - 1)^{-1} \sum_{j \neq i}^{q} b_j,$$

where the coefficient estimates are taken from the fitted first-degree Scheffé polynomial [i.e., (17) without the β_0 term].

Cornell [5] presents designs for fitting the models of (17) and (18) by showing that since the parameters there can be expressed as linear functions of the parameters in Scheffé's models, then designs that are appropriate for fitting the Scheffé models can also be used for estimating the parameters in Cox's polynomial.

For modeling an extreme change in the response as the value of $x_i \to 0$, Draper and St. John [13] suggest that inverse terms $(1/x_i)$ be included in the Scheffé polynomial. An extreme change in the response might be the erratic performance of a two-stroke combustion engine when the proportion of oil in the fuel mixture tends to zero, and it is of interest to be able to detect this with our model.

For a system containing q components and for detecting extreme changes in the response as $x_i \rightarrow 0$, some of the equations suggested are

$$\eta = \sum_{i=1}^{q} \beta_i x_i + \sum_{i=1}^{q} \beta_{-i} x_i^{-1},$$

$$\eta = \sum_{i=1}^{q} \beta_i x_i + \sum_{i<j}^{q}\sum \beta_{ij} x_i x_j + \sum_{i=1}^{q} \beta_{-i} x_i^{-1},$$

$$\eta = \sum_{i=1}^{q} \beta_i x_i + \sum_{i<j}^{q}\sum \beta_{ij} x_i x_j + \sum_{i<j<k}^{q}\sum\sum \beta_{ijk} x_i x_j x_k + \sum_{i=1}^{q} \beta_{-i} x_i^{-1}.$$

It is assumed that the value of x_i never reaches zero, but could be very close to zero, that is, $x_i \rightarrow \epsilon_i > 0$, where ϵ_i is some extremely small quantity defined for each application of these models. When, according to the design, some of the blends have $x_i = 0$, then in order to include the term x_i^{-1} in the model, one adds a small positive amount, say c_i, to each value of x_i. For most problems a value of c_i would be somewhere between 0.02 and 0.05, and ideally one would let $c_1 = c_2 = \cdots = c_q = c$, according to Draper and St. John.

ADDITIONAL TOPICS

Marquardt and Snee [23] present several statistics, F, R^2, and the adjusted R_A^2 for use when testing different hypotheses involving the model parameters. For the Scheffé model $E(y) = \sum_{i=1}^{q} \beta_i x_i$, the hypothesis that states the response does not depend on the mixture components is equivalent to the hypothesis $H_0: \beta_i = \beta$ for all i. The test of this hypothesis is performed using the ratio

$$F = \frac{\sum_{u=1}^{N} (\hat{y}_u - \bar{y})^2/(q-1)}{\sum_{u=1}^{N} (y_u - \hat{y}_u)^2/(N-q)} = \frac{\text{mean square regression}}{\text{mean square residual}},$$

where $\hat{y}_u$ is the predicted response at the uth experimental setting using the fitted model and $\sum_{u=1}^{N}(y_u - \hat{y}_u)^2$ is the sum of the squares of the residuals at the N experimental settings.

Model reduction resulting from setting up contrasts among some of the terms in the Scheffé models and using the contrasts as terms in a derived model form is discussed by Park [26] and Cornell [7, Chap. 5]. How to measure the rate of change (or slope) of the surface along the component axes in order to learn more about the shape of the surface is outlined by Cornell and Ott [10]. Designs that are appropriate when including process variables (such as cooking time and cooking temperature) in mixture experiments are presented by Hare [20]. Reparametrized model forms containing process variables with mixture components are covered in Cornell [7, Chap. 3].

Finally, the use of the symmetric simplex designs suggested by Murty and Das [25] and the fitting of standard polynomial forms for the sequential exploration of mixture surfaces is discussed by Goel and Nigam [16].

CONCLUDING REMARKS

The main difference between mixture experiments and the standard regression-type problem is that mixtures consist of ingredients whose proportionate values are restricted to be nonnegative and sum to unity whereas the values of the concomitant variables in the regression study are allowed to take on negative values and vary over wide ranges. Another property of a mixture experiment is that the measured response, which is a function of the ingredients, is assumed to depend only on the relative proportions of the ingredients present in the mixture and not on the amount of the mixture.

Research on mixture designs and models, as represented by the number of papers that have appeared in the statistical literature, is a relatively new activity. Almost all of the theory and methodology that has emanated from the statistical community has surfaced since 1960. A complete bibliography, through 1979, is listed in Cornell [7].

References

[1] Becker, N. G., (1968). *J. R. Statist. Soc.*, **30**, 349–358. (The introduction of models which are homogeneous of degree one for studying additive effects of the components.)

[2] Becker, N. G., (1970). *Biometrika*, **57**, 329–338.

[3] Becker, N. G., (1978). *Aust. J. Statist.*, **20**, 195–208.

[4] Cornell, J. A. (1973). *Technometrics*, **15**, 437–455. (A complete review of nearly all of the published statistical papers on mixture designs and models from 1958 to 1973.)

[5] Cornell, J. A. (1975). *Technometrics*, **17**, 24–35.

[6] Cornell, J. A. (1979). *Technometrics*, **21**, 95–106.

[7] Cornell, J. A. (1981). *Experiments with Mixtures: Designs, Models, and the Analysis of Mixture Data*. Wiley, New York.

[8] Cornell, J. A. and Good, I. J., (1970). *J. Amer. Statist. Ass.*, **65**, 339–355. (Considers the blending of categories of components.)

[9] Cornell, J. A. and Gorman, J. W. (1978). *Biometrics*, **34**, 251–263.

[10] Cornell, J. A. and Ott, L. (1975), *Technometrics*, **17**, 409–424.

[11] Cox, D. R. (1971). *Biometrika*, **58**, 155–159.

[12] Draper, N. R. and Lawrence, W. E. (1965). *J. R. Statist. Soc. B*, **27**, 450–465, 473–478.

[13] Draper, N. R. and St. John, R. C. (1977). *Technometrics*, **19**, 37–46.

[14] Goel, B. S. (1980a). *Biom. J.*, **22**, 345–350.

[15] Goel, B. S. (1980b). *Biom. J.*, **22**, 351–358.

[16] Goel, B. S. and Nigam, A. K., (1979). *Biom. J.*, **21**, 277–285.

[17] Gorman, J. W. (1970). *J. Quality Tech.*, **2**, 186–194.

[18] Gorman, J. W. and Hinman, J. E. (1962). *Technometrics*, **4**, 463–487. (A lucid presentation on the use of simplex-lattice designs and Scheffé's polynomials.)

[19] Hare, L. B. (1974). *Food Tech.*, **28**, 50–62.

[20] Hare, L. B. (1979). *Technometrics*, **21**, 159–173.

[21] Kenworthy, O. O. (1963). *Ind. Quality Control*, **19**, 24–26.

[22] Kurotori, I. S. (1966). *Ind. Quality Control*, **22**, 592–596. (The introduction of pseudocomponents for exploring a subregion of the simplex space where the subregion is defined by the placing of lower bounds on the component proportions.)

[23] Marquardt, D. W. and Snee, R. D. (1974). *Technometrics*, **16**, 533–537.

[24] McLean, R. A. and Anderson, V. L. (1966). *Technometrics*, **8**, 447–454. (Presents an algorithm for locating the vertices of a constrained subregion of the simplex.)

[25] Murty, J. S. and Das, M. N. (1968). *Ann. Math. Statist.*, **39**, 1517–1539.

[26] Park, S. H. (1978). *Technometrics*, **20**, 273–279.

[27] Quenouille, M. H. (1959). *J. R. Statist. Soc. B*, **21**, 201–202.

[28] Saxena, S. K. and Nigam, A. K. (1977). *Technometrics*, **19**, 47–52.

[29] Scheffé, H. (1958). *J. R. Statist. Soc. B*, **20**, 344–360. (The first paper to present the simplex-lattice designs and the corresponding polynomial models for mixture experiments.)

[30] Scheffé, H. (1963). *J. R. Statist. Soc. B*, **25**, 235–263.

[31] Snee, R. D. (1971). *J. Quality Tech.*, **3**, 159–169.

[32] Snee, R. D. (1973). *Technometrics*, **15**, 517–528. (A lucid presentation of several model forms as well as ways of analyzing mixture data.)

[33] Snee, R. D. (1975). *Technometrics*, **17**, 149–159.

[34] Snee, R. D. (1979). *Chem. Tech.*, **9**, 702–710.

[35] Snee, R. D. and Marquardt, D. W. (1974). *Technometrics*, **16**, 399–408.

[36] Snee, R. D. and Marquardt, D. W. (1976). *Technometrics*, **18**, 19–29.

[37] Thompson, W. O. and Myers, R. H. (1968). *Technometrics*, **10**, 739–756.

(DESIGN OF EXPERIMENTS
FACTORIAL EXPERIMENTS
KIEFER–WOLFOWITZ PROCEDURE
LATTICE DESIGNS
LINEAR REGRESSION
PREDICTION
RESPONSE SURFACE DESIGNS
STEEPEST DESCENT, METHOD OF)

J. A. CORNELL

MIXTURE METHOD

The mixture method is one of several fundamental tools used in constructing efficient exact algorithms for generating repeated realizations of random variables on a computer (*see* GENERATION OF RANDOM VARIABLES). Its key feature is that it generates with high probability a realization from a simple distribution that is close to the desired one.

The mixture method is applicable and useful for both continuous and discrete (probability) distributions. Like other tools (e.g., acceptance–rejection method, inverse-distribution method), it may be used alone or in combination with other tools. Like almost all methods currently used for generating random variables, it is designed to use uniform random numbers, which are available and efficiently generated on most computers.

The mixture method is based on a representation of the (cumulative) probability distribution function or, more simply, the (discrete or continuous) probability density function $f(\cdot)$, from which realizations are desired, in terms of a mixture

$$f(\cdot) = \sum_{i=1}^{n} p_i \cdot f_i(\cdot), \tag{1}$$

where $f_i(\cdot)$ (> 0) denote density functions that are elements of the mixture, p_i $(p_i \geqslant 0, \sum_{i=1}^{n} p_i = 1)$ denote the mixture weights, and the positive integer n denotes the number of mixture elements (*see* MIXTURE DISTRIBUTIONS). The method obtains realizations from $f(\cdot)$ by generating realizations from $f_i(\cdot)$ with probability p_i. Specifically, for each desired realization X from $f(\cdot)$, the generating portion of the mixture method first generates a (integer) realization $I \sim P(I = i) = p_i$ of the mixture element identifier according to the discrete distribution $\{p_i, i = 1, 2, \ldots, n\}$, and then generates a realization from the selected element $f_i(\cdot)$ using some convenient method. That is, for each X desired from $f(\cdot)$, the following two steps are executed:

M1. (Choose at random an element of the mixture.)

Generate $I \sim P(I = i) = p_i$,

$i = 1, 2, \ldots, n$.

M2. (Generate at random a variable from the selected element of the mixture.)

Generate and return $X \sim f_I(\cdot)$.

It is readily apparent that the variable X generated by this method has the desired distribution: its density defined by its generation is just

$$\sum_{i=1}^{n} P(I = i) \cdot f_i(\cdot) = \sum_{i=1}^{n} p_i \cdot f_i(\cdot),$$

which from (1) is just the desired density $f(\cdot)$.

Whereas the mixture method will yield realizations X with the desired distribution for any mixture (1), only carefully chosen mixtures will provide efficient algorithms. The overall goals in making this choice are that (in step **M1**) the integer variable $I \sim P(I = i) = p_i$, $i = 1, 2, \ldots, n$, is efficiently generated, and that (in step **M2**) those densities $f_i(\cdot)$ corresponding to the large p_i are simple ones (easy to generate from). Two common choices for these $f_i(\cdot)$ are the uniform density and the triangular ("sum of two independent uniforms") density (*see* RECTANGULAR DISTRIBUTIONS).

Two main classes of mixture method algorithms can be distinguished: special-purpose algorithms and general-purpose algorithms. These are discussed separately in the next two sections.

SPECIAL-PURPOSE ALGORITHMS

Most implementations of the mixture method have been special-purpose algorithms: algorithms hand-fitted to the shape of the distribution $f(\cdot)$. Many clever special-purpose mixture algorithms have been proposed for various of the common families of probability distributions, notably the normal, t, χ^2, F, beta, gamma, Cauchy, Poisson, geometric, and binomial. The number of mixture elements is chosen to be small, typically five or less. The integer variable I is generated most simply when $n = 2$ merely by comparing a uniform (0, 1) variable with p_1. When $n > 2$, the integer variable may be generated by a simple search procedure such as sequential or binary search, or for large n more efficiently by one of the several discrete-variable generating techniques (e.g., Walker's Method*, Indexed-search Method).

The algorithm of Kinderman and Ramage [3] is one implementation of the mixture method for the normal distribution and provides a good illustration. The choice of mixture in this algorithm is:

$p_1 = 0.884$;

$f_1(\cdot)$ = a scaled version of the triangular (sum-of-two-independent-uniforms) density on $|x| \leqslant 2.22$.

$p_2 = 0.089$;

$f_2(\cdot)$ = remainder density (generated in three pieces by the acceptance–rejection method).

$p_3 = 0.027$;

$f_3(\cdot)$ = the normal tail $\phi(x) \cdot I(|x| > 2.22)$.

The density $f_1(\cdot)$ is the simple one, and realizations from it are generated most of the time ($p_1 = 0.884$). The densities $f_2(\cdot)$ and $f_3(\cdot)$ are residual ones from which variables are generated much less efficiently, but realizations are generated from these only $p_2 = 0.089$ and $p_3 = 0.027$ of the time, respectively. The triangular subdensity $p_1 \cdot f_1(\cdot)$ is shown in Fig. 1.

A common design strategy for special-purpose mixture algorithms is illustrated by this example. First choose a simple density $f_1(\cdot)$ that is close to the desired density $f(\cdot)$, then choose p_1 as large as possible under the restriction that $p_1 f_1(\cdot) \leqslant f(\cdot)$. One may also choose another simple density $f_2(\cdot)$ close to the density

$$\frac{f(\cdot) - p_1 \cdot f_1(\cdot)}{1 - p_1},$$

then choose p_2 as large as possible under the restriction that $p_2 \cdot f_2(\cdot) \leqslant f(\cdot) - p_1 \cdot f_1(\cdot)$, and so on, up to the selected number of mixture densities.

GENERAL-PURPOSE ALGORITHMS

Several general-purpose implementations of the mixture method have been developed in recent years. These are algorithms intended for application to a wide class of distributions. They compute as an initial overhead and use during the generation of realizations a stored table of constants that tailor the method to the desired distribution. General-purpose mixture methods for discrete distributions include Walker's method*, the in-

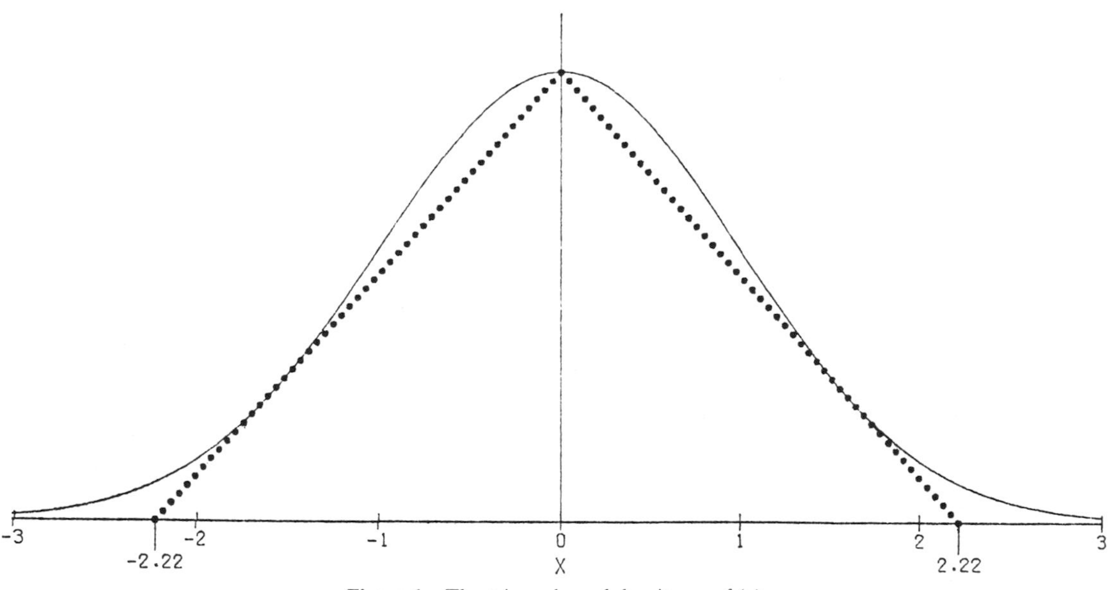

Figure 1 The triangular subdensity $p_1 \cdot f_1(\cdot)$.

dexed-search method, and Marsaglia's table method*. There are also several general-purpose mixture methods, some under current development, that are potentially applicable to broad classes of *continuous* distributions: the rectangle-wedge-tail method [11], the ziggurat method [12], the alias-rejection-mixture method [7], and the kernel method [8]. Further development of these toward the goal of general applicability to continuous distributions is needed.

The *equiprobable mixture method* underlies almost all general-purpose methods proposed to date. In these applications of the mixture method the number n of mixture densities is chosen to be relatively large, the weights are chosen to be the *same* (equiprobable mixture), and the vast majority of the mixture densities are typically chosen to be uniform (either continuous or discrete). Unlike the special-purpose applications of the mixture method, in these applications the efficiency of generating $I \sim p_i$ is attained not by choosing n to be small, but rather by choosing the p_i to all be the same, so that I is a discrete uniform random variable, easily generated on a computer. For example, Marsaglia's table method [9] for discrete distributions uses an equiprobable mixture for each place of a decimal (or other base) expansion of a probability, and uses degenerate (one-point) densities as elements of the mixtures. Another example for discrete distributions is Walker's table method [5, 14], which uses one equiprobable mixture with two-point densities as the elements of the mixture.

ACCEPTANCE-COMPLEMENT METHOD

The recent *acceptance-complement method* [6] allows one to obtain efficient sampling from the mixture even when the probability p_1 of the simple element $f_1(\cdot)$ is not large. Preparations for this method include:

1. Choose a mixture with two elements:

$$f(\cdot) = p_1 \cdot f_1(\cdot) + p_2 \cdot f_2(\cdot).$$

2. Choose a dominating density $h^*(\cdot) \geqslant p_2 \cdot f_2(\cdot)$ that bounds the subdensity $g_2(\cdot) = p_2 \cdot f_2(\cdot)$.

To generate random variables, the following two steps are performed *once* per variable generated:

AC1. Prepare for the acceptance-complement comparison by generating the (independent) random variables X and U.

Generate $X \sim h^*(\cdot)$.

Generate $U \sim$uniform $(0, 1)$, independent of X.

AC2. Acceptance-complement comparison.

If $U \leqslant g_2(X)/h^*(X)$, then return (accept) X.

Otherwise, generate and return Y, independent of (X, U), from the density $f_1(\cdot)$.

Flexibility is afforded by a choice of both the two-element mixture $f(\cdot) = p_1 \cdot f_1(\cdot) + p_2 \cdot f_2(\cdot)$ and the dominating density $h^*(\cdot)$. As with the acceptance–rejection method of Von Neumann [13], to which it is related, the acceptance–complement method has good efficiency when the dominating density $h^*(\cdot)$ can be chosen to be a density from which random variables are easily generated, and the ratio $g_2(X)/h^*(X)$ is not difficult to compute. The latter guideline can be relaxed through comparing U first to simple bounding functions (e.g., constants) for the ratio $g_2(X)/h^*(X)$.

The acceptance-complement method has been implemented as a special-purpose method hand-tailored to specific distributions, for example to the Cauchy distribution* [6]. It has also been used as an ingredient in some of the equiprobable mixture-type general-purpose methods discussed in the preceding section. Marsaglia [12] and Kronmal et al. [7] have suggested its use in general-purpose methods and illustrated its use specifically for the normal distribution. As with other general-purpose methods, its routine implementation for broad classes of continuous distributions has not yet been accomplished.

HISTORY AND LITERATURE

The mixture method was suggested by Marsaglia [10]. Good accounts of this method and of other methods of random variable generation are provided in Kennedy and Gentle [2], Ahrens and Dieter [1] and Knuth [4].

References

[1] Ahrens, J. H. and Dieter, U. (1973). *Non-Uniform Random Numbers*. Institut für Math. Statistik, Technische Hochschule, Graz, Austria.
[2] Kennedy, W. J., Jr., and Gentle, J. E. (1980). *Statistical Computing*. Marcel Dekker, New York.
[3] Kinderman, A. J. and Ramage, J. G. (1976). *J. Amer. Statist. Ass.*, **71**, 893–896.
[4] Knuth, D. E. (1981). *The Art of Computer Programming*, Vol. 2: *Seminumerical Algorithms*, 2nd ed. Addison-Wesley, Reading, MA.
[5] Kronmal, R. A. and Peterson, A. V., Jr. (1979). *Amer. Statist.*, **33**, 214–218.
[6] Kronmal, R. A. and Peterson, A. V., Jr. (1981). *Amer. Statist. Ass.*, **76**, 446–451.
[7] Kronmal, R. A., Peterson, A. V., Jr., and Lundberg, E. D. (1979). *Proc. ASA Computing Section, 1978*, 106–110.
[8] Kronmal, R. A., Larsen, M. P., and Peterson, A. V., Jr. (1982). Eleventh International Biometric Conference, Toulouse, France, September 6–11, 1982. (The kernel-plus-acceptance-complement method: a general-purpose method for the computer generation of random variables from a continuous distribution.)
[9] Marsaglia, G. (1963). *Commun. ACM*, **6**, 37–38.
[10] Marsaglia, G. (1961). *Ann. Math. Statist.*, **32**, 894–898.
[11] Marsaglia, G., MacLaren, M. D. G., and Bray, T. A. (1964). *Commun. ACM*, **7**, 4–10.
[12] Marsaglia, G. (1982). Personal communication.
[13] Von Neumann, J. (1951). *Appl. Math. Ser.*, **12**, 36–38.
[14] Walker, A. J., (1977). *ACM Trans. Math. Software*, **3**, 253–256.

(GENERATION OF RANDOM VARIABLES
MARSAGLIA'S TABLE METHOD
RANDOM NUMBERS
SIMULATION
WALKER'S METHOD)

ARTHUR V. PETERSON
RICHARD A. KRONMAL

MIZUTANI DISTRIBUTION *See* GENERALIZED HYPERGEOMETRIC DISTRIBUTIONS

MÖBIUS FUNCTION

The Möbius function μ is defined on integers recursively as $\mu(1)$, $\mu(n) = 0$ if n is divisible by the square of a prime, and $\mu(n) = (-1)^r$ if n is a product of r distinct primes. The main properties of a Möbius function is that it is multiplicative and, moreover

$$\sum_{d \mid n} \mu(d) = \begin{cases} 1 & (n = 1) \\ 0 & (n > 1). \end{cases}$$

Here the summation is carried over all divisors d of n. Applications of the Möbius function occur in probabilistic number theory* [1].

Reference

[1] Kubilius, J. P. (1964). "Probabilistic methods in the theory of numbers." *American Mathematical Society Translations of Mathematics Monographs*, Vol. 11. American Mathematical Society, Providence, RI.

(PROBABILISTIC NUMBER THEORY)

MODAL BLOCK ALGORITHM *See* CLASSIFICATION

MODE *See* MEAN, MEDIAN, MODE, AND SKEWNESS

MODEL CONSTRUCTION: SELECTION OF DISTRIBUTIONS

The simplest parametric inference model assumes a random sample $X_1, \ldots, X_n$ on a parent random variable (rv) X with distribution function (df) F, where F is a member of a parametric class $\mathscr{F} = \{F_\theta : \theta \in \Omega\}$ of distribution functions, and $\theta = (\theta_1, \ldots, \theta_p)'$ is a p-component vector. When $\mathscr{F}$ is a known parametric class of distributions there are available in classical parametric inference many techniques for making inferences about the components of θ. The inferences made on the parameters depend on the parent family $\mathscr{F}$ used. Therefore, it is important that the parent family be selected with care.

In general, all available information should be used in selecting a parametric class $\mathscr{F}$. Information may be available from different sources. Sometimes it may be possible to deduce a model from a scientist's understanding of the phenomena producing the sample. Mathematical results, such as the central limit theorem* or Poisson process*, can sometimes be used to deduce a reasonable model. However, in many cases we shall have to use the information in the sample itself to validate a model—techniques for this purpose are *data based* methods, to which this discussion will be limited.

Given a sample, we consider the problem of selecting one of a collection of k families $\mathscr{F}_1, \ldots, \mathscr{F}_k$ of distribution functions as the best-fitting family. A selection procedure or rule assigns each sample $X_1, \ldots, X_n$ to exactly one of the families $\mathscr{F}_1, \ldots, \mathscr{F}_k$. For a particular rule, let P_j denote the probability that the class $\mathscr{F}_j$ is selected when it is the correct family, and let $w_1, \ldots, w_k$ denote nonnegative numbers that measure the relative utility of correctly choosing the individual families. For a given set of weights $w_1, \ldots, w_k$, we consider one selection rule better than another if it gives a larger value to the weighted sum $w_1 P_1 + \cdots + w_k P_k$.

Our approach to constructing a selection rule is as follows. A selection statistic S_j is first computed for each class, $j = 1, \ldots, k$. We then use the following rule:

Select F_j for which

$$w_j S_j = \max\{w_1 S_1, \ldots, w_k S_k\}. \quad (1)$$

In this article we discuss three types of selection statistics: optimal invariant, suboptimal invariant, and maximum likelihood selection statistics. The invariant rules are applicable only for continuous parent random variables. The maximum likelihood rules can, in principle, be used for both continuous and discrete distributions. *See also* DISCRIMINANT ANALYSIS.

INVARIANT SELECTION RULES

The material of this section is more accessible to readers familiar with the topic of invariance as given in Lehmann [13] or Fraser [8]. (*See also* INVARIANCE CONCEPTS IN STATISTICS.)

Suppose that a transformation g applied to the parent random variable X transforms the class $\mathscr{F}_j$ of distributions onto itself for $j = 1, \ldots, k$. If this transformation is applied to each sample member, the problem of selecting $\mathscr{F}_j$ for the transformed values is the same as for the untransformed values. Thus it is natural to consider a selection statistic S_j that is invariant with respect to the transformation g.

Suppose that there is a group G of g transformations under which F_j is invariant. Every statistic that is invariant under G can be expressed as a function of a maximal invariant statistic, say $m_j(x_1, \ldots, x_n)$. Suppose first that these statistics are all the same or equivalent for the k classes. Then it is natural to base the selection of a family $\mathscr{F}_j$ on the maximal invariant statistic, say $m(x_1, \ldots, x_k)$. We use as selection statistic S_j the value of the density of the maximal invariant statistic with respect to the same dominating σ-finite measure for every one of the k classes. With this choice of selection statistic used in the rule (1), the weighted sum $w_1 P_1 + \cdots + w_k P_k$ is a maximum for any invariant rule. This result is closely related to results for most powerful invariant tests for separate families of distributions. These rules are thus called optimal invariant selection (OIS) rules. See Lehmann [13], Hajek and Sidak [10], Quesenberry and Starbuck [15], and, especially, Hogg et al.

[11]. The weights w_j can be chosen to represent the relative utility of correct decisions for the various distribution classes or the prior probabilities of the classes, in which case the weighted sum maximized is the total probability of correct selection. In practice, we often use equal prior probabilities of $1/k$ for each class.

Important groups of transformations in practice include scale parameter transformations,

$$G_S = \{ g(x) = ax, a > 0 \};$$

location-scale transformations,

$$G_{LS} = \{ g(x) = ax + b, a > 0, -\infty < b < \infty \};$$

and scale-shape transformations,

$$G_{SS} = \{ g(x) = ax^b, a > 0, b > 0 \}.$$

Suppose first that $\mathscr{F}_j$ is a scale parameter class of distributions, that is, that $\mathscr{F}_j$ is invariant under the group G_S of transformations for $j = 1, \ldots, k$; and let f_j denote the density function of that member of $\mathscr{F}_j$ that has scale parameter θ (say) equal to one. A scale invariant selection statistic for family j is

$$S_j = \int_0^\infty f_j(\lambda x_1, \ldots, \lambda x_n)\lambda^{n-1}\, d\lambda. \quad (2)$$

If $\mathscr{F}_j$ is a location-scale family for $j = 1, \cdots, k$, let f_j denote the density function of the distribution in this class with scale parameter one and location parameter zero. A location-scale invariant selection statistic for this family is given by

$$S_j = \int_0^\infty \int_{-\infty}^\infty f_j(\lambda x_1 - \mu, \ldots, \lambda x_n - \mu) \times \lambda^{N-2}\, d\mu\, d\lambda. \quad (3)$$

If $\mathscr{F}_j$ is a scale-shape family, that is, a family invariant with respect to the transformation group G_{SS}, then a scale-shape selection statistic is given by

$$S_j = \int_0^\infty \int_0^\infty f_j(\gamma x_1^\lambda, \ldots, \gamma x_n^\lambda)\gamma^{n-1}\lambda^{n-2} \times (x_1 \ldots x_n)^\lambda\, d\gamma\, d\lambda, \quad (4)$$

where f_j is the density of the family with both scale and shape parameters equal to one.

To illustrate the use of invariant selection statistics, the scale invariant procedure will be given for the failure distributions of Table 1. Note that the shape parameters (α, β, and σ) for the gamma*, Weibull*, and lognormal* densities are assumed known. For most problems it is easier to compute the logarithm of the selection statistic than the statistic itself, and for this reason we give the logarithms of the scale invariant selection statistics for the distributions of Table 1 in Table 2.

To apply the scale invariant selection procedure we compute the values of $\ln S_j$ given in Table 2 and select the appropriate distribution. These particular selection procedures were studied in Kent and Quesenberry (K-Q) [12], with $w_1 = \cdots = w_k = 1/k$, and

Table 1 Densities of Failure Distributions

Name	Symbol	Density
Exponential $\theta > 0$, all densities	$E(\theta)$	$\theta^{-1}\exp(-x/\theta)$
Gamma $\alpha > 0$, known	$G(\theta, \alpha)$	$\theta^{-\alpha}[\Gamma(\alpha)]^{-1}x^{\alpha-1}\exp(-x/\theta)$
Weibull $\beta > 0$, known	$W(\theta, \beta)$	$\frac{\beta}{\theta}\left(\frac{x}{\theta}\right)^{\beta-1}\exp\{-(x/\theta)^\beta\}$
Lognormal $\sigma > 0$, known	$\mathrm{LN}(\theta, \sigma)$	$\frac{1}{\sigma x\sqrt{2\pi}}\exp[-\{\ln(x/\theta)\}^2/2\sigma^2$

Table 2 Optimal Invariant Selection Statistics

Family	Logarithm of S
$E(\theta)$	$\ln[\Gamma(n)] - n \cdot \ln\left(\sum_{i=1}^{n} x_i\right)$
$G(\theta, \alpha)$ α known	$-n \cdot \ln[\Gamma(\alpha)] + (\alpha - 1)\left(\sum_{i=1}^{n} \ln x_i\right)$ $+\ln[\Gamma(n\alpha)] - n\alpha \cdot \ln\left(\sum_{i=1}^{n} x_i\right)$
$W(\theta, \beta)$ β known	$(n-1) \cdot \ln(\beta) + \ln[\Gamma(n)]$ $+(\beta - 1)\left(\sum_{i=1}^{n} \ln x_i\right)$ $-n \cdot \ln\left(\sum_{i=1}^{n} x_i^{\beta}\right)$
$\mathrm{LN}(\theta, \sigma)$ σ known	$-(n-1) \cdot \ln\left(\sigma\sqrt{2\pi}\right)$ $-\frac{1}{2}\ln n - \sum_{i=1}^{n} \ln x_i - \left[\sum_{i=1}^{n} \ln^2 x_i - \left(\sum_{i=1}^{n} \ln x_i\right)^2 \Big/ n\right] \Big/ (2\sigma^2)$

the formulas of Table 2 as well as some of the numerical results to be given in the Example are from that paper.

In some problems one or more of the classes of distributions $\mathscr{F}_j$ will involve parameters other than those related to the invariant selection statistic. For example, the gamma, Weibull, and lognormal distributions of Table 1 involve shape parameters denoted by α, β, and σ, respectively. In these cases the scale invariant selection statistics are functions of the additional parameters and in order to compute the selection statistics these parameters will be replaced by estimates. In general, the resulting selection statistics and procedures are no longer optimal, or necessarily invariant. However, in many problems this approach gives very good procedures that do turn out to be invariant. These procedures are here called suboptimal invariant selection (SIS) rules.

As an example we substitute maximum likelihood estimates for α, β and σ in the gamma, Weibull, and lognormal selection statistics of Table 2. This procedure will be illustrated in the Example. Siswadi and Quesenberry [17] recommended this suboptimal procedure over the optimal scale-shape invariant procedure on the basis of computational ease and relatively good performance in terms of the total probability of correct selection.

MAXIMUM LIKELIHOOD SELECTION RULES

When the parameter vector θ is completely known, it is natural to use the likelihood function $L_j(x_1, \ldots, x_n; \theta)$, say, as the selection statistic, in view of the Neyman–Pearson lemma*. Since θ is not known here, the maximum likelihood estimate $\hat{\theta}$ will be substituted for it to obtain the maximum likelihood selection (MLS) statistic:

$$S_j = \sup_{\theta} L_j(x_1, \ldots, x_n; \theta),$$
$$= L_j(x_1, \ldots, x_n; \hat{\theta}). \tag{5}$$

The MLS rule consists of using (5) in the rule (1). When conditions are satisfied to assure convergence in probability of $\hat{\theta}$ to θ, then for n large, this MLS procedure should be a good approximation to the rule based on known θ that maximizes $w_1 P_1 + \cdots +$

Table 3 ML Selection Statistics for Failure Distributions

Distribution	$\ln S$
$E(\theta)$	$-n \ln \hat{\theta} - \sum_{1}^{n} \frac{x_j}{\hat{\theta}}$
$G(\theta, \alpha)$	$-n[\hat{\alpha} \ln \hat{\theta} + \ln\{\Gamma(\hat{\alpha})\}]$ $+(\hat{\alpha} - 1)\sum_{1}^{n} \ln x_j - \sum_{1}^{n} \frac{x_j}{\hat{\theta}}$
$W(\theta, \beta)$	$n \ln \hat{\beta} - n\hat{\beta} \ln \hat{\theta}$ $+(\hat{\beta} - 1)\sum_{1}^{n} \ln x_j - \sum_{1}^{n} (x_j/\hat{\theta})^{\hat{\beta}}$
$\mathrm{LN}(\theta, \sigma)$	$-n \ln(\hat{\sigma}\sqrt{2\pi}) - \sum_{1}^{n} \ln x_j$ $-\sum_{1}^{n} \frac{[\ln(x_j/\hat{\theta})]^2}{2\hat{\sigma}^2}$

$w_k P_k$. Cox [4] has observed that the MLS procedure (for a location-scale problem) can be considered an approximation to the Bayes rule with equal prior probabilities for families. MLS rules are often, in fact, invariant with respect to the group G of transformations that define the OIS procedures of the last section. When this is so, generally the OIS procedures are superior. However, in some problems, the rules are actually the same; in many problems the OIS procedure is difficult to find and to compute, whereas the MLS procedure is relatively easy to find and compute, and gives weighted sums reasonably close to the best possible. For these reasons, Dumonceaux et al. [6] recommended the MLS procedure and Dumonceaux and Antle [5] applied this approach to select between lognormal and Weibull distributions. Bain and Englehardt [2] used an MLS procedure to select between gamma and Weibull distributions. Siswadi and Quesenberry [17] considered an MLS procedure for selecting among Weibull, lognormal, and gamma distributions with type I censored data.

In practice it is often easier to compute and compare $\ln S_j$ than S_j itself. The $\ln S_j$'s for the distributions of Table 1 are given in Table 3. The values of these selection statistics will be given in the Example.

Example. To illustrate the application of the SIS and MLS procedures, we consider selecting one of the four distributions of Table 1 to fit the yarn-strain data of Table 6 of K-Q. This data was reported in 1970 by Picciotto, who presented 22 sets of observations from experiments testing the tensile fatigue characteristics of a polyester/viscose yarn to study the problem of warp breakage during weaving. The experiment consisted of placing 100 samples of yarn into a 10-station testing apparatus that subjected the yarn to 80 cycles per minute of a given strain level. The cycle at which the yarn failed (cycles-to-failure) was recorded. Table 4 gives the frequency distribution for one of these samples at a given strain level (2.3%). Also, the ML estimates of the parameters of the distributions and of the SIS selection statistics reported in K-Q, as well as the $\ln S$ for MLS rules are given in Table 4.

The SIS method picks the gamma as the best fitting distribution while the MLS method prefers the Weibull distribution. However, there is a virtual tie between the gamma and Weibull distributions by both methods. The lognormal distribution is ranked third in both cases, and the exponential is last. The SIS and MLS methods agree very well for this example. As a final check we have drawn in Fig. 1 a histogram of the frequency distribution given in Table 4 and the four ML estimating densities: viz. $E(222.0)$, $G(99.2, 2.24)$, $W(247.9, 1.60)$, and $\text{LN}(174.7, 0.77)$. These graphs are in agreement with the conclusions stated.

Table 4 Grouped Data and Numerical Results for Example

Cycles-to-Failure-Data					
Interval	Frequency	Interval	Frequency	Interval	Frequency
0–100	21	300–400	12	600–700	1
100–200	32	400–500	4	700–800	0
200–300	26	500–600	3	800–900	1

		$\ln S$	
Distribution	Parameter Estimates	Scale Invariant	Maximum Likelihood
$E(\theta)$	$\hat{\theta} = 222.0$	-641.64	-640.26
$G(\theta, \alpha)$	$\hat{\theta} = 99.2, \hat{\alpha} = 2.24$	-627.03	-625.25
$W(\theta, \beta)$	$\hat{\theta} = 247.9, \hat{\beta} = 1.60$	-627.05	-625.20
$\text{LN}(\theta, \sigma)$	$\hat{\theta} = 174.7, \hat{\sigma} = .77$	-633.41	-631.75

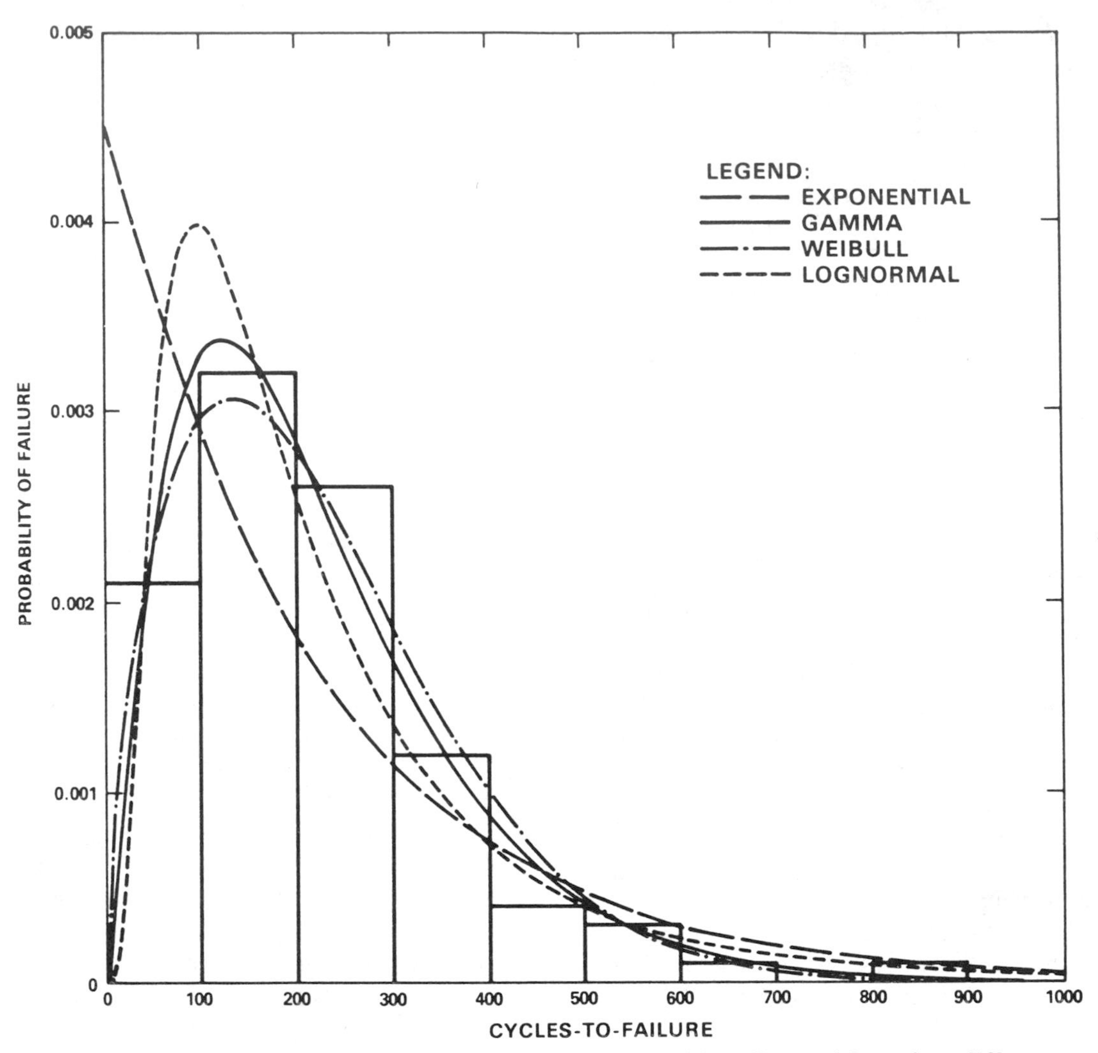

Figure 1 Histogram and ML densities for yarn data. (Reproduced from Kent and Quesenberry [12] by permission of the American Statistical Association.)

REMARKS ON OTHER TECHNIQUES

There are other approaches to the selection of distributions in addition to those based on invariant and maximum likelihood rules described. Some of these are graphical techniques such as the total time on test graphs of Barlow and Campo [3] for choosing failure distributions. Probability plots* are often used to aid in the selection of distributions in many contexts (see, e.g., Gnanadesikan [9] and Quesenberry et al. [16]). Nelson [14] considers hazard plotting* for censored data*.

A general approach is as follows. Let an omnibus composite goodness-of-fit* statistic (Anderson–Darling*, chi-squared, etc.) be computed from the sample for each class of distributions. Then we can choose that family that gives, say, the smallest value of the test statistic. If the P-values*—observed significance levels—of the test statistics can be computed, at least approximately, then the P-values can be compared and used as an index to select a distribution with the largest P-value, even when different goodness-of-fit statistics are computed for different families. Selection based on goodness-of-fit statistics in these ways in general will not give very efficient selection rules. See Dyer [7] for further discussion and some numerical comparisons for particular applications of some rules of this type for the location-scale or scale parameter families for $k = 2$.

Atkinson [1] poses a method that imbeds the rival distributions in a larger class with an additional parameter and selects by making tests on the extra parameter. Volodin [18] uses a similar general strategy to discriminate between gamma and Weibull distributions.

Summary

While we have considered explicitly only univariate random sample problems, the invariant and maximum likelihood methods apply in principle to problems including multivariate distributions, censored samples, and concomitant variables models. In practice, applications for these problems often lead to intractable mathematics or difficult computing problems. We hope that the methods discussed here will be helpful in providing some avenues of approach to problems of particular interest to readers.

Acknowledgment

We express appreciation to the American Statistical Association* for permission to reproduce some of the discussion, the entries of Tables 2 and 4 and Fig. 1 from the paper K-Q [12] in *Technometrics*.

References

[1] Atkinson, A. C. (1970). *J. R. Statist. Soc.*, **B32**, 323–345.

[2] Bain, L. J. and Engelhardt, M. (1980). *Commun. Statist. A*, **9**, 375–381.

[3] Barlow, R. E. and Campo, R. (1975). *Reliability and Fault Tree Analysis: Theoretical and Applied Aspects of System Reliability and Safety Assessment.* Society for Industrial and Applied Mathematics, Philadelphia, pp. 451–481.

[4] Cox, D. R. (1961). *Proc. 4th Berkeley Symp. Math. Statist. Prob.*, **1**, 105–123.

[5] Dumonceaux, R. and Antle, C. E. (1973). *Technometrics*, **15**, 923–926.

[6] Dumonceaux, R., Antle, C. E., and Haas, G. (1973). *Technometrics*, **15**, 19–27.

[7] Dyer, A. R. (1973). *J. Amer. Statist. Ass.*, **68**, 970–974.

[8] Fraser, D. A. S. (1957). *Nonparametric Methods in Statistics*. Wiley, New York, Secs. 2.3, 3.7.

[9] Gnanadesikan, R. (1977). *Methods for Statistical Data Analysis of Multivariate Observations*. Wiley, New York.

[10] Hajek, J. and Sidak, Z. (1967). *Theory of Rank Tests*. Academic Press, New York.

[11] Hogg, R. V., Uthoff, V. A., Randles, R. H., and Davenport, A. S. (1972). *J. Amer. Statist. Ass.*, **67**, 597–600.

[12] Kent, J. and Quesenberry, C. P. (1982). *Technometrics*, **24**, 59–65.

[13] Lehmann, E. L. (1959). *Testing Statistical Hypotheses*. Wiley, New York.

[14] Nelson, W. (1972). *Technometrics*, **14**, 945–966.

[15] Quesenberry, C. P. and Starbuck, R. R. (1976). *Commun. Statist. A*, **5**, 507–524.

[16] Quesenberry, C. P., Whitaker, T. B., and Dickens, J. W. (1976). *Biometrics*, **32**(4), 753–759.

[17] Siswadi and Quesenberry, C. P. (1982). *Nav. Res. Logist. Q.*, **29**, 557–569.

[18] Volodin, I. N. (1974). *Theory Prob. Appl.*, **19**, 383–393.

(DISCRIMINANT ANALYSIS
EXPLORATORY DATA ANALYSIS
SELECTION PROCEDURES)

C. P. QUESENBERRY

MODELING, STATISTICAL AND PROBABILISTIC *See* STATISTICAL AND PROBABILISTIC MODELING

MODELS I, II, AND III

Model I is another name for the *fixed-effects model.* It is a linear model in which

$$\text{observed value} = (\text{linear function of unknown parameters}) + \text{residual},$$

the residuals being mutually independent random variables with zero expected values. It is also known as the *parametric* or *systematic model*.

There is general agreement for the terminology used to describe Models I and II;

see, for example, Johnson and Leone [3, Chap. 13], Ostle and Mensing [4, Chap.10], and Snedecor and Cochran [5, Chap. 10]. These terms appear to have been first introduced by Churchill Eisenhart in 1947 [2]. Model II is another name for the *random-effects* or *components of variance model* (*see* VARIANCE COMPONENTS), in which

observed value
  = constant
    + (linear function of random variables)
    + residual.

In this model, the "random variables" are independent of the residuals and, like the residuals, have zero mean and are independent. However, the same value of a "random variable" can be a common component of several data values. (*See* FIXED-, RANDOM-, AND MIXED-EFFECTS MODELS, where these concepts are discussed for normally distributed residuals in more detail.)

Model III is sometimes used as another name for the mixed-effects model, in which [4, Chap. 10]

observed value
  = (linear function of
    unknown parameters)
    + (linear function of random variables)
    + residual.

Dunn and Clark [1, Chap. 9] depart from the preceding tradition in restricting Model II to random effects from an infinite population of possible values, and in defining Model III as a random-effects model in which the random effects are drawn from a finite population of possible values.

References

[1] Dunn, O. J. and Clark, V. A. (1974). *Applied Statistics: Analysis of Variance and Regression.* Wiley, New York.

[2] Eisenhart, C. (1947). *Biometrics*, **3**, 1–21.

[3] Johnson, N. L. and Leone, F. C. (1977). *Statistics and Experimental Design in Engineering and the Physical Sciences*, Vol. 2, 2nd ed. Wiley, New York.

[4] Ostle, B. and Mensing, R. W. (1975). *Statistics in Research*, 3rd ed. Iowa State University Press, Ames, IA.

[5] Snedecor, G. W. and Cochran, W. G. (1967). *Statistical Methods*, 6th ed. Iowa State University Press, Ames, IA.

(ANALYSIS OF VARIANCE
FIXED-, RANDOM-,
  AND MIXED-EFFECTS MODELS
GENERAL LINEAR MODEL
ONE-WAY CLASSIFICATION
VARIANCE COMPONENTS)

MODIFIED BISERIAL CORRELATION

See BISERIAL CORRELATION

MODIFIED HERMITE POLYNOMIALS

See CHEBYSHEV–HERMITE POLYNOMIALS

MODIFIED NORMAL DISTRIBUTIONS

Although this term might be applied generically to *any* distribution obtained by altering ("modifying") a normal distribution, it is more particularly used to refer to a class of distributions introduced by Romanowski [1, 2]. The alternative name—*modulated* normal distribution—distinguishes the class more distinctly.

These distributions are, in fact, compound (mixture*) distributions obtained by ascribing the power function distribution, with PDF $f_T(t) = (a+1)t^a$ for $0 \leqslant t \leqslant 1$ and $a \geqslant -1$, to $\sigma^{-2} \times$ (variance) of a normal distribution with zero mean.

The CDF is

$$F_x(x) = \frac{a+1}{\sigma\sqrt{2\pi}} \int_{-\infty}^{x} \int_0^1 t^{a-1/2} \times \exp\left\{-\tfrac{1}{2} y^2 (t\sigma^2)^{-1}\right\} dt\, dy.$$

The variance is $\sigma^2(a+1)/(a+2)$; the kurtosis is

$$\beta_2 = 3(a+2)^2 \{(a+1)(a+3)\}^{-1}.$$

The distribution is symmetrical about zero.

Special cases include:

$a = 0$ (equinormal*)

$a = \frac{1}{2}$ (radiconormal*)

$a = 1$ (lineonormal*)

$a = 2$ (quadrinormal*)

Some relevant tables can be found in Romanowski [3].

References

[1] Romanowski, M. (1964). *Bull. Géod.*, **73**, 195–216.

[2] Romanowski, M. (1969). *Metrologia*, **4**(2), 84–86.

[3] Romanowski, M. and Green, E. (1965). *Bull. Géod.*, **78**, 369–377.

MODIFIED (OR CURTAILED) SAMPLING PLANS

Modified sampling is a type of sequential sampling procedure. Usually the term has been used when sampling by attributes under either a binomial or a hypergeometric sampling plan in a quality control* sampling inspection setting (described in the following sections). Modified sampling implies that items are sampled one at a time, terminating when the ultimate decision is, in fact, already determined. In practice the phrase *curtailed sampling* is in more common usage than *modified sampling*. The basic idea can be applied to any nonnegative random variable whether discrete or continuous.

SINGLE SAMPLING AND THE BINOMIAL

Suppose that a production line process produces items a fraction p of which are defective. If $p = p_0$ (or less), the process is regarded as operating satisfactorily and no corrective action is necessary. On the other hand, if the fraction defective is as poor as $p = p_1 > p_0$, improvements are essential and it is desirable to uncover that state of affairs. To make a decision, the sampling plan (n, c) is used. This means that a random sample of n units are observed, the number of defective units x in that sample is counted, and it is concluded that $p \leqslant p_0$ only if $x \leqslant c$.

One way to determine the plan is to find the minimum n and the accompanying c such that

$$\Pr(X \leqslant c)\begin{cases} \geqslant 1 - \alpha_0 & \text{if} \quad p = p_0 \\ \leqslant \beta_1 & \text{if} \quad p = p_1, \end{cases} \qquad (1)$$

where X is a binomial random variable and α_0, β_1 are preselected small probabilities. For easy methods of finding such a plan see a monograph by Guenther [1]. This can also be done on a computer.

Regardless of the method used to determine n and c, sampling can be curtailed. Specifically, inspect items one at a time, stop sampling, and conclude that the process is operating

A. Unsatisfactorily as soon as $c + 1 = d$ defectives are observed, or

B. Satisfactorily as soon as $n - c$ nondefectives are observed,

whichever occurs first. The sampling plan is said to be *semicurtailed* if (**B**) is replaced by

B′. Satisfactorily as soon as n items are inspected without finding d defectives.

If sampling is curtailed or semicurtailed, the number of observations required to reach a decision, say W, is a random variable which can assume values from $c + 1$ to n. The probability distribution of W is easily derived [1, Exercises 1.15, 1.16]. Of particular interest is the expected value of W, say $E(W)$, sometimes called the *average sample number** (ASN). Letting $b(x; n, p)$ be the binomial probability density function and

$$E(r; n, p) = \sum_{x=r}^{n} b(x; n, p),$$

then if sampling is curtailed,

$$E(W) = \frac{c+1}{p}\left[E(c+2; n+1, p)\right] + \frac{n-c}{1-p}\left[1 - E(c+1; n+1, p)\right]. \qquad (2)$$

The corresponding formula for semicurtailed sampling is

$$E(W) = \frac{c+1}{p}\left[E(c+2; n+1, p)\right] + n\left[1 - E(c+1, n, p)\right]. \qquad (3)$$

SINGLE SAMPLING AND THE HYPERGEOMETRIC

If a lot of N items contains $k = k_0$ (or less) defectives, it is regarded as acceptable. On the other hand, if $k = k_1 > k_0$, the lot is unacceptable, and it is desirable to uncover that fact. As in the binomial case the decision is based upon a sampling plan (n, c). Now x is the observed number of defectives in a simple random sample of size n.

One way to determine the plan is to find the minimum n and accompanying c such that

$$\Pr(X \leqslant c)\begin{cases} \geqslant 1-\alpha_0 & \text{if} \quad k = k_0 \\ \leqslant \beta_1 & \text{if} \quad k = k_1, \end{cases} \tag{4}$$

where X is a hypergeometric* random variable and α_0, β_1 are preselected small probabilities. For easy methods of finding such a plan see Guenther [1]. This can also be done on a computer.

As with the binomial, sampling can be curtailed by following (**A**) and (**B**) or semicurtailed by replacing (**B**) by (**B**′). Letting $p(N, n, k, x)$ be the hypergeometric density and

$$P(N, n, k, r) = \sum_{x=0}^{r} p(N, n, k, x),$$

the counterparts of (2) and (3) are

$$\begin{aligned} E(W) = {} & \frac{(c+1)(N+1)}{k+1} \\ & \times\left[1 - P(N+1, n+1, k+1, c+1)\right] \\ & + \frac{(n-c)(N+1)}{N+1-k} \\ & \times P(N+1, n+1, k, c) \end{aligned} \tag{5}$$

and

$$\begin{aligned} E(W) = {} & \frac{(c+1)(N+1)}{k+1} \\ & \times\left[1 - P(N+1, n+1, k+1, c+1)\right] \\ & + nP(N, n, k, c). \end{aligned} \tag{6}$$

Again, the probability density of W is easily derived [1, Exercise 1.30].

CURTAILING WITH THE POISSON

For some material coming from a production line it is more appropriate to count defects per unit rather than defectives. This may be the case when the product is measured by the square yard (e.g., cloth, roll roofing) or is an assembled unit (e.g., television set, refrigerator). A model frequently used for computing probabilities associated with the number of defects per unit is the Poisson*.

Let μ be the expected number of defects per unit. If $\mu = \mu_0$ (or less), the process is operating satisfactorily. On the other hand, if $\mu = \mu_1 > \mu_0$, it is essential to discover this and take appropriate action. The decision can be made using a sampling plan (n, c). This implies that a random sample of n units is inspected, the total number of defects Y is observed, getting y, and it is concluded that $\mu \leqslant \mu_0$ only if $y \leqslant c$.

One way to determine the plan is to find the minimum n and the accompanying c such that

$$\Pr(Y \leqslant c)\begin{cases} \geqslant 1-\alpha_0 & \text{if} \quad \mu = \mu_0, \\ \leqslant \beta_1 & \text{if} \quad \mu = \mu_1, \end{cases} \tag{7}$$

where Y is a Poisson random variable with expected value $n\mu$ and α_0, β_1 are preselected small probabilities. For easy methods of finding such a plan, again see Guenther [1].

Regardless of the method used to determine n and c, sampling can be curtailed. Now termination is possible only if too many defects are counted, that is, as soon as $\sum_{i=1}^{w} x_i \geqslant d = c+1$, where $x_1, x_2, \ldots, x_w$ are the observed number of defects in the first w units inspected. Letting $p(x; \mu)$ be the Poisson density function and

$$E(r; \mu) = \sum_{x=r}^{\infty} p(x; \mu),$$

then [1, Exercise 1.40]

$$E(W) = n - \sum_{j=1}^{n-1} E(d; j\mu). \tag{8}$$

If only the fraction of the last unit inspected when the dth defect is found is used in the count, a somewhat simpler result is obtained. Now

$$E(W) = n[1 - E(d; n\mu)] + (d/\mu)E(d+1; n\mu) \quad (9)$$

and only two Poisson sums are required. This can be obtained from Hald's [2] formula (12.1.4) by replacing μ with $n\mu$. It can also be derived by using the waiting time distribution for a Poisson process*. (See, e.g., Hogg and Craig [3, Remark, pp. 104 and 105]. Their density of W is the one needed for (9) if $w < n$. Our W requires the density for $w = n$, that is, $\Pr(W = n)$. This is the same as their $\Pr(W > n) = 1 - E(d; n\mu)$. From this partly continuous, partly discrete density, the expectation (9) is easily obtained.)

DOUBLE SAMPLING WITH THE BINOMIAL AND HYPERGEOMETRIC

The decision for the kind of problem proposed in the preceding sections can be based on a double sampling* plan. Usually this requires two sample sizes n_1, n_2 and two acceptance numbers c_1 and c_2. A first sample of size n_1 is inspected and x, the observed number of defectives, is determined. If $x \leqslant c_1$, the decision is acceptance, while if $x \geqslant c_2 + 1$, the decision is rejection. On the other hand, if $c_1 + 1 \leqslant x \leqslant c_2$, a second sample of size n_2 is inspected and y, the observed number of defectives in the second sample, is counted. Now the decision for acceptance is made only if $x + y \leqslant c_2$. As with single sampling, plans that satisfy two power or operating characteristic (OC) conditions can be found. (See the monograph by Guenther [1].) Here one would certainly prefer to use a computer.

Curtailing or semicurtailing can be used in both samples; formulas for the expected number of observations have been derived. For the situation in which semicurtailed sampling is used in both samples, Guenther [1, pp. 34–35] gives these expected values for both the binomial and the hypergeometric.

Miscellaneous Comments

Some further discussion of curtailed sampling can be found in Chapters 2 and 12 of Hald's book [2].

Although curtailing or semicurtailing may not on the average save much sampling when quality is good ($p = p_0$, $k = k_0$, $\mu = \mu_0$), the reduction may be considerable when quality is poor ($p = p_1$, $k = k_1$, $\mu = \mu_1$).

References

[1] Guenther, W. C. (1977). *Sampling Inspection in Statistical Quality Control*, Monographs and Courses No. 37. Griffin, London. (Intermediate-level book on sampling inspection.)

[2] Hald, A. (1981). *Statistical Theory of Sampling Inspection by Attributes*. Academic Press, New York. (Mathematically, a moderately difficult book containing extensive information about attribute sampling, 57 pages of tables, and 13 pages of references. Good reference book.)

[3] Hogg, R. B. and Craig, A. T. (1978). *Introduction to Mathematical Statistics*, 4th ed. Macmillan, New York. (Upper division and first-year graduate text in mathematical statistics.)

(ACCEPTANCE SAMPLING
AVERAGE SAMPLE NUMBER
BINOMIAL DISTRIBUTION
DOUBLE SAMPLING
HYPERGEOMETRIC DISTRIBUTION
POISSON DISTRIBUTION
QUALITY CONTROL, STATISTICAL
SAMPLING PLANS)

WILLIAM C. GUENTHER

MODIFIED POWER SERIES DISTRIBUTION

Let X be a discrete random variable with probability distribution

$$P(X = x) = \frac{a(x)(g(\theta))^x}{f(\theta)}, \qquad x \in T, \quad (1)$$

where T is a subset of the set of nonnegative

integers; $a(x) > 0$; $g(\theta)$ and $f(\theta)$ are positive, finite, and differentiable. The class of distributions given by (1) has been called by Gupta [7] modified power series distributions (MPSD), also denoted MPSD($g(\theta)$, $f(\theta)$).

If $g(\theta)$ is invertible, (1) reduces to Patil's [34] generalized power series distribution*, and if, in addition, T is the entire set of nonnegative integers, it reduces to the power series distribution first given by Noack [31]. Hence the MPSD class includes, among others, the binomial*, the negative binomial*, the Poisson*, and the logarithmic series* distributions; it contains the generalized negative binomial distribution (GNBD) [19], the generalized Poisson distribution (GPD) [2], the generalized logarithmic series distribution (GLSD) [20], the lost game distribution (LGD), the distribution of the number of customers served in a busy period [3], and their truncated forms. These generalized distributions do not belong to the power series class and hence the MPSD class properly contains the class of power series distributions. A truncated MPSD is also a MPSD in its own right.

Generalized Negative Binomial Distribution

$$P(X = x) = \frac{n\Gamma(n+\beta x)}{x!\,\Gamma(n+\beta x - x + 1)} \frac{\{\theta(1-\theta)^{\beta-1}\}^x}{(1-\theta)^{-n}},$$

$x = 0, 1, 2, \ldots$; $0 < \theta < 1$; $|\theta\beta| < 1$, $\beta = 0$ or $\beta \geqslant 1$; $g(\theta) = \theta(1-\theta)^{\beta-1}$; $f(\theta) = (1-\theta)^{-n}$.

Generalized Poisson Distribution

$$P(X = x) = \frac{\lambda_1(\lambda_1 + \lambda_2 x)^{x-1}}{x!} \frac{(\theta e^{-\lambda_2\theta})^x}{e^{\lambda_1\theta}},$$

$x = 0, 1, 2, \ldots$; $\theta\lambda_1 > 0$, $\theta\lambda_2 < 1$; $g(\theta) = \theta e^{-\lambda_2\theta}$; $f(\theta) = e^{\lambda_1\theta}$.

Generalized Logarithmic Series Distribution

$$P(X = x) = \frac{\Gamma(x\beta)}{x\Gamma(x)\Gamma(x\beta - x + 1)} \frac{\theta^x(1-\theta)^{\beta x - x}}{-\ln(1-\theta)},$$

$x = 1, 2, 3, \ldots$; $0 < \theta < 1$, $\beta \geqslant 1$, $0 < \theta\beta < 1$; $g(\theta) = \theta(1-\theta)^{\beta-1}$; $f(\theta) = -\ln(1-\theta)$.

Lost-Game Distribution

$$P(X = x) = \frac{\binom{2x-a}{x}}{2x-a} \frac{a\{\theta(1-\theta)\}^x}{\theta^a},$$

$x = a, a+1, \ldots$; $0 < \theta < \frac{1}{2}$, $a > 1$; $g(\theta) = \theta(1-\theta)$; $f(\theta) = \theta^a$.

Number of Customers Served in a Busy Period

$$P(X = x) = \frac{(xk + x - 2)!}{x!\,(xk-1)!} \frac{\left[\theta/(\theta+1)^{k+1}\right]^x}{\theta/(\theta+1)},$$

$x = 1, 2, 3, \ldots$; $k \geqslant 1$, $\theta = \rho/k$, where ρ is the traffic intensity; $g(\theta) = \theta/(\theta+1)^{k+1}$; $f(\theta) = \theta/(\theta+1)$.

For the MPSD class, we discuss moments, negative and factorial moments*, cumulants*, and moment and probability-generating functions*, and their relationships. Maximum likelihood* and minimum variance unbiased estimators* are developed and specialized. Finally, some characterizations and miscellaneous results are described.

MOMENTS AND GENERATING FUNCTIONS OF MPSD

Let μ_r' = rth moment about the origin, μ_r = rth moment about the mean, $\mu^{[r]}$ = rth factorial moment, and κ_r = rth cumulant; the following results have been established by Gupta [9], Gupta and Singh [4], Gupta [6], Kumar and Consul [28], and Gupta [18].

$$E(X) = \frac{g(\theta)}{f(\theta)} \frac{f'(\theta)}{g'(\theta)}.$$

$$\mu_{r+1}' = \frac{g(\theta)}{g'(\theta)} \frac{d\mu_r'}{d\theta} + \mu_r'\mu_1'.$$

$$\mu_{r+1} = \frac{g(\theta)}{g'(\theta)} \frac{d\mu_r}{d\theta} + r\mu_2\mu_{r-1}.$$

In particular,

$$\mathrm{Var}(X) = \frac{g(\theta)}{g'(\theta)} \frac{d\mu_1'}{d\theta}.$$

$$\mu^{[r+1]} = \frac{g(\theta)}{g'(\theta)} \frac{d\mu^{[r]}}{d\theta} + \mu^{[r]}\mu^{[1]} - r\mu^{[r]}.$$

$$\kappa_{r+1} = \frac{g(\theta)}{g'(\theta)} \sum_{j=1}^{r} \binom{r-1}{j-1} \mu_{r-j}' \frac{d\kappa_j}{d\theta} - \sum_{j=2}^{r} \binom{r-1}{j-2} \mu_{r+1-j}' \kappa_j.$$

$$\mu_r = \sum_{y=0}^{\infty} \sum_{i=0}^{\infty} a(y+i) \frac{(y+i)!}{i!} \times \frac{(g(\theta))^{y+i}}{f(\theta)} S(r, y),$$

where $S(r, y)$ is the Stirling number* of the second kind.

$$\mu^{[r]} = \sum_{i=0}^{\infty} \frac{a(r+i)(r+i)!}{i!} \frac{(g(\theta))^{r+i}}{f(\theta)}.$$

The moment-generating function is

$$M_X(t) = \sum_{s=0}^{\infty} \frac{t^s}{s!} \sum_{y=0}^{\infty} \sum_{i=0}^{\infty} \binom{y+i}{i} \frac{y!\,a(y+i)}{f(\theta)} \times (g(\theta))^{y+i} S(s, y).$$

Let $M(r,k) = E(X+k)^{-r}$; $r = 1, 2, \ldots$. Then

$$M(r,k) = \frac{1}{f(\theta)(g(\theta))^k} \int_0^{\theta} M(r-1,k) \times g'(\theta) f(\theta) (g(\theta))^{k-1} d\theta,$$

$$M(1,k) = E\left(\frac{1}{X+k}\right) = \frac{\int_0^{\theta} g'(\theta) f(\theta) (g(\theta))^{k-1} d\theta}{f(\theta)(g(\theta))^k}.$$

Negative moments have been employed in deriving the bias of maximum likelihood estimators* (see Kumar and Consul [28] and Gupta [18]).

Suppose θ can be written as a function of $g(\theta)$, say $\psi(g(\theta))$, by means of Lagrange's formula. Then the probability generating function* of X [6] is

$$G_X(t) = \frac{f\psi(tg(\theta))}{f\psi(g(\theta))}.$$

The preceding results have proved useful in obtaining the moments, factorial and negative moments, cumulants, and generating functions for the generalized distributions described earlier.

ESTIMATION

In this section, we develop the maximum-likelihood estimators* (MLE) and the minimum variance unbiased (MVU) estimators of θ or some parametric function $\phi(\theta)$ of θ for the MPSD or its truncated version. The MVU estimators of the probability function of the MPSD will also be discussed.

Maximum Likelihood Estimation

Let $\overline{X}$ be the sample mean based on a random sample $X_1, X_2, \ldots, X_N$ from an MPSD $(g(\theta), f(\theta))$ with mean $\mu(\theta)$. The likelihood equation for θ (see Gupta [8]) is

$$N \frac{g'(\theta)}{g(\theta)} \left(\overline{X} - \mu(\theta)\right) = 0. \tag{2}$$

The solution of (2) yields

$$\overline{X} = \mu(\hat{\theta}). \tag{3}$$

If $\mu(\theta)$ is invertible, the MLE of θ, obtained by inverting (3), is given by $\hat{\theta} = \mu^{-1}(\overline{X})$. If $\mu(\theta)$ is not invertible, one may solve (3) iteratively using the Newton–Raphson method*.

The bias of $\hat{\theta}$ is given by

$$b(\hat{\theta}) = -\frac{1}{2N\mu_2^2(\theta)} \frac{g(\theta)}{g'(\theta)} \times \left[\mu_3(\theta) + \frac{g(\theta)g''(\theta) - (g'(\theta))^2}{(g'(\theta))^2} \mu_2(\theta)\right], \tag{4}$$

where $\mu_2(\theta)$ and $\mu_3(\theta)$ are the second and the third central moments of the MPSD. From (4), a necessary and sufficient condition for $\hat{\theta}$ to be unbiased is

$$\mu_2(\theta) = c\exp\left\{-\int\left[\frac{g''(\theta)}{g'(\theta)} - \frac{g'(\theta)}{g(\theta)}\right]d\theta\right\}, \tag{5}$$

where c is a constant independent of θ. For generalized power series distributions, the MLE is unbiased only for the Poisson distribution [8]. The asymptotic variance of $\hat{\theta}$ is given by

$$\operatorname{Var}(\hat{\theta}) = \frac{g(\theta)}{g'(\theta)} \Big/ \left(N\frac{d\mu}{d\theta}\right). \tag{6}$$

For a one-one function $\phi = \omega(\theta)$ of θ, the MLE of ϕ is given by $\hat{\phi} = \omega(\hat{\theta})$. The bias and the asymptotic variance of ϕ are

$$b(\hat{\phi}) = -\frac{1}{2N\mu_2^2}\frac{g(\theta)}{g'(\theta)}\frac{d\phi}{d\theta} \times\left[\mu_3 - \mu_2\left\{1 - \frac{g(\theta)g''(\theta)}{g'(\theta)^2} + \frac{g(\theta)}{g'(\theta)}\frac{d^2\phi}{d\theta^2}\Big/\frac{d\phi}{d\theta}\right\}\right], \tag{7}$$

$$\operatorname{Var}(\hat{\phi}) = \left\{\frac{g(\theta)}{g'(\theta)}\right\}\left(\frac{d\phi}{d\theta}\right)^2 \Big/ (N\mu_2(\theta)), \tag{8}$$

respectively.

The MLE of θ along with its bias and the asymptotic variance for some special MPSDs such as GNBD, GPD, and their decapitated versions have been derived in Gupta [8]. Kumar and Consul [28] developed recurrence relations for the negative moments of the MPSD and its displaced and decapitated forms and hence obtained the bias and the variance of the MLE of θ for certain MPSDs. Negative moments have also been utilized by Gupta [18] for obtaining asymptotic expressions for the bias and the variance of the MLE $\hat{\theta}$ of θ for the LGD. Similar expressions are derived for the distribution of the number of customers served in a busy period in an $M/E_k/1$ queue (see Gupta [5] and QUEUEING THEORY).

Minimum Variance Unbiased Estimation*

The MVU estimators for the MPSD are developed in the cases (a) when its range is known and (b) when its range is unknown. Let

$$I^r = \{r, r+1, r+2, \ldots\},$$

where r is a nonnegative integer, and let T in (1) be such that $T \subseteq I^0$.

RANGE KNOWN. Let $X_1, X_2, \ldots, X_N$ be a random sample from (1) and let $Z = \sum_{i=1}^N X_i$ be the sample sum. Then Z is sufficient and complete for θ (*see* SUFFICIENCY; COMPLETENESS). The distribution of Z is also an MPSD given by

$$P(Z = z) = b(z, N)(g(\theta))^z/(f(\theta))^N, \quad z \in D_n, \tag{9}$$

where

$$D_n = \left\{z \,\middle|\, z = \sum_{i=1}^N X_i,\ X_i \in T,\ i = 1, 2, \ldots, N\right\} \subseteq I^0.$$

The following theorem gives a necessary and sufficient condition for $\phi(\theta)$ to admit a unique MVU estimator [12, 29].

Theorem. There exists an essentially unique unbiased estimator of $\phi(\theta)$ with minimum variance if and only if $\phi(\theta)\{f(\theta)\}^N$ is analytic at the origin and has an expansion of the form

$$\phi(\theta)\{f(\theta)\}^N = \sum_{z \in E_n} c(z, N)(g(\theta))^z, \tag{10}$$

where $c(z, N) \neq 0$ for $z \in E_n \subseteq I^0$ and $E_n \subseteq D_n$. When $\phi(\theta)$ is MVU estimable, the estimate is given by

$$\psi(z, N) = \begin{cases} \dfrac{c(z, N)}{b(z, N)}, & \text{if } z \in E_n, \\ 0, & \text{otherwise.} \end{cases} \tag{11}$$

Using this theorem, MVU estimators of θ and $\phi(\theta)$ have been derived by Gupta [14, 20], Jani [24–26], and Kumar and Consul

[29] for some MPSDs and their left truncated versions when the truncation point is known. The MVU estimators for bivariate and multivariate versions of the MPSD have been considered by Shoukri [36] and Patel [33]. (*See also* MULTIVARIATE POWER SERIES DISTRIBUTIONS.)

RANGE UNKNOWN. For this case, Kumar and Consul [29] and Jani [24] have developed MVU estimators of r^m and $\phi(\theta)$ and derived results for some left truncated MPSDs. See also Jani [22, 23] and Patel and Jani [32].) The MVU estimators for truncated versions of multivariate MPSD are developed by Patel [33].

MVU ESTIMATION FOR THE PROBABILITIES OF MPSD. The MVU estimator for the probability $P(X = x)$, $x \in T$, of the MPSD has been developed by Kumar and Consul [29], Jani [24], and Gupta and Singh [17] and is given by

$$\widehat{P(X = x)} = P(X = x \mid Z = z) = \frac{a(x)b(z - x, N - 1)}{b(z, N)}, \quad z \in (N - 1)[T] + x.$$

MVU estimators of $P(X = x)$ for certain special MPSDs are developed by these authors.

CHARACTERIZATIONS

In this section, we present characterizations of the MPSD in damage models*, reliability theory*, and renewal processes*.

Length-Biased Distributions

Suppose X has the MPSD $(g(\theta), f(\theta))$ given by (1). Define a random variable Y_1 with probability function

$$P(Y_1 = y) = \frac{yP(X = y)}{\mu(\theta)}. \quad (12)$$

Then Y_1 has a length-biased distribution. Such distributions arise in life-length studies (see Gupta [14]). Then $E(Y_1 - 1) = E(X)$ if and only if X is Poisson [9]; $E(Y_1) = 1 + 2E(X)$ if and only if X is geometric [14].

Forward and Backward Recurrence Times in Renewal Processes

In fatigue studies* let X represent the number of cycles to failure that measures the life of a component operating in a system. The sequence of component life lengths forms a renewal process. At any time t, let U_t and V_t be the backward and the forward recurrence times in this process. For large values of t, the distribution of U_t or V_t is given by

$$P(Y = y) = P(X > y)/\mu(\theta), \quad (13)$$

where $\mu(\theta) = E(X)$. Let X have an MPSD $(g(\theta), f(\theta))$. Then, X has a geometric distribution if and only if $E(Y) = E(X)$ for all θ in a set I such that the values $g(\theta)$ fill a nondegenerate interval (see Gupta [10]).

Ratio of Variance and Mean

Let X have a MPSD $(g(\theta), f(\theta))$ and $\psi(\theta) = \mathrm{Var}(X)/E(X)$. Then [13]

$$\frac{f'(\theta)}{f(\theta)} = c\frac{g'(\theta)}{g(\theta)}\exp\left\{\int \psi(\theta)\frac{g'(\theta)}{g(\theta)}\,d\theta\right\}, \quad (14)$$

where c is a constant. The following characterizations are available for the geometric and the Poisson distributions:

X has a geometric distribution* if and only if $\psi(\theta) = 1 + \mu(\theta)$.

X has a Poisson distribution if and only if $\psi(\theta) = 1$.

Correlation Between Numbers of Objects of Two Types

Let X have a binomial distribution with parameters N and p. Suppose N has MPSD $(g(\theta), f(\theta))$. The correlation* coefficient ρ between X and $N - X$, the numbers of objects of two types, is given by (15). We have the following characterizations in terms of ρ:

$\rho = 0$ if and only if N has a Poisson distribution [11].

Let $g(\theta)$ be a monotonic increasing function in a subspace ω of the parameter space Ω. Then $\rho \gtrless 0$ according as $\ln f(\theta)$ is convex or concave in ω with respect to the function $g(\theta)$. If $g(\theta)$ is a decreasing function of θ in ω, then $\rho \gtrless 0$ according as $\ln f(\theta)$ is concave or convex in ω with respect to the function $g(\theta)$ [21, 35].

Cumulants

A discrete probability distribution is a MPSD if and only if the recurrence relation

$$\kappa_{r+1} = \frac{g(\theta)}{g'(\theta)} \frac{\partial \kappa_r}{\partial \theta}$$

between its cumulants κ_{r+1} and κ_r, $r = 1, 2, \ldots,$ holds [24].

Miscellaneous

MISCLASSIFICATION. Jani and Shah [26] considered a situation where the observation corresponding to $x = 1$ of an MPSD is misclassified as $x = 0$ with probability α, $0 \leqslant \alpha \leqslant 1$. They derived recurrence relations for raw and central moments of the misclassified MPSD and developed MLEs for α and θ. (See also Lingappaiah and Patel [30]).

APPLICATIONS IN GENETICS. Let X and Y denote the number of boys and girls, respectively, in a family with N children. Of interest is the correlation ρ between the random variables X and $Y = N - X$ when X has a binomial distribution with parameters N and p, and N itself is regarded as a random variable with a MPSD [11]:

$$\rho = \frac{(pq)^{1/2}[f(\theta)E'_\theta(N) - f'(\theta)]}{[pf(\theta)E'_\theta(N) + qf'(\theta)]^{1/2} \times [qf(\theta)E'_\theta(N) + pf'(\theta)]^{1/2}}, \quad (15)$$

where $E(N)$ is the expected value of N and prime denotes differentiation with respect to θ. For the case $p = q = \frac{1}{2}$, the forms of $f(\theta)$ are characterized for $\rho \gtrless 0$. A table of expressions for ρ is given for certain MPSDs. (See also Janardan [21] and Rao [35].)

RESOLUTION OF MIXTURES. Abu-Salih [1] considered the resolution of a mixture of observations from two MPSDs and used the method of maximum likelihood to identify the population of origin of each observation and to estimate the parameters of that population.

TAIL PROBABILITIES. Jani and Shah [25] derived an integral expression for the tail probabilities of an MPSD in terms of absolutely continuous distributions. Some MPSDs are given as examples.

References

[1] Abu-Salih, M. S. (1980). *Rev. Colomb. Mat.* **19**, 197–208.

[2] Consul, P. C. and Jain, G. C. (1973). *Technometrics*, **15**, 791–799.

[3] Daniels, H. E. (1961). *J. R. Statist. Soc. B*, **23**, 409–413.

[4] Gupta, P. L. and Singh, J. (1980). *Statistical Distributions in Scientific Work*. Vol. 4, C. Taillie, G. P. Patil, and B. A. Baldessari, eds. Reidel, Dordrecht, pp. 189–195.

[5] Gupta, P. L. (1982). *Commun. Statist. A*, **11**, 711–719.

[6] Gupta, P. L. (1982). *Math. Operat. Statist.*, **13**, 99–103.

[7] Gupta, R. C. (1974). *Sankhyā B*, **36**, 288–298; **37**, 255 (erratum, 1975). (This paper introduces the class of modified power series distributions. Examples and some characterizations are also provided. Intermediate level. Essential reading.)

[8] Gupta, R. C. (1975). *Commun. Statist. A*, **4**, 689–697. (Maximum likelihood estimators are developed. Expressions for bias and asymptotic variance are developed along with some characterizations.)

[9] Gupta, R. C. (1975). *Commun. Statist. A*, **4**, 761–765.

[10] Gupta, R. C. (1976). *Scand. J. Statist.*, **3**, 215–216.

[11] Gupta, R. C. (1976). *Sankhyā B*, **38**, 187–191.

[12] Gupta, R. C. (1977). *Commun. Statist. A*, **6**, 977–991. (Minimum variance unbiased estimators are developed for the MPSD and its truncated versions.)

[13] Gupta, R. C. (1977). *Math. Operat. Statist.*, **8**, 523–527.

[14] Gupta, R. C. (1979). *Commun. Statist. A*, **8**, 601–607.

[15] Gupta, R. C. (1979). *Commun. Statist. A*, **8**, 685–697.

[16] Gupta, R. C. (1981). *Statistical Distributions in Scientific Work*. Vol. 4, C. Taillie, G. P. Patil, and B. A. Baldessari, eds. Reidel, Dordrecht, pp. 341–347.

[17] Gupta, R. C. and Singh, J. (1982). *Math. Operat. Statist.*, **13**, 71–77. (MVU estimators for the probabilities of the MPSD are developed using the Rao–Blackwell theorem.)

[18] Gupta, R. C. (1984). *J. Statist. Plan. Inf.*, **9**, 55–62.

[19] Jain, G. C. and Consul, P. C. (1971). *SIAM J. Appl. Math.*, **21**, 501–513. (A new distribution called the generalized negative binomial distribution is presented. This is an example of an MPSD.)

[20] Jain, G. C. and Gupta, R. P. (1973). *Trab. Estadistica*, **24**, 99–105.

[21] Janardan, K. G. (1981). "Correlation between the numbers of two types of children in a family with the Markov–Pólya survival model." *Tech. Report No. 81-16*. Institute for Statistics and Applications, Dept. Mathematics and Statistics, University of Pittsburgh.

[22] Jani, P. N. (1977). *Sankhyā B*, **39**, 258–278.

[23] Jani, P. N. (1978). *J. Indian Statist. Ass.*, **16**, 41–48.

[24] Jani, P. N. (1978). *Metron*, **36**, 173–185.

[25] Jani, P. N. and Shah, S. M. (1979). *Metron*, **37**, 75–79.

[26] Jani, P. N. and Shah, S. M. (1979). *Metron*, **37**, 121–136.

[27] Kemp, A. W. and Kemp, C. D. (1968). *J. R. Statist. Soc. B*, **30**, 160–163.

[28] Kumar, A. and Consul, P. C. (1979). *Commun. Statist. A*, **8**, 151–166.

[29] Kumar, A. and Consul, P. C. (1980). *Commun. Statist. A*, **9**, 1261–1275.

[30] Lingappaiah, G. S. and Patel, I. D. (1979). *Gujarat Statist. Rev.*, **6**, 50–60.

[31] Noack, A. (1950). *Ann. Math. Statist.*, **21**, 127–132.

[32] Patel, S. R. and Jani, P. N. (1977). *J. Indian Statist. Ass.*, **15**, 157–159.

[33] Patel, S. R. (1979). *Metrika*, **26**, 87–94.

[34] Patil, G. P. (1962). *Ann. Inst. Statist. Math.*, **14**, 179–182.

[35] Rao, B. R. (1981). *Commun. Statist. A*, **10**, 249–254.

[36] Shoukri, M. M. (1982). *Biom. J.*, **24**, 97–101.

(LOGARITHMIC SERIES DISTRIBUTION
MULTIVARIATE POWER SERIES DISTRIBUTIONS
NEGATIVE BINOMIAL DISTRIBUTION
POISSON DISTRIBUTIONS
POWER SERIES DISTRIBUTIONS)

RAMESH C. GUPTA
RAM C. TRIPATHI

MODULUS TRANSFORMATION

A family of transformations of data proposed by John and Draper [1] to obtain approximate normality from symmetric long-tailed distributions:

$$y^{(\lambda)} = \begin{cases} \operatorname{Sign}(y)\left[\dfrac{(|y|+1)^{\lambda}-1}{\lambda}\right], & \lambda \neq 0, \\ \operatorname{Sign}(y)\left[\log(|y|+1)\right], & \lambda = 0. \end{cases}$$

Here sign(y) equals 1 if $y > 0$ and -1 if $y < 0$.

Observe that if $\lambda < 0$, $y^{(\lambda)}$ is restricted to the interval $[\lambda^{-1}, -\lambda^{-1}]$. This family of transformations is monotonic, continuous at $\lambda = 0$ and applicable in the presence of negative values. It is closely related to the transformations proposed by Box and Cox*.

Reference

[1] John, J. A. and Draper, N. R. (1980). *Appl. Statist.*, **29**, 190–197.

(BOX AND COX TRANSFORMATION)

MOIVRE, ABRAHAM DE *See* DE MOIVRE, ABRAHAM

MOIVRE–LAPLACE THEOREM (GLOBAL)

A version of this theorem, sometimes known as the integral (or global) Moivre–Laplace theorem and a particular case of the central limit theorem*, can be stated as follows.

Let

$$P_n(k) = \binom{n}{k} p^k q^{n-k},$$

$$0 < p < 1, \quad q = 1 - p, \quad k = 0, 1, \ldots, n,$$

the $P_n(k)$ being binomial* probabilities,

$$P_n(a, b] = \sum_{a < x \leqslant b} P_n\left(np + x\sqrt{npq}\right).$$

Then

$$\sup_{-\infty < a < b \leqslant \infty} \left| P_n(a, b] - \left(\sqrt{2\pi}\right)^{-1} \int_a^b e^{-x^2/2}\, dx \right|$$

$\to 0$ as $n \to \infty$.

An alternative version is: if $S_n = X_1 + \cdots + X_n$ is the sum of n independent Bernoulli random variables* with common parameter p, then

$$\sup_{-\infty < a < b \leqslant \infty} \left| P\left\{ a < \frac{S_n - ES_n}{\sqrt{\mathrm{Var}(S_n)}} \leqslant b \right\} - \frac{1}{\sqrt{2\pi}} \int_a^b e^{-x^2/2}\, dx \right| \to 0$$

as $n \to \infty$. The last result implies a more familiar form of the theorem: For any $0 \leqslant A < B \leqslant \infty$,

$$P(A < S_n \leqslant B) - \left[\Phi\left(\frac{B - np}{\sqrt{npq}} \right) - \Phi\left(\frac{A - np}{\sqrt{npq}} \right) \right] \to 0$$

as $n \to \infty$, where $\Phi(\cdot)$ is the cumulative standardized normal distribution function

$$\Phi(x) = \int_{-\infty}^{x} \frac{1}{\sqrt{2\pi}} e^{-t^2/2}\, dt.$$

(DE MOIVRE, ABRAHAM
LAPLACE, PIERRE SIMON
LIMIT THEOREM, CENTRAL
NORMAL DISTRIBUTION)

MOIVRE–LAPLACE THEOREM (LOCAL)

If $\{X_n\}$ is a sequence of binomial* random variables with parameters (n, p) then

$$\sqrt{n} \left| \Pr(X_n = r) - \frac{1}{\sqrt{\{np(1-p)\}}} \phi\left(\frac{r - np}{\sqrt{\{np(1-p)\}}} \right) \right|$$

$\to 0$ as $n \to \infty$, where $\phi(y) = (2\pi)^{-1/2} \exp(-\frac{1}{2}y^2)$ (the standardized normal density).

A generalization of this result to lattice distributions* is given by Gnedenko [1].

Reference

[1] Gnedenko, B. V. (1962). *The Theory of Probability*, 4th ed., translated from the Russian by B. D. Seckler. Chelsea, New York, Section 43, Chap. 8.

(BINOMIAL DISTRIBUTION
LIMIT THEOREM, CENTRAL)

MOMENT GENERATING FUNCTION

See GENERATING FUNCTIONS

MOMENT MATRIX

For an n-dimensional random variable $\mathbf{X} = (X_1, \ldots, X_n)$ with a joint cumulative distribution function $\Phi(\mathbf{x})$, the moment matrix is an $n \times n$ matrix whose (i, j)th element is

$$\int x_i x_j \, d\Phi(\mathbf{x}), \qquad \begin{array}{l} i = 1, \ldots, n, \\ j = 1, \ldots, n. \end{array}$$

This matrix is positive definite and symmetric. It is related to the variance-covariance matrix*.

MOMENT PROBLEM

Let F be the distribution function of some random variable X and suppose moments

$\mu_n = \mathscr{E}X^n = \int_{-\infty}^{\infty} x^n \, dF(x)$ exist for all $n \geqslant 1$. The problem of characterizing F from its set of moments $\{\mu_n\}$ is a moment problem; in its full generality, it considers a set of constants $\{\mu_0 = 1, \mu_1, \mu_2, \ldots\}$ and asks if they can be moments of a distribution function. In order to make a more precise statement, we first take note of a couple of facts.

First, it is possible to find two distinct distribution functions *that have the same set of moments*. For example, suppose Z is a standard normal random variable. Then $Y = \exp(Z)$ has a lognormal distribution* with density function

$$f(x) = (2\pi)^{-1/2} x^{-1} \exp\{-(\ln x)^2/2\}, \qquad x > 0,$$

and zero elsewhere. For $|\alpha| \leqslant 1$, set

$$f_\alpha(x) = f(x)\left[1 + \alpha \sin(2\pi \ln x)\right].$$

Then $\{f_\alpha : |\alpha| \leqslant 1\}$ is an infinite collection of densities each of which has the same set of moments as f does. (This interesting example is due to Heyde [4].) Consequently, in general, moments do not determine a distribution function uniquely.

Second, moments of any random variable X necessarily satisfy certain conditions. For example, if $\beta_\nu = \mathscr{E}|X|^\nu$ is the absolute moment of order ν, then $(\beta_\nu)^{1/\nu}$ is an increasing function of ν. This is the well-known Liapunov's inequality*. Similarly, the quadratic form

$$\mathscr{E}\left(\sum_{i=1}^{n} X^{\alpha_i} t_i\right)^2 = \sum_i \sum_j \mathscr{E}X^{\alpha_i + \alpha_j} t_i t_j \geqslant 0,$$

so that $\det(\mathscr{E}X^{\alpha_i + \alpha_j}) \geqslant 0$, yielding a relation between moments of various orders of X. Thus the term moment problem is associated in general with two types of problems.

TYPE I. Given a sequence $\{\mu_n, \; n = 0, 1, 2, \ldots; \; \mu_0 = 1\}$ of constants, find necessary and sufficient conditions under which $\{\mu_n\}$ is a set of moments of some distribution function F.

The solution of a problem of type I depends on the *spectrum* of F, which is defined to be the set of all real numbers x for which $F(x + \epsilon) - F(x - \epsilon) > 0$ for any $\epsilon > 0$. (Any such point x is a point of increase of F.) If the spectrum of F is a finite interval $[a, b]$ then Hausdorff [3] gives a necessary and sufficient condition under which $\{\mu_n\}$ is a set of moments of F. In the case when the spectrum of F is $[0, \infty)$, the problem has been solved by Stieltjes [7], and when F has spectrum on $(-\infty, \infty)$, the solution has been given by Hamburger [2]. Unfortunately, the conditions are too complex and largely uncheckable to be of any practical use (see Shohat and Tamarkin [6, Chap. 1] for details). This has led to a search for simple sufficient conditions that can be checked.

TYPE II. Given that $\{\mu_n\}$ is a set of moments of some distribution function F, when do $\{\mu_n\}$ determine F uniquely? What are distribution functions with moments $\{\mu_n\}$?

Once again no complete solution for the uniqueness problem is known. For example, when the spectrum of F is $[a, b]$, the moments can be used to define an analytic characteristic function* and hence the distribution F is uniquely determined by $\{\mu_n\}$. This is not usually the case when F is concentrated in $[0, \infty)$ or in $(-\infty, \infty)$. The most general sufficient condition for the unique determination of a distribution is due to Carleman (1926), who showed that $\{\mu_n\}$ determines F uniquely if the series $\sum_{n=1}^{\infty} (\mu_{2n})^{-n/2}$ diverges. It is known that this condition is not necessary (*see also* CARLEMAN'S CRITERION).

The nonuniqueness of F, given $\{\mu_n\}$, is closely connected to the fact that the characteristic function of F is not analytic at the origin. The distribution functions F that have analytic characteristic functions are precisely those for which the moment generating function* exists. (For necessary and sufficient conditions under which a characteristic function is analytic, see Lukacs [5, Section 7.2].) In the case of the lognormal distribution, as mentioned earlier, moments of all orders exist and are given by

$$\mu_n = \mathscr{E} Y^n = \exp(n^2/2).$$

A simple application of the well-known ratio test shows that the series $\sum_{n=0}^{\infty} t^n e^{n^2/2}/n!$ diverges for $t \neq 0$ so that the moment generating function of the lognormal distribution does not exist. Thus the set of moments $\{\mu_n\}$ does not determine F uniquely.

The problem of moments has a long and glorious history and has led to the development of several branches of mathematics. Earliest references date back to the work of Chebyshev* (1873). The term moment problem, however, appears to occur for the first time in the work of Stieltjes in 1894–1895 in which he introduced the concept of the Stieltjes integral. Stieltjes' work was extended by Hamburger in 1920–1921, closely followed by Hausdorff's work in 1923. Several excellent monographs give an overview of the subject; we refer, in particular, to Akhiezer [1] and Shohat and Tamarkin [6].

References

[1] Akhiezer, N. I. (1965). *The Classical Moment Problem*. Oliver and Boyd, Edinburgh, Scotland.

[2] Hamburger, H. (1920, 1921). *Math. Ann.*, **81**, 235–319; **82**, 120–164.

[3] Hausdorff, F., (1923). *Math. Z.*, **16**, 220–248.

[4] Heyde, C. C. (1963). *J. R. Statist. Soc. B*, **25**, 392.

[5] Lukacs, E. (1970). *Characteristic Functions*. Hafner, New York.

[6] Shohat, J. A. and Tamarkin, J. D. (1960). *The Problem of Moments*. Mathematical Survey **1**, American Mathematical Society, Providence, RI.

[7] Stieltjes, T. J. (1894–1895). *Ann. Fac. Sci. Toulouse*, **8J**, 1–122; **9A**, 5–47.

(CARLEMAN'S CRITERION
CHARACTERISTIC FUNCTIONS
GENERATING FUNCTIONS)

V. K. ROHATGI

MOMENT-RATIO DIAGRAMS

Moment-ratio diagrams show the values of the moment ratios $\mu_3/\mu_2^{3/2}$ and μ_4/μ_2^2 for certain classes of distribution (μ_r denotes the rth central moment).

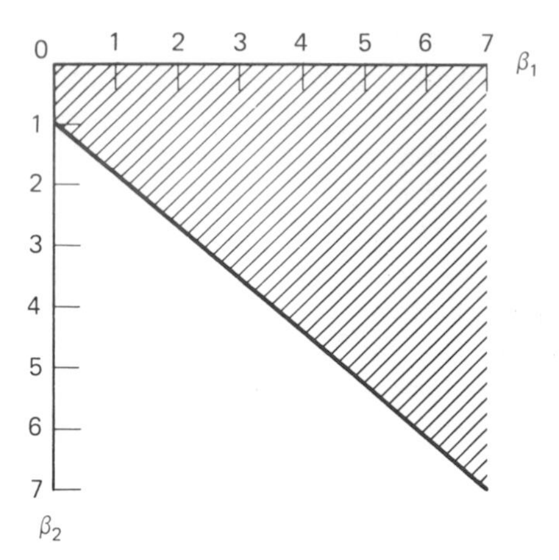

Figure 1 Traditional moment-ratio diagram.

The classical form of the diagram shows values of the ratios

$$\alpha_3^2 = \beta_1 = \mu_3^2/\mu_2^3, \qquad \alpha_4 = \beta_2 = \mu_4/\mu_2^2.$$

It is traditionally drawn with β_1 as abscissa and β_2 as ordinate, but "upside-down"—that is, with increasing values of β_2 corresponding to downward motion (see Fig. 1).

Since, for all distributions,

$$\beta_2 - \beta_1 - 1 \geqslant 0,$$

the shaded region in the figure is not relevant. It is often called the *impossible region*.

In order to reduce the proportion of wasted space, Craig [1] suggested plotting values of $\delta = (2\beta_2 - 3\beta_1 - 6)/(\beta_2 + 3)$ and β_1. This gives the "impossible region" shown in Fig. 2. Note that $\delta = 2$ corresponds to $\beta_2 = \infty$. This arrangement is especially convenient for the Pearson system*.

Both kinds of diagram suffer from the defect that information on the sign of μ_3 is lost. When it is important to retain this

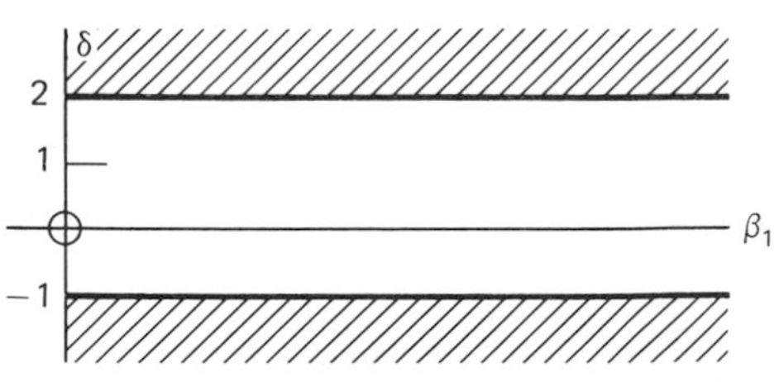

Figure 2 Craig's moment-ratio diagram [1].

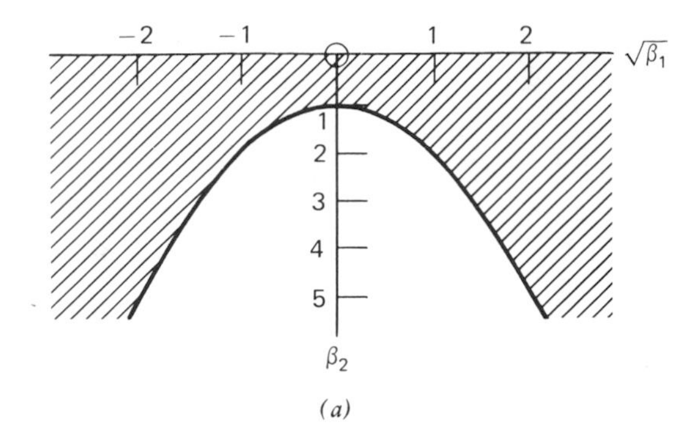

(a)

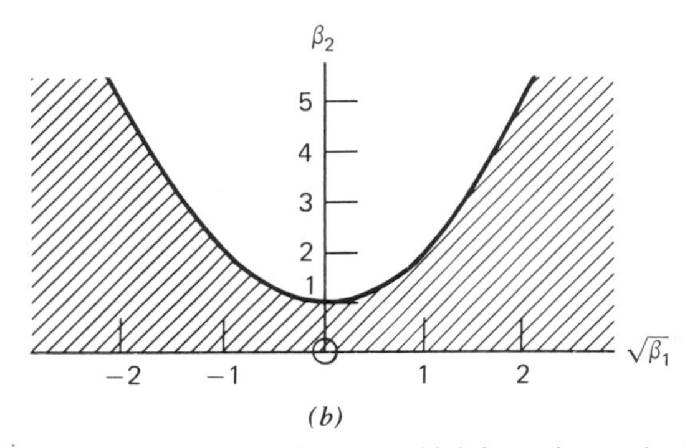

(b)

Figure 3 Moment-ratio diagrams with information retained.

information β_1 (or α_3^2) is replaced by $\sqrt{\beta_1}$ (or α_3) $= \mu_3/\mu_2^{3/2}$. (The notation of $\sqrt{\beta_1}$ to include the sign of μ_3 is also traditional.) This leads to diagrams of the form shown in Figs. 3*a* and 3*b*, constructed according to whether the traditional upside-down presentation is retained or not.

Although the shape of a distribution is not determined by the values of $\sqrt{\beta_1}$ and β_2, these moment ratios provide a useful way of exhibiting relationships among distributions in a system. This is especially the case when there is a one-to-one correspondence between ($\sqrt{\beta_1}$, β_2) values and members of the sytem as for the Pearson and Johnson systems. The representation is also helpful even when this is not so, for example, in power transformations* of gamma distributions* [2] and Burr distributions* [3].

References

[1] Craig, C. C. (1936). *Ann. Math. Statist.*, **7**, 16–28.

[2] Johnson, N. L. and Kotz, S. (1972). *Biometrika*, **59**, 226–229.

[3] Rodriguez, R. N. (1977). *Biometrika*, **64**, 129–140.

(FREQUENCY CURVES, SYSTEMS OF
JOHNSON'S SYSTEM OF DISTRIBUTIONS
MOMENT RATIOS
PEARSON'S DISTRIBUTIONS)

MOMENT RATIOS

If the random variables X and Y are linked by the linear mathematical relationship

$$Y = a + bX,$$

then the central moments* of X and Y are

linked by the relationship

$$\mu_r(Y) = b^r \mu_r(X).$$

In particular $\mu_2(Y) = b^2 \mu_2(X)$ and, if $b > 0$,

$$\frac{\mu_r(Y)}{\{\mu_2(Y)\}^{r/2}} = \frac{\mu_r(X)}{\{\mu_2(X)\}^{r/2}}.$$

The ratio $\mu_r / \mu_2^{r/2}$ is called a *moment ratio* or a *standardized moment*. Its value is preserved under increasing linear transformations. Such transformations change the location and scale parameters of the corresponding distribution, but leave the shape unchanged. For this reason the ratios are often called *shape-factors* although they do not necessarily define the shape uniquely. (*See also* MOMENT-RATIO DIAGRAMS.)

The notations

$$\alpha_r = \mu_r / \mu_2^{r/2} \quad \text{and} \quad \beta_{r-2}^{(r-2)/2} = \mu_r / \mu_2^{r/2}$$

are used for these quantities.

For much theoretical work it is convenient to use *cumulant ratios* $k_r / k_2^{r/2}$ rather than moment ratios. (The first s such ratios and the first s moment ratios—$\mu_3 \mu_2^{-3/2}, \dots,$ and $\mu_{s+2} \mu_2^{-(s+2)/2}$—have a one-to-one relationship.)

The sample moment ratios $\sqrt{b_1}$ and b_2 for data $x_1, \dots, x_n$ are defined by

$$\sqrt{b_1} = m_3 / m_2^{3/2}, \qquad b_2 = m_4 / m_2^2,$$

where m_r is the rth sample moment given by

$$m_r = \sum_i (x_i - \bar{x})^r / n, \qquad \bar{x} = \sum_i x_i / n.$$

The statistics $\sqrt{b_1}$ and b_2 were suggested by Karl Pearson* for judging whether sets of data departed from normality (*see* DEPARTURES FROM NORMALITY, TESTS FOR for further discussion). Pearson and Hartley [3, Tables 34B and 34C] tabulate 5 and 1 percent points of $\sqrt{b_1}$ and b_2 under normal distribution assumptions for $25 \leqslant n \leqslant 5000$; this edition contains corrections by Pearson [2] that do not appear in earlier editions. Patel and Read [1, Section 5.7] list expressions for the moments of $\sqrt{b_1}$ and b_2 under normality and provide further references.

References

[1] Patel, J. K. and Read, C. B. (1982). *Handbook of the Normal Distribution*. Dekker, New York.

[2] Pearson, E. S. (1965). *Biometrika*, **52**, 282–285.

[3] Pearson, E. S. and Hartley, H. D. (1966). *Biometrika Tables for Statisticians*, 3rd ed., Vol. 1. Cambridge University Press, London.

(CUMULANTS
DEPARTURES FROM NORMALITY, TESTS FOR
MOMENT-RATIO DIAGRAMS
MOMENTS)

MOMENTS

Let X be a random variable with cumulative distribution function $F_X(\cdot)$. If the expected value* of X^k

$$EX^k = \int x^k \, dF_X(x), \qquad k = 1, 2, \dots \tag{1}$$

exists, it is called the kth moment of X (about zero) or the kth *crude* moment of X. The kth moment of the random variable $X - EX$ is called the kth *central* moment*. It is computed using the formula

$$E(X - E(X))^k = \int (x - E(X))^k \, dF_X(x).$$

The kth moment of the random variable $|X|$ is called the kth *absolute* moment of X. (*See also* CUMULANTS; EXPECTED VALUE; FACTORIAL MOMENTS; STANDARD DEVIATION; VARIANCE.)

For absolutely continuous* distributions (1) becomes $\int x^k f_X(x)\,dx$, where $f_X(x)$ is the probability density; for (atomic) discrete distributions, taking on values a_i with probabilities p_i $(i = 1, 2, \dots)$, the kth moment is given by $\sum_{i=1}^{\infty} a_i^k p_i$.

Let $X_1, \dots, X_n$ be random variables with the joint distribution function $F_{X_1, \dots, X_n}(\cdot)$.

The quantity

$$\mu(k_1, \ldots, k_n) = \int \cdots \int x_1^{k_1} \cdots x_n^{k_n} \, dF_{X_1, \ldots, X_n}(x_1, \ldots, x_n) = E(X_1^{k_1} \ldots X_n^{k_n}),$$

where $k_i \geqslant 0$ and $\sum_{i=1}^n k_i = k$, is called a *mixed moment of the variables $X_1, \ldots, X_n$ of order k*. The mixed moment of order 2 for two random variables $X - E(X)$ and $Y - E(Y)$ is $E[(X - E(X))(Y - E(Y))]$ and is the *covariance** of X and Y.

MOMENTS, METHOD OF *See* METHOD OF MOMENTS

MOMENTS, PARTIAL

The nth partial (or incomplete) moment of a random variable X over the set $A \subset R$ is given by

$$\int_A x^n f(x) \, dx$$

if X is continuous with density f, and by

$$\sum_{x \in A} x^n p(x)$$

if X is discrete with probability function p. If $A = R$, then the integral (sum) above represents the usual complete moment of X, computed about zero. If the integral (sum) does not converge, the nth partial moment does not exist. Each set A gives rise to a partial moment of X; in the following discussion A is assumed to be of the form $(-\infty, y]$ if X is continuous. In the discrete case, the sum extends over all x from the range of X that satisfies $x \leqslant y$. This nth partial moment will be denoted by $E_{-\infty}^y[X^n]$. Partial moments of this form often occur in applications (some will be shown later) and can be used to compute partial moments over other sets.

CALCULATION TECHNIQUES

For some distributions, partial moments may be calculated directly.

Example. Let X have a beta distribution* with parameters $\alpha > 0$ and $\beta > 0$. Then for $y \in [0, 1]$,

$$E_{-\infty}^y[X^n] = \int_0^y x^n \frac{\Gamma(\alpha + \beta)}{\Gamma(\alpha)\Gamma(\beta)} x^{\alpha - 1}(1 - x)^{\beta - 1} dx = \frac{\Gamma(\alpha + \beta)}{\Gamma(\alpha)} \frac{\Gamma(\alpha + n)}{\Gamma(\alpha + \beta + n)} \times F_{\text{be}}(y \mid \alpha + n, \beta),$$

where $F_{\text{be}}(\cdot \mid \alpha + n, \beta)$ is the cumulative distribution function (cdf) of a beta distribution with parameters $\alpha + n$ and β.

For some distributions, especially discrete ones, it is easier to determine partial factorial moments, which may then be converted to partial moments as in the complete case.

Example. Let X have a Poisson distribution* with parameter $\lambda > 0$. Then

$$E_{-\infty}^y[X(X-1) \ldots (X - n + 1)] = \lambda^n F_{\text{poi}}(y - n \mid \lambda),$$

where F_{poi} represents the cdf of a Poisson distribution. For $n = 2$, clearly

$$E_{-\infty}^y[X^2] = E_{-\infty}^y[X(X-1)] + E_{-\infty}^y[X] = \lambda^2 F_{\text{poi}}(y - 2 \mid \lambda) + \lambda F_{\text{poi}}(y - 1 \mid \lambda).$$

For the normal distribution, recursive relationships can be developed to compute partial moments [4]. Let X be distributed as $N(\mu, \sigma^2)$ and let Z be distributed as $N(0, 1)$. Using integration by parts, it can be shown that for $n \geqslant 2$,

$$E_{-\infty}^y[Z^n] = -y^{n-1}\phi(y) + (n-1)E_{-\infty}^y[Z^{n-2}],$$

where ϕ is the standard normal density. Also, by definition,

$$E_{-\infty}^y[Z^0] = \Phi(y)$$

and

$$E_{-\infty}^y[Z] = -\phi(y),$$

where Φ represents the standard normal cdf.

Also

$$E_{-\infty}^{y}\left[X^{n}\right] = E_{-\infty}^{y}\left[(\mu + \sigma Z)^{n}\right]$$
$$= \sum_{i=0}^{n}\binom{n}{i}\mu^{i}\sigma^{n-i}E_{-\infty}^{y^{*}}\left[Z^{n-i}\right],$$

where $y^{*} = (y - \mu)/\sigma$.

The partial moment generating function* of X is given by

$$M_{-\infty}^{y}(t) = E_{-\infty}^{y}\left[e^{tX}\right]$$
$$= \begin{cases} \int_{-\infty}^{y} e^{tx} f(x)\,dx, & \text{continuous,} \\ \sum_{x \leqslant y} e^{tx} p(x), & \text{discrete.} \end{cases}$$

To obtain a partial moment generating function over a set of the form $(a,b]$, observe that

$$M_{a}^{b}(t) = E_{-\infty}^{b}\left[e^{tX}\right] - E_{-\infty}^{a}\left[e^{tX}\right]$$
$$= M_{-\infty}^{b}(t) - M_{-\infty}^{a}(t).$$

Partial moments can be obtained from $M_{-\infty}^{y}(t)$ by expanding in powers of t:

$$E_{-\infty}^{y}\left[X^{n}\right] = \left[\frac{d^{n}}{dt^{n}} M_{-\infty}^{y}(t)\right]_{t=0}.$$

Example. Let X have a gamma distribution* with parameters $\alpha > 0$ and $\beta > 0$. Then for $y > 0$,

$$M_{-\infty}^{y}(t) = \int_{0}^{y} e^{tx}\left\{\beta^{\alpha}x^{\alpha-1}e^{-\beta x}/\Gamma(\alpha)\right\}dx$$
$$= \frac{\beta^{\alpha}}{(\beta - t)^{\alpha}} F_{\text{gam}}(y \mid \alpha, \beta - t),$$

where $F_{\text{gam}}(\cdot \mid \alpha, \beta - t)$ is the cdf of a gamma distribution with parameters α and $\beta - t$.

RELATIONSHIP TO CONDITIONAL MOMENTS

The conditional expectation of a function $h(X)$ of the random variable X over a set $A \subset R$ can be written as

$$E\left[h(X) \mid X \in A\right]$$
$$= \begin{cases} \int_{A} h(x) f(x)\,dx \Big/ \int_{A} f(x)\,dx, & \text{continuous,} \\ \sum_{x \in A} h(x) p(x) \Big/ \sum_{x \in A} p(x), & \text{discrete,} \end{cases}$$
$$= E_{A}\left[h(X)\right]/P(X \in A),$$

where $E_{A}[h(X)]$ represents the partial expectation of $h(X)$ over the set A. (*See also* CONDITIONAL PROBABILITY AND EXPECTATION.)

Example. Assume X is distributed as $N(0, \sigma^{2})$ and find $E[X|X < 0]$.

From above, $E[X \mid X < 0] = 2E_{-\infty}^{0}[X]$. Using the recursion formula, $E_{-\infty}^{0}[X] = -\sigma\phi(0) = -0.3989\sigma$.

Hence $E[X \mid X < 0] = -0.7978\sigma$.

Example. Assume X has a $N(\mu, \sigma^{2})$ distribution. Then $|X|$ has a folded normal distribution (*see* FOLDED DISTRIBUTIONS). To find $E(|X|)$, observe that

$$E(|X|) = E(X|X > 0)P(X > 0)$$
$$- E(X|X < 0)P(X < 0)$$
$$= E_{0}^{\infty}\left[X\right] - E_{-\infty}^{0}\left[X\right].$$

From the recursion formula,

$$E_{-\infty}^{0}\left[X\right] = -\sigma\phi(\mu/\sigma) + \mu\Phi(-\mu/\sigma).$$

Also,

$$E_{0}^{\infty}\left[X\right] = E\left[X\right] - E_{-\infty}^{0}\left[X\right]$$
$$= \sigma\phi(-\mu/\sigma) + \mu\Phi(\mu/\sigma).$$

Substitution and simplification gives

$$E(|X|) =$$
$$\sigma\sqrt{2/\pi}\exp\left\{-\tfrac{1}{2}\theta^{2}\right\} - \mu\left[1 - 2\Phi(\Theta)\right],$$

where $\Theta = \mu/\sigma$.

Example. Let W have a normal distribution with parameters μ and σ, truncated to the right at τ. Then

$$E(W) = E_{-\infty}^{\tau}[X]/\Phi[(\tau - \mu)/\sigma],$$

where X has a $N(\mu, \sigma^2)$ distribution.

USES OF PARTIAL MOMENTS

Partial moments are used extensively in decision theory* and in Bayesian point estimation [3] (*see* BAYESIAN INFERENCE). They occur in sequential analysis* in the context of optimal stopping-rule* problems [1]. The optimal use of screening variables in classification* problems under a variety of loss functions involves partial moments [2].

Example. A vendor wanting to sell a good must decide in advance how much of the good to order. Suppose that the vendor makes a gain of d dollars on each unit of the good sold but suffers a loss of c dollars on each unit ordered but not sold. Assume that the demand for the good, X, has density f and cdf F. The vendor's expected profit is

$$(c + d)E_{-\infty}^{\alpha}[X] - (c + d)\alpha F(\alpha) + d\alpha,$$

where α is the amount the vendor orders. The quantity α^* that maximizes the vendor's expected profit is given by $F(\alpha^*) = d/(d + c)$, and the expected profit based on α^* is $(c + d)E_{-\infty}^{\alpha^*}[X]$.

References

[1] De Groot, M. H. (1970). *Optimal Statistical Decisions*. McGraw-Hill, New York.

[2] Marshall, A. W. and Olkin, I. (1968). *J. R. Statist. Soc. B*, **30**, 407–435.

[3] Raiffa, H. and Schlaifer, R. (1961). *Applied Statistical Decision Theory*. Harvard University Press, Boston.

[4] Winkler, R. L., Roodman, G. M., and Britney, R. R. (1972). *Manag. Sci.*, **19**, 290–296.

(ABSOLUTE MOMENTS
CONDITIONAL PROBABILITY AND EXPECTATION
FACTORIAL MOMENTS
FOLDED DISTRIBUTIONS
TRUNCATED DISTRIBUTIONS)

RAMONA L. TRADER

MOMENTS, TRIGONOMETRIC

The sth trigonometric moment of a distribution over the range $0 \leqslant X \leqslant 2\pi$, is the expected value* of $\exp(isX)$, s an integer. Regarded as a function of a continuous s, it is the characteristic function* of X. For the case for which it is defined, it is

$$\tau_s = E[\sin sX] + iE[\cos sX].$$

The trigonometric moments $\tau_1, \tau_2, \ldots,$ define the distribution of X uniquely [2]. For a recent application, see Gupta [1].

If X is a random variable, the expected values of trigonometric functions such as

$$\sin^{\alpha} X \cos^{\beta} X$$

are also called trigonometric moments. When X is an angle, they arise naturally in such contexts as Fourier and periodogram analysis* and distributions on spheres and hyperspheres.

References

[1] Gupta, D. (1978). *Metron*, **36**, 119–130.

[2] Hurwitz, A. (1903). *Math. Ann.*, **57**, 425–446; **59**, 553.

(CIRCULAR NORMAL DISTRIBUTION
DIRECTIONAL DISTRIBUTIONS)

MOMENT TEST OF NORMALITY *See* D'AGOSTINO'S TEST OF NORMALITY

MONOTONE LIKELIHOOD RATIO

Let $f(x;\theta)$ denote the probability density function of a family of distributions, either discrete or absolutely continuous, where θ is a parameter taking values in an interval Θ. The family is said to have a *monotone likelihood ratio* (MLR) if, based on a random sample $(x_1, \dots, x_n)$ from a member distribution there exists a statistic $T = T(x_1, \dots, x_n)$ such that the ratio

$$\frac{L(\theta' : x_1, \dots, x_n)}{L(\theta'' : x_1, \dots, x_n)} = \frac{L(\theta')}{L(\theta'')}$$

of likelihood functions is either nonincreasing in T or nondecreasing in T for every $\theta' < \theta''$. Frequently $x_1, \dots, x_n$ are independent, and then $L(\theta : x_1, \dots, x_n) = \prod_{i=1}^{n} f(x_i; \theta)$.

Example 1. One-Parameter Exponential Families*. Let

$$f(x;\theta) = A(\theta)B(x)\exp[C(\theta)d(x)], \quad -\infty < x < \infty.$$

Then

$$L(\theta')/L(\theta'') = \exp\left[\{C(\theta') - C(\theta'')\}\sum_{i=1}^{n} d(x_i)\right],$$

which is nonincreasing (respectively, nondecreasing) in $T = \sum_{i=1}^{n} d(x_i)$ whenever $\theta' < \theta''$ if $C(\cdot)$ is nondecreasing (respectively, nonincreasing) in θ.

Example 2. Uniform Distribution. Let

$$f(x;\theta) = \begin{cases} \theta^{-1}, & 0 < x < \theta, \\ 0, & \text{otherwise}, \end{cases}$$

and let $x_{(n)} = \max(x_1, \dots, x_n)$. Then whenever $\theta' < \theta''$,

$$\frac{L(\theta')}{L(\theta'')} = \begin{cases} (\theta''/\theta')^n, & 0 < x_{(n)} < \theta', \\ 0, & \theta' \leqslant x_{(n)} < \theta'', \end{cases}$$

which is nonincreasing in $x_{(n)}$.

A necessary and sufficient condition for the family $f(x;\theta)$ to have MLR in x holds if $\partial^2 \log f(x;\theta)/\partial\theta^2$ exists; it is that this derivative be nonnegative for all θ and all x [1, Section 3.13]. If θ is a location parameter, so that $f(x;\theta) = g(x - \theta)$, then the family f has MLR in x if and only if $-\log g$ is convex [1 Section 8.2].

The MLR property is useful in developing uniformly most powerful (UMP) tests (*see* HYPOTHESIS TESTING and NEYMAN–PEARSON LEMMA). Suppose that $x_1, \dots, x_n$ is a random sample from the parent distribution with density $f(x;\theta)$, where θ lies in an interval I, and suppose that the family $\{f(x;\theta), \theta \in I\}$ has MLR in $T = T(x_1, \dots, x_n)$. Then [2, Section IX.3.2] in testing the null hypothesis H_0 that $\theta \leqslant \theta_0$ against the alternative H_1 that $\theta > \theta_0$,

1. If the MLR is nondecreasing in T and if t exists such that

   $$\Pr[T < t \mid \theta = \theta_0] = \alpha,$$

   then the critical region $\{(x_1, \dots, x_n): T(x_1, \dots, x_n) < t\}$ is that of a UMP size α test of H_0 vs. H_1.
2. If the MLR is nonincreasing in T and if t' exists such that

   $$[\Pr T > t' \mid \theta = \theta_0] = \alpha,$$

   then the critical region $\{(x_1, \dots, x_n): T(x_1, \dots, x_n) > t\}$ is that of a UMP size α test of H_0 vs. H_1.

An application of (2) to Example 2 leads to the UMP size α critical region

$$\{(x_1, \dots, x_n) : x_{(n)} > \theta_0(1 - \alpha)^{1/n}\}$$

in testing $H_0 : \theta \leqslant \theta_0$ vs. $H_1 : \theta > \theta_0$.

When certain composite hypothesis-testing problems remain invariant under some group of transformations, the Neyman–Pearson lemma may be used with the data expressed in terms of a maximal invariant statistic v_n, say, and the parameters in terms of a maximal invariant τ, say, of interest. If the family of resulting densities has MLR in v_n, then a UMP test of $H_0 : \tau \leqslant \tau_0$ vs. $H_1 : \tau > \tau_0$ will exist. See Example 4 of INVARIANCE CONCEPTS IN STATISTICS for an illustration.

References

[1] Lehmann, E. L. (1959). *Testing Statistical Hypotheses*. Wiley, New York.

[2] Mood, A. M., Graybill, F. A., and Boes, D. C. (1974). *Introduction to the Theory of Statistics*. McGraw-Hill, New York.

(HUNT–STEIN THEOREM
HYPOTHESIS TESTING
INVARIANCE CONCEPTS IN STATISTICS
NEYMAN–PEARSON LEMMA)

MONOTONE RELATIONSHIPS

Prior information regarding the parameters of a statistical model frequently pertains to the "shape" of a parameter set and is quantified by restricting the relative magnitudes of these parameters. For example, in ballistics, penetration probabilities are monotone functions of muzzle velocities; in bioassay*, the probability of a specific response to a drug is assumed to be a monotone function of the dosage level; in insurance, the force of mortality* is assumed to increase with age at least after 30; in biology, certain parametric models, such as the Poisson* or generalized Poisson, are fit to data partly because of their shape (i.e., unimodal); in reliability*, failure rates are assumed to be increasing, decreasing, or U-shaped.

Statistical techniques that take advantage of prior information regarding the shape of a parameter set have been widely researched. A comprehensive account and extensive bibliography of this research up to about 1971 can be found in Barlow et al. [1] and surveys of some of the more recent work are given in refs. 11, 13, and 14. (*See also* ISOTONIC INFERENCE and ORDER RESTRICTED INFERENCES.)

Taking a monotone relationship into account in a statistical analysis should increase the efficiency of the analysis significantly, provided of course that the relationship actually holds. Perhaps the most familiar example is the problem of testing the equality of the means of two normal populations with known variances based on independent samples from these populations. If it is known that one of the means is larger than the other, then a one-sided test (i.e., one that takes the order restriction into account) is more powerful than the two-sided test which ignores the order restriction. In fact, the one-sided test is uniformly most powerful in this example (*see* HYPOTHESIS TESTING). Additional studies of the power of order restricted tests are described in ref. 1 (see, in particular, Section 3.4).

REDUCTION OF ERRORS OR EXPECTED ERRORS; LEAST SQUARES

Another use of prior knowledge about a monotone relationship to increase the efficiency of a statistical analysis is by reduction of errors or expected errors in estimates. There are several interesting results along these lines. Suppose we have independent samples from each of k populations indexed by the parameters $\theta_1, \theta_2, \ldots, \theta_k$ and suppose we have prior information that these parameters satisfy a monotone relation such as $\theta_1 \leqslant \theta_2 \leqslant \cdots \leqslant \theta_m \geqslant \cdots \geqslant \theta_k$ $(1 \leqslant m \leqslant k)$. In many such problems there exist "basic" estimates, $\hat{\theta}_1, \hat{\theta}_2, \ldots, \hat{\theta}_k$, which, because of sampling variability, may not satisfy the desired relation and must be modified or "smoothed." For example, if $\theta_1, \theta_2, \ldots, \theta_k$ are the means of k normal populations, then the basic estimates are the sample means and the smoothed estimates are called the *isotonic regression*. The results about errors compare the error or expected error in the smoothed estimates to those in the basic estimates.

The isotonic regression of $\hat{\theta}$ is a least-squares* projection of $\hat{\theta}$ onto the subset D of Euclidean k-space consisting of all points that satisfy the monotone relationship. We denote this point by $\bar{\theta} = (\bar{\theta}_1, \bar{\theta}_2, \ldots, \bar{\theta}_k)$. The point $\bar{\theta}$ solves

$$\min_{g \in D} \sum_{i=1}^{k} (\hat{\theta}_i - g_i)^2 w_i, \qquad (1)$$

where $w_1, w_2, \ldots, w_k$ are positive weights. The weight w_i is usually proportional to the relative precision of $\hat{\theta}_i$ as an estimator of θ_i, and in many problems w_i is the size of the sample from the ith population. (A number of algorithms for computing $\hat{\theta}$ are given in Chapters 1 and 2 of ref. 1.)

Suppose $\bar{\theta}$ solves (1) for some set of positive weights $w_1, w_2, \ldots, w_k$. Then

$$\sum_{i=1}^{k} (\hat{\theta}_i - g_i)^2 w_i \geqslant \sum_{i=1}^{k} (\hat{\theta}_i - \bar{\theta}_i)^2 w_i + \sum_{i=1}^{k} (\bar{\theta}_i - g_i)^2 w_i$$

for all $g \in D$. (This is Theorem 1.3 in ref. 1.) The sum on the left is the square of a weighted distance from $\hat{\theta}$ to g. In particular, this holds for $g = \theta$, and we refer to the sum on the left as the *total square error* in θ. The second term on the right when evaluated at $g = \theta$ is the total square error in $\bar{\theta}$. Thus this error is larger for $\hat{\theta}$ than for $\bar{\theta}$, a rather remarkable result as it speaks to error and not expected error.

The order restricted estimate $\bar{\theta}$ has an even stronger error reducing property for a linear order such as $\theta_1 \leqslant \theta_2 \leqslant \cdots \leqslant \theta_k$ in that

$$\sum_{i=1}^{k} \phi(\bar{\theta}_i - \theta_i) w_i \leqslant \sum_{i=1}^{k} \phi(\hat{\theta}_i - \theta_i) w_i$$

for every convex function $\phi(\cdot)$ (cf. Malmgren [6]). In particular, taking $\phi(x) = |x|$, we see that $\bar{\theta}$ reduces the total absolute error. By taking $\phi(x) = |x|^p$ for $p \geqslant 1$, we find

$$\left[\sum_{i=1}^{k} |\bar{\theta}_i - \theta_i|^p w_i\right]^{1/p} \leqslant \left[\sum_{i=1}^{k} |\hat{\theta}_i - \theta_i|^p w_i\right]^{1/p}.$$

Letting $p \to \infty$, we obtain

$$\max_{1 \leqslant i \leqslant k} |\bar{\theta}_i - \theta_i| \leqslant \max_{1 \leqslant i \leqslant k} |\hat{\theta}_i - \theta_i| \quad (2)$$

so that $\bar{\theta}$ also reduces the largest error. Among other things, (2) is generalized to partial orders in Robertson and Wright [9].

The estimates $\bar{\theta}_1, \bar{\theta}_2, \ldots, \bar{\theta}_k$ are, in general, biased. In fact, for the linear order $\theta_1 \leqslant \theta_2 \leqslant \cdots \leqslant \theta_k$ and the least-squares problem

$$\bar{\theta}_1 = \min_{1 \leqslant j \leqslant k}\left(\sum_{i=1}^{j} \hat{\theta}_i w_i \Big/ \sum_{i=1}^{k} w_i\right),$$

so that if the basic estimates $\hat{\theta}_i$; $i = 1, 2, \ldots, k$ are unbiased then $\bar{\theta}_1$ is the minimum of k unbiased estimates. In light of this, it is rather surprising that for the normal means problem and the linear order $\theta_1 \leqslant \theta_2 \leqslant \cdots \leqslant \theta_k$,

$$E\left[(\hat{\theta}_i - \theta_i)^2\right] \geqslant E\left[(\bar{\theta}_i - \theta_i)^2\right].$$

Thus the mean square error is smaller for $\bar{\theta}_i$ than for $\hat{\theta}_i$ for $i = 1, 2, \ldots, k$ (see Lee [5]).

GENERAL ESTIMATES

Estimates that minimize $[\sum_{i=1}^{k} |\hat{\theta}_i - g_i|^p]^{1/p}$ for $1 \leqslant p \leqslant \infty$ are considered in refs. 2, 15, and 16. The solutions are not, in general, unique. However, for $p = \infty$ (estimates that minimize $\max_{1 \leqslant i \leqslant k} |\hat{\theta}_i - g_i|$), one solution θ', lying in the "center" of the solution set, has a norm-reducing property, which implies that

$$\max_{1 \leqslant i \leqslant k} |\theta'_i - \theta_i| \leqslant \max_{1 \leqslant i \leqslant k} |\hat{\theta}_i - \theta_i|. \quad (3)$$

Suppose $X_{i1}, X_{i2}, \ldots, X_{in_i}$ denote the items of the sample from the ith population. Estimates that minimize

$$\sum_{i=1}^{k} \sum_{j=1}^{n_i} |X_{ij} - g_i| w_i \quad (4)$$

subject to the order restriction are considered in refs. 3, 7, 8, 10, 12. (This is not, in general, equivalent to minimizing $\sum_{i=1}^{k} (|\hat{\theta}_i - g_i| w_i)$ for some basic estimates $\hat{\theta}_1, \hat{\theta}_2, \ldots, \hat{\theta}_k$). Of course, the solutions are not necessarily unique. A general class of such problems is considered in ref. 9; there it is shown that $\max_{1 \leqslant i \leqslant k} |\bar{\theta}_i - \theta_i| \leqslant \max_{1 \leqslant i \leqslant k} |\hat{\theta}_i - \theta_i|$ where $\bar{\theta}$ is an order restricted solution to (4) and $\hat{\theta}$ is the collection of medians of the samples (i.e., $\hat{\theta}$ is an unrestricted solution).

Order restricted estimates have been criticized both because they "oversmooth" and because they are not sufficiently smooth (*see* GRADUATION). The first criticism is based on the fact that the smoothing process averages values in $\hat{\theta}$ regardless of the strength of the evidence in $\hat{\theta}$ that the monotone relationship does not hold. This suggests that a Bayesian approach where the prior assigns "most" of its mass to D would be of interest. A less restrictive way to characterize a monotone relationship that addresses this criticism in some problems is considered in ref. 4.

The second criticism is based on the fact that $\bar{\theta}$ is usually constant over several subsets of $\{1, 2, \ldots, k\}$ and has the shape of a step function. In many practical problems, investigators feel that θ does not have such a form. Estimates obtained by combining the ideas of splines* (i.e., smooth estimates) and isotonic regression are considered in [17]. The evidence that order restricted estimates are closer to the true θ than are the basic estimates leads one to feel that if additional smoothing is required, then one might do well to smooth $\bar{\theta}$ rather than $\hat{\theta}$.

References

[1] Barlow, R. E., Bartholomew, D. J., Bremner, J. M., and Brunk, H. D. (1972). *Statistical Inference Under Order Restrictions*. Wiley, New York.

[2] Barlow, R. E. and Ubhaya, V. A. (1971). *Proceedings of the Symposium on Optimizing Methods in Statistics*. Academic Press, New York, pp. 77–86.

[3] Cryer, J. D., Robertson, T., Wright, F. T., and Casady, R. J. (1972). *Ann. Math. Statist.*, **43**, 1459–1469.

[4] Dykstra, R. L. and Robertson, T. (1983). *J. Amer. Statist. Ass.*, **78**, 342–350.

[5] Lee, C. (1981). *Ann. Statist.*, **9**, 686–688.

[6] Malmgren, E. (1972). "Contributions to the estimation of ordered parameters." Ph.D. thesis. Department of Statistics, University of Iowa, Iowa City, IA.

[7] Robertson, T. and Waltman, P. (1968). *Ann. Math. Statist.*, **39**, 1030–1039.

[8] Robertson, T. and Wright, F. T. (1973). *Ann. Statist.*, **1**, 422–432.

[9] Robertson, T. and Wright, F. T. (1974). *Ann. Statist.*, **2**, 1302–1307.

[10] Robertson, T. and Wright, F. T. (1975). *Ann. Statist.*, **3**, 350–362.

[11] Robertson, T. and Wright, F. T. (1980a). *Proceedings of Symposia in Applied Mathematics*, Vol. 23. American Mathematical Society, Providence, RI, pp. 55–71.

[12] Robertson, T. and Wright, F. T. (1980b). *Ann. Statist.*, **8**, 645–651.

[13] Sager, T. W. (1983). *Commun. Statist. A*, **12**, 529–558. (An interesting survey of the estimation of modes and isopleths.)

[14] Snijders, T. A. B. (1979). *Asymptotic Optimality Theory for Testing Problems with Restricted Alternatives*. Mathematical Center Tracts 113, Mathematisch Centrum, Amsterdam.

[15] Ubhaya, V. A. (1974a). *J. Approx. Theory*, **12**, 146–159.

[16] Ubhaya, V. A. (1974b). *J. Approx. Theory*, **12**, 315–331.

[17] Wright, I. W. and Wegman, E. J. (1980). *Ann. Statist.*, **8**, 1023–1035.

Bibliography

Brunk, H. D. (1975). *Ann. Prob.*, **3**, 1025–1030. (Defines a pseudometric on the collection of σ-lattices of measurable subsets of a probability space. L_p norms of differences of conditional p-means of the same function given two σ-lattices are bounded by distances between these σ-lattices in the pseudometric.)

Dykstra, R. L. (1982). *J. Amer. Statist. Ass.*, **77**, 621–628. (Considers maximum likelihood estimation of two stochastically ordered distributions when the data include censored observations.)

Dykstra, R. L. and Robertson, T. (1982). *Ann. Statist.*, **10**, 708–716. (An efficient algorithm for computing the isotonic regression in two or more independent variables.)

Leurgans, S. (1981). *Ann. Statist.*, **9**, 905–908. (The only linear combinations of the order statistics having the Cauchy mean value property are the arithmetic mean, the percentiles, and the weighted midranges.)

Robertson, T. and Wright, F. T. (1982). *Ann. Statist.*, **10**, 1234–1245. (Several notions of the degree of conformity to a trend are developed.)

Robertson, T. and Wright, F. T. (1983). *Biometrika*, **70**, 597–606. (An approximation for the "level probabilities" for unequal weights.)

Sackrowitz, H. and Strawderman, W. (1974). *Ann. Statist.*, **2**, 822–828. (The maximum likelihood estimators of ordered binomial parameters are, in general, inadmissible.)

Wright, F. T. (1982). *Ann. Statist.*, **10**, 278–286. (Considers a monotone regression estimate obtained by grouping observations. The resulting estimator has an asymptotic normal distribution and its rate of convergence is faster than that of the usual isotonic regression.)

(GRADUATION
ISOTONIC INFERENCE

ORDER RESTRICTED INFERENCE
SPLINES)

TIM ROBERTSON

MONOTONICITY AND POLYTONICITY, COEFFICIENTS OF *See* POLYTONICITY AND MONOTONICITY, COEFFICIENTS OF

MONTE CARLO METHOD (DATA ADJUSTING) *See* EDITING STATISTICAL DATA

MONTE CARLO METHODS

Monte Carlo methods are those in which properties of the distributions of random variables are investigated by use of simulated random numbers. The methods, aside from the collection of data, are similar to the usual statistical methods in which random samples are used in making inferences concerning actual populations. Generally in applications of statistics, a model is used to simulate some phenomenon that has a random component. In Monte Carlo methods, on the other hand, the object of the investigation is a model itself, and random or pseudo-random events are used to study the model.

Often in application of Monte Carlo methods, the problem being studied does not have an explicit random component; however, in these cases a deterministic parameter of the problem is expressed as a parameter of some random distribution and then that distribution is simulated. An example of this type is the famous Buffon needle problem*, in which the value of π is estimated by repeatedly tossing a needle onto a surface on which parallel lines are drawn and determining the proportion of tosses in which the needle crosses a line. (See Perlman and Wichura [13] for a discussion of this problem.) During World War II and immediately thereafter, Monte Carlo methods were extensively used in studying deterministic problems (primarily solutions of differential equations) arising in work on the atomic bomb by Fermi, von Neumann, Ulam, and Metropolis (see Metropolis and Ulam [10]). The name Monte Carlo (from the casino in Monaco) for these methods dates from that period.

Monte Carlo methods are often used by statisticians to investigate distributional problems that are mathematically intractable, such as evaluation of distribution functions or moments of a distribution (*see* ref. 6 and COMPUTERS AND STATISTICS). Monte Carlo is also widely used in robustness* studies of statistical procedures. The method in this case involves simulating observations from an alternative distribution and computing from these observations the statistics for the procedure in the usual way. From the empirical distributions for the statistics obtained in this manner, the robustness of the ordinary statistical procedure can be evaluated.

The idea of simulating a process to study its properties is so basic and straightforward that it is not possible to say who first used such methods. An early documented use of a Monte Carlo method was by the American statistician Erastus Lyman De Forest in 1876 (see Stigler [17]). In a study of smoothing (adjustment) of a time series*, De Forest proposed a statistic based on the residuals and the fourth differences of the original series to assess the adequacy of the smoothing. He derived an expression for the asymptotic standard deviation of the statistic, but then sought empirical verification of the expression. Using the inverse cdf method (*see* GENERATION OF RANDOM VARIABLES), on 100 cards of equal size, he wrote $F^{-1}(u_i)$ for $u_i = 0.005, 0.015, \dots, 0.995$, where F is the normal cumulative distribution function. These cards he placed in a box and drew out one at a time to simulate a normal process, and from the values of the statistic computed from these artificial samples, he obtained partial confirmation of the expression he had derived. Although there were some flaws in De Forest's reasoning and his simulation* study was rather primitive, the study was a contribution to Monte Carlo methodology. Another important early contribution was made by "Student" [18] while studying the distributions of the correlation* coefficient and of the t statistic (see also Pearson [12] and GOSSET, WILLIAM SEALY). Student used actual biometric data to simulate realizations of normally distributed random variables. Other early uses of Monte Carlo methods by statistical research workers are described in Teichroew [19].

Since the introduction of the digital computer, the random numbers used in Monte Carlo studies are most often generated by the computer (*see* GENERATION OF RANDOM VARIABLES), and this facility has led to the widespread use of the Monte Carlo technique. In the period from 1978 to 1982, Monte Carlo methods were used in the research reported in approximately 30% of the articles in the *Journal of the American Statistical Association**. The role of Monte Carlo methods has become similar to that of experimentation in the natural sciences, and the need for proper, careful conduct and reporting of Monte Carlo experimentation has been emphasized [8]. A good example of extensive and well-planned use of Monte Carlo is reported in Andrews et al. [1], a study conducted at Princeton University of alternative point estimators of location in symmetric distributions with heavy tails (*see* HEAVY-TAILED DISTRIBUTIONS). In such distributions (e.g., Cauchy) the ordinary sample mean is not a good estimator because its variance is large or even infinite. The distributions of many of the other estimators studied are quite intractable, and hence Monte Carlo methods were used to estimate their moments and other properties. The Monte Carlo studies made extensive use of variance reducing methods (see the section Variance Reducing Methods), but even so these studies consumed many hours of computer time. While the Monte Carlo results are interesting in their own right, one of the important uses of Monte Carlo is to identify quickly promising statistical methods worthy of further study.

Another somewhat different Monte Carlo method is the *Monte Carlo test* used in hypothesis testing*. In this method, introduced by G. A. Barnard [2], data sets are simulated under the null hypothesis* and the proportion yielding values of the test statistic more extreme than the observed value is used in the decision rule. The number of simulated data sets required for a given level of significance and power is smaller than intuition might suggest, and for commonly used test-operating levels is often as low as 99 (see refs. 3 and 9).

As an example of Monte Carlo tests, Besag and Diggle [3] considered the problem of testing for spatial pattern in the locations of 65 pine trees in a square plot (*see* SPATIAL DATA ANALYSIS). The distances between all pairs of trees were used as the basis of the test of a null hypothesis of uniform distribution over the plot. A chi-square test* comparing these 2,080 distances with the distribution of the distance between two randomly selected points in the square is not valid because of the lack of independence of the intertree distances. A statistic of the same form as a chi-square goodness-of-fit* statistic was used, but instead of the decision being based on the chi-square distribution, the decision was based on values of this statistic computed from 99 simulations of 65 points randomly distributed in the square. As it turned out, the statistic computed from the distances between the pine trees was very close to the median of the 99 simulated values; hence no evidence was provided by this statistic for the rejection of the null hypothesis. (If the statistic were assumed to have a chi-square distribution, it would be significant at the 0.01 level.)

MONTE CARLO METHODS AND EVALUATION OF AN INTEGRAL

Every quantitative result of a Monte Carlo experiment is essentially an estimate of the value of some integral. Since the integral

$$\theta = \int_0^1 f(u)\,du$$

(if it exists) is the expectation of $f(U)$, where U is a uniformly distributed random variable over $(0, 1)$, the value of the integral can be estimated unbiasedly by

$$(1/n)\sum_{i=1}^{n} f(U_i),$$

where $U_1, U_2, U_3, \ldots, U_n$ are a random sample from a $U(0, 1)$ distribution. The variance of the estimate of the integral is

$$(1/n)\int_0^1 [f(u) - \theta]^2 \, du.$$

This method of estimating the integral is *crude Monte Carlo*, or *sample-mean Monte Carlo*. An alternate method, *hit-or-miss Monte Carlo*, is to obtain random points in the rectangular region bounded by vertical lines at 0 and 1 and by horizontal lines at $\min[f(u)]$ and $\max[f(u)]$. If $f(u) \geqslant 0$, the proportion of points falling below the curve defined by $f(u)$ multiplied by the area of the rectangle, plus $\min[f(u)]$, is an unbiased estimate of the integral.

In most cases, the integral being evaluated is a complicated multiple integral. If the integral is simple enough to be evaluated analytically or by a good method of numerical integration*, Monte Carlo methods should not be used since an unnecessary random error would be introduced thereby (*see also* n-DIMENSIONAL QUADRATURE).

VARIANCE REDUCING METHODS

Since Monte Carlo experimentation involves a statistical estimation problem, the same desiderata and techniques of estimation apply. The most commonly required properties for an estimator are unbiasedness* and minimum (or small) variance. Much Monte Carlo research has been concerned with reducing the variance of the estimator. The most important variance reduction technique consists of avoidance of the use of random components to the extent possible, by analytic reduction of the problem. For example, if the integral can be evaluated analytically over part of the range of integration, the Monte Carlo method should be restricted to the remainder of the range. Another illustration of such avoidance is crude Monte Carlo instead of hit-or-miss Monte Carlo. In the latter method, a random ordinate is compared to the function evaluated at a random abscissa yielding a binomial process. In crude Monte Carlo, the randomness in the ordinate is avoided. The variance of crude Monte Carlo is never greater than that of hit-or-miss Monte Carlo (see Hammersley and Handscomb [5, p. 54]).

The other variance reduction techniques in Monte Carlo are essentially the same as those familiar in sampling: probability sampling, stratified sampling, and use of auxiliary variables (see SAMPLING). Unequal probability sampling is called *importance sampling** in the literature on Monte Carlo, and its use, as well as that of stratification, is quite straightforward. The methods of auxiliary variables are far richer in possibilities for variance reduction and can be used in one form or another in almost any Monte Carlo study on nontrivial problems.

An important method of variance reduction uses antithetic (i.e., negatively correlated) variates. If X and Y are negatively correlated unbiased estimators of θ, a linear combination of X and Y exists that has less variance than either individually. In simple problems, antithetic variates are readily available. For example, in the Monte Carlo integration of $\int_0^1 f(u)\,du$ above, U_i and $1 - U_i$ are useful antithetic variates. Although the use of U and $1 - U$ from a basic uniform random number generator is a common antithetic variate method, it does not always yield negatively correlated pairs of variates. For example, in the well-known log-sine-cosine transformation that yields normal variates from uniform ones, antithetic uniform variates will yield nonnegatively correlated normal variates.

Regression* methods can also be used to good advantage in reducing the variance of Monte Carlo estimates when there are auxiliary variates available. This method is also used in ordinary sampling techniques and, as in those applications, the special case of zero intercept (the ratio estimator) is an important one. A simple regression estimator with a preassigned slope is also often used and, in Monte Carlo, the regressor in this method is called a *control variate*.

The *conditional expectation* of Y given X may be used to achieve impressive reductions in the variance of a Monte Carlo estimator. (This is essentially a regression method in which the regression is known.)

$$\mathrm{Var}(Y) = \mathrm{Var}\{E(Y|X)\} + E\{\mathrm{Var}(Y|X)\} \geqslant \mathrm{Var}\{E(Y|X)\}.$$

Therefore, while the expectation of Y is the same as the expectation over X of $E(Y|X)$, the variance of the latter is less.

If the objective of the Monte Carlo is comparison of positively correlated random variables or of similar stochastic systems, use of common random numbers will often improve the precision. The method of common random numbers can be extended even further when, as frequently occurs, it is desired to compare the variances of unbiased estimators of some parameter. As an example, suppose two estimators T_1 and T_2 are unbiased for τ, and it is desired to estimate $\mathrm{Var}(T_1) - \mathrm{Var}(T_2)$ by Monte Carlo. Now

$$\begin{aligned}\mathrm{Var}(T_1) - \mathrm{Var}(T_2) &= E(T_1^2) - \tau^2 - E(T_2^2) + \tau^2 \\ &= E(T_1^2) - E(T_2^2).\end{aligned}$$

Since T_1 and T_2 are very likely positively correlated, the Monte Carlo estimate of $E(T_1^2 - T_2^2)$ formed by taking the mean of the differences of the squared values should have less variance than the difference in the estimates of $\mathrm{Var}(T_1)$ and $\mathrm{Var}(T_2)$.

Another variance reduction method useful when the objective is to estimate a variance involves decomposition into independent variables. Suppose it is desired to estimate $\mathrm{Var}(X)$ and X can be expressed as $Y + Z$, where Y and Z are independent. If $\mathrm{Var}(Z)$ is known and Y can be simulated, the Monte Carlo study should be conducted on Y rather than on X, since the variance of Y is smaller and its variance can be estimated with less variance. A good example of this method (given in Simon [16]) is in estimating the variance of the median M of a sample of size p drawn from a normal distribution with mean 0 and variance 1. If $M_1, M_2, \ldots, M_n$ are the medians of n samples of size p, an estimate of the variance is $(1/n)\sum_{i=1}^n M_i^2$. An improved estimator can be developed by expressing M_i as $\bar{X}_i + (M_i - \bar{X}_i)$ where $\bar{X}_i$ is the ith sample mean. By the sufficiency-ancillarity theorem (*see* ref. 16 and BASU THEOREMS), $\bar{X}_i$ and $(M_i - \bar{X}_i)$ are independent. The estimate of the variance of $(M_i - \bar{X}_i)$, $(1/n)\sum_{i=1}^n (M_i - \bar{X}_i)^2$, which has less variance than $(1/n)\sum_{i=1}^n M_i^2$, can then be added to the known variance of $\bar{X}_i$ to get a better estimate of the variance of the median.

Decomposition into a sum of conditionally independent variates can also be effective in reducing the variance of an estimator of variance. An important example is given by Relles [14]. The problem is to estimate the variance of a scale and location invariant function ϕ of a t-variable. For $i = 1, 2, \ldots, n$, let X_i be distributed as a normal random variable with mean 0 and variance 1, let S_i be such that νS_i^2 has a chi-square distribution with ν degrees of freedom, let all X_i and S_i be mutually independent, and let $Y_i = X_i / S_i$. Also, let X and S be vectors of length n containing the X_i and S_i, respectively. Now define the following quantities:

$$U = S(S'S)^{-1}S'X,$$
$$w = S'X/(S'S).$$

The vector of fitted values U is independent of the vector of residuals $X - U$ and the sufficiency-ancillarity theorem can be used to show that $Y_i - w$ is independent of w, given S (see ref. 16). Since ϕ is scale and location invariant, $E\{\phi(Y_i)\} = 0$ and so $\mathrm{Var}\{\phi(Y_i)\} = E\{\phi^2(Y_i)\}$. Because of conditional independence of $Y_i - w$ and w (hence of $\phi(Y_i) - w$ and w),

$$E\{\phi^2(Y_i)|S\} = E\{[\phi(Y_i) - w]^2 | S\} + E\{w^2 | S\}.$$

Conditional on S, w is normally distributed with mean 0 and variance $1/S'S$, and $\nu S'S$ is distributed as chi-square with $n\nu$ degrees of freedom. So, taking expectations with respect to S yields

$$E\{\phi^2(Y_i)\} = E\{\phi(Y_i) - w\}^2 + \nu/(n\nu - 2).$$

The conclusion is that the average of observations on $\{\phi(Y_i) - w\}^2$ plus the fixed quantity $\nu/(n\nu - 2)$ can be used instead of the average of observations on $\phi^2(Y_i)$ to estimate the variance of $\phi(Y_i)$. The former estimator has smaller variance than the latter. Depending on ϕ, the relative difference in the variances may be extremely large.

The value of a variance reducing method depends on the actual reduction in computations (or in computer time), since the variance can be reduced just by increasing the sample size. If such a method halves the variance at the expense of a fourfold increase in computer time, the method is of no value since enlarging the sample size would be just as effective. With the availability of very cheap computing, the value of efforts to achieve variance reduction for simple problems may be moot, since extremely large sample sizes, even to the size of the period of the random number generator, are possible. (Of course, simple problems could probably be solved analytically, and Monte Carlo would not be appropriate.) The larger problems being studied by Monte Carlo continue to require very significant amounts of computer time and consideration of variance reduction techniques is by no means irrelevant.

Variance reducing techniques (especially if they are thought to be clever) are called swindles (*see* MONTE CARLO SWINDLE).

SIMULATION

Although the essence of Monte Carlo methods is simulation*, this latter term has generally come to denote experimentation on more complex models and, in particular, on models that involve a time variable. In the usual Monte Carlo methods, the observations are independent and identically distributed; in simulation, the observations often are neither. An example of a simulation rather than Monte Carlo method of study is in a queueing problem. Suppose in a manufacturing process that arrivals of parts to a work station follow a Poisson distribution* whose parameter varies depending on the time of day. The time required for processing the part at the work station follows (say) a gamma distribution*. The parts go through a number of work stations and are combined with other parts along the way. At each station there is a nonzero probability of the assemblage's becoming defective. Simulation may be used to estimate the length of time required to produce a finished part, to estimate the number of parts produced per unit of time, to study the queue lengths, to explore alternative work station configurations, and so on. Observations during simulation of the system characteristics, such as queue lengths, are not independent. Estimation of variance or higher moments is made difficult by this lack of independence; much of the research in simulation methodology has been devoted to this problem. Also, because of the lack of independence, variance reducing methods are often not available and when it is possible to use them, they must be used extremely carefully.

In addition to *discrete event simulation*, which is what the simulation described earlier is called, the term simulation refers to other techniques, often using analogue devices, to study continuous time and motion problems (e.g., the flight of aircraft). These techniques, however, do not use statistical methods to any great extent nor are continuous simulation methods used often in statistical research.

Literature

Although a number of good articles and texts deal explicitly with Monte Carlo methods and simulation, much of the best literature is in publications concerned with other problems, such as Andrews et al. [1]. Studies that use Monte Carlo often result in advances in the technique, in the discovery of new swindles, etc., and the reports of such studies contribute substantially to the literature on Monte Carlo methods. A large proportion of the articles in such journals as *Journal of the Royal Statistical Society**, *Series C*, *Journal of the American Statistical Association*, *Technometrics**, and others on statistical theory and methods contain discussions of Monte Carlo studies. Articles in these journals also occasionally deal directly with the methodology, but such articles are more likely to be found in *Journal of Statistical Computation and Simulation**, or in *Communications in Statistics**: *Simulation and Computation*. Journals devoted to the physical sciences also often contain articles reporting on the use of Monte Carlo methods. An early collection of papers on various aspects of Monte Carlo methods is in the symposium proceedings edited by Meyer [11]. Good accounts of Monte Carlo methods and the underlying theory are provided in Hartley [6] and in books by Hammersley and Handscomb [5], Hengartner and Theodorescu (in German) [7], and Rubinstein [15]. Variance reduction is discussed extensively in Hammersley and Handscomb [5, 55–74], Hartley [6], and Simon [16].

Results of simulation studies are often reported in journals on the biological sciences, industrial engineering, management science*, and systems analysis. *Management Science*, *Operations Research**, and *Naval Logistics Research Quarterly** are core journals for the technique of discrete event simulation. The books by Fishman [4] and Rubinstein [15] provide good general discussions of simulation.

References

[1] Andrews, D. F., Bickel, P. J., Hampel, F. R., Huber, P. J., Rogers, W. H., and Tukey, J. W. (1972). *Robust Estimates of Location: Survey and Advances*. Princeton University Press. Princeton, NJ.

[2] Barnard, G. A. (1963). *J. R. Statist. Soc. B*, **25**, 294.

[3] Besag, J. and Diggle, P. J. (1977). *Appl. Statist.*, **26**, 327–333.

[4] Fishman, G. S. (1978). *Principles of Discrete Event Simulation*. Wiley, New York.

[5] Hammersley, J. M. and Handscomb, D. C. (1964). *Monte Carlo Methods*. Methuen, London.

[6] Hartley, H. O. (1977). *Statistical Methods for Digital Computers*. Wiley, New York, pp. 16–34.

[7] Hengartner, W. and Theodorescu, R. (1977). *Einführung in die Monte-Carlo-Methode*. Carl Hanser Verlag, Munich.

[8] Hoaglin, D. C. and Andrews, D. F. (1975). *Amer. Statist.*, **29**, 122–126.

[9] Marriott, F. H. C. (1979). *Appl. Statist.*, **28**, 75–77.

[10] Metropolis, N. and Ulam, S. (1949). *J. Amer. Statist. Ass.*, **44**, 335–341.

[11] Meyer, H. A., ed. (1956). *Symposium on Monte Carlo Methods*. Wiley, New York.

[12] Pearson, E. S. (1938). *Biometrika*, **30**, 210–250.

[13] Perlman, M. D. and Wichura, M. J. (1975). *Amer. Statist.*, **29**, 157–163.

[14] Relles, D. A. (1970). *Technometrics*, **12**, 499–515.

[15] Rubinstein, R. Y. (1981). *Simulation and the Monte Carlo Method*. Wiley, New York.

[16] Simon, G. (1976). *Appl. Statist.*, **25**, 266–274.

[17] Stigler, S. M. (1978). *Ann. Statist.*, **6**, 239–265.

[18] "Student" (W. S. Gosset) (1908). *Biometrika*, **6**, 1–25 and 302–310.

[19] Teichroew, D. (1965). *J. Amer. Statist. Ass.*, **60**, 27–49.

(COMPUTERS AND STATISTICS
GENERATION OF RANDOM VARIABLES
IMPORTANCE SAMPLING
MONTE CARLO SWINDLE
SIMULATION
STRATIFIED SAMPLING)

JAMES E. GENTLE

MONTE CARLO SWINDLE

Experimental sampling or simulation* is often used to solve a problem that is analytically intractable. When the original problem can be replaced by a simpler one that can be shown to have the same answer, the simulation is distinguished as a *Monte Carlo swindle*. A *swindle* involves a probabilistic change in the sampling as a result of problem reformulation or decomposition. It may, but need not, rely on analytical tools such as those used to reduce the amount of computation and/or increase the precision in the results. Generally, the result of a swindle is a total or average; it is the quantities being totaled that distinguish a swindle from simple experimental sampling.

A swindle may be formulated to use auxiliary variables, as in the antithetic variable swindle (*see* MONTE CARLO METHODS), or it may involve control variables. For example, to determine the distribution of a random variable, say Y, a swindle may rely on the known distribution of another random variable X. The Monte Carlo then estimates the joint distribution of X and Y and uses the known marginal distribution of X to calculate the unknown marginal distribution of Y. Notice that this swindle involves sampling from a probability distribution that is different from simply sampling from the distribution of Y alone. The relative efficiency of this swindle can be considerable [2].

Another example is a simple form of the "location swindle" for comparing variances of location estimators (*see* MONTE CARLO METHODS). The full location and scale swindle can be extended to apply to random variables whose distribution is expressible as a ratio, say $X_i = Z_i / Y_i$, where the distribution of $\mathbf{Z} = (Z_1, \ldots, Z_n)$ is known. Then, conditional on $\mathbf{Y}$, the distribution of $\mathbf{X}$ is completely known, and the quantities

$$a = \mathbf{Z}^t\mathbf{Y}/\mathbf{Y}^t\mathbf{Y},$$

$$b = \left\{(\mathbf{Z} - a\mathbf{Y})^t(\mathbf{Z} - a\mathbf{Y})/(n-1)\right\}^{1/2},$$

and the configuration

$$\mathbf{C} = ((X_1 - a)/b, \ldots, (X_n - a)/b)$$

are independent. Thus this swindle involves functions of the configuration, instead of the sample, and uses information about its conditional distribution to achieve savings in both computation and accuracy in the estimates of the target parameters. Applications of swindles of this sort to estimate variances of location and scale estimators and percent points of distributions are described in refs. 1, 3, 4, and 6.

A list of references for Monte Carlo swindle may be found in Simon [5] and in MONTE CARLO METHODS.

Acknowledgment

The author wishes to thank J. W. Tukey for his advice on the presentation of this article.

References

[1] Andrews, D. F., Bickel, P. J., Hampel, F. R., Huber, P. J., Rogers, W. H., and Tukey, J. W. (1972). *Robust Estimates of Location: Survey and Advances*. Princeton University Press, Princeton, NJ.

[2] Fieller, E. C. and Hartley, H. O. (1954). *Biometrika*, **41**, 494–501.

[3] Gross, A. M. (1973). *Appl. Statist.*, **22**, 347–353.

[4] Relles, D. A. (1970). *Technometrics*, **12**, 499–515.

[5] Simon, G. (1976). *Appl. Statist.*, **25**, 266–274.

[6] Tukey, J. W. (1979). In *Robustness in Statistics*, Robert Launer and Graham Wilkinson, eds. Academic Press, New York, pp. 75–102.

(MONTE CARLO METHODS
SIMULATION)

KAREN KAFADAR

MONTMORT, PIERRE RÉMOND DE

Born: October 27, 1678, in Paris, France.

Died: October 7, 1719 (of smallpox) in Paris, France.

Contributed to: probability theory.

The second of three sons of François and Marguerite Rémond, who were of the nobility, Montmort traveled widely in Europe in his youth. He came under the influence of Father Nicholas de Malebranche, with whom he studied religion, philosophy, and mathematics. He succeeded his elder brother as canon of Nôtre-Dame but resigned in 1706 to marry and settle down at the country estate of Montmort, which he had bought with the fortune his father had left him. His marriage was a happy one, and during this simple and retired life he set to work on the theory of probability. In 1708, the first edition of ref. 3 appeared, "... where with the courage of Columbus he revealed a new world to mathematicians ..." according to Todhunter [5]. At the time, Montmort was aware of, and partly motivated by, the work by the Bernoullis* (reviewed in 1705 and 1706) on the book that was to be published posthumously in 1713 as Jacob (James) Bernoulli's *Ars Conjectandi*, under the editorship of Jacob's nephew Nicolaus (Nicholas). This publication in turn was motivated by the appearance of Montmort's work. Nicolaus and Montmort evolved an extensive and fruitful technical correspondence, some of which is included [together with a single letter from Jean (John) Bernoulli] as the fifth part of the substantially expanded second edition of Montmort's treatise [3], also published in 1713. It is clear from the correspondence that the mathematical influence of the Bernoullis (not to mention their contributions) on the second edition, was substantial. Montmort was piqued by De Moivre's* *De Mensura Sortis* (the Latin precursor of the *Doctrine of Chances*), which appeared in 1711 and which he regarded as plagiaristic [5]. It was, in fact, quite scathing in attacking his own first edition [1]; Montmort retaliated with an *Avertissement* in his second edition. Contrary to popular opinion, the breach was never properly healed [5, p. 102; 2].

The value of Montmort's work is partly in his scholarship. He was well-versed in the work on chance* of his predecessors (Pascal*, Fermat*, Huygens*), met Newton* on one of a number of visits to England, corresponded with Leibnitz, but remained on good terms with both sides during the strife between their followers. His friendship with Brook Taylor led to the latter's facilitating the publication of ref. 4 on the summation of infinite series, an element of Montmort's mathematical interests which enters into his probability work and distinguishes it from the earlier purely combinatorial problems arising out of enumeration of equiprobable sample points. Although ref. 3 to a large extent deals with the analysis of popular gambling* games, it focuses on the mathematical properties and is thus written for mathematicians rather than gamblers. The Royal Society elected Montmort a Fellow in 1715 and the *Académie Royale des Sciences* made him an associate member (as he was not a resident of Paris) the following year.

An extensive summary of the whole of ref. 3 is given in ref. 5; Todhunter thus accords him a very substantial place in the history of the subject. Montmort's best-known contribution to elementary probability is a result connected with the problem of matching pairs (in connection with the card games *Rencontre*, *Treize*, and Snap), in which n distinct objects are assigned a specific order, while n matching objects are assigned random order. The probability u_n of at least one match, where A_i is the event that there is a match in the ith position, is, in modern textbook analysis:

$$\Pr\left(\bigcup_{i=1}^{n} A_i\right) = \sum \Pr(A_i) - \sum \Pr(A_i \cup A_j) + \sum \Pr(A_i \cup A_j \cup A_k) \cdots = \sum_{j=1}^{n} \frac{(-1)^{j-1}}{j!},$$

of which the limit as $n \to \infty$ is $1 - e^{-1}$. This is thought to be the first occurrence of an exponential limit in the history of probability. Montmort's general iterative procedure for calculating u_n from $nu_n = (n-1)u_{n-1} + u_{n-2}$ is, on the other hand, based on a conditional probability argument (given in a commentary by Nicolaus Bernoulli) according to the outcome at the first position.

Montmort also worked with Jean Bernoulli on the problem of points* considered by Pascal and Fermat (for players of unequal skill). He worked with Nicolaus on the problem of duration of play in the gambler's ruin problem, possibly prior to De Moivre, and at the time the most difficult problem solved in the subject. Finally, in a letter of September 9, 1713, Nicolaus proposed the following problems to Montmort:

> *Quatrième Problème:* A promises to give an *écu* to B, if with an ordinary die he obtains a six with the first toss, two *écus* if he obtains a six with the second toss. . . . What is the expectation of B?
>
> *Cinquième Problème:* The same thing if A promises B *écus* in the progression 1, 2, 4, 8, 16, ...

It is clear that the St. Petersburg Paradox*, as subsequently treated by Daniel Bernoulli in 1738 is but an insignificant step away. In his reply Montmort indicates the solution to Montmort's problems and describes them as being of no difficulty.

References

[1] David, F. N. (1962). *Games, Gods, and Gambling: The Origins and History of Probability and Statistical Ideas from the Earliest Times to the Newtonian Era.* Griffin, London, Chap. 14. (Strong on biographical detail.)

[2] Hacking, I. (1974). "Montmort, Pierre Rémond de." *Dictionary of Scientific Biography*, Vol. 9, C. C. Gillispie, ed. pp. 499–500.

[3] Montmort, P. Rémond de. (1713). *Essay d'Analyse sur les Jeux de Hazard*, 2nd ed. Quillau, Paris (Reprinted by Chelsea, New York, 1980. Some copies of the second edition bear the date 1714. The first edition appeared in 1708.)

[4] Montmort, P. Rémond de. (1720). "De seriebus infinitis tractatus." *Philos. Trans. R. Soc.*, **30**, 633–675.

[5] Todhunter, I. (1865). *A History of the Mathematical Theory of Probability*. Macmillan, London, Chapter VIII. (Reprinted by Chelsea, New York, 1949 and 1965.)

Bibliography

Fontenelle, (1721). "Éloge de M. de Montmort." *Histoire de l'Académie Royale des Sciences pour l'Année 1719*, pp. 83–93. (The chief biographical source cited in refs. 1 and 5.)

Maistrov, L. E. (1974). *Probability Theory: A Historical Sketch.* Academic Press, New York. (Translated and edited from the Russian edition (1967) by S. Kotz. Chapter III Section 2 alludes to Montmort's work.)

Taylor, Brook. (1793). *Contemplatio Philosophica*, William Young, ed. London. (Contains part of Montmort's correspondence with Taylor.)

(BERNOULLIS, THE
GAMBLING, STATISTICS IN
MATCHING PROBLEM, THE
PROBABILITY, HISTORY OF (OUTLINE)
PROBLEM OF POINTS
ST. PETERSBURG PARADOX
SUMMATION OF SERIES)

E. SENETA

MOOD MEDIAN TEST *See* BROWN-MOOD TEST

MOOD SCALE TEST *See* MOOD'S DISPERSION TEST

MOOD'S DISPERSION TEST

Mood [7] introduced a two-sample nonparametric test for dispersion as a competitor to the two-sample F-test* for dispersion (i.e., the normal theory test based on the ratio of two sample variances). Let $x_i (i = 1, \ldots, m)$ and y_i $(i = 1, \ldots, n)$ denote independent random samples from the random variables X and Y, respectively. The null hypothesis states that X and Y have the same distribution. The pooled sample observations $(x_1, \ldots, x_m, y_1, \ldots, y_n)$ are ordered from smallest to largest. Suppose that k classes of distinct values occur in this ranking scheme. (Each class is formed by all observations having a particular value.) Let t_j denote the number of observations in the jth class of values ranked from below. Then $\sum_{i=1}^{k} t_i = m + n = N$ is the total number of observations in the pooled sample. Let a_j and b_j denote the number of x's and y's, respectively, in the jth class (i.e., $a_j + b_j = t_j$ for $j = 1, \ldots, k$, $\sum_{i=1}^{k} a_i = m$ and $\sum_{i=1}^{k} b_i = n$). Also let $S_j = \sum_{i=1}^{j} t_i$ and $a_0 = b_0 = t_0 = S_0 = 0$. In the presence of ties, Mood's dispersion test statistic is given by

$$M = \sum_{j=1}^{k} a_j \phi_j,$$

where ϕ_j is the arithmetic mean of the t_j rank functions associated with the conceptual collection of the t_j values in the jth class if there had been no ties. In particular,

$$\phi_j = t_j^{-1} \sum_{I = S_{j-1}+1}^{S_j} \left(I - \frac{N+1}{2} \right)^2,$$

where $[I - (N + 1)/2]^2$ is the rank function of Mood's dispersion test associated with the pooled sample's Ith-ordered observation. Under the null hypothesis and given a specific configuration of ties (i.e., $t_1, \ldots, t_k$), the mean and variance of M are given [3] by

$$E_0(M) = \frac{m(N^2-1)}{12},$$

$$\operatorname{var}_0(M) = \frac{mn(N+1)(N^2-4)}{180} - \frac{mn}{180N(N-1)} \sum_{j=1}^{k} t_j(t_j^2-1) \times \left[t_j^2 - 4 + 15(N - S_j - S_{j-1})^2\right].$$

The subtracted nonnegative expression in $\operatorname{var}_0(M)$ has been termed *Mielke's adjustment*. Mood [7] established the large sample normality of M and obtained the mean and variance of M under the null hypothesis when there are no ties (i.e., $t_j = 1$ for $j = 1, \ldots, k = N$). The type of alternatives that Mood's dispersion test detects quite adequately is when the distribution of X is either far more dispersed or far more concentrated than the distribution of Y but X and Y have about the same median. In contrast, the type of alternative that Mood's dispersion test detects very poorly is when the distribution of X differs from the distribution of Y by a simple shift in location.

Limiting Pitman efficiencies* of Mood's dispersion test relative to locally most powerful tests* and to the Ansari–Bradley dispersion test [1] for scale alternative detection (*see* ANSARI–BRADLEY W-STATISTICS) are obtained for selected origin-symmetric distributions. The origin-symmetric distributions are the double exponential (*see* LAPLACE DISTRIBUTION), logistic*, normal*, uniform*, and seven t-distributions*. Let B denote Ansari–Bradley's dispersion test statistic. Then the rank function associated with B is $|I - (N+1)/2|$, in the notation corresponding to M.

Furthermore, Ansari–Bradley's dispersion test is asymptotically equivalent to Siegel–Tukey's test* [9]. The limiting Pitman efficiencies of both Mood's and Ansari–Bradley's dispersion tests relative to locally most powerful tests [denoted by $e(M,T)$ and $e(B,T)$] and of Mood's dispersion test relative to Ansari–Bradley's dispersion test [denoted by $e(M,B)$] are presented in Table 1 for the specified origin-symmetric distributions.

An obvious feature of Table 1 is that Mood's dispersion test is an asymptotically optimum test for the Cauchy distribution* since $t(2)$ denotes the t distribution with two degrees of freedom. Incidentally, in the notation corresponding to M, a test statistic with rank function $|I - (N+1)/2|^v$, where $v > 0$, is an asymptotically optimum test for the distribution having a density function given [4] by

$$f(x) = \tfrac{1}{2} v^{-1/v} (1 + |x|^v/v)^{-(v+1)/v}.$$

An inspection of Table 1 indicates that Ansari–Bradley's dispersion test appears to perform better than Mood's dispersion test

Table 1 Limiting Pitman Relative Efficiencies for Scale Alternative Detection

Distribution	$e(M,T)$	$e(B,T)$	$e(M,B)$
Double exponential	0.868	0.750	1.157
Logistic	0.874	0.732	1.194
$t(1)$	0.924	0.986	0.938
$t(2)$	1.000	0.937	1.067
$t(3)$	0.982	0.876	1.122
$t(4)$	0.956	0.831	1.151
$t(5)$	0.933	0.797	1.170
$t(10)$	0.865	0.715	1.209
$t(20)$	0.818	0.665	1.230
Normal, $t(\infty)$	0.760	0.608	1.250
Uniform	0.000	0.000	1.667

only when distributions with substantially heavier tails than the Cauchy distribution are encountered. As distributions with progressively lighter tails are encountered, both Mood's and Ansari–Bradley's dispersion tests become increasingly less effective relative to the locally most powerful test. While Mood's dispersion test is superior to Ansari–Bradley's dispersion test for the uniform distribution, they are both useless relative to the limiting locally most powerful test.

Further, the effectiveness of tests based on either M or B in detecting dispersion alternatives requires that the distributions of X and Y have the same location. Alternative tests have been suggested by Hollander [2] and Moses [8] that accommodate this requirement. If the location requirement is satisfied, then the tests based on either M or B are usually far more sensitive than these alternative tests in detecting dispersion alternatives.

A disturbing observation pertaining to a class of commonly used permutation* statistics which includes M, B, and the test statistic suggested by Hollander [2] as special cases (*see* HOLLANDER EXTREME TEST) is that the underlying analysis space of statistics within this class is a complex nonmetric space [5, 6]. As a consequence, inferences based on statistics in this class are misleading, since they are inconsistent with the conceptual intent of most subject matter investigators (i.e., an investigator assumes that the inferential results of a test procedure correspond to a familiar Euclidean geometry when this assumption is almost always false). Since an alternative class of permutation statistics that eliminates this geometry problem exists, the routine use of this class is advocated. This class and the previously mentioned class of commonly used permutation statistics are both contained in an even broader class of statistics (*see* MULTIRESPONSE PERMUTATION PROCEDURES). The analogues of M, B, and the test statistic suggested by Hollander [2] in the alternative class are based on the same rank functions associated with the originally defined statistics [5, 6].

References

[1] Ansari, A. R. and Bradley, R. A. (1960). *Ann. Math. Statist.*, **31**, 1174–1189. (Introduces the Ansari–Bradley dispersion test.)

[2] Hollander, M. (1963). *Psychometrika*, **28**, 395–403. (Describes a nonparametric dispersion test which accommodates for location differences.)

[3] Mielke, P. W. (1967). *Technometrics*, **9**, 312–314. (Provides adjustment for ties with Mood's dispersion test statistic's variance.)

[4] Mielke, P. W. (1972). *J. Amer. Statist. Ass.*, **67**, 850–854. (Describes a class of dispersion tests and investigates their asymptotic properties.)

[5] Mielke, P. W., Berry, K. J., Brockwell, P. J., and Williams, J. S. (1981). *Biometrika*, **68**, 720–724. (Notes a geometric problem with a class of nonparametric tests which includes Mood's dispersion test.)

[6] Mielke, P. W., Berry, K. J., and Medina, J. G. (1982). *J. Appl. Meteorol.*, **21**, 788–792. (Amplifies on geometric problem with class of nonparametric tests which includes Mood's dispersion test.)

[7] Mood, A. M. (1954). *Ann. Math. Statist.*, **25**, 514–522. (Introduces Mood's dispersion test.)

[8] Moses, L. E. (1952). *Psychometrika*, **17**, 239–247. (Introduces a combinatorial type dispersion test that accommodates location differences.)

[9] Siegel, S. and Tukey, J. W. (1960). *J. Amer. Statist. Ass.*, **55**, 429–444. (Introduces Siegel–Tukey's dispersion test.)

(ANSARI–BRADLEY *W*-STATISTICS
F-TESTS
HEAVY-TAILED DISTRIBUTIONS
HOLLANDER EXTREME TEST
MULTIRESPONSE PERMUTATION PROCEDURES
RANK TESTS
SCALE TESTS
SIEGEL–TUKEY TEST)

PAUL W. MIELKE, JR.

MOORE–PENROSE INVERSE

Given an $m \times n$ matrix $\mathbf{B}$, define $\mathbf{B}^-$ (an $n \times m$ matrix) satisfying

$$\mathbf{B}\mathbf{B}^-\mathbf{B} = \mathbf{B}.$$

This matrix $\mathbf{B}^-$ is usually referred to as the generalized inverse* of $\mathbf{B}$. The matrix $\mathbf{B}^-$ usually is not unique. If, however, $\mathbf{B}^-$ satisfies the three following conditions:

$$\mathbf{B}^{-}\mathbf{B}\mathbf{B}^{-} = \mathbf{B}^{-},$$
$$(\mathbf{B}\mathbf{B}^{-})' = \mathbf{B}\mathbf{B}^{-},$$
$$(\mathbf{B}^{-}\mathbf{B})' = \mathbf{B}^{-}\mathbf{B},$$

where $\mathbf{B}'$ is the transpose of $\mathbf{B}$, then it is unique and is called the Moore–Penrose inverse or the pseudo-inverse.

It has been used extensively in the theory of linear regression analysis (see, e.g., Seber [1]).

Reference

[1] Seber, G. A. F. (1977). *Linear Regression Analysis.* Wiley, New York.

(GENERALIZED INVERSES)

MORAL CERTAINTY

In modern English, the adjective *moral* refers primarily to the distinction between right and wrong. But eighteenth and nineteenth century authors on probability often used the word in a different sense. They used *moral certainty*, for example, to mean very high probability.

In order to understand the origins of this usage, we must examine the Latin origins of the word moral. The Latin noun *mos* means a custom or a practice of a person or a people. The plural of *mos* is *mores*, and since the Romans considered it virtuous to follow the customs of one's ancestors, *mores* was used to translate into Latin the Greek word ἤθη, which means ethics. Cicero (106–143 B.C.) coined the adjective *moralis* in order to translate ἠθικός, or "ethical." Later Latin writers followed Cicero in using *moralis* to mean ethical, but because of the more general meaning of *mos*, they also used *moralis* (and the abverb *moraliter*) in other ways.

In the seventeenth and eighteenth centuries, *moralis* and *moraliter* were widely used in discussions of argument and evidence. Certain kinds of evidence were called moral evidence (*evidentia moralis*). Such evidence could make something "morally certain" (*moraliter certum*) or "morally impossible" (*moraliter impossibile*). Originally, moral evidence was the evidence provided by arguments that appeal to knowledge of human nature or to the desire to conform to human custom. Testimony, since its credibility rests on the knowledge that people are usually honest, is moral evidence. It is morally certain that a mother loves her child. Everyone else's believing in God is a moral argument for my believing in God. Gradually, however, the category of moral argument was broadened to include all arguments that are only usually reliable. The adjective "moral" was then explained by saying that something is morally certain if we can take it for granted in deciding on our own actions. Descartes (1596–1650), for example, described moral certainty as the kind of certainty that is adequate for practical life.

One factor contributing to the broadening of the category of moral argument was its position in a widely accepted trichotomy. Philosophers and theologians distinguished three kinds of argument—metaphysical, physical, and moral. Metaphysical arguments were based on self-evident principles. Physical arguments were based on laws of nature, or, more generally, on invariable experience. The category of moral argument, being more vaguely defined, tended to include everything that was left over.

The terms *moral certainty* and *moral impossibility* were first used in a technical sense in probability theory by James Bernoulli (*see* BERNOULLIS, THE). In Part IV of his *Ars Conjectandi* (published posthumously in 1713), Bernoulli defined these terms by saying that a thing is morally certain if its probability is so close to complete *certainty* that the difference cannot be noticed and morally impossible if its probability is only as great as this unnoticeable difference. He suggested that judicial authorities should establish a convention fixing the limits of moral certainty at a probability of, say, 0.99 or 0.999.

The adjective *moral* acquired yet another use in probability theory in the mid-eighteenth century when Gabriel Cramer and Daniel Bernoulli developed the idea of expected utility. (*See* UTILITY; SAINT PETERSBURG PARADOX.) In a 1728 letter in French to Nicholas Bernoulli, which Daniel published in his 1738 paper on utility, Cramer used "moral value" (*valeur morale* in French) for what we now call utility and "moral expectation" (*espérance morale* in French) for what we now call expected utility. This use of *moral* was based, of course, on the idea that the utility of money is its value in relation to a person's *mores*, his customs and behavior.

The broad interpretation of *moral* that we find in Descartes, Bernoulli, and Cramer was widely accepted in the eighteenth and nineteenth centuries. Even the dictionary of the French Academy in the nineteenth century defined *moral certainty* as a "certainty founded on probabilities as strong as one can have in ordinary matters of life." Yet many authors during this period also used *moral* in narrower ways. The philosopher Leibniz (1646–1716), for example, used *morally necessary* and *morally impossible* to refer to actions that are right and wrong, respectively. This usage was also followed by Kant and by many theologians.

The use of *morally certain* to mean highly probable seems to be rare in modern English. Yet some dictionaries still sanction it, along with a number of other interpretations.

Literature

For some early examples of the use of *moral certainty* see *Index Scolastico-Cartesian*, by Etienne Gilson (originally published in Paris in 1912, reprinted by Burt Franklin, New York, 1964). Descartes's definition of moral certainty is in Section 205 of Part IV of his *Principiorum Philosophiae*. For Leibniz's views on moral necessity, see Robert Merrihew Adams's paper on pp. 243–283 of *Leibniz: Critical and Interpretative Essays* (Michael Hooker, ed., University of Minnesota Press, Minneapolis, 1982). For a general discussion of the relation between mathematics and the "moral sciences" in the eighteenth century, see Lorraine J. Daston's paper on pp. 287–309 of *Epistemological and Social Problems of the Sciences in the Early Nineteenth Century*. (H. N. Jahnke and M. Otte, eds. Reidel, Dordrecht, Holland, 1981).

(CHANCE
PROBABILITY, HISTORY OF
WEIGHT OF EVIDENCE)

GLENN SHAFER

MORAN DAM *See* DAM THEORY

MORBIDITY

Morbidity is a term used in epidemiological studies to describe sickness in human populations. It differs from concepts such as fertility* and mortality (*see* RATES) in that morbidity rates have to be strictly defined in order to avoid ambiguity. Morbidity can be disease-specific or based on a group of diseases; in this context measurements are restricted to patients who have presented themselves to a physician (therapist, nurse, doctor's aide) with a clinical complaint, and the presence in the population of hidden morbidity that cannot be measured is an unknown factor. The problem of measuring morbidity (as well as of detecting persons who are infected with communicable diseases) is thus rendered more difficult. Long-term studies involve problems in data collection* such as those associated with under-reporting and follow-up*.

A different arena in which the term *morbidity* appears is that of manpower planning* and the compilation of labor statistics. There the number of *morbid days* refers to time lost from work other than for vacations and is included in measuring productivity and economic indicators. We do not consider this field further here.

Disease-specific illness is defined in the United States by the Center for Disease Control (CDC) for most endemic diseases, and morbidity rates are based on these definitions. As an illustration, in an outbreak of diphtheria in San Antonio, Texas, in 1970 (see Marcuse and Grand [3]), reported cases were labeled as diphtheria if the patient had either (1) symptoms attributable to diphtheria and a culture confirmation *or* (2) strictly defined symptoms suggestive of diphtheria "plus either a tonsillar or pharyngeal membrane or a major complication of diphtheria," again strictly defined. Strict definitions of this kind are necessary to ensure as uniform a reporting procedure as possible and to safeguard against subjective differences between physicians (or between hospitals) in deciding if each patient is a case or not. For some psychiatric illnesses such as depression, such strict definitions are much harder to apply, and the measurement of morbidity becomes more subjective (*see* PSYCHOLOGY, STATISTICS IN). For the reliability of laboratory screening tests in measuring morbidity, *see* MEDICAL DIAGNOSIS, STATISTICS IN.

The two principal measures are *prevalence* and *incidence*. A key factor in measurement, and a complicating one, is that people become sick at different points in time and remain ill for different lengths of time. In a given city or region or state, disease-specific morbidity is measured by:

$$\text{point prevalence} = \frac{\begin{pmatrix}\text{number of cases at}\\ \text{a particular moment}\end{pmatrix}}{\begin{pmatrix}\text{number in population}\\ \text{at that moment}\end{pmatrix}}.$$

$$\text{period prevalence} = \frac{\begin{pmatrix}\text{number of cases during}\\ \text{a specified time period}\end{pmatrix}}{\begin{pmatrix}\text{number in population}\\ \text{at midpoint of period}\end{pmatrix}}.$$

$$\text{incidence} = \frac{\begin{pmatrix}\text{number of } \textit{new} \text{ cases}\\ \text{during a specified time period}\end{pmatrix}}{\begin{pmatrix}\text{number in population}\\ \text{at midpoint of period}\end{pmatrix}}.$$

Where immunity from disease is a factor, the population is usually defined to include only those members at risk (*see* EPIDEMIOLOGICAL STATISTICS). Prevalence is useful in measuring the (momentary) health of a community and hence the adequacy of the community's health services, but it is affected by the duration of illness. The incidence rate is a better estimate of the probability of the disease occurring in particular groups of people. By restricting the numerator to new cases only, the incidence rate gives a useful picture of the progress of an epidemic, for which rates will be recorded on a weekly basis, say. Prevalence and incidence rates are frequently measured specific to sex, age, racial, socioeconomic, and sometimes geographic groups. In the San Antonio diphtheria outbreak, for example, new cases occurred at no more than 2 per week for the first six months of 1970, increased to 6 per week in July, and peaked at 22 in one week in August before declining gradually. The incidence rate was highest in children 10–14 years old, was 4, 46, and 49 per 1,000,000 for Anglos, blacks, and Mexican-Americans, respectively, and was 5.6, 25.2, and 69.9 per 1,000,000 among the upper 25%, middle 50% and lower 25% by socioeconomic status, respectively (as measured by census tracts) [3].

The severity of illness is measured also according to CDC definitions, symptoms being defined as mild, moderate or severe, for example. For life-threatening diseases, the severest endpoint, mortality, is measured by the *case fatality rate*, defined as

$$\text{case fatality rate} = \frac{\begin{pmatrix}\text{number of deaths from the}\\ \text{disease during specified period}\end{pmatrix}}{\begin{pmatrix}\text{number of cases at the}\\ \text{beginning of the period}\end{pmatrix}}.$$

Comparison studies of disease between geographic regions and points in time involve adjusted measure rates (e.g., for age if the age distributions between regions differ substantively). For discussion see Fox et al. [1] and MacMahon and Pugh [2] (*see also* VITAL STATISTICS).

References

[1] Fox, J. P., Hall, C. E., and Elveback, L. R. (1970). *Epidemiology: Man and Disease*. Macmillan, New York.

[2] MacMahon, B. and Pugh, T. F. (1970). *Epidemiology: Principles and Methods*. Little, Brown, Boston.

[3] Marcuse, E. K. and Grand, M. G. (1973). *J. Amer. Med. Ass.*, **224**, 305–310.

(CLINICAL TRIALS
EPIDEMIOLOGICAL STATISTICS
FOLLOW-UP
HEALTH STATISTICS, NATIONAL CENTER FOR
HISTORICAL STUDIES
LIFE TABLES
VITAL STATISTICS)

CAMPBELL B. READ

MORGAN–PITMAN TEST *See* PITMAN–MORGAN TEST

MORGENSTERN–FARLIE SYSTEM *See* FARLIE–GUMBEL–MORGENSTERN DISTRIBUTION

MORTALITY *See* ACTUARIAL STATISTICS, LIFE; CLINICAL TRIALS; COMPETING RISKS; DEMOGRAPHY; FORCE OF MORTALITY; LIFE TABLES; RATES; SURVIVAL ANALYSIS; VITAL STATISTICS

MOSES TEST *See* HOLLANDER EXTREME TEST

MOSS

MOSS is the acronym for minimum orthogonal sum of squares. In assessing distributional properties of observed multivariate data by means of probability plots* a useful procedure is to fit a straight line to the scatter of the points in the p-dimensional space by minimizing the sum of squares of perpendicular deviations of the points from the line and to compute the value of the obtained minimum orthogonal sum of squares.

The linear principal component analysis* is an algorithm for fitting the MOSS line and computing the MOSS value. See, e.g., Gnanadesikan [1] for more details.

Reference

[1] Gnanadesikan, R. (1977). *Methods for Statistical Data Analysis of Multivariate Observations*. Wiley, New York.

(LEAST SQUARES
PROBABILITY PLOTTING)

MOST APPROXIMATIVE METHOD

In this, a modification of the minimax method*, one assumes that the intervals (A_i, B_i) containing the respective residual errors r_i are known and seeks to determine for each unknown (say x) the smallest interval (I, S) containing that unknown (i.e., an interval such that for every value of x less than I or greater than S, one or more of the residuals r_i lie outside the given interval). This modification of the minimax method* (*method of minimum approximation*) was proposed by Goedseels [2] for use when the limits of the errors are known; otherwise he advocated use of the method of minimum approximation. Many of the properties of that method hold also for the *most approximative method*. Further results concerning the most approximative method and generalizations thereof have been given by Goedseels [3–5], Tits [7], and Alliaume [1]. The method was used extensively by Goedseels and some of his associates at the Catholic University of Louvain and the Royal Observatory of Belgium, but it has seen little use elsewhere. Further information about it and its applicability was given by Harter [6].

References

[1] Alliaume, M. (1927). *Ann. Soc. Sci. Bruxelles*, **A47**(1), 5–14, 60–68.

[2] Goedseels, P. J. E. (1909). *Théorie des Erreurs d'Observation*, 3rd ed. Charles Peeters, Louvain, Belgium; Gauthier-Villars, Paris.

[3] Goedseels, P. J. E. (1910). *Ann. Soc. Sci. Bruxelles*, **34**(2), 257–287.

[4] Goedseels, P. J. E. (1911). *Ann. Soc. Sci. Bruxelles*, **35**(1), 351–368.

[5] Goedseels, P. J. E. (1925). *Exposé Rigoreux de la Méthode des Moindres Carrés*. Ceuterick, Louvain, Belgium.

[6] Harter, H. L. (1974–1976). *Int. Statist. Rev.*, **42** (1974), 147–174, 235–264, 282; **43** (1975), 1–44, 125–190, 269–278; **44** (1976) 113–159. (A comprehensive survey of the literature on the method of least squares and various alternatives, including the most approximative method.)

[7] Tits, L. (1912). *Ann. Soc. Sci. Bruxelles*, **36**(2), 253–263.

(MINIMAX METHOD)

H. LEON HARTER

MOST POWERFUL TEST *See* HYPOTHESIS TESTING

MOST PROBABLE NUMBER

The *most probable number* is the maximum likelihood estimate* of the density of organisms in a liquid when several samples of the liquid are examined for the presence or absence of growth without direct counting of organisms. One basic assumption made is that the organisms are randomly distributed in the liquid so that the number of organisms in a sample Y for an appropriately scaled measure of dilution x follows a Poisson distribution* with mean λx, where λ is the density of organisms in the liquid. Therefore the probability that y organisms are in a sample is given by

$$P(Y = y) = e^{-\lambda x}(\lambda x)^y / y!, \quad y = 0, 1, \ldots .$$

Another basic assumption is that each sample exhibits growth if it contains at least one organism. Under these assumptions the probability of a sterile sample, i.e., $P(Y = 0)$, is $\theta = \exp(-\lambda x)$. For a single dose x, n independent samples are observed. Then the number s of samples observed to be sterile has a binomial distribution* with parameters n and θ, so the maximum likelihood estimate of θ is $\theta = s/n$. The corresponding maximum likelihood estimate of λ is the most probable number $\hat{\lambda} = -[\ln(s/n)]/x$, where ln denotes natural logarithm.

More generally, suppose that several dilutions are studied independently with n_i samples at the ith dilution, s_i of which are sterile, $i = 1, 2, \ldots, k$. In this case the likelihood function is a product of k binomial functions and reaches a maximum for $\lambda = \hat{\lambda}$, the most probable number. There is no explicit formula for $\hat{\lambda}$ for $k > 1$, but iterative procedures have been developed for its calculation where $\hat{\lambda}$ is the solution of the equation

$$\sum_{i=1}^{k} \frac{(n_i - s_i)x_i e^{-\hat{\lambda} x_i}}{(1 - e^{-\hat{\lambda} x_i})} = \sum_{i=1}^{k} s_i x_i .$$

The estimator $\hat{\lambda}$ was named the most probable number by McCrady [10]. Cochran [1] discussed its experimental background and calculation. He also gave guidelines for the design of serial dilution assays. Peto [11] developed the same model to describe the invasion of microorganisms. He presented an iterative method for the calculation of the most probable number and the estimator of its variance. Mather [9] also discussed experimental design and presented iterative calculations for the most probable number (the format commonly used for quantal bioassays) for the situation in which log λ is a linear function of x. Support for Mather's formulation was given by Epstein [4] from the perspective of extreme value theory. Both Peto and Mather presented tables to aid in the calculation of the most probable number. Finney [5] also presented iterative calculations for λ in the bioassay context. For a fuller discussion of quantal bioassays, *see* BIOASSAY, STATISTICAL METHODS IN *and* QUANTAL RESPONSE ANALYSIS.

The most probable number was compared with several other estimators for the same model by Cornell and Speckman [2]. A detailed illustration of the iterative calculations required to compute the most probable number is also displayed in their article based on data presented by Edington et al. [3]; the most probable number can be calculated for any spacing of doses and has desirable large-sample properties. It also performed well in small-sample studies they conducted. They suggested the simple method of partial totals presented by Speckman and Cornell [13] as an appropriate alternative for small samples for equally spaced doses. The Fisher (see Fisher and Yates [6, Table VIII 2]) and Spearman–Kärber (see Johnson and Brown [7]) methods were suggested as alternatives regardless of the sample size for dosages with equally spaced logarithms. The latter is discussed for a variety of models in the entry KÄRBER METHOD. Bayesian estimation for this same model was discussed by Petrasovits and Cornell [12].

Koch and Tolley [8] summarized the literature on most probable number analysis and extended its use to examinations of trends in most probable numbers and to comparisons of most probable numbers between experiments. They incorporated separate most probable number calculation from dilution series or microbial experiments, with corresponding variance estimation, into linear categorical data analysis. Weighted least-squares* procedures were developed and applied to a series of experiments leading to comparisons of extinction patterns of bacterial populations under different conditions. This article by Koch and Tolley is a comprehensive presentation of the history and current status of most probable number analysis.

References

[1] Cochran, W. G. (1950). *Biometrics*, **6**, 105–116.

[2] Cornell, R. G. and Speckman, J. A. (1967). *Biometrics*, **23**, 717–737.

[3] Edington, C. W., Epler, J. L., and Regan, J. D. (1962). *Genetics*, **47**, 397–406.

[4] Epstein, B. (1967). *Biometrics*, **23**, 835–839.

[5] Finney, D. J. (1964). *Statistical Methods in Biological Assay*, 2nd ed. Hafner, New York, Section 21.5.

[6] Fisher, R. A. and Yates, F. (1963). *Statistical Tables for Biological, Agricultural, and Medical Research*. 6th ed. Oliver and Boyd, Edinburgh, Scotland.

[7] Johnson, E. A. and Brown, B. W., Jr. (1961). *Biometrics*, **17**, 79–88.

[8] Koch, G. G. and Tolley, H. D. (1975). *Biometrics*, **31**, 59–92.

[9] Mather, K. (1949). *Biometrics*, **5**, 127–143.

[10] McCrady, M. H. (1915). *J. Infect. Dis.*, **17**, 183–212.

[11] Peto, S. (1953). *Biometrics*, **9**, 320–335.

[12] Petrasovits, A. and Cornell, R. G. (1975). *Commun. Statist.*, **4**, 851–862.

[13] Speckman, J. A. and Cornell, R. G. (1965). *J. Amer. Statist. Ass.*, **60**, 560–572.

(BIOASSAY, STATISTICAL METHODS IN
BIOSTATISTICS
KÄRBER METHOD
QUANTAL RESPONSE ANALYSIS)

RICHARD G. CORNELL

MOST STRINGENT TEST

The basic ideas of statistical testing are presented in HYPOTHESIS TESTING. As explained there, when both the null hypothesis H_0 and the alternative hypothesis H_1 specify the distribution of the data completely, there is a best test of H_0 against H_1. The best size α test has the property that no other test with probability at most α of rejecting H_0 falsely when the data are from the distribution specified by H_0 has larger probability of rejecting H_0 correctly when the data are from the distribution specified by H_1. In many cases, however, the hypotheses are composite and there is no test that is uniformly best for all distributions specified by H_1. That is, there is no size α test that has the largest possible probability of rejecting H_0 for each distribution specified by H_1.

When there is no best test, the criterion of *stringency* can be applied to choose between competing tests. Stringency requires that the test have minimal shortcoming over the alternative. The *shortcoming* of a test at a distribution F in H_1 refers to the difference between its probability of rejection under F and the largest possible probability of rejection by any test against F. The most stringent test has minimal shortcoming. The definitions of shortcoming and most stringent test are most easily understood in the context of an example.

Example. Suppose that H_0 specifies that the probability p of heads for a coin is 0.5 and H_1 specifies that $p \neq 0.5$. No test of these hypotheses has the largest possible probability of rejection (i.e., largest possible power) for every $p \neq 0.5$. The test that rejects when c or fewer heads are observed in n tosses is best for any $p < 0.5$, and the test that rejects when $n - c$ or more heads are observed is best for any $p > 0.5$, where c depends on the size α of the test. To fix the ideas, consider the case $n = 10$ and $c = 2$, which corresponds to $\alpha = 0.055$. The power function of each of the two best one-sided tests for this case is drawn in Fig. 1. Since each of these tests is best against one side of $p \neq 0.5$, the maximum probability of rejection by any test of H_0 at a particular p in H_1 is given by the maximum of these two power curves. The maximum of these curves is the *envelope power curve*. The shortcoming of any test of H_0 against H_1 is the maximum difference between its power curve and the envelope power curve. The most stringent test is the one that minimizes this maximum difference. From Fig. 1, it can be seen that the shortcoming of the best test for $p < 0.5$ and the shortcoming of the best test for $p > 0.5$ are both 1. This shortcoming can be decreased only by using a test that protects against both $p < 0.5$ and $p > 0.5$ by rejecting H_0 for both small and large values of x. If we allow only nonrandomized tests (i.e., tests for which the decision to accept or reject depends only on the number of heads

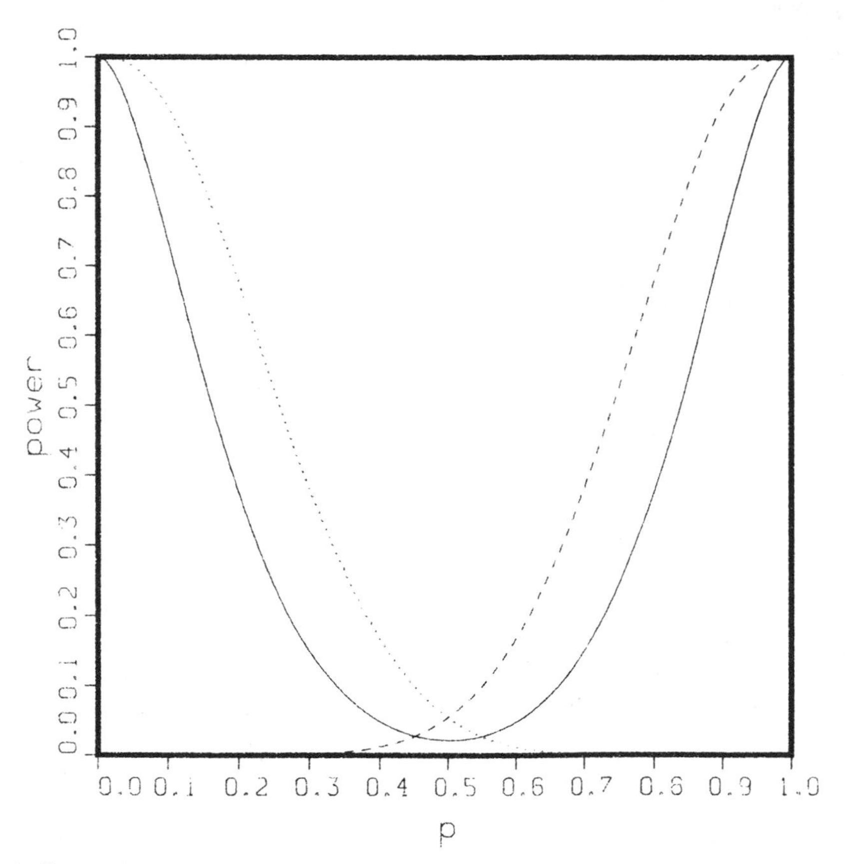

Figure 1 Power of tests of $p = 0.5$ for $n = 10$, $\alpha = 0.055$. The dotted line indicates the best test against $p < 0.5$; the dashed line, the best test against $p > 0.5$; and the solid line, the most stringent test against $p \neq 0.5$.

observed), then there are only a finite number of two-sided tests of size at most α. Consideration of each of these tests in turn shows that the most stringent test of size at most 0.055 rejects when either $x \leqslant 1$ or $x \geqslant 9$. This test has size 0.023. Its maximum shortcoming, which occurs at $p = 0.20$ and $p = 0.80$, is 0.30.

Most stringent tests exist under weak conditions. Their use is limited, however, because they are often difficult to determine. Methods that sometimes lead to a most stringent test include invariance* arguments and minimax* reasoning. These approaches are discussed in Lehmann [1].

References

[1] Lehmann, E. L. (1959). *Testing Statistical Hypotheses*. Wiley, New York, Section 8.5.

(HYPOTHESIS TESTING
MINIMAX TESTS)

DIANE LAMBERT

MOVER–STAYER MODEL

This model was introduced by Blumen et al. [1] as a model for industrial mobility of labor and further considered by Goodman [2].

It is defined by an m-state Markov chain (*see* MARKOV PROCESSES) with transition probabilities of the form

$$\pi_{ij} = (1 - s_i)p_j, \qquad i \neq j = 1, \ldots, m,$$
$$\pi_{ii} = (1 - s_i)p_i + s_i, \qquad i = 1, \ldots, m,$$

where $\{p_k\}$ is a probability distribution and

$$1 - s_i \geqslant 0, \qquad (1 - s_i)p_i + s_i \geqslant 0 \qquad \text{for all} \quad i = 1, \ldots m.$$

The conditional probabilities of state change are given by

$$P_{ij} = \pi_{ij}/(1 - \pi_{ii}) = p_j/(1 - p_i).$$

This model thus serves as an early example of quasi-independence* in contingency tables for $i \neq j$ (with missing or excluded diagonal). Estimation procedures for this model were considered by Morgan and Titterington [3]. In this article, various applications of this model in geology*, meteorology*, ethnology, and psychiatry are discussed.

References

[1] Blumen, I., Kogan, M., and McCarthy, P. J. (1955). In *Cornell Studies of Industrial and Labor Relations*, Vol. 6: *The Industrial Mobility of Labor as a Probability Process*. Cornell University, Ithaca, NY.

[2] Goodman, L. A. (1961). *J. Amer. Statist. Ass.*, **56**, 841–868.

[3] Morgan, B. J. T. and Titterington, D. M. (1977). *Biometrika*, **64**, 265–269.

(CONTINGENCY TABLES
INDUSTRY, STATISTICS IN
MARKOV PROCESSES
QUASI-INDEPENDENCE)

MOVING AVERAGES

In statistical studies, a series of data arranged by chronological order is a time series*. For example, the records of daily mean temperatures and barometer pressure in a given locality are time series. Generally, data in a time series are equally spaced. The values assumed by a variable at time t may or may not embody an element of random variations, but in the majority of cases such an element is present, if only as an error of measurement. The variations in the data with time may be relatively smooth and orderly, or they may be rather complex and without apparent pattern.

It is convenient to consider that the variations in time are produced by superimposed sinusoidal waves of various amplitudes, frequencies, and phases. The Fourier theorem states that no matter how complicated these variations in the series may be they can be always adequately approximated by the superposition of a number of simple component periodic functions [2]. The amplitudes,

Table 1 Illustration of Smoothing a Time Series by Means of a Typical Moving Average

Time Series Original Values	Moving Average Weights					Smoothed Values
378	−0.073					
371	0.294	−0.073				
395	0.558	0.294	−0.073			387.76
413	0.294	0.558	0.294	−0.073		426.25
487	−0.073	0.294	0.558	0.294	−0.073	475.05
499		−0.073	0.294	0.558	0.294	499.55
498			−0.073	0.294	0.558	503.09
525				−0.073	0.294	
552					−0.073	

frequencies, and phases of these functions are generally changing constantly with time.

The main purpose of applying a moving average to a time series is to reduce the amplitude of high-frequency oscillations in the data without significantly affecting the low-frequency components. Above some high-frequency or for certain frequency bands, depending on the property of the moving average, this reduction is complete for all practical purposes.

The assumption on which the use of moving averages is justified is that high-frequency oscillations or those corresponding to certain frequency bands are either random errors or are of no significance to the particular type of analysis of the data to be performed after its application.

A moving average consists of a sequence of fractional values called *weights*. The moving-average estimate corresponding to an observation X_t is computed from observations X_{t-m} through X_{t+m} as follows:

$$X_t^a = \sum_{j=-m}^{m} w_j X_{t+j} = w_{-m} X_{t-m} + \cdots + w_0 X_t + \cdots + w_m X_{t+m}, \tag{1}$$

where w_j are the weights, w_0 being the principal weight for symmetric moving averages, and $w_{-j} = w_j$ for all j. The *span* of the moving average, given by the number of observations used in computing it, equals $2m + 1$ in equation (1).

The operation of applying a moving average to a time series, also known as *smoothing* (*see* GRADUATION), is a special case of a broader general process of filtering, a concept brought into the field of time-series analysis from electrical engineering.

An application of a moving average is given in Table 1. The weights are cumulatively cross-multiplied by the adjacent values in the time series, and the resulting product is entered opposite the time-series values multiplied by the principal weight. Then the weights are moved down one time increment (data interval) along the time series and the cross-multiplication is repeated to obtain a second smoothed value. This process is repeated until all the time-series data have been used. The sum of the weights of a moving average determines the ratio of the mean of the output or smoothed series to the mean of the original series. It is generally desirable to leave the mean of the series unchanged. Thus the sum of the weights of most moving averages equals one.

BASIC PROPERTIES OF MOVING AVERAGES

A moving average transforms an original time series $\{X_t\}$, the input, into a smoothed time series $\{X_t^a\}$, the output. The three basic properties of moving averages or smoothing functions are: (a) scale preservation, (b) superposition, and (c) time invariance.

A moving average L preserves scale if an amplification of the input $\{X_t\}$ by a given scale factor α results in the amplification of the output by the same factor, that is,

$$L(\alpha\{X_t\}) = \alpha L\{X_t\}. \tag{2}$$

The superposition principle states that if two time series $\{X_t\}$ and $\{Y_t\}$ are added together and given as the input to the moving average L, then the output will be the sum of the two series that would have resulted from using the two initial inputs to the L transformation separately. Hence

$$L(\{X_t\} + \{Y_t\}) = L\{X_t\} + L\{Y_t\}. \tag{3}$$

These two first properties express the fact that a moving average is a linear operator. The time-invariant property states that if two inputs $\{X_t\}$ and $\{X_{t+\tau}\}$ to the moving average L are the same except for a relative time lag τ, then the outputs will also be the same except for the displacement τ. That is, if $\{X_t^a\} = L\{X_t\}$ then,

$$\{X_{t+\tau}^a\} = L\{X_{t+\tau}\} \qquad \text{for all } \tau. \tag{4}$$

This means that the smoothing function or moving average always responds in the same manner no matter what time in history a given input is presented to the moving average.

CONSTRUCTION OF MOVING-AVERAGE FORMULAS

Moving averages are applied to time series when it is assumed that the systematic component, sometimes called signal or trend*, is a smooth function of time that cannot be closely approximated by a single function over the entire span of the series. A smooth function can be approximated well by a polynomial of a fairly low degree over some span of time, but not over the entire range. The polynomial that approximates the systematic part in one interval may not be the same in other intervals. Therefore, the assumption of smoothness is a *local* property whereas the assumption of a polynomial signal or trend is a *global* property that concerns the entire time interval [1].

A common use of moving averages in economic and social data is to eliminate from each series not only irregular fluctuations but also seasonal variations. Thus the smoothed series is not intended to *replace* the original data but to *supplement* them. This differs from smoothing made to physical observations in order to eliminate errors of measurement. It also differs from smoothing intended to estimate the "universe" from a sample such as that known as *graduation**, which is used for mortality tables by actuaries.

Depending on the purposes of the moving averages, several formulas have been constructed. The estimation of weights can be done by fitting polynomials or from "summation" formulas.

Moving Averages by Fitting Polynomials

The calculation of the weights of a moving average by fitting polynomials is extensively treated in refs. 3 and 5. To generate the set of weights of a moving average, the span of the average and the degree of the polynomial must be chosen in advance.

For a given span, say $2m+1$, and a polynomial of degree p, not greater than $2m$, the coefficients are calculated by least squares and the midordinate of the fitted polynomial is used to estimate the smoothed middle value [5].

Suppose that the terms of the moving average are $X_{t-m}, \dots, X_t, \dots X_{t+m}$ and that the polynomial that will give the smoothed values is

$$f_t(j) = \alpha_0 + \alpha_1 j + \alpha_2 j^2 + \cdots + \alpha_p j^p,$$
$$j = -m, \dots, m. \tag{5}$$

The smoothed value of the middle term X_t is $f_t(0)$, that is, α_0. The estimates of the α's, denoted by a's, can be obtained by least squares*, solving the equations

$$\frac{\partial}{\partial a_i} \sum_{j=-m}^{m} (X_{t+j} - a_0 - a_1 j - \cdots - a_p j^p)^2$$
$$= 0, \qquad i = 0, 1, \dots, p, \tag{6}$$

which gives the normal equations

$$a_0 \sum_{j=-m}^{m} j^i + a_1 \sum_{j=-m}^{m} j^{i+1} + \cdots + a_p \sum_{j=-m}^{m} j^{i+p} = \sum_{j=-m}^{m} j^i X_{t+j}, \quad i = 0, 1, \ldots, p. \quad (7)$$

By symmetry, the sum of any odd power of j is zero. Then, for even i, the coefficients of $a_1, a_3, \ldots$ in equation (7) are zero, and for odd i, the coefficients of $a_0, a_2, \ldots,$ in equation (7) are zero. Solving equation (7) for a_0,

$$a_0 = \sum_{j=-m}^{m} w_j X_{t+j}, \quad (8)$$

where $w_j = w_{-j}$. The weights depend only on m and p, but not on the X's. For a given degree p of the fitted polynomial, the variance of the smoothed series decreases with increasing span, and, for a given span, the variance goes up with increasing p.

Moving Averages from Summation Formulas

During the last century and beginning of this one, actuaries looked at the problem of obtaining moving-average weights that, when fitted to second- or third-degree polynomials, would fall exactly on those polynomials and when fitted to stochastic, nonmathematical data, would give smoother results than can be obtained from the weights obtained by fitting polynomials by the method of least squares (*see* ACTUARIAL STATISTICS).

Actuaries obtained moving-average weights from *summation* formulas, which are extensively discussed in refs. 3–5 and are based on the following principle.

Let Δ denote the operation of differencing (*see* FINITE DIFFERENCES, CALCULUS OF); $\Delta X_t = X_{t+1} - X_t$. Let $[2m+1]X_t$ symbolize the sum of $2m+1$ consecutive terms of which X_t is the middle one. Then it is possible to find combinations of these operations of differencing and summation that, when differences of above a certain order are neglected, merely reproduce the functions operated on. That is,

$$f\{\Delta, [\ \ ]\}X_t = X_t + \text{high differences.} \quad (9)$$

The smoothed value X_t^a is then

$$X_t^a = f\{\Delta, [\ \ ]\}X_t. \quad (10)$$

Two well-known moving averages that are correct to third differences in the sense that this order of differences is not affected by the procedure, are the Spencer's* 15- and 21-point formulas. The Spencer's 15-term moving average can be obtained by first calculating

$$X_t^* = \tfrac{1}{4}(-3X_{t-2} + 3X_{t-1} + 4X_t + 3X_{t+1} - 3X_{t+2}), \quad (11)$$

then averaging, with equal weights, five successive X_t^*'s, next averaging four successive terms of the resulting series, and finally averaging four successive terms of this last series. Hence

$$X_t^a = \tfrac{1}{320}[4][4][5]X_t^*. \quad (12)$$

The final weights are symmetric with $(w_{-7}, \ldots, w_0)$, which is equal to

$$\tfrac{1}{320}(-3, -6, -5, 3, 21, 46, 67, \underline{74}), \quad (13)$$

where $\underline{74}$ is the principal (middle) weight. The set of weights (13) gives a smooth graduation because the weight diagram is itself a smooth curve. Recognition of the fact that the smoothness of the resulting graduation depends directly on the smoothness of the weight diagram led to Henderson's Ideal formula, which minimizes the sum of squares of the third differences of the smoothed curve. The fulfillment of this criterion is equivalent to making the sum of the squares of the third differences of the set of weights of the moving averages a minimum.

The reduction in the variance of the third differences of the smoothed series is a function of the set of weights of the filter. In effect, assuming that the original series $\{X_t\} = \{F_t + U_t\}$, where F_t is the systematic component and U_t is the random part, such that $E(U_t) = 0$ and

$$E(U_t U_s) = \begin{cases} 1, & t = s \\ 0, & t \neq s \end{cases}$$

and that the smoothed series is $\{X_t^a\} = \sum_j w_j \{X_{t+j}\}$, then

$$\operatorname{var} \Delta^3\{X_t^a\} = E\left(\Delta^3\{X_t^a\} - E\Delta^3\{X_t^a\}\right)^2 = E\left(\Delta^3 \sum_j w_j\{U_{t+j}\}\right)^2. \tag{14}$$

By applying the third differences, solving the squares and the summation and applying the mathematical expectation E, equation (14) becomes

$$\operatorname{var} \Delta^3\{X_t^a\} = \sum\left(\Delta^3 w_j\right)^2. \tag{15}$$

The lack of smoothness is given by the sum of squares of the third differences of the weights. To calculate $\sum(\Delta^3 w_j)^2$ for a particular set of weights, the set must be considered infinite, where to the right and the left of the actual values there is an infinite number of zeros. Thus, for example, the $\sum(\Delta^3 w_j)^2 = 12/169$ if the w_j's are all equal to $1/13$ and the $\sum(\Delta^3 w_j)^2 = 1/72$ if the w_j's are those of the well-known centered 12-term moving average (the first and last weights are equal to $\frac{1}{24}$ and the eleven remaining weights are equal to $\frac{1}{12}$). The latter gives a smoother curve than the simple 13-term moving average.

If the span of the average is $2m - 3$, the general expression of Henderson's Ideal formula [4] for the nth term is

$$315\{(m-1)^2 - n^2\}\{m^2 - n^2\} \times \frac{\{(m+1)^2 - n^2\}\{(3m^2 - 16) - 11n^2\}}{8m(m^2-1)(4m^2-1)(4m^2-9)(4m^2-25)}. \tag{16}$$

To derive a set of 15 weights from this formula, 9 is substituted for m and the values are obtained for each n from -7 through 7. The sum of squares of the third differences of these weights is 12% smaller than that obtained from the Spencer's 15-term formula.

There is a relationship between smoothing by fitting polynomials by least squares and smoothing from summation formulas. In the least-squares formula, all deviations between observed and fitted values are assumed equally weighted and thus the sum of squares of the errors is made a minimum. In the summation formulas, the deviations between observed and fitted values are not equally weighted and if different weights are applied then the sum of squares of the deviations is made a minimum [4]. The Henderson's moving averages give the same results as if weighted least squares had been used, where the weights are those that give the "smoothest" possible curve, the latter in the sense that the sum of squares of the third differences is made a minimum.

References

[1] Anderson, T. W. (1971). *The Statistical Analysis of Time Series*. Wiley, New York.

[2] Jenkins, G. M. and Watts, D. G. (1968). *Spectral Analysis and Its Applications*. Holden-Day, San Francisco.

[3] Kendall, M. G. and Stuart, A. (1966). *The Advanced Theory of Statistics*, Vol. 3. Hafner, New York.

[4] Macaulay, F. R. (1931). *The Smoothing of Time Series*. National Bureau of Economic Research, New York.

[5] Whittaker, E. and Robinson, G. (1924). *The Calculus of Observations: A Treatise on Numerical Mathematics*. Blackie, London.

(ACTUARIAL STATISTICS
AUTOREGRESSIVE–INTEGRATED MOVING AVERAGE (ARIMA) MODELS
AUTOREGRESSIVE MOVING AVERAGE (ARMA) MODELS
FINITE DIFFERENCES, METHOD OF
GRADUATION
TIME SERIES)

ESTELA BEE DAGUM

MOVING-OBSERVER SAMPLING METHOD

This is often used in transformation studies, when the subjects of a survey are in motion. For example, if it is required to estimate the number of pedestrians on one block in a city, the observer walks the length of the block and records the number of people he passes and the number who pass him. The observer *then* walks back again (at the same speed) and repeats the process. Subtracting the number of people who overtake him from those he passes and averaging the results obtained from the two "journeys" (back and forth) yields an estimate of the average number of people on the block during the time at which the sampling was performed.

For applications to vehicle flow and more detailed analysis, see refs. 1 or 2.

References

[1] Kish, L. (1965). *Survey Sampling*. Wiley, New York.

[2] Yates, F. S. (1971). *Sampling Methods for Censuses and Surveys*. Griffin, London.

(SAMPLING PLANS
SURVEY SAMPLING)

MOVING SUMS (MOSUM)

The monitoring of a sequence of observations in order to decide whether they are realizations of identically distributed random variables is a common problem in various situations of applied statistics, two of which will be discussed in this article:

a. Testing the constancy of a regression relationship over time.

b. Continuous sampling inspection.

The basic model is $Y_t = E[Y_t] + U_t$, U_t being independently distributed according to $N(0, \sigma^2)$, that is, normal with mean zero and variance σ^2. In (b), $E[Y_t]$ is the target value μ_0, which may be known or unknown. In (a), let $E[Y_t]$ be $\mathbf{z}_t'\boldsymbol{\beta}$, $\mathbf{z}_t$ being the k-vector of regressor variables observed at time t and $\boldsymbol{\beta}$ the k-vector of regression coefficients. A further parameter is σ^2, which might be known in situations of type (b).

The model used for (b) can be seen as a simplified version of that used for (a). In both situations, the aim of the analysis is to determine whether the assumption of constancy of the respective model parameters is violated. Therefore, the test statistics and test procedures can be treated in common. In most control situations, however, the tests are performed during the continuing process, i.e., on the basis of a growing amount of data. This contrasts with (a), where the test usually is performed after all data have been obtained (i.e., as a means of checking the adequacy of the model that is the basis of the analysis).

MOSUM TESTS

The crucial assumption of the model is that its parameters ($\boldsymbol{\beta}$ or μ_0, σ^2) are constant, at least within the period of observation. For checking whether this assumption is fulfilled, tests can be designed that check whether the error terms U_t follow the random pattern assumed in the model. This can be done by looking at the residuals* between the observations y_t and their estimates $\hat{Y}_t$ obtained from the fitted model. The least-squares residuals have a nondiagonal and singular covariance matrix; moreover, residuals based on the whole sample will be less sensitive to changes than residuals calculated only from information up to the respective time. More suitable are the so-called *recursive residuals*

$$W_t = \frac{Y_t - \mathbf{z}_t'\mathbf{b}_{t-1}}{\sqrt{1 + \mathbf{z}_t'(\mathbf{Z}_{t-1}'\mathbf{Z}_{t-1})^{-1}\mathbf{z}_t}},$$

$$t = k+1, \dots, T; \quad (1)$$

here, $\mathbf{Z}_{t-1}' = (\mathbf{z}_1, \dots, \mathbf{z}_{t-1})$ and $\mathbf{b}_{t-1}$ are the least-squares estimates of $\boldsymbol{\beta}$ based on the observations obtained up to time $t-1$. Under the assumption of constancy of the parameters the recursive residuals are independently distributed $N(0, \sigma^2)$. Nonconstant $\boldsymbol{\beta}$

or μ_0, however, imply nonzero expectation, and nonconstant σ^2 causes heteroscedasticity* and autocorrelation of the recursive residuals. These properties can be used to design suitable test procedures for testing the constancy.

The *moving sums* of recursive residuals (*mosum statistics*) are defined as

$$M_t = \sigma^{-1} \sum_{r=t-\kappa+1}^{t} W_r, \qquad t = k + \kappa, \ldots, T. \quad (2)$$

Under constancy, the vector $\mathbf{M}$ of mosum statistics is distributed according to $N(\mathbf{0}, \mathbf{V})$, the elements of $\mathbf{V}$ being $v_{t,t+i} = \kappa - |i|$ for $|i| \leqslant \kappa$ and 0 otherwise. These moments still apply if σ is substituted by its estimate s, which is determined to be the square root of the average of the squared recursive residuals, the distribution of $\mathbf{M}$, however, being normal only in the asymptotic case. Related test statistics are the moving sums of squared recursive residuals (*mosum-sq statistics*) [11]

$$\mathrm{MS}_t = \sigma^{-2} \sum_{r=t-\kappa+1}^{t} W_r^2, \qquad t = k + \kappa, \ldots, T, \quad (3)$$

which can be shown to follow marginally a chi-squared distribution with κ d.f. and $\mathrm{cov}[\mathrm{MS}_t, \mathrm{MS}_{t+i}] = 2v_{t,t+i}$. In the case of unknown variance, the mosum-sq statistics can be modified suitably.

As long as the assumption of constancy is not violated, the deviations of the mosum and mosum-sq statistics from their expectations will be nonsystematic, whereas after a violation, systematic deviations will be observed. Therefore, the test whether significant deviations are present can be used as a means to detect nonconstancy of the parameters.

TESTING A REGRESSION MODEL FOR CONSTANCY OVER TIME

Let the regression model be fitted to a sample of fixed size T. To test whether the model parameters $(\boldsymbol{\beta}, \sigma^2)$ are constant, several procedures are available in the literature [11]. Most of these [6–9, 14] can be considered as part of the category of *overfitting* (i.e., overparameterization of the model, which necessitates checking whether the additional parameters contribute to the explanation of the data). A different approach is to check simultaneously the cumulative sums of the recursive residuals (*cusum statistics*) [5, 18]. Two shortcomings, however, are inherent in the definition of cusum statistics. First, an increasing number of observations contribute to the statistics, so that the relative weight of observations after a violation of constancy decreases with increasing time of the onset of the violation. Second, to give each of the individual tests the same type I error probability, the critical limits must form a parabolic curve.

Both of these drawbacks are avoided when mosum statistics are substituted for the cusum statistics in the test procedure [10]. For testing against a two-sided alternative, the mosum statistics M_t, $t = k + \kappa, \ldots, T$ are compared with critical limits $\pm c\sqrt{\kappa}$. The constant c must be chosen so that the overall null hypothesis (i.e., $E[M_t] = 0$ for all $t = k + \kappa, \ldots, T$) is erroneously rejected at a given significance level α. Exact critical limits can be derived only from the multivariate distribution of the M_t. Neglect of the correlations (cf. the use of Bonferroni's* or Šidák's [19] inequalities) leads to conservative limits. For large values of κ (i.e., large correlations) the use of Hunter's inequality [12] improves the limits considerably [4]. Bonferroni-type critical limits can be easily derived by use of the corresponding marginal distribution; Hunter-type critical limits are based on the bivariate distribution so that the lack of tables might cause additional difficulties [20, 21]. Hunter-type individual significance levels are given in ref. 4. Monte Carlo estimates for c of the exact critical limits can be found in ref. 11.

An analogous test procedure can be based on mosum-sq statistics. The power of the test procedures in various situations of nonconstancy is discussed in ref. 11. Simulation* results indicate trends of the power depending on the extent of nonconstancy, κ, T, and the time of onset of nonconstancy. In the case of a shifting intercept, the power of the mosum test dominates that of all other tests including the cusum test. With a shifting variance, the cusum-sq test and the mosum-sq test are comparable in power and superior to the mosum and cusum test. In the case of superimposition of the shifting variance upon the nonconstancy of the intercept, the power of the mosum-sq and cusum-sq test are additionally increased; the power of the mosum test decreases slightly, whereas that of the cusum test collapses. Consequently, the combined application of the mosum, cusum, and mosum-sq (or cusum-sq) tests can serve to distinguish the cases of nonconstant mean, nonconstant variance, and their simultaneous presence.

The method can be applied in quality control* when a limited series of observations is to be checked [2, 3]. For example, when the number of samples per day is fixed in advance; α is the proportion of false alarms per day. More commonly, however, the control is to be performed during the continuing process.

CONTINUOUS SAMPLING INSPECTION

The situation can be sketched as follows. Let X be a normally distributed random variable with $\text{var}[X] = \sigma_x^2$. Independent samples $x_{t1}, \ldots, x_{tn}$ of size $n \geqslant 1$ are drawn at $t = 1, 2, \ldots$ in order to check whether $E[X] = \mu_0$, μ_0 being the target value. This can be done on the basis of the mosum statistics (2): Denoting Y_t as the mean value of $X_{t1}, \ldots, X_{tn}$, the recursive residuals turn out to be $W_t = Y_t - \mu_0$ with $\text{var}[W_t] = \text{var}[Y_t] = \sigma_x^2/n = \sigma^2$. When μ_0 is unknown, in (1), $\mathbf{z}_t'\mathbf{b}_{t-1} = \bar{y}_{t-1}$. The control limit parameter c now must be chosen in accordance with a given average run length* (ARL).

A control technique based on mosum statistics was discussed in ref. 16 for the first time. Modifications such as for the case of unknown variance σ^2 and for the control of σ^2 are treated in ref. 3. Devices are the moving-sum (average) charts.

Average Run Length

The usual indicator of the performance of a control technique is the ARL, the expected value of the run length; apart from its dependence on the parameters of the control procedure, it is determined by the disturbance $E[X] - \mu_0 = \delta\sigma$. The derivation of the run-length distribution for the mosum statistics is a difficult task: neither Brownian motion* approximation nor Wald-type arguments (*see* SEQUENTIAL ANALYSIS) are applicable. For many purposes, however, the ARL can be taken as a suitable characteristic of the run-length distribution.

For the ARL, bounds can be derived from a formula given in ref. 13 for weighted sums of independent and identically distributed normal random variables; the ARL of the one-sided mosum procedure with control limit $c\sqrt{\kappa}$ is contained in the interval

$$\frac{\kappa}{1 - P_\kappa(h_\delta)} \leqslant \text{ARL} \leqslant \kappa + \frac{P_\kappa(h_\delta)}{P_{\kappa-1}(h_\delta) - P_\kappa(h_\delta)}, \tag{4}$$

where $h_\delta = c\sqrt{\kappa} - \delta\kappa$ and $P_\kappa(x) = \Pr[M_{\kappa+1} < x, \ldots, M_{2\kappa} < x]$. For moderate and large values of κ, the evaluation of (4) is difficult. For example, the interval for $\kappa = 2$, $c = 3$, and $\delta = \sqrt{2}/2 \doteq 0.71$ is given by $170.2 \leqslant \text{ARL} \leqslant 180.3$.

An upper bound of the ARL can be derived from a further formula given in ref. 13 for the case of moving sums of independent but arbitrarily distributed random variables. The control procedure underlying this formula, however, is slightly modified; it starts by comparing $w_1/\sigma, \ldots, (w_1 + \cdots + w_{\kappa-1})/\sigma$ with the control limit before checking the mosum statistics M_t, $t \geqslant \kappa$. The ARL of this procedure is bounded above by

$$\text{ARL} \leqslant \kappa/(1 - P_\kappa^*), \tag{5}$$

where $P_\kappa^* = \Pr[M_\kappa < c\sqrt{\kappa}\,]$. For normally distributed variables, $P_\kappa^* = \Phi(c - \delta\sqrt{\kappa})$, with Φ denoting the distribution function of the standard normal variable. For moderate and large values of κ, the use of (5) for the mosum procedure leads to rather crude bounds (e.g., for $c = 3.29$ and $\delta = 0.5$, bounds of 335 and 137 are obtained for $\kappa = 5$ and 20, respectively, corresponding Monte Carlo estimates being 115.2 and 55.9 with respective standard deviations of 9.8 and 4.1 [3]). For small values of κ, the bounds underestimate the ARL of the mosum procedure. So, for $\kappa = 2$, $c = 3$, and $\delta = \sqrt{2}/2$, (5) gives 87.9, which is not within the limits evaluated from (4). Monte Carlo estimates for the ARL of the mosum procedure are given in refs. 3 and 16.

A bound for $\text{ARL}_\pm$ of the two-sided mosum procedure can be derived from the ARL values of the corresponding one-sided procedures:

$$\text{ARL}_\pm \geqslant (\text{ARL}_+^{-1} + \text{ARL}_-^{-1})^{-1}; \quad (6)$$

on the right-hand side events in which both bounds, the lower and the upper, are crossed within the same run are counted twice. These events become rare for large values of c and κ.

Comparison

Candidates to be compared with the mosum technique are the Shewhart $\bar{x}$-chart and the cusum (V-mask) technique (*see* CUMULATIVE SUM CONTROL CHARTS). From a formal point of view, the $\bar{x}$-chart is a special mosum procedure ($\kappa = 1$), whereas the V-mask technique can be seen as the simultaneous application of mosum statistics with $\kappa = 1, 2, \ldots$. Concerning the ease of operation, the mosum technique might be judged comparable to the $\bar{x}$-chart.

The behavior of the methods in terms of the ARL can be summarized [17] as follows: Large disturbances ($\delta > 2.5$) are more quickly detected by $\bar{x}$-charts than by the V-mask and the mosum technique; the reverse is true for small disturbances. Given a fixed value ARL_0 for $\delta = 0$, the (ARL, δ)-curves of the $\bar{x}$-chart are completely determined, but those of the V-mask and mosum technique depend on one parameter. As a design criterion, the parameter can be chosen in such a way that, given ARL_0, a minimal value ARL_1 for $\delta = \delta_1$ is achieved. By means of numerical optimization, it is shown in ref. 1 that, for practically relevant parameter values, under all V-masks with given ARL_0, the optimal one is that for which the reference value k is $\delta\sigma/2$. The minimum, however, is rather flat. According to Monte Carlo estimates [15], ARL-values for the mosum technique that are optimal in this sense are very similar to the corresponding V-mask results.

Robustness versus nonnormality of the process is discussed in ref. 3.

Due to the lack of a feasible analytical or numerical approach to the determination of the ARL or other statistical properties of the mosum technique, it is not surprising that the number of available results is much smaller than for the cusum technique. This may be the reason that, in spite of the simplicity of its application, the mosum control technique is not widely used in practice.

References

[1] Bauer, P. and Hackl, P. (1984). In *Frontiers in Quality Control*, H.-J. Lenz, G. B. Wetherill, and P-Th. Wilrich, eds., Physica-Verlag, Würzburg, Germany, pp. 199–207.

[2] Bauer, P. and Hackl, P. (1978). *Technometrics*, **20**, 431–436.

[3] Bauer, P. and Hackl, P. (1980). *Technometrics*, **22**, 1–7.

[4] Bauer, P. and Hackl, P. (1982). "The Application of Hunter's Inequality in Simultaneous Testing." *Tech. Rep. 11/82*, Dept. of Statistics, University of Economics, Vienna, Austria.

[5] Brown, R. L., Durbin, J., and Evans, J. M. (1975). *J. R. Statist. Soc. B.*, **37**, 149–192.

[6] Chernoff, H. and Zacks, S. (1964). *Ann. Math. Statist.*, **35**, 999–1018.

[7] Farley, J. U. and Hinich, M. J. (1970). *J. Amer. Statist. Ass.*, **65**, 1320–1329.

[8] Farley, J. U., Hinich, M. J., and McGuire, T. W. (1975). *J. Econometrics*, **3**, 297–318.

[9] Ferreira, P. A. (1975). *J. Amer. Statist. Ass.*, **70**, 370–374.

[10] Hackl, P. (1978). In *Models and Decision Making in National Economies*, J. M. L. Jansen, L. F. Pau,

and A. Straszak, eds. North-Holland, Amsterdam, pp. 219–225.
[11] Hackl, P. (1980). *Testing the Constancy of Regression Models over Time*. Vandenhoeck and Ruprecht, Göttingen, Germany.
[12] Hunter, D. (1976). *J. Appl. Prob.*, **13**, 597–603.
[13] Lai, T. L. (1974). *Ann. Statist.*, **2**, 134–147.
[14] Quandt, R. E. (1958). *J. Amer. Statist. Ass.*, **53**, 873–880.
[15] Reinoehl, C. (1982). "Optimale Parameter der MOSUM-Technik." *Technical Report 3/82*, Dept. of Statistics, University of Economics, Vienna.
[16] Roberts, S. W. (1959). *Technometrics*, **1**, 239–250.
[17] Roberts, S. W. (1966). *Technometrics*, **8**, 411–430.
[18] Schweder, T. (1976). *J. Amer. Statist. Ass.*, **71**, 491–501.
[19] Šidák, Z. (1967). *J. Amer. Statist. Ass.*, **62**, 626–633.
[20] Stoline, M. R. (1983). *J. Amer. Statist. Ass.*, **78**, 367–370.
[21] Worsley, K. J. (1982). *Biometrika*, **69**, 297–302.

(AVERAGE RUN LENGTH
CONTROL CHARTS
CUMULATIVE SUM CONTROL CHARTS
QUALITY CONTROL, STATISTICAL
SAMPLING PLANS)

PETER HACKL

MRLF *See* MEAN RESIDUAL LIFE FUNCTION

MTTF *See* MEAN TIME TO FAILURE

MUIRHEAD'S THEOREM

This theorem is a generalized form of inequality between different means. It states that, for any n positive numbers $u_1, \ldots, u_n$,

$$\sum_j \prod_{i=1}^n u_{j_i}^{\alpha_i} \geqslant \sum_j \prod_{i=1}^n u_{j_i}^{\beta_i}$$

if and only if

$$\sum_{i=1}^k \alpha_i \geqslant \sum_{i=1}^k \beta_i, \qquad k = 1, 2, \ldots, n-1,$$

$$\sum_{i=1}^n \alpha_i = \sum_{i=1}^n \beta_i,$$

where the vector $\boldsymbol{\alpha}$ *majorizes* the vector $\boldsymbol{\beta}$ and $\sum_{(j)}$ denotes summation over all permutations $(j_1, j_2, \ldots, j_n)$ of the first n integers $1, 2, \ldots, n$.

Dividing both sides of the inequality by $n!$, we see that we get an inequality between two *generalized means*. The well-known inequality—arithmetic mean $\geqslant$ geometric mean—follows by taking

$$\alpha_1 = 1, \qquad \alpha_i = 0, \qquad i > 1,$$

$$\beta_i = n^{-1}, \qquad i = 1, 2, \ldots, n.$$

The inequality—geometric mean $\geqslant$ harmonic mean—is obtained by taking

$$\alpha_i = 0, \qquad i = 1, 2, \ldots, n-1, \qquad \alpha_n = -1;$$

$$\beta_i = -n^{-1}, \qquad i = 1, 2, \ldots, n.$$

(ARITHMETIC MEAN
GEOMETRIC MEAN
HARMONIC MEAN)

MULTICOLLINEARITY

The term multicollinearity has been used in statistics in situations in which one variable is very nearly a linear combination of other variables. Thus it is not a generalization of the mathematical term collinearity, which refers to the property of several points being on the same straight line. Some have used the terms interchangeably; Morrison [8, pp. 271–273] refers to two highly correlated variables as *nearly collinear* in that one variable is nearly a linear transformation of the other. He illustrates a possible consequence of this in regression* analysis, where two such variables are used as predictors. He considers multicollinearity to be the case in which a multiple correlation of one variable with another is near unity, so that one variable is nearly a linear transformation of the others. By these definitions, *collinearity* is a special case of *multicollinearity*, since ordinary correlation is a special case of multiple correlation*. Multicollinearity is the term that generally has been used, particularly in the econometrics* literature. Goldberger [5, p. 80] defines multicollinearity as "the situation which arises when some of all of the explanatory variables are so highly correlated one with another that it becomes very difficult, if not impossible, to disentangle their influences and obtain a reasonably precise estimate of their [separate] effects."

In the example of multicollinearity given by Morrison, the exchange rate of dollars in German marks (x_1) and in Swiss francs (x_2) were used as independent variables in a regression analysis to predict the price of gold. The exchange rates were correlated 0.95 and were thus said to be nearly collinear. Although the exchange rates were correlated 0.87 and 0.84 with the dependent variable, the regression coefficients were not significant using the Scheffé multiple comparison* test. The overall regression equation and the regression equations using each x variable separately were all highly significant. This is obvious from the high correlations with the price of gold. Morrison also displays the confidence ellipsoid, a long narrow ellipse that does not include the point (0, 0) although it intersects both the x and y axes. This indicates that within 95% confidence limits either β_1 or β_2, but not both, can be zero. Morrison states that the effect of the high correlation is to cause confusion when both x variables are used together. His recommendation in this case is to use only one of the variables.

Although this advice may be satisfactory in a prediction situation, it is not a satisfactory answer when the aim is to understand the influences of independent variables. The difficulty, as stated by Theil [9], is that "the data do not really enable the analyst to distinguish the effects of the variables on the dependent variable; at least, these data do not enable him to do so with any real precision, and the standard errors of the coefficients will therefore be large." A similar situation arises in nonorthogonal analysis of variance where it is sometimes difficult to distinguish between the effects of two different factors. The problems in the ANOVA* context have been discussed by Appelbaum and Cramer [1].

The identification of multicollinearity can be quite simple when one anticipates the problem. On many occasions multicollinearity may be suspected when one encounters the occurrence of one of its possible conseqences. Since multicollinearity is defined as the near linear dependence of one variable on a set of other variables, the multiple correlations of each variable with every other variable in an appropriate set can be used as a guide. Thus in regression analysis a very high multiple correlation among the independent variables is a definite indicator of multicollinearity. This can be examined by computing the inverse of the correlation matrix among the independent variables. The squared multiple correlation of the ith variable with all the others is given by $1 - 1/r^{ii}$. A more efficient method of examining linear dependencies involves the successive multiple correlations of variable 2 with variable 1, 3 with 1, and 2, 4 with 1, 2, and 3, etc. These coefficients are readily computed as a by-product of a regression analysis using the Cholesky (or square root) method of computation (*see* LINEAR ALGEBRA, COMPUTATIONAL). They can also be obtained easily when the method of solution involves orthogonalization or use of the "sweep" operator. In the Cholesky method of triangularization of a correlation matrix, each squared diagonal element is of the form $1 - R_{i \cdot p}^2$, where $R_{i \cdot p}$ is the multiple correlation of the ith variable with the earlier variables. If multicollinearity exists, one of these multiple correlations will be close to one indicating a linear dependency involving some of the earlier variables. Computer programs should always check these correlations since continuation of the computation will inevitably lead to loss of accuracy. As a rule, the number of leading zeros in $1 - R^2$ is equal to the number of lost significant figures if computation is continued. Stepwise regression* computer programs typically have a tolerance placed on $1 - R^2$ or its square root to prevent the loss of accuracy in computation.

It has sometimes been suggested that the determinant of the correlation matrix may be used as an indicator of multicollinearity. Since the determinant is equal to the product of the stepwise $1 - R^2$ values mentioned earlier, the determinant may be small even though no multiple correlation is large. This measure cannot be recommended.

Undetected multicollinearity has a number of possible consequences. The most serious of these is *computational error*. A perfect linear dependency among the independent variables may cause a division by zero $(1 - R^2)$ in the initial stage of computation. Typically the division will be by a very small number due to round off error* in computation. The result may be the loss of many significant figures and meaningless solutions, possibly a negative sum of squares. This is less likely to occur with computer programs that use orthogonalization methods of computation or that use double precision arithmetic. Problems of this sort are well known to users of statistical computer programs. Well-designed computer programs will give a warning message, possibly indicate the nature of the linear dependency, and continue the computation with the deletion of the offending variable.

There are typically several signals of multicollinearity. The *standard errors* of one or more regression coefficients may be *very large*, resulting in very wide confidence intervals for the coefficients. Many or all of the coefficients will be nonsignificant contributors to the regression equation even though the individual variables may be highly correlated with the dependent variable. This was the case in the example considered by Morrison. Another indicator is the presence of *regression coefficients* that are *very large*, even when standardized. Standardized coefficients are usually between -1 and $+1$; a coefficient of 2 or more is typically an indicator of multicollinearity. Often they will occur in pairs of opposite sign, and coefficients as large as 1000 are not uncommon.

Although some may find the consequences of multicollinearity confusing, they are easily understood if one considers the models in various tests of significance (see Cramer [3]). The key idea is that the t tests* customarily given by regression computer programs compare the model involving all the x variables with a model in which one of the x variables has been omitted. In the presence of multicollinearity, such a variable may be redundant even though it may be highly correlated with the y variable. In an extreme situation, if one includes several x variables that are highly correlated with the y variable as well as their sum, every one of the x variables will be redundant and nonsignificant although all of them will provide excellent prediction of the y variable.

Multicollinearity does not invariably indicate the presence of redundant variables. Psychological statisticians are familiar with the concept of the *suppressor variable*, which is highly correlated with another dependent variable but uncorrelated with the dependent variable. Such a variable can substantially increase the multiple correlation when combined with a variable that is only modestly correlated with the dependent variable. Indeed Cramer [4] has contrived an example in which two x variables have an almost perfect correlation with each other while each have a virtually zero correlation with y. Despite this extreme multicollinearity, both variables have highly significant regression coefficients, and the multiple correlation is equal to one. This situation is very simply explained geometrically if the variables are represented by three vectors in a plane. The two $\mathbf{x}$ vectors can have a small angle between them indicating a high correlation, while both may be at nearly right angles to $\mathbf{y}$ indicating low correlations. Since all three vectors are in a plane, the $\mathbf{y}$ vector must be a linear combination of the $\mathbf{x}$ vectors. Variables such as these will have partial correlations with the y variable, when the other x variables are partialed out. In the example considered by Cramer, the partial correlations were virtually equal to one. Obviously, one would not wish to delete either variable despite the high multicollinearity.

As Leser [7] has noted, "multicollinearity is serious when emphasis lies on the estimation of individual parameters in the relationship, but less serious when the objective of prediction of the dependent variable is stressed." One can easily eliminate the multicollinearity in the data by transforming the variables so that they are uncorrelated. This has no effect on the predictions and eliminates any computational problems. Even though regression coefficients may have large standard errors, the predictions may be very stable. This will always be the case when variables responsible for the multicollinearity do not reduce the multiple correlation when they are omitted form the analysis. Indeed, if one uses well-conditioned computational methods such as orthogonalization, the effects of multicollinearity in such situations is inconsequential. Where there is interest in the regression coefficients, the methods of ridge regression* [6] are often used as an alternative to deleting redundant variables. This results in biased estimates of the coefficients, although their mean square errors* will be smaller than otherwise. Deleting variables may have the same effect. Latent root regression* is still another way of detecting and eliminating multicollinearity. A linear dependency among the x variables will necessarily result in a latent root (eigenvalue) of zero. Although the value of a small latent root is not as directly related to regression analysis as the multiple correlation, this method provides an alternative way of identifying multicollinearities and determining whether the x variables involved are redundant.

Distinctions between fixed and random regression models are important. Willan and Watts [10] discuss the effects of multicollinearity on prediction. As already noted, one can transform to uncorrelated variables without affecting the predictions. Willan and Watts' concern is really with optimum allocation. They show, that if one can choose the x values as in a fixed regression model, the efficiency or prediction can vary greatly. Similarly, Goldberger [5, p. 80] claims that "Multicollinearity is a property of the sample data and not of the population." Clearly a random sample from a population with highly correlated variables will typically yield highly correlated variables in the sample. In a fixed regression situation, the presence of multicollinearity is not dependent on the population, but on how the data is selected.

Example 1. In the example given by Morrison, the correlation matrix is

$$\begin{array}{c} \\ x_1 \\ x_2 \end{array} \begin{array}{ccc} x_1 & x_2 & y \\ \left[\begin{array}{l} 1.0 \\ 0.9538 \end{array}\right. & \begin{array}{l} 0.9538 \\ 1.0 \end{array} & \left.\begin{array}{l} 0.8740 \\ 0.8418 \end{array}\right] \end{array}$$

The standardized regression coefficients are given by

$$\beta_1 = 0.787, \qquad \beta_2 = 0.091,$$

with standard errors both equal to 0.330 and t statistics of

$$t_1 = 2.39, \qquad t_2 = 0.275,$$

while the multiple correlation equals 0.8744. This example does not exhibit any of the serious distortions of multicollinearity since the correlation between the x variables is only about 0.95. The multiple correlation is virtually identical to the correlation between x_1 and y, indicating that x_2 is completely redundant. It must be remembered that the t-tests for the regression coefficients test the additional predictive value of the variable in question. Either variable by itself is obviously highly predictive, and these t-tests are not relevant to this question.

Example 2. The adverse effects of multicollinearity will generally be apparent only when the multiple correlation among the x variables is as great as 0.99. Consider the correlation matrix below for various values of r_{12}, the correlation between x_1 and x_2.

$$\begin{array}{c} \\ x_1 \\ x_2 \end{array} \begin{array}{ccc} x_1 & x_2 & y \\ \left[\begin{array}{c} 1.0 \\ r_{12} \end{array}\right. & \begin{array}{c} r_{12} \\ 1.0 \end{array} & \left.\begin{array}{c} 0.50 \\ 0.49 \end{array}\right] \end{array}$$

Table 1 gives values of R, β_i, σ_β, and t_i for various values of r_{12}. The regression coefficients are of opposite sign and become increasingly large, even though both x_1 and x_2 have virtually the same correlation with y. With r_{12} equal to 0.99 or 0.999, the standard errors of the regression coefficients are large and neither t value is significant. The multi-

Table 1

r_{12}	R	β_i	σ_β	t_i
0.99	0.501	0.75	1.25	0.60
		− 0.25	1.25	− 0.20
0.999	0.543	5.25	3.83	1.37
		− 4.75	3.83	− 1.24
0.9999	0.863	50.25	7.29	6.89
		− 49.75	7.29	− 6.83

ple correlations are only slightly larger than the correlation between x_1 and y. We may say that either x variable is redundant. With r_{12} equal to 0.9999, the situation is somewhat different. Although both standardized regression coefficients are very large, both are highly significant. Neither variable is redundant since the multiple correlation is increased to 0.863.

References

[1] Appelbaum, M. I. and Cramer, E. M. (1974). *Psychol. Bull.*, **81**, 335–343.

[2] Belsley, D. A., Kuh, E., and Welsch, R. (1980). *Regression Diagnostics: Identifying Influential Data and Sources of Collinearity*. Wiley, New York (extensive references).

[3] Cramer, E. M. (1972). *Amer. Statist.*, **26** 26–30.

[4] Cramer, E. M. (1974). *Mult. Behav. Res.*, **9**, 241–243.

[5] Goldberger, A. S. (1968). *Topics in Regression Analysis*. Macmillan, New York.

[6] Hoerl, A. E. and Kennard, R. W. (1970). *Technometrics*, **12**, 55–67.

[7] Leser, C. E. V. (1969). *Economic Techniques and Problems*, Griffin's Statistical Monographs and Courses, **20**. Griffin, London.

[8] Morrison, D. F. (1983). *Applied Linear Statistical Methods*. Prentice-Hall, Englewood Cliffs, NJ.

[9] Theil, H. (1971). *Principles of Econometrics*. Wiley, New York.

[10] Willan, A. R. and Watts, D. G. (1978). *Technometrics*, **20**, 407–412.

(COLLINEARITY
LATENT ROOT REGRESSION
MULTIPLE CORRELATION COEFFICIENT
RIDGE REGRESSION)

ELLIOT M. CRAMER

MULTIDIMENSIONAL CENTRAL LIMIT THEOREMS

The *multidimensional central limit theorem* (CLT) is generally regarded as a generic name applied to any theorem giving convergence in distribution to the multivariate normal distribution* for a sum of an increasing number of random vectors. Results of this kind hold under far-reaching circumstances and, as with the one-dimensional case, give the multivariate normal distribution its central place in the theory of probability and statistics (*see* MULTINORMAL DISTRIBUTION).

Comparatively little attention has been given to the central limit behavior of random vectors in Euclidean space $\mathbb{R}^d$, $d > 1$, since the Cramér–Wold device (see following discussion) enables much theory to be obtained from straightforward extension of results for the one-dimensional case.

Let $\mathbf{X}_n$ and $\mathbf{X}$ be random (column) vectors in $\mathbb{R}^d$ and suppose that $\mathbf{t}'\mathbf{X}_n \xrightarrow{d} \mathbf{t}'\mathbf{X}$ as $n \to \infty$ for each vector $\mathbf{t}$ in $\mathbb{R}^d$. Here the prime denotes transposition and $\xrightarrow{d}$ denotes convergence in distribution. Then, in terms of characteristic functions*

$$Ee^{iu\mathbf{t}'\mathbf{X}_n} \to Ee^{iu\mathbf{t}'\mathbf{X}}$$

as $n \to \infty$ for each real u. Taking $u = 1$ and recalling that $\mathbf{t}$ is arbitrary, we have that $\mathbf{X}_n \xrightarrow{d} \mathbf{X}$ from the continuity theorem for characteristic functions. This is the Cramér–Wold device and its application leads, for example, to the following classical form of the CLT for sums of independent random vectors in $\mathbb{R}^d$.

Theorem 1. Let $\mathbf{X}_1, \mathbf{X}_2, \ldots$ be independent d-dimensional random vectors such that $E\mathbf{X}_i = \mathbf{0}$ and $E\mathbf{X}_i\mathbf{X}_i' = \mathbf{V}_i$. Suppose that as $n \to \infty$

$$n^{-1}\sum_{i=1}^{n} \mathbf{V}_i \to \mathbf{V} \neq 0$$

and for every $\epsilon > 0$,

$$n^{-1}\sum_{i=1}^{n} E\left(\|\mathbf{X}_i\|^2 I\left(\|\mathbf{X}_i\|^2 > \epsilon n^{1/2}\right)\right) \to 0, \quad (1)$$

where I denotes the indicator function and

$\|\mathbf{X}\| = (\mathbf{X}'\mathbf{X})^{1/2}$ is the Euclidean norm of the vector $\mathbf{X}$. Then the random vector $n^{-1/2}(\mathbf{X}_1 + \cdots + \mathbf{X}_n)$ converges in distribution as $n \to \infty$ to the d-dimensional normal distribution with zero mean vector and covariance matrix $\mathbf{V}$.

If the random vectors $\mathbf{X}_i$ in Theorem 1 are identically distributed, then the Lindeberg condition* (1) is automatically satisfied.

It should be noted that Theorem 1 contains a central limit result for the multinomial distribution*. Let $(Y_1^{(m)}, \ldots, Y_r^{(m)})$, $m = 1, 2, \ldots, n$, be independent and identically distributed random vectors having at most one coordinate different from zero and such that for $m = 1, 2, \ldots, n$ and $j = 1, 2, \ldots, r$,

$$P(Y_j^{(m)} = 1) = p_j, \quad P(Y_j^{(m)} = 0) = 1 - p_j,$$
$$P(Y_1^{(m)} = 0, \ldots, Y_r^{(m)} = 0)$$
$$= q = 1 - p_j - \cdots - p_r.$$

Then

$$(Z_1, Z_2, \ldots, Z_r)' = \sum_{m=1}^{n} (Y_1^{(m)}, Y_2^{(m)}, \ldots, Y_r^{(m)})'$$

has a multinomial distribution. We readily find from Theorem 1 that

$$\left([np_1(1-p_1)]^{-1/2}(Z_1 - np_1), \ldots, [np_r(1-p_r)]^{-1/2}(Z_r - np_r)\right)'$$

converges in distribution as $n \to \infty$ to the normal distribution $N(\mathbf{0}, \boldsymbol{\Sigma})$ where $\boldsymbol{\Sigma} = (\sigma_{ij})$ with

$$\sigma_{ii} = 1, \qquad i = 1, 2, \ldots, r,$$
$$\sigma_{ij} = -[p_i p_j / \{(1-p_i)(1-p_j)\}]^{1/2},$$
$$i, j = 1, 2, \ldots, r, \quad i \neq j$$

(e.g., see Fisz [5, Theorem 6.13.2, p. 235; see also Theorem 6.13.1]).

In general, however, norming by scalars is not appropriate in higher dimensions. Any such norming by scalars must have the same order of magnitude as the maximum of the norming constants for the one-dimensional components. The limit will then be degenerate for all components whose norming constants are of lower order asymptotically than the maximum.

The following simple example illustrates the problem. Let $\{X_i, Y_i, i = 1, 2, \ldots\}$ be independent random variables with zero means and finite second moments. Let $s_n^2(X) = \sum_{k=1}^n EX_k^2$, $s_n^2(Y) = \sum_{k=1}^n EY_k^2$ and suppose that

$$s_n(X)/s_n(Y) \to 0,$$
$$s_n^{-1}(X) \sum_{k=1}^{n} X_k \xrightarrow{d} N(0,1) \quad \text{and}$$
$$s_n^{-1}(Y) \sum_{k=1}^{n} Y_k \xrightarrow{d} N(0,1)$$

as $n \to \infty$, $N(0, 1)$ denoting the unit normal law. Then, if

$$S_n = \sum_{k=1}^{n} (X_k \cos\phi + Y_k \sin\phi,$$
$$- X_k \sin\phi + Y_k \cos\phi)'$$

for fixed $\phi \neq 0 \pmod{\pi/2}$, it is clear that one has to rotate back $\phi \pmod{\pi/2}$ and normalize componentwise in order to obtain a nondegenerate two-dimensional limit.

A general approach to the problem of norming requires operator normalization; a comprehensive discussion of this approach has been provided by Hahn and Klass [7]. For independent random vectors $\mathbf{X}_i$, $i = 1, 2, \ldots$ in $\mathbb{R}^d$, they obtained necessary and sufficient conditions for the existence of linear operators A_n such that $A_n \sum_{i=1}^n \mathbf{X}_i$ converges in distribution to the standard multivariate normal law in $\mathbb{R}^d$. In an earlier paper, Hahn and Klass [6] dealt with the case of independent and identically distributed random vectors using matrix normalization; in this case, no centering difficulties arose.

In the case of finite variances the sufficiency part of the result of Hahn and Klass [7] specializes to the following generalization of Theorem 1 (see, e.g., Bhattacharya and Ranga Rao [2, Corollary 18.2, p. 183]).

Theorem 2. Let $\mathbf{X}_{n1}, \ldots, \mathbf{X}_{nk_n}$ be independent d-dimensional random vectors such that $E\mathbf{X}_{ni} = \mathbf{0}$ and $E\|\mathbf{X}_{ni}\|^2 < \infty$, $1 \leqslant i \leqslant k_n$.

Write $\mathbf{V}_n = k_n^{-1}\sum_{j=1}^{k_n} E(\mathbf{X}_{nj}\mathbf{X}'_{nj})$ and suppose that $\mathbf{T}_n$ is a symmetric positive definite matrix satisfying $\mathbf{T}_n^2 = \mathbf{V}_n^{-1}$, $n \geqslant 1$. Then, if $k_n \to \infty$ and for every $\epsilon > 0$,

$$k_n^{-1} \sum_{j=1}^{k_n} E\left(\|\mathbf{T}_n\mathbf{X}_{nj}\|^2 I\left(\|\mathbf{T}_n\mathbf{X}_{nj}\| > \epsilon k_n^{1/2}\right)\right) \to 0,$$

where I denotes the indicator function, then $k_n^{-1/2}\mathbf{T}_n\sum_{j=1}^{k_n}\mathbf{X}_{nj}$ converges in distribution to the standard d-dimensional normal law.

This result is a multidimensional version of the (sufficiency part of the) classical Lindeberg–Feller theorem*. An extension to Hilbert space has been provided by Kandelaki and Sazonov [10].

For rate of convergence results in the case of $\mathbb{R}^d$-valued random vectors see the book of Bhattacharya and Ranga Rao [2] and the monograph by Sazonov [12]. Emphasis in research on this topic has been on the order of approximation; explicit numerical bounds on the error of approximation that are valid for finite samples are notable for their absence from the literature.

As with the one-dimensional case, there are many generalizations of Lindeberg–Feller type results and general theories for convergence to infinitely divisible laws have been developed in a variety of settings (*see* INFINITE DIVISIBILITY). For a discussion in the case of Hilbert space–valued random variables see Laha and Rohatgi [11]; for Banach space–valued random variables, see Araujo and Giné [1]. A comprehensive treatment of the problem in the very general setting of locally compact groups is provided by Heyer [9].

Invariance principles* and functional central limit theorems* can also be interpreted as multidimensional central limit results. These results are ordinarily formulated in the setting of a metric space such as the space of continuous functions on $[0, \infty)$ or $[0, 1]$ and convergence of finite-dimensional projections (giving limit results in Euclidean space of arbitrary dimension) can be deduced readily. Indeed, the convergence of finite-dimensional distributions is a necessary but not sufficient requirement for convergence in such general settings. A concise perspective of this subject and its history can be obtained from the Introduction to Csörgö and Révész [4]; see also the review article by Heyde [8].

Comments on the Literature

Multidimensional central limit theorems have been little discussed in texts, even at the graduate level, but an elementary discussion has been provided by Breiman [3, Chap. 8]. With the exception of the text of Fisz [5] and the more advanced text of Laha and Rohatgi [11], the following references are at the level of research monographs and papers.

References

[1] Araujo, A. and Giné, E. (1980). *The Central Limit Theorem for Real and Banach Valued Random Variables.* Wiley, New York.

[2] Bhattacharya, R. N. and Ranga Rao, R. (1976). *Normal Approximations and Asymptotic Expansions.* Wiley, New York.

[3] Breiman, L. (1969). *Probability and Stochastic Processes wtih a View Towards Applications.* Houghton Mifflin, Boston.

[4] Csörgö M. and Révész, P. (1981). *Strong Approximations in Probability and Statistics.* Academic Press, New York.

[5] Fisz, M. (1963). *Probability Theory and Mathematical Statistics.* Wiley, New York.

[6] Hahn, M. G. and Klass, M. J. (1980). *Ann. Prob.*, **8**, 262–280.

[7] Hahn, M. G. and Klass, M. J. (1981). *Ann. Prob.*, **9**, 611–623.

[8] Heyde, C. C. (1981). *Int. Statist. Rev.*, **49**, 143–152.

[9] Heyer, H. (1977). *Probability Measures on Locally Compact Groups.* Springer-Verlag, Berlin.

[10] Kandelaski, N. N. and Sazonov, V. V. (1964). *Theor. Prob. Appl.* **9**, 38–46.

[11] Laha, R. G. and Rohatgi, V. K. (1979). *Probability Theory.* Wiley, New York.

[12] Sazonov, V. V. (1981). In *Lecture Notes in Mathematics*, No. 879: *Normal Approximation—Some Recent Advances.* Springer, Berlin.

(CONVERGENCE OF SEQUENCES OF RANDOM VARIABLES
LIMIT THEOREM, CENTRAL

LIMIT THEOREMS
LINDEBERG CONDITION
LINDEBERG–FELLER THEOREM
MULTINORMAL DISTRIBUTION)

C. C. HEYDE

MULTIDIMENSIONAL CONTINGENCY TABLES

Simple 2×2 contingency tables* are the precursors of general multidimensional contingency tables. Typically, a 2×2 contingency table arises in either comparing two independent samples with respect to a dichotomous response or studying the nature of association* between two dichotomous responses. The model for the former case relates to a product of two binomial distributions* while in the latter case, we have a multinomial distribution*. From the point of view of statistical inference*, in the first case, the null hypothesis of interest is the homogeneity* of the two samples, while in the second case, it relates to the lack of association between the two attributes. For both cases, an exact test (*see* FISHER'S EXACT TEST) [9], conditional in nature, is available; it employs the conditional distribution of the cell frequencies given the marginals. A randomization procedure [29] enables one to use the exact signficance level, while continuity corrections to justify use of the asymptotic chi-square distribution* for moderate sample sizes have been considered by Yates [30] and others. For a good account of these developments, *see* CATEGORICAL DATA, CHI-SQUARE TESTS, CONTINGENCY TABLES, and CONTINUITY CORRECTIONS.

For a proper interpretation of multidimensional contingency tables we introduce the concepts of *factors* and *responses*. The term *factor* denotes an experimental characteristic that can be controlled while *response* denotes a characteristic associated with the experimental outcome that may not be controlled in the given experimental setup. Both factors and responses are categorical in nature; these categories may be ordered in some way or they may even be purely categorical. In the former case, they are called structured; in the latter case, *unstructured*. In the 2×2 contingency table, for the first case, we have a single factor (population 1 or 2) and a single dichotomous response, while in the second case, we have two dichotomous responses. In multidimensional contingency tables, we may have more than one factor and/or response, and these may even be polychotomous. For the classical Bartlett [3] data, reported under the entry CONTINGENCY TABLES, we have two structured factors, each at two levels and an unstructured (binary) response. Grizzle et al. [15] considered another useful example relating to a $4 \times 4 \times 3$ table with two factors: hospital $(1, 2, 3, 4)$ and surgical procedure (A, B, C, D), both unstructured, and a response (severity of the "dumping syndrome," an undesirable sequel of surgery for duodenal ulcer), structured, with three levels: none, slight, and moderate. Thus we have a total of 16 subsamples; and for each subsample, a trinomial distribution.

Suppose now that the various combinations of the different factors constitute a set I $(= \{i\})$ of cardinality k $(\geqslant 1)$, so that we have k independent samples of sizes $n_1, \dots, n_k$, respectively, drawn from k populations indexed by the elements of I. To incorporate possible *incomplete contingency tables*, we may allow the provision that not all possible combinations of the different factors need be members of the set I. Thus, if there are l $(\geqslant 1)$ factors and the mth factor has a_m $(\geqslant 1)$ levels, for $m = 1, \dots, l$, then $k \leqslant a_1 \dots a_l = a^*$, say. Also, suppose that the various combinations of the levels of the different responses constitute a set J $(= \{j\})$ of cardinality q. Typically, $q = s_1 \dots s_r$, where r $(\geqslant 1)$ is the number of responses and $s_1, \dots, s_r$ (all $\geqslant 2$) denote the number of categories for the individual responses. In this setup, $j = (j_1, \dots, j_r)$, where $1 \leqslant j_m \leqslant s_m$, for $m = 1, \dots, r$. Here also, to incorporate incomplete tables, we may allow some particular combinations of the response categories as unobservable, so that $q \leqslant s_1 \dots s_r$. To introduce the probability model for the table just described, for each $i \in I$, we denote the cell probability for the jth cell by π_{ij}, for $j \in J$. Then, pertaining to the multifactor, multiresponse model, the multidimensional contingency table relates to the matrix $\{n_{ij}, j \in J, i \in I\}$ of observed cell frequencies, and rests on the (product multinomial) probability law:

$$\prod_{i \in I} \left\{ \left(n_i! \Big/ \prod_{j \in J} n_{ij}! \right) \prod_{j \in J} \pi_{ij}^{n_{ij}} \right\}, \qquad (1)$$

where

$$\sum_{j \in J} \pi_{ij} = 1, \qquad \sum_{j \in J} n_{ij} = n_i, \qquad \forall i \in I. \quad (2)$$

For the 2×2 contingency tables, in the first case, we have $k = 2$, $r = 1$, $s_1 = q = 2$, while in the second case, $k = 1$, $r = 2$, $s_1 = s_2 = 2$, and $q = 4$. For the two-dimensional $b \times c$ contingency table (with $b \geqslant 2$, $c \geqslant 2$), if we have a single factor with b categories and a response with c categories, then $k = b$, $r = 1$, $s_1 = q = c$, while if we have two responses having b and c categories, respectively, then $k = 1$, $r = 2$, $s_1 = b$, $s_2 = c$, and $q = bc$. In either case, under the null hypothesis (of homogeneity* of the b populations or the independence of the two responses), the conditional distribution of the n_{ij}, given the marginals, is independent of the π_{ij}, and hence the Fisher exact test procedure is applicable (*see* FISHER'S EXACT TEST). Also, if $\pi_{i\cdot} = \sum_{j \in J} \pi_{ij}$ and $\pi_{\cdot j} = \sum_{i \in I} \pi_{ij}$, then for the two response model, measures of association* are based on the quantities $(\pi_{ij}/(\pi_{i\cdot}\pi_{\cdot j}) - 1)$, $1 \leqslant i \leqslant b$, $1 \leqslant j \leqslant c$. Other hypotheses of interest include the homogeneity of the marginals (i.e., $\pi_{i\cdot} = \pi_{\cdot i}$, $i = 1, \ldots, b$) when $b = c$.

Formulation of plausible statistical hypotheses or suitable measures of association* (or interactions) in higher-dimensional contingency tables has been an area of fruitful research during the past three decades. The formulation may depend on the number of factors and/or responses as well as on other considerations. For example, in the three-dimensional contingency table relating to three responses (i.e., in the model (1), $k = 1$, $r = 3$, $q = s_1 s_2 s_3$ with $s_1 \geqslant 2$, $s_2 \geqslant 2$, $s_3 \geqslant 2$), if we rewrite the π_{ij} as $\pi_{j_1 j_2 j_3}$ with $j = (j_1, j_2, j_3) \in J = \{j : 1 \leqslant j_i \leqslant s_i,\ i = 1, 2, 3\}$, then, we may have a variety of hypotheses of interest. Let us write

$$\pi_{j_1 j_2 \cdot} = \sum_{j_3 = 1}^{s_3} \pi_{j_1 j_2 j_3},$$

$$\pi_{j_1 \cdot\cdot} = \sum_{j_2 = 1}^{s_2} \sum_{j_3 = 1}^{s_3} \pi_{j_1 j_2 j_3}.$$

(The other marginal probabilities are hypotheses defined in a similar manner). Then we may formulate the following:

a. The hypothesis of complete independence of the three responses:

$$H_0^{(1)}: \quad \pi_{j_1 j_2 j_3} = \pi_{j_1 \cdot\cdot}\pi_{\cdot j_2 \cdot}\pi_{\cdot\cdot j_3}, \qquad \forall j \in J. \qquad (3)$$

b. The hypothesis of independence of the first two responses and the last one:

$$H_0^{(2)}: \quad \pi_{j_1 j_2 j_3} = \pi_{j_1 j_2 \cdot}\pi_{\cdot\cdot j_3}, \qquad \forall j \in J, \quad (4)$$

and a similar formulation for any other pair with one left out.

c. The hypothesis of no partial association between the first two response variables, given the third one,

$$H_0^{(3)}: \quad \pi_{j_1 j_2 j_3} = \pi_{j_1 \cdot j_3}\pi_{\cdot j_2 j_3}/\pi_{\cdot\cdot j_3}, \qquad \forall j \in J, \qquad (5)$$

and a similar formulation for any two response variables, given the other.

d. The hypothesis of pairwise independence,

$$H_0^{(4)}: \begin{cases} \pi_{j_1 j_2 \cdot} = \pi_{j_1 \cdot\cdot}\pi_{\cdot j_2 \cdot}, \\ \pi_{j_1 \cdot j_3} = \pi_{j_1 \cdot\cdot}\pi_{\cdot\cdot j_3}, \\ \pi_{\cdot j_2 j_3} = \pi_{\cdot j_2 \cdot}\pi_{\cdot\cdot j_3}, \end{cases} \qquad \forall j \in J. \qquad (6)$$

e. The hypothesis of no second-order interaction among the three response variables. One possible version of this hypothesis is

$$H_0^{(5)}: \quad \pi_{j_1 j_2 j_3} = \frac{\pi_{j_1 j_2 \cdot}\pi_{\cdot j_2 j_3}\pi_{j_1 \cdot j_3}}{\pi_{j_1 \cdot\cdot}\pi_{\cdot j_2 \cdot}\pi_{\cdot\cdot j_3}}, \qquad \forall j \in J. \qquad (7)$$

In each of these testing problems, one may obtain suitable estimators of the cell probabilities in (1) subject to the constraint imposed by the null hypothesis. These may be the maximum likelihood estimators*, the minimum chi-square* estimators, the modified minimum chi-square estimators, or any other estimators having an asymptotically normal distribution with a dispersion matrix attaining the Cramér–Rao limit. These are termed the BAN (best asymptotically normal) estimators (*see* ASYMPTOTIC NORMALITY). Thus BAN estimators of the parameters in the model (1), under the appropriate null hypothesis, can be incorporated to provide the estimates of the expected frequencies for the different cells, and the usual goodness-of-fit* test statistic (based on the discrepancies between the observed and the estimated expected frequencies) can be used to test the null hypothesis. Under the appropriate null hypothesis, this test statistic will have asymptotically a chi-square distribution with degrees of freedom equal to $q - 1$ $-$ (number of independent constraints imposed on the parameters by the null hypothesis); *see* CHI SQUARE TESTS. In some specific cases, the Fisher exact test procedure can also be extended to the higher-dimensional case. Other formulations of the hypothesis of no second-order interactions of note are those of Roy and Kastenbaum [25] and Bhapkar and Koch [4], among others. These are based on suitable contrasts in the log-probabilities (i.e., $\log \pi_{j_1 j_2 j_3}$) and their natural BAN estimators. From the computational point of view, often they may appear to be quite cumbrous, though their asymptotic chi-square distribution theory remains intact. Other notable contributions to the statistical analysis of interactions (of various orders) in multidimensional contingency tables have been made by Plackett [21], Goodman [10, 11], Darroch [8], Birch [4, 5], and Altham [1] among others. In this context, analysis of categorical data by linear models, developed by Grizzle et al. [15] and extended further by the North Carolina School, plays a vital role. The likelihood-ratio* approach, mostly developed by Goodman and the school led by him, is also worth mentioning. Note that the total degrees of freedom* [10, 11] $(q - 1)$ for the goodness-of-fit statistic may be partitioned into various components (accounting for the variation due to the main effects and interactions of various orders) for which the corresponding likelihood-ratio-type chi-square statistics provide appropriate test statistics (*see* PARTITION OF CHI-SQUARE). In the case of three-dimensional contingency tables with one factor and two response variables (i.e., in (1), $k \geqslant 2$, $r = 2$, $s_1 \geqslant 2$, $s_2 \geqslant 2$, and $q = s_1 s_2$), average partial association of the two response variables, given the factor, can be formulated in a meaningful way, and the log-linear model or the likelihood-ratio model may then be adapted to test for partial association. The case of four- or higher-dimensional contingency tables conceptually presents no difficulty, but may involve greater computational complications. There will be a larger class of plausible hypotheses and the problem of defining higher-order interactions in a convenient manner (capable of simple analysis schemes) may become more involved. Often, among the various possible formulations, any single one may fail to qualify as the best one. However, the log-linear and the likelihood-ratio approaches still provide workable solutions in such cases.

In the multidimensional contingency tables, if the factors and/or responses are structured, then more informative procedures can be based on some nonparametric methods (arising out of grouped data models). For some simple cases, some of these tie-adjusted nonparametric procedures have been discussed in Grizzle and Williams [16], Sen [26], (*see also* CHI-SQUARE TESTS and RANK TESTS FOR GROUPED DATA). As mentioned earlier, model (1) may apply even to some incomplete tables. For such tables, in view of the structural zero probabilities for certain cells, in the formulation of the hypotheses in (3)–(7), we may need to replace the marginal probabilities by appropriate positive constants and then redefine the hypothesis of independence or no interaction in an equivalent way. For such incomplete tables, with such equivalent formulations, tests for quasi-independence and interactions (based on BAN estimators, likelihood-ratio-type statistics and log-linear models) are discussed in detail in Bishop et al. [7, Chap. 5] (1975) and in Haberman [17, Chap. 7].

References

[1] Altham, P. M. E. (1970). *J. R. Statist. Soc. B*, **32**, 63–73.

[2] Altham, P. M. E. (1970). *J. R. Statist. Soc. B*, **32**, 395–407.

[3] Bartlett, M. S. (1935). *J. R. Statist. Soc. Suppl.* **2**, 248–252.

[4] Bhapkar, V. P. and Koch, G. (1968). *Biometrics*, **24**, 567–594.

[5] Birch, M. W. (1964). *J. R. Statist. Soc. B* **26**, 313–324.

[6] Birch, M. W. (1965). *J. R. Statist. Soc. B* **27**, 111–124.

[7] Bishop, Y. M. M., Fienberg, S. E., and Holland, P. W. (1975). *Discrete Multivariate Analysis: Theory and Practice*. MIT Press, Cambridge, MA.

[8] Darroch, J. N. (1962). *J. R. Statist. Soc. B.*, **24**, 251–263.

[9] Fisher, R. A. (1922). *J. R. Statist. Soc.*, **85**, 87–94

[10] Goodman, L. A. (1963). *J. R. Statist. Soc. B*, **25**, 179–188.

[11] Goodman, L. A. (1964). *J. Amer. Statist. Ass.*, **59**, 319–322.

[12] Goodman, L. A. (1968). *J. Amer. Statist. Ass.*, **63**, 1091–1131.

[13] Goodman, L. A. (1971). *J. Amer. Statist. Ass.*, **66**, 339–344.

[14] Grizzle, J. E. (1967). *Amer. Statist.*, **21**, 28–32.

[15] Grizzle, J. E., Starmer, C. F., and Koch, G. G. (1969). *Biometrics*, **25**, 489–504.

[16] Grizzle, J. E., and Williams, O. D. (1972). *Biometrics*, **28**, 137–156.

[17] Haberman, S. J. (1979). *Analysis of Quantitative Data*, Vol. 2. Academic Press, New York.

[18] Ku, H. H. and Kullback, S. (1968). *J. Res. Natl. Bur. Stand. Sec. B.*, **72**, 159–199.

[19] Mantel, N. (1970). *Biometrics*, **26**, 291–304.

[20] Odoroff, C. L. (1970). *J. Amer. Statist. Ass.*, **65**, 1617–1631.

[21] Plackett, R. L. (1962). *J. R. Statist. Soc. B*, **24**, 162–166.

[22] Plackett, R. L. (1964). *Biometrika*, **51**, 327–337.

[23] Read, C. B. (1977). *Commun. Statist. A*, **6**, 553–562.

[24] Roy, S. N. (1957). *Some Aspects of Multivariate Analysis*. Asia Publishing House, Calcutta, India.

[25] Roy, S. N. and Kastenbaum, M. A. (1956). *Ann. Math. Statist.*, **27**, 749–757.

[26] Sen, P. K. (1968). *Sankhyā A*, **30**, 22–31.

[27] Stuart, A. (1953). *Biometrika*, **40**, 105–110.

[28] Stuart, A. (1955). *Biometrika*, **42**, 412–416.

[29] Tocher, K. D. (1950). *Biometrika*, **37**, 130–144.

[30] Yates, F. (1934). *J. R. Statist. Soc. Suppl.*, **1**, 217–235.

(ASSOCIATION, MEASURES OF
CATEGORICAL DATA
CHI-SQUARE TESTS
CONTINGENCY TABLES
FISHER'S EXACT TEST
GOODNESS OF FIT
LIKELIHOOD RATIO TESTS
PARTITION OF CHI-SQUARE)

P. K. SEN

MULTIDIMENSIONAL SCALING

In this entry we summarize the major types of multidimensional scaling (MDS), the distance models used by MDS, the similarity data analyzed by MDS, and the computer programs that implement MDS. We also present three brief examples. We do not discuss experimental design, interpretation, or the mathematics of the algorithms. The entry should be helpful to those who are curious about what MDS is and to those who wish to know more about the types of data and models relevant to MDS. It should help the researcher, the statistical consultant, or the data analyst who needs to decide if MDS is appropriate for a particular set of data and what computer program should be used.

For a more complete, but still brief, introduction to MDS, the reader should turn to Kruskal and Wish [6]. A complete discussion of the topics covered here as well as of experimental design, data analysis, and interpretive procedures can be found in Schiffman et al. [14]. An intermediate-level mathematical treatment of some MDS algorithms is given in Davison (1983). An advanced treatment of the theory of MDS, illustrated with innovative applications, is presented by Young and Hamer [21]. Reviews of the current state of the art are presented by Young (1984a; 1984b). Multidimensional scaling is related to principal components analysis*, factor analysis*, cluster analysis, and numerical taxonomy; the reader is referred to the appropriate entries in this encyclopedia, along with the SCALING and PROXIMITY DATA entries.

OVERVIEW OF MULTIDIMENSIONAL SCALING

Multidimensional scaling (MDS) is a set of data analysis techniques that display the structure of distance-like data as a geometrical picture. It is an extension of the procedure discussed in SCALING.

MDS has its origins in psychometrics, where it was proposed to help understand people's judgments of the similarity of members of a set of objects. Torgerson [18] proposed the first MDS method and coined the term, his work evolving from that of Richardson [11]. MDS has now become a general data analysis technique used in a wide variety of fields [14]. For example, the book on theory and applications of MDS by Young and Hamer [21], presents applications of MDS in such diverse fields as marketing*, sociology*, physics*, political science*, and biology. However, we limit our examples here to the field with which the author is most familiar, psychology*.

MDS pictures the structure of a set of objects from data that approximate the distances between pairs of the objects. The data, which are called similarities, dissimilarities, distances, or proximities, must reflect the amount of (dis)similarity between pairs of the objects (*see* MEASURES OF SIMILARITY, DISSIMILARITY AND DISTANCE). In this article we use the term *similarity* generically to refer to both similarities (where large numbers refer to great similarity) and to dissimilarities (where large numbers refer to great dissimilarity).

In addition to the traditional human similarity judgment, the data can be an "objective" similarity measure (the driving time between pairs of cities) or an index calculated from multivariate data (the proportion of agreement in the votes cast by pairs of senators). However, the data must always represent the degree of similarity of pairs of objects (or events).

Each object or event is represented by a point in a multidimensional space. The points are arranged in this space so that the distances between pairs of points have the strongest possible relation to the similarities among the pairs of objects. That is, two similar objects are represented by two points that are close together, and two dissimilar objects are represented by two points that are far apart. The space is usually a two- or three-dimensional Euclidean space, but may be non-Euclidean and may have more dimensions.

MDS is a generic term that includes many different specific types. These types can be classified according to whether the similarities data are qualitative (called nonmetric MDS) or quantitative (metric MDS). MDS types can also be classified by the number of similarity matrices and the nature of the MDS model. This classification yields classical MDS (one matrix, unweighted model), replicated MDS (several matrices, unweighted model), and weighted MDS (several matrices, weighted model). We discuss the nonmetric/metric and the classical/replicated/weighted classifications in the following sub-sections.

Classical MDS

The identifying aspect of *classical MDS* (CMDS) is that there is only one similarity matrix. Table 1 is a matrix of similarity data suitable for CMDS; it contains the flying mileages between 10 American cities. The cities are the "objects," and the mileages are the "similarities." An MDS of these data gives the picture in Fig. 1, a map of the

Figure 1 CMDS of flying mileages between 10 American cities.

relative locations of these 10 cities in the United States. This map has 10 points, one for each of the 10 cities. Cities that are similar (have short flying mileages) are represented by points that are close together, and cities that are dissimilar (have large mileages) by points far apart.

Generally, CMDS employs Euclidean distance to model dissimilarity. That is, the distance d_{ij} between points i and j is defined as

$$d_{ij} = \sqrt{\sum_a^r (x_{ia} - x_{ja})^2},$$

where x_{ia} specifies the position (coordinate) of point i on dimension a. The distance can also be defined according to the Minkowski model:

$$d_{ij} = \sqrt[p]{\sum_a^r |x_{ia} - x_{ja}|^p},$$

where the value of p ($\geqslant 1$) is set by the investigator.

For either definition of distance there are n points, one for each of the n objects. There are also r dimensions, where the value of r is determined by the investigator. The coordinates x_{ia} are contained in the $n \times r$ matrix **X**. Using matrix algebra, the Euclidean model can be defined as

$$d_{ij} = \left[(\mathbf{x}_i - \mathbf{x}_j)(\mathbf{x}_i - \mathbf{x}_j)'\right]^{1/2},$$

where $\mathbf{x}_i$ is the ith row of **X** and contains the r coordinates of the ith point on all r dimensions. A simple matrix expression for the Minkowski model is not possible. For both models, the distances d_{ij} are contained in the $r \times n$ symmetric matrix **D**. Finally, the similarities s_{ij} are contained in the matrix **S**, also $n \times n$.

METRIC CMDS. The first major CMDS proposal [18] was metric (i.e., the similarities had to be quantitative). Torgerson's development required the data to be at the ratio level of measurement, although this was soon generalized to the interval level [8]. While the data could contain random error, this early type of MDS required that the data be dissimilarities (not similarities), complete (no missing values), and symmetric (the dissimilarity of objects I and J had to equal that of objects J and I). These CMDS proposals also required the distance model to be Euclidean. The flying mileage example is metric CMDS because flying mileages are at the ratio level of measurement.

For metric CMDS, the distances **D** are determined so that they are as much like the dissimilarities **S** as possible. There are a variety of ways in which "like" is strictly defined, but a common one is a least-squares* definition. In this case, we define

$$l\{\mathbf{S}\} = \mathbf{D} + \mathbf{E},$$

where $l\{\mathbf{S}\}$ is read "a linear transformation of the similarities." If the measurement level is ratio, then the linear transformation has a zero intercept, but can be nonzero when the level is interval. If the data are similarities, the slope of the transformation is negative; if dissimilarities, it is positive.

In the preceding equation, **E** is a matrix of errors (residuals) that in the least-squares optimization situation, we wish to minimize. Since the distances **D** are a function of the coordinates **X**, the goal of CMDS is to calculate the coordinates **X** so that the sum of squares of **E** is minimized, subject to suitable normalization of **X**. We also need to calculate the best linear transformation $l\{\mathbf{S}\}$. Torgerson's method does not actually minimize the sum of squares of **E**, nor do ALSCAL or MULTISCALE. The KYST, MINISSA, and SMACOF programs do.

Table 1 Flying Mileages Between 10 American Cities

Atlanta	Chicago	Denver	Houston	Los Angeles	Miami	New York	San Francisco	Seattle	Washington, DC	
0	587	1212	701	1936	604	748	2139	2182	543	Atlanta
587	0	920	940	1745	1188	713	1858	1737	597	Chicago
1212	920	0	879	831	1726	1631	949	1021	1494	Denver
701	940	879	0	1374	968	1420	1645	1891	1220	Houston
1936	1745	831	1374	0	2339	2451	347	959	2300	Los Angeles
604	1188	1726	968	2339	0	1092	2594	2734	923	Miami
748	713	1631	1420	2451	1092	0	2571	2408	205	New York
2139	1858	949	1645	347	2594	2571	0	678	2442	San Francisco
2182	1737	1021	1891	959	2734	2408	678	0	2329	Seattle
543	597	1494	1220	2300	923	205	2442	2329	0	Washington, DC

These programs are all discussed in the Computer Programs section.

NONMETRIC CMDS. The second major CMDS proposal [5, 15] was nonmetric. That is, the data could be at the ordinal level of measurement (*see* ORDINAL DATA). In addition, the data **S** could be either complete or incomplete, symmetric or asymmetric, and similarities or dissimilarities.

These nonmetric CMDS proposals extended the distance model to the Minkowski case and generalized the relation between similarities and distances. They enable defining

$$m\{\mathbf{S}\} = \mathbf{D} + \mathbf{E},$$

where $m\{\mathbf{S}\}$ is read "a monotonic transformation of the similarities." If **S** is actually dissimilarities then $m\{\mathbf{S}\}$ preserves order, whereas if **S** is similarities, it reverses order. Thus, for nonmetric CMDS, we need to solve for the monotonic (order-preserving) transformation $m\{\mathbf{S}\}$ and the cordinates **X**, which together minimize the sum of squares of the errors **E** (after normalization of **X**). This exact problem is solved by the MINISSA, KYST, and SMACOF programs (discussed in the final section) while ALSCAL and MULTISCALE solve closely related problems.

The nonmetric optimization represents a much more difficult problem to solve than the metric problem and is an important breakthrough in multidimensional scaling. In fact, nonmetric CMDS is the first example of using quantitative models to describe qualitative data that belongs to the approach discussed by Young [19] (*see* QUALITATIVE DATA, STATISTICAL ANALYSIS).

It is reassuring to know that when we degrade the flying mileages (Table 1) into ranks of flying mileages, and then submit the ranks to nonmetric CMDS, the map that results is indistinguishable from that shown in Fig. 1.

Replicated MDS

The next major development, *replicated MDS* (RMDS), permitted the analysis of several matrices of similarity data simultaneously [7]. There are m matrices $\mathbf{S}_k$, one for each subject k, $k = 1, \ldots, m$.

RMDS uses the same distance models as CMDS, but uses them to describe several similarity matrices rather than one. With RMDS, the matrix of distances **D** is determined so that it is simultaneously like all the similarity matrices $\mathbf{S}_k$.

For metric RMDS, the least-squares definition of "like" is

$$l_k\{\mathbf{S}_k\} = \mathbf{D} + \mathbf{E}_k,$$

where $l_k\{\mathbf{S}_k\}$ is the linear transformation of the kth similarity matrix $\mathbf{S}_k$ which best fits the distances **D**. The data may be similarities or dissimilarities and may be at the ratio or interval levels, just as in metric CMDS. The analysis minimizes the sum of the squared elements in all error matrices $\mathbf{E}_k$, subject to normalization of **X**.

For nonmetric RMDS, we minimize the several $\mathbf{E}_k$ in

$$m_k\{\mathbf{S}_k\} = \mathbf{D} + \mathbf{E}_k,$$

where $m_k\{\mathbf{S}_k\}$ is the monotonic transformation of the similarity matrix $\mathbf{S}_k$ which is a least-squares fit to the distances in matrix **D**. The data may be similarities of dissimilarities, just as in CMDS.

Note that for RMDS each linear or monotonic transformation l_k or m_k is subscripted, letting each data matrix $\mathbf{S}_k$ have a unique linear or monotonic relation to the distances **D**. Since k ranges up to m, there are m separate linear or monotonic transformations, one for each of the m dissimilarity matrices $\mathbf{S}_k$. This implies that RMDS treats all the matrices of data as being related to each other (through **D**) by a systematic linear or monotonic transformation (except for a random error component). The KYST and SMACOF programs minimize the sum of squares of $\mathbf{E}_k$, while ALSCAL and MULTISCALE solve other closely related problems. In psychological terms, RMDS accounts for differences in the ways subjects use the response scale (i.e., differences in response bias).

Jacobowitz [4] used RMDS to study the way language develops as children grow to

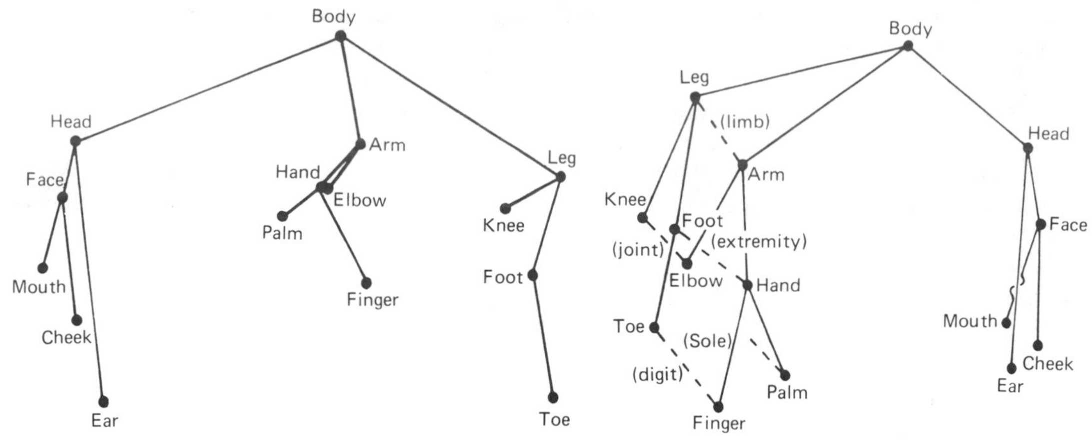

Figure 2 (a) RMDS of children's similarity judgments about 15 body parts; (b) RMDS of adults' similarity judgments about 15 body parts.

adulthood. In his experiment he asked children and adults to judge the similarity of all pairs of 15 parts of the human body. The judges were five-, seven-, and nine-year-olds, and adults. There were 15 judges at each age. Four separate RMDS analyses were done, one for each age group.

The RMDS results for the five-year-olds are shown in Fig. 2*a*, and for the adults in Fig. 2*b*. The analysis located the points in the space, but did not draw the lines. The lines were drawn by Jacobowitz to interpret the psycholinguistic structure that people have for body-part words. Jacobowitz theorized that the structure would be hierarchical. We can see that it is. He further theorized that the structure would become more complex as the children become adults. This theory is also supported, since the adults' hierarchy also involves a classification of corresponding arm and leg terms. (In Fig. 2*b* the corresponding terms are linked by dashed lines, the implied classification terms are shown in parentheses, and the word *sole*, which was not a stimulus, is shown in the position that we would predict it to be in if the study were repeated.)

Weighted MDS

The next major MDS development, *weighted MDS* (WMDS), generalized the distance model so that several similarity matrices $\mathbf{S}_k$ could be assumed to differ from each other in systematically nonlinear or nonmonotonic ways. Whereas RMDS only accounts for individual differences in response bias, WMDS incorporates a model to account for individual differences in the fundamental perceptual or cognitive processes that generate the responses. For this reason, WMDS is often called *individual differences scaling* (INDSCAL) and is often regarded as the second major breakthrough in multidimensional scaling.

WMDS invokes the following definition of weighted Euclidean distance:

$$d_{ijk} = \sqrt{\textstyle\sum_a^r w_{ka}(x_{ia} - x_{ja})^2}\,, \qquad w_{ka} \geqslant 0,$$

which, in matrix algebra is

$$d_{ijk} = \left[(\mathbf{x}_i - \mathbf{x}_j)\mathbf{W}_k(\mathbf{x}_i - \mathbf{x}_j)'\right]^{1/2},$$

where $\mathbf{W}_k$ is a $r \times r$ diagonal matrix. The diagonal values, which must not be negative, are weights for subject k on each of the r dimensions.

WMDS is appropriate for the same type of data as RMDS. However, RMDS generates a single distance matrix $\mathbf{D}$, while WMDS generates m unique distance matrices $\mathbf{D}_k$, one for each data matrix $\mathbf{S}_k$. The distances $\mathbf{D}_k$ are calculated so that they are all as much like their corresponding data matrices $\mathbf{S}_k$ as possible. For metric WMDS, the least-squares problem is

$$l_k\{\mathbf{S}_k\} = \mathbf{D}_k + \mathbf{D}_k\,,$$

and for nonmetric WMDS, the problem is

$$m_k\{\mathbf{S}_k\} = \mathbf{D}_k + \mathbf{E}_k .$$

Thus, for WMDS, we need to solve for the matrix of coordinates $\mathbf{X}$, the m diagonal matrices of weights $\mathbf{W}_k$, and the m transformations m_k or l_k. We wish to do this so that the sum of squared elements in all error matrices $\mathbf{E}_k$ is minimal subject to normalization contraints on $\mathbf{X}$ and $\mathbf{W}_k$.

Neither of the two most commonly used computer programs solve either of the problems defined. (These programs and others are discussed in the last section.) The INDSCAL program, by Carroll and Chang [1], provided the first metric WMDS solution. However, it optimizes the fit of scalar products to a transformation of the data. The ALSCAL program, by Takane et al. [17] (see also Young and Lewyckyj [22] and Young et al. [24], provided the first and still the only algorithm to incorporate both nonmetric and metric solution to WMDS and optimize the fit of squared distances to the data. In fact, ALSCAL is still the only algorithm to provide the user with nonmetric and metric solutions to the CMDS, RMDS, and WMDS situations discussed, and it is regarded as the third major breakthrough in multidimensional scaling.

The MULTISCALE algorithm by Ramsay [10] provided the first metric WMDS solution to optimize the preceding index (it fits distances to the data). Finally, the SMACOF algorithm [2, 3] and its associated program [16], which is still under development, will more than likely be the first program to be able to fit distances to the data so that the sum of squares of $\mathbf{E}$ is strictly minimized, where the distances may be CMDS, RMDS, or WMDS distances, and where the transformation may be metric or nonmetric.

While WMDS incorporates the RMDS

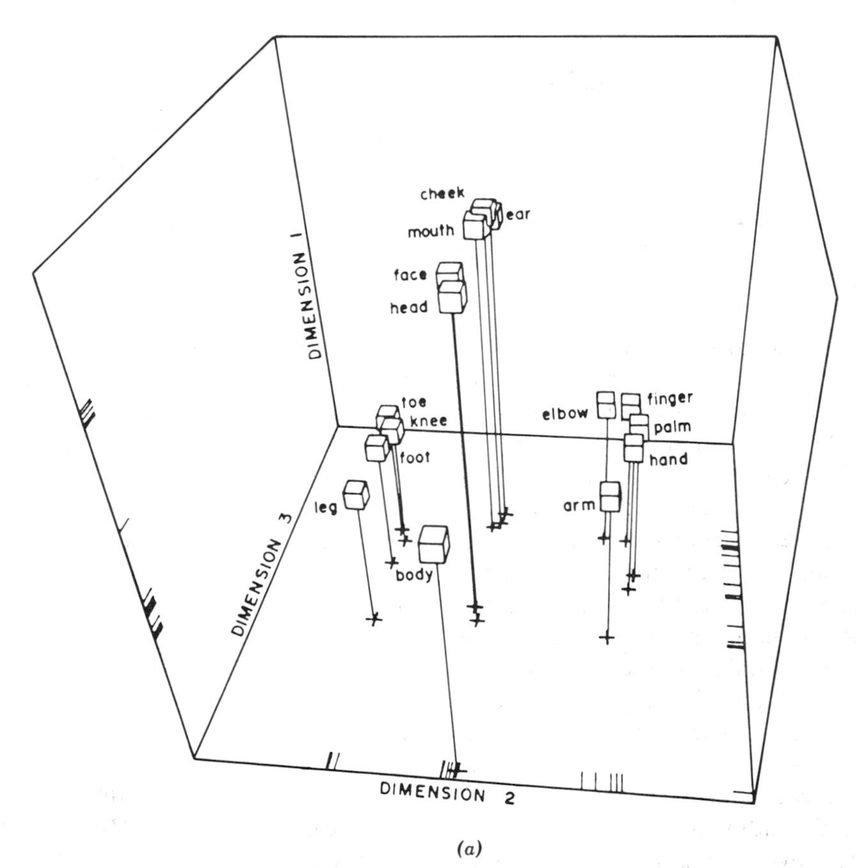

(a)

Figure 3 (a) WMDS of children's and adults' similarity judgments about 15 body parts.

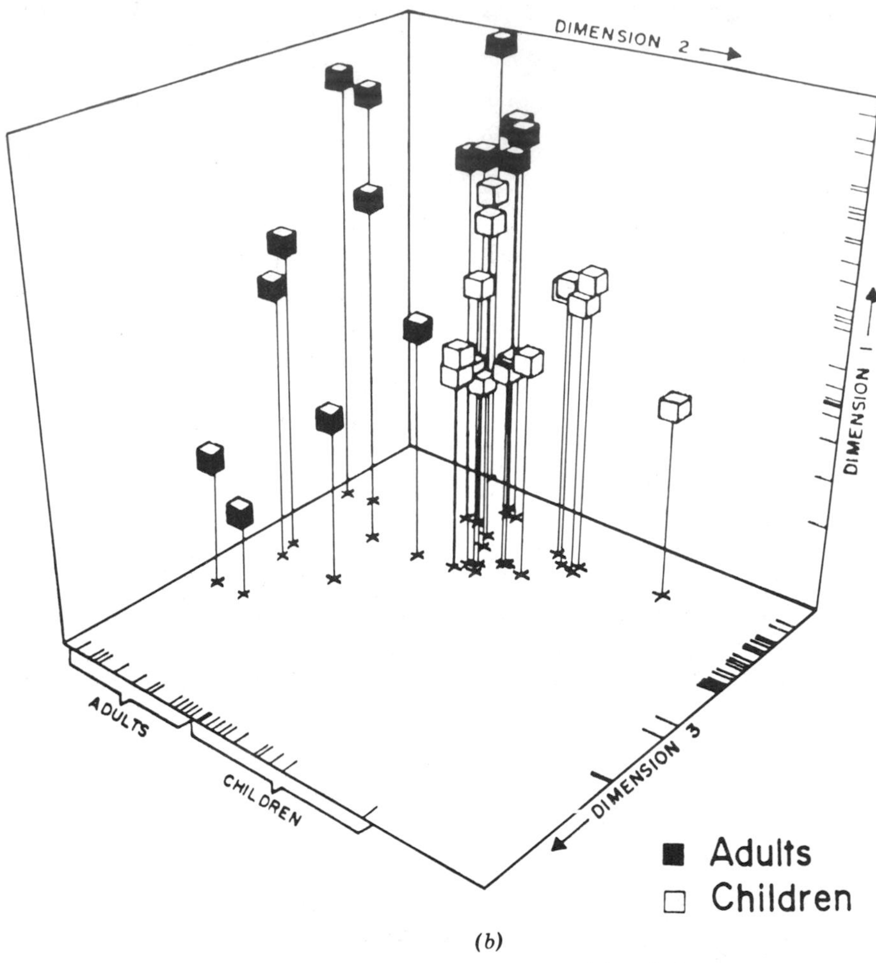

(b)

Figure 3 (b) Subject weights from WMDS analysis.

notion of individual differences in response bias (via m_k and l_k), the important aspect of WMDS is that it provides specific parameters for individual variation in cognitive or perceptual processes. These parameters are the weights. The weights are interpreted as the importance, relevance, or salience of each dimension to each individual. A large weight means that the dimension is important to the individual, a small weight means the dimension is unimportant. If the similarity matrices correspond to experimental conditions, say, rather than to people, the interpretation is that the weights reflect the importance of each dimension in the various experimental conditions.

The Jacobowitz data already discussed provide a nice example of WMDS. An analysis of the 15 five-year-olds together with the 15 adults provided the results displayed in Fig. 3. In Fig. 3*a*, we see that the stimulus structure is the anticipated hierarchy. In Fig. 3*b*, which is the weight space, we see that the children and adults occupy different parts of the space, showing that the children and adults have different cognitive structures for parts of the body.

COMPUTER PROGRAMS

Several computer programs have become a significant part of the MDS discipline (*see* STATISTICAL SOFTWARE). These programs,

Table 2 Characteristics of Several MDS Computer Programs

	MINISSA	KYST-2a	INDSCAL	ALSCAL-83	MULTISCL-2	SMACOF-1b
Similarity	Yes	Yes	Yes	Yes	Yes	Yes
Asymmetric	No	Yes	Yes	Yes	Yes	Yes
Missing	Yes	Yes	No	Yes	Yes	Yes
Two-way	Yes	Yes	No	Yes	Yes	Yes
Measurement	N	MN	M	MN	M	MN
Model	C	CR	WO	CRWO	CRWO	CRW
Fit	D	D	P	S	DL	D
Algorithm	L	L	L	L	M	L
Converge	No	No	Yes	Yes	No	Yes
Stimuli	100	100	dyn	dyn	50	dyn
Matrices	dyn	dyn	dyn	dyn	100	dyn
Elements	4950	4000	dyn	dyn	15000	dyn
Dimensions	10	6	10	6	10	dyn

and several of their characteristics, are listed in Table 2. A complete reference for each program is given in the bibliography.

The first four rows of Table 2 refer to the type of data each program can analyze—specifically, whether each program can analyze similarity data in addition to dissimilarity data, asymmetric data in addition to symmetric data, data with missing elements in addition to data without, and two-way in addition to three-way data.

The next two rows of Table 2 refer to the types of analyses each program can provide. The Measurement row refers to whether the program can provide only nonmetric analyses (N), only metric analyses (M), or both (MN). The Model row refers to whether the program can provide analyses that are classical (C), replicated (R), weighted (W), or other types (O).

The next three rows of Table 2 refer to several aspects of the iterative algorithm employed by each program. The Fit row refers to the aspect of the model that is fit to (a transformation of) the data (D indicates distances, P scalar products, S squared distances, and L log distances). The Algorithm row indicates whether the program is a least-squares program (L) or a maximum likelihood program (M). The Converge row shows whether the algorithm is convergent (each iteration must improve the fit index being optimized) or not.

The final four rows specify the maximum size problem that can be analyzed by each program. Some programs place specific limits on the number of stimuli, matrices, dimensions, or total number of data elements. These limits are indicated by a number. Other programs are dynamic and place no limit. These are indicated as "dyn."

The ALSCAL-84, MULTISCALE-2, and SMACOF-1B programs are the current state-of-the-art programs. Of the programs listed in Table 2, ALSCAL-84 [23] is the most flexible, fitting the widest range of models to the widest range of data. The ALSCAL algorithm is convergent (which is desirable) and is faster than MULTISCALE but slower than SMACOF. ALSCAL is the only MDS program currently available in major statistical systems (SAS and SPSS) and is the easiest program to use. However, the algorithm optimizes the fit of squared distances to the dissimilarities, which is not the most desirable optimization criterion. ALSCAL is descriptive, having no inferential aspects.

MULTISCALE-2 [10] has the unique feature that it is based on the maximum likelihood principle. Of the programs listed in Table 2, it is the only one that enables statistically based significance tests and that can be used for inferential purposes. MULTISCALE provides the user with a selection of models that is smaller than that provided by ALSCAL, but larger than that provided by SMACOF. However, of the three programs, MULTISCALE is the least flexible in the types of data that can be analyzed, and the slowest. Also it has a nonconvergent algorithm.

SMACOF-1B [16], clearly the fastest of these three programs, optimizes the fit of distances to dissimilarities by a convergent algorithm. The algorithm [2, 3] is the simplest and most elegant of any program listed in Table 2. It fits the CMDS, RMDS, and WMDS models and is as flexible as ALSCAL in the types of data it can analyze. However, SMACOF is currently under active development; it is difficult to use and is not available in any statistical package. When fully mature, SMACOF will be the program of choice.

References

[1] Carroll, J. D. and Chang, J. J. (1970). *Psychometrika*, **35**, 238–319. (A key paper: Provides the first workable WMDS algorithm, and one that is still in very wide use. Generalizes singular value (Eckart–Young) decomposition to N-way tables.)

[2] deLeeuw, J. (1977). In *Recent Developments in Statistics*, J. R. Barra et al., eds. North-Holland, Amsterdam. (Advanced mathematical paper that proposes the SMACOF algorithm and proves its convergence. Difficult but elegant.)

[3] deLeeuw, J. and Heiser, W. J. (1977). In *Geometric Representations of Relational Data*, J. C. Lingoes, ed. Mathesis Press, Ann Arbor, MI. (Continues the work published in the preceding reference.)

[4] Jacobowitz, D. (1973). "Development of semantic structures." Unpublished Ph.D. dissertation. University of North Carolina at Chapel Hill.

[5] Kruskal, J. B. (1964). *Psychometrika*, **29**, 1–27; 115–129. (Completes the second major MDS breakthrough started by Shepard by placing Shepard's work on a firm numerical analysis foundation. Perhaps the most important paper in the MDS literature.)

[6] Kruskal, J. B. and Wish, M. (1977). *Multidimensional Scaling*. Sage Publications, Beverly Hills, CA. (Very readable and accurate brief introduction to MDS that should be read by everyone wanting to know more.)

[7] McGee, V. C. (1968). *Multivar. Behav. Res.*, **3**, 233–248.

[8] Messick, S. J. and Abelson, R. P. (1956). *Psychometrika*, **21**, 1–17.

[9] Ramsay, J. O. (1982). *J. R. Statist. Soc. A*, **145**, 285–312. (Foundations for one aspect of the current state of the art. Introduces hypothesis testing into the MDS framework, providing statistical tests to help decide on the appropriate dimensionality and model.)

[10] Ramsay, J. O. (1982). *Multiscale II Manual*. Department of Psychology, McGill University, Montreal, Canada. (Very high-quality user's guide to the program based on the preceding reference.)

[11] Richardson, M. W. (1938). *Psychol. Bull.*, **35**, 659–660.

[12] Roskam, E. E. *MINISSA Standard Version*. Nijmegen Mathematics–Psychology Department, University of Nijmegen, Nijmegen, Holland. (The MINISSA user's guide.)

[13] SAS Institute. (1980). *SAS Supplemental Library User's Guide*. SAS Institute, Cary, NC.

[14] Schiffman, S. S., Reynolds, M. L., and Young, F. W. (1981). *Introduction to Multidimensional Scaling*. Academic Press, New York.

[15] Shepard, R. N. (1962). *Psychometrika*, **27**, 125–140; 219–246. (Started the second major MDS breakthrough by proposing the first nonmetric algorithm. Intuitive arguments placed on firmer ground by Kruskal.)

[16] Stoop, I. and de Leeuw, J. (1982). *How to Use SMACOF-IB*. Department of Data Theory, University of Leiden, The Netherlands. (A complete user's guide.)

[17] Takane, Y., Young, F. W., and de Leeuw, J. (1977). *Psychometrika*, **42**, 7–67. (The third major MDS breakthrough. Combined all previous major MDS developments into a single unified algorithm.)

[18] Torgerson, W. S. (1952). *Psychometrika*, **17**, 401–419. (The first major MDS breakthrough.)

[19] Young, F. W. (1981). *Psychometrika*, **46**, 357–388. (A readable overview of nonmetric issues in the context of the general linear model and components and factor analysis.)

[20] Young, F. W. (1984). *Research Methods for Multimode Data Analysis in the Behavioral Sciences*, H. G. Law, C. W. Snyder, J. Hattie, and R. P. MacDonald, eds. (An advanced treatment of the most general models in MDS. Geometrically oriented. Interesting political science example of a wide range of MDS models applied to one set of data.)

[21] Young, F. W. and Hamer, R. M. (1984). *Theory and Applications of Multidimensional Scaling*. Erlbaum Associates, Hillsdale, NJ. (The most complete theoretical treatment of MDS and the most wide-ranging collection of applications.)

[22] Young, F. W. and Lewyckyj, R. (1979). *ALSCAL-4 user's guide*, 2nd ed. Data Analysis and

Theory Associates, Carrboro, NC. (A brief ALSCAL user's guide.)

[23] Young, F. W. and Lewyckyj, R. (1983). In *SAS Institute Supplemental Library User's guide*. SAS Institute, Cary, NC. (Complete ALSCAL user's guide. One aspect of the current state of the art.)

[24] Young, F. W. Takane, Y., and Lewyckyj, R. (1980). *Amer. Statist.* 117–118. (An abstract.)

(CLASSIFICATION
COMPONENTS ANALYSIS
FACTOR ANALYSIS
GRAPH-THEORETIC CLUSTER ANALYSIS
HIERARCHICAL CLASSIFICATION
MEASURES OF SIMILARITY, DISSIMILARITY, AND DISTANCE
PROXIMITY DATA)

FORREST W. YOUNG

MULTILINEAR FORMS *See* FORMS, BILINEAR

MULTINOMIAL COEFFICIENTS

The coefficient of $y_1^{\alpha_1} y_2^{\alpha_2} \dots y_k^{\alpha_k}$ in the expansion of $(y_1 + y_2 + \cdots + y_k)^n$ is

$$n! \Big/ \left(\prod_{i=1}^{k} \alpha_i! \right), \qquad \sum_{i=1}^{k} \alpha_i = n.$$

This is called a *multinomial coefficient* and is denoted by

$$\binom{n}{\alpha_1, \alpha_2, \dots, \alpha_k}.$$

(MULTINOMIAL DISTRIBUTIONS)

MULTINOMIAL DISTRIBUTIONS

In this article, the multinomial distribution (MD), multivariate multinomial distribution (MMD), negative multinomial distribution (NMD), and other related distributions are considered. The MD is a generalization of the binomial distribution, and NMD is the generalization of the negative binomial distribution. The MD arises in categorical data* analysis. Situations in which the NMD is useful occur in the study of "accident proneness" of individuals.

DEFINITION AND STRUCTURE OF THE MD

The s-dimensional MD with parameters n and $\mathbf{p}' = (p_1, p_2, \dots, p_s)$ is defined by the joint probability function (pf) given by

$$f_{X_1, X_2, \dots, X_s}(x_1, x_2, \dots, x_s) = n! \prod_{i=1}^{s} (p_i^{x_i}/x_i!), \tag{1}$$

$x_i = 0, 1, 2, \dots, n$ $(i = 1, 2, \dots, s_i)$, where $\sum_{i=1}^{s} x_i = n$, $0 < p_i < 1$, with $\sum_{i=1}^{s} p_i = 1$. Since for the MD $\sum_{i=1}^{s} p_i = 1$ and $\sum_{i=1}^{s} x_i = n$, the pf given by (1) is often written as

$$f_{X_1, X_2, \dots, X_s}(x_1, x_2, \dots, x_s) = \frac{n!\,(1 - \sum_1^{s-1} p_i)^{(n - \sum_1^{s-1} x_i)}}{(n - \sum_{i=1}^{s-1} x_i)!} \left[\prod_{i=1}^{s-1} \left(\frac{p_i^{x_i}}{x_i!} \right) \right]$$

Here we shall assume the pf given by (1).

Genesis

Consider a series of n independent trials. Suppose each trial can result in only one of s mutually exclusive events A_i, with probability p_i, $i = 1, 2, \dots, s$ where $\sum_{i=1}^{s} p_i = 1$. Further, let the r.v. X_i represent the number of occurrences of event A_i, $i = 1, 2, \dots, s$. Then the joint distribution of $(X_1, X_2, \dots, X_s)$ is given by (1).

Property 1. If $(X_1, X_2, \dots, X_s)$ have the MD (1) then the *marginal distribution* of any subset of $(X_1, X_2, \dots, X_s)$ is also an MD. Further, the *conditional distribution* of any subset of $(X_1, X_2, \dots, X_s)$, given the remaining X_j's, is an MD. In particular, (a) X_i has the binomial distribution* with parameters n and p_i, (b) (X_i, X_j) has an MD with parameters n, p_i, p_j, and (c) the conditional distribution of X_i, given $X_j = x_j$, is binomial with parameters $n - x_j$ and $p_i/(1 - p_j)$.

Property 2. The probability-generating function* of the MD is given by

$$G(t_1, t_2, \ldots, t_s) = \left(\sum_{i=1}^{s} t_i p_i\right)^n.$$

The joint factorial moments* of the MD are

$$\mu_{(r_1, r_2, \ldots, r_s)} = E\left[X_1^{(r_1)} X_2^{(r_2)} \cdots X_s^{(r_s)}\right] = n^{(\Sigma r_i)} \prod_{i=1}^{s} p_i^{r_i}.$$

In particular,

$$E(X_i) = np_i, \qquad \operatorname{Var}(X_i) = np_i(1 - p_i),$$
$$\operatorname{Cov}(X_i, X_j) = -np_i p_j,$$
$$E\left[X_i \mid X_j = x_j\right] = (n - x_j) p_i (1 - p_j)^{-1}$$

(linear regression).

Property 3. The probability inequalities

$$\Pr\left[\bigcap_{i=1}^{s} (X_i \leqslant a_i)\right] \leqslant \prod_{i=1}^{s} \Pr(X_i \leqslant a_i),$$

due to Mallows [24], are valid for the MD for any values of $a_1, a_2, \ldots, a_s$. Thus the MD belongs to the negative quadrant dependent class considered by Lehmann [22].

Property 4. The modes of the multinomial distribution are discussed in Finucan [10]. Roughly, the modes are located near the expected value point. For details, see Johnson and Kotz [17, Chap. 11].

Property 5. If $X_1, X_2, \ldots, X_s$ are independent Poisson variables with parameters $\lambda_1, \lambda_2, \ldots, \lambda_s$, respectively, then the conditional distribution of $(X_1, X_2, \ldots, X_s)$, given $\sum_{i=1}^{s} X_i = n$, is multinomial with parameters n and $\mathbf{p}$, where $p_i = \lambda_i / \sum_1^s \lambda_i$, $i = 1, 2, \ldots, s$. This property is useful in many applications of the Poisson distribution*. Estimation of the parameters of the Poisson distribution using this property is discussed in Bol'shev [8]. Another application of this property for obtaining a characterization of the Poisson distribution using a splitting model appears in Rao and Srivastava [39] and Ratnaparkhi [40]. A characterization of the multinomial distribution is considered in Janardan [16].

Property 6. The concepts of stochastic majorization* of a random vector $\mathbf{X}$ and Schur family in the parameter vector, say $\boldsymbol{\theta}$, of the multivariate distributions of $\mathbf{X}$, are discussed in Nevius et al [28]. The related concepts of Schur convexity, Schur concavity, and their applications in the study of probability inequalities and inferential problems are discussed in Marshall and Olkin [25]. For applications of these concepts in the study of the multinomial distribution, we refer the reader to Alam [2], Marshall and Olkin [25, Chap. 11], Olkin [30], Perlman and Rinott [36], Rinott [41], and Wong and Yue [55].

Property 7. If $X_1, X_2, \ldots, X_s$ have the MD with parameters n and $\mathbf{p}$, where $\mathbf{p}$ itself is a random vector having the Dirichlet distribution*, then the resulting mixture distribution* is known as the *compound multinomial distribution* [15]. Mosimann [26] found this compound MD to be useful in the analysis of pollen data in paleoecological studies.

Property 8. If $X_1, X_2, \ldots, X_s$ have the MD with parameters n and $\mathbf{p}$ where n itself is a random variable having the logarithmic distribution* with parameter θ, then the resulting mixture distribution is the *multivariate modified logarithmic distribution*. For details, see Patil and Bildikar [34].

Property 9. If $X_1, X_2, \ldots, X_s$ have the MD given by

$$f_{X_1, X_2, \ldots, X_s}(x_1, x_2, \ldots, x_s) = \frac{n!}{x_0! \prod_{i=1}^{s} x_i!} x_0^{p_0} \cdot \prod_{i=1}^{s} x_i^{p_s},$$

where $x_0 = (n - \sum_{i=1}^{s} x_i)$ and $p_0 = 1 - \sum_{i=1}^{s} p_i$ and if n itself is a random variable having the negative binomial distribution* with parameters k and p, then the resulting mixture distribution is the s-*variate negative multinomial distribution* with parameters k

and $\theta_i = qp_i/(1 - qp_0)$, $(i = 1, 2, \ldots, s)$, where $q = 1 - p$.

Property 10. Let $X_1, X_2, \ldots, X_s$ have the MD given by (1). Let $(X_1, X_2, \ldots, X_s)$ denote the realization of the corresponding multinomial experiment. Then $X^2 = \sum_{i=1}^{s}[(x_i - np_i)^2/(np_i)]$ has approximately the chi-squared distribution with $(s - 1)$ degrees of freedom [35]. This property is found to be useful in statistical analysis, in particular, in categorical data* analysis (*see* CHI-SQUARE TESTS *and* CONTINGENCY TABLES).

Property 11. The MD arises as a special case of: the multivariate power-series distribution* and the sum-symmetric power-series distribution (see Patil [33]); the multivariate Pólya distribution (see Steyn [45]).

ESTIMATION OF PARAMETERS OF MD

Let $X_1, X_2, \ldots, X_s$ have an MD given by (1). If n and s are known, then the maximum likelihood* estimates of $p_i (i = 1, 2, \ldots, s)$ are given by $\hat{p}_i = f_i.n$, where f_i is the observed frequency of A_i. The problem of simultaneous confidence regions for $\mathbf{p}$ was studied by Quesenberry and Hurst [37] and Goodman [12]. A sequential estimation of the parameters of the multinomial distribution appears in Bhat and Kulkarni [7]. Assuming that $p_i = 1/s$ $(i = 1, 2, \ldots, s)$, the maximum likelihood estimator of s was obtained by Lewontin and Prout [23].

APPROXIMATIONS

The computation of exact probabilities for the MD is difficult. A number of approximations that could be useful for this purpose are considered by Bennett [6], Hoel [14], Johnson and Young [18], Lancaster and Brown [21], Rüst [42], Studer [47], and Vora [50]. The relationship between the sums of the multinomial probabilities and multiple integrals was considered by Olkin and Sobel [31] and Stoka [46]. Improvements to the chi-squared approximation, recorded as property 10 of the MD were suggested by Wise [52, 53] and Hoel [14]. Approximation to the moments and the distribution of the likelihood-ratio statistics for goodness of fit of the MD are discussed by Smith et al. [44].

APPLICATIONS OF THE MD

The MD has many applications in statistical data analysis. In particular, it is prominently used in situations where the experimental data need to be considered as multiple categories of events (e.g., in categorical data analysis). Another important application of the MD is in Maxwell–Boltzmann statistical thermodynamics.

OTHER DISTRIBUTIONS RELATED TO THE MD

The compound multinomial distributions are recorded as properties **7** and **8** of the MD. The truncated multinomial distribution (arising due to the unobservability of certain A_i's considered in the genesis of the MD) and the related estimation problem is considered by Asano [3], Batschelet [5], and Geppert [11].

The following related distributions also often arise in practice.

Bivariate Binomial Distribution (BBD)

Consider a two-way cross-classified contingency table corresponding to two characters, say c_1 and c_2, observed for each individual in a population. Let $p_{10}, p_{01}, p_{11}, p_{00}$ denote the probabilities that an individual possesses c_1 but not c_2, c_2 but not c_1, both c_1 and c_2, and neither c_1 nor c_2, respectively. Take a sample of n individuals from the selected population. Let X_1 and X_2 denote the numbers of individuals in the random sample possessing characters c_1 and c_2, respectively. Then (X_1, X_2) has the *bivariate binomial distribution* with parameters n, p_{10}, p_{01}, and p_{11}.

Its joint probability function is given by

$$f_{X_1,X_2}(x_1,x_2) = \sum_{j=0}^{\min(x_1,x_2)} \left[\frac{n!\, p_{11}^j p_{10}^{x_1-j} p_{01}^{x_2-j} p_{00}^{n-x_1-x_2+j}}{j!\,(x_1-j)!\,(x_2-j)!\,(n-x_1-x_2+j)!} \right],$$

for $x_i = 0, 1, 2, \ldots, n$ $(i = 1, 2)$; $0 < p_{10} < 1$, $0 < p_{01} < 1$, $0 < p_{11} < 1$, $0 < p_{10} + p_{01} + p_{11} < 1$, $p_{00} = 1 - p_{10} - p_{01} - p_{11}$.

Property 1. The probability-generating function of the BBD is given by

$$g(t_1 t_2) = (p_{00} + p_{10}t_1 + p_{01}t_2 + p_{11}t_1t_2)^h.$$

Property 2. The moments of the BBD are

$$E(X_i) = np_i, \qquad V(X_i) = np_i(1 - p_i), \qquad i = 1, 2$$

and

$$\operatorname{Cov}(X_1, X_2) = n(p_{11} - p_1 p_2),$$

where $p_1 = p_{10} + p_{11}$, $p_2 = p_{01} + p_{11}$.

Property 3. If (X_1, X_2) has the BBD with parameters n, p_{10}, p_{01}, and p_{11}, then X_i has the binomial distribution with parameters n and p_i, $(i = 1, 2)$, where p_i is defined as in property **2**.

Property 4. The BBD with parameters n, p_{10}, p_{01}, and p_{11} tends to the bivariate Poisson distribution with parameters λ_1, λ_2, and λ_{12} as $n \to \infty$, $p_{10} \to 0$, $p_{01} \to 0$ and $p_{11} \to 0$ such that $np_{10} \to \lambda_1$, $np_{01} \to \lambda_2$, $np_{11} \to \lambda_{12}$, $0 < \lambda_i < \infty$, $i = 1, 2$, $0 < \lambda_{12} < \infty$.

For details of the BBD see Aitken and Gonin [1] and Capobianco [9]. An extension of the BBD, which arises in the study of $2 \times s$ contingency tables, is called the multivariate binomial distribution; for details, see Wishart [54].

MULTIVARIATE MULTINOMIAL DISTRIBUTION (MMD)

The MMD distribution arises as an extension of the BBD in the analysis of p-way cross-classified contingency tables*. Thus the MMD is the joint distribution of the p multinomial distributions arising in such tables. For details, see Wishart [54] and Steyn [45]. Tallis [48, 49] studied a multinomial distribution where each X_i $(i = 1, 2, \ldots, s)$ has the same marginal distribution and $\operatorname{corr}(X_i, X_j) = \rho$, for all i and j such that, $i \neq j$, $i = 1, 2, \ldots, s$, $j = 1, 2, \ldots, s$. For further details, see Tallis [48, 49]. The MD arising through stochastic matrices* appears in Gyires [13].

NEGATIVE MULTINOMIAL DISTRIBUTION (NMD)

Definition and Structure of the NMD

The negative multinomial distribution as mentioned in property **9** of the MD arises as a mixture distribution*; alternately, as an s-variate extension of the negative binomial distribution. The s-dimensional NMD with parameters k and $\mathbf{p}' = (p_1, p_2, \ldots, p_s)$ is defined by the joint pf

$$f_{X_1,X_2,\ldots,X_s}(x_1, x_2, \ldots, x_s) = \frac{\Gamma(k + \sum_{i=1}^s x_i)}{\Gamma(k)\prod_{i=1}^s x_i!}\, p_0^k \cdot \left(\textstyle\prod_{i=1}^s p_i^{x_i}\right), \qquad (2)$$

for $x_i = 0, 1, 2, \ldots$ $(i = 1, 2, \ldots, s)$, $0 < k < \infty$, $0 < p_i < 1$,

$$\sum_{i=1}^s p_i < 1, \qquad p_0 = 1 - \sum_{i=1}^s p_i.$$

Genesis

Let X_i $(i = 1, 2, \ldots, s)$ be independent random variables, X_i having the Poisson distribution with parameter $\beta\lambda_i$. If β has a gamma distribution* with parameters α and k, then the resulting mixture distribution of $(X_1 X_2, \ldots, X_s)$ is a negative multinomial with parameters k and $\mathbf{p}$, where $p_i = \lambda_i/[\alpha + \sum_{j=1}^s \lambda_j]$, $i = 1, 2, \ldots, s$. For further details, see Bates and Neyman [4] and Neyman [29].

Property 1. The probability generating function of the NMD is given by

$$G(t_1, t_2, \ldots, t_s) = p_0^k \left(1 - \sum_{i=1}^{s} t_i p_i\right)^k.$$

The joint factorial moments are

$$\mu_{(r_1, r_2, \ldots, r_s)} = E\left[\prod_{i=1}^{s} X_i^{(r_i)}\right]$$
$$= \left(k + \sum_{1}^{s} r_i - 1\right)^{(\Sigma r_i)} \prod_{i=1}^{s} \left(\frac{p_i}{p_0}\right)^{r_i}.$$

In particular,

$$E[X_i] = kp_i / p_0,$$
$$\operatorname{Var}(X_i) = kp_i(p_i + p_0)/p_0^2,$$
$$\operatorname{Cov}(X_i, X_j) = kp_i p_j / p_0^2.$$

For details regarding the cumulants of the NMD, see Wishart [54].

Property 2. The marginal distribution of any subset of $(X_1 X_2, \ldots, X_s)$ is again negative multinomial. In particular, the distribution of X_i is negative binomial. Therefore, the NMD is sometimes referred to as the *multivariate negative binomial distribution*.

Property 3. A special case of the negative multinomial distribution with parameters k and $\mathbf{p}$, k a positive integer, is known as the s-variate Pascal distribution. It arises as a waiting-time distribution. For details, see Sibuya et al. [43].

Property 4. If $X_1 X_2, \ldots, X_s$ have the NMD with parameters k and $\mathbf{p}$, where $\mathbf{p}$ itself is a random vector having the Dirichlet distribution, then the resulting mixture distribution is known as the *compound negative multinomial distribution*. Mosimann [27] found this compound distribution to be useful in the analysis of pollen data in paleoecological studies.

Property 5. See property 9 of the MD.

Property 6. The s-variate NMD with parameters k and $\mathbf{p}$ tends to the joint distribution of s independent Poisson variables with parameters $\boldsymbol{\lambda}' = (\lambda_1, \lambda_2, \ldots, \lambda_s)$ as $k \to \infty$ and $p_i \to 0$ such that $kp_i \to \lambda_i$, $i = 1, 2, \ldots, s$, $0 < \lambda_i < \infty$.

Property 7. The NMD arises as special case of each of: the multivariate power-series distribution and the sum-symmetric power-series distribution (see Patil [33]; the multivariate Pólya distribution (see Steyn [45]).

For more details regarding the NMD, see Sibuya et al. [43]. Expressions for the sums of NM probabilities have appeared in Khatri and Mitra [20]. Methods for obtaining the tail probabilities of the NMD are considered in Joshi [19].

ESTIMATION OF PARAMETERS OF NMD

Let $X_1, X_2, \ldots, X_s$ have the s-variate NMD with parameters k and $\mathbf{p}$. The estimation of functions of parameters $\mathbf{p}$ when k is known is discussed by Sibuya et al. [43] and Patil [32]. The maximum likelihood equations for k and $\mathbf{p}$ are given in Johnson and Kotz [17] and Sibuya et al. [43].

APPLICATIONS OF NMD

The use of NMD in the study of "accident proneness" is discussed in Bates and Neyman [4] and Neyman [29]. Other situations where the NMD is useful occur in inverse sampling*.

OTHER DISTRIBUTIONS RELATED TO NMD

The use of the compound NMD is mentioned in property 4 of the NMD. The bivariate negative binomial distribution which arises as the mixture of the bivariate Poisson distribution is considered by Wishart [54] and Wiid [51]. The s-variate extension of the bivariate negative binomial distribution is known as the multivariate negative binomial distribution; for details, see Steyn [45].

References

[1] Aitken, A. C. and Gonin, H. T. (1935). *Proc. R. Soc. Edinburgh*, **55**, 114–125.

[2] Alam, K. (1970). *Ann. Math. Statist.*, **41**, 315–317.

[3] Asano, C. (1965). *Ann. Inst. Statist. Math. Tokyo*, **17**, 1–13.

[4] Bates, G. E. and Neyman, J. (1952). *Univ. Calif. Publ. Statist.*, **1**, 215–253.

[5] Batschelet, E. (1960). *Biom. Zeit.*, **2**, 236–243.

[6] Bennett, R. W. (1962). *Aust. J. Statist.*, **4**, 86–88.

[7] Bhat, B. R. and Kulkarni, N. V. (1966). *J. R. Statist. Soc. B*, **28**, 45–52.

[8] Bol'shev, L. N. (1965). *Theor. Prob. Appl.*, **10**, 446–456.

[9] Capobianco, M. F. (1964). Ph.D. Thesis. Dept. of Mathematics, Polytechnic Institute of Brooklyn, Brooklyn, NY.

[10] Finucan, H. M. (1964). *Biometrika*, **51**, 513–517.

[11] Geppert, M. P. (1961). *Biom. Zeit.*, **3**, 55–67.

[12] Goodman, L. A. (1965). *Technometrics*, **7**, 247–254.

[13] Gyires, B. (1981). *Statistical Distributions in Scientific Work*, Vol. 4, C. Taillie, G. P. Patil, and B. A. Baldessari, eds. Reidel, Boston, pp. 231–242.

[14] Hoel, P. G. (1938). *Ann. Math. Statist.*, **9**, 158–165.

[15] Ishii, G. and Hayakawa, R. (1960). *Ann. Inst. Statist. Math. Tokyo*, **12**, 69–80.

[16] Janardan, K. G. (1974). *Scand. Actuarial J.*, **1**, 58–62.

[17] Johnson, N. L. and Kotz, S. (1969). *Discrete Distributions*. Wiley, New York.

[18] Johnson, N. L. and Young, D. H. (1960). *Biometrika*, **47**, 463–469.

[19] Joshi, S. W. (1974). *Ann. Inst. Statist. Math. Tokyo*, **27**, 95–97.

[20] Khatri, C. G. and Mitra, S. K. (1968). *Tech. Rep. 1/68*, Indian Statistical Institute, Calcutta, India.

[21] Lancaster, H. O. and Brown, T. A. I. (1965). *Aust. J. Statist.*, **7**, 40–44.

[22] Lehmann, E. L. (1966). *Ann. Math. Statist.*, **37**, 1137–1153.

[23] Lewontin, R. C. and Prout, T. (1956). *Biometrics*, **12**, 211–223.

[24] Mallows, C. L. (1968). *Biometrika*, **55**, 422–424.

[25] Marshall, A. W. and Olkin, I. (1979). *Inequalities: Theory of Majorization and Its Applications*. Academic Press, New York.

[26] Mosimann, J. E. (1962). *Biometrika*, **49**, 65–82.

[27] Mosimann, J. E. (1963). *Biometrika*, **50**, 65–82.

[28] Nevius, S. E., Proschan, F., and Sethuraman, J. (1977). *Ann. Statist.*, **5**, 263.

[29] Neyman, J. (1963). *Proceedings of the International Symposium on Discrete Distributions, Montreal*, pp. 1–14.

[30] Olkin, I. (1972). *Biometrika*, **59**, 303–307.

[31] Olkin, I. and Sobel, M. (1965). *Biometrika*, **52**, 167–179.

[32] Patil, G. P. (1965). *Sankhyā A*, **28**, 225–238.

[33] Patil, G. P. (1968). *Sankhyā B*, **30**, 335–336.

[34] Patil, G. P. and Bildikar, Sheela (1967). *J. Amer. Statist. Ass.*, **62**, 655–674.

[35] Pearson, K. (1900). *Philos. Mag., 5th Ser.*, **50**, 157–175.

[36] Perlman, M. D. and Rinott, Y. (1977). "On the Unbiasedness of Goodness of Fit Tests." Unpublished manuscript referred to ref. 25).

[37] Quesenberry, C. P. and Hurst, D. C. (1964). *Technometrics*, **6**, 191–195.

[38] Rao, C. R. (1957). *Sankhyā*, **18**, 139–148.

[39] Rao, C. R. and Srivastava, R. C. (1979). *Sankhyā Ser. A.*, **41**, 124–128.

[40] Ratnaparkhi, M. V. (1981). *Statistical Distributions in Scientific Work*, Vol. 4, C. Taillie, G. P. Patil, and B. A. Baldessari, eds. D. Reidel, Boston, 357–363.

[41] Rinott, Y. (1973). *Israel J. Math.*, **15**, 60–77.

[42] Rüst, H. (1965). *Zeit. Wahrscheinlichkeitsth. verw. Geb.*, **4**, 222–231.

[43] Sibuya, M., Yoshimura, I., and Shimizu, R. (1964). *Ann. Inst. Statist. Math. Tokyo*, **16**, 409–426.

[44] Smith, P. J., Rae, D. S., Manderscheid, R. W., and Silbergeld, S. (1981). *J. Amer. Statist. Ass.*, **76**, 737–740.

[45] Steyn, H. S. (1951). *Ned. Adad. Wet. Proc., Ser. A*, **54**, 23–30.

[46] Stoka, M. I. (1966). *Studii Cercetari Mat.*, **18**, 1281–1285.

[47] Studer, H. (1966). *Metrika*, **11**, 55–78.

[48] Tallis, A. M. (1962). *J. R. Statist. Soc. B*, **24**, 530–534.

[49] Tallis, A. M. (1964). *J. R. Statist. Soc. B*, **26**, 82–85.

[50] Vora, S. A. (1950). Ph.D. Thesis. University of North Carolina, Chapel Hill, NC.

[51] Wiid, A. J. B. (1957–58). *Proc. R. Soc. Edinburgh Sect. A*, 65, 29–34.

[52] Wise, M. E. (1963). *Biometrika*, **50**, 145–154.

[53] Wise, M. E. (1964). *Biometrika*, **51**, 277–281.

[54] Wishart, J. (1949). *Biometrika*, **36**, 47–58.

[55] Wong, C. K. and Yue, P. C. (1973). *Discrete Math.*, **6**, 391–398.

[56] Young, D. H. (1967). *Biometrika*, **54**, 312–314.

(BINOMIAL DISTRIBUTION
DIRICHLET DISTRIBUTION
MULTINOMIAL COEFFICIENTS
MULTIVARIATE DISTRIBUTIONS
MULTIVARIATE POWER-SERIES DISTRIBUTIONS
NEGATIVE BINOMIAL DISTRIBUTION
POISSON DISTRIBUTIONS)

M. V. RATNAPARKHI

MULTINOMIAL PROBIT: MULTINOMIAL LOGIT

The problem of modeling relationships between a categorical* response variable Y and a set of regressor variables $\mathbf{x}' = (x_1, x_2, \ldots, x_p)$ occurs in bioassay*, epidemiology, transportation theory, econometrics*, and in many other socioeconomic areas. When the dependent variable is binary*, both probit (*see* QUANTAL RESPONSE ANALYSIS) and logit* models have been found to be extremely useful. Multinomial* logit (MNL) and multinomial probit (MNP) models have been developed more recently to analyze polytomous* response variables. These qualitative response models can be classified into two distinct categories depending on whether the dependent variable has an ordered or unordered structure. The nature of the independent variables also offers many possible combinations.

ORDERED MODELS

We start with an example from bioassay introduced by Gurland et al. [10] to extend the binary probit model. Suppose a dosage x of an insecticide is administered to an insect and as a consequence the insect either dies, becomes moribund, or stays alive. The response variable Y is defined to take on values 0, 1, or 2 depending on whether the insect is alive, moribund, or dead. In order to construct a model relating the distribution of Y to the dosage level x, we assume the existence of an unobservable continuous random variable Z that measures the level of poisoning of the insect at dosage level x and two real numbers $\alpha_1 < \alpha_2$ such that $Y = 0$, 1, or 2 depending on whether $Z \in (-\infty, \alpha_1]$, $(\alpha_1, \alpha_2]$, or (α_2, ∞), respectively. We further assume that at dosage level x, Z is distributed normally with mean $\beta_0 + \beta_1 x$ and variance σ^2. Then the distribution of Y given by

$$\begin{aligned}
\Pr[Y = 0 \mid x] &= \Pr[Z \leqslant \alpha_1 \mid x] \\
&= \Phi(\gamma_1 - \delta x), \\
\Pr[Y = 1 \mid x] &= \Pr[\alpha_1 < Z \leqslant \alpha_2 \mid x] \\
&= \Phi(\gamma_2 - \delta x) - \Phi(\gamma_1 - \delta x), \\
\Pr[Y = 2 \mid x] &= \Pr[Z > \alpha_2] \\
&= 1 - \Phi(\gamma_2 - \delta x),
\end{aligned}$$

specifies the model, where $\gamma_1 = (\alpha_1 - \beta_0)/\sigma$, $\gamma_2 = (\alpha_2 - \beta_0)/\sigma$, $\delta = \beta_1/\sigma$, and Φ is the standard normal cumulative distribution function. Ordered models usually are applicable when the values of the qualitative dependent variable may be assumed to correspond to intervals of values of a latent continuous variable. In general, suppose the response variable Y takes values $y_1, y_2, \ldots, y_k$ on some scale, where $y_1 < y_2 \cdots < y_k$. We assume the existence of a latent continuous variable Z such that the event $[Y = y_i]$ is observed when $Z \in (\alpha_{i-1}, \alpha_i]$, $i = 1, 2, \ldots, k$; $\alpha_0 = -\infty$, $\alpha_k = \infty$, and $\alpha_0 < \alpha_1 < \alpha_2 \cdots < \alpha_k$. If the conditional distribution of $Z - \boldsymbol{\beta}'\mathbf{x}$ for a given vector of regressor variables $\mathbf{x}$ does not depend on $\mathbf{x}$, then

$$\begin{aligned}
&\Pr[Y = y_i \mid \mathbf{x}] \\
&\quad = \Pr[Z \in (\alpha_{i-1}, \alpha_1] \mid \mathbf{x}] \\
&\quad = \Psi(\alpha_i - \boldsymbol{\beta}'\mathbf{x}) - \Psi(\alpha_{i-1} - \boldsymbol{\beta}'\mathbf{x}),
\end{aligned}$$

where $\Psi(\cdot)$ may be chosen to be any appropriate cumulative distribution function as the distribution of $Z - \boldsymbol{\beta}'\mathbf{x}$ for given $\mathbf{x}$. A multinomial response model is called MNP or MNL of the ordered type depending on whether $\Psi(\cdot)$ is chosen to be normal or logistic*. The possibility of developing other models using alternative distribution functions is apparent.

UNORDERED MODELS

When the polytomous variable does not have an ordered structure, the existence of a scalar-valued latent variable becomes hard to justify. This may happen if multiple aspects of the response are used to classify an individual. Such problems are quite common in the analysis of economic choice variables and route choice problems in transportation studies. Unordered MNL models have been studied by Cox [5], McFadden [15], and Domencich and McFadden [7] while MNP models in the unordered setting have been investigated by Ashford and Sowden [4], Daganzo [6], and others.

When the response variable Y takes k distinct values $y_1, y_2, \dots, y_k$ and p regressor variables $\mathbf{x}' = (x_1, x_2, \dots, x_p)$, then the MNL model is specified by

$$\Pr[Y = y_i \mid \mathbf{x}] = \frac{\exp[\alpha_{0i} + \boldsymbol{\beta}_i'\mathbf{x}]}{\sum_{i=1}^{k}\exp[\alpha_{0i} + \boldsymbol{\beta}_i'\mathbf{x}]}, \quad i = 1, 2, \dots, k,$$

where $\alpha_{01} + \boldsymbol{\beta}_1'\mathbf{x}$ is assigned the value 0 for all $\mathbf{x}$ for identifiability* of the parameters.

McFadden [15] derived this model by maximizing stochastic utility functions associated with the categories of a multinomial response variable while a similar analysis was given by Marschak [14] for the binary logit case. Let $U_i(\mathbf{x})$ $i = 1, 2, \dots, k$ be the stochastic utility functions associated with the k states of the response variable for an individual with characteristics $\mathbf{x}$, and let $U_i(\mathbf{x}) = \mu_i(\mathbf{x}) + \epsilon_i$, where $\mu_i(\mathbf{x})$ is the nonstochastic component and ϵ_i is the random element distributed independently of $\mu_i(\mathbf{x})$. If we assume that

$$[Y = y_i \mid \mathbf{x}] \equiv \left[U_i(\mathbf{x}) = \max_{1 \leqslant j \leqslant k} U_j(\mathbf{x})\right]$$

and ϵ_i, $i = 1, 2, \dots, n$, are independently identically distributed random variables with $\Psi(u) = \exp[-\exp(-u)]$ as their distribution function, then

$$\Pr[Y = y_i \mid \mathbf{x}] = [e^{\mu_i(\mathbf{x})}] \Big/ \sum_{i=1}^{k} e^{\mu_i(\mathbf{x})}, \quad i = 1, 2, \dots, k.$$

If we further specify $\mu_1(\mathbf{x}) = 0$ and $\mu_i(\mathbf{x}) = \alpha_{0i} + \boldsymbol{\beta}_i'\mathbf{x}$ for $i \geqslant 2$, we obtain the MNL model. It is obvious that other specifications of $\mu_i(\mathbf{x})$ are possible. One deficiency of the model is that the ratio of the probabilities of any two categories does not depend on other categories. This property of the model is related to the assumption of the independence and the choice of the extreme value distribution* of the errors. To overcome this deficiency, McFadden has developed a variation of the model known as *generalized extreme value model* where a pair of errors has Gumbel's type-B bivariate distribution. The details of this model and further extensions of it have been discussed in Amemiya [1] and McFadden [16].

The MNP model may be similarly specified by the stochastic utility maximization method. If we assume that the errors have a normal multivariate distribution with mean $\mathbf{0}$ and variance-covariance matrix $\boldsymbol{\Sigma}$, then

$$\Pr[Y = y_i \mid \mathbf{x}] = \Pr[U_i(\mathbf{x}) > U_j(\mathbf{x}), \quad j = 1, 2, \dots, k, j \neq i \mid \mathbf{x}]$$

where $\mathbf{U}(\mathbf{x}) = [U_1(\mathbf{x}), \dots, U_k(\mathbf{x})]$ has a multivariate normal distribution with mean $(\mu_1(\mathbf{x}), \dots, \mu_k(\mathbf{x}))$ and variance covariance matrix $\boldsymbol{\Sigma}$.

The MNP model thus specified is complicated and the computational problems of estimating the parameters when there are more than three categories are quite formidable. Further details are discussed in Daganzo [6] and Manski and McFadden [16]. Note that binary logit* and probit models are special cases of MNP and MNL models in ordered as well as unordered cases, but the ordered multinomial models are quite distinct and unrelated to the unordered ones.

DISCRIMINANT ANALYSIS AND MNL MODEL

In discriminant analysis*, the objective is to classify an individual with characteristics $\mathbf{x}$ into one of k categories denoted by $y_1, y_2, \ldots, y_k$, where $\mathbf{x}$ is assumed to have a density f_i if the individual belongs to category y_i. Suppose the prior probability of an individual belonging to category y_i is π_i where $\pi_i > 0$ and $\sum_{i=1}^{k} \pi_i = 1$. In Bayesian discriminant analysis, the decision to classify an individual into one of the categories is based on using the posterior probabilities $\Pr[Y = y_i \mid \mathbf{x}]$, $i = 1, 2, \ldots, k$ and a loss matrix, describing consequences of making wrong decisions (*see* BAYESIAN INFERENCE). Although the objective of the discriminant analysis is quite different from that of the MNL models, it is interesting to note that if we assume that the f_i's are normal density functions with mean vector $\mathbf{0}$ and a common variance-covariance matrix $\boldsymbol{\Sigma}$, then $\Pr[Y = y_i \mid \mathbf{x}]$ has precisely the same structure as in the MNL model.

Amemiya [1] has given an excellent review of qualitative response models in which he has also discussed the possible use of the discriminant analysis formulation of the MNL model.

MULTIVARIATE MODEL

A multivariate model with discrete dependent variables may be treated as a univariate model with the number of categories at most equal to the product of the number of categories for the individual variables. The specification of the probability of the categories usually takes into account the underlying multivariate structure. If the individual variables are ordered and latent random variables are assumed to have a multivariate normal distribution, then the model may be specified as an ordered multivariate MNP model. Such a bivariate probit model with binary response variables was used by Ashford and Sowden [4]. But a similar model using the multivariate logistic distribution is not useful, as this implies that the correlation coefficient between any pair of latent random variables is $\frac{1}{2}$ [12]. Because of this, MNL models of the unordered type usually are formulated in the multivariate case.

MNL models are also used in the analysis of contingency tables* and are very closely related to log-linear models. For an $r \times s$ contingency table, let p_{ij} denote the probability for the ith row, jth column of the table. If we assume $p_{ij} = \exp(m_{ij})/d$, where $d = \sum_{i=1}^{r} \sum_{j=1}^{s} \exp(m_{ij})$, and choose $m_{11} = 0$ for identifiability, then $\log[p_{ij} \mid p_{11}] = m_{ij}$. An extensive literature exists on the parametrization of m_{ij} into suitable components to denote different effects and the considerable arsenal of methodology developed for analyzing log-linear models may be used to allow simultaneous analysis of all these logits and estimation of the parameters. Haberman [11], Fienberg [9], and Nerlove and Press [17] have discussed the interrelationship between logit and log-linear models in two-dimensional and multidimensional tables with combinations of ordered as well as unordered structures of the underlying latent variables.

PARAMETER ESTIMATION

Suppose observations are grouped into k categories corresponding to values of regression vectors $\mathbf{x} = (x_1, x_2, \ldots, x_p)$. Then the likelihood function is of the product multinomial form for MNP as well as MNL models and the parameters are usually estimated by maximizing the likelihood function, although minimum χ^2 methods* may also be used when the number of observations per cell is reasonably large. In all these cases, explicit estimators cannot be derived, as the estimating equations are nonlinear functions of the parameters. Instead iterative schemes using the Newton–Raphson method* or other numerical methods for solving nonlinear equations are usually employed. Convergence is usually rapid, but infinite estimators are possible. In MNL models of the unordered type, Cox [5] has suggested the use of weighted regression* methods. The asymptotic properties of maximum likelihood, minimum X^2 as well as regression estimators are, in general, well understood, as they belong to the family to BAN estimators (*see* LARGE-SAMPLE THEORY), but small sample properties of these estimators, especially in nonlinear situations, are not well understood. Anderson [2] and Anderson and Blair [3] have discussed estimation procedures for the MNL model under an alternate sampling scheme called "separate sampling" where **x**'s are sampled for given values of the categorical response variable. The maximum likelihood method using the discriminant analysis formulation of MNL models has also been used by Efron [8] and others to estimate the parameters.

Daganzo [6] and Manski and McFadden [13] have reported computational methods related to estimation problems in MNP models. But a satisfactory computational procedure still needs to be developed.

For logistic regression* models, computer software packages are available in SAS, BMDP, and GLIM*. Details of these programs with their respective capabilities have been discussed in Wijesinka et al. [18].

References

[1] Amemiya, T. (1981). *J. Econ. Lit.*, **19**, 1483–1536.

[2] Anderson, J. A. (1972). *Biometrika*, **59**, 19–35.

[3] Anderson, J. A. and Blair, V. (1982). *Biometrika*, **69**, 123–136.

[4] Ashford, J. R. and Sowden, R. R. (1970). *Biometrics*, **26**, 535–546.

[5] Cox, D. R. (1970). *The Analysis of Binary Data*. Methuen, London.

[6] Daganzo, C. (1979). *Multinomial Probit—The Theory and Its Application to Demand Forecasting*. Academic Press, New York.

[7] Domencich, T. A. and McFadden, D. (1975). *Urban Travel Demand—A Behavioral Analysis*. North-Holland, New York.

[8] Efron, B. (1975). *J. Amer. Statist. Ass.*, **70**, 892–898.

[9] Fienberg, S. F. (1977). *The Analysis of Cross-Classified Categorical Data*. MIT Press, Cambridge, MA.

[10] Gurland, J., Lee, I., and Dahm, P. A. (1960). *Biometrics*, **16**, 382–397.

[11] Haberman, S. J. (1979). *Analysis of Qualitative Data*, Vol. 2. Academic Press, New York.

[12] Johnson, N. L. and Kotz, S. (1972). *Distributions in Statistics: Continuous Multivariate Distributions*. Wiley, New York.

[13] Manski, C. F. and McFadden, D. (1981). *Structural Analysis of Discrete Data*. MIT Press, Cambridge, MA.

[14] Marschak, J. (1960). *Stanford Symposium on Mathematical Methods in Social Sciences*, K. Arrow, ed. Stanford University Press, Stanford, CA, pp. 319–339.

[15] McFadden, D. (1974). *Frontiers in Econometrics*, P. Zarembka, ed. Academic Press, New York. pp. 105–142.

[16] McFadden, D. (1981). *Structural Analysis of Discrete Data*, C. F. Manski and D. McFadden, eds. MIT Press, Cambridge, MA.

[17] Nerlove, M. and Press, S. J. (1973). *Rep. No. R1306-EDA/NIH*, Rand Corporation, Santa Monica, CA.

[18] Wijesinka, A., Begg, C. B., and McNeil, B. J. (1982). *Tech. Rep. No. 2622*, Dept. of Biostatistics, Sidney Farber Cancer Institute, Boston.

Bibliography

Finney, D. J. (1971). *Probit Analysis*. Cambridge University Press, Cambridge, England.

Plackett, R. L. (1974). *The Analysis of Categorical Data*. Griffin, London.

(BIOASSAY, STATISTICAL METHODS IN
ECONOMETRICS
QUANTAL RESPONSE ANALYSIS)

B. B. BHATTACHARYYA

MULTINOMIAL SAMPLING *See* NATURALISTIC SAMPLING

MULTINORMAL DISTRIBUTIONS

INTRODUCTION

The historical developments leading to a definition of the multivariate normal distribution are discussed by Hilary Seal [35]. The earliest attempts are credited to Bravais [6], and Schols [34]. Francis Galton* [12], analyzing the correlation* in bivariate data, considered the structure of a bivariate normal* density function. Assuming contours of equal density to be concentric ellipses, he developed, with the help of J. D. H. Dickson, the formula for the probability density function in a form we use to this day. The original diagram from Galton's paper is reproduced in the article CORRELATION. It was left to Edgeworth [11] to attempt a four- and higher-dimensional extension of the normal distribution. However, Karl Pearson* [30] appears to have been the first to introduce the "modern" form of the multivariate normal density function.

The present-day multivariate analyst considers the distribution from a variety of perspectives. So far, no single definition can be given to accommodate these different views. In the univariate case, the random variable Z with $E(Z) = 0$, $\mathrm{Var}(Z) = 1$ has the probability density function $(2\pi)^{-1/2}\exp(-z^2/2)$, $-\infty < z < \infty$. The multidimensional extension is given by considering $Z_1, \ldots, Z_p$, which are independent $N(0, 1)$ random variables. Then their joint probability density function (pdf) can be written as

$$(2\pi)^{-p/2}\exp(-\mathbf{z}'\mathbf{z}/2), \qquad \mathbf{z} \in R^p.$$

In this case, we write $\mathbf{Z} \sim N_p(\mathbf{0}, \mathbf{I})$. This can be generalized using the following definition.

Definition 1. A p-dimensional random vector $\mathbf{X}$ is said to have a nonsingular p-dimensional multivariate normal distribution if the joint pdf of the elements of $\mathbf{x}$ is

$$\begin{aligned} f(x_1, \ldots, x_p) &= (2\pi)^{-p/2}|\mathbf{\Sigma}|^{-1/2} \\ &\quad \times \exp\{-(\mathbf{x}-\boldsymbol{\mu})'\mathbf{\Sigma}^{-1}(\mathbf{x}-\boldsymbol{\mu})/2\}, \\ &\qquad \mathbf{x} \in R^p. \end{aligned}$$

This is denoted by $\mathbf{X} \sim N_p(\boldsymbol{\mu}, \mathbf{\Sigma})$.

When $\mathbf{\Sigma}$ is of rank $r < p$, the vector $\mathbf{X}$ has the *singular* normal distribution. It is possible to transform $\mathbf{X}$ to $\mathbf{Y} = \mathbf{AX} + \mathbf{b}$, with $\mathbf{A}$ an $r \times p$ matrix of rank r; then $\mathbf{Y}$ has the *nonsingular* multivariate normal distribution in r-dimensions. This leads to the following definition.

Definition 2. (Srivastava and Khatri [37]) A p-dimensional random vector $\mathbf{X}$ is said to have the multivariate normal distribution $N_p(\boldsymbol{\mu}, \mathbf{\Sigma})$ if $\mathbf{X}$ has the same distribution as $\mathbf{Y} = \boldsymbol{\mu} + \mathbf{DZ}$, where $\mathbf{D}$ is a $p \times r$ matrix of rank r, $\mathbf{\Sigma} = \mathbf{DD}'$, and $\mathbf{Z} \sim N_r(\mathbf{0}, \mathbf{I})$. Here r is called the *rank* of the distribution of $\mathbf{X}$. In what follows the rank of $\mathbf{\Sigma}$ is taken to be p.

The distribution is also defined by using one of its main characteristic properties (see property **4** in the Characterizations section).

ELEMENTARY PROPERTIES

Most of the basic properties are derivable from the moment-generating function* (mgf) $\exp[\mathbf{t}'\boldsymbol{\mu} + \mathbf{t}'\mathbf{\Sigma}\mathbf{t}/2]$. From this it follows that $E(\mathbf{X}) = \boldsymbol{\mu}$, $\mathrm{Var}(\mathbf{X}) = \mathbf{\Sigma}$. If $\mathbf{\Sigma}$ is positive definite, then there exists a nonsingular transformation that standardizes $\mathbf{X}$ to $N_p(\mathbf{0}, \mathbf{I})$. From the mgf it can be seen that all the third moments around $\boldsymbol{\mu}$ are zero. The fourth moment is

$$\begin{aligned} E\{(X_i - \mu_i)(X_j - \mu_j)(X_k - \mu_k)(X_l - \mu_l)\} \\ = \sigma_{ij}\sigma_{kl} + \sigma_{ik}\sigma_{jl} + \sigma_{il}\sigma_{jk}, \end{aligned}$$

where σ_{ij} is the covariance between X_i and X_j. Several other results follow directly from the mgf:

1. If $\mathbf{Y} = \mathbf{AX} + \mathbf{b}$, $\mathbf{A}$ $(r \times p)$, $\mathbf{b}$ $(r \times 1)$ being constants, then $\mathbf{Y} \sim N_r(\mathbf{A}\boldsymbol{\mu} + \mathbf{b}, \mathbf{A\Sigma A}')$.

2. If $\mathbf{X}$ is partitioned into subvectors $\mathbf{X}_1$ $(q \times 1)$, $\mathbf{X}_2$ $[(p-q) \times 1]$, then, defining partitions of $\boldsymbol{\mu}$, $\mathbf{\Sigma}$ appropriately, it is possible to see that the marginal distribution of $\mathbf{X}_1$ is $N_q(\boldsymbol{\mu}_1, \mathbf{\Sigma}_{11})$, and of $\mathbf{X}_2$ is $N_{p-q}(\boldsymbol{\mu}_2, \mathbf{\Sigma}_{22})$. In particular, it follows that each element of $\mathbf{X}$ has a univariate normal distribution. Note here that the marginal normality of the elements of $\mathbf{X}$ does not ensure their joint normality. Consider, for example, for $p = 2$, $f(x_1, x_2) = \frac{1}{2}[\phi_1(x_1, x_2) + \phi_2(x_1, x_2)]$ with ϕ_i being a standard bivariate normal density* function having the correlation coefficient ρ_i. Each of the marginals is a univariate normal distribution, while $f(x_1, x_2)$ is not the bivariate normal density function.

3. The random vectors $\mathbf{X}_1, \mathbf{X}_2$, in the normal case, are independent if and only if their covariance matrix* $\mathbf{\Sigma}_{12}$ is zero.

4. If $\mathbf{Y}_i$ are independent $N_p(\boldsymbol{\mu}_i, \mathbf{\Sigma}_i)$, for $i = 1, 2$, then $\mathbf{Y}_1 + \mathbf{Y}_2$ is $N_p(\boldsymbol{\mu}_1 + \boldsymbol{\mu}_2, \mathbf{\Sigma}_1 + \mathbf{\Sigma}_2)$.

The exponent $(\mathbf{X} - \boldsymbol{\mu})'\mathbf{\Sigma}^{-1}(\mathbf{X} - \boldsymbol{\mu})$ in Definition **1** has the χ^2 distribution* with p degrees of freedom.

CONDITIONAL DISTRIBUTIONS

If $\mathbf{X}$ is partitioned $\mathbf{X}_1$ $(q \times 1)$, $\mathbf{X}_2$ $[(p-q) \times 1]$, then the conditional distribution of $\mathbf{X}_1$ given $\mathbf{X}_2 = \mathbf{x}_2$ is $N_q[\boldsymbol{\mu}_1 + \mathbf{\Sigma}_{12}\mathbf{\Sigma}_{22}^{-1}(\mathbf{x}_2 - \boldsymbol{\mu}_2), \mathbf{\Sigma}_{11} - \mathbf{\Sigma}_{12}\mathbf{\Sigma}_{22}^{-1}\mathbf{\Sigma}_{21}]$; a similar result holds for the conditional distribution of $\mathbf{X}_2$ given $\mathbf{X}_1 = \mathbf{x}_1$. From the conditional distribution, it follows that the regression* of $\mathbf{X}_1$ on $\mathbf{X}_2$ is linear. The conditional covariance matrix, on the other hand, is constant (homoscedastic). The matrix $\mathbf{\Sigma}_{12}\mathbf{\Sigma}_{22}^{-1}$ is the *matrix of partial regression coefficients* of $\mathbf{X}_1$ on $\mathbf{X}_2$. If $q = 1$, then $E(X_1 \mid \mathbf{X}_2 = \mathbf{x}_2)$ is of the form $\alpha + \sum_{i=2}^{p} \beta_i x_i$, the so-called general linear model*. In this case, the scalar measure of correlation between X_1 and $\mathbf{X}_2$ is the *multiple correlation* coefficient*; its value is given by $R = (\boldsymbol{\sigma}_{12}'\mathbf{\Sigma}_{22}^{-1}\boldsymbol{\sigma}_{21})^{1/2}/\sigma_{11}^{1/2}$.

In the conditional distribution of $\mathbf{X}_1$ given $\mathbf{X}_2 = \mathbf{x}_2$, the coefficient of correlation between any two elements of $\mathbf{X}_1$ is the *partial correlation coefficient**. On the other hand, the nonzero characteristic roots* of the matrix $\mathbf{\Sigma}_{11}^{-1}\mathbf{\Sigma}_{12}\mathbf{\Sigma}_{22}^{-1}\mathbf{\Sigma}_{21}$ are the *canonical correlations* (*see* CANONICAL ANALYSIS) between $\mathbf{X}_1$ and $\mathbf{X}_2$. In all these cases, the correlations are measures, in some sense, of the association* between the sets of random variables under consideration.

PROBABILITY INEQUALITIES

Probability inequalities involving the multivariate normal distribution have been found to be useful in inferential problems. Tong [42] has discussed such inequalities in some detail. If $\mathbf{X} \sim N_p(\mathbf{0}, \mathbf{\Sigma})$, $\sigma_{ij} = l_i l_j (\sigma_{ii}\sigma_{jj})^{1/2}$, $l_i^2 \leqslant 1$, Sidak [36] and Khatri [25] have shown that for any set of positive constants $c_1, c_2, \dots, c_p$,

$$P\{|X_i| \geqslant c_i,\, i = 1, 2, \dots, p\} \geqslant \prod_{i=1}^{p} P\{|X_i| \geqslant c_i\},$$

$$P\{|X_i| \leqslant c_i,\, i = 1, 2, \dots, p\} \leqslant \prod_{i=1}^{p} P\{|X_i| \leqslant c_i\}.$$

If $\mathbf{X} \sim N_p(\mathbf{0}, \mathbf{\Sigma})$, and writing $\Phi[\mathbf{y} \mid \mathbf{\Sigma}] = P\{X_i \leqslant y_i,\ i = 1, 2, \dots, p \mid \mathbf{\Sigma}\}$, it is possible to prove (Srivastava and Khatri [37]) that:

1. If $\mathbf{\Sigma}^*, \mathbf{\Sigma}$ are two positive definite matrices with $\sigma_{ii}^* = \sigma_{ii}$, $\sigma_{ij}^* \leqslant \sigma_{ij}$, then $\Phi[\mathbf{y} \mid \mathbf{\Sigma}^*] \leqslant \Phi[\mathbf{y} \mid \mathbf{\Sigma}]$ for all $\mathbf{y}$;
2. If $\rho_{ij} = \rho$, for all $i < j$ in $\mathbf{\Sigma}$, where ρ_{ij} is the correlation coefficient between X_i and X_j, then $\Phi(\mathbf{y} \mid \rho)$ is a monotonic increasing function of ρ and $\Phi(\mathbf{y} \mid \rho)$ is greater than, equal to or less than $\prod_{i=1}^{p} P\{X_i < y_i\}$ according as ρ is positive, zero or negative;
3. If $\mathbf{\Sigma}$ has $\sigma_{ij} \geqslant 0$ for all $i < j$, then $\Phi(\mathbf{y} \mid \mathbf{\Sigma}) \geqslant \prod_{i=1}^{p} P\{X_i \leqslant y_i\}$. Further extensions to these and other results are to be found in Das Gupta et al. [9].

EVALUATION OF MULTIVARIATE PROBABILITIES

Let $\mathbf{X} \sim N_p(\mathbf{0}, \mathbf{R})$, $\mathbf{R}$ being the correlation matrix*. The usual problem considered is the determination of $P\{X_1 \geqslant h_1, X_2 \geqslant h_2, \dots, X_p \geqslant h_p\}$. Kendall [24] has represented this probability as

$$\sum{}^* \prod{}^* \left\{ \frac{\rho_{ij}^{n_{ij}}}{n_{ij}!} \right\} \prod_{i=1}^{p} H_{n_i - 1}(h_i)\phi(h_i),$$

the so-called called tetrachoric series*. Here $\sum^*$ is a multiple summation over n_{ij}, $\prod^*$ a multiple product over $1 \leqslant i \leqslant j \leqslant p$; $H_n(x)$

is the Chebyshev–Hermite polynomial* of nth degree, and $\phi(x)$ is the standard normal density function. Convergence problems associated with this and other related series representations of this probability are examined by Harris and Soms [19].

A more interesting problem is the determination of the *orthant probability** $\Phi[\mathbf{0} \mid \mathbf{R}]$, given by the integral

$$(2\pi)^{-p/2}|\mathbf{R}|^{-1/2} \times \int_0^\infty \cdots \int_0^\infty \exp\left(\frac{-\mathbf{x}'\mathbf{R}^{-1}\mathbf{x}}{2} \right) d\mathbf{x}.$$

The following special cases should be mentioned:

1. If p is odd, then
$$2\Phi[\mathbf{0} \mid \mathbf{R}] = 1 - \sum P(E_i) + \sum_{i<j} P(E_i E_j) - \cdots,$$
where E_i is the event $\{X_i \geqslant 0\}$. If $p = 3$, then $\Phi[\mathbf{0} \mid \mathbf{R}]$ reduces to
$$\tfrac{1}{2} - \tfrac{1}{4}\pi^{-1}(\cos^{-1}\rho_{12} + \cos^{-1}\rho_{13} + \cos^{-1}\rho_{23})$$
(see David [10]).
2. If $\rho_{ij} = \rho$, then
$$2L_p(\rho) = 1 - \frac{p}{2} + \binom{p}{2} L_2(\rho) - \cdots + \binom{p}{p-1} L_{p-1}(\rho),$$
where $L_p(\rho) = \Phi[\mathbf{0} \mid (1-\rho)\mathbf{I} + \rho \mathbf{1}\mathbf{1}']$.
3. Steck [38] has expressed $\Phi[\mathbf{h} \mid \mathbf{R})$ when $p = 3$ in terms of probabilities involving independent univariate normal variables.
4. If $p = 4$, Moran [27] has given
$$\Phi(\mathbf{0} \mid \mathbf{R}) = \frac{1}{16} + \frac{S_1}{8\pi} + \frac{S_2}{4\pi^2} + \cdots,$$
where $S_1, S_2, \ldots$ are sums of products of the correlations in $\mathbf{R}$. For the case of $\rho_{12} = \rho_{34} = \alpha$, $\rho_{13} = \rho_{24} = \beta$, and $\rho_{14} = \rho_{23} = \alpha\beta$ with $|\alpha| \leqslant 1$, $|\beta| \leqslant 1$, Cheng [7] has given a simple representation for the orthant probability as
$$\tfrac{1}{16} + \tfrac{1}{4}\pi^{-1}\left[\sin^{-1}\alpha + \sin^{-1}\beta + \sin^{-1}(\alpha\beta)\right] + \tfrac{1}{4}\pi^{-2}\left[(\sin^{-1}\alpha)^2 + (\sin^{-1}\beta)^2 - (\sin^{-1}\alpha\beta)^2\right].$$

Johnson and Kotz [22, Chap. 35] have a detailed discussion of these results.

TRUNCATED DISTRIBUTIONS

Obviously, the truncation in a multidimensional distribution can be done in several ways. Essentially, what one examines is the (conditional) distribution of $\mathbf{X}_2$ given $\mathbf{a} < \mathbf{X}_1 < \mathbf{b}$. General results are complicated in the multidimensional case. Truncation* of the type $\sum_{j=1}^{p} a_j x_j \geqslant d$ can be reduced to the same form by appropriately transforming the variables. Other truncations, considered in the literature, include the "elliptical" truncation due to Tallis [41]. Here the random variables in $\mathbf{X}$ are constrained to lie inside the ellipse $a < \mathbf{X}'\mathbf{R}^{-1}\mathbf{X} \leqslant b$. In the standard bivariate normal distribution*:

> If X_1 is truncated at h, then $f(x_1, x_2) = \phi(x_1, x_2)/[1 - \Phi(h)]$. Hence the correlation in the truncated distribution is given by $\rho_T = \rho[\rho^2 + (1-\rho^2)/\mathrm{Var}(X_1)]^{-1/2}$.
>
> If X_1 is truncated at a and X_2 at b, both on the left, then $f(x_1 x_2) = \phi(x_1, x_2)/Q(a, b)$, where $Q(a,b) = P\{X_1 \geqslant a,\ X_2 \geqslant b\}$. In this case, Gajjar and Subrahmaniam [13] have given the expression for ρ_T. Table 1 summarizes the effects of the degree of truncation and ρ on $Q(a,b)$.

Problems of estimation in the bivariate case have been presented in the literature. (See Johnson and Kotz [22, Chap. 36]). A study of the behavior of the sample correlation coefficient and its transforms in the truncated bivariate normal distribution has been presented by Subrahmaniam and Gajjar [39].

Table 1 Q(a, b) for Various Values of (a, b) and ρ

		ρ					
a	b	− 0.90	− 0.50	− 0.05	0.05	0.50	0.90
0.5	0.5	0.000	0.036	0.089	0.101	0.163	0.245
0.0	0.5	0.011	0.081	0.147	0.161	0.226	0.296
− 0.5	0.5	0.063	0.146	0.207	0.219	0.272	0.307
0.0	0.0	0.071	0.166	0.242	0.257	0.333	0.428
− 0.5	0.0	0.203	0.273	0.338	0.352	0.418	0.488
− 0.5	− 0.5	0.383	0.419	0.471	0.484	0.546	0.628
− 1.5	0.5	0.242	0.261	0.285	0.290	0.305	0.308
− 1.5	0.0	0.433	0.442	0.464	0.469	0.490	0.499
− 1.5	− 0.5	0.624	0.627	0.643	0.647	0.671	0.691
− 1.5	− 1.5	0.866	0.866	0.870	0.871	0.884	0.910
− 2.5	− 2.5	0.987	0.987	0.987	0.987	0.988	0.990

CHARACTERIZATIONS

Several characterizations available for the univariate normal distribution have analogs in the multinormal case. However, the proofs as well as the conditions under which such analogs hold are more complicated. Some important results are:

Property 1. The independence of the sample mean $\bar{\mathbf{X}}$ and variance matrix $\mathbf{S}$ characterizes the multinormal distribution.

Property 2. If $\mathbf{X}_1, \mathbf{X}_2$ are independent p-vectors, then their sum is multinormal if and only if each of the vectors $\mathbf{X}_1, \mathbf{X}_2$ is multinormal.

Property 3. Ghurye and Olkin [14] have generalized the Darmois–Skitovich theorem*: Consider $\mathbf{X}_1, \dots, \mathbf{X}_n$, n independent p-dimensional random vectors, and let $\mathbf{A}_1, \dots, \mathbf{A}_n, \mathbf{B}_1, \dots, \mathbf{B}_n$ be $p \times p$ nonsingular matrices. If $\mathbf{W}_1 = \sum_{i=1}^{n} \mathbf{A}_i \mathbf{X}_i$, $\mathbf{W}_2 = \sum_{i=1}^{n} \mathbf{B}_i \mathbf{X}_i$ are independent, then the $\mathbf{X}_i$ are normally distributed. It should be noted that if $\mathbf{A}_i$ (or $\mathbf{B}_i$) is zero, then $\mathbf{X}_i$ can be arbitrary. On the other hand, if $\mathbf{A}_i$ is singular, then the corresponding vector $\mathbf{X}_i$ is partly normal.

Property 4. The most important characterization is that $\mathbf{X}$ is multinormal if and only if every linear combination of the elements of $\mathbf{X}$ is univariate normal. This characterization is used by some authors to define a multivariate normal distribution. (See, e.g., Rao [33, Chap. 8].)

Kagan et al. [23] have considered other characterizations of the multinormal distribution.

ESTIMATION

Let $\mathbf{X}_1, \mathbf{X}_2, \dots, \mathbf{X}_N$ be a random sample of size N from $N_p(\boldsymbol{\mu}, \boldsymbol{\Sigma})$ with $N > p$. Then the maximum likelihood* estimators of $\boldsymbol{\mu}$ and $\boldsymbol{\Sigma}$ are given by

$$\bar{\mathbf{X}} = \frac{1}{N} \sum_{i=1}^{N} \mathbf{X}_i, \qquad \hat{\boldsymbol{\Sigma}} = \frac{1}{N} \mathbf{A}$$

with

$$\mathbf{A} = \sum_{i=1}^{N} (\mathbf{X}_i - \bar{\mathbf{X}})(\mathbf{X}_i - \bar{\mathbf{X}})'.$$

The estimator for $\boldsymbol{\Sigma}$ can easily be modified to obtain an unbiased estimator $\mathbf{S} = \mathbf{A}/n$, where $n = N - 1$.

Although these estimators are easily obtained and their properties are well established, they are not optimal in a decision-theoretic sense since they are inadmissible (*see* ADMISSIBILITY). Based on a sum of squared errors loss function*

$$L(\boldsymbol{\mu}, \hat{\boldsymbol{\mu}}, \boldsymbol{\Sigma}) = (\boldsymbol{\mu} - \hat{\boldsymbol{\mu}})' \boldsymbol{\Sigma}^{-1} (\boldsymbol{\mu} - \hat{\boldsymbol{\mu}}),$$

James and Stein [21] showed that the estimator

$$\hat{\boldsymbol{\mu}}^* = (1 - c/\overline{\mathbf{X}}'\mathbf{S}^{-1}\overline{\mathbf{X}})\overline{\mathbf{X}} \qquad \text{with}$$

$$c = (p-2)/[N-(p-2)]$$

has a smaller expected loss (risk*) than $\overline{\mathbf{X}}$ for $p \geqslant 3$; hence $\overline{\mathbf{X}}$ is not admissible for $p \geqslant 3$. Similar types of estimators for $\boldsymbol{\Sigma}$ have been developed by James and Stein [21] and more recently by Olkin and Selliah [29] and Haff [18]. A general discussion of decision-theoretic estimation (including Bayes, empirical Bayes* and minimax*) for the parameters of the multinormal distribution can be found in Muirhead [28, Chap. 4] and Berger [4].

When sets of observations from multinormal data are missing or incomplete, special estimation problems arise. In the bivariate case, suppose the incomplete sample is $(x_1, x_2, \ldots, x_n, x_{n+1}, \ldots, x_N)$ and $(y_1, y_2, \ldots, y_n)$ with mean vector (μ_1, μ_2), common variance σ^2, and correlation coefficient ρ. The maximum likelihood estimators can be obtained by writing the likelihood function as a product of the likelihood of the x's with the conditional likelihood function of the y's given the x's. The estimators are then the solution to the following four equations [8]:

$$\hat{\mu}_1 = \bar{x}^*, \qquad \hat{\mu}_2 = \bar{y} - \hat{\rho}(\bar{x} - \bar{x}^*),$$

$$\hat{\rho} = \frac{S_{12}}{N\hat{\sigma}^2 - (S_1^{*2} - S_1^2)},$$

$$\hat{\sigma}^2 = \left(S_1^{*2} - S_1^2 + \frac{S_1^2 + S_2^2 - 2\hat{\rho}S_{12}}{1 - \hat{\rho}^2}\right)(N+n)^{-1},$$

with

$$\bar{x}^* = \sum_{i=1}^{N} \frac{x_i}{N}, \quad \bar{x} = \sum_{i=1}^{n} \frac{x_i}{n}, \quad \bar{y} = \sum_{i=1}^{n} \frac{y_i}{n},$$

$$S_1^2 = \sum_{i=1}^{n} (x_i - \bar{x})^2, \quad S_2^2 = \sum_{i=1}^{n} (y_i - \bar{y})^2,$$

$$S_1^{*2} = \sum_{i=1}^{N} (x_i - \bar{x}^*)^2,$$

$$S_{12} = \sum_{i=1}^{n} (x_i - \bar{x})(y_i - \bar{y}).$$

The authors show that in $[-1, 1]$ there is exactly one root having the same sign as S_{12}, which is the solution to the cubic equation

$$f(\hat{\rho}) = n(S_1^{*2} - S_1^2)\hat{\rho}^3 - (N-n)S_{12}\hat{\rho}^2 + [N(S_1^2 + S_2^2) - n(S_1^{*2} - S_2^{*2})]\hat{\rho} - (N+n)S_{12} = 0.$$

This real root is the unique MLE of ρ.

Anderson [1] has considered a similar pattern of missing data and obtained the MLEs with $\sigma_1^2 \neq \sigma_2^2$.

Hocking and Smith [20] have developed estimation procedures for more general situations of estimating parameters for incomplete data* from the p-variate multinormal distributions.

TESTS OF HYPOTHESES AND CONFIDENCE REGIONS

The following results are useful in deriving the sampling distributions for statistics related to the multivariate normal:

Property 1. If $\mathbf{Z}$ is $N_p(\mathbf{0}, \boldsymbol{\Sigma})$, then the quadratic form $\mathbf{Z}'\boldsymbol{\Sigma}^{-1}\mathbf{Z}$ is distributed as χ^2 with p degrees of freedom.

Property 2. If the $p \times p$ positive definite matrix $\mathbf{A}$ can be written as $\sum_{\alpha=1}^{m} \mathbf{Z}_\alpha \mathbf{Z}'_\alpha$, where $\mathbf{Z}_1, \ldots, \mathbf{Z}_{(m)}$ are independent $N_p(\mathbf{0}, \boldsymbol{\Sigma})$, then the distribution of the elements in $\mathbf{A}$ is the *Wishart distribution** with m degrees of freedom and covariance matrix $\boldsymbol{\Sigma}$. This is written as $\mathbf{A} \sim W_p(m, \boldsymbol{\Sigma})$ with the subscript denoting the dimension of $\mathbf{A}$.

Property 3. If $\mathbf{Z} \sim N_p(\mathbf{0}, \boldsymbol{\Sigma})$ and $\mathbf{A} \sim W_p(m, \boldsymbol{\Sigma})$ with $\mathbf{Z}$ and $\mathbf{A}$ independently distributed, then $\mathbf{Z}'(\mathbf{A}/m)^{-1}\mathbf{Z}$ is said to have *Hotelling's* T_m^2 *distribution* with m degrees of freedom (*see* HOTELLING'S T^2).

Analogous to the independence of the sample mean and variance in the univariate normal case, here $\overline{\mathbf{X}}$ and $\mathbf{S}$ are independently distributed with $\overline{\mathbf{X}} \sim N_p(\boldsymbol{\mu}, \boldsymbol{\Sigma}/N)$ and $\mathbf{S} \sim W_p(n, \boldsymbol{\Sigma}/n)$.

With the covariance matrix $\boldsymbol{\Sigma}$ known, property **1** can be used to show that $\sqrt{N}(\overline{\mathbf{X}} - \boldsymbol{\mu})$ is $N_p(\mathbf{0}, \boldsymbol{\Sigma})$; hence $N(\overline{\mathbf{X}} - \boldsymbol{\mu})'\boldsymbol{\Sigma}^{-1}(\overline{\mathbf{X}} - \boldsymbol{\mu}) \sim \chi^2$ with p degrees of freedom. From this result, tests of hypotheses and confidence regions for $\boldsymbol{\mu}$ can be developed. To test $H_0: \boldsymbol{\mu} = \boldsymbol{\mu}_0$, we use the critical region

$$N(\overline{\mathbf{X}} - \boldsymbol{\mu}_0)'\boldsymbol{\Sigma}^{-1}(\overline{\mathbf{X}} - \boldsymbol{\mu}_0) \geqslant \chi^2_{p,\alpha},$$

where $\chi^2_{p,\alpha}$ denotes the upper $100\alpha\%$ point of a χ^2 distribution with p degrees of freedom. Based on $\overline{\mathbf{X}}$, a $100(1-\alpha)\%$ confidence region for $\boldsymbol{\mu}$ is $N(\boldsymbol{\mu} - \overline{\mathbf{X}})'\boldsymbol{\Sigma}^{-1}(\boldsymbol{\mu} - \overline{\mathbf{x}}) \leqslant \chi^2_{p,\alpha}$, which is the surface and interior of an ellipsoid centered at $\overline{\mathbf{X}}$.

Using property **3**, it follows that

$$N(\overline{\mathbf{X}} - \boldsymbol{\mu})'\mathbf{S}^{-1}(\overline{\mathbf{X}} - \boldsymbol{\mu}) \sim T^2_n.$$

From this result, tests concerning $\boldsymbol{\mu}$ can be developed when $\boldsymbol{\Sigma}$ is unknown using the inequality

$$N(\overline{\mathbf{X}} - \boldsymbol{\mu})'\mathbf{S}^{-1}(\overline{\mathbf{X}} - \boldsymbol{\mu}) \geqslant T^2_{n,\alpha}.$$

The $100(1 - \alpha)\%$ confidence region for $\boldsymbol{\mu}$ is given by

$$N(\boldsymbol{\mu} - \overline{\mathbf{X}})'\mathbf{S}^{-1}(\boldsymbol{\mu} - \overline{\mathbf{X}}) \leqslant T^2_{n,\alpha}.$$

The relationship

$$(n - p + 1)\mathbf{T}^2/(np) = F_{p,n-p+1}$$

simplifies these calculations, since percentiles of the F-distribution are readily available. These results can be extended to tests and confidence regions for the mean vectors of two populations.

Other tests of hypotheses (e.g., discriminant analysis*, tests of equality for k mean vectors, MANOVA, equality of covariance matrices, canonical correlations) are based on the joint distribution of the characteristic roots derived from independent Wishart distributions. A detailed discussion is given in Pillai [31, 32].

COMPLEX ANALOGS

The complex multivariate normal distribution was introduced by Wooding [43]. Goodman [16] has studied its properties in the context of time-series* analysis.

Definition 3. Consider a complex random vector $\mathbf{Z} = \mathbf{X} + i\mathbf{Y}$ in which both $\mathbf{X}$ and $\mathbf{Y}$ are p-dimensional random vectors. Let $E(\mathbf{X}) = \boldsymbol{\mu}_1$, $E(\mathbf{Y}) = \boldsymbol{\mu}_2$, $\mathrm{Var}(\mathbf{X}) = \mathrm{Var}(\mathbf{Y}) = \boldsymbol{\Sigma}_{11}$, $\mathrm{Cov}(\mathbf{X}, \mathbf{Y}) = \boldsymbol{\Sigma}_{12}$, a skew symmetric matrix. The vector $\mathbf{Z}$ is said to have the p-dimensional complex normal distribution if the joint probability density function of the elements of $\mathbf{Z}$ is given by

$$\begin{aligned} f(z_1, z_2, \ldots, z_p) &= \pi^{-p}|\boldsymbol{\Sigma}_z|^{-1} \\ &\quad \times \exp\left\{-(\mathbf{z}^* - \boldsymbol{\mu}_z^*)'\boldsymbol{\Sigma}_z^{-1}(\mathbf{z} - \boldsymbol{\mu}_z)\right\}, \end{aligned}$$

where $\mathbf{a}^*$ is the complex conjugate of $\mathbf{a}$, $\boldsymbol{\mu}_z = \boldsymbol{\mu}_1 + i\boldsymbol{\mu}_2$ and $\boldsymbol{\Sigma}_z = \boldsymbol{\Sigma}_1 + i\boldsymbol{\Sigma}_2$ is a Hermitian positive definite matrix. In this case, we write $\mathbf{Z} \sim CN_p(\boldsymbol{\mu}_z, \boldsymbol{\Sigma}_z)$.

Other definitions of the complex normal distribution are given in the literature. (See, e.g., Srivastava and Khatri [37] and Goodman [16, 17].)

Miscellany

CENTRAL LIMIT THEOREM. The multivariate normal distribution arises in practice as a limiting joint distribution of several statistics. This is the consequence of the multidimensional central limit theorem* (CLT), which can be stated:

Let $\mathbf{X}_1, \mathbf{X}_2, \ldots$ be a sequence of independent, identically distributed random variables in p-dimensions. Let $E(\mathbf{X}_\alpha) = \boldsymbol{\mu}$, $\mathrm{Var}(\mathbf{X}_\alpha) = \boldsymbol{\Sigma}$. Then the sampling distribution of the statistics $\sqrt{N}[\mathbf{X}_N - \boldsymbol{\mu}]$ is asymptotically $N_p(\mathbf{0}, \boldsymbol{\Sigma})$. An example of the application of the multivariate CLT is the result that if $\mathbf{A} \sim W_p(\nu, \boldsymbol{\Sigma})$, then $[\mathrm{Vec}(\mathbf{A}) - \nu\,\mathrm{Vec}(\boldsymbol{\Sigma})]/\sqrt{\nu}$ is asymptotically $N_{p^2}[\mathbf{0}, (\mathbf{I} + \mathbf{K})(\boldsymbol{\Sigma} \otimes \boldsymbol{\Sigma})]$, where $\mathbf{K}$ is a $p^2 \times p^2$ commutative matrix and $\otimes$ denotes a Kronecker product. For details, we refer to Muirhead [28, p. 90].

LIMITING FORM OF MULTINOMIAL DISTRIBUTION. The multivariate normal distribution arises also as the limiting form of the multinomial distribution*. Thus if $X_1, \ldots, X_{k+1}$ have a (joint) multinomial distribution with parameters $n, p_1, p_2, \ldots, p_{k+1}$, where $\sum_{i=1}^{k+1} p_i = 1$, then the random vector $\mathbf{U} = \sqrt{n}\,(\hat{\mathbf{p}} - \mathbf{p})$ has asymptotically the $(k + 1)$-dimensional singular normal distribution with mean vector zero and covariance matrix $(1/n)[\boldsymbol{\Delta}_p - \mathbf{p}\mathbf{p}']$, where $\boldsymbol{\Delta}_p = \operatorname{diag}(p_1, \ldots, p_{k+1})$. (See Bishop et al. [5, Chap. 14].)

ENTROPY. Let $f(\mathbf{x})$ be the probability density function of a p-dimensional random variable $\mathbf{X}$. The quantity $-\int f(\mathbf{x}) \ln f(\mathbf{x})\, d\mathbf{x}$ is the *entropy** of the distribution. The problem is one of maximizing the entropy subject to given expectation vector $\boldsymbol{\mu}$ and variance matrix $\boldsymbol{\Sigma}$ of $\mathbf{X}$. Rao [33, Chap. 8] has shown that the solution is obtained by choosing $f(\mathbf{x})$ as the $N_p(\boldsymbol{\mu}, \boldsymbol{\Sigma})$ density function. The entropy of the distribution is shown by him to be $(p/2)\ln 2\pi + p/2 + \frac{1}{2}\ln|\boldsymbol{\Sigma}|$.

TESTS OF MULTINORMALITY. The assumption of multivariate normality underlies most of the standard techniques used for estimation and testing in the analysis of multivariate data. Multivariate normality can be assessed by univariate techniques for evaluating marginal normality or multivariate techniques for judging joint normality. These results are discussed in Chapter 5 of Gnanadesikan [15]. (*See also* MULTIVARIATE NORMALITY, TESTING FOR *and* MARDIA'S TEST FOR MULTINORMALITY.)

MULTINORMAL AS A STABLE DISTRIBUTION. The multinormal distribution belongs to the type of distributions called stable. (*See* MULTIVARIATE STABLE DISTRIBUTIONS.)

Example. While several examples are available in the literature to illustrate the application of the multinormal distribution, Table 2 is intended to exemplify a complex design situation. The reader is referred to Anderson [2, Chap. 8] for other examples. The data refer to the yield of five varieties of barley in two consecutive years under six different location conditions. Thus any pair in parentheses is bivariate normal with mean vector $\boldsymbol{\mu}_{ij}$, variance matrix $\boldsymbol{\Sigma}$, $i = 1, 2, \ldots, 6$, $j = 1, 2, \ldots, 5$. The problem is a simple example of MANOVA, where one is trying to test for the equality of varietal, location effects when the observation is a vector rather than a scalar. For details of the analysis, we refer to Chapter 8 of Anderson [2].

References

[1] Anderson, T. W. (1957). *J. Amer. Statist. Ass.*, **52**, 200–204.

Table 2 Two-Way Classified Bivariate Normal Data

		Varieties				
		M	S	V	T	P
Location	UF	$\binom{81}{81}$	$\binom{105}{82}$	$\binom{120}{80}$	$\binom{110}{87}$	$\binom{98}{84}$
	W	$\binom{147}{100}$	$\binom{142}{116}$	$\binom{151}{112}$	$\binom{192}{148}$	$\binom{146}{108}$
	M	$\binom{82}{103}$	$\binom{77}{105}$	$\binom{78}{117}$	$\binom{131}{140}$	$\binom{90}{130}$
	C	$\binom{120}{99}$	$\binom{121}{62}$	$\binom{124}{96}$	$\binom{141}{126}$	$\binom{125}{76}$
	GR	$\binom{99}{66}$	$\binom{89}{50}$	$\binom{69}{77}$	$\binom{89}{62}$	$\binom{104}{89}$
	D	$\binom{87}{68}$	$\binom{77}{67}$	$\binom{79}{67}$	$\binom{102}{92}$	$\binom{96}{94}$

[2] Anderson, T. W. (1958). *An Introduction to Multivariate Statistical Analysis*. Wiley, New York.

[3] Basu, D. (1956). *Sankhyā*, **17**, 221–224.

[4] Berger, J. O. (1980). *Statistical Decision Theory: Foundations, Concepts, and Methods*. Springer-Verlag, New York.

[5] Bishop, Y. M. M., Fienberg, S. E., and Holland, P. W. (1975). *Discrete Multivariate Analysis: Theory and Practice*. MIT Press, Cambridge, MA.

[6] Bravais, A. (1846). *Mém. Acad. R. Sci. Inst. France*, **9**, 255–332.

[7] Cheng, M. C. (1969). *Ann. Math. Statist.*, **40**, 152–161.

[8] Dahiya, R. C. and Korwar, R. M. (1980). *Ann. Statist.*, **8**, 687–692.

[9] Das Gupta, S., Eaton, M. L., Olkin, I., Perlman, M. D., Savage, L. J., and Sobel, M. (1972). *Proc. 6th Berkeley Symp. Math. Statist. Prob.*, **2**, 241–264.

[10] David, F. N. (1953). *Biometrika*, **40**, 458–459.

[11] Edgeworth, F. Y. (1892). *Philos. Mag., Ser. 5*, **34**, 190–204.

[12] Galton, F. (1886). *Proc. R. Soc. Lond.*, **29**, 42–63. (There is an appendix by J. D. Hamilton Dickson on pp. 63–72.)

[13] Gajjar, A. V. and Subrahmaniam, K. (1978). *Commun. Statist. B*, **7**, 455–477.

[14] Ghurye, S. G. and Olkin, I. (1962). *Ann. Math. Statist.*, **33**, 533–541.

[15] Gnanadesikan, R. (1977). *Methods for Statistical Data Analysis of Multivariate Observations*. Wiley, New York.

[16] Goodman, N. R. (1963). *Ann. Math. Statist.*, **34**, 152–177.

[17] Goodman, N. R. (1963). *Ann. Math. Statist.*, **34**, 178–180.

[18] Haff, L. R. (1979). *Ann. Statist.*, **7**, 1264–1276.

[19] Harris, B. and Soms, A. P. (1980). *J. Multivariate Anal.*, **10**, 252–267.

[20] Hocking, R. R. and Smith, W. B. (1968). *J. Amer. Statist. Ass.*, **63**, 159–173.

[21] James, W. and Stein, C. (1961). *Proc. 4th Berkeley Symp. Math. Statist. Prob.*, **1**, 361–379.

[22] Johnson, N. L. and Kotz, S. (1972). *Distributions in Statistics: Continuous Multivariate Distributions*. Wiley, New York.

[23] Kagan, A., Linnik, Y. V. and Rao, C. R. (1972). *Characterization Problems of Mathematical Statistics*. Wiley, New York.

[24] Kendall, M. G. (1941). *Biometrika*, **32**, 196–198.

[25] Khatri, C. G. (1967). *Ann. Math. Statist.*, **38**, 1853–1867.

[26] Kshirsagar, A. M. (1972). *Multivariate Analysis*. Marcel Dekker, New York.

[27] Moran, P. A. P. (1956). *Proc. Cambridge Philos. Soc.*, **52**, 230–233.

[28] Muirhead, R. J. (1982). *Aspects of Multivariate Statistical Theory*. Wiley, New York.

[29] Olkin, I. and Selliah, J. B. (1977). *Statistical Decision Theory and Related Topics*, Vol. II, Gupta and Moore, eds. Academic Press, New York, 313–326.

[30] Pearson, K. (1896). *Philos. Trans. R. Soc. Lond. A*, **187**, 253–318.

[31] Pillai, K. C. S. (1976). *Can. J. Statist.*, **4**, 157–184.

[32] Pillai, K. C. S. (1977). *Can. J. Statist.*, **5**, 1–62.

[33] Rao, C. R. (1965). *Linear Statistical Inference and its Applications*. Wiley, New York. (Second edition, 1973.)

[34] Schols, Ch. M. (1875). *Verh. Nederl. Akad. Wet.*, **15**, 1–75.

[35] Seal, H. L. (1967). *Biometrika*, **54**, 1–24.

[36] Sidak, Z. (1967). *J. Amer. Statist. Ass.*, **62**, 626–633.

[37] Srivastava, M. S. and Khatri, C. G. (1979). *An Introduction to Multivariate Statistics*. North-Holland, Amsterdam.

[38] Steck, G. P. (1958). *Ann. Math. Statist.*, **29**, 780–800.

[39] Subrahmaniam, K. and Gajjar, A. V. (1980). *J. Multivariate Anal.*, **10**, 60–77.

[40] Subrahmaniam, K. and Subrahmaniam, K. (1973). *Multivariate Analysis: A Selected and Abstracted Bibliography*. Marcel Dekker, New York.

[41] Tallis, G. M. (1963). *Ann. Math. Statist.*, **34**, 940–944.

[42] Tong, Y. L. (1980). *Probability Inequalities in Multivariate Distributions*. Academic Press, New York.

[43] Wooding, R. A. (1956). *Biometrika*, **43**, 212–215.

(BIVARIATE NORMAL DISTRIBUTION
CORRELATION
DISCRIMINANT ANALYSIS
GENERAL LINEAR MODEL
HOTELLING'S T^2
HOTELLING TRACE
LAMBDA CRITERION, WILKS'S
MAHALANOBIS D^2
MULTIDIMENSIONAL CENTRAL LIMIT THEOREM
MULTIPLE CORRELATION
MULTIPLE LINEAR REGRESSION
MULTIVARIATE ANALYSIS
MULTIVARIATE ANALYSIS OF VARIANCE
MULTIVARIATE DISTRIBUTIONS
MULTIVARIATE NORMALITY, TESTING FOR

NORMAL DISTRIBUTION
PILLAI TRACE
WISHART DISTRIBUTION)

S. Kocherlakota
K. Kocherlakota

MULTIPHASE SAMPLING

Multiphase sampling means that the sampling is conducted in two or more parts or phases. We consider a single universe (population), initially unstratified, for simplicity of illustration. A frame that comprises N sampling units is assumed. A first-phase sample of n units is randomly selected by use of this frame. A survey is conducted on these n units to obtain information on a variate X that can be used either to classify the n units into, say, K strata, or to assist in the estimation of another variate Y.

The second-phase sampling is undertaken by subsampling the set of n first-phase units. This second phase will be conducted by random selection of m_i units in each of the established strata or by a single selection of m units out of the n units in the first-phase sample. In either case, more detailed investigation, which may include fieldwork, will collect information on a variate Y for each of the units in the selected subsample (a single variate Y is discussed here although most surveys obtain information about a number of different Y variates).

Note that the same set of sampling units is dealt with throughout multiphase sampling, whereas, by contrast, in multistage sampling the sampling units are nested or hierarchically arranged: primary, secondary, tertiary, etc., units are established, depending on the number of stages in the multistage sampling design (*see* STRATIFIED MULTISTAGE SAMPLING). Desired information is collected on the lowest stage units in a single-phase operation.

In studying the application of sampling theory to survey problems, one is confronted by the many uses of the prefix *multi-*. In addition to the term discussed here, we have multistratification, multitemporal, multivariate, multinomial, and multistage as already noted. Perhaps, multistage and multiphase are most frequently confused. It is possible, of course, to combine multistage and multiphase techniques in complex survey designs (see Kish [7, 12.1C]; also Jessen [6, 10.9]). Most applications of multiphase sampling involve only two phases in the sampling plan for which the terms two-phase sampling or double sampling* are often used as descriptors.

The multiphase approach may be taken when it is convenient and cheap to collect certain information in the first phase from a relatively large sample [9]. For example, sometimes the first-phase information can be obtained from available files, census* records, aerial photographs, or satellite images so that fieldwork is not needed to obtain this first-phase information. The information so obtained may be used for stratification of the universe or in estimation of a total or average of a variate Y.

For the theoretical developments in multiphase sampling, see Neyman [8] and Watson [10]. Watson credits W. G. Cochran [2] for the "mathematical basis of the method". Neyman presented the technique for use in developing the stratification by sampling in the first phase. On the other hand, Cochran concentrated on the use of the first-phase information in the estimation process for the Y total or $\overline{Y}$.

It must be noted that the use of the first-phase sampling for stratification has received little practical application. Thus most applications have been in estimation with the employment of ratio*, regression, and difference estimators (the latter arising when the regression slope b is assumed or determined to be equal to one).

While mail sampling has limited application, it is of interest to note that mail sampling with subsampling of the nonrespondents* may be viewed as a special form of multiphase sampling for stratification. The first mailing (or a combination of two or more mailings) divides the universe of N potential respondents into N_1 respondents and N_2 nonrespondents, but N_1 and N_2 are unknown numbers. The n_1 respondents to the mailing to a sample of n addresses can be used to estimate the weight N_1/N by forming n_1/n. There are $n_2 = n - n_1$ nonrespondents and these are subsampled by selection, of, say, m_2 out of the n_2. Contact is then made by telephone or direct interview with the m_2 units. Since the information obtained by telephone or by direct interview may not be comparable in quantity of quality with that obtained by mail, a subset, say m_1, of the n_1 respondents should also be contacted by the same method as the m_2 set in order that comparison may be made of the methods of data collection*.

It is necessary to consider when multiphase sampling should be used; a decision to employ it will depend on the costs of obtaining the desired information at the first phase and at the second phase, and on the respective variances and the covariances of X and Y. The alternative is to undertake a single sampling of n_0 units to obtain the Y information. Kish [7, pp. 440–443] gives an excellent opening discussion of this decision situation. For the regression estimation case with double sampling, Cochran [2] presents a figure (12.1) that shows the relative efficiency of double vs. single sampling in terms of the ratio of the cost per unit in the second phase c'' to the cost per unit in the first phase c' and the finite population correlation R between y_i and x_i for the N units in the frame. Three curves are plotted to show (1) equal precision for double and single sampling, (2) a 25% increase in precision for double sampling, and (3) a 50% increase in precision for double sampling. Jessen [6] has extended the Cochran-type chart to include greater increases in efficiency for double sampling (Fig. 10.1) and also gives a useful table (10.1) presenting the same type of information. A useful formula is

$$R^2 > \frac{4c''c'}{(c'' + c')^2},$$

which shows what value of R is required to make double sampling preferred over single sampling when the costs per unit of sampling in the first and second phases are known.

The availability of satellite-recorded information has made possible a variety of new applications in sampling (*see* GEOGRAPHY, STATISTICS IN). Multiphase or double sampling has been used to describe correctly some of these applications. In other cases, when the X information is obtained for the entire universe, as it often is, by use of satellite data, the proper description returns to regression estimation [6, p. 331]. Even so, the information is still obtained in two phases. Organizations involved in these satellite applications have been US-NOAA, USDA, US-AID, UN-FAO, various universities, and the Environmental Research Institute of Michigan (ERIM). An example is provided by Allen et al. Of particular interest is the use of the repeated passes of the Landsat on an 18-day cycle so that the multitemporal change in crops can be utilized. Discriminant analysis* and clustering techniques are used to separate and classify areas of the Earth scene. Higher-resolution source materials or field observations are then used for the second phase and combined with the satellite-derived classification for regression estimation.

Both Kish [7] and Yates [11] note that multiphase sampling can be extended to three or more phases. Neither author, however, gives an example. A really complex example is given by Jessen [6] in which he describes six phases. The descriptions given are brief; interested readers will need to refer to Jessen [5] in order to better understand this complex example. The example is of immense practical interest, however, since it is concerned with the estimation of the total tonnage of a tree crop (oranges) in a major producing area of Florida. No difficulties arise in estimation when extensions are made to three or more phases, but the variance formulas become more complex with increase in the number of phases. Another extension of double sampling for sampling the same universe on repeated occasions (multitemporal) has been described by Cochran [2, 12.9–12.11]. Interesting optimization considerations, first examined by Jessen [3], are developed by Cochran for retaining the initial sample of n units or the dropping of old units and adding new units to the sample on succeeding occasions.

References

[1] Allen, J. A., Latham, J. S., Colwell, J. E., Reinhold, R., and Jebe, E. H. (1982). "Monitoring the Changing Areal Extent of Irrigated Lands of the Gefara Plain, Libya". In *Proceedings of the International Symposium on Remote Sensing of the Environment, 19–25, January 1982, Cairo, Egypt*, Vol. II. Environmental Research Institute of Michigan, Ann Arbor, MI, pp. 1119–1126.

[2] Cochran, W. G. (1977). *Sampling Techniques*, 3rd ed. Wiley, New York.

[3] Jessen, R. J. (1942). *Iowa Agric. Expt. Statist. Res. Bull. 304*,

[4] Jessen, R. J. (1955). *Biometrics*, **11**, 99–109.

[5] Jessen, R. J. (1972). In *Statistics Papers in Honor of G. W. Snedecor*, T. A. Bancroft, ed. ISU Press, Ames, IA, pp. 145–165. (See also ref. 4.)

[6] Jessen, R. J. (1978). *Statistical Survey Techniques*. Wiley, New York.

[7] Kish, L. (1965). *Survey Sampling*. Wiley, New York.

[8] Neyman, J. (1938). *J. Amer. Statist. Ass.*, **33**, 101–116.

[9] UN Statistics Office. (1950). *The Preparation of Sampling Survey Reports*, UN Series C, No. 1. UN, New York. Also (1964) Series C, No. 1, Rev. 2. (See also ref. 7, p. 440.)

[10] Watson, D. J. (1937). *J. Agri. Sci.*, **27**, 474–483.

[11] Yates, F. (1971). *Sampling Methods for Censuses and Surveys*, 3rd ed. Griffin, London.

(DOUBLE SAMPLING
FINITE POPULATIONS, SAMPLING FROM
NONRESPONDENTS, SUBSAMPLING THE
SAMPLING PLANS
STRATIFIED MULTISTAGE SAMPLING
SURVEY SAMPLING)

EMIL H. JEBE

MULTIPLE COMPARISONS

Introductory statistics courses typically include as an important topic the confidence interval* for the mean of a normal distribution*. The interval involves Student's t-distribution* and can be formally described as follows. For a sample of n independent observations $Y_1, \ldots, Y_n$, each distributed according to a normal distribution with mean μ and variance σ^2, a $100(1-\alpha)\%$ confidence interval for μ is given by

$$\bar{y} - t_{n-1}^{\alpha/2} \frac{s}{\sqrt{n}} < \mu < \bar{y} + t_{n-1}^{\alpha/2} \frac{s}{\sqrt{n}}, \quad (1)$$

where

$$\bar{y} = \frac{1}{n} \sum_{i=1}^{n} y_i, \qquad s^2 = \frac{1}{n-1} \sum_{i=1}^{n} (y_i - \bar{y})^2, \quad (2)$$

and $t_{n-1}^{\alpha/2}$ is the upper $\alpha/2$ percentile point of a t-distribution with $n-1$ degrees of freedom (i.e., $\Pr[T_{n-1} > t_{n-1}^{\alpha/2}] = \alpha/2$).

Suppose now that the investigator has two sets of data. For example, consider the plasma bradykininogen levels in a group of normal control patients and in a group of patients with active Hodgkin's disease. For the control patients $\bar{y}_1 = 6.08$, $s_1 = 1.36$, $n_1 = 23$, and for the Hodgkin's patients $\bar{y}_2 = 4.24$, $s_2 = 1.30$, $n_2 = 17$. Application of (1) to the two data sets with the appropriate t critical values ($t_{22}^{0.025} = 2.07$, $t_{16}^{0.025} = 2.12$) gives the 95% confidence interval (5.49, 6.67) for μ_1 and the 95% confidence interval (3.57, 4.91) for μ_2. Each interval has probability 0.95 of including its corresponding true population mean.

But what is the probability that both intervals simultaneously contain their respective population means? The answer to this question depends on the degree of probabilistic dependence between the two data sets. If the two data sets come from totally separate experiments, so that the data sets are independent, then the probability of joint coverage is $0.95 \times 0.95 = 0.9025$. If there is dependence, the probability can be greater or less than this. The smallest it can be is 0.90. This lower bound comes from the elementary probability inequality

$$\Pr[A_1 \cap A_2] \geqslant 1 - \Pr[A_1^c] - \Pr[A_2^c], \quad (3)$$

where A_i^c is the complement of A_i (*see* BONFERRONI INEQUALITIES).

The joint probability indicates that the investigator should not be too smug about his or her 95% confidence intervals. The probability of making at least one error is certainly greater than the 0.05 claimed for each interval and could be as high as 0.10. While this increase in the probability of an error may not seem so disastrous, the situation deteriorates as the number of different confidence intervals increases. For five independent confidence intervals, the probability of at least one error is $1 - (0.95)^5 = 0.2262$ and for ten intervals, it is 0.4013.

The goal of the *multiple comparisons* procedures is to control this joint probability. Specifically, suppose an investigator makes a number of statistical statements (viz., confidence intervals or tests) that comprise a family $\mathscr{S} = \{S_f, f \in \mathscr{F}\}$. The number of elements in $\mathscr{F}$ is usually finite but can be infinite. Let N_w be the number of incorrect statements in $\mathscr{S}$. A multiple comparisons procedure aims at controlling N_w over all possible data configurations by requiring $P[N_w > 0] \leqslant \alpha$.

Before delving into an overview of the field of multiple comparisons, one cautionary note should be sounded. As the size of the family $\mathscr{F}$ increases, the statistician has to become more and more conservative in order to maintain $P[N_w > 0] \leqslant \alpha$. The statistical tests have to be weaker and the confidence intervals broader. Therefore, the investigator and the statistician must use judgement on how to balance between not falsely rejecting any null hypotheses and not having the power to reject any when they are false. For confidence intervals, the probability of coverage must be judged against the lengths of the intervals.

While the preceding introductory remarks have been couched in terms of separate confidence intervals for normal means, the most frequent application of multiple comparisons methods is in comparing k populations. The investigator typically wants to know if the k populations under study are all the same or whether they differ. If they are not all the same, he or she usually wants to know which ones differ. The techniques designed to answer this question in the case of comparing the means of normal populations are reviewed in the first section. Analogous distribution-free* rank procedures are discussed in the section on Nonparametric Tests. When the individual observations are binary, the problem becomes one of comparing binomial proportions, and this topic is covered in the section on Categorical Data.

Confidence intervals and tests on regression coefficients and regression surfaces are described in the final section on Regression Analysis.

For those who seek additional reading, O'Neill and Wetherill [36] and Krishnaiah [29] give surveys of the field. Miller [33] includes his 1966 treatise on multiple comparisons [31] and his 1977 review article covering developments in the field from 1966–1976 [32]. (*See also* MULTIVARIATE MULTIPLE COMPARISONS.)

A subject akin to multiple comparisons is *ranking* and selection*. However, the goal of ranking and selection procedures is somewhat different from that of multiple comparisons methods. Their aim is to select a subset of the populations having the highest parameter values with the probability of a correct selection exceeding a prescribed level (*see* RANKING PROCEDURES *and* SELECTION PROCEDURES).

COMPARISON OF NORMAL MEANS

Consider k samples from k different normal distributions (*see* ONE-WAY CLASSIFICATION). Specifically, let Y_{ij}, $i = 1, \dots, k$; $j = 1, \dots, n_i$, be independently distributed according to $N(\mu_i, \sigma^2)$. The problem is to test the null hypothesis $H_0: \mu_1 = \cdots = \mu_k$ and, if H_0 is rejected, to decide which means differ, or it is to construct simultaneous confidence intervals on the $\binom{k}{2}$ mean differences $\mu_i - \mu_{i'}$.

The oldest multiple comparisons procedure for the testing problem is the *Fisher protected least significant difference (LSD)* *test*. It rejects the null hypothesis H_0 if the F statistic for the one-way analysis of variance* is significant, i.e.,

$$F = \frac{\nu}{k-1} \cdot \frac{\sum_{i=1}^{k} n_i(\bar{y}_{i\cdot} - \bar{y}_{\cdot\cdot})^2}{\sum_{i=1}^{k}\sum_{j=1}^{n_i}(y_{ij} - \bar{y}_{i\cdot})^2} > F^{\alpha}_{k-1,\nu}, \tag{4}$$

where

$$\bar{y}_{i\cdot} = \frac{1}{n_i}\sum_{j=1}^{n_i} y_{ij}, \quad \bar{y}_{\cdot\cdot} = \sum_{i=1}^{k}\sum_{j=1}^{n_i} y_{ij} \Big/ \sum_{i=1}^{k} n_i, \tag{5}$$
$$\nu = \sum_{i=1}^{k}(n_i - 1),$$

and $F^{\alpha}_{k-1,\nu}$ is the upper α percentile point of the F-distribution with $k-1$ and ν degrees of freedom. If H_0 is rejected, then all pairwise comparisons of means are handled as nonsimultaneous single comparisons based on the t-distribution. The initial F-test "protects" all the subsequent multiple comparisons that employ the "least" large critical value, namely, an ordinary t percentile point.

This test continues to be used in practice because of its ease and simplicity. However, if, for example, one population mean differs markedly from the others that are all the same, the F statistic will be significant, and subsequently there may be false rejections among the equal means because of the multiple comparisons being made.

The investigator or statistician can always use *Bonferroni t-tests or intervals*. The Bonferroni t confidence intervals are

$$\mu_i - \mu_{i'} \in \bar{y}_{i\cdot} - \bar{y}_{i'\cdot} \pm t_{\nu}^{\alpha/2c} s\left(\frac{1}{n_i} + \frac{1}{n_{i'}}\right)^{1/2}, \tag{6}$$

where

$$s^2 = \frac{1}{\nu}\sum_{i=1}^{k}\sum_{j=1}^{n_i}(y_{ij} - \bar{y}_{i\cdot})^2, \tag{7}$$

$t_{\nu}^{\alpha/2c}$ is the upper $\alpha/2c$ percentile point of the t-distribution with ν degrees of freedom, and $c = \binom{k}{2}$ or whatever number of comparisons are being made. The unusual t percentile points can be obtained from computer programs, some programmable calculators, Dunn [4], Moses [34], Miller [33], or other sources. By the Bonferroni inequality for c events [see relation (3)], the probability that the c intervals [relation (6)] are all correct is greater than or equal to $1 - \alpha$. The null hypothesis H_0 is rejected if some intervals [relation (6)] do not include zero, and these means are declared different.

While the lengths of the intervals (6) can be improved on by the specialized Tukey-type range procedures for this problem, the Bonferroni inequality can be applied in more complicated situations in which specialized techniques don't exist. Very slight improvement in the lengths of the intervals can be obtained through use of the probability inequality for multivariate t-distributions due to Šidák [39] (see Dunn [6]). A refined Bonferroni inequality due to Hunter [25] can also lead to improvement, but the method is computationally complicated.

Scheffé [38] gave a multiple comparisons interpretation to the F statistic (*see* SCHEFFÉ'S SIMULTANEOUS COMPARISON PROCEDURE). He established that the projection of the ellipsoid confidence region for the linearly independent set of mean differences onto the vector $\mathbf{c} = (c_1, \ldots, c_k)$, where $\sum_{i=1}^{k} c_i = 0$, is

$$\sum_{i=1}^{k} c_i\mu_i \in \sum_{i=1}^{k} c_i\bar{y}_{i\cdot} \pm ((k-1)F^{\alpha}_{k-1,\nu})^{1/2} s\left(\sum_{i=1}^{k}\frac{c_i^2}{n_i}\right)^{1/2}. \tag{8}$$

These intervals for different $\mathbf{c}$ are called *Scheffé (S-) intervals*. The probability that the intervals are simultaneously correct for all possible $\mathbf{c}$ subject to $\sum_{i=1}^{k} c_i = 0$ is the same as the probability associated with the confidence ellipsoid, namely, $1 - \alpha$. The linear combinations $\sum_{i=1}^{k} c_i\mu_i$ with $\sum_{i=1}^{k} c_i = 0$ are called *contrasts** and include the pairwise mean differences as special cases. If the F-test is significant [see relation (4)], there are contrasts whose intervals [relation (8)] do not contain zero, which is the null value under H_0. In almost all instances the confidence intervals [relation (8)] are longer than the corresponding Bonferroni- and Tukey-type intervals for pairwise mean differences. For more general contrasts, the Scheffé intervals tend to be shorter than the corresponding Tukey-type intervals. The Bonferroni intervals for general contrasts can be shorter or longer than the Scheffé intervals depending on how many contrasts are being considered.

For a balanced one-way analysis of variance where $n_i \equiv n$, $i = 1, \ldots, k$, the *Tukey studentized range* (*T*-) *intervals* are

$$\mu_i - \mu_{i'} \in \bar{y}_{i'}. \pm q^{\alpha}_{k,v} s n^{-1/2}, \tag{9}$$

where $q^{\alpha}_{k,v}$ is the upper α percentile point of the studentized range* distribution for k means with v degrees of freedom. Harter [21, 22] provides the best tables of $q^{\alpha}_{k,v}$, and some of these are reproduced in Miller [33]. The probability that the $\binom{k}{2}$ intervals (9) for all $i \neq i'$ are simultaneously correct is exactly $1 - \alpha$. This follows from the fact that the maximum pairwise mean difference is precisely the range. Note that (9) uses the factor $(1/n)^{1/2}$ rather than the more standard $(2/n)^{1/2}$ given by the standard deviation of the difference, which is used in (6) and (8) when $n_i \equiv n$. Any pair of means whose corresponding confidence interval [see (9)] does not cover zero is declared significantly different. The intervals [relation (9)] can be extended to encompass contrasts as well.

Tukey coined the phrase "wholly significant difference (*WSD*)" for the term $q^{\alpha}_{k,v} s(1/n)^{1/2}$ in (9) to contrast it with the *least significant difference (LSD)* term $t^{\alpha/2}_{v} s(2/n)^{1/2}$ used in nonsimultaneous testing.

The Tukey studentized range procedure gives the shortest simultaneous confidence intervals for pairwise mean differences when it is applicable, that is, when the experimental design is balanced. In the unbalanced case where the n_i are not all equal, there are approximate Tukey-type procedures.

The *Spjøtvoll–Stoline T′-method* gives the intervals [40]

$$\mu_i - \mu_{i'} \in \bar{y}_{i}. - \bar{y}_{i'}. \pm q'^{\alpha}_{k,v} s \max\left\{ \frac{1}{\sqrt{n_i}}, \frac{1}{\sqrt{n_{i'}}} \right\}, \tag{10}$$

where $q'^{\alpha}_{k,v}$ is the upper α percentile point of the studentized augmented range distribution for k means with v degrees of freedom. Tables of $q'^{\alpha}_{k,v}$ are available in Stoline [43].

The *Hochberg GT2-intervals* [24] are

$$\mu_i - \mu_{i'} \in \bar{y}_{i}., - \bar{y}_{i'} \pm |m|^{\alpha}_{c,v} s \left(\frac{1}{n_i} + \frac{1}{n_{i'}} \right)^{1/2}, \tag{11}$$

where $c = \binom{k}{2}$ and $|m|^{\alpha}_{c,v}$ is the upper α percentile point of the studentized maximum modulus* distribution for c means with v degrees of freedom. Hahn and Hendrickson [20] give general tables of $|m|^{\alpha}_{k,v}$ for $k = 1(1)6(2)12, 15, 20$, and these appear in Miller [33]. For use with this technique, Stoline and Ury [45] and Ury et al. [49] give special tables of $|m|^{\alpha}_{c,v}$ with $c = \binom{k}{2}$ for $k = 2(1)20$ and $v = 20(2)50(5)80(10)100$, respectively.

Finite intersection procedures proposed earlier by Krishnaiah [28] are related to the intervals (11) and (12). Also, the tables of the studentized largest chi-square distribution* by Armitage and Krishnaiah [1] provide critical points for the studentized maximum modulus by taking the square root of the entries in the table with one degree of freedom for the numerator.

Gabriel [13] suggested combining separate confidence intervals based on the critical constant $|m|^{\alpha}_{c,v}$.

Intervals with a more complicated factor multiplying $q^{\alpha}_{k,v} s$ have been proposed by Genizi and Hochberg [15].

Ury [48] compared the relative lengths of (10) and (11) and found the T' method to be superior for mildly imbalanced designs whereas the GT2 method is superior for greater imbalance. However, these intervals, as well as the Genizi–Hochberg intervals, are always longer than the intervals proposed much earlier by Tukey [47] and Kramer [27].

The *Tukey–Kramer intervals* are simply Tukey's intervals [see (9)] with the scalar factor equal to the standard deviation of an unbalanced mean difference adjusted by $\sqrt{2}$ for the studentized range, i.e.,

$$\mu_i - \mu_{i'} \in \bar{y}_i. - \bar{y}_{i'}. \pm q^{\alpha}_{k,v} s \left[\frac{1}{2} \left(\frac{1}{n_i} + \frac{1}{n_{i'}} \right) \right]^{1/2}. \tag{12}$$

Another interpretation of (12) is that n in (9) has been replaced by the harmonic mean of n_i and $n_{i'}$.

Over the years the Tukey–Kramer intervals have been neglected because it was not known whether the probability of simultaneous coverage is always greater than or equal to $1 - \alpha$. Probability inequalities guarantee this for the Spjøtvoll–Stoline, Hochberg, and Genizi–Hochberg intervals. However, Monte Carlo work by Dunnett [9] indicates that the Tukey–Kramer intervals are conservative (i.e., coverage probability $\geqslant 1 - \alpha$) in the cases considered, and recent theoretical work by Hayter [23] proves this to be true in general. Earlier Kurtz [30] and Brown [3] had established this for $k = 3$ and $k = 3, 4, 5$, respectively.

For a detailed comparison of all the aforementioned procedures, the reader is referred to Stoline [44]. Stoline recommends the general use of the Tukey–Kramer intervals.

For the numerical example in the introduction, there was actually a third group of patients with inactive Hodgkin's disease. For this group $\bar{y}_3 = 6.51$, $s_3 = 1.63$, and $n_3 = 27$ after one very high value was trimmed. The pooled standard deviation s equals 1.46. The Tukey–Kramer 95% confidence intervals for the three mean comparisons are

$$\mu_1 - \mu_2 \in 1.84 \pm 1.12 = (0.72, 2.96),$$

$$\mu_1 - \mu_3 \in -0.43 \pm 1.00 = (-1.43, 0.57), \tag{13}$$

and

$$\mu_2 - \mu_3 \in -2.27 \pm 1.09 = (-3.36, -1.18).$$

One can conclude that the active Hodgkin's disease patients (group 2) differ from the controls (group 1) and the patients with inactive Hodgkin's disease (group 3) and there is no demonstrable difference between the latter two groups. The constant $q^{0.05}_{3,64}/\sqrt{2} = 2.40$ was used in computing the intervals (13). The corresponding Bonferroni and Scheffé critical constants, $t^{0.05/6}_{64} = 2.46$ and $(2F^{0.05}_{2,64})^{1/2} = 2.51$, respectively, are slightly larger but would lead to the same conclusion. If the three intervals were treated separately in nonsimultaneous fashion, the critical constant would be $t^{0.025}_{64} = 2.00$.

All the intervals mentioned in this section are based on the assumption that σ^2 is the same for all samples. For techniques when this assumption is violated the reader is referred to Tamhane [46] and Dunnett [10].

The methods of this section can be extended to the two-way classification and higher-way designs.

A slightly different problem is where the investigator wants to compare k experimental populations against a single standard control population. Dunnett [7, 8] has pioneered the work in this area, and a description of the techniques can be found in Miller [33].

With the exception of Fisher's protected LSD test, all of the testing procedures described in this section have the property that the comparison between any two means is unaffected by the comparisons between, and with, the other means. Procedures with this property have been labeled "simultaneous test procedures (STP)" by Gabriel [12]. There are more general *multiple comparisons procedures (MCP)* that proceed in a stepwise fashion. The critical constant at each stage depends on the number of means begin compared. Test procedures of this type can increase the power for individual comparisons, but they lack any confidence interval interpretation. (For details, *see* MULTIPLE RANGE AND ASSOCIATED TEST PROCEDURES.)

Work has been done on multiple comparisons procedures from a Bayesian decision-theoretic point of view. For references and a description of these techniques, *see* k-RATIO t-TESTS, t-INTERVALS, AND POINT ESTIMATES FOR MULTIPLE COMPARISONS.

NONPARAMETRIC TESTS

All of the techniques described in the preceding section depend on the assumption of normally distributed data for their distribution theory. Considerable effort has been expended by a number of statisticians in creating rank test analogs to these procedures that are distribution-free.

Consider still the one-way classification*, but in this setting the Y_{ij}, $i = 1, \ldots, k$, $j = 1, \ldots, n_i$, are only assumed to be independently distributed according to the continuous distributions F_i, $i = 1, \ldots, k$. The problem is to test the null hypothesis H_0: $F_1 = \cdots = F_k$ and, if H_0 is rejected, to decide which populations differ. The tests considered here are constructed primarily for detecting shift alternatives.

The classical nonparametric test for this hypothesis is the *Kruskal–Wallis rank test*, which rejects H_0 if

$$H = \frac{12}{N(N+1)} \sum_{i=1}^{k} n_i \bar{R}_{i\cdot}^2 - 3(N+1) \quad (14)$$

exceeds its critical value, which for large samples ($n_i \geqslant 5$) can be taken to be $\chi_{k-1}^{2\alpha}$, the upper α percentile point of a χ^2 distribution with $k-1$ degrees of freedom (*see* KRUSKAL–WALLIS TEST). In expression (14)

$$N = \sum_{i=1}^{k} n_i,$$

$$R_{ij} = \text{rank of } Y_{ij} \text{ in } Y_{11}, \ldots, Y_{kn_k}, \quad (15)$$

$$\bar{R}_{i\cdot} = \frac{1}{n_i} \sum_{j=1}^{n_i} R_{ij}.$$

Nemenyi [35] used a Scheffé-type projection argument on (14) to suggest that F_i and $F_{i'}$ be declared different if

$$|\bar{R}_{i\cdot} - \bar{R}_{i'\cdot}| > (\chi_{k-1}^{2\alpha})^{1/2} \left[\frac{N(N+1)}{12} \right]^{1/2} \times \left(\frac{1}{n_i} + \frac{1}{n_{i'}} \right)^{1/2}. \quad (16)$$

The Bonferroni critical value in this case would be $g^{\alpha/(2c)} = t_{\infty}^{\alpha/(2c)}$, the upper $\alpha/(2c)$ percentile point of a unit normal distribution where $c = \binom{k}{2}$. Usually $g^{\alpha/(2c)} < (\chi_{k-1}^{2\alpha})^{1/2}$, so testing the $\binom{k}{2}$ mean rank differences with the Bonferroni critical value substituted into (16) produces a more powerful multiple comparisons test (see Dunn [5]).

The asymptotic covariance structure of the mean rank vector $(\bar{R}_{1\cdot}, \ldots, \bar{R}_{k\cdot})$ is of a form such that the analog of the Tukey–Kramer procedure should be valid asymptotically (see Miller [33, p. 171]). This procedure would reject $F_i = F_{i'}$ if

$$|\bar{R}_{i\cdot} - \bar{R}_{i'\cdot}| > q_{k,\infty}^{\alpha} \left[\frac{N(N+1)}{12} \right]^{1/2} \times \left[\frac{1}{2} \left(\frac{1}{n_i} + \frac{1}{n_{i'}} \right) \right]^{1/2}, \quad (17)$$

where $q_{k,\infty}^{\alpha}$ is the upper α percentile point of the studentized range distribution for k means with infinite degrees of freedom. The right-hand side of the inequality in (17) should be smaller than its counterpart in (16) with $(\chi_{k-1}^{2\alpha})^{1/2}$ or $g^{\alpha/2c}$.

For the bradykininogen levels example, the three group mean ranks are

$$\bar{R}_{1\cdot} = 38.33, \quad \bar{R}_{2\cdot} = 15.94, \quad \bar{R}_{3\cdot} = 42.63. \quad (18)$$

Averaged ranks were used for ties, and the one very high value in group 3 was not trimmed for this rank analysis. At the 5% significance level the Tukey–Kramer type procedure defined by (17) gives

$$|\bar{R}_{1\cdot} - \bar{R}_{2\cdot}| = |38.33 - 15.94| > 14.81,$$
$$|\bar{R}_{1\cdot} - \bar{R}_{3\cdot}| = |38.33 - 42.63| < 13.03, \quad (19)$$
$$|\bar{R}_{2\cdot} - \bar{R}_{3\cdot}| = |15.94 - 42.63| > 14.25.$$

Thus group 2 (active Hodgkin's) is judged to be significantly different from groups 1 (controls) and 3 (inactive Hodgkin's) with the latter two groups being judged alike. This conclusion agrees with the normal theory analysis. The constant $q_{3,\infty}^{0.05}/\sqrt{2} = 2.34$ was used in (19). The corresponding critical constants $g^{0.05/6} = 2.39$ and $(\chi_2^{20.05})^{1/2} = 2.45$ for the Bonferroni and Nemenyi procedures, respectively, are slightly larger. The nonsimultaneous critical constant would be $g^{0.025} = 1.96$.

Steel [41, 42] and Dwass [11] independently proposed an alternative ranking scheme. For comparing populations i and i' the usual two-sample Wilcoxon rank statistic is computed using just the observations in these two samples. Its value is then compared with the critical point appropriate to the maximum of $\binom{k}{2}$ interrelated Wilcoxon rank statistics under H_0. For details, see the original articles or Miller [33].

An advantage of the Steel–Dwass ranking is that the comparison between populations i and i' is not contaminated by observations from other populations. Also, confidence intervals for the differences in location parameters can be constructed, whereas this is not feasible with the Kruskal–Wallis ranking. A disadvantage of the Steel–Dwass ranking is that $\binom{k}{2}$ rankings are required to perform the test rather than one large ranking. Also, tables of critical values are available only for the equal sample size case $n_i \equiv n$. Koziol and Reid [26] have proved that these two different methods of ranking are asymptotically equivalent under H_0 or under sequences of alternatives tending to H_0. For alternative hypotheses where $k - 1$ of the populations are identical and one differs from the rest, Oude Voshaar [37] has shown that, for the test based on (17) with $n_i \equiv n$, the probability of rejecting the equality of some identical pair can exceed α, but numerical work indicates that the probability does not exceed α by very much.

The problem where k populations are compared to a single control population has also been studied from the nonparametric point of view (see Miller [33]).

For the situation where there is pairing between the observations as in a two-way classification, Hollander and Nemenyi proposed the use of signed rank statistics, but these are not quite distribution-free when considered jointly (see Miller [33]). Multiple comparisons tests can also be derived from the Friedman rank test for the two-way classification (*see* FRIEDMAN'S CHI-SQUARE TEST).

Multiple comparisons analogs to the sign test* and the Kolmogorov–Smirnov* test exist as well. The interested reader is referred to Miller [33].

CATEGORICAL DATA

When the individual response is quantal or binary (i.e., 0 or 1), the mean becomes a proportion. Not surprisingly, there are multiple comparisons techniques for comparing k proportions.

Consider first the case of k proportions $\hat{p}_i = r_i/n_i$, where the r_i are independently binomially distributed with parameters p_i and sample sizes n_i, $i = 1, \ldots, k$, respectively. For testing the null hypothesis $H_0: p_1 = \cdots = p_k$ the customary *Pearson chi-square statistic* is

$$X^2 = \sum_{i=1}^{k} \frac{(r_i - n_i\hat{p})^2}{n_i\hat{p}(1-\hat{p})}, \qquad \hat{p} = \sum_{i=1}^{k} r_i \Big/ \sum_{i=1}^{k} n_i.$$

Asymptotically X^2 has a χ^2 distribution with $k - 1$ degrees of freedom (*see* CHI-SQUARE TESTS *and* CONTINGENCY TABLES). Scheffé-type projections of the X^2 statistic suggest declaring p_i and $p_{i'}$ unequal if

$$|\hat{p}_i - \hat{p}_{i'}| > \left(\chi^{2\,\alpha}_{k-1}\right)^{1/2}\left[\hat{p}(1-\hat{p})\left(\frac{1}{n_i} + \frac{1}{n_{i'}}\right)\right]^{1/2}. \tag{20}$$

The Tukey–Kramer and Bonferroni critical constants $q^{\alpha}_{k,\infty}/\sqrt{2}$ and $g^{\alpha/2c}$, where $c = \binom{k}{2}$, respectively, are smaller than $(\chi^{2\,\alpha}_{k-1})^{1/2}$. These critical constants are the same as those used in conjunction with the rank analysis (16)–(17).

While (20) is designed for binary data*, it can be illustrated by dichotomizing the bradykininogen data. The median for the combined data from all three groups is 5.73. Categorizing observations according to whether they fall above or below the combined median leads to the following 2×3

contingency table:

	Groups 1	2	3
> 5.73	14	3	17
< 5.73	9	14	11

(21)

The observed proportions falling above the combined median are $\hat{p}_1 = 0.61$, $\hat{p}_2 = 0.18$, and $\hat{p}_3 = 0.61$. By (20) with $\hat{p} = 1/2$ and $q_{3,\infty}^{0.05}/\sqrt{2}$ replacing $(\chi_2^{2\,0.05})^{1/2}$,

$$\begin{aligned} |\hat{p}_1 - \hat{p}_2| &= |0.61 - 0.18| > 0.37, \\ |\hat{p}_1 - \hat{p}_3| &= |0.61 - 0.61| < 0.33, \\ |\hat{p}_2 - \hat{p}_3| &= |0.18 - 0.61| > 0.36, \end{aligned} \tag{22}$$

which gives the same conclusion as for the normal theory and rank analyses. Nemenyi [35] proposed the use of median tests; for more details about them, the reader is referred to Miller [33].

The preceding test procedure does not lead to simultaneous confidence intervals on the differences $p_i - p_{i'}$. To obtain these, one needs to base the approach on the *Goodman statistic* [17]

$$Y^2 = \sum_{i=1}^{k} \frac{(r_i - n_i\tilde{p})^2}{n_i\hat{p}_i(1 - \hat{p}_i)}, \quad \tilde{p} = \sum_{i=1}^{k} \hat{w}_i\hat{p}_i \Big/ \sum_{i=1}^{k} \hat{w}_i,$$
$$\hat{w}_i^{-1} = \hat{p}_i(1 - \hat{p}_i)/n_i.$$

The Y^2 statistic (*see* GOODMAN'S Y^2) leads to the intervals

$$p_i - p_{i'} \in \hat{p}_i - \hat{p}_{i'} \pm \left(\chi_{k-1}^{2\,\alpha}\right)^{1/2} \times \left[\frac{\hat{p}_i(1 - \hat{p}_i)}{n_i} + \frac{\hat{p}_{i'}(1 - \hat{p}_{i'})}{n_{i'}}\right]^{1/2} \tag{23}$$

for $i \neq i'$, which have asymptotic probability greater than or equal to $1 - \alpha$ of being simultaneously correct. The intervals (23) could be shortened by substituting the Bonferroni critical constant $g^{\alpha/2c}$ with $c = \binom{k}{2}$ in place of $(\chi_{k-1}^{2\,\alpha})^{1/2}$.

Rather than being independent, the r_i may be the observed frequencies in the categories $i = 1, \ldots, k$ for a sample of size n from a multinomial distribution with probabilities $p_1, \ldots, p_k$, $\sum_{i=1}^{k} p_i = 1$. The *Gold confidence intervals* [16] for the differences in the probabilities are

$$\begin{aligned} p_i - p_{i'} &\in \hat{p}_i - \hat{p}_{i'} \\ &\pm \left(\chi_{k-1}^{2\,\alpha}\right)^{1/2}\left[\hat{p}_i(1 - \hat{p}_i) + 2\hat{p}_i\hat{p}_{i'} + \hat{p}_{i'}(1 - \hat{p}_{i'})\right]^{1/2}. \end{aligned} \tag{24}$$

Goodman [18] correctly points out that the Bonferroni critical constant $g^{\alpha/(2c)}$ with $c = \binom{k}{2}$ usually improves the intervals.

There are suitable simultaneous confidence intervals for testing a goodness-of-fit* null hypothesis $H_0: p_i = p_i^0$, $i = 1, \ldots, k$, where the p_i^0 are specified. Also, there are multiple comparisons procedures for testing independence in an $r \times c$ contingency table and for testing the equality of cross-product ratios. For references and details, the reader is referred to Miller [33].

REGRESSION ANALYSIS

The type of multiple comparisons made in regression analysis differs from that described in the first three sections. Instead of comparing pairs of sample values, one against another, the intent is to compare a number of sample values against their unknown mean values.

The simple linear regression* model is

$$Y_i = \alpha + \beta x_i + e_i, \qquad i = 1, \ldots, n, \qquad e_i \text{ independent } N(0, \sigma^2). \tag{25}$$

A common problem is to construct a confidence region for α and β. A confidence ellipsoid based on the F statistic is often referred to, but it is hardly ever drawn in practice. Instead, the Scheffé-type intervals

$$\begin{aligned} \alpha &\in \hat{\alpha} \pm \left(2F_{2,n-2}^{\alpha}\right)^{1/2}s \times \left(n^{-1} + \bar{x}^2/\textstyle\sum_{i=1}^{n}(x_i - \bar{x})^2\right)^{1/2}, \\ \beta &\in \hat{\beta} \pm \left(2F_{2,n-2}^{\alpha}\right)^{1/2}s \times \left(1/\textstyle\sum_{i=1}^{n}(x_i - \bar{x})^2\right)^{1/2}, \end{aligned} \tag{26}$$

which arise from projecting the F confidence ellipsoid onto the coordinate axes, are used. In (26),

$$\hat{\alpha} = \bar{y} - \hat{\beta}\bar{x},$$

$$\hat{\beta} = \sum_{i=1}^{n} (x_i - \bar{x})(y_i - \bar{y}) \Big/ \sum_{i=1}^{n} (x_i - \bar{x})^2, \tag{27}$$

$$s^2 = \frac{1}{n-2} \sum_{i=1}^{n} (y_i - \hat{\alpha} - \hat{\beta}x_i)^2.$$

The intervals (26) can be shortened by substituting the Bonferroni critical constant $t_{n-2}^{\alpha/4}$ for $(2F_{2,n-2}^{\alpha})^{1/2}$. Both sets of intervals have probability greater than or equal to $1 - \alpha$ of simultaneously covering α and β.

If the regression model is reparametrized as

$$Y_i = \alpha + \beta(x_i - \bar{x}) + e_i, \tag{28}$$

then the intervals

$$\alpha \in \hat{\alpha} \pm |m|_{2,n-2}^{\alpha} s/\sqrt{n},$$

$$\beta \in \hat{\beta} \pm |m|_{2,n-2}^{\alpha} s \big/ \left\{\textstyle\sum_{i=1}^{n}(x_i - \bar{x})^2\right\}^{1/2} \tag{29}$$

can be used. With this parametrization

$$\hat{\alpha} = \bar{y},$$

$$\hat{\beta} = \sum_{i=1}^{n} (x_i - \bar{x})(y_i - \bar{y}) \Big/ \sum_{i=1}^{n} (x_i - \bar{x})^2, \tag{30}$$

$$s^2 = \frac{1}{n-2} \sum_{i=1}^{n} (y_i - \hat{\alpha} - \hat{\beta}(x_i - \bar{x}))^2.$$

The critical constant $|m|_{2,n-2}^{\alpha}$, which is the upper α-percentile point of the studentized maximum modulus* distribution for two means with $n-2$ degrees of freedom, is always smaller than $(2F_{2,n-2}^{\alpha})^{1/2}$ or $t_{n-2}^{\alpha/4}$. Good tables of $|m|_{k,v}^{\alpha}$ are available in Hahn and Hendrickson [20]; these are reproduced in Miller [33]. The probability is exactly $1 - \alpha$ that α and β are contained in the intervals (29).

A request is sometimes made of the statistician for a confidence interval for the value of the regression function at x. If just one x is specified, then an interval based on Student's t-distribution can be used. However, if x is unspecified and is, instead, a generic term for any value of the independent variable, then a confidence band on the whole regression function is needed.

The classical *Working–Hotelling–Scheffé confidence band** is

$$\alpha + \beta x \in \hat{\alpha} + \hat{\beta}x \pm (2F_{2,n-2}^{\alpha})^{1/2} s \times \left[\frac{1}{n} + \frac{(x - \bar{x})^2}{\sum_{i=1}^{n}(x_i - \bar{x})^2}\right]^{1/2}. \tag{31}$$

The parameters and estimates in (31) correspond to those for model (25). These intervals follow from projections of the F confidence ellipsoid for (α, β) (see Miller [31]). The probability is exactly $1 - \alpha$ that the confidence intervals (31) are simultaneously correct for all values of x, $-\infty < x < +\infty$.

Since the expressions (31) as functions of x are hyperbolas, the confidence band is seldom actually drawn in practice. The *Graybill–Bowden confidence band* [19]

$$\alpha + \beta(x - \bar{x}) \in \hat{\alpha} + \hat{\beta}(x - \bar{x}) \pm |m|_{2,n-2}^{\alpha} s \times \left[\frac{1}{\sqrt{n}} + \frac{|x - \bar{x}|}{\left\{\sum_{i=1}^{n}(x_i - \bar{x})^2\right\}^{1/2}}\right] \tag{32}$$

utilizes straight-line segments and is therefore much easier to draw. The parameters and estimates in (32) are based on the model (28). The intervals (32) come from projections of the rectangular confidence region (29) for (α, β), and the probability is exactly $1 - \alpha$ that the intervals are simultaneously correct for all x.

Straight-line confidence bands of uniform width over a finite x range have also been given by Gafarian [14] and Bowden and Graybill [2].

Figure 1 exhibits 10 pairs of bilirubin production levels measured by a standard old method and an easier new method. The least-squares* estimates (30) under model (28) are $\hat{\alpha} = 15.00$, $\hat{\beta} = 0.97$, and $s = 2.06$. The corresponding 95% simultaneous confi-

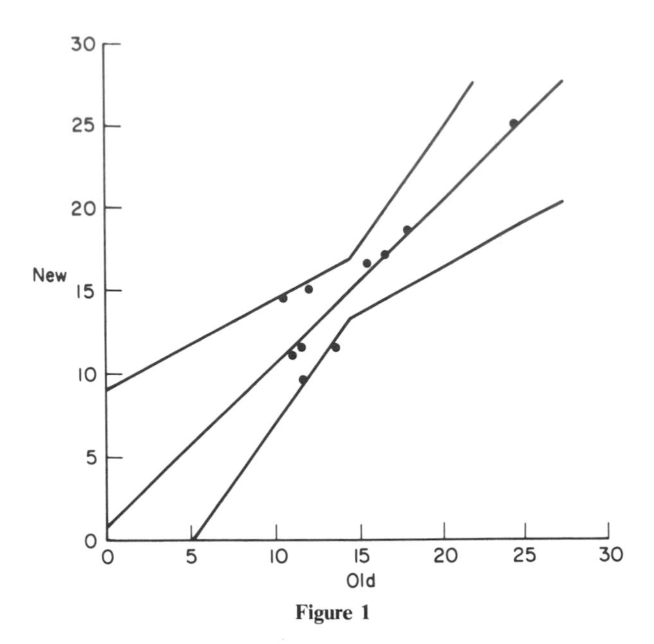

Figure 1

dence intervals (29) with $|m|_{2,8}^{0.05} = 2.72$ and $\sum_{i=1}^{n}(x_i - \bar{x})^2 = 166.58$ are

$$\begin{aligned} \alpha &\in 15.00 \pm 1.77 = (13.23, 16.77), \\ \beta &\in 0.97 \pm 0.43 = (0.54, 1.40). \end{aligned} \tag{33}$$

The estimated regression line $0.95 + 0.97x$ and the Graybill–Bowden linear-segment confidence band (32) are drawn in Fig. 1. The Working–Hotelling–Scheffé confidence band with $(2F_{2,8}^{0.05})^{1/2} = 2.99$ or $t_8^{0.05/4} = 2.75$ would consist of a pair of hyperbolas that are slightly farther from the regression line at $\bar{x} = 14.48$ and slightly closer at $x = 5$ or 25.

The prediction problem is to predict the value of a future observation Y at a specified value x^0 of the independent variable, and the calibration* (sometimes called discrimination) problem is to estimate the value of the independent variable x that gave rise to the dependent variable value y^0. Simultaneous confidence intervals are available when more than one x^0 or y^0 are specified. For details on these techniques, the reader is referred to Miller [33].

Many of the preceding ideas and methods extend to the multiple linear regression* situation (see Miller [33]).

References

[1] Armitage, J. V., and Krishnaiah, P. R. (1964). "Tables for the Studentized Largest Chi-Square Distribution and Their Applications." *Techn. Rep. No. ARL 64-188*, Aerospace Research Laboratories, Wright-Patterson Air Force Base, Ohio.

[2] Bowden, D. C., and Graybill, F. A. (1966). *J. Amer. Statist. Ass.*, **61**, 182–198.

[3] Brown, L. D. (1979). "A Proof That Kramer's Multiple Comparisons Procedure for Differences Between Treatment Means Is Level-α for 3, 4, or 5 Treatments." Manuscript, Cornell University, Ithaca, NY.

[4] Dunn, O. J. (1961). *J. Amer. Statist. Ass.*, **56**, 52–64.

[5] Dunn, O. J. (1964). *Technometrics*, **6**, 241–252.

[6] Dunn, O. J. (1974). *Commun. Statist.*, **3**, 101–103.

[7] Dunnett, C. W. (1955). *J. Amer. Statist. Ass.*, **50**, 1096–1121.

[8] Dunnett, C. W. (1964). *Biometrics*, **20**, 482–491.

[9] Dunnett, C. W. (1980). *J. Amer. Statist. Ass.*, **75**, 789–795.

[10] Dunnett, C. W. (1980). *J. Amer. Statist. Ass.*, **75**, 796–800.

[11] Dwass, M. (1960). In *Contributions to Probability and Statistics, Some k-Sample Rank-Order Tests.* I. Olkin et al., eds. Stanford University Press, Stanford, CA, pp. 198–202.

[12] Gabriel, K. R. (1969). *Ann. Math. Statist.*, **40**, 224–250.

[13] Gabriel, K. R. (1978). *J. Amer. Statist. Ass.*, **73**, 724–729.

[14] Gafarian, A. V. (1964). *J. Amer. Statist. Ass.*, **59**, 182–213.

[15] Genizi, A., and Hochberg, Y. (1978). *J. Amer. Statist. Ass.*, **73**, 879–884.

[16] Gold, R. Z. (1963). *Ann. Math. Statist.*, **34**, 56–74.

[17] Goodman, L. A. (1964). *Ann. Math. Statist.*, **35**, 716–725.

[18] Goodman, L. A. (1965). *Technometrics*, **7**, 247–254.

[19] Graybill, F. A., and Bowden, D. C. (1967). *J. Amer. Statist. Ass.*, **62**, 403–408.

[20] Hahn, G. J., and Hendrickson, R. W. (1971). *Biometrika*, **58**, 323–332.

[21] Harter, H. L. (1960). *Ann. Math. Statist.*, **31**, 1122–1147.

[22] Harter, H. L. (1969). In *Order Statistics and Their Use in Testing and Estimation*, Vol. 1: *Tests Based on Range and Studentized Range of Samples from a Normal Population*. Aerospace Research Laboratories. (Available from Superintendent of Documents, U.S. Government Printing Office, Washington, DC 20402.)

[23] Hayter, A. J. (1984). *Ann. Statist.*, **12**, 61–75.

[24] Hochberg, Y. (1974). *J. Multivariate Anal.*, **4**, 224–234.

[25] Hunter, D. (1976). *J. Appl. Prob.*, **13**, 597–603.

[26] Koziol, J. A. and Reid, N. (1977). *Ann. Statist.*, **5**, 1099–1106.

[27] Kramer, C. Y. (1956). *Biometrics*, **12**, 307–310.

[28] Krishnaiah, P. R. (1965). *Ann. Inst. Statist. Math.*, **17**, 35–53.

[29] Krishnaiah, P. R. (1979). In *Developments in Statistics*, Vol. 2, P. R. Krishnaiah, ed. Academic Press, New York. pp. 157–201.

[30] Kurtz, T. E. (1956). "An Extension of a Multiple Comparison Procedure." Ph.D. dissertation. Princeton University, Princeton, NJ.

[31] Miller, R. G., Jr. (1966). *Simultaneous Statistical Inference*. McGraw-Hill, New York.

[32] Miller, R. G., Jr. (1977). *J. Amer. Statist. Ass.*, **72**, 779–788.

[33] Miller, R. G., Jr. (1981). *Simultaneous Statistical Inference*, 2nd ed. Springer, New York.

[34] Moses, L. E. (1978). *Comm. Statist. B*, **7**, 479–490.

[35] Nemenyi, P. (1963). "Distribution-free Multiple Comparisons." Ph.D. dissertation. Princeton University, Princeton, NJ.

[36] O'Neill, R. T. and Wetherill, B. G. (1971). *J. R. Statist. Soc. B*, **33**, 218–241.

[37] Oude Voshaar, J. H. (1980). *Ann. Statist.*, **8**, 75–86.

[38] Scheffé, H. (1953). *Biometrika*, **40**, 87–104.

[39] Šidák, Z. (1967). *J. Amer. Statist. Ass.*, **62**, 626–633.

[40] Spjøtvoll, E., and Stoline, M. R. (1973). *J. Amer. Statist. Ass.*, **68**, 975–978.

[41] Steel, R. G. D. (1960). *Technometrics*, **2**, 197–207.

[42] Steel, R. G. D. (1961). *Biometrics*, **17**, 539–552.

[43] Stoline, M. R. (1978). *J. Amer. Statist. Ass.*, **73**, 656–660.

[44] Stoline, M. R. (1981). *Amer. Statist.*, **35**, 134–141.

[45] Stoline, M. R., and Ury, H. K. (1979). *Technometrics*, **21**, 87–93.

[46] Tamhane, A. C. (1979). *J. Amer. Statist. Ass.*, **74**, 471–480.

[47] Tukey, J. W. (1953). "The Problem of Multiple Comparisons." Mimeographed notes. Princeton University, Princeton, NJ.

[48] Ury, H. K. (1976). *Technometrics*, **18**, 89–97.

[49] Ury, H. K., Stoline, M. R., and Mitchell, B. T. (1980). *Commun. Statist. B*, **9**, 167–178.

(BONFERRONI INEQUALITIES AND INTERVALS
CONFIDENCE INTERVALS AND REGIONS
GOODMAN'S Y^2
k-RATIO t-TESTS, t-INTERVALS AND POINT ESTIMATES FOR MULTIPLE COMPARISONS
MULTIPLE RANGE AND ASSOCIATED TEST PROCEDURES
MULTIVARIATE MULTIPLE COMPARISONS
RANKING PROCEDURES
SCHEFFÉ'S SIMULTANEOUS COMPARISON PROCEDURE
SELECTION PROCEDURES
TUKEY'S SIMULTANEOUS COMPARISON PROCEDURE
WORKING–HOTELLING–SCHEFFÉ CONFIDENCE BANDS)

R. MILLER

MULTIPLE CORRELATION COEFFICIENT

Although ideas related to multiple correlation appear in Galton's [7] discussion of anthropometric data, and in even earlier works of Bravais [3] and Edgeworth [4], it was G. U. Yule [30] who fitted a plane to a swarm of points using the method of least squares* and thus launched the techniques of multiple regression and multiple correlation. Actually K. Pearson* [24], Yule's professor and mentor, visualized a very basic role for the multiple correlation coefficient when he advocated maximizing it, instead of the least squares, as the general principle for fitting a plane of regression.

The multiple correlation coefficient $\mathscr{P}_{1\cdot(23\ldots p)}$ between jointly distributed variables X_1 and $(X_2, X_3, \ldots, X_p)$ is the maximum product moment correlation between X_1 and any linear combination $\sum_2^p a_j X_j$ of $X_2, X_3 \ldots X_p$. If

$$\mathbf{\Sigma} = (\sigma_{ij}) = \begin{pmatrix} \sigma_{11} & \boldsymbol{\sigma}'_{12} \\ \boldsymbol{\sigma}_{12} & \mathbf{\Sigma}_{22} \end{pmatrix}$$

is the covariance matrix of $(X_1, X_2, \ldots, X_p)$, then

$$\mathscr{P}_{1\cdot(23\ldots p)} = \underset{a_2, \ldots, a_p}{\mathrm{Max}} \left\{ \mathrm{Corr}\left(X_1, \sum_2^p a_j X_j \right) \right\}$$
$$= \left\{ \boldsymbol{\sigma}'_{12} \mathbf{\Sigma}_{22}^{-1} \boldsymbol{\sigma}_{12} / \sigma_{11} \right\}^{1/2}. \quad (1)$$

An important property of $\mathscr{P}^2$ is that, in the multiple linear regression of X_1 on $X_2, \ldots, X_p$, $\mathscr{P}^2$ is the proportion of the total variation due to regression (*see* MULTIPLE LINEAR REGRESSION, where numerical examples are given under normality assumptions). Let $\mathbf{X}' = (X_1, X_2, \ldots, X_p)$ be distributed according to a p-variate normal distribution with covariance matrix $\mathbf{\Sigma}$ and let $\mathbf{X}_1, \mathbf{X}_2, \ldots \mathbf{X}_N$ be a random sample from this population (*see* MULTINORMAL DISTRIBUTION). Then the sample multiple correlation coefficient $R_{1\cdot(23\ldots p)}$ between X_1 and $(X_2, X_3, \ldots, X_p)$ is the analogous quantity $\{\mathbf{s}'_{12}\mathbf{S}_{22}^{-1}\mathbf{s}_{12}/s_{11}\}$,$^{1/2}$ where $\mathbf{S}$ is the sample covariance matrix partitioned similarly to the aforementioned $\mathbf{\Sigma}$. In multiple linear regression* analysis $R^2_{1\cdot(23\ldots p)}$ is known as the *coefficient of determination**.

SAMPLING DISTRIBUTION

Let $R = R_{1\cdot(23\ldots p)}$ be the multiple correlation coefficient of a sample of size N from a p-variate normal population with $\mathscr{P} = \mathscr{P}_{1\cdot(23\ldots p)} = 0$. Then

$$\{(N-p)/(p-1)\}\{R^2/(1-R^2)\} \sim F(p-1, N-p) \quad (2)$$

(*see* F-DISTRIBUTION). That is, in the case of independence the distribution of R^2 is related by (2) to that of the variance ratio with $(p-1, N-p)$ degrees of freedom. Equivalently,

$$T^2 = (n-2)R^2/(1-R^2)$$

then has a Hotelling's T^2-distribution*; when $p = 2$ this reduces to the square of Student's t-statistic (with $n-2$ degrees of freedom) for testing for independence in a bivariate normal distribution*.

In 1928, R. A. Fisher* [5] observed that "of the problems of exact distribution of statistics in common use that of the multiple correlation is the last to have resisted a solution." The problem is important because of the "practical necessity" for significance tests and "the great theoretical interest owing to the close connection which must exist between it and the simple correlation coefficient on the one hand, and on the other to the form already obtained from the uncorrelated material." He then interpreted R^2 as a projection, the cosine of an angle in the sample space, and using the characteristic geometric method, obtained the density of R^2:

$$f_{R^2}(x) = \frac{(1-\mathscr{P}^2)^{n/2}(1-x)^{n-p-1/2}}{\Gamma\left(\frac{n-p-1}{2}\right)\Gamma\left(\frac{n}{2}\right)} \times \sum_{j=0}^{\infty} \left[\frac{(\mathscr{P}^2)^j x^{(p-1)/2+j-1}\Gamma\left(\frac{n}{2}+j\right)}{j!\,\Gamma\left(\frac{p-1}{2}+j\right)} \right], \quad (3)$$

where $n = N - 1$. Particular cases of these results were earlier discussed by Yule (1907) and Isserlis (1915).

The exact non-null distribution of R^2 can now be variously derived and expressed in several equivalent forms, e.g., see Garding [8], Gurland [10], Moran [18], and Wilks [29]. The most elegant approach is based on an ingenious representation originally due to Wijsman [28] but often and erroneously attributed to others. Using the method of random orthogonal transformations he introduced earlier, Wijsman shows that

$$\tilde{R}^2 \sim \{(\rho\chi_{N-1} + Z)^2 + \chi^2_{p-2}\}/\chi^2_{N-p}, \quad (4)$$

where $\tilde{R}^2 = R^2/(1-R^2)$, $\tilde{\rho}^2 = \rho^2/(1-\rho^2)$; Z, χ_{N-1}, χ_{p-2} and χ_{N-p} are independently distributed unit normal, and chi variables (*see* CHI DISTRIBUTION) with indicated degrees of freedom, and $\sim$ denotes equivalence in distribution. The representation can be readily used, as Gurland [10] did, to obtain the characteristic function* of $\tilde{R}^2$. It can then be inverted to construct the density of $\tilde{R}^2$. Among various forms of the density of R^2, two are noteworthy. The first (due to Fisher) is

$$f_{R^2}(x) = \frac{(1-\mathscr{P}^2)^{(N-1)/2} x^{(p-3)/2}(1-x)^{(N-p)/2-1}}{B\left(\left(\frac{p-1}{2}\right),\left(\frac{N-p}{2}\right)\right)} \times {}_2F_1\left(\frac{N-1}{2},\frac{N-1}{2};\frac{p-1}{2};\mathscr{P}^2 x\right), \quad (5)$$

where ${}_2F_1$ is the Gauss hypergeometric function. The second form (due to Gurland) is

$$f_{R^2}(x) = \sum_{j=0}^{\infty} b_j f_j(x), \quad (6)$$

where $f_j(\cdot)$ is the density of a beta* variable with parameters $((p-1)/2+j, (N-p)/2)$, and

$$b_j = \frac{\Gamma\left(\frac{N-1}{2}+j\right)}{j!\,\Gamma\left(\frac{N-1}{2}\right)} \frac{(\mathscr{P}^2)^j}{(1-\mathscr{P}^2)^{(n-1)/2}}. \quad (7)$$

Equivalently, the density of $R^2/(1-R^2)$ is that of the ratio X/Y of two independent variables X and Y, such that

$$X \sim \chi^2_{p-1+2K}, \qquad Y \sim \chi^2_{N-p},$$

where K has a negative binomial distribution* with density proportional to $(\mathscr{P}^2)^k(1-\mathscr{P}^2)^{(N-1)/2}$ for $k = 0, 1, 2, \ldots$ [9, Sec. 6.9]. Gurland and Milton [11] also consider convergence characteristics of these series.

CUMULANTS

The jth moment of R about 0 is given by Banerjee [2] as

$$\mu_j'(R) = \frac{(1-\mathscr{P}^2)^{(N-1)/2}\Gamma\left(\frac{p+j-1}{2}\right)}{\Gamma\left(\frac{N-1+j}{2}\right)} \times D^j {}_2F_1\left(\frac{N-1+j}{2},\frac{N}{2};\frac{p-1}{2};\mathscr{P}^2\right), \quad (8)$$

where D denotes the operator $\frac{1}{2}\mathscr{P}^3(\partial/\partial\mathscr{P})$. From this, or otherwise, the mean and variance of R^2 are seen to be

$$E(R^2) = \mathscr{P}^2 + \frac{p-1}{N-1}(1-\mathscr{P}^2) - \frac{2(N-p)}{N^2-1}\mathscr{P}^2(1-\mathscr{P}^2) + O(N^{-2}), \quad (9)$$

and

$$\operatorname{Var}(R^2) = \begin{cases} \dfrac{4\rho^2(1-\rho^2)^2(N-p)^2}{(N^2-1)(N+3)} + O(N^{-2}) & \text{if } \rho \neq 0 \\ \dfrac{2(p-1)(N-p)}{(N-1)^2(N+1)} + O(N^{-2}) & \text{if } \rho = 0. \end{cases} \quad (10)$$

Moschopoulos and Mudholkar [19] obtain and use the m.g.f. of $T = -\log(1-R^2)$ to derive asymptotic expansions for the first four cumulants of T up to $O(N^{-4})$. The corresponding expressions for the cumulants of R^2 are obtained from these using Taylor expansions*.

ASYMPTOTIC DISTRIBUTIONS

If $\mathscr{P} \neq 0$ or 1, then the asymptotic distribution of $\sqrt{N}(R^2 - \mathscr{P}^2)/\{2\mathscr{P}(1 - \mathscr{P}^2)\}$ is $N(0, 1)$ as $N \to \infty$. However, if $\mathscr{P} = 0$ then, as $N \to \infty$, NR^2 is asymptotically noncentral chi-square* with $(p - 1)$ d.f. and noncentrality parameter δ^2.

Fisher's z-transform of the ordinary correlation coefficient plays an important role in bivariate correlation analysis (*see* FISHER'S z-TRANSFORMATION). In the present case, let $Z = \tanh^{-1}R$ and $\zeta = \tanh^{-1}\mathscr{P}$. Then as $N \to \infty$ the asymptotic distribution of $\sqrt{N}(Z - \zeta)$ is $N(0, 1)$, provided that $\mathscr{P} \neq 0$. If $\sqrt{N}\mathscr{P}$ is fixed then, as Gajjar [6] shows, $\sqrt{N}Z$ is asymptotically a noncentral chi-variable with $(p - 1)$ d.f. and noncentrality parameter $N\zeta^2$.

APPROXIMATIONS

In view of the practical importance of R^2 and the intricate nature of its exact distribution, several approximations for its non-null distribution have appeared in the literature. Tiku [27], Khatri [14], and Gurland [10] present moments-based F and noncentral-F* approximations for $\tilde{R}^2$. Lee [15] carefully studies these and constructs new approximations tied to the normal distribution by applying the Geary–Fieller reasoning to certain power-transforms of the numerator and denominator of Wijsman's representation of $\tilde{R}^2$. Lee also examines Fisher's z-transform of R and finds it inadequate; but see the editor's note accompanying his paper. Recently, Moschopoulos and Mudholkar [19] constructed a Gaussian approximation to the non-null distribution by applying the Wilson–Hilferty reasoning to $T = -\log(1 - R^2)$.

INFERENCE

The practical importance of R^2 stems from its role in multiple regression. As the coefficient of determination, it is used to assess the goodness of the regression fit and also to test the independence between X_1 and $(X_2, X_3, \ldots X_p)$, i.e., to test $H_0 : \mathscr{P}^2 = 0$. The best invariant test (see ref. 21) of H_0 rejects it when $R^2 \geqslant$ constant, where the critical constant is determined using the F-distribution in (2). For admissible, Bayes and minimax properties of this test, see references in Giri [9].

The sample multiple correlation R^2 is the maximum likelihood estimator of $\mathscr{P}^2$, but it is not unbiased. The best unbiased estimation of $\mathscr{P}^2$ is discussed in Olkin and Pratt [22]. For methods of constructing confidence intervals for $\mathscr{P}^2$, see Moschopoulos and Mudholkar [19] and Biometrika Tables [23]. The relevant percentiles of R^2 appear in Lee [16] and *Biometrika Tables for Statisticians* [23].

NON-NORMAL CASE

The theory of multiple correlation has been extended in two non-normal directions. The nonparametric version, the *multiple rank correlation**, is proposed and studied by Moran [17]. However, much of the recent work is targeted at replacing the underlying multivariate normal distribution by spherical and elliptically contoured distributions. Much of the normal theory results carry over, at least in structural form, to these more general models. For example the relation (2) between R^2 and the F-distribution remains valid under the spherical model, and the asymptotic distributions described earlier undergo very little change under the elliptical models. For a review see Muirhead [20, 21].

Literature

Multiple correlation being a very basic topic, it is discussed in most monographs devoted to multivariate analysis*, e.g. see Anderson [1], Giri [9], Muirhead [21], and Srivastava and Khatri [25]. More recent references appear in Subrahmaniam and Subrahmaniam

[26]. An excellent summary of results is given in Johnson and Kotz [13, Chap. 32].

References

[1] Anderson, T. W. (1958). *An Introduction to Multivariate Analysis*. Wiley, New York.

[2] Banerjee, D. P. (1952). *J. Ind. Soc. Agric. Statist.*, **4**, 88 – 90.

[3] Bravais, A. (1846). *Acad. R. Sci. Inst. France*, **9**, 256–332.

[4] Edgeworth, F. Y. (1892). *Philos. Mag., 5th Series*, **34**, 194–204.

[5] Fisher, R. A. (1928). *Proc. R. Soc. Lond. A*, **121**, 654–673.

[6] Gajjar, A. V. (1967). *Metron*, **26**, 189–193.

[7] Galton, F. (1888). *Proc. R. Soc. Lond.*, **45**, 135–145.

[8] Garding, L. (1941). *Skand. Aktuarietidskr.*, **24**, 185–202.

[9] Giri, N. C. (1977). *Multivariate Statistical Inference*, Academic Press, New York.

[10] Gurland, J. (1968). *J. R. Statist. Soc. B*, **30**, 276–283.

[11] Gurland, J. and Milton, R. C. (1970). *J. R. Statist. Soc. B*, **32**, 381–394.

[12] Isserlis, L. (1917). *Philos. Mag., 6th series*, **34**, 205–220.

[13] Johnson, N. L. and Kotz, S. (1970). *Distributions in Statistics. Continuous Univariate 2*. John Wiley, New York, Chap. 32.

[14] Khatri, C. G. (1966). *Ann. Inst. Statist. Math.*, **18**, 375–380.

[15] Lee, Y. S. (1971). *J. R. Statist. Soc. B*, **33**, 117–130.

[16] Lee, Y. S. (1972). *Biometrika*, **59**, 175–189.

[17] Moran, P. A. P. (1948). *Biometrika*, **35**, 203–206.

[18] Moran, P. A. P. (1950). *Proc. Camb. Philos. Soc.*, **46**, 521–522.

[19] Moschopoulos, P. G. and Mudholkar, G. S. (1983). *Commun. Statist. Comp. Simul.*, **12**, 355–371.

[20] Muirhead, R. J. (1980). In *Multivariate Statistical Analysis*, R. P. Gupta, ed. North-Holland, New York.

[21] Muirhead, R. J. (1982). *Aspects of Multivariate Statistical Theory*. Wiley, New York.

[22] Olkin, I. and Pratt, J. W. (1958). *Ann. Math. Statist.*, **29**, 201–2ll.

[23] Pearson, E. S. and Hartley, H. O., eds. (1972). *Biometrika Tables for Statisticians*, Vol. II, Cambridge University Press, New York.

[24] Pearson, Karl (1920). *Biometrika*, **13**, 25 and 45.

[25] Srivastava, M. S. and Khatri, C. G. (1979). *An Introduction to Multivariate Statistics*. North-Holland, New York.

[26] Subrahmaniam, S. and Subrahmaniam, K. (1973). *Multivariate Analysis: A Selected and Abstracted Bibliography*. Marcel Dekker, New York.

[27] Tiku, M. L. (1966). *J. Ind. Soc. Agric. Statist.*, **18**, 4–16.

[28] Wijsman, R. A. (1959). *Ann. Math. Statist.*, **30**, 597–601.

[29] Wilks, S. S. (1932). *Ann. Math. Statist.*, **3**, 196–203.

[30] Yule, G. U. (1897). *J. R. Statist. Soc.*, **60**, 3.

[31] Yule, G. U. (1907). *Proc. R. Soc. Lond. Ser. A*, **79**, 182–193.

(COEFFICIENT OF DETERMINATION
CORRELATION
FISHER'S *z*-TRANSFORMATION
F-TESTS
MULTINORMAL DISTRIBUTION
MULTIPLE LINEAR REGRESSION
MULTIVARIATE ANALYSIS)

GOVIND S. MUDHOLKAR

MULTIPLE CRITERIA DECISION MAKING (MCDM)

Multiple criteria decision making (MCDM) has become one of the fastest growing fields of inquiry in operational sciences since the early seventies. A recent MCDM bibliography listed about 2000 published works that have appeared during the past decade or two. The word *multiple* identifies the major concern and focus of this burgeoning field: it is that multiple criteria rather than a single criterion characterize human choice and decision making.

It has become increasingly less satisfactory to view the world around us in a unidimensional way and to use only a single criterion when judging what we see. Humans always compare, rank, and order the objects of their experience with respect to *criteria of choice*. But only in very simple, straightforward, or routine situations can we assume that a *single* criterion of choice will be fully satisfactory.

We may pick the *largest* apple from a basket (criterion of size), the *cheapest* brand of beer (price), the *highest* salary offer (dollar amount), or the *shortest* route home (distance). But often we worry whether the largest apple is the sweetest, the juiciest, the most aromatic, or the freshest. We may be concerned not only with our beer's price but also with its taste, caloric content, carbonation, and alcoholic content. We agonize about whether the highest salary offer is the one that also promises the highest rate of salary increase, whether it is accompanied by generous fringe benefits, and whether the job provides comfortable working conditions or stimulates sufficient interest and provides a challenge.

In a very definitive sense we can talk about decision making *only* if at least two criteria are present. If only one criterion exists, mere measurement and search suffice for making a choice. For example, if you are asked to select the largest apple from a basket, the tallest man from a team, the heaviest book from a shelf, are you engaged in decision making? Or is it sufficient to measure with respect to the criterion in question and search for the "maximal" alternative? This reasoning does not imply that measurement and search are simple and easy activities.

In many real situations, decision criteria are not perfectly measurable but are loaded with uncertainty*, imprecision, and fuzziness (*see* FUZZY SET THEORY). For example, we might not be able to measure precisely the size of the apples from which the largest is to be selected. In such a case, we could make the wrong decision due to approximate or imprecise measurement. As long as we have only one criterion, decision theory*, with its loss and risk function approach, would allow us to deal with the situation. However, the uncertainty and risk measurement in multidimensional situations, *multivariate risk theory*, has not yet been fully developed. One possible multidimensional theory of risk has been advanced by Colson and Zeleny; a number of utility-related treatments can be found in the work of Keeney and Raiffa.

We speak about multiple *criteria*. What are criteria? Criteria are measures, rules, and standards that guide human choice and decision making. A number of different types of these guiding measures can be considered. We shall introduce only three basic criterion types: *attributes*, *objectives*, and *goals*.

Attributes refer to descriptors of objective reality. A person might be described in terms of height, weight, coloring, age, or wealth. Other attributes might be more subjectively colored, e.g., intellect, beauty, figure, and social status. One can choose any attribute or attributes as criteria of choice or decision making. A theory dealing specifically with the aggregation* of attributes into a single criterion of "utility function" is designated *multiattribute utility theory* (*MAUT*).

Objectives represent *directions* of improvement or preference along individual attributes or complexes of attributes. There are only two directions: more and less (i.e., maximize and minimize). Thus height in itself is an attribute, but finidng the tallest among the alternatives, or maximizing height, is an objective. It is at this point in the decision process that the decision maker's needs and desires enter. That is, will the decision maker choose to maximize height, minimize age, or maximize amiableness? Thus an attribute becomes an objective when it is assigned (by a human) a purpose, a direction of desirability or improvement. "To maximize horsepower" is an objective directing the search along the attribute horsepower. Such MCDM methodologies as *multiobjective programming* or *compromise programming* are designed to assist in resolving a conflict among a number of incommensurable objectives or objective functions.

Goals are specific values, or levels defined in terms of either attributes or objectives and determined a priori. They can be precise, desired levels of attainment as well as more fuzzily delineated or vague ideals. "Maximizing gas mileage" is a well-stated *objective* in the search for an automobile. "Achieving gas mileage of 26 miles per gallon" is a clearly stated *goal* indicating a specific reference value for that objective. Thus goals refer quite unambiguously to particular target levels of achievement that can be defined in terms of both attributes and objectives. The most common methodology specifically designed to deal with the attainment of goals is referred to as *goal programming*.

A basic solution concept applicable in each of the methodologies utilized in the MCDM process is the concept of *nondominance*, or *Pareto optimality*. The nondominance solution concept, usually stated as the Pareto principle or Pareto optimality* principle, postulates that a solution B is dominated by solution A if by moving from B to A we improve at least one objective function and worsen no others. Solution A is nondominated if there is no other solution that would improve at least one objective and not worsen any other.

The usefulness of nondominated solutions in MCDM methodology can be illustrated by examining some of the advantages of finding them:

1. Multiple objectives are often incommensurate, both qualitative and quantitative, and carry different weights of importance. This leads to a complex problem of trade-off evaluation using the decision maker's utility or preference function. Reliable construction of a utility function may, however, be too complex, unrealistic, or impractical. The set of nondominated solutions then provides a meaningful step forward under such conditions of relative ignorance.

2. If more is always preferable to less, then any solution that maximizes the utility function of a *rational* decision maker must be nondominated: If more is preferred to less, then only higher or equal utility may be derived from the increased levels of corresponding attributes or criteria of choice.

3. If N consists of only a relatively small number of solutions or alternatives of choice, there is no need to search for the decision maker's utility function. Consequently it makes sense to explore the set of all feasible decision alternatives X and to characterize its subset N of all nondominated solutions before engaging in the assessment of u. It is not wise to gather and process all the information needed for utility assessment without finding the approximate size of N first.

4. The set of nondominated alternatives can be useful in dealing with more complicated types of X, e.g., discrete point sets or nonconvex sets of feasible alternatives.

Thus we note that the nondominance solution concept forms a recurring basis for most MCDM solution methodologies. For specific, detailed solution processes for each of the MCDM methodologies mentioned—multiattribute utility theory, multiobjective programming, compromise programming, and goal programming—refer to the Bibliography.

Bibliography

Cochrane, J. L. and Zeleny, M., eds. (1973). *Multiple Criteria Decision Making*. University of South Carolina Press, Columbia, SC. (The first substantial reference to MCDM as an organized field of inquiry. This is a classic volume containing original contributions of some 40 researchers and thinkers who met in South Carolina in 1972. Still a source of fresh and challenging research and applications ideas related to MCDM.)

Cohon, J. L. (1978). *Multiobjective Programming and Planning*. Academic Press, New York. (Well-written monograph concentrating on the methodology of linear multiobjective programming and its applications, especially in the public and governmental sectors.)

Colson, G. and Zeleny, M. (1979/1980). *Uncertain Prospects Ranking and Portfolio Analysis Under the Conditions of Partial Information*. Oelgeschlager, Gunn and Hain Publishers, Cambridge, MA. [Highly technical monograph introducing the multidimensional concept of risk and its measurement via the so-called portfolio ranking vector (PRV)].

Ignizio, J. P. (1976). *Goal Programming Extensions*. Heath, Lexington, MA. (Specialized monograph dealing with preemptive version of goal programming accompanied by some examples and extensions.)

Keeney, R. L. and Raiffa, H. (1976). *Decisions with Multiple Objectives: Preferences and Value Tradeoffs*. New York: Wiley, New York. (An extensive monograph dealing with MAUT and especially concerned with the aggregation of multiple attributes into a single utility suprafunction. A well-written book delving into the earlier MCDM approach and thinking.)

Lee, S. M. (1972). *Goal Programming for Decision Analysis*. Auerbach Publishers, Philadelphia, PA. (One of the earliest monographs concerned with the theory and applications of preemptive version of goal programming. Obviously outdated by now, but still a good example of the early MCDM thinking and writing.)

Starr, M. K. and Zeleny, M., eds. (1977). *Multiple Criteria Decision Making*. North-Holland, Amsterdam. (A stimulating and original research reference establishing the trends and concepts dominating MCDM during the eighties. A source of research topics for the MCDM newcomers.)

Zeleny, M. (1974). *Linear Multiobjective Programming*. Springer-Verlag, New York. (The original monograph and extension of linear multiobjective programming together with multicriterion simplex method and its computer code.)

Zeleny, M. (1975). *Multiple Criteria Decision Making: Kyoto 1975*. Springer-Verlag, New York. (High quality and high visibility. Proceedings of the MCDM conference in Kyoto. Contributions by Charnes and Cooper, Marschak, Hammond, Polak, Zadeh, Rapoport, Haimes, and many others.)

Zeleny, M. (1982). *Multiple Criteria Decision Making*. McGraw-Hill, New York. (The first comprehensive textbook covering MCDM as a whole. Accompanied by a number of examples and instructor's manual of real-life cases, this book has become a standard in the operations research–management sciences literature.)

Zeleny, M. (1984). *MCDM—Past Decade and Future Trends*. JAI Press, Greenwich, Conn. (A decennial festschrift volume commemorating the 10 years of MCDM since the South Carolina Conference in 1972. This is a source book for MCDM research, teaching, and applications. It contains reviews and summaries of individual MCDM methodologies as well as challenges and scientific projections into the future.)

(DECISION THEORY
FUZZY SET THEORY
RISK MEASUREMENT, FOUNDATIONS OF
RISK THEORY
UTILITY)

MILAN ZELENY

MULTIPLE DECISION PROCEDURES

A statistical decision problem in which there are only a finite number of possible decisions is called a multiple decision problem or a multidecision problem. Any decision function or decision rule that might be used in such a problem is called a multiple decision procedure (*see also* DECISION THEORY).

Problems of testing statistical hypotheses form an important special class of multiple decision problems in which there are only two possible decisions: accept or reject the null hypothesis. A simple example of a problem with three decisions is one in which θ is a real-valued parameter and it must be decided, on the basis of some statistical data, which of the following three possibilities is correct: $\theta < \theta_0$, $\theta_0 \leqslant \theta \leqslant \theta_1$, or $\theta > \theta_1$, where θ_0 and θ_1 are given numbers ($\theta_0 < \theta_1$).

Another type of multiple decision problem is one in which it must be decided which of k populations ($k \geqslant 2$) has the largest mean, based on random samples from each of the populations. In a variant of this problem, the means of the k populations must be ranked from largest to smallest (*see* RANKING PROCEDURES and SELECTION PROCEDURES).

The parameter space Ω in a multiple decision problem is not restricted in any way and might contain either just a finite number of possible values of the parameter θ or an infinite number. In every multiple decision problem, an appropriate loss $L(\theta, d)$ must be specified for every value of $\theta \in \Omega$ and every possible decision d.

(DECISION THEORY)

MORRIS H. DEGROOT

MULTIPLE DECREMENT TABLES

Suppose there is a community subject to a single decrement C_1. To fix ideas, this decrement could be death, and in conformity with accepted practice (*see* LIFE TABLES), the probability of death between ages x and $x + dx$ may be written $\lambda(x)\,dx + o(dx)$, so that the probability of surviving to age $x + 1$, given that the community member has survived to exact age x, is

$$\exp\left\{-\int_x^{x+1} \lambda(t)\,dt\right\} = \frac{S(x+1)}{S(x)}, \quad (1)$$

where $S(x) = \Pr\{X > x\}$ is the survival distribution [3, Chap. 3].

Let us assume that instead of the single decrement C_1, there are k mutually exclusive decrements $C_1, C_2, \ldots C_k$ that act independently, each subject to its own instantaneous probability of decrement (approximately) $\lambda_i(x)\,dx$ $(i = 1, 2, \ldots k)$. Then the total probability of decrement at age x is

$$\sum_{i=1}^{k} \lambda_i(x)\,dx,$$

and, in an obvious notation,

$$\begin{aligned}\frac{S(x+1)}{S(x)} &= \exp\left\{-\sum_{i=1}^{k}\int_x^{x+1}\lambda_i(t)\,dt\right\}\\ &= \prod_{i=1}^{k}\frac{S_i(x+1)}{S_i(x)}\\ &\equiv \prod_{i=1}^{k} p_i^*(x, x+1) \equiv p^*(x, x+1)\\ &\equiv 1 - q^*(x, x+1). \qquad (2)\end{aligned}$$

The single decrement survival distribution (1) has become the multiple decrement survival distribution (2). The foregoing, expressed in terms of multiple causes of death, is in essentially the same form that Cournot uses in his elementary probability text of 1843 [2, pp. 317–321].

Now consider how to estimate the probability of the ith decrement at age x last birthday, $q_i^*(x, x+1)$, in the given community. Write N for the number of individuals in the community aged exactly x and d_i for the number of decrements before age $x + 1$ from cause i. Then, conditional on N, the probability of the observed decrements (proportional to the likelihood of the realization) is the multinomial

$$\frac{N!}{(\prod_{i=1}^{k} d_i)!\,(N - \sum_{i=1}^{k} d_i)!} \times \prod_{i=1}^{k}\{q_i^*(x, x+1)\}^{d_i} \times \left\{1 - \sum_{i=1}^{k} q_i^*(x, x+1)\right\}^{N - \sum d_i}$$

and the maximum likelihood estimate of the required probability is

$$\hat{q}_i^*(x, x+1) = \frac{d_i}{N}, \qquad i = 1, 2, \ldots k$$

[3, paragraph 12.2].

Example. It is instructive to illustrate this by an example given by Böhmer [1]. Age x is 48 and $N = 285$ policies of a life insurance company that may be decremented by (a) death, (b) disability, or (c) lapse. Hence $k = 3$, and it is assumed that the group can be augmented by new cases during the year of age. We suppose the year to be subdivided into seconds and instead of passing directly from age x to age $x + 1$, we rewrite (2) as

$$p^*(x, x+1) = \prod_{i=1}^{3}\prod_{j=1}^{M} p_i^*(x + t_{j-1}, x + t_j),$$
$$t_0 = 0, \quad M = 315{,}360. \qquad (3)$$

As pointed out in ref. 3 (paragraph 6.7), if there are no decrements from t_{j-1} to t_j, the multiplicand is unity and it is only the decrements that contribute to the probability estimate. These decrements and the three increments which, like N, become part of the conditioning, are ordered as follows:

Type of Decrement	No. Before Decrement	No. After Decrement
Death	285	284
Lapse	284	283
Increment	283	284
Death	284	283
Disability	283	282
Increment	282	283
Disability	283	282
Increment	282	283
Death	283	282
Death	282	281

Death $\quad \hat{p}_1^*(48, 49) = \frac{284}{285}\cdot\frac{283}{284}\cdot\frac{282}{283}\cdot\frac{281}{282} = 0.9860$

Disability $\quad \hat{p}_2^*(48, 49) = \frac{282}{283}\cdot\frac{282}{283} = 0.9929$

Lapse $\quad \hat{p}_3^*(48, 49) = \frac{283}{284} = 0.9965$

This laborious estimation procedure has become practicable in an age of electronic computers [12]. The summary *occurrence/exposure* estimate of $q_i^*(x, x+1)$, e.g.,

$$\tilde{q}_1^*(48, 49) = \frac{4}{285 - \frac{1}{2}(3-3)} = 0.0140$$

used by today's demographers [7; 3, paragraph 12.4] is much simpler and may not be further from the truth than the previous value [9].

The foregoing estimate of $p_1^*(x, x+1)$ can be written in the form

$$\frac{N-j}{N} = \prod_{l=1}^{j} \frac{N-l}{N-l+1} = \prod_{l=1}^{j} \left(1 - \frac{1}{N-l+1}\right),$$

and the corresponding cumulative instantaneous death rate at age x, that is $\int_x^{x+1} \lambda_i(t)\, dt$, as

$$-\sum_{l=1}^{j} \ln\left(1 - \frac{1}{N-l+1}\right) \simeq \sum_{l=1}^{j} \frac{1}{N-l+1} \quad \text{for } N \text{ large.}$$

Successive values of this plotted against a hypothetical cumulative hazard rate should result in an approximate straight line. Nelson [10] has utilized such a procedure to good effect [3, paragraph 7.7].

It will have been noticed that relation (2) depends explicitly on the independence of the causes of decrement. Although dependence of two causes is not equivalent to probabilistic dependence there is a *prima facie* case for caution in declaring the opposite. Thus, in the foregoing illustration, the policyholder who let his policy lapse by failure to pay a due premium was presumably not near death or disability so far as these hazards can be foreseen. Lapse is thus likely to be probabilistically dependent on the other two causes of decrement, and relation (2) should be reconsidered. This difficulty had been mentioned in the controversies of the 1760s and 1870s [12], but to this day actuaries have not attempted to resolve it.

In the 1950s statisticians began to study multiple causes of failure under the title *competing risks** [12]. They naturally thought in terms of a joint survival distribution function

$$S_{1,\dots,k}(x_1, x_2, \dots, x_k) = \Pr\left\{\bigcap_{i=1}^{k} (X_i > x_i)\right\}$$

with marginal survival distribution functions

$$S_i(x_i) = S_{1\dots k}(0, \dots, x_i, \dots, 0), \quad i = 1, 2, \dots k$$

[3, paragraph 9.3]. There is now no assumption of independence of the random variables X_i $(i = 1, 2, \dots k)$. The observable random variable, failure because of one of the k causes, is

$$X = \min(X_1, X_2, \dots X_k),$$

and the overall survival function is

$$S_x(x) = \Pr\{X > x\} = \Pr\left\{\bigcap_{i=1}^{k} (X_i > x)\right\} = S_{1\dots k}(x, x, \dots x).$$

The instantaneous hazard rate at "age" x from cause of failure C_i in the presence of all other causes is then defined as

$$h_i(x) = \lim_{\theta \to 0} \frac{1}{\theta} \Pr\left\{(x < X_i < x + \theta) \bigcap_{j \neq i}^{k} (X_j > x) \,\middle|\, \bigcap_{j=1}^{k} (X_j > x)\right\}$$
$$= -\frac{1}{S_x(x)} \times \frac{\partial S_{1\dots k}(x_1, \dots, \dots x_k)}{\partial x_i}\Bigg|_{\bigcap_{i=1}^{k}(x_i = x)},$$

which is not necessarily the same as our previously utilized

$$\lambda_i(x) = -\frac{1}{S_i(x)} \frac{dS_i(x)}{dx} = -\frac{d \ln S_i(x)}{dx}.$$

If we define an index I such that

$$I = \begin{cases} i & \text{if } X_i = X, \\ 0 & \text{otherwise,} \end{cases}$$

the distribution function of time of failure

for C_i is

$$Q_i^*(x) = \Pr\{(X \leqslant x) \cap (I = i)\}$$
$$= \int_0^x h_i(t) S_x(t)\, dt,$$

where the asterisk denotes that failure from cause C_i is being considered in the presence of all causes C_j $(j = 1, 2, \dots k)$. The hazard rate of this distribution is

$$h_i(x) = \frac{1}{S_x(x)} \frac{dQ_i^*(x)}{dx}.$$

Noting that

$$\int_0^\infty h_i(t) S_x(t)\, dt \neq 1,$$

we have

$$S_i^*(x) = \Pr\{\text{failure time} > x \mid \text{eventual failure from cause } C_i\}$$
$$= \int_x^\infty h_i(t) S_x(t)\, dt \Big/ \int_0^\infty h_i(t) S_x(t)\, dt$$

with hazard rate

$$\lambda_i^*(x) = -\frac{d \ln S_i^*(x)}{dx}$$

[3, paragraph 9.4]. Observe that $h_i(x) = \lambda_i^*(x)$ when the X_j's are independent, but $S_i(x) \neq S_i^*(x)$. However, there is a theorem [3, paragraph 9.6] that corresponding to any $S_i(x)$, there is an equivalent collective with independent failure hazard rates.

In general, then, the multiple decrement functions for nonindependent causes of failure are not estimable from observations of X and I. This is why an approach through hazard rates seems to be more fruitful than one through survival distribution functions [11]. On the other hand, if the survival distribution function $S_{1 \dots k}(x_1, x_2, \dots x_k)$ is given explicitly, the other functions can be calculated and the hazard rates estimated even when the causes of failure are dependent. As a simple example, suppose $k = 2$ and

$$S_{12}(x_1, x_2)$$
$$= S_1(x_1) S_2(x_2)$$
$$\times \left[1 + \theta\{1 - S_1(x_1)\}\{1 - S_2(x_2)\}\right],$$
$$|\theta| \leqslant 1,$$

and

$$S_1(x) = e^{-x} = S_2(x);$$

then $S_X(x)$, $h_x(x)$, $h_i(x)$, $Q_i^*(x)$, and the survival distribution with independent hazards are easily derived [3, paragraph 9.8].

Another feature of the statistician's approach is the incorporation of concomitant variables* in the model to compare different treatments affecting time to failure. With the usual restriction on concomitant variables that these should be measured before the treatments are given, the hazard rate, including a $1 \times s$ vector $\mathbf{z}' = (z_1, z_2, \dots z_s)$ of concomitant variables depending on r parameters β_j $(j = 1, 2, \dots r)$, may be written

$$\lambda(t; \mathbf{z}) \equiv \lambda(t; \mathbf{z}; \boldsymbol{\beta}).$$

The survival function is then

$$S(t; \mathbf{z}) = \exp\left[-\int_0^t \lambda(u; \mathbf{z})\, du\right],$$

and the likelihood for a sample of N items (or individuals), with τ_j the time at which item j entered the study and t_j the time when it was last observed, is

$$L = \prod_{j=1}^N \lambda(t_j; \mathbf{z})^{\delta_j} \frac{S(t_j; \mathbf{z}_j)}{S(\tau_j; \mathbf{z}_j)},$$

where

$$\delta_j = \begin{cases} 1 & \text{if item } j \text{ failed at } t_j \\ 0 & \text{if item } j \text{ continued at } t_j \end{cases}$$

[3, paragraph 13.3]. A monograph devoted to this model, which has been successful in medical research, is that of Kalbfleisch and Prentice [8]. If the t_j are modified so that they apply within m ranges of time during each of which t_{ij} $(i = 1, 2, \dots, m)$ has an exponential distribution*, L becomes the likelihood of a log-linear model* of multivariate contingency [6] and can be analyzed numerically by methods described in Fienberg [4] or Haberman [5].

References

[1] Böhmer, P. E. (1912). *Rapp. 7me Congr. Int. Actuariel*, **2**, 327–46.

[2] Cournot, A. (1843). *Exposition de la Théorie des Chances et des Probabilités*. L. Hachette, Paris. (Reprinted n.d. by Bizzarri, Rome.)

[3] Elandt-Johnson, R. C. and Johnson, N. L. (1980). *Survival Models and Data Analysis*. Wiley, New York.

[4] Fienberg, S. E. (1978). *The Analysis of Cross-Classified Categorical Data.* MIT Press, Cambridge, MA.

[5] Haberman, S. J. (1978/79). *Analysis of Qualitative Data*, Vols. I and II. Academic Press, New York.

[6] Holford, T. R. (1980). *Biometrics*, **36**, 299–305.

[7] Jensen, U. F. and Hoem, J. M. (1982). "Multistate life table methodology: A probabilist critique" In *Multidimensional Mathematical Demography*, K. C. Land and A. Rogers, eds., Academic Press, New York.

[8] Kalbfleisch, J. D. and Prentice, R. L. (1980). *The Statistical Analysis of Failure Time Data*. Wiley, New York.

[9] Miller, R. B. and Hickman, J. C. (1983). *Scand. Actuarial J.*, 77–86.

[10] Nelson, W. (1972). *Technometrics*, **14**, 945–966.

[11] Prentice, R. L., Kalbfleisch, J. D., Peterson, A. V., Jr., Flournoy, N., Farewell, V. T., and Breslow, N. E. (1978). *Biometrics*, **34**, 541–554.

[12] Seal, H. L. (1977). *Biometrika*, **64**, 429–39, (1981). *Mitt. Ver. schweiz. Versich.-Math.*, **81**, 167–175.

(ACTUARIAL STATISTICS—LIFE
CONCOMITANT VARIABLES
DEMOGRAPHY
LIFE TABLES
SURVIVAL ANALYSIS)

H. L. SEAL

MULTIPLE INDICATOR APPROACH

Data analyses often proceed on the assumption of absolutely no measurement errors*, even where measurement has been problematic. If the response to the question, Are all variables perfectly measured? is no, this should be followed by the query, What are the *sources* of measurement error and how can they be brought explicitly into the equation system? If one writes an equation linking "true" and "measured" values of a variable, there must be a *substantive* rationale justifying it. For indirect measurement, it is useful to postulate a causal model connecting unmeasured variables to their indicators, often requiring a complex "auxiliary" measurement theory to supplement the substantive theory [1, 3].

Consider the models of Figs. 1 and 2, in which the conceptual or "true" variables are represented by X_i and their measured indicators Y_j are assumed to be *effects* of these X_i. In Fig. 1, adapted from Costner and Schoenberg [4], industrial development X_1 is measured by three indicators: GNP/capita (Y_1), energy consumption/capita (Y_2), and labor diversity (Y_3). Industrial development is assumed to affect political development X_2, which is measured by four indicators: executive functioning (Y_4), party organization (Y_5), power diversification (Y_6), and an

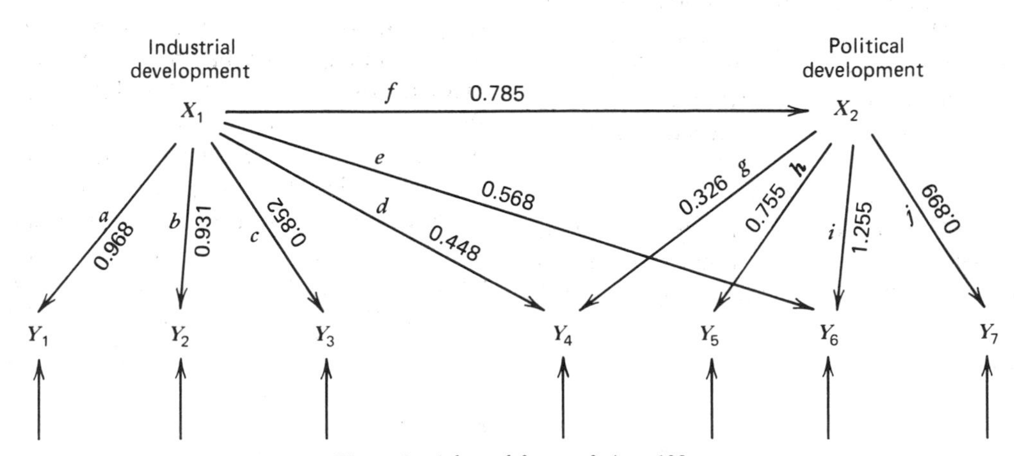

Figure 1 Adapted from ref. 4, p. 198.

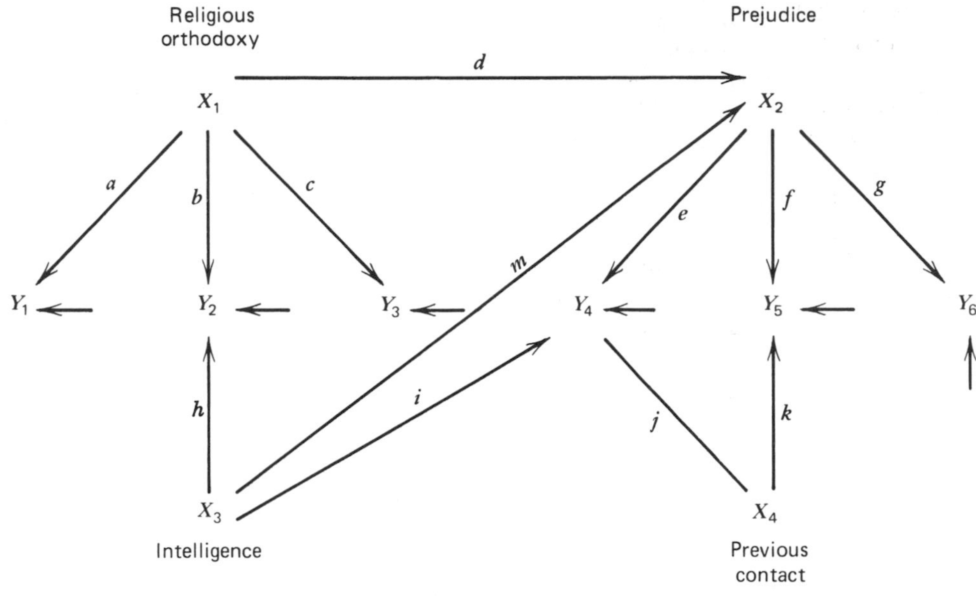

Figure 2

index of political representation (Y_7). (See Olsen [11] for a discussion of these indices.) Complicating the model, however, is the assumption that industrial development X_1 may also affect Y_4 and Y_6, two of the indicators of political development. The intercorrelations for these seven indicator variables are provided in Table 1, and numerical estimates of the path coefficients $a, b, c, \ldots, j$ (standardized beta weights) have been inserted in Fig. 1.

In Fig. 2 we assume the researcher is interested in the effects of religious orthodoxy, X_1, on prejudice, X_2, but two other measured variables, intelligence, X_3, and previous contact, X_4, also affect some of the indicator variables Y_j, as diagrammed. This type of situation, involving nonrandom measurement errors caused by uncontrolled and unmeasured variables, is very common in the social sciences.

Table 1 Intercorrelations Among Indicators Y_i Used in Fig. 1

	Y_1	Y_2	Y_3	Y_4	Y_5	Y_6	Y_7
Y_1	1.00	0.95	0.83	0.66	0.56	0.45	0.67
Y_2		1.00	0.83	0.70	0.54	0.38	0.66
Y_3			1.00	0.62	0.54	0.38	0.61
Y_4				1.00	0.47	0.45	0.60
Y_5					1.00	0.64	0.64
Y_6						1.00	0.67
Y_7							1.00

In general, the approach involves constructing causal models containing combinations of measured and unmeasured variables. In the simplest cases, as in Figs. 1 and 2, the variables of substantive interest are measured by indicators assumed to be effects of combinations of the unmeasured variables, in which case confirmatory factor analysis* may be used to obtain estimates of the path coefficients linking the true substantive variables (*see* PATH ANALYSIS). In other instances, indicator variables may be *causes* of some substantive variables, as in Fig. 3, in which case Jöreskog's more general analysis of covariance* structures may be used [6]. Both procedures yield maximum likelihood* (ML) estimates, but have the disadvantage of being sensitive to specification errors and tend to distribute these errors throughout the system. Therefore, exploratory search procedures for locating troublesome assumptions are also needed.

The models of Figs. 1 and 2 are confirmatory factor analysis models with certain relationships among specified factors assumed a priori. If constant coefficients, linear equations, and independent observations are assumed throughout, confirmatory factor analysis provides ML estimates, either using standardized variables or by arbitrarily selecting one indicator of each variable as a reference indicator and setting its coefficient equal to unity, permitting the other coefficients to remain unstandardized.

If we let the matrix $\mathbf{\Phi}$ represent the matrix correlations among the unmeasured X_i, the $\mathbf{\Phi}$ matrices for Figs. 1 and 2 are, respectively,

$$\underset{(2\times 2)}{\mathbf{\Phi}_1} = \begin{pmatrix} 1 & f \\ f & 1 \end{pmatrix},$$

$$\underset{(4\times 4)}{\mathbf{\Phi}_2} = \begin{bmatrix} 1 & d & 0 & 0 \\ d & 1 & m & 0 \\ 0 & m & 1 & 0 \\ 0 & 0 & 0 & 1 \end{bmatrix}.$$

We may also construct the respective factor-loading matrices $\mathbf{\Lambda}$ by inserting the proper symbols for the nonzero coefficients linking the measured Y_i to the unmeasured X_i, as follows:

$$\underset{(7\times 2)}{\mathbf{\Lambda}_1} = \begin{bmatrix} a & 0 \\ b & 0 \\ c & 0 \\ d & g \\ 0 & h \\ e & i \\ 0 & j \end{bmatrix},$$

$$\underset{(6\times 4)}{\mathbf{\Lambda}_2} = \begin{bmatrix} a & 0 & 0 & 0 \\ b & 0 & h & 0 \\ c & 0 & 0 & 0 \\ 0 & e & i & j \\ 0 & f & 0 & k \\ 0 & g & 0 & 0 \end{bmatrix}.$$

If we place a sufficient number of constraints (usually zero assumptions) on the elements of $\mathbf{\Phi}$ and $\mathbf{\Lambda}$, the unrestrained parameters may then be estimated. The population matrix $\mathbf{\Sigma}$ of correlations (or covariances) among indicators may then be expressed as

$$\mathbf{\Sigma} = \mathbf{\Lambda\Phi\Lambda}' + \mathbf{\Psi},$$

where $\mathbf{\Psi}$ is a diagonal matrix of unique disturbances (represented diagrammatically by short arrows).

If the equation system is identified, one may calculate the sample indicator correlation (or covariance) matrix $\mathbf{S}$, obtaining ML estimates $\hat{\mathbf{\Phi}}$, $\hat{\mathbf{\Lambda}}$, and $\hat{\mathbf{\Psi}}$ by minimizing the function $F = \frac{1}{2}[\log|\mathbf{\Sigma}| + \operatorname{tr}(\mathbf{S\Sigma}^{-1})]$, under the assumption of multivariate normal distributions of the factors and the disturbance terms. The estimate $\hat{\mathbf{\Sigma}}$ may then be compared with the obtained matrix $\mathbf{S}$ and a series of likelihood ratio tests* made to compare the relative adequacies of nested models. Large chi-square values relative to degrees of freedom suggest a poor fit and the need to use additional free parameters. In the case of Fig. 1, Costner and Schoenberg [4] judge this model to provide a much more satisfactory fit than others they consider; therefore, they conclude that the estimates given in Fig. 1 are appropriate.

As implied, not all indicators will be effects of unmeasured ones. Jöreskog's [6, 7] analysis of covariance structures provides a merging of structural-equation* and factor-analytic models, handling combinations of cause-and-effect indicators, as illustrated in Fig. 3, where our interest might be in the relationship between anxiety level η_i and learning η_2. In addition to two effect indicators of anxiety (Y_1 and Y_2), perhaps a self-report (a physiological response), and two of learning (Y_3 and Y_4), perhaps two different tests, an experimenter may have attempted to infer anxiety by manipulating it, using different test instructions ($X_1 = \xi_1$) and classroom conditions ($X_2 = \xi_2$). Or, very frequently in nonexperimental social research, we "measure" some subjective state (such as anxiety or frustration) in terms of some stimulus variable (such as the level of unemployment).

By taking cause indicators such as X_1 and X_2 as perfectly measured exogenous variables and suitably redefining the other variables, it is often possible to represent a multiple-indicator model in terms of the equa-

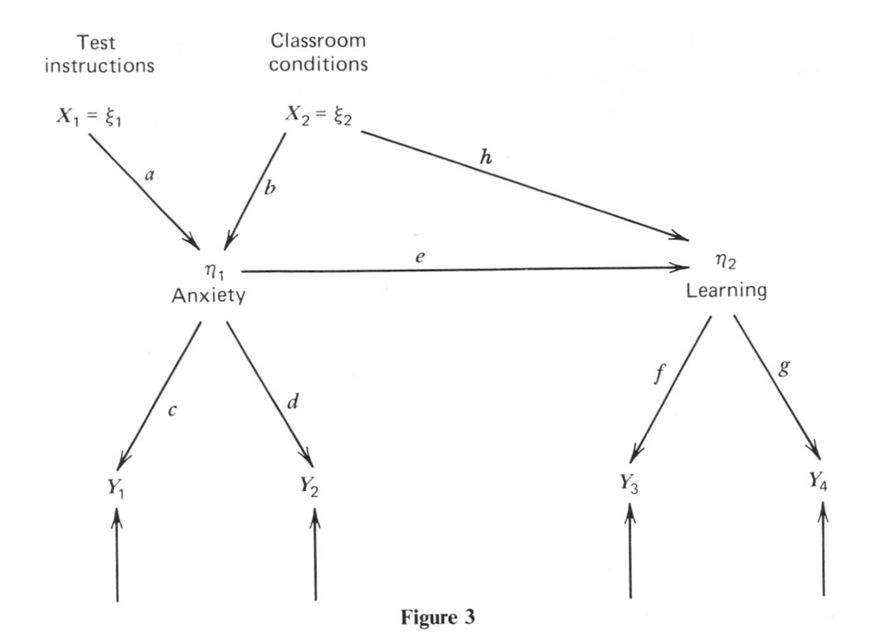

Figure 3

tion system

$$\beta\eta = \Gamma\xi + \zeta, \quad (1)$$

$$\mathbf{X} = \nu + \Lambda_X\xi + \delta, \quad (2)$$

$$\mathbf{Y} = \mu + \Lambda_Y\eta + \epsilon, \quad (3)$$

where ζ $(m \times 1)$, δ $(q \times 1)$, and ϵ $(p \times 1)$ are vectors respectively representing the errors in the equations, measurement errors in exogenous variables, and measurement errors in endogenous variables (*see* ECONOMETRICS).

Equation (1) represents the set of structural (causal) equations linking the m true endogenous variables η to a set of n true exogenous variables ξ, some of which may be cause indicators such as X_1 and X_2 in Fig. 3. In Fig. 3, equation (2) representing the measurement errors in exogenous variables has been made trivial by setting $X_1 = \xi_1$ and $X_2 = \xi_2$. Equation (3) represents the set of equations for each of the effect indicators Y_i as measures of the η_i, here the two theoretical variables, anxiety and learning. The β and Γ matrices in equation (1) are coefficient matrices for the structural equations, and the matrices Λ_x and Λ_y are factor loading matrices. The matrix Σ becomes much more complex, but ML methods may again be used to provide estimates and goodness-of-fit* tests to evaluate alternative (nested) models. The LISREL* series of programs are readily available to users [8].

In models of this type there are likely to be a large number of unknown parameters relative to empirical information. The system may then be underidentified (*see* IDENTIFIABILITY). Although necessary conditions for identification may be stated for the confirmatory factor-analysis model, the general necessary and sufficient conditions have not been determined. In specific models, however, it is often possible to determine whether the coefficients can be estimated by solving for them in terms of the empirically obtained correlations [3]. In general, the greater the complexity of the model with respect to sources of measurement biases, the more indicators needed to yield identified models. Since most measurement-error models are exploratory, substantially overidentified systems are preferred.

As a cautionary note, measurement-error models are likely to be misspecified. ML procedures will be superior to less efficient approaches only insofar as specification errors are relatively minor. ML methods have the disadvantages of obscuring these errors and dispersing their effects across the entire model, making it difficult to locate the faulty assumptions. Costner and Schoenberg [4] and Burt [2] suggest working with submodels to locate such errors. For instance, two-indicator submodels can locate sources of nonrandom error that crosscut unmeasured variables, with three-indicator submodels helping to locate nonrandom errors confined to a single variable.

One limitation of these procedures is the assumption of linear relationships between true and measured variables. Alternative assumptions allowing for satiation or threshold effects may be more appropriate. For instance, if the theoretical variable is a utility measured indirectly by an objective variable (e.g., money), it is unrealistic to assume linearity. Or one may use multiplicative or other nonadditive measurement-error models, allowing for variable slopes connecting indicators and true values, with these slopes related to other variables in one's theory [10].

References and Further Reading

[1] Blalock, H. M. (1968). In *Methodology in Social Research*, H. M. Blalock and A. B. Blalock, eds. McGraw-Hill, New York, Chap. 1. (A general orientation written for social scientists.)

[2] Burt, R. S. (1976). *Sociol. Meth. Res.*, **5**, 3–52.

[3] Costner, H. L. (1969). *Amer. J. Sociol.*, **75**, 245–263.

[4] Costner, H. L. and Schoenberg, R. (1973). In *Structural Equation Models in the Social Sciences* (Seminar), A. S. Goldberger and O. D. Duncan, eds. Chap. 9. (This book also contains a number of additional relevant chapters.)

[5] Hauser, R. M. and Goldberger, A. S. (1971). In *Sociological Methodology 1971*, H. L. Costner, ed. Jossey-Bass, San Francisco, Chap. 4.

[6] Jöreskog, K. G. (1970). *Biometrika*, **57**, 239–251.

[7] Jöreskog, K. G. (1973). In *Structural Equation Models in the Social Sciences* (Seminar), A. S. Goldberger and O. D. Duncan, eds. Chap. 5.

[8] Jöreskog, K. G. and Sörbom, D. (1978). *LISREL IV: A General Computer Program for Estimation of a Linear Structural Equation System by Maximum Likelihood Methods*. National Education Resources, Chicago. (LISREL programs can be obtained from the Educational Testing Service.)

[9] Long, J. S. (1976). *Sociol. Meth. Res.* **5**, 157–206. (Excellent expository treatment of Jöreskog's approach with numerous references.)

[10] Namboodiri, N. K., Carter, L. F., and Blalock, H. M. (1975). *Applied Multivariate Analysis and Experimental Designs*. McGraw-Hill, New York. (Also discusses additional types of measurement-error models in Chaps. 12 and 13.)

[11] Olsen, M. E., (1968). *Amer. Sociol. Rev.* **33**, 669–712.

(ECONOMETRICS
FACTOR ANALYSIS
GROUP TESTING
LISREL
MEASUREMENT ERROR
PATH ANALYSIS
POLITICAL SCIENCE, STATISTICS IN
PSYCHOLOGICAL TESTING THEORY
PSYCHOLOGY, STATISTICS IN
SOCIOLOGY, STATISTICS IN
STRUCTURAL EQUATIONS MODELS)

H. M. BLALOCK

MULTIPLE LINEAR REGRESSION

Multiple linear regression is the name given to a generalization of the theory and techniques of simple linear regression* for situations where there are at least two explanatory variables (*see* REGRESSION COEFFICIENTS). Let

$$f(X_1, X_2, \dots, X_k) = b_0 + b_1 X_1 + \cdots + b_k X_k$$

denote the multiple regression equation, where $X_1, X_2, \dots, X_k$ are k explanatory variables related to a response variable Y.

Data for a multiple regression situation can be thought of as a cloud of points $(Y_i, X_{i1}, X_{i2}, \dots, X_{ik})$, $i = 1, \dots, n$, in $(k+1)$-dimensional space.

The quantities b_i, $i = 0, 1, \dots, k$, are the *regression coefficients*. Individually these can be meaningless, but collectively they can play an important role in predicting a value of the response variable Y. To study their joint properties and in general to investigate multiple regression problems, it is helpful to employ matrix notation.

Let $\boldsymbol{\phi}$ denote a null matrix, $\mathbf{A}'$ the transpose of a matrix $\mathbf{A}$, $\mathbf{A}^{-1}$ the inverse of a nonsingular matrix $\mathbf{A}$, and $\mathbf{Y}'$ denote the row vector of response observations, or response random variables.

Let $\epsilon(\mathbf{Z})$ denote the matrix of the expected value of the variables in the matrix $\mathbf{Z}$. Let

$$\mathbf{X} = \begin{bmatrix} 1 & X_{11} & \cdots & X_{1j} & \cdots & X_{1k} \\ 1 & X_{21} & \cdots & X_{2j} & \cdots & X_{2k} \\ \vdots & \vdots & & \vdots & & \vdots \\ 1 & X_{i1} & \cdots & X_{ij} & \cdots & X_{ik} \\ \vdots & \vdots & & \vdots & & \vdots \\ 1 & X_{n1} & \cdots & X_{nj} & \cdots & X_{nk} \end{bmatrix},$$

$$\mathbf{b} = \begin{bmatrix} b_0 \\ b_1 \\ \vdots \\ b_j \\ \vdots \\ b_k \end{bmatrix}.$$

In multiple linear regression it is assumed that $\epsilon(\mathbf{Y}) = \mathbf{X}\boldsymbol{\beta}$; that is, the ith observation Y_i has as its expectation, a known linear combination of parameters $\beta_0, \beta_1, \ldots, \beta_k$. If the Y_i's are independent with common variance σ^2, then in matrix notation these assumptions can be expressed as $\mathrm{Cov}[\mathbf{Y}] = \sigma^2\mathbf{I}$, where $\mathbf{I}$ is an $n \times n$ identity matrix, and $\mathrm{Cov}[\mathbf{Y}]$ denotes the $n \times n$ covariance matrix of $\mathbf{Y}$. The (i,i)th element of $\mathrm{Cov}[\mathbf{Y}]$ is the variance of Y_i and the (i,j)th element equals the covariance of Y_i and Y_j, $i \neq j$.

Denote by $\mathbf{E}$ the vector of differences $\mathbf{E} = \mathbf{Y} - \epsilon(\mathbf{Y})$. The elements of $\mathbf{E}$ are called *residuals* (*see* RESIDUAL ANALYSIS). $\mathbf{Y}$ can then be written

$$\mathbf{Y} = \mathbf{X}\boldsymbol{\beta} + \mathbf{E}, \qquad \epsilon(\mathbf{E}) = \boldsymbol{\phi}.$$

Much of the significance of the matrix approach rests in the fact that many statistical problems can be phrased in the context of this one general model (*See* GENERAL LINEAR MODEL).

If the variables X_i are selected in such a way that $\mathbf{X'X}$ is nonsingular, then regression methods are essentially based on the following results:

1. $\mathbf{b} = \hat{\boldsymbol{\beta}} = (\mathbf{X'X})^{-1}\mathbf{X'Y}$ is the least-squares* vector for estimating the parameters in $\boldsymbol{\beta}$.
2. $\mathbf{b} = (\mathbf{X'X})^{-1}\mathbf{X'Y}$ is the maximum likelihood* estimator of $\boldsymbol{\beta}$ when $\mathbf{Y}$ is distributed as n-variate normal.
3. $\epsilon(\mathbf{b}) = \boldsymbol{\beta}$.
4. $\mathrm{Cov}(\mathbf{b}) = \sigma^2(\mathbf{X'X})^{-1}$.
5. If $\mathbf{Y}$ is distributed normally, then $\mathbf{b}$ is a $(k+1)$-variate normal.
6. Under normality, the maximum likelihood estimator of σ^2 can be expressed as $\tilde{\sigma}^2 = \mathbf{Y}'[\mathbf{I} - \mathbf{X}(\mathbf{X'X})^{-1}\mathbf{X}']\mathbf{Y}/n$.
7. $\epsilon(\tilde{\sigma}^2) = (n-k-1)\sigma^2/n$; hence an unbiased estimator of σ^2 is given by $\hat{\sigma}^2 = \mathbf{Y}'[\mathbf{I} - \mathbf{X}(\mathbf{X'X})^{-1}\mathbf{X}']\mathbf{Y}/(n-k-1)$.
8. If $\mathbf{Y}$ is normal, then $(n-k-1)\hat{\sigma}^2/\sigma^2 = \mathbf{Y}'[\mathbf{I} - \mathbf{X}(\mathbf{X'X})^{-1}\mathbf{X}']\mathbf{Y}/\sigma^2$ is chi-square with $(n-k-1)$ degrees of freedom.
9. $\mathbf{b}$ and $\hat{\sigma}^2$ are independent.
10. Furthermore, even without the normality assumption, the following result holds: For any linear combination $\mathbf{h}'\boldsymbol{\beta}$, the linear unbiased estimate with minimum variance is $\mathbf{h}'\hat{\boldsymbol{\beta}} = \mathbf{h}'(\mathbf{X'X})^{-1}\mathbf{X'Y} = \mathbf{h'b}$ (*see* GAUSS–MARKOV THEOREM).

Under normality $(\mathbf{h}'\hat{\boldsymbol{\beta}} - \mathbf{h}'\boldsymbol{\beta})/s_{\mathbf{h'b}}$ is distributed as a Student t-variable with $(n-k-1)$ degrees of freedom, where $s_{\mathbf{h'b}} = (\hat{\sigma}^2\mathbf{h}'(\mathbf{X'X})^{-1}\mathbf{h})^{1/2}$. To test the hypothesis $\mathbf{h}'\boldsymbol{\beta} = a$, an appropriate test statistic is $t = (\mathbf{h}'\hat{\boldsymbol{\beta}} - a)/s_{\mathbf{h'b}}$. To set a $100(1-\alpha)\%$ confidence interval on $\mathbf{h}'\boldsymbol{\beta}$, one reads the tabulated $t_{\alpha/2}$ value from the Student t tables and then computes the numerical values of the interval limits by the formulas $\mathbf{h}'\hat{\boldsymbol{\beta}} \pm t_{\alpha/2}s_{\mathbf{h'b}}$.

To illustrate these concepts, suppose $n = 7$ observation triplets constitute a set of data:

Y	X_1	X_2
13	4	1
7	5	3
2	2	3
15	6	1
8	3	2

Suppose we wish to fit the plane $f(X_1, X_2) = b_0 + b_1X_1 + b_2X_2$ to this data. In matrix form, we have

$$\mathbf{Y} = \begin{bmatrix} 13 \\ 7 \\ 2 \\ 15 \\ 8 \end{bmatrix}, \quad \mathbf{X} = \begin{bmatrix} 1 & 4 & 1 \\ 1 & 5 & 3 \\ 1 & 2 & 3 \\ 1 & 6 & 1 \\ 1 & 3 & 2 \end{bmatrix}, \quad \mathbf{b} = \begin{bmatrix} b_0 \\ b_1 \\ b_2 \end{bmatrix},$$

$$\boldsymbol{\beta} = \begin{bmatrix} \beta_0 \\ \beta_1 \\ \beta_2 \end{bmatrix}, \quad \mathbf{X'X} = \begin{bmatrix} 5 & 20 & 10 \\ 20 & 90 & 37 \\ 10 & 37 & 24 \end{bmatrix},$$

$$(\mathbf{X'X})^{-1} = \begin{bmatrix} 5.103 & -0.710 & -1.032 \\ -0.710 & 0.129 & 0.097 \\ -1.032 & 0.097 & 0.323 \end{bmatrix},$$

$$\mathbf{X'Y} = \begin{bmatrix} 45 \\ 205 \\ 71 \end{bmatrix}, \quad \hat{\boldsymbol{\beta}} = \begin{bmatrix} 10.871 \\ 1.387 \\ -3.710 \end{bmatrix},$$

$$\hat{\sigma}^2 = \frac{0.839}{2} = 0.419.$$

To test the hypothesis $H_0 : \beta_1 = \beta_2$, one can express H_0 as $\mathbf{h}'\boldsymbol{\beta} = 0$ where $\mathbf{h}' = [0, 1, -1]$. For the example

$$s^2_{\mathbf{h'b}} = \hat{\sigma}^2[0, 1, -1] \begin{bmatrix} 5.103 & -0.710 & -1.032 \\ -0.710 & 0.129 & 0.097 \\ -1.032 & 0.079 & 0.323 \end{bmatrix} \times \begin{bmatrix} 0 \\ 1 \\ -1 \end{bmatrix} = 0.1081,$$

$t = \mathbf{h'b}/s_{\mathbf{h'b}} = 5.097/0.3288 = 15.5$, which implies rejection of H_0.

To test $\beta_1 = 0$, one uses $\mathbf{h}' = [0, 1, 0]$, in which case $s_{\mathbf{h'b}} = 0.2326$ and $t = 5.96$.

Among the many useful concepts associated with regression is that of the multiple correlation* coefficient R^2, defined to be the proportion of the corrected total sum of squares explained by regression.

$$R^2 = \frac{\text{sum of squares due to regression}}{\text{total sum of squares corrected for the mean}}$$

$$= \frac{\sum_{i=1}^n (Y_i - \bar{Y})^2 - \sum_{i=1}^n [Y_i - f(X_{i1}, \dots, X_{ik})]^2}{\sum_{i=1}^n (Y_i - \bar{Y})^2}.$$

An important aspect of multiple regression analysis involves seeking the best subset of a set of explanatory variables. Consider a situation with four predictor variables. $n = 18$ experimental units have been selected, and the data for them is displayed in Table 1.

Table 1 Data: Four Predictors and One Response

X_1	X_2	X_3	X_4	Y_5	X_1	X_2	X_3	X_4	Y_5
2.5	20	7	0	1.3	1.0	22	7	6	0.9
2.0	15	13	4	0.1	0.8	20	10	5	0.7
4.6	35	5	8	1.9	3.0	17	12	0	0.8
0.6	17	11	3	0.3	3.2	28	8	4	1.6
0.2	9	4	10	0	0.5	13	1	7	0.7
1.6	14	9	4	0.3	3.4	17	5	2	1.1
2.7	30	13	5	1.4	0.9	8	3	3	0.2
2.9	25	8	1	1.5	4.4	31	14	6	1.0
3.8	18	2	9	1.0	2.6	26	13	8	0.5

Table 2 exhibits the sample correlations among the variables as computed from the data in Table 1.

For a situation with k predictor variables there exist $2^k - 1$ different regressions of Y on subsets of the predictors. For large values of k, then many possible subsets exist, and running all possible regressions would be prohibitive. For our example $k = 4$ and 15 regressions exist. Displayed in Table 3 are the R^2 values corresponding to these 15 regressions. Notice that among single-variable predictor regressions X_3 is the worst; R^2 for X_3 alone is 0.003. This agrees with the -0.056 correlation value for X_3 and Y. Among all two-variable possibilities, X_2 and X_3 would be the best according to the R^2 criterion. A combination of X_2, X_3, and X_4 explains 95% of the variability in Y. So we see that X_3 in conjunction with other predictors is important but by itself is poor. According to the R^2 criterion, bringing in X_1 along with X_2, X_3, and X_4 is of little value.

Several methods are available for selecting a good subset of predictor functions without

Table 2 Correlations

	X_1	X_2	X_3	X_4	Y
X_2	0.711	1			
X_3	0.193	0.416	1		
X_4	−0.052	0.098	−0.264	1	
Y	0.708	0.788	−0.056	−0.145	1

Table 3 Variables in the Model and Their Corresponding R^2 Values

X_3	0.003	X_1, X_4	0.513	X_1, X_3, X_4	0.567
X_4	0.021	X_1, X_3	0.540	X_1, X_2, X_4	0.701
X_1	0.501	X_1, X_2	0.665	X_1, X_2, X_3	0.820
X_2	0.621	X_2, X_4	0.671	X_2, X_3, X_4	0.950
X_3, X_4	0.031	X_2, X_3	0.799	X_1, X_2, X_3, X_4	0.952

doing all $2^k - 1$ regressions. One of these methods is *stepwise regression**. It is often used when k is large. We illustrate by applying it to the $k = 4$ data situation described earlier.

We start with X_2 because the correlation between X_2 and Y is greater than the correlation between X_i and Y for $i = 1, 3, 4$. The t-ratio for the slope in the regression of Y on X_2 is 5.12, which suggests we keep X_2 and try pairs (X_2, X_i), $i = 1, 3, 4$. $R^2 = 0.799$ for the pair (X_2, X_3). The t-ratio for X_3 is -3.64 and for X_2 is 7.70. Both are sufficiently large t-ratios, so we keep both X_2 and X_3. If either had been (in absolute value) less than an arbitrary but prechosen t value (depending on the degrees of freedom), then the predictor corresponding to the small t-value would have been deleted and other pairs investigated. Since both X_2 and X_3 are kept in the model, we next try X_2, X_3, X_4 and X_1, X_2, X_3. Since $R^2 = 0.950$ is maximum among all R^2 associated with X_2, X_3 and one other variable, we calculate t-ratios for the coefficients in the three predictor models including X_2, X_3, and X_4. All three t-ratios are large, hence we next try regressing Y on X_1, X_2, X_3, and X_4. At this point, the t-ratio for X_1 is only 0.84. Thus we delete X_1 and choose X_2, X_3, and X_4 as the "best" subset.

Curvilinear regression can be viewed as multiple regression where some predictor variables are functions of others. A special case is *polynomial regression*. For polynomial regression in one variable, the regression function takes the form

$$f_k(X) = b_0 + b_1 X + b_2 X^2 + \cdots + b_k X^k.$$

To illustrate curvilinear regression models, let us fit successively higher-degree polynomials to the $n = 22$ data points displayed in Table 4. For each fit, the R^2, the MSE (the mean squared error*), and the t-ratio for regression coefficients will be studied until an adequate fit of the data is realized.

The least-squares staight line is $f_1(X) = 11.2 + 3.54X$ with $R^2 = 0.849$, MSE = 6.309, and slope t-ratio equal to 10.6.

To fit a quadratic, one lets $X_2 = X^2$. The **X′X** and **X′Y** matrices become

$$\mathbf{X'X} = \begin{bmatrix} n & \Sigma X_i & \Sigma X_i^2 \\ \Sigma X_i & \Sigma X_i^2 & \Sigma X_i^3 \\ \Sigma X_i^2 & \Sigma X_i^3 & \Sigma X_i^4 \end{bmatrix}, \quad \text{and}$$

$$\mathbf{X'Y} = \begin{bmatrix} \Sigma Y_i \\ \Sigma X_i Y_i \\ \Sigma X_i^2 Y_i \end{bmatrix}.$$

The quadratic fit of the data is not much better than the straight line fit. $f_2(X) = 10.7 + 4.04X - 0.0872X^2$. $R^2 = 0.850$, MSE = 6.602, and the t-statistic for testing whether $\beta_2 = 0$ is only $t = -0.34$. Notice that R^2 stayed almost the same while the MSE increased.

The cubic fit is, however, much better. $f_3(X) = 2.96 + 18.7X - 6.45X^2 + 0.752X^3$

Table 4

X_i	Y_i	X_i	Y_i	X_i	Y_i
1.0	15.9	2.9	21.9	4.4	25.0
1.5	18.0	1.3	20.3	3.7	23.1
2.0	20.5	0.5	8.9	2.2	20.6
2.6	21.1	5.2	30.8	3.1	22.1
3.9	23.8	0.7	13.2	3.3	22.3
5.0	29.4	1.8	19.0	4.2	20.4
4.7	26.4	4.9	27.5	5.3	34.9
0.3	9.5				

and the t-ratios for the β_i's are now all significant. R^2 is now 0.961, and the MSE has decreased to 1.827.

Since a cubic explains about 96.1% of the variability in the response, this leaves little room for improvement. The quartic equation is $f_4(X) = 4.99 + 12.9X - 2.10X^2 - 0.428X^3 + 0.105X^4$, with $R^2 = 0.964$ and the MSE reduced to 1.767. The t ratios have all decreased with only β_1 near significance.

Notice how very different the coefficients are in the different regression equations. There may even be sign changes depending on the presence or absence of a higher-power term. This is due, at least in part, to the presence of multicollinearity* in the powers of X as predictors of Y. Multicollinearity means that some of the columns of the **X** matrix are highly correlated.

The cubic equation seems adequate for these several reasons, and a plot of the data supports this.

The application of regression techniques may run into snags. For example, the columns of the **X** matrix may be mathematically dependent, in which case $\mathbf{X'X}$ is singular. This may happen because of the model chosen or the data fit, or it may be that the predictor variables are dependent. Regardless of the reason, when $\mathbf{X'X}$ is singular, least-squares regression estimates of $\boldsymbol{\beta}$ cannot be obtained from the matrix equation $\mathbf{X'X}\hat{\boldsymbol{\beta}} = \mathbf{X'Y}$.

If the $\mathbf{X'X}$ matrix is nonsingular but near singular, there are multicollinearity difficulties. It may be computationally difficult to obtain an accurate inverse and even if the inverse is computed precisely, the estimate of the elements in $\boldsymbol{\beta}$ may have large standard errors. Multicollinearity among predictors may in fact give estimates of β_i's that have the wrong sign and/or are off by more than an order of magnitude. These and related problems have given rise to a host of new regression techniques. Some of these techniques strive to incorporate into the analysis prior information relative to the parameters. Ridge regression* is a term used to describe some of the methods. Whereas ordinary least squares leads (under satisfied assumptions) to unbiased but often unrealistic estimates, ridge regression estimates are biased, but often more realistic. In situations where multicollinearity exists, ridge regression estimates tend to have much smaller mean squared error than do the least-squares estimates (*see* MULTICOLLINEARITY; *see also* LATENT ROOT REGRESSION).

Bibliography

Allen, D. M. and Cady, F. B. (1982). *Analyzing Experimental Data by Regression*. Wadsworth, Belmont, CA.

Carmer, E. M. (1972). *Amer. Statist.*, **26**, 26–30.

Chatterjee, S. and Price, B. (1977). *Regression Analysis by Example*. Wiley, New York.

Cook, R. D. (1977). *Technometrics*, **19**, 15–18.

Daniel, C., Wood, F. S., and Gorman, J. W. (1980). *Fitting Equations to Data: Computer Analysis of Multifactor Data*, 2nd ed. Wiley, New York.

Draper, N. and Smith, H. (1981). *Applied Regression Analysis*, 2nd ed. Wiley, New York.

Dutka, A. F. and Ewen, F. J. (1971). *J. Quality Tech.*, **3**, 149–155.

Furnival, G. M. (1971). *Technometrics*, **13**, 403–408.

Furnival, G. M. and Wilson, R. W., Jr. (1974). *Technometrics*, **16**, 499–512.

Graybill, F. A. (1976). *Theory and Application of the Linear Model*. Duxbury, Belmont, CA.

Goldberger, A. S. (1964). *Econometric Theory*. Wiley, New York.

Gunst, R. F. and Mason, R. L. (1980). *Regression Analysis and Its Application*. Marcel Dekker, New York.

Hocking, R. R. (1976). *Biometrics*, **32**, 1–49.

Hoerl, A. E. and Kennard, R. W. (1970). *Technometrics*, **12**, 55–67.

Hoerl, A. E. and Kennard, R. W. (1970). *Technometrics*, **12**, 69–82.

Jaech, J. L. (1966). *Ind. Quality Control*, **23**, 260–264.

Kleinbaum, D. G. and Kupper, L. L. (1978). *Applied Regression Analysis and Other Multivariate Methods*. Duxbury, Belmont, CA.

Lamotte, L. R. and Hocking, R. R. (1970). *Technometrics*, **12**, 83–94.

Lindley, D. V. (1968). *J. R. Statist. Soc. B*, **30**, 31–53.

Marquardt, D. W. and Snee, R. D. (1975). *Amer. Statist.*, **29**, 3–20.

Mendenhall, W. (1968). *Introduction to Linear Models and the Design and Analysis of Experiments*. Wadsworth, Belmont, CA.

Mosteller, F. and Tukey, J. W. (1977). *Data Analysis and Regression*. Addison-Wesley, Reading, MA.

Neter, J. and Wasserman, W. (1974). *Applied Linear Statistical Models*. Richard D. Irwin, Georgetown, Ontario, Canada.

Obenchain, R. L. (1977). *Technometrics*, **19**, 429–439.

Schilling, E. G. (1974). *J. Quality Tech.*, **6**, 74–83.

Weisberg, S. (1980). *Applied Linear Regression*. Wiley, New York.

Williams, E. J. (1959). *Regression Analysis*. Wiley, New York.

Younger, M. S. (1979). *A Handbook for Linear Regression*. Duxbury, North Scituate, MA.

(CURVE FITTING
ELIMINATION OF VARIABLES
GAUSS–MARKOV THEOREM
GENERAL LINEAR MODEL
LATENT ROOT REGRESSION
LINEAR REGRESSION
MEAN SQUARE ERROR
MULTICOLLINEARITY
MULTIPLE CORRELATION
MULTIVARIATE ANALYSIS OF VARIANCE
REGRESSION COEFFICIENTS
REGRESSION POLYNOMIALS
REGRESSION VARIABLES, SELECTION OF
RESIDUAL ANALYSIS
RIDGE REGRESSION
STEPWISE REGRESSION)

ROBERT HULTQUIST

MULTIPLE MATRIX SAMPLING

Multiple matrix sampling, or matrix sampling as the procedure is frequently referred to, is a sampling procedure in which the characteristics of a complete data matrix are estimated from the characteristics of a randomly selected sample of the entries in that matrix. More specifically and given a matrix of scores consisting of K columns and N rows, a random sample of scores from this matrix is obtained by sampling at random k columns from K columns, n rows from N rows, and taking only those scores common to the n rows and k columns selected. The results obtained from this matrix sample may be used to estimate unbiasedly values of the parameters of the $N \times K$ matrix. The typical procedure, however, is one taking multiple samples and using the combined results to estimate values of parameters of interest; one example is depicted in Fig. 1, where an X indicates a score in the sample. Columns have been sampled randomly and without replacement, and all columns have been sampled—with the same true for rows. (It should be noted that the column and row numbers are arbitrary. For example, the 1, 2, and 3 column numbers denote the first, second, and third columns selected randomly.) The reader interested in retracing the statistical developments in multiple matrix sampling is referred to the following influential publications: Hooke [3, 4], Johnson and Lord [5], Wilks [14], Lord and Novick [6], Shoemaker [8], Sirotnik [11], Sirotnik and Wellington [12], and Wellington [13].

Although multiple matrix sampling may

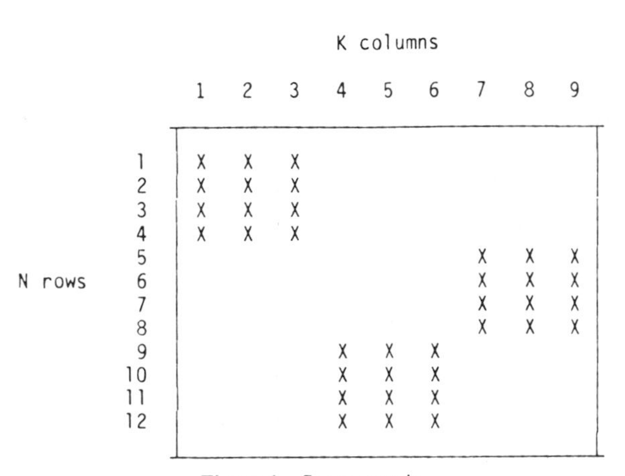

Figure 1 Score matrix.

be used effectively in a variety of contexts, applications to date are found most frequently within the bailiwick of education and specifically in the areas of achievement testing and program evaluation (*see* PSYCHOLOGICAL TESTING THEORY). Within this context, the abstract description of multiple matrix sampling acquires more meaning if, in place of K columns and N rows, are substituted, respectively, test items and examinees. Here parameters commonly of interest are the mean test score (where the test score for examinee i is the sum of his K item scores), the variance of test scores, the relative frequency distribution of total test scores in the examinee population, the mean item score for each item, and the differences among mean test scores on two or more occasions (e.g., mean post-test score minus mean pretest score). To illustrate the strategy here, assume a 60-item test and 100 students enrolled in a given program. One way to implement multiple matrix sampling is by randomly dividing the 60-item test into four 15-item subtests, randomly dividing the 100 students into four subgroups of 25 students each, and assigning one subtest to each subgroup. Here each student in the program is tested (although a sample of students could have been used instead) with different students taking different subtests. Defining t as the number of subtests, k as the number of items per subtest, n as the number of examinees to which each subtest is administered, and a given sampling plan* by $(t/k/n)$, this particular sampling plan may be denoted as (4/15/25). When $k = 60$ and $N = 100$, other sampling plans are possible as, for example (2/30/50), (5/12/20), and (10/6/10), when items and examinees are sampled exhaustively and without replacement. Determining the relative merits of alternative sampling plans for a particular examinee population and item domain is done routinely by manipulating values of t, k, and n within the equation for the standard error of estimate for the estimator or estimators of interest and using estimated values for all parameters in those equations.

Within the context of achievement testing, multiple matrix sampling is a statistical tool that makes possible the implementation of domain-referenced achievement testing, where the domain of interest is that (frequently large) item domain collectively operationalizing those skills which students should acquire as a function of program participation. This tool is particularly well-suited to evaluating the effectiveness of an instructional program where the focus is on estimating the performance of the entire group of students and not differentiating among individual students (*see* EDUCATIONAL STATISTICS). The rationale linking multiple matrix sampling, domain-referenced achievement testing, and program evaluation is presented in detail by Shoemaker [9, 10] and provides further clarification to comments made by Lord in PSYCHOLOGICAL TESTING THEORY.

When considering the use of multiple matrix sampling, it is important to weigh both its advantages and limitations. Although both are described here in the context of education and achievement testing, they are more broadly applicable. The primary advantages of using multiple matrix sampling are:

1. Reduction in Testing Burden: Because each examinee tested responds to only a subset of items from the complete test or item domain, the testing time (burden) is markedly less than that required to administer all items to a given examinee. Additionally, it is not necessary to test all program participants.

2. More Comprehensive Testing for a Given Amount of Testing Burden: For a given amount of testing time per student, group performance may be assessed over a larger item domain using multiple matrix sampling because the amount of testing burden is a function of subtest length not size of the test or item domain.

3. Reduced Standard Error of Estimate: Estimates of group performance obtained through multiple matrix sampling will estimate parameters more efficiently (have a reduced standard error of estimate) than will other sampling procedures collecting the same number of observations, where one observation is defined as the score obtained by one student on one item. For example, two alternative sampling strategies are item sampling (where a subset of items selected randomly from the item domain is administered to all examinees) and examinee sampling (in which a subgroup of examinees selected randomly from the examinee population is administered all items). If the same number of observations is collected under item sampling, examinee sampling, and multiple matrix sampling—and simple random sampling is used, the standard error of estimate will be less using multiple matrix sampling.

4. Potential Political Asset: For those program evaluations voluntary in nature, participation of various sites and personnel may be increased through using multiple matrix sampling because less student testing time will be required and concomitantly there will be less disruption of the daily classroom routine.

5. Results Less "Test Specific": The use of multiple matrix sampling in an evaluation decreases the likelihood that the obtained results are a function of the particular test used, because different students are administered different subtests. When student achievement is both pretested and posttested, typically all subtests are assigned randomly to examinees at each testing time.

Multiple matrix sampling has some limitations, the primary ones being:

1. More Complex Testing Procedure: With multiple matrix sampling, the evaluator must contend with an assortment of subtests, each of which must be assigned individually to those students selected for testing. There is the added possibility of different sets of instructions for each subtest and, most certainly, more guidelines will be necessary for all personnel involved.

2. Best Suited for Group Assessment: Multiple matrix sampling is best suited for assessing group performance and not the performance of individual participants in a program. The reason here is two-part: (a) not all students participating in the program may be tested and (b) those students selected for testing respond to only a single subtest. This limitation is relevant particularly to those testing programs providing results to multiple users—only one of which may be the program evaluator.

3. Potential Context Effect: Standardized achievement tests frequently are used to assess levels of student achievement in program evaluations with obtained results compared to those associated with a normative population. Here the normative distribution to which these results are compared was one generated by administering the complete standardized test to all examinees. Herein lies a potential problem when multiple matrix sampling is used to estimate group performance on a standardized achievement test; the problem is labeled generally the context effect. When items from a standardized achievement test are divided into subtests for administration via multiple matrix sampling, the subsequent results being contrasted wtih those obtained from the population on which the test was normed, the assumption is made that an examinee's response to an item in a subtest and his or her response to the same item when administered the complete standardized test are one and the same. To the extent that this is not true, a context effect is said to be present in the results obtained from multiple matrix sampling. However, results to date suggest that generally the context effect is minimal.

The reader interested in using multiple matrix sampling can find the necessary equations for a variety of estimators and their associated standard errors in Shoemakcr [8, 9] and Sirotnik and Wellington [12] among others. To illustrate the kinds of equations found there, the estimator for the mean test score is

$$\mathrm{EST(MEAN)} = K\sum_{i}^{N}\sum_{j}^{K}\frac{X_{ij}}{tkn}, \qquad (1)$$

where X_{ij} refers to the score obtained by

examinee i on item j. The estimator for the variance of test scores is

$$\text{EST(VAR)} = \frac{K(N-1)}{N} \times \left[d_4 - d_3 + (K-1)d_2 - (K-1)d_1 \right], \quad (2)$$

where

$$d_1 = \frac{S_1 - S_2 - S_3 + S_4}{tkn(tkn - k - n + 1)}, \quad (3)$$

$$d_2 = (S_2 - S_4)/\left[tkn(k-1) \right], \quad (4)$$

$$d_3 = (S_3 - S_4)/\left[tkn(n-1) \right], \quad (5)$$

$$d_4 = S_4/(tkn), \quad (6)$$

$$S_1 = \left(\sum_i^N \sum_j^K X_{ij} \right)^2, \quad (7)$$

$$S_2 = \sum_i^N \left(\sum_j^K X_{ij} \right)^2, \quad (8)$$

$$S_3 = \sum_i^K \left(\sum_j^N X_{ij} \right)^2, \quad (9)$$

$$S_4 = \sum_i^N \sum_j^K X_{ij}^2, \quad (10)$$

when the number of items is the same for all subtests and the number of examinees is also the same for all subtests. If items are assigned through simple random sampling and without replacement to subtests and subtests are assigned randomly to examinees, the standard error of estimate of that defined by equation (1) is estimated by the square root of

$$\begin{aligned} \text{VAR}\left[\text{EST(MEAN)} \right] &= K(tknN)^{-1} \\ &\times \left[d_1 \{ NK(tkn - k - n + 1) - tkn(N-1)(K-1) \} \right. \\ &+ d_2 \{ NK(k-1) - tkn(K-1) \} \\ &+ d_3 \{ NK(n-1) - tkn(N-1) \} \\ &\left. + d_4(NK - tkn) \right] \end{aligned} \quad (11)$$

In addition to estimating individual parameters, it is possible to estimate the entire normative frequency distribution of test scores that would have been obtained by testing all examinees on all items. For example, if items were scored dichotomously, the negative hypergeometric distribution might be used (*see* HYPERGEOMETRIC DISTRIBUTIONS). Other distribution functions used frequently are the family of Pearson curves—particularly the type I distribution. Examples of using selected Pearson curves (*see* PEARSON'S DISTRIBUTIONS) to estimate normative distributions (and contrasting the obtained results with those of the negative hypergeometric distribution) are given by Brandenburg and Forsyth [1, 2].

References and Further Reading

[1] Brandenburg, D. C. and Forsyth, R. A. (1974). *Educ. Psychol. Meas.*, **34**, 3–9.

[2] Brandenburg, D. C. and Forsyth, R. A. (1974). *Educ. Psychol. Meas.*, **34**, 475–486.

[3] Hooke, R. (1956). *Ann. Math. Statist.*, **27**, 55–79.

[4] Hooke, R. (1956). *Ann. Math. Statist.*, **27**, 80–98.

[5] Johnson, M. C. and Lord, F. M. (1958). *Educ. Psychol. Meas.*, **18**, 325–329.

[6] Lord, F. M. and Novick, M. R. (1968). *Statistical Theories of Mental Test Scores*. Addison-Wesley, Reading, MA, Chap. 11, pp. 234–260.

[7] Shoemaker, D. M. (1973). *J. Educ. Meas.*, **10**, 211–219.

[8] Shoemaker, D. M. (1973). *Principles and Procedures of Multiple Matrix Sampling*. Ballinger, Cambridge, MA.

[9] Shoemaker, D. M. (1975). *Rev. Educ. Res.*, **45**, 127–147.

[10] Shoemaker, D. M. (1980). *Educ. Evaluation Policy Anal.*, **4**, 37–49.

[11] Sirotnik, K. (1975). In *Evaluation in Education: Current Applications*, W. J. Popham, ed. McCutchan, Berkeley, CA.

[12] Sirotnik, K. and Wellington, R. (1977). *J. Educ. Meas.*, **14**, 343–399.

[13] Wellington, R. (1977). *Psychometrika*, **41**, 375–384.

[14] Wilks, S. S. (1962). *Mathematical Statistics*. Wiley, New York.

(EDUCATIONAL STATISTICS
PSYCHOLOGICAL TESTING THEORY
PSYCHOLOGY, STATISTICS IN)

DAVID M. SHOEMAKER

MULTIPLE RANGE AND ASSOCIATED TEST PROCEDURES

Having taken independent observations y_{ij} ($i = 1, \ldots, k$; $j = 1, \ldots, n_i$) from k normal populations $N(\mu_i, \sigma_i^2)$, an experimenter wants to know which pairs of means are unequal. Testing their equality two at a time at the conventional level (e.g., $\alpha = 0.05$) does not adequately protect against type I error (incorrect rejection), even if k is as small as 3 or 4, and *multiple comparison* procedures* (MCPs) are advocated. See Dunnett [10], Games [15], O'Neill and Wetherill [23], Thomas [32], Spjøtvoll [30], Chew [5], Stoline [31], and Miller [22] for survey articles.

MCPs are appropriate only for comparing unstructured populations or treatments (e.g., k experimental corn varieties). They are not appropriate if the k treatments correspond to k levels of a quantitative factor (e.g., temperature), are made up of combinations of two or more factors, or are such that it is possible a priori to construct meaningful orthogonal contrasts among them. See Chew [6].

MCPs can be grouped into two categories: (1) *simultaneous test procedures* (STPs) where the same critical value is used to test all pairs of means, if $n_i = n$ and $\sigma_i^2 = \sigma^2$; and (2) sequential or *stepwise multiple comparison procedures* (SMCPs), where the critical value depends on the number of sample means that fall between the two being tested after the k means are arranged in rank order.

SIMULTANEOUS TEST PROCEDURES

1. *Fisher's protected least significant difference method* is identical with the Student's t-test*, except that it requires a significant F-test* for the equality of all k means before individual paired differences may be tested. (None of the other STPs and SMCPs have this prerequisite.) Two sample means $\bar{y}_i$ and $\bar{y}_j$ will be declared to be significantly different at the $100\alpha\%$ level if their absolute difference exceeds

$$\text{LSD} = t_{\nu,1-(\alpha/2)}\sqrt{s^2\left(n_i^{-1} + n_j^{-1}\right)}\ ,$$

where $t_{\nu,\gamma}$ is the (100γ)th percentile of Student's t-distribution with ν degrees of freedom (d.f.) and s^2 (with ν d.f.) is the pooled estimate of σ^2. This method is readily extended to general contrasts by simply replacing the standard error of a paired difference by that of the contrast.

2. The *Fisher–Bonferroni method* (FBM) also uses Student's t-test to compare each pair of means but the significance level α is reduced to α/m, where m is the total number of contrasts we wish to make. Tables of these unusual percentiles of the t-distribution* are given in Bailey [1] and Games [16].

3. *Tukey's honestly significant difference* (HSD) *method* tests all pairs against $W_k = q_{k,\nu,1-\alpha}\sqrt{s^2/n}$, where $q_{k,\nu,1-\alpha}$ is the $100(1-\alpha)$th percentile of the distribution of the Studentized range of k means with ν d.f. (Fisher's LSD $= W_2$.) For $n_i \neq n_j$, Dunnett [11] recommends replacing n by the harmonic mean* of n_i and n_j; for both $n_i \neq n_j$ and $\sigma_i \neq \sigma_j$, see Dunnett [12]. For an extension to covariance analysis, see Bryant and Paulson [3]. (*See also* TUKEY'S SIMULTANEOUS TEST PROCEDURE.)

4. In *Scheffé's method*, the critical value for testing any two means is

$$S = \sqrt{(k-1)F_{k-1,\nu,1-\alpha}(2s^2/n)}\ ,$$

where $F_{k-1,\nu,\gamma}$ is the (100γ)th percentile of the F-distribution with $(k-1)$ and ν d.f. Both Tukey's and Scheffé's methods can be extended to general contrasts, with Tukey's method being more powerful for paired comparisons and Scheffé's method being superior for general contrasts. (*See also* SCHEFFÉ'S SIMULTANEOUS TEST PROCEDURE.)

To illustrate the preceding methods, we will use data from an experiment (completely randomized design* with $n_i = 10$ replicates) conducted at the U.S. Horticultural Research Laboratory, Orlando, Florida, to compare the infectivity of $k = 6$ biotypes of a certain nematode on citrus seedlings. The following are the ranked average dry-stem weights in grams for the six biotypes.

Biotypes	1	2	3	4	5	6
Means	11.11	10.76	10.68	9.76	7.94	6.25

The analysis of variance gives $s^2 = 7.6793$ with $\nu = 6 \times 9 = 54$ d.f. and $F = 4.84 > F_{5,54,0.95} = 2.39$. The results (with $\alpha = 0.05$) are as follows, where means within parentheses are not significantly different. In the FBM, we take $m = k(k-1)/2 = 15$.

Method	LSD	Grouping of Means	No. of Significant Pairs
Fisher	2.49	(1, 2, 3, 4), (4, 5), (5, 6)	7
FBM	3.82	(1, 2, 3, 4, 5), (4, 5, 6)	3
Tukey	3.67	(1, 2, 3, 4, 5), (4, 5, 6)	3
Scheffé	4.69	(1, 2, 3, 4, 5,), (2, 3, 4, 5, 6)	1

5. *Dunnett's method* handles the special case where one of the k treatments is a control and we wish to compare it only with each of the remaining $(k-1)$ treatments. A treatment differs from control if their difference exceeds $d_{\alpha,k^*,\nu}\sqrt{2s^2/n}$, where $k^* = (k-1)$ is the number of treatments other than the control. Values of $d_{\alpha,k^*,\nu}$ are tabulated for both one- and two-sided alternatives.

In our example, the first treatment was a control (the seedlings were not inoculated). A one-sided alternative is appropriate since the nematode can only have a deleterious effect or none at all. The 5% LSD is $2.29\sqrt{2(7.6793)/10} = 2.84$. Only biotypes 5 and 6 differed from control, as in Fisher's LSD method.

Robson [25] extends the method to balanced incomplete block designs (*see* BLOCKS, BALANCED INCOMPLETE). Dudewicz et al. [7] discuss optimum allocation of experimental units if the variances are unequal and unknown. For a nonparametric method, see Hollander and Wolfe [18] and Levy [20].

STEPWISE MULTIPLE COMPARISON PROCEDURES

A. To apply the *Newman–Keuls* multiple range test* (NKMRT), arrange the means in order of magnitude and declare two means p apart [with $(p-2)$ means in between] to be different if they differ by more than $W_p = q_{p,\nu,1-\alpha}\sqrt{s^2/n}$. (Tukey's HSD $= W_k$ and since $W_p < W_k$ for $p < k$, the NKMRT will tend to give more significant pairs than the HSD test.) For the example in paragraph 4 with $\sqrt{s^2/n} = 0.8763$, we have

p	2	3	4	5	6
$q_{p,54,.95}$	2.84	3.41	3.75	4.00	4.18
W_p	2.49	2.99	3.29	3.50	3.67

The order of testing is step-down; that is, we start with $p = k$, followed by $p = k-1, \ldots, 2$. Two means that are not significantly different are underlined, including all intermediate means. No further testing is made between any two means already underscored by the same line. The NKMRT groups the means into (1, 2, 3, 4, 5) and (5, 6), with four significant pairs (16, 26, 36, 46).

B. *Duncan's multiple range test** (DMRT) is a slight modification of the NKMRT with α in W_p replaced by $\alpha_p = 1-(1-\alpha)^{p-1}$, the rationale being that that is the protection one gets in the usual tests of $(p-1)$ orthogonal contrasts among p means. The critical value for testing two ranked means p apart is $R_p = q_{p,\nu,1-\alpha_p}\sqrt{s^2/n}$. For our example, we have

p	2	3	4	5	6
0.05_p	0.0500	0.0975	0.1426	0.1855	0.2262
$q_{p,54,1-0.05_p}$	2.84	2.98	3.08	3.15	3.21
R_p	2.49	2.61	2.70	2.76	2.81

The means are grouped into (1, 2, 3, 4), (4, 5), and (5, 6), as in Fisher's method. Since $R_p < W_p$ for $p > 2$, the DMRT will tend to give more significant differences than the NKMRT. If the DMRT (or NKMRT) is performed only after a significant F-test, Shaffer [29] recommends that the largest difference between the two means be tested against R_{k-1} (or W_{k-1}) in order to "increase the probability of detecting differences without changing the control of Type I error."

If $\bar{y}_i$ and $\bar{y}_j$ are correlated with covariance $c\sigma^2$, replace s^2/n by $s^2(n_i^{-1} + n_j^{-1} - 2c)/2$. For an extension to covariance analysis, see Bryant and Bruvold [4]. (For Duncan's latest views on MCPs, *see* k-RATIO t-TESTS, t-INTERVALS AND POINT ESTIMATES FOR MULTIPLE COMPARISONS.)

C. The critical value in *Welsch's GAPA test* [34] for testing two ranked means p apart is $G_p = C_p\sqrt{s^2/n}$. For $\nu = 54$ d.f. and $k = 6$, $C_p = 3.47, 3.82, 3.97, 4.17$ approximately, and $G_p = 3.01, 3.30, 3.44, 3.44, 3.61$ for $p = 2, \ldots, 6$, respectively. The means are grouped into (1, 2, 3, 4, 5) and (5, 6), with four significant pairs. The test is more powerful than Tukey's HSD. Welsch also has a step-up SMCP, where ranked means 2 apart are tested first, followed by means 3 apart, etc.

D. *Ryan's* test* [27] uses the Student's t but at a level $\alpha_p = \alpha/[k(p-1)/2]$ in testing 2 ranked means p apart. If $k = 6$ and $\alpha = 0.05$, $\alpha_p = 0.0033$, 0.0042, 0.0056, 0.0083, and 0.0167 for $p = 6, \ldots, 2$, respectively. No tables are provided.

E. Some other SMCPs are Peritz's test (unpublished but described in Einot and Gabriel [13], Section 1.8], Tukey's wholly significant difference (WSD) method and the Marcus, Peritz, and Gabriel [21] test (MPGT). Peritz's test is a modification of the NKMRT and Ryan test to make the NKMRT have a certain *closure* property. An algorithm for performing this tedious test is given in Begun and Gabriel [2]. The critical value of Tukey's WSD method is the average of the critical values of Tukey's HSD and NKMRT. The MPGT is an improved stepwise version of the Dunnett test. (For MCPs with categorical data* and regression, see MULTIPLE COMPARISONS.)

ERROR RATES: COMPARISON AMONG THE MCPS

Ryan [26] and Federer [14] define the following three type I error rates (ERs), following Tukey [33]: (1) α_c = ER per comparison = expected ratio of the number of incorrect rejections to the total number of nonsignificant comparisons tested; (2) α_e = ER per experiment = expected number of incorrect rejections per experiment where the null hypothesis H_0 is true; and (3) α_w = ER experimentwise = expected ratio of the number of experiments with at least one incorrect rejection to the total number of experiments with H_0 true.

From Bonferroni's inequality*, $\alpha_w \leqslant \alpha_e$ (approximately equal for small α_e). If each experiment has m comparisons, $\alpha_e = m\alpha_c$. Thus we can ensure that $\alpha_w \leqslant 0.01$ (say) if we make m comparisons at $\alpha_c = 0.01/m$ each, but this is inefficient if $\alpha_w \geqslant 0.10$ or if m is large. The rates α_e and α_w also depend on whether H_0 is the complete null hypothesis H_{0c} (all k means equal) or the partial null hypothesis H_{0p} (at least 2 means equal). The maximal α_e (or α_w) is the maximum of α_e (or α_w) under all partial null hypotheses.

Ryan [26] and Federer [14] discuss the choice of the type of ER. In general, α_c is appropriate when all comparisons are equally important and one wrong rejection does not affect the validity or usefulness of the other comparisons from the same experiment. The rates α_e and α_w do not distinguish between an experiment with 2 treatments and one with 10, say, treatments, where it will be easier to make one or more wrong rejections if H_0 is true. The rate $\alpha_w = 0.05$ is thus more stringent than $\alpha_c = 0.05$ for $k \geqslant 3$, and one should use $\alpha_w \geqslant 0.10$ for $k \geqslant 3$ where one would use $\alpha_c = 0.05$ (see Hartley [17, p. 49]).

Several Monte Carlo* studies have been made to compare the various MCPs. "Clearly, the different methods are not comparable when each one is studied with a probability of say, five percent of the kind of Type I errors discussed in its original presentation" [13]. The Bonferroni method uses α_e, while Tukey's, Scheffé's, Dunnett's, Welsch's, Ryan's, and Peritz's tests use α_w. The ER in Fisher's protected LSD and the NKMRT is α_w only under the complete null hypothesis H_{0c}. Duncan's ER is per degree of freedom.

Einot and Gabriel [13] compared the per pair powers of the MCPs for fixed maximal α_w by adjusting Duncan's α_p to $1-(1-\alpha)^{(p-1)/(k-1)}$. The NKMRT was omitted because the maximal α_w cannot be controlled if $k > 3$. In descending order of power, the results were Peritz > Ryan > Duncan > Tukey > Scheffé. However, the Monte Carlo study (for $k = 3$–5) showed that the power differences were small, and Tukey's HSD was recommended for its simplicity, extension to general contrasts, and the availability of corresponding simultaneous confidence bands. Using all-pairs power as criterion, however, Ramsey [24] found the increase in power in the Peritz test could be substantial, with the more convenient Welsch test not far behind.

More basic than the choice of an MCP is the question of whether MCPs are relevant. In a comment on Ramsey [24], Gabriel wondered about the real purpose of multiple comparisons. In the discussion following O'Neill and Wetherill [23], Plackett viewed the subject of multiple comparisons as being essentially artificial. In fact, one may even ask if significance testing is relevant. In actual experiments, it is inconceivable that the k true means will be (exactly) equal. It is more logical to assume a priori that all means are unequal and conduct the experiment either to estimate just how unequal they are or for ranking* and selection* purposes.

References

[1] Bailey, B. J. R. (1977). *J. Amer. Statist. Ass.*, **72**, 469–478.

[2] Begun, J. and Gabriel, K. R. (1981). *J. Amer. Statist. Ass.*, **76**, 241–245.

[3] Bryant, J. L. and Paulson, A. S. (1976). *Biometrika*, **63**, 631–638.

[4] Bryant, J. L. and Bruvold, N. T. (1980). *J. Amer. Statist. Ass.*, **75**, 874–880.

[5] Chew, V. (1976). *HortScience*, **11**, 348–357.

[6] Chew, V. (1976). *Proc. Fla. State Hort. Soc.*, **89**, 251–253.

[7] Dudewicz, E. J., Ramberg, J. S., and Chen, H. J. (1975). *Biom. Zeit.*, **17**, 13–26.

[8] Duncan, D. B. (1955). *Biometrics*, **11**, 1–42.

[9] Dunnett, C. W. (1964). *Biometrics*, **20**, 482–491.

[10] Dunnett, C. W. (1970). In *Statistics in Endocrinology*, J. W. McArthur and T. Colton, eds. MIT Press, Cambridge, MA, pp. 79–103.

[11] Dunnett, C. W. (1980). *J. Amer. Statist. Ass.*, **75**, 789–795.

[12] Dunnett, C. W. (1980). *J. Amer. Statist. Ass.*, **75**, 796–800.

[13] Einot, I. and Gabriel, K. R. (1975). *J. Amer. Statist. Ass.*, **70**, 574–583.

[14] Federer, W. T. (1964). (In lecture notes on the "Design and Analysis of Experiments," given at Colorado State University, Fort Collins, Colorado, July 13–August 7, 1964.)

[15] Games, P. A. (1971). *Amer. Educ. Res. J.*, **8**, 531–565.

[16] Games, P. A. (1977). *J. Amer. Statist. Ass.*, **72**, 531–534.

[17] Hartley, H. O. (1955). *Commun. Pure and Appl. Statist.*, **8**, 47–72.

[18] Hollander, M. and Wolfe, D. A. (1973). *Nonparametric Statistical Methods*. Wiley, New York.

[19] Keuls, M. (1952). *Euphytica*, **1**, 112–122.

[20] Levy, K. J. (1980). *Amer. Statist.* **34**, 99–102.

[21] Marcus, R., Peritz, E., and Gabriel, K. R. (1976). *Biometrika*, **63**, 655–660.

[22] Miller, R. G., Jr. (1981). *Simultaneous Statistical Inference*, 2nd ed. Springer, New York. (The only book of its kind.)

[23] O'Neill, R. and Wetherill, G. B. (1971). *J. R. Statist. Soc. B*, **33**, 218–250. (Contains discussion and about 250 references classified into 15 categories.)

[24] Ramsey, P. H. (1978). *J. Amer. Statist. Ass.*, **73**, 479–485. (Comment and rejoinder on pp. 485–487.)

[25] Robson, D. S. (1961). *Technometrics*, **3**, 103–105.

[26] Ryan, T. A. (1959). *Psychol. Bull.*, **56**, 26–47.

[27] Ryan, T. A. (1960). *Psychol. Bull.*, **57**, 318–328.

[28] Scheffé, H. (1953). *Biometrika*, **40**, 87–104.

[29] Shaffer, J. P. (1979). *J. Educ. Statist.*, **4**, 14–23.

[30] Spjøtvoll, E. (1974). *Scand. J. Statist.*, **1**, 97–114.

[31] Stoline, M. R. (1981). *Amer. Statist.*, **35**, 134–141.

[32] Thomas, D. A. H. (1973). *The Statist.*, **22**, 16–42.

[33] Tukey, J. W. (1953). "The Problem of Multiple Comparisons." Mimeographed notes. Princeton University, Princeton, NJ.

[34] Welsch, R. E. (1977). *J. Amer. Statist. Ass.*, **72**, 566–575.

Bibliography

Bradu, D. and Gabriel, K. R. (1974). *J. Amer. Statist. Ass.*, **69**, 428–436.

Johnson, D. E. (1976). *Biometrics*, **32**, 929–934.

Jolliffe, I. T. (1975). In *Applied Statistics*, R. P. Gupta, ed. North-Holland, New York.

Krishnaiah, P. R., Mudholkar, G. S., and Subbaiah, P. (1980). In *Analysis of Variance*, Vol. 1: *Handbook of Statistics*, P. R. Krishnaiah, ed. North-Holland, New York.

Lehmann, E. L. and Shaffer, J. P. (1977). *J. Amer. Statist. Ass.*, **72**, 576–578.

Shaffer, J. P. (1977). *Biometrics*, **33**, 293–303.

Shirley, E. A. C. (1979). *Appl. Statist.*, **28**, 144–151.

(CONFIDENCE INTERVALS AND REGIONS
k-RATIO *t*-TESTS, *t*-INTERVALS AND POINT ESTIMATES FOR MULTIPLE COMPARISONS
MULTIPLE COMPARISONS
ONE-WAY CLASSIFICATION
RANKING PROCEDURES
SCHEFFÉ'S SIMULTANEOUS TEST PROCEDURE
SELECTION PROCEDURES
t-TESTS
TUKEY'S SIMULTANEOUS TEST PROCEDURE)

VICTOR CHEW

MULTIPLE-RECORD SYSTEMS (DATA)

A multiple-record system is a framework for the estimation of the size of a population from the combined data for two or more sources of information. The population elements are usually human beings or events pertaining to them such as births, deaths, accidents, crimes, diagnoses of diseases, experiences with a product or service, and other phenomena for which accurate enumeration by a single source may not be economical or feasible. More generally, the population elements under consideration can be animals, objects, or attributes observed in a laboratory since the statistical principles underlying estimation from multiple-record systems are analogous to those for CAPTURE-RECAPTURE METHODS (see El-Khorazaty et al. [7] for discussion).

Dual record systems involving two information sources have been applied extensively in demography*. For example, vital rates for births and deaths might be estimated from the combination of civil registration systems with a periodic sample survey* covering the same time period and area (see Coale [3]). Also, this framework enables some evaluation of the extent of completeness of each information source. The usual dual record system estimator for population size is

$$\hat{N} = n_1 n_2 / n_{11}, \quad (1)$$

where n_j is the number of events identified by the jth system for $j = 1, 2$ and n_{11} is the number of events identified by both systems. As noted in Chandrasekar and Deming [2], usage of this estimator is justified under the following assumptions:

1. No target coverage bias (i.e., all events recorded by each source correspond to the population of interest).
2. No correlation bias (i.e., the identification of an event by one source is statistically independent of its identification by the other source).
3. No matching bias (i.e., all events identified by both sources are included in n_{11}).

Violations of assumptions (1–3) can produce downward or upward bias in $\hat{N}$. More specifically, the population size is underestimated when there is positive correlation between the two sources and an excess of erroneous matches (i.e., n_{11} is too large); and it is overestimated when there is overcoverage (i.e., the n_j are too large). The impact of correlation bias can sometimes be reduced by partitioning the population elements into relatively homogeneous strata for which the results from equation (1) are added together. Issues concerning matching bias are primarily a matter of data quality and thus can be difficult to resolve in many applications. Some strategies for dealing with the underlying *record linkage problem** are given in DeGroot et al. [4], Deming and Glasser [5], DuBois [6], Madigan and Wells [9], and Tepping [11]. Another consideration is that $\hat{N}$ can be unstable for situations where n_{11} tends to be small, and so other estimators may be preferable to it (*see* CAPTURE-RECAPTURE METHODS). For more complete discussion of the dual record system and its assumptions, see El-Khorazaty et al. [7] and Marks et al. [10].

A good example of a multiple-record system with more than two sources is given in Wittes et al. [13]. It is concerned with the use of five sources to estimate the number of infants with a specific congenital anomaly for a target area and time period. These were hospital obstetric records, hospital inpatient records, Department of Public Health records, Department of Mental Health records, and school records. From this information, many potential estimates are available; and the choice among them depends on what assumptions are considered appropriate. One general strategy of interest is the use of *log-linear models** to describe the structure in the 2^5 *incomplete contingency table** for the cross-classification of the five sources and to obtain the implied estimates for the missing cell for noninclusion by all sources and the total population. The capabilities of such analysis are illustrated for this example in Bishop et al. [1, Chap. 6] and Koch et al. [8]. Finally, an example involving both multiple sources and stratification is given in Wittes [12].

References

[1] Bishop, Y. M. M., Fienberg, S. E., and Holland, P. W. (1975). *Discrete Multivariate Analysis: Theory and Practice*. MIT Press, Cambridge, MA.

[2] Chandrasekar, C. and Deming, W. E. (1949). *J. Amer. Statist. Ass.*, **44**, 101–115.

[3] Coale, A. J. (1961). "The Design of an Experimental Procedure for Obtaining Accurate Vital Statistics." International Population Conference, New York, 372–375.

[4] DeGroot, M. H., Feder, P. I., and Goel, P. K. (1971). *Ann. Math. Statist.*, **42**, 578–593.

[5] Deming, W. E. and Glasser, G. J. (1959). *J. Amer. Statist. Ass.*, **54**, 403–415.

[6] DuBois, N. S. D. (1969). *J. Amer. Statist. Ass.*, **64**, 163–174.

[7] El-Khorazaty, M. N., Imrey, P. B., Koch, G. G., and Wells, H. B. (1977). *Int. Statist. Rev.*, **45**, 129–157.

[8] Koch, G. G., El-Khorazaty, M. N., and Lewis, A. L. (1976). *Commun. Statist. A*, **5**, 1425–1445.

[9] Madigan, F. C. and Wells, H. B. (1976). *Demography*, **13**, 381–395.

[10] Marks, E. S., Seltzer, W., and Krotki, K. J. (1974). *Population Growth Estimation: A Handbook of Vital Statistics Measurement*. The Population Council, New York.

[11] Tepping, B. J. (1968). *J. Amer. Statist. Ass.*, **63**, 1321–1332.

[12] Wittes, J. T. (1974). *J. Amer. Statist. Ass.*, **69**, 93–97.

[13] Wittes, J. T., Colton, T., and Sidel, V. W. (1974). *J. Chronic Dis.*, **27**, 25–36.

(CAPTURE-RECAPTURE METHODS
POPULATION OR SAMPLE SIZE ESTIMATION
RECORD LINKAGE/MATCHING SYSTEMS)

GARY G. KOCH

MULTIPLE SAMPLING

This is an acceptance sampling* procedure in which successive samples of predetermined sizes $n_1, n_2, \ldots, n_k$ are taken. After each sample is taken a decision is made, based on the observations so far available whether to accept, reject, or proceed to the next sampling stage. The special case $k = 2$ is discussed in detail in DOUBLE SAMPLING. If each sample size is $1(n_1 = n_2 = \cdots n_k = 1)$, we have a (truncated) sequential sampling* situation.

Although multiple sampling is most commonly used when attributes ("defective," "nonconforming," etc.) rather than variables ("length," "tensile strength," etc.) are measured, it can be used in the latter situation also.

(ACCEPTANCE SAMPLING
DOUBLE SAMPLING
MULTIPHASE SAMPLING
QUALITY CONTROL, STATISTICAL
SAMPLING PLANS
SEQUENTIAL SAMPLING)

MULTIPLE TIME SERIES

Multiple time series is the study of probability models and methods of data analysis that represent and summarize the relations between several time series*. The theory of multiple time series is usually developed as an extension of the theory of a scalar time series.

When a variable, denoted Y, is observed at successive (usually equispaced) times t, the series of observations $Y(t)$, $t = 0, \pm 1, \ldots$ is called a time series, or more precisely a *univariate time series*. When the value $Y(t)$ at time t is a vector of real (or complex) numbers, the time series is called a *multiple* or *multivariate* time series. Time-series analysts seem to have used *multiple* in preference to *multivariate* to describe the case of vector-valued observations.

Univariate time-series analysis seeks to model $Y(t)$ as a function of the time variable t by representing $Y(t) = \mu(t) + Z(t)$, where $\mu(t) = E[Y(t)]$ is the *mean value function* and $Z(t)$ is the fluctuation function. The theory of models for time series starts with the assumption that the time series $Y(t)$ has been preprocessed so that we may assume $\mu(t) = 0$.

The basic problem of the *theory* of time-series analysis is to study the dependence (or correlation) structure of zero mean time series. The time series $Y(t)$ is called *covariance stationary* if there exists a sequence $R(v)$, $v = 0, \pm 1, \ldots$ (called the *covariance* function) such that the covariance kernel

$$K(s,t) = E[\{Y(s) - \mu(s)\}\{Y(t) - \mu(t)\}]$$

satisfies, for all t,

$$K(t + v, t) = R(v).$$

Note that $K(t,t) = R(0) = \text{Var}[Y(t)]$. We call

$$\rho(v) = R(v)/R(0), \qquad v = 0, \pm 1, \ldots$$

the *correlation function* of the covariance stationary time series $Y(t)$.

The *spectral density function* of a univariate covariance stationary time series $Y(t)$ is the Fourier transform (*see* INTEGRAL TRANSFORMS) of its correlation function

$$f(\omega) = \sum_{v=-\infty}^{\infty} \exp(-2\pi i v\omega)\rho(v),$$
$$-0.5 \leqslant \omega \leqslant 0.5.$$

This definition implicitly assumes that $\rho(v)$ is summable; the process of transforming the time series to fulfill this assumption is a basic problem of empirical time-series analysis that is too complicated to explain in a short article. The variable ω is interpreted as *frequency*. The Fourier transform of the covariance function is also used and is called the *unnormalized spectral density function*.

In estimating $\rho(v)$ and $f(\omega)$ one may want to choose estimators $\hat{\rho}(v)$ and $\hat{f}(\omega)$ which, in addition to statistical criteria, possess the following mathematical properties:

$$\rho(0) = 1, \qquad \rho(-v) = \rho(v),$$

$\rho(v)$ is positive definite in the sense that

$$\sum_{i,j=1}^{n} c_i c_j^* \rho(i-j) \geqslant 0$$

for any integer n, and complex coefficients $c_1, \ldots, c_n$ (c^* denotes the conjugate of the complex number c);

$$\int_{-0.5}^{0.5} f(\omega)\,d\omega = 1, \quad f(-\omega) = f(\omega), \quad f(\omega) \geqslant 0.$$

The spectral density function $f(\omega)$ provides a *spectral representation* of $\rho(v)$ in the sense that

$$\rho(v) = \int_{-0.5}^{0.5} \exp(2\pi i v\omega) f(\omega)\,d\omega,$$
$$v = 0, \pm 1,, \ldots .$$

The *spectral distribution function* $F(\omega)$ is defined by

$$F(\omega) = 2\int_0^{\omega} f(\omega')\,d\omega', \qquad 0 \leqslant \omega \leqslant 0.5.$$

To study the relation between two time series $Y_1(t)$ and $Y_2(t)$, it is convenient to *stack* the values observed at time t into a 2×1 vector $\mathbf{Y}(t)$ with transpose $\mathbf{Y}^*(t) = [Y_1(t), Y_2(t)]$. An asterisk on a vector or matrix denotes its complex-conjugate transpose. We call $\mathbf{Y}(t)$, $t = 0, \pm 1, \ldots$ a *multiple time series*.

In defining correlation and spectral density matrices of a multiple time series, we consider an $r \times 1$ vector $\mathbf{Y}(t)$ with transpose $\mathbf{Y}^*(t) = [Y_1(t), \ldots, Y_r(t)]$, where $Y_1(\cdot), \ldots, Y_r(\cdot)$ are r time series whose linear relationships one seeks to model.

A zero mean multiple time series is *covariance stationary* if there exists a sequence of matrices $\mathbf{R}(v)$, $v = 0, \pm 1, \ldots$ [called a *covariance matrix function*] such that the covariance matrix kernel

$$\mathbf{K}(s,t) = E\left[\mathbf{Y}(s)\mathbf{Y}^*(t)\right]$$

satisfies for all t

$$\mathbf{K}(t+v,t) = \mathbf{R}(v).$$

The (h, j)th element of $\mathbf{R}(v)$ is denoted

$$R_{hj}(v) = \operatorname{Cov}\left[Y_h(t+v), Y_j(t)\right].$$

The function $R_{jj}(v)$ is a *covariance* function; for $h \neq j$, $R_{hj}(v)$ is a *cross-covariance* function. The function

$$\rho_{hj}(v) = R_{hj}(v)/\{R_{hh}(0)R_{jj}(0)\}^{1/2}$$

is a *cross-correlation* function. The function $\rho_{jj}(v) = R_{jj}(v)/R_{jj}(0)$ is a *correlation* function [or even an *autocorrelation* function to emphasize its distinction from a cross-correlation function]. The matrix function $\boldsymbol{\rho}(v)$ with (h, j)th entry equal to $\rho_{hj}(v)$ is the *correlation matrix* function.

The matrix Fourier transform

$$\mathbf{f}(\omega) = \sum_{v=-\infty}^{\infty} \exp(-2\pi i\omega v)\boldsymbol{\rho}(v), \qquad -0.5 \leqslant \omega \leqslant 0.5,$$

is called the *spectral density matrix*. For each ω, $\mathbf{f}(-\omega) = \mathbf{f}^*(\omega)$ and $\mathbf{f}(\omega) \geqslant 0$ (in words, $\mathbf{f}(\omega)$ is a non-negative definite matrix).

For $h \neq j$, the (h, j)th entry

$$f_{hj}(\omega) = \sum_{v=-\infty}^{\infty} \exp(-2\pi i v\omega)\rho_{hj}(v)$$

is called the *cross-spectral density*; it is a complex-valued function whose real (Re) and imaginary (Im) parts are denoted

$$c_{hj}(\omega) = \operatorname{Re} f_{jj}(\omega), \qquad \text{the } \textit{co-spectrum},$$

$$q_{hj}(\omega) = -\operatorname{Im} f_{hj}(\omega), \qquad \text{the } \textit{quadrature spectrum}.$$

An alternative way of expressing $f_{hj}(\omega)$ is to write it in polar form

$$f_{hj}(\omega) = G_{hj}(\omega)\exp 2\pi i\phi_{hj}(\omega),$$

where

$$G_{hj}(\omega) = \left\{c_{hj}^2(\omega) + q_{hj}^2(\omega)\right\}^{1/2},$$

$$2\pi\phi_{hj}(\omega) = \tan^{-1}\{-q_{hj}(\omega)/c_{hj}(\omega)\};$$

the arctangent function is defined as a function of two variables and has range $-\pi$ to π. The function $G_{hj}(\omega)$ is the *cross-amplitude spectrum*, and $\phi_{hj}(\omega)$ the *phase spectrum*. The *coherence spectrum* $W_{hj}(\omega)$ is defined by

$$W_{hj}(\omega) = |f_{hj}(\omega)|^2/f_{hh}(\omega)f_{jj}(\omega).$$

The letter W is chosen in honor of Norbert Wiener*, who first introduced these concepts in his celebrated 1930 paper "Generalized Harmonic Analysis" [21].

To interpret the various spectra one can define for a multiple time series, consider a 2×2 spectral density matrix

$$\mathbf{f}(\omega) = \begin{bmatrix} f_{11}(\omega) & f_{12}(\omega) \\ f_{21}(\omega) & f_{22}(\omega) \end{bmatrix};$$

using the basic matrix operation *sweep* fundamental to regression analysis one can transform $\mathbf{f}(\omega)$ to

$$\begin{bmatrix} f_{11}^{-1}(\omega) & f_{11}^{-1}(\omega)f_{12}(\omega) \\ f_{21}(\omega)f_{11}^{-1}(\omega) & f_{22}(\omega) - f_{21}(\omega)f_{11}^{-1}(\omega)f_{12}(\omega) \end{bmatrix},$$

whose entries provide regression analysis in the frequency domain, as we now explain.

One can write $Y_2(t)$ as a sum of $\hat{Y}_2(t)$, which is the linear function of the time series $Y_1(\cdot)$ best approximating $Y_2(t)$ in the sense of minimum mean square error and a residual $\tilde{Y}_2(t)$. We write $\hat{Y}_2(t)$ as the output of a linear filter:

$$\hat{Y}_2(t) = \sum_{s=-\infty}^{\infty} b(s) Y_1(t-s).$$

The coefficients $b(s)$ are determined by the normal equations*, for all u,

$$\begin{aligned} E[Y_2(t) Y_1(u)] &= E[\hat{Y}_2(t) Y_1(u)] \\ &= \sum_{s=-\infty}^{\infty} b(s) E[Y_1(t-s) Y_1(u)], \end{aligned}$$

which we write as

$$R_{21}(t-u) = \sum_{v=-\infty}^{\infty} b(s) R_{11}(t-s-u).$$

Therefore, for all v (letting $v = t - u$)

$$\frac{\sigma_2}{\sigma_1} \rho_{21}(v) = \sum_{s=-\infty}^{\infty} b(s) \rho_{11}(v-s),$$

where $\sigma_1^2 = R_{11}(0)$, $\sigma_2^2 = R_{22}(0)$. We obtain a formula for $b(s)$ by first finding a formula for

$$B(\omega) = \sum_{s=-\infty}^{\infty} b(s) \exp(-2\pi i \omega s),$$

the *regression transfer function*. Replacing correlations by their spectral representations, one obtains the fundamental formula for the transfer function of the filter transforming $Y_1(\cdot)$ into $\hat{Y}_2(\cdot)$:

$$B(\omega) = \frac{\sigma_2}{\sigma_1} f_{21}(\omega) f_{11}^{-1}(\omega).$$

The gain and phase spectra $G_{21}(\omega)$ and $\phi_{21}(\omega)$ provide measures of the gain and phase of the filter with transfer function $B(\omega)$.

The residual time series $\tilde{Y}_2(t)$ has spectral density proportional to

$$\begin{aligned} f_{22}(\omega) &- f_{21}(\omega) f_{11}^{-1}(\omega) f_{12}(\omega) \\ &= f_{22}(\omega)\{1 - W_{12}(\omega)\}. \end{aligned}$$

The coherence spectrum plays the role of the squared correlation coefficient. At the frequencies ω where $W_{12}(\omega)$ is close to 1, there is a close linear fit between the two time series. Estimation of coherence can be very delicate in practice. If one is unaware of how to take proper care, one can spuriously conclude that the coherence is zero (see Parzen [11] and Cleveland and Parzen [6]).

To estimate correlation functions and spectral density matrices from a sample $\mathbf{Y}(t)$, $t = 1, \ldots, T$ of a zero mean covariance stationary multiple time series one first estimates $\mathbf{R}(v)$ by the *sample covariance matrix* function

$$\mathbf{R}_T(v) = \frac{1}{T} \sum_{t=1}^{T-v} \mathbf{Y}(t+v) \mathbf{Y}^*(t), \qquad v = 0, \ldots, T-1,$$

Its elements are denoted

$$R_{hj,T}(v) = \frac{1}{T} \sum_{t=1}^{T-v} Y_h(t+v) Y_j(t).$$

For purposes of visual examination of the estimators, one usually prints a sequence of matrices (which we write out for the case $r = 2$):

$$\begin{bmatrix} \rho_{11,T}(0) & \rho_{12,T}(0) \\ \rho_{21,T}(0) & \rho_{22,T}(0) \end{bmatrix} \begin{bmatrix} \rho_{11,T}(1) & \rho_{12,T}(1) \\ \rho_{21,T}(1) & \rho_{22,T}(1) \end{bmatrix} \ldots,$$

where

$$\rho_{hj,T}(v) = R_{hj,T}(v) / \{R_{hh,T}(0) R_{jj,T}(0)\}^{1/2}$$

is the *sample correlation function*. The matrix $\boldsymbol{\rho}_T(v) = \{\rho_{hj,T}(v)\}$ is the *sample correlation matrix* of lag v. The sample spectral density matrix is defined by

$$\mathbf{f}_T(\omega) = \sum_{|v|<T} \exp(-2\pi i \omega v) \boldsymbol{\rho}_T(v).$$

The sample unnormalized spectral density matrix is defined by

$$\frac{1}{T} \left\{ \sum_{s=1}^{T} \exp(-2\pi i \omega s) \mathbf{Y}(s) \right\} \times \left\{ \sum_{t=1}^{T} \exp(-2\pi i \omega t) \mathbf{Y}(t) \right\}^*.$$

$\mathbf{f}_T(\omega)$ might be considered a "natural" estimator of $\mathbf{f}(\omega)$, but it is *not* consistent. One approach to forming consistent estimators of $\mathbf{f}(\omega)$ is by smoothing $\mathbf{f}_T(\omega)$. However this approach does not lead easily to determining optimal orders of smoothing. Fitting autoregressive (AR), moving average (MA), and autoregressive-moving average (ARMA)* schemes to a multiple time series not only provides spectral estimators but also directly provides solutions to the prediction problem and the testing for white noise problem.

The time series $\mathbf{Y}(t)$, $t = 0, \pm 1, \ldots,$ is called *white noise** when the random vectors $\{\mathbf{Y}(t)\}$ are independent and identically distributed (and thus obey the model of traditional multivariate analysis). White noise is interpreted as a hypothesis that there is no dependences over time between the variables. A fully nonparametric test of the hypothesis of white noise could be based on a test of $H_0 : \boldsymbol{\rho}(v) = 0$ for $v \neq 0$. It is difficult to develop a statistical test of H_0 with good properties directly without specifying an alternative hypothesis. Two important ways of specifying alternative hypotheses are the multivariate moving average* model MA(q) and autoregressive model AR(p) to be defined below.

A multiple zero mean stationary time series $\mathbf{Y}(t)$ is said to obey an ARMA(p,q) scheme [or *autoregressive-moving average scheme* of orders p and q] if $\mathbf{Y}(t)$ satisfies the representation

$$\begin{aligned} \mathbf{Y}(t) &+ \mathbf{A}_p(1)\mathbf{Y}(t-1) + \cdots \\ &+ \mathbf{A}_p(p)\mathbf{Y}(t-p) \\ &= \boldsymbol{\epsilon}(t) + \mathbf{B}_q(1)\boldsymbol{\epsilon}(t-1) + \cdots \\ &+ \mathbf{B}_q(q)\boldsymbol{\epsilon}(t-q), \end{aligned}$$

where $\boldsymbol{\epsilon}(t)$ is a zero mean Gaussian white noise multiple time series with covariance matrix $\boldsymbol{\Sigma} = E[\boldsymbol{\epsilon}(t)\boldsymbol{\epsilon}^*(t)]$.

The matrix polynomials

$$\mathbf{G}_p(z) = \mathbf{I} + \mathbf{A}_p(1)z + \cdots + \mathbf{A}_p(p)z^p,$$

$$\mathbf{H}_q(z) = \mathbf{I} + \mathbf{B}_1(1)z + \cdots + \mathbf{B}_q(q)z^q$$

are assumed to satisfy the invertibility conditions: all the zeros of $\log\det \mathbf{G}_p(z)$, and all the zeros of $\log\det \mathbf{H}_q(z)$, lie outside the unit circle in the complex z-plane. The ARMA model is expressed in operator form

$$\mathbf{G}_p(L)\mathbf{Y}(t) = \mathbf{H}_q(L)\boldsymbol{\epsilon}(t),$$

where L is the backward shift operator defined by $L\mathbf{Y}(t) = \mathbf{Y}(t-1)$.

A finite parameter ARMA(p,q) can be fit to a multiple time series either as an exact model or as an approximating model. Infinite order AR and MA representations of a multiple time series $Y(\cdot)$ can be shown to exist under suitable conditions; the *spectral density matrix is bounded above and below* means that there exists positive constants c_1 and c_2 such that

$$c_1\mathbf{I} \leqslant \mathbf{f}(\omega) \leqslant c_2\mathbf{I} \qquad \text{for all } \omega$$

where $\mathbf{A} \leqslant \mathbf{B}$ for matrices $\mathbf{A}$ and $\mathbf{B}$ means that $\mathbf{B} - \mathbf{A}$ is a positive definite matrix; *log det spectra is integrable* means that the following quantities are finite:

$$\int_{-0.5}^{0.5} \log\det \mathbf{f}(\omega)\, d\omega = \log\det \boldsymbol{\Sigma}_\infty$$

where $\boldsymbol{\Sigma}_\infty$ is the covariance matrix of $\tilde{\mathbf{Y}}(t) = \boldsymbol{\epsilon}(t)$, the errors of prediction of $\mathbf{Y}(t)$ given $\mathbf{Y}(t-1), \mathbf{Y}(t-2), \ldots,$ using infinite memory minimum mean square error linear prediction.

When the spectral density matrix is bounded above and below, the multiple time series possesses an AR(∞) representation

$$\mathbf{G}_\infty(L)\mathbf{Y}(t) = \boldsymbol{\epsilon}(t).$$

When log det spectra is integrable, the multiple time series possesses an MA(∞) representation

$$\mathbf{Y}(t) = \mathbf{H}_\infty(L)\boldsymbol{\epsilon}(t).$$

Each representation has its applications. The AR(∞) representation is convenient for expressing formulas for the horizon h predictor of $Y(t+h)$ given $Y(t), Y(t-1), \ldots;$ the MA(∞) representation is convenient for expressing the mean square prediction error.

When the multiple time series $\mathbf{Y}(\cdot)$ obeys an ARMA(p,q) scheme, it has AR(∞) and MA(∞) representations with

$$\mathbf{G}_\infty(z) = \mathbf{H}_q^{-1}(z)\mathbf{G}_p(z),$$

$$\mathbf{H}_\infty(z) = \mathbf{G}_p^{-1}(z)\mathbf{H}_q(z).$$

A nonparametric approach to modeling multiple time series is provided by directly estimating $\mathbf{G}_\infty(z)$ and $\mathbf{H}_\infty(z)$.

From an MA(∞), the spectral density matrix $\mathbf{f}(\omega)$ is obtained by

$$\mathbf{f}(\omega) = \mathbf{H}(e^{-2\pi i\omega})\boldsymbol{\Sigma}\mathbf{H}^*(e^{-2\pi i\omega})$$

From an AR(∞), the inverse spectral density matrix is obtained by

$$\mathbf{f}^{-1}(\omega) = \mathbf{G}^*(e^{-2\pi i\omega})\mathbf{\Sigma}^{-1}\mathbf{G}(e^{-2\pi i\omega}).$$

When a multiple time series can be assumed to have zero means and is covariance stationary with spectral density matrices bounded above and below, an effective way of estimating spectra, testing for white noise, forming minimum mean square error linear predictors, and estimating time domain (transfer function) models relating time series is to fit AR(p) schemes.

To estimate the coefficients of an AR(p), there are fast algorithms available based on the Yule–Walker equations expressing the fact that an AR(p) can be interpreted as a finite memory predictor. Equations for the coefficients of the predictor are derived by regarding $\mathbf{G}_p(L)\mathbf{Y}(t)$ as the prediction errors of the minimum mean square error linear predictor of $\mathbf{Y}(t)$ given $\mathbf{Y}(t-1), \ldots, \mathbf{Y}(t-p)$; the projection theorem in Hilbert space states that the prediction errors are orthogonal to the conditioning variables, that is,

$$E\left[\{\mathbf{G}_p(L)\mathbf{Y}(t)\}\mathbf{Y}^*(t-k)\right] = 0, \qquad k = 1, \ldots, p.$$

Further, the prediction error covariance matrix

$$\begin{aligned}\mathbf{\Sigma}_p &= E\left[\{\mathbf{G}_p(L)\mathbf{Y}(t)\}\{\mathbf{G}_p(L)\mathbf{Y}(t)\}^*\right] \\ &= E\left[\{\mathbf{G}_p(L)\mathbf{Y}(t)\}\mathbf{Y}^*(t)\right].\end{aligned}$$

Consequently, one can determine $\mathbf{A}_p(j)$, $j = 1, \ldots, p$ by solving [with $\mathbf{A}_p(0) = \mathbf{I}$]

$$\sum_{j=0}^{p} \mathbf{A}_p(j)\mathbf{R}(k-j) = 0, \qquad k = 1, \ldots, p, \qquad (I)$$

and then determine $\mathbf{\Sigma}_p$ by

$$\mathbf{\Sigma}_p = \sum_{j=0}^{\infty} \mathbf{A}_p(j)\mathbf{R}(-j). \qquad (II)$$

We call (I) and (II) the population Yule–Walker equations; the *sample Yule–Walker equations* for estimators $\mathbf{A}_p(j)$ and $\mathbf{\Sigma}_p$ are obtained by substituting an estimator $\mathbf{R}_T(v)$ for $\mathbf{R}(v)$.

The problem of *model order identification* develops methods for determining the orders p and q of approximating ARMA(p,q) schemes. AR(p) order determining criteria have been given by Akaike [1] and Parzen [13]. Approaches to ARMA model identification are given by Akaike [1, 2], Box and Tiao [4], and Tiao and Box [17].

The theory of statistical analysis of multiple time series was pioneered by Quenouille [16], Whittle [19, 20], and Tukey [18]. The basic textbooks are Hannan [8], Brillinger [5], and Priestley [15]. An excellent account of the engineering contributions is Kailath [9]. Autoregressive estimators of spectral density matrices of multiple time series were emphasized by Parzen [12]. The periodic autoregression approach to multiple time series spectral estimation is discussed by Newton [10]. The theory of ARMA parameter estimation is reviewed by Anderson [3]. Some analogies between multivariate analysis* and regression analysis and multiple time series analysis are discussed in Brillinger [5] and Parzen and Newton [14].

An excellent illustrative example of multiple time series model identification is given by Cooper and Wood [7].

References

[1] Akaike, H. (1974). A new look at the statistical model identification. *IEEE Trans. Aut. Control*, **AC-19**, 716–722.

[2] Akaike, H. (1976). In *Advances and Case Studies in System Identification*, R. Mehra and D. G. Lainiotis, eds. Academic Press, New York and London.

[3] Anderson, T. W. (1980). In *Directions in Time Series*, D. R. Brillinger and G. C. Tiao, eds. Institute of Mathematical Statistics, Hayward, California, pp. 49–59.

[4] Box, G. E. P. and Tiao, G. C. (1977). *Biometrika*, **64**, 355–365.

[5] Brillinger, D. R. (1981). *Time Series: Data Analysis and Theory*. Holden Day, San Francisco.

[6] Cleveland, W. S. and Parzen, E. (1975). *Technometrics*, **17**, 167–172.

[7] Cooper, D. M. and Wood, E. F. (1982). *J. Time Ser. Anal.*, **3**, 153–164.

[8] Hannan, E. J. (1970). *Multiple Time Series*. Wiley, New York.

[9] Kailath, T. (1974). *IEEE Trans. Inf. Theory*, **IT-20**, 145–181.

[10] Newton, H. J. (1982). *Technometrics*, **24**, 109–116.

[11] Parzen, Emanuel (1967). *Proc. Berkeley Symp. Math. Statist. Prob.*, Vol. 1, L. LeCam and J. Neyman, eds. University of California Press, Berkeley, CA, pp. 305–340.

[12] Parzen, Emanuel (1969). In *Multivariate Analysis*, Vol. II, P. Krishnaiah, ed. Academic Press, New York, pp. 389–409.

[13] Parzen, Emanuel (1977). In *Multivariate Analysis*, Vol. IV, P. Krishnaiah, ed. North-Holland, Amsterdam, pp. 283–295.

[14] Parzen, Emanuel and Newton, H. J. (1979). In *Multivariate Analysis*, Vol. V, P. Krishnaiah, ed. North-Holland, Amsterdam, pp. 181–197.

[15] Priestley, M. B. (1981). *Spectral Analysis and Time Series*. Academic Press, London.

[16] Quenouille, M. H. (1957). *The Analysis of Multiple Time Series*. Griffin, London.

[17] Tiao, G. C. and Box, G. E. P. (1981). *J. Amer. Statist. Ass.* **76**, 802–816.

[18] Tukey, J. W. (1959). In *Probability and Statistics*, The Harold Cramér Volume, U. Grenander, ed. Wiley, New York.

[19] Whittle, P. (1953). *J. R. Statist. Soc. B*, **15**, 125–139.

[20] Whittle, P. (1963). *Biometrika*, **50**, 129.

[21] Wiener, N. (1930). *Acta Math.*, **55**, 117–285.

(AUTOREGRESSIVE MOVING AVERAGE (ARMA) MODELS
MOVING AVERAGE
TIME SERIES)

EMANUEL PARZEN

MULTIPLICATION PRINCIPLE

This is a basic counting principle used in elementary probability theory and usually stated as follows:

> If one operation can be performed in any of n_1 ways, and a second operation can be performed in any of n_2 ways, the operations can be combined (the second immediately following the first) in $n_1 n_2$ ways.

A more abstract formulation is

> If a set A has n_1 elements and a set B has n_2 elements, then the Cartesian product $A \times B$ has $n_1 n_2$ elements.

(COMBINATORICS)

MULTIRESPONSE PERMUTATION PROCEDURES

Multiresponse permutation procedures (MRPP) provide analyses compatible with the natural Euclidean space of response measurements being investigated [11, 13]. For example, suppose a finite population (Ω) of N artifacts ($\omega_1, \ldots, \omega_N$) associated with a given archaeological investigation is partitioned into g distinct groups ($S_1, \ldots, S_g$) of classified artifact types along with an excess group (S_{g+1}) consisting of unclassified artifacts (S_{g+1} may be empty). The response measurements associated with each artifact (ω_I) are the r location coordinates ($x_{1I}, \ldots, x_{rI}$) within the archaeological site being investigated (r may be 2 or 3, depending on the inclusion of the depth coordinate). The investigator wishes to know if different groups of artifact types are physically separated from one another (*see* ARCHAEOLOGY, STATISTICS IN). MRPP yield a statistical basis for deciding if such a physical separation between groups exists. (The artifacts of the finite population, $\Omega = \{\omega_1, \ldots, \omega_N\}$, in this example may of course be replaced by any type of object where each object may involve any collection of r response measurements [10, 11].) Let $n_i \geqslant 2$ be the number of classified artifacts in group S_i ($i = 1, \ldots, g$), let $K = \sum_{i=1}^{g} n_i$, and let $n_{g+1} = N - K \geqslant 0$ be the number of remaining unclassified artifacts in the excess group S_{g+1}. Let $\Delta_{I,J}$ be a symmetric distance function value of the location coordinates associated with artifacts ω_I and ω_J. While ordinary Euclidean distance may appear to be the only sensible choice of $\Delta_{I,J}$ in the present archaeological context, the consequence of this intuitive choice will be shown to have serious ramifications on many of the most commonly used statistical tests. The test statistic underlying MRPP is given by

$$\delta = \sum_{i=1}^{g} C_i \xi_i ,$$

where

$$\xi_i = \binom{n_i}{2}^{-1} \sum_{I<J} \Delta_{I,J} \Psi_i(\omega_I) \Psi_i(\omega_J)$$

is the average distance function value for all distinct pairs of artifacts (objects) within group S_i $(i = 1, \ldots, g)$, $\sum_{I<J}$ is the sum over all I and J such that $1 \leqslant I < J \leqslant N$, $\Psi_i(\omega_I)$ is 1 if ω_I belongs to S_i and 0 otherwise, $C_i > 0$ $(i = 1, \ldots, g)$, and $\sum_{i=1}^{g} C_i = 1$. The recommended choice of C_i is the simple proportion of classified artifacts in group S_i (i.e., $C_i = n_i / K$ for $i = 1, \ldots, g$). The null hypothesis for MRPP assigns an equal probability to each of the

$$M = N! \Big/ \left(\prod_{i=1}^{g+1} n_i! \right)$$

possible distinct allocations of the N artifacts (objects) to the $g + 1$ groups. If the different groups of artifacts are separated from one another physically and the distance function is Euclidean distance, then a concentration of response measurements (location coordinates) within groups would occur (i.e., the values of ξ_i would be small). Thus a small value of δ indicates a concentration of response measurements within the g groups [2, 10, 11].

The symmetric distance function $(\Delta_{I,J})$ defines the structure of the MRPP analysis space. This structure should be compatible with the natural Euclidean space of the response measurements being investigated [11, 13] (*see* MEASURES OF SIMILARITY, DISSIMILARITY AND DISTANCE). Consider the class of symmetric distance functions given by

$$\Delta_{I,J} = \left[\sum_{k=1}^{r} (x_{kI} - x_{kJ})^2 \right]^{v/2},$$

where $v > 0$. The MRPP analysis space is nonmetric when $v > 1$ (the triangle inequality property of a metric space fails) [9, 11, 13, 14] and is metric when $v \leqslant 1$ (a distorted metric space when $v < 1$). Since $v = 1$ yields the only nondistorted metric space (a Euclidean space), only $v = 1$ provides an analysis space that is compatible with the natural Euclidean space of the response measurements. While the validity of a permutation test is not affected by these geometric considerations, the rejection region (power characteristics) of almost any test is highly dependent on the underlying geometry (i.e., a permutation test will surely be misleading if its rejection region is incomprehensible). As is subsequently demonstrated, this concern involves some of the most commonly used statistical tests (e.g., one-way analysis of variance* and Mann–Whitney–Wilcoxon tests*). As a further example, suppose that (x_{1I}, x_{2I}, x_{3I}) denotes the rectangular coordinates of the Ith of N points on the edge of a unit sphere centered at the origin. The symmetric distance function for this example is the arc length in radians given by

$$\Delta_{I,J} = \cos^{-1}(x_{1I}x_{1J} + x_{2I}x_{2J} + x_{3I}x_{3J}).$$

Except that two rather than three dimensions are involved, an analogous symmetric distance function is given for points on the edge of a unit circle centered at the origin.

The choice of C_i governs the efficiency of the MRPP statistic. Both Mantel and Valand [6] and Mielke et al. [12] suggested an inefficient choice given by $C_i = \binom{n_i}{2} / \sum_{j=1}^{g} \binom{n_j}{2}$. The recommended efficient choice is the simple group size proportion given by $C_i = n_i / K$ [8]. O'Reilly and Mielke [15] introduced the present form of the MRPP statistic and mentioned a further inefficient choice given by $C_i = 1/g$. Since the permutation tests based on the two-sample t and one-way analysis of variance statistics are special cases of MRPP when $C_i = (n_i - 1) / (K - g)$, this efficient choice was, in effect, introduced by Fisher [4].

Since small values of δ indicate a concentration of response measurements within the g groups, the probability statement under the null hypothesis for MRPP given by $P(\delta \leqslant \delta_0)$ is the P-value* for δ_0 (an observed value of δ). For situations when M is small (say less than a million), an efficient algorithm has been developed to calculate the exact P-value [1]. When M is larger (as with most applications of MRPP), a method of moments* approximation requires calculation of the mean, variance and skewness of δ (denoted by μ_δ, σ_δ^2, and γ_δ) under the null hypothesis for MRPP. Efficient algorithms for calculating μ_δ, σ_δ^2, and γ_δ exist [7, 8, 12]. The standardized statistic given by $T = (\delta - \mu_\delta)/\sigma_\delta$ is fitted to the Pearson-type III distribution by setting the skewness parameter of that distribution equal to γ_δ [10]. The Pearson type-III random variable has mean zero, variance one, and is characterized by its skewness parameter [5, 10] (*see* PEARSON'S DISTRIBUTIONS). Empirical evidence suggests that this method of moments approximation for *P*-values is quite good [9, 13].

Preliminary findings indicated circumstances under the null hypothesis for MRPP when the asymptotic distribution of $N(\delta - \mu_\delta)$ is nondegenerate and nonnormal with γ_δ being substantially negative [7, 8, 12]. Also under the null hypothesis for MRPP, theorems now exist that:

1. Prescribe conditions when the asymptotic distribution of $N^{1/2}(\delta - \mu_\delta)$ is normal [15].
2. Specify the nondegenerate nonnormal asymptotic distribution of $N(\delta - \mu_\delta)/\mu_\delta$ for the univariate case [3].

Mielke et al. [13] show that the two-sample *t*, one-way analysis of variance, Wilcoxon–Mann–Whitney and many other well-known nonparametric tests are special univariate cases of MRPP when $N = K$, $C_i = (n_i - 1)/(N - g)$, and $\Delta_{I,J} = (x_{1,I} - x_{1J})^2$. Because the symmetric distance function for these special cases is squared Euclidean distance, the associated MRPP analysis space is nonmetric and is not compatible with the natural Euclidean space of the response measurements (this observation is very disturbing). Empirical power comparisons between nonparametric tests involving Euclidean distance and squared Euclidean distance as the symmetric distance function suggest that distinct advantages are attained with Euclidean distance [9, 11].

Permutation techniques closely related to MRPP also exist for analyzing univariate matched-pairs* data and multivariate randomized blocks data [9, 14] (*see* BLOCKS, RANDOMIZED COMPLETE). These related permutation techniques include the one-sample *t*, randomized blocks analysis of variance, sign*, Wilcoxon signed-ranks*, Friedman* two-way analysis of variance, Spearman rank correlation*, Spearman footrule, Cochran's Q*, and many other tests as special univariate cases. *See also* PERMUTATION PROCEDURES.

Selected applications of MRPP have involved disciplines such as archaeology*, climatology (*see* METEOROLOGY, STATISTICS IN), and weather modification* [2, 10, 13, 16].

References

[1] Berry, K. J. (1982). *Appl. Statist.*, **31**, 169–173. (Contains an efficient method for finding exact MRPP *P*-values.)

[2] Berry, K. J., Kvamme, K. L., and Mielke, P. W. (1983). *Amer. Antiq.* **48**, 547–553. (Provides an application of MRPP in anthropology.)

[3] Brockwell, P. J., Mielke, P. W. and Robinson, J. (1982). *Aust. J. Statist.*, **24**, 33–41. (Asymptotic nonnormal distribution of MRPP statistic is given for the univariate case.)

[4] Fisher, R. A. (1925). *Metron*, **5**, 90–104. (Implicitly introduces an efficient version of the MRPP statistic.)

[5] Harter, H. L. (1969). *Technometrics*, **11**, 177–187. (Gives description and excellent tables for Pearson type-III distribution.)

[6] Mantel, N. and Valand, R. S. (1970). *Biometrics*, **26**, 547–558. (Introduces an early version of MRPP with inadequate distributional assumptions.)

[7] Mielke, P. W. (1978). *Biometrics*, **34**, 277–282. (Demonstrates asymptotic nonnormality property of MRPP.)

[8] Mielke, P. W. (1979). *Commun. Statist. A*, **8**, 1541–1550. Errata: *A*, **10**, 1795; *A*, **11**, 847. (Introduces an efficient version of MRPP with asymptotic nonnormality property.)

[9] Mielke, P. W. and Berry, K. J. (1982). *Commun. Statist. Theor. Meth.* **11**, 1197–1207. (A variation of MRPP for handling matched pairs and power comparisons.)

[10] Mielke, P. W., Berry, K. J., and Brier, G. W. (1981). *Monthly Weather Rev.*, **109**, 120–126. (Pearson type-III approximation suggested and an application of MRPP in climatology provided.)

[11] Mielke, P. W., Berry, K. J., Brockwell, P. J., and Williams, J. S. (1981). *Biometrika*, **68**, 720–724. (Makes initial mention of geometric problem associated with many well-known statistical techniques and presents power comparisons for a new class of nonparametric tests.)

[12] Mielke, P. W., Berry, K. J., and Johnson, E. S. (1976). *Commun. Statist. A*, **5**, 1409–1424. (Peculiar distributional characteristics noted with an inefficient version of MRPP.)

[13] Mielke, P. W., Berry, K. J., and Medina, J. G. (1982). *J. Appl. Meteor*, **21**, 788–792. (Amplifies

geometric problem with many commonly used statistics and presents weather modification application of MRPP.)

[14] Mielke, P. W. and Iyer, H. K. (1982). *Commun. Statist. Theor. Meth.*, **11**, 1427–1437. (Variation of MRPP for analyzing randomized block experiments.)

[15] O'Reilly, F. J. and Mielke, P. W. (1980). *Commun. Statist. A*, **9**, 629–637. (Describes conditions when asymptotic distribution of studentized MRPP statistic is normal.)

[16] Wong, R. K. W., Chidambaram, C., and Mielke, P. W. (1983). *Atmos.-Ocean* **21**, 1–13. (Gives an application of MRPP in weather modification.)

(ANALYSIS OF VARIANCE
BLOCKS, RANDOMIZED COMPLETE
DISTRIBUTION-FREE METHODS
GEOMETRY IN STATISTICS
MATCHED PAIRS
MEASURES OF SIMILARITY, DISSIMILARITY, AND DISTANCE
PERMUTATION TESTS)

PAUL W. MIELKE, JR.

MULTISERVER QUEUES

The theory of multiserver queues is much more complex than the theory of single-server queues. Although the ideas involved are essentially the same, the extension to several servers requires different methods of analysis.

A typical queueing system consists of a group of s servers and a waiting room of size r (*see* QUEUEING THEORY for a general description). Customers arriving when all servers are engaged form a waiting line and are subjected to specific queue discipline and service mechanism (in a full availability system, no queue is formed if at least one server is free). The waiting time is the time between the instant of arrival and the instant at which a customer is admitted to a server (when $r = 0$, no waiting is possible, and customers arriving when the system is blocked are lost).

The primary stochastic characteristics of a queueing system are specified by the *input process* (flow of incoming customers) and the *service process* (the service time being the length of time a customer spends with a server). The principal objects of interest are the following stochastic processes*.

1. $Y = (Y_t, 0 \leqslant t < \infty)$, where Y_t is a random variable (RV) representing a number of customers in the system at time t. The queue length (number of waiting customers) is $\max(0, Y_t - s)$.
2. $W = (W_t, 0 \leqslant t < \infty)$ or $(W_n, n = 0, 1, \dots)$, where W_t (or W_n) is the waiting time of a customer arriving at time t (or of the nth customer).
3. $(B_n, n = 1, 2, \dots)$, where B_n is the duration of the nth busy period (time interval when all—or some—servers are busy).
4. Output process describing departures from the system after completion of service.

The following are the most typical multiserver queues.

MARKOVIAN QUEUES

The process Y is assumed to be a time-homogeneous Markov chain (*see* MARKOV PROCESSES) with matrix $\mathbf{P}(t) = (p_{ij}(t))$ of transition probabilities:

$$p_{ij}(t) = \mathrm{pr}(Y_t = j \mid Y_0 = i),$$
$$t \geqslant 0, \quad i, j = 0, 1, \dots, r + s,$$

satisfying the (forward) Kolmogorov equation

$$d\mathbf{P}(t)/dt = \mathbf{P}(t)\mathbf{Q}, \qquad \mathbf{P}(0) = \mathbf{I},$$

where $\mathbf{Q} = (q_{ij})$ is the *intensity matrix* with $\mathbf{QI} = \mathbf{0}$, $(q_{ii} = -q_i)$, and $q_{ij}h$ being the probability of transition from i to j during a short time interval of length h. In applications, $\mathbf{Q}$ is specified by the structure of the system (e.g., group input and batch service are represented by q_{ij} for $j > i$ and $j < i$, respectively). In particular, for the birth-and-death process* with coefficients λ_i and μ_i:

$$q_{i\,i+1} = \lambda_i, \qquad q_{i\,i-1} = \mu_i,$$
$$q_i = \lambda_i + \mu_i, \qquad q_{ij} = 0 \qquad \text{otherwise.}$$

The explicit time-dependent solutions for $p_{ij}(t)$ are usually too involved, so the equilibrium (steady state, ergodic) solution is of practical interest:

$$\lim_{t\to\infty} p_{ij}(t) = p(j) = \mathrm{pr}(Y_t = j),$$

which can be found from the matrix equation: $\mathbf{p}\cdot\mathbf{Q} = \mathbf{0}$.

For the birth-and-death process, the explicit form of $\mathbf{p}$ is available (in terms of λ_i and μ_i); moreover,

$$M_t = Y_t + \int_0^t [\mu(Y_\tau) - \lambda(Y_\tau)]\,d\tau, \qquad t \geqslant 0$$

form a martingale*, with $\mathbb{E}M_t = \mathbb{E}Y_0$.

In Markovian queues, the waiting time W of a customer can be regarded as the first entrance time to the subset $(0, 1, \dots, s-1)$ for a (modified) queue length process and its (complementary) distribution function (when conditioned on the initial state $i \geqslant s$):

$$W^c(t) = \mathrm{pr}(W > t)$$

satisfies a (backward) Kolmogorov equation. Similarly, the busy period B initiated by the state i (1 or s, in particular) is the first entrance time to the state $i-1$ (for the original chain).

Markovian queues are best known and have received considerable attention in the literature (see refs. 9, 23, 34, 38, 46, 47).

Examples listed here are classical and are variants of the $M/M/s$ system (Poisson input with intensity λ, exponential service with mean $1/\mu$) with

$$\lambda_i = \lambda, \qquad \mu_i = \min(i, s)\cdot\mu,$$

$\rho = \lambda/(s\mu)$ (traffic intensity) and $A = \lambda/\mu$ (traffic, measured in erlangs).

Strict Order Service ($r = \infty$)

The equilibrium solution exists iff $\rho < 1$, and is given by

$$p(j) = p(0)A^j/j! \qquad \text{for } j < s,$$
$$= p(0)A^j/(s!\,s^{j-s}) \qquad \text{for } j \geqslant s,$$

with constant $p(0)$ determined from normalization $\mathbf{p}\cdot\mathbf{1} = 1$. The waiting time distribution is

$$W^c(t) = W^c(0)e^{-(1-\rho)s\mu t}, \qquad t \geqslant 0,$$

where

$$W^c(0) = p(s)(1-\rho)^{-1}$$

is the classical Erlang delay formula (see, e.g., Cooper [9, p. 72]. The mean waiting time $M = \mathbb{E}W$ is

$$M = W^c(0)/[s\mu(1-\rho)]$$

and (the Little formula—see Cooper [9, p. 156] and Little [27]):

$$\mathbb{E}Y_t = \lambda(M + 1/\mu).$$

The distribution of the busy period (initiated by state s) has the same form as for the system $M/M/1$ (with μ replaced by $s\mu$), and its mean is again M. Moreover, the output from the queue (in equilibrium) is also Poisson with intensity λ (Burke Theorem; see Cooper [9, p. 141] and Brémaud [4, p. 123]).

Random Service ($r = \infty$)

Expressions for the transform of the waiting time distribution are available, but it is of interest that the probability of delay $W^c(0)$ and the mean waiting time M are the same as those for the strict order service.

Loss System ($r = 0$)

The equilibrium solution is given by the celebrated Erlang distribution:

$$p(j) = \frac{A^j}{j!} \Big/ \sum_{k=0}^{s} \frac{A^k}{k!}, \qquad 0 \leqslant j \leqslant s$$

with mean $A[1 - p(s)]$, where $p(s)$ is the classical Erlang loss formula (see Cooper [9, p. 5]).

Infinite Number of Servers ($s = \infty$)

The equilibrium solution exists always and is given by the Poisson distribution* with mean A.

Finite System

Finite waiting room and finite number servers, subject to various queue disciplines, received considerable attention in the literature; see refs. 9 and 23.

NON-MARKOVIAN QUEUES

The general approach consists of modifying a queueing process (by selecting a subprocess or by enlarging the original process) in such a way as to produce an auxiliary Markov process*. In the most typical situations the input process (X_n, $n = 0, 1, \ldots$) is of the regenerative type, where X_n is the instant of arrival of the nth customer, with interarrival times $T_n = X_n - X_{n-1}$ being independent identically distributed RVs with d.f. U and mean $1/\lambda$. The service process (L_n, $n = 0, 1, \ldots$) has also iid RVs representing the service time of the nth customer, with d.f. F and mean $1/\mu$. The input and the service processes are assumed independent, and the strict order service with $r = \infty$ is considered. The process Y is Markovian iff both U and F are exponential distributions*.

(a) For exponential service times and arbitrary input ($GI/M/s$ system) the process (Y_n, $n = 0, 1, \ldots$), where $Y_n = Y(X_n -)$ is the number of customers met by the nth customer on arrival, constitutes an (imbedded) discrete parameter Markov chain. Explicit expressions for the steady-state probabilities $p(j) = \text{pr}(Y_n = j)$ are available. The waiting-time distribution (in equilibrium) has the same form as for the $M/M/s$ system, with ρ replaced by β in formulas, where $0 < \beta < 1$ is the unique solution of the characteristic equation

$$\beta = U^*\left[(1 - \beta)\mu s\right]$$

with U^* denoting the Laplace–Stieltjes transform of the d.f. U. Expressions for the busy period distribution are also known. See references 9, 23, 46, and 50.

(b) For Poisson input and arbitrary service times ($M/G/s$ system), the supplementary variable technique considers enlargement to random vectors $(Y_t, L_t^1, \ldots, L_t^s)$, where L_t^i are the elapsed times of customers being served at time t. For details of waiting-time analysis, see refs. 3, 9, 23, 24, 46, and 50.

The probability of delay $W^c(0)$ depends now on the form of distribution F. In contrast, in the loss system ($r = 0$), the famous result asserts that the probability of loss is independent of the form of F (and is given by the Erlang loss formula); see also ref. 5. Furthermore, the equilibrium distribution of customers present at time t and the distribution of customers met on arrival coincide.

(c) The general system $GI/G/s$ presents formidable difficulties. Kiefer and Wolfowitz [22] obtained an s-dimensional integral equation (which generalizes the Lindley equation for $s = 1$; see "single-server queue") and found that the steady-state waiting time distribution exists when $\rho < 1$. The complete solution for the waiting time distribution in $GI/G/s$ was obtained by F. Pollaczek [32, 33] by the method of several complex variables. Representing the waiting time W_n of the nth customer in the form

$$W_n = \max(0, \min(T_{n1}, \ldots, T_{ns})),$$

where T_{nv} ($v = 1, \ldots, s$) are time intervals between the arrival of the nth customer and terminations of the last s services, Pollaczek showed that the generating function

$$\Phi(\alpha, z) = \sum_{n=0}^{\infty} z^n \mathbb{E}\left(e^{-\alpha W_n} \mid t_{0v}, v = 1, \ldots, s\right)$$

(for fixed initial conditions t_{0v} at $n = 0$) can be expressed in terms of a solution of a system of s simultaneous integral equations. Simplification can be obtained when the Laplace–Stieltjes transforms of service time distributions F are rational functions of the argument (in particular, when F is exponential). Excellent accounts of the Pollaczek theory were given by P. Le Gall [25] and by Cohen [7]; see also the brief summary in ref. 48. Important simplifications and extensions were obtained by de Smit in a series of papers [42, 43].

OTHER METHODS

Queueing literature is enormous in volume and in scope, and ranges from theoretical works using sophisticated mathematical techniques to works describing complex systems encountered in practice. References given here represent a cross section of the literature and contain mostly books, many of which have large bibliographies. In particular, ref. 9 is an excellent introduction, ref. 7 has a very informative up-to-date survey of the field; ref. 13, 41, and 44 contain proceedings; of great value is the bibliographical list [39] containing several hundred items. Some of the books on stochastic processes which contain discussions of queueing theory are listed in part B.

The following are brief comments on some methods used in the study of multiserver queues; no claim to completeness is made, however.

(a) In its development, Queueing Theory proceeded from studies of individual systems toward general consideration of types of stochastic processes and formulation of laws and methods applicable to sufficiently wide classes of systems. A successful formulation of a general theory was proposed by V. E. Beneš [1, 2]. L. Takács stressed the role of fluctuation theory and combinatorial methods [50, 51, 66]. N. U. Prabhu examined the Wiener–Hopf technique [34, 63]. J. Keilson investigated properties of distributions, especially stochastic monotonicity [20, 21, 60]; see also ref. 57. J. Th. Runnenburg pointed out advantages of the method of collective marks [37]; see also refs. 9 and 23. J. W. Cohen [7] developed a theory of derived Markov chains. Stochastic integrals* (introduced by R. Fortet) were discussed by P. Le Gall [25, 26]. W. Whitt [53] examined the connection between counting processes and queues; see also ref. 45. Point processes and martingales* in queueing theory were discussed by P. Brémaud [4]. Semi-Markov processes, regenerative processes, weak convergence approach, and others, received considerable attention in the literature; see refs. 3, 6, 52, 63, 64. Many investigations were devoted to specific topics of interest such as various forms of duality, the role of the Little formula [27, 35], "phase-type distributions" [28], output processes [14], heavy traffic effects [18], the insensitivity property [5], optimization problems [4, 24]), approximation theory (see [10], [31]), etc.

(b) On the other extreme, special systems with specified structural properties received considerable attention in the literature (especially in Operations Research* studies). Special queue disciplines, priorities, scheduling, time sharing, buffer systems, time-dependent input and service, limited availability systems, variable number of servers are just a few illustrative examples. Works of I. Adiri, P. Naor, M. Yadin, M. Hofri, and E. G. Coffman, L. Kleinrock, M. Neuts, M. F. Ramalhoto, D. P. Gaver, A. Descloux, R. L. Disney, R. V. Evans and others should be mentioned here; a very informative survey of this field is in refs. 9, 23. For priorities see ref. 19, and for applications to teletraffic see refs. 9, 24, 46; 23 for applications to computers *see also* NETWORKS OF QUEUES.

(c) Numerical computations and simulation* methods are of growing importance at present; see refs. 17, 23, 24, and 29. Statistical analysis of queues has also been discussed in the literature; see ref. 11 for the survey. Estimation of parameters in queueing models (ML* method), testing of hypotheses concerning basic queueing assumptions (input and service processes, Markov property, customer behavior), sampling and other techniques have been used. Of special interest have been estimation of probabilities (of delay, of loss, etc.) and of averages (the number in the system, the waiting time, the busy period, etc.); see refs. 11, 12, 13, 16, 24, 36, 46, and 65. Statistical infrerence problems in Markovian queues were treated in ref. 55 and in more general stochastic processes in Ref. 61. Social and psychological effects of queueing were discussed in ref. 40.

References

A. THEORY AND APPLICATIONS (* DENOTES ADVANCED MATHEMATICAL TEXT)

[1]* Beneš, V. E. (1963). *General Stochastic Processes in the Theory of Queues*. Addison-Wesley, Reading, MA.

[2] Beneš, V. E. (1965). *Mathematical Theory of Connecting Networks and Telephone Traffic*. Academic Press, New York.

[3]* Borovkov, A. A. (1976). *Stochastic Processes in Queueing Theory*. Springer, Berlin.

[4]* Brémaud, P. (1981). *Point Processes and Queues, Martingale Dynamics*. Springer, New York.

[5] Burman, D. Y. (1981). *Adv. Appl. Prob.*, **13**, 846–859.

[6] Cohen, J. W. (1976). *On Regenerative Processes in Queueing Theory*. Springer, New York.

[7]* Cohen, J. W. (1982). *The Single-Server Queue*, rev. ed. North-Holland, Amsterdam.

[8] Connolly, B. W. (1975). *Lecture Notes on Queueing Systems*. Ellis Horwood, Chichester, England.

[9] Cooper, R. B. (1981). *Introduction to Queueing Theory*, 2nd ed. North-Holland, Amsterdam.

[10] Cosmetatos, G. P. (1976). *Operat. Res. Quart. I*, **27**, 615–620; supplement, **28**, 596–597 (1977).

[11] Cox, D. R. (1965). pp. 289–316 in reference 44.

[12] Cox, D. R. and Smith, W. L. (1961). *Queues*. Methuen, London; Wiley, New York.

[13] Cruon, R., (1967). *Queueing Theory*. The English University Press, London.

[14] Daley, D. J. (1976). *Adv. Appl. Prob.*, **8**, 395–415.

[15] Gnedenko, B. V. and Kovalenko, I. N. (1968). *Introduction to Queueing Theory*. Israel Program for Scientific Translations, Jerusalem, Israel.

[16] Gross, D. and Harris, C. M., (1974). *Fundamentals of Queueing Theory*. Wiley, New York.

[17] Iglehart, D. and Shedler, G. S. (1980). *Regenerative Simulation of Response Times in Networks of Queues*. Springer, New York.

[18] Iglehart, D. and Whitt, W. (1970). *Adv. Appl. Prob.*, **2**, 150–177, 355–369.

[19] Jaiswal, N. K. (1968). *Priority Queues*. Academic Press, New York.

[20] Keilson, J. (1965). *Green's Function Methods in Probability Theory*. Griffin, London.

[21] Keilson, J. (1965). pages 43–71 in ref. 44.

[22]* Kiefer, J. and J. Wolfowitz (1955). *Trans. Amer. Math. Soc.*, **78**, 1–18.

[23] Kleinrock, L. (1975). *Queueing Systems (I and II)*. Wiley, New York.

[24] Kosten, L. (1973). *Stochastic Theory of Service Systems*. Pergamon Press, Oxford.

[25] Le Gall, P. (1962). *Les Systèmes avec ou sans Attente et les Processus Stochastiques*. Dunod, Paris.

[26] Le Gall, P. (1974). *Stoch. Processes Appl.*, **2**, 261–280.

[27] Little, J. D. C. (1961). *Operat. Res.*, **9**, 383–387.

[28] Neuts, M. F. (1975). In *Liber Amicorum Prof. Em. H. Florin*, Dept. of Mathematics, University of Louvain, Belgium, pp. 173–206.

[29] Neuts, M. F. (1981). In *Matrix-Geometric Solutions in Stochastic Models—An Algorithmic Approach*. The John Hopkins University Press, Baltimore.

[30] Newell, G. F. (1971). *Applications of Queueing Theory*. Chapman-Hall, London.

[31] Newell, G. F. (1973). *Approximate Stochastic Behavior of n-Server Service Systems with Large n*. Springer, New York.

[32]* Pollaczek, F. (1961) *Theorie Analytique des Problèmes Stochastiques Relatifs a un Groupe de Lignes Téléphoniques avec Dispositif d'Attente*. Gauthier-Villars, Paris.

[33]* Pollaczek, F. (1965). In ref. 44, pp. 1–42.

[34] Prabhu, N. U. (1965). *Queues and Inventories*. Wiley, New York.

[35] Ramalhoto, M. F. et al. (1981). *A Survey on Little's Formula*. Nota no. 23, Centro de Estatistica, Universidade de Lisboa, Lisbon, Portugal (to be published).

[36] Riordan, J. (1962). *Stochastic Service Systems*. Wiley, New York.

[37] Runnenburg, J. Th. (1965). pp. 399–438 in reference 44.

[38] Saaty, T. L. (1961). *Elements of Queueing Theory*. McGraw-Hill, New York.

[39] Saaty, T. L. (1966). *Nav. Res. Logist. Quart.* **13**, 447–476.

[40] Saaty, T. L. (1967). pp. 205–214 in ref. 13.

[41] Shlifer, E. (ed.) (1975). *Proceedings XX TIMS*, Jerusalem. Academic Press, New York.

[42] de Smit, J. H. A. (1971). "Many-Server Queueing Systems." Thesis. University of Delft, The Netherlands.

[43] de Smit, J. H. A. (1975). pp. 555–558 in Ref. 41.

[44] Smith, W. L. and W. E. Wilkinson eds. (1965). *Congestion Theory*. University of North Carolina Press, Chapel Hill, NC.

[45] Sonderman, D. (1979). *Adv. Appl. Prob.*, **11**, 439–455.

[46] Syski, R. (1960). *Introduction to Congestion Theory in Telephone Systems*. Oliver and Boyd, Edinburgh.

[47] Syski, R. (1965). pp. 170–227 in ref. 44.

[48] Syski, R. (1967). pp. 33–60 in ref. 13.

[49] Syski, R. (1975). pp. 507–508, 547–554 in ref. 41.

[50] Takács, L. (1962). *Introduction to the Theory of Queues*. Oxford University Press, New York.

[51] Takács, L. (1965). pp. 337–398 in ref. 44.

[52] Whitt, W. (1974). *Adv. Appl. Prob.*, **6**, 175–183.

[53] Whitt, W. (1981). *Adv. Appl. Prob.*, **13**, 207–220.

B. BOOKS ON STOCHASTIC PROCESSES, WITH APPLICATIONS TO QUEUEING

[54] Bharucha-Reid, A. T. (1960). *Elements of the Theory of Markov Processes and Their Applications*. McGraw-Hill, New York.

[55] Billingsley, P. (1961). *Statistical Inference for Markov Processes*. University of Chicago Press, Chicago.

[56] Billingsley, P. (1979). *Probability and Measure*. Wiley, New York.

[57] van Doorn, E. (1981). *Stochastic Monotonicity and Queueing Applications of Birth-and-Death Processes*. Springer, New York.

[58] Feller, W. (1966). *An Introduction to Probability Theory and Its Applications*, Vols. 1 and 2. Wiley, New York.

[59] Karlin, S. and H. M. Taylor (1975). *First Course in Stochastic Processes*, 2nd ed. Academic Press, New York.

[60] Keilson, J. (1981). *Markov Chain Models—Rarity and Exponentiality*. Springer, New York.

[61] Liptser, R. S. and Shiryayev, A. N. (1978). *Statistics of Random Processes*. Springer, New York.

[62] Parzen, E. (1962). *Stochastic Processes*, Holden-Day, San Francisco.

[63] Prabhu, N. U. (1980). *Stochastic Storage Processes*. Springer, New York.

[64] Rolski, T. (1981). *Stationary Random Processes Associated wtih Point Processes*. Springer, New York.

[65] Ross, S. M. (1980). *Introduction to Probability Models*, 2nd ed. Academic Press, New York.

[66] Takács, L. (1967). *Combinatorial Methods in the Theory of Stochastic Processes*. Wiley, New York.

Added in Proof: Connections with boundary problems in complex analysis were developed into a powerful tool for queueing theory in the book *Boundary Value Problems in Queueing System Analysis*, by J. W. Cohen and O. J. Boxma, North-Holland, Amsterdam, 1983.

(MARKOV PROCESSES
NETWORKS OF QUEUES
OPERATIONS RESEARCH
QUEUEING THEORY
STOCHASTIC PROCESSES)

R. SYSKI

MULTISTAGE SAMPLING *See* STRATIFIED MULTISTAGE SAMPLING

MULTISTAGE TESTS *See* SEQUENTIAL ANALYSIS

MULTISTATE COHERENT SYSTEMS

One inherent weakness of traditional reliability theory* (*see* COHERENT STRUCTURE THEORY) is that the system and the components are always described just as functioning or failed. Fortunately, this is being replaced by a theory for multistate systems of multistate components. This enables one in a power generation system, for instance, to let the system state be the amount of power generated, or, in a pipeline system, the amount of oil running through a crucial point. In both cases, the system state may be measured on a discrete scale. References 1, 4 and 8 initiated the research in this area in the late seventies. Here we summarize, starting from two recent papers [2, 7].

Let the set of states of the system be $S = \{0, 1, \ldots, M\}$. The $M + 1$ states represent successive levels of performance ranging from the perfect functioning level M down to the complete failure level 0. Furthermore, let the set of components be $C = \{1, 2, \ldots, n\}$ and the set of states of the ith component be S_i $(i = 1, \ldots, n)$, where $\{0, M\} \subseteq S_i \subseteq S$. Hence the states 0 and M are chosen to represent the endpoints of a performance scale that might be used for both the system and its components.

If X_i $(i = 1, \ldots, n)$ denotes the state or performance level of the ith component and $\mathbf{x} = (x_1, \ldots, x_n)$, it is assumed that the state ϕ of the system is given by the structure function $\phi = \phi(\mathbf{x})$. A series of results in multistate reliability theory can be derived for the following systems.

Definition 1. A system is a *multistate monotone system* (MMS) iff its structure function ϕ satisfies:

1. $\phi(\mathbf{x})$ is nondecreasing in each argument.

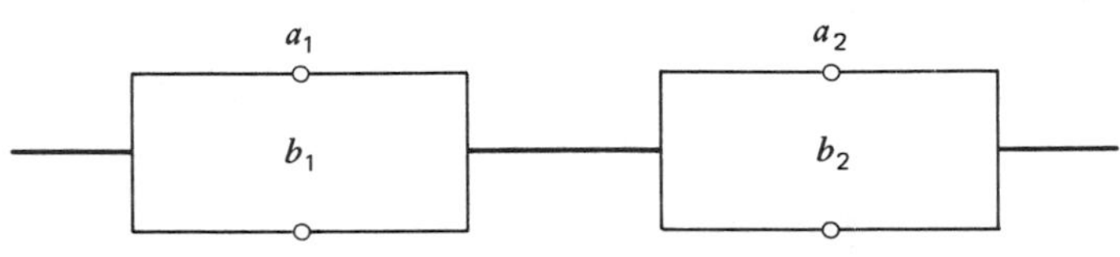

Figure 1 An MMS network.

Table 1. State of System in Fig. 1

Component 2	3	0	2	3
	1	0	1	2
	0	0	0	0
		0	1	3
		Component 1		

2. $\phi(\mathbf{0}) = 0$ and $\phi(\mathbf{M}) = M$ ($\mathbf{0} = (0, \ldots, 0)$, $\mathbf{M} = (M, \ldots, M)$).

As a simple example of an MMS consider the network of Fig. 1. Here component 1 (2) is the parallel module of the branches a_1 and b_1 (a_2 and b_2). Let $x_i = 3$ ($i = 1,2$) if two branches work and 1 (0) if one (no) branch works. The state of the system is given in Table 1. State 1 is a critical one both for each component and the system as a whole in the sense that the failing of a branch leads to the 0 state. In binary theory the functioning state comprises the states $\{1,2,3\}$ and hence just a rough description of the system's performance is possible.

DETERMINISTIC PROPERTIES OF MULTISTATE SYSTEMS

We start by generalizing each of the concepts "minimal path set" and "minimal cut set" in coherent structure theory*. In the following $\mathbf{y} < \mathbf{x}$ means $y_i \leqslant x_i$ for $i = 1, \ldots, n$ and $y_i < x_i$ for some i.

Definition 2. Let ϕ be the structure function of an MMS and let $j \in \{1, \ldots, M\}$. A vector $\mathbf{x}$ is a *minimal path (cut) vector to level* j iff $\phi(\mathbf{x}) \geqslant j$ and $\phi(\mathbf{y}) < j$ for all $\mathbf{y} < \mathbf{x}$ ($\phi(\mathbf{x}) < j$ and $\phi(\mathbf{y}) \geqslant j$ for all $\mathbf{y} > \mathbf{x}$). The corresponding *minimal path (cut) sets to level j* are given by $C^j(\mathbf{x}) = \{i \mid x_i \geqslant 1\}$ ($D^j(\mathbf{x}) = \{i \mid x_i < M\}$).

For the structure function tabulated in Table 1, the minimal path (cut) vectors for instance, to level 2 (1), are (3, 1) and (1, 3) ((3, 0) and (0, 3)).

We now impose further restrictions on the structure function ϕ. The following notation is needed:

$$(\cdot_i, \mathbf{x}) = (x_1, \ldots, x_{i-1}, \cdot, x_{i+1}, \ldots, x_n),$$
$$S^0_{i,j} = S_i \cap \{0, \ldots, j-1\},$$
$$S^1_{i,j} = S_i \cap \{j, \ldots, M\}.$$

Definition 3. Consider an MMS with structure function ϕ satisfying

1. $\min_{1 \leqslant i \mid n} x_i \leqslant \phi(\mathbf{x}) \leqslant \max_{1 \leqslant i \leqslant n} x_i$. If in addition $\forall i \in \{1, \ldots, n\}$, $\forall j \in \{1, \ldots, M\}$, $\exists(\cdot_i, \mathbf{x})$ such that
2. $\phi(k_i, \mathbf{x}) \geqslant j$, $\phi(l_i, \mathbf{x}) < j$ $\forall k \in S^1_{i,j}$, $\forall l \in S^0_{i,j}$, we have a *multistate strongly coherent system* (MSCS).
3. $\phi(k_i, \mathbf{x}) > \phi(l_i, \mathbf{x})$ $\forall k \in S^1_{i,j}$, $\forall l \in S^0_{i,j}$, we have a *multistate coherent system* (MCS).
4. $\phi(M_i, \mathbf{x}) > \phi(0_i, \mathbf{x})$, we have a *multistate weakly coherent system* (MWCS).

All these systems are generalizations of a system introduced in ref. 4. The first one is presented in ref. 7, whereas the two latter for the case $S_i = S$ ($i = 1, \ldots, n$) are presented in ref. 6. When $M = 1$, all reduce to the established binary coherent system (BCS). The structure function

$$\min_{1 \leqslant i \leqslant n} x_i \qquad \left(\max_{1 \leqslant i \leqslant n} x_i\right)$$

is often denoted the multistate series (parallel) structure.

Now choose $j \in \{1, \ldots, M\}$ and let the states $S^0_{i,j}$ ($S^1_{i,j}$) correspond to the failure (functioning) state for the ith component if a binary approach had been applied. Condition **2** of Definition **3** means that for all components i and any level j, there shall exist a combination of the states of the other components, $(\cdot_i, \mathbf{x})$, such that if the ith component is in the binary failure (functioning) state, the system itself is in the corresponding binary failure (functioning) state. Loosely speaking, modifying ref 2, condition **2** says that every level of each component is relevant to the same level of the system. Condition **3** says that every level of each component is relevant to the system, whereas condition **4** simply says that every component is relevant to the system.

For a BCS, one can prove the following practically very useful principle: Redundancy at the component level is superior to redundancy at the system level, except for a parallel system where it makes no difference. Assuming $S_i = S$ ($i = 1, \ldots, n$), this is also true for an MCS, but not for an MWCS.

We now mention a special MSCS type. Introduce the indicators ($j = 1, \ldots, M$)

$$I_j(x_i) = 1 \quad (0) \qquad \text{if} \quad x_i \geq j \quad (x_i < j)$$

and the indicator vector

$$\mathbf{I}_j(\mathbf{x}) = (I_j(x_1), \ldots, I_j(x_n)).$$

Definition 4. An MSCS is said to be a *binary-type multistate strongly coherent system* (BTMSCS) iff there exist binary coherent structures ϕ_j, $j = 1, \ldots, M$ such that its structure function ϕ satisfies

$$\phi(\mathbf{x}) \geq j \Leftrightarrow \phi_j(\mathbf{I}_j(\mathbf{x})) = 1$$

$$\text{for all} \quad j \in \{1, \ldots, M\} \quad \text{and all } \mathbf{x}.$$

Choose again $j \in \{1, \ldots, M\}$ and let the states $S_{i,j}^0$ ($S_{i,j}^1$) correspond to the failure (functioning) state for the ith component if a binary approach is applied. By the preceding definition ϕ_j will uniquely determine from the binary states of the components the corresponding binary state of the system. The MMS of Fig. 1 is an MSCS, but not a BTMSCS. In ref. 7 it is shown that if all ϕ_j are identical, the structure function ϕ reduces to the one suggested in ref. 1. Furthermore, it is indicated that most of the theory for a BCS can be extended to a BTMSCS.

PROBABILISTIC PROPERTIES OF MULTISTATE SYSTEMS

We now concentrate on the relationship between the stochastic performance of the system and that of the components. Let X_i denote the random state of the ith component and let

$$\Pr(X_i \leq j) = P_i(j), \qquad \bar{P}_i(j) = 1 - P_i(j);$$

$$i = 1, \ldots, n; \quad j = 0, \ldots, M.$$

P_i represents the *performance distribution of the ith component*. Now if ϕ is a structure function, $\phi(\mathbf{X})$ is the corresponding random system state. Let

$$\Pr(\phi(\mathbf{X}) \leq j) = P(j), \qquad \bar{P}(j) = 1 - P(j);$$

$$j = 0, \ldots, M.$$

P represents the *performance distribution of the system*. We also introduce the *performance function of the system*, h, defined by

$$h = E\phi(\mathbf{X}).$$

We obviously have that

$$h = \sum_{j=1}^{M} \bar{P}(j-1).$$

Hence, for instance, bounds on the performance distribution of the system automatically give bounds on h.

We briefly illustrate how coherent structure theory* bounds are generalized to bounds on the performance distribution of an MMS of associated components. First we give the crude bounds

$$\prod_{i=1}^{n} \bar{P}_i(j-1) \leq \bar{P}(j-1) \leq 1 - \prod_{i=1}^{n} P_i(j-1).$$

Next we give bounds based on the minimal path and cut vectors. For $j \in \{1, \ldots, M\}$ let

$$\mathbf{y}_r^j = (y_{1r}^j, \ldots, y_{nr}^j), \qquad r = 1, \ldots, n_j,$$

$$\mathbf{z}_r^j = (z_{1r}^j, \ldots, z_{nr}^j), \qquad r = 1, \ldots, m_j$$

be the system's minimal path (cut) vectors to level j and

$$C^j(\mathbf{y}_r^j), \qquad r = 1, \ldots, n_j,$$

$$D^j(\mathbf{z}_r^j), \qquad r = 1, \ldots, m_j,$$

the corresponding minimal path (cut) sets to level j. Then

$$\prod_{r=1}^{m_j} \left[1 - \Pr\left(\bigcap_{i \in D^j(\mathbf{z}_r^j)} (X_i \leq z_{ir}^j) \right) \right]$$

$$\leq \Pr[\phi(\mathbf{X}) \geq j]$$

$$\leq 1 - \prod_{r=1}^{n_j} \left[1 - \Pr\left(\bigcap_{i \in C^j(\mathbf{y}_r^j)} (X_i \geq y_{ir}^j) \right) \right],$$

$$j = 1, \ldots, M.$$

These are simplified in the case of independent components.

As a simple application of the crude bounds, consider the system of Fig. 1. Let the probability of a branch working be p, and assume that branches within a component work independently whereas the two components are associated. Then

$$2p^4 + \{1-(1-p)^2\}^2 \leqslant h \leqslant 3-(1-p)^4-2(1-p^2)^2.$$

For $p=0$ and $p=1$, we get the obvious results, whereas for $p=\frac{1}{2}$, $\frac{11}{16} \leqslant h \leqslant \frac{29}{16}$.

Almost all efforts on multistate systems theory have been concentrated on mathematical generalizations of the traditional binary theory. This research has, moreover, been quite successful. One key area where much research remains is the development of appropriate measures of component importance. Finally, there is a need for several convincing case studies demonstrating the practicability of the generalizations introduced. We know that some are under way.

References

[1] Barlow, R. E. and Wu, A. S. (1978). *Math. Operat. Res.*, **3**, 275–281.

[2] Block, H. W. and Savits, T. S. (1982). *J. Appl. Prob.*, **19**, 391–402.

[3] Block, H. W. and Savits, T. S. (1982). "Continuous Multistate Structure Functions." *Tech. Rep. No. 82-27*, Dept. of Mathematics and Statistics, University of Pittsburgh. (This paper initiates the research on multistate monotone systems with a continuous state space.)

[4] El-Neweihi, E., Proschan, F., and Sethuraman, J. (1978). *J. Appl. Prob.*, **15**, 675–688.

[5] Funnemark, E. and Natvig, B. (1985). "Bounds for the Availabilities in a Fixed Time Interval for Multistate Monotone Systems", *Adv. Appl. Prob.*, **17**, (to appear). (This paper generalizes all existing bounds from binary theory to multistate monotone systems.)

[6] Griffith, W. (1980). *J. Appl. Prob.*, **17**, 735–744.

[7] Natvig, B. (1982). *Adv. Appl. Prob.*, **14**, 434–455.

[8] Ross, S. M. (1979). *Ann. Prob.*, **7**, 379–383.

(COHERENT STRUCTURE THEORY
RELIABILITY, PROBABILISTIC)

B. NATVIG

MULTISTRATIFIED SAMPLING

Stratified sampling is one of the sampling designs often used to obtain statistical information. It involves the formation of strata (i.e., the stratification of the sampling units) and requires the following operations:

1. Choice of the stratification variable(s).
2. Choice of the number L of strata.
3. Determination of the way in which the range of the stratification variable(s) is to be divided in order to define the strata.

Apart from stratification, the design requires:

4. The choice of the total sample size n.
5. The choice of the sample size to be taken from the hth stratum, n_h, for $h = 1, \ldots, L$.

The procedures for carrying out these five activities depend on whether one is interested in obtaining information about a single variable (*unistratified sampling*) or two or more variables (*multistratified sampling*). Here we concentrate on the multivariate problem (for the univariate problem see, e.g., Cochran [4, Chap. 5]). In practice, the design operations are generally made in the numerical order presented. However, for simplicity, we consider them in reverse order. The first section is devoted to the sample allocation problem (point 5), the second section to sample size determination (4), and the third section to the stratification problem (1–3).

Throughout we assume independent simple random samples are obtained within each stratum, although the discussion could be generalized to more complex stratified designs. Also, we assume the parameters to be estimated are (without loss of generality) the K population means $\theta_1, \ldots, \theta_K$ of a certain set of variables $X_1, \ldots, X_K$. Let N be the population size, N_h the size of the stratum h, $\bar{x}_{k,h}$ the sample mean of variable X_k for stratum h, $\hat{\theta}_k$ a linear unbiased estimator of θ_k using stratified random sampling, $\sigma^2_{k,h}$ the variance of variable X_k within stratum h defined with divisor $N_h - 1$, and

$W_h = N_h/N$. Then $\hat{\theta}_k$ would be given by

$$\hat{\theta}_k = \sum_{h=1}^{L} W_h \bar{x}_{k,h}, \qquad k = 1, 2, \ldots, K,$$

and its variance by

$$V(\hat{\theta}_k) = \sum_{h=1}^{L} W_h^2 (N_h - n_h) \frac{\sigma_{k,h}^2}{N_h n_h},$$

$$k = 1, \ldots, K. \quad (1)$$

SAMPLE ALLOCATION

Assume that L, n, and the stratification are given. The problem to be discussed is how to distribute n among the L values $n_1, \ldots, n_L$. Several options are available.

Proportional Allocation

We could distribute n according to the size of each stratum, setting $n_h = W_h n$ for $h = 1, \ldots, L$. This is *proportional allocation* and with it equation (1) reduces to, say, $V_p(\hat{\theta}_k)$, with

$$V_p(\hat{\theta}_k) = \frac{N - n}{Nn} \sum_{h=1}^{L} W_h \sigma_{k,h}^2,$$

$$k = 1, \ldots, K.$$

Optimum Allocation (Minimizing a Loss of Information Measure)

Note that *Neyman's allocation** (see Cochran [4, p. 97]) would *not* be an optimum solution to the multivariate allocation problem since it would give, in general, K different values for every n_h. The first value of n_h would be equal to $N_h \sigma_{1,h} n/(N_1 \sigma_{1,1} + \cdots + N_L \sigma_{1,L})$, obtained from minimizing $V(\hat{\theta}_1)$; ... ; and the K'th from minimizing $V(\hat{\theta}_K)$. A similar problem arises when minimizing a variance for a fixed cost or the cost for a fixed variance (see Cochran [4, pp. 95–97]). The nature of the conflict is clear; an allocation that may be optimum for one variable may not be for another.

In the multivariate case, an alternative to proportional allocation is the procedure due to Geary [10], which minimizes a total relative loss of information measure, and which is an extension of Neyman's allocation (see Dalenius [6] for details). An additional form of allocation that uses nonlinear programming* will be presented in the subsection Proportional Allocation. In that case, the sample size is determined simultaneously with $n_1, \ldots, n_L$ (see also Cochran [4, pp. 120–121], Chakravarthy [1] and Ghosh [11]).

SAMPLE SIZE DETERMINATION

Proportional Allocation

Now assume that L and the stratification are both given and that $n_h = W_h n$, where n is to be determined. In the univariate problem, n may be calculated assuming the density of the estimator of the mean is approximately normal and defining a maximum length for a confidence interval* (other equivalent procedures include the definition of a fixed cost or a fixed variance).

In the multivariate problem, the above procedure could be carried out for each one of the K variables. A problem is that (as for n_h in Neyman's allocation) K different values for n could be obtained. If the biggest one is used, all desired precision levels would be satisfied. However, the precision gain of some estimators may not be substantial (because of the bigger sample size) and one may decide to sacrifice precision of some estimates (namely of those with corresponding bigger values for n) in order to have a smaller sample size.

Optimum Allocation (Minimizing the Sample Size)

Now we mention a procedure due to Kokan [17] and Chatterjee [3] for determination of n without imposing a priori a given sample allocation. The problem may be set up as: Minimize $n = n_1 + \cdots + n_L$, subject to $V(\hat{\theta}_k) \leqslant \epsilon_k$ for $k = 1, \ldots, K$, and $0 \leqslant n_h \leqslant N_h$ for $h = 1, \ldots, L$, where ϵ_k is the desired precision level of the estimate of θ_k, chosen according to the importance of X_k and the purpose to which $\hat{\theta}_k$ is to be put. (Alternatively, a cost function could replace $n = n_1 + \cdots + n_L$, which is the formulation used by Kokan and Chatterjee.) The solution may be found by nonlinear programming; an algorithm is given in Chatterjee [2] based on the method of Zoutendijk [24] (see also Dalenius [5, 6] and Kokan and Khan [18]).

THE STRATIFICATION PROBLEM

In the first and second sections, we assumed a stratification was given. We now turn to the stratification problem. Several reasons motivate stratification of units in a sampling design: for example, facilitating of administrative work, defining domains of study, or gaining in the precision of the estimates. The first two considerations are less amenable to a mathematical solution, so we comment on procedures that concentrate on precision*.

Regarding the choice of stratification variables (point 1) past information on the variables of interest or of proxies may be used in the stratification procedure. (For a discussion on the effects of this in the univariate case, see Dalenius [6] and Cochran [4, pp. 131–132].) Now we proceed to the description of how the information on the stratification variables (which to ease notation we also denote by $X_1, \ldots, X_K$) is to be used to form the strata (point 3), and how to determine an appropriate value for the number of strata L (point 2). Again, we distinguish according to the sample allocation to be used.

Proportional Allocation

For the univariate problem, Dalenius [6] assumes the probability density function of the variable of interest, say $f(x)$, is continuous and sets up the stratification problem as finding the strata boundaries $x^{(1)}$, $x^{(2)}, \ldots, x^{(L-1)}$ in order to minimize the variance of the estimator, subject to $x^{(0)} < x^{(1)} < \cdots < x^{(L-1)} \leqslant x^{(L)}$, where $x^{(0)} = \min[x]$ and $x^{(L)} = \max[x]$. Dalenius [6, p. 175] comments on the implications of the assumption on $f(x)$ and shows that $x^{(1)}, \ldots, x^{(L-1)}$ must satisfy the simultaneous equations $x^{(h)} = (\mu_h + \mu_{h+1})/2$ for $h = 1, \ldots, L-1$, where μ_h is the mean of stratum h. The solution to the above system may be found by a procedure analogous to that employed by Dalenius and Hodges [7, 8], which uses the cubic root of the cumulative (see Jarque [16, pp. 44–45], Singh [21], and Ekman [9] for details).

The preceding result is most useful when $K = 1$. When the number of variables of interest is 2, Ghosh [12] and Sadasivan and Aggarwal [20] consider minimizing the determinant of the variance–covariance matrix of the estimators and in that case, their result could be applied. We now discuss procedures useful when K is larger.

In the multivariate stratification problem, there are K variances that should be considered since a stratification optimum for one variable may be inadequate for another. So the first step is to define numerically manageable stratification criteria which contemplate the multivariate nature of the problem. For this we base our discussion on Jarque [15] (see also Hagood and Bernert [14] and Golder and Yeomans [13]).

Assume the procedure of Dalenius is applied individually to the variables $X_1, \ldots, X_K$ and denote the K optimum univariate stratifications by $S_1^*, \ldots, S_K^*$. Each of the S_k^* gives respectively a lower bound for $V_p(\hat{\theta}_k)$. Denote these *lower bounds* by $V^*(\hat{\theta}_1), \ldots, V^*(\hat{\theta}_K)$ and define by $V_S(\hat{\theta}_k)$ the variance of $\hat{\theta}_k$ obtained when using stratification S. In general, there will not exist a stratification S that attains simultaneously the K lower bounds. Define a *measure of closeness* between S and the optimum univariate stratification S_k^*, say $d_k(S) = V_S(\hat{\theta}_k) / V^*(\hat{\theta}_k)$. Observe that $d_k(S)$ is the reciprocal of the efficiency of stratification S, so $d_k(S) \geqslant 1$, and that stratifications with low values for $d_1(S), \ldots, d_K(S)$ would be preferred.

To obtain an efficient stratification, say S^*, a criterion suggested in Jarque [15, p.

164] is to find the stratification that minimizes $F(S) = d_1(S) + \cdots + d_K(S)$. $F(S)$ is a quadratic scale invariant function and there are clustering algorithms readily available for its minimization, for instance, Ward's [23] algorithm (see Sparks [22]) or the k-Means algorithm* (see MacQueen [19]).

A more general criterion is to use $G(S) = \Psi_1 d_1(S) + \cdots + \Psi_K d_K(S)$, where the Ψ_k are given weights. For instance, we could set $\Psi_k = V^*(\hat{\theta}_k)/V^T(\hat{\theta}_k)$, where the $V^T(\hat{\theta}_k)$ are target variances, and choose to minimize $G(S)$. Other functional forms involving $d_1(S), \ldots, d_K(S)$ may be used. For example, we may want to minimize $D(S) = (\mathbf{d}(S) - \mathbf{1})'\mathbf{\Lambda}(\mathbf{d}(S) - \mathbf{1})$, where $\mathbf{d}(S) = (d_1(S), \ldots, d_K(S))'$, $\mathbf{1}$ is a $K \times 1$ vector of ones, and $\mathbf{\Lambda}$ is a $K \times K$ matrix of constants. Yet another (generalized variance*) criterion is to minimize the determinant of the variance-covariance matrix $\mathbf{\Sigma}$ of $\hat{\theta}_1, \ldots, \hat{\theta}_K$.

All these criteria are reasonable, but for a choice based on computational ease, $F(S)$ and $G(S)$ have the advantage over $D(S)$ and the generalized variance approach. In a numerical exercise using real data (see Jarque [15, pp. 165–167]), the S that minimized $F(S)$ also minimized $D(S)$ with $\mathbf{\Lambda} = \mathbf{I}$. If the choice is based on a single criterion, the stratification minimizing it would clearly always be preferred. However, two stratifications S_1 and S_2 may be such that, say $F(S_1) < F(S_2)$. S_2 still may be preferred due to a more desirable set of values of the variances.

A further comment refers to the choice of the number of strata $L(2)$. In the univariate case, for $L > 6$, little variance reduction would be obtained (see Cochran [4, p. 133]). In the multivariate case, the function

$$I(L) = \sum_{k=1}^{K} \frac{V_S(\hat{\theta}_k : L)}{V(\hat{\theta}_k : 1)}$$

may serve to determine L, where $V_S(\hat{\theta}_k : L)$ denotes $V_S(\hat{\theta}_k)$ with L strata and $V(\hat{\theta}_k : 1)$ is $V(\hat{\theta}_k)$ when using a simple random sample of size n. In Jarque [15] the value of $I(L)$ for different procedures using real data was computed and with $L > 6$ there was still a significant reduction in variance, suggesting the need of $L > 6$ in multivariate situations. In general, a criterion to determine L is to plot $I(L)$ and set the value of L as that beyond which little change in $I(L)$ is obtained.

Optimum Allocation

The preceeding subsection referred to the stratification problem when using proportional allocation. The empirical study in Jarque [15, p. 168] showed that the optimum stratification for that allocation also had a good performance for other allocations, motivating its use in survey practice.

Now again concentrate on the case where desired precision levels have been established for each estimate. In the subsection on Optimum Allocation, we assumed a stratification was given. Yet the procedure described in Kokan [17] and Chatterjee [3] could be generalized. In the first stage, alternative stratifications would be found, and in the second stage the nonlinear programming solution would be applied to each. Then the stratification satisfying the desired precision levels with the minimum sample size could be chosen as optimum. Regarding the determination of L in the optimum allocation case, a third stage could be incorporated, repeating the two stages for each value of L within a given range. Then the optimum L (and corresponding stratification) would be the one satisfying the restrictions with minimum sample size.

In general, the choice between the use of results for proportional allocation or optimum allocation would depend on the computational facilities available. For instance, the computational expense of carrying out the procedure of the final subsection may be substantial, leading one to prefer using the results of the subsection on Proportional Allocation.

References

[1] Chakravarthy, I. M. (1955). *Sankhyā*, **14**, 211–216.

[2] Chatterjee, S. (1966). *Tech. Rep. No. 1*, Dept. of Statistics, Harvard University, Cambridge, MA.

[3] Chatterjee, S. (1968). *J. Amer. Statist. Ass.*, **63**, 530–534.

[4] Cochran, W. G. (1963). *Sampling Techniques*. Wiley, New York.

[5] Dalenius, T. (1953). *Skand. Aktuarietidskr.*, **36**, 92–102.

[6] Dalenius, T. (1957). *Sampling in Sweden. Contributions to the Methods and Theories of Sample Survey Practice*. Almqvist & Wiksell, Stockholm.

[7] Dalenius, T. and Hodges, J. L. (1957). *Skand. Aktuarietidskr.* **3–4**, 198–203.

[8] Dalenius, T. and Hodges, J. L. (1959). *J. Amer. Statist. Ass.*, **54**, 88–101.

[9] Ekman, G. (1959). *Ann. Math. Statist.* **30**, 219–229.

[10] Geary, R. C. (1949). *Technical Series*.

[11] Ghosh, S. P. (1958). *Calcutta Statist. Ass. Bull.*, **8**, 81–90.

[12] Ghosh, S. P. (1963). *Ann. Math. Statist.*, **34**, 866–872.

[13] Golder, P. A. and Yeomans, K. A. (1973). *Appl. Statist.*, **22**, 213–219.

[14] Hagood, M. J. and Bernert, E. H. (1945). *J. Amer. Statist. Ass.*, **40**, 340–344.

[15] Jarque, C. M. (1981). *Appl. Statist.*, **30**, 163–169.

[16] Jarque, C. M. (1982). "Contributions to the Econometrics of Cross-Sections," Ph.D. dissertation. Faculty of Economics, Australian National University, Canberra, Australia.

[17] Kokan, A. R. (1963). *J. R. Statist. Soc. A*, **126**, 557–565.

[18] Kokan, A. R. and Khan, S. (1967). *J. R. Statist. Soc. B*, **29**, 115–125.

[19] MacQueen, J. (1967). *Proc. 5th. Berkeley Symp. Math. Statist. Prob.*, **1**, 281–297. University of California Press, Berkeley, CA.

[20] Sadasivan, G. and Aggarwal, R. (1978). *Sankhyā C*, **40**, 84–97.

[21] Singh, R. (1975). *Sankhyā C*, **37**, 109–115.

[22] Sparks, D. N. (1973). *Appl. Statist.*, **22**, 126–130.

[23] Ward, J. H. (1963). *J. Amer. Statist. Ass.*, **58**, 236–244.

[24] Zoutendijk, G. (1959). *J. R. Statist. Soc. B*, **21**, 338–355.

(NEYMAN ALLOCATION
SAMPLING PLANS
STRATIFIED SAMPLING)

CARLOS M. JARQUE

MULTITRAIT-MULTIMETHOD MATRICES

A multitrait-multimethod matrix is a correlation between measurements obtained when each of a number of traits is measured by each of a number of methods. The use of a correlation* matrix of this type was suggested by Campbell and Fiske [4] for investigating the validity of tests as measures of psychological constructs (*see* PSYCHOLOGICAL TESTING THEORY). Two types of validity were considered. *Convergent validity* was regarded as the extent to which a test correlates with different measures of the same construct. *Discriminant validity* was the extent to which a test does not correlate with measures of different constructs.

A multitrait-multimethod matrix provided by Taylor [11] in a study on attitudes of 320 workers is shown in Table 1. Three traits, Attitude to Supervisors ($T1$), Attitude to Co-workers ($T2$), and Attitude to Work ($T3$) were each measured by three methods, Taylor's projective technique ($M1$), a Likert rating scale ($M2$), and an Osgood rating scale ($M3$). The combination of the ith trait and the kth method is denoted by T_iM_k in the row and column headings in Table 1.

Campbell and Fiske [4] proposed the subjective assessment of convergent and discriminant validity by inspection of various sets of elements of the multitrait-multimethod matrix. In Table 1, the symmetric diagonal blocks, with elements $r(T_iM_k, T_jM_k)$, are the *heterotrait-monomethod* blocks while the nonsymmetric off-diagonal blocks, with nondiagonal elements $r(T_iM_k, T_jM_l)$, $i \neq j \neq k \neq l$, are the *heterotrait-heteromethod* blocks. The diagonals of the heterotrait-heteromethod blocks, with elements $r(T_iM_k, T_iM_l)$, are the *validity diagonals*. Campbell and Fiske [4] specified four requirements for a multitrait-multimethod matrix:

1. The elements of the validity diagonal should be substantial.
2. Any element of validity diagonal should be larger than all other elements in the

Table 1 A Multitrait-Multimethod Matrix

	T_1M_1	T_2M_1	T_3M_1	T_1M_2	T_2M_2	T_3M_2	T_1M_3	T_2M_3	T_3M_3
T_1M_1	1.00	0.55	0.55						
T_2M_1	0.55	1.00	0.57						
T_3M_1	0.55	0.57	1.00						
T_1M_2	0.85	0.46	0.48	1.00	0.57	0.52			
T_2M_2	0.51	0.73	0.50	0.57	1.00	0.58			
T_3M_2	0.52	0.51	0.76	0.52	0.58	1.00			
T_1M_3	0.77	0.43	0.40	0.81	0.44	0.42	1.00	0.51	0.45
T_2M_3	0.50	0.71	0.48	0.47	0.69	0.42	0.51	1.00	0.48
T_3M_3	0.46	0.42	0.69	0.42	0.41	0.70	0.45	0.48	1.00

corresponding row and column of its heterotrait-heteromethod block, i.e., $r(T_iM_k, T_iM_l) > r(T_iM_k, T_jM_l)$ and $r(T_iM_k, T_iM_l) > r(T_jM_k, T_iM_l)$.

3. Any element of a validity diagonal should be larger than all the nondiagonal elements in the corresponding row of the heterotrait-monomethod block to its side and all nondiagonal elements in the corresponding column of the heterotrait-monomethod block above it, i.e., $r(T_iM_k, T_iM_l) > r(T_iM_k, T_jM_k)$ and $r(T_iM_k, T_iM_l) > r(T_jM_l, T_iM_l)$.
4. The same pattern should be exhibited by the nondiagonal elements of all heterotrait-monomethod blocks as well as all heterotrait-heteromethod blocks.

The first requirement is indicative of convergent validity and the last three of discriminant validity. The multitrait-multimethod matrix in Table 1 meets all four requirements. In particular, the last requirement is met since all nondiagonal elements in both the monomethod and heteromethod blocks are more or less equal.

Hubert and Baker [6, 7] gave some possible measures of the extent to which the Campbell–Fiske requirements are met. Permutation tests on these measures involving reassignment of row and column headings of the multitrait-multimethod matrix were suggested.

A number of models have been tried when analyzing multitrait-multimethod matrices (cf. Schmitt et al. [9]). Examples are: a restricted factor analysis* model (e.g. Jöreskog [8, Table 8]), three-mode* factor analysis (e.g. Tucker [12] and Bentler and Lee [1, Section 6]) and multiple battery factor analysis (e.g. Browne [2, Section 5]). These models do not provide information concerning the extent to which the Campbell–Fiske requirements are satisfied.

In an investigation of a number of empirical multitrait-multimethod matrices, Campbell and O'Connell [5] observed that method effects on correlation coefficients tend to be multiplicative rather than additive. Factor analysis models, however, do not imply a multiplicative method effect, and doubts were expressed by Campbell and O'Connell [5, pp. 424–425] as to their appropriateness for multitrait-multimethod matrices.

A direct product model that does imply multiplicative method effects and insight concerning the extent to which the Campbell–Fiske conditions are met was proposed by Swain [10]. This model imposes constraints on standard deviations and also implies equality of all elements of any validity diagonal. These unwanted restrictions may be eliminated [3] by regarding each observed variable as consisting of a common component and a unique component and adopting a direct product structure for the correlation matrix of the common components only.

Suppose that there are t traits and m methods under consideration and P is the $mt \times mt$ population correlation matrix for observed scores. This model is

$$P = D_\zeta\left(P_M \otimes P_T + D_\eta^2\right)D_\zeta,$$

Table 2 Parameter Estimates

Trait Correlations $\hat{P}_T$				Method Correlations $\hat{P}_M$			
T_1	1.00			M_1	1.00		
T_2	0.61	1.00		M_2	0.89	1.00	
T_3	0.56	0.66	1.00	M_3	0.87	0.86	1.00

Indices of Communality $\hat{\zeta}$

T_1M_1	T_2M_1	T_3M_1	T_1M_2	T_2M_2	T_3M_2	T_1M_3	T_2M_3	T_3M_3
0.96	0.91	0.91	1.00	0.93	0.93	0.93	0.90	0.88

where P_M is an $m \times m$ nonnegative definite matrix with unit diagonals, P_T is a $t \times t$ nonnegative definite matrix with unit diagonals, and D_ζ and D_η are $mt \times mt$ diagonal matrices. In this model, a diagonal element of D_ζ represents an *index of communality* or correlation coefficient between an observed score and its common component, a diagonal element of D_η represents the ratio of a unique component standard deviation to a common component standard deviation, and the common component correlation matrix is the direct product $P_M \otimes P_T$. Because P has unit diagonals D_η may be regarded merely as a function of D_ζ:

$$D_\eta = \left(D_\zeta^{-2} - I\right)^{1/2}.$$

Nondiagonal elements of P_M are referred to as "method correlations" and those of P_T as "trait correlations" as they are similarity indices with the mathematical properties of correlation coefficients.

The Campbell–Fiske requirements are met by the common component correlation matrix $P_M \otimes P_T$ rather than by the observed variables correlation matrix P. The second and fourth requirements are always met, while the first requirement implies that the method correlations must be substantial; the third requirement is satisfied if all method correlations are larger than all trait correlations.

Maximum multivariate normal likelihood estimates of D_ζ, P_T, and P_M, obtained from the correlation matrix of Table 1, are shown in Table 2. All trait correlations are close in magnitude as are all method correlations. Since the method correlations are substantial and are all greater than all trait correlations, the Campbell–Fiske requirements are satisfied by the estimated common component correlation matrix $\hat{P}_M \otimes \hat{P}_T$.

References

[1] Bentler, P. M. and Lee, S. Y. (1979). *Brit. J. Math. Statist. Psychol.*, **32**, 87–104.

[2] Browne, M. W. (1980). *Brit. J. Math. Statist. Psychol.*, **33**, 184–199.

[3] Browne, M. W. (1984). *Brit. J. Math. Statist. Psychol.*, **37**, 1–21.

[4] Campbell, D. T. and Fiske, D. W. (1959). *Psychol. Bull.* **56**, 81–105.

[5] Campbell, D. T. and O'Connell, E. J. (1967). *Multivariate Behav. Res.*, **2**, 409–426.

[6] Hubert, L. J. and Baker, F. B. (1978). *Multivariate Behav. Res.*, **13**, 163–179.

[7] Hubert, L. J. and Baker, F. B. (1979). *Brit. J. Math. Statist. Psychol.*, **32**, 179–184.

[8] Jöreskog, K. G. (1974). In *Contemporary Developments in Mathematical Psychology*, R. C. Atkinson et al., eds. W. H. Freeman, San Francisco, pp. 1–56.

[9] Schmitt, N., Coyle, B. W., and Saari, B. B. (1977). *Multivariate Behav. Res.*, **12**, 447–478.

[10] Swain, A. J. (1975). "Analysis of Parametric Structures for Variance Matrices." Ph.D. thesis. University of Adelaide, Australia.

[11] Taylor, T. R. (1983). "A Multivariate Approach to the Prediction of Behavior Towards Attitude Objects." National Institute for Psychological Research. *Tech. Rep.* in preparation.

[12] Tucker, L. R. (1967). *Multivariate Behav. Res.*, **2**, 139–151.

(GROUP TESTING
PSYCHOLOGICAL TESTING THEORY
PSYCHOLOGY, STATISTICS IN)

M. W. Browne